Peter van Oosterom

Siyka Zlatanova

Elfriede M. Fendel

Geo-information for Disaster Management

Peter van Oosterom
Siyka Zlatanova
Elfriede M. Fendel
(Editors)

Geo-information for Disaster Management

With 516 Figures

Professor Dr. Peter van Oosterom
Dr. Siyka Zlatanova
Elfriede M. Fendel

Delft University of Technology
OTB Research Institute for Housing, Urban and Mobility Studies
Section GIS Technology
Jaffalaan 9
2628 BX Delft
The Netherlands

Cover illustration: On 26 December 2004 a magnitude 9 earthquake generated a devastating Tsunami causing many casualties in countries around the Indian ocean. A model prediction of this phenomenon is overlaid with in situ sea level measurements of the Envisat satellite altimeter recorded 3:15 hours after the event. More details on this unique observation can be found in the article of Ambrosius et al. in this book. The original data was kindly provided by Remko Scharroo/NOAA and rendered by an artist.

Library of Congress Control Number: 2005920463

ISBN 3-540-24988-5 **Springer Berlin Heidelberg New York**

Springer is a part of Springer Science+Business Media
springeronline.com

Printed in The Netherlands

Cover design: Erich Kirchner
Production: Luisa Tonarelli
Typesetting: Camera-ready by Elfriede M. Fendel
Printing: Krips bv, Meppel
Binding: Litges + Dopf, Heppenheim

Printed on acid-free paper 30/2132/LT – 5 4 3 2 1 0

Foreword

One thing was sure when we started to organize this symposium last year: there will always be more (natural or man-made) disasters all over the world. Of course, no one could have anticipated the huge disaster of the Tsunami on 26 December 2004 in Asia, just a few months before the symposium. The cover of this book contains an impression of the situation 3 hours and 15 minutes after Tsunami (thanks to Remko Scharroo, NOAA for providing us with these satellite altimetry images; more details can be found in the paper on page 323-336). This extremely disastrous event did give the organizers mixed feelings. Of course, very deep sadness, but also the reinforced belief that geo-information (technology) must be used to optimize the disaster management (both before and after situations).

Researchers, developers, users and geo-information providers all participated in 'The First International Symposium on Geo-information for Disaster Management', Delft, the Netherlands, 21-23 March 2005. This created the most appropriate atmosphere for work and discussions between different professionals. During the symposium two basic types of presentations could be recognized: the ones with more focus on research and development of geo-information technology, and the others with more focus on the practical needs and solutions for users and managers in disaster management. The observation one could make is that these two basic types are often mixed in a presentation that contains both (parts of) users needs and technology solutions.

Geo-information technologies offer a variety of opportunities to aid management and recovery in the aftermath of industrial accidents, road collisions, complex emergencies, earthquakes, fires, floods and similar catastrophes. These context-aware technologies can provide access to needed information, facilitate the interoperability of emergency services, and provide high-quality care to the public.

Disaster management depends on large volumes of accurate, relevant, on-time geo-information that various organizations systematically create and maintain. This information may be described in catalogues and made available through Geo-Information Infrastructures, such as Infrastructure for Spatial Information in Europe (INSPIRE), based on ISO, CEN, and OpenGIS standards. While the semantics of geo-information might be clear to the producer, formal semantics are seldom available. This complicates real-time machine processing in support of disaster management.

Disaster management poses significant challenges for data collection, data management, discovery, translation, integration, visualization and communication based on the semantics of the heterogeneous (geo-) information sources with differences in many aspects: scale/resolution, dimension (2D or 3D), classification and attributes schemes, temporal aspects (up-to-date-ness, history, predictions of the future), spatial reference system used, etc.

For this reasons the Section GIS Technology at the OTB Research Institute for Housing, Urban and Mobility Studies at the Delft University of Technology took the initiative to organize the First Symposium on Geo-information for Disaster Management. This First Symposium focused primarily on the response and secondarily on the relief phases of Disaster Management encouraging a wide discussion on systems and requirements for use of geo-information under time and stress constraints and unfamiliar situations, environments and circumstances.

The organizers of the First symposium believe: the initiated discussion between technology developers (software and hardware), disaster management bodies, information providers, developers of standards and users will accelerate the development of advanced context-aware technologies for disaster management.

Recognizing the importance of disaster management issues, several universities, international organizations and vendors have taken the initiative to make this symposium an annual event, which will be organized in different continents. Three follow-up symposiums are already planned: India (2006), Canada (2007) and China (2008). These symposia are already officially approved as events of the International Society for Photogrammetry and Remote Sensing (ISPRS)

Goal and Objectives

The fundamental goal of the Symposium was to tackle disaster management problems in their entirety, considering: a. technology (both software and hardware applicable for Disaster Management), b. user requirements for geo-information (both management and mobile users), and c. information providers (data and standards). Therefore, during the Symposium the following aspects were addressed: 1. the state-of-the-art in Disaster Management, 2. a review of tools, software, existing geo-information sources, organizational structures and methods for work in crisis situations, 3. an outline of the drawbacks in current use, discovery, integration and exchange of geo-information, and 4. some suggestions for future research directions.

Conference topics

In order to reach the goal and objectives described above the Symposium focused on the following topics, which are reflected on by the papers included in this book:

- User Needs, Requirements and Technology Developments
- Data Collection and Data Management
- Data Integration and Knowledge Discovery
- End-User Environments
- Positioning and Location-Based Communication
- Information Systems for Specific DM Applications

Paper Selection Process

The Symposium on Geo-information for Disaster Management is a refereed symposium. The papers were submitted as extended abstracts and reviewed by at least two members, but usually three or four members, of the Scientific Program Committee. Nearly 170 full abstracts were submitted. The 22 best were selected for an oral presentation in a plenary session, next another 49 good papers were selected for an oral presentation in a parallel session. Finally, from the 50 submitted poster presentations, 27 submissions were selected for publication in this book. The authors of the selected abstracts were asked to submit a long paper for oral presentations or a short paper for poster presentations. In the preparation of the final papers, the authors had to consider the comments of the reviewers[1].

[1] Every effort has been made to ensure that the advice and information in this book is true and accurate. However, neither the publisher nor the authors, nor the editors can accept any legal responsibility or liability for any errors or omissions that may be made. The information content of papers collected in this volume is the sole responsibility of the respective authors.

Acknowledgement

Any conference takes considerable organization, and this one – being the first – especially. The editors would like to express their appreciation and gratitude towards the members of the Scientific Program Committee for completing the review process in time.

Such an event cannot be organized without the support of the international scientific community and the ICT-industry. A special thanks goes to the sponsors: GIN, RWS-AGI, Intergraph, Bentley, ESRI, and TU Delft/GDMC and the supporting organizations: ISPRS, EuroSDR, AGILE, and UNOOSA for recognizing the importance of this symposium.

This symposium is the result of the research program 'Sustainable Urban Areas' (SUA) carried out by Delft University of Technology.

Peter van Oosterom, Siyka Zlatanova and Elfriede M. Fendel
January 31, 2005

Table of Contents

User Needs, Requirements and Technology Developments

Plenary Contributions

Oral Contributions

Poster contributions

Data Collection and Data Management

Plenary Contributions

Oral Contributions

Poster contributions

Data Integration and Knowledge Discovery

Plenary Contributions

Oral Contributions

Poster Contributions

End-User Environments

Plenary Contributions

Oral Contributions

Oral Contributions

Poster Contributions

Information Systems for Specific DM Applications

Plenary Contributions

Oral Contributions

Poster Contributions

Author Index

Programme Committee

Chair
Peter van Oosterom
Co-chairs
Christian Heipke
Jonathan Li
Hardy Pundt
David Stevens
Peter Woodsford
Sisi Zlatanova
Members
Rune Aasgard
Ben Ale
Orhan Altan
Costas Armenakis
Michael Batty
Jaap Besemer
Roland Billen
Lars Bodum,
Antonio Camara
Jun Chen
Volker Coors
Oscar Custers
Naser El-Sheimy
Andrea Fabbri
Elfriede M. Fendel
Ben Gorte
Dorota Grejner-Brzezinska
Michel Grothe
Daniel Holweg
Qingshan Jiang
Jitske de Jong
Pieter Jonker
Allison Kealy
Tjeu Lemmens
Rainer Malaka
Mark Millman
Chris Parker
Friso Penninga
Wilko Quak
Alias Abdul Rahman
Javier Ramos
Orvind Rideng
Henk Scholten
Wolfgang Steinborn
Jantien Stoter
Manolis Stratakis
Theo Tijssen
Edward Verbree
Marian de Vries
Jinling Wang
Pietro Zanarini

Local Organizing Committee

The Local Organizing Committee consists of staff members of the Section GIS Technology of the OTB Research Institute for Housing, Urban and Mobility Studies, Delft University of Technology.
Elma Bast-Gast
Elfriede M. Fendel
Tjeu Lemmens
Peter van Oosterom
Friso Penninga
Wilko Quak
Axel Smits
Theo Tijssen
Edward Verbree
Marian de Vries
Sisi Zlatanova

Orchestra: Developing a Unified Open Architecture for Risk Management Applications

Alessandro Annoni[1], Lars Bernard[1], John Douglas[2], Joseph Greenwood[3], Irene Laiz[4], Michael Lloyd[5], Zoheir Sabeur[4], Anne-Marie Sassen[6], Jean-Jacques Serrano[2] and Thomas Usländer[7]

[1] Joint Research Centre, European Commission, TP262, I-21020 Ispra (VA), Italy.
Email: {lars.bernard|alessandro.annoni@jrc.it}
[2] BRGM, 3, avenue Claude-Guillemin - BP 6009 - 45060 Orléans Cedex 2, France.
Email: j.douglas@brgm.fr
[3] Research and Innovation, Ordnance Survey, Romsey Road, Southampton, SO16 4GU, United Kingdom.
Email: Joe.Greenwood@ordnancesurvey.co.uk
[4] BMT Cordah Limited, 7 Ocean Way, Ocean village. Southampton SO14 3TJ, United Kingdom.
Email: Zoheir.sabeur@bmtcordah.com
[5] AMRIE, 20-22 Rue Du Commerce, 1000, Belgium.
Email: amrie@info.org
[6] Atos Origin, sae. C, Albarracin 25, 28037 Madrid, Spain.
Email: anne-marie.sassen@atosorigin.com
[7] Fraunhofer IITB, Fraunhoferstr. 1, 76131 Karlsruhe, Germany.
Email: Thomas.Uslaender@iitb.fraunhofer.de

Abstract

Due to organizational and technological barriers, actors involved in the management of natural or man-made risks cannot cooperate efficiently. In an attempt to solve some of these problems, the European Commission has made "Improving risk management" one of its strategic objectives of the IST program. The integrated project Orchestra is one of the projects that recently started in this area. The main goal of Orchestra is to design and implement an open service oriented software architecture that will improve the interoperability among actors involved in multi-risk management. In this paper we will describe the goals of Orchestra and explain some of the key characteristics of the project. These are:

- The chosen design process of the Orchestra Architecture.
- How to further improve geospatial information and standards for dealing with risks
- How ontologies will be used to bring interoperability from a syntactical to a semantical level.

The paper ends with two examples demonstrating the benefits of the Orchestra Architecture. One is in the area of coastal zone management, and the other is related with managing earthquake risks.

1 The Current Situation in Risk Management

Increasing numbers of natural disasters have demonstrated to the European Commission and the Member States of the European Union the paramount importance of the natural hazards subject for the protection of the environment and the citizens. The flooding experienced throughout central Europe in August 2002 is the most recent example of the damage caused by unforeseen weather driven natural hazards. The summer of 2003 clearly showed the growing problem of droughts in Europe including the Forest Fires in Portugal with more than 90,000 ha of burnt areas. There is strong scientific evidence of an increase in mean precipitation and extreme precipitation events on the one hand and water shortages for certain regions on the other hand which implies that weather driven natural hazards may become more frequent.

The European Spatial Development Perspective and new EC Regional Policy regulations emphasize the need of a better spatial planning and require new tools for the impact assessment of regional developments on natural and technological risks and vice versa. Now the new regulations for renewed Structural Funds and instruments for the period 2007-2013, adopted by the EC on 14 July 2004, foresee specific measures for "developing plans and measures to prevent and cope with natural risks".

The different types of risks (affecting the territory of the EU) need to be better addressed by an integrated approach to risk management. Prevention, preparedness and response, three major phases of risk management, usually involve a vast range of sectoral institutions and organizations at various administrative levels with different systems (monitoring, forecasting, warning, information, etc.) and services. Unfortunately, exchange of relevant information needed for dealing with risks is often limited to a raw data exchange level and true efficiency, in most cases, is hindered by administrative and legal boundaries as well as a lack of interoperability on the technical side.

Because natural and man-made disasters are not limited by administrative boundaries, cross-border aspects also need to be carefully considered and as a consequence major efforts are required to harmonize data and making services interoperable.

The application of different policies, procedures, standards and the lack of interoperability of systems, result in problems related to an efficient data management and information delivery, all critical elements of Risk Management. Interoperability based on standardization will help close the gap between the different actors and is also expected to stimulate the development of an operational European service market supported by the sharing of procedures, interfaces and resources.

A substantial portion of IT expenditure in Europe supports the maintenance of thousands of legacy systems, the vast majority of which were not designed to work together. Recent events have underscored the need to be able to consolidate information from disparate systems to support citizen protection and security, disaster management, criminal justice, and other missions, crossing pan-European agency boundaries and extending into national, state and local government areas. One of the most urgent and important challenges currently facing governments is to get these systems to interoperate and share information.

2 The Orchestra Project

In an attempt to solve some of these problems, the European Commission has made "Improving risk management" one of its strategic objectives of the IST program. The integrated project Orchestra is one of the projects that recently started in this area. The main goal of Orchestra is to design and implement an open service oriented software architecture that will improve the interoperability among actors involved in multi-risk management.

In order to realize this goal, the key objectives for the project are the following:

- To design an open service-oriented architecture for risk management that links spatial and non-spatial information services. In this context Orchestra will provide input to INSPIRE and GMES (see below).
- To develop the service infrastructure for deploying risk management services.

- To develop thematic services that are useful for various multi-risk management applications based on the architecture.
- To validate the Orchestra architecture and thematic services in a multi-risk scenario.
- To provide software standards for risk management applications. In particular, the de facto standard of OGC and the standards of ISO and CEN are envisaged to be influenced.

The Orchestra project started in September 2004. Currently the focus of the work is on understanding user needs, system requirements and an assessment of useful technologies. This is considered the necessary for input for design decisions for the Orchestra architecture. In the following we will first explain the process used to create the architecture, the so-called Orchestra Reference Model.

3 The Orchestra Reference Model

The architectural process of Orchestra is based on the principles of the following international standards:

- The Reference Model for Open Distributed Processing (ISO/IEC 10746 RM-ODP) is used for the structuring of ideas and documentation.
- The OpenGIS Service Architecture (especially ISO/DIS 19119) is used for the taxonomy of the Orchestra services.

3.1 Mapping of RM-ODP Viewpoints

RM-ODP is an international standard for creating open, distributed processing systems. It provides an overall conceptual framework for building distributed systems in an incremental manner. The Orchestra architectural process uses the RM-ODP viewpoints for the structuring of ideas and their documentation. The mapping of the viewpoints to Orchestra is indicated in **Table 1**. As the Orchestra deployment will have the nature of a loosely-coupled distributed system based on operational services rather than a distributed application based on computational objects, in Orchestra the "computational viewpoint" is referred to as the "service viewpoint".

Viewpoints	Mapping to Orchestra	Usage example
Enterprise	Reflects the analysis phase in terms of the system and the user requirements as well as the technology assessment	Use case description of a geo-processing service
Information	Covers the conceptual model of all kinds of information with their thematic, spatial, temporal characteristics as well as their meta-data.	UML class diagram defining the information elements that are used by the geo-processing service
Computational (referred to as Service Viewpoint)	Covers the Orchestra services that enable syntactical and semantic interoperability and administration across system boundaries	UML specification of the geo-processing service
Engineering	Covers the mapping of the Orchestra service specifications to the chosen service infrastructure	Mapping of the UML specification to WSDL
Technology	Covers the technological choices of the service infrastructure and the operational issues of the infrastructure.	Usage of W3C Web Services and UDDI

Table 1. Mapping of the RM-ODP Viewpoints to Orchestra

The Orchestra Reference Model covers all five viewpoints in the following manner:

- The analysis phase is described as part of the Enterprise Viewpoint.
- The design phase encompasses the harmonized specification of the Information and Service viewpoint resulting from requirements of the Enterprise viewpoint. The result is the Orchestra architecture that is, by definition, a platform-neutral specification according to the requirements of ISO/DIS 19119 (i.e. specification in UML).
- The Orchestra architecture does not cover the Engineering and Technology viewpoints.
- The aspects of the Engineering and Technology viewpoints are combined in one or more process steps. Each step represents one mapping to a specific service infrastructure (e.g. W3C Web Services) and leads to a platform-specific Orchestra Implementation Specification.

3.2 Compliance with the OpenGIS Service Architecture

The Orchestra architecture is a "simple service architecture" and shall use the service taxonomy of the Open GIS Architecture (ISO/DIS 19119). Thus, Orchestra will provide human interaction services (e.g. catalogue or map viewers), model/information management services (e.g. feature and map access services, query support services), workflow/task services (e.g. service chaining support), processing services (e.g. statistical calculation), communication services (e.g. web services) and system management services (e.g. authorization support).

3.3 Analysis and Design Process

The Orchestra architecture is being designed in an iterative way recognizing the fact that both the requirements of the system and end users and the technological progress in the IT market and in IT standardization have a dynamic nature and cannot be completely caught in a one-shot design. Thus, a global iteration cycle between the analysis and the design phase of the architecture is foreseen (see Fig. 1).

A **consolidation process** in-between ensures that, at a defined point in time, there is a common understanding of the system requirements, the user requirements and an assessment of the current technology as a foundation to design the Orchestra architecture.

System requirements encompass all aspects that today prevent interoperability between systems. They are expressed in terms of architectural properties that a system should follow in order to improve the exchange, sharing and using of information across system boundaries.

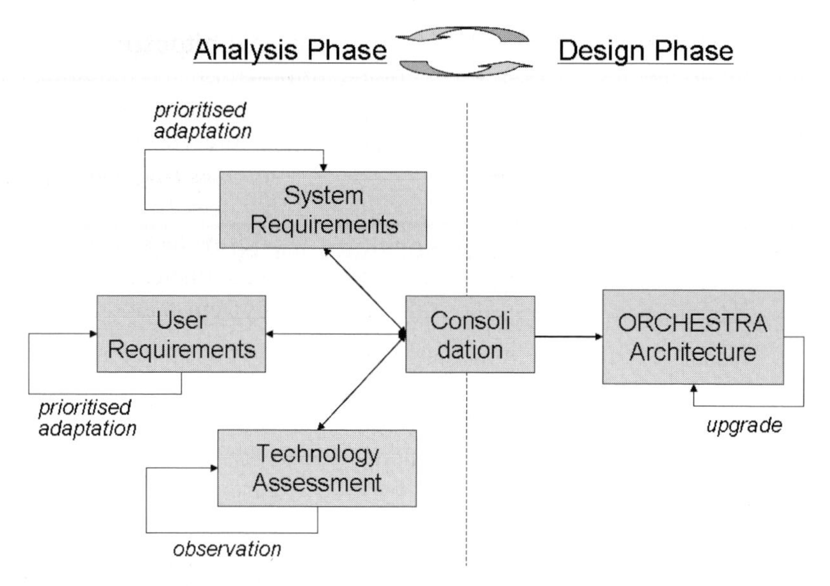

Fig. 1. Dynamic Orchestra Architectural Process

User requirements, on the one hand, represent the view of the end user of the Orchestra system that is being specified independent of the technological foundation of the system. It is expressed in terms of end-user services, information presentation and availability requirements and non-functional aspects such as trustworthiness of information, response time or quality of service. On the other hand, they comprise the view of the system engineer that will build thematic services (e.g. for flood or fire forecast). Here, user requirements are expressed in terms of the domain-specific information model (ontology) that has to be supported by the principal information management capabilities of the Orchestra platform.

Both the system and the user requirements are dynamic in the sense that they will be prioritized and adapted in local iteration cycles due to the result of the consolidation process.

Technology assessment is also a continuous process. Orchestra aims at building the architecture on top of technologies, tools and products that are either standard approaches or have proven to be successful in solving interoperability problems in deployed use-cases.

The dynamic nature of the input factors of the Orchestra architecture naturally leads to an iterative architectural design process. Various but controlled upgrades of the Orchestra architecture will be required to adapt the architecture to the changing needs.

In the following two sections of this paper we will highlight two important technologies for Orchestra: geographic information and ontologies for improvement of semantic interoperability. We will explain what Orchestra will contribute to the state of the art in these fields.

4 Geographic Information Services for Risk Management

Geographic information is involved in all phases of risk management processes - from prevention to immediate reaction. The geospatial aspects may be explicit (e.g. topographic maps providing background information) or hidden (tables about population distribution in an affected area). In the same way either dedicated tools are used to analyze or incorporate geospatial aspects (e.g. the usage of a GIS by a GI expert) or the information is integrated via interoperable GI components or GI services in a specific risk management application. Orchestra focuses on the latter case.

The transition from closed and monolithic GIS to open and interoperable GIS introduces a paradigm shift in the development of GIS-standards. Standards do not any more focus on formats for the file exchange of geodata but are directed to the specification of service interfaces. These services provide geo-information that are processed according the user's query and their specific needs, e.g. a specific part of the cadastre, a fastest route from *A* to *B*, a thematic map on the current ozone pollution etc.

4.1 Interoperable GI Services

On-going activities to realize interoperable GI services currently mainly focus on encoding and accessing geographic information in an interoperable manner. The existing (partly draft) ISO standards and OGC specifications cover encoding and accessing geographic information to achieve interoperable GI services. Thus following the OGC service taxonomy this work falls mainly into the category model/information management services. However, specifications to define interoperable GI services to process spatio-temporal information (processing services) need still to be researched and there is an urgent need to define the respective specifications to support risk management applications as intended by Orchestra. Examples of functionalities that need to be covered by these geo-processing services are:

- Weighted combination of different geo-information layers to support assessment and spatial decisions.

- Spatial and topological operators (buffer, generalization, etc.).
- Feature extraction services, e.g. to operate on remote sensing data.
- Geostatistical operators to extract key values or aggregate spatial information.

Moreover Orchestra will investigate on interoperability of GI services with spatio-temporal simulation models to ease the incorporation of e.g. flooding or weather forecast simulations in risk management applications.

Interoperable GI services are one cornerstone for the successful implementation of Spatial Data Infrastructures (SDI). SDI evolve all over Europe on institutional, regional and national levels. In 2001 the European Commission started the INSPIRE initiative (INfrastructure for SPatial InfoRmation in Europe; http://inspire.jrc.it/home.html) to streamline national SDI developments with respect on supporting European environmental policies. The initiative intends to trigger the creation of a European Spatial Data Infrastructure (ESDI) that delivers to the users integrated geospatial information services from various distributed sources, crossing institutional and political boundaries. The target users of INSPIRE include policy-makers, planners and managers at European, national and local level and the citizens and their organizations.

4.2 European Directive on Infrastructure for Spatial Information in Europe (INSPIRE)

Following 3 years of intensive collaboration with Member States experts and stakeholder consultation, the European Commission has adopted on the July 2004 a proposal for a Directive establishing an infrastructure for spatial information in the Community (INSPIRE) (COM(2004) 516 final).

The adoption of the proposal marks an important step on the way forward to a European-wide legislative framework that helps in achieving an ESDI. This proposal does not only address policy related issues concerning the development of an ESDI but also dedicates three chapters to the technical requirements that have to be fulfilled by the member states to establish the ESDI. These three chapters are on Metadata, Interoperability of spatial data sets and services, and Network services. Under these chapters the proposal list general requirements on these issues as well as it formulates the requirement to adopt appropriate implementing rules.

During the INSPIRE preparatory phase (2005-2006) the ORCHESTRA project will provide input towards the drafting as well as the piloting of the INSPIRE implementing rules in the risk management domain. The first input can be expected on the topic of the INSPIRE network services.

The establishment of an ESDI will represent significant added value for - and will also benefit from - other Community initiatives such as Council Regulation (EC) No 876/2002 of 21 May 2002 setting up the Galileo Joint Undertaking (OJ L 138, 28.5.2002) and Global Monitoring for Environment and Security (GMES): Establishing a GMES capacity by 2008 (COM(2004) 65 final). In order to exploit the synergies between these initiatives, Member States should consider using the data and services resulting from Galileo and GMES as they become available, in particular those related to the time and space references from Galileo.

5 The Use of Ontologies for Improved Semantic Interoperability

If there are many distributed and heterogeneous sources of information and application services to be used then the key is interoperability between them. This needs to work on at least three levels; the syntax level, the structural level and the semantic level. To date most standard work has focused on the syntax of communicating between disparate software services, for example the Web Map Service standard of the OGC, which specifies the syntax for communicating between the mapping client and the service. Work has also concentrated on the structure of the data used in those interactions such as Simple Feature Specification of the OGC, which allows structured data to be shared between systems. These two aspects make the physical connectivity and exchange of data possible and enable a distributed architecture to be realized.

However the content of the information in the distributed systems being connected may be referring to substantially different things and consequently not interoperate at the level of its semantics. For example in the risk management domain the term 'bank' in different data sources could mean 'a steep natural incline' or 'a business establishment in which money is kept'. One set of data is clearly suitable for use in a flood modeling application and the other is not! This is further complicated in multiple risk management scenarios where the information with different terminology from multiple information communities must be interpreted together.

In order to resolve this semantic interoperability when sharing data between systems the meaning of the terms must be stated so that the computer can resolve the differences. This can be done in ontology, which in a computer science setting is a "specification of a conceptualization". It formally describes a set of concepts and the relationships that hold between them in a given context in a logical manner so the semantics can be interpreted by machines. The W3C have defined the OWL (Web Ontology Language) standard for the representation of ontologies to describe the semantics of disparate resources on the network.

The approach in Orchestra will be to define ontologies for common domains of risk management such as flooding or forest fire and the common concepts of risk management that bind these domains. The rich high-level formal semantics within these ontologies should be capable of representing the abstract concepts and relations such as the causes, propagators and effects of risks. The simpler semantics of individual data sources and applications are represented in their own ontologies and are then mapped to the rich high level semantics of these ontologies This allows heterogeneous and abstract semantics of different data sources and applications to be interpreted through the richer semantics of the domain. By having a common semantics to which data and services are ultimately mapped it improves the selection and comparison of data sources based on their content and it becomes possible for the machine to infer whether two data sources are comparable or suitable for use in an application service. The formal representation of the common concepts of risk management also provide a common semantics to support cross domain query in multi risk management. For example the data about the results of forest fire can act as an input to understanding the increased risk to flooding due to mass vegetation changes.

To allow the ontologies to be used to bind together the different data sources and services of the Orchestra architecture they need to be mapped to the common meta information model and be processed by the core services. The project will define the mapping of the ontologies to the meta information model of the Orchestra architecture to which the different services and data sources will also be referenced. This common structure for using meta information will be used to combine the different services by providing the relationship to the syntax and structural description of information. Semantic web-services will need to be incorporated into the architecture as part of the core services to expose and manipulate the ontologies. These at a minimum include ontology management and storage services, inference services to infer the semantic equivalence between different data sources and services and query brokering services. These are required to transform the semantics into syntax and structural connectivity to enable access to the different data sources.

Through the meta information model and the services of the Orchestra architecture Orchestra will seek to provide a framework in which semantic interoperability can start to be achieved in addition to and integrated with the syntactical and structural interoperability. The key benefits to using formal and explicit semantics lie in the area of selecting and combining disparate information from a variety of sources to be used for a particular risk management task. With a functional semantic layer the differing uses of terminology across different risk management systems is no longer a barrier to the productive sharing of information and approaches. The appropriate use of data for a given application service or function can also be matched to ensure the end results are consistent. In an open architecture this allows multiple organization to share information and collaborate as the formal semantic enables plurality of views particularly in multi-risk scenarios where different information communities must work together.

6 Examples of the Use of Orchestra

The Orchestra architecture will not take the form of a ready to use application for risk management. To the contrary, it should be considered as a collection of services, tools and methodologies that so-called system users can use to develop risk management applications for end users. It is our vision that chains of cooperating services will become reality and that system users can add value to already existing services. They can for instance combine services for one risk with services for another risk, and that way develop a multi-risk management application. There will be less need for reinvention of the wheel, and more effort can be spend on creative solutions for specific problems.

The Orchestra architecture will be generic in the sense that it should provide basic functionality that is useful in applications for any kind of risk (natural or technological), also in cross border situations. Based on this architecture, the project will develop a few thematic services that are useful in various risk management applications, such as a cadastre service, a weather service, etc.

A large part of the project will be dedicated to the validation of the Orchestra architecture and thematic services. We will now describe two possible scenarios for the use of Orchestra. The first one is related to coastal zone management, and the second one to earth quake risk management.

6.1 An Example of the Use of Orchestra for Risk Management at Coastal Zone

The study of the environmental risks which may be generated from maritime transport activity in coastal European waters requires the establishment of advanced modeling techniques and evaluation benchmarks. The support of a generic evaluation of the environmental risks within an open information system infrastructure is key for improving the risk management methodology.

In Orchestra, various numerical modeling applications could be deployed to assess the environmental risks induced by ship traffic activity at coastal zones. This can be efficiently achieved through a generic sharing of common toxicity, traffic networks, coastal zone environmental databases and numerical modeling kernels. The generic sharing of data information is formalized through advanced knowledge modeling at coastal zone, including standard data and meta-data access methods. Maritime transport activity could induce the multiple risk of introducing anti-foulants, ballast waters, oil and chemical spills, and atmospheric emissions into coastal waters and coastal zones. Ship traffic spatial networks in European coastal waters can be geographically identified, updated in time and shared within common databases (Figure 2). The generic access to such geospatial information within Orchestra will lead to rapid risk management of ship traffic in European coastal waters.

The engendered multiple risks can be predicted through multi-modeling the fate of anti-foultants such as TBT, alien organisms in ballast waters and SO_x/NO_x gases in the atmosphere following on ship traffic network within a region ship traffic network of interest. The various discharges of ballast waters, anti-foulants and gas emissions could lead to quantifiable risks of multiple orders around coastal zone environments, including health risks. For instance, TBT anti-foulants exposure may lead to marine species genetic disorders; ballast waters could introduce alien marine organisms and diseases and; gas emissions cause poor air quality and respiratory problems around the same zone of interest. Oil spills could not only damage marine organisms, but also damage commercial activities such as fishing and tourism, and hence contribute to raising unemployment levels in coastal working communities.

The above mentioned risks could themselves induce secondary risks on local economies at coastal zone, i.e. fishing and leisure industries, and may contribute to unemployment levels within coastal working communities.

The various types of risks are established as probability of occurrences of the so-called Predicted Environmental Concentration (PEC) and compared to critical exposures; the so-called No Effect Concentration (NEC) value. Spatial hazard maps can therefore be predicted within coastal zone and updated according to changes in environmental conditions and ship traffic activity with time (Figure 3). Hazards maps will lead to the dynamic evaluation of multi-risks and their probability of exceeding with respect to EU and international statutory criteria. These are with respect to air and water quality standards but also ecosystem toxicity thresholds. The economic indicator thresholds are much more challenging during the evaluation exercise but can still be implemented as additional information layers within the Orchestra model applications if historical economic indicators are available for the regions of study.

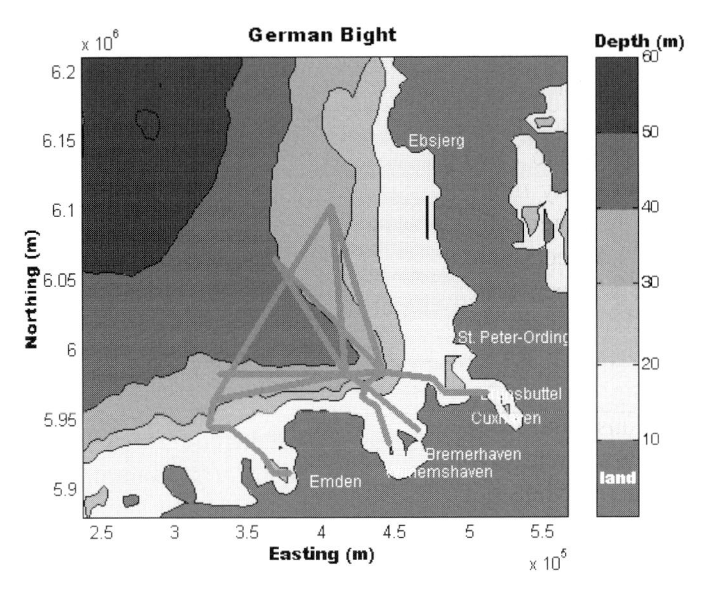

Fig. 2. Illustrative ship traffic network in the German Bight

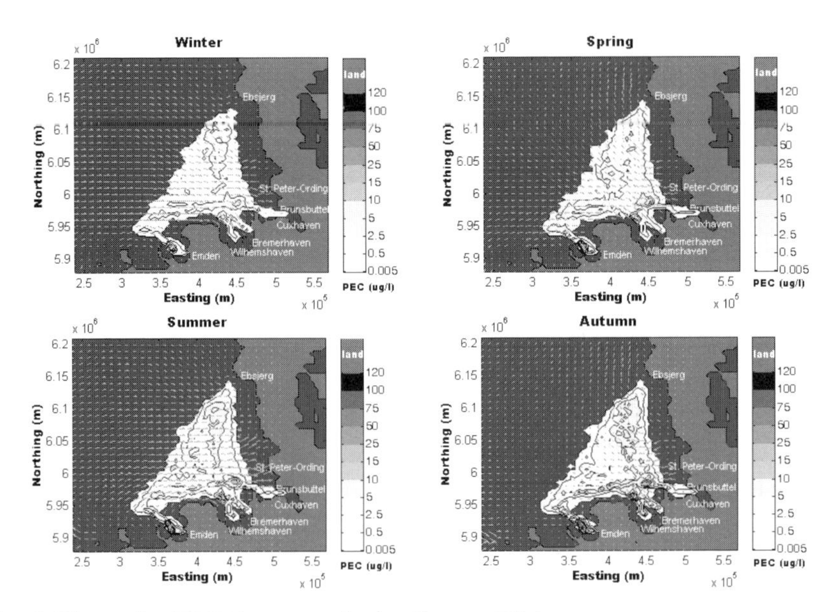

Fig. 3. Illustrative TBT Exposures in the German Bight

6.2 An example of the Use of Orchestra for Earthquake Risk Management

Earthquakes can occur in border regions (both national and international), e.g. the 1976 Friuli earthquake sequence that caused damage in Italy and in Slovenia. They can induce other potentially damaging events, such as landslides (e.g. the 2001 El Salvador earthquake where the biggest single lost of life was caused by a landslide triggered by the earthquake) or fires (e.g. the 1923 Kanto Plain earthquake where the fires following the earthquake caused many more causalities than the actual earthquake shaking). Because of the potentially catastrophic nature of large earthquakes, which can affect all aspects of society, their correct management relies on close cooperation between different organizations at different levels. For these reasons earthquake risk management is a good example of the type of risk that is of particular interest to Orchestra. Earthquake risk management is an interesting contrast to coastal zone management because earthquakes are a rapid-onset hazard (the strong shaking that causes damage during an earthquake usually lasts for less than one minute), although the effects can last for decades, whereas coastal zone hazards are usually slow-onset hazards. In addition, the main technique for managing earthquake risk is to reduce the vulnerability of the elements at risk by using better construction techniques whereas coastal zone risk management also includes efforts to reduce the actual hazard (a step that is impossible for earthquakes). For these and other reasons it is a big challenge to develop an architecture that is appropriate for both these risks (and others).

Orchestra will help to improve the management of earthquake risk in Europe by facilitating easier data exchange between involved parties. For example, the estimated level of seismic hazard often does not exactly match at national and international borders due to differences in input data and methodology, whereas in reality earthquakes do not respect human borders. Similar problems can occur for building vulnerability assessments in different countries due to differences in the way this is performed and because of the difficulty in accessing vital information. Therefore the assessed earthquake risk across borders is difficult to compare and consequently it is difficult assign priorities to its management. This leads to a non-optimal allocation of resources and therefore waste and higher real risks in some regions than expected.

7 Concluding Remarks

The Orchestra project (www.eu-orchestra.org) started in September 2004 and will run until August 2007. The following organizations are involved in the project:

- Atos Origin, Spain
- European Commission – DG Joint Research Centre, Italy
- Hochschule fuer Technik und Wirtschaft des Saarlandes, Germany
- Open Geospatial Consortium (Europe) Limited, United Kingdom
- BRGM, France
- Ordnance Survey, United Kingdom
- Fraunhofer IITB, Germany
- ARC Seibersdorf research GmbH, Austria
- Eidgenoessische Technische Hochschule Zuerich, Switzerland
- Intecs, Italy
- DATAMAT S.p.A., Italy
- TYPSA, Spain
- BMT Cordah Limited, United Kingdom
- The Alliance of Maritime Regional Interests in Europe, Belgium

The project will work together closely with two other Integrated Projects in the field called WIN (http://www.win-eu.org) and OASIS (www.oasis-fp6.org). WIN will concentrate more on organizational issues relevant for improved interoperability in risk management and OASIS focuses on crisis management. The three projects will use the same architectural principles and make their results interoperable. These results will be provided as input to INSPIRE and to GMES (Global Monitoring for Environment and Security, www.gmes.info).

References

Craglia M, Annoni A, Smith RS, Smits P (2002) Spatial Data Infrastructures: Country Reports- GINIE – EUR 20428 EN http://wwwlmu.jrc.it/ginie/doc/SDI_final_en.pdf

Craglia M, Annoni A, Klopfer M, Corbin C, Pichler G, Smits P (2003) GINIE Book: Geographic Information in the Wider Europe. GINIE http://wwwlmu.jrc.it/ginie/doc/ginie_book.pdf

European Commission (2003) "INSPIRE State of Play Reports- Spatial Data Infrastructures in Europe : State of play 2002", http://inspire.jrc.it/state_of_play.cfm

European Commission (2003) “INSPIRE State of Play Reports- Spatial Data Infrastructures in Europe : State of play 2002”, http://inspire.jrc.it/state_of_play.cfm

European Commission (2004) Proposal for a Directive “establishing an infrastructure for spatial information in the Community (INSPIRE)” COM(2004) 516 final. (http://inspire.jrc.it)

Nebert D (ed.) (2001)“Developing Spatial Data Infrastructures: The SDI Cookbook, Version 1.1”, Global Spatial Data Infrastructure, Technical Committee

Smits P (ed) (2002) INSPIRE Architecture and Standards Position Paper. (http://inspire.jrc.it)

Laser Scanning Applications on Disaster Management

Andrea Biasion, Leandro Bornaz and Fulvio Rinaudo

Politecnico di Torino - Dipartimento di Georisorse e Territorio,
C.so Duca degli Abruzzi, 24, 10129 Torino, Italy.
Email: andrea.biasion@polito.it; leandro.bornaz@polito.it;
fulvio.rinaudo@polito.it

Abstract

The recent upgrading of laser scanning devices has led to a set of new surveying techniques for civil engineering and environmental analysis.

The terrestrial laser scanner allows complete and dense 3D digital models of the surface of any object to be reconstructed. This is very useful for natural hazards and risk assessment where morphological investigation is a starting point to evaluate stability properties. In the case of disaster management, a 3D model is useful to acquire and monitor emergency situations. Field acquisitions, with laser scanner devices are, in addition, very fast and they are made in safe conditions.

However, particular attention must be paid during the acquisition, processing and modeling phases.

- LIDAR data often include elevated noise (usually gross errors and outliers) that has to be removed with opportune techniques before starting the 3D registration and modeling.
- In addition, 3D contexts usually have very complex shapes that cannot be recorded with sufficient resolution with a single scan. For this reason, two or more scans must be taken from different points of view of the same object, in order to eliminate shaded areas.
- To obtain the final 3D model of the object it is therefore necessary to align and geo-reference the single scans using suitable registration techniques.
- The integration of DDSM geometric data (e.g. the point cloud acquired by using the laser scanner) with image radiometric data allows a new concept of the Solid Image to be obtained.

- Solid images can be used in many surveying applications to determine, in real time, the position of any point in a 3D reference system, using a normal PC or to carry out correct three-dimensional measurements (lines, areas, volumes, angles, etc.), by just selecting some points on the image.
- In addition, with the Solid Image, even unskilled users can easily plot profiles, sections and plans using simple drawing functions, and can generate stereo models and realistic 3D models.

An application example of these new surveying techniques in disaster management is shown.

1 Laser Scanner Instruments

In recent years a new category of instruments has been introduced in the field of surveying. These instruments can acquire portions of land and objects of various shapes and sizes in a quick, cheap and safe way. These instruments, based on laser technology, are commonly known as terrestrial laser scanners.

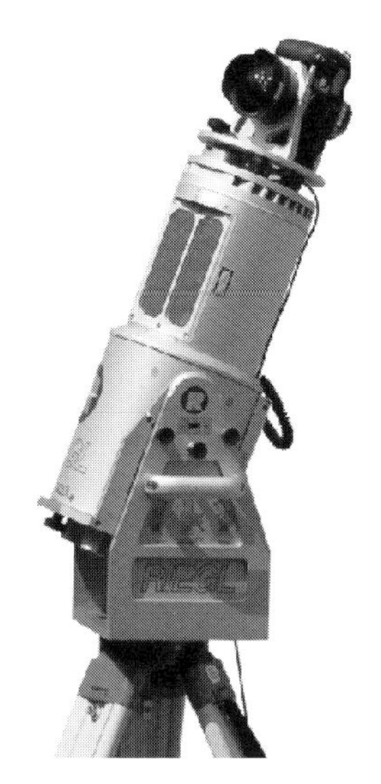

Fig. 1. Laser scanner RIEGL LMZ 420i

While laser scanner instruments based on the triangulation principle and with high degrees of precision (less than 1 mm) have been widely used since '80s, TOF (Time Of Flight) instruments have only been developed for metric survey applications in the last 5 years. In this paper we consider these kind of instruments.

TOF technology based laser scanners can be considered as highly automated topographic total stations. They are usually made up of a laser and of a set of mechanisms that allow the laser beam to be directed in space, according to the object that is being surveyed.

The distance between the centre of the instrument and a generic point is measured, on a known direction, therefore the X, Y and Z coordinates of the measured point can be computed for each recorded distance-direction. Laser scanners allow millions of points to be recorded in a few minutes.

Thanks to these characteristics, laser scanning can be used in a variety of fields:

- Topography and Mine surveys;
- Architecture and building surveys;
- Archaeology and cultural heritage surveys;
- Monitoring and Civil engineering;
- City modelling;
- Tunnel surveys;
- Virtual reality.

As far as disaster management is concerned, a large portion of natural objects can be acquired in various situations. Compared with other "classical" survey techniques, such as topographic or photogrammetric techniques, laser scanner instruments offer considerable advantages, as will be shown in the paper.

2 Laser Scanner Survey

Acquisition is the first step in laser scanner survey techniques. To obtain a valuable final result, this phase should be correctly planned and executed. As acquisition is the only action carried out on the field, a real time data check is suggested.

Compared with other survey techniques, such as topography or photogrammetry, the laser scanner survey offers some advantage, such as:

- short survey sessions in which it is possible to record a great number of points;
- the data treatments and elaboration phases can be conducted in separate places and at different times, after the survey;

- the availability of new informatics instruments allows a great deal of information about the surveyed object to be extracted quite simply;
- night acquisitions can be conducted (laser beams do not require light to be reflected);
- unmanned survey stations can be installed for multi-temporal acquisitions.

When the object has a complex shape or when a single scan cannot record the whole object, a series of scans must be performed. This series has to be correctly planned to avoid hidden areas.

As previously mentioned, terrestrial laser scanners can be considered as highly automatic motorized total stations.

While with total stations the operator directly chooses the points to be surveyed, laser scanners randomly acquire a dense cloud of points. The operator only selects the portion of the object he would like to acquire and the density of the points he desires in the scan (either the angular step of the scan in vertical and horizontal planes or the average metric step on the object). Once these initial values have been chosen, the acquisition is completely automatic.

3 The Treatment of Laser Scanner Data

3.1 Laser Scanner Data

The result of a laser survey is a very dense point cloud (also called DDSM – Dense Digital Surface Model). For each point of the model the spherical coordinates d, α, β (distance, zenith and azimuth angle) and sometimes the reflectivity values are first obtained, then the Cartesian coordinates X, Y, and Z.

As this set of points is acquired in a completely arbitrary way, while certain parameters are selected by the operator, it is necessary to manage this data in a critical and reasonable way. Particular attention must be paid to the quality of the original data.

The precision in the determination of spherical coordinates (used by the instruments), is almost the same for each scanner model. The accuracy of the distance measurement ranges from 5 mm to 25 mm, according to the laser class used.

3.2 Laser Scanner Data Treatment

Laser scanner data treatment consists of a set of actions that are necessary to obtain the correct digital model of an object, starting from a set of point clouds.

This set of actions can be divided into 2 different steps:

1. the pre-treatment (or preliminary treatment) of the laser data;
2. the final product creation.

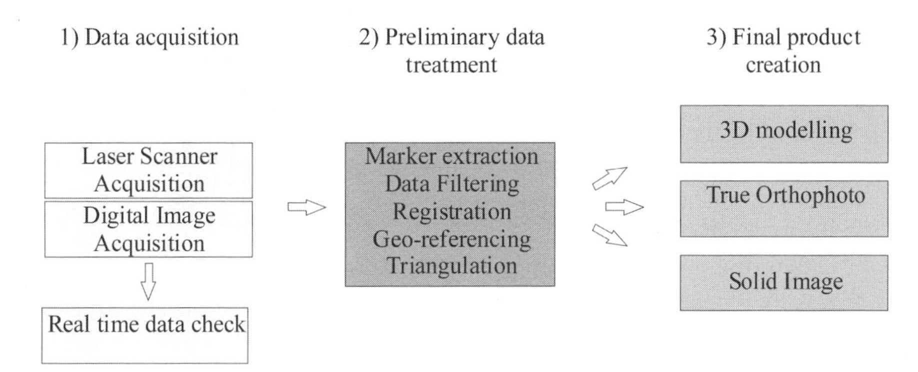

Fig. 2. Laser scanner data workflow

By "preliminary data treatment" we mean all the operations that are directly carried out on the point cloud, such as, for example, data filtering (noise reduction), point clouds registration, geo-referencing and multiple scans triangulation operations.

These operations include:

1. automatic high reflectivity point collection;
2. data noise reduction;
3. point cloud alignment and/or georeferencing;
4. laser scanning triangulation.

The result of these procedures is a complex, "noise free" point cloud (without any outliers, gross or systematic errors) and this is the correct starting point for the second stage of the laser data treatment: 3D modeling, orthoprojection and solid image generation.

3D modeling consists of a set of operations that, starting from any point cloud, allows a surface model of the object to be formed.

While there is a huge range of different products on the market to carry out solid modelling, just a few software packages for correct preliminary treatment of terrestrial laser scanner data can be found.

The true orthophoto and the solid image are new products in the surveying field that can be made by integrating a 3D laser scanner digital model with classical photogrammetric information.

3.3 Automatic High Reflectivity Point Collection

Modern laser scanner devices record, for each acquired point, the direction of the laser beam (horizontal and vertical angles), the measured distance and the reflectivity values (the energy reflected by the measured point). This set of information makes it possible to calculate the 3D coordinates of each point using simple geometric equations.

The reflectivity value is connected to the type of material that makes up the object, an aspect that can be of fundamental importance in the analysis and development of automatic algorithms or in a preliminary classification of the materials.

One of the possibilities offered by the knowledge of this set of information is the opportunity of automatically registering (or georeferencing) two adjacent point clouds. To do this, it is sufficient to arrange some high reflectivity stickers (markers) on the object during the scan.

When the laser beam strikes the markers, the recorded reflectivity value is very high and is usually much higher than the one recorded on natural points (for example, rock, wood etc). If this simple property is used, it is possible to automatically identify the position of the markers inside the 3D model acquired with the laser scanner.

The markers should however be suitably sized and arranged on the object if a correct determination of their position have to be obtained. Their position and size have to be planned considering the laser-object distance, the used angular resolution and the average inclination of the measuring directions.

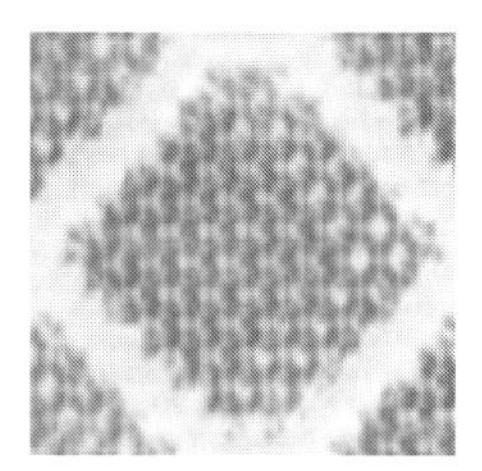

Fig. 3. The markers

3.4 Data Noise Reduction

One of the fundamental preliminary operations of the treatment of terrestrial laser data is data filtering. In fact, the data acquired by laser scanner devices always have noise that is lower than the tolerance of the used instruments.

Noisy data prevent a correct interpretation of the details of an object from being made. In order to obtain a "noise free" model of the object, it is necessary to use specific algorithms that are able to reduce or eliminate, as much as possible, the acquisition errors that can be found in the point clouds. Many algorithms have been developed by the authors in recent years in order to reduce the noise (e.g. the robust median estimator).

The implemented algorithms are also able to remove any scattered points that do not belong to the object (vegetation, cars in movement, people, urban furnishings etc.).

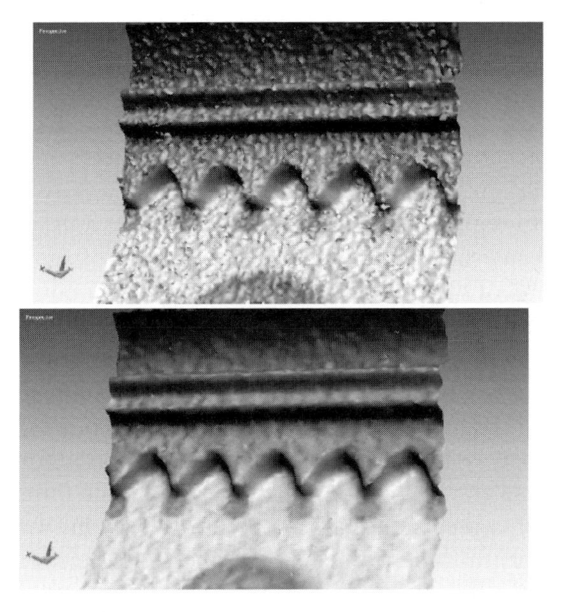

Fig. 4. 3D model of a noisy and a filtered point cloud

This phase introduces a residual error in the point cloud, due to the point position changes introduced by the algorithms. This error has been estimated by the authors, according to experiences acquired over the years, to be about 10 mm.

3.5 Point Cloud Alignment and/or Georeferencing

In the previously mentioned case, when a single scan is not enough to completely describe the object, more than one scan has to be made. In these cases each scan has its own reference system: the reconstruction of the 3D model of the surveyed object requires the registration of the scans in a single (local or global) reference system. This phase can be performed in an interactive way through the identification of the homologous points inside the overlapping portion of two adjacent scans.

Once the homologous points (at least three) have been collected, a simple 6 parameter transformation (3 rotations and 3 translations) can be estimated and all the points of a scan can be moved into the reference system of a scan that has been assumed as the dominant.

If this simple approach is followed, all the scans which describe an object can be made to refer to a unique coordinate system (e.g. the first scan coordinate system or an external system defined by at least three points).

3.6 Laser Scanner Triangulation

Very large objects and/or complex shapes of objects require multi-scan registration.

If one tries to register all the scans using the previously described algorithms, unexpected and unacceptable deformation of the final 3D model occurs at the end of the process. If 8 scans are simply connected with 30% overlapping over an area façade of 40 m, the discrepancies tested on a set of check points will be of about 2 m! The registration of a long series has the same problems which occur during the photogrammetric triangulation of a single strip. In these cases, a series of control points, referring to an external reference system, must be provided (using a minimum overlap of at least 30%. One control point must be provided each three adjacent scans in order to reach a precision that is compatible with the range accuracy of the laser scanner). In order to reach the expected precision, a rigorous adjustment must be performed. Each scan is considered as a separate object: tie points (homologous unknown points) and control points (homologous known points) are considered with different weights (w = 1 for the tie points and w = 2 for the control points).

All the fitting parameters can be estimated and all the points of the scans transformed into the external reference system.

3.7 Management of the Digital Images and Creation of the Solid Image

A digital photo image can be considered, with a good approximation, as a central perspective of an acquired object. If the internal and external orientation parameters of the camera are known, it is possible to establish the direction in the space of each point of the object which is represented by a pixel in the image.

Using a dense digital surface model (DDSM) of the acquired object, each pixel (and therefore each direction in the space) can be associated to the distance value between the perspective centre and the point of the object which is represented by the pixel itself. In this way, each pixel can represent the 3D position of the corresponding point of the object in an appropriate reference system

The integration of the DDSM geometric data with the radiometric data of the image allows the new concept of the solid image to be obtained [Dequal, 2003].

Solid images can be used in architectural, archaeological and land surveying applications to determine, in real time, the position of any point in a 3D reference system, using a common PC or to carry out correct 3D measurements (lines, areas, volumes, angles, etc.) by just selecting some points on the image.

Oriented images and the related DDSM allow one to automatically produce artificial stereoscopic pairs and 3D color model.

4 An Application: The Trappistes Falaise

In the morning of November 29 2003, a 600 m^3 rock fall partially destroyed the Trappistes tunnel (Vallais region, near Martigny, in Switzerland) that was originally built to protect the road against landslides, and one person was killed. The installation of a sophisticated monitoring system, named GUARDAVAL allowed the road to be quickly opened again. At the same time, geologists were asked to evaluate the risk of other similar massive rock falls occurring in the area. It was clearly shown that similar massive events were very unlikely to occur again, but the detachment of small rock portions had to be prevented. Among the various interventions that were chosen, one was the reinforcement of the stability of those portions through anchor gears.

The understanding of the spatial organization of geological structures is a key approach in the analysis of rock fall risks and their mitigation. As on-site field investigation is still the basic way of providing the necessary data to describe the discontinuity pattern, classical tools, such as stereo-pairs are often not sufficient to appreciate the propagation of fractures in a rock mass, or how they intersect with each other, generating blocks that can be unstable.

Fig. 5. The Trappistes falaise after the rockslide

Today, 3D modeling computer software allows complex surfaces to be built, their intersections to be computed and realistic 3D volume models to be produced.

The building and analysis of a 3D geological model has been instrumental in helping the geologists evaluate the possibility of another massive event occurring, the relative instability of the rock slabs in the detachment niche and the positioning of the anchors.

A collaboration between the Politecnico di Torino – DIGET, and CREALP (Centre de Recherche sur l'Environnement ALPin) has led to an high resolution laser scanning of the cliff being obtained. A 3D topographic surface was built using 1.4 million points, which offers sufficient accuracy for the main fractures to be seen and evaluated.

There are several advantages, in practical terms, and concerning the safety of the personnel:

- classical geomechanical surveys made with geological compasses are carried out by specialized personnel, with alpinist experience, who run certain risks when working on the instable walls;

- laser scanner productivity is higher than with any other traditional survey instrument (it can acquire millions of points in a short time), throughout the whole day (night acquisitions are possible);
- when necessary, a permanent station can be placed for multi-temporal continuous acquisitions;
- a brief survey is necessary, but most of the elaboration of the data can be done in a safe place;
- the laser scanner data (millions of point clouds) give a complete description of the object, compared with traditional survey instruments, that can reproduce only certain particulars.

The principal purposes of the laser scanner survey were:

- to help the geologists evaluate whether another massive event would occur, the relative instability of the rock portions in the detachment niche and the positioning of the anchors;
- to verify multi-temporal capability of rockslide movements with actual laser scanner technologies;
- solid Image application for geomechanics parameter determination (DIP and DIP direction).

4.1 The Survey

The survey was conducted on the side of the valley opposite the rockslide, because of the limited width of the valley. Laser scans were taken from three different points (shown in blue – fig. 6), to avoid hidden areas, for a complete description of the wall.

Four points are shown in red where high – reflectivity markers were placed. The coordinates of these points, that have been placed in stable places, have been acquired with GPS instruments, to create a reference system correctly oriented to the North. The other high – reflective markers placed on the wall and used only for the registration of the three scans, and for local movement control are shown in green. To evaluate the multi-temporal movements, the survey was carried out twice. The instrument used is a RIEGL LMS-Z420 laser scanner with the following characteristics:

- laser class 1 (safe for the human eye), and class 3R (more power for a longer range);
- maximum acquisition range: 250 m for class 1 and 1000 m for class 3R;
- distance measurement accuracy: ±5 mm (class 1), ±15 mm (class 3R);
- acquisition speed: 6600 – 10000 points/sec.;
- acquisition window: 360° (horizontal) x 80° (vertical);

– angular resolution: 10 mgon.

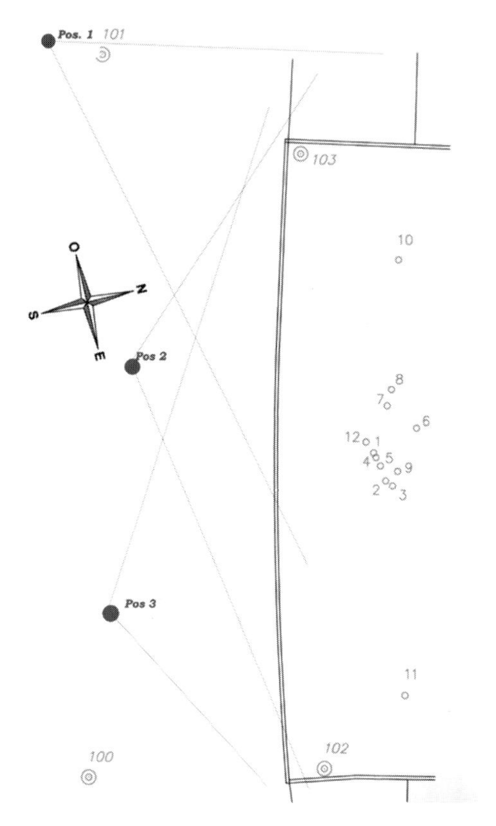

Fig. 6. Survey plan

4.2 Data Treatment

The previously described data treatment operations have been performed by using a software developed by the authors called LSR 2004. The same software allows the production and the management of Solid Images.

The 3D multi-temporal modelling was performed by using commercial software, and was used to:

– evaluate any variation in volume of the rock mass;
– obtain information about the exposure and slope orientation and about the principal joint family. It's also possible to extract information from the 3D model, such as contour lines, sections and profiles.

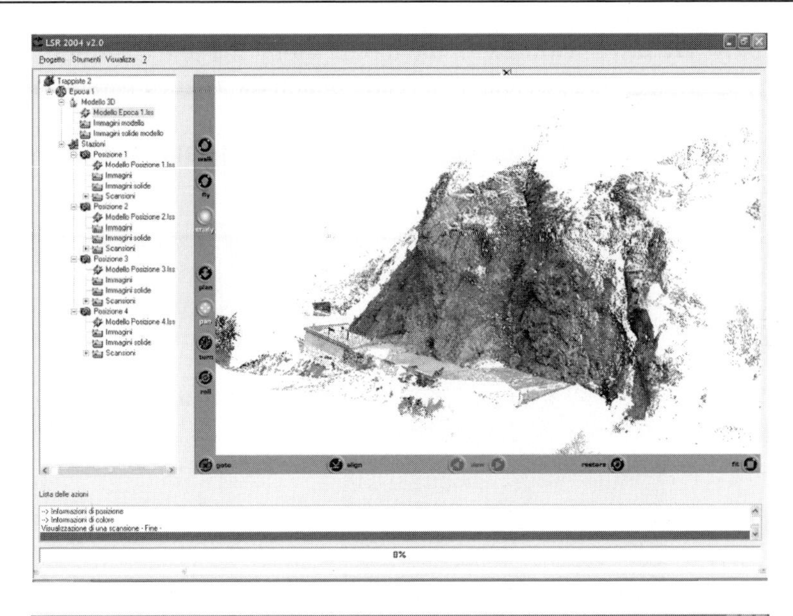

Fig. 7. Data treatments: pre – treatments and modeling

4.3 Solid Image

The extraction of the geomechanical parameters, such as distance measurements, angles, DIP and DIP direction proved to be very useful in this case study. By just clicking with the mouse along the joint families on the raw aspect image, it is possible to directly obtain the needed paramctcrs.

The accuracy of these measurements does not take into consideration the noise introduced by modeling, as the treated point cloud is used in the Solid Image.

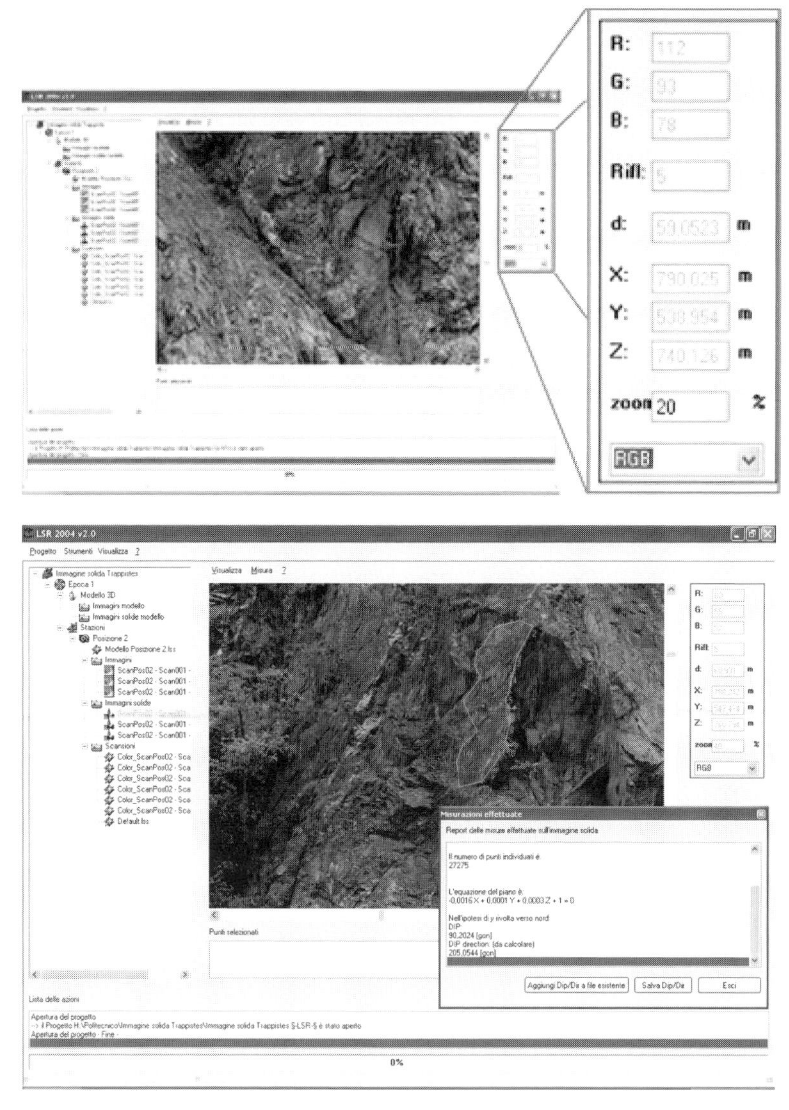

Fig. 8. Solid image software and application

5 Conclusions

Solid image is an economic and user-friendly alternative of the well known 3D realistic models usually generated by commercial software. Solid image preserve all the radiometric and photo-interpretation powerful of a photographic image and can be managed also by using well-known software (e.g. Adobe Photoshop).

This case study shows that Solid image can be very helpful to extract all the required information in order to study stability problems in cliffs and rock falls hazards.

In the future the possibility of exporting the block model of the fault to stability simulation software should help to increase to the degree of confidence in the evaluation of cliff stability and rock fall risks.

Laser scanner shows its best performances when an intelligent integration with digital photogrammetry is conceived.

References

Biasion A, Bornaz L, Rinaudo F (2004) Monitoraggio di eventi franosi in roccia con tecniche laser-scanner, Atti convegno nazionale SIFET, Chia Laguna (CA) 22-24/09/2004

Bornaz L, Dequal S (2003) The solid image: A new concept and its applications - ISPRS Commission V, WGV/4 – Ancona, 01-03, 2003

Big Brother or Eye in the Sky? Legal Aspects of Space-Based Geo-Information for Disaster Management

Frans G. von der Dunk

Space Law Research, International Institute of Air and Space Law,
Faculty of Law, Leiden University, Steenschuur 25, 2311 ES Leiden,
the Netherlands.
Email: F.G.vonderDunk@law.leidenuniv.nl

Abstract

Amongst the methods of gathering geo-information inter alia for disaster management purposes, the use of satellites is a particularly interesting one in view of their global coverage. In this area a number of recent interesting developments have taken place, such as the establishment of a Charter on Space and Major Disasters, and the rapidly evolving plans for GMES and GEOSS.

These developments raise a number of legal issues, related to state responsibility, state liability, and the respective roles of intergovernmental organizations and commercial and/or private entities in this regard. These issues cannot be easily solved, since the international legal environment for the use of space data for disaster management purposes from the other end offers a rather fragmented picture. Some rules and principles exist on the international level, often however not well-defined and leaving room for conflicting interpretations. Others are confined to certain national territories, certain types of activities or even certain types of natural or legal persons. In still other cases no specific rules or principles can be found to be applicable at all.

1 Introduction

Amongst the methods of gathering geo-information inter alia for disaster management purposes, the use of satellites (remote sensing; earth observation) is a particularly interesting one in view of its global coverage. Also, in this area a number of recent interesting developments have taken place, such as the establishment of a Charter on Space and Major Disasters, and the rapidly evolving plans for Global Monitoring for the Environment and Security (GMES) and the Global Earth Observation System of Systems (GEOSS).

These developments raise a number of legal issues, which should be tackled and, largely, solved if the possibilities of satellites to contribute to disaster management is to prosper as comprehensively as possible. This will partly determine whether the capabilities of satellite earth observation in such cases would make a satellite an "Eye in the Sky", capable of seeing where more conventional methods cannot look as efficiently and thus greatly contributing to mankind's wellbeing in such specific cases as disaster management, or whether it will turn out to be more of a "Big Brother", such capabilities being used predominantly against the interest of human beings rather than to help them.

The paper will present a first overview of the legal issues involved, how they might interfere with the ultimate objectives of disaster management and what might be done in respect of the most problematic gaps and overlaps in the legal regime(s) at a relatively short notice. In doing so, the paper will refer as appropriate in particular cases to the topic of disaster management, or relate more general conclusions to that specific topic.

2 Satellite Earth Observation for Disaster Management: The Role of Space Law

By way of starting point it should be clear that in view of the novelty of international disaster management and in particular the uses of satellite information in such a context, as of yet little dedicated and focused legal regulation exist. As a consequence, the international legal environment for the use of space data for disaster management purposes offers a rather fragmented picture. Some rules and principles exist on the international level, often however not well-defined and leaving room for conflicting interpretations. Others are confined to certain national territories, certain types of activities or even certain types of natural or legal persons. In still other cases no specific rules or principles can be found to be applicable at all.

The analysis thus represents an effort to address the most salient legal aspects of such operations from a more general, international space law-perspective. It is divided into four sub-themes as they arise under general space law:
1. international responsibility,
2. international liability,
3. the role of intergovernmental organizations and
4. the role of private entities.

Prior to that, a brief description of satellite earth observation, in particular of GMES and the Charter on Space and Major Disasters, will be provided.

The observation of the earth, including its oceans and atmosphere, has long been viewed as one of the major benefits of human space activities. Its applications have widened, in particular over the last two decades, from largely strategic ('spying') ones to a wide variety of uses: for agricultural development, mining purposes, industrial and urban development, environmental controls, monitoring of arms and disarmament treaties up to and including, with the rapidly growing level of resolution of the data available on the open market, also geographical information systems (GIS).

Increasingly, also the use of satellite data in the context of major disasters, whether natural (earthquakes, floods, typhoons) or man-made (oil spills, landslides, refugees) became of interest. Though only rarely capable of avoiding the disaster altogether, such data could be used pre-disaster to mitigate its catastrophic effects by their warning capability, and post-disaster by rapid damage assessment and enhancement of the efficiency of rescue and damage-mitigation operations, as well as longer-term recovery and rehabilitation. For the sake of this paper, 'disaster management' is viewed as comprising both categories of disaster-related activities.

The United Nations Declaration on Principles Relating to Remote Sensing of the Earth from Outer Space, Resolution 41/65[1], adopted by consensus on 3 December 1986, sets out the main legal principles in this regard. Thus, Principle X states the need for states to convey relevant information regarding threats to the "Earth's natural environment" to other states concerned; and even more to the point, in a similar vein, Principle XI calls upon states in the possession of relevant information regarding natural dis-

[1] Res. 41/65; Official Records of the General Assembly, Forty-first Session, Supplement No. 20 (A/41/20 and Corr. 1); UN Doc. A/AC.105/572/Rev.1, at 43; 25 ILM 1334 (1986).

asters to duly inform other states concerned[2]. Principle VIII furthermore describes the envisaged role of the United Nations itself in this context: it "shall promote international cooperation, including technical assistance and coordination in the area of remote sensing"[3].

It should be noted, that Resolutions of the UN General Assembly are not binding legal documents per se; however, they may over time evolve into reflections of customary international law. In the case of Resolution 41/65, the fact that it was adopted by consensus is surely conducive to such a development. Moreover, UN General Assembly Resolutions carry considerable weight already in the political/moral sphere; hence states will be rather prudent in avoiding to be seen as neglecting the principles set out in such a Resolution. In any case, these principles provided a solid point of departure for the development of further legal rules and principles once the time would be ripe.

With a view to disaster management in particular, however, since then more stringent and/or legally binding rules and obligations were not developed. The inherent international and sovereignty-sensitive aspects, coupled to the enormous investment necessary for operational satellite systems, caused states with earth observation capacities (basically until then a handful of developed states) generally speaking to be weary in taking upon their shoulders any (additional) obligations potentially further complicating their satellite operations.

With the increasing onslaught, visibility and global character (at least in terms of effects, such as in the case of refugees or aid programs) of major disasters over the last decades, no longer confined moreover to the developing world, the understanding grew that such an attitude would in the end be counter-productive. Furthermore, to the extent developing nations still bore the brunt of disasters, it is particularly noteworthy that leading developing nations such as India, China, Indonesia and Brazil acquired indige-

---

[2] The full text of Principle X runs as follows: "Remote sensing shall promote the protection of the Earth's natural environment. To this end, States participating in remote sensing activities that have identified information in their possession that is capable of averting any phenomenon harmful to the Earth's natural environment shall disclose such information to States concerned."
The full text of Principle XI runs as follows: "Remote sensing shall promote the protection of mankind from natural disasters. To this end, States participating in remote sensing activities that have identified processed data and analyzed information in their possession that may be useful to States affected by natural disasters, or likely to be affected by impending natural disasters, shall transmit such data and information to States concerned as promptly as possible."

[3] Principle VIII, Res. 41/65.

nous satellite earth observation capabilities, entering the club of 'haves' in terms of space.

These developments resulted not so much (yet) in the development of a distinct legal regime for disaster management or even the use of satellite data in that context – that still seems a bridge too far at this moment. It did result, though, in two institutional developments in particular. The most visible results of the changing paradigms thus are twofold: the development, on the one hand, of the GMES-concept, and on the other hand, of the Charter on Space and Major Disasters.

3 Global Monitoring for the Environment and Security (GMES)

The first-mentioned development was initiated by the two pre-eminent European international organizations in terms of space: the European Space Agency (ESA) and the European Union as represented by the European Commission. These two organizations had over the 1990's grown closer together in defining the interests and policies of 'Europe' when it came to space, and such closer co-operation had resulted already in the first joint project, for a European Global Navigation Satellite System 'Galileo', planned to be operational as of 2008[4].

As a consequence inter alia of a few oil spill disasters, highly visible and political discussions concerning the Kyoto Protocol and some major human tragedies in the Balkans and elsewhere, ESA and the Commission started considering, roughly along the lines of their co-operation on Galileo, to arrive at a coherent system of providing satellite observation data for a broad range of purposes: environmental but also others as long as related to both civil and 'military'/political security of the peoples of Europe. This became the concept of GMES: "Global Monitoring for Environment and Security (GMES) is a joint endeavour by ESA and the European Commission to establish an independent capability for global monitoring, in support of European environment and security goals. GMES is envisioned as a complete decision-support system for use by the public and policymakers, enabling the acquisition, interpretation and distribution of

4 See on Galileo e.g., the author's *Quis vadit cum vobis*, Galileo? – Institutional Aspects Of Europe's Own Satellite Navigation System, in *Proceedings of the Forty-Sixth Colloquium on the Law of Outer Space* (2004), 360-70; Liability for Global Navigation Satellite Services: A Comparative Analysis of GPS and Galileo, 30 *Journal of Space Law* (2004), 129-67; Of Co-operation and Competition: GALILEO as a Subject of European Law", in *Legal Aspects of the Future Institutional Relationship between the European Union and the European Space Agency* (2003), 47-64.

all useful information related to the environment, risk management and the natural resources. It represents a vital part of Europe's contribution to issues affecting the global environment and the safety of the Earth. Making GMES happen involves improving current deficiencies in European information gathering by better coordinating existing information gathering resources situated on the ground. It also means optimizing the use of current and future Earth Observation systems – whose unique perspectives provide a whole new dimension of information about the Earth"[5].

In other words, the first 'phase' consists essentially of an inventory of the types of satellite data currently available to Europe for those purposes, with the idea to identify gaps – whether in types of data, in terms of geographical or thematic coverage, or in time: many existing satellite earth observation operations were developed as one-off operations, with little planning in place for the after-satellite life of the data or continuation of generation of data after the end-of-life of the satellite. Once the gaps would be identified, in a second phase thus decisions would have to be taken as to where it would be possible and (technologically, politically, financially, socially) feasible to develop instruments and/or satellites to fill such gaps.

It is clear that a major benefit of GMES, already to some extent in the first phase but certainly in the second phase, would be in the area of disaster management. There is no doubt that disaster management presents one of the key drivers and key applications for GMES, as part of both 'Environment' and civil 'Security'.[6] In view of the coherence of the participating entities, the dedication of the two entities as well as the member states behind them, the technical and operational expertise of ESA and the legislative machinery and political clout of the EU, it might be expected that GMES may come to represent a solid contribution to the future global capabilities to manage major disasters, possibly raising such operations to a fundamentally higher level. Yet, how large those benefits would be, would only (and likely gradually) become clear once any second phase of GMES will become implemented. It may be noted here, that the GMES Services Element, agreed upon by ESA as its contribution to GMES's first phase in

---

5 See *http://www.esa.int/export/esaEO/SEMV343VQUD_environment_0.html*.

6 Thus, for example currently amongst the GMES Services initiated under ESA guidance the following are considered: EO based risk information services for forest fire and flood management led by Astrium (France) (called "Risk-EOS"); water pollution risk and soil sealing maps for water management and soil protection, led by InfoTerra GmbH (Germany) (called "SAGE"); Real time Ocean Surveillance for Environment and Security covering oil pollution and water quality and led by Alcatel (France) (called "ROSES"); and working with the humanitarian community to improve access to maps, satellite imagery and geographic information led by Infoterra UK) (called "RESPOND").

November 2001, is to run for five years; the capacity for GMES to be an "operational system" is envisaged as of 2008[7].

Meanwhile, the European development of the GMES concept has also triggered non-European governmental earth observation satellite operators to consider the importance of enhancing global use of satellite data for disaster management. This has led to current discussions on ensuring that as between those various satellite operators as much co-operation in the area of (global) disaster management would be accommodated as was politically and economically possible. Currently, these discussions have given rise to the concept of Global Earth Observation System of Systems (GEOSS), the idea being that unavoidably a number of systems will continue to be operated separately for some time to come, but that those systems should at least find a way to work together in the relevant areas of access to, exchange of, and interpretation of data in a sort of rudimentary system – a 'system of systems'.

4 The Charter on Space and Major Disasters

A more immediate development in the area of disaster management and potential involvement of satellite data therein concerns the establishment of the Charter on Space and Major Disasters[8], which focuses directly and exclusively on the mitigation of major disasters and their harmful effects, without creating any new international bureaucratic and cumbersome institutional layer.

The Charter was established by a number of leading space agencies with operational remote sensing capabilities, initiated by ESA and the French space agency CNES in 1999 as a follow-up to the UNISPACE III Conference, where the potential of earth observation in the context of major disasters was prominently discussed. The Canadian Space Agency CSA, the US National Oceanic and Atmospheric Administration (NOAA) and the Indian Space Agency ISRO, and most recently (in July 2003) the Argentine National Commission on Space Activities CONAE joined, so that the charter currently counts six partners[9]. The International Charter, declared formally operational on 1 November 2000, aims at providing a unified system of space data acquisition and delivery to those affected by natural or man-made disasters. Each member agency has committed resources to

---

7 See *http://esamultimedia.esa.int/docs/GMES_Newsletter_1.pdf*, at p. 2.

8 The full name is Charter On Cooperation To Achieve The Coordinated Use Of Space Facilities In The Event Of Natural Or Technological Disasters.

9 See *http://www.disasterscharter.org/main_e.html.*

support the provisions of the Charter and thus is helping to mitigate the effects of disasters on human life and property: ESA provides data from ERS and Envisat, CNES from the SPOT satellites, CSA from the Radarsat satellites, ISRO from the IRS satellites, NOAA from the POES and GOES satellites and CONAE from SAC-C.

Article 6.1 of the Charter stipulates in this respect that requests to adhere to the Charter may be made by any space system operator or space agency with access to space facilities which agrees to contribute to the commitments made by the parties under Article IV; therefore, it is a de facto prerequisite for membership to the Charter to possess capability to operate satellite systems or at least of doing so in the near future. Those space facilities are not necessarily limited to earth observation satellites or instruments; "space systems for observation, meteorology, positioning, telecommunications and TV broadcasting or elements thereof such as onboard instruments, terminals, beacons, receivers, VSATs and archives" are also included[10]. Indeed, for example GOES and POES are meteorological satellites.

Upon request by a "beneficiary body", the member agencies acquire the data of the area affected by the disaster from their satellites, process the data so as to create useful images, analyze them further if necessary, and distribute the resulting information free of charge to those states affected by the disaster via associated bodies. Only so-called "authorized users" can activate the Charter: a state affected by a disaster who wishes to access relevant data needs to contact either one of the "associated bodies"[11] or one of the "co-operating bodies"[12] acting in partnership with an associated body.

Indeed, the Charter has been able to assist the countries affected by disaster rather promptly, though constant efforts are made to bring the time of response down even further. For instance, in the case of a flood occurring in Toulouse in 2002, the image was available 38 hours after the request to activate the Charter and just 14 hours after image acquisition. As per De-

[10] Art. I, Charter on Space and Major Disasters.

[11] An "associated body" is "an institution or service responsible for rescue and civil protection, defense and security under the authority of a State whose jurisdiction covers an agency or operator that is a party to the Charter"; Art. 5.2.

[12] Cooperating bodies includes the European Union, the UN Bureau for the Coordination of Humanitarian Affairs and other recognized national or international organizations with which the parties may have cause to cooperate in pursuance of the Charter. A "cooperating body" does not operate a space system but acts in partnership with the an associated body which does; see Art. 3.5.

cember 2004, the Charter had been activated a total of 58 times for various disasters all around the world[13].

By way of further examples and with reference to one particular satellite data provider, over the past year and a half data from the SPOT satellites (through SPOTImage) have been used in the following cases [14]:

- April 2003, Italy: volcanic eruption on the island of Stromboli
- April 2003, India: violent storm in the state of Assam, leaving thousands of people homeless and killing about 30 people
- April 2003, Argentina: floods devastating the whole province of Santa Fe and neighboring regions, causing 60,000 inhabitants to be evacuated
- May 2003, Turkey: earthquake (6.1 on the Richter scale) killing hundreds of people and causing extensive damage in the state of Bingöl
- May 2003, Algeria: very violent earthquake at Boumerdes (6.6 on the Richter scale) killing more than 2,200 people and injuring 9,000
- July-August 2003, France: Var, Corse and the Alpes Maritimes departments suffering from gigantic forest fires
- July-August 2003, Portugal: gigantic forest fires
- December 2003, France: Gard and Bouches-de-Rhône departments: floods causing 30,000 inhabitants to be evacuated and killing 7 people
- January 2004, Iran: 6.7 magnitude earthquake in Bam and environments: more than 30,000 people being killed, 50,000 injured and 100,000 homeless.

It is clear therefore, that the Charter can be of help in a large number of rather varying events; not only in developing but certainly also in developed countries.

5 Legal Issue # 1: International Responsibility

Firstly, under current space law states are responsible for "national activities in space", even if conducted by "non-governmental entities", and for ensuring these activities are in conformity with the law[15]. This raises issues of the extent to which there are relevant legal obligations pertaining to

---

13 See *http://www.disasterscharter.org/disasters_e.html.*

14 See *http://www.spotimage.fr/html/_167_210_214_215_.php.*

15 Art. VI, Treaty on Principles Governing the Activities of States in the Exploration and Use of Outer Space, including the Moon and Other Celestial Bodies (hereafter Outer Space Treaty), London/Moscow/Washington, done 27 January 1967, entered into force 10 October 1967; 610 UNTS 205; TIAS 6347; 18 UST 2410; UKTS 1968 No. 10; Cmnd. 3198; ATS 1967 No. 24; 6 ILM 386 (1967).

the duty to provide information versus the right of ownership over data, including intellectual property rights and data protection issues, since only non-conformity with such clear legal obligations could raise the issue of state responsibility.

Here, it is noteworthy to recall that Principle XI of Resolution 41/65 calls upon states in the possession of relevant information regarding natural disasters to duly inform other states concerned[16]. Being an elaboration of an even more general principle of 'good neighborliness' amongst states, this would be seen by many as representing customary law, and hence a binding legal obligation.

However, the mere existence of the Charter – which of course restates such 'intentions' – at the same time makes clear that such an obligation can only move from the realm of theory to that of practical relevance if, indeed, mechanisms and procedures are provided for. Even with the Charter itself being a binding document, though not on the state but on the agency level, 'violations' of the obligations contained in it Charter would be difficult to define, let alone to determine the proper 'reparation' to address any state responsibility. To begin with, the Charter itself does not provide so much for 'rights', either of victim states or individual victims, but for obligations of a still rather general nature. Also, it would be difficult to imagine what forms of 'reparation', what 'sanctions' should follow in case of any 'violation' of a relevant obligation.

The above, as a consequence, certainly leaves any existing protection of copyrights, patents or other potentially applicable intellectual property rights (such as trademarks) in tact, so that in a sense any obligation to provide data provides an exception to the right of a copyright holder to not allow use of their data as such or only allow it against certain fees – a right effectively waived by the Charter for the limited purposes which it deals with. Therefore also, at this point inherent difficulties in applying copyrights, usually requiring some 'intellectual effort' or 'creativity' directly at the level of data generation need not be dealt with as these are still predominantly dealt with at the national level, each state largely maintaining its sovereignty in dealing with those issues by legal means.

The GMES and GEOSS concepts steer completely clear from such difficult questions, as they provide for informal – though highly relevant and, hopefully, beneficial – institutional co-operation mechanisms. Again, any general duty of 'good neighborliness' even as applied to space in the context of Resolution 41/65 would be too vague to distil any clear-cut obligations for the purpose of state responsibility and questions of eventual reparation.

---

[16] See *supra*, …

6 Legal Issue # 2: International Liability

Secondly, along similar lines under current space law states are specifically liable for damage caused by space objects[17]. Such damage, however, is generally seen as focusing on physical damage caused by direct impact of a satellite; it remains rather doubtful – to say the least – whether liability can be apportioned under the Outer Space Treaty or the Liability Convention for damage caused by absence of or non-access to certain data, conversely caused by certain data and their usage. Further in detail, issues arise as to what kind of liability – absolute or fault – would apply, what level of compensation could be expected, whether any waivers of liability would, could or should apply, and appropriate procedures to effectively and fairly arrive at dispute settlements on liability issues.

As to the Charter, services are provided on a "best efforts" basis, implying that Charter members will take all necessary measures in rendering aid but do not guarantee successful results. A specific provision in the Charter clearly waives the liability of satellite operators called upon to provide data under the Charter: "The parties shall ensure that associated bodies which, at the request of the country or countries affected by disaster, call on the assistance of the parties undertake to: (…) confirm that no legal action will be taken against the parties in the event of bodily injury, damage or financial loss arising from the execution or non-execution of activities, services, or supplies arising out of the Charter"[18]. So the member agencies would assume no liability arising from the Charter service. Death cases are also subject to the waiver of liability, even though it is not stipulated specifically in the above clause.

This waiver of liability, however, does not comprehensively solve the problem. Firstly, since the Charter is concluded among the partner agencies but not with all the potential crisis victims, the waiver of liability is not mutually agreed upon. Wherever the victim of a crisis is not one of the countries to which the Charter partners belong, the one-sided waiver of liability raises questions as to its validity.

---

[17] Art. VII, Outer Space Treaty, and Artt. I(c), II, III, Convention on International Liability for Damage Caused by Space Objects (hereafter Liability Convention), London/Moscow/Washington, done 29 March 1972, entered into force 1 September 1972; 961 UNTS 187; TIAS 7762; 24 UST 2389; UKTS 1974 No. 16; Cmnd. 5068; ATS 1975 No. 5; 10 ILM 965 (1971); provide for liability of a "launching State" respectively the "launching States" for any damage caused by a space object launched by such state, as procured by such state, or as launched from the territory or facility of such a state.

[18] Art.5.4, Charter.

Furthermore, the Charter provides for a waiver of liability only concerning cases arising between the affected country and the Charter partners. It does not mention, for instance, cases arising from potential liability of value-added service providers with respect to Charter partners or states affected by disaster. The Charter does not stipulate whether a state affected by the disaster can bring action against value-added service providers directly, in case these are somehow involved in the damage being caused.

This finally raises issues regarding the so-called 'Good Samaritan' principle, a principle known in various national jurisdictions which essentially means that a person who injures another in imminent danger while attempting to aid him, and who is then sued by the aided one, will not be charged with contributory negligence unless the rescue attempt is an unreasonable one or the rescuer acts unreasonably in performing the attempted rescue[19]. Its purpose is to prevent people from being unduly reluctant to help a stranger in need, for fear of legal repercussions should they make some mistake in doing so.

The Good Samaritan doctrine has been used widely in different jurisdictions throughout the world. In Canada and the United States, it is incorporated by means of specific acts. The principle is also reflected in different national laws in European countries. If the aid worker has worsened the condition of the imperiled person, many techniques are available to assess the rescuer's conduct: from mitigation of damages in Dutch law to the presumption of a low standard of care in French and English law. Since the Good Samaritan principle is incorporated into domestic law of many countries, it is considered to reflect customary international law. What it means in the context of the International Charter, and whether its main criteria and parameters are overruled by it, remains an issue to be dealt with in further detail, however.

The further development of the law on such issues will be followed with particular interest by the EU and ESA in their on-going GMES-related activities. For example, whether liability for wrongful information could effectively be waived in the case of usage of GMES-derived information, might be considerably clarified when, in the case of the Charter, actual disputes have arisen, and have to be adjudicated, on such issues. To what extent moreover, in view also of the limitation of the Good Samaritan doctrine to cases of aid, such legal outcomes would apply to other types of usage of GMES data (e.g. on security-related issues), is yet to be analyzed as well.

---

[19] See *http://pa.essortment.com/goodsamaritanl_redg.htm.*

7 Legal Issue # 3: The Role of Intergovernmental Organizations

Thirdly, the role of intergovernmental organizations in this area results in some particular ramifications of (state) responsibility and liability, due to the fact that under international space law these organizations are given a kind of secondary status. This issue may be relevant in the context of disaster management since, for instance, ESA is one of the founding fathers and parties of the Charter on Space and Major Disasters, GMES is a project jointly initiated by ESA and by the Commission on behalf of the European Union, and both these organizations consequently are also involved in the discussions on GEOSS.

As to state responsibility under the Outer Space Treaty, it also applies "when activities are carried on in outer space (...) by an international organization"[20], though in such cases jointly with the responsibility of the international organization itself. Further to this provision, states are to resolve "any practical questions arising in connection with activities carried on by international intergovernmental organizations"[21]. As a consequence, effectively the member states of neither ESA nor the EU can hide behind those organizations in case any of their activities in the context of the Charter or of GMES would violate applicable rules of international law; it is their duty and responsibility to ensure that the organizations themselves do not undertake any such violating activities.

As to liability, intergovernmental organizations equally enjoy a similar secondary status. This status effectively allows them to act as liable entities, and a similar construction under the Registration Convention allows them to themselves register satellites and exercise concomitant competencies over them[22]. ESA actually enjoys the relevant status, as it has deposited relevant Declarations in respect of both Conventions, but the EU does not: in the latter case, any liability would revert directly to the individual member states to the extent of course qualifying as "launching States' under the relevant Articles.

But even in the case of ESA, there is a subsidiary liability for the member states: in case ESA is not able to satisfy any justified claim for com-

---

[20] Art. VI, Outer Space Treaty.

[21] Art. XIII, Outer Space Treaty.

[22] See respectively Art. XXII, Liability Convention, and Art. VII, Convention on Registration of Objects Launched into Outer Space (hereafter Registration Convention), New York, done 14 January 1975, entered into force 15 September 1976; 1023 UNTS 15; TIAS 8480; 28 UST 695; UKTS 1978 No. 70; Cmnd. 6256; ATS 1986 No. 5; 14 ILM 43 (1975).

pensation under the Liability Convention within six months, the individual member states (to the extent again qualifying as "launching States") are held to jointly compensate the damage concerned[23].

When it comes to the Charter or GMES therefore, to the extent that responsibility and/or liability would arise, the secondary status of the intergovernmental organizations involved may cause problems of efficient and proper handling of disputes. How, for example, will proper reparation in case of violations of international obligations be guaranteed when the relative responsibility of the organizations respectively their member states is not clearly outlined? Similar issues might arise with respect to compensation for any damage caused, where the risk of delay is almost inherent in the provision of Article XXII(3) of the Liability Convention that only after six months of non-payment by the relevant intergovernmental organization individual member states may be called upon to compensate.

8 Legal Issue # 4: The Role of Private Entities

Fourthly, focusing on possible involvement of commercial and/or private entities in relevant activities, apart from aforementioned issues relating to access to data, privacy and data protection (which are of special relevance in a commercial environment), particular issues arise on the point of (national) licensing and certification, and the way law handles the public goods-versus-fair competition dichotomy. These themes will not be further explored here, as this would obviously be beyond the scope of the paper, but are nevertheless of primary importance for the current topic and will therefore have to be taken into due consideration.

The issue of state responsibility under Article VI of the Outer Space Treaty as discussed supra strongly impacts upon this issue, as such responsibility also arises for privately conducted space activities. Hence, it points to the need for national legislation to implement on a national level any international obligations relevant; from this perspective it may be pointed out that so far only a handful of states have actually established such more or less comprehensive space legislation. This concerns, in varying degrees of detail and elaboration, the United States, Norway, Sweden, the United Kingdom, Russia, South Africa, the Ukraine, Australia and Brazil, as well as Hong Kong as Special Administrative Region within the People's Republic of China. In addition, states such as Argentina, Canada, France and Japan have important pieces of national legislation in place which come

[23] See Art. XXII(3)(a), Liability Convention.

close to, but do not yet really provide for proper legislative control over private space activities conducted within their respective jurisdictions[24]. A still larger number of states, however, so far remain without any transparent and coherent domestic legal means to control and monitor relevant private activities. A lot of work remains to be done in this area.

As a consequence, the legal implications for example for SPOTImage, the French private company involved in Charter operations, may be difficult to analyze or even ascertain. As for GMES, at this point it is not at all clear whether, and if so, at what level private entities might be involved in its operations. It may be pointed out here, however, that GMES is very often referred to as the 'second Galileo', primarily because it equally is a co-operation project of the EU and ESA. Further to that, one can not rule out that, once GMES would call for additional operational satellite systems or operations, the example of a Public-Private Partnership (PPP) as currently being established for Galileo, where a private operator of the satellite system would be supervised by a public entity, would be aimed for in this context as well.

In any case, any involvement of private entities, either currently in the context of the Charter, or in future in the context of GMES, would result in bringing a number of legal regimes into the picture either trying to ensure private activities will be as beneficial to society at large as possible, or actively stimulate their involvement by protecting their interests – such as intellectual property rights regimes allowing companies to market their inventions and operations. Such legal regimes are largely developed at national levels, and moreover generally without any specific space application in mind. This warrants thorough further analysis as to the gaps and overlaps, inconsistencies and further problems created by such an extended and complex legal environment.

9 Concluding remarks

In conclusion, a rudimentary (certainly as far as the international level is concerned) legal framework can be discerned providing a very limited set of parameters to disaster management activities involving satellite data. Whilst at the national level, in particular when it comes to private sector-involvement, much more detailed regimes sometimes do apply, questions automatically arise as to the consequences of any such applicability to such

[24] See e.g. the author's Heeding the Public-Private Paradigm: Overview of National Space Legislation around the World, in *2004 Space Law Conference Papers Assembled* (2004), 20-34.

an international, as well as specific space-related issue as the use of satellite data for disaster management.

The Charter and, as a next-generation step, GMES may be seen as major milestones in the international domain to address this highly important area of global relations, but they remain principally at the level of institutional practical co-ordination, co-operation and efficiency-enhancement. Considerable analysis would be required on the issue of how these new developments – the Charter, GMES but more in general any use of satellite data for such activities of clear benefit to humanity as disaster management – may, would or should result in new complementary legal developments. Actually, as of yet 'legal practice' is hardly in existence, which does not even allow us a solid evaluation of the various consequences of applicability (or not) of the legal regimes which do exist – intellectual property rights, licensing, liability and so on.

How should for example, liability (civil/financial as much as criminal) be approached and dealt with, in an internationally-harmonised fashion, in the difficult and different context of disaster management using satellite data? How should the balance between the right of owners of the relevant intellectual property to do with the data as they see fit and the interests of actual or potential victims or victim states be established? Is there a proper way to prevent abuse of data by free-riders? These and many other questions will have to be solved if one is to ensure satellites, from a disaster management perspective, will indeed act like "Eyes in the Sky", rather than as "Big Brothers".

ICT for Environmental Risk Management in the EU Research Context

Karen Fabbri and Guy Weets

Directorate General Information Society and Media, Unit ICT for Transport and the Environment, European Commission BU 31 5/83 ; 1049 Brussels, Office: avenue de Beaulieu 31, B1160 Brussels, Belgium. Email: karen.fabbri@cec.eu.int; guy.weets@cec.eu.int

Abstract

The management and mitigation of natural and man-induced risks is a topic of growing worldwide concern, especially in regions where population densities and the frequency of extreme events are increasing. Risk management, being a highly multidisciplinary activity, has many facets that include the advancement of thematic and applied research, the integration and deployment of new and existing technologies, the provision of usable services for the citizens, and the need to adequately consider aspects of risk perception and communication. The EC is actively supporting research on risk management through out its programs for Research and Technological Development. The availability of high quality, rapidly accessible and secure geo-information is the basis of sound decision making for risk management and disaster prevention In particular, scientific advances in the field of information and communication technology have contributed to interoperability and harmonization of geo-spatial information, and the integration of in situ sensor networks and satellite communication for disaster alert systems.

1 European Research in the Field of Risk Management

The overall aim of European research policy is to strengthen the European research community in order to promote scientific excellence and innovation to advance knowledge and understanding, and to support the implementation of related European policies. The European Commission (EC), and in particular, the Directorates-General for Research and Information Society Technologies, have been supporting research on natural and technological disasters since the early 1980's through the EC's successive Framework Programs for Research and Technological Development (RTD).

Multinational and interdisciplinary research in this field has addressed floods, landslides, avalanches, forest fires, earthquakes, volcanic eruptions, industrial hazards, and overall risk management issues. European Union (EU) research has lead to the development of methods and technologies for improved hazard forecasting and monitoring; risk assessment, management and mitigation. In parallel, research efforts have also focused on the use of Information and Communication Technologies (ICT) in support of risk and crisis management. The objective being to improve preparedness and response to major crises, and improve the safety of rescue teams, thereby reducing economic losses and the toll on human lives.

Advances have been made particularly in the areas of flood forecasting, the design of earthquake resistant structures, forest fire hazard mapping and suppression techniques, volcanic eruption monitoring, landslide and avalanche hazard assessment, and industrial process safety related to the implementation of the Seveso II Directive. On the crisis management side, research results have lead to the improvement of situational awareness, as well as better decision support tools in support of the whole command chain.

2 FP6 Research

Currently, research under the Sixth Framework Program (2003-2006) focuses on a more holistic approach in which hazard vulnerability and risk assessment are addressed in an integrated manner with the aim of mitigating the environmental, social and economic effects of natural disasters. Where relevant, but in particular for floods, and storms, research is to address climate variability issues and the potential impacts of climate change. Ongoing ICT research is focusing on (i) the improvement of I interoperability of civil protection equipment allowing for joint intervention in case of major disaster, (ii) the development of harmonized geo-spatial information for the maintenance of vulnerability information and allowing the management of inter-related risks (including domino effects), (iii) developing new approaches for the deployment of in-situ sensor networks and bridging the gap between in situ and remote sensing observation in the context of GMES (described below).

3 EU-MEDIN

The EC has launched two disaster-related initiatives. Firstly, the Euro-Mediterranean Disaster Information Network (EU-MEDIN), which has the aim of disseminating and networking the activities of the European disaster research community. In particular by promoting strategies for the harmonization of methods and the integration of data, disseminating European research results, stimulating new integrated and applied research and common frameworks for risk assessment; and acting as a central node for research results including methods, tools, case studies, datasets, and recommendations. (See :http://www.eu-medin.org/)

4 GMES

Secondly, the EC is a co-founder of the Global Monitoring of Environment and Security (GMES) initiative, in co-operation with the European Space Agency (ESA). The objective of GMES is to build a European capacity for global monitoring of the environment and security by 2008. The initiative aims to integrate environmental observation data from all available sources, satellite and in situ, while taking account of EU policies and data standardization methods. GMES research activities support user-driven development of services targeted to provide reliable, timely and independent environmental information to European stakeholders. The "security" component of GMES explicitly addresses natural disasters. GMES is a major European contribution to the global ad-hoc Group on Earth Observation (GEO), of which the EC is one of the four co-chairs. GEO's primary task is to develop a 10-year Implementation Plan for a coordinated and sustainable Global Earth Observation System of Systems (GEOSS) presented at the global Earth Observation Summit in February 2005. One of the nine GEO societal benefit issues is "Reducing loss of life and property damage due to natural and human-made disasters". European Research policy strongly supports both initiatives, GEO and GMES. (See: http://www.gmes.info/)

5 ICT for Risk Management

Currently, all 25-member states of the EU have very different information systems, operational procedures and communication systems that do not talk to each other. Since 2000 significant R&D activities have addressed these issues and include the development of a software platform that improves interoperability between civil protection systems as well as enhancing contingency and continuity planning in the public and private sectors.

The intention is to design, validate and demonstrate a generic integrated risk management open system to support the whole chain of activities, from initial assessment to post-disaster recovery. Special attention is given to operations of European civil protection organizations in the case of large-scale natural and industrial disasters across all 25 EU member states.

The RTD activities are divided into two sub-activities: the first one covering the monitoring planning, forecasting, preparedness and recovery phases; the second dealing mainly with the alert and the response phase.

5.1 Risk Management and Forecasting

Until recently, risk management components were developed independently by a vast range of institutions and organizations. In addition, the exchange of relevant information needed for dealing with risks is too often hindered by administrative and legal boundaries, as well as lack of interoperability on the technical side. The recently endorsed INSPIRE initiative (see http://www.ec-gis.org/e-esdi/) aims at harmonizing geo-spatial information across the EU. This new Directive offers a unique opportunity for a major overhaul in disaster preparedness, contingency planning as well as community involvement in risk reduction.

Current projects are working on developing service architectures for risk management based on "open" standards. This approach should dramatically reduce the cost of building and maintaining risk management applications, and it will also allow for the appropriate handling of systemic risk and domino effects.

5.2 Crisis Management Operations

Initial investigations revealed that civil protection organizations appeared not to benefit as much as other professionals from the new developments in ICT. Many such public authorities are still poorly equipped, and since this market is rather small, providers companies are slow to invest in up-to-date and cost effective equipment and applications. This is one of the reasons why the EU ultimately decided to invest in this field. It was also perceived as an opportunity to improve equipment ***interoperability***, in order to allow different emergency actors, possibly belonging to different regional or national European authorities, to work jointly in case of large scale and cross-boarder disasters.

Crisis management operations are based on a three-level architecture concept: the "coordination and command centre" supported by "function-specific control rooms", the "mobile command centres" and the "crew" in the field. The function-specific control rooms host the local management and interface with auxiliary functions (technical and scientific support, short term forecasting, meteorological office, emergency health care, public utilities, damage assessment, etc.).The reporting is made to the upper level, that is, the coordination centre, which is located on the premises of the local, regional or national government authorities, depending on the size of the crisis, with a secure link to the crisis centre of the EC in case of major disaster.

Several RTD projects populated this concept with a full range of advanced applications and emergency management tools that were tested and validated by end-users. These elements are being integrated in Emergency Information Systems over a communication infrastructure supporting voice and data exchange through a robust messaging service.

A key objective of the current ICT projects is the creation of a civil crisis management methodology and an interoperable command and control infrastructure. The current approach is building on NATO experience and adapts successful methods initially developed for military applications to the needs of civil protection agencies. The projects will examine the effectiveness of an EU network of trans-national headquarters with the capability of dealing with more than one trans-regional crisis at any one time, and integrate applications into a full fledged C4I (Command Control Communication Computer and Intelligence) for Civil protection operations.

Future developments should address the need for better integration of the wide range public safety communications and evolve into a network-centric crisis management framework (see http://www.cordis.lu/ist/so/risk-management/).

5.3 Security Research

The Preparatory Action on 'Enhancement of the European industrial potential in the field of Security Research 2004-2006', which constitutes the Commission's contribution to the wider EU agenda for addressing key security challenges facing Europe today, will clearly benefit from the research results and standardisation efforts made in the field of crisis management operation.

6 Conclusions

Although there is still a need to further our understanding of the underlying processes leading to natural and technological hazards, there is particular urgency for the improve the integration of hazard assessment with local *vulnerability* issues in order to fully assess the potential risks to society. It is only once this integrated chain (*hazard-vulnerability-risk*) is taken into account, that we can start to better predict the environmental, social and economic consequences of disasters; and hence evaluate alternative mitigation measures that can be economically justified and socially acceptable. Research on disaster reduction should also move forward in the context of multiple (systemic) risks, firstly, to deal with domino and indirect effects and secondly, to bridge the gap between science and real life disaster prevention and management requirements. For all this to be achieved, sizeable efforts will be required for the harmonization of assessment techniques and the integration and standardization of geo-spatial data and analytical tools. In addition, future research must focus on the integration of a wide range of sensors (in-situ, airborne, and satellite) to provide optimal and cost effective information awareness for risk and emergency management. There is also a need to promote the inter-operability of secure and dependable public safety communication systems, including ad-hoc broadband networks for emergency operations and alert networks.

Airborne Passive Microwave Radiometry for Emergency Response

Roland B. Haarbrink[1] and Anatoly M. Shutko[2]

[1] Miramap, ESA/ESTEC (EUI-PP), Keplerlaan 1, P.O. Box 299, 2200 AG Noordwijk, the Netherlands.
Email: rhaarbrink@miramap.com

[2] Institute of Radioengineering and Electronics, Russian Academy of Sciences, 1, Vvedensky Square, City of Friazino, 141190 Moscow, Russia/Center for Hydrology, Soil Climatology and Remote Sensing, Alabama A&M University, P.O. Box 1208, Normal AL, USA.
Email: ashutko@ms.ire.rssi.ru; anatoli.shutko@email.aamu.edu

Abstract

In the event of a natural or terrorist disaster, the key to rescue and recovery operations is timely information presented in standardized data formats. The quicker the response, the higher the likelihood that lives would be saved and property damage would be minimized. Geospatial information should also be tailored to the needs of first responders. The addition of a multi-band Airborne Passive Microwave Radiometry system to existing optical remote sensing solutions with a wireless data downlink capability has an enormous potential for preventing or assisting in disaster operations. In crisis mode, this proven technology can deliver critical geospatial updates to assist first responders. Timely soil moisture and depth to water table maps over large areas can be created to monitor water barriers, to manage dangerously high water tables, and to prevent flooding disasters.

1 Introduction

The international community is raising the level of investigations to detect areas of water seepage through water barriers and to reveal areas with dangerously high groundwater level. An independent study [12] performed by the Dutch Institute of Health and Environment (RIVM) indicates that 549 km (15%) of the primary water barriers in the Netherlands do not meet specifications and that 1217 km (35%) of them has possible issues. The same study also indicates that the financial damage of flooding would lead up to 2570 million Euros for each kilometer of water barrier. The problem is illustrated by the following two incidents (Fig. 1).

On 26 August 2003 the water barrier at Wilnis broke through and flooded an urban area. 1500 persons had to be evacuated, it caused material damage worth millions of Euros and only partially covered by insurance companies to buildings, roads and yards, and it caused psychological damage to the victims. The cause of the break was drought that weakened the barrier.

On 27 January 2004 the water barriers in Stein caused 500 persons to be evacuated and halted canal transportation to some large chemical plants. The levee was deformed, but stood strong using sand bags. The cause of the threat was a broken water pipe that saturated the barrier.

Fig. 1. Water Barriers Under Pressure at Wilnis (left) and Stein (right)

2 Airborne Passive Microwave Radiometry

Optical remote sensing technologies including photography, lidar and multispectral imagery are not able to produce underground soil moisture maps, simply because their wavelengths are not long enough to penetrate the ground. Thermal infrared imagery was used to detect water seepage and to locate weak water barriers, but the measurement were based on temperature changes rather than directly on soil moisture changes, so this technology was not robust enough to monitor water barriers. Other people have tried active radar technologies, but SAR and IFSAR measurements heavily depend on the surface texture, roughness and vegetation, and radar systems usually must be operated from expensive jet aircraft [5, 8, 15].

Passive Microwave Radiometry (PMR) technology on the other hand makes it possible to produce very detailed, geo-referenced maps showing the soil moisture variation at surface within first few centimeters to decimeters in grams per cubic centimeter (g/cc), and the depth to shallow water table down to several meters. PMR is based on measurements of naturally emitted radiation of the earth in the millimeter to decimeter range of wavelengths. Within these bands, the land surface radiation is primarily a function of the free water content in soil, but it is also influenced by other parameters, such as shallow groundwater, above ground vegetation biomass, salinity and temperature of open water, where the sensitivity is a function of the wavelength.

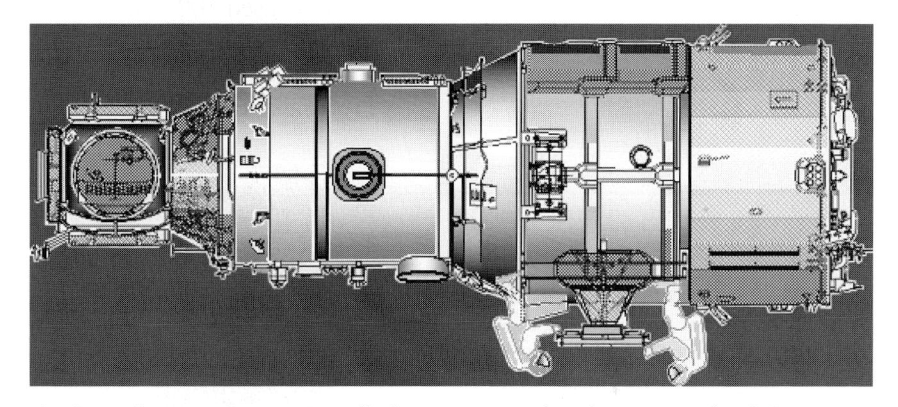

Fig. 2. Radiometer System Installation on International Space Station [7]

Since the first developments of space borne PMR systems for environmental monitoring, the technology has proven successful in the measurement of surface and subsurface soil moisture. For the first time in space exploration history, the Soviet Cosmos 243 spacecraft was launched with four single beam radiometers in 1968 [21]. The Priroda spacecraft module with four channel scanning and non-scanning microwave radiometer systems was attached to Space Station Mir in 1996 [6], and an 8 beam L-band (21-cm) radiometer is planned to be installed on the International Space Station in 2007 by the Russian Academy of Sciences (RAS) for soil moisture maps of large areas, for salinity maps of oceans, for monitoring the thermal flow of water in the ocean, and for estimating the ocean-land-atmosphere energy dynamics (Fig. 2). The European Space Agency (ESA) is also preparing the Soil Moisture and Ocean Salinity (SMOS) mission to be launched in 2007.

When mounting the same type of radiometers onboard a light aircraft or autonomous helicopter (Fig. 3) and using dedicated data modeling software, the spatial resolution of the map products increase dramatically from tens of kilometers to a few meters compared to the space borne PMR data. This increase of detail is essential for mapping potential hazard areas where water seepage occurs or for producing soil moisture maps in case of an emergency.

Fig. 3. Autonomous Helicopter with Microwave Radiometers for Emergency Mapping [1]

Through the years, many microwave radiometric remote sensing studies of the environment were performed for scientific research and application from aircraft and satellites. Leading scientists have conducted airborne PMR test flights and have developed microwave radiometric methodologies to determine the sea surface state, water surface temperature, geothermal sources, moisture profiles, integral water content in soil, water table depth, and the penetration of electromagnetic waves through clouds and vegetation [2, 4, 9, 10, 11, 13, 14, 16, 18, 19, 20, 22, 23, 24, 25, 26]. Multi-band radiometer methodologies using joint collection and processing of 0.8, 2, 5 and 21 cm channel data from scanning and push broom radiometer systems and using a-priori information on the soil composition have proven successful in producing water seepage maps and depth to shallow water table maps. The principle of multi-band PMR to generate these maps is illustrated in Figure 4.

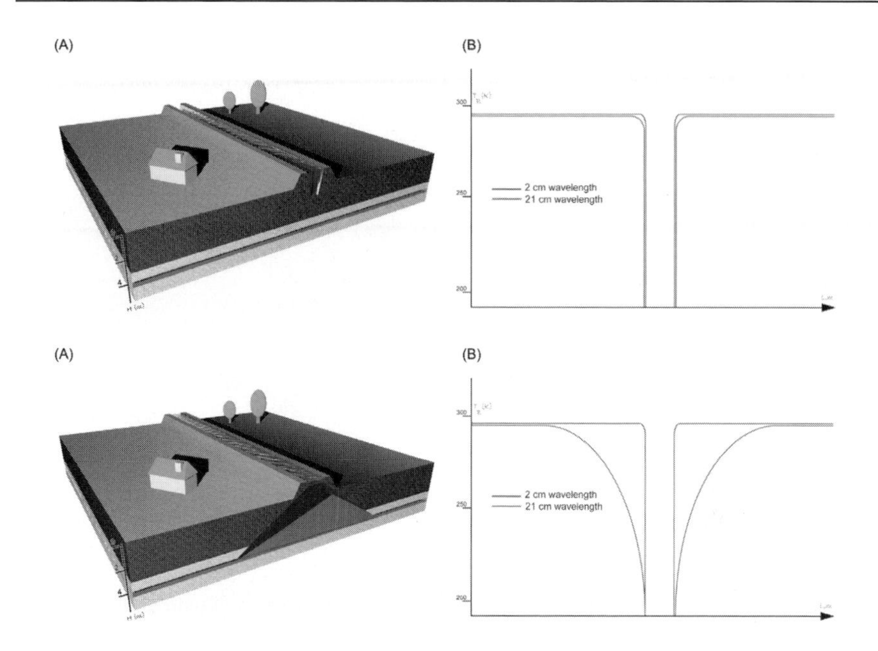

Fig. 4. Principle of Multi-Band PMR to Detect Water Seepage [21]

When collecting PMR data from a light aircraft or autonomous helicopter, the sensor sensitivity to different environmental parameter changes is recorded and analyzed. The datasets are then compared to each other, validated with corresponding field measurements, and the maps are generated through thematic interpretation with special retrieval algorithms. Onboard GPS equipment is used to geo-reference the data, so that the digital maps can be overlaid with existing vector databases, aerial photography and elevation models of the area.

The radiometer data can be transmitted from aircraft to ground processing office through a wireless downlink capability to speed up the time from collection to data delivery in case of an emergency. The resulting soil moisture maps or depth to ground water table maps showing potential hazard areas would be made available within a few hours of collection and can be used in any GIS software package. An example of map showing soil moisture changes is given below in Figure 5.

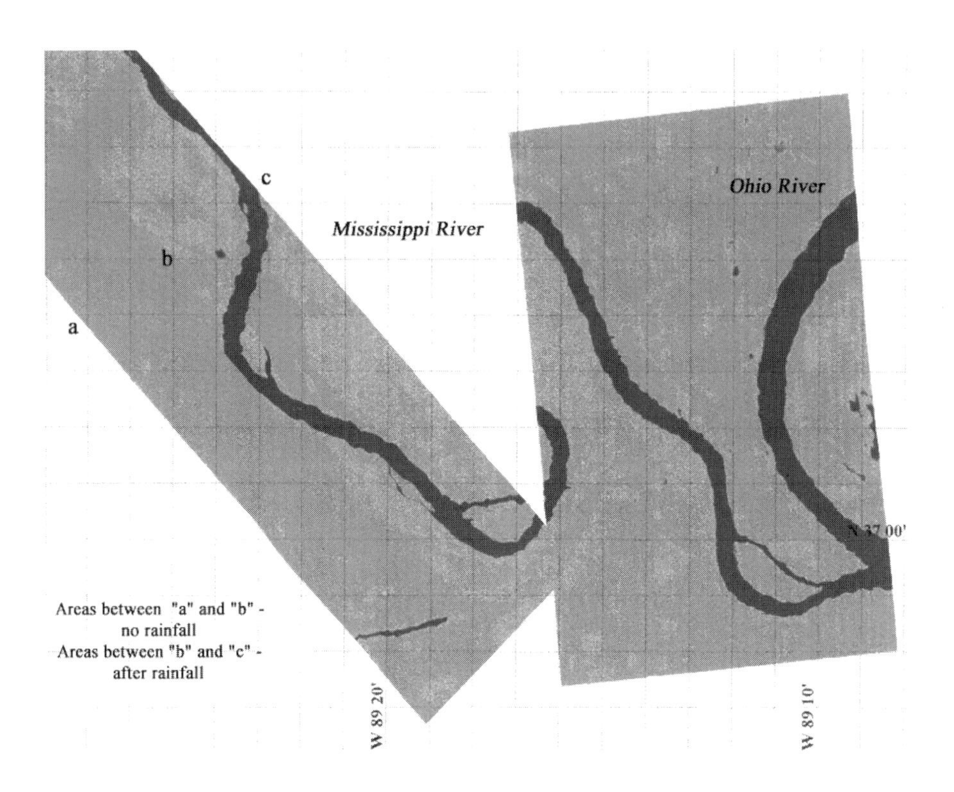

Fig. 5. Map Showing Soil Moisture Changes [17]

Although traditional ground sample techniques to monitor water barriers would always be needed to track local instabilities of water barriers and ground water, as well as to calibrate PMR sensors to the soil composition of a specific area, airborne PMR has these benefits:

Safety: Airborne PMR is a contact less technology that does not weaken water barriers under very high water pressure.

Actuality: Collection of large areas at a time while providing emergency response services with same-day aircraft availability and same or next day data delivery.

Reliability: Continuous multi-channel measurements are covering all parts of the target area instead of a few samples on the ground. Airborne PMR measurements are first of all directly related to changes in soil moisture and to a lesser extent to changes in water temperature.

Flexibility: Airborne PMR can be flown day and night under almost any weather condition such as fog and clouds. Optical systems using wavelengths shorter than 0.8 cm are considerably influenced by the atmosphere such as fog, clouds and rain.

Cost-efficiency: Thousands of soil moisture measurements instead of a few samples a day increase the cost-efficiency. For larger areas, the costs to monitor water barriers could be reduced by five to ten times compared to traditional ground sample technologies by development of Geo-Information Monitoring System that includes remotely sensed data, in-situ measurements and mathematical modeling of spatial-temporal changes of physical parameters within the area of observation [3].

3 Conclusion

The addition of a multi-band PMR system to existing optical remote sensing solutions has an enormous potential for preventing or assisting in disaster operations. In crisis mode, this proven technology can deliver critical geospatial updates to assist first responders. Timely soil moisture and depth to water table maps over large areas can be created to monitor water barriers, to manage dangerously high water tables, and to prevent flooding disasters.

4 Epilogue

The airborne PMR solution is also an excellent tool to monitor temperature increase of land, boundaries of a forest fire in cloudy conditions, changes in salinity and temperature of water surfaces, presence of oil slicks, polluted harbors, on-ground snow melting, and ice on water surfaces, roads and runways.

Miramap is promoting the science, development and practical applications of the PMR technology, while offering affordable collection and processing systems to produce joint soil moisture data, digital photography, lidar elevation data and thermal imagery for emergency response mapping. Miramap is supported by the European Space Agency, the Russian Academy of Sciences, and the Alabama A&M University.

Acknowledgement

The authors are thankful to their partners and colleagues N. Eldering, B. Naulais, J. Eldering, A. Khaldin, E. Novichikhin, I. Sidorov, A.

Chukhlantsev, S. Golovachev, E. Reutov, V. Pliushchev, V. Krapivin, T. Coleman and F. Archer for their valuable contribution to Passive Microwave Radiometry developments and practical use.

References

[1] Archer F, Shutko A, Coleman T (2004) Microwave Autonomous Copter System. Special Flier, Center for Hydrology, Soil Climatology and Remote Sensing, AAMU
[2] Chukhlantsev AA, Golovachev SP, Shutko AM (1989) Experimental study of vegetable canopy microwave emission. Adv Space Res, vol 9, no 1, pp 317-321
[3] Chukhlantsev AA, Golovachev SP, Krapivin VF, Shutko AM, Coleman TL, Archer F (2004) A remote sensing-based modeling system to study the Aral-Caspian water regime. Proc 25th Asian Conference on Remote Sensing, Thailand, vol 1, pp 506-511
[4] Golovachev SP, Reutov EA, Chukhlantsev AA, Shutko AM (1989) Experimental investigation of microwave emission of vegetable crops. Izvestia VUZ'ov, Radiofizika, vol 32, pp 551-556 (in Russian, English translation in University Digest, Radiophys Quantum Electron)
[5] Haarbrink RB (2003) High Altitude LIDAR To Enhance GeoSAR System Performance. Proc ISPRS Workshop 3D Reconstruction From Airborne Laserscanner and InSAR Data in Dresden
[6] Jackson TJ, Hsu AY, Shutko AM, Tishchenko Y, Petrenko B, Kutuza B, Armand N (2002) Priroda microwave radiometer observations in the Southern Great Plains. Int J Remote Sensing, vol 23 (2), pp 231-248
[7] Khaldin A, Abliazov V (2004) Radiometer System Installation on International Space Station. Flier by Special Design Bureau, Institute of Radioengineering and Electronics, Russian Academy of Sciences
[8] Kutuza B, Shutko AM, Pliushchev V, Ramsey III E, Logan B, DeLoach S, Haldin A, Novichikhin E, Sidorov I, Nelson G (2000) Advantages of synchronous multispectral SAR and microwave radiometric observations of land covers from aircraft platforms. Proc EUSAR, Munich, 5 pp
[9] Macelloni G, Paloscia S, Pampaloni P, Ruisi R (1994) Spaceborne microwave radiometry of land surfaces. In: Lurie, Pampaloni and Shiue (ed) Microwave Instrumentation and Satellite Photogrammetry for Remote Sensing of the Earth. Proc SPIE 2313, pp 129-135
[10] Macelloni G, Paloscia S, Pampaloni P, Ruisi R (2001) Airborne multifrequency L- and Ka-band radiometric measurements over forests. IEEE Trans Geosci Remote Sensing, vol 39, pp 2507-2513
[11] Macelloni G, Paloscia S, Pampaloni P, Santi E (2003). Global scale monitoring of soil and vegetation using active and passive sensors. Int J Remote Sensing, vol 24, pp 2409-2425

[12]Milieu- en Natuurplanbureau (2004) Risico's in bedijkte termen. RIVM, Bilthoven (in Dutch)

[13]Mkrtchan FA, Reutov E, Shutko AM, Kostov K, Michalev M, Nedeltchev N, Spasov A, Vichev B (1988) Experiments in Bulgaria for determination of soil moisture in the top one-meter using microwave radiometry and a priori information. Proc IGARSS, pp 665-666

[14]Pampaloni P, Paloscia S (1985) Experimental relationships between microwave emission and vegetation features. Int J Remote Sensing, vol 6, pp 315-323

[15]Pampaloni P, Sarabandi K (2004) Microwave remote sensing of land. Radio Science Bulletin, vol 308, pp 30-48

[16]Pampaloni P (2004) Microwave radiometry of forests. Waves in Random Media, vol 14, pp 275-298

[17]Pliushchev VA, Sidorov IA, Malinin AM, Biriukov ED, Shutko AM, DeLoach S (2000) Airborne complex for pre-fire condition detection in forested areas by means of microwave radiometry. Science Consuming Technologies, vol 1, no 1, pp 48-53

[18]Reutov EA, Shutko AM (1986) Determination of the soil water content using microwave radiometry and a priory information. Soviet Journal of Remote Sensing, vol 5 (1), pp 100-106

[19]Reutov EA, Shutko AM (1987) Determination of moisture content of not uniformly moistened soils with a surface transition layer using microwave multi channel radiometry. Soviet Journal of Remote Sensing, vol 6 (1), pp 98-103

[20]Shutko AM (1982) Microwave radiometry of lands under natural and artificial moistening. IEEE Transactions on Geoscience and Remote Sensing, vol GE-20, pp 18-26

[21]Shutko AM (1986) Microwave Radiometry of Water Surface and Grounds. Nauka/Science Publ House, Moscow (In Russian, English translation available in: FTD-ID (RS) T-0584-89, Microfiche NR: FTD 89C-000798 L, 1989)

[22]Shutko AM (1987) Remote sensing of waters and land via microwave radiometry (The principles of method, problems feasible for solving, economic use). Proc Study Week on Remote Sensing and Its Impact on Developing Countries, Pontifical Academy of Sciences, Vatican City, pp 413-441

[23]Shutko AM (1992) Soil-vegetation characteristics at microwave wavelengths. In: Chapter 5, TERRA-1: Understanding the Terrestrial Environment, The Role of the Earth Observations from Space, Taylor & Francis, London-Washington DC, pp 53-66

[24]Shutko AM, Haldin AA, Novichikhin EP, Milshin AA, Golovachev SP, Grankov AG, Mishanin VG, Jackson TJ, Logan BJ, Tilley GB, Ramsey III EW, Pirchner H (1995) Microwave radiometers and their application in field and aircraft campaigns for remote sensing of land and water surfaces. Proc IGARSS, pp 734-735

[25]Shutko AM, Haldin A, Novichikhin E, Yazerian G, Chukhray G, Vorobeichik E, Agura V, Kalashnik S, Sarkisjants V, Sklonnaja N, Logan B, Ramsey III E (1997) Application of microwave radiometers for wetlands and estuaries

monitoring. Proc 4th International Conference on Remote Sensing for Marine and Coastal Environment, vol 1, pp 553-561

[26]Shutko AM (1997) Remote sensing of soil moisture and moisture related parameters by means of microwave radiometry: instruments, data, and examples of application in hydrology. In: Sorooshian, Gupta, Rodda (ed) NATO ASI Series, Global Environmental Change, vol 46, Springer, pp 263-273

Flood Vulnerability Analysis and Mapping in Vietnam

Hoang Minh Hien, Tran Nhu Trung, Wim Looijen and Kees Hulsbergen

Correspondence: H2iD, Geomatics Business Park, Voorsterweg 28, 8316 PT Marknesse, the Netherlands.
Email: hulsbergen@h2id.nl

Abstract

Acknowledging the disastrous implications of Climate Change for socio-economic development, the public-private institute of the Geomatics Busines Park has since 2002 worked on a new concept coined Extreme Events Engineering and Monitoring (EEEM). In EEEM, the word 'Engineering' stands for all types of logic reasoning and activities as long as these help people to firmly reduce their vulnerability to extreme events.

We see EEEM as a 'bi-polar' concept, aiming to deeply integrate a long chain of analyses and activities. On the 'high end' of the EEEM chain, Earth Observation plays an important role. Down at the 'lower end', EEEM firmly rests in the mud in the form of grass roots based evacuation and relief plans as well as long term land use improvements.

Between both poles of the EEEM chain, the so-called Integrated Vulnerability Analysis (IVA) forms the central defence strategy against this 'axis of peril'.

Vietnam is very vulnerable for natural disasters such as flooding, especially during such extreme events as overflowing rivers and excessive local rainfall during typhoons. This paper describes an initial application of EEEM at the vulnerable city of Hue, which was struck in 1999 by extremely heavy rainfall and extensive flooding of the Perfume River.

1 Extreme Events Engineering & Monitoring (EEEM)

The Geomatics Business Park (GBP) is a young and fast growing Public-Private set-up in Marknesse, the Netherlands, with some 20 institutes and companies. The scientific and commercial mission of the GBP is to generate added value by integrating Earth Observation techniques with Water- and Land related Infrastructure expertise in a 'One stop shopping centre' for a very wide range of clients. Within the R&D program of the GBP a new concept emerged some three years ago: 'Extreme Events Engineering and Monitoring' (EEEM).

Acknowledging the disastrous imminent implications of Climate Change for world wide socio-economic developments, EEEM poses a new approach in which Earth Observation (EO) plays an important role. Until recently, potentially useful EO applications in the domain of natural disaster management (e.g. flooding) were not even considered, simply because the hi-tech area of EO was in practice too distant from the more down-to-earth oriented Flood Disaster Management and Relief Operations.

The ambition of EEEM is to help effectively bridging this gap. At the start, defining basic user requirements plays a key role, since the potential of EO is rapidly increasing and thus is sometimes beyond the working knowledge of flood disaster management. Equally true, in EO circles the potential needs of flood disaster management are not always readily clear.

'Extreme Events' can roughly be categorized in the following groups:

- Short or (extremely) long duration
- Multiple or singular
- Local, regional or global
- Water-, land-, air-related or a combination
- Man-made or natural.

Some examples of extreme events are illustrated below.

1.2 Water Related Extreme Events

Water related extreme events include droughts, floods, typhoons and accelerated sea level rise.

Fig. 1. Glacier Perito Moreno breaks down in 2003 for the first time in 16 years in Lago Argentino. Credits: AP Photo/Guillermo Gallardo-Telam

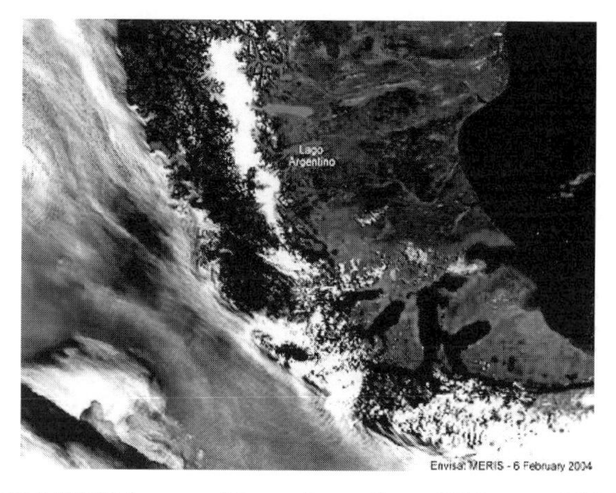

Fig. 2. ENVISAT-MERIS image of Lago Argentino of February 2004 © ESA

Drought, which is often aggravated by human action, may affect very large areas for months, even years, and thus has a serious impact on regional food production, often reducing life expectancy for entire populations and economic performance of large regions.

Floods come in several types. River Floods originate from winter and spring rains, coupled with snow melt, and torrential rains from decaying tropical storms and monsoons. Coastal Floods are generated by winds from intense off-shore storms and tsunamis. Urban Floods are exacerbated by intense urbanization, increasing runoff up to six times what would occur on natural terrain. Flash Floods can occur within minutes after excessive rainfall, a dam or levee failure, or an emergency release of water from a reservoir filled to capacity.

Typhoons (or tornados) can cause enormous damage due to high wind speeds and accompanying rainfall while devastating waves batter coastlines, ships, offshore structures and harbors.

Accelerated sea level rise is one of the slowest but world-wide felt and serious extreme events. Low lying countries, such as the Netherlands and large parts of Vietnam, are extremely vulnerable for sea level rise due to global warming. Under this global warming ocean water volumes expand, ice caps melt and sea level rises.

Fig. 3. Etna eruption October 2002 © ESA

1.3 Land and Air Related Extreme Events

Land and air related extreme events include volcanic eruptions, earthquakes, fires and landslides. Persons and properties on the ground near erupting volcanoes or near the epicenters of earthquakes are under threat. Ash clouds from major volcano eruptions endanger aircraft and airport operations over distances of thousands of kilometers. The damage done by earthquakes to physical infrastructure disrupts public life.

Fig. 4. Earthquake Bam, December 2003 © Space Imaging

Fires are caused by human activities or by natural phenomena such as lightning or volcanoes. They cause loss of human life and personal property, economic upsets, and changes in regional and global atmospheric composition and chemistry. Moreover, forest fires add to the risk of deforestation in terms of accelerated run-off and flooding.

Landslides threaten settlements, structures that support transportation, natural resources and tourism. They cause considerable damage to highways, railways, waterways and pipelines. They commonly coincide with other extreme events such as earthquakes, volcanic eruptions, and flash floods caused by heavy rainfall.

In any of these cases, extreme events deserve much focus since they:

- Are beyond imagination for most people
- Are increasingly outside traditional 'design conditions'
- Cause massive damage and prolonged social disruption
- Are far beyond local coping capacity, threatening development.

1.4 What Can Be Done with EEEM and Who Are Beneficiaries?

Climate Change is inevitably associated with the occurrence of more extreme events. The IPCC in its Third Assessment Report concludes that floods and periods of drought will more frequently occur. However, too often in spatial planning the impact of (increased future frequency of) extreme events is not taken into account when an area is developed. With growing urbanization and expansion of settlements especially in fertile floodplains, disastrous effects of floods will increase. These serious effects can be minimized if extreme events are taken into account during the planning and engineering process, and if extreme events are monitored as a basis for understanding and better forecasting.

1.5 Engineering

To better be prepared for, adapt to and cope with the expected increasing frequency and severity of future extreme events, it is clear that the associated spatial planning, the changing boundary conditions, the design and the construction of the physical infrastructure, residential areas and other buildings in disaster prone areas should be based on:

- Historical information on the occurrence of extreme events
- Information (estimates) on future occurrence of extreme events
- Detailed vulnerability information for such areas; the concept of vulnerability is further discussed in Section 4.

This means that in the process of planning new works a proper (engineering) analysis has to be made focusing on the production of the above type of information.

This implies a major difficulty especially for very large scale developments, since the total time duration between first concept and commissioning of these grand works often exceeds a decade. During this relatively long time the hydraulic boundary conditions (which to a large extent define the engineering design) themselves may change appreciably, precisely because of Climate Change. It is clear that through this specific type of uncertainty the (financial) risk for the entire project may be difficult to estimate. This asks for an even stronger focus on monitoring, understanding, modeling and predicting quantitative characteristics of future extreme events.

1.6 Monitoring

Areas susceptible to extreme events, as well as the extreme events themselves, should be closely monitored. Main goals of this monitoring are:

- Understanding of the physical processes of extreme events
- Building up a statistical data base for predictions
- Better insight into existing weak spots, and
- Identification of potential possibilities for interventions to better cope with the effects of these events in the future.

Requirements for monitoring and its analysis include the following:

- Information should be easy to understand and use

- Easy access to data sources and documentation
- Easy integration of extreme event information into other systems

Monitoring activities should be designed to facilitate:

- Early warning of extreme events
- Estimation of area, intensity and duration of extreme events
- Identification of confidence level for the occurrence of extreme events
- Planning for immediate relief and long term management for extreme event mitigation
- Education of the beneficiaries of extreme event information on how to interpret the information

1.7 Beneficiaries

Several 'levels' of beneficiaries are identified:

- Primary policy makers at the national level and within international organizations; also the associated researchers
- Policy makers at provincial level; further consultants, relief agencies, researchers and insurance companies; also trade, transport, and engineering companies
- Local policy makers at implementation level, and also local producers such as farmers, suppliers, builders and water managers
- But most of all, the final beneficiaries of EEEM are the people in the field facing the threats of extreme events every day

2 Flooding Disasters in Vietnam

Vietnam is one of the most disaster prone countries in the world. Hazards include floods, typhoons, inundations, droughts, and forest fires. Vietnam's average per capita income is $ 500 per year. Central Vietnam, one of the poorest regions with 7.5 million inhabitants, is more vulnerable than other regions. Over the last 25 years, especially flooding disasters form a growing threat, killing hundreds of people each year. Many locations are hit by disasters while they have not yet recovered from the consequences of the previous year's disaster. The recent accumulation of natural disasters in Vietnam indicates that an abnormal climate trend is in progress.

Examples are the large floods of 1996, hitting all regions of the country, causing heavy loss of lives and property, with an estimated total damage of

$ 720 million; this was considered the biggest economic loss ever in Vietnam. On November 2, 1997 the typhoon Linda landed in Bac Lieu province on the southern tip of Vietnam, claiming 3,000 lives. In late 1997 and early 1998 a severe drought occurred in most of the Central Provinces. In addition, five big tropical storms consecutively hit the Central Provinces, causing heavy rainfall and flooding. The province of Thua Thien Hue and neighboring provinces were struck by devastating floods in November 1999, killing some 700 people. Maximum rainfall in the capital Hue City reached 1,384 mm during a single day, causing immense havoc in this UNESCO World Heritage site. In 2000 and 2001 the Mekong River delta suffered from particularly large floods. It was the most severe flooding of the last 70 years, in terms of extremely early time of occurrence, deep inundation, total inundated area, and inundation period. Almost 500 people were killed; total economic loss was $ 280 million.

In the near future frequency and severity of extreme events (such as devastating river flooding, possibly combined with local torrential rains and typhoon related storm surges in the lower river branches and lagoons) are expected to increase due to Climate Change.

The Vietnamese Disaster Management Unit (DMU) has national responsibility for mitigating of and responding to natural disasters in general. The DMU is responsible for the day-to-day organization of disaster relief through information provision and relief operations. In planning and implementing strategies and measures for natural disaster mitigation in Vietnam, the DMU plays a central role. The DMU operates under the Central Committee of Flood and Storm Control (CCFSC) which is directly positioned under the central government. The powerful Ministry of Agriculture and Rural Development (MARD) is charged with the Vice Chairmanship, while 15 other Ministries and some important central Departments and Services are represented as Members in the CCFSC.

Strategies and measures for disaster mitigation in Vietnam include:

- Development of policies on problems of flood and inundation
- Improving integrated management systems, including establishment of laws and regulations, and assigning responsibilities to concerned ministries and agencies
- Planning and improving structural and non-structural measures
- Improving flood warning and forecasting methods
- Paying attention to some specialized non-structural measures, for example development of flood and inundation prone area maps; land use planning; flood response options

3 EEEM Activities in Vietnam

In 2003, two EEEM fact finding missions to Hanoi and Hue focused on user needs and on the specific user environment, in terms of physical, technical, societal and institutional conditions. These missions were effective thanks to open and very constructive talks with the DMU and to the fact that the missions were embedded in two broad and long term projects: the Vietnamese-Netherlands Integrated Coastal Zone Management (VNICZM), and the Coastal Cooperative Program (CCP, Netherlands Ministry of Transport, Public Works and Water Management). General findings of this reconnaissance phase include:

- Flood control is of utmost importance to Vietnam, especially in the Red River Basin (Hanoi), the Huong (Perfume) River Basin (Hue City) and in the Mekong Delta.
- Flood preparedness is needed in Vietnam and can help save many lives
- Flash floods are a reoccurring phenomenon in Vietnam; there is no system in place for flash flood warning
- The people of Vietnam have learned to live with floods but timely information on timing, duration, inundation depth and spatial extent of flood is required
- Factual, quantitative and effective (mapped) information on the vulnerability of flood-prone areas is missing
- Urban planning and planning of waterworks does not take vulnerability of potentially flooded areas into account
- Near-real-time rainfall estimates are required; in combination with vulnerability maps they can be used to set up flood early warning systems
- Ground stations are needed for immediate and cost effective Earth Observation data usage
- The speed of high-water flood waves descending from mountainous regions to low-land areas is mostly unknown
- Capacity building in flood modeling and flood management, Earth Observation and GIS applications is needed
- Earth Observation and GIS activities are mostly carried out at Universities and at government institutions in Hanoi, not by consultancies and not in remote areas
- Accurate Digital Elevation Models are required
- Combination of existing GIS layers with new Earth Observation derived information is needed

Based on these first findings, a follow-up phase of this EEEM project was assigned under the Netherlands' NIVR-GO program, which runs from August 2004 to June 2005. During this consecutive phase, the use of Earth

Observation and GIS data in a context of flood disaster warning, mitigation and relief will be analyzed and demonstrated during two visits to Vietnam. Geographical focus is on two river systems: the Huong (Perfume) River near Hue City in Central Vietnam and the Mekong River in South Vietnam. Specifically, the concept of Flood Vulnerability Mapping (now under development) will be addressed.

A 'first-order' Flood Vulnerability Map will be composed using data from Landsat7, ERS, ENVISAT, RADARSAT and ASTER. The great flood of November 1999, which inundated large parts of the lower Huong (Perfume) River area, has been relatively well documented in terms of ground truth, especially with respect to time dependent inundation depth and damage. During a Workshop in April 2005 in Vietnam the findings of this project will be communicated and evaluated with the DMU together with local and provincial users and international experts.

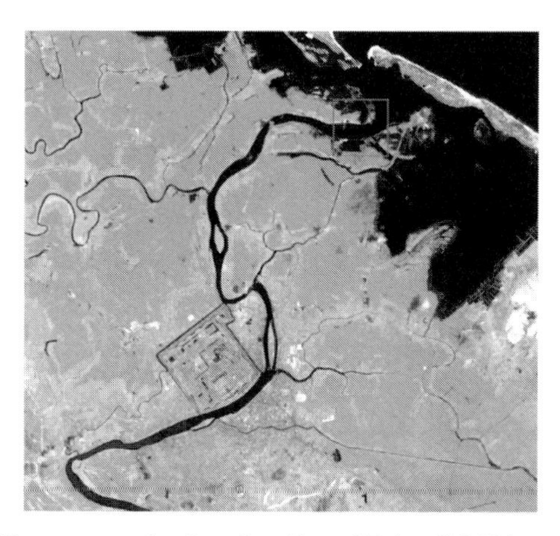

Fig. 5. The City of Hue as seen by Landsat 7 on 17 April 2003

4 Need for Flood Vulnerability Analysis

Water in Vietnam has always been a basic fact of life and of death as well. More than 2000 rivers and streams wind through the mountainous and hilly landscape, originally covered with dense forests. They formed a mostly narrow alluvial coastal plain. Two major river systems import water from upstream countries: the Red River in the north and the Mekong River in the south. These rivers have formed large delta plains in Vietnam, scattered with agricultural settlements and major cities amidst a myriad of river branches and channels. In between these two large river deltas, the third major region of Vietnam can be characterized as a relatively narrow strip of high mountains steeply sloping towards the sea: 'Central Vietnam' with smaller cities like Hue.

Seasonal floods are a basic part of life in Vietnam. Historically, floods were accepted as a given fact. A flood is thought to be perceived as a 'disaster' if and when awareness emerges that its devastating effects may be curbed, mitigated, and perhaps even prevented, through better understanding of its causes and through a gradual development of means of intervention. At that moment the notion of 'flood disaster management' is born.

Three different strategies for river flood management have gradually been developed for the three main regions of Vietnam. In the Red River delta the main line is "Dike protection", including watershed forest protection, upstream reservoir building, flood diversion schemes, dike system improvement, river bed dredging, and providing emergency spillways. In Central Vietnam the main line is "Active preparedness, mitigation, and adaptation", including improvement of drainage channels with concrete linings, changing crop patterns, river bed dredging, setting up warning systems, supporting local preparedness, and planning multi-purpose upstream reservoirs. In addition for the central coastal area this line includes mangrove planting, fisher village repositioning, sea dike improvement, stabilization of tidal inlets, evacuation planning, and building fishing boat shelters. Finally in the Mekong delta the main line is "Living together with floods", including building embankments for the protection of residents and crops, changing crop patterns, setting up evacuation areas, elevated house construction methods, and resettlement.

These official flood management strategies are embedded in a legal framework. Planning and implementation of these strategies is supported by a number of disaster prevention and mitigation projects. Some of these projects are supported by UNDP and other donors.

An issue in preparing effective and broadly supported flood mitigation strategies is formed by the need to find a clear and effective ranking mechanism among all these plans. Which project must be done first? And why that particular project? What will be its costs? And, much more difficult to define, what will be the benefits, but also: what might be the (unforeseen) longer term environmental drawbacks? In other words: to which of the many national, regional and local flood mitigating needs should the scarce available resources (funding and qualified personnel) be allocated in order to generate maximum benefit on short, medium and long term?

A convincing answer to this important issue is hard to find. Especially so since we deal with a 'moving target' with double uncertainty. First, the vulnerability of people, economy and ecology is changing faster than ever due to the vast economic growth rate (economic developments tend to concentrate in low lying areas, preferably next to rivers and coastlines). Second, because the climate together with its extreme events is changing 'as we speak', in a direction which at this moment is difficult to tell with precision. Yet, finding an answer, even by way of approximation, would be quite rewarding especially if such answer can be generated using a transparent methodology. Such clarity could greatly strengthen the commitment of all parties involved, because the goal and the way towards it would be clear. This would make it easy and even attractive for the stakeholders to accept the method and therefore support the needed actions.

A potentially successful approach towards finding the sought method of ranking is through a 'Vulnerability Analysis'. What we mean by 'Vulnerability Analysis' (or VA) is a well structured and systematic method of research which may be described as 'defining the links in a chain of (mostly uncontrolled and/or unwanted) processes through which an external hazard is affecting ourselves and/or our environment'. This could just seem a very lengthy way to express that we 'want to understand what happens to us'. But VA is more than that. A major characteristic of VA is that it almost 'automatically' comes up with suggestions as to What We Must Do To Reduce Vulnerability. And even more than that, VA also tells us (though not entirely automatically) which weak factor in the chain of events will, after being repaired, yield maximum benefit for the least effort. This comes close to what was our goal from the start.

As an example of how VA in general may work, let us first briefly consider the issue of traffic accidents. In Vietnam, 1000 people are killed each month through accidents in street traffic. So there is good reason to classify everyone on the road as being 'vulnerable to the hazard of traffic'.

Now suppose we are responsible to reduce the killings, in other words 'to reduce this specific vulnerability'. We could approach this problem in a structured way by starting to make a thorough analysis of each and every lethal accident ('extreme event') in as much detail as possible. This should include (and this is crucially important) all thinkable possible adverse circumstances, both physical and mental, which somehow might have played a role in causing this particular person to be a traffic victim. (It could be especially effective to also investigate people who have only been slightly injured, and have therefore luckily survived: the 'nearly-killed').

By doing this, one deliberately closes in on all possible 'weak points' in the (physical, social and mental) process of street traffic. Central to successfully identifying these weak points is continuously asking oneself questions like: "Might the victim have survived if he/she had….." or "Might the victim have survived if this or that 'external' physical circumstance had been a little more….." Next, these 'weak points' form the onset for ever so many potential points of intervention and improvement. One simple example: A car driver is killed after his car frontally hit a roadside tree. The list of logical qucstions (and onc out of many possible interventions) might include:

- Was he drunk? (Reduce allowed blood alcohol promillage)
- Did he speed? (Make artificial bumps)
- Did he wear safety belts? (Prescribe & check using safety belts)
- Did he phone? (Forbid using the phone while driving)
- Did he avoid hitting a child? (Teach traffic rules at school)
- Did his brakes fail? (Let older cars have yearly check-ups)
- Was the road too close to the tree? (Change road design rules)

To be done for all individual accidents. As can be seen, both physical and non-physical potential interventions are generated in an almost automatic way. The next logical questions concern benefits, costs and implementability, leading to the sought ranking order in the long list.

A first, simple, preliminary but quite useful result of such traffic accidents VA is a road map with so-called 'black spots': locations with an extremely high concentration of lethal accidents. Depending on the particular case, even without further analysis, some conspicuous warnings signaling "black spot" could be erected a little upstream of the spots at hand ('early warning system'). These might already do good work, at the expense of only a very small effort. More effort, but also with a higher yield, might be put into taking some other measures. So far for the traffic example.

Coming back now to the flood disaster issue, one can perhaps see some strong similarities with the traffic victims' issue. Both types of hazard, though completely different in character, show some striking parallels in their systematic analysis approach, especially in the need to understand what the physical and mental detailed processes of the natural phenomena and the potential victims really are. But also in the way to determine a full scope of potential ways of mitigating, escaping, protecting, and diverting the well defined partial risks. Finally also in the possibility to sort out and develop a ranking order of 'best practices'. Just as in the example of traffic hazards, it is possible in the flooding hazard domain to regard various consecutive stages or levels of accuracy.

In a first stage of flood VA, it would already be a great help if good 'vulnerability maps' were available identifying 'black spots' by way of a crude 'early warning system'. Just as in the traffic accident issue, local knowledge of the actual historic flooding phenomena is necessary. Also, 'near flood disasters' are worth while to be analyzed. Methods should be prepared to monitor any new floods in detail.

With over 2000 rivers in Vietnam, usage of Earth Observation methods clearly is a must. Apart from monitoring the inundated areas during floods, numerical hydraulic models of river and inundation behavior should be developed to improve the understanding of the relevant phenomena. These should be calibrated through field measurement campaigns. The models should be fed with data about rainfall and produce information about run-off, groundwater flows, and also show the elevating effect of typhoons on water levels along the coast and in lagoons, and their reducing effects on river drainage capacity.

Among all the various ingredients which together make up the vulnerability of an area (including its inhabitants) to flooding risks, lack of awareness is probably the most underestimated, 'mental' factor. We have provisionally identified several 'levels of problem ownership' which hopefully will make the local people less vulnerable in several consecutive steps, through applying a range of typically 'non-physical' interventions.

5 Geo Information for Flood Vulnerability Analysis

In EEEM the role of Earth Observation and GIS is important but it is not the solution to the problem. However, Earth observation and GIS can assist in deriving part of the required information. Mapping vulnerability is one item which can be resolved using a combination of Earth Observation data, GIS data and local knowledge, even at scales 1:50,000. The local topography of the terrain can quickly be derived from stereo satellite imagery at resolutions of 1 to 15m. Deriving land use information from multi temporal satellite imagery is a straight forward activity.

Existing GIS data, e.g. road network and river network, can be overlaid with the Earth Observation derived information to present a visualization of the terrain and to analyze and simulate various scenarios. In combination with e.g. hydrological forecasting models, the impact of potential extreme events can be determined and used in early warning systems. These can help prepare the local population to take immediate action in response to the extreme events.

Given the availability of EO and GIS, it is equally important to educate and train local (government) people on the use of the information, on the integration of that information into the decision-making process, and to integrate the Earth Observation derived information with other information, such as demographic information and social-economic information. All these different types of information should be integrated in hazard maps, hazard plans, hazard manuals, evacuation plans etc.

Besides that, Earth Observation and GIS based information should be readily available, i.e. in real-time, since immediate actions in case of extreme events are required.

The EEEM activities in Vietnam deal mainly with producing flood vulnerability maps of the area around the city of Hue, and of the area between Can Tho and Chau Doc in the Mekong River delta. The approaches used in both instances are similar: an already occurred flood is taken as a reference and radar data from ERS, ENVISAT and RADARSAT is used to determine the flood extent.

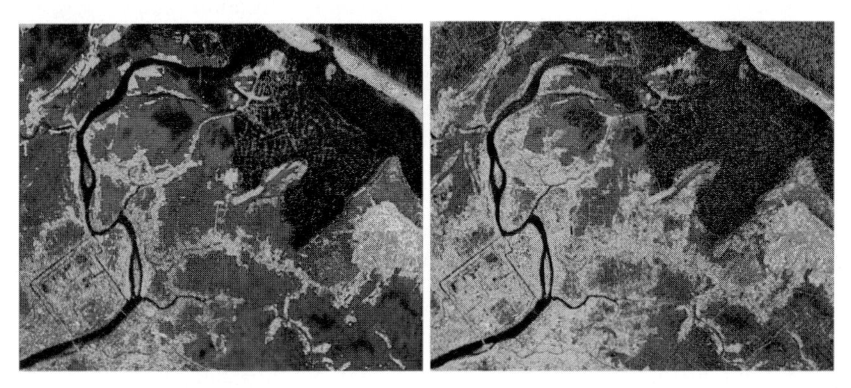

Fig. 6. RADARSAT images of Hue taken on 6 November 1999 (left) and 15 November 1999 (right) on top of a Landsat7 image of 19 September 1999. The sea is in the top right hand corner. Light grey indicates non-flooded areas. Dark grey and black indicate flooded areas. The 700 m by 700 m inner imperial city was temporarily inundated to a depth of 2 m according to eyewitnesses.

Furthermore, a basic land use map is produced on the basis of Landsat 5 and 7 data. The land use classes are: agriculture, water, forest, urban and other. To determine flood vulnerability, information on the topography is required. To that end data from the Shuttle Radar Topography Mission (SRTM) is used at 90m grid size. For the area around the city of Hue a Digital Elevation Model (DEM) at 15m grid size was derived using ASTER data.

These three types of data are combined to derive a flood vulnerability map of the particular area. For the area around Hue a separate analysis is carried out to identify land use changes between 1991 and 2003 in order to assess the influence of the floods in land use and to study forest changes as a possible indicator for increased flooding upstream the Huong River.

To raise awareness among local and provincial authorities a GIS simulation is developed that enables the visualization of floods and the effects on people, land and properties.

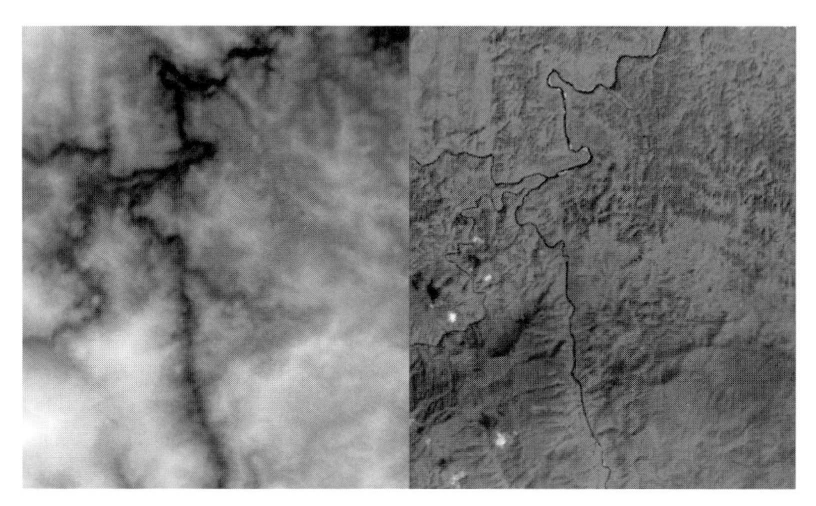

Fig. 7. ASTER derived DEM (left) and original ASTER image (right) of the area near Hue City of February 2002

One of the problems associated with the extensive low-lying area in the Mekong River delta is formed by the small height differences. In the absence of accurate and up-to-date information on the topography the FLI-MAP laser altimetry system by Fugro, the Netherlands, will be used to demonstrate the potential of this technique to derive DEMs with vertical resolutions of up to 5cm. First, a pilot project is envisaged early 2005 in the Red River delta near Hanoi. The laser altimetry technique has successfully been applied in the Netherlands, South Africa and the USA to map dykes, railways and power lines very accurately.

The province of Ha Giang in North Vietnam is frequently suffering from flash floods due to very heavy rainfall. Here, a flash flood risk map is constructed using topographic data from the SRTM mission. Using existing GIS data on soil properties in combination with slope and aspect information provides information on the possible vulnerability for flash floods. Furthermore, rainfall estimates are derived from low resolution meteorological satellite data. Together, these sources of information will assist the people of the Ha Giang province to better prepare for flash floods emergencies and to save human lives and properties.

Acknowledgements

The VNICZM offices in Hanoi and Hue are thanked for their support during field work and for the exchange of information. The moral support from RIKZ, especially through Dr Robbert Misdorp is greatly acknowledged. Last but not least, the authors would like to thank the Foundation Geomatics Business Park, for their initial support to develop EO and GIS based information products, and the Netherlands Agency for Aerospace Programs (NIVR) for their support to develop and demonstrate the flood vulnerability analysis and mapping in Vietnam.

For more information on EEEM please consult www.eeem.nl.

Geo Information Breaks through Sector Think

Stefan Diehl[1] and Jene van der Heide[2]

[1] Public Aid 'Gelderland Midden' region, Department of Disaster Recovery and Large-Scale Emergency Management, Beekstraat 39, PO Box 5364, 6902 EJ Arnhem, the Netherlands.
Email: stefan.diehl@hvdgm.nl

[2] Municipality of Arnhem, Department Geo Information, Van Oldenbarneveldtstraat 90, PO Box 9200, 6800 HA Arnhem, the Netherlands.
Email: jene.van.der.heide@arnhem.nl

Abstract

In case of disaster recovery an enormous information flow arises between administrative and operational groups, which during ' times of peace' would not exist in this form, or if so, at a much more modest scale. This complex structure of administrative tiers and operational services is served by as many, or perhaps even a larger number of (geo) information systems. The use of geo information and geographic information systems for combining, analyzing and visualizing data at this point has not taken sufficient root in the disaster recovery structure. If geo information is used at all, it is within the individual organization, and often it is not possible to share data with other partners in the chain.

The central problem as defined for our paper thus reads: *"What should be the focus of the Public Order and Safety Sector in order to establish an adequate geo information facility?"*

Further to this central problem, we shall deal with the following derived questions in this paper:

- What is the project result of *Veiligheidsnet/Geo-Informatie (Safety Net/Geoinformation,* referred to in this paper by its Dutch name*)*?
- What is a geo information facility?
- Why is *Veiligheidsnet/Geo-Informatie* a success and how can others learn from it?

The output of the *Veiligheidsnet/Geoinformatie* project is 'VNET'. VNET is a system that can be used by chain partners within as well as outside of the Province of Gelderland to communicate with each other via maps in the event of a disaster. VNET can also be used for consulting geo information during 'times of peace', in other disaster management stages. Examples would be work processes of the pro-action, prevention or preparation departments.

In addition to the intended project output there has also been some spin-off. The collaboration between the different chain partners (regional as well as cross-regional) in the area of geo-information has improved.

One has become aware of the necessity of consolidating geo information within the scope of disaster recovery and there is an increased awareness of the critical factors for success, the absence of which interferes with a further development of the use of geo information.

Our definition of a geo information facility is the following: "Opening up and making accessible the information that indicates the relative and the absolute situation of objects and occurrences." The supply of geo information has five critical factors for success: social, technological, organizational, economic and political.

The single most important pillar on which the *Veiligheidsnet/Geo-Informatie* project rests is Collaboration. It has been demonstrated that chain partners from different disciplines can work together (the social and organizational aspect) in order to arrive at a functionally and technologically sound solution. The project can serve as an example for the organization of the geo information supply throughout the entire Public Order and Safety Sector.

The Public Order and Safety Sector must be aware of the following aspect, if it wants to establish an adequate geo information facility in the future:

- sector think hampers effective collaboration;
- municipalities have insufficient access to geo information within the scope of disaster recovery;
- there is no adequate political steering that would enable a method of exchange;
- the 'polder model' is neither effective nor efficient in bringing about a solid geo information facility.

The experience gained through VNET indicates that the above aspects can be overcome and that the central proposition can actually be refuted. It is a matter of smart organization.

Preface

Geo information is crucial in adequate disaster recovery. Close collaboration between the geo information departments of the municipalities of Arnhem and Renkum, the Province of Gelderland and Public Aid 'Gelderland Midden' has resulted in a higher level of geo information knowledge with the regional fire brigade as well as a higher level of knowledge on disaster recovery in municipalities and the province. On the shop floor one looked at the problem with the supply of information by better facilitating the geo information supply. This enhances the quality of disaster recovery.

In the Netherlands, the use of geo information is currently viewed as offering great opportunities for the Public Order and Safety Sector. Evaluations of recent emergencies demonstrated that a sound information position is crucial to all partners in the chain. This is underpinned in national and local policy. The *Veiligheidsnet/Geo-Informatie* project has shown that chain partners from different disciplines can work together to arrive at a functional and technological solution. In our country there are various initiatives to this effect.

The conclusion that there is political support for the idea of incorporating the use of geo information in the organization of disaster recovery and that, moreover, there are no technological impediments, prompt us to raise the question why it is that these matters have not been properly arranged yet in the Netherlands.

All ingredients for an optimum supply of geo information for the benefit of Dutch disaster recovery organizations are there: there is money, political support, knowledge and expertise, the technology is up to it and the Veiligheidsnet/Geo-Informatie project has proven that problems of a social and organizational nature can be overcome.

It is just a matter of getting it done.

1 Introduction

In the Netherlands, aid services, municipalities, provinces and other parties involved practice with disaster scenarios every year. Sometimes, an actual disaster occurs. Each year, situations occur in the Netherlands where the process of disaster recovery and crisis management is started off. This practical experience shows that high quality information is needed to allow for the adequate recovery of disasters. Said information is usually tied to a particular location, which is why we can call it geo information. This geo information is not automatically available for each partner in the chain, especially not in case of an emergency. This is not, for that matter, a new phenomenon in the Public Order and Safety Sector (Dutch acronym: OOV). Which is all the more remarkable as municipalities and Provinces are making considerable progress with their geo information facilities.

Public Aid 'Gelderland Midden' is convinced of the strategic importance of geo information in the organization of disaster recovery and crisis management. The geo information facility must be completed so that information will become available to the different partners in the chain faster.

The Province of Gelderland has the ambition to set up a safety network, called *Veligheidsnet*, that will support the parties involved in the event of a disaster or an emergency, as well as in the daily work processes. What is special about *Veiligheidsnet*, is the pivotal role that geo information plays in it. Herein below, said partial project shall be referred to as *Veiligheidsnet/Geo-Informatie*. Apart from widely known communication vehicles such as the telephone, fax machine and face-to-face communication, the *Veiligheidsnet* vehicle focuses on communication via maps. Having relevant geo information available and accessible during disasters, disturbances of the public order and at times of peace will help the stakeholders to act more adequately. The objective of the project *Veiligheidsnet/Geo-Informatie* is: *the realization and implementation of a system that allows for the exchange of geographic and administrative data between the different partners in the chain.* From this objective it naturally follows that there will also be an investigation into opportunities for disclosing the data that the different chain partners have available under normal circumstances. At this point, two prototypes are available, which are being tested during the disaster practice in the different municipalities.

Starting from the *Veiligheidsnet/Geo-Informatie* project, this paper shall pay attention to the results of the project on the one hand and the context of the *Veiligheidsnet/Geo-Informatie* project on the other. We hope that this will lead to a further collaboration and sharing of knowledge between the parties involved, and that eventually this will lead to a better supply of information during disasters, incidents as well as the day-to-day work processes.

1.1 The Use of Geo Information

In disaster recovery an enormous flow of data arises between administrative and operational groups, which during ' times of peace' would not exist in this form, or if so, at a much more modest scale. As disaster recovery involves a large number of chain partners (see annexes 2 and 3), and seeing how each partner in the chain has its own data system and architecture in place via which the information can be opened up, the streamlining of said data flow in the event of a disaster is incredibly difficult. Or so it seems.

The Netherlands have coped with a number of considerable disasters, of which the fireworks disaster in Enschede is by far the largest in scale. Twenty-two people lost their lives, 947 were injured and an entire neighborhood was destroyed. The evaluation of this disaster clearly demonstrated that crisis management mainly entails information management, including the difficulties that arise in terms of the supply of information.

*Quote from the Oosting Committee research report 'Vuurwerkramp Enschede' (*Enschede Fireworks Disaster*), February 2001:* "The recovery of disasters takes place within a complex administrative/organizational culture. First of all, there is the tiered system of the internal administration. The basis of the recovery of disasters rests with the municipality' (...) 'The system of the distribution of administrative responsibilities over the different administrative tiers goes hand in hand with a system of different operational services. Together they constitute a complex structure, which, however, will only be activated the extraordinary circumstances of a disaster.'

This complex structure of administrative tiers and operational services is served by as many, or perhaps even a larger number of (geo) information systems. Tat this point in time, the use of geo information and geographic information systems for combining, analyzing and visualizing data at has not taken sufficient root in the disaster recovery structure. If geo information is used at all, it is within the own organization, and often it is not possible to share data with other partners in the chain.

1.2 Problem Definition

The central problem as defined for our paper reads: "What should be the focus of the Public Order and Safety Sector in order to establish an adequate geo information facility?"

From this central problem, we shall deal with the following derived questions in this paper:

- What is the project result of *Veiligheidsnet/Geo-Informatie*?
- What is a geo information facility?
- Why is *Veiligheidsnet/Geo-Informatie* a success and how can others learn from it?

1.3 Structure of the Paper

In this paper we shall give an account of the *Veiligheidsnet/Geoinformatie* project. The context of the project, however, is equally important. Our reason for writing the present paper is to demonstrate what should be the focus of the Public Order and Safety Sector if it wants to establish an adequate geo information facility, and that a project such as *Veiligheidsnet/Geo-informatie* shows that there are plenty of opportunities for doing so.

In Section 2 we shall introduce five factors that are critical to the success of the geo information facility. In that same Section we shall establish the correlation between these five critical factors for success and the output of the *Veiligheidsnet/Geo-informatie* project as well as the broader context of the Public Order and Safety Sector. A proper understanding of the term geo information facility is required if one is to gain a thorough understanding of the project and its context.

We shall then home in on the output of the *Veiligheidsnet/Geo-informatie* project: VNET.

In Section 3 we shall set out the objective and parameters governing the project, as well as its design, output, architecture and costs and benefits. In Section 4 a conclusion and some recommendations will follow.

2 The Supply of Geo Information in a Wide Perspective

In the Netherlands, there is much interest in geo information. In all administrative echelons we can see interesting initiatives developing. At the level of central government, there is the policy document called 'Ruimte voor Geo-Informatie' (*Room for Geo Information*)[1]. In Provinces we see collaboration with municipal projects such as DURP [2](Digital Exchange of Spatial Planning Initiatives) being launched successfully. In municipalities we see how more and more geo information can be opened up via the Intranets, while on the Internet more and more interactive topographical materials become available.

The project *Veiligheidsnet/Geo-Informatie* sees to an enhanced supply of geo information during disasters and calamities but also during the day-to-day work processes. But what exactly do we mean when we talk about the supply of geo information and which aspects are pivotal in terms of the

---

[1] For further information please refer to: www.ruimtevoorgeoinformatie.nl

[2] For further information please refer to: http://www.vrom.nl/pagina.html?id=7406

success of the supply of geo information? In our Introduction you have already read that thus far the Sector of Public Order and Safety has not made adequate arrangements for the supply of geo information. Which aspect of supplying geo information should one home in on in order to turn this around?

2.1 Definition and Aspects

Our definition of a geo information facility is the following: "Opening up and making accessible the information that indicates the relative and the absolute situation of objects and occurrences."

The supply of geo information has five aspects[3]. Each of these individually are critical success factors:

1. Social;
2. Technological;
3. Organisational;
4. Economic;
5. Political.

One example of the social aspect is the presence of 'keen' staff. Staff that are enthusiastic and willing to make the effort to 'tear down walls' and that are not afraid to think out of the box. The technological aspect concerns the defining of the technological parameters for disclosing and managing geo information. In fact, we are talking about the required hardware and software. The organizational aspects concerns the way in which you can establish a decisive project organization with all the different organizations or parts thereof that are involved, but also how you can organize the eventual work of the project organization in a structured manner. The arranging for and covering of financial requirements and making manifest the costs and benefits are all part of the economic aspect. And finally you will need political and administrative commitment to be able to set up the geo information facility.

Based on these five aspects we can determine which particular area of the geo information supply in the Public Order and Safety Sector (OOV) deserves more attention, and subsequently in which of the areas the project *Veiligheidsnet/Geo-Informatie* proves to be successful.

---

[3] Based on presentation by B. ten Brinke, *Organisatie en GIS*, 2000

2.2 The Aspects Applied to the Public Order and Safety Sector

Embedding the use of geo information for the sake of disaster recovery is not something that can be arranged overnight. All the parameters for a correct geo information facility must be fulfilled. During the realization of the project one came across a number of issues that needed to be resolved before a further integration of geo information could took place.

2.2.1 Politics

The Home Office noted that the information supply within the scope of disaster recovery is not what it should be. Quote (from the Crisis Management Policy Plan 2004-2007[4]) 'The communication and exchange of information between the crisis partners is becoming a more and more crucial bottleneck. The different parties involved in the crisis all have different information systems, architectures and protocols in place.'

This means that there definitely is the political commitment to properly arrange the exchange of information during a disaster. It turns out to be difficult, however, to succeed in giving shape to this commitment. What is lacking is a policy on the use of geo information in disaster recovery that gets nationwide support. The political arena must be more directive in case of the geo information supply between the organizations involved.

One must lay down requirements for the quality level of the geo information facility. There is much to be gained from stimulating the market and informing the partners within the chain on the importance of a sound geo information planning and the use of geo information.

2.2.2 Organization and the Social Aspect

The use of geo information per safety region and/or chain partner is fragmented. Also, there are massive difference in terms of geo information knowledge per safety region and between chain partners themselves. Disaster recovery is a combined effort of multiple organizations. This is governed by law. So then how can a good geo information facility between multiple organizations be realized?

---

[4] To be downloaded here: http://www.minbzk.nl/contents/pages/9070/definitievetk-versiecrisisbeheersingsdu-eindversie-25juni2004.pdf

There proved to be quite a few obstacles to overcome in the OOV Sector and sector-focused working and sector think still prevail[5]. The public aid services, provinces and municipalities are strongly inclined to approach matters from their own perspective. Now this is only logical, in part anyway. Still, if you need to work together you must organize matters from a different perspective. This is where the social and organizational aspect simultaneously come into play. Information is the flaw in disaster recovery. The typically Dutch compartmentalization and sector operation have brought this flaw to light and made the system vulnerable.

Quote (Regional Disaster Recovery Policy Plan 2005-2008): '*Given the conclusion that the matching of information is the flaw in crisis management, we are placing an emphasis on a multiple discipline approach to enhance the flow of information' (...) 'It will be necessary to embed the use of geo information in the organization of disaster recovery' (...) 'chain partners need to agree on their data being used across the different disciplines' (...) 'chain partners are advised to digitalize their analogous data and to make them available to the other partners' (...) 'we need to invest in knowledge and education'*.

This contrasts sharply with the successful results of municipalities and provinces, which have already organized their data facilities fairly well (horizontal attuning) and on whose priority lists the opportunities this offers feature prominently[6]. In disaster recovery, municipalities are responsible for many processes (see annex 1). In many of the processes, geo information plays an important role. This does not imply that municipalities have actually worked out their geo information facility in case of disaster recovery for 100%. It appears that in actual practice, the geo information staff have not discovered the area of disaster recovery yet. And it is this very connection that has been an important factor contributing to the success of the *Veiligheidsnet/Geo-Informatie* project. Apart from organizing and arranging the horizontal attuning, geo staff of municipalities should also arrange for the vertical attuning.

---

[5] Also see the following articles: GIS magazine, # 7, volume 2. and Vi-Matrix, # 5, volume 12

[6] See for instance: gis.pagina.nl

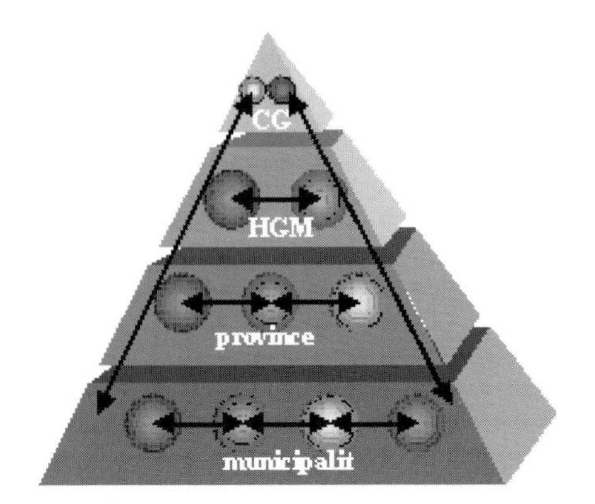

Fig. 1. The Attuning Pyramid

2.2.3 Technology

In Dutch OOV projects the system is usually the focal point. This 'system think' leads to the wrong discussions. Technology is a condition that needs to be fulfilled, but it is only a means to an end and not an end per se. The discussion that is necessary concerns which data must be available to the partners in the chain (process think) at what point in time, and what would be their required quality.

The technological aspect is often yet unnecessarily seen as a bottleneck. Technology is basically a non-issue. The bottleneck is much more the smart organization (the actual organizational and the social aspect) than it is the search for and implementation of a technological solution.

One good example of this is the C2000 communications system for public aid services in the Netherlands. This was a project that was characterized by huge budget and term overruns. It still remains to be seen whether the eventual outcome is in line with the original expectations.

A further problem is that such system think often has a paralyzing effect on the further development of a geo infrastructure to be used for disaster recovery. Parties often adopt an expectant attitude because they find themselves facing systems that are 'closed'. An example of this would be the data system used for the incident room called the Integrated Incident Room System (GMS being the Dutch acronym). The purpose of this system is to improve the information position of the operator. The system offers good functionality, yet since there are no open standards and also since it was placed within the shielded police domain, the shared incident room gets partly isolated as well.

Another problem is that one tends to await developments on a national level. One must let go of the 'system think'. We should not be creating a 'disaster recovery system' but rather a method based on OpenGIS principles, where multiple system types can tap into.

In the Netherlands one major problem is the absence of a nationwide communication infrastructure that can be used in disaster recovery for the exchange of data. The infrastructure available is the Nationaal Noodnet (the National Emergency Net, telephony/fax) and the C2000 system (radiotelephones/walkie-talkies). These do not provide sufficient possibilities as to the opening up of geo information. This is exactly where the challenge lies: to expand the infrastructure in the short term and to make it sufficiently broad.

2.2.4 Economic

Finally, there is the economic aspect of the geo information facility. Politicians will make available sufficient financial means to really achieve good results. These financial means, however, are not used efficiently enough. Efficiency would improve hugely from increasing the collaboration. Good ideas should be used and fleshed out in a broader context. This can only be done if the organizational and the social aspect are not arranged in a sector context.

In short: the OOV Sector has trouble setting up a sound geo information facility because of the compartmentalization and because it proves difficult to collaborate on a cross-organization level. The main bottleneck for organizing a good geo information facility is the social and organizational aspect.

2.3 The Aspects Applied to the Veiligheidsnet/Geo-Informatie Project

Veiligheidsnet/geo-informatie is a pilot project. After the initiation stage it is time to anchor the results in the organizations. According to the Nolan stages[7], the *Veiligheidsnet/Geo-Informatie* project is currently in between the phases of initiation and contagion. The project did, however, provide an opening to the stages of control and integration (see Figure 2).

[7] See: www.infocratie.nl for further information on the Nolan stages

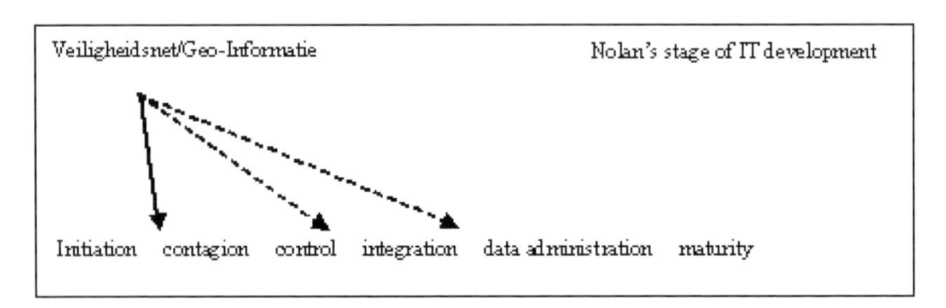

Fig. 2. Nolan's stages of IT development applied to Veiligheidsnet/geo-information

The success of *Veiligheidsnet/Geo-Informatie* lies in the fact that it has yielded sound results fast, with multiple parties being involved. It was especially the way in which the social aspect was fleshed out that contributed to this success. The project was carried out by young, ambitious people who were not (yet) 'infected' with sector think. The staff members were especially knowledgeable in the field of geo information. This provided a fresh outlook on the issues at hand and it yielded creative ideas for shaping the geo information facility. Translating this to the five aspects of geo information supply, we see that the social aspect was an important factor for success. The other aspects were arranged for the duration of the projects, however, they will be put under some pressure after the *Veiligheidsnet/Geo-Informatie* is completed.

The project output of the *Veiligheidsnet/Geo-Informatie* project is an application called 'VNET'.

In the next section we shall set out the objective and parameters governing VNET, as well as its design, output, architecture and costs and benefits.

3 The VNET System

Thanks to the upping of the collaboration between the geo departments of the Arnhem and Renkum municipalities, the Province of Gelderland and the Public Aid Service 'Gelderland Midden', to name one contributing factor, one has grown aware of the enormous possibilities offered by geo information within the scope of disaster recovery. A number of joint projects contributed to this growing awareness as well.

In this section we shall discuss the objective and parameters governing , as well as its design, its architecture, its initial output and its costs.

3.1 Objective

The objective of the project *Veiligheidsnet/Geo-Informatie* is the realization and implementation of a system that allows for the exchange of geographic and administrative data between the different partners in the chain. From this objective it naturally follows that there will also be an investigation into opportunities for disclosing the data the different chain partners have available in 'times of peace'.

VNET shows how there are no technological impediments that hamper a sound exchange of geo information between all partners in the chain. Once the pilot shall be completed it should be possible to bring it into production and to let the chain partners join in.

3.2 Parameters

Disaster recovery involves many different parties that all use a large variety of data systems. It is important to let these systems communicate with each other. In this context, two important parameters were defined for the *Veiligheidsnet/Geo-Informatie* project:

- The system must be based on the OpenGIS principle, so that data from different distributed sources can be combined fast and easy. Therefore, the system must operate independent of the sources.
- The system must be scaleable without this entailing enormous costs or structural changes to the software and hardware. The system should be set up in such a way that other public aid regions in the Netherlands are able to use the system as well.

3.3 Project Design

The project *Veiligheidsnet/Geo-informatie* is in the hands of four chain partners: The Province of Gelderland, the municipalities of Arnhem and Renkum and the public aid service 'Gelderland Midden'. Nieuwland Automatiseringen, established in Wageningen, was commissioned to realize the technological aspects of the project.

To the project the RAD (Rapid Application Development) method was applied.

This means that no lengthy preliminary studies are conducted and no bulky requirement programs are written, but that the processes are short and that results are gained fast.

The total project turnover time is eighteen months. For the functional and technical design and the technical realization six months were taken. The system is tested for 6 months in trials.

3.4 Project Output

The system provides functionality for the exchange of information between the command post in ‘the field’, the regional centre of co-ordination (RCC), the different municipal co-ordination centers (GCCs), the provincial co-ordination centre (PCC) and further partners in the chain. Each chain partner can log on to the system and get its own layer. Within this layer, dots, lines and surfaces can be drawn. This is a set of symbols developed for VNET. No (inter)national set of symbols is available to be used in disaster recovery. To these dots, lines and surfaces attribute data can be added.

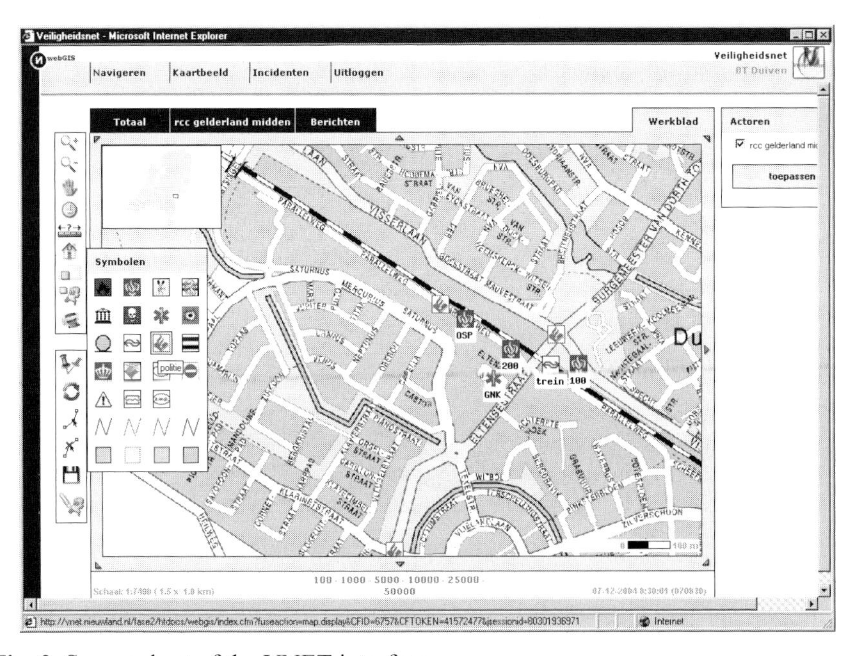

Fig. 3. Screen short of the VNET interface

The chain partners can contact each other with requests for information or a request to take certain decisions. This is referred to as a standard order, or a ‘sitrep’, which is short for a ‘situation report’.

The chain partners are information of any new sitrep or standard order that has become available or of any new symbols place don the map. This is done via the 'push principle': one is actively offered the information.

Furthermore, all data flows are logged on, so that for the evaluation of a disaster (exercise) it can be retrieved exactly who did what and when it was done. All parties have access to background materials such as aerial photos, the Grootschalige Basiskaart Nederland (GBKN, the Large Scale Basic Map of the Netherlands) and further topographic maps. Various functionalities are available, such as tools for measuring, or for navigation on postcode, address or XY co-ordinate. A map archive is available for filing important maps and retrieving them at a later stage.

3.5 Architecture

The classic method of integration on a data level is both costly and time-consuming. This is one of the reasons why the OpenGIS principle was taken for a starting point, as it expressly starts from the idea of data integration at a middleware level. Initially these are just geo data, but later on the platform should also be suited for the exchange of e.g. 3D animations, photos, video streaming and so on. Also it should support 'on-site' work processes (see Figure 4).

3.5.1 Technology

For this project, WebGIS was used, which is a product developed by Nieuwland Automatisering from Wageningen[8]. A number of functionalities were added to WebGIS, to make it suitable for disaster recovery. The system now complies with all the parameters.

The underlying sources are (geo)databases, map servers, streaming media and search engines. It should be possible to reach these sources and get answers to questions. This is realized via ODBC/JDBC, webservers or native calls. It should at the very least be possible to basically reach all sources via online technologies (TCP/IP). This Figure is a rendition of the geo connector. This is a communications layer that makes the software situated above it independent of the underlying sources and that is fully based on open standards. This communication layer provides the knowledge as to which source system is to be consulted in case of a particular question and as to how these sources are constructed. The programming layer consists of small building blocks built up according to a modular

[8] For further information, please refer to the Nieuwland Automatisering website: www.nieuwland.nl

principle. The presentation layer of application is built up out of these small building blocks, which can be reused. This makes for very easy maintenance and also ensures that new applications can be realized fast. We have actually experienced this first hand.

The software is based on Internet standards such as Java/J2EE or NET. An important portion of the software is used for making the data and setting ready for configuration by the required user applications via an administration layer.

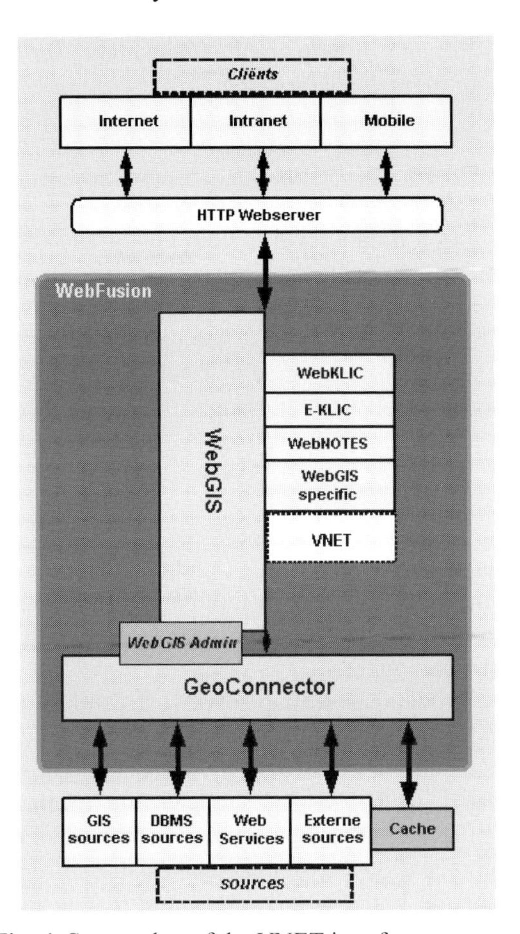

Fig. 4. Screen shot of the VNET interface

3.5.2 Security

Various distributed and possibly sensitive sources are deployed for disaster recovery. These are served to users via online technology. This is where the issue of security comes in. During the pilot phase the communication is done place via Internet, where https and certificates ensure the required level of security. If the project output is taken into production it will be possible to opt for VPN connections, the use of direct SDSL line or, alternatively, there may be a dedicated data net for disaster recovery in the future, which would make security even tighter. Apart from this it is wise to execute VNET redundantly.

For the joint disclosing of data it is necessary to chart the different privacy-related laws and bylaws with which the chain partners must comply. Think of the Data Protection Act and the Police Files Act. For instance: can the ambulance operator pass on to the police that they need to protect themselves carefully as the person injured in a stabbing incident is HIV positive? This consequences and directions for solutions are still to be established.

3.5.3 Data

For VNET we distinguish between three types of data: (1) situational-durational data, (2) operations-related process data and (3) reference data. In the event of a disaster, the situational-durational data are added in VNET. Examples of such situations are roadblocks, disaster locations and toxic clouds. Considering their very nature, situation-durational data are not available prior to the occasion. VNET does include a standard set of signs that can be used to add frequent data fast. By operations-related process data we mean the data that are managed by the individual partners in the chain. Municipalities and provinces especially have a multitude of data available. This may range from population data to information on the sewerage. These data are important in answering specific questions. Take the following question, for instance: How many people live in the area to be evacuated? The operations-related process data are to be of an adequate quality and must be accessible in compliance with the OpenGIS standard. As municipalities and provinces manage the most geo information of all chain partners by far, a close tie-up with the development of geo information in disaster recovery is mandatory. This is one of the things that contributed to the success of VNET. Finally, we mentioned reference data. These are the topographical backgrounds (top10vector, top25, top50,

GBKN, ..) used as a reference to which the situational-durational and operations-related process data are added.

The shared use of data is highly recommended. We sometimes refer to these as distributed sources, see also Section 3.2. Data should remain close to their original source. As for the topographical backgrounds: as it stands, these are purchased by all the individual parties that use them. This is of course a somewhat odd situation, as the organizations have already paid for the topographical maps from the Topografische Dienst Kadaster (*Topographic Service for the Netherlands*) via the regular tax levies and the usage by other governmental bodies should be free of charge.

3.6 The Initial Output

Over the past year, the project participants have worked hard to realize the project's objective and to deliver the project output. In the last four months of 2004 there have been extensive trials of the system. The project has not been completed yet, but time has come to start taking stock. The question we need to ask is: 'what is the concrete yield of the project?'. The benefits cannot be expressed in a monetary value. The yield mainly concerns improvement in quality. It was quite a boost for the project team that the Landelijk Beraad Rampenbestrijding (the National Committee for Disaster Recovery) awarded the project the innovation prize for disaster recovery 2004. This really highlights the fact that we are heading in the right direction in our efforts to research and establish the matters at hand in an innovative manner.

The VNET application has been realized. In the event of a disaster, chain partners within as well as outside of the Province of Gelderland can communicate with each other via maps. VNET can also be used for consulting geo information during 'times of peace', in other stages of disaster management. Examples would be work processes of the pro-action, prevention or preparation departments.

In addition to the intended project output there has also been some spin-off. The collaboration between the different chain partners (regional as well as cross-regional) in the area of geo-information has improved.

One has become aware of the necessity of consolidating geo information within the scope of disaster recovery and there is an increased awareness of the critical factors for success the absence of which interferes with the further development of the use of geo information.

3.7 Costs

In September 2004, the first prototype of the system was released (release 1.0). This release was used during a number of trials. In these trials a number of aspects that required improvements were found, and these improvements were in fact made. In the months of November and December 2004 the system was used extensively. The system was even put into operation during two (near disaster) incidents. During these two months a list of requirements and needs was drawn up, which will be implemented early 2005 (release 2.1).

The development of release 1.0 cost about 22,500 Euro, the further development for release 2.0 cost 12,500 Euro. Implementing the list of requirements and needs for release 2.1 is estimated to cost approximately 5,000 Euro.

Apart from these development costs a WebGIS license was required.

4 Conclusions and Recommendations

We can conclude that the project *Veiligheidsnet/Geo-Informatie* can serve as a good example of close collaboration between chain partners. The project is in sharp contrast with its wider context. The supply of geo information in the OOV Sector proves to be difficult to organize. We see the same thing surfacing in the problem central to this paper.

The central problem as defined for our paper reads: "What should be the focus of the Public Order and Safety Sector in order to establish an adequate supply of geo information?"

The Public Order and Safety Sector must be aware of the following aspect, if it wants to establish an adequate geo information facility in the future:

- sector think hampers effective collaboration;
- municipalities have insufficient access to geo information within the scope of disaster recovery;
- there is no adequate political steering that would enable a method of exchange;
- the polder model is neither effective nor efficient in bringing about a solid supply of geo information.

If we set off these reasons against the *Veiligheidsnet/Geo-informatie* project we see that it is the actual weaknesses of organizing the geo information facility in a broader context that are the strengths of the project

Veiligheidsnet/Geo-Informatie. This can be explained by the fact that this is a project with a limited scope, both in geographic and in organisational terms. Nevertheless, the innovative application offers valuable insights that can be used in the Netherlands as well as in an international context.

VNET allows for the exchange of geographic and administrative data between the different chain partners in the event of a disaster. The architecture on which this application is based relies on OpenGIS principles, is scaleable and web based. This implies that parties that wish to use VNET do not have to purchase any software that should be able to communicate with the hardware and software already available. VNET is not yet ready. It is a pilot project. We are now facing the challenge of anchoring the application in the organization of disaster recovery.

Annex 1

The process of disaster recovery involves a large number of chain partners. In the event of disaster recovery and large-scale and/or exceptional actions, the Fire Brigade is in charge of the process as a whole. Per working process the process responsibility rests with the following parties:

Process
Process Responsible

1. Alerting
   municipality
2. Source and effect control
   fire brigade
3. Advice and information
   municipality
4. Alerting the population
   municipality
5. Clearance and evacuation
   police
6. Fencing off disaster area
   police
7. Traffic control
   police
8. Maintaining the legal order
   police
9. Decontaminating people and animals
   GHOR
10. Decontaminating vehicles and infrastructure
    fire brigade
11. Collecting contaminated goods
    municipality
12. Preventative public health and medical/environmental measures
    GHOR
13. Medical aid chain
    GHOR
14. Relief and care
    municipality

15. Registration of victims
    municipality
16. Identification of fatal casualties
    police
17. Funeral arrangements
    municipality
18. Observations and measurements
    fire brigade
19. Giving directions
    police
20. Making accessible and clearing up
    fire brigade
21. Care/logistics of disaster recovery staff
    fire brigade
22. Primary needs victims
    municipality
23. Criminal investigation
    police
24. Psycho-social aid and care
    GHOR
25. Damage registration
    municipality
26. Communications
    fire brigade
27. Environment
    municipality
28. Follow-up care
    municipality
29. Reporting
    municipality

GHOR: Medical Aid with Incidents and Disasters

Annex 2

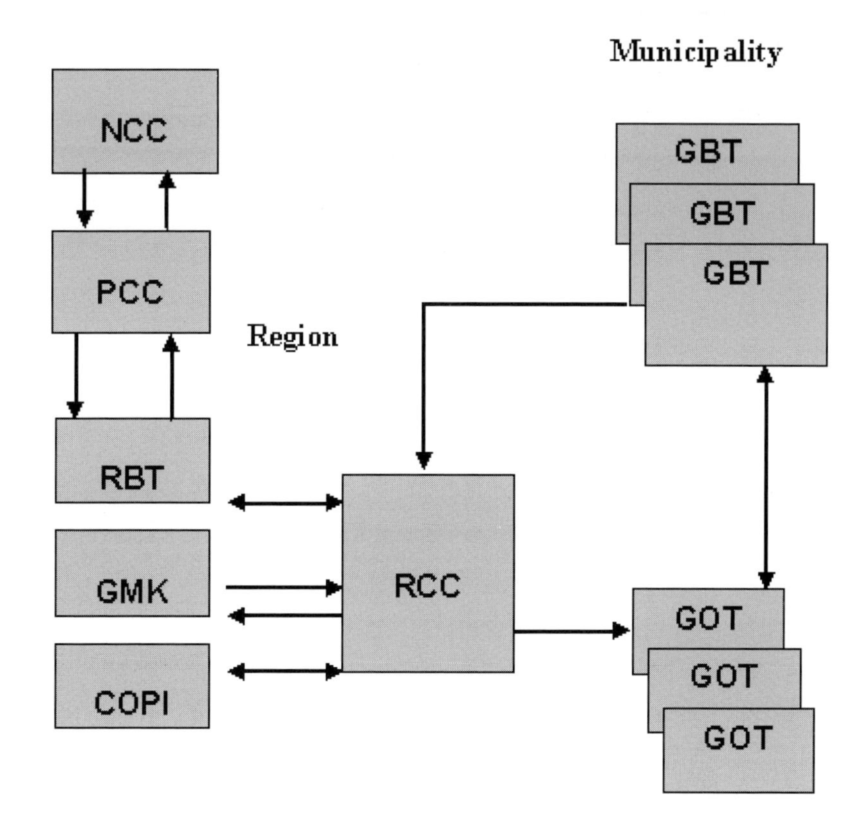

NCC:	National Co-ordination Centre
PCC:	Provincial Co-ordination Centre
RBT:	Regional Policy Team
GMK:	Municipal Emergency Room
COPI:	Command Centre Scene of the Incident
RCC:	Regional Co-ordination Centre
GBT:	Municipal Policy Team
GOT:	Municipal Operational Team

Annex 3

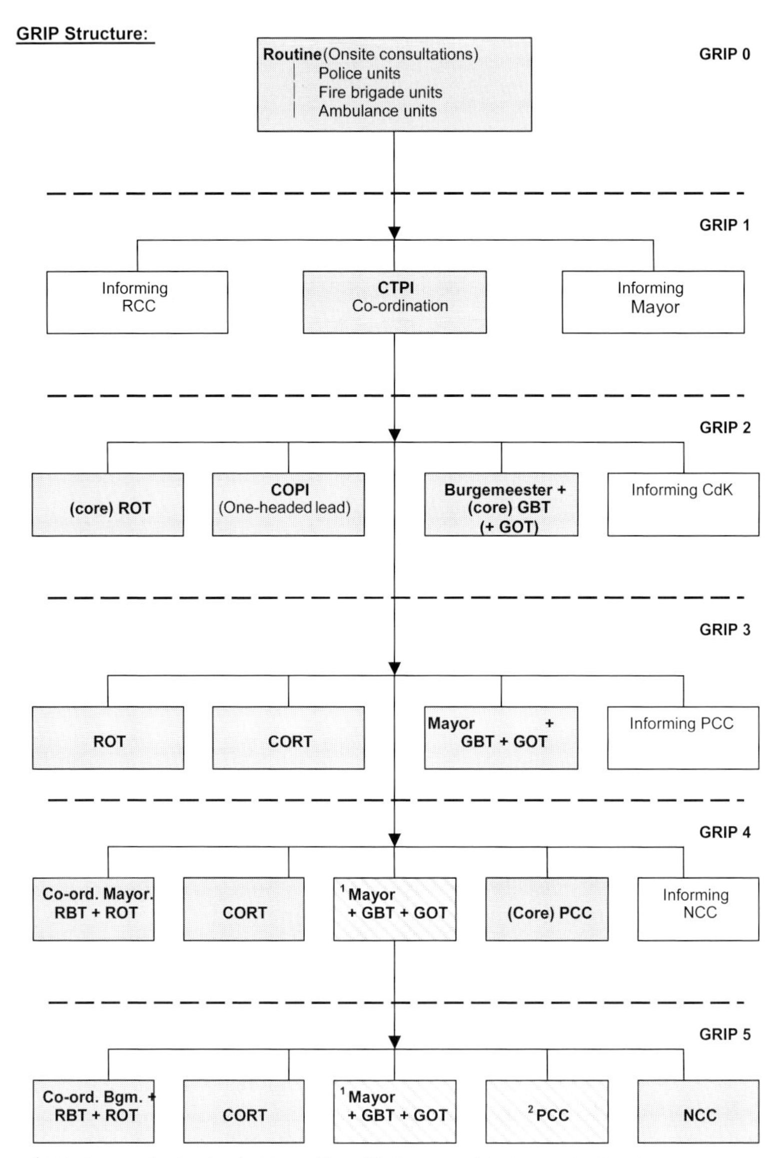

[1] Main focus policy-forming decision-making shifts from a municipal to a regional level.
[2] Main focus policy-forming decision-making on a national level.

Accurate On-Time Geo-Information for Disaster Management and Disaster Prevention by Precise Airborne Lidar Scanning

Steffen Firchau and Alexander Wiechert

TopoSys GmbH, Obere Stegwiesen 26, 88400 Biberach, Germany.
{s.firchau, a.wiechert}@toposys.com

Abstract

TopoSys GmbH manufactures and operates high-end fiber based airborne Lidar sensor systems which also include a for spectral optical line scanner. Airborne Lidar sensor systems are capable of playing a dominant role at data acquisition for disaster prevention and disaster management, especially the systems of TopoSys which focus on high precise, high dense mapping.

Obviously disaster management could only be as good as the information it has got for its decisions. The data need to be on-time, precise, accurate and actual. The presentation shows concepts how to use Lidar sensor systems for on-time data acquisition. Basic products like high resolution RGB and CIR ortho images or elevation models and their usage for disaster management are being presented as well as derived products like shore line or vectorized flooded areas. The sketch of a disaster data acquisition plan (DDAP) is shown which integrates real-time data acquisition e.g. for the use for police, rescue teams or relief organizations and post processed data for e.g. insurances, services or activities for prevention.

In addition to concepts for data acquisition for disaster management, ideas for data acquisition for disaster prevention like regular survey and change monitoring of critical areas are being presented.

A view on the future is also given how airborne Lidar sensor systems will be developed further to better serve the needs of efficient autonomous on-time data collection.

1 Background

Flood disasters in the last years, especially the flooding in Central and Eastern Europe in 2002, have shown an increasing demand for precise, accurate and actual data of the surface. From the time of the first flood disasters and the availability of Airborne Lidar Scanners as a new technology, most of the areas that were involved in high water events or are at high risk have been scanned by Airborne Lidar sensor systems to generate digital elevation models (DEM).

In the Laser scanning technology the distance between an aircraft and the earth's surface is measured by determining the traveling time of a light pulse. To be able to generate a digital model of the topography, it is necessary to accurately record the position and altitude of the aircraft, as well as the angle of every emitted measuring beam. The differentiation of the first and last light echo following a single measuring pulse permits identifying and determining further properties of the terrain.

The first echo is reflected by the ground surface, hence by treetops, high-voltage transmission lines or roof ridges, the last echo mostly by the land underneath. A Digital Surface Model (DSM) is generated by the selection of the first echo. The last echos are the starting point for generating the Digital Terrain Model (DTM).

Besides the Laserscanner the system is equipped with an optical line scanner that acquires spectral data simultaneously. In combination with the elevation data the raw spectral data that is acquired in four spectral channels can be rectified and processed to true ortho images in true color (RGB) and color infrared (CIR).

DSM, DTM, true ortho images in RGB and CIR are the standard products of an Airborne Lidar Scanner but a variety of application-oriented products complements the range of products. The described applications for disaster management and disaster prevention are focusing on river modeling and hydraulic analyses, coastline protection and change detection for example in areas at risk of earthquakes, landslides or avalanches. For disaster prevention the creation of a highly accurate and up-to-date data base is of fundamental importance whereby in case of an incidence of a disaster rapid availability - ideally in real-time - of elevation and image data is most important.

2 TopoSys Falcon

The easiest way to differentiate between the Laserscanner systems is a classification according to the concept of beam deflection. In contrast to oscillating and rotating mirrors used for beam deflection by most of the system manufacturer, TopoSys Falcon deflects the laser beams through a mechanical fixed array of fibers that are firmly aligned to a permanent view angle of 14.3°. Falcon's laser pulse rate of approximately 83 kHz is not coupled to flight height, viewing angle or number of resulting echoes. Therefore the laser data of the Falcon is ideally suitable for the generation of highly precise and dense elevation models.

A narrow view angle as given at the Falcon guarantees high penetration into dense vegetation. Especially for the generation of digital terrain models (DTM) it is of great importance to have reflections of the ground. The measurement of both first echo (FE) and last echo (LE) enables the calculation of the DTM from last echo reflections and in addition the calculation of the DSM from the first echo. Particularly for hydraulic simulations high precise digital terrain models without any information about vegetation and buildings are sought-after.

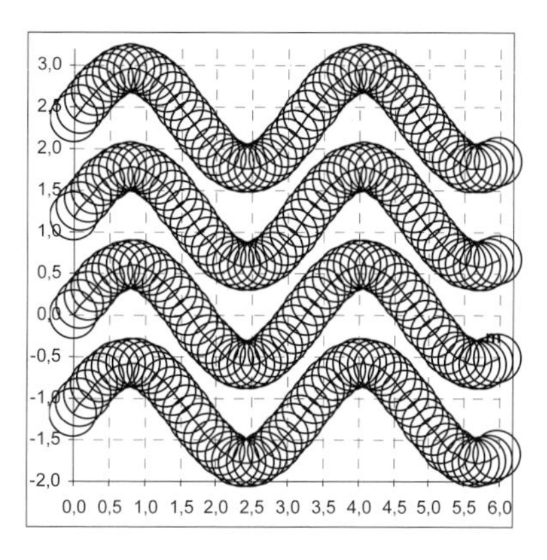

Fig. 1. Scan Pattern Falcon

Due to the system concept the high overlap of measurements in flight direction is another advantage of the Falcon system. 83,000 pulses per second and the fiber array with 127 fibers result in a scan rate of 653 Hz in flight direction. With an averaged flying speed of 65 m per second adjacent scans are displaced by just 10 cm (figure 1). Erroneous measurements caused by birds or clouds can be detected without difficulty. In contrast to other system's single measurements Falcon's multiple overlapping measurements enhance the reliability of data enormously.

Besides the advantages in the reliability of data the accuracy in edge detection increases what is of high importance for the survey of dikes.

In addition to the large overlap of measurements in flight direction due to the system concept of laser fiber scanners the swing mode was introduced in 2003 (figure 1). The scan pattern on the ground has been optimized and the coverage of the penetrated surface has been improved; small longish objects (e.g. walls) oriented parallel to the flight direction are measured reliably by the Laserscanner.

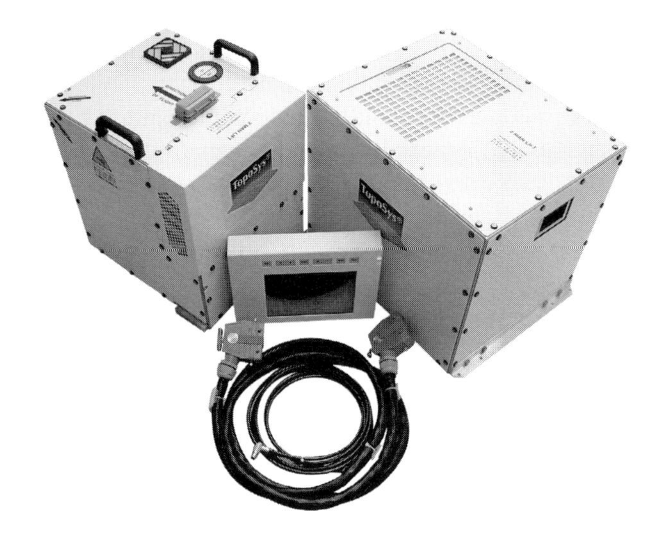

Fig. 2. Airborne Lidar Scanner Falcon II

For many applications Falcon's integrated line scanner is not only a pleasant supplement. Image data recorded simultaneously with the Lidar range measurements frequently simplify the interpretation of elevation data. The RGB/NIR line scanner is an optical scanner which captures image data digitally in four spectral channels. Due to automatic control of exposure and aperture the line scanner needs hardly any manual settings. Image data of an optical line scanner is recorded and processed strip wise. Ortho rectification is done with the help of a digital surface model derived by the Laserscanner. Only the use of a digital surface model including vegetation and buildings guarantees error-free ortho rectification of the image data where all objects are correctly positioned [Pflug et al., 2004]. The output images are georeferenced RGB and CIR true-ortho images and perfectly suited for discrimination of sealed and non-sealed surfaces and as input for geographical information systems [Schnadt and Katzenbeisser, 2004].

Data processing of the Falcon system is easy and done with a software package especially developed for the Falcon system. The modular software package comprises laser and image data processing as well as specific modules for data analysis, flight planning and quality control. Depending on the level of accuracy and the product first results of a survey flight can be provided within a few hours after landing.

3 High Precision Topographic Base Data

The detailed and precise survey of earth's surface and topography as it is done by airborne Lidar sensor systems permits an efficient methodology to create highly accurate base datasets. Additional advantages within the creation of a base dataset with a Lidar system like Falcon result from the capability to store image data at the same time with the elevation data. Particularly image data such as RGB and CIR true ortho images provide assistance with the interpretation of elevation data. As result of the first-time survey of an area digital elevation models with raster widths of at least 2 meters are available. Such a complete area-wide base dataset is already accessible for some of the German states. In most of them the data acquisition is carried out at the moment and area-wide elevation models will be available in the next couple of months.

Due to flood protection for some of the most important watercourses or for areas at risk high precise elevation models with raster widths up to 0.50 meters are available. In the past only cross sections surveyed by terrestrial surveying (tachymetry) have been taken into account for the simulation of high water incidences. In the meantime cross sections can be easily extracted from high precise digital elevation models. If necessary the cross sections can be complemented with terrestrial surveyed underwater cross sections. Simple 1D hydraulic simulations can be performed with those cross sections. For much more complex hydraulic simulations as done for 2D simulations digital elevation models from laserscanning are the input dataset. Not only accuracy, up-to-dateness and rapid data acqusition but also the total costs of this method are advantageous.

For the simulation of high water scenarios different types of elevation models can be calculated depending on the way of processing:

- Elevation models of the surface from first echo data, that describe roofs, treetops, power lines, etc.
- Elevation models of the terrain from last echo data including building and vegetation or excluding buildings and vegetation

- Difference models from first and last echo, which give exact information about shapes of buildings and vegetation

For the hydraulic simulation itself usually elevation models without buildings and vegetation are used. The other elevation models are useful to derive information about the surface roughness and to define roughness parameters for hydraulic calculations. Due to restrictions of the maximum number of points as input dataset for hydraulic simulation software packages the dataset has to be thinned out. Even a relatively small area of 10 km² consists of 10,000,000 points at a raster width of 1 meter. This huge dataset and a lot of more additional information have to be processed by the software package.

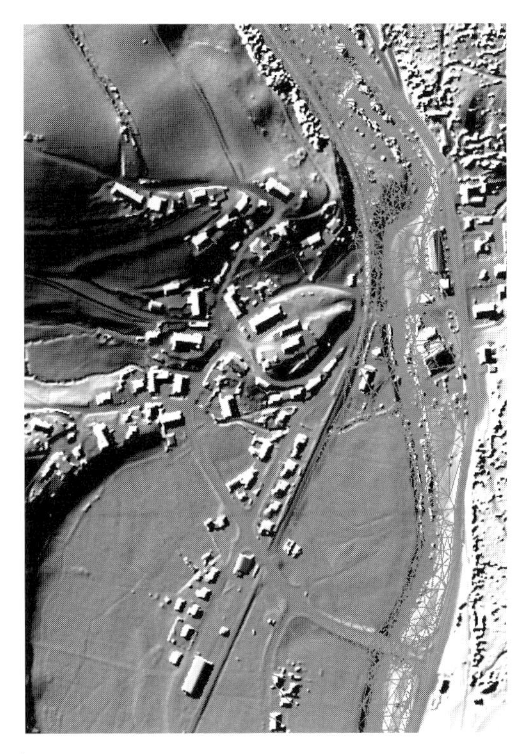

Fig. 3. TIN of a river area

For this reason purpose-built TINs (Triangular Irregular Networks) are calculated. TINs are reduced to 2-5% of the original number of points without loss of accuracy. They contain all important information of the surface including important information of edges. The calculation of TINs is usually based on digital terrain models. As a result of hydraulic simulations maps with areas at high risk in case of flooding are generated.

Mapping is done in different levels of probability for high water incidences. Effects of constructions on future high water incidences can be simulated.

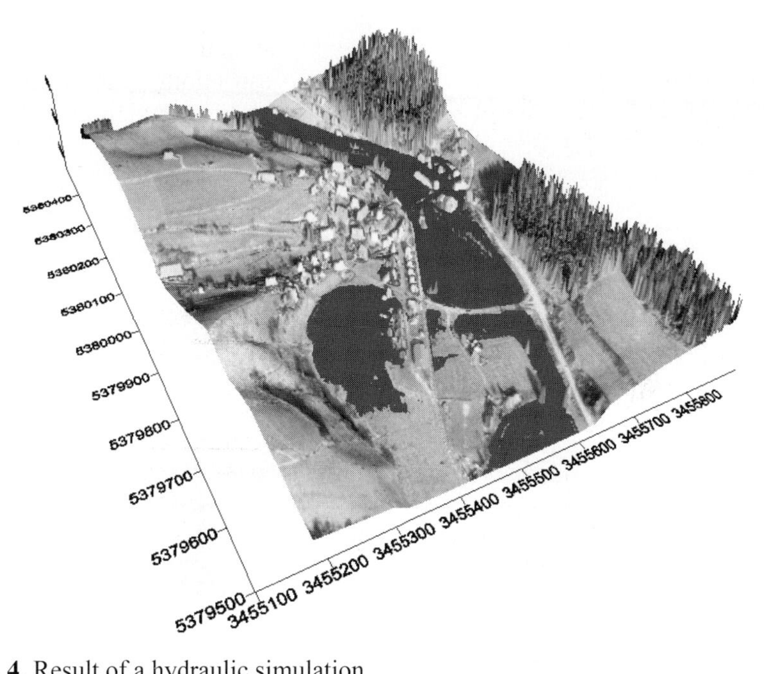

Fig. 4. Result of a hydraulic simulation

Besides preventive measures for flooding disasters elevation and image data can also be used as base dataset in areas that are at high risk for earthquakes. Also in alpine sceneries that are at risk of landslides and avalanches the first-time acquisition of a base dataset is recommendable. Furthermore the repetition of the survey at regular intervals is advantageous, e.g. at intervals of several months or years. Regular surveys can detect changes on the surface. As a result of so-called monitoring applications and preventive measures initiated after change detection, catastrophes can be avoided at some times.

The calculation of difference models is an important and helpful tool. Particularly with elevation data changes in the terrain can be detected easily. Besides, the visualization of detected changes is simple and demonstrative. Difference models between first and last echo have been mentioned above. However difference models between surface and terrain models can give the heights of buildings and vegetation as an absolute or relative height. But the most potential type is a difference model between surveys at particular times, e.g. differences in terrain between the present and the previous year.

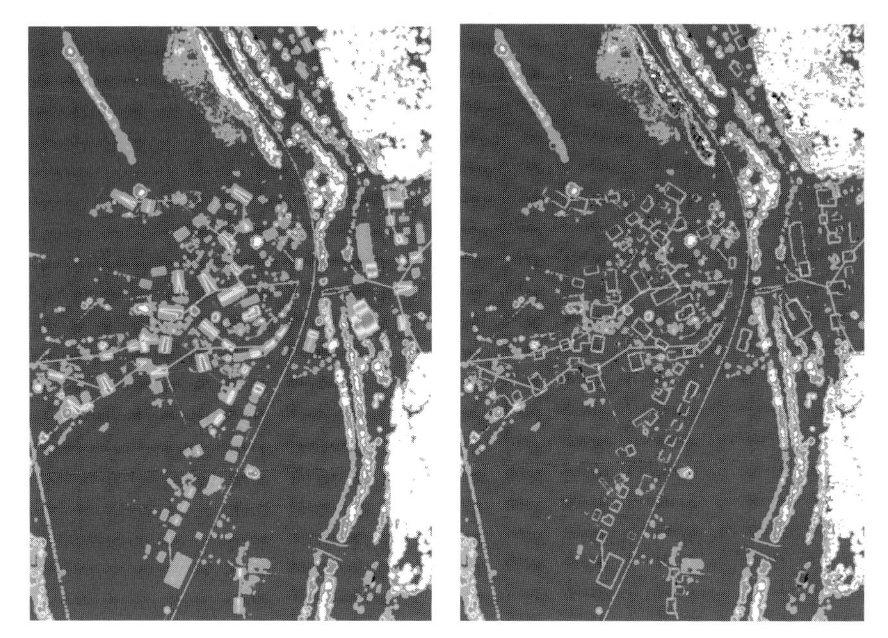

Fig. 5. Difference Models DSM-DTM (left) and FE-LE (right)

For reliable change detection the elevation models have to be adjusted to each other with the assistance of fixed objects (reference data). Furthermore the reference points should not be located within the area of interest. Difference models are calculated after the adjustment of the elevation models in location and height. Land subsidence of a few centimetres can be detected without difficulty in the calculated difference models. The detection of horizontal changes and movements is also possible.

4 Real-Time Data Acquisition

In case of an abruptly and unexpectedly occurring disaster happening in areas where no base dataset is available the situation is completely different. The fast operation of a Lidar system in disaster areas can be the solution in such circumstances. On one hand the system has to be on call quickly and on the other hand fast data processing must be possible. The way of data processing depends on the disaster operation and area. The results of the processed data should be available in various types.

The Lidar sensor system is usually mounted in an airplane all the time and is ready for operation anytime. Therefore the time for the system being ready for use in the disaster area only depends on the distance from the base airport to the area to be surveyed. Another precondition is an advance warning to put the crew into stand-by. Without advance warning the reaction is extended by 3 hours. In worst case the Lidar system can be in operation in the disaster area inside Europe within 12 hours, without advance warning within 15 hours. If there is an advance warning for a particular area, e.g. thunderstorm or flood warnings, the system can be transferred in those areas. The reaction time for data acquisition in case of disaster is minimized from many hours to several minutes. For operations outside Europe the Lidar sensor system is usually packed, transferred and integrated in an airplane or helicopter on site.

Depending on shape and size of the area to be surveyed it has to be decided whether data acquisition is done by airplane or helicopter. If the area has the shape of a rectangular area the survey with an airplane is chosen. Data acquisition can be done with a small number of turns. This results in a better performance (scanned area per hour) in contrast to surveys done by helicopter. Surveys of longish areas such as the survey of whole coastal sections with a maximum width of less than 500 metres should be performed by helicopter. Main reason for operating with helicopters is the large amount of turns that are necessary for the survey of longish areas. In contrast to the airplane turns flown by helicopters are allowed to have a much narrower radius. The flight time without data acquisition can be reduced and the survey capacity during one flight increases. The operation with helicopters has even more advantages: Reduced flying altitudes above ground and slow flight velocities result in higher point densities and better accuracies of the elevation models. Besides, areas can be flown at clouded sky with low flying altitudes adapted to the altitude of the clouds.

In case of flood disasters the water-land boundary is of particular interest. Airborne Lidar surveys can provide the water-land boundary at specific dates during inundation. All the flooded areas including water levels can be reconstructed afterwards with the assistance of the water-land boundary and the base elevation dataset of the whole area. For the acquisition of the water-land boundary survey flights along the boundary are required. For such a survey flight it is not possible to make a detailed flight planning in advance (before take-off) like it is usually done for survey flights of large areas and data acquisition for base datasets. Instead of the mentioned flight planning data acquisition of the water-land boundary can be done by contact flight with a helicopter. In addition to the elevation data CIR true ortho images support the digitalization of the water-land boundary.

For areas that are at high risk of subsidence, landslides or avalanches the area-wide acquisition of the surface is preferable. Due to the topography and narrow valleys in high mountains the survey often has to be performed by helicopters.

As a result of the surveys high precise elevation models and true ortho images are calculated. These datasets can be analyzed by experts and actions can be taken. As rough rule-of-thumb it can be stated that the time for calculating elevation models and true ortho images is equivalent to the duration of the survey flight. Generally laser and image data acquisition can be performed within 4 hours of flight before landing is required. If the data is processed immediately after landing another 4 hours later processing of elevation and image data is finished and data can be analyzed. Intermediate data is available even earlier.

The raw processing time of the elevation models would be faster and could almost be performed in real-time but at a first processing stage the reconstruction of the flight path is necessary. For the precise reconstruction of the flight path navigation data of the complete survey flight must be available. Currently the accuracy of the captured real-time solution of the navigation systems is not precise enough to complete the processing of the data real-time during the survey flight.

5 Outlook

The main future demand we see for high density Lidar data while less density dense and less precise data will be provided by satellites. Also we feel that highly integrated multiple sensor platforms will play a dominant role in the upcoming mapping projects. While Lidar scanner systems based on mirror technology seems to have reached their top, the fiber scanners are almost at the very beginning of their development and offer huge potential to serve future needs of the market.

Next to the mentioned Falcon Lidar system for future applications it is intended to do data acquisition with further developed systems that can be operated in unmanned aerial vehicles. The captured data is transferred via radio link to the base station where data processing can be performed in real-time.

Speeding up the data processing and automated object extraction are additional topics which will become more and more relevant over the next years to have the data in time.

References

Pflug, M., Rindle, P., Katzenbeißer, R., 2004. True-Ortho-Bilder mit Laser-Scanning und multispektralem Zeilenscanner. Photogrammetrie, Fernerkundung, Geoinformation, 3, pp. 173-178.

Schnadt, K., Katzenbeißer, R., 2004. Unique Airborne Fiber Scanner Technique for Application-Oriented Lidar Products. International Archives of Photogrammetry, Remote Sensing and Spatial Information Sciences, Vol. XXXVI – 8/W2

Methodology for Making Geographic Information Relevant to Crisis Management

Anders Grönlund

Lantmäteriet Emergency and Public Safety, National Land Survey of Sweden, SE-801 82 Gävle, Sweden.
Email: anders.gronlund@lm.se

Abstract

The following questions will be addressed in this contribution:

1. Spatial data for risk management - Risk or opportunity?
2. Is spatial data needed in risk management systems?
3. What is the use of spatial data in risk management system?
4. Is there a difference in the way a decision maker must think before, under and after?

Alan Leidner is the chief of Citywide GIS, New York City Department. He has experience from the 9 11 terrorist attack on the World Trade Centre and says in the introduction to *Confronting Catastrophe, a GIS handbook:* "No other technology allows for the visualisation of an emergency or disaster situation as effectively as GIS. By placing the accurate physical geography of disaster event on a computer monitor, and then align other relevant features, events, conditions or threats with that geography, GIS lets police, fire, medical and managerial personnel make decisions based on the data they can see and judge for themselves. This visualised information can be of critical relevance to a disaster manager: the size and direction of wildfire perimeters, the location of broken levees or of hazardous chemical spill release points, or the whereabouts of surviving victims inside a bombed building. GIS can be a matter of life and death."

1 Responsibilities

In Sweden we have a crisis management decree that is led by Swedish Emergency Management Agency (SEMA). The Swedish crisis manage-

ment system consists of government authorities divided into 6 different areas. National Land Survey of Sweden (NLS) is responsible for co-operating the sector geographic information. The sector has for the moment 6 members, NLS, Swedish Meteorological and hydrological institute (SMHI), Geological survey of Sweden (SGU), Swedish Maritime Administration (SMA), Swedish geotechnical institute (SGI) and Swedish National Road Administration (SNRA). One of the big aims for the sector is to make geographic information relevant for crisis management work.
To fulfill the objectives in the decree the sector is working with several different projects in which we want to point out three important main ways:
1. Understanding of the use of geographic information
2. Geo support including the need of geographic information
3. New classification of existing data

2 Benefits for the Society

As a decision-maker, an emergency services coordinator or head of security you must have access to correct and up-to-date information for your planning before you take decisions, during a crisis situation, and for following up the impact of crisis situations on society. In our concept called KRIS-GIS® we have a method to train decision-makers to request the information that is really needed. When joint decisions involving several persons are to be taken, it is vital that all speak the same technical language. This is certainly something we have all experienced in our own every day activities. For crisis management it is absolutely crucial that no misunderstanding can arise in interpreting concepts, symbols etc. as the decisions that are taken based on them can involve life or death.
Doing experiments to test how geographic information can be used for simulating a series of events, such as flooding, with the help of 3D tools is a part of the concept. A test module is being developed for training planners and decision-makers who use the crisis management system to choose the correct type of support for taking decisions before, during and after serious emergencies.

Method support as will be described later in this document helps all involved to reach a common understanding of a situation.

2.1 Collection

Purpose, usually everyone have a purpose for doing something. When you collect data you have a purpose. If someone wants to use that data and that someone has another purpose there could be a problem. When using data in a GIS it is often the attributes attached to the object that is of importance, not the object itself. Example, someone ask you if you have data of real estates. You say yes. When that someone uses your data he or she wants the geometry and that's missing. In the Swedish Real Estate system there is data of 3,5 million real estate. There is a lot of attributes but no geometry.

The starting point is the risk and vulnerability analysis that must be carried out by all local authorities and government authorities. The risk and vulnerability analysis hopefully shows what data and information is needed before a crisis situation appears.

There are some municipalities in Sweden that have a strategy for using GIS in the daily work. They say that geographic information is a part of all information. They also have a central storage of the data and the administrations in a municipality deliver data. After that all administrations can use that data. One of the problems is the documentation of the data, metadata (see 2.4 Quality)

2.2 The Need of Geographic Information

What sort of information will be needed? What sort of information do we have and what additional information must we have access to and how can we get it? The need of geographic information (GIBB) is a method that the Swedish armed forces use. If you want to make a decision you need information. GIBB is a matrix which helps you identify the type and content of the information that is required.What kind of information depends of the questions that surrounding the decision. If you want an answer to the questions then you probably must classify your data in new ways. The classification has to be made by experts in certain fields. After that you take your newly classified data in a GIS and make analysis. Now you have to talk to experts again and ask them about what answers you can have from the analysis. Hopefully you have the right answers and can make a good decision.

2.3 Data Management

If there is a need for spatial data in risk management there is also a need to help the decision makers be aware of the meaning of the spatial data.

There is two ways, education and exercises or exercises and education. Which way to go depends? If you want someone to learn or understand how important spatial data is you must educate them and exercise can be a good method. The understanding of the use of geographic information is relatively low. If you want someone to learn or understand the importance of spatial data you must educate them and an exercise ca is a good method.

You can have an exercise with the aim to let the decision-makers know what data is available in the municipality.

In Sweden we have started a study in geo support. Many countries have geo support or geo cells within the armed forces. We want to see if there are needs for a civilian geo support. The geo support should contain of 4-5 experts in GIS and a network of experts in different fields e.g. geology, meteorology and hydrology. Then if someone wants to make a decision because of a risk they can ask the geo support for help. Important is that the user tell the purpose and the geo support provide the product. Reports and studies show that municipalities have difficulties in using there own data or information when it comes to crisis. Even if they use GIS in their daily work it's not sure that they know how to use GIS in a crisis situation.

2.4 Quality

The data that exists in the society today are collected with certain purpose. Most of the data that is wanted for crisis management already exist. That mean that the data is collected with a certain purpose. Now that we want to use the same data but we have a different purpose there must be a way to relay on the data.

Metadata for a certain dataset is needed and that creates a new problem. What kind of metadata?

Is that documentation from the collector showing, purpose, organisation or a full description as in ISO 19100? Metadata is something thing that you must have in a standardized way. That is the only way to know if the data is usable for my new purpose. The problem in Sweden today is that there is only a few organization that have metadata with there data. The municipalities can't se any reason for metadata because they know there own data.

We have a standard ISO 19115 about metadata but a standard in that form is not of any use in the municipality. The GIS should in the future read a metadata file and tell the user how good the analysis is.

The sector geographic information research and development in 2004(R&D) evaluates the relationship between quality and classification in specific information and how the results can be interpreted. Studies of water flow have been carried out along the central sections of Eskilstunaån and the terrain elevation model that was used was not accurate enough to show, with certainty, whether there is a risk for flooding or land slides. During 2004, a more accurate terrain model, produced from data collected using airborne laser scanning, have been produced and be used for new water flow studies. Comparative analyses of the results of the two studies are planned to begin during January 2005. During 2005 studies of the transportation of hazardous goods are also planned to be started.

2.5 Standardization

In Sweden we have a organization called Swedish standards institute (SIS). A lot of spatial data producers are member in SIS. SIS is making standards for data exchange including conceptual model, application schemas and XML-schemas. Main focus for the standards is data exchange. ISO 19100 is the standard that Sweden will use. How useful is that standard for crisis information?

INSPIRE, is that the answer to all questions? Maybe, but there is a long way to go and still it is the best step forward in many years. As it says in the introduction to INSPIRE: “Spatial information can play a special role in this new approach because it allows information to be integrated from a variety of disciplines for a variety of uses. A coherent and widely accessible spatial description of the Community territory would deliver the requisite framework for coordinating information delivery and monitoring across the Community.

Spatial information may also be used to produce maps, which are a good way of communicating with the public. Unfortunately, the technical and socio-economic characteristics of spatial information make the problems of coordination, information gaps, undefined quality and barriers to accessing and using the information particularly acute”.

2.5.1Cartography

From Elements of cartography 6 ed: “One of the most important concerns for future cartographers will be standards. Increasingly, local units will feed data up to central (state, national, global) coordinating organizations. These organizations will be responsible for setting standards and facilitating data distribution. Standards are needed for data quality, data exchange, hardware and software interoperability, and data collection procedures. Knowledge of data models, features, attributes, and data set lineage are examples of concepts that cartographers must now learn."

Data is stored in databases and the visualization depends on a certain purpose. Many organisations have agreed on the symbols but not how to visualize the symbols. Using cartography you want to communicate something with someone, usually geographic information and crisis information as an overlay. There is several decisions a cartographer must make when it comes to visualize crisis information.

- What kind of classification is needed?
  - What about the size for visualization?
- What kind of selection must be done?
- How shall thc symbolization bc donc?
  - What shape shall be used?
  - What about the orientation
  - Can all colours be used?
  - Can all values of colour be used?
  - Shall there be certain patterns?

All these questions must have an answer other it's difficult to have situation awareness. In Sweden some authorities have decided what symbols should be used for a certain event. They also have what the meaning of different colours, for example red for danger.

2.6 Semantics

When we speak we use some words that have different meanings in different branches. What's the difference between a street and a road? Synonymous or homonymous are often used and that can be confusing talking to other branches. Do we mean the same thing?

The sector geographic information has a concept which is a way of finding functional forms for crisis management support and, therefore, requires close co-operation between all involved parties. This is important as similar information can exist in several databases.

2.7 Agreements

NLS has agreements with the municipalities about the use of NLS data. Use for crisis management is still a blank leaf in the agreement and therefore need to be looked into urgently.

Outsourcing of different parts of work in the municipality makes it difficult to have relevant information in a crisis situation. Many municipalities outsource all operation of IT to other organizations and some of the organizations are abroad. What agreements are taken in account so that the information is available in crisis situations? For example Force majeure. The consequence can be that the supply of information stops due to Force majeure.

3 Conclusions

There is a lot of information out in the society. There is in some situations an information overflow. If there is the decision makers must know what relevant information is? Most of the data and information that is of use in a crisis situation comes from different administrations daily work. Then the word relevant can mean knowledge. Knowledge about what data and information is available in the society. Relevant can also mean that the data and information is good enough for analysis.

If we mean different things but using the same word it's hard to have good situation awareness. In the sector geographic information we use GIBB to see what information is needed. Another problem is that the decision maker can't ask the right questions. We don't want the decision maker to tell us what data or information that he or she needs. We want the decision maker to tell us why he or she wants that data or information.

If we don't have standards for metadata, cartography it will be difficult to have situation awareness. For example the letter H means, landing place for helicopter, dog patrol and homeless. If we don't have situation awareness it is difficult to make the right decisions.

Municipalities must do risk and vulnerability analysis in the same way so that the county administration can compare and see if there is some risks that can effect other risks.

If a decision maker wants to use geographic information in a crisis situation he or she must know want question they must ask so that they have the right answers and make the right decisions. If the decision makers can see what decisions he or she probably has to take then experts in different fields can find relevant information. So co-operation and co-ordination is of great importance. Co-operation in collecting data and also have metadata for the data and co-ordination in what different administrations in the municipality wants and needs.

If we can achieve parts of this then GIS and analysis with GIS could be really useful.

The Value of Gi4DM for Transport & Water Management

Michel J.M. Grothe, Harry C. Landa and John G.M. Steenbruggen

Ministry of Transport, Public Works and Water Management, Department of Geo-Information and ICT, P.O. Box 5023, 2600 GA, the Netherlands.
Email: m.j.m.grothe@agi.rws.minvenw.nl; h.c.landa@agi.rws.minvenw.nl; j.g.m.steenbrugen@agi.rws.minvenw.nl

Abstract

The tasks and responsibilities of the ministry of Transport, Public Works and Water management in the Netherlands are focused on water management (water quality and water quantity) and traffic and transport over waterways, roads, rail and in the air. In the past several calamities have taken place in those sectors: heavy car collisions due to intense fog nearby the city of Breda (1972, 1990), pollution of Rhinewater due to fire in Sandoz-plant (1986), airplane crashes near Schiphol and Eindhoven (1994, 1995), derailment of chloride trains in Delfzijl and Kijfhoek (2000, 1986), river floods (1993, 1995), and recently some dike collapses due to heavy drought and rain (2003, 2004). Besides managing these large(r) calamities daily traffic incident management along the main roads and waterways is a major task of the Ministry as well. At all those occasions the need for (spatial) information and supporting information systems is large. That need has only grown because the attention has also shifted to the proactive and preventive phase in the calamity control and incident management.

1 Introduction

The directorate-general for Public Works & Water Management as part of the Dutch ministry of Transport, Public Works & Water management (V&W) is since 1798 responsible for maintaining and administering the main roads and waterways in the Netherlands. These tasks include protection of the country against floods from both the rivers and the sea. V&W is an organization having 15,000 employees, an annual budget of approx. € 7 billion (US $ 9 billion), and more than 200 offices throughout the country. Accurate and up-to-date geo-information has always been a necessity for administering the main water- and road networks of the Netherlands.

A new Dutch government policy has urged V&W to deliver more value for money on the same budget and to simultaneously reduce its employees. To achieve these goals, the organization is moving from a decentralized approach to a centralized steering model using uniform working models and organization-wide standards.

The department of Geo-information and ICT (AGI) is responsible for providing the organization with the IT and the (geospatial) information needed for its tasks. The department of Geo-information and ICT (AGI) is working with V&W to meet the challenge of reducing ICT (information and communications technology) costs considerably. The strategy to meet this challenge is built on the principles of uniform working models, open standards, server-based computing and central data hosting and maintenance. In this paper we will give an overview of the use of GI in Disaster Management within the ministry of Transport, Public Works and Water in the Netherlands. Attention is given to the following subjects:

- A brief introduction to the theoretical background of disaster management.
- An actual status description of DM and geo-information within V&W in terms of strategy, organization, and working methods.
- A substantive description of used computer systems and information technology in this field and the role an position of geo-information in the information and application architecture.
- Insight in the underlying renewed GeoData Infrastructure based on ISO/OGC web services and metadata standards - Geoservices and Location Based Services - and adopted application architecture and organization.
- A descriptive case in flood management.

2 Theoretical Background of Disaster Management

Any discussion of disasters depends on, and often references, a common understanding of disaster taxonomy. In (Green. W. et al., 2002) three classes are described as the highest order range of disaster events (shown in Table 1): natural disasters, human systems failures, and conflict based disasters.

Class	Distinguishing Characteristics
Natural Disaster	any event which reaches the definition of a disaster, which results from natural forces, and in which human intervention is not the primary causation of those forces
Human Systems Failure	any event which reaches the definition of a disaster and which results from significant human failure in any portion of a systems definition of the event, including input, process, and output – this may include events which wholly involve the built environment or which initiate events that are otherwise natural in their action
Conflict Based Disaster	any event which reaches the definition of a disaster and which results from internal conflict within a nation or external conflict directed at it, including not only the obvious threats of war, revolution, and terrorism, but also politically, racially, or economically based civil disorder, as well as internal state sponsored terrorism, genocide, and ethnic cleansing

Table 1. Natural disasters, human systems failures, and conflict based disasters

Natural disasters (i.e. earthquakes, floods) pose a threat to population, its goods and the environment. Urban areas are particularly vulnerable not only because of the concentration of population but due to the interplay that exists between people, buildings, and technological systems. Disasters pose a threat to sustainable development as they have the potential to destroy decades of investment and effort, and cause the deviation of resources intended for primary tasks such as education, health and infrastructure. United Nations has made a call to put the issue of urban disaster management high on national agendas as there is a steady increase in urbanization. According to HABITAT (1996), in 1975 approximately 38% of the world's population lived in urban areas and they estimate that this figure will reach 60% by the year 2025.

The scope of disaster management can be best understood by means of a tri-dimensional matrix (see Figure 1) describing the three types of elements involved (levels of government, management phase, implementation measure) and the resulting range of possible implementation strategies (Masser *et al*, 2000). Implementation strategies consist of structural measures which involve the modification of the environment (i.e. re-strengthening/demolition of buildings, construction of dykes/drainage) while non-structural measures involve activities such as coordination and communication (i.e. emergency drills, warning systems). Disaster management involves a cycle which should consist of an organized effort to mitigate against, prepare for, respond to, and recover from a disaster (FEMA, 1998).

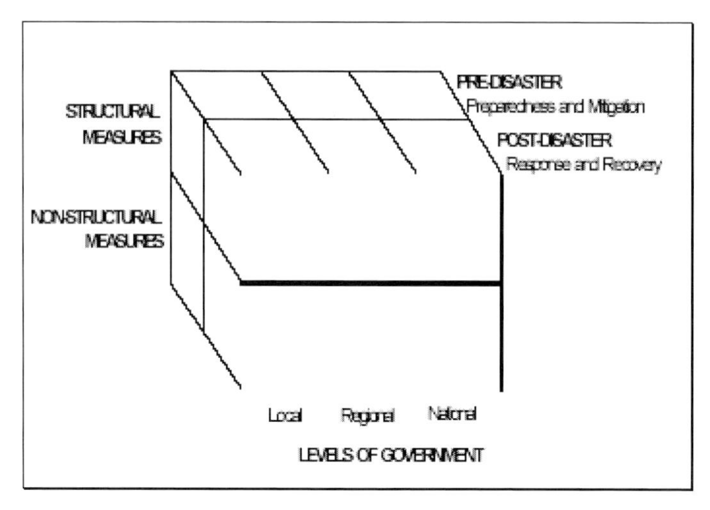

Fig. 1. Disaster Management (adapted from Tobin, 1997)

The following definitions describe each of the phases of this cycle:

- *Response* refers to activities that occur during and immediately following a disaster. They are designed to provide emergency assistance to victims of the event and reduce the likelihood of secondary damage.
- *Recovery* constitutes the final phase of the disaster management cycle. Recovery continues until all systems return to normal or near normal.
- *Mitigation* relates to pre-activities that actually eliminate or reduce the chance or the effects of a disaster. Mitigation activities involve the assessment of risk and reducing the potential effects of disasters as well as post disaster activities to reduce the potential damage of future disasters.
- *Preparedness* consists of planning how to respond in case an emergency or disaster occurs and working to increase resources available to respond effectively.

Informed decisions are a pre-requisite for the formulation of successful mitigation, response, preparedness and recovery strategies. To a large extent, however, successful strategies depend on the availability of accurate information presented in an appropriate and timely manner. Information is also important as it increases the transparency and accountability of the decision-making process and it can therefore contribute to good governance. Initiatives such as the Global Disaster Information Network (GDIN) have recently been launched to increase awareness of the importance and the value of disaster-related information "the right information, in the right format, to the right person, in time to make the right decision" (GDIN, 2001).

3 Status Description of DM and GI within V&W

3.1 V&W and Safety & Security Policy

In the Netherlands the ministry of Transport, Public Works and Water management (V&W) has a number of tasks and responsibilities in the field of safety and security. The following domains are distinguished:

- Traffic and transportation by road, rail, water and air: examples are security on Schiphol airport, (inter-)national aviation security, social security in public transport, safety on main roads and rivers and transport of dangerous goods.
- Physical infrastructure: examples are tunnels, bridges, rail shunting emplacements and underground tubes.
- Water management: examples are protection against high water levels of rivers and sea (e.g. dikes and retention areas) (water defense), level management of surface waters in relation to fresh water intake, dumping of cooling water and/or inland shipping (water quantity) and oil and chemicals spill detection and recovery (water quality)

The primary policy tasks of V&W are within these domains and the safety and security policy is an integrated part of those tasks. The ministry has formulated a generic view on safety and security which consists of the following four core elements:

- To strive for a permanent improvement of safety and security
- Within the above-mentioned domains a process of ongoing improvement has been implemented. The process is focusing on permanent reduction of (the chance on) killed and injured people and the (chance on) social disruption. It is evident that in the process of permanent improvement SMART policy objectives and milestones have to be set in order to make progress explicit.
- To weigh measures explicitly and transparently
- V&W operates in an area of constant tension in which trade-offs between economy, environment and security have to be made. Other policy aims then safety and security must explicitly be assessed, weighed and exchanged against (the chance on) killed and injured and/or material damage. V&W prefers a transparent view on those assessments in order to allow for clear political choices. This is achieved by assembling parcels of measures with varying ambitions by which the positive and negative consequences on costs, economy, environment and security become obvious.

- To be prepared for inevitable risks
- Risks are inevitable: even though it is the aim to improve working procedures and results to the max and to achieve better safety situations there will always be a change on risk or misfortune. We don't control nature and above that it is humans nature to create riskful situations. Within this context we not always provide for sufficient financial and technical means to prevent for unsafe situations. These so called inevitable risks oblige V&W to take constantly into account the possibility of an accident or calamity after which it is best to minimize the consequences of such an accident or calamity as much as possible. A useful and perhaps most important method is to practice for such events; this also allows for building and testing the quality of organizational and communicational structures and procedures.
- To obtain and maintain a safety culture within the ministry
- A necessary condition to achieve permanent improvement is to explicitly manage the aspect of safety and security. This can be obtained by the installation of dedicated units within the organization, by explicit reservation of funds and by agenda setting. In this way awareness is created within the organization.

To transform this view into more practical and concrete safety and policies the concept of "the safety chain" (see figure 2). This chain is composed of five phases: pro-action, prevention, preparation, response and recovery.

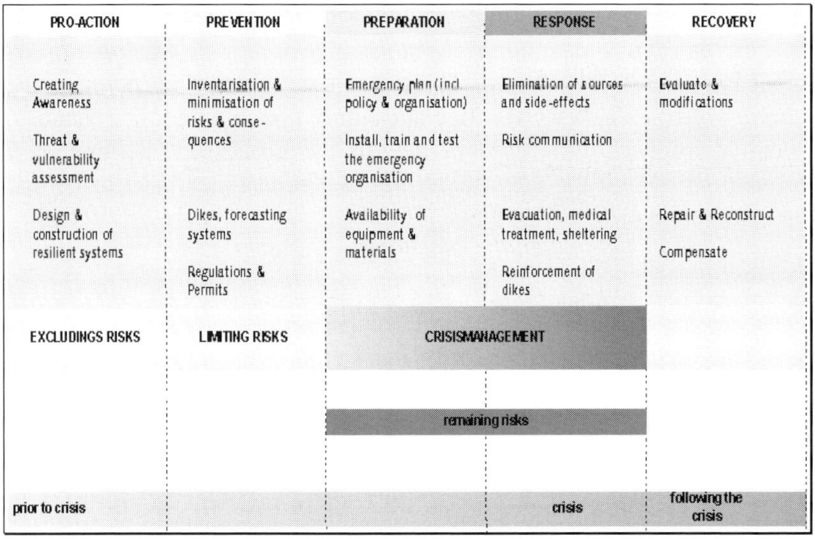

PRO-ACTION	PREVENTION	PREPARATION	RESPONSE	RECOVERY
Creating Awareness	Inventarisation & minimisation of risks & consequences	Emergency plan (incl. policy & organisation)	Elimination of sources and side-effects	Evaluate & modifications
Threat & vulnerability assessment		Install, train and test the emergency organisation	Risk communication	
Design & construction of resilient systems	Dikes, forecasting systems Regulations & Permits	Availability of equipment & materials	Evacuation, medical treatment, sheltering Reinforcement of dikes	Repair & Reconstruct Compensate
EXCLUDINGS RISKS	LIMITING RISKS	CRISISMANAGEMENT		
		remaining risks		
prior to crisis			crisis	following the crisis

Fig. 2. The safety chain

3.2 V&W and Crisis Management

Since the end of 1999 V&W has grouped its activities in the field of crisis management in the departmental coordination centre crisis management (V&W-DCC). This centre builds on the strategic objective of crisis management: a decisive V&W-organization that functions administratively, organizationally and operational in a coherent way during incidents and calamities. From the central objective the following points have been derived:

- Development and implementation of crisis management policy: development of policy on planning, education, training and all other areas related to crisis management.
- Preparing of emergency plans in such a way that a common insight arises in the tasks and responsibilities of the organizational services concerned, insight in the dangers where the organizational service concerned is responsibly for and the insight in relevant networks.
- Training and practicing staff so that people concerned with regard to knowledge, skills and experience are well prepared to crises.
- Embedding in the organizational context. The departmental structure and working methods must be coupled to the external network. During crisis circumstances the partners in the network must be able to find each other and know where they can count on.

3.3 Organization

The generic safety and security policy of V&W is developed by the program board safety & security; the policy for and the operational crisis management as such is the responsibility of the departmental coordination centre crisis management (DCC) (see figure 3).

The DCC is the central office for all Directorate-Generals (DGs) and the central services within V&W. In case of a crisis a V&W alarm number is available. A protocol has been developed in which has been described clearly who's is responsible for what en who informs who if there is something at hand. This is summarized in network cards for all domains; a comprehensive overview which describes the organizational context of the organization (see figure 4). Onto these network cards the actors are added for each domain. These network analyses and competence diagrams together with de description of the internal organization and checklists for decision-making form the handbook of crisis management of V&W.

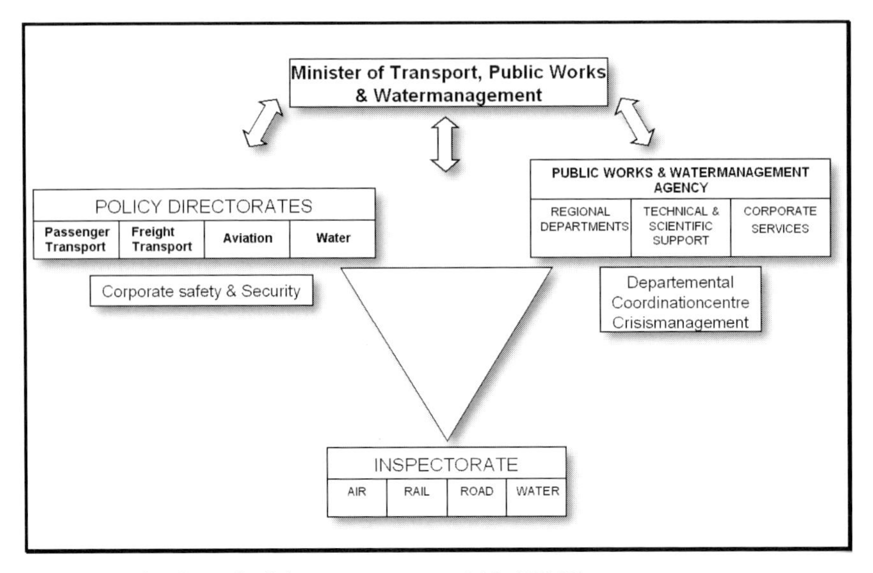

Fig. 3 Organization of crisis management within V&W

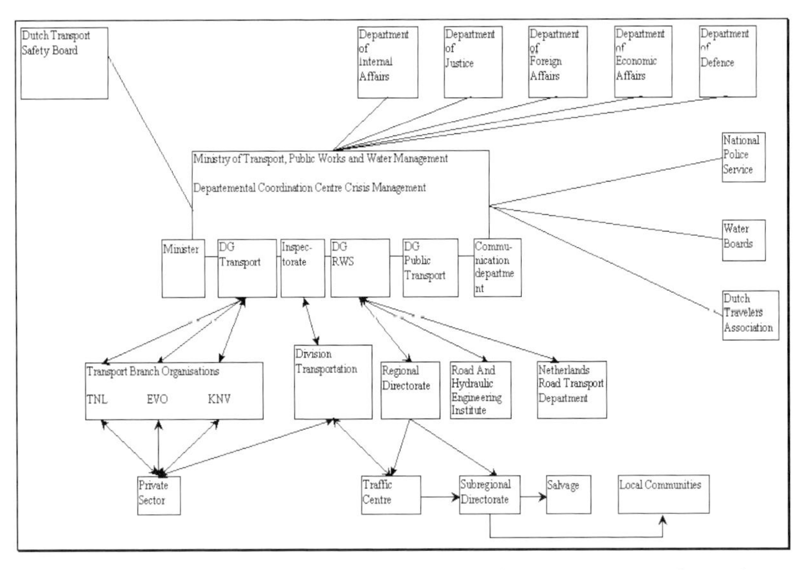

Fig. 4. Operational organization network during crisis events on main roads

3.4 Working Methods

At a strategic level the topics of safety, security and crisis management are being developed in a number of ways. Examples are the internalization of external costs, development of performance indicators for measuring policy effectiveness, developing future scenarios in order to review safety and security plans, developing uniform and accepted models for risk assessment and last but not least working on public confidence. Each of these themes is analyzed on several aspects. The economic aspect is considering safety and security as an economic problem; a central question is for example "how much money does it cost to protect “X” capital goods or “Y” lives. The aspect of spatial planning approaches security and crisis management as spatial problem with questions such as "where can I plan houses or industrial areas and where definitely not". If security is considered as an administrative problem questions on "who has which responsibilities and mandates and are they properly addressed". The fourth and more generic approach consider security and crisis management as a social problem and aims at developing socially acceptable security level/dangers as parameters for decision-making and aims at a more integrated approach in which aforesaid aspects are taken along.

4 (Geographic) Information Systems for Disaster Management

Within V&W there are several information systems that provide information for DM. Two more or less evident information systems for DM within V&W will be outlined here in more detail Infraweb and HIS from a geo-information infrastructure point of view. Infraweb is an information system for the input, storage, analysis, presentation, distribution and workflow processing of calls, incidents and calamities at the waterways and main roads in the Netherlands. Infraweb is a web based system hat is available public and consists of several modules: calls, communication, mapping, logbook, reporting, links and agents and maintenance. There are system coupling with the systems IVS’90 for monitoring shipping traffic and WVO-info for water quality spill management. HIS is the High Water information System that has three objectives:

- Perform impact analysis of water dikes during the preparation phase of a high water situation;
- Support communication during a high water situation;
- Perform the monitoring status of dikes during a high water situation through relevant information.

Both HIS and Infraweb have a geographical database, module and interface. The role of geo-information in these information systems for Disaster Management is evident. Geo-information is the integrator of data and information. Both are stand alone information systems with specific communication interfaces to other information systems. The distribution of geo-information is based on the simple concept of data duplication, especially concerning geographic core data (topographic maps, aerial photos, et cetera). Each of the information systems has it's own geodatabase which is regularly updated with new updates of core geodatasets. Geographical core data exchange between these systems is based on data duplication through physical media (see figure 5).

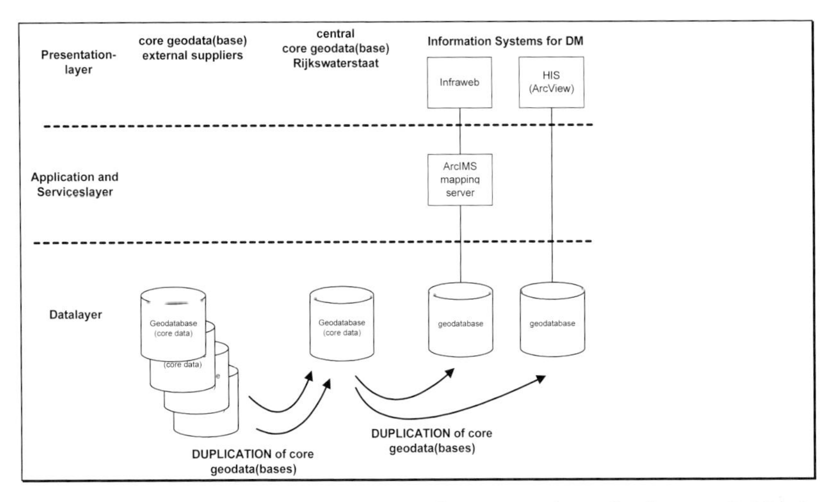

Fig. 5. Distribution of geo-information from central geodatabase to DM information systems

It is expected that geographical data exchange in the near future will be based on direct and online access at the core database of the supplier. In figure 6 the data duplication situation and online access situation is shown. Online access will be based on open interface standards of ISO/OGC.

For the above mentioned information systems Infraweb and HIS this implies that their future geographical core data distribution will be based on these open interfaces. This is part of the newly build geo-infrastructure of Rijkswaterstaat.

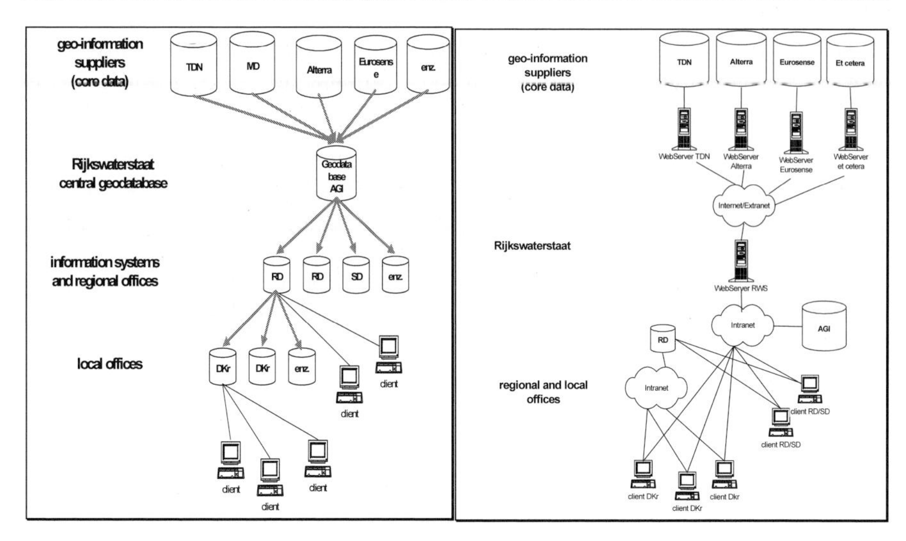

Fig. 6. Distribution of geo-information within Rijkswaterstaat now and in the near future

5 GeoData Infrastructure Rijkswaterstaat

In this chapter we will outline the renewed GeoData Infrastructure based on ISO/OGC web services and metadata standards and adopted application architecture and infrastructure.

5.1 Towards a New GeoData Infrastructure

As water management was crucial to survive in the low-parts of the Netherlands the management of the water systems was already conducted in a sophisticated and organized way in the middle ages. There are e.g. map series of thematic maps of the water system on a scale of 1 to 19:000 (15 sheets) dated 1611. V&W has thus a long and standing tradition of mapping and geospatial data processing. V&W, like many similar organizations, moved through the consequent stages in which the process of geospatial data management was automated. First the mapping process was automated using mostly CAD and automated drawing techniques. Then stand-alone desktop GI-Systems were introduced, first mainly for more complicated geo-processing task like modeling and analyzing data for policy making. When desktop GIS became more lightweight and easier to use and access, the use of GIS further spread throughout the organization. As it was relatively easy to develop add-ons, scripts and applications, and there was no strict policy for application development and maintenance, many smaller and larger GIS-applications popped-up. Because less emphasis was laid on data management and data distribution and application maintenance was not embedded within the organization this lead to a sub-optimal situation in which software maintenance was expensive and data management cumbersome.

At an early stage the responsible professionals recognized the drawbacks of the very bottom-up approach in which geospatial data processing was managed and developed. Single-use GIS applications were replaced by multi user client-server type solutions but the bottom up steering (and funding) of development still blocked a more structural approach.

A new Dutch government policy has urged V&W in the last years to deliver "more value" on the same budget and to simultaneously reduce its number of employees. To achieve these goals, the organization is moving from a decentralized approach to a centralized steering model using uniform working models and organization-wide standards. When the board of directors, alerted by the rising costs of IT and geo-data management, recognized the problems in this field, a centralized steering model for IT was put in place. This makes it now possible to work on Enterprise GIS: geodata once and securely managed and accessible for any worker who needs it any time and anywhere it is needed. In this paper components of this centralized solution are discussed.

Geospatial data management at V&W moves to a new stage in which the demands of today can be met. Key factors in this geospatial data management policy are centralization of geo-data in centrally managed geodatabases, the use of open standards, the exchange of Windows-clients for small-footprint browser-based web-clients and the use of mobile technology which is seamlessly connected to the main GDI.

About ten years ago, during the rapid bottom-up development of GIS within the organization there was an awareness that some form of standardization was necessary in order to be able to exchange data and share applications. By that time the only way to achieve this was to standardize on a vendor and ESRI-software was selected to provide this "GIS-standard" for V&W. Now there comes a need for open standards in order not to have a "vendor lock-in" and to be able to communicate with other organizations. In 2004 V&W made the decision to base development on the Open Geospatial Consortium standards. Especially the OGC Services Architecture was adopted. A geo-information infrastructure based on the OpenGeospatial Consortium (OGC) Services Architecture has been established using both open source software (OSS) and proprietary vendor components. This infrastructure has already enabled broad geo-information sharing throughout the organization and has proven to be cost effective. Expected future developments include feature services and the implementation of a transactional web feature service for mobile clients.

The use of open standards makes it now feasible to use open source software. For some developments open source software proves to be competitive with commercial based software and is successfully deployed. Since the whole of the organization is now connected with a reliable broad-band network, data can be stored in centralized databases and other centrally managed data stores. As data sharing and interoperability becomes more important and open standards are required a transition is made from file-based storage e.g. in coverage files of shape files to storage in a database which is geo-enabled. V&W uses both the ESRI SDE-database application and Oracle Spatial for this purpose.

5.2 OGC-Based Web Services Infrastructure Rijkswaterstaat

Within the "Geoservices" project an OGC-based services architecture infrastructure has been implemented. The most important functional element of the OGC-based web services architecture is the concept of "publish-find-bind". This powerful concept works as follows (see figure 7). The data manger publishes geo-information in the form of maps and served by an Internet map server in a registry. The registry is known as the OGC Web Catalogue Services (WCS). The user of an application has a find button, for access to the registry and to answer the questions: which map servers are available and which maps are available? After the maps are found in the catalogue database, the user can bind the maps through the web mapping user interface of the application. The OGC web services concept is based on a distributed systems concept. Data is directly accessible at the source database through an OGC-based services interface. Sharing geo-information in an open and transparent way is what this is all about. The architecture offers users of geo-information an easy way to access, publish, find and bind geo-information through Internet technology.

The adopted software architecture is characterized by a three layers architecture: presentation layer, services layer and data layer (see figure 8). The architecture is modular and scalable. The kernel of the architecture is the services layer. The modularity of the services layer is illustrated by the fact that several different software components for the Java platform as well as the Apache/PHP platform. For basic web mapping services OGC Web Map Server, Web Feature Server and Web Coverage Server Rijkswaterstaat uses the Minnesota MapServer (MS4W) of the University of Minnesota (Open Source Software). The Web Catalogue Service is the Open

Source Software (OSS) product Degree from the German vendor LatLon. Degree is Java based. A Dutch company Geodan offers a geocoder service based on the Application Services Providing (ASP) business model, to gazetteer. An underlying location database offers access to all addresses, postcodes and municipalities in the Netherlands.

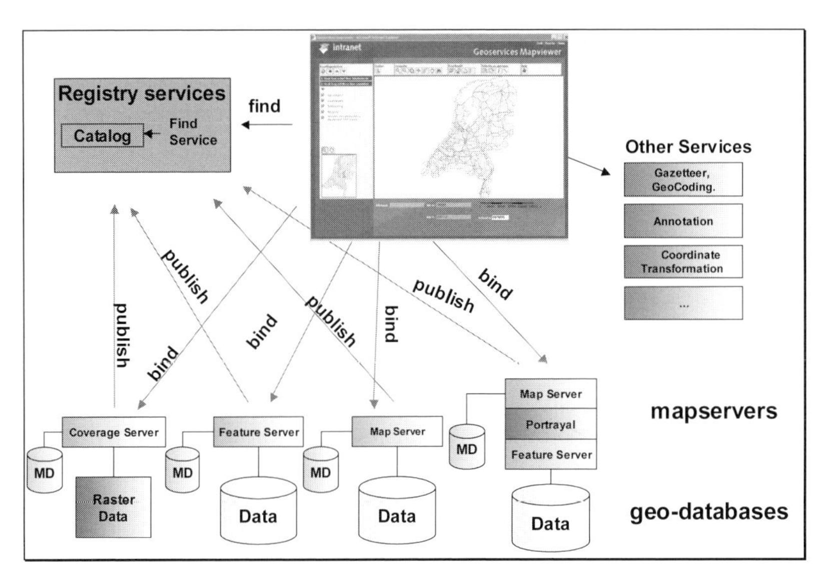

Fig. 7. The concept of publish-find-bind

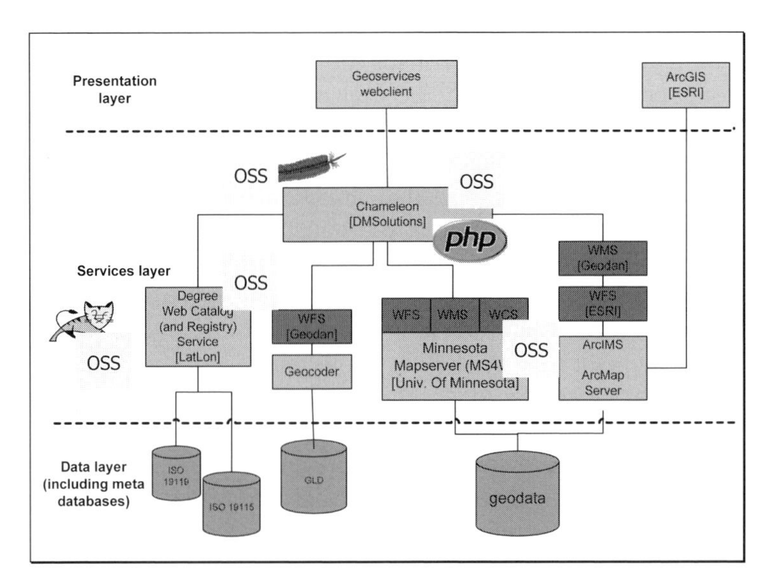

Fig. 8. Adopted OGC-based software infrastructure

The client software platform is the Canadian product Chameleon, based on a PHP server-side scripting language (no plug-ins needed). Chameleon is a OSS product management by the Canadian company DMSolutions. Furthermore for mapping ArcIMS from ESRI is used as well, especially for web mapping in combination with the ESRI's ArcGIS desktop client software. This geo-information infrastructure based on the OGC web services architecture has been established using both open source software products and COTS components. This infrastructure has already enabled broad geo-information sharing throughout the organization and has proven to be cost effective.

5.3 Location Based Services

To carry out the day-to-day activities many Rijkswaterstaat employees are out in the field-inspecting infrastructure, checking permits and regulations etc. To supply these workers with hand-held mobile computers with GPS location and a wireless connection to the office network proved to be very productive. Apart from the time saved also the quality of the decisions made increased because of the information available on the spot. In a pilot project mobile computers running a web browser were connected to several distributed databases and able to do vector editing with WFS-T in these databases all through OGC standard protocols (see figure 9). Expected future implementations include the implementation of a transactional web feature service for mobile clients.

Location-based Services (LBS) are information services, accessible through devices such as mobile phones, PDA's, tablet and laptops. Their essence is the ability to locate a user or an object in space, and to use this to make information processes aware of the location dimension. Location-based services include car and personal navigation, field work, point-of-interest search, track and trace, emergency services, fleet management and entertainment. In general there are two main service categories (Grothe en Steenbruggen, 2002):

Pull services: when one self is requesting for some information, e.g. 'Where is the nearest cash machine?'

Push services: when one is involved in some kind of service without actively being part of it, e.g. when one is being traced down (Tracking Services) by somebody else, who is using a service such as 'Friend Finder.' or 'Advertising'.

The importance and value of location in mobile services has been underlined in many studies. According to Durlacher's report (Durlacher, 2001) the 'killer' mobile applications will be those that utilize the key characteristics of the mobile channel: location-specificity, personalization, and immediacy. In terms of map services location-specificity means that the maps in use are from the area of interest, either defined by some positioning method or user-defined (using e.g. addresses, place names or coordinates). Personalization can be regarded as a tailored product for a specific task (e.g. vehicle navigation) or individually personalized content according to the user needs and taking the terminal specific characteristics into account. Immediacy means that the data content is actual. Most of the time applications will use a combination of static (e.g. topographic maps) and dynamic (e.g. traffic info, water levels) information

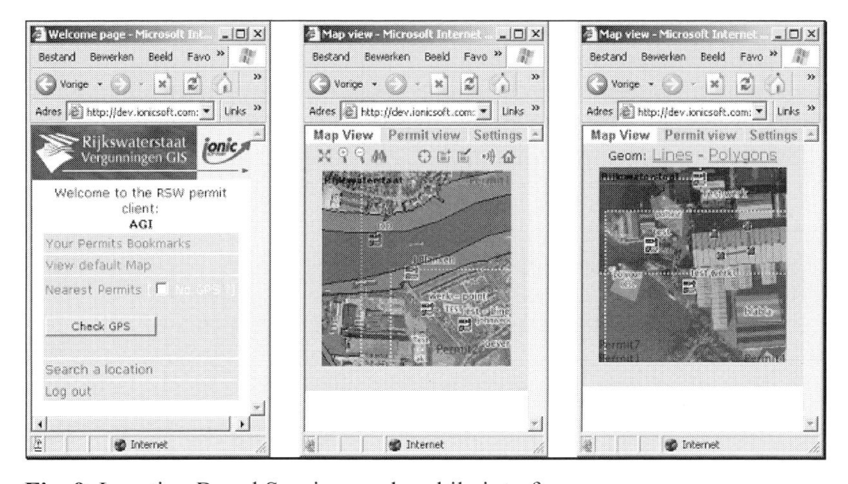

Fig. 9. Location Based Services and mobile interfaces

Cell Broadcast (CB) is an important 'new' push LBS technology with high potential for disaster management. CB uses a separate channel within the GSM and provides a medium for sending real-time messages to all mobile phones within an given geographic area of a mobile network. The position of an active mobile phone is tracked by the operator and thus always known on cell-id. The Netherlands is divided by approximate 5000 cells. Cell Broadcast has not been implemented in the Netherlands. The main reason is that the technology is network specific. There are 5 operators active and an operator can only reach his own subscribers. This is a important barrier. The Ministry of Economic affairs has started a project to create an general platform for this barrier. This results in an project where there will be an infrastructure available that the government can use to send important public message to specific areas to all mobile phones. In case of an disaster messages can be sent to inform relevant areas to take adequate actions (Steenbruggen, 2004).

6 DM Scenario[1] for OGC-Compliant Services Architecture

6.1 Introduction

Information systems for Disaster Management will benefit from these new developments in an evolutionary way. These benefits can best be shown by outlining a DM scenario. The scenario is an as realistic as possible attempt to demonstrate the usefulness of geographical web services based on the ISO/OGC web services architecture. This scenario consists of a number of scenes in which flood defence is simulated and disaster management in the Netherlands is illustrated (see figure 10). It is a fictive simulation of a ship crash at the Rhine river near Nijmegen in the Netherlands during a critical high water situation.

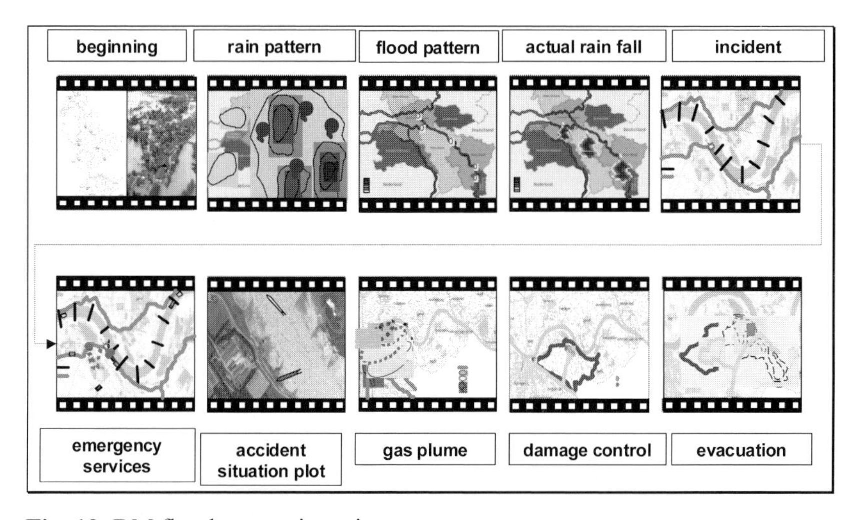

Fig. 10. DM floods scenario script

6.2 DM Scenario; the Beginning

It is Monday morning when Willem van der Gaag, crisis coordinator of the V&W Institute for Inland Water Management and Waste Water Treatment

[1] This scenario is a developed for a OGC web services simulation by V&W Department of Geo-Information and ICT, Twijnstra en Gudde Management Consultants and the project organizations Viking, NOAH and HIS.

(RIZA), starts up his computer and inspects the water information system of the rivers Rhine and Maas. He knows, today the water level is expected to rise to alarming heights. Yesterday at the national news showed the first alarming pictures of floods by excessive rainfall in NordRhein Westfalen and the Eiffel (Germany). Combined with extreme high rise water levels of the rivers caused by melting water and rainfall in Switzerland and part of the East of France this will lead to floods in the rivers basins in the Netherlands. RIZA is responsible for the river water level monitoring and discharge (water transport per m3 per second) at Lobith and Borgharen, the locations where respectively Rhine and Maas enter the Netherlands. Besides water quantity, RIZA also monitors water quality. RIZA calculates the expected water rise and water level. As soon as the actual water transport level in combination with the expected time of transport exceeds the defined limits, RIZA will alarm the regional and local water managers. In case of an expected flood situation, the local fire fighting and police department will be informed as a precaution.

Willem van de Gaag is checking the situation in Lobith using his geographical information system (GIS); the maps show the situation of the water system Rhine at the location Lobith. He also checks the Rheinatlas (http://www.iksr.org) where he explores the previously predefined flood risk areas.

Scene 1 Mapping the Weather Forecast

Willem is evaluating the water situation in Lobith. He predicts the water expected water levels in the next 48 hours. In order the have realistic water predictions he needs information about the local weather situation in the water system of Rhine and Maas. Therefore he accesses the database of the Royal Netherlands Meteorological Institute (KNMI). While the water rise prediction system calculates the expected water levels, Willem is inspecting the weather maps of KNMI with his GIS. He displays the rain fall maps of the last three days and the rainfall predictions of the next three days. He also takes a look at the maps of the actual and expected wind directions and wind speed. These maps show an increasing rainfall in the river basin area. When he receives the prediction outcomes, he knows that Wednesday morning Lobith is expecting a high water transport level with an expected maximum peak Wednesday evening.

This information leads to the a 'flood warning message' to the regional and local water managers and emergency services (local fire fighting and police departments). This message is an early warning that a flood crisis is to be expected. The message consist of all relevant information including the maps and is offered through geographical web services technology.

Especially the map with the location where the Rhine meets the Waal, Netherrhine and IJssel is showing transparently information of the alarming flood situation. Even taking the uncertainty of the predictions into account, a two-day flood situation is expected with even a chance of extreme floods. Through email, fax, phone and geographical web services all responsible and involved parties are informed.

Scene 2 High Water Rise

In Arnhem the regional crisis manager Joost Hilhorst receives the incoming flood warning message and is going to measure the 'exact' water rise level per kilometer section. This means that for each kilometer at the waterway the expected water level and discharge is calculated. As long the calculated levels remain under the level of the dikes along the waterway, no floods will occur. However with strong winds and river surges this could lead to floods. Especially at locations with weaker dike stability this can have impacts. The dike management information system of Rijkswaterstaat will inform the regional crisis manager about the status and stability of the dikes. The weak dike spots are mapped together with the detailed, predicted water levels for each river section and located on a risk map. Joost Hilhorst puts forward the risk map during a meeting with other flood experts. They decide that it is time to express 'Warning phase 1'. The decision is in particular based on the fact that the map shows a maximum of 0.8 meters water rise in de Waalbandijk in the Ooijpolder and 0.4 meters at the lock in this dike. Joost and his colleagues also know that during renewal of the road deck the asphalt at the dike road is removed the last week. This even makes the spot more vulnerable.

Scene 3 Extreme Rainfall

Tuesday morning Joost Hilhorst is inspecting the latest weather forecasts. These show increased possibility of extreme rainfall in the Rhine water system in Gelderland and Noord Limburg. The weather radar maps show heavy rainfall patterns with strong winds near Wesel and Nijmegen. Messages of water floods in cellars of houses and tunnels in the region give an strong indication that the water level at the waterways will increasingly and the emergency services (especially fire fighting units) will be busy in next hours. At the same time the weather forecast shows stormy weather expected on Wednesday evening.

Because of the water level and waves due to the expected stormy weather Joost and his colleagues expect that the transport possibilities at the Waal river will be limited to zero at Wednesday afternoon. A warning to professional shipping companies is send out; between Wednesday 1200 hours and Friday evening 1200 hours ship traffic will closed at the Waal corridor between Lobith and Weurt. The margin of the dike stem level for each dike section remains under the critical levels and therefore restricted ship traffic is still possible. The warning message to the ships is accompanied by a map with detailed restrictions.

For the critical situation at the Waalbandijk in the Ooijpolder it is decided to raise the dike stem level with an extra layer of sandbags. Wednesday morning a contractor of Rijkswaterstaat will place a double row of sandbags (40 cm) at a length of 30 meters next to the lock at the Waalbandijk.

Scene 4 Incident: Shipping Accident

The same Wednesday morning 10.00 hours a ship incident message arrives at the emergency room of the regional Fire Fighting Unit. At the location of river section or 'kilometerraaij' 854 two ships are crashed. The captain of one of the ships, loaded with coal, indicates that he maneuvered his ship into a dike in order to avoid the crash. According to the incident protocol Henk Groen at the emergency room collects the necessary information from the captain and tries to figure out if any human victims are involved, whether there is fire or a leak with chemicals or gasoline. The captain of the Rhine vessel loaded with coal is confused and with the second ship there is no contact. Henk Groen decides to contact the river information services system in order to obtain information about the actual traffic at the incident location. At the same time he warns emergency services including a diving team. Boat units of Rijkswaterstaat and police are ordered to the incident spot as well. Henk can track and trace the mobile units on the computer.

The incident spot is marked in the system and because of the expected impact the crisis situation is scaled upwards to the Dutch GRIP Level 2; GRIP level 2 means that coordination of the calamity is done by local government. First, there is a possibility of a breach in the dike at the spot of the large coal vessel during a flood. Second, it is expected that this ship restricts the water transport with strong a chance of water damming. Henk Groen views the municipality map and informs the crisis coordinator of the municipality. At the same time an inter local warning is dispatched to the surrounding municipalities in case the incident accumulates to GRIP 3 level.

Scene 5 Crisis coordination

After a couple of minutes the first police surveillance unit is near the incident spot. Through their C2000 communication system the unit informs the emergency room about the local situation. However, the police surveillance is confronted with the road block of sand and other maintenance materials for the renewal of the asphalt layer. The unit is 300 meters away from the incident spot and continues on foot. At first visual inspection they do not observe fire or any other alarming activity. The emergency room decides to order the contractor to clear the road. Also the voluntary fire fighting unit of the municipality of Ubbergen arrives at the road block and can not access the incident spot. The police unit on foot discovers a small sandy track in the direction of the stranded coal vessel. This small track is not on the map of the emergency room and Henk Groen asks for a more detail roadmap of the local road network manager. It seems that there is a small track that gives access 300 meters downstream the incident location. The emergency room informs and guides the diving team from Nijmegen to this spot.

The Ammoniac Tanker and its Gas Plume

Then by phone the emergency room Gelderland-Zuid in Nijmegen receives the information about the second ship from the River Information Centre from Rotterdam. There is a strong indication that the second ship might be the ammoniac tanker Diana from Düsseldorf. Up till know it was not possible to reach the ship by phone. According the GPS-based River Information System the last location of the ship is 100 meters downstream the stranded coal vessel. This indicates that the tanker is damaged. Henk Groen immediately informs the mobile rescue units via C2000 to perform observations at the Waalbandijk of a possible gas leak from the ammoniac tanker. This warning is also communicated with the regional crisis centre of the security region of Gelderland Zuid, that became operational during the last couple of hours. The potential of an gas plume calamity makes this incident an inter municipal incident and therefore an Dutch GRIP 3 level incident.

Scene 7 Gas Plume Evacuation Scenarios

Martin Slootsdijk, being responsible information manager at the regional crisis centre Gelderland-Zuid, immediately investigates the wind maps, the actual ship movements at the Waal. Next he calculates a plume using a integrated plume model of TNO-FEL in case of an gas explosion at the ammoniac tanker. The resulting gas plume maps are overlayed with the maps of housing and inhabitants, economic activities, cattle and the road network in order to get an indication of numbers of persons and cattle for evacuation. Because of the fact that the tanker still moves downstream he also copies the plumes to downstream locations and recalculates the number of evacuees. It is obvious that the tanker should be stopped in order to avoid plumes at larger dense populated city of Nijmegen. He alarms the water police and Rijkswaterstaat to stop the tanker for driving downstream.

Scene 8 Prevention

In the mean time at the crisis centre of Rijkswaterstaat East in Arnhem the crisis team is working on a strategy to avoid further damage at the dike by the Rhine vessel. Repairing the damage is one activity of strategy, the other is trying to avoid further damage because of the fact that the ship's position is on top of the water direction. This might cause in combination with the heavy rainfall and strong winds an additional water rise of 0.15 m as is calculated by one of the information systems. The safety margin becomes very narrow. At the same time there a possibility that the coal ship is flooding away, causing more damage at the dike.

In order the prevent damage the most actual information about industrial activities (dangerous goods that might have environmental impact) and heritage sites (especially monuments) in the area that are threatened by possible floods are collected. At the Risk Map of the Netherlands it is shown that one industrial site is located in the area; it is an stone/concrete factory. Through the permit database it is shown that that the factory has permits for storage of oil spill depot. The crisis centre orders these goods to be removed. The monuments database (or KICH-portal) shows that several monuments are located in the area. However, it is decided that no further actions are necessary to prevent damage at goods in these monuments (all private properties).

Scene 9 High Water Evacuation

In the mean time at the regional crisis centre of Gelderland-Zuid the crisis team is gathering the necessary information to see whether evacuation of people and cattle will become inevitable considering the potential gas plume and possibility of floods. At that moment a message from the water police arrives at the centre. There has been communication with the ammoniac tanker. The captain of the tanker has confirmed that his gas tank pressure level is ok and he and his crew did not observe any visual damage at the gas tanks. It is decided that the ammoniac tanker will be sent for further visual inspection to the coal harbor of Weurt.

In the mean time several calculations are performed in order to determine the size of an possible evacuation of people, cattle and goods. Through model based calculation with the HIS system it is shown large parts of the Ooijpolder will be flooded. In a period of one hour after a major dike collapse the water will reach the village of Ooij. However the inhabitants will need 3 hours to leave through the southern and eastern route. Ooij has 4 elderly homes with 25 inhabitants, three public archives of the Municipality of Ubbergen that need special attention.

6.3 Scenario Synthesis

Besides the end of the fictive incident, because that is a matter of an exercise in collecting and handling information and decision making, this scenario shows that the traditional instruments such as phone and email for handling crisis situations changes through the use of geographical web services. By plotting information on the map and offering online access to distributed geographic databases, these information services increase speed of handling during crisis situations. Especially, information management can benefit from OGC-compliant web services oriented architectures in crisis situations that have the following characteristics,:

- When different organizations with different responsibilities are involved;
- When organizations are located at different locations;
- When organizations use different distributed geographical information systems and geodatabases.

There are risks in geo web services architectures using for DM infrastructures as well. One of the main risks s the reliability and availability of the Internet and mobile communication infrastructures. The communication network infrastructure is an essential component of such a services oriented architecture with online data access at the point of supply. The alarm organization in the Netherlands (police, fire departments and) use their own specific communication network (C2000), but for other government organizations that have a role in DM are depending on a stabile network infrastructure.

At the same time the (core data) supplier has the responsibility of offering online data services 7 days a week and 24 hours a day. This means that clear service level agreements are necessary.

A strategy of uniform working models, open standards, server-based computing and central data hosting and maintenance is feasible and has many advantages. OGC standards are matured so these enable the construction of an enterprise geo-infrastructure based on open standards. Advantages for the organization are a robust data management, the widespread availability of geo-information and low threshold for data sharing.

7 Final remarks

Setting up successful disaster data/information infrastructure networks is one of the greatest challenges faced by the geo-information community. This requires considerable effort and commitment by the various users and providers of information as well as clarity on the rights and obligations of all the parties involved. To provide a reliable information infrastructure, the availability of different building blocks is crucial. For example, to provide successful location based services, the availability of location technology, (wireless) communication infrastructure, geo-information and geoservices is crucial.

Supplying decision-makers with raw geographic information is a very common practice which has generally yielded negative results. Decision-makers are often provided with maps of hazard zones, population density, urban growth and land-use, on which to base their decisions. But, as little attempt had been made to integrate the data sets, the implications of the information can be not immediately visible, causing confusion and in the worst cases, the formulation of inadequate policies. Simply stated, decision makers require information that allows them to establish the cost-benefit implications of the various strategies that can be implemented. Consequently, loss forecasts are essential as they establish the amount of damage that can be avoided by investing a given amount of financial resources. The production of loss forecasts requires additional datasets as well as further steps of data integration.

The value of geo information for disaster management within V&W is evident: hind – and forecasting, scenario-analysis and not in the last place operational management in case of events are well proven applications. For the near future focus will be on trend analysis, development and implementation of new location based services, training, simulation and quality management of information and services.

References

Durlacher Research Ltd, Eqvitec Partners Oy (2001) UMTS Report: An Investment Perspective. http://www.durlacher.com/downloads/umtsreport.pdf

Federal Emergency Management Agency – FEMA (1998), Introduction to Mitigation Independent Study Course. IS 393. USA

Global Disaster Information Network (GDIN), http://www.gdin-international.org/about_policy.html

Green WG III, Mc Ginnis SR (2002) Thoughts on the higher order taxonomy of disasters, Notes on the Science of Extreme Situations, Paper No. 7

Grothe M, Steenbruggen JGM (2002) Altijd en overal (geo)informatie binnen handbereik. Bouwstenen en toepassingen van LBS, in Geo-nieuws 1-2002

Masser I, Montoya L (2002) GIS in Urban Disaster Management. In City Development Strategies. May 2002

Montoya L (2002) GIS and Remote Sensing in Urban Disaster Management, 5th AGILE Conference on Geographic Information Science, Palma (Balearic Islands, Spain) April 25th-27th 2002

Steenbruggen JGM, Grothe M (2004) Location Based Services gepositioneerd als informatiedienst!, in: GIN 2004-6

Steenbruggen JGM (2004) Cell Broadcast, een nieuwe locatiegebonden informatiedienst! in: GeoNieuws 3-2004

Tobin G, Monts B. (1997) Natural Hazards: Explanation and Integration. The Guilford Press. USA.

A Case Study in Multiagency GIS for Managing a Large-Scale Natural Disaster

Russ Johnson

Public Safety Industry, Fire/EMS/Disaster Management/Homeland Security, ESRI, 380 New York Street, Redlands, CA, 92373-8100 USA..
Email: russ_johnson@esri.com

Abstract

This paper will outline how 15 government agencies in California worked together using GIS to plan for, develop mitigation plans and respond to the catastrophic fires in Southern California in the fall of 2003. The focus of the presentation will detail how several different agencies and the private sector created a shared GIS data base to collaborate on how to provide protection, prevention and response to a pending wildfire catastrophe. Examples of how innovative uses of GIS and mobile GIS were used to assist first responders will be highlighted.

This project resulted in creating a common vision of the problem, identification of priority problems that required shared resources and cooperative efforts. GIS was used throughout the incident to planning, support public information and provide the framework for high level briefings including the President Bush.

1 Introduction

This paper will provide a best practice case study of multiple agencies and the private sector using geographic information system (GIS) technology to manage a complex and hazardous situation. The paper will describe how agencies worked together, created a common vision and common understanding, and integrated a plan for a potential catastrophic event.

2 Hazardous Pre-event Conditions, Multiple Participating Agencies

In October 2003, wildfires in Southern California burned more than 750,000 acres, destroyed 2,232 homes, and killed 14 people. More than 6,000 firefighters battled the infernos. Yet the consequences of these fires could have been worse had it not been for the high level of prevent planning, preparedness, mitigation, and response between the multiple government agencies and private agencies tasked with wildfire response. This cooperative planning was supported by GIS technology—providing government officials and private agencies a shared vision of the problem and an understanding of priority actions required to reduce the consequences of this event to life, property, and natural resources. This case study will examine lessons learned and how GIS technology provided powerful tools for preparedness, response, recovery, and analysis.

In October 2003, Southern California forests were extremely dry and vulnerable when the fires struck. Years of dry conditions and insect infestations killed hundreds of thousands of trees. The areas affected by the tree mortality and high fire threat were densely populated forest communities. The forested areas are within the San Bernardino National Forest, a series of mountains that range between 2,500 and 11,000 feet elevation.

Recognizing the potential threats from this massive vegetation mortality problem, representatives from the state, local, and federal government, as well as private companies and volunteer organizations, collaborated to analyze the occurrence of bark beetles, dead and dying trees, and associated fire risk in the San Bernardino and Riverside County, California, mountains. Together, they formed the Mountain Area Safety Taskforce (MAST).

Agencies participating in MAST included

- Federal
  - U.S. Forest Service—San Bernardino National Forest
  - Natural Resources Conservation Service
- State
  - California Department of Forestry and Fire Protection
  - California Department of Transportation (Caltrans)
  - California Highway Patrol
- County
  - County of San Bernardino
  - County Office of Emergency Services
  - San Bernardino County Fire Department
  - San Bernardino County Roads Department

  - San Bernardino County Sheriff's Department
  - San Bernardino County Solid Waste Management
  - Riverside County Board of Supervisors
  - Riverside County Department of Transportation
  - Riverside County Executive Office
  - Riverside County Transportation and Land Management Agency
  - Riverside County Fire Department
  - Riverside County Flood Control
  - Riverside County Office of Emergency Services
  - Riverside County Sheriff's Department
  - Riverside County Waste Management
- Local
  - Arrowbear Lake Fire Department
  - Big Bear City Fire Department
  - Big Bear Lake Fire Protection District
  - Crest Forest Fire Protection District
  - Running Springs Fire Department
  - Fern Valley Water District
  - Idyllwild Fire Protection District
  - Idyllwild Water District
  - Lake Hemet Water District
  - Pine Cove Water District
- Energy Companies and Private Organizations
  - Southern California Edison
  - Bear Valley Electric Company
  - NEXTEL
  - Pine Cove Homeowners' Association
- Volunteer Organizations
  - Inland Empire Fire Safe Alliance
  - Angelus Oaks Fire Safe Council
  - Arrowhead Communities Fire Safe Council
  - Bear Valley Fire Safe Council
  - Lytle Creek Fire Safe Council
  - Mill Creek Canyon Fire Safe Council
  - Mountain Rim Fire Safe Council
  - Wrightwood Fire Safe Council
  - San Bernardino National Forest Association
  - Mountain Communities Fire Safe Council

The MAST groups were organized under the Incident Command System (ICS). This is now a requirement in the United States under the National Incident Management System that conforms to United States Presidential Directive #5:

Homeland Security Presidential Directive/HSPD-5 Subject: Management of Domestic Incidents Purpose

To enhance the ability of the United States to manage domestic incidents by establishing a single comprehensive national incident management system.

For additional information visit
http://www.whitehouse.gov/news/releases/2003/02/20030228-9.html.

The Incident Command System is an emergency management organization that brings multiple agencies into a command and general staff with positions focused on work production. The organizations identified a group of unified commanders and staff necessary to deal with the issues surrounding the tree mortality problem.

The primary objective of this organizational structure was to develop an integrated multiagency plan. This involved developing evacuation plans, clearing forests of potential tree hazards from routes into and out of the mountains, and providing emergency planning and hazard mitigation information to the public. In addition, the plan provided for a multiparticipant (yet single) voice as a means of securing federal funding to combat the problems that threatened the communities. This unified front included concerned state and local governments officials working with federal legislators to combat the growing fire threats. The structure and directives also included a program aimed at reducing materials that would potentially act as fuel for a fire threat and creating fuel breaks. This required an extensive planning and logistical effort for the removal of dead standing trees, the reduction of fuel on the ground, and the creation of defensible space or cleared area around homes and developed areas. These developed areas included campgrounds, parks, commercial buildings, communications, and government facilities.

Finally, as part of developing a multiagency plan, it was important to explore commercial use or disposal options for waste wood products, and to identify and develop plans for ensuring long-term forest and natural resource sustainability.

3 Applying GIS for Preplanning

In developing the initial list of priorities and action plans, it was agreed that analyzing the potential wildfire threat and identifying mitigation requirements constituted the highest priority.

The participating agencies quickly recognized the necessity of using a central information technology hub for many reasons; the response to this information need was to establish a comprehensive GIS that would provide optimized information automation, accuracy, maintenance, and dissemination. Moreover, a GIS information base would provide a better method for communication and collaboration among the many agencies and a common framework for shared decision support. Simply put, GIS provided a single solution to analyze, visualize, and understand problems and opportunities for every type of emergency management variable. The MAST organization overcame an important hurdle in creating a Memorandum of Understanding (MOU) that broke down the institutional barriers in exchanging data. Through the MOU, the MAST agencies agreed to assemble GIS data into one shared database, which would be used for modeling, analysis, resource management, and response planning. Thus, when the fires ignited, MAST had already considered evacuation and emergency response plans for the area and had thousands of maps and models in hand.

While this paper revolves around preparing and responding to a large-scale emergency, the real story is about people and organizations uniting across traditional boundaries and departments and how GIS technology facilitated this endeavor.

The MAST organizations shared the geography and the risk associated with Southern California fire hazards, but none of them could address the mitigation requirements on their own. Their joint efforts were rooted in a common goal of reducing the region wide risk of a major fire and minimizing impact on mountain communities should one occur.

ESRI joined MAST to assist in providing GIS services. Specifically, ESRI helped establish and manage the MAST GIS Center as well as hosting the associated MAST Public Information Web site at www.calmast.org. The center provided technical resources for MAST including helping to aggregate disparate data from dozens of participating organizations. The Public Information Web site provides the general public with information about vegetation mortality and related fire hazard conditions in the Southern California area and provides access to fire prevention, emergency preparedness, and wildfire recovery guidance and programs maintained by the MAST organizations and their member agencies. The Web site was enabled with interactive GIS mapping, which allowed residents to enter their address and see where they reside in relationship to the hazardous areas, potential safe refuge areas, shelters, and other important geographic information.

The MAST effort recognized the need to continue to engage and grow the sphere of concerned disciplines and organizations. Public sector participants who contributed data and expertise included federal agencies such as the San Bernardino National Forest and the Natural Resources Conservation Service. At the state level, participants included the California Departments of Forestry and Fire Protection, Transportation, Parks and Recreation, Fish and Game, Highway Patrol, and Office of Emergency Services. County-level participation came from the San Bernardino County and Riverside County Offices of Emergency Services, Sheriff's Departments, Fire Departments, Transportation Departments, and more. Local participation included fire departments from Big Bear City, Arrowbear Lake, Running Springs, Crestline, Idyllwild, Lake Hemet, and more. At a regional level, the Department of Environmental Quality came onboard.

Private sector partners in the project provided expertise, materials, and equipment to the MAST GIS Center and the Web site. Hewlett–Packard (HP) provided computer equipment, DigitalGlobe gave satellite imagery, Southern California Edison contributed its utility data and expertise, NEXTEL provided its network data, and Geographic Data Technology provided geographic data.

4 Building Data and Applications

As one would expect with so many participants, data came in various projections, often in different formats, and frequently without reliable metadata. To build a comprehensive database, the MAST database manager converted the data to common formats, resolved projection issues, and tracked down metadata for a data set or multiple data sets to determine the most recent data for the MAST mission.

Data layers included utilities, terrain, imagery, population, critical facilities, water systems, vegetation, historical fire occurrence, police stations, fire stations, schools, hospitals, roads, and so forth. Vegetation mortality had previously been mapped using fixed-wing aircraft, with area ecologists looking out the window to estimate the extent and severity of the problem often using hand held global positioning system (GPS) devices. To bring these data up-to-date and to make them less subjective, the MAST GIS Center obtained newly collected 20-meter resolution imagery from Spot 2 and Spot 3 satellites and 61-centimeter resolution multispectral QuickBird imagery from DigitalGlobe. Imagery analysis revealed that San Bernardino and Riverside Counties of California, affected by years of drought and bark beetle infestation, contained an estimated one million weakened (or dead) trees.

After building the database, MAST set out to develop GIS applications that could be used in the event of a fire. For example, vegetation treatment priorities were created factoring in such variables as vegetation mortality, population, roads, and utility infrastructure at risk. Fire dispersion modeling based on vegetation types, age class, mortality, proximity to developments, and natural resource values were conducted to identify critical values at risk. The modeling was done using a federally developed modeling program named FARSITE. The FARSITE Fire Area Simulator is a program for personal computers that simulates the growth and behavior of a fire as it spreads through variable fuel and terrain under changing weather conditions. It includes surface and crown fire spread and intensity, transition to crown fire, and spotting models. FARSITE has been used to project the growth of ongoing wildfires and prescribed fires and in planning activities for suppression, prescribed fire, prevention, and fuel assessment. Applications were also developed to support a comprehensive, multiyear reforestation management plan including prioritization of dead tree removal. In addition, evacuation and fire response planning were refined.

5 Responding to the San Bernardino County, California Wildfires

In the fall of 2003, hot, dry winds tore through Southern California. Soon, a series of fires started in several locations. As the fires grew, the MAST GIS Center was able to immediately start supporting the fire suppression efforts. This support included providing incident commanders with a common operational picture to support decision making and enhance communication. GIS specialists from the various responding agencies, as well as volunteers from ESRI, staffed seven incident command posts and operations centers, making maps to support the firefighting efforts. In addition to the maps from the MAST central database, many new maps were generated on-the-fly based on individual requests. As GIS became an integral tool for combating the firestorms, emergency response personnel began to understand firsthand the powerful capability that GIS could provide.

Some of these maps included

- Fire spread prediction maps
- Smoke plumes and affected vulnerable populations
- Mountain tops suitable for radio repeaters
- Areas designated by combinations of slope, flammable vegetation, aspects, and canyons that would be impossible to protect safely by firefighters
- Critical protection priorities
- Work assignments
- Flight routes and air hazards
- Facilities
- Evacuation centers

As the fires began to grow and became large and complex, National Incident Management teams were activated to provide command and control expertise. These teams consist of highly trained experts in command, safety, public information, logistics, planning, operations, financial management, and a host of other disciplines necessary to organize, manage, and control very complicated disasters. These teams used GIS for operational incident management. In compliance with the Incident Command System component of NIMS, GIS is used to

- Provide Incident Action Plan Maps—These maps include divisions, branches, incident facilities, transportation routes, air routes, command posts, staging areas, logistical facilities, and so forth. They provide public safety personnel with information on where they are assigned to work, objectives, and other information necessary to carry out their assignment.
- Provide Incident Prediction—Predictions of modeling plumes, blasts, fire spread, floods, and so forth, are provided for incident commanders to anticipate incident spread and deployment of public safety personnel in order to contain or control the incident.
- Provide Public Information—Maps and analyses are provided for news media to inform the public of the status of the incident, road closures, shelters, and so forth. Incident commanders often use maps during interviews to explain and detail their efforts and current status.

At the height of the fires, GIS specialists provided near real-time, integrated mapping support to nine firefighting incident command centers and firefighters. The advanced planning by MAST facilitated the effective evacuation of tens of thousands of displaced mountain community citizens, directing them along predetermined paths to planned evacuation facilities. These well-coordinated community evacuations had been preplanned and the order to begin each evacuation was based upon predetermined "trigger" points. In addition, law enforcement personnel had preplanned who would coordinate evacuations and where to eliminate confusion. These efforts are critically important in fast-moving wildfires, with fire trucks trying to advance to an incident and residents trying to evacuate an incident.

The San Bernardino County Sheriff's Department and other local police agencies used GIS software for a host of other activities and responses to the fast moving wildfires. GIS-generated maps were used to help monitor traffic, display fire perimeters, assess damage, patrol evacuated areas, plan community reentry, and much more. The department also used GIS for crime scene management because of the suspicion that the fire was started by an arsonist.

6 Providing Diverse Analysis and Mapping

Maps of the fire progression combined current fire perimeter from GPS devices to provide a historical record of fire location by operational period. Incident commanders used facility maps to show staging areas, drop points, base camps, and command posts. The mountainous terrain demanded extensive use of aircraft, so maps showing air routes and air hazards were generated. Maps helped decision makers effectively allocate resources such as deploying crews to protect utility assets and mountain populations. GIS specialists generated maps of the most hazardous locations and areas that were undefendable for firefighters, factoring in current weather conditions, ingress and egress routes, terrain, and highly flammable vegetation. They also mapped safety areas into which firefighters could flee if the flames turned their way.

Each morning, GIS teams at the MAST GIS Center and from incident command centers, as well as various fire response GIS contractors, generated both digital and paper maps for firefighter briefings. Every 12 hours, a new updated incident action plan is developed that takes into account progress made and fire spread conditions from the previous action plan and firefighter efforts. Maps are a key component of these plans. Each area of the fire is divided and each division has personnel assigned with key objectives for resource protection and fire line construction. Other maps include travel plans, firefighter drop points, hazardous areas, and logistical support facility locations. The maps also help make crews aware of the context of the fire and help them formulate strategies and tactics. Fire crews took maps into the field that showed such critical features as dozer lines, water sources, and staging areas and proposed fire line construction plans.

During large-scale disasters, politicians and other high-level government officials are often compelled to visit the incident first-hand to assess overall impact and to provide support. To effectively carry out these visits, it is important for emergency responders to provide a comprehensive situation overview and to have access to affected areas and affected citizens. During the Southern California wildfires siege, numerous elected officials constantly visited the incident. Some of these officials included U.S. President George W. Bush, California Governor Arnold Schwarzenegger, California Secretary of Agriculture Ann M. Veneman, and numerous other state and local politicians and officials.

In a separate briefing for President George W. Bush, with former Governor Gray Davis and Governor Schwarzenegger in attendance, GIS-produced hard-copy maps showed fire extent, property damage, fire lines, fire perimeters, and vegetation mortality. Meanwhile, online maps at www.calmast.org helped to disseminate timely information to the public, evacuation centers, and journalists.

In addition to providing maps to fire staff, law enforcement, and government officials, maps provided valuable information to displaced citizens. The thousands of evacuated people were given highly detailed maps on a daily basis to illustrate where the fires had progressed and which homes and neighborhoods had been destroyed or saved.

GIS-generated map books were distributed to crew leaders and firefighters who responded from outside the area to assist local first responders and firefighters.

These map books contained information and maps such as

- Areas of Safe Refuge
- Staging Areas
- Potential Incident Command Posts
- Community Protection Priorities
- Hazardous Areas That Were Deemed Unprotectable
- Critical Facilities
- Transportation Routes and Road Networks

These maps assisted firefighters from outside the area to become quickly oriented and to understand protection priorities and unsafe areas quickly with high-quality and detailed maps.

Moderate Resolution Imaging Spectroradiometer (MODIS) was used when conditions made it difficult for helicopters to fly and record perimeters. These conditions existed when winds were extreme and/or smoke made it difficult to fly through or observe the location of fire perimeters. MODIS data is collected from two low-orbiting (800 km) satellites called Terra and Aqua that transmit data twice a day.

Some of these data sets are calibrated radiance, geolocation fields, clouds mask, aerosol, and precipitable water. The MODIS satellites observe earth through 36 bands with three different resolutions (1 pixel = 250 m, 500 m, or 1 km [at nadir, 6 km at edge]). MODIS data collected for the wildfires was one-kilometer pixels that exceeded 1,000 degrees. These points would collectively demonstrate a close approximation of the fire perimeters. This data would be approximately six hours old when collected and posted.

7 Utilizing Data Visualization

At daily briefings, 3D GIS technology provided firefighters with views of the terrain by displaying digital elevations of the earth's surface and high-resolution imagery draped over the digital elevation models to create a realistic view of the affected area that could be navigated, toured, and flown through. During the briefing, GIS provided a virtual tour of the fire perimeters, potentially affected areas, and the overall progress of the events. Firefighters were updated with topographic features, such as ridges, that might be suitable for constructing a fire line well in advance of the fire. This type of orientation and contingency planning saved time and created a common understanding among fire crews of what actions were needed at very specific locations. Firefighters requested this type of technology briefing for all of the command posts.

8 Pinpointing New Fire Lines and Creating Contingency Plans Using Mobile GIS

Mobile GIS helped emergency responders prepare for the potential spread of fire to other communities not yet affected by the spreading fires.

The San Bernardino County Fire Department used mobile GIS on handheld devices to create digital maps and contingency plans for the Big Bear community in the event that the Southern California Old Fire scorched its way past firefighter perimeters. Staff went to Big Bear City, one of the communities potentially threatened by the Old Fire, as a contingency detachment. They were tasked with addressing the possibility that the Old Fire might spread past the adjacent Lake Arrowhead community (a community that was being impacted by the fire) and toward Big Bear Lake, a heavily populated resort community in the San Bernardino National Forest. This GIS staff operated out of a mobile computing laboratory (at the

Big Bear Incident Command Post) featuring state-of-the-art hardware, software, plotters, signage equipment, and other technologies.

In the aftermath of the contingency planning, the new Big Bear maps were brought back to the Incident Command Post for final fire response efforts, reentry operations, and rebuilding work.

A second group also using the services of the GIS unit was the public information officers, who are responsible for handling public and media inquiries. Digital and paper fire maps showing fire perimeter, fire progression, fire break, property data, and more helped inform affected communities as to what was going on and support the media in their reporting efforts. The third group consisted of fire and GIS experts who used digital three-dimensional maps to present the latest status of the fire, its containment, and final plans to senior government officials. Digital maps that were made for operations, for example, only had to be altered slightly for the information group. GIS made these types of alterations possible in just minutes.

9 GIS in San Diego County, California

On October 25, 2003, a human-caused (arson) fire began near the mountain town of Ramona in San Diego County, California. Known as the Cedar Fire, the speed and ferocity of the blaze required thousands of firefighters and other emergency responders. Much of the GIS-based emergency response to the San Bernardino County fires described above took place in a similar fashion in San Diego County.

For the San Diego fires, the California Department of Forestry and Fire Protection deployed a GIS trailer and staff to San Diego to support the CDF GIS team members. The trailer became a 24-hour information resource, providing firefighters and others with continuous accurate mapping and analysis. GIS team members produced hundreds of detailed maps integrating multiple information sources from field observations and infrared camera-equipped helicopters. Similar to the Grand Prix, Old, and other Southern California fires, the San Diego Fire morning briefings included full-size map plots showing fire progression, fire containment, bulldozer lines, and other critical data. GIS enabled effective allocation of crews, bulldozers, engines, aircraft, and other assets.

10 Post Fire Analysis for the San Bernardino County Fires

Helped by a change in the weather, firefighters eventually gained the upper hand on all of the Southern California wildfires. By that time, they had consumed approximately 700,000 square miles. The MAST GIS lab quickly stepped up to support many government efforts to map destroyed homes and analyze debris flow risk.

One of the immediate needs was to determine structure loss and structure damage. Field reconnaissance began, which frequently involved assessment teams walking the burn perimeter with GPS receivers to determine the extent of the fire. The MAST team created several damage assessment applications that involved use of a Tablet PC to conduct windshield surveys of damage assessment. This allowed agencies to quickly understand which and to what extent homes and structures were damaged. More than 2,000 homes were destroyed and several thousand homes were damaged. Field personnel also used GIS land parcel data to record damage and store it in the GIS database. By clicking on a particular parcel, a simple form appeared that allowed the field inspector to categorize structures as destroyed, damaged, or unaffected. This form also allowed users to provide additional comments to further clarify the extent of damage and capture digital pictures linked to the particular land record parcel.

This triage information was then downloaded at the end of each day to the central GIS database. The results of this data collection could then be displayed on the map using GIS with color-coded results. More detailed information could be obtained by clicking on the parcel record. Parcel records contain tax assessment information and property values. Quick estimates of financial loss could be calculated within the GIS application. The use of these tools made damage assessment data collection more efficient, saved time, and allowed further analysis of the fire behavior and structural loss much easier.

In addition to the information collected for damage assessment, other beneficial analysis could be conducted. By viewing the structural damage on the map with other GIS data layers (fire perimeters, burn intensity, topography, surrounding vegetation types, vegetation mortality, age of structures, construction types etc.), it became possible to understand what structure features and physical features performed worst (or best) during a wildfire.

Many other teams immediately set out to assess the effects of the fires on vegetation and soil and analyze the increased risk of runoff and flooding until the burned areas recovered. Specifically, the Federal Emergency Mapping Agency (FEMA) initiated the development of its Postfire Advisory Flood Hazard Maps to assist homeowners and state, local, and federal agencies in recognizing and addressing the increased flood risks. To begin to model and understand potential flood and debris movement, burn intensity had to be identified. The increased incidence of catastrophic wildfires in the western United States and the encroachment of development into fire prone ecosystems have created a critical need for methods to quantify potential hazards posed by debris flows produced from burned watersheds. Debris flows are one of the most hazardous consequences of rainfall on recently burned hill slopes, and agencies are beginning to use GIS tools to determine both the probability and magnitude of such destructive events from individual drainage basins. One of the critical elements in calculating debris movement is burn intensity. To create burn intensity maps, current imagery was collected. Using imagery, GIS could analyze which landscapes had burned with high, moderate, or low intensity. With this information, the United States Geological Survey Service (USGS) used slope, soils, and other data to evaluate potential debris flow risks in the high burn intensity areas. These USGS maps show preliminary assessments of the probability of debris flow and estimates of peak discharges that could be generated by debris flows issuing from various burned areas. Specifically, models were created to estimate flow in response to various intensities of rainstorms.

On December 25, 2003, rainfall on exposed slopes generated debris flow and floods from basins throughout the burned area. These events led to the deaths of 16 people in Waterman Canyon and Devore Canyon. In this event, too, emergency personnel used GIS to help guide rescue and recovery operations and gain a better understanding of the tragedy. The public safety officials responsible for managing the flood recovery used the MAST GIS Center to provide the required information, analysis, and mapping to manage the floods.

After the flood incidents had been controlled and the search, rescue, and recovery efforts were over, county officials implemented an emergency notification system. Using the GIS streets data, addresses, and emergency telephone database, a notification application was put into operation. Small residential areas or polygons that could be affected by debris movement and floods were predetermined and mapped. Remote sensors measuring rainfall were placed on vulnerable slopes that had experienced high-intensity fire. Using GIS, flood models were run for these vulnerable slopes in order to better understand and predict the potential of debris movement. From these models, thresholds were established for emergency notification to potentially affected citizens. Based on a volume of rainfall within a specified time frame (i.e., one-half inch of rain within a one-hour time period), the appropriate phone numbers (from the address/telephone database) within the premapped areas would be called with an emergency evacuation message.

11 Counting the Costs

In early November 2003, President Bush had already declared five Southern California counties federal disaster areas. In the two months following that declaration, $176 million in disaster relief was approved. Disaster relief and recovery efforts were managed by 17 federal agencies and 35 state departments and agencies staffed by 481 FEMA personnel and 411 personnel from other agencies.

These costs are only the beginning of the toll of the Southern California wildfires. Burn assessment and landslide mitigation efforts continue to this day. Public health officials are considering the long-term effects of smoke exposure on local children. In addition to the environmental and health repercussions, the economic impact to mountain communities will be felt for years. However, in these efforts, the MAST program continues to play an important role and serve as an excellent example of what can be accomplished when agencies work across traditional boundaries to best accomplish shared goals.

12 Concluding Remarks

The 2003 Southern California wildfire incidents represent a best practice for multiple government agencies collectively preparing, collaborating, sharing information, and responding to a major emergency event. Looking to the future, the MAST operation can act as a model of how public and private agencies can work together on disasters as they occur. More importantly, these lessons can assist all agencies looking for methods to prepare and mitigate homeland security threats in the wake of the events of September 11, 2001.

Agencies must share information, plan together, develop mitigation strategies, and become jointly prepared. GIS is a tool that can facilitate this requirement. GIS is able to take complex problems and analyze all of the related issues and provide answers in graphic form that can be effectively communicated and understood. This provides various agencies with the ability to create a common vision, common language, and common understanding of the problem as well as prioritize needs and the appropriate mitigation requirements.

The focus and requirements of crisis management at the local community level today include

1. Improving continuity of government/operations
2. Empowering employees with personal workplace disaster plans
3. Evaluating infrastructure vulnerability
4. Reducing vulnerabilities
5. Improving intelligence coordination
6. Strengthening individual and community preparedness for all disasters
7. Continuing building our capability to respond to nuclear, biological, and chemical events
8. Developing greater capability to rescue victims from collapsed structures
9. Supporting regional and state efforts in terrorism preparedness and response
10. Improving operational coordination to disasters at large venues

GIS is a technology that can support all of these requirements. As people with different missions begin to work together and form relationships, the possibility for multiple agency planning and preparedness becomes greater, and GIS can play a critical role in facilitating this. Using GIS for homeland security, emergency and disaster management is another way to leverage an investment that has multiple benefits.

User-Oriented Provision of Geo-Information in Disaster Management: Potentials of Spatial Data Infrastructures considering Brandenburg/ Germany as an Example

Petra Köhler

Geo Forschungs Zentrum Potsdam (GFZ), Data & Computing Center, Telegrafenberg A3, 14473 Potsdam, Germany.
Email: p.koehler@gfz-potsdam.de

Abstract

Actual and high-quality data and information – in the most cases spatial data or geo-information (GI) - are the foundation for decision making in disaster management especially in situations where losses are to be minimized and lives are to be saved. Data and technology suppliers are talking about the magic thing: the "user" or "end user". But, providing innovative products hoping the user will be positive about and spend a lot of money for means more than asking "who is the user?" and "where can he be found?".

Anyhow, those products shall support information flows and working processes, often without having the time or chance to reconsider and discard decisions. First of all, data and information products must be available and usable, i.e. they must be known and available in that form and time they are needed. Technological solutions must be application-oriented, aiming at fulfilling the user's needs in his specific environment.

These facts are taken up by the Special Interest Group "Disaster Management" under the umbrella of the "Geodaten-Infrastruktur Brandenburg (GIB)" in one of the eastern states of Germany. In the GIB framework a network of administrative players, data and software suppliers, scientific institutions and users of geographic information is developed and cooperations and reconciliations are established to build up a local Spatial Data Infrastructure (SDI). The initiative aims at the interdisciplinary and cross-institutional availability and usability of spatial data, its opening for various working fields and the establishment of an information and communication platform for administration, industry, science and society according to GI.

In the following the potentials of GI, geographic information technology (IT) and SDI for disaster management are identified. First cognitions and results as well as implementation approaches under involvement of the "users" are presented to provide application-oriented support to the effective disaster prevention and coping.

1 Introduction: Spatial Data Infrastructure - What is it?

1.1 GI and IT as Resources in Disaster Management

Geo-information provide an enormous variety of content and know-how that can be used in the context of disaster management. Reference data mirror facts and circumstances on the earth's surface. They provide a basis for spatial referencing of thematic data which give an account of issues in the context of environment, infrastructures, statistics etc.

Topographic maps are used for micro-scale demonstrations whereas satellite images play an important role visualizing macro-scale processes and changes of the landscape. With their increasing resolution they among others enable the visual detection of damages due to an extreme flood event. Linked with this valuable reference information thematic data like land use data, utilities maps and demographic distributions actually provide essential information concerning the characteristic of appearances, facts and relations and thus act as input for modeling, scenarios and simulations on the one hand and on the other hand as foundation for planning and coordination of measures and in this way for decision support.

Modern information technology is seen as driving engine while using spatial data and geographic information as resource in disaster management. The increasing storage and processing capacity, increasing networking of actors and resources via internet and effective communication channels lead to complex and seminal solutions. Geographic information systems (GIS), spatial enabled databases, internet map services and modules for mobile data acquisition are some examples bearing high potentials for disaster management.

The usage of common architectures and established standards for data and interfaces is the decisive precondition for the cross-institution and integrating use of data. International standardization of data and services ensures openness and flexibility of applications and the integration of complementary tools to cope with sophisticated questions and processes.

Standardization associations like the Open Geospatial Consortium (OGC), the International Standardization Organization (ISO) and the World Wide Web Consortium (W3C) provide corresponding specifications that are more and more picked up in application development.

1.2 Spatial Data Infrastructure: Framework for the Opening of Spatial Data

To ensure suitable conditions using spatial data as foundation for complex applications in any field of work the establishment of a framework is required which arranges and warrants the availability and usability of spatial data as well as of spatial information technologies. Spatial data infrastructures bring out such frameworks and according to Nebert (2001) are defined as

„base collection of technologies, policies and institutional arrangements that facilitate the availability of and access to spatial data. The SDI provides a basis for spatial data discovery, evaluation, and application for users and providers within all levels of government, the commercial sector, the non-profit sector, academia and by citizens in general".

Thus, various components at different levels are to be implemented to build up a spatial data infrastructure. The *organizational level* comprises networking and interactions between administration, industry, science and society aiming at the coordination and sustainment of the SDI's institutionalization and the realization of the relating concepts and measures. The *political and legal level* defines access and use conditions, pricing models etc. and anchor the infrastructure concept in the principles and statutes of the institutional framework. At the *data level* integration and the generation of transparent and comprehensible offers are to be provided to open GI for application and to ensure the development of a "geo-information market". The *technological level* aims at the introduction of standardized interfaces and the securing of interoperability of heterogeneous applications. According to that, the implementation of organizational structures, the definition of responsibilities, the documentation and publication of data sources and services considering the introduction of standards and approved concepts and the unification of access and usage rights become apparent as essential actions to realize a sustainable spatial data infrastructure.

2 Disaster Management as Focusing Field of Work

Natural and man-made disasters like extreme floods, earthquakes, forest fires, hurricanes and terrorist attacks are events of high hazard potential. Leading to damage or destruction of property, infrastructure, health and life they cause impacts that are to be prohibited. This is the goal of disaster management: Disaster management comprises different stages in a cycle starting with the mitigation and the preparation for an expected disastrous event, going on with its real occurrence and the taking of measures to prohibit and reduce its impacts. Actions of recovery subsequently pass again into the stages of mitigation and preparation. In the following disaster management means the totality of measures referring to mitigation, preparedness, response and recovery (FEMA 2001).

2.1 General Requirements in Disaster Management

Each of the stages of disaster management requires the orientation on given and potential events and situations and the making of decisions. The foundation for decision making is funded information derived from actual and high-quality data. Managing, conditioning, analyzing and processing as well as presenting such data and information depend on information technological applications, e.g. data warehouses, information systems, dispatching and coordinating systems and tools like mobile devices.

The definition of tasks, the operation of processes and the formulation of significant results of analyses, modeling and simulations many times trace back to scientific algorithms and procedures. Disaster research in particular is aiming at understanding the origin and the occurrence of extreme events by natural and man-made disasters and reproducing their impacts. New and scientific substantiated procedures, information products, services and technologies for an improved disaster management are developed in this environment of research and development (R&D).

To ensure well-structured measures previous to, during and following disastrous events, the knowledge of suitable resources and their characteristics like features, availability and usability as well as possible interactions is required. Their documentation and evidence by metadata (descriptive "data about data" and information platforms like WWW-portals is of high importance but also embedding disaster management and the different actors and decision makers in effective and superior organizational structures.

Thus, the general requirements of disaster management comprise

- disaster relevant data and information products,

- disaster relevant methods and procedures,
- disaster relevant tools and technologies and a
- suitable organizational environment

and can be summarized as the availability of the right information as foundation for decision making at the right time at the right place and in the right form.

2.2 Barriers

The variety of data sets and information nearly comes along with the variety of data suppliers, responsibilities and access regulations. In Germany most of the available (spatial) data is held by the communities and by federal and national agencies. Central „shops“ distributing cross-institution products hardly do not exist which results in enormous expenditures in obtaining data on the part of users and clients. Scarcely transparent and comprehensible offers, inconsistent access and use constraints enforce that effect. Moreover, varying data structures and proprietary formats complicate the integration and common usage of data.

Even R&D organizations and enterprises produce and hold manifold data stocks. In science, up to now data and solutions have been exchanged primarily between research institutions, but the dissemination of results and information products to non-research target groups is pursued increasingly. The development of disaster management applications by IT industry many times is rather geared to technological innovation than to the users' requirements. The lack of early dialogue between industry and user results in the lack of consideration of given organizational and financial basic conditions and of customization of state-of-the-art products. Thus, incompatibility of systems and solutions as well as isolated applications for the realization of complex scenarios enforce the problem of difficult integration and usability of disaster relevant data and information within the scope of multi-actor decision making.

3 User-Oriented Support of Operational Disaster Management: Methods of Resolution inside the SDI of Brandenburg

3.1 Spatial Data Infrastructure Brandenburg and Disaster Management

In Brandenburg, one of the eastern federal states of Germany, the initiative „Geodaten-Infrastruktur Brandenburg (GIB)“ was founded in 2001 fostering the implementation of the Brandenburg SDI (KÖHLER 2003). In 2004 the Brandenburg Parliament resolved upon the constitution of an interministerial committee and the formulation of a master plan as basis for the realization of the local spatial data infrastructure.

Its potentials are to be exploited for different branches and among others for disaster management. Thus, a workshop dealing with „GIS und spatial data in disaster protection in Brandenburg“ for the first time brought together responsible players of the Ministry of Interior, of the communities' disaster management as well as GIS experts, data and software suppliers and GeoForschungsZentrum Potsdam (GFZ) as local research institution. Expected added values by spatial data, modern information technology and the establishment of a common infrastructure were demonstrated by presentations. Furthermore, working groups examined weak points of present situations and formulated first methods of resolutions.

3.2 Necessary Actions

The discussions exposed calls for actions at different levels: First of all a lack of communication between the players in local operational disaster management was stated. This affects both the communication between communities and state, between communities themselves and between particular agencies and staff positions. The dissemination and exchange of disaster relevant data and information do not follow present possibilities and software systems and solutions supporting manifold processes and workflows, e.g. information systems or complex dispatching and management systems, do hardly exist in control centers. The availability and existing access points to digital data sources in many cases are unknown and the acquisition is too laborious. These facts hamper processes and coordination especially in time-critical decision needs.

The resulting demands become apparently: Fostering of coordination and communication between all participating groups and provision of disaster relevant data and information as foundation for decision support and under consideration of modern information and communication technology.

Following measures are to be realized to guarantee support of processes and workflows and effective disaster prevention and coping:

- Organizational measures: Introduction of well-defined organizational structures with involvement of operational disaster managers; analysis of working processes and the specific actors' requirements; determination of unified information flows.
- Measures to ensure optimized availability and usability of spatial data: Establishment of transparent data offers; introduction of suitable usage conditions for disaster coping; unification of data and information management.
- Technological measures: Implementation of services providing disaster relevant data and information; development of application-oriented prototypes; usage of approved architectures and standards.

In this context the initiation of pilot projects and the transfer from research and development to application play an essential role. Institutions of disaster research like GFZ Potsdam and the Center for Disaster Management and Risk Reduction Technology (CEDIM) as well as networks of GI industry are valuable partners.

3.2.1 Special Interest Group "Disaster Management"

To realize these approaches a Special Interest Group (SIG) Disaster Management was founded subsequent to the workshop described above. As part of the GIB initiative it consists of representatives of the Ministry of Interior, the local fire departments, the communities' GIS experts, the geodetic survey and of GFZ Potsdam. Its goal: optimization of access to and usage as well as exchange of disaster relevant information. Common procedures in data and information management shall be implemented and arrangements on spanning information and communication be reached. Existing gaps are to be detected and methods of resolution developed and realized by pilot projects and test beds. Thus, real application orientation and know-how transfer is expected. Cooperation with scientific projects and neighboring regions and moreover the orientation on national and international activities are aspired to guarantee information exchange, flexibility and interoperability. In detail the tasks comprise the

- examination of concrete scenarios referring to potential hazardous events including the interactions of responsible actors at different levels and the usage of spatial data, software applications etc.,
- formulation of a catalogue of requirements according to organizational structures and the usage of spatial data and information technology,

- deduction of optimizing strategies and recommendations,
- organizational and technological realization of methods of resolution in pilot projects and the
- reconcilement with further developments in the context of GIB, eGovernment und similar initiatives in close-by (federal) states etc.

These activities result in the locking up of a master plan aiming at a common and unified technical and organizational infrastructure as foundation for the development and operation of control centers and staff positions in disaster management.

3.2.2 Current Works

The analysis of the users' requirements must be the foundation for further activities and measures, looking at specific surroundings and working processes. During the first workshop requirements on organizational and IT aspects were already addressed. Moreover, a survey was started by GeoForschungsZentrum Potsdam and Ministry of Interior to consolidate the workshops' results: It covers both existing and required data and technological resources in authorities, fire departments, control centers etc. including detailed queries on characteristics, features and problems.

At a second event different services, systems and applications were presented to demonstrate existing solutions for data and information dissemination and sharing. Currently, the SIG deals with the analysis and evaluation of such procedures and systems. Afterwards a concept will be formulated to initiate first steps relating to real implementations. Thereby, the connection to the Brandenburg SDI initiative bears a lot of advantages: The implementation of standardized web services to provide data and metadata to the public is seen as essential foundation for more complex scenarios in disaster management.

Furthermore, there are projects planned linking operational disaster coping with disaster research. Risk analyses as future results of different scientific investigations are examples for valuable output that can be provided in form of maps, web service, software modules etc.

4 Conclusions

Spatial data and geographic information technologies bear numerous potentials for the protection of and coping with natural and man-made disasters. However, manifold organizational groupings and responsibilities, proprietary data structures and incompatible systems and solutions hamper the effective exploitation of these resources. The establishment of spatial data infrastructures as frameworks ensuring the optimized availability and usability of spatial data and information is required to steer and optimize processes and work flows under participation of administration, industry, science and citizens.

Following added values are expected under the consideration of these approaches:

- development of complimentary organizational structures and networks
- initiation of pilot projects and transfer of know-how and technology from R&D to application
- supply of application-oriented data and information products as well as of interoperable software tools and services.

The usage of approved standards on the one hand and the involvement of the users during the preparation stage of the particular infrastructure modules already is the precondition for the approach's success. Finally, the result is the optimization of working processes in all the stages of the disaster management cycle.

References

Federal Emergency Management Agency (2001) Information Technology Architecture, Version 2.0 - The road to e-FEMA (Volume 1). Washington, http://www.fema.gov/pdf/library/it_vol1.pdf

Köhler P (2003): Geodateninfrastruktur Brandenburg: Organisatorische und praktische Umsetzung. In: Bernard L et al. (ed) Geodaten- und Geodienste-Infrastrukturen - von der Forschung zur praktischen Anwendung, Beiträge zu den Münsteraner GI-Tagen. Münster, pp 151-160

Nebert D (ed) (2001) Developing Spatial Data Infrastructures: The SDI Cookbook. http://www.gsdi.org/pubs/cookbook/cookbook0515.pdf

PEGASUS: A Future Tool for Providing Near Real-Time High Resolution Data for Disaster Management

Nicolas Lewyckyj and Jurgen Everaerts

Flemish Institute for Technological Research (Vito), Center for Remote Sensing & Earth Observation Processes (TAP), Boeretang 200, B-2400 Mol, Belgium.
Email: nicolas.lewyckyj@vito.be

Abstract

Today, geo-information acquired by remote sensing techniques is more and more used for the management of all the phases of a crisis or disaster situation. Information is generally provided either by spaceborne or by airborne instruments. However, both technologies suffer some limitations. It is thought that a High Altitude Long Endurance (HALE) Unmanned Aerial Vehicle (UAV) can combine advantages of both approaches and reduce to the minimum the disadvantages. The Flemish Institute for Technological Research (Vito) has therefore decided to develop a system combining a HALE UAV with some state-of-the-art high resolution sensors for Earth Observation in the framework of the PEGASUS project. The present paper describes the project concept, the payload definition and some possible applications in the framework of disaster management.

1 Introduction

Remote Sensing (RS) is traditionally performed by either airborne or spaceborne systems, each having distinct advantages and disadvantages. Airborne systems operate mostly around several hundreds of meters for helicopters and up to 2 to 10 km above sea level for airplanes. They offer great flexibility, short response times and are able to generate very high resolution data (typically few cm). They are however expensive to operate and their operation is strongly limited by air traffic control constraints and by weather conditions. The coverage is also limited as typical swaths are about a few kilometers. Furthermore, the instruments are subjected to severe vibrations as the aircraft are operating in an unstable part of the atmosphere. Due to these turbulences, it is difficult to navigate the aircraft along the planned lines. State-of-the-art Inertial Measurement Unit (IMU) and differential GPS systems are therefore a must to allow a correct and accurate geo-referencing of the image during post-processing in a short turn-around time.

Spaceborne systems offer a very stable platform and allow global coverage (e.g. SPOT Végétation produces daily a global coverage of the earth at 1km ground pixel size). Although the resolution of the satellites has improved significantly (e.g. IKONOS provides panchromatic images with a 1m ground resolution), it is still considerably inferior to the spatial resolution of airborne systems. Moreover, due to orbital mechanics, satellites are either limited to focus always at the same area of the earth or to pass over certain regions at regular time intervals. In the latter case, regions suffering of frequent cloud cover require long acquisition periods (up to more than a year) before an area is completely imaged.

During the last years, satellite images are more and more used for disaster management as the International Charter "Space and Major Disasters" (Béquignon 2004) offers the possibility to obtain free images. However it is clear that the current situation is still insufficient: decision makers do not receive geographical information with adequate spatial resolution and/or in due time. Furthermore, there is no real database of high spatial resolution images (like IKONOS or Quickbird) that can be used for a precise estimation of the damage consequent to the disaster (e.g. earthquake).

2 The PEGASUS Project

As an alternative to conventional airborne and spaceborne platforms, the PEGASUS (Policy support for European Governments by Acquisition of information from Satellite and UAV-borne Sensors) project will use an unmanned aerial vehicle (UAV). It is thought that solar High Altitude Long Endurance (HALE) Unmanned Aerial Vehicle (UAV) can combine advantages of both approaches and reduce to the minimum the disadvantages. This initiative is officially supported by the Flemish Government (Belgium) that decided to fund the acquisition of the platform in June 2004.

The solar HALE UAV platform (also called stratellite) will be able to operate at middle European latitudes (52° North) between 14 and 20 km altitude for weeks to months without landing. The system will be powered by a combination of solar energy cells and batteries. In a later phase, batteries may be replaced by fuel cells.

Despite its aptitude to cover large areas (up to 125.000 km^2 on a yearly basis), dedicated missions (e.g. limited continuous survey of small areas with a high information update rate) are definitely within the scope of the project. This can be very useful for disaster prediction, detection, evaluation, management and recovery phases.

Four high resolution instruments will equip the stratellite. The instruments selection has been performed on the basis of the user community requirements (Fransaer et al. 2004). A multispectral digital camera will be developed first. Later on, a laser altimeter, a thermal camera and a Synthetic Aperture Radar instrument will complete the suite of sensors. The synergy of the sensors will ensure the system to provide all weather, day and night, high quality data. The data will be directly downlinked to a mobile ground station and forwarded to a central Processing and Archiving Facility (PAF). After standard processing, the collected data will be available for the stakeholders in near real-time via internet. The system will deliver data and information that are directly usable for programs like the Global Monitoring for Environment and Security (GMES) as well as for medium and small scale mapping applications.

The HALE UAV will be controlled via an autonomous mobile ground station operating within a radius of 150 km. The trajectories will be preprogrammed up to 3 days in advance but direct control of the platform by an operator/pilot is also foreseen. This will allow a great flexibility of the system on an operational basis. The UAV carrier itself can also be transported in containers and the take-off takes typically few hours. Rapid deployment of the whole system will thus be possible if required.

The system is set to fly its first demonstration tests in 2005, in Belgium. Later on it is foreseen to develop a fleet of HALE UAVs hovering above the whole of Europe.

2.1 Why solar High Altitude Long Endurance (HALE) UAVs ?

The use of Unmanned Aerial Vehicle (UAV) offers two main advantages as compared to conventional Aircraft:

- the unmanned character allows very long uninterrupted missions (e.g. relay for telecommunication),

- the platforms may hover above dangerous areas avoiding the sacrifice of people in order to gather crucial information (e.g. for disaster management purpose like in Chernobyl in 1986).

These aspects were directly understood by military services that originally developed the concept of UAV. The use of Unmanned Aerial Vehicles however grows nowadays drastically within the civil community. Actually about 400 different officially registered UAVs exist, scaling from few centimeters (micro UAV) up to 70m wingspan (Helios prototype or Global Hawk) or to more than 200 meters (stratospheric balloons). The range of weights encountered for UAV starts form several grams and go up to tons and the altitudes reached by unmanned platforms vary between several cm up to nearly 31 km above sea level (the Helios prototype holds the world altitude record for non-rocket powered aircraft by flying to 96,863 ft in August 2001).

According to EURO-UVS (2003), the majority of UAVs is still in a development stage, but nearly 30 of them are already market ready. Beside the military and security aspects, they can be used for telecommunication and for earth observation (mapping, vegetation, risk management (fire, earthquakes, inundations, …), environmental monitoring, local monitoring (archeology, detection of unauthorized oil spills, leakage in pipes, high voltage cable survey), etc.…).

Most UAV's are aerodynamic, i.e. generating lift by moving (airplane or flying wing design) but aerostatic systems (blimps or balloons) are currently also in development. The latter ones rely on very large volumes (200.000 m^3 or more) of helium to provide lift, which allows very large payload to be carried. However, steering such large volumes requires a large amount of power and such carriers are more sensitive to their environment (e.g. high wind speeds). According to Küke (2000) these systems are likely to be an order of magnitude more expensive than aerodynamic systems.

The choice of using a High Altitude platform was directed by two major elements:

- for safety reasons, the platforms should affect the current air traffic (up to 12 km altitude) as little as possible,
- in the low stratosphere (between 14 en 25 km) wind speeds are lower and turbulences limited, what confers a better stability to the UAV.

It was therefore opted for a UAV able to remain between 14 and 20 km altitude. To fulfill the user requirements, the platform should be able to survey large area and/or to be available rapidly for emergency situations when required (e.g. forest fires in the Mediterranean area). The platform should therefore be able to move relatively rapidly (aerodynamic concept).

Furthermore, in order to restrict to the minimum the number of take-offs and landings of the UAV (crossing the air traffic corridors) and in order to optimize the image acquisition costs, it was opted for long endurance missions (weeks to months continuous survey). Finally, small light weight UAVs were preferred to large ones for both safety (in case of loss of control) and for engineering (physical constraints on the structure of the carrier) reasons. These latter aspects insure also better maneuverability.

The long endurance character implies the solar character of the UAV as nuclear power supply is technologically also possible (Graham-Rowe 2003) but not an acceptable due to the risks involved in case of a mishap. The solar character offers also the advantage to be environmental friendly (totally free of polluting emissions).

In the specific context of crisis management, HALE platforms are certainly of greatest interest as they are easily dirigible and able to cover a large area on the one hand or able to remain above a dedicated area for a long period on the other hand. The High Altitude character of the system allows for efficient mission planning (above air traffic control). Moreover, unlike manned aircraft, no time is wasted for flight clearance, transfer flights or crew replacement and data acquisition can be performed continuously for very long periods.

3 Characteristics of the Payload and Possible Applications

Weight, power consumption and volume are the keywords constraining the design of the payload (instruments and auxiliary systems). The payload of the solar HALE-UAV that will be used within the framework of the PEGASUS project is limited to a maximum weight of 2 to 3 kg, unless the solar HALE UAV is scaled up. The power supply for the instruments is about 50 watts. The rest of the power generated by the solar cells is used by the flight systems itself. The volume constraint is probably the most difficult one to deal with: the different sensors should be designed to fit within the limited volume and the irregular shape of the aircraft's fuselage. In the future, the UAV platform should evolve in a manner that it will be able to carry heavier payloads and provide more power.

As previously mentioned, four different kinds of sensors are foreseen at the present moment: a Multispectral Digital Camera (MDC), a LIght Detection And Ranging sensor (LIDAR), a digital thermal camera and a Synthetic Aperture Radar (SAR). The sensors to be developed are based on well known and proven technology. This allows to reduce significantly the time effort as well as the costs for development. The different instruments will be defined and developed in sequence allowing a possible response to changes in the market. Costs for data processing will also be significantly reduced by designing the instruments to the constraints of the requirements, but not better (data have to conform to the applications' requirements but not exceed them, what implies that the instruments will deliver data that are "just good enough").

Knowing that the images will be processed in a semi-automatic way and made available via internet within one hour, decision makers will be able to rapidly access very high precision and high accuracy geographical information. This point is critical for a performing management of crisis and disaster situations. Moreover, because all the data acquired by the solar HALE UAV will be stored in a huge database, a rapid evaluation of the damages will be possible by comparing the situation before and after the disease.

3.1 The Multispectral Digital Camera (MDC)

The Multispectral Digital Camera will provide images in up to 10 narrow spectral bands (10 nm individual band width) in the visual and near-infrared spectrum (400 – 1000 nm). From an altitude of 20 km, the ground pixel size will be 15 to 20 cm. The instrument will be composed by 12000 pixel wide line CCD arrays (see Reulke 2003 for an overview of available sensor technology) providing a swath width of 1800 to 2400 m.

A position accuracy of 15 cm will be guaranteed by the combination of a high-grade position and orientation systems, the forward oversampling (pushbroom system flying at low speed) and the use of ground control points. The Multispectral Digital Camera is the first instrument to be implemented.

3.2 LIDAR

Covering the same swath as the MDC, the LIDAR instrument will provide elevation information with a point density between 1 point per 2-4 m^2. The pulse repetition frequency is set to 15 kHz, which will produce a point density of 1 point per 2.5 m^2in the best case. The instrument will record the first and last reflected pulse, and the intensity of the reflected pulses. Higher point densities useful for detailed city mapping (e.g. Noble and Nix 2003) can be obtained by multiple overpasses over the same area. Combining that information with images of the Multispectral Digital Camera, orthophoto can be produced. Moreover, LIDAR data will allow a statistical improvement of a Digital Surface (or Terrain) Models, e.g. coastal zone or flood plane mapping.

Being an active instrument, the main challenges in the design of the LIDAR are the power required and the limited mass in which this power has to be dissipated (5 kg maximum for the upscaled model of the solar HALE UAV).

LIDAR data can typically be used to evaluate damages subsequent to earthquakes (e.g. building destruction,…) especially in remote locations difficult to access.

3.3 Thermal Digital Camera

The thermal digital camera will operate in two thermal infrared bands (SWIR : 3-5 μm and LWIR : 8-12 μm). The respective spatial resolutions will be 1.13m for the SWIR region and 2.25 m for the LWIR region (depending on the wavelength). The lower resolution obtained for the thermal camera as compared to the resolution of the MDC is due to the fact that, from a design point of view, it must be possible to exchange different sensors without changing the structure and the aerodynamic of the carrier. For that reason, the aperture of the thermal camera is chosen to be the same as for the Multispectral Digital Camera. This directly affect the focal length and the resolution. Furthermore, to cover the same swath as the Multispectral Digital Camera, the SWIR sensor will be a line array of 1 600 pixels, and the LWIR sensor will have an 800 pixels wide line.

Thermal images can be used for numerous applications. The most known in the framework of disaster management is obviously the forest fire detection. However, thermal images allow also "night vision" enhancing thereby the efficiency of rescue-teams during the night. Furthermore, such high resolution allows the accurate detection of heat-losses or even the present of moisture in the ground (if not too deep). Survey of embankments or levees is thus possible, permitting to avoid breaks, and thus flooding. Finally people search by rescue teams (e.g. boot sinking) is also possible.

3.4 SAR

The Synthetic Aperture Radar adds an all weather, day-and-night capability to the sensors suite. Aimed at environmental and security applications (oil spills, flooding, …), it will operate at short wavelength (X-band). A preliminary study has shown that a 2.5 m ground resolution over a 4.5 km swath is achievable, with a 3 kHz pulse repetition frequency. Being an active sensor, the SAR instrument requires also more power. Before the development of the SAR instrument starts, an evaluation of the necessary and available power for the instrument during night time will therefore be performed. Weight constraints may also be a serious challenge.

The possibility offered by this active sensor is critical for disaster management in areas with important cloud coverage: the SAR instrument allows to "see through the clouds".

3.5 Auxiliary Payload

The auxiliary payload is composed by an GPS/INS system for position and attitude registration and by a data transmission system (S- or X-band). These systems will be common for all instruments (including for the navigation). Attitude determination in real time is only required to support the image acquisition.

4 Conclusions

Once operational, the PEGASUS system will provide the decision makers with geographical information with very high spatial, spectral and temporal resolution. The data will be less expensive than spaceborne or airborne systems and will be available on internet very rapidly. The synergy of sensors will allow an all weather, 24-hours a days survey and the compact character of the whole system will allow for rapid deployment and great flexibility of the trajectories.

The HALE UAV flying between 14 and 20 km above see level will offer the advantages associated with both airborne and spaceborne systems, without suffering their disadvantages. However the limitations with respect to the weight of the payload require some technological development, especially for the active sensors.

The system is set to fly its first demonstration tests in 2005, in Belgium. Later on it is foreseen to develop a fleet of HALE UAVs hovering above the whole of Europe.

References

Béquignon J (2004) The International Charter "Space and Major disasters", Proc of the United Nations international Workshop on the Use of Space Technology for Disaster Management, Munich, Germany, October 18-22 2004. Also

available on URL:
http://www.zki.caf.dlr.de/media/download/unoosa_workshop_presentations/05_pres_session03_chair-wade/13_UNOOSA-DLR_Bequignon_ESA.ppt

Fransaer D, Vanderhaeghen F, Everaerts J (2004) PEGASUS: Business Plan for a Stratospheric Long Endurance UAV System for Remote Sensing, Proc ISPRS congress, Istanbul, Turkey, July 2004

Graham-Rowe D (2003) Nuclear-powered drone aircraft on drawing board, New Scientist, 22 February 2003

Küke R (2000) HALE Aerostatic Platforms, ESA Study Contract Report, contract 13243/98/NL/JG

Noble P, Nix M (2003) 3D North Sydney – Precise 3D database for Retrieval and Visualization. In Fritsch (ed.) Protogrammetric Week 2003, Wichmann Verlag, Heidelberg, Germany

Reulke R (2003) Film-based and Digital Sensors – Augmentation or Change in Paradigm? In Fritsch (ed) Protogrammetric Week 2003, Wichmann Verlag, Heidelberg, Germany

Disaster Management: The Challenges for a National Geographic Information Provider

Chris Parker and Mark Stileman

Ordnance Survey, Romsey Road, Maybush, Southampton, United Kingdom, SO16 4GU.
Email: chris.parker@ordnancesurvey.co.uk

Abstract

Effective disaster management requires getting the right information (often geographically related) to the right place at the right time. Minimising response times to incidents is therefore critical. Ordnance Survey's Mapping for Emergencies unit addresses this requirement through out-of-hours incident support. Additionally, a Pan-government agreement now gives a wide range of central government organisations direct access to a suite of Ordnance Survey's products, allowing better preparedness in the planning phases of disaster management.

Recent and anticipated (often disruptive) developments in spatial databases, GPS, wireless, mobile and computing technologies have changed, and will continue to change, the way in which geographic information (GI) can be collected, maintained, analysed, integrated and delivered to the end user in disaster management and other domains. GI is increasingly part of the information mainstream. These developments have changed the role of Ordnance Survey from being the nation's map maker to being the geographic information provider to the nation, with a substantial role in developing a geographic framework in which both geographic and related information can be efficiently integrated, exchanged and understood.

These developments prompt a number of research challenges within the domain of disaster management. Three are considered and illustrated:

- The application of user-centred design techniques to user requirements and behaviours in order to identify where GI adds value in contributing to task efficiency
- Considerations of database modelling in 3 and 4 dimensions
- Exploring GI portrayal with new technology

1 Introduction

This paper considers the role and challenges for a national GI provider, Ordnance Survey, in meeting the GI needs of the disaster management domain. It considers:

- The nature and scope of disaster management
- Ordnance Survey's historic support in meeting mapping and data requirements for emergency response and crisis management
- Recent developments in providing government with direct access to GI
- The development of the Digital National Framework™ (DNF®) as a consistent framework within which to use and link GI and related information

The paper then considers the research challenges for Ordnance Survey against a background of developments in disruptive technologies and their impacts on the future characteristics of GI. The nature of these research challenges with respect to disaster management is then discussed with special reference to work on understanding user needs and behaviours, modelling data in 3 and 4 dimensions and exploring means of information portrayal and visualisation afforded by new technology.

2 Disaster Management

According to the Cabinet Office Civil Contingencies Secretariat (CCS) (www.ukresilience.info/handling.htm) the management of disasters, whether man-made or natural, involves:

- Risk assessment
- Risk prevention or mitigation
- Preparation including contingency planning, training, and exercising
- Emergency response and recovery

Disaster management may require multiple, governmental and non governmental agencies, to work with multiple scenarios, across administrative, organisational, linguistic and domain boundaries, at local, regional, national and/or international levels, utilising information drawn from numerous diverse sources and systems. As an example of this complexity, summarising from information on the Cabinet Office CCS UK resilience website (previously cited), 28 organisations within Lead Government Departments in England are responsible for handling over 40 types of specified crises. In addition, local government administrations, emergency services, utilities and voluntary organisations are likely to be locally involved with critical roles at operational, tactical and strategic levels.

Actors and decision makers involved in disaster management require geographic information, often integrated with other information, in order to carry out their tasks more effectively and efficiently. Paraphrasing Dr Robert MacFarlane, visiting Fellow to the Civil Contingencies Secretariat Emergency Planning College, the challenge in disaster management is:

"…to get the right resources to the right place at the right time…to provide the right information to the right people to make the right decisions at the right level at the right time."

3 Ordnance Survey and Disaster Management

Ordnance Survey, Great Britain's national mapping agency, was created in 1791 as a military organisation in order to better prepare Britain's defences against the posed threat of Napoleonic invasion. The British Government instructed that a map of the south coast be drawn up. The first map depicting Kent, the most vulnerable county to invasion, is shown in Figure 1 below.

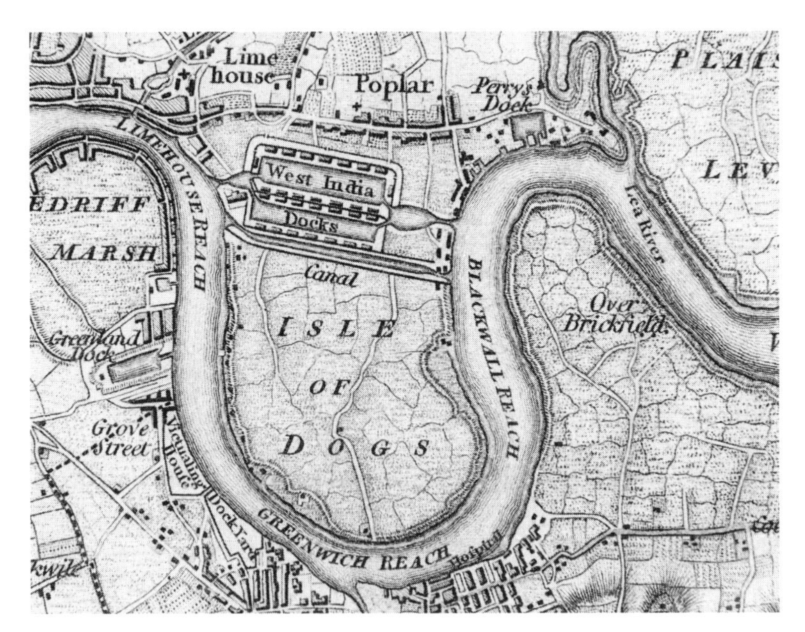

Fig.1. Extract from the first Ordnance Survey map of Kent, finalised in 1801

From that time until the present day Ordnance Survey maps and GI have been used in support of various disaster management tasks, whether man-made or the result of natural hazards.

Ordnance Survey is now an independent government department and Executive Agency, reporting directly to a government minister. Established as a public sector Trading Fund since 1999, it now enjoys greater commercial flexibility and greater responsibility, existing with no subsidy from the taxpayer. Specific work required in the national interest is carried out, at cost, for the Government under a National Interest Mapping Services Agreement (NIMSA) contract.

Largely in response to the Lockerbie disaster on 21 December 1988, when an American airliner was blown up over this small Scottish town, Ordnance Survey established its Mapping for Emergencies unit as an out-of-hours service to the emergency services and crisis managers. Examples of response include:

- Provision of mapping, geographic information and customised products
- Query response and logistical support for:
  - missing persons searches
  - mountain rescue
  - accident site investigation
  - geographic profiling
  - incident room support
  - mission planning

3.1 Use of GI in Managing a Crisis

When the foot-and-mouth disease outbreak started in 2001, the Department of Environment, Food and Rural Affairs (Defra), as the Lead Government Agency, needed to find a quick method of sharing information about the location of infected properties to inform both the public and other agencies involved in managing the crisis. To do this, Defra created and delivered a geographic information system (GIS) application with access to a wide range of Ordnance Survey mapping and digital data. This allowed a range of maps to be produced. Small-scale maps highlighted regional spread patterns to Cabinet Office decision makers and maps of livestock populations and movements for epidemiologists. Mid-scale maps identified land use to the army teams for burial sites and large-scale maps identified farm buildings and infrastructure for cleansing teams. Digital GI held in a GIS allowed:

- Flexibility in updating and maintaining currency in information in a rapidly changing situation
- Rapid electronic distribution across the country for display and printing
- Easy-to-read maps for public access through a website that received over 600,000 hits per day at the height of the outbreak

The GIS developed was used throughout the Department in supporting helpline queries, identifying locations and restrictions and issuing movement licences. This GI is now used throughout Defra to aid emergency planning and strategic approaches to man-made and natural disaster management.

4 Requirements of GI in Disaster Management

The foot-and-mouth crisis provides a good illustration of the information management requirements of various actors in performing their respective disaster management tasks effectively. If disasters are to be managed effectively then each actor or group of actors responsible for carrying out a task or set of tasks, needs to carry out those tasks as efficiently and effectively as possible; not just in isolation, but as part of a system of networked activity and information flows. Since 70–80% of information is resolvable to geographic location then the nature and characteristics of GI, and the way in which it is used, is paramount in managing crises effectively.

Providing the actors with the right GI, in the right place at the right time in order that they can make the right decisions, at the right level, at the right time are the requirements of any GI customer, but the implications of failing to meet these needs may be catastrophic within the domain of disaster management. The challenge for the GI provider is to deliver information to the actors and decision makers in a way that enables their expertise to be more effectively applied in carrying out their task responsibilities. This needs to be achieved without burdening them with GI management issues such as locational accuracy, data currency, data integration, semantic ambiguity, device, network, information portrayal and communication considerations.

Those working in the domain of disaster management require timely, location-based information, integrated with other information, presented in the most appropriate way to carry out the task at hand.

5 Role of a National GI Provider in Meeting Requirements for Disaster Management

Since everything happens somewhere, what, then, is the role of a national GI provider in meeting the above requirements for information within the domain of disaster management?

5.1 Access to National GI

One role is to provide better access to national GI by ensuring government departments and agencies have better access to Ordnance Survey products and services directly. In support of this, a Pan-government agreement launched in 2003 gives over 500 British government departments and agencies access to a wide range of digital map data. This is a significant step towards ensuring that common standards are consistently used for decision making.

5.2 A Consistent Geographic Framework for Integrating Information and Processes

According to Andrew Pinder, former Government e-Envoy:

"Geography is one of the key common frameworks that will enable us to link information together and boost efficiency in government."

A second role for a national GI provider is to contribute to a geographic framework for integrating information and processes.

Underpinning most information is location (Murray and Shiell 2004) and therefore geography potentially provides the common denominator to link disparate information sources required for effective disaster management across many organisations. Whilst technology now makes it easy to collect information of all kinds, store it and use it in many different ways, Murray and Shiell point out its very ease of use, the limits and priorities of different organisations, and the lack of effective standards provide barriers to seamless information exchange between agencies and actors and hence effective decision making and response.

Murray and Shiell describe the DNF, launched in response to the need to provide a vision for better GI exchange and integration which:

"...provides a permanent, maintained and definitive geographic base to which information with a geo-spatial content can be referenced" (Ordnance Survey 2000) and incorporates:

"...a set of enabling principles and operational rules that underpin and facilitate the integration of georeferenced information from multiple sources."

These principles are that:

- Data should be collected at the highest resolution, whenever economically feasible, once only and then reused
- Information captured should be located to the national geodetic referencing framework for the United Kingdom and Ireland based on the European Terrestrial Reference System 1989 (ETRS89). This enables real-time transformation and positioning information, linking location

data to the wider European framework and allowing data exchange with other European countries
- Features are uniquely referenced by a unique identifier, identifying the feature and source using the namespace concept in XML, allocated by a registry
- Composite features are created from existing features wherever possible
- The existence of data created is registered within a central registry
- Such information may subsequently be used to meet analysis and multi-resolution publishing requirements
- The DNF should incorporate and adopt existing de facto and de jure standards whenever they are proven and robust
- GI from any source can be associated and integrated in a 'plug-and-play' manner

The DNF operates within a service orientated architecture where technical documents, standards and guidelines and a searchable directory of datasets are made available as a web service, implemented using existing metadata standards.

Based on DNF principles, OS MasterMap® was developed as a single database. Nationally maintained and consistent, it comprises over 400 million objects with unique reference numbers (called TOIDs) for all features, providing a common denominator for disparate datasets held within the public and private sectors in Great Britain (Figure 2).

This provides the means to accurately georeference uniquely identified geographic features in a coherent and consistent georeferencing framework in order to integrate disparate datasets, underpinned by location and geography, - a very necessary set of characteristics of GI in the information age.

Whilst information requirements vary across the domain of disaster management, according to role and task responsibilities, the main challenges, common to all, are obtaining the right information as rapidly as possible in order to take the right decisions and actions whilst minimising response times. This means improving locational information and being able to provide that information to mobile devices.

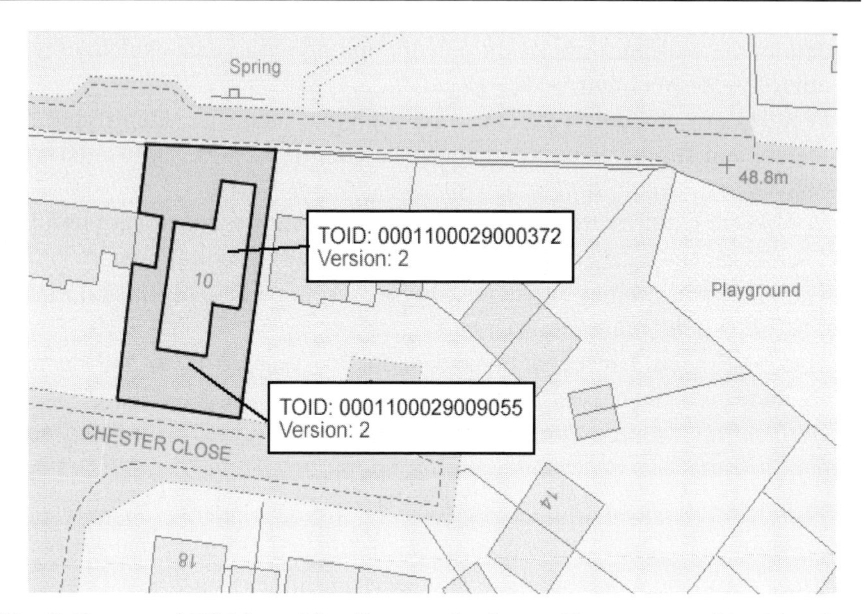

Fig. 2. Excerpt of OS MasterMap Topography Layer. Every geographic object has a unique TOID® for data association

For example, the Dumfries and Galloway Police now have access to OS MasterMap over an Intranet-based GIS, allowing their officers to quickly and accurately pinpoint the location of emergency services and to identify incident patterns to more effectively allocate resources. This has resulted in a 70% saving over their previous system. The National Crime and Operations Faculty (NCOF) at Centrex® operates as a national resource in partnership with police forces across the country. A key role is assisting in missing person searches, where experience has shown the first two hours of any search are critical. Using GIS has enabled NCOF to decide search areas in minutes rather than days, ensuring rapid mobilisation of teams and prioritising of resources. Paper-based searches are now replaced by GIS with access to digital GI, including OS MasterMap and the ability to coordinate searches remotely using mobile devices such as PDAs, laptops and mobile phones. The Greater Manchester Ambulance Service (GMAS) is now using their GI for predictive analyses, plotting where and when incidents are likely to occur. Within the Shropshire Fire and Rescue Service GI is held in a cab-based GIS, allowing rapid coordination with the command and control centre.

By providing direct access to its information through the Pan-government agreement and local authority service level agreements, and by providing objects with unique identifiers within in a national geographic framework, Ordnance Survey is providing the means by which information can be more easily integrated, providing a framework for more effective risk assessment, emergency planning and response through more effective decision making and enabling faster response times to incidents.

6 Research Challenges for a National GI Provider in the Disaster Management Domain

If the requirements of disaster management actors and decision makers are: to have the right information communicated in the right way at the right time, what, then, are the research challenges of a national GI provider in trying to meet these requirements?

Over the last 10–15 years technological advances offered by GPS, spatial databases, the Internet, wireless and mobile devices have revolutionised the way location-based data is collected, stored, maintained, analysed and delivered to the user. The map is now just one expression of the spatial database. These advances have brought GI to the information mainstream and are enabling its increasingly pervasive use as a fundamental ingredient to effective decision making and task management. They have enabled a change in Ordnance Survey's role from that of the nation's map maker to that of fundamental provider of the nation's geographic reference framework, whilst maintaining its role in providing national map coverage. These developments are enabling a move from a product centric to a database centric business, from which products and services can be derived.

6.1 Developments in Disruptive Technology

Developments in disruptive technology[1] will continue to offer new ways by which (geographic) information is captured, analysed and used. In a review of future societal, consumer, global and technological trends and their impacts on GI, a near future disruptive technological vision is offered by Robin Manning, Research Foresight Manager at BT® (Parker in press). This is summarised in table 1.

[1] Disruptive technologies are those that, when combined with socio-economic change, fundamentally alter the way tasks and activities are carried out and often fundamentally alter the way in which society, businesses and individuals operate. They often accelerate the rate of change. For example, the information age exists on the trading of knowledge made possible by the disruptive technologies of the computer and Internet.

Information stored locally Miniaturisation of sensors and ultra simple computing (smart dust) and miniaturisation of sensors, processors, storage capacity (1 bit per 20 atoms!) and wireless will allow information to be sensed and stored on the device and in the network.	***Semantic encoding*** The meaning of the information will be held on the device in the network. Distributed processing will allow intelligence to be held in the network. Moving towards device to device understanding of information.
Wireless technology Bluetooth®, WiFi but also, software radio, Ultra Wide Band wireless, embedded wireless, optical wireless. Near-field communications and radio frequency identification. Tag Technology: remotely powered short- range tags, battery powered long- range tags. In space: high altitude platforms, micro satellites.	***Open source software (and hardware?)*** Such as Linux™ for business, governments, desktops and home? Embedded Linux within mobile devices. Where next? Open source applied to hardware, chips?
Will enable:	
Pervasive ubiquitous computing Enabled through: "Pin head web server", wearable computing, sensing and sensor networks.	***Ad hoc networking*** E.g. Mesh Radio, Cybiko Games Peer to peer networking E.g. Napster®, Bluetooth file swap.
Sensor networks and smart environments Smart environments rich in processors, tags, data stores, sensors and communicators with trillions of processors. Networked everything (almost): cameras, players, TV, car entertainment, navigation devices, kitchen and white goods, remote controllers, tools, mobiles, mobile games, laptop/PDAs, medical equipment, industrial plant. Non-electronic goods connected by RFID tags. Networked people through wearable computing and bio-electronics.	***Disappearing computers & new interfaces*** Haptics, Biosensing & Biometrics (Fingerprints, iris scan, facial recognition, galvanic skin response, heart rate, muscle movements, brain activity) Everyday objects (fridges, microwave cookers) become computing interfaces. Computers disappear, computing becomes ambient.
Federated processing through Grid computing Increasing bandwidth and sharing computing resources through networked super computers using computing and storage as a utility. Better security, utilisation and lower costs?	

Table 1. A Disruptive ICT Vision (after Robin Manning, BT)

6.2 Impacts of Future Trends on Characteristics of GI

Technology will enable virtually everything (objects and people) to be identified, tagged, sensed and monitored. Miniaturisation will allow processing on the device and geographic intelligence to be held on the network. Whilst increased bandwidth and processing power will allow complex distributed analyses across trusted super computing environments. The following impacts on GI are suggested (Parker in press):

The need to accurately georeference uniquely identified geographic features in a coherent and consistent georeferencing framework in order to integrate disparate datasets, underpinned by location and geography, will be a necessary characteristic of GI in the information age.

GI will need to be more accurate, more current, available in near real time, and contain greater attribution. The developing market for location-based services (LBS) based on wireless and mobile devices will require the right information to be delivered when, where and how the user wants it, customised to meet their context and user profile. The value of GI for LBS and other services will be context, profile and time dependent. Whilst in the past we have used GI explicitly as paper or digital data, increasingly it is likely to be implicit, embedded and integrated, often invisible behind the service being delivered. Delivery of GI as a service rather than as a product (analogous to receiving the telephone number rather than the telephone directory) will start to prevail. Whilst topology and geometry will be of paramount importance for complex spatial analyses, the results may be portrayed as text, voice, sound, haptics and/or colour rather than traditional cartography.

In a spatial database GI is no longer confined to a 2-D planimetric representation of the world. Increasingly, applications – such as building information modelling (BIM), which joins up data in the architectural, engineering and construction domain, and contingency planning applications, for example – will require true 3-D data models, or the means to derive them. For many applications, including LBS, tasks and decisions are bounded by events and, increasingly, the temporal (4th dimensional) aspects of GI will be required.

Finally, GI is starting to be modified or created by the user within a reference framework, perhaps under the guidelines of some information commons (Onsrud 1998). See, for example, community authoring with OS MasterMap at www.urbantapestries.net.

These trends in GI reflect the GI requirements of the disaster management community as well. In a database-centric environment the overarching challenge for the national GI provider is no longer the design of mapping products but the design of the whole process of delivery of GI to the user as part of mainstream information delivery. The map is now merely one means of portraying the results of a query on the spatial database.

6.3 Implications for GI Research

Given the requirements of the disaster management community (to communicate the right information in the right way at the right time), the impacts of future technological trends on the characteristics of GI, and the application of DNF principles in managing the collection and integration of GI efficiently, the research challenges for this GI provider in meeting the needs for disaster management could be stated as:

- Understanding users' needs and behaviours
- Capturing data once and:
  - integrating with many databases
  - using at many scales
  - modelling in several dimensions
  - personalising in many ways for many different users
- Automating as far as is possible the data capture, data manipulation and data delivery processes

The remainder of this paper considers the nature of these research challenges with special reference to work on understanding user needs and behaviours, modelling data in 3- and 4-D, and exploring means of information portrayal and visualisation with new technology.

6.4 Understanding User Needs and Behaviours

Within the disaster management domain actors and decision makers use GI to carry out tasks and activities. They are not interested in GI as an end in itself but as a means to an end in making the right decision at the right time in order to carry out tasks or activities effectively and efficiently. GI is one ingredient required to make an informed decision in order to carry out a task. Other ingredients may include resources, costs and time. Users want Information to make informed decisions based on an answer to their question. GI adds value when it meets those conditions; when it contributes to informed decision making, allowing more effective execution of tasks and activities and the saving of time, costs, resources and lives. GI loses value if it adds nothing to the decision making or when there are barriers to accessing the right information, at the right time. Examples of such barriers could include time taken to manipulate the required data, inappropriate scales, poor accuracy or over-complex symbology.

Determining what, when and how to portray the GI that will *make a difference* is a design issue. In determining how to ensure GI adds value within the disaster management domain we can learn from the approaches of user-centred design in the consumer world (for example, Marzano 1998) and from human factors integration within the defence research environment. Whilst technological advances have (and will continue to) made data collection easier, contributing to an explosion of potential information availability, they have made the information management task more complex, hence the need for information management frameworks, standards and governing principles. We live in a world of high information complexity. As information consumers, however, we want clarity and simplicity. Within the disaster management domain users want the answer to their question, the whole answer and nothing but the answer! Anything else complicates the picture and adsorbs precious time. According to Aarte and Marzano (2003) high complexity requires high design in order to provide clarity and simplicity. For GI providers this means removing information complexity by applying the principles of user-centred design to the design of product and information service delivery; to look with the eyes of the consumers and users, to think with them. Just as user-centred design thinking is applied to the production of a PDA, laptop, car, toothbrush or computer game to ensure usability, so, as providers of GI, we have to design the delivery of the information service to the user appropriate to their requirements. This is crucial within the disaster management domain.

The UK Ministry of Defence Human Factors Integration Defence Technology Centre is applying human factors and ergonomics in order to ensure defence systems with a human interface, including information systems, maximise overall effectiveness by considering the capabilities, characteristics and limitations of the human operators. The approach is to design the system around the people rather than requiring the people to adapt or work around poor designs and working environments that make operation difficult or even dangerous. In summarising their research they ask: Can this person, with this training, do these tasks, to these standards under these conditions? (Goom 2003). The question is just as relevant in understanding the GI needs of disaster management actors.

Our user needs research focuses on *what, when, where* and *how* GI is important in people's tasks and decision-making processes. A three-stage research approach identifies, firstly, the elements key to decision making for a given task, secondly, the participants' experience of carrying out a given task and, thirdly, user behaviour whilst performing a task. Take the task of wayfinding, for example. Existing research indicates that key elements to decision making in wayfinding tasks are landmarks, orientation and action. For GI to offer value in this context it needs to assist orientation, reduce complexity and increase participants' confidence. In the LBS4ALL project (www.lbs4all.org), which is exploring LBS aids to navigation for visually impaired people and older people, interviews indicate that useful GI in this context is that which reduces uncertainty and risk.

Finally, by focussing on user behaviour in critical incidents while travelling, the ValuedLBS project (www.lboro.ac.uk/research/esri/lbs/) aims to identify useful GI; that is, *what* information at what *time* will *make a difference*.

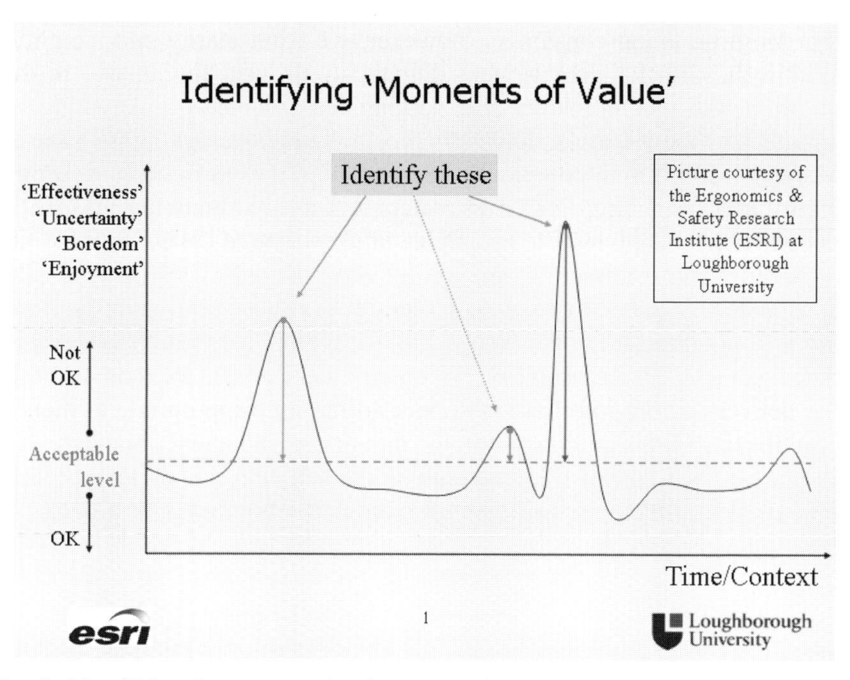

Fig. 3. Identifying "moments of value" over time and context in identifying usefulness of GI

Figure 3 illustrates "moments of value" where the right GI would, in this case, reduce risk and uncertainty. Whilst this example is taken from wayfinding research in the LBS context, the approach in identifying moments of value could also apply to where GI might increase effectiveness, decision making and response times within the disaster management context. The challenge is then to identify when location and GI provides a good enough indication of context to indicate when the service (and its delivery mechanism) is most valuable to the user (May, pers comm).

6.5 Capturing Data Once, Integrating with Many Databases – Semantic Interoperability

Effective disaster management is dependent on the ability to share the right information with the right organisations at the right time. Some of the barriers to effective use of shared information can be removed by reducing the time and cost of information processing, which is currently dependent on significant human intervention and manual input. Research at Ordnance Survey is looking at how to encode the meaning of different data sources so that machines can intelligently process data, and combine it with other data to be used across applications in a semantic web environment. Currently, the meaning of our data is implicit only to humans, not computers, which leads to ambiguities of interpretation, inefficiencies in processing, limitations in automated data sharing and interoperability limited to the syntactic and structural levels. By developing web ontology language (OWL)-based ontologies, explicit meaning to the terms used in our data can then be shared and interpreted by machines and humans within an information framework within which different organisations can collaborate using Ordnance Survey and other information (Greenwood 2003). Developing these ideas within an information infrastructure for risk management, as part of the EC 6th framework programme, IST ORCHESTRA project (http://www.eu-orchestra.org/index.shtml) is described at this symposium in Sassen et al (2005).

6.6 Capturing Data at One Resolution, Providing it at Many Resolutions – Generalisation

Within the disaster management domain GI is required at the resolution appropriate to the levels at which decisions and tasks are being carried out, be this at strategic, tactical or operational, or at all of these levels. The challenge for a national GI provider is to be able to capture data at the most appropriate resolution demanded by national circumstance but to provide it at the level appropriate to the task at hand. At Ordnance Survey our generalisation research is working towards the holy grail of automating and customising the generalisation process, which will derive medium- and small-scale maps from a single large-scale digital database, OS MasterMap. In addition, the aim is to develop a strategy and the appropriate tools to build on-demand generalisation applications, running as automatically as possible. This work is described further in Regnauld (2003).

6.7 Capturing Data Once and Modelling in Several Dimensions

Developments in spatial database and visualization technologies are stimulating user awareness of the potential uses of topologically structured digital data beyond 2-D planimetric views. Disaster management occurs in the real world in all its dimensions where there are requirements not only for conventional 2-D expressions of the spatial database but also 2.5, 3 and 4 dimensions as well. For example, flood modelling, blast, and pollution plume modelling, mission rehearsals, contingencies and evacuation planning. Better ways of modelling the real world to reflect potential user requirements in the disaster management domain are needed. These include better data models that can be more effectively used to create required products or views of the data.

The research poses a number of questions. What are the features that need to be considered for modelling the diverse aspects of the real world? What are the theoretical data models that allow the identified features to be modelled effectively? What are the practical issues involved with the use of the data models to handle live data? What generic data framework could encompass the full range of features?

6.7.1 Feature-Based Digital Mapping in 3-D Space

We are working collaboratively to explore a framework for feature-based digital mapping in 3-D space (Slingsby, Longley and Parker, in press). The design issues that are considered important and which we try to address are:

- The need for a data repository that can store data in a compact and a flexible form
- The need to be able to provide 3-D geometrical views of urban data
- The need to be able to hold spaces, both internal and external, to buildings in the same framework
- The need to have a framework that can be updated in an incremental fashion
- The provision for pedestrian accessibility in the model
- The need to cope with alternative conceptualisations of real-world features

Our model develops the 2-D digital database of OS MasterMap in the conceptualisation of features, accessibility and 3-D geometry.

Conceptualisation of Features

In OS MasterMap prescribed building unit extents are not consistently based on 2-D geometrical considerations. Features are based on simple 2-D geometries and have no concept of hierarchies of features (for example, rooms, flats, storeys, buildings and land parcels). Our model provides for the concept of features being sets of geometrical elements where each geometrical element may be part of one or more features and where they are grouped either as a simple list (non derivable features) or according to criteria (derivable features). This approach aims to provide a flexible way in which the same geometrical database can be used for different conceptualisations of real-world features.

Accessibility

Pedestrian access connectivity is integral to our framework, which ensures that fully connected customised pedestrian networks that pass seamlessly through and between buildings can be extracted from the model. Attributes on access points will indicate different access permissions, which correspond to the spaces between access points. This information can be used for pedestrian modelling, accessibility and evacuation analyses, way-finding and routing applications. Pedestrian access is seen as integral to the model because the urban environment is one designed for people to work and move about in.

3-D Geometry

In our framework we have extracted the urban environment as a set of connected floor spaces, between which may be walls (barriers) and access points. Rather than the requirements for large amounts of height data, we use rules associated with pedestrian accessibility or surface characteristics to reconstruct a 3-D geometry. More height data can be added in the form of spot heights and these are used to improve the 3-D geometry. Support for 3-D representation of roof morphologies and support for non-vertical walls could be added with the addition of more attributes and 2.5-D rules. For a fuller explanation of the model, see Slingsby, Longley and Parker (in press).

6.7.2 Spatio-Temporal Considerations

Disasters are dynamic events (or a series of dynamic events) that occur at particular locations over time. The management of disasters also requires actions at particular locations over time and therefore the incorporation of time as the fourth dimension in GI is important for effective disaster management. A spatio-temporal data model should facilitate:

- An understanding of the rules that govern real-world change
- Explanation of the state of the world at the current time (now) and previous times
- Predictions of the state of the world in the future

Worboys and Hornsby (2004) have been working on a well founded model that encompasses both geographic objects and events and have applied this thinking to modelling time aspects of Ordnance Survey's Integrated Transport Network™ (ITN) Layer within OS MasterMap. Their model, the Geospatial Event Model (GEM) is based on three principle components: Geospatial objects, events and their settings. Figure 4 shows these components and their interactions.

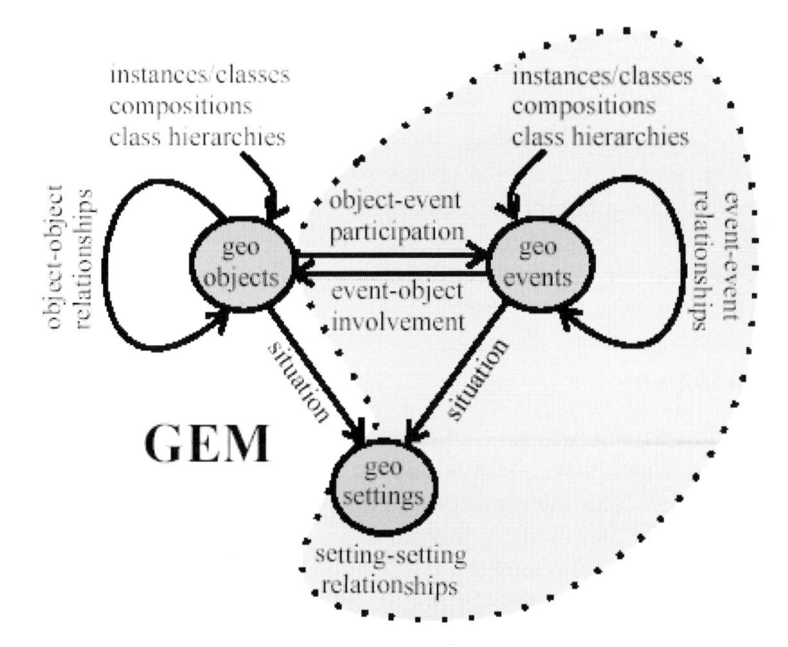

Fig. 4. The GEM model: Objects, events and their interaction (from Worboys and Hornsby 2004)

Their model allows the geo-spatial object model (objects and their references to spatial locations) to be extended with a dynamic component (events and their spatio-temporal references). The GEM introduces several new areas of analysis, including event modelling, spatio-temporal setting representation, event-event relationships and object-event relationships.

These components allow a structured form of analysis of dynamic domains with applications in disaster management, particularly the response and recovery phases. With reference to the work with the OS MasterMap ITN Layer, further work includes taking the OWL-based ontology developed through to database development, consideration of visualisation issues and domain-specific demonstrator case studies.

6.8 Exploring Information Portrayal with New Technology

Actors and decision makers in the disaster management domain are not necessarily well versed in map-based information. Providing the right information means considering how GI might be best portrayed to the user, given their profile and context. This need not necessarily be a map. Given the developments in spatial databases, GPS and other location technologies, mobile devices and wireless networks, and other, perhaps more intuitive, ways of providing the right GI information at the location of the user's task. The examples below illustrate ways this is being explored at Ordnance Survey.

6.8.1 Sound Stage

The sound stage (figure 5) is an internal environment that demonstrates how GI may be received by the user in the external, real-world environment.

Figure 5. Sound stage: As the user moves across the map floor (representing the real- world environment) equipped with a headset and PDA linked to an ultrasonic receiving unit, position is tracked by the ultrasound network made up of a single transmitter and an array of transducers installed overhead (emulating GPS). Software developed by the Mobile Bristol – http://www.mobilebristol.com – project is used to define the application area within the ultrasound network's effective range and, through a simple interface, allows multimedia content to be embedded at specific locations or zones corresponding with features appearing on the map floor. For example as the user walks past the pub feature audio visual information is received on the PDA screen

It illustrates how, by using GPS-enabled mobile devices, the user's location can be determined, allowing only the information relevant to the user at that location to be received as text, sound and images. Based on analysis of a spatial database, the user receives only the GI relevant to them at that location in the mode they want, which may not necessarily be a map.

6.8.2 Jaguar – Personal LBS Guide

Jaguar is a personal LBS application that uses large-scale geographic data to determine the user's context (figure 6). Driven by the user's current location, Jaguar can be demonstrated on a range of different mobile devices such as PDAs and mobile phones. A position fix is used to obtain data from a remote spatial database containing third party data associated with features in OS MasterMap. Context awareness will be an important driver to personal LBS usage and can in part be mediated by Ordnance Survey data.

The context of the user can then be determined from items such as:

- A *safety* rating (currently derived from fictional data emulating crime probability by polygon)
- Proximity to points of interest (such as shops)
- Proximity to virtual graffiti posted by other users, linked to map features
- How fast the user is moving (on foot or in a car)

When the location sensitivity of mobile devices becomes better than 10-m accuracy, a wide range of compelling applications built on data associated with OS MasterMap become possible.

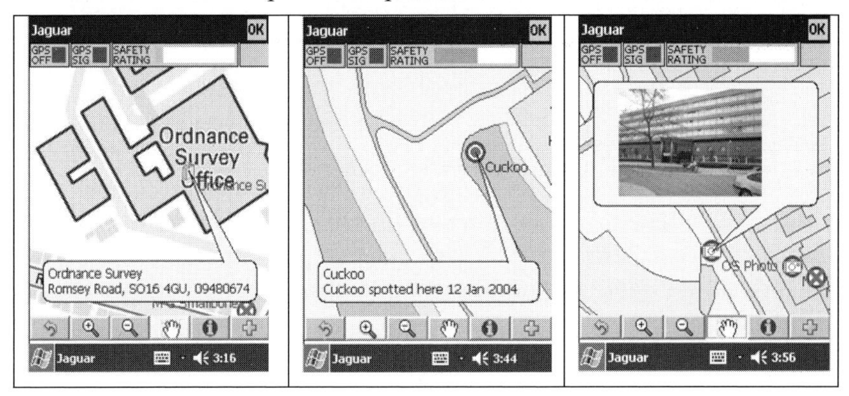

Fig. 6. User interface to Jaguar showing virtual graffiti and safety application

6.8.3 Augmented Reality and Magic Windows

Augmented reality (AR) describes a real-world view superimposed with computer-generated graphics, generating a composite view made up of real and virtual elements. Thus the user's perception of his/her surroundings is "augmented".

AR has massive implications for applications in many areas, including disaster management and personal navigation. Conventional 2-D mapping requires a certain degree of skill and spatial awareness to make use of properly, and reports indicate a large proportion of the population lack this skill. AR techniques obviate the need for map-reading skills since virtual labels and wayfinding information can be attached to real-world objects appropriate to the user's location (figure 7). The technology exists to produce magic window type applications where the information a user requires is determined by their location, the direction in which their device is pointing and the requirements of their task at hand (figure 8).

Fig. 7. Augmented reality. GI when, where and how the user wants it

Fig. 8. A virtual 1:1 map that can be seen only through a magic window: the screen of a hand-held device such as a tablet PC or PDA

Both of the above examples use position fixing to query a remote spatial database and related data in order to provide the appropriate GI to that uscr's location and contcxt – an important stcp in providing the right information at the right time in the right place and filtering out unnecessary information.

7 Conclusions

Disaster management is a complex domain and effective disaster management, from risk assessment and mitigation through to response and recovery, requires the right information to be delivered to the right people at the right time to make the right decisions at the right level at the right time. Since all events happen somewhere, at sometime, GI forms the backbone of the information requirement. Since many organisations are involved, data sources required are many and varied. Information sharing and integration is therefore of paramount importance.

Whilst developments in technology have made data capture easier, the total information space has become more complex and the task of data management harder. This has also changed the nature and demands placed on GI. The challenge for Ordnance Survey as a national GI provider is to provide a robust GI framework to manage that complexity, based on sound principles and standards within which GI and other information can be related and shared. The DNF provides the basis for GI data sharing in the UK and direct access to OS MasterMap, based on DNF principles, through a Pan-government agreement, giving lead government departments with responsibilities for disaster management direct access to over 400 million uniquely referenced geographic features, allowing better planning and a reduction in response times to events.

If the right information is to be provided when and where the disaster management user wants it, then significant challenges remain to be solved. These include:

- Understanding user needs and behaviours in order to identify what, when, where and how GI adds value to the task – identifying moments of value – by applying techniques of user-centred design and human factors
- Enabling the meaning of GI to be understood by humans and machines through the development of ontologies and semantic reference systems
- Generalising information appropriate to the user's task,
- Developing spatial and spatio-temporal database structures that represent the dimensionality and event-based nature of the real world
- Using continual developments in technology to provide the right information to the user appropriate to their task (personalising information according to context), at their location, at that time.

Acknowledgement

The authors are grateful for the assistance of James Wardroper and colleagues in Research & Innovation, Ordnance Survey for the preparation of this paper.

References

Aarts E, Marzano S (2003) The new everyday view on ambient intelligence. 010 Publishers

Annoni A, Bernard L, Douglas J, Greenwood J, Laiz I, Lloyd M. Sabeur Z, Sassen AM, Serrano JJ, Usländer T (2005) Orchestra: Developing a Unified Open Architecture for Risk Management Applications. Proc First Symposium on Geographic Information for Disaster Management. University of Delft, Springer-Verlag, The Netherlands

Goom M (2003) Overview of human factor integration. Human Factors Integration Defence Technology Centre accessed at: http://www.hfidtc.com/public/pdf/1%20%20GOOM%20%20OVERVIEW%20OF%20HFI.pdf

Greenwood J (2003) Sharing feature based geographic information – a data model perspective. Ordnance Survey, Proc 7th International Conference on GeoComputation

Manning R (2004) Disruptive technologies. Internal presentation, BT, Adastral Park, UK. 1 July
Marzano S (1998) Creating value by design: thoughts. Royal Philips Electronics, V & K Publishing, Blaricum, The Netherlands
May A (2004) Personal communication. 1 October 2004. Ergonomics & Safety Research Institute (ESRI), University of Loughborough, United Kingdom
Murray K, Shiell D (in press) A Framework for geographic information in Great Britain
Onsrud HJ (1998) The Tragedy of the information commons. In: Policy issues in modern cartography. Elsevier Science, pp 141–158
Ordnance Survey (2000) The Digital National Framework – Consultation paper 2000/1 [no longer available online – but can be obtained from the author]
Parker CJ (in press) Research challenges for a geo-information business
Regnauld N (2003) Algorithms for the amalgamation of topographic data. Proc 21st International Cartographic Conference, Durban, South Africa
Slingsby AD, Longley PA and Parker CJ (in press) A New framework for feature-based digital mapping in three-dimensional space
Warboys MF and Hornsby K (2004) From objects to events: GEM, the geo-spatial event model. In: Egenhofer MJ, Freska C, Miller HJ (eds) Geographic Information Science, LNCS 3234, Proc Third international Conference, Adelphi, MD, USA, Springer – Verlag Berlin and Heidelberg, pp 327-343

Websites visited

http:/www.ukresilience.info/handling.htm
http:/www.lbs4all.org
http://www.mobilebristol.com
http:/www.urbantapestries.net
http://www.hfidtc.com/
http://www.lboro.ac.uk/research/esri/lbs/
http://www.eu-orchestra.org/index.shtml

CNES Research and Development and Available Software in the Framework of Space-Images Based Risk and Disaster Management

Hélène Vadon and Jordi Inglada

Centre National d'Etudes Spatiales (CNES), 8 avenue Edouard Belin, 31401 Toulouse Cedex 9, France.
Email: helene.vadon@cnes.fr

Abstract

CNES has been involved for four years in the so called International Charter "space and major disasters". In this framework, both software development and research activities have been carried out, which aim at testing the usefulness of space based images for risk and disaster management, and at improving the space image based products deliverable to the end users.

Space images provide unique spatial coverage and potentially high site revisit opportunities. In case of disaster (like fire, flood, earthquake…), comparing images acquired before and after the event is the usual way to extract information about the spatial extension and the magnitude of the disaster. But to perform this comparison, two steps are necessary: co-registration of images and geo-referencing on the terrain. The typical technical problems are the following: First the images are not always geo-referenced, which means not fully superposable to a map, and even when this is the case, geo-referencing is never perfect, because orbital and attitude data are not known with a sufficient accuracy. Second, because we deal with unforeseen events, thus using in the urgency whatever image is available, we may be obliged to use and compare images from different satellite instruments, when no adequate image pair is available from a single satellite.

The paper develops the technical issues of co-registration and geo-referencing. It analyses the impact of the local DEM (Digital Elevation Model) quality. It considers both cases of similar and non similar images, like for example an optic image and a radar one.

The paper addresses the present status of CNES software for risk and disaster management, pointing out its unique features and also the still missing parts. The current research directions are also presented.

1 Introduction

CNES, the French national space agency, has been involved for many years in geometrical quality assessment of satellite systems. In order to fulfill the requirements related to this task, it has developed a set of robust and efficient engineering tools, related to satellite geometry. In parallel, it has been involved for about four years in the so called International Charter « Space and Major Disasters ». This activity has led to research and development activities, which also have led to complementary development of specific tools, in particular in the field of co-registration of images from different instruments, which is normally not required for traditional satellite image quality assessment, and also of change detection and object extraction.

Although those two domains, geometrical Image Quality assessment and change detection between images, seem to be independent from each other, they are not distinct in reality for two main reasons. First, efficient change detection can only be achieved with very high quality image relative geometrical modeling. Change detection and damage assessment algorithms require very accurate co-registration between images. Second, geocoding is a key issue for risk management, because it is required as soon as one wishes to directly relate the changes to the land cover. A very good knowledge of the absolute satellite instrument geometry is also required for any efficient remote sensing image post processing.

2 Satellite Imagery Utilization for Disaster Management

Using satellite images in case of disaster management has obviously many advantages.

First, the spatial extension of images is usually rather large, which is very important if one wants to get an overview of the whole region. The observed area will be typically 10km*10km for very high resolution (0.6m) optical imaging systems, 60km*60km for high resolution ones (2.5m), 100 km for a 20m resolution radar satellite, 500km for a scanSAR type radar image, and 1000km for a Vegetation type imaging satellite.

Second, satellite images are available anywhere in whatever the local ground situation is: no local telecommunication network is used, no local facility or personnel are required. Damage assessment may be performed from a remote place. For example, in the framework of the Charter « Space and Major Disasters » applied to the recent Bam earthquake, in Iran, damage assessment has been performed from Germany, and only relevant information has been sent to the local Civil Protection Authorities.

Third, there are so many different satellites in operation and such a dense image archive than one should be able to find, for every case, the most appropriate one to use. For example, in case of flooding, with very dense cloud coverage, one will select radar systems, which are able to image whatever the atmospherically conditions are. In the case of earthquake damages in a city, one will rather select high-resolution optical systems. Moreover, recent research work performed at CNES has shown that it is possible, to a certain extend, to evaluate damages from two images acquired by different satellite systems, like optical and radar one (Inglada et al., 2003).

However, there are of course also drawbacks using satellite images. Their cost is always high, but this can be compensated by international agreements such as CNES, ESA and RADARSAT corporation, for quick and free delivery of images in case of environmental disaster.

In a more technical point of view, the work to extract relevant information from space images is quite complex: in general, one image is extracted from the archive, first, to represent the past surface reference state and, second, the system is programmed urgently to obtain the new state. One hopes, looking at differences, to be able to detect damaged areas. But this operation, rather easy to explain, is difficult to implement. First, the images to be compared must be perfectly registered, and if possible georeferenced. With typical image localization accuracy of ten to twenty meters, this operation is usually performed manually, at least partially. In case of urgency like for disaster, every manual operation is a limitation, because the personnel is not available day and night.

Another difficulty of using satellite images for disaster management is purely operational. Most of the satellite follow sun synchronous orbits, leading to a given revisit period, which is typically a few days. Therefore, one has no choice for the time to acquire the post-event image. Moreover, there is only a little chance that one finds in the archive a pre-event acquisition with exactly the same conditions, like the same incidence angle. The consequence is that a Digital Elevation Model (DEM) will have to be used to avoid local deformation between the images. And precise DEMs are not available everywhere.

Finally, extracting relevant information related to damage from image differences is also a challenge, as will be shown later in the paper, but this is not a specific drawback of satellite systems, this is rather related to the limited resolution of satellite systems compared to airborne ones.

3 Satellite Based Disaster Images Map Projection

3.1 Co-Registration and Geo-Referencing of Images

One must differentiate between the two concepts: co-registration and geo-referencing.

Co-registration is mandatory for appropriate superposition of images, which is a pre-requisite for surface change detection. But it is possible to achieve accurate co-registration without accurate geo-referencing. In this case, one can project the processed resulting image (for example the change detection image) using a standard earth geometrical model like an ellipsoid, if no DEM is available, or using a DEM if available. This projection may be cartographic or geographic using any given map projection. Of course, in case a standard ellipsoid is used, the images are not truly geo-referenced (we call them "level 2" in our terminology, "level 1" corresponds to images in the acquisition geometry and "level 3" corresponds to the geo-referenced images). But this intermediate product might still be usable, i.e. one will be able to localize in parallel the pixels in the original images, after projection on the ground, and their homologues in the result image (like for example changes image). This can be simply achieved by recognition of details in the original images, which corresponds to finding land marks.

Geo-referencing, if accurately performed, solves the problem of image to map superposition. Geo-referencing is the process of re-sampling the raw image in such a way that it becomes registered to a given map. This allows building of any kind of map-projected database of images from different sensors or satellite systems. Accurate geo-referencing process implies that is has been possible to associate every line, column coordinate of the original satellite image with a geographic (latitude, longitude) or cartographic (X,Y) coordinate on the ground. In the geo-referenced cartographic image **(Fig. 1)**, every image point of coordinates (b,a) is associated to the cartographic map projection coordinate (x,y). In this figure, r is the interval, in meters, between two image points (the "resolution").

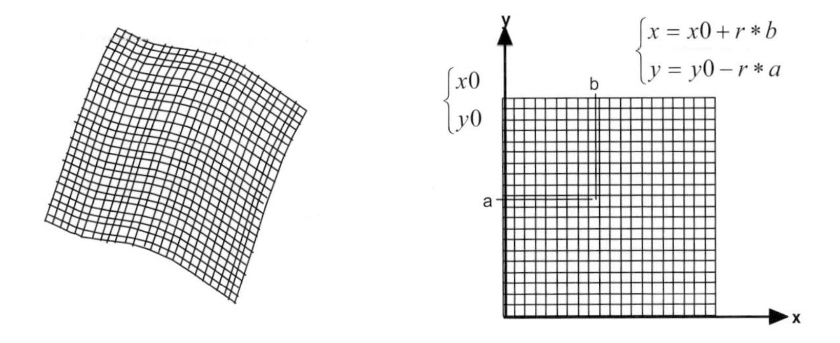

Fig. 1. Raw image geometry (left) and geo-referenced image geometry (right)

Different strategies can be considered when using co-registration and geo-referencing for change detection (in this paper, we only consider automatic processes):

– The satellite system from which the images are acquired provides a very precise geometrical model, and an accurate DEM is available over the area: in this case, images may be independently geo-referenced, and damage assessment may be performed on those geo-referenced images, leading to a directly interpretable result image. One must be careful that combination of errors may double the relative co-registration error. For example, a 5m rms image localization error, which is indeed a very good accuracy, may lead for some local terrain areas to a 10m relative positioning error, which, if applied to a 2.5 resolution image, might represents 4 pixels! This is totally unacceptable for change detection algorithms, which require a superposition better than one pixel.
– The satellite systems provide a rather good geometrical model, which is, however, not sufficient for processing the geo-referenced images directly. In this case, one must combine co-registration with geo-referencing. This is the most common case. Two alternatives can be considered
  1. Co-registration is applied first and geo-referencing afterwards. This is possible when both images are delivered with a geometrical model. Co-registration consists of finding homologous points in two images. Applying a bundle block adjustment based has proved to be always more efficient than trying to remove geometrical residuals by fitting to them a polynomial function. After the bundle block adjustment, better geometrical models will be available which in turn will allow a more accurate geo-referencing. This option is the best option for images coming from the same satellite.

2. Geo-referencing of the images is performed first, independently for each of them. In a second step, a compensation for residual relative positioning errors is achieved by a local computation of misregistrations, followed by re-sampling of one image to the other. The reference image will be chosen as the best geo-referenced one. After the first step, knowing accuracy of both satellite images and the DEM, it is possible to derive the maximum local misregistration error. Misregistration computation is limited to a small area, which is not the case in the option 1, where homologous points may be very far away in one and in the other image, being in raw geometry.

3.2 Co-Registration Process Description and Accuracy

CNES, as a space agency, is responsible for image quality assessment of its own optical satellite systems (Spot and Helios), and also evaluates image quality of other satellites. Therefore, it has been involved for a long time in research and operational software development in the field of co-registration algorithms. As a matter of fact, computation of local shifts between images, which is a mandatory step of co-registration, is one of key issue of geometrical quality assessment work: it is required for inter-band geometrical model computation, as well as absolute positioning accuracy determination. Co-registration is a two steps process: first computation of the local shifts, second computation of image interpolation. The second step, the interpolation process, is not developed in the paper.

3.2.1 Co-Registration of Similar Images

1. Optical to optical images

As mentioned above, a geometrical disparity assessment is generally the first step of the process. The traditional way of performing disparity measurement is to use linear correlation. The correlation-based shift computation process is robust to image noise, and the measured disparity accuracy on similar images has proven to be very high (Vadon and Massonnet, 2000), to a few hundredths of a pixel. This accuracy is amply sufficient for any post-processing algorithm, among which the change detection ones. Therefore, when using optical images only, the geo-referencing process is far more critical than the co-registration one.

2. Radar to radar images

Disparity measurement between two radar images may be achieved different ways. If one wishes to co-register multi-look images, the correlation technique may be used the same way it is used for optical images. But the speckle noise is less similar when the acquisition incidence angles become different. The accuracy of such a process is poor. On the other hand, radar images geometry is very simple to model. Comparing this model to the real local shifts computed by correlation, and extracting from the real shifts only the average values, one is able to reconstruct a very precise disparity map between the two images (Massonnet 1994).

Disparity measurement between single look complex images, based on complex correlation, will provide good results only if both images have similar speckle characteristics, which implies they have been acquired in interferometric conditions. But even in this case, knowing that the accuracy achieved when using a geometrical model combined with a rough mutli-look image correlation is sufficient, we would have no reason for correlating large complex images instead of small multi-look images, in the scope of co-registration.

3.2.2 Multi-Sensor Images

Multi-sensor image co-registration is an emerging technique, on which CNES has performed R&D work recently, and for which the algorithms have just been introduced in its operational disparity measurement software (Inglada and Giros, 2004a). This technique of co-registering very different images, although less efficient for similar images than the one based on the correlation, will however be very useful in many applications, among which natural disaster management: it may happen that only two images are available, one before and one after the disaster event, and that one of them is a radar one whereas the other is an optical one.

Optical to radar geometrical disparity measurement is a specific case of the larger domain of disparity measurement between images acquired in different spectral bands. It is clear that linear correlation technique is not appropriate in this case, being based on the research of the position of homologous points with the maximum correlation rate. The underlying assumption is that the two "similar" areas (small areas around the estimated homologous points) have a radiometric linear (ax+b) relation. And this assumption is totally wrong in those cases.

The idea then is to replace the correlation rate by another radiometric similarity measurement, based on the statistics of the local radiometry. This technique has been successfully used in the medical image community. The homologous points will be the ones which maximize statistical dependence between the local radiometric distributions. There are different possible functions to be used, but their common point is that all of them work on local (around the points assumed to be homologous) histograms. A consequence is that computation is much slower than for correlation, but numerical results show that it is possible to achieve a sub-pixel accuracy. A limitation of this technique is that, because it is based on local statistics, the sub-images on which the histograms are computed must be larger than for correlation (typical 55*55 pixels windows for statistical measurements to be compared to typical 13*13 pixels windows in the case of correlation). A consequence of this is that, although it is possible to obtain a sub-pixel accuracy for the shift measurement, their high spatial frequencies will not be accessible.

3.3 Geo-Referencing Process Description and Accuracy

3.3.1 Geo-Referencing of Optical Images

Remote sensing images are acquired with instruments mounted on satellites. Therefore, the geometrical modeling is a combination of time dependant satellite ephemeris (position and velocity) and attitude (roll, pitch and yaw angles, **Fig. 2**), and instrument internal geometry and elementary detectors viewing directions. The model may be delivered with the data or not. The model is generally delivered only with raw geometry images. If the user gets an already geo-referenced product, the model is clearly no more useful. When a geometrical model is delivered, it may take different forms. For example, it is expressed physically with SPOT products (ephemeris, attitude sampled data), and as polynomial functions with the IKONOS images.

As far as localization accuracy is concerned, it depends on the image resolution. To provide orders of magnitude, it is of 30m for Spot5 2.5 images, and will be 12m for the Pleiades 0.70m resolution ones. In all cases, geo-referencing without ground control points is not very precise, leading to more than 10 pixels shifts. And this is usually no sufficient for a direct superposition to a precise map.

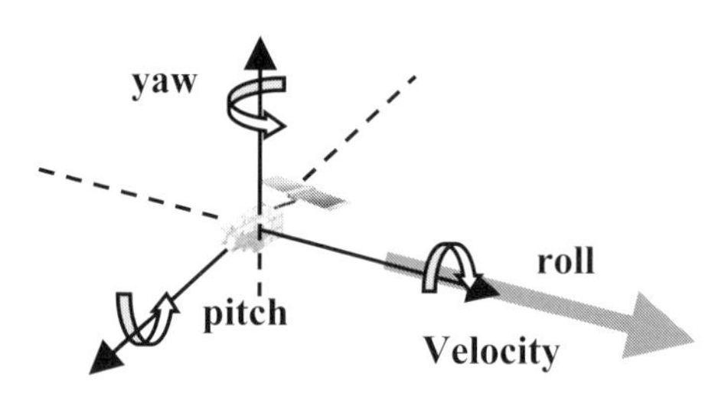

Fig. 2. Definition of the satellite attitude angles

We will now consider the impact of localization errors. The impact on image position and velocity restitution errors is shown in **Figure 3**, and the impact of roll, pitch and yaw restitution error is shown in **Figure 4**. Both figures are representative of a case of CCD line of detectors based instrument. Similar effects might be drawn about the impact of errors in the derivatives of the position (velocity) and attitude angles restitution.

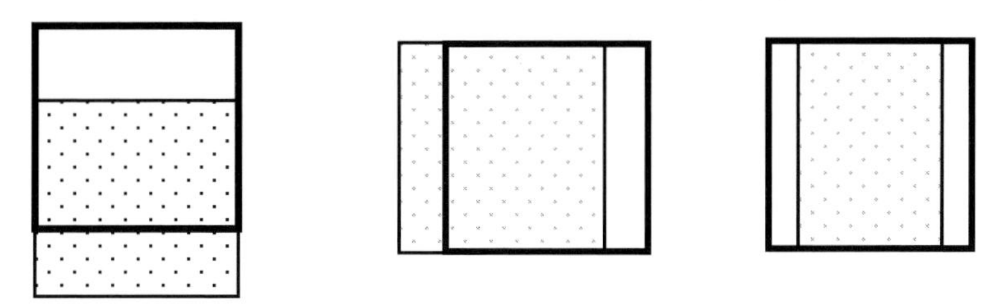

Fig. 3. Impact of position restitution errors (in X, Y and Z or along track, across track and altitude)

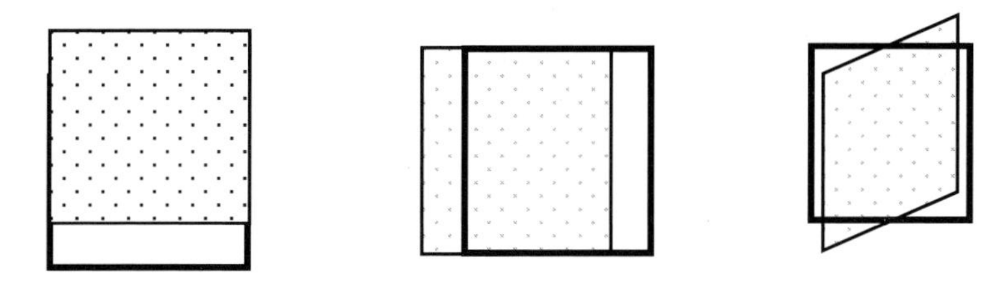

Fig. 4. Impact of attitude restitution errors

The above-mentioned effects are global but, on relief areas, they will be additional local errors, depending on the local elevation error. Those DEM related errors are proportional to the altitude error, as shown in **Figure 5**.

As a conclusion, we can say that geo-referencing errors will not only be seen as global translations or low frequency shifts, but will also lead to local artefacts. Those local artefacts will be proportional to the DEM errors, which explains why precise relief knowledge is strategic in the geo-referencing process. Furthermore, the local translation induced by a DEM error is proportional to the resolution. A 2m resolution image, processed with the same DEM as a 10m resolution image, will exhibit 5 times more translation, in terms of pixel shift.

That is why the approach for geo-referencing has always been a manual one, an operator taking many ground control points on the image and their homologous points in the images, in order to build a model fitting best to those points.

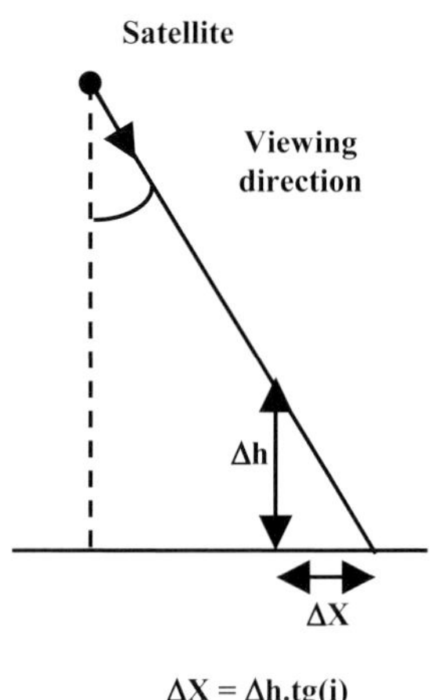

Fig. 5. Impact of an elevation error Δh on the image

An alternative is to build a world-wide geo-referenced image database, and to use this already geo-referenced database for all future geo-referencing process. The advantage is that the process can become automatic, the drawback being the cost of such a database.

To provide an idea of the magnitude of DEM related errors, let's take a 10m error in the relief model, an image at a 1 m resolution, and a viewing incidence angle of 10°. The shift in X, due to the DEM error, will be 10m*tg(10°)=1.77m. This value is negligible in a 10m resolution image, but with a high resolution one (1m), this represents 1.77 pixels, which will decrease the geo-referencing accuracy, hence any change detection algorithm accuracy (except if the second image is an optical image also and has be acquired at exactly the same incidence).

3.3.2 Geo-Referencing of Radar Images

Radar image acquisition principle is very different from optical one, more particular the image geometry is not sensitive to the satellite attitude. Therefore, we will not encounter problems such as the low-frequency deformations observed in the process of geo-referencing optical images. On the other hand, the errors in the DEM impact the geo-referencing process in a similar way as they do for optical images, but this time with a Δx error (in horizontal positioning) proportional to Δh/tg(i), i being the incidence angle. Because most of the radars work at incidence angles around or more than 45°, we can say that the impact of a Δh error in elevation will at most lead to a Δh meters error in horizontal positioning. The remarks we had for optical images, on the relation of the localization error (in pixel) with the image resolution remains valid for radar images. Therefore, for the same reasons than in optics, the knowledge of the local elevations is strategic for accurate geo-referencing.

4 Change Detection and Information Extraction

Information extractable from the analysis of a couple of remote sensing images is traditionally categorized into two types: geometrical, for example surface movement measurement, and radiometrical, for example change detection.

As far as change detection is concerned, CNES has implemented the following algorithms:

- Intensity ratio, at pixel level
- Mutual information, on a local area
- Local morphological gradient average direction.
- Comparison of the probability density distance (before the disaster/ after it). This method has proved to be well suited to radar / radar images change detection (Inglada et al, 2003), when the images are acquired at different incidences.

Besides change detection, it is also possible to follow different approaches. In particular, one can think of another kind of information, which may be extracted independently from 2 images, and compared afterwards: the object recognition. There are many advantages of using objects instead of pixels in the framework of disaster management. Damage assessment is easier, because the knowledge of a change in an area in the image is replaced by the knowledge of a change of an object, which has a meaning to the end user. Moreover, recognizing objects in the images, independently from comparison between those objects in different images, leads to additional information on the ground land cover.

CNES has been working for the last years on automatic man-made object recognition in high-resolution images (Inglada and Giros 2004b). Object detection is understood as finding the smallest rectangular area in the image, which contains the object. The algorithms developed are based on learning methods. In order to build a system, which is independent on the object to be recognized, we use a supervised leaning approach based on support vector machines.

With the use of 10 object classes (isolated buildings, paths, cross-roads, bridges, highways, suburbs, wide roads, narrow roads, roundabouts), and a learning process using 150 examples per class, we have built the confusion matrix of the recognition system. This matrix figures out the percentage of good detection, as well as the percentage of false alarm. Conclusion of this first study has shown that it is possible to obtain about 80% of good detection.

One of the drawbacks of this technique is that it is very slow, in terms of computer processing time, and therefore it is difficult to build extensive tests, in particular if one wishes to analyze images such as Spot5 very high resolution ones, which dimension are 24000*24000 pixels. This is why one of the research directions is the pre-processing of the images using totally different techniques, which will pre-select the candidate areas in the images. This pre-selection should ideally select more candidates than there are really on the ground, the final process input images being only composed of those candidates. The method used is based on pre-conscious user models.

Another way of detecting changes between images of the same spectral band is to use the coherence information. Coherence is a criterion that exists both when comparing radar to radar images, in which case it is the interferometric coherence, and optical to optical ones, in which case we talk about correlation rates. The coherence information has already proven its add-on value on real case disaster management, like during the Bam earthquake where a fault opening, unknown locally, has been discovered on the coherence image.

5 CNES Software for Risk Management

CNES has initiated the development of software, aimed at being operationally able to produce damage maps from remote sensing images. The main objectives of this software were to

- Superpose images acquired at different dates from any satellite instrument, radar or optical
- Detect changes between any couple of images
- Produce damage maps, adapted to the final user needs

The development environment has been chosen to be "Python", which is an interpretated object-oriented language, portable on various hardware (UNIX, Linux, and Windows). One advantage of this environment is also that it is possible, and easy, to automatically build a Man Machine Interface to the various applications.

With this implementation, the interfaces to the basic algorithms are described in the XML language, with the following parameters: physical path of the software application, algorithm parameters and on-line help (man).

The software includes basic components from various areas, such as

- Geometrical modeling (optic and radar, such as direct and inverse localization, co-localization, geo-referencing grids computation)
- Shift computation (similar and non similar images)
- Re-sampling
- DEM building
- Object recognition
- Change detection

This software is not fully implemented yet, but should hopefully be at the end of the year.

6 CNES Research Program Related to Risk

On going research work related to disaster management is in the following directions:

- Continuation of the research on object extraction, aiming at improving the computer time by introducing a two pass method: the first pass will pre-select potential candidates, and will be based on a very efficient process in terms if computer processing time. The second will refine the list and will be based on the learning algorithm on which CNES has already worked.
- Implementation of optical to radar image geometrical models. This process should allow direct co-registration of an optical image and a radar image, without geo-referencing them first. This will allow, as is already done for optical to optical images co-registration, a global bundle block adjustment with refinement of both optical and radar geometrical models.
- Continuation of work on radar stable detectors.
- Development of algorithms for optic / radar change detection.

References

Inglada, Giros (2004a) "On the possibility of multi-sensor image registrarion", IEEE Transactions on Geoscience and Remote Sensing, vol. 42, no. 10, pp 2104-2120, October

Inglada, Giros (2004b) "Automatic man-made object recognition in high resolution remote sensing images". IGARSS

Inglada et al (2003) "Lava flow Mapping during the Nyiragongo January 2002 eruption over the City of Goma in the Frame of the International Charter Espace and Major Disasters", IGARSS

Massonnet (1994) "Giving an Operational Status to SAR Interferometry", in First Workshop on ERS-1 Pilot Projects, ESA, Toledo, Spain, 22-24 June 1994, pp 379-382

Vadon, Massonnet (2000) "Earthquake displacement fields mapped by very precise correlation: Complementarity with radar interferometry", IGARSS

A Decision Support System for Preventive Evacuation of People

Kasper van Zuilekom[1], Martin van Maarseveen[1] and Marcel van der Doef[2]

[1] University of Twente, Faculty of Engineering, Center for Transport Studies, P.O. Box 217, 7500 AA Enschede, the Netherlands.
Email: k.m.vanzuilekom@utwente.nl;
m.f.a.m.vanmaarseveen@utwente.nl

[2] Directorate General of Public Works and Water Management, the Road and Hydraulic Engineering Institute, P.O. Box 5044, 2600 GA Delft, the Netherlands.
Email: m.r.vddoef@dww.rws.minvenw.nl

Abstract

As a densely populated country in a delta the Netherlands have to be very considered about flooding risks. Up to 65% of its surface is threatened by either sea or rivers. The Dutch government has started a research project 'Floris' (Flood Risk and Safety in the Netherlands) to calculate the risks of about half of the 53 dike-ring areas of The Netherlands. This project has four tracks: (1) determining the probability of flooding risks of dike-rings areas; (2) the reliability of hydraulic structures; (3) the consequences of flooding and (4) coping with uncertainties.

As part of the third track, the consequences of flooding, the Ministry of Transport, Public Works and Water Management has asked the University of Twente to develop a Decision Support System for analyzing the process of preventive evacuation of people and cattle from a dike-ring area.

This Support System, named Evacuation Calculator (EC), determines the results of several kinds of traffic management in terms of evacuation progress in time and traffic load. The EC makes a distinction between four types of traffic management scenarios: (1) reference; (2) nearest exit; (3) traffic management; (4) out-flow areas. The scenarios one and two represent a situation where no traffic management or limited traffic management is present. Scenario three (traffic management) calculates an optimal traffic management (given the model assumptions). Within the fourth scenario the user has the freedom to adjust the scenarios by (re)defining out-flow areas. In this way the user has the possibility to adapt to local possibilities and restraints. The limited data need and efficient algorithms in the EC make it possible to model large-scale problems.

Targets in the EC development were twofold: (1) a safe estimate of the evacuation time and (2) to support the development of an evacuation planning. These targets are met by the development of scenarios with specific and well defined objectives. Optimization methods were developed to solve the problems and meet the objectives.

The classical framework of transport planning is used as a basis, but with extensions:

- Trip generation: a broad range of traffic categories are defined. For each category has there own departure rate in time.
- Trip distribution: the core of the EC. The objectives of the scenarios are determining the distribution. The evacuation time is calculated.
- Traffic assignment: visualization of the traffic flows.

The paper will describe the structure of the EC, its objective functions and problem solving techniques. Furthermore a case study of dike-ring Flevoland is presented.

1 Introduction

Water plays a key role in the safety of the Netherlands. Up to 65% of its area, an area in which many of the economic activities take place, is threatened by either sea or rivers. It is a condition that needs permanent attention. Moreover, the country has to cope with serious consequences of environmental changes. The climate is changing as a result of pollution and use of fossil energy. Temperatures are expected to go up, rainfall will increase in intensity and frequency, and eventually sea level will rise. At the same time the soil will sink because of gas and salt extraction. All these factors together make it more difficult to protect the Netherlands against flooding, despite dikes and hydraulic structures.

In view of these problems the Dutch government has started the research project 'Floris' (Flood Risk and Safety in the Netherlands). This project has four tracks: (1) determining the probability of flood risks; (2) the reliability of hydraulic structures; (3) the consequences of flooding, and (4) coping with uncertainties. As part of the third track, the consequences of flooding, the Dutch Ministry of Transport, Public Works and Water Management has asked the Centre for Transport Studies of the University of Twente to develop a method for describing and analyzing the process of preventive evacuation of people and cattle from a dike-ring area. The method has been implemented in a Decision Support System (DSS) called the Evacuation Calculator. Primarily, the DSS will be used for an ex ante evaluation of the process of preventive evacuation for some 26 of the 53 dike-rings[1] of the Netherlands.

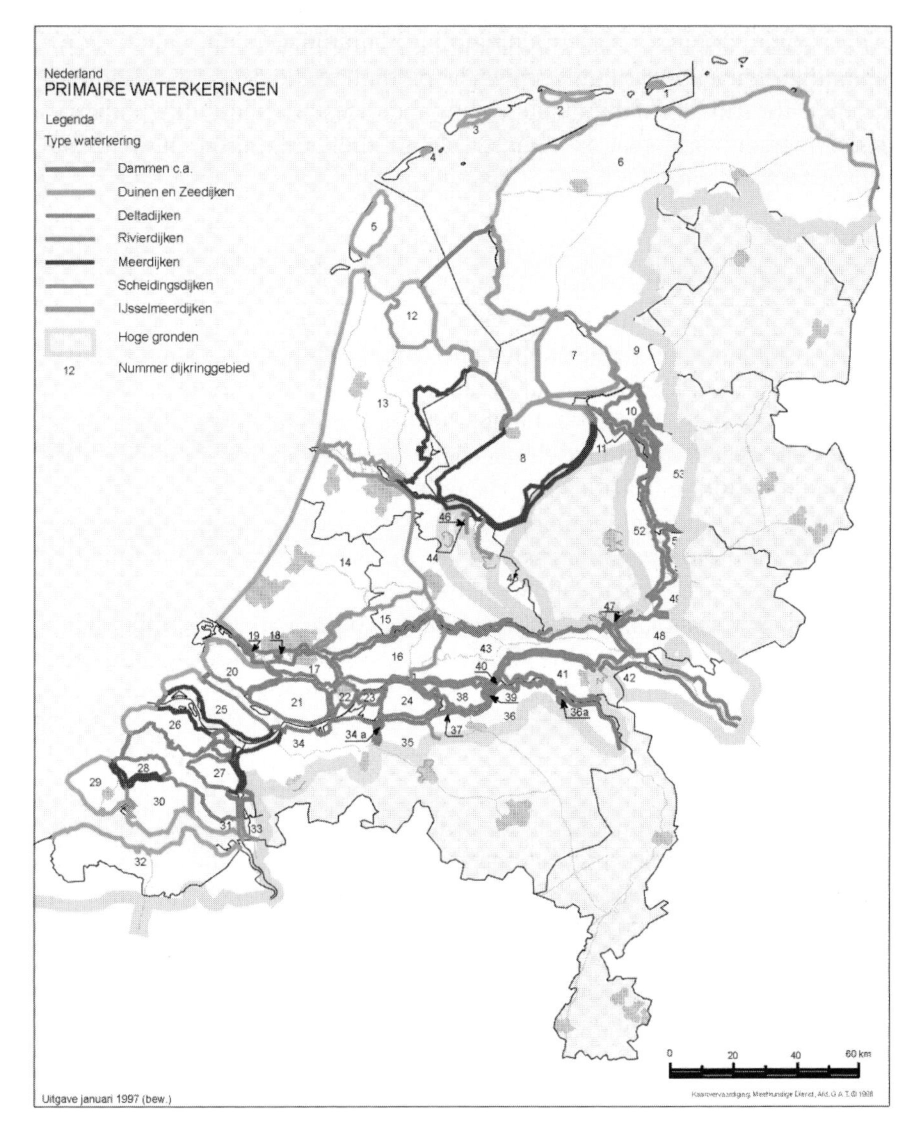

Fig. 1. The dike-rings of the Netherlands

[1] A dike-ring is an area which is protected against flooding by a system of dikes and hydraulic structures.

The key issue in this respect is the progress of the evacuation with an emphasis on the total time span needed for preventive evacuation. An additional benefit of the DSS is that it can be very helpful in the design of efficient strategies for organizing the evacuation in a specific dike-ring area within the framework of setting up an evacuation plan.

This paper discusses the evacuation problems in general and those of preventive evacuation of dike-rings in particular. In detail a method is specified for trip distribution and routing that uses efficiently the potential of the network and that is transferable to the application of a traffic management scheme.

2 Evacuation of People

There is an increasing interest for modeling evacuations. Studies have been initiated by risk analysis of nuclear power plants (Sheffi, 1982) and hurricanes (Hobeika, 1985; Urbiana, 2001). There are many causes that require an evacuation. These can be natural phenomena as extreme weather conditions (hurricanes, heavy rainfall, wildfires caused by drought), springtide and geological phenomena (earth quakes, volcanism, tsunami), but also human activities as industrial accidents, failure of hydraulic structures, accidents with transports of hazardous goods and attacks by terrorists. Expansion of human activities to vulnerable areas increases the impact of extreme circumstances. There are great differences in the predictability of time, location, scale and outcome of the dangerous situation.

In the Dutch situation dangerously high water levels of rivers can be predicted several days in advance. Although it is uncertain if and when the dike-ring will be flooded there will be enough grounds to start preparations. The aim of precautionary action is to reduce the risk and the consequences. One of the possibilities of reducing risk and consequences is the preventive evacuation of the dike-ring. It is important that the preventive evacuation is well organized, efficient and will need a minimum of time in order to avoid casualties. An accurate estimate of the evacuation time is helpful in determining the start of the evacuation. It implies that the decision can be taken as late as possible, at a moment where there is a more precise picture of the threat. A superfluous evacuation should be avoided. The crisis team needs to find a balance between an early decision (where the organization is not critical, casualties are unlikely, but an evacuation could be redundant) or a late decision (where the organization is critical and casualties could happen).

The whole process of a preventive evacuation can be outlined in a timeline.

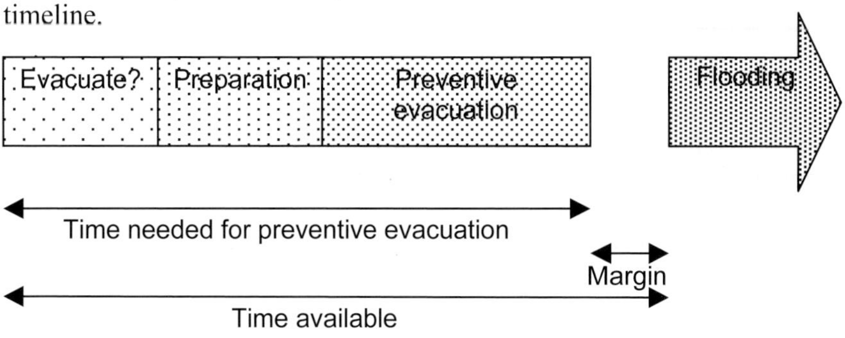

Evacuate? Preparation Preventive evacuation

Flooding

Time needed for preventive evacuation

Margin

Time available

Fig. 2. Time line of the evacuation

From the point of view of the evacuee the whole process looks like: (1) organization of the departure; (2) departure from home; (3) travelling in the direction of a safe area; (4) leaving the danger area through one of its exits[2]; (5) continuation of the journey to the destination in the safe area.

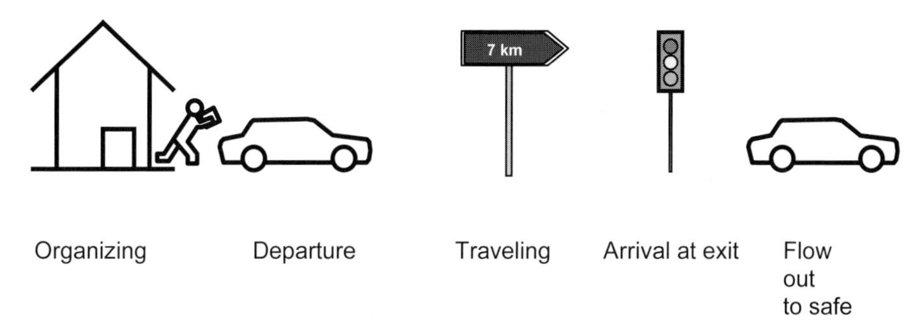

Fig. 3. Phases of the evacuation as seen by the evacuee

With a preventive evacuation there is neither actual flooding nor immediate threat. It is assumed that traffic behavior is normal and that the usual assumptions for modeling behavior are applicable. In cause of an actual flooding behavior will change from the 'normal' state to flee behavior. In the latter situation it is uncertain whether the usual assumptions are applicable.

---

[2] In general the dike-rings in the Dutch situation have several roads (exits) to surrounding areas.

2.1 The Abilities of the Authorities during (Threat of) a Disaster

In the situation of a disaster or a threat of disaster the capabilities of the authorities are enlarged. Depending on the size of the area a coordinator is assigned. The coordinator could be the mayor, a coordinating mayor (if several municipalities are involved), the province or the Ministry of the Interior. The authorities are entitled to take all the necessary actions within the restrictions of the constitution.

For the organization of a preventive evacuation it not only means that the enforcement of an evacuation is allowed, but also that any action to speedup the process and to increase the efficiency is permitted. This could mean: enforcement of time of departure, choice of exit and route to the exit.

2.2 The Process of Decision Making

A disaster plan as an evacuation plan is one of the aspects of the whole process of decision making during the threat of flooding. Authorities like municipalities, province and the central government are involved. Other functional organizations as the polder-board, the department of water management and environmental affairs of the province, the directorate general of public works and water management are involved. The evacuation plan is one of the preparatory plans that form the basis of the final approach. See for this process the figure below (Martens, 2002, as mentioned in Boetes, 2003).

During a critical situation the crisis team will go through three process steps: (1) judgment of the situation; (2) formulation of a plan; (3) judgment of the functioning of the chosen approach. The quality of the chosen approach depends, partly, on available resources, well prepared plans, procedures and commitments about the organizational structure (Boetes, 2003).

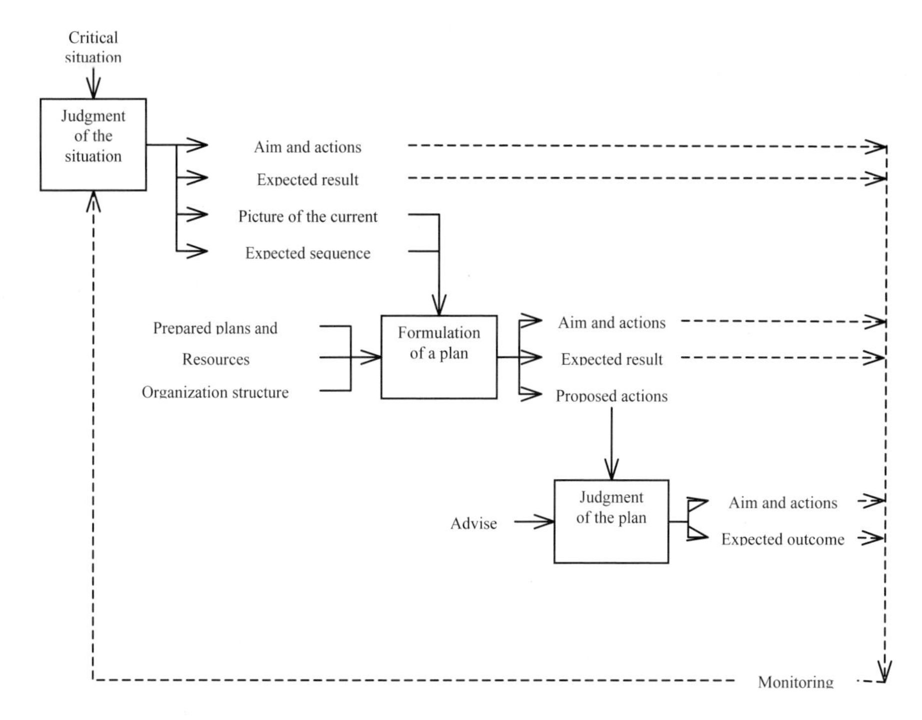

Fig. 4. The process of decision making under threat of flooding (Martens, 2002)

2.3 Modeling of a Preventive Evacuation

During a preventive evacuation there is a process of matching supply and demand as in normal traffic situations, although the setting in case of a preventive evacuation is quite specific. The matching of supply and demand can be modeled from a 'What if' or a 'How to' approach (Russo, 2004).

In a 'What if' approach a situation (or scenario) is modeled and the results of the model are analyzed. Stepwise the situation is adjusted until no further improvements seem possible. The final result is interpreted and translated in to an evacuation plan. The final result depends on the interpretation and adjustments of the modeler. The quality of the result is, by lack of a formal objective function, unclear. It is possible to use detailed and complex models in this situation. The modeler will focus on those aspects of the model that are important for the problem.

In a 'How to' approach the result is determined by the objective function, the constraints and structure of the model. Not in all cases an optimal solution can be guaranteed (due to local optima). The objective function, constraints and solving techniques could limit the complexity of the model. The focus on the objective function can overshadow other difficult quantifiable objectives.

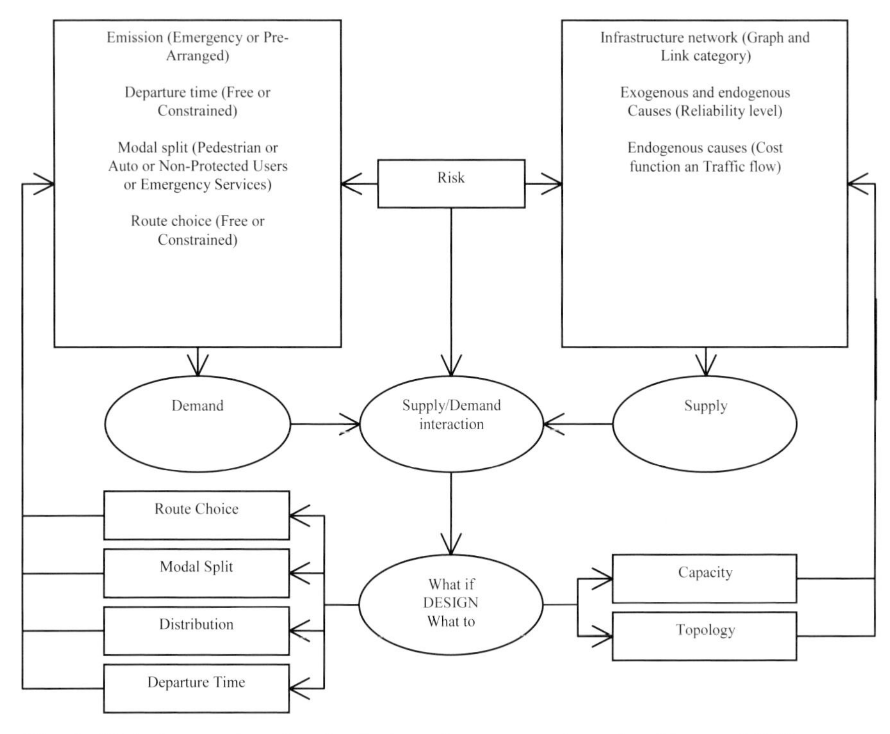

Fig. 5. Global procedure for the design of an evacuation plan (Russo, 2004)

Cova and Johnson (Cova, 2003) describe a procedure to eliminate crossing of routes and minimize the weavings on crossing, where extra distance to the safe area is allowed. They choose for this approach knowing that in many, urban, networks the crossings determine the capacity of the network. Elimination of crossing traffic and reducing weaving is a logical next step. In the approach of Cova and Johnson the distribution and route choice are the determining factors for the objective function.

Sheffi (Sheffi, 1985) handles the problem of the simultaneous trip distribution and assignment in general. Evacuation can be seen as such a type of problem. Goal is to find that distribution and assignment where a system optimum is obtained. Every change in the final solution, distribution or assignment, will affect the objective function. Sheffy proves this problem can be solved by using existing techniques and a modest adjustment of the network. The equilibrium assignment is used in a network were there is only one, spanning, destination. Chiu (Chiu, 2004) uses this pragmatic solution. Implicitly a perfect control of destination and route choice is assumed. The solution should be considered as a best-case solution that guides to (sub) optimal solutions with more realistic constrains.

For the 'Floris' project the focus is on a conservative and realistic estimate of the evacuation time together with a proposal for the traffic management during the evacuation. Efficiency of the method in terms of data handling and computing time is of importance as about 26 dike-rings will be investigated. From this perspective a 'How to' approach that uses the capabilities of the crisis team to influencing the traffic flows is preferable. The solution for this problem is found in a method that focuses on the trip distribution.

3 Formulation of the Methods

In the situation of a preventive evacuation there are many uncertainties and inaccuracies: (1) the number of people and cattle in the dike-ring during the threat of flooding; (2) the number of cars and trucks involved; (3) the time of departure; (4) the state of the network at the time of evacuation; (5) the route choice. Experiences under threat of hurricanes show a large discrepancy between expected and actual departure rate. During hurricane Opal people left their homes about three hours later than the slowest estimate of response rate (Alabama, web).

As a result of this it is not functional to focus on a maximum of model accuracy. A model with a limited complexity is appropriate for this situation. Sensitivity analysis is helpful to determine the critical processes.

Four different scenarios are developed:

1. Traffic management; within the capabilities of the crisis team a 'How to' model will suggest an efficient organization of the preventive evacuation. The inhabitants are directed to specific exits of the dike-ring.
2. Reference; this is a scenario where the inhabitants of the dike-ring are free in their choice of the exit.
3. Nearest exit; in this scenario the evacuees will go to there nearest exit of the dike-ring, regardless capacity and use of this exit.
4. Flow off areas; in this scenario inhabitants are directed to specific exits.

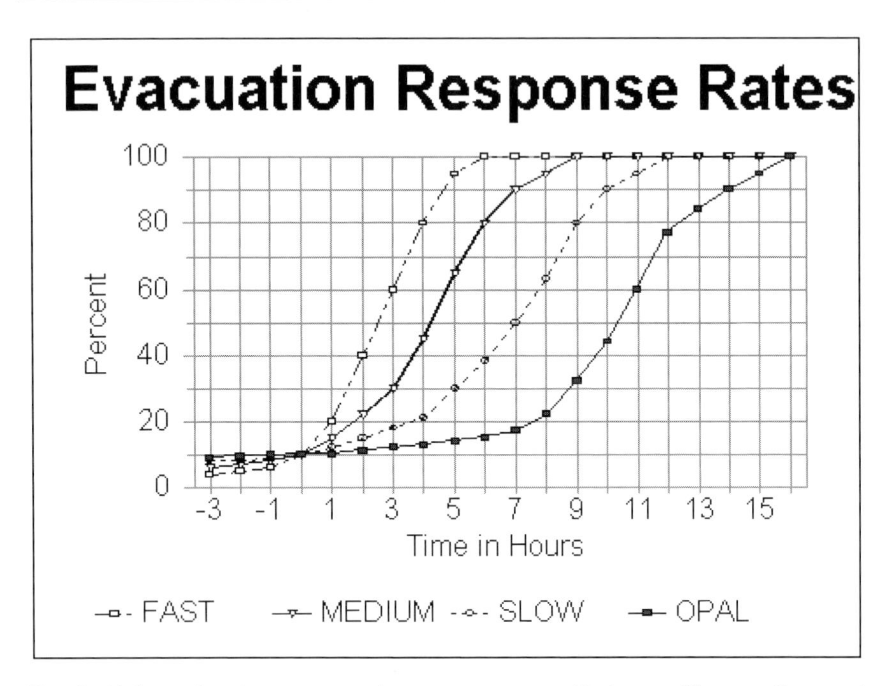

Fig. 6. Alabama hurricane evacuation response rates. Estimates (fast, medium and slow) and actual response rate during hurricane Opal (Alabama, web)

3.1 The Traffic Management Method

The capabilities for influencing behavior are important constrains for an evacuation plan as for the development of an evacuation method.
Possibilities for manipulation are:

- Time of departure. By means of information and direct orders the time of departure can be influenced.
- Trip distribution. It is possible to instruct to go to a specific exit.
- Mode of travel. In general people with a car available will use their car. For people without own means of transport the authorities will be responsible for supplying public transport.
- Route choice. By means of information and instructions it will be possible to guide the traffic.

The number of evacuees cannot be influenced.

The purpose of the evacuation plan can be defined as: a distribution and routing of the evacuees in such a way that the evacuation time is short and the possibilities of the network are utilized efficiently while the necessary traffic management can be realized.

Crossing streams of traffic is a source of waiting times and disturbances and should be avoided. Diverging traffic introduces a choice problem for drivers and the local traffic management. Diverging of traffic is not allowed in the method. In the actual implementation of an evacuation plan it can be introduced in specific situation. For the time being pure converging flow of traffic is assumed. This makes introduction of one-way traffic (reverse laning or contra flow (Urbina, 2003)) to increase capacity possible.

Using pure converging traffic flows delays at crossings are negligible. The traffic volume at the exits will be the highest (as a result of the converging flows) as a result of this it is likely that the exit will be the bottleneck. It is assumed that it is possible to assign a capacity to an exit that is appropriate for the route to the exit.

These assumptions lead to the following conceptual model:

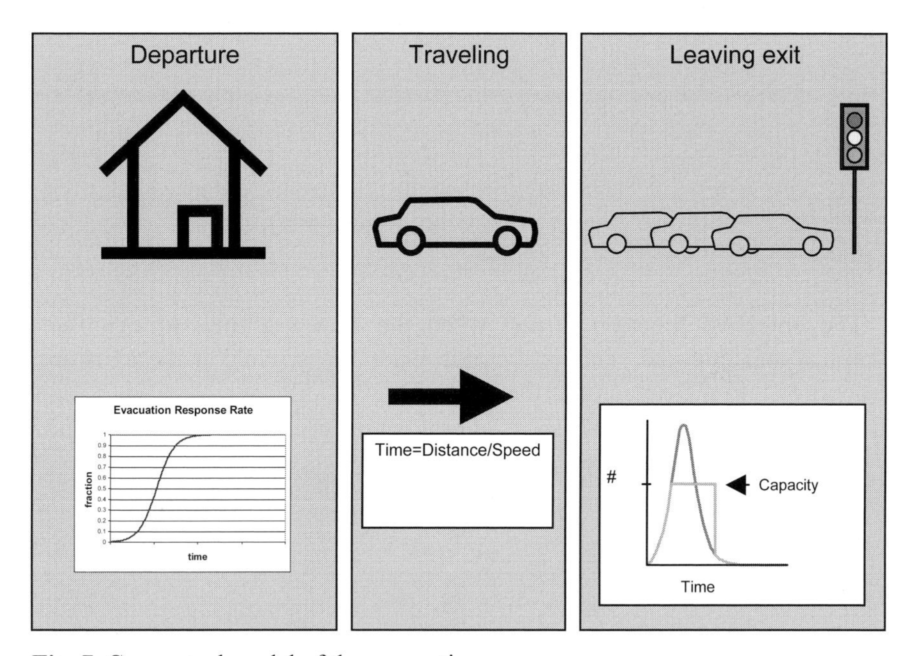

Fig. 7. Conceptual model of the evacuation

The task is now to create a trip distribution where the traffic flows are converging and the capabilities of the exits are used well. For the time being we consider the whole of traffic which will leave the dike-ring. When the trip distribution is determined it is possible to create the distribution in time by using the evacuation response rate and the arrival time at the exit.

Let P_i in Person Car Units, PCU, be the trip production by evacuees from zone i.

The evacuation time is determined by the last car leaving the dike-ring. The objective function is the defined by:

minimize(maximum(flow out time)).

Where:

- All evacuees will leave the dike-ring.
- The traffic flow to the exits is efficient.

This objective function suits a preventive evacuation. In situations where the urgency is high and casualties not avoidable this objective function is not valid. In these types of situations it is of great interest to limit the casualties given the time to flooding.

The flow out time U_j [hours] of an exit j is determined by the arrivals at the exit A_j [PCU/hour] and the capacity C_j [PCU/hour] of the exit:

$$U_j = \frac{A_j}{C_j} \quad (1)$$

$$where:$$

$$\sum_i P_i = \sum_j A_j$$

The objective function is met when the flow out time of exits are identical and minimal. This is the case when the arrivals at the exits are proportional to the capacity of the exits:

$$A_j = T\frac{C_j}{\sum_j C_j} \quad (2)$$

$$where:$$

$$T = \sum_i P_i$$

Every distribution of the productions with these attractions will match the objective function, but will not necessarily result in efficient traffic flows to the exits.

Let the distance traveled from origin i to destination j along the shortest path be z_{ij}.

Let the number of trips from origin i to desination j be T_{ij}.

Then the total vehicle distance is defined as the weighted sum of trips and distance traveled: $\sum_i \sum_j z_{ij} T_{ij}$

By minimizing the total vehicle distance, given de productions and attractions, unnecessary vehicle distances are avoided. This problem is known as the classic transportation problem:

$$\min\left(\sum_i \sum_j z_{ij} T_{ij}\right) \tag{3}$$

$$\begin{aligned} &where: \\ &\sum_j T_{ij} = P_i \\ &\sum_i T_{ij} = A_j \\ &T_{ij} \geq 0 \end{aligned}$$

The transportation problem needs an initial OD-matrix. In the implementation of the Evacuation Calculator the trips form i to j are proportional to the production and attraction:

$$T_{ij} = \frac{P_i \cdot A_j}{T} \tag{4}$$

When the resulting OD-matrix is loaded to the network using an All-Or-Nothing assignment the traffic flows are convergent.

F_{tk} is the fraction of evacuee category k who will leave in time interval t. In the implementation the user is free to define the fraction for each F_{tk} or by using a logistic function:

$$F_{tk} = \frac{1}{1+\exp\left(a_k\left(t-b_k\right)\right)} - \frac{1}{1+\exp\left(a_k\left(\left(t-1\right)-b_k\right)\right)} \tag{5}$$

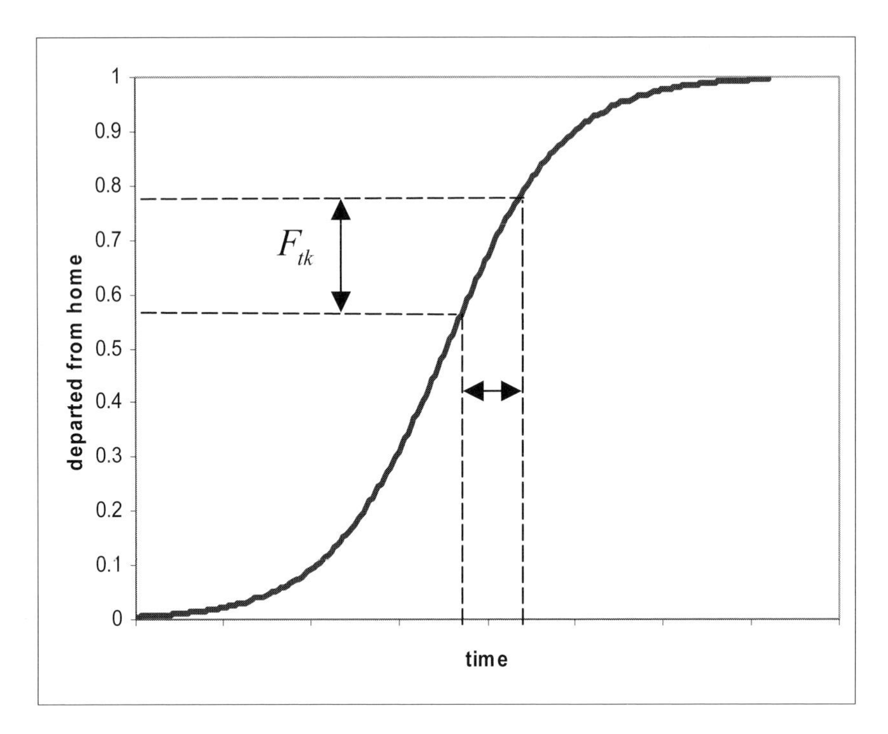

Fig. 8. Use of the logistic function for *Ftk*

The OD-matrix per time interval is now determined by:

$$T_{ijt} = \sum_k T_{ijk} F_{tk} = \sum_k T_{ij} \frac{P_{ik}}{P_i} F_{tk} \tag{6}$$

After departure from home the vehicles will arrive after r_{ij} time intervals at the exit. The travel time is dependent on the distance from origin to exit, z_{ij} [km], and the average speed, $\bar{v}$ [km/h], in the dike-ring and the number of time intervals in an hour, I:

$$r_{ij} = Int\left(\frac{z_{ij} I}{\bar{v}}\right) + 1 \tag{7}$$

Now it is possible to calculate the arrivals, ARR_{jk} [PCU] at exit j for a time interval t:

$$ARR_{jtk} = \sum_i T_{ijk} F_{t-r_{ij},k} \tag{8}$$

$$ARR_{jt} = \sum_k ARR_{jtk}$$

The vehicles that arrive at the exit will leave the dike-ring for as far capacity allows. In a time interval the number of vehicles that leave the exit are limited to.

$$\frac{C_j}{I} \tag{9}$$

The flow out DEP_{jt} at exit j for time interval t depends on the available traffic (delayed and just arrived) and the capacity of the exit:

$$DEP_{jt} = \min\left(ARR_{jt} + DEL_{j,t-1}, \frac{C_j}{I} \right) \tag{10}$$

$DEP_{j,t-1}$ are those vehicles that could not pass in earlier time interval(s) using exit j. For the first time interval there are no delayed vehicles:

$$DEL_{j,0} = 0 \tag{11}$$

Vehicles that cannot pass in time interval t will be delayed and will use a later time interval:

$$DEL_{jt} = \max\left(ARR_{jt} + DEL_{j,t-1} - DEP_{jt}, 0\right) \tag{12}$$

The number of departing vehicles for category k is calculated by assuming that at arrival the categories are spread homogeneous over all vehicles. A fraction of all delayed and just arrived vehicles will eventual leave the dike-ring eventually:

$$DEPRATIO_{jt} = \frac{DEP_{jt}}{ARR_{jt} + DEL_{j,t-1}} \tag{13}$$

The departures for category k in time interval t are now:

$$DEP_{jkt} = DEPRATIO_{jt}\left(ARR_{jkt} + DEL_{jk,t-1}\right) \tag{14}$$

The delayed vehicles for category k in time interval t are:

$$DEL_{jkt} = ARR_{jkt} + DEL_{jk,t-1} - DEP_{jkt} \tag{15}$$

Central unit of measurement in the model is the PCU. The number of evacuee in PCU Q_k can be transformed into the number of persons (or cattle) N_k by using the occupancy degree μ_k and the PCU-value of the used type of vehicle PCU_k :

$$N_k = Q_k \frac{\mu_k}{PCU_k} \tag{16}$$

For each exit and all exits together the resulting output of the model is:

- Arrivals and flow out [PCU/time interval].
- Arrivals [PCU/time interval] for all categories together and for each category.
- Flow out [PCU/time interval] for all categories together and for each category.
- Flow out [number/time interval] for all categories together and for each category.

The resulting OD-matrix (whole evacuation or per time interval) is available for (dynamic) assignment.

The method has some relationship to the first stages of the classic one-mode traffic model:

- The trip end calculation. Where trip production is determined by the social economic data of the zones. There is a difference in calculation of the attractions. Here the capacities of the exits are leading.
- The trip distribution. Special for this method is the minimization of total vehicle distance.

The results (graphs and OD-matrix) are available in seconds. Assignment of the OD-matrix give a better insight in the resulting traffic flows.

The Traffic management scenario uses the 'How to' approach. The other implemented scenarios (Reference, Nearest exit and Flow out areas) use the 'What if' approach.

3.2 The Reference Scenario

The differences between the Reference scenario and the Traffic management scenario are:

- Trip end calculation: the user is free in setting the (relative) attraction of the exits. In general these will be determined by the traffic volumes on the exits on normal working days.
- Trip distribution: the distribution is identical to the initial distribution of the Traffic management scenario. There is no minimization of total vehicle distance.

3.3 The Nearest Exit Scenario

Here the differences with the Traffic management scenario are:

- Trip end calculation: the attractions are not explicit chosen, but are a result of the distribution.
- Trip distribution: all productions of an origin are allocated to the nearest exit.

$$t_{ij} = P_i \tag{17}$$

$$\text{for that } j \text{ where:}$$

$$z_{ij} = \min\left(z_{ij} \forall j\right)$$

$$t_{ij} = 0 \text{ for all other } j$$

3.4 The Flow Off Scenario

The Traffic management scenario will result in sets of more or less independent flow off areas which use one or more exits. It is not likely that this flow off areas can be transferred to an evacuation plan without change. Geographic, jurisdictional and other local constraints will make adjustments necessary. With the Flow out scenario the user is free in defining sub areas with one or more exits. Within every sub area the trip end calculation and distribution is solved with the Traffic management method: (1) attractions proportional to capacities and (2) minimizing the total vehicle distance given the productions and attractions.

The minimization of the total vehicle distance for all flow out areas is solved by manipulating the distance matrix z_{ij} before the actual minimization. Those relations that are not part of an flow out area are flagged with an extreme large value. The further procedures are identical to the Traffic management scenario. The deviation to the Traffic management scenario is:

- Trip distribution: manipulation of the distance matrix z_{ij}.

$$
\begin{aligned}
& z_{ij} = \infty \forall i \in D_m \wedge j \notin D_m \\
& z_{ij} = \infty \forall j \in D_m \wedge i \notin D_m \\
& \text{where:} \\
& D_m = \text{the set of } i \text{ and } j \text{ defining sub area } m
\end{aligned}
\tag{18}
$$

4 Case Study of Flevoland

Flevoland is one of the larger dike-rings in The Netherlands. Its surface is about 98 square kilometers. With about 258 thousand inhabitants (102 thousand houscholds) thc arca is not very densely populated, at least to Dutch standards. The two larger cities are Almere and Lelystad. Others are considerably smaller.

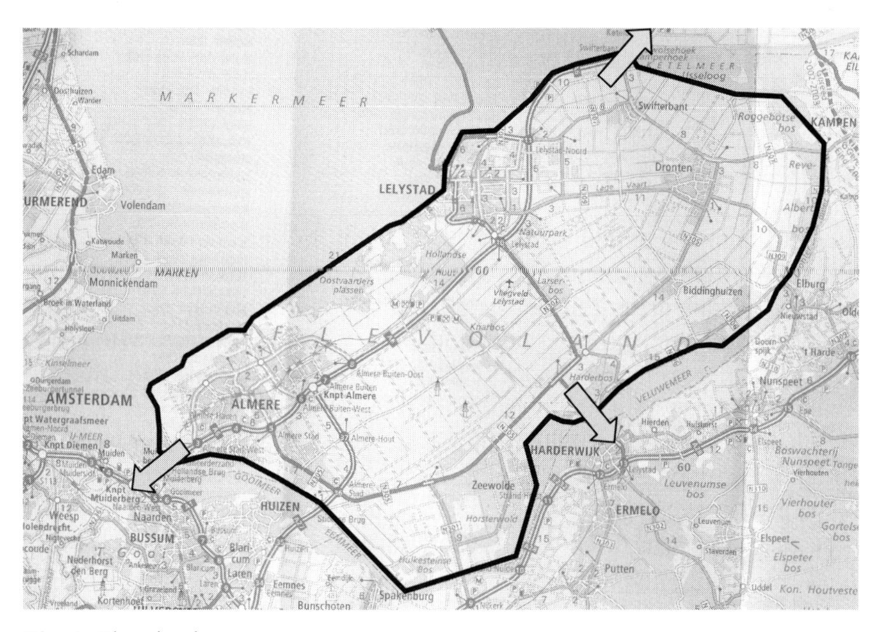

Fig. 9. Flevoland

The Network

For demonstration purpose we use just three of the eight exits[3]. The capacities of the exits [PCU/hour] are set to 6600 for the south-west exit, 4300 for the north-east exit and 1500 for the south east exit. This leads to a total capacity of 12400 [PCU/hour].

Trip End Calculation

We define two categories: (1) person cars (2) people who need assistance or public transport to leave the dike-ring.
We assume that[4]:

- All passenger cars will leave the dike-ring. Passenger cars have a PCU of 1.
- 10% of the inhabitants in the age of 35 to 64 and 50% of those in the age of 65 and more will use a bus. Busses will transport on average 20 passengers. A bus has a PCU-value of 2.

This will result in 93630 PCU for the category passenger cars and 2079 PCU for the category of bus users, 95708 PCU in total. The evacuation will need at least seven and a half hours (95708/12400).

Departure Rates

For each category the departure rate is defined with a logistic curve. The characteristics of these curves are:

- Cars: 50% will have been departed in 5 hours, 90% in 6 hours.
- Bus users: 50% in 8 hours, 90% in 9 hours.

For both categories it implies that 40% of the population will leave home in one hour. During the peak hours about 38000 PCU will enter the network. This exceeds the capacity of the exits three times (38283/12400).

---

[3] With eight exits Flevoland has a relative large capacity of exits.

[4] Up till now the trip end calculation, departure rate as the average speed in the dike-ring are not yet defined in the Floris research project. Further research on these topics is needed.

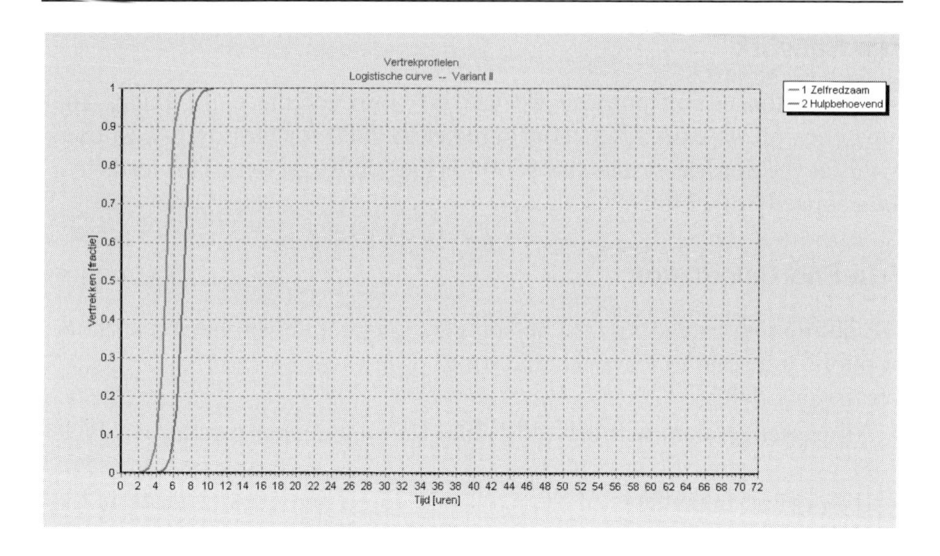

Fig. 10. The departure rate in the Flevoland Case.

Average Speed

The average speed in the dike-ring is set to 25 km per hour.

Results of the Evacuation Calculator

In the Reference scenario all three exits have the same relative weight. This will result in overload of the south-east exit particularly as this exit has the least capacity. With the Reference scenario the dike-ring is empty in 25½ hours. During 4 hours the maximum flow out of 12400 PCU/hour is reached. The bottle neck is at 4:30 till 25:15 at its maximum capacity, other exits are 9:15 and 12:00 hour below their capacity. Clearly this is a worst case scenario that can be improved easily.

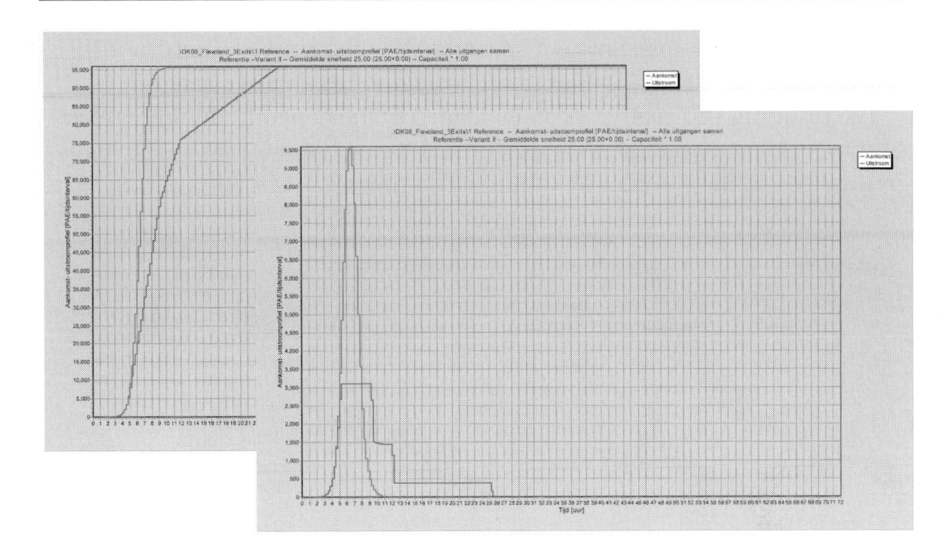

Fig. 11. Evacuated PCUs in time per time interval and cumulative for the Reference method.

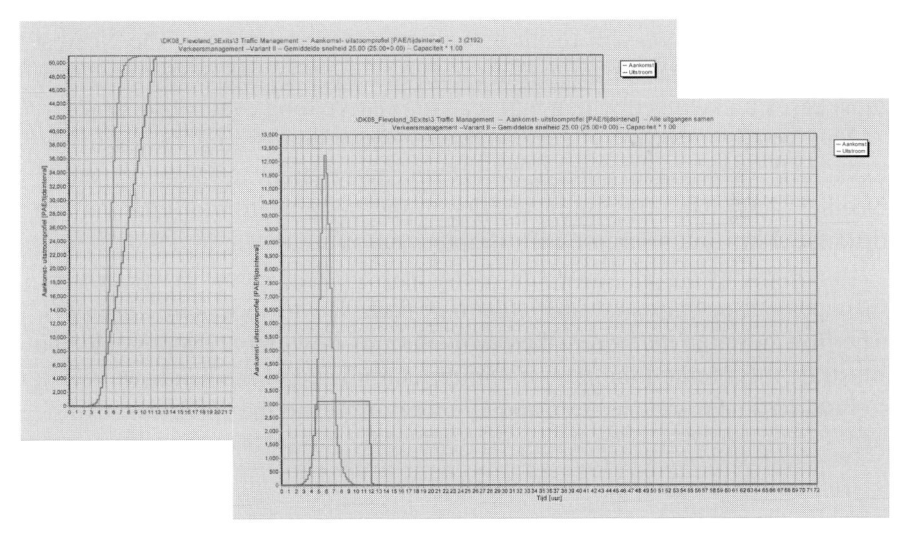

Fig. 12. Evacuated PCUs in time per time interval and cumulative for the Traffic management method.

The Nearest exit scenario performs better than the Reference scenario, but the evacuation time is still high. The dike-ring is empty after 20 hours. Bottle neck is still the south-east exit. At the north-east exit the load drops below capacity at 9:15, at the south-west exit at 12:00 hour.

With the Traffic management scenario the dike-ring is empty at 12:15 hour. From 4:45 till 11:45 all exits are loaded at maximum capacity.

The resulting OD-matrices can be loaded to the network. In this case an All-Or-Nothing assignment is performed using OmniTrans[5]. The plots give an idea of the complexity of the traffic flows.

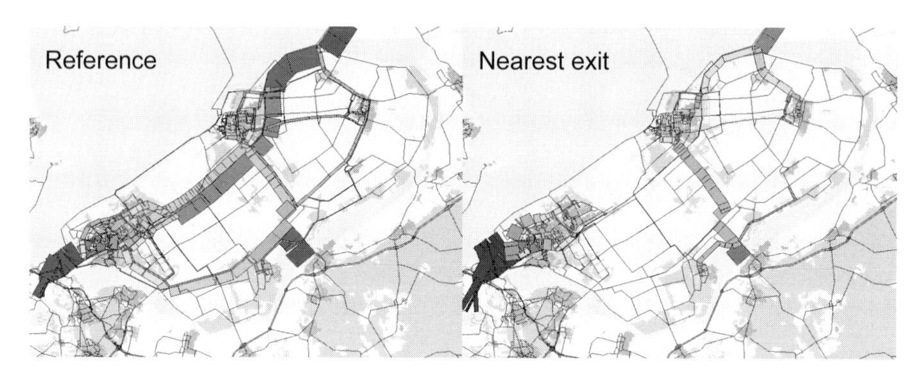

Fig. 13. All-Or-Nothing assignment of the methods Reference and Nearest exit

Typical for the Reference method is the crossing traffic flows. These crossing traffic flows will result in waiting times and delays. The Nearest exit method will lead to converging traffic flows. The use of the exits is in most cases unbalanced which will lead to relative long evacuation times.

The Traffic management method will show converging routes to the exits. All exits have similar workload in relation to there capacity. Needless long trips are eliminated due to the minimization of total vehicle distance. Flow out areas are identifiable.

In some aspects the flow out areas that are suggested by the Traffic management scenario are difficult to realize due to local circumstances. The flow out area method makes evaluation of flow out areas possible. To illustrate this method the city of Almere is assigned to the south-west exit, Lelystad to the north-east exit and all other origins to the south-east exit.

[5] OmniTrans (www.OmniTrans-Internation.com) is an application for traffic planning.

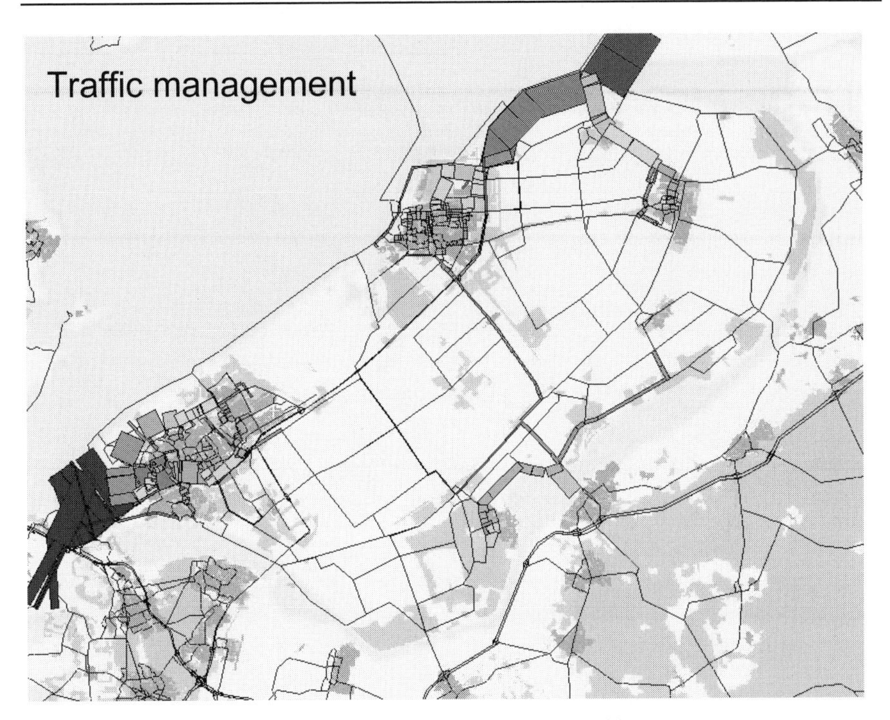

Fig. 14. All-Or-Nothing Assignment of the method Traffic management

The result has hardly any impact on the south-west exit, but shifts traffic from the north-east to the south-east exit. The total evacuation time will be 18:45 minutes. If for some reason this would be preferable special attention should be paid to the routes to the south-east exit. Reverse laning and special attention to the merging traffic near the exit is necessary.

Defining the dike-ring and running the methods, including the All-Or-Noting assignment, will take a few hours. For a re-run of a method several seconds is needed for the Evacuation Calculator and some minutes for the AON assignment. This is made possible by using a network for the whole of the Netherlands[6]. All dike-rings are defined within this network. This makes the data handling uncomplicated. The Evacuation Calculator will create a directory with data for each dike-ring.

[6] The network is part of the NRM (New Regional Modal) which is developed by the AVV Transport Research Center of the Ministry of Transport, Public Works and Water Management.

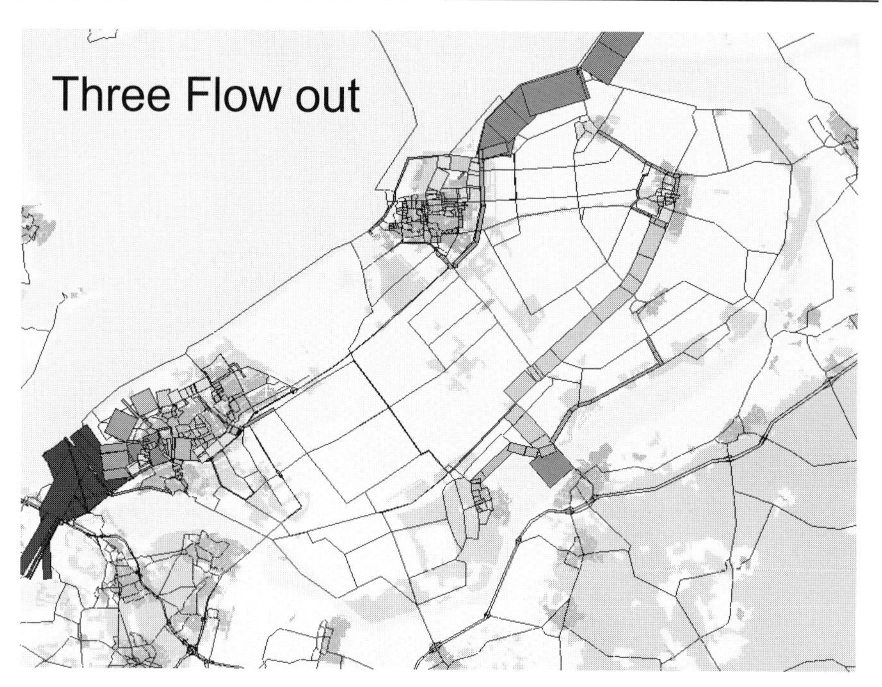

Fig. 15. All-Or-Nothing assignment of the flow out method

5 Conclusions

The developed Traffic management model for preventive evacuation follows a 'How to' approach. It produces more or less independent outflow areas which will assist in the development of an evacuation plan. The scenarios Reference and Nearest exit are especially useful in determining a safe estimate of the evacuation time.

The model is relatively simple in terms of complexity, data need, preparation and run time. Its accuracy is appropriate for this situation where many of the process steps can have very different outcomes.
For a better understanding of the capabilities of the Evacuator Calculator further research is needed with respect to dynamic macro (or micro) assignment.

Preventive evacuations of dike-rings in the Netherlands are quite rare. As a consequence important behavioral parameters in the model are not yet based on actual data. It is suggested to develop a monitoring program for in case a crisis situation occurs. Elements of this monitoring program should be at least: (1) the trip generation for the several categories; (2) the departure rates for the categories; (3) speed and traffic volumes during the evacuation; (4) supporting activities of police, fire brigade and army; (5) guiding information (radio and television); (6) perception of the situation by the evacuees.

References

Alabama, web http://www.sam.usace.army.mil/hesdata/Alabama/altranspage.htm

Boetes E, Brouwers N, Martens S, Miedema B, Vemde R van(2002), Evacuatie bij hoogwater: informatie voor een verantwoord besluit tot evacuatie, scriptie vierde jaargang MCDM (Master of Crisis and Disaster Management), Netherlands Institute for Fire and Disaster Management (NIBRA) & The Netherlands School of Government (NSOB), 2002

Chiu YC (2004) Traffic Scheduling Simulation and Assignment for Area-Wide Evacuation., 7th Annual IEEE Conference on Intelligent Transportation Systems (ITSC 2004).; Washington D.C., 2004

Cova JT, Johnson JP (2003) A network flow model for lane-based evacuation routing., Transportation Research Part A, Volume 37, page 579-604, Elsevier Science Ltd., 2003

Hobeika AG, Jamei B (2001) MASSVAC: A model for calculating evacuation times under natural disaster., Emergency Planning, Simulation Series 15/23, 1985

Urbina E, Wohlson B (2003) National Review of hurricane evacuation plans and policies: a comparison and contrast of state practices, Transportation Research Part A, vol 37, pp 257-275, Elsevier Science Ltd

Martens S (2002) Wat maakt een operationeel leider competent; Orrientatie op de competenties van operationeel leiders, scriptie vierde jaargang MCDM (Master of Crisis and Disaster Management), Netherlands Institute for Fire and Disaster Management (NIBRA) & The Netherlands School of Government (NSOB), 2002

Russo F, Vitetta A (2004) Models for evacuation analysis of an urban transportation system in emergency conditions., 10th World Conference on Transport Research (WCTR 2004), Istanbul, 2004

Sheffi Y (1985) Urban Transportation Networks: Equilibrium Analysis with Mathematical Programming Methods., Prentince-Hall Inc., Englewood Cliffs, New Jersey, 1985

Sheffi Y, Mahmassani H, Powell WB (1982) A Transportation Network Evacuation Model., Transportation Research, Part A, Volume 16A, No. 3, page 209-218, Pergamon Press, 1982

Considering Elevation Uncertainty for Managing Probable Disasters

Georgios A. Achilleos

Lab. of Space Geometric Representation, National Technical University of Athens, 2 Kitsou Str., Athens 11522, Greece.
Email: ageorgea@central.ntua.gr

Abstract

The existence of elevation errors in the Digital Elevation Models (DEMs) usually is ignored, during spatial analysis of risk assessment and disaster management problems. As a result, conclusions are extracted, decisions are taken and actions are designed and executed, while the problem is examined on a wrong point of view. This paper describes the attempt to introduce a new model, the DEEM (Digital Elevation Error Model), which incorporates elevation uncertainty and accompanies a DEM uniquely. The use of an uncertain DEM, combined with a probabilistic "soft" decision approach, eliminates the risk of taking decisions that do not imply to the real problem's basis. Research has shown deviations existing in results, up to 20-50% for volume measurements, area measurements, definition of boundaries, visibility calculation, etc., from those of a "hard" decision approach. The absence of an integrated GIS, able to manage data uncertainty, forces for by-pass approaches of the problem but not the appropriate ones.

1 Introduction

The insufficiency of binary logic and "hard" decision making to represent and manage spatial data in set with their spatial errors, comprises a remarkable difficulty for widely accepted risk assessment and disaster management procedures [Burrough 1986, Fisher 1993, Soulantzos 2001, Veis and Agantza 1989, Zadeh 1965]. The uncertainty management theory, constitutes an important assistant and contributor towards this direction. Data are accompanied by their errors and are managed through the GIS Spatial Analysis modules, aiming to reach a result with known uncertainty [Burrough 1986, Cressie 1991, Davis 1973, Fisher 1993, Fisher 1995, Soulantzos 2001, Wenzhong et al 1994, Wenzhong and Tempfli 1994, Zadeh 1965].

This research shows that the problems are serious, as it concerns cases consisting risk applications. This is due to the fact that estimating risk through an uncertain spatial database, leads to the dispute of results. Therefore, this forces the designers to undertake a Risk Level within the project design, usually based on the confidence the data provide.

Most of the disaster management applications make use of elevation data, kept in DEMs. These are uncertain elevation models and this uncertainty influences the application's results. Incorporating elevation uncertainty within the elevation models, introduces a new concept, the Uncertain DEM (UDEM), which is composed from the DEM and its DEEM (Digital Elevation Error Model).

The description of the DEEM is presented in this paper. Further on, the impact of this model in volume calculation and in visibility analysis is approached, in order to understand the size of the problem.

2 Uncertain DEMs

2.1 The Concept of an UDEM

The meaning of a UDEM, composes a hybrid data structure of uncertain altimetry. This structure leads to the incorporation of elevation data uncertainty management, within a GIS, although most of the GISs, do not include modules for uncertain data management. This uncertainty is propagated to the results of the analysis.

The UDEM, provides the ability for choosing the confidence level (C.L.) to work with, and this C.L. is usually compatible with the data accuracy. Simultaneously, a risk level is introduced into the results, in case these are applied in practice.

The UDEM is defined as hybrid, due to the fact that by choosing a C.L., a conventional DEM is defined automatically, fully compatible to any GIS management.

2.2 The DEEM

The DEEM is a data structure model, incorporating the uncertainty contained by the elevation information [Achilleos 2002a]. The DEEM follows exactly the same file structure as the DEM that accompanies and has a "personal" elevation error value for every elevation value in the DEM. In combination with the DEM (Figure 1), the DEEM is based on the acceptance of a normal distribution to describe these elevation errors. This distribution covers the 99.9% of data within the range of values (Figure 2) (see Eq. 2.1) [Cressie 1991, Davis 1973]:

$$\bar{x} \pm 3 \cdot \sigma_x \tag{2.1}$$

$$P\,(H < \bar{x} + 3 \cdot \sigma_x) = \sim 100\%$$

$$P\,(H > \bar{x} - 3 \cdot \sigma_x) = \sim 100\%$$

$$P\,(H < \bar{x} - 3 \cdot \sigma_x) = \sim 0\%$$

$$P\,(H > \bar{x} + 3 \cdot \sigma_x) = \sim 0\%$$

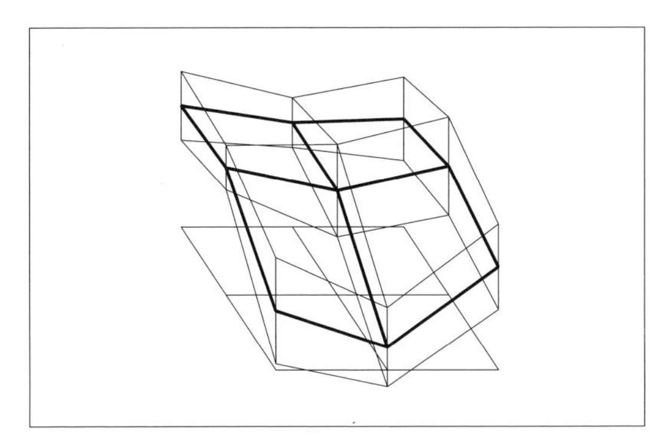

Fig. 1. Uncertain DEM and its DEEM

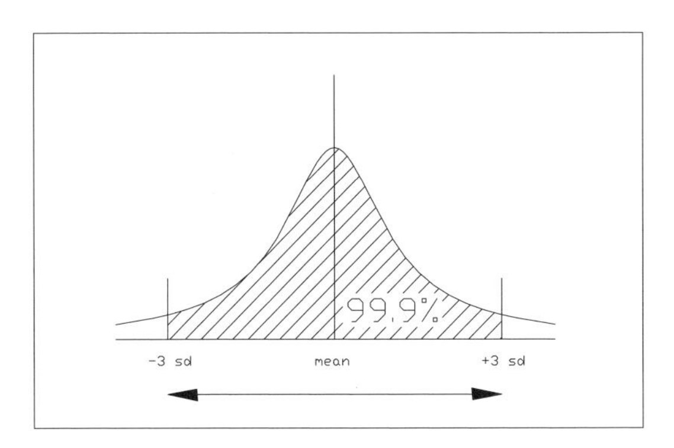

Fig. 2. Probability Function

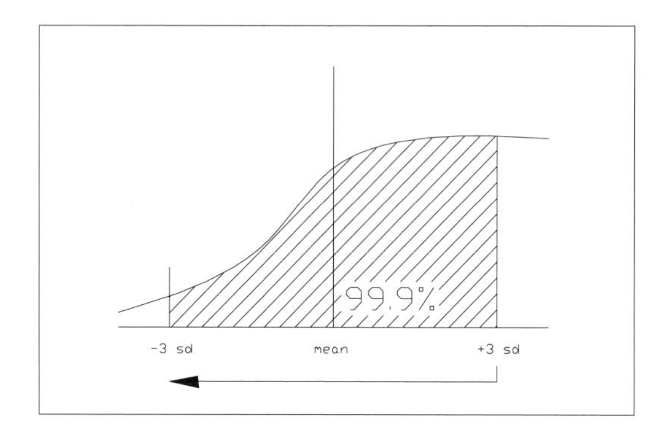

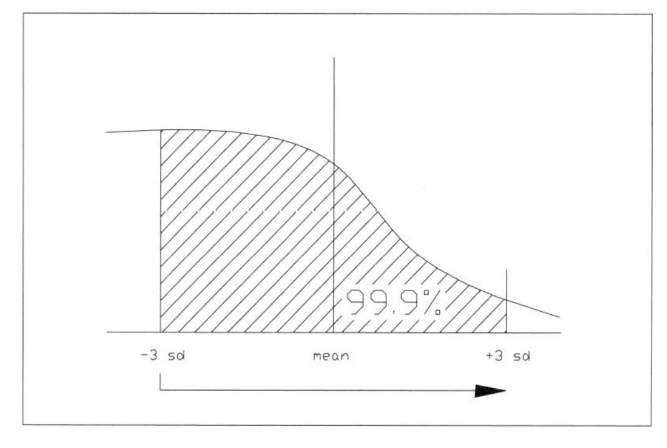

Fig. 3. Cumulative Probability Function (a, b)

The probability function, giving the bell shaped graph (Figure 2), is (see Eq. 2.2), while the cumulative probability function, giving the S shaped graph (Figure 3 (a), (b)), is (see Eq. 2.3):

$$f(x) = \frac{1}{\sqrt{2 \cdot \pi}} \cdot \frac{1}{\sigma} \cdot e^{\left\{-\frac{(x-\bar{x})^2}{2 \cdot \sigma^2}\right\}} \tag{2.2}$$

$$F(x) = \int_{\infty}^{x} f(x) = \int_{\infty}^{x} \frac{1}{\sqrt{2 \cdot \pi}} \cdot \frac{1}{\sigma} \cdot e^{\left\{-\frac{(x-\bar{x})^2}{2 \cdot \sigma^2}\right\}} \tag{2.3}$$

2.3 From an UDEM to a DEM

The transition from an UDEM to a DEM, is based on the selected C.L. (as well as the selected risk level). The critical value of the elevation, according to the C.L., is calculated from the cumulative probability function, as the value for which the Riemann integral with that elevation value as a limit, gives an area equal to the C.L. The determination of this critical value, takes place for every pixel of the DEM, resulting a new DEM that represents the selected C.L. A selection of a different C.L. leads to a result of a different DEM.

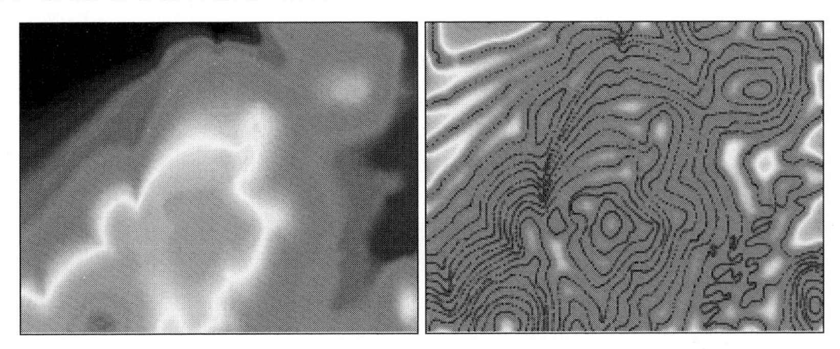

Fig. 4. DEM and its DEEM

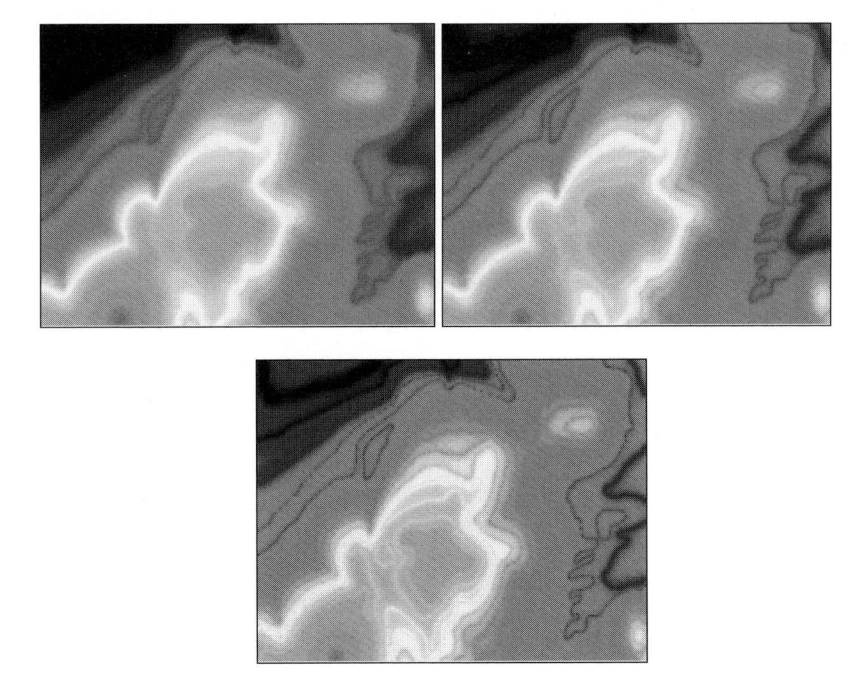

Fig. 5. DEM for C.L. 75%, 90% and 99%

Figures 4 and 5 (Figures 4, 5 above) present in order, the DEM, its DEEM and the DEMs for the C.L.s 75%, 90% and 99%.

3 Risk Assessment and Disaster Management through Uncertainty

3.1 Importance of a Risk Assessment

The term "risk assessment", introduces the meaning of danger calculation and if possible, avoidance (much more serious, in case it concerns human lives).

A large amount of actions taken on environment, influence and complicate human lives. Many of these actions put these human lives in danger, such as construction projects (dams, nuclear power stations, etc.), dangerous materials transportation (nuclear waste, dangerous gases and liquids, etc.), immediate actions services planning (fire observation network, ambulance and police station allocation, etc.) and many more. Risky planned projects and actions containing low level of safety and hygiene, do not incorporate requirements for development.

The meaning of uncertainty, influencing this attempt for risk estimation and assessment, covers a variety of factors. Some of these factors are: data, models, conditions, relationships between conditions, etc. The basic component of planning is the set of data that is used. This set of data, usually contains stochastic errors, and the only available element is an estimation of their distribution. Due to this fact, the incorporation of uncertainty within the procedure of planning is unavoidable.

Therefore, the problem is not the avoidance of errors and the introduced uncertainty, but their incorporation within the procedure.

3.2 The Influence of Using Uncertain Terrains in Assessment

The use of UDEMs as a representation of terrains, influences the risk assessment analysis. Therefore, there are contestations concerning the results, especially by specialists believing on soft decision approaches.

The conventional analysis, using a deterministic approach, comes to a decision importantly different from that of the uncertainty theory.

Obviously, determining geometrical characteristics, which may or may not have a certain hypostasis and their determination is based on a DEM, demands the consideration of any elevation error included. The influence is immediately observed in area and volume calculations, site selections, zone and boundaries definitions, 3D calculations, etc.

This paper, covers two case studies of analysis for environmental health risk assessment and disaster management: case study 1 concerning capacity volume calculations of reservoirs behind dams and case study 2 concerning visibility analysis.

4 CASE STUDIES

4.1 Capacity Volume Calculations

The certain case study covers a fictitious preliminary study of an artificial reservoir, resulting from a dam construction. The aim is the estimation of the total capacity of the 3D solid the water will fill. Further on, the selection of the certain position and specifications for the project, are compared with other alternative places and different specifications, in order to achieve the final place selection and dam dimensioning and reinforcement.

The elevation data used for the analysis come from topographic 1:20.000 scale maps. The contour interval is 5m while the accuracy is ±2.5m [Richardus 1973, Robinson 1994, Yoeli 1984]. The elevation level of the dam top and its overflow position is 340m.

The DEM with the dam's selected axis are presented in Figure 6 (Figure 6 (left)). Binary analysis provides the area that stands higher than 340m (Figure 6 (right)).

Using a membership function (classified in Figure 7), researches can estimate the area which presents probability 75% to be altimetricaly higher than the 340m. The same area can be estimated for a probability of 90% or 99% (Figures 8 right and left, Figure 9). It can be realized that the greater the C.L. is, the greater the proportion of the study area which is altimetricaly higher than the 340m. This increament of the area, introduces points for water overflow (water "escape" points), which arithmetically increase with the increament of the C.L.

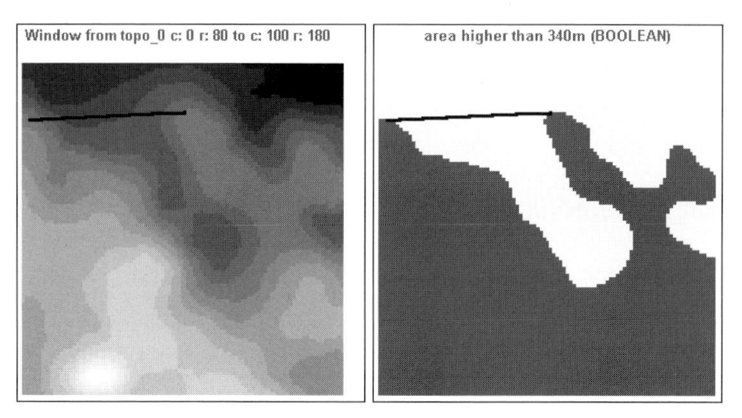

Fig. 6. DEM (left) and Area Elevations > 340m (Binary Result) (right)

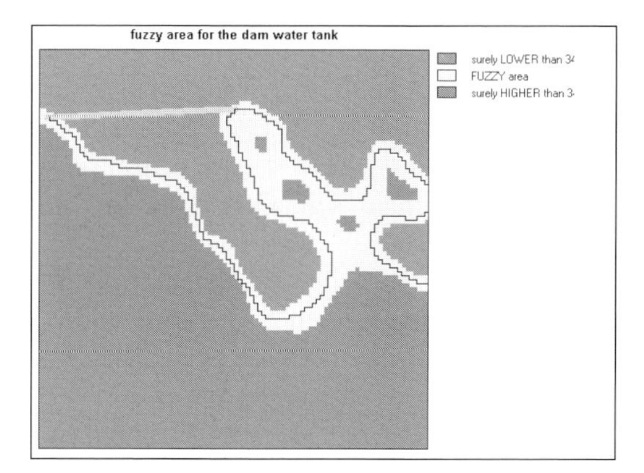

Fig. 7. Fuzzy Area (midtone: surely < 340m, dark: surely > 340m, white:FUZZINESS)

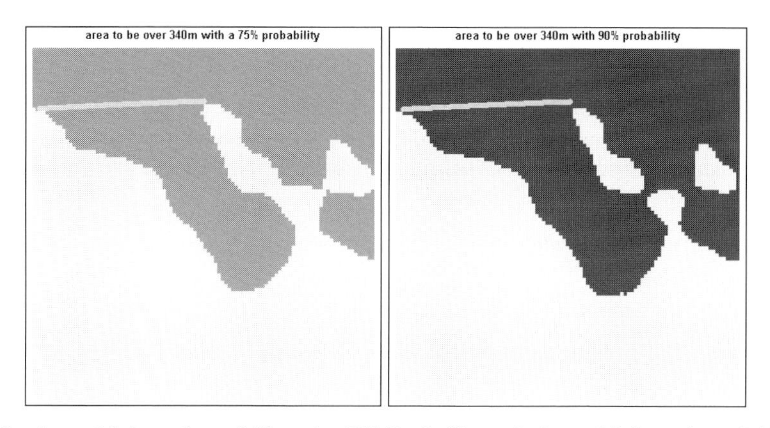

Fig. 8. Area higher than 340m (p>75%) (left) and Area higher than 340m (p>90%) (right)

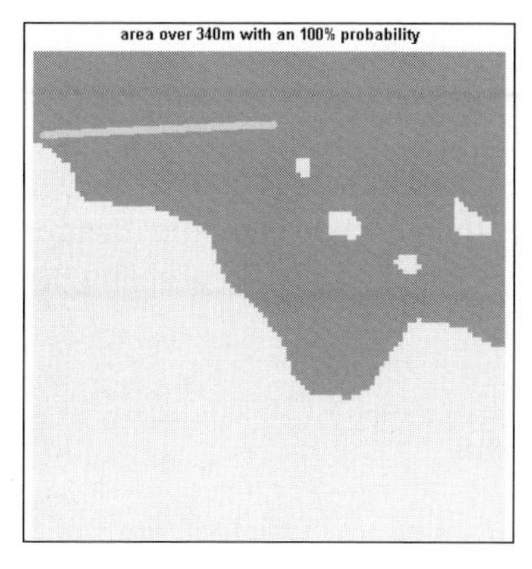

Fig. 9. Area higher than 340m (p>99%)

The volume calculations are based on the geometric volume estimation of solid shapes. This is the product of the shape's base multiplied by its height. In the certain case, the height of the solid shape is considered to be the mean elevation difference, presented in the area of the reservoir (base).

This approach is applied due to the problem shown, of the volume underestimation. This problem is a result of the fact that many points of the area, with elevations lower than the dam top (340m), are included in the horizontal area of the base, for certain values of the C.L. These points give negative elevation differences, which influence seriously the final calculation of the reservoir's capacity.

The results for the area and the capacity (volume) of the reservoir are presented in Table 1 (Table 1). The observed deviations of the water tank capacity are within the range of 9.9% to 42.5%.

C. L. [%]	AREA [1000*m2]	MEAN ELEVATION [m]	ELEVATION DIFFERENCE [m]	SOLID SHAPE VOLUME [1000*m3]	DEVIATION [%]
BIN	3547.5	280.83	59.17	209905.5	----
75	3897.5	286.97	53.03	230615.0	9,9
90	4267.5	292.81	47.19	252507.9	20,3
99	5055.0	303.98	36.02	299104.3	42,5

Table 1. Results for Area and Volume

It is obvious that a false calculation or a high data inaccuracy leads to a high risk for a dam failure or a large water amount overflow. These possible scenarios put into danger the human lives and the environment behind the dam. The risk assessment analysis works better, taking into account that data are inaccurate and thus, specialists are better prepared for a disaster management situation.

The researcher can estimate the size of the disaster the managers have to face in case of an emergency. The overflowed water amount or the total water flow beyond the dam, in case it collapses, can be estimated and thus, the area that will be flooded and set into danger can be defined. The situation can then be less dangerous and probably the area can be protected more actively.

4.2 Visibility Analysis

Visibility analysis in cases of immediate actions services planning (i.e. fire observation network, army observatories), is a strategic procedure, considering that these cases are time sensitive cases. In case of an emergency situation, action must be taken immediately to avoid setting into danger the environment and its habitats [Fisher 1993, Fisher 1995].

Visibility calculation is one of the basic procedures existing mostly in every GIS. Furthermore, as this procedure is strength related to the man's everyday, being one of his five senses, attention should be paid in order to ensure an accurate calculation.

The designed algorithm GAVOS, uses the elevations from a DEM and their variations from a DEEM, to calculate an estimation of the probability for which a TargetPoint is visible from a ViewPoint. The best-fitted Line of Sight (LoS) (Figure 10), giving the maximum probability of visibility existence, is selected. The probability for line AB to exist, is the product of all intermediate probabilities (see Eq. 4.1):

$$PA, Pi_1, Pi_2, ..., Pi_{15}, PB \qquad (4.1)$$
$$P = P(A) \cdot P(i_1) \cdot P(i_2) \cdot ... \cdot P(i_{15}) \cdot P(B)$$

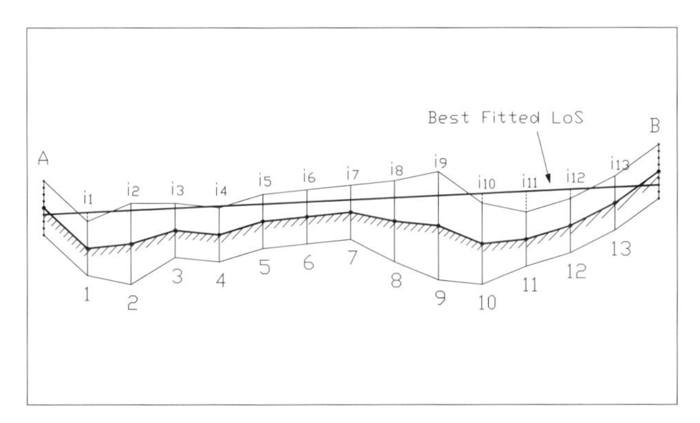

Fig. 10. Best Fitted LoS

GAVOS algorithm is still under research, as a part of a broader schedule, concerning UDEMs and their influence to DEM products. A short investigation concerning the elevation error size alteration within a DEM, shows that the probability two points to be visible to each other can vary from 100% to less than 30% (elevation errors from 2m to 6m respectively) (Table 2).

Elevation Error	Maximum Probability to See
2.00	96.40
3.00	78.21
4.00	53.12
5.00	40.71
6.00	27.16

Table 2. Variation of Visibility Existence Probabilities

5 Remarks - Comments

Obviously, it is purposive to embody uncertainty within the data sets and further on, within the calculation procedures, in order that possible rough situations to be abstained during the implementation of any project. Further more, the autocorrelation concerning this uncertainty, should also be embodied within the calculation procedures, as this autocorrelation performs an important role in the results' fuzziness.

The ideal condition could be a GIS, able to embody uncertainty, autocorrelations and cross-correlations, in order to execute analyses and estimations, without complicated and failure sensitive procedures.

This ascertainment is not only a result of the research concerning the capacity of water tanks or the visibility analysis. It is a broader ascertainment of research incorporating the area and volume calculations of certain magnitudes, line definitions, etc. through a UDEM [Achilleos 2002a, Fisher 1993, Fisher 1995].

The main problem, concerning these cases, is the definition of the uncertainty (thus, the definition of the DEEM [Achilleos 2002a, Li 1992, Li 1993, Mikhail 1978, Ostman 1987]). This problem can be approached in different ways. An approach, defines the DEEM through the procedure of the DEM generation, where the elevation error, is calculated through objective steps [Achilleos 2002a].

The deviations presented during the application of the uncertain approaches, may lead to wrong estimations, to re-selection of position, to the selection of a different design solution, to the project total budget failure and also to possible problems of the project safety and further on, the project operation.

Similar case studies mentioned in the beginning of this article (i.e. dangerous materials transportation, dangerous materials leakage, etc.) need to be studied under a similar point of view in order to embody data uncertainty within the procedure.

REFERENCES

Achilleos G (2002a) Study of Errors in the Procedures of Digitizing Contour Lines and Generating Digital Elevation Models, PhD Thesis, Lab. of Geography & Spatial Analysis, School of Rural & Surveying Engineers, National Technical University of Athens, Athens (greek language).

Burrough PA (1986) Principles of Geographical Information Systems for Land Resources Assessment, Oxford University Press, Great Britain.

Carter JR (1988) Relative Errors Identified in USGS Gridded Models, Proceedings of Ninth International Symposium on Computer - Assisted Cartography, Baltimore, Maryland, AUTOCARTO 9.

Cressie NAC (1991) Statistics for Spatial Data, John Wiley and Sons, Inc., New York.

Davis JC (1973) Statistics and Data Analysis in Geology, John Wiley & Sons, Inc., USA.

Fisher PF (1993) Algorithm and Implementation Uncertainty in Viewshed Analysis, International Journal of Geographical Information Systems, Vol. 7, No. 4, pp. 331-347.

Fisher PF (1995) An Exploration of Probable Viewsheds in Landscape Planning, Environment and Planning B: Planning and Design, Vol. 22, pp. 527-546.

Li Z (1992) Variation of the Accuracy of Digital Terrain Models with Sampling Intervals, Photogrammetric Record, 14 (79), April 1992, pp. 113-128.

Li Z (1993) Mathematical Models of the Accuracy of Digital Terrain Model Surfaces Linearly Constructed from Square Gridded Data, Photogrammetric Record, 14 (82), pp. 661-674.

Mikhail E (1978) Panel Discussion: The Future of DTM (presented at the ASP DTM Symposium, May 9-11, 1978, St.Louis, MO), Photogrammetric Engineering & Remote Sensing, Vol. 44, No. 12, pp 1487-1497.

Ostman A (1987) Accuracy Estimation of Digital Elevation Data Banks, Photogrammetric Engineering & Remote Sensing, Vol. 53, No. 4, pp. 425-430.

Richardus P (1973) The Precision of Contour Lines and Contour Intervals of Large- and Medium- Scale Maps, Photogrammetria, 29, pp.81-107.

Robinson GJ (1994) The Accuracy of Digital Elevation Models Derived from Digitised Contour Data, Photogrammetric Record, 14(83), pp. 805-814.

Soulantzos G (2001) Fuzzy Logic in certain Location Problems, Journal of Hellenic Association of Rural & Surveying Engineers (greek language).

Veis G, Agantza – Mbalodimou A (1989) Error Theory and Least Squares Methods, Semester Teaching Notes, School of Rural & Surveying Engineers, National Technical University of Athens, Athens (greek language).
Wenzhong S, Ehlers M, Tempfli K (1994) Modeling and Visualizing Uncertainties in Multi-Data-Based Spatial Analysis, Proceedings of EGIS / MARI' 94, Paris, pp. 454-464.
Wenzhong S, Tempfli K (1994), Modelling Positional Uncertainty of Line Features in GIS, ASPRS/ACSM, Annual Convention & Exposition Technical Papers, Nevada, pp. 696-705.
Yoeli P (1984) Error - Bands of Topographical Contours with Computer and Plotter (Program Koppe), Geo-Processing, 2, pp. 287-297.
Zadeh LA (1965) Fuzzy Sets, Information and Control, 8: 338-353.

Emergency Preparedness System for the Lower Mekong River Basin: A Conceptual Approach Using Earth Observation and Geomatics

Ferdinand Bonn, Guy Aubé, Claire Müller-Poitevien and Goze Bénié

Centre d'applications et de recherches en télédétection (CARTEL), Université de Sherbrooke, Sherbrooke, Québec, Canada, J1K2R1.
Email: Ferdinand.Bonn@usherbrooke.ca

Abstract

With the aid of the Canadian Space Agency and the Canadian geospatial industry (Ærde, HCL, Strata 360), CARTEL is working in close collaboration with the Mekong River Commission (MRC) and the Cambodian Red Cross to establish a Flood Emergency Response System in Cambodia. Earth Observation (EO) and Geomatics generally insists on spatial distribution and the accessibility of the health centres, food warehouses, flood safe areas as well as the spatial distribution of the disease according to the changing factors of the physical environment and planning and the management of public health in order to improve quality of the decision-making process. The strategy is centered on the analysis of the needs for the managers of the Mekong River Commission and of the Cambodian Red Cross. It also focuses on the questions and data relating to the follow-up in space and time of the flood events and its effects on the local communities. This study takes into account the three dimensions of security: vulnerability, preparedness, response. The methodology includes the following stages: (1) analysis of the needs of the Mekong River Basin managers such as health, transportation, safe areas, infrastructures at risk, food security; (2) design of the diagram of the emergency response system including the identification and the conception of georeference data and spatial analysis functionalities; (3) design and development of a database and metadata; (4) development of vulnerability maps by using multi-date EO imagery (RADARSAT-1, aerial photography, high resolution optical imagery) combined with the historical and topographic data.

1 Introduction

In recent years, it has become a pressing issue for governments to improve human security in emergency response situations. Some of the recent initiatives using space technologies to assist in disaster management are the Disaster Emergency Logistic Telemedicine Advanced Satellites System developed with support of ESA, the International Charter on Space and Major Disasters signed by CNES, CONAE, CSA, ESA, ISRO, NOAA, and the UN, the Mozambique Flood Information System developed with the support of the German Aerospace Center (DLR), the Environmental Monitoring Information Network in Bangladesh under development by RADARSAT International (RSI), the GIS-Based Flood Information System (AWRA, 2003), the National Urban Search and Rescue Response System (FEMA, 2003), the Real-Time Emergency Management via Satellite (REMSAT, MDA, 2001), the UN International Strategy for Disaster Reduction, etc.

When natural disasters occur, such as storms or floods, population's relocalisation and its management constitute a complex problem for health and security. These disasters mainly strike unprivileged populations, who reside in high-risk areas, such as along the banks of rivers or in non- protected coastal zones. As an example, water treatment facilities are often destroyed by floods, which bring drinking water contamination by bacteria, viruses, parasites and toxic substances. Rice fields, fisheries and aquaculture infrastructures are also damaged, causing food insecurity. In 2000 in Cambodia, 3,500,000 people were affected by floods (347 deaths, 7,068 houses destroyed, 347,107 ha of rice fields damaged). But droughts, which can also occur in this area, are an equally important threat to human security and they can disrupt the food security in the area.

2 The Lower Mekong Basin

Located in SE Asia, the Mekong River is 4 800 km long, flowing north-south. It drains a large, multinational watershed of 795 000 km^2 shared by China, Myanmar, Laos, Thailand, Cambodia and Vietnam (figure 1). The Mekong River Commission (MRC) has been established in order to harmonize management, environmental issues and water use between the different countries involved. Subjected to a monsoon climate, with the rainy season occurring from May to November, the Mekong usually enters a flooding stage from July to December. The overall hydrology of the Mekong is relatively complex, especially in the lower part of its basin, located in Cambodia and Vietnam.

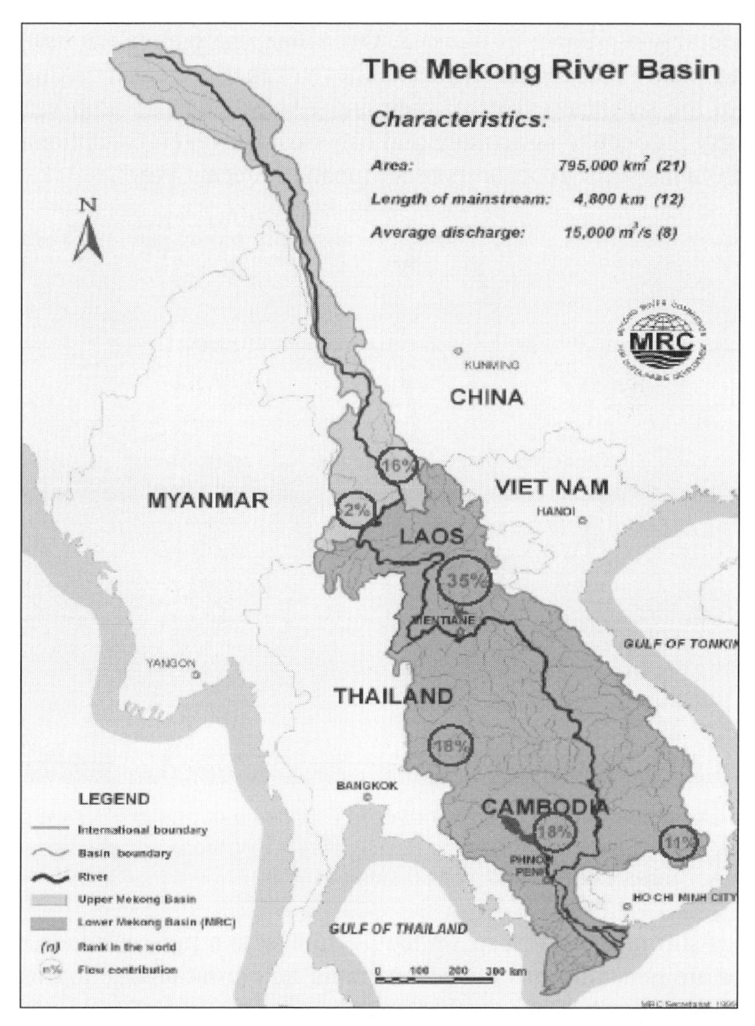

Fig. 1. Overview of the Mekong river basin. Percentages in circles refer to basin proportions in the different countries. (modified from Mekong River Commission)

The Cambodian Floodplain plays an important hydrological role in the regulation and dampening of the floodwave (Fujii et al., 2003). Especially, the great lake of Cambodia (Tonle Sap), which is a dead end tributary of the Mekong, fills and drains according to the floods and the tidal cycles in the delta, located in Vietnam.

The flooding process is part of regular life in the lower Mekong basin. It provides irrigation for crops, water for the fisheries and navigation facilities for the communication network. The water area therefore varies considerably as shown in figure 2. Over time, the population has adapted its way of life to these conditions and even benefits from it, taking advantage of the seasonal cycles of high and low water to develop its original and very productive agriculture and fisheries. However, exceptional floods and droughts seem to occur more frequently in recent years.

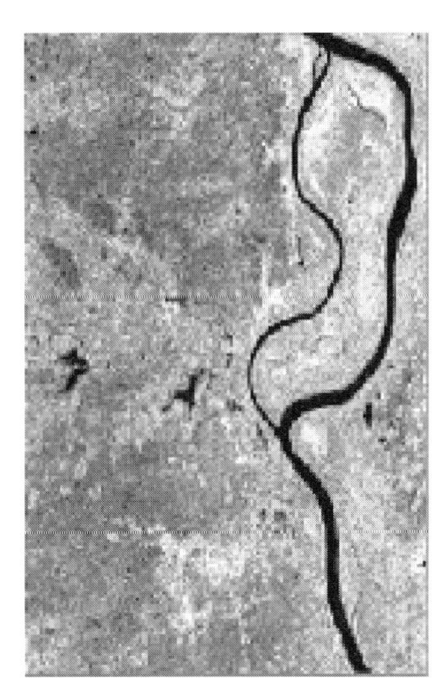
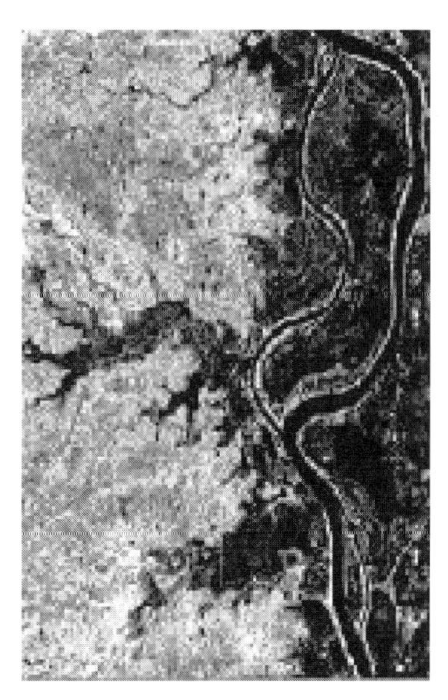

Fig. 2. A section of the Tonle Sap river in Cambodia during dry season (left) and flood season (right) as observed by RADARSAT-1. Water areas appear in black. (© CSA, courtesy of Hatfield Consultants Ltd.)

It is still difficult to explain their origin with a high certitude, but two causes are generally mentioned: the rapid deforestation due to illegal logging and agricultural expansion, which reduces the water flow retarding capacity in the basin, and the climate change which produces an increase of sea level, higher tides and more frequent typhoons.

3 Objectives of the Project

As many developing countries having suffered from natural disasters but also from long periods of war, Cambodia lacks a good infrastructure in updated maps, in geospatial information and in communications. The Cambodian Red Cross (CRC), which is the major national agency for emergency planning and intervention, is taking its rescue and mitigation decisions by relying on the experience of the people in the field and on the existing geographical data in the country. The former have a good knowledge of the local situation and have also an extended experience of living in emergency situations, but they usually lack the more general overview at regional or national levels, which may be required for planning interventions in the case of a large crisis. The existing geographical information is usually scattered, outdated and not organized in a flexible or standardized format. It is therefore difficult for the authorities to plan a structured approach in emergency preparedness and mitigation based on spatially distributed information.

The objectives of the present project are, therefore, to develop a model of an emergency response system based on geomatics, and to install it as a prototype at the MRC, more specifically inside its Flood Management and Mitigation Programme (FMMP).

4 Methodology

4.1 Evaluation of User Needs

The CRC and the FMMP of the MRC need an operational decision support system based on geographic information in order to be able to plan their operations for population rescue, food distribution, safe areas locations and health services.

After having worked in close cooperation with the CRC in order to identify its geospatial information needs, these needs can be summarized by a list of questions requiring answers of the "where?" type, requiring a geographic location answer. These questions and information needs can be divided in five subcategories:

- location of and access to health services
- location of and access to flood safe areas
- needs related to evacuation roads and waterways
- inventory and location of infrastructures at risk
- access to food security

The CRC also provided our team with the adequate information about the existing approaches, its past experience in emergency situations and some examples of local actions taken. We jointly decided to develop a prototype GIS based emergency response system for the district of Kândal, located close to Phnom Penh, for easier validation in the field.

4.2 Analysis of Similar Systems

The second step of the approach was to look at other systems existing elsewhere in the world. These systems vary in complexity, from relatively simple flood risk maps based on digital topography and past flood extensions observed by satellite to more complex systems relying on an integrated multi-risk management system. There are two Canadian examples of these cases: a) the 1997 Red River flood in Manitoba which used RADARSAT-1 data in a near real time mode for mapping flood extent and planning rescue and protection activities in the field (Bonn and Dixon, 2004) b) the REMSAT system, developed by MDA and Telesat with the support of CSA, ESA and the telecom industry (MDA 2001). This system has shown it efficiency during the severe forest fires that occurred in British Columbia, Canada in 2003.

The European Commission has also contributed to the development of similar initiatives under the EU-MEDIN program, where risk assessment, mitigation and geographic information are strongly integrated (Ghazi, 2003). The important choices that we had to make needed to take into account the capability of the end user to make the system work in an environment that had its technological limitations: frequent power failures, limited computing facilities, possible disruption of communications.

4.3 Development of the Conceptual Model

The conceptual model used in the system is based on the principles developed by Bénié (2003) for the health care aspects and presented in figure 3. It is basically a system that uses a series of feed-back loops for the decision making support system. It includes steps for validation and adjustments with the real world situation, and provides constantly updated information to the decision makers.

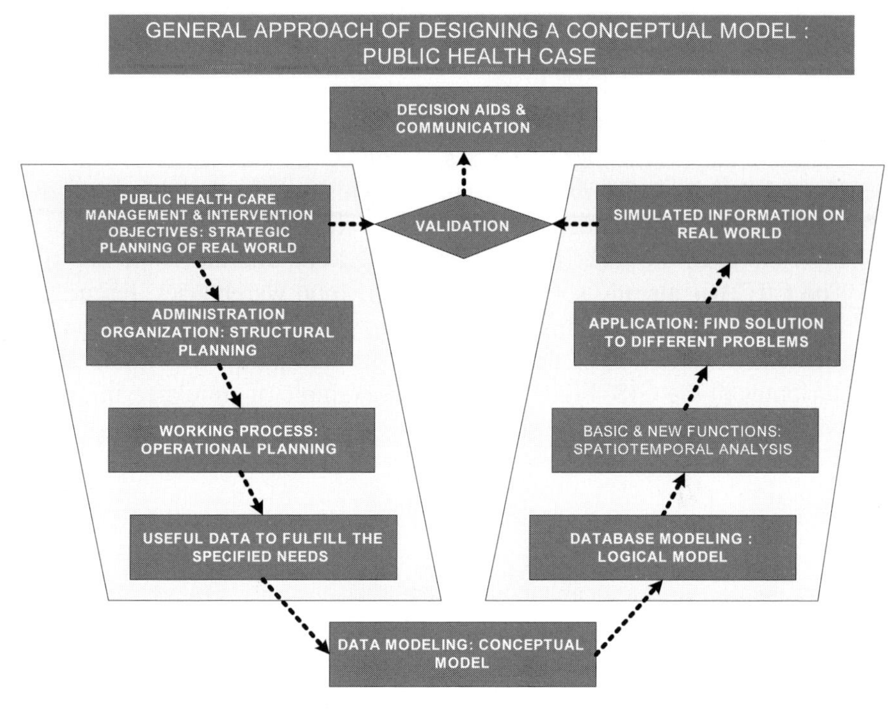

Fig. 3. Conceptual structure of a geographic information system for public health (Bénié & al., 2003)

This step of the conceptual model development is often overlooked in GIS developments, but it is a crucial one to ensure a smooth functioning of the whole system. The system should be able to answer smoothly and efficiently to a series of queries such as:

- Where are the nearest Health Centres from the Flood zones?
- Where is the nearest Health Centre from the village X?
- What is the Health Center capacity? The occupation rate?
- What is the shortest way from village X to Health Center X?
- What are the villages at less than X km from a Health Centre?
- What are the villages located in a possible cholera zone? Malaria zone?
- Does the hospital X possess an emergency kit?
- Where are the safe sites in case of floods?
- Where are the evacuation roads for village X?
- Etc.

4.4 Implementation of a Prototype

After this conceptual work, we have chosen to develop and implement a prototype of the system for two vulnerable communes located close to Phnom Penh Peam Oknha Ong & Kaoh Reah, district of Lveam Aem, Kandal Province, Cambodia) (figure 4). This will be the pilot area for the prototype to be implemented and tested by the CRC. This area is located in the vicinity of the junction between the Mekong and the Tonle Sap rivers. The CRC has already some inventories of food warehouses, rescue facilities and safe areas for this region, but they are not yet mapped and documented. Constitution of the prototype will incorporate this information as attributes in the GIS. Figure 5 shows an example of simulated information on food warehouses in Cambdodia.

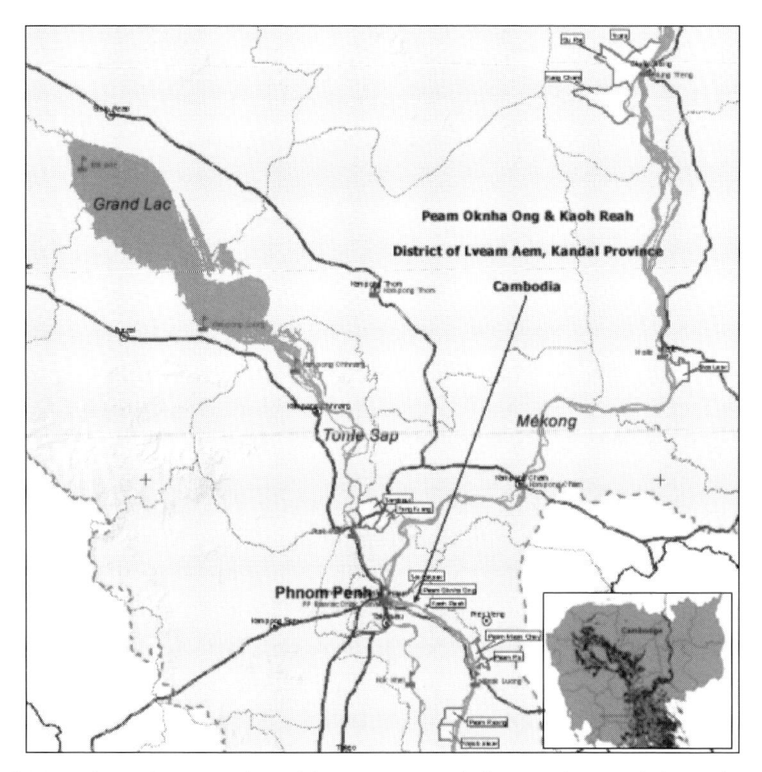

Fig. 4. Location of two vulnerable communes (pilot area), Kandal Province, SE of Phnom Penh (modified from Mekong River Commission and Cambodia Red Cross)

warehouses_cambodia.dbf

Wh_no	Province	District	Sub_dist	Village	Org	Build_alt	1floor_alt	2floor_alt	Coating	Food_type	Food_kg	Water_l	Km_nat_rd
23	Phnom Penh	1201	120108	Phum 8	WFP	4	3	0	Bricks	Rice	12000	3000	0,5
24	Kampong Cham	808	80807	Krang Mkak	WFP	10	3,2	6,2	Wood	Rice	50000	0	3
25	Kandal	810	81013	Khpob Kraom	CRC	6	4,2	0	Bricks	Rice	10000	2500	20
26	Kandal	811	81104	Ta Khmau	OXFAM	4	3	0	Wood	Wheat	20000	5000	4,3
27	Kandal	802	80201	Kandal Leu	WFP	5	3,7	0	Wood	Rice	5000	1200	3
28	Prey Veng	1403	140307	Kampong Trabaek	AAH	6	3,8	0	Bricks	Dry Fruits	2000	0	10
29	Kampong Cham	307	30109	Cheung Chhrok	WFP	3	4	0	Bricks	Dry Fish	3000	1300	2.3

Fig. 5. Simulated Geospatial Information table on food warehouses in the study area (Aubé, 2004)

This information is then completed with the geographic location and incorporated as attributes in the GIS, with links to the appropriate queries.

5 Preliminary Results

These results are at their very early stages and have not yet been validated in the field or by the end users, these steps being planned for the second half of 2005 and through 2006.

5.1 Conceptual Model Structure

The conceptual model is structured in classes (level 1), subclasses (levels 2 to 5) and entities (level 6). Figure 6 gives an approximation of this structure.

5.2 Attribute Tables

The attribute tables such as the one presented in figure 5 are linked to the entities and can be mapped or retrieved through queries resulting from spatial analysis. The type of questions to be asked to the system are quasi endless. Typically they can be spelled as: what is the shortest (or the safest) route from any place to the nearest safe zone, hospital or food warehouse.

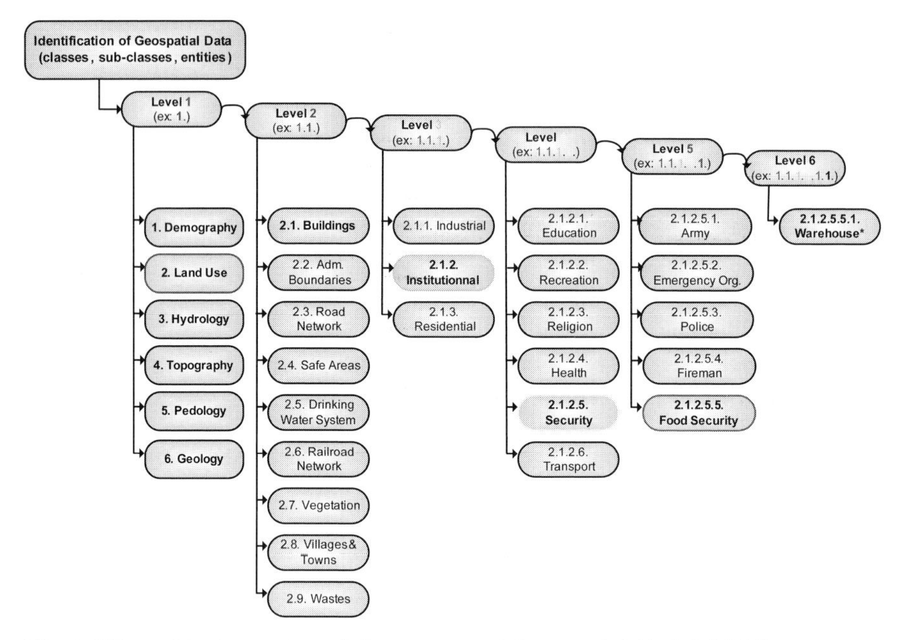

Fig. 6. Hierarchical structure of the conceptual data model (Aubé, 2004)

5.3 User Interface

The user interface is designed to be as friendly as possible, and in some aspects it looks like a web interface.

6 Conclusions

The development of the MERS (Mekong Emergency Response System) for the MRC and the CRC is intended at this stage to be a demonstration project to be experimented by the end users on a very local basis. If successful, it will be enlarged and integrated in the Flood Management and Mitigation Program of the MRC, with the CRC acting as the major end user. In its present stage, it is more a preparedness system than a response system, therefore the title of this paper. It has, however, the potential to evolve into a response system when it will be integrated in the FMMP. This may however require additional functionalities such as links with the flood forecasting systems and weather data, but its structure could allows this future flexibility.

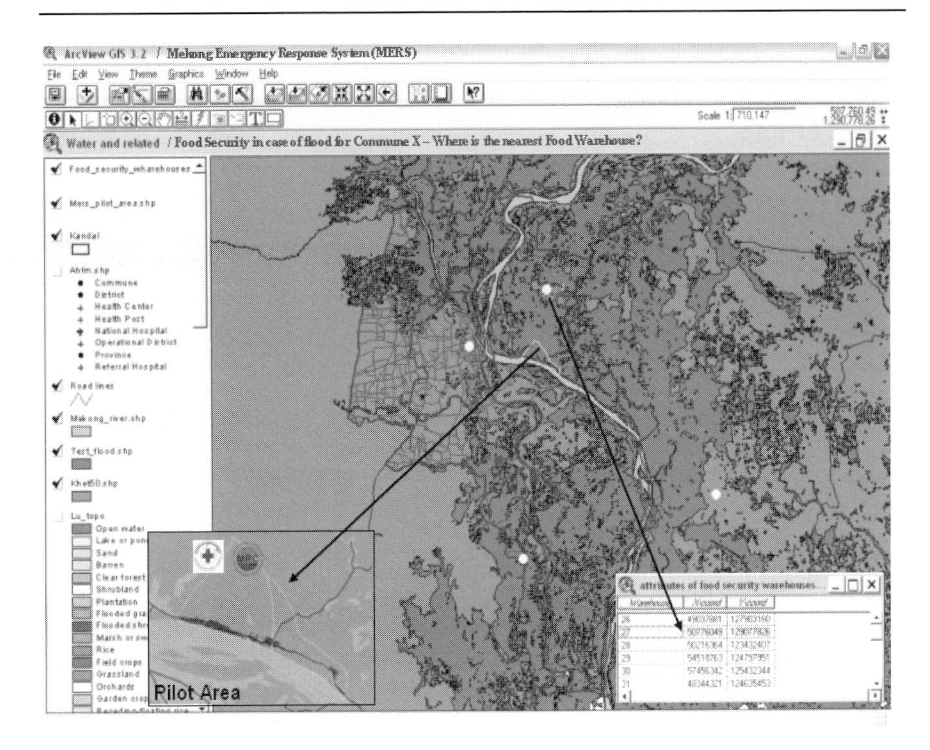

Fig. 7. Example of the MERS prototype user interface and possible outputs (Aubé, 2004)

Acknowledgements

The authors would like to thank the Canadian Space Agency for the support of this project through its EOADP program, Pierre Dubeau, Andy Dean and Tom Boivin from Hatfield Consultants Ltd., ,Valter Blasevic and Bill Kemp from Strata 360, Ulf Hedlund and Manitaphone Mahaxay from MRC for their useful advice as well as Dr Sam Ath from CRC for the user requirements and Michel Pomerleau for the ongoing programming.

References

Aubé G (2004) Système d'intervention d'urgence dans le basin du Fleuve Mékong: développement d'un modèle conceptuel de données appliqué à la sécurité et à la santé publique lors d'inondations (Kândal, Cambodge). Mémoire soumis, MSc en géographie, Université de Sherbrooke, 128 p

AWRA, American Water Resources Association, (2003) Development of a GIS-Based Flood Information System for Floodplain Modeling and Damage Calculation. www.awra.org/jawra/papers/J98063.html

Bénié GB et al (2003) Notes de cours de SIG. Département de Géographie et Télédétection, Université de Sherbrooke, Sherbrooke

Bonn F, Dixon R (2004) Monitoring flood extent and forecasting excess runoff leading to increased flood risk with RADARSAT-1 data. Natural Hazards, in press

FEMA, Federal Emergency Management Agency (2003) National Urban Search and Rescue (US&R) Response System. Washington, www.fema.gov/usr/

Fujii H, Garsdal H, Ward P, Ishii M, Morishita K, Boivin T (2003) Hydrological Roles of the Cambodian Floodplain of the Mekong River. International Journal of River Basin Management, 1:3, pp 1-14

Ghazi A (2003) Disaster research and EU-MEDIN. EU-MEDIN workshop, Thessaloniki, Greece, 26-27 May 2003 http://www.eu-medin.org

MDA (2001) Real-Time Emergency Management via Satellite - REMSAT. Richmond, BC, www.remsat.com

Framing Spatial Decision-Making and Disaster Management in Time

Bernard Cornélis

Scientific and Public Involvement in Risk Allocations Laboratory (SPIRAL) & Haute École Charlemagne
Rue Richard Orban, 35 – B-4257 Berloz, Belgium.
Email: cornber@yahoo.com

Abstract

Indubitably, information and communication technologies, amongst which geographic information systems, can help in the management of disasters. Yet, there are just one element which has to interact with the other ingredients. Hence, their presence should be well-thought in order to avoid hindering the return to a "normal" situation, or worse, enhancing the effects of the disasters. By taking a spatial information science perspective, this contribution broadens the debate from technical issues to conceptual ones. It first identifies decision-making as being a major topic of interest for both disaster management and geographic information. After defining the concepts of 'spatial decision', 'disaster', 'risk', 'crisis' and the purpose of 'plans', this paper puts forward a framework for considering these concepts and the decision-making processes in time. By combining, in this perspective, the different concepts, a typology of disaster situations can be established and hence corresponding technological solutions or needs can be pointed out.

1 Introduction

When two fields of activities meet, a certain number of crossroads can be defined, especially if they are multidisciplinary in nature. Usually, in this type of confrontation, the standpoints taken originate from one of the fields. In the case of disaster management and geographic information, one of the meeting-points is decision-making. For the former, it is an intrinsic characteristic of the management activity (Laudon and Laudon 2002). While for the latter, it is a widely claimed purpose of the field's actions (see for example UCGIS 1996).

Decision-making is a field of investigation in itself. Hence, looking at our convergence point through the lens of decision science gives relief to the bridges linking both domains of activity. Furthermore, it is enhanced by the integration of time through the conceptual definition of disaster and of the decision-making processes.

1.1 Realities in Presence

The focus of interest of actors working in *disaster management* is centered on the reduction of the impacts of an event, which is expected, happening, or over. This activity can take different forms such as co-ordination, prevention, mitigation, response, or reaction. According to the role of the actors involved (civil protection, emergency medical aid, public order forces, press and media, politicians, civil servants, the people and private companies), the means of actions and the objectives will differ. Furthermore, their need of information, and subsequently of spatial information and of information technology, is variable just as is the role of space in their activities.

For historical reasons, most people working in the field of *geographic information* focus their interest on data and technology. Their activities both at the research and at the application levels spread over topics ranging from spatial data acquisition, handling, analysis, to data quality assessment, data mining, data infrastructures, and from distributed computing, interoperability, user interfaces design, to real-time processing, inferencing for decision support, enhancement of problem-solving environments... (Karimi and Blais 1996). With the development of the geographic information science approach (Goodchild 1992; Mark 1999), several themes such as public participation and ontological foundation of geographic information are being investigated (Craig et al. 2002; Mark et al. 2000).

In the field of *decision science*, some are working on the making of decision, others are focusing on the decision itself or on the effective outcome of a judgment also referred to as 'choice'. In addition, some are looking at decisions made by individuals, or by or within organizations, or by individuals as member of organizations (Simon et al. 1986). Moreover, the decision environment presents situations which can be stable or certain, risky, and uncertain, either in the outcomes or in their processes. Another way to qualify decisions is by looking at their frequency and at the level of definition of the decision problem. The terms 'structured', 'semi-structured', and 'unstructured' decisions are defined on these criteria (Turban and Aronson 1998). Combined, these classifications of the decision field give a hint to the complexity of dealing with decisions. Not to mention that cognition definitions place decisions among the higher cognitive processes (Costermans 1998; Reed 1999). So to apprehend decision realities, models have been established. The 'decisional fountain' model in its temporal dimension is the lens used here to give depth to the link between disaster management and geographic information.

1.2 Some Definitions

In contrast to the abundant literature on decisions, decision support systems and spatial decision support systems, the concept of '*spatial decision*', although widely used, is barely or rarely defined (Cornélis 1998; 2002a; 2002b; van Herwijnen and Rietveld 1999). For the purpose of this contribution, it is simplistically considered as being a decision using spatial information.

'*Disasters*' are defined as being occurrences having ruinous results, or causing widespread destruction and distress. Some of them called cataclysms or catastrophes are sudden, while others such as droughts are long-standing events. Some are spatially limited such as bombing while others are widespread like epidemic (- 1996; Dauphiné 2001; UNEP IE 1998). It should be pointed out that it is only after the failure of the systems' protection capacity that the occurrence of an event is qualified of 'disaster'. Before that, potential threats or events fall into what is called 'risk'.

Some authors (such as Stirling 1999) reserve the use of the term 'risk' to designate potential events for which the outcomes are well defined and for which there exist a firm basis for probabilities. In a broader sense – used in this paper –, '*risk*' is defined as being the potentiality of an event characterized by a likelihood within a certain timeframe together with its magnitude or consequences for the systems and their objects. In a risk analysis perspective, the last characteristic is often expressed as being the vulnerability of the system to a certain event (Boroush et al. 1998; Desroches et al. 2003; Haimes 2004). Both type of characteristics can be spatially defined as can attest the literature in the geographic information field (Agumya and Hunter 1999; Chen et al. 2001; Church and Cova 2000; Cornélis and Billen 2001; Cova 1999; Cova and Church 1997; Dowd 2003; Frank et al. 2000; Gueremy 1987; Hewitt III 1993; Husdal 2001; Lackey 1998; Lawson and Williams 2000; Manche 1997; Myaux et al. 1997; Pidd et al. 1996; Pidd et al. 1997; Prathumchai et al. 2001; Rampini et al. 1995; Rejeski 1993; Rubio and Bochet 1998; Steinberg 1992; Thumerer et al. 2000).

Disasters, whatever their origin (natural, technological or civil), often create crisis situations (Barthélemy and Courrèges 2004) and hence require crisis management or, in other words, to decide special measures to solve the problems caused by the crisis. Whatever the system it applies to (individuals, organizations, local communities, society, environment,…), a '*crisis*' is generated by an event with impact(s) threatening prior values or the realization of some objectives. It requires extra resources, usually not in the hand of the victim(s), to get back to a "normal" situation. Furthermore, for the different actors involved, the incertitude level is high due to the multiplicity of perspectives in the response, to the time constraints linked both to the evolution of the situation and to the time necessary to implement the responses (Zwetkoff 2000).

To be prepared for the realization of risks, emergency plans are elaborated. The purpose of these *plans* is to reduce the incertitudes (through the development of detection and prevision activities), the inconsistencies (by promoting the efficiency through optimal and continuous adaptation of the means to the aims) and the diversity (by making compatible the conflicting rationalities of the actors). Whether the plans end up in legal texts or not, they can be evaluated according to the objectives, the means used, the outputs and the outcomes for the different actors involved or not. This evaluation requires several simulation techniques. Without evaluation and sometimes even with, the plans are just symbolic responses legitimizing the governing body (Clarke 1999). It should also be pointed out that an outdated emergency plan can be more harmful then having no plan at all (Charbonneau 1992).

2 Time Framing

Just like referencing phenomenas in space favors a better understanding, their referencing in time also reveals part of the picture. Time is often represented in linear or cyclical ways. These representations allow the visualization of the temporal relations events or their different phases have with one another (simultaneity, succession, recurrence, frequency, length, time shifts,…). Spatio-temporal representations and analysis are still being explored nowadays (Bailly and Beguin 2001; Gutiérrez Puebla 2001; Péguy 2001; Pereira 2002; Piveteau 1995). The concepts defined earlier are refined here with their temporal dimension.

2.1 Disaster, Risk and Crisis

When presenting disaster and risk graphically in time, Dauphiné (2001) presents them as being respectively ‘punctual’ and ‘permanent’ in time. This representation has been enhanced (Figure 1) by introducing a variability in the representation of risk (mild grey below the time line), by giving a temporal dimension to disaster, by introducing crisis in the picture and by scaling the magnitude of the different elements (Y-axis). As can be seen in this illustration, a crisis is not necessarily associated with a disaster nor is the opposite. A disaster can occur even when the risk is low. Note that the biggest crisis are not necessarily associated with a major disaster. The media transmitted information, the perception of the people, the political agendas can initiate crisis.

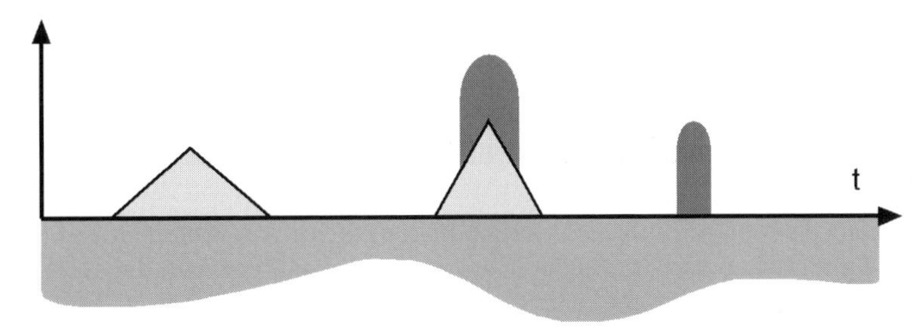

Fig. 1. Temporal representation of disaster (light grey triangles), risk (mild grey) and crisis (dark grey area) presenting three situation types

The introduction of the temporal dimension to characterize disaster induces that a triggering event set off the disaster, hence defining an initial moment (d_i). The disaster can also be defined by its highest point (d_h) and by its ending one (d_e). The evolution patterns and the time lags between these particular temporal points (d_{i-h}, d_{i-e}, d_{h-e}) further enhance the characterization of disaster in time. Within the disaster period, there are a couple of specific moments which should be included to get a complete view. The moment when the premises are becoming observable (d_o) should be differentiated from the moment the disaster is becoming obvious and visible to the people (d_v). d_o and d_v are usually almost simultaneous, especially for small d_{i-e}, while for long d_{i-e} the observations might be questioned and not recognized as being the premises of a major disaster. The elapsed time between d_i and d_o (d_{i-o}) corresponds to the latent period of the disaster (Figure 2). Mitigation, response, and reaction are only possible after d_o and often only happen after d_h.

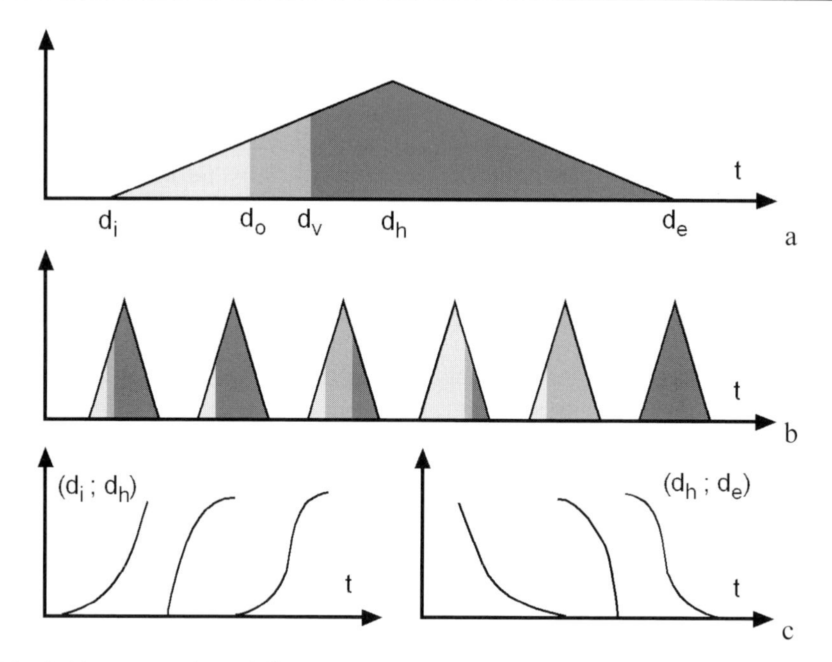

Fig. 2. Representation of disaster on a time line: **a**) the different moments of a disaster (the light grey correspond to the latent period; the mild grey to the awareness raising; and the dark grey to the reaction period); **b**) variations in the succession of d_o and d_v; **c**) examples of variations in the evolution patterns of (d_i; d_h) on the left and of (d_h; d_e) on the right

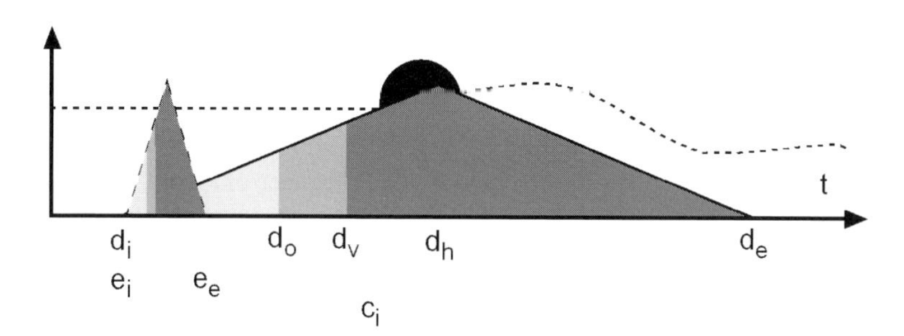

Fig. 3. Example of a time representation of a disaster with its triggering event and a crisis the disaster generated. The small triangle represents the triggering event, the big one, the disaster and its internal temporality, the black sun represents the crisis and the doted line the crisis threshold level

The temporal representation of figure 1 and 2 can be further enhanced by focusing on the triggering event and by defining its initial moment (te_i), highest point (te_h), ending point (te_e), and its observable (te_o) and visible

(te_v) moments. Here also the temporal and evolution patterns should be considered (as in figure 2). The temporal relation between disaster and triggering event as well as the causal relation between one type of disaster/triggering event and another should be considered. The crisis too can be defined in time. The initial moment of a crisis (c_i) corresponds either to the point in time when a threshold level of magnitude has been crossed or when the disaster reaches a cumulative effect greater than the adaptation capacity of the system. The threshold level of magnitude can be plotted in time, just like risk it has a fluctuating profile (Figure 3).

2.2 Decision-Making Processes

The *decisional fountain* model originated from a global reflection on the different facets of decisions. It is a generic multi-dimensional model which can be used for screening decision-making processes (Cornélis and Brunet 2002; Cornélis and Viau 2000). Implicitly, this conceptual model integrates time through its five successive phases (Figure 4): documentation (DO), decisional analysis (DA), decision-taking (DT), decision implementation (DI) and decision evaluation (DE). Obviously, the DI and DE phases can not be performed if the DT has not occurred, just like a decision can not be taken without some kind of DO and DA. The analogy with the fountain comes from the DI temporal characteristic of decisions: some effects (droplets) can be immediate, while others can impact a lot later.

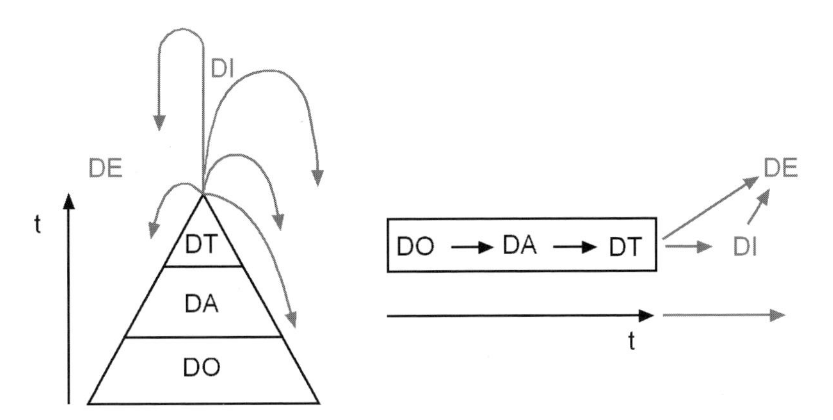

Fig. 4. The decisional fountain model and its temporal dimension. Left its 'traditional' view with time arrows, right its transposition in a time-line view

When taking a closer look at the time-line view, the temporal pattern of the decision-making process shows potential decisional situations (Figure 5). These situations can be classified in two families: the strict sequence (DO-DA-DT-DI-DE) and the weak sequence (DO-DA-DT)-DI-DE in which the DO-DA-DT sequence is a strict one. In emergency situations, time allocated is reduced and is spent accordingly between the phases.

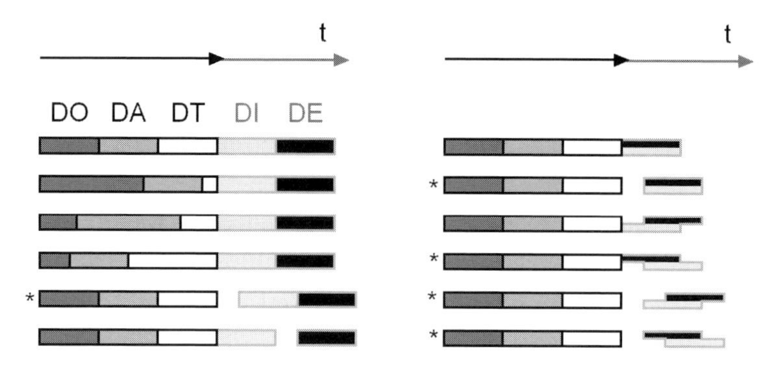

Fig. 5. Examples of time-line representations of the decisional fountain phases for strict sequences (left) and for weak sequences (right). * indicates decision-making processes which can correspond to the emergency planning activity. The blank space between (DO-DA-DT) and DI-DE stands for the lapse of time before the decision is implemented

Each phase of the decisional fountain model can be integrated in time in order to grasp the evolution of each of its constituting elements. For example, by zooming on the decision space in the DT phase, the evolution of the preferred alternatives during the making of decision can be plotted by analogy to Lund time-geography approach. Hence, the "trajectory" of a decision-maker's preference for an alternative or for a set of alternatives can be plotted. The final decision is determined by the 'position' within the decision space of the preferences at the time the decision is to be taken (Figure 6). Although this approach should be explored in more details, a few remarks can be stated. First, the decision space is not fixed as it is for the geographical space in Hägerstrand approach (Chardonnel 1999). Then, in a decision process analysis, what is important is to identify the constraints which might be in the path of reaching certain decisions. Furthermore, the connection with the other elements of decisional processes should ideally be made clear. Nevertheless, the premise that individuals have goals is shared by the analysis of decision-making processes. So, the follow up of a decision-maker's preferences and goals can be sketched on such diagrams.

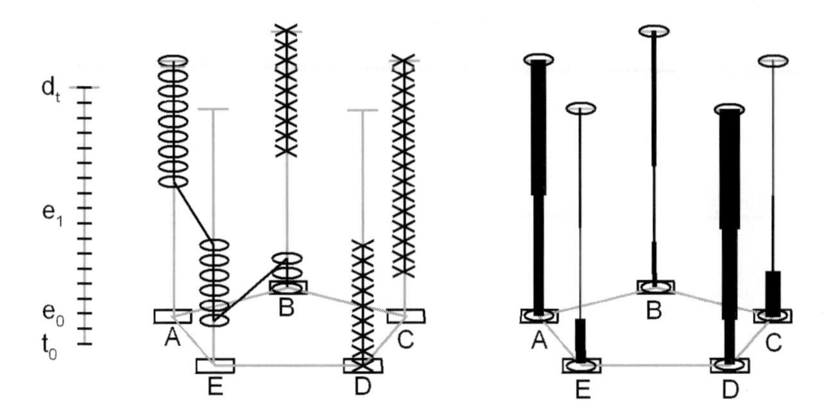

Fig. 6. Representation of discrete decisional spaces integrated in time. In the left scale representing time, t_0 stands for the moment the decisional problem is addressed; e_0 and e_1 for events influencing the preferences; and d_t the time of taking the final decision. In the left diagram, ellipses represent the selection and crosses the elimination of the different alternatives. The grey parts of the alternative's time lines represent alternatives for which the decision-maker is 'neutral' or is not expressing himself. On the right diagram, the preferences and their evolutions are sketched for a situation different from the left setting. These ordered preferences could express selection or elimination preferences

3 Implications for Disaster Management and Geographic Information Solutions

Just like the spatial dimension is important to better understand disaster and its triggering events, to better prepare disaster responses, the temporal dimension and the different decision processes are part of the full picture. By combining, for example, the temporal set of decisional fountains and the classification of disaster types in their various dimensions, different scenarios can be identified for the operational, tactical or strategic levels. These scenarios point to information technology requirements or developments, spatial and non-spatial data needs, model development, and even to the limits of scientific knowledge and beyond. This prospect opens up to other considerations related to philosophical, political or societal questions (see for example Scanlon 1988). Which risks are our societies ready to accept? Which spatial decisions should be considered by the disaster management plans (proportional, egalitarian, radical or fluctuating spatial repartition)? Which events should be monitored? These of course go far beyond the scope of this short introductory contribution, but if we, as citizen of the world, omit to consider them, the true power of geographic information science will turn out to be nothing more than a device to hide the responsibility and to legitimize decisions taken by a few of us.

References

- (1996) Disaster dictionary. http://www.disasterrelief.org/Library/Dictionary/

Agumya A, Hunter GJ (1999) Translating uncertainty in geographical data into risk in decisions. In: Shi W, Goodchild MF, Fisher PF (eds) Proceedings of the International Symposium on Spatial Data Quality '99, The Hong-Kong Polytechnic University, Hong-Kong, pp 574-584

Bailly A, Beguin H (2001) Introduction à la géographie humaine. Armand Colin, Paris

Barthélemy B, Courrèges P (2004) Gestion des risques: Méthode d'optimisation globale. Éditions d'Organisation, Paris

Boroush M, Garant R, Davies T (1998) Understanding risk analysis. American Chemical Society, Washington, DC

Charbonneau S (1992) La gestion de l'impossible: La protection contre les risques techniques majeurs. Economica, Paris

Chardonnel S (1999) Du temps, de l'espace et des populations... Pour une géographie intégrant le temps. In: Emplois du temps et de l'espace: Pratiques des populations d'une station touristique de montagne, Université Joseph Fourier - Grenoble 1, Grenoble, pp 35-73

Chen K, Blong R, Jacobson C (2001) MCE-RISK: integrating multicriteria evaluation and GIS for risk decision-making in natural hazards. Environmental Modelling & Software 16:387-397

Church RL, Cova TJ (2000) Mapping evacuation risk on transportation networks using a spatial optimization model. Transportation Research Part C 8:321-336

Clarke L (1999) Mission improbable: Using fantasy documents to tame disaster. The University of Chicago Press, Chicago

Cornélis B (1998) Managing decision in spatial decision support systems. Proceedings of the 1st AGILE conference

Cornélis B (2002a) Defining the concept of 'spatial decision'. Working Paper ULg

Cornélis B (2002b) A spatial perspective on decision-making methods and processes. In: Mateu J, Montes F (ed) Spatial Statistics through Applications, vol 13. WIT Press, Southampton, pp 1-19

Cornélis B, Billen R (2001) La cartographie des risques et les risques de la cartographie. In: Hupet P (ed) Risque et systèmes complexes - Les enjeux de la communication, vol 2. PIE-Peter Lang, Bruxelles, pp 207-222

Cornélis B, Brunet S (2002) A policy-maker point of view on uncertainties in spatial decisions. In: Shi W, Fisher PF, Goodchild MF (eds) Spatial data quality, Taylor & Francis, London, pp 168-185

Cornélis B, Viau AA (2000) Decision processes with regard to drought monitoring and mitigation. In: Vogt JV, Somma F (eds) Drought and drought mitigation in Europe, vol 14. Kluwer Academic Publishers, Dordrecht, pp 279-290

Costermans J (1998) Les activités cognitives - Raisonnement, décision et résolution de problèmes. De Boeck Université, Bruxelles

Cova TJ (1999) GIS in emergency management. In: Longley PA, Goodchild MF, Maguire DJ, Rhind DW (eds) Geographical information systems, vol 2 - Management issues and applications. John Wiley & Sons, Inc., New York, pp 845-858

Cova TJ, Church RL (1997) Modeling community evacuation vulnerability using GIS. International Journal of Geographical Information Science 8:763-784

Craig WJ, Harris TM, Weiner D (ed) (2002) Community participation and geographic information systems. Taylor & Francis, London

Dauphiné A (2001) Risques et catastrophes: Observer-spatialiser-comprendre-gérer. Armand Colin, Paris

Desroches A, Leroy A, Vallée F (2003) La gestion des risques: principes et pratiques. Hermes Science, Paris

Dowd PA (2003) The assessment and analysis of financial, technical and environmental risk in mineral resource exploitation. In: Fabbri AG, Gaál G, McCammon RB (eds) Deposit and geoenvironmental models for resource exploitation and environmental security, vol 80. Kluwer Academic Publishers, Dordrecht, pp 187-211

Frank WC, Thill J-C, Batta R (2000) Spatial decision support system for hazardous material truck routing. Transportation Research Part C 8:337-359

Goodchild MF (1992) Geographical information science. International Journal of Geographical Information Systems 6(1):31-45

Gueremy P (1987) Principes de cartographie des risques inhérents à la dynamique des versants. Travaux de l'Institut de Géographie de Reims 69:5-41

Gutiérrez Puebla J (2001) Escalas espaciales, escalas temporales. Estudios Geográficos LXII:89-104

Haimes YY (2004) Risk modeling, assessment, and management. John Wiley & Sons, Inc., Hoboken, New Jersey

Hewitt III MJ (1993) Risk and hazard modeling. In: Goodchild MF, Parks BO, Steyaert LT (ed) Environmental modeling with GIS, Oxford University Press, Oxford, p 317

Husdal J (2001) Can it really be that dangerous? Issues in visualisation of risk and vulnerability. www.husdal.com/mscgis/

Karimi HA, Blais JARR (1996) Current and future directions in GISs. Computers, Environment and Urban Systems 20(2):85-97

Lackey RT (1998) Fisheries management: integrating societal preference, decision analysis, and ecological risk assessment. Environmental Science & Policy 1(4):329-335

Laudon KC, Laudon JP (2002) Management information systems: Managing the digital firm. Prentice-Hall, Inc., Upper Saddle River, NJ

Lawson AB, Williams FLR (2000) Spatial competing risk models in disease mapping. Statistics in Medicine 19:2451-2467

Manche Y (1997) Propositions pour la prise en compte de la vulnérabilité dans la cartographie des risques naturels prévisibles. Revue de Géographie Alpine 85(2):49-62

Mark DM (1999) Geographic information science: Critical issues in an emerging cross-disciplinary research domain. Workshop report

Mark DM, Egenhofer MJ, Hirtle SC, Smith B (2000) Ontological foundations for geographic information science. http://www.ucgis.org/priorities/research/research_white/2000%20Papers/emerging/ontology_new.pdf

Myaux J, Ali M, Felsenstein A, Chakraborty J, de Francisco A (1997) Spatial distribution of watery diarrhoea in children: Identification of "risk areas" in a rural community in Bangladesh. Health & Place 3(3):181-186

Péguy C-P (2001) Espace, temps, complexité: vers une métagéographie. Belin, Paris

Pereira GM (2002) A typology of spatial and temporal relations. Geographical Analysis 34:21-33

Pidd M, de Silva FN, Eglese RW (1996) A simulation model for emergency evacuation. European Journal of Operational Research 90(3):413-419

Pidd M, Eglese RW, de Silva FN (1997) CEMPS: A prototype spatial decision support system to aid in planning emergency evacuations. Transactions in GIS 1(4):321-334

Piveteau J-L (1995) Temps du territoire. Éditions Zoé, Carouge-Genève

Prathumchai K, Honda K, Nualchawee K (2001) Drought risk evaluation using remote sensing and GIS: a case study in Lop Buri Province. 22nd Asian Conference on Remote Sensing

Rampini A, Binaghi E, Carrara P, Antoninetti M, Tomasoni R, Tryfonopoulos D, Gonzalez FA, Mazzetti A (1995) FIREMEN: a knowledge-based decision support system for fire risk evaluation in Mediterranean environment. In: Benciolini GB, Roli F, Wilkinson GG (ed) Il telerilevamento ed i sistemi informativi territoriali nella gestione delle risorse ambientali, vol EUR 16330 IT. Office for Official Publications of the European Communities, Luxembourg, pp 79-87

Reed SK (1999) Cognition - Théories et applications. De Boeck Université, Bruxelles

Rejeski D (1993) GIS and risk: a three-culture problem. In: Goodchild MF, Parks BO, Steyaert LT (eds) Environmental modeling with GIS, Oxford University Press, Oxford, pp 318-331

Rubio JL, Bochet E (1998) Desertification indicators as diagnosis criteria for desertification risk assessment in Europe. Journal of Arid Environments 39:113-120

Scanlon J (1988) Winners and losers: some thoughts about the political economy of disaster. International Journal of Mass Emergencies and Disasters 6(1):47-63

Simon HA, Dantzig GB, Hogarth R, Plott CR, Raiffa H, Schelling TC, Shepsle KA, Thaler R, Tversky A, Winter S (1986) Decision making and problem solving. In: Research briefings 1986: Report of the research briefing panel on decision making and problem solving, National Academy Press, Washington DC, pp 1-16

Steinberg J (1992) La cartographie synthétique des risques naturels et technologiques en milieu urbain. Bullletin de l'Association des Géographes Français 5:456-464

Stirling A (1999) On science and precaution in the management of technological risk. European Commission, Brussels and Luxembourg

Thumerer T, Jones AP, Brown D (2000) A GIS based coastal management system for climate change associated flood risk assessment on the East coast of England. International Journal of Geographic Information Systems 14(3):265-281

Turban E, Aronson JE (1998) Decision support systems and intelligent systems. Prentice-Hall, Inc., London

UCGIS (1996) Research priorities for geographic information science. Cartography and Geographic Information Systems 23(3):115-127

UNEP IE (1998) Hazard identification and evaluation in a local community. United Nations Publication, Paris

van Herwijnen M, Rietveld P (1999) Spatial dimensions in multicriteria analysis. In: Thill J-C (ed) Spatial multicriteria decision making and analysis, Ashgate Publishing Ltd, Aldershot, pp 77-99

Zwetkoff C (2000) Vers un nouveau mode de gestion de la sécurité alimentaire ? In: Dioxine: de la crise à la réalité, Éditions de l'Université de Liège, Liège, pp 85-107

Disaster Monitoring Based on Portable Terminal for Real-Time RADARSAT-1 Data Acquisition

Olga Gershenzon

R&D Center ScanEx, Russian Federation 22/5 L'va Tolstogo, 119021, Moscow, Russia.
Email: info@scanex.ru

Abstract

R&D Center ScanEx (www.scanex.ru) is leading manufacturer of personal ground stations and terminals, i.e. antenna systems for receiving, storing and processing Earth observation images. Their unique features compared to the traditional systems for gathering information about the Earth from space are as follows: affordable price, compactness, a technology on the basis of a standard PC, ease of the operation, a unified technology of data storage, processing, and image thematic analysis. All of these leads to cheaper data, simpler data acquisition technology, and quicker access time for the widest possible range of users. A personal ground station is the unique means to enable users to receive images of the Earth from space directly at their PCs. UniScan™ ground station (in stationary and mobile modifications) by R&D Center ScanEx is flexible solution for receiving information from wide range of Earth observation satellites: Terra/Aqua, IRS-1C,1D,P6, RADARSAT-1, EROS A1 and others. R&D Center ScanEx offers the complete chain for RS data acquisition, storage and processing on the base of its own hardware and software solutions. Affordable price of ground stations makes it possible to install such equipment not only for national remote sensing Centers, but at research, education organizations. The main possibility that gives UniScan™ technology is access to RS data in real-time mode – data is ready to analysis after 15-30 minutes after data reception. This is too important for such RS data applications as disaster management (flooding, fires monitoring).

R&D Center ScanEx in cooperation with Antrix Corp. offers integrated solution for acquisition information from recently launched IRS-P6 satellite by affordable price which includes hardware, software and 1,000 minutes time for data downlink. IRS-P6 data with resolution of 5.8, 23 and 56 m and short revisit period (5, 24 and 5 days correspondingly) make this data valuable for very different RS data applications.

1 Introduction

We consider remote sensing terminals (RS terminals) – compact, affordable, simple for operation systems for acquisition images of the Earth from space- as significant stage in realization of the idea of decentralization, speeding-up and simplification of access to RS data.

When we need one image for definite date for limited area of interest we can use service of centralized archives even if time of data delivery and data cost are not convenient for us. Centralized archives we understand as several global archives where RS data are received, processed and stored. Usually it takes much time for delivery ordered image to end user, and long chain of RS data distributors, companies offering service for value-added products very complicates access to RS data and increases its cost. But in case we want to integrate images from space into on-line decision-making system cost of images and time of its delivery are crucial parameters.

Really RS terminals and personal ground stations (PGS) for RS data acquisition and processing (by analogy with personal computers) enable to use images from space just after 15-30 minutes after data reception on a place of decision-making. Such approach is in principle important both on the level of governmental bodies, regional administrations and large industrial corporations.

In particular this caused wide implementation of RS terminals on the base of UniScan™ technology in frames of Ministry of Natural Resources of Russian Federation (Moscow, Gelendzhik, Irkutsk, Ekaterinburg, Yakutsk, Yuzhno-Sakhalinsk - see figure 1), in the interests of regional administrations (Belgorod, Chita, Salekhard – see figure 1) and large industrial companies (Atyrau, Priozersk – “KazGeoCosmos” Company, figure 1). In spite of difference in the status and organizational form of above-mentioned users the list of problems that they resolve is varied only in geographic features of regions. This list includes following applied tasks: detecting of wildfires and forecast of their development, water pollution monitoring (thermal pollutions and oil spills), forests monitoring (illegal felling, consequences after wildfires, etc.), control over compliance with the licensing agreements in natural resources development, control over illegal fishing, monitoring of regional infrastructure development, rapid and objective damage assessment after natural disasters and others.

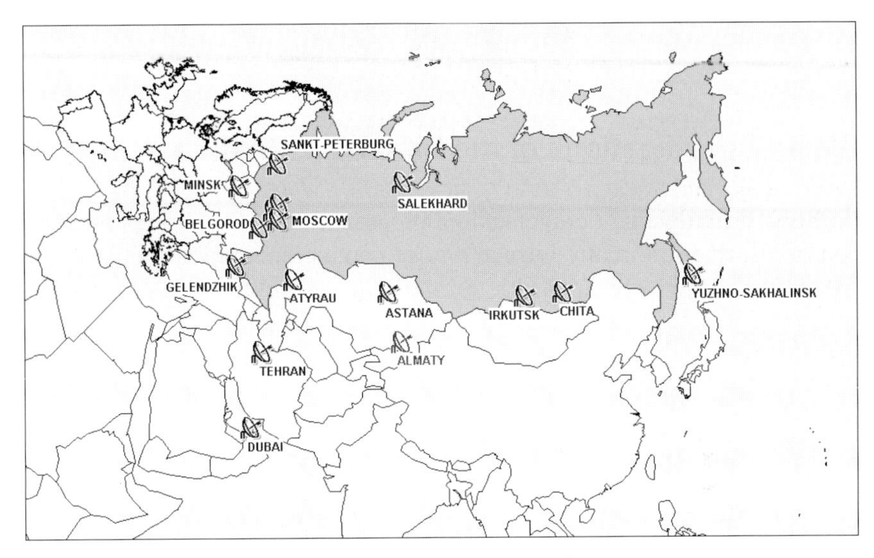

Fig. 1. Operational UniScan™ ground stations network

2 What is UniScan™ Ground Station and RS Terminals?

What do we mean when we say Personal Ground stations and terminals? We mean antenna systems for receiving, storing and processing Earth observation images. Their unique features compared to the traditional systems for gathering information about the Earth from space are as follows: affordable price, compactness, a technology on the basis of a standard PC, ease of the operation, a unified technology of data storage, processing, and image thematic analysis. All of these leads to cheaper data, simpler data acquisition technology, and quicker access time for the widest possible range of users. RS terminal is ground station intended for receiving information especially from one or two satellites for specific applications. A personal ground station is more unified means to enable users to receive images of the Earth from space directly at their PCs from different Earth Observation satellites for solving wide range of tasks. UniScan™ ground station (in stationary and mobile modifications) by R&D Center ScanEx is flexible solution for receiving information from satellites: Terra/Aqua, IRS-1C/D, IRS-P6, RADARSAT-1, EROS A and others.

UniScan™ station is designed for receiving and processing images transmitted from low-orbiting Earth satellites in X-band with data rates up to 120 Mbps (in near future up to 160 Mbps). Such data rate allows to transmit detailed images with the spatial resolution of several meters.

3 Main Specifications of the UniScan™ Station

UniScan™ hardware is universal and programmable. It provides for reception of information in any format, whose parameters are within the following limits:

Parameter	Range
Carrier frequency	8.0 ... 8.4 GHz
Digital data rate	2.5 ... 120 Mbps (QPSK) 1.25 … 60 Mbps (BPSK)
Modulation	BPSK, QPSK, SQPSK

The users would not need any hardware modifications to adapt the station for a new satellite and format, only some additional software will be required provided that:

- format parameters are within the limits indicated above;
- power of the satellite transmission is sufficient to be received by the particular antenna system.

Sich-1M, Monitor-E No 1, IRS-P5, IRS-P6, RADARSAT-2, SPOT 4 and 5, EROS A, Envisat-1 satellites and many others transmit (or will transmit) information within the UniScan™ range of parameters.

UniScan™ is designed in two configurations:

- mesh-reflector antenna, 3.6 m in diameter with a 3-axis rotating support (UniScan™-36);
- solid-reflector antenna, 2.4 m in diameter with a 2-axis "X – Y" rotating support (UniScan™-24)

The UniScan™ consists of:

- the antenna system;
- the control unit;
- computer interface boards;
- an Intel Pentium IV based PC (one or two, see below);
- the software;
- a set of connecting cables.

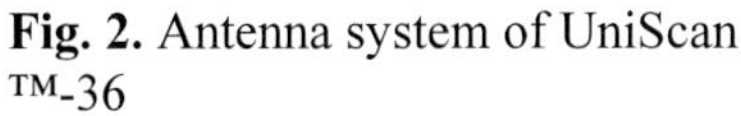

Fig. 2. Antenna system of UniScan ™-36

Fig. 3. Antenna system of UniScan ™-24

The ingest interface boards comprise Viterbi decoders and FPGAs for the frame synchronization and in-line processing (differential decoding, simple formatting, etc.).

PCs for the UniScan™ station are capable of ingesting, logging and temporary storage of the relevant large data streams. If a station is intended for parallel data reception via two radio downlinks (as is the case with the IRS-1C/1D satellites), it is supplied with two PCs connected into a local network and two PC interfaces.

R&D Center ScanEx offers the complete chain for RS data acquisition, storage and processing on the base of its own hardware and software solutions. Affordable price of ground stations makes it possible to install such equipment not only for national remote sensing Centers, but at research, education organizations.

The system is equipped with specialised software for:

- the control of acquisition and data recording to a PC hard disc (ScanReceiver®);
- the preliminary processing (data formatting, geolocation and calibration). Definite structure of this software depends on station configuration, i.e. list of data to be acquired;
- the visualisation and analysis of images (ScanMagic®, one license);
- cataloguing images (ScanEx Catalogue Manager®);
- additional thematic image processing (ScanEx Image Processor®, one license).

All software is designed for MS Windows.

Data recording and storage is implemented in formats developed by the R&D Center ScanEx. They preserve all of the metadata received from a satellite and the metadata added in the course of reception.

These formats are fully supported by the delivered software. It is also possible to obtain low-level products in standard formats such as

- MODIS: MOD 01, MOD 02, MOD 03 (Terra); MYD 01, MYD 02, MYD 03 (Aqua);
- IRS (LISS-3, PAN): FAST;
- RADARSAT: RADARSAT Level 0, Level1 in CEOS SAR formats.

R&D Center ScanEx in cooperation with Antrix Corp. offers integrated solution for acquisition information from recently launched IRS-P6 satellite by affordable price which includes hardware, software and 1,000 minutes time for data downlink. IRS-P6 data with resolution of 5.8, 23 and 56 m and short revisit period (5, 24 and 5 days correspondingly) make this data valuable for very different RS data applications.

4 Disaster Monitoring Using RADARSAT-1 Data

In June 2004 the "RADARSAT-1 Reception License Agreement" between Canadian Space Agency, RADARSAT Int. and R&D Center ScanEx was signed.

Radar data can be effectively used for different tasks in disaster management such as flooding monitoring, oil spills monitoring on shelf, sea ice monitoring. In addition, capabilities for Terra/Aqua MODIS data acquisition can be installed on RADARSAT-1 terminal. This gives a possibility of regular monitoring wildfires in near-real time mode for effective land use. Also MODIS data can be used in addition to radar data, for example, for flooding and sea ice monitoring.

As known the shelf of northern seas becomes the place of concentration of huge oil resources. Development of oil reservoirs on the shelf can impact on water and seaside environment. That is why regular ecological monitoring of the shelf becomes too important at present time.

Regular monitoring of vast squares of the seas is difficult problem. One of the efficient ways to resolve this issue is to use images of the Earth from space. The most effective information source for these purposes can be SAR (Synthetic Aperture Radar) data from RADARSAT-1 satellite.

At present the RADARSAT-1 program (www.rsi.ca) is the most reliable radar satellite mission in the world. It provides end users with the data from 8 and up to 100 m spatial resolution within a swath from 50 and up to 500 km wide. A unique feature of the SAR imaging as compared with optical instruments is its independence from a natural illumination (i.e. time of the day and season) and cloudiness, which is absolutely transparent for this type of sensing. In case of emergency the program offers a unique opportunity to place a request 29 hours prior to the actual time of acquisition with near 100% guarantee. Data can be used for analyzing after 1-1.5 hours after its acquisition.

To receive satellite images in real-time mode and get them ready for processing and analysis is the most important task for disaster management. This approach reduces the risk of potential environmental damage to the lands, forests, offshore and as well as to fragile coastal ecosystems. Portable RADARSAT-1 terminal by R&D Center ScanEx is easy for installation (both in stationary and mobile variants) and to operate, and equipped with all necessary software for data reception and near real-time processing. This is turnkey up-to-date affordable solution for disaster management. Four such terminals will be put into operation within Russia and Kazakhstan during year 2004 (see figure 1). The circle of visibility of these stations is about 2,000 km from the place of installation that gives a possibility to observe vast regions around a place of installation of a RADARSAT-1 terminal.

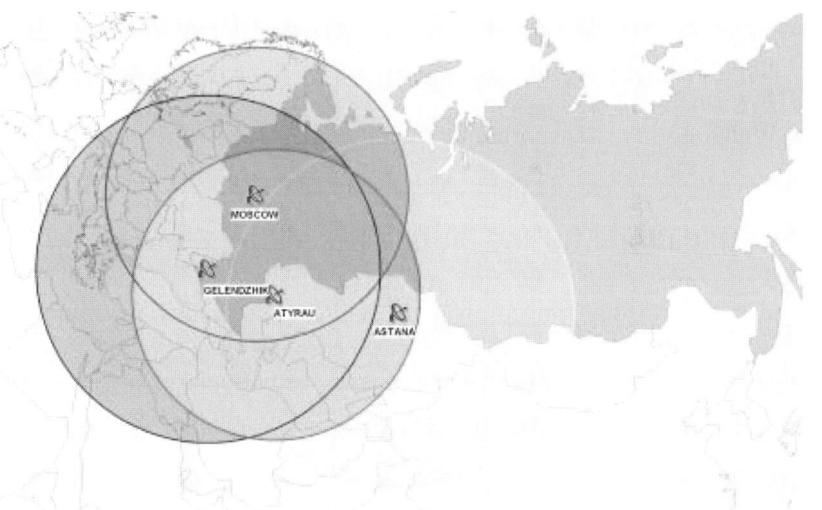

Figure 4. Network of RADARSAT-1 ground stations based on RADARSAT-1 terminal by R&D Center ScanEx

5 Mobile Receiving Station Based on the UniScan™ Technology

The mobile station is implemented in two configurations: on the base of UniScan™-24 and UniScan™-36 stations. In technical terms, the system is small and relatively light with the weight about 400 kg or 1,000 kg and a 2.4 m or 3.6 m antenna dish respectively. The system can be set up in a new location in 2 to 3 hours. Once the imagery is received, the advanced software for data reception, storage and processing enables the operator to analyse the situation and give relevant recommendations within a period of time as short as 30 minutes. For best results, the received images should be combined with background information such as local cartographic/GIS layers or geolocated archive imagery from Resurs-O1, Meteor-3M, Landsat, IRS or Terra satellites.

Unique features of the R&D Center ScanEx mobile receiving station:

- a simplified small-size antenna which enables reliable reception at satellite elevations starting from 5-10° with the radius of the reception zone up to 1,500 – 3,000 km;
- rapid operational deployment (2…3 hours) without the use of any additional equipment;
- low cost compared to similar stations by other manufacturers;
- high mobility (transported by a lightweight truck with a trailer).

6 Applications for the Mobile Receiving Station

The mobile system, in comparison with the stationary receiving centers may be the most efficient solution for the following:

- real-time data reception in emergency situations to enhance the efficiency of rescue operations and damage assessment;
- real-time data acquisition for the regions, not covered by the stationary receiving centers, to create imagery archives;
- deployment prior to the completion of a full-scale Earth observation center;
- creation of a national Earth observation infrastructure in developing countries;
- demonstration of Earth observation capabilities at international exhibitions and aerospace shows;
- specific Earth observation tasks in remote regions for a limited period of time (e.g. dealing with accidents and natural disasters, prospecting for oil, gas and other mineral resources, pipeline construction etc.).

Fig. 4. Indoor equipment of mobile groundstation

Fig. 5. Mini laboratory on the base of mini trailer

The basic complete set of the mobile receiving station consists of the same as UniScan™, but besides this includes a dismountable platform for antenna system installation and auxiliary equipment for a truck-mounted UniScan™-36 system.

By customer request, the following may be supplied in addition to the basic set: a light truck, a trailer equipped as a mobile laboratory, an autonomous petrol-based generator (220 V output, enables consumption of up to 2 kW). The particular components of the set are to be detailed in the mobile station delivery contract.

7 Conclusion

The main possibility that gives UniScan™ technology and terminals on its base is access to RS data in real-time mode – data is ready to analysis after 15-30 minutes after data reception. This is too important for RS data applications for any dynamic phenomena such as wildfires and their sites detection, oil slicks detection and control of their motion; crop quality on day by day (or at least a week by week) basis; control of water pollution on regular basis and so on.

User Requirements for a Mobile Disaster Documentation System

Sven H. Leitinger[1], Christian Scheidl[2], Stefan Kollarits[2] and Johannes Hübl[3]

[1] Salzburg Research Forschungsgesellschaft mbH, Jakob-Haringer Str. 5/III, 5020 Salzburg, Austria.
Email: sven.leitinger@salzburgresearch.at

[2] PRISMA solutions EDV-Dienstleistungen GmbH Klostergasse 18, 2340 Mödling, Austria.
Email: {christian.scheidl, stefan.kollarits}@prisma-solutions.at

[3] Institute of Mountain Risk Engineering, University of Natural Resources and Applied Life Sciences, Peter Jordanstraße 82, 1190 Wien, Austria.
Email: johannes.huebl@boku.ac.at

Abstract

According to a successful disaster management the analysis of the disaster is needed. For this purpose the documentation of the disaster is necessary. Up to now many studies deal with different aspects of mobile systems for the disaster management. They are developed for disaster management activities of public safety units, but there are no applications for the documentation process during and after a disaster. In this paper we describe an approach for a mobile disaster documentation system. The main focus lies on the user requirements of the different user groups. The disaster documentation focuses mostly on information about a disaster and possible activities undertaken by disaster and public safety experts. The documentation structure should use a common language based on a standard terminology. A standardized documentation structure would help to harmonize the information basis, accessibility, and better integration in spatial decision-making processes.

1 Introduction

The alpine regions have a high potential for mountain hazards and risks, such as avalanches, debris flows or mountain torrents. In order to successfully react in the case of a disaster, we need a careful hazard and risk analysis. The documentation of disasters in the response and recovery phase represents an important input for the risk analysis (Cutter, 2003, Leitinger, 2004). Up to now the documentation of the disasters has been done in the form of sheets, analog maps and images at the event site. Hübl et al. (2002) describe this process and the scientific and technical background of the documentation in the DOMODIS-Handbook.

The current state of geographic information systems (GIS) can provide decision makers with the information they need to manage a disaster. In the last years researchers and the industry start to develop mobile GIS for field workers of public safety organizations. These mobile applications can be also used in the process of documentation. The operators responsible for the documentation of a disaster face with different user requirements for such systems.

In this short paper, we describe an approach for a mobile disaster documentation system. We start with an overview of related work of mobile applications in the field of disaster management. Than we explain the user requirements for the documentation process and at the end we conclude this paper with some directions for our further research.

2 Related Work

Zerger (2003) describes several applications of a GIS used for disaster decision support. These systems focus on technical specifications of a GIS software system, spatial data themes, spatial data capture techniques, modeling of the hazard and its spatial extent and cartographic presentation of results. Only a few studies deal with mobile applications for the disaster management decision support. These applications are designed to support mainly public safety organizations (fire brigades, civil protection, public authorities) during the response phase of the emergency management. Zlatanova and Holweg (2004) discuss a mobile system which uses 3D geodata for different clients in the emergency management. In this approach the requirements to the supporting system with the respect to different user groups (decision makers, field workers, desktop and web clients) are described. Furthermore the study concentrates on the specific data requirements for the diverse clients. Another study in this field was done by Erharuyi and Fairbairn (2003) which focuses on scheduling and information management issues for the field workers and the interoperability among databases and GIS which should enable real-time mobile data processing. According to them, the emergency field personnel should not only be able to navigate geo-information and services provided by the central office, but also gather process, upload and renew information residing in the central database. This application was tested in a study case of oil spill management in the Niger Delta of Nigeria. Horz and Schmidt (2003) describe a mobile map-based information management service for early warning and disaster management activities. In this approach, the mobile end-users can view, manage, annotate and communicate geo-data in the field. The field workers can add additional geo-referenced information to the maps. The system, called MoMoSat, uses secure real-time connections over satellites and WLAN enabling the communication between mobile units and the primary data store secure real-time. Meissner et al. (2002) present the requirements and technology needed for the communication process and the information systems used during the response and recovery phases. Their focus lies on the design and architecture of the network, and on the configuration, scheduling and data management issues.

Above mentioned studies deal with several different topics such as the data and user requirements, system architecture, information management and communication technologies of mobile applications for the disaster management. They focus basically on the needs of public safety organizations. This work can serve as a theoretical basis for our approach where we aim at developing a mobile disaster documentation system. The following section examines the user requirements of the documentation process, which differ from the requirements of public safety organizations.

3 User Requirements for the Documentation Process

In general terms, documentation is any communicable material (such as text, video, audio, etc., or combinations thereof) used to explain some attributes of an object, system or procedure (Wikipedia, 2004). Disaster documentation focuses mostly on information about a disaster and possible activities undertaken by disaster and public safety experts. The main responsibility of the disaster experts is to study the reasons for a disaster, the disaster consequences, and to document the security procedures. These experts are very seldom present at the site where the disaster occurred. On the other hand, the public safety teams are immediately informed about the disaster and actively involved in the response phase. This documentation should have an appropriate structure adequate for public safety teams and it should permit a rapid context formation, and information extraction and retrieval. It should enable them to monitor the development and consequences of a disaster, and to collect the relevant data. A mobile documentation application that enables an efficient documentation process shall fulfill the following requirements:

- different level of education (general knowledge - expert knowledge),
- different level of experience (practioners - theorists),
- objective data interpretation,
- invariability in time and space,
- interpretation instead of objective recordings (matter of socialization in case of catastrophic events),
- different terminologies (dialect, languages, data semantic).

The documentation structure should use a common language based on a standard terminology. Depending on the chronological and thematic level of detail, the documentation structure should consist of two parts; the event documentation and event notification. The lowest level of detail can be recorded with the help of the event notification independently of the educational level of the expert. The level of detail that can be acquired by an expert should depend on the experience and the knowledge of the expert. The structure should be standardized and usable by all experts involved in the disaster management activities. A standardized documentation structure would help to harmonize the information basis, accessibility, and better integration in spatial decision-making processes. The cornerstones of a uniform documentation structure are the spatial level of detail (what do I have to observe?), the chronological level of detail (when do I have to observe?), the applied level of detail (who has to observe?), and the thematic level of detail (what has to be recorded?). In order to take an advantage of such a uniform documentation structure we have to reconsider the current documentation processes. Disaster process phenomena should be recorded based on a detailed guideline as discussed by Kienholz et al. (1995), and be usable to public safety organizations.

In our current project GUSTAV we plan to develop a mobile disaster documentation application which will be extended by an event notification structure. The project is financed by the Austrian Ministry for Transport, Innovation and Technology. Special attention will be devoted to the realization of an effective collection of disaster phenomena by public safety experts. By means of such an event notification combined with a uniform documentation structure advantage of new tools, like mobile GIS applications, can be gained. Co-authors of this paper are currently working on a transalpine event notification guideline within the Interreg IIIB project

DIS-ALP. The cooperating partners on the project are the Institute of Geography of the University of Bern (GIUB) and the Centre for Naturals Hazards and Risk Management (CENAR, BOKU).

4 Conclusion and Further Work

In this short paper we briefly overviewed several studies of mobile applications for the decision support in the disaster management. These applications present different research areas of mobile GIS and serve as the basis for our further development of a wireless disaster documentation system. We concentrated on the user requirements for such application. Our discussion provided in this paper is in an initial phase of the development and needs additional research. In our future work we will consider the usability requirements and execute interviews with possible users of the application. The idea would be to interview the public safety as well as disaster experts and register their needs and suggestions for the functionalities of the application. We will also study trends and future development of a mobile GIS in the field of disaster management, and possible integration of geo-data needed within the documentation process. Within GUSTAV project we aim to design an software architecture for a mobile disaster documentation system.

References

Cutter SL (2003) GI Science, Disasters, and Emergency Management. Transactions in GIS, vol 7 (4), pp 439-445

Erharuyi N, Fairbairn D (2003) Mobile Geographic Information Handling Technologies to support Disaster Management. Geography, vol 88 (4), pp 312 - 318

Horz A, Schmidt D (2003) MOMOSAT - Mobile GIS Collaboration in Early Warning and Disaster Management. In: 4. Forum Katastrophenvorsorge - Extended Abstracts, German Committee for Disaster Reduction, pp 23-26

Hübl J, Kienholz H, Loipersberger A (eds) (2002) DOMODIS - Documentation of Mountain Disasters. Internationale Forschungsgesellschaft Interpraevent, Schriftenreihe 1, Handbuch 1, Klagenfurt, Austria

Kienholz H, Krummenacher B (1995) Symbolbaukasten zur Kartierung der Phänomene. Miteilung des Bundesamtes für Wasser und Geologie Nr. 6, BUWAL, BWG, online located at: http://www.umwelt-schweiz.ch/buwal/shop/files/pdf/phpN7vLeT.pdf

Leitinger SH (2004) Comparision of GIS-based Public Safety Systems for Emergency Management. In: Fendel E, Rumor M (eds) Proceedings of UDMS '04, Chioggia, Italy

Meissner A et al. (2002): Design Challenges for an Integrated Disaster Management Communication and Information System. The first IEEE workshop on Disaster Recovery Networks, New York, June 2002

Wikipedia The Free Encyclopedia. online located at: http://en.wikipedia.org/wiki/Documentation

Zlatanova S, Holweg D (2004) 3D Geo-Information in Emergency Response: A Framework. In: Proceedings of the Fourth International Symposium on Mobile Mapping Technology (MMT'2004), Kunming, China, pp 6

Zerger A, Smith DI (2003) Impediments to using GIS for real-time disaster decision support. Computers, Environment and Urban Systems, vol 27 (2), pp. 123-141

Use of Photogrammetry, Remote Sensing and Spatial Information Technologies in Disaster Management, especially Earthquakes

Orhan Altan

Faculty of Civil Engineering. Department of Geodesy and Photogrammetry, Istanbul Technical University, 36626 Ayazaga-Istanbul/ Turkey.
Email: oaltan@itu.edu.tr

Abstract

With each passing day, catastrophe risk for urban regions of the world is increasing. One of these catastrophes is the earthquake and recent events in Northridge and Kobe were typical examples of what can happen when a major earthquake strikes directly under a densely populated area. Mega cities created by the rapid urbanization and development in unsafe areas led to far greater losses experienced in the past. In order to reduce the property losses after an earthquake a quick repair process is a major task. This process must be based on detailed plans for rebuilding or strengthening procedures of the buildings. The major damage loss is caused by earthquakes.

Geodetic science plays an important role in the earthquake research. By means of long-term measurement, deformations caused by the breakage of the earth crust caused by the moving plates can be examined. Photogrammetry and Information System techniques are new tools in the earthquake research. Terrestrial photogrammetric methods have been used for the first time to document the damages after an earthquake in Friaul, Italy. There are many attempts to use photogrammetry, remote sensing and information sciences in the earthquake damaged areas. Some of them are related with the earthquake prediction, long and short term, some of them is related to the damage recording and assessment. A similar study to this research is the work after the Kobe earthquake.

In all these studies they claim of data collection as well as before or after an earthquake. Earlier earthquakes revealed problems in the processes of documenting and analyzing the building damage that occurred due to earthquake disasters which demanded much effort in terms of time and man power. The main difficulties appeared because analogue damage assessments created a great variety of unstructured information that had to be put in a line to allow further analysis. Apart from that, documentation of damage effects was not detailed and could only be carried out on the spot of a disaster.

1 Introduction

With each passing day, catastrophe risk for urban regions of the world is increasing. One of these catastrophes is the earthquake and recent events in Northridge and Kobe were typical examples of what can happen when a major earthquake strikes directly under a densely populated area. Megacities created by the rapid urbanization and development in unsafe areas led to far greater losses experienced in the past. In order to reduce the property losses after an earthquake a quick repair process is a major task. This process must be based on detailed plans for rebuilding or strengthening procedures of the buildings. The aim of this paper is to present a different approach in the monitoring, documentation and analyze the damages in the buildings after an earthquake. At the end an example of using remotely sensed data in an earthquake is also given.

2 Photogrammetry and GIS in the Earthquake Research

Geodetic science plays an important role in the earthquake research. By means of long-term measurement, deformations caused by the breakage of the earth crust caused by the moving plates can be examined. Photogrammetry and Information System techniques are new tools in the earthquake research. Terrestrial photogrammetric methods have been used for the first time to document the damages after an earthquake in Friaul, Italy (Foramitti, 1980). A similar study to this research is the work after the Kobe earthquake (Kiremidjan and King, 1995). In this study they declare that the information system is an essential tool in the earthquake research.

They evaluate the damages of the building by using their own computer program developed in the 80's for a research project in Zagrep University (Anicic and Radic, 1990). The use of an expert system for the evaluation of earthquake damages based on expert systems using evaluation tables is research currently under investigation (Papnoni, Tazir and Gavarini, 1989). The research work at Karlsruhe University "Strong Earthquake", Germany, sponsored by the German Science Foundation (DFG) use data acquired at a smaller scale.

3 Data Aquisition by Photogrammetry

Photogrammetry is an efficient tool in monitoring of spatial objects with respect to location, form and shape. Its main advantage to other measuring techniques lies in the fact that the measurement is done on the images and indirect measuring possibility opens the users of this method a wide range of application possibilities. So the recorded images contain a great extend of information so that many of the detailed acquisition of deformation can be done afterwards. To establish the deformation of a building from its complex details, three-dimensional coordinates of characteristic points related to the structure of the building must be known. In order to measure them, these points must be projected at least in two images.

With known camera calibration parameters (interior orientation parameters) the unknown 3D-object coordinates (XYZ) can be computed by measuring their image coordinates of the object (in this case of the building) points. Their values can be determined with an adjustment procedure (bundle adjustment). The faulty measurements will be eliminated by this way and a precise measuring capability can be reached. In order to relate the determined XYZ coordinates to an overall coordinate system, control points with known coordinates are used.

Today in addition to so called classical, analog ways of photogrammetric data handling, digital methods are also used. This enables an automation of data processing by means of image analysis and matching techniques. In this context 3D-object reconstruction techniques, classification or image detection and their integration into a deformation analysis procedure using information system technology can also be used.

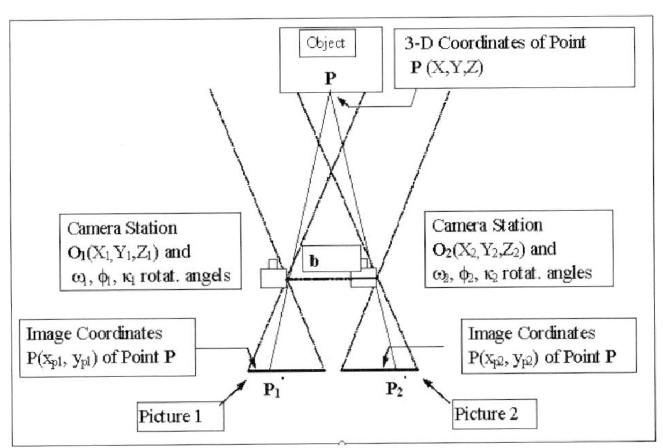

Fig. 1. General case photogrammetric data acquisition

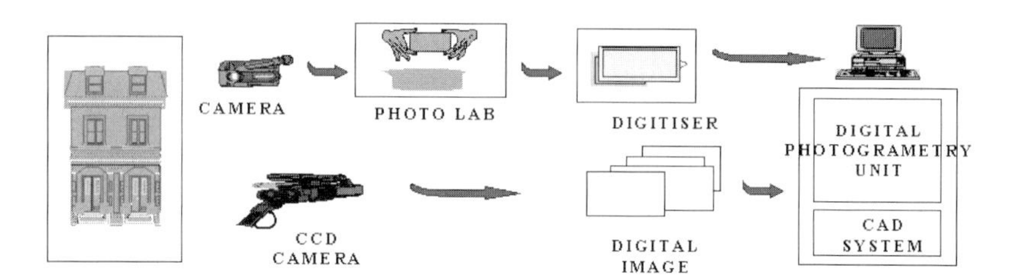

Fig. 2. Data flow in photogrammetric evaluation process

In order to determine the deformation of a building as a whole system, the 3D-shape of building must be reconstructed. This reconstruction procedure must be based on the determined coordinate of the building characteristic points. As the damaged buildings after an earthquake is a potential danger for the investigators and also passengers the first determination must be based on the points on thc facade of the building.

Secondly no prior measurements of these buildings are available in order to relate the deformed values to. In this investigation the following way was chosen. At least two plumb lines hanging down on the facade of the building define a vertical plane. Based on the assumptions that the lowest points of the building can be considered stable, and the facade is build, as in most cases, vertically, all deviations can be related to this vertical plane and to the coordinate system, which can be defined for the specific case (see Külür, 1998).

Fig. 3. Digital photo of the damaged schoolhouse obtained by KODAK DCS 200 and plumb lines used as scale information

This approach allows a quick documentation. Using the digital technology, the photographic processing has disappeared and on line registration can be used. In the research work a digital camera by KODAK (DCS 200) has been used. This allows to load the pictures directly to the laptop computer at site and to begin immediately with the processing of the gained data. The damaged high school building in DINAR was photographed with this camera. The processing and evaluation of the images was done with a photogrammetric software package PICTRAN (Schewe, 1995). The result is the coordinates of the characteristic points defining the movement of the building. These characteristic points are on the 11 different axes of the high-school building. They are also chosen on the heights of the stories of the building (in this case 4). So by means of axes and stories as quasi-heights, a systematic grid is placed on the facade of the building.

From these coordinates displacement values are calculated according to the following formulas;

Relative Height Differences: $H_l(i) = Z(i) - Z(i\text{-}1)$

Absolute Depth Differences: $D_a(i) = Y(i) - Y(1)$

Relative Depth Differences: $D_r(i) = Y(i) - Y(i\text{-}1)$

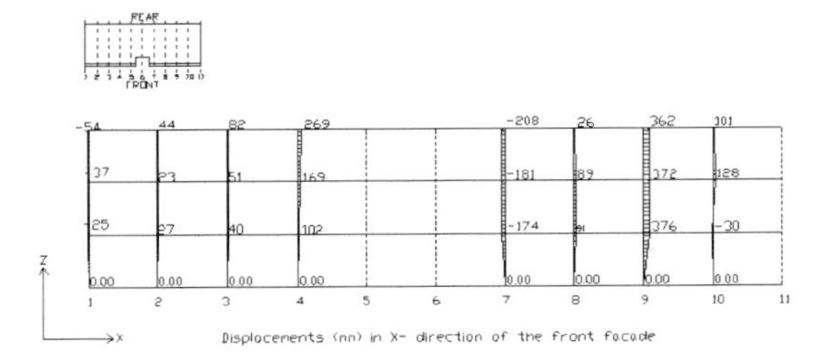

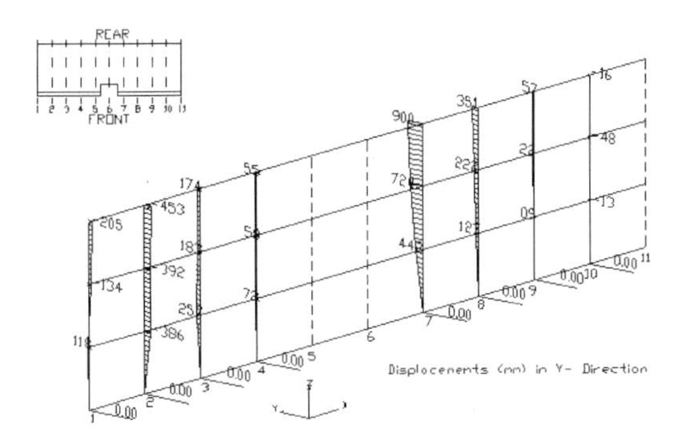

Fig. 4. Deviations from the vertical plane

The relative displacement in one direction of two overlaying stories is calculated from their displacement values as;

$\Delta_i = d_i - d_{i-1}$

where d_i and d_{i-1} are the displacements of a column in i. and i-1. story. If we denote with h_i the height of these two adjacent stories, then the relative displacements of the point in question must be compared with two maximum values calculated as;

$(\Delta_i)_{max} / h_i \leq 0.0035$

$(\Delta_i)_{max} / h_i \leq 0.02 /R$

Here the coefficient R is the load reduction factor of the building with respect to the structural system and natural response coefficient of the building.

4 Information Systems and Query (Analysis)

Geographic Information Systems (GIS) consist of computer hardware, software and geographic data. They are designed to efficiently capture, store, update, manipulate and display all forms of geographically referenced information (Bill and Fritsch, 1991). Spatial data is obtained in digital form, rearranged, analyzed according to several querying parameters and later presented in the form of either alphanumeric or graphical displays. Information systems can compile a great number of descriptive and geometric data, which is stored in a database.

The units in the geographic information systems are objects, which can be described by quantitative and qualitative components. Here the mentioned object should be regarded as a unit that exists in nature and that can be categorized geographically, physically and descriptively. Location and shapes of three-dimensional objects are defined as units with single meaning determined by point coordinates.

As mentioned above, a geographic information system differs from CAD or AM/AF systems in the joint management and administration of its geometric and thematic data. These thematic data are called as attributes. Objects are grouped according to their common characteristics or attributes. The data can be acquired by any kind of measurement technique, eg. geodetic and photogrammetric measurement techniques, digitiser, scanner, CCD-cameras and satellite images.

4.1 Linking Photogrammetry and Information Systems

The main goal of the research project is the combination of the two methods described in the foregoing paragraphs. After the acquisition of the geometric deformations with photogrammetric methods, this information has to be analyzed. Based on this analysis different conclusions can be drawn, e.g. Evaluations concerning the stability of the building, decisions, whether it can be rebuilt or sold are torn down. Such decisions depend on many factors. An information system can provide an easy access to the stored data. In order to store the data efficiently and allow the access to it, a data model has to be established. To this end, the damage assessment sheets of the ITU, as well as international coding schemes have been analyzed. Based on this study, a detailed data model has been put up, including the structural and damage related aspects of several building types (Volz, 1998). This model has been implemented in the GIS-product ArcView. Using the programming language Avenue, some of the analysis procedures could be automated; e.g. the determination of damage degrees. These functions help to accelerate the analysis considerably.

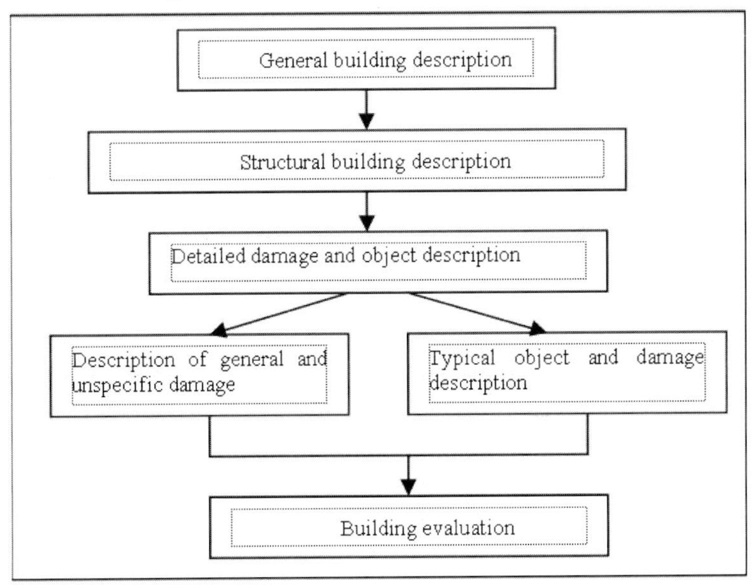

Fig. 5. The object levels of data model according to the object class principle

4.2 Description of the Methods Used in Earthquake Damage Acquisition and Analysis

The method used in Dinar for the purpose of damage acquisition is developed by a joint research of the Department of Civil Engineering of the Middle East University in Ankara and the "ITU" working group "Earthquake Engineering". This concept is developed upon request of the Ministry of Building and Housing. Basically the acquisition consists of different parts, which was grouped later on for a building. The main part of this damage report is the "report", where all important information, attributes for building description, damage acquisition and analysis, are listed.

The total damage description is divided in different components. Besides analyzing the damage for building in discussion, the building condition in a case of an earthquake is also considered. The geotechnical properties of the ground degree of the settlement of the basement are also considered. The essential description of the building is done by means of object class principle. The objects are described in different hierarchical level, by means of geometrical and semantically data see Figure 6.

The highest level of object class is the building itself, which consists of construction elements, objects of the second level. At these construction elements one can observe damage marks, e.g. cracks which are the elements of a further level. The structuring of the data within the GIS-products Arc-VIEW is done in form of tables, which was constructed by means of the relational database dBase.

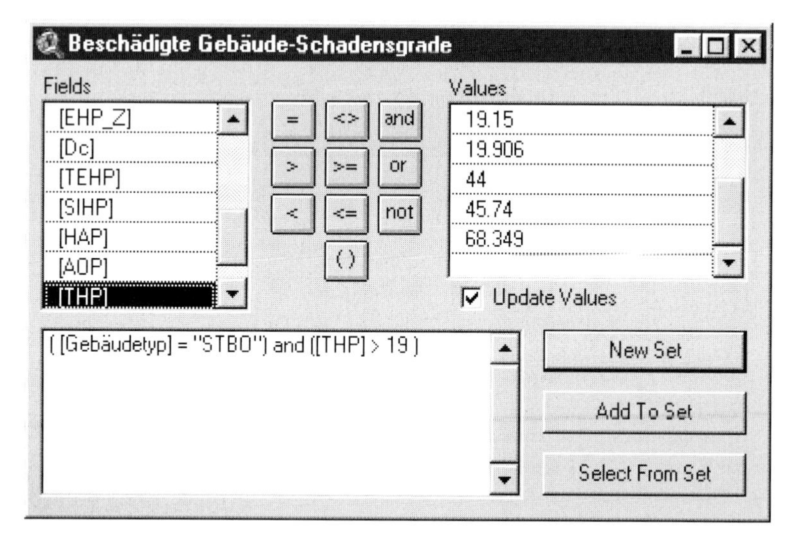

Fig. 6. The "Query-Builder" in ArcVIEW

The input of data was realized by forms in MS Access. It was made easy by means of default values. Figure 6 shows such a form. There a question of buildings of the type "reinforced concrete whose damage points exceed 19 is listed.

4.3 Data Acquisition by means of Remote Sensing

One of the major problems in Earthquake research is to predict and timely warn these natural catastrophes. Earthquake prediction by existing ground based facilities are not fully reliable and the recent unpredicted earthquakes in the last years (Iran, Morocco, Turkey) point the need for scientific progress in solving this problem and in employing additional evidence for earthquake prediction. Russian scientists and engineers have recently proposed the concept of a geo-space system for prediction and monitoring earthquakes and other natural and man-made catastrophes, which is based on a system capable of monitoring precursors of earthquakes in the ionosphere and magnetosphere of the Earth and using these precursors to make short-term forecast of Earthquakes.

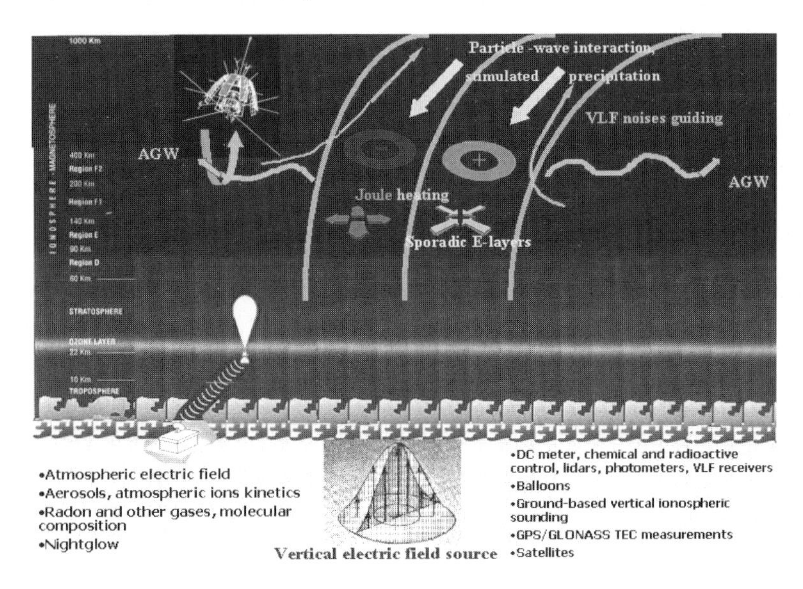

Fig. 7. Seismo-ionospheric coupling model schematic presentation (Sergey, 2004)

The colleagues from Russia are investigating in recent research projects the interaction between ionosphere's F layer variations and different variations occurring in circumterrestrial environment (atmosphere, ionosphere and magnetosphere) associated with seismic activity, and detected by means of ground base and satellite monitoring. They proclaim that the obtained results lead that the inospheric precursors of Earthquakes are really existing phenomena and the developed procedures for their detection offer a possibility of their practical utilization in systems for warning earthquakes and for making short term forecasts. (Sergey, 2003).

4.4 Analysis by means of GIS

The essential part of a GIS is its analyzing capability. The fundamental analyzing methods are used in selection of objects according of typical descriptive or spatial criteria, their aggregation, etc. So one can easily define and locate all buildings with definite damage descriptions and this can be very helpful for possible repair process.

Starting from the sample data, the first item was to design a concept for the automatic damage analysis and assessment. This was done for the construction element "reinforced concrete column". Basically the damage analysis must start from the lowest object level, as the damage of an upper level element has to be calculated from the sum of investigated levels, at first. A single crack can be described by many parameters like, length, width, trajectory, position etc. According to these parameter values, the crack influences the damage grade of the specific element. For the total assessment of one element, other damage indicators must also be considered. In case of a column, besides the investigation of the cracks, the deflection of the column from vertical line, the visibility of the steel reinforcement or the falling off the concrete mantle should be investigated. After the consideration of all these parameters the damage grade of a column can be calculated. After the completion of the damage grades (e.g. "Without damage", "slight damage", "middle damage" and "strong damage") for all the construction elements the total damage grade of a building can be calculated. This procedure reflects the basic idea of the damage determination used in earthquake analysis.

5 Conclusions

The documentation and analysis method explained in the above paragraphs reflect the results of a short-term study. In the preliminary evaluations so far made, the results displayed great differences from the manually obtained ones. In the current prototype, the building model is very detailed - further investigations have to focus on a thorough investigation of this data model in order to point out ways to accelerate the acquisition process. In general the use of an information system allows for the integrated documentation and analysis of earthquake damages and can be extended for a wider use as a catastrophe management system in general.

The main problem for further investigation is the acquisition of detailed object geometry by means of photogrammetric methods. In order to accelerate this procedure there exist appropriate tools. For instance the use of exemplary model in digital form for specific building types and to complete it by individual damage attributes has to be investigated. The completion of photogrammetric acquisition could lead to an automatic object extraction so that the digitization effort can be minimized.

In this preliminary work the marking of control points on the facade as hanging plumb lines with weights was a job with some risks. In further study these hanging down from the windows of the building should be replaced by mobile control-point systems.

Another very important and useful application could be the "Compass-Vulcan Program" for creation of space-born systems providing short-term prognosis of natural and man-maid catastrophes, including earthquake forecasting. For this purpose the combined efforts not only the interest of the scientist's also international programs are necessary.

References

.... (1996) Dinar Earthquake Damage Evaluation Report, (in Turkish), Turkish Earth-quake Foundation Publications, Istanbul

Aksoy A (1995) Kurzer Überblick über die Vermessungsarbeiten in der Türkei, Neue Technologien in der Geodäsie, Altan, Lucius (eds), Istanbul, pp 1-10

Altan O, Fritsch D, Sester M (1997) Antrag zum Projekt Dokumentation und Analyse von Erdbebenschäden mittels Geo-Informationssystemen, (unpublished), Stuttgart, pp 1-14

Anicic D, Radic R (1990) Computerised Assessment of Earthquake Damage, Earthquake Damage Evaluation and Vulnerability Analysis of Building Structures, International Network of Earthquake Engineering Centres (INEEC). Series on Engineering Aspects of Earthquake Phenomena, Oxon, S.1-19. (UB: 4B 247)

Bill R, Fritsch D (1991) Grundlagen der Geo-Informationssysteme, Bd. 1, Wichmann Verlag

Foramitti H (1980) Erdbebeneinsatz der terrestischen Photogrammetrie in Friaul, ISPRS XIV Congress, Comm. V, vol XXIII, part B5, pp 191-299

Kiremidjian AS, King,S (1995) An integrated earthquake damage and loss methodology through GIS, VII. International Conference on Soil Dynamics and Earthquake Engineering (SDEE 95), (UB: 4B 1252), Boston, p 664

Külür S, (1998) Kalibrierung und Genauigkeitsuntersuchung eines digitalen Bildaufnahmesystems, PFG

Schewe H (1995) A PC - Based System for Digital Close-Range Photogrammetry, Proceedings of the First Turkish - German Joint Geodetic Days, Altan, Gründig (eds), Istanbul, p 255-261

Sergey P, Kirill B (2003) Ionospheric Precursors of Earthquakes, Springer Verlag

Sergey P, Kirill B (2004) "Compass-Vulcan" Space system for Monitoring of Natural and Man-Made Catastrophes, lecture notes at the seminar in Bosporus Universoti, August

Tazir TZH, Gavarini C (1989) AMEDEUS, A KBS for the Assessment of Earthquake Damaged Buildings, Report of the IABSE Colloquium on Expert Systems in Civil Engineering, Bergamo, pp 141-150

Volz S (1998) Versuch zur Optimierung der Dokumentation und Analyse von Erdbeben-schäden und Gebäuden mittels eines Geo-Informationssystems am Fallbeispiel der Stadt Dinar/Türkei, Diploma Thesis, University of Stuttgart

The 26 December 2004 Sumatra Earthquake and Tsunami Seen by Satellite Altimeters and GPS

Boudewijn Ambrosius[1], Remko Scharroo[2], Christophe Vigny[3], Ernst Schrama[1] and Wim Simons[1]

[1] Delft University of Technology, Delft Institute of Earth Observation and Space Systems, Kluyverweg 1, 2629 HS Delft, the Netherlands.
Email:{b.a.c.ambrosius; e.j.o.schrama; w.j.f.simons}@lr.tudelft.nl

[2] NOAA Laboratory for Satellite Altimetry, Silver Spring, Maryland, USA.
Email: remko.scharroo@noaa.gov

[3] Laboratoire de Geologie, École Normale Supérieure (ENS), Paris, France.
Email: vigny@mailhost.geologie.ens.fr

Abstract

On 26 December 2004 a strong earthquake with an epicenter west of the coast of Sumatra generated a tsunami in the Indian Ocean. The earthquake had a magnitude of 9, which makes it a rare event since earthquakes greater than magnitude 8.5 have occured about once every 10 years since 1900 according to [5]. The last time a tsunami was generated by a magnitude 9 earthquake was on Good Friday 1964 off the coast of Alaska. The Sumatra earthquake and the tsunami that followed caused many victims in the countries surrounding the Indian Ocean. At the time of writing the death toll stands at 225000, which is an incredible number that is changing by the day. This article discusses two unique scientific aspects related to the earthquake and the tsunami. Four satellite altimeters picked up the traveling wave in the Indian Ocean, evidenced by the fact that a tsunami model matches the satellite observations. Early GPS observed displacements at a few sites in the Sumatra, Thailand, Malaysia region confirm the predictions of a geophysical model.

1 Introduction

Despite all human tragedy that resulted from the Indian Ocean earthquake and tsunami on 26 December 2004, the event is also unique from a geophysical research perspective. Two different techniques are discussed in this paper. The first concerns the observation of the traveling tsunami by four satellite altimeter systems that crossed the Indian Ocean in the first 9 hours after the earthquake. The second is an observation of horizontal GPS site displacements as a result of stress released in the Earth's crust and upper mantle. The latter observations are part of a cooperative research project between the TU Delft and partners in Indonesia, Thailand and Malaysia. The motivation of this paper is to shortly discuss both remarkable observations, which have been the source for public attention in national and international media. It will be shown that relatively simple modeling techniques can greatly help in the interpretation of both satellite altimeter and GPS observations.

2 The Tsunami Captured by Satellite Altimeters

The tsunami generated by the Sumatra earthquake is a shallow water wave, also called a gravity wave, that is well described by corresponding equations found in for instance [2]. Gravity waves travel with a speed of about the square root of $g.H$ (here g is 9.81 m/s^2, H is the ocean depth). For example, for a 4 km deep ocean we find a wave speed of about 720 kilometer per hour. As soon as the tsunami reaches shallow seas the wave speed significantly drops and the height of the wave increases. This phenomenon was widely reported in the media directly after the tsunami hit the shores of several countries around the Indian Ocean and the Andaman Sea. Some amateur videos show a decrease of the water level before the wave crest hits the beach. In other cases the videos show no retreat of the shoreline and the crest of the wave directly hits the beach. Whether or not the water level first drops depends on the shape of the wave front as it leaves the tsunami generation point.

The Delft University of Technology and NOAA maintain a common radar altimeter database system known as RADS [6]. This system was set-up in such a way that geophysical data records containing the 1-per-second satellite altimeter data including numerous instrumental and geophysical corrections are organized in a consistent manner. The tools that come with the RADS database system allow one to efficiently search large volumes of data. One essential data mining tool is the collinear track analysis program, hereby it is possible to overlay passes observed by any altimeter over previous passes observed in earlier cycles. Each altimeter system that we use for sampling the tsunami over flies a ground track pattern that repeats itself after a certain period. For the TOPEX/Poseidon and Jason-1 altimeter systems the length of a repeat cycle is about 10 days, see also [1]. This also means that ground tracks of both systems are separated by 315 km at the equator, which makes it difficult to observe the full extent of a traveling tsunami with a single orbiting altimeter system.

Despite the temporal and spatial sampling patterns of satellite altimetry, four altimeters captured the height profile of the tsunami in the deep ocean. In our opinion, these observations are unique and it is only possible because all four radar altimeters happened to observe sea level height profiles over the Indian Ocean during the first few hours after the earthquake, which took place on Sunday, 26 December 2004 at 00:58:53 UTC, see [3]. TOPEX/Poseidon and Jason-1 mapped the Indian Ocean about 2 hours after the tsunami started, as shown in Plates 1 and 2. The top part of each plate shows the results of a tsunami prediction model, which essentially solves the shallow water equations starting with an initial condition that matches the generation point west of the coast of Sumatra. In the bottom part of Plate 1 and 2 one can see the evolution of the relative sea level observed by either TOPEX/Poseidon or Jason-1 on 26 December 2004 relative to the sea level observed in the previous cycle. These sea level anomalies are presented by the black line in the graph under the map. Both TOPEX/Poseidon and Jason-1 appear to observe a change in the sea level that matches well the shaded height profile representing the tsunami model prediction run by NOAA, see [8].

Plate 3 shows a similar situation for the Envisat altimeter system, which observed the event 3:15 hours after the initiation of the tsunami. In this case it is also obvious that the model prediction matches the satellite observation. In Plate 4 the same is shown for the US Navy GFO altimeter system which mapped the Indian ocean about 8:50 hours later. In this figure the wave field is smeared out and it becomes more difficult to detect a tsunami with an orbiting altimeter system.

It should be emphasized that the observations of the satellite altimeters only refer to sea level anomaly differences between the altimeter cycles during the event and previous cycles. Moreover, the observations are only made along altimeter tracks. They do not observe the full extent of the tsunami model predictions shown in Plates 1 to 4. Unfortunately, this knowledge was not always been picked up by news media that were very eager to show the color images stating that "it" was observed by a satellite (sic).

The first satellite altimeter results show that the wave height, here represented as wave amplitude and not crest to bottom height difference as is sometimes reported, decreases over time. Two hours after the earthquake the peak wave height as observed by the altimeters in Plate 1 and 2 is about 60 cm in the ocean, in Plate 3 (3:15 hours after the event) it is reduced to 40 cm and in Plate 4 (8:50 hours later) the signal is reduced to as little as 5 to 10 centimeter. The accuracy of the sea level observation by orbiting altimeter systems is about 3 cm and today this is not a limiting factor for observing this event. However, it was pure coincidence that 3 out of the 4 altimeter systems were able to make the observation. They were at the "right" place at the "right" place, which is a delicate statement where the press officers (for instance at the TU Delft) directly inserted the text "as far as this can be said for this tragic event".

Several texts that appeared in the media also included a caveat emptor to the effect that the satellite altimeter observations are not suited for a real-time tsunami alert system. A network of buoys and bottom pressure recorders, so-called tsunameters, are more suited for such an alert system. The operation of such a network, together with the seismic observation of the epicenter and the magnitude of the underlying earthquake is a separate issue beyond the scope of this paper.

3 Crustal Site Displacements Observed with GPS

The Global Positioning System (GPS) allows to compute station coordinates in a global reference frame with millimeter accuracy, provided that suitable software, accurate satellite positions and clock information distributed by i.e. the International GPS Service (IGS) are incorporated in the analysis, see [4]. A well-known application of GPS is to observe the (relative) motions of tectonic plates.

The South East Asia region is a tectonically complex region where several plates are colliding with relative speeds of up to 10 cm per year. In Europe there are also such plate motions, except that the relative velocities are only about 1 to 1.5 centimeter per year. This explains why there are frequent earthquakes and volcanic eruptions along the plate boundaries in South East Asia and that big off-shore earthquake events are capable of initiating a tsunami. From a scientific point of view, it is important to understand the mechanism of an earthquake, and to map all high-risk areas. Prediction of earthquakes with any geophysical modeling or observation tool is at present not demonstrated.

The December 26 earthquake has resulted in deformations of the 2 involved tectonic plates in the SE Asia region. We will show that significant crustal deformations have been observed up to 1000 km distance from the epicenter of the earthquake. The reason for such large scale displacements is that stress has built up over a period as long as hundred years. In this area the stress results from the submerging of the Australian plate in a North Easterly direction under the Sundaland plate, which forms the largest part of SE Asia. During the earthquake, this stress has been released in a matter of minutes. The motion of one plate beneath another is called subduction, which is a slow and non-continuous process that is accompanied with stress release in the form of earthquakes. These events frequently occur below the ocean along the Indonesian archipelago which spans more than 5000 km, and they may initiate tsunamis. The crustal displacements as a result of earthquakes are a result of the fact that the Earth's lithosphere and mantle behave like a visco-elastic body.

Seismic observations of earthquakes are used on a routine basis to determine the epicenter and the magnitude of the event. Less well known is the precise surface deformation pattern as a result of an earthquake. In case of the Sumatra earthquake, the displacements at the plate boundaries are difficult to observe since they are located on the ocean floor. The epicenter displacements can be inferred with the help of a geophysical model that is calibrated with remotely observed GPS displacement vectors. In this case we rely on post-earthquake GPS observations in the area at sites that were previously observed.

Since 1994 TU Delft (DEOS) participates in and organizes GPS observation campaigns in SE Asia. These activities are coordinated with colleagues at universities and institutes in France, Indonesia, Malaysia, Singapore and Thailand. Analysis of the many GPS observations in SE Asia has resulted in a kinematical model which contains motions that are determined with accuracies up to 1 millimeter per year. Since 1997 a focal point of the research activities of DEOS is the area around the city of Palu on the island of Sulawesi in Indonesia, which is located on top of a fault system that poses an acute natural hazard for the 300,000 living there. This threat has become very evident, because less than a month after the mega-thrust earthquake event near Sumatra, the city was hit on 24 January 2005 by a magnitude 6.3 earthquake. The two events are most probably not related, because Palu is almost 3000 km away from the 26 December 2004 magnitude 9.0 earthquake's epicenter.

An important project to the tsunami event is the EC-ASEAN "SE Asia: Mastering Environmental Research with GEodetic Space Techniques" (SEAMERGES) project. The goal of this EU funded project is to transfer knowledge in the research of high-risk areas that could be affected by earthquakes, land slides and extreme subsidence, floods and even tsunamis. The use of geodetic techniques with the support of space technology is one of the drivers of the SEAMERGES project.

In a report issued by the SEAMERGES project shortly after the disaster in SE Asia, it was predicted by a model described in [7] that co-seismic displacements (those occurring during the earthquake) closest to the earthquake could be as large as several tens of centimeters and even of the order of a few meters towards the epicenter. Figure 5 is taken from [7] and it shows the co-seismic motions as predicted by a geophysical model in the region as a result of stress release of the Sumatra earthquake. The deformation vectors shown in this figure are the consequence of the earthquake and they are all pointing towards the epicenter gaining length as they approach the center. The deformations even extend to Thailand. The island of Phuket at 750 km from the epicenter could been shifted by at least 15 cm and the shift at Bangkok, at a distance of 1400 km, is predicted at 2 cm.

While most of the recent GPS data gathered after the event is not yet processed, the first GPS observed displacements confirm the model predictions: a permanent GPS station of the IGS in Singapore (which was installed in 1997 by the Faculty of Aerospace Engineering at the TU Delft) was already confirmed to have moved by 2 cm. A permanent GPS site of Indonesian partners in the SEAMERGES project in Medan, Sumatra, was confirmed to have moved by more than 15 cm.

Post-seismic displacements are the consequence of a material property of the Earth's crust and mantle and have already been confirmed by SEAMERGES partners in Malaysia. Also here the co-seismic displacements can be determined because a number of GPS stations were operational before, during and after the earthquake. A first GPS result from the Department of Survey and Mapping Malaysia (DSMM), which is one of the SEAMERGES partners, can be seen in Figure 6. It shows the change of position of the Malaysian permanent station ARAU, about 500 km SW of Phuket. This figure demonstrates the co-seismic displacement around 26 December 2004 (day 361 in the plot). The station displacements are given in the north, east and vertical directions. A jump of about 3 cm to the south and 15 cm to the west is clearly visible. It is possible to obtain this kind of information within a day of the actual GPS measurements. Differences observed between the predicted (for ARAU: 6 cm to the south, 10 cm to the west, taken from [7]) and the actual motions (illustrated in Figure 6) are important input parameters to constrain geophysical models. These models can also be further improved, and enable a better understanding of the (future) earthquake mechanism and the (possible) regions affected.

4 Conclusions

Four satellite altimeter systems demonstrated the presence of a tsunami in the deep Indian Ocean. In the deep ocean the wave height (crest to trough difference) of such a wave is about 1.2 meters while the spatial wavelength appears to be 200 km. The Sumatra earthquake has resulted in a stress release that caused deformations of at least 10 cm at stations 750 km away from the earthquake and 2 cm in a GPS station in Singapore 1100 km from the epicenter. In both cases relatively simple models explain the observations, although it should be remarked that our results are very preliminary and that more time is required to interpret and validate all observations and models.

The first scientific results were presented during a special seminar in the framework of the SEAMERGES project in Kuala Lumpur, Malaysia on 16 February 2005. The scientific implications of the tsunami and the earthquake of 26 December 2004, were presented at the European Geosciences Union (EGU) assembly, 24-29 April 2005 Vienna, Austria.

Acknowledgements

The SEAMERGES project (http://www.deos.tudelft.nl/seamerges) is part of the ASEAN-EU University Network Program (AUNP). It is a joint initiative by the European Union (EU) and the ASEAN University Network (AUN) aiming to improve co-operation between higher education institutions in European Union Member States and ASEAN countries, and to promote regional integration within ASEAN countries.

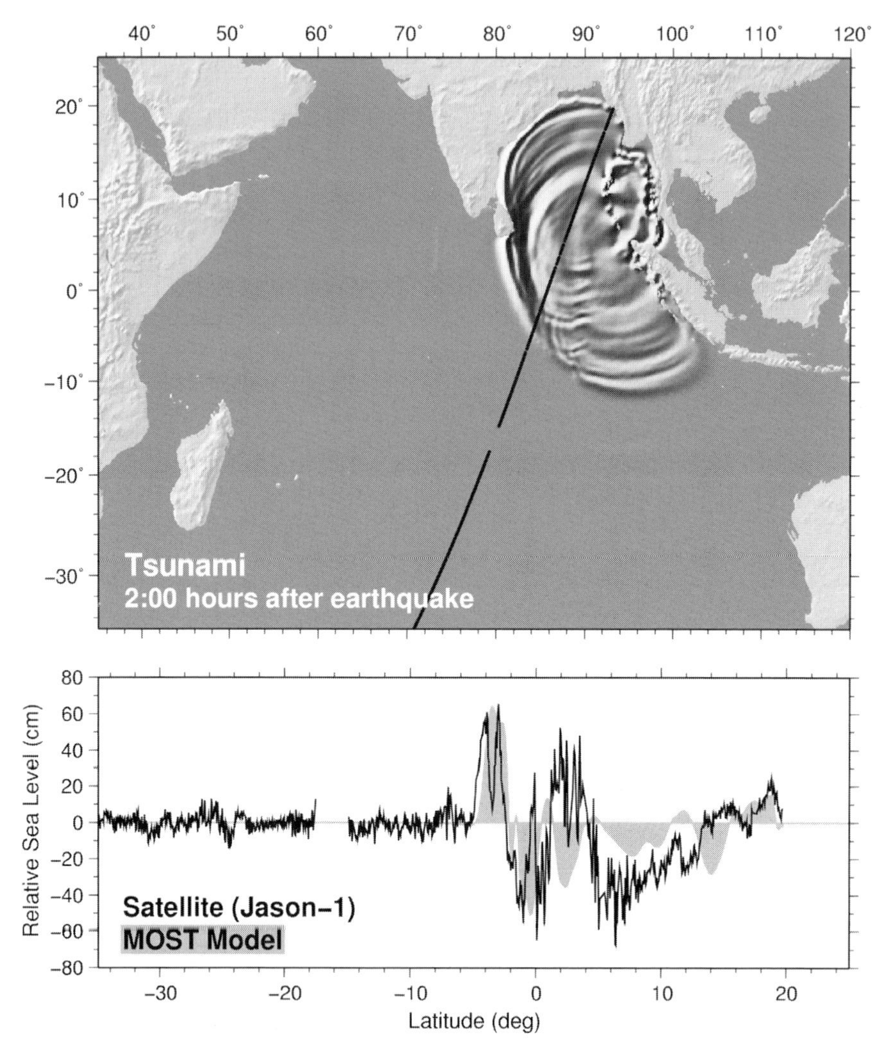

Fig. 1. The 26 December 2004 tsunami seen by the JASON-1 altimeter jointly operated by NASA and CNES. The ground track intersects the equator 2 hours after the Earthquake on 26 December 2004. The upper map shows the modeled tsunami and the satellite track along which altimeter data was used to obtain sea level anomalies. The bottom graph shows the observed relative sea level in black, the shaded line shows the sea level predicted by NOAA's tsunami model, see [8].

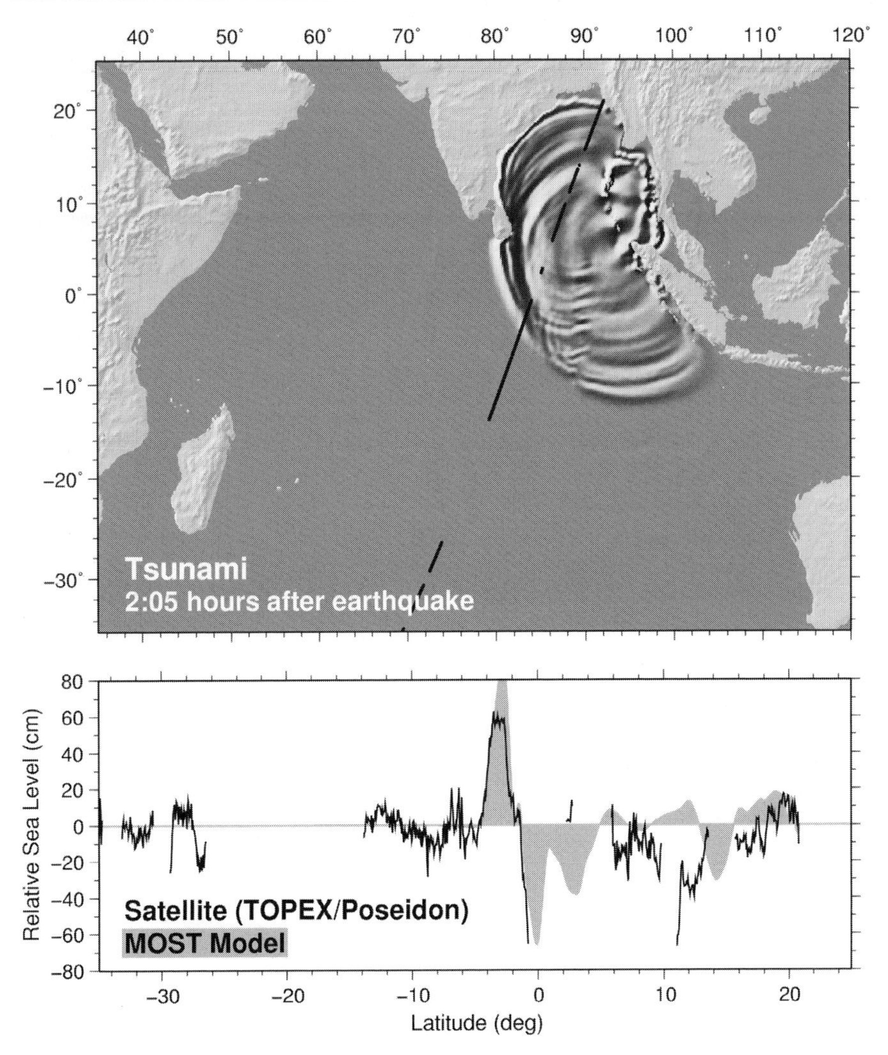

Fig. 2. The tsunami seen by the TOPEX/Poseidon altimeter operated by NASA at 2 hours and 5 minutes after the Earthquake on 26 December 2004. This shows the largest recorded gravity wave seen by any of the four altimeters, the maximum observed crest to trough height is about 1.2 meter. For details see figure 1.

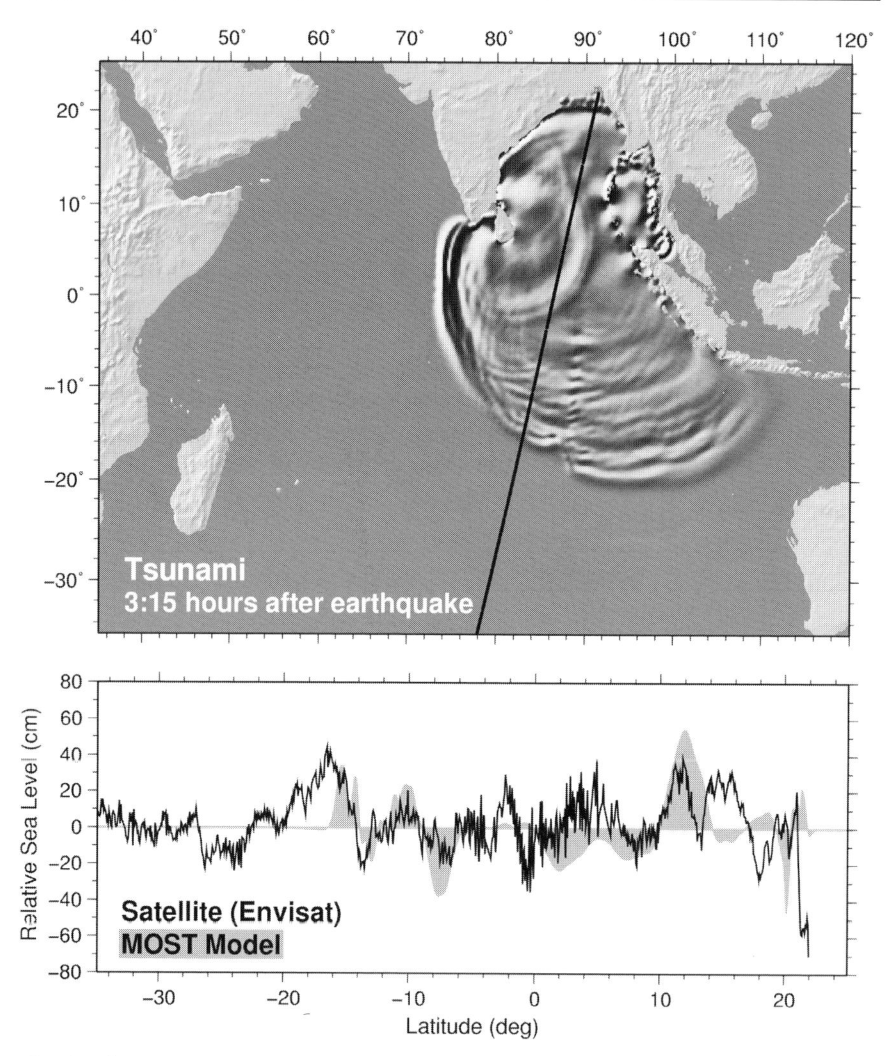

Fig. 3. The tsunami seen by ESA's Envisat altimeter 3 hours 15 minutes after the Earthquake on 26 December 2004. For details see figure 1

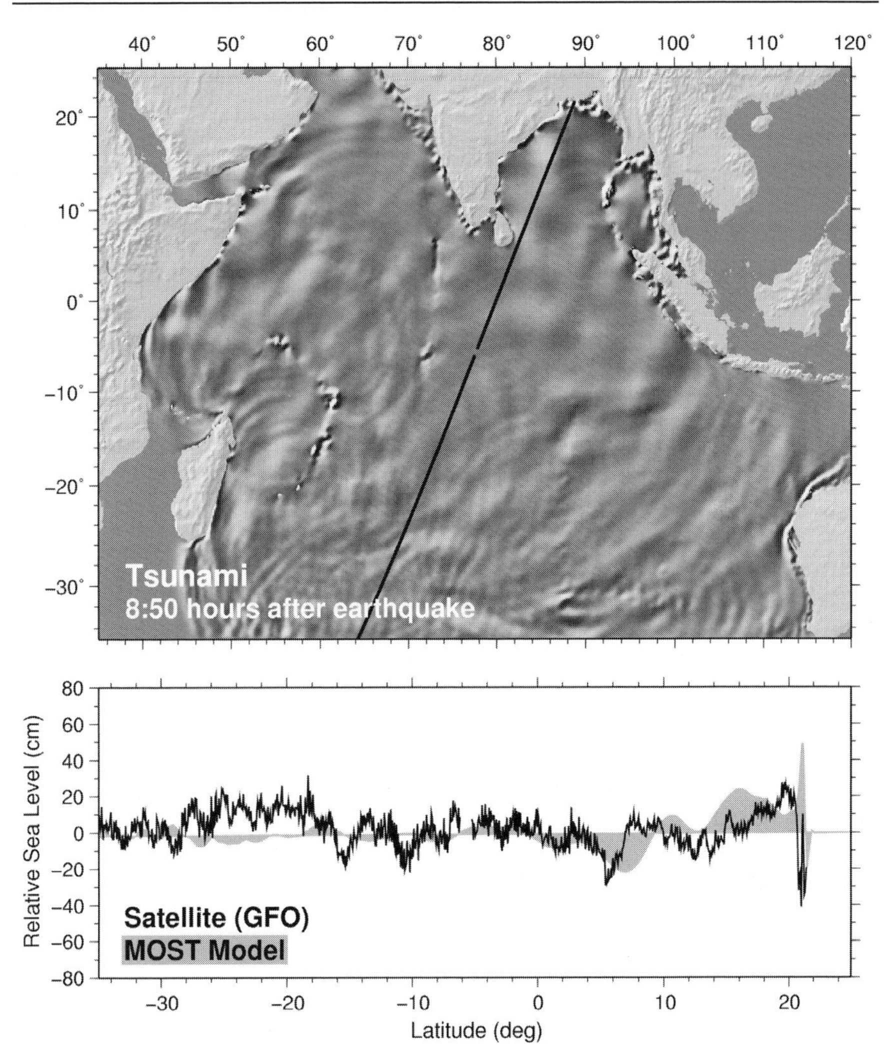

Fig. 4. The tsunami seen by the Geosat Follow-on (GFO) altimeter (operated by the US Navy) at 8 hours 50 minutes after the Earthquake on 26 December 2004. For details see figure 1

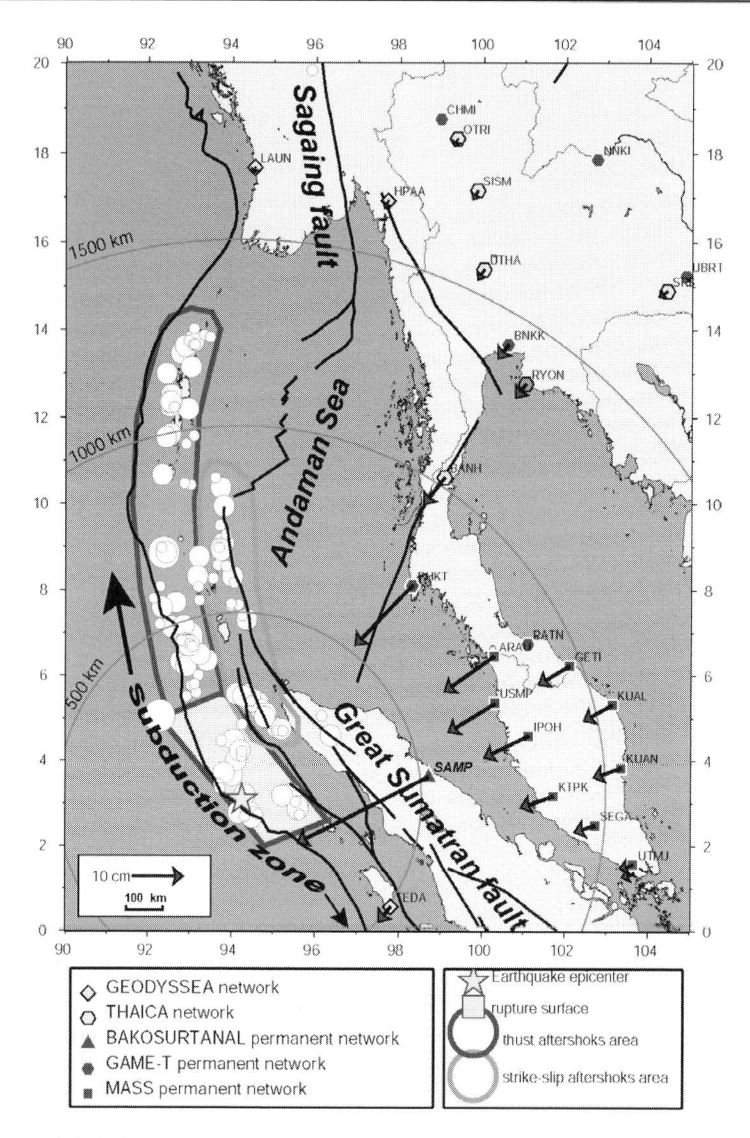

Fig. 5. Modeled deformations predicted for GPS stations in the region, see [7]. The modeled displacements are the result of the Sumatra earthquake with a magnitude of 9. All aftershocks are shown until 4 January 2005. The actual displacements are verified at three stations in Medan in Sumatra, in Singapore, and at station ARAU in Malaysia

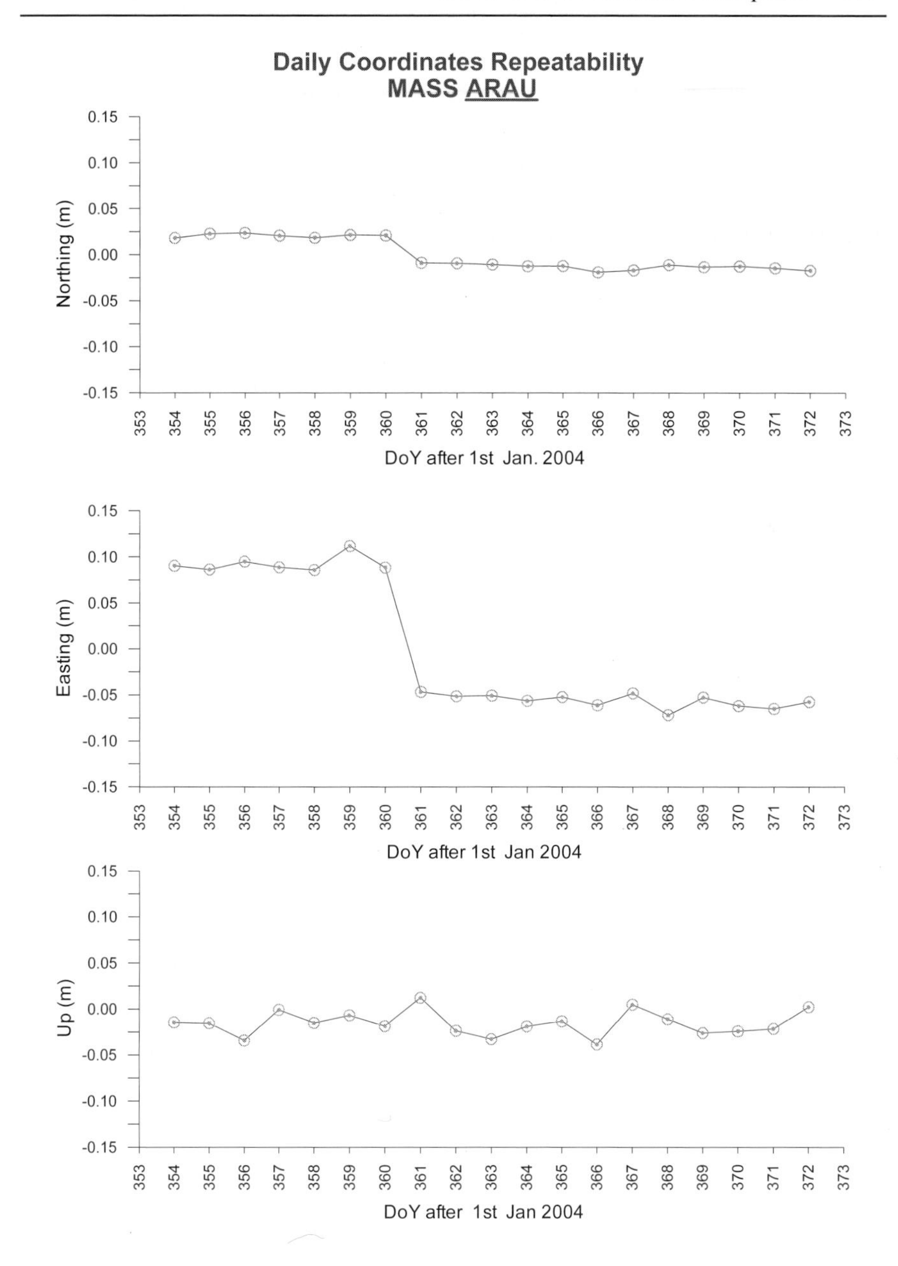

Fig. 6. Changes in daily positions (coordinate repeat-abilities) related to the 26 December 2004 earthquake near Sumatra. This figure was provided by DSMM/Malaysia and it shows the co-seismic motion of the permanent GPS station (ARAU) in Malaysia, which is located more than 750 km away from the earthquake's epicenter

References

[1] Fu LL, Cazanave A (2001), Satellite Altimetry and Earth Sciences, A Handbook of Techniques and Applications, International Geophysics Series, Vol. 69, Academic Press
[2] Gill A (1982) Atmosphere-Ocean Dynamics, Academic Press
[3] http://earthquake.usgs.gov/eqinthenews/2004/usslav/
[4] http://igscb.jpl.nasa.gov/
[5] http://neic.usgs.gov/neis/eqlists/10maps_world.html
[6] http://www.deos.tudelft.nl/altim/rads/literature.shtml
[7] http://www.deos.tudelft.nl/seamerges/docs/Banda_aceh.pdf
[8] http://www.pmel.noaa.gov/tsunami/indo_1204.html

Near-Real Time Post-Disaster Damage Assessment with Airborne Oblique Video Data

Norman Kerle[1], Rob Stekelenburg[2], Frank van den Heuvel[3] and Ben Gorte[3]

[1] International Institute for Geo-Information Science and Earth Observation (ITC), Hengelosestraat 99, P.O. Box 6, 7500 AA Enschede, the Netherlands.
Email: kerle@itc.nl (Corresponding author)

[2] InnoStack, Ommerbos 51, 7543 GG Enschede, the Netherlands.
Email: mail@innostack.com

[3] Delft University of Technology, Delft Institute of Earth Observation and Space Systems (DEOS), Kluyverweg 1, 26 29 HS Delft, the Netherlands.
Email: {f.a.vandenHeuvel; b.g.h.gorte}@lr.tudelft.nl

Abstract

Natural and man-made disasters lead to challenging situations for the affected communities, where comprehensive and reliable information on the nature, extent, and the consequences of an event are required. Providing timely information, however, is particularly difficult following sudden disasters, such as those caused by earthquakes or industrial accidents. In those situations only partial, inaccurate or conflicting ground-based information is typically available, creating a well-recognized potential for satellite remote sensing to fill the gap. Despite continuous technical improvements, however, currently operational, non-classified, space-based sensors may not be able to provide timely data. In addition, even high spatial resolution satellites (< 1m) are limited in their capacity to reveal true 3D structural damage at a level of detail necessary for appropriate disaster response in urban areas.

Uncalibrated oblique airborne imagery, both video and photography, is typically the first data type available after any given disaster in an urban setting, usually captured by law enforcement or news agencies. In this study we address the use of video data for systematic, quantitative, and near-real time damage assessment, using video and auxiliary data of an industrial disaster in 2000 in Enschede, the Netherlands, and of Golcuk, Turkey, acquired after the 1999 Kocaeli (or Marmara) earthquake.

We focus in particular on texture-based damage mapping based on both empirical and more generic, geometric indicators. Data-specific attributes included color indices and edge characteristics, while the data-independent approach included rotation invariant Local Binary Pattern and contrast operators (LBP/C). In an earlier step of the project, an interface was created to allow the near-real time processing of video streams, and, depending on positional information encoded with the data, their combination with auxiliary data such as maps or pre-disaster image data. Here we further investigated the potential of the available imagery for 3D reconstruction of the disaster area. Correspondences between consecutive video frames were established automatically by feature point tracking and used for the estimation of the coordinates of the terrain points as well as the camera parameters. Furthermore, we quantitatively assessed the quality of the reconstruction based on the data available. The ultimate goal of the project is to establish a versatile processing platform that supports extraction of such information as well as damage mapping, but also partial 3D reconstruction and integration of pre-event GIS and other auxiliary data.

1 Introduction

Natural disasters result in mortality and property damage that exceeds the response and recovery capabilities of the affected area. Any meaningful and organized response to such events is reliant on detailed information on the nature, scale and spatial distribution of the consequences. This information also has to be made rapidly available to all response forces involved, as the overall cost of a disaster, both in terms of economic damage and fatalities, depends on how quickly the event is responded to, and how efficiently response activities are managed. However, the response to virtually every extensive, in particular urban, disaster in recent years was delayed by a slow inventory of the event's consequences. This is true for disasters in developing countries (e.g. following earthquakes in Koceali [Turkey; 1999], Gujarat [India; 2001], or Bourmerdes [Algeria, 2004]), but also developed ones (e.g. following the 1995 Kobe, Japan, earthquake). This results from a shortage of reliable information coming from the disaster site, difficulties in access, disruptions of power and communication lines, and organizational reasons, such as insufficient preparedness and poor coordination between the different response parties.

The potential of geoinformatics, in particular GIS and remote sensing technology, for all phases of the disaster management cycle, that is hazard and risk assessment, disaster prevention and preparedness, in-disaster monitoring, and post-disaster relief and reconstruction, has been extensively described (e.g. Walter 1994; Showalter 2001; Kerle and Oppenheimer 2002). Whether active or passive, pan or multispectral, or high- or low-resolution data are most suitable is entirely dependent on the hazard or disaster type under consideration. However, one rule applicable to all types is the temporal constraint in the syn-event and post-disaster relief phases. While for all pre-event phases, as well as during reconstruction, time is not typically a critical parameter, it is the most significant constraint during the immediate disaster response, to an extent that speed takes precedence over information detail or quality.

The first image data likely to be available after a disaster in an urban setting are oblique airborne videos or photographs captured by the news media or law enforcement agencies. In particular the acquisition of video data poses no technical challenges, giving this sensor type a valuable versatility unmatched by space-based devices. The imagery acquired, however, differs substantially from standard air- and spaceborne remote sensing, both in principal purpose and data characteristics, and, therefore, established image analysis methods can only be used within limits. Such surveys tend to (i) be unplanned in terms of data acquisition scheme, (ii) use an unstable platform, (iii) focus on highly damaged areas at the expense of complete coverage, (iv) produce oblique, uncalibrated data with comparatively low spatial resolution, and (v) suffer from frequent cutting and zooming, resulting in spatially discontinuous data streams with scales changes between and within frames. Commonly used devises are analogue or digital BetaCams (360 and 720 pixels per line, respectively) or HDTV (High Definition Television) with 1920 pixels per line. Law enforcement agencies are also increasingly using sophisticated cameras that acquire infrared or thermal imagery to document the aftermath of disasters.

1.1 Research Objectives

The goal of this project is to establish a comprehensive damage assessment methodology that works on any type of post-disaster airborne oblique video data. The system would facilitate (i) automatic and generic structural building damage mapping, if applicable using user-specified training areas, (ii) extraction of any encoded GPS/IMU data and georegistration of the video data, and (iii) integration with reference imagery if available. Accounting for the different types of video/TV cameras in use, the system should (iv) be scalable in terms of sophistication and level of processing. Additionally, it should (v) support partial 3D reconstruction of the damaged area, and display of the data in 3D, with potential superimposing of damage and pre-event reference data. In this paper we focus on the 3D reconstruction aspects, as well as texture-based damage detection.

In the following sections we briefly introduce the two disaster sites considered, and discuss the scientific value of video data by reviewing studies that used them for disaster response and damage mapping. We also discuss technical developments from other disciplines that appear applicable for this task.

2 Study Areas and Data Used

2.1 The Enschede Fireworks Disaster

On 13 May 2000, a series of explosions occurred at a fireworks factory located within a residential area in Enschede, the Netherlands, killing 22 people and severely damaging or destroying nearly 500 buildings. Areas close to the factory suffered heavily, both from the pressure waves as well as fires, while damage severity rapidly declined towards the furthest affected structures, approximately 1 km from the explosion site.

Acquisition of video imagery, the principal data source used for this paper, was initiated approximately two hours after the disaster by the Dutch National Police Aviation Branch, and was repeated on the following days and augmented by high-resolution vertical aerial photographs 12 days later. Video data were captured with a Sony Interline Transfer camera with an ½ inch CCD chip (6.4 × 4.8 mm) at 25 frames per second, written to an analogue VHS tape, and later digitised at a resolution of 720 × 540 pixels. For this paper, a sequence of 761 frames (30.4 sec), during which no zooming occurred, was manually selected. A sample frame is shown in Fig. 1. Thermal imagery was also collected, using an Agema Thermovision 1000 with a 8-12 μm spectral band.

The vertical aerial photos were not used for the actual damage assessment, which was instead based on ground-based surveys. Disaster response was also hindered by outdated map material. Incidentally, a planned aerial survey of Enschede at a scale of 1:18,000 was carried out just 4 hours before the disaster occurred, providing reference data that could have been, but were not, used in the disaster response phase.

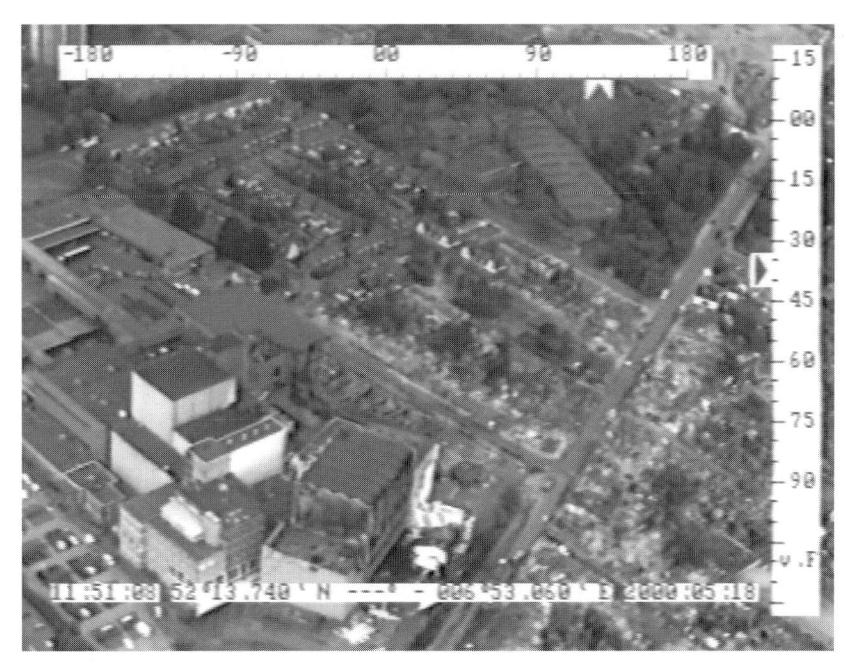

Fig. 1. An image from the sequence used in the experiment. We previously automatically extracted the encoded GPS, azimuth and inclination information for frame footprint estimation (see Kerle and Stekelenburg 2004). Note that all image data used are originally in color

2.2 The 1999 Marmara Earthquake

A devastating earthquake (Magnitude 7.4) occurred on 17 August 1999 close to Kocaeli in the north-western part of Turkey, along the North Anatolian Fault Zone. It resulted in widespread and extensive damage in a highly populated and industrialized region. More than 15,000 people are estimated to have died, and some 40,000 building collapsed or were heavily damaged by the event. Golcuk was one of the most damaged towns in the region with the number of fatalities and destroyed buildings exceeding 5,000 and 2,300, respectively. Because of the expanse of the affected arca, damage evaluation was slow. The principal types of building damage observed were damage due to foundation failure and poor detailing, soft story and intermediate story collapse, and pancake collapse. Mortality was particularly high in pancaked buildings.

Aerial video imagery of Golcuk was acquired by a media agency on the day of the earthquake. Total footage was approximately five minutes in length, with a resolution of 720 × 576 lines. In addition, data from a detailed ground-based damage inventory were used to assess the accuracy of the video-based damage mapping.

3 Previous Use of Video Data for Damage Assessment

Video data have been used in scientific research since the 1980s, where they were primarily used in resource management and agricultural applications (for a review of early studies see Mausel et al. 1992). To our knowledge, video data were first applied for damage assessment in a study by Everitt et al. (1989), in this case to map wildfires with mid-infrared videos. Since then a number of video-based methods have been developed for damage assessment. In addition, work in other disciplines, in particular computer science, machine vision and robotics, has resulted in techniques that appear also well suited to the damage assessment problem.

The 1995 Kobe earthquake sparked a range of efforts to apply video and TV data on urban disasters, leading to several qualitative and quantitative studies. Early studies were entirely qualitative, focusing on visual damage mapping based on oblique images. Ogawa et al. (1999) and Ogawa and Yamazaki (2001) analyzed individual HDTV images of post-earthquake Kobe, and mapped damage into four different categories. The approach was largely emulated by Hasegawa et al. (2000a; 2000b). Visual analysis can be of use for detailed assessment of clearly different damage types. However, it is impractical for large areas, time consuming and subjective, and results are difficult to map and integrate with auxiliary spatial information.

The visual analysis approaches were superseded by more automated methods based on training data, for which the extensive, and clearly visible structural damage sustained in Kobe appeared well suited. Hasegawa et al. (1999) extracted color and edge characteristics of differently damaged wooden buildings from HDTV training sets, to map different damage classes based on thresholds, though with limited success. Mitomi et al. (2000) tested the concept on damage data of the 1999 Kocaeli and 1999 Chi-Chi (Taiwan) earthquakes. Judging the success, or comparing the applicability of the approach in different settings, is not possible, since no absolute accuracy assessments were carried out. One clear limitation of the method is the frame-specific training, which requires separate adjustment for all frames. This was partially overcome in later work by the same authors (2002), where damage mapping was based on a maximum likelihood classification, as well as a more generic texture analysis based on entropy and Angular Second Moment (ASM). As before, however, the results were not evaluated quantitatively (see also Yamazaki 2001, for a detailed review).

Extensive research has been carried out in 3D virtual model construction of urban areas, especially for tourism and heritage conservation applications (e.g. Haala and Brenner 1999; Pollefeys et al. 2004). The parameters of such models, in terms of completeness, detail and realism, are governed by the application. Though the time limitations of a post-disaster situation rarely allow for the involved computations required in 3D model creation, even partial models of a disaster site can be of value. Depending on the quality and detail, they allow collapsed or heavily damaged houses to be identified. Jang et al. (2004) created 3D models of cultural heritage sites in Japan from video data collected by a remote controlled helicopter. A similar system, integrating air- and ground-based imagery and ground control data, was developed by Otani et al. (2004), to model buildings in 3D, as well as to use imagery to texturize the models. Recent work at the Katholieke Universiteit Leuven (Belgium) and the University of North Carolina at Chapel Hill (US) has developed algorithms to use uncalibrated transverse video streams for detailed textured surface reconstruction (e.g. Pollefeys et al. 2004).

Instead of completely rendered models, Weindorf et al. (1999) extracted wireframe models of pre- and post-disaster buildings from reference aerial photographs, and aerial videos, respectively. Unlike the mono-temporal information extracted in the above studies, the authors derived sufficient information for a per-building change detection.

Integration of multiple datasets, but also spatial positioning of damage extracted from mono-temporal imagery, is reliant on georegistered datasets. Such information can be encoded in the video by a linked GPS/IMU. In an earlier stage of this study we extracted GPS and camera inclination and azimuth information from video data (see Fig. 1), and used it to calculate approximate frame footprints (Kerle and Stekelenburg 2004). Extensive research has also gone into mosaicing of video sequences. Such series cover larger areas and are easier to reference, especially where ground control is scarce, or extensive damage hinders identification of known points (see for example Kumar et al. 2000).

In principle, methods that can be applied to standard photographic data can also be used on video imagery. To account for data-specific processing needs, a range of commercial software tools has been developed. They can be used to extract individual frames (e.g. VideoCap Pro, PowerDirector Pro), to produce 3D models (e.g. PhotoModeller, ImageModeler), to time/GPS-stamp video data (e.g. Scenalyzer) and to create mosaics and orthorectify video data (e.g. RavenView by Observera).

4 Methodology

4.1 Video-Based 3D Reconstruction

A 30 second VHS video sequence (761 frames) of the Enschede site, during which no apparent zooming occurred, was manually chosen. The processing was carried out in MatchMover® (RealViz; www.realviz.com), which is capable of tracking 3D camera data and motion from uncalibrated video sequences, thereby allowing 3D reconstruction of manually or automatically generated feature points. Naturally, the success of the method depends on the characteristics of the data available. Critical factors are the quality of the imagery, the motion of the camera, and the characteristics of the scene related to the presence of distinct features in the imagery. The following procedure was followed: Step (i) extracts 2D points suitable for tracking, i.e. with good contrast in all directions, using an interest operator. New points are also being extracted during tracking, as points are occasionally lost during tracking, e.g. when they 'leave' the image. Step (ii) tracks extracted features by matching between consecutive frames. In step (iii) the interior and exterior camera parameters are calculated. In the current experiment the only interior orientation parameter was the (fixed) focal length. The principal point was assumed to be in the centre of the frame and lens distortion was neglected. This simplification implies the introduction of minor systematic errors in the 3D reconstruction. The derived exterior orientation parameters comprise the position and orientation of each frame in space, from which the position and orientation parameters of all frames was calculated. This included the flying height in meters relative to the surface. Additionally, a surface is interpolated from the 3D coordinates of the tracked points, which was then georeferenced using GPS information extracted from selected video frames.

4.2 Texture-Based Damage Detection

The Kobe damage areas used as training sets by Hasegawa et al. (2000b) were relatively locally homogenous. This suggests that texture segmentation algorithms may be suitable to detect such areas without the need for frame-specific training areas and continuous parameter adaptation to account for illumination differences. Many different texture classification methods have been devised (for a comparative review see Randen and Husoy 2002). We adopted the method proposed by Ojala and Pietikäinen (1999), who used Local Binary Patterns and contrast (LBP/C) to segment synthetic texture pattern as well as natural scenes into homogenous areas based on information on the spatial structure of local texture. The LBP method uses a small number of parameters, simple image features and has shown promising results for natural scenes. The original LBP is calculated from a 3 × 3 matrix, while C is a measure of contrast (Fig. 2).

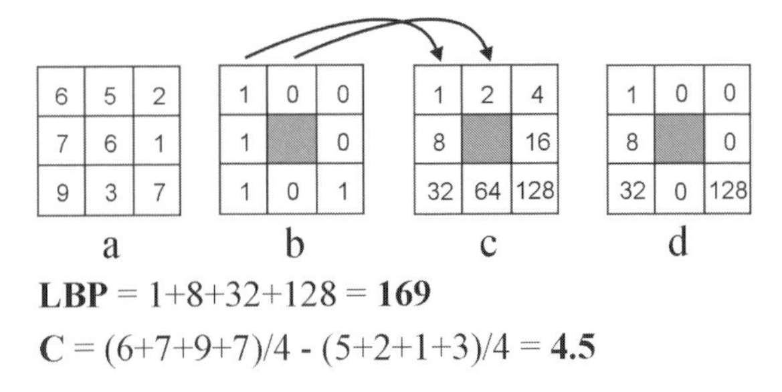

Fig. 2. Calculation of LBP and C for a 3 × 3 matrix. **a** shows the original image data, **b** whether a pixel is larger (1) or smaller (0) than the center pixel, **c** the pixel weights, and **d** the product of **b** and **c**, which is then summed to derive the LPB. From Ojala and Pietikäinen (1999)

This LBP can be made rotational invariant by classifying the 256 possibilities into 36 unique rotation patterns (Pietikäinen et al. 2000), as shown in Fig. 3. Lucieer (2004) used variance VAR instead of C, as well as a multiscale approach with different matrix sizes.

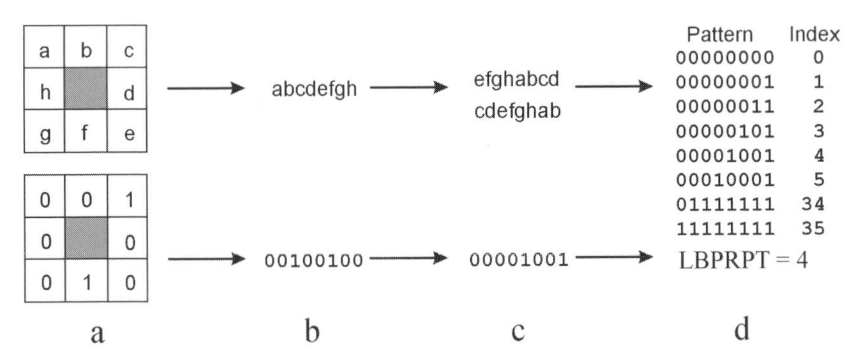

Fig. 3. Rotation invariant LBP. The 256 possibilities are classified into 36 unique rotation patterns. After Pietikäinen et al. (2000)

We have implemented the methodology of Ojala and Pietikäinen (1999), which consists of three steps: (i) hierarchical splitting, (ii) agglomerative merging and (iii) pixelwise classification, where the latter is a refinement of the merging stage. Two descriptors (LBP/C, LBP/VAR or other pairs) are calculated for the image and their distribution is approximated by a two-dimensional histogram.

Pietikäinen and Ojala (2002) concluded that the distribution of local spatial patterns and contrast plays a critical role in successful segmentation. Given the limited success of the LBP descriptor (see *Results*), we also implemented a similar feature descriptor called HVD (Horizontal Vertical Diagonal lines). The algorithm searches for differences between the centre and outer pixel in a N × N matrix, and is implemented as multi-scale:

X	•	X	•	X
•	x	x	x	•
X	x	O	x	X
•	x	x	x	•
X	•	X	•	X

For example, a 5 × 5 matrix calculates Abs(O-x) + Abs(O-X), thereby considering both the 3 × 3 and 5 × 5 scale.

5 Results

5.1 3D Reconstruction

Applying the automated procedure outlined in Sect. 4.1 resulted in the 3D reconstruction of 1533 points and the position and orientation parameters of all 761 frames. Figure 4b shows a view on the reconstructed points with one of the frames projected onto it. In each frame at least 100 points were tracked. The focal length was estimated to be 1701 pixels or 15.12 mm (75.6 mm if translated to a frame format of 32 × 24 mm). The flying height above the terrain was calculated to a range between 360 and 395 m (Fig. 6). The pixel size on the ground would be approximately 0.2 m for a vertical view. Due to the oblique viewing direction the ground pixel size varies between ca. 0.3 m and several meters. Residuals are computed by the software for each point at each epoch, as shown for one point in Fig. 5, and are on the order of 1 pixel. The quality of the 3D reconstruction is not only a function of the precision of the localization of the tracked point in image space, but also strongly depends on the number of frames in which the point is tracked, in combination with the amount and direction of the movement of the camera. In other words, the accuracy of the 3D position of a point strongly depends on the base-height ratio achieved by tracking its position in the imagery. From the information above we conclude that the accuracy of the 3D coordinates is at the meter level.

Fig. 4. a Sample frame with the track line of one point through the sequence superimposed. **b** partially reconstructed surface with the frame shown in Fig. 1 draped on top

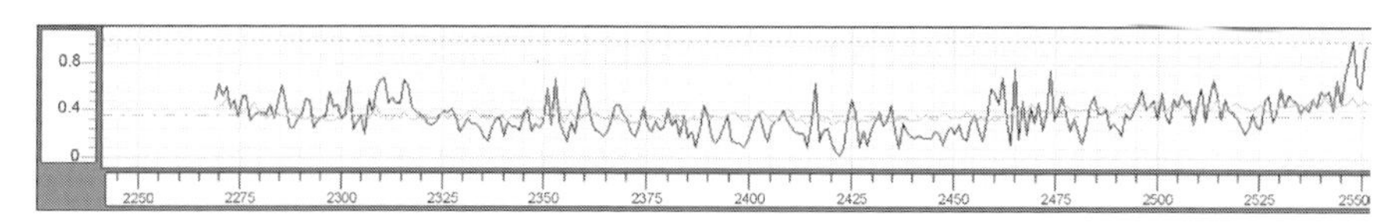

Fig. 5. Residual plot for one of the tracked points, showing residuals of less than 1 pixel

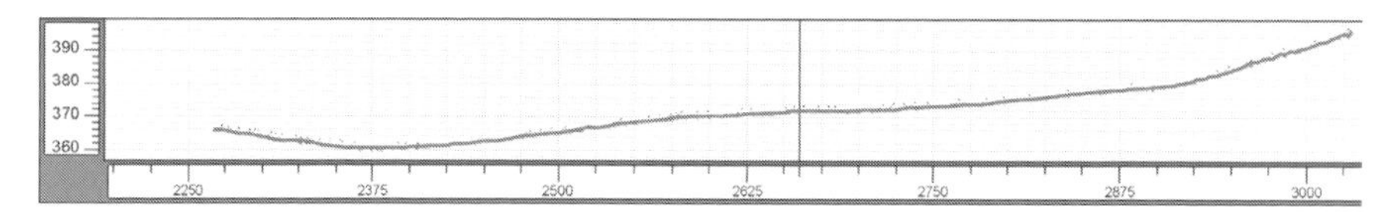

Fig. 6. Flying height during the 761 frames varying between 360 and 395 meter

5.2. Texture Segmentation and Local Binary Patterns

Both the original and the rotation invariant LBP/C algorithms were first tested on synthetic pattern, and then applied to video frames of Gulcuk and Enschede. Similar to the studies by Ojala, Pietikäinen and colleagues, the operators worked well on our artificial texture patterns. However, these test patterns are characterized by near perfect homogeneity within individual textures, and are thus far easier to process than natural scenes.

Difficulties encountered with those data also raise another question: what exactly constitutes urban disaster damage? It is a subjective concept, and, naturally, highly variable for different disaster types. Earthquakes tend to produce well-constrained damage patterns, from houses tilting or slumping as a result of liquefaction, structural damage ranging from cracks to internal damage compromising structural integrity, soft-story or pancake-collapse, or complete collapse. Industrial accidents can be more complicated. Even if only man-made disasters resulting in widespread structural damage are considered (excluding, for example, toxic gas discharge), damage patterns can vary widely. The pressure waves and fires caused by the Enschede explosions caused highly non-uniform damage, but also resulted in buildings close to the blast site simply disappearing. One benchmark for urban structural damage assessment based only on post-event imagery is visual detection: if damage can not be identified clearly, either because it is entirely internal, or because the entire damaged structure has disappeared, no algorithm will be of use. For the latter case, only integration of pre-disaster information can help.

An additional complication is the comparatively low resolution and poor quality (e.g. poor interlacing) of video data. As stated above, ground pixel size can be several meters, leading to substantial averaging of damage pattern. Figure 7 shows comparative results for Golcuk: (a) is a damage map based on multi-treshold processing (hue, variance, edge and edge variance), with an overall accuracy of 86% (Ozisik and Kerle 2004); (b) shows the LBP/C (5 × 5) implementation for the same scene. Although the method shows promise, especially for better quality data, at this point the less generic training-based method perform more reliably.

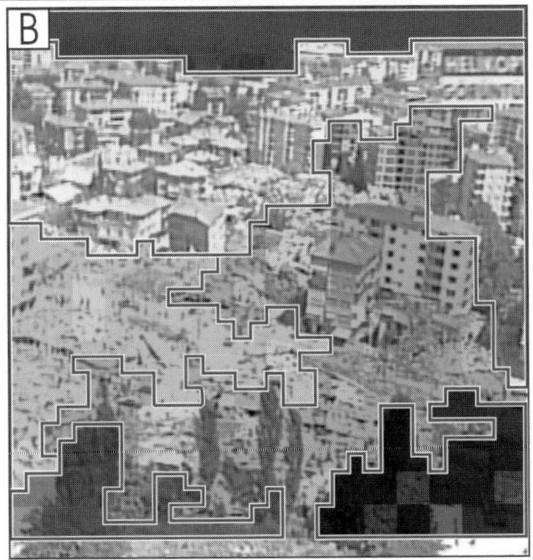

Fig. 7. Processed scene of Golcuk. **a** Based on color and geometric characteristics of training data, and **b** using LBP/C (5 × 5 matrix). See text for discussion

In both our artificial textures and natural images the rotation invariant LBP/C (3 × 3 matrix), which only has 36 data bins, worked better than the original version (256 × 8 bins), though it is still very dependent on threshold parameter values. We also implemented the LBP/VAR, and found that it outperformed the contrast-base approach for the smaller matrix sizes. Finally, we also implemented the HVD method described above, with a performance matching that of the LBP. However, further quantitative comparison of the methods on different image types is required.

6 Conclusions

The aim of this study was to investigate the utility of uncalibrated airborne video data for post-disaster urban damage assessment. Focus was on partial 3D reconstruction of such an area, as well as texture-based segmentation of those frames, with the aim of identifying homogenous damaged areas. A part of the Enschede disaster area was successfully reconstructed from 30 seconds of uncalibrated video data. The quality of the reconstruction is limited, due to the limitations inherent in video data recorded in VHS format, in combination with an unfavorable imaging configuration. However, the results are encouraging, and more research is required to assess fully the potential of this type of data for 3D reconstruction, and subsequent integration with auxiliary spatial data. Currently, we are investigating the possibilities for an increase in spatial resolution as well as the geometric quality of the reconstruction. Even when the helicopter has circled the object to be reconstructed, it is unlikely that features can be tracked during the whole sequence as their appearance in the imagery changes considerably with the viewing angle. Therefore, a quality loss in the average viewing direction with at least a factor two can be expected. This factor can be considerably higher for points tracked in a limited number of frames or tracked in a part of the sequence where the movement of the helicopter is minimal.

We also tested a texture segmentation approach based on rotation invariant LBPs and contrast/variance on both artificial textures and natural scenes. Although results showed improvement over previously used LBP algorithms, especially for the disaster scenes success was limited. All results showed great sensitivity to parameter settings. At this point, previous damage assessment results for the Golcuk (Turkey) site, based on training-specific image characteristics, still outperforms more generic segmentation methods. An approach based on horizontal, vertical and diagonal lines (HVD) matched the performance on artificial texture patterns, but could not improve results for natural scenes. Both the LBP and HVD approaches are still very dependent on threshold parameter values for the splitting and merging operations.

We next plan to explore the potential of multiscale segmentation, including color information. Furthermore, the research will continue towards making the method more robust, i.e. less dependent on initial parameters, matrix sizes etc. The focus here will be on a multiscale version of LBP/VAR, that will also incorporate color to distinguish between natural (trees, sky, etc.) and man-made features.

The review of previous studies, as well as results presented here, demonstrates that video data, regardless of type and availability of auxiliary information, have a strong potential to aid in disaster response, little of which has been realized to date. Given the limitations of space-based data types for damage assessment, we feel that video data can be developed as a principal data source for urban damage assessment. The methods tested here appear to be realistically implementable for standard or more sophisticated video data, such as obtained by the media or law enforcement agencies. Furthermore, a system could be implemented that establishes a radio or satellite link to a processing facility, and that allows near-real time access to the information, or remote processing. For example, Wright and El-Sheimy reported on such a system for forest fire detection, which has also previously been established for radar-based flood monitoring in China.

Acknowledgments

The video imagery of the Enschede disaster site was made available by the Enschede Police's TOL-Team. We particularly thank Hans Kamperman for his support. The Golcuk video data were made available by Show TV.

References

Everitt JH, Escobar DE, Davis MR (1989) Mid-Infrared video: A possible tool for thermal analysis of wildfires. Geocarto International 4: 39-45

Haala N, Brenner C (1999) Extraction of buildings and trees in urban environments. Isprs Journal of Photogrammetry and Remote Sensing 54: 130-137

Hasegawa H, Aoki H, Yamazaki F, Matsuoka M, Sekimoto I (2000a) Automated detection of damaged buildings using aerial HDTV images. Proceedings of the International Geoscience and Remote Sensing Symposium. Honolulu, USA

Hasegawa H, Aoki H, Yamazaki F, Sekimoto I (1999) Attempt for automated detection of damaged buildings using aerial HDTV images. Proceedings of the 20th Asian conference on remote sensing. Hong Kong

Hasegawa H, Yamazaki F, Matsuoka M, Sekimoto I (2000b) Extraction of building damage due to earthquakes using aerial television images. Proceedings of the 12th world conference on earthquake engineering. Auckland, New Zealand

Jang HS, Lee JC, Kim MS, Kang IJ, Kim CK (2004) Construction of national cultural heritage management system using RC helicopter photographic surveying system. Proceedings of the XXth ISPRS Congress. Istanbul, Turkey

Kerle N, Oppenheimer C (2002) Satellite remote sensing as a tool in lahar disaster management. Disasters 26: 140-160

Kerle N, Stekelenburg R (2004) Advanced structural disaster damage assessment based on aerial oblique video imagery and integrated auxiliary data sources. Proceedings of the XXth ISPRS Congress. Istanbul, Turkey

Kumar R, Samarasekera S, Hsu S, Hanna K (2000) Registration of highly-oblique and zoomed in aerial video to reference imagery. Proceedings of the 15th International Conference on Pattern Recognition. Barcelona, Spain

Lucieer A (2004) Uncertainties in segmentation and their visualisation. ITC/Utrecht

Mausel PW, Everitt JH, Escobar DE, King DJ (1992) Airborne videography - current status and future perspectives. Photogrammetric Engineering and Remote Sensing 58: 1189-1195

Mitomi H, Matsuoka M, Yamzaki F (2002) Application of automated detection of buildings due to earthquakes by panchromatic television images. Proceedings of the 7th National Conference on Earthquake Engineering. Boston, MA

Mitomi H, Yamzaki F, Matsuoka M (2000) Automated detection of building damage due to recent earthquakes using aerial television images. Proceedings of the 21st Asian Conference on Remote Sensing. Taipei, Taiwan

Ogawa N, Hasegawa H, Yamaguchi Y, Matsuoka M, Aoki H (1999) Earthquake damage survey methods based on airborne HDTV, photography and SAR. Proceedings of the 5th U.S. Conference on Lifeline Earthquake Engineering. Seattel, USA

Ogawa N, Yamazaki F (2001) Photo-interpretation of building damage due to earthquakes using aerial photographs. Proceedings of the 12th World Conference on Earthquake Engineering. Auckland, New Zealand

Ojala T, Pietikäinen M (1999) Unsupervised texture segmentation using feature distributions. Pattern Recognition 32: 477-486

Ojala T, Pietikäinen M, Maenpaa T (2002) Multiresolution gray-scale and rotation invariant texture classification with local binary patterns. IEEE Transactions on Pattern Analysis and Machine Intelligence 24: 971-987

Otani H, Aoki H, Yamada M, Ito T, Kochi N (2004) 3D Model measuring system. Proceedings of the XXth ISPRS Congress. Istanbul, Turkey

Ozisik D, Kerle N (2004) Post-earthquake damage assessment using satellite and airborne data in the case of the 1999 Kocaeli earthquake, Turkey. Proceedings of the XXth ISPRS Congress. Istanbul, Turkey

Pietikäinen M, Ojala T, Xu Z (2000) Rotation-invariant texture classification using feature distributions. Pattern Recognition 33: 43-52
Pollefeys M, Gool LV, Vergauwen M, Verbiest F, Cornelis K, Tops J, Koch R (2004) Visual modeling with a hand-held camera. International Journal of Computer Vision 59: 207-232
Randen T, Husoy J (2002) Filtering for Texture Classification : a comparative study. IEEE Transactions on pattern analysis and machine intelligence 24: 971-987
Showalter PS (2001) Remote sensing's use in disaster research: a review. Disaster prevention and management 10: 21-29
Walter LS (1994) Natural hazard assessment and mitigation from space: the potential of remote sensing to meet operational requirements. Proceedings of the Natural hazard assessment and mitigation: the unique role of remote sensing. Royal Society, London, UK
Weindorf M, Voegtle T, Baehr H-P (1999) An approach for the detection of damages in buildings from digital aerial information. In: Wenzel F, Lungu D and Novak O (eds) Vrancea Earthquakes: Tectonics, Hazard and Risk Mitigation. Dordrecht, Netherlands Kluwer Academic Press: 341-348
Yamazaki F (2001) Applications of remote sensing and GIS for damage assessment. Proceedings of the 8th International Conference on Structural Safety and Reliability. Newport Beach, USA

Abilities of Airborne and Space-Borne Sensors for Managing Natural Disasters

Mathias J.P.M. Lemmens

Section GIS Technology, OTB Research Institute for Housing, Urban and Mobility Studies, TU Delft, P.O. Box 5030, 2600 GA Delft, the Netherlands.
Email: t.lemmens@otb.tudelft.nl

Abstract

Mankind is putting increasing pressure on the one and only Earth his race has to share, not only by sheer numbers but also, unfortunately, wasteful lifestyle. Those numbers and that lifestyle are causing many areas to become rapidly more vulnerable to a wide range of technological, environmental and natural hazards. It is generally agreed that the availability of proper geo-data is crucial for the entire management cycle of disasters. The Geo-information technology community has produced and continues to produce floods of airborne, space-borne and other accurate, timely and detailed data. All these data are well suited as sound foundation for disaster management. The Geo-information technology community has also developed and continues to develop sophisticated technology to process and analyse the data in order for it to arrive at the right information at the right time and to disseminate it to the right persons. The present paper provides backdrops on the abilities of airborne and space-borne sensors for managing natural disasters.

1 Introduction

Disasters are of all ages. When they strike, more often than not they strike suddenly and ruthlessly. The damage disasters cause fills us with deep awe for the unimaginable forces natural phenomena are able to release. Annual economic loss associated with natural disasters doubled every decade of the last half century. Economic loss resulting from natural hazards even tripled in the course of the nineties and now amounts to an annual one million millions Euro. The Christmas 2004 tsunami disaster is unprecedented in terms of the amount of sudden loss of life, injured victims and demolished constructions. However, in terms of economic losses, this natural disaster, which did overwhelmingly affect nearly every coastal zone of the Indian Ocean, seems to be of a quite modest extent. Indeed, the ratio of actual damage and financial losses in the poor regions of the world is much more profound than in the richer ones. This seems to be an unavoidable social phenomenon, notwithstanding its unfairness.

Together with the Yangtze River summer flood of 1998 and the 2001 and 2003 earthquakes hitting Gujarat, India, and Bam, Iran, respectively, the Christmas 2004 tsunami disaster shows that management of natural disasters is urgent and should be given high priority on the international agenda. Disaster management involves many diverse activities. These activities can be grouped into five subsequent stages. The first stage concerns *assessment*: inventorising the sensitivity of the region to certain types of catastrophes. In this stage the risks and danger for human life and environment are determined. The second stage involves *mitigation*. When the risks and danger are known, one may start to take provisions to make the region less vulnerable to the occurrence of the catastrophes to which the area is sensitive. Proper land use planning and management, and taking strengthening provisions are the actions to be carried out here. In the third stage – *preparedness* – planning of emergency aid and development of scenarios and monitoring systems are central together with the establishment of early warning systems.

The urgency of the activities in the above three stages is often hard to understand, because they have to be carried out when the sky is still seemingly cloudless. Nothing has happened yet; there is no urgency to force authorities to put effort into these activities. Given the necessary financial resources and all the other priorities many countries are facing, authorities will thus often feel no drive to come into action. This is apparently different from the fourth and fifth stage. Now, the catastrophe has actually struck. Thus authorities cannot afford to keep their hands crossed; the gravity of the catastrophe creates an obligation to respond rapidly. The fourth stage – *response* – is the most dramatic stage. The catastrophe caused unthinkable human suffering and environmental damage. Rescue teams will attempt to save lives, injured people will be cured and nursed, and relief will be offered to sufferers by food support and provisional housing. This stage is world news for a few weeks and given high priority by all news stations all over the world. The dramatic images displayed during these days are sometimes all that the general public will remember of the disaster for many years. When the general public has gone back to the order of the day, the fifth stage arrives: *recovery*. In the recovery stage, actions are undertaken so that survivors can, in the foreseeable future, pick up their daily lives again. This stage thus mainly consists of revitalization and reconstruction. Houses are rebuilt; roads and railroads as well as other works of infrastructure are repaired. An important although often neglected part of this stage is strategic development; this means tackling the question of how to prevent the area from future disasters in order to secure a safe and sustainable future. An essential part of this tail stage is thus that it acts as a driving force and fosterer for starting and keeping vivid the initial stage: assessment.

Prior to, during and after a disaster, taking appropriate action is critical. For that the right information should be available at the right time to the right persons. The appropriateness of information depends on:

- the type of disaster one wants to cope with
- which of the above five stages the information is needed for
- the geographical scale of the disaster.

2 Type of Disaster

Disaster risks stem from sudden energy release in one of the three basic environmental compartments: air, water and land. Table 1 gives examples of natural events, which may take place in each of these compartments and their consequences. Note that all of the mentioned events may result in one or more consequences as indicated.

Air	Water	Land	Consequences
Cyclones	Sea Floods	Earthquakes	Loss of human lives
Windstorms	River Floods	Volcanic activity	Epidemiological diseases
Lack of rainfall (drought)	Tsunami	Biological Plagues (e.g. locusts)	Destruction of buildings
Abundance of rainfall (flood)		Forest Fires	Destruction of land
		Land slides	Pollution

Table 1. Compartments and examples of natural events

The extent of disasters may vary widely over space and time.

2.1 Spatial Extent

The spatial extent of the effects of a natural phenomenon may vary largely from a local neighborhood to the entire globe. More specifically, the spatial extent of a natural disaster may be of:

- spot size (e.g. volcanic activity)
- local (e.g. earthquakes)
- zonal (e.g. cyclones, sea floods, river floods)
- regional (e.g. drought, forest fires)
- continental (e.g. drought, plagues)
- global (e.g. sea level rising)

Many civilizations all over the world flourish as a result of being located on river deltas. To keep all feet dry in the lowlands, two enemies have to be combated: flood threat from the sea and flood threat from the river(s). On the other hand, water is a vital source of life. To stay on friendly terms with your biggest enemy requires an almost divine level of management. In addition to their preference for residing in river deltas, the human race is also inclined to settle in earthquake-sensitive and volcanic-active areas. To prevent human beings from death or injury early-warning systems have been developed based on advanced geo-information technology. For example, Murai and Araki (2003) developed a new, advanced method for earthquake prediction using GPS network triangles.

2.2 Temporal Extent

When considering the temporal extent it is appropriate to distinguish two stages, which correspond to the above pre-disaster stage and post-disaster stage. The first one - the culmination period - involves the pre-disaster period in which the destructive forces are culminated. Some disasters are caused by a slow process of culmination of forces over a longer time (e.g. centuries or millennia). For example for an earthquake the culmination period may take centuries, whilst for a cyclone this period may be counted in terms of weeks. The second stage concerns the occurrence of the event and is determined by the abruptness of energy release. An earthquake may come overnight and the released energy may strike hard, sudden and short. A cyclone may harass a zone over an entire continent for several days.

The three dimensions of appearance - spatial extent, culmination period and abruptness of strike - determine which type(s) of geo-data are the most fit for use. For example volcanic activity has a very small spatial extent and will strike abruptly. To monitor properly a natural phenomenon with these characteristics, permanent observation of the spot is vital. Because the spatial extent is of spot size the use of a ground-based station suites the best. Over the last five years ground-based infrared monitoring systems have been designed and tested at several active volcanoes (Harris, et al., 2003). The thermal time series measurements obtained at each of these active volcanoes allow for timing for the onset of eruptive events.

2.3 Observable Quantities

The above brings us to an important issue: which observable quantities are essential for arriving at an optimal solution in the most cost-effective way? Such essential parameters should be observable or at least be derivable in a sufficient accurate way from observable quantities by using computer models. For economical reasons, the number of parameters should also be few as possible, preferable just one. For monitoring of volcanic activity, temperature is such an essential observable. Continuously recorded thermal data provide information on the thermal evolution of a volcano. From proper analysis of time series of thermal data important features can be derived such as event frequencies, repose time, ejection velocities and time of event onsets (Harris, et al., 2003). River flood management would benefit a lot when information on the amount of dropped rainfall together with the amount of forecasted rainfall and rate of river water level rise provide information about the risk of flood, and which measures are to be taken to tackle the threat in the best way, such as strengthening of dikes, evacuation or willingly flooding of areas of little economical value in order to save densely populated regions where activities of high economical value take place. Such a scenario requires that in previous stages a bunch of data are collected such as land use and the three-dimensional topography in the format of a Digital Elevation Model (DEMs) of high accuracy and high resolution. Airborne Lidar is a very suited data acquisition technology for the generation of DEMS of river areas, urban conglomerates and coastal zones. It is sometimes underestimated that one of the prerequisites to keep the disaster management system effective is that the database has to be maintained on a permanent or semi-permanent basis.

In addition it is necessary that in the pre-disaster stage, the phenomenon has been understood so well that proper models can be determined and transferred to analytical models, which can be digitally handled by computer. In this modeling process, remotely sensed data play a key role for arriving at an understanding of the physical behavior of disasters. For example, Villegas (2004) reports on using Landsat_TM imagery for understanding the eruption of the Nevado del Ruiz Volcano, Colombia on 13th

November 1985. With 23,000 people being killed it was the fourth worst volcanic disaster in recorded history.

3 Sensor Technology

The last decade has thrown up many changes in the way we collect information of the earth surface. From the technological point of view, the main driving forces behind these changes have been the development of new sensors and the ICT revolution. Examples of such sensors are high-resolution airborne and space-borne digital cameras, the Orbiting Positioning Systems (GPS and Glonass), Laser Altimetry (airborne and terrestrial Lidar), and Interferometric SAR, which enable the collection of geo-data with a level of detail, accuracy, automation and collection rate never before seen. The accuracy and level of detail of the geo-data produced by these sensors, not only support the national economy in fields like agriculture, forestry, mining, water and coastal management, marine fisheries and sustainable development, but they are also very useful for disaster management purposes. This progress is accompanied by two other technological revolutions: computer technology and telecommunication in general, and more specifically the rapid growth of the internet and its use. These developments are still rapidly expanding. Table 2 and 3 provide values of the basic features of important space-borne remote sensing observation systems. The choice of a data type or mix of data types heavily depends on the disaster management stage, for which the data is used, and the spatial extent and temporal extent of the disaster type. A geo-data type, of which the importance can hardly be overlooked, is the base-map. The function of a base-map is twofold. First of all the base-map acts as the geometric infrastructure to which all other geo-data sets are referenced in order to match a certain spot or feature in the data set to the proper location in the real world. A second function of the base map is related to its thematic contents. Topography, buildings and so on provide information on the pre-disaster state of the real world. By comparing the base-map with post-disaster data, an inventory of the damage can be made and decisions on proper actions being taken. The presence of a base-map accompanied by proper use may thus rescue people and save economical valuable goods. It is a general observation that combination of data sets is crucial when examining remotely sensed data.

Satellite	GR (pan)	GR (ms)	SS/SW	SR (ms)
Aster	-	VNIR 5m,SWIR 30m, TIR 90m	60x60km	VNIR, SWIR, TIR
Orbview 3	1m	4m	8x8km	B, G, R, NIR
Quickbird	61cm	2.45m	16.5km	B, G, R, NIR
SPOT 2 & 4	10m	20m	60x60km	G, R, NIR, SWIR
SPOT 5	2,5m & 5m	10m, 20m, 1km	60x60km	G, R, NIR, SWIR
Ikonos	1m	4m	11.3km	B, G, R, NIR
Landsat TM	15m	30m, TIR 60m	170x183km	6 bands +TIR
Resourcesat-1 LISS IV	5m	5m	23,9km	G, R, NIR, SWIR
CBERS 1	20m	20m	113km	B, G, R, NIR

Table 2., Spectral and spatial characteristics of important high resolution space borne earth observations sensors. GR: Ground Resolution, pan: panchromatic, ms: multispectral, SS/SW Scene Size or Swath Width, SR: Spectral Resolution, NIR Near-Infrared, SWIR: Short-Wave Infrared, TIR: Thermal Infrared. (Source: GIM International, 2004, vol. 18, nr. 7)

Satellite	Nadir RT	Off-nadir RT	Stereo
Aster	16d	Var	Y/IT (64s)
Orbview 3	15d	3d	Y/IT+AT
Quickbird	1 - 3.5d	-	Y/IT+AT
SPOT 2 & 4	26d	1-3d	Y/AT
SPOT 5	26d	1-3d	Y/AT
Ikonos	-	3d	Y/IT
Landsat TM	16d	N	N
Resourcesat-1 LISS IV	24d	5d	Y/AT(5d)
CBERS 1	26d	5d	Y/AT(5d)

Table 3. Viewing characteristics of important high resolution space-borne earth observations sensors. RT: Revisit Time, IT: In-Track Stereo, AT: Along-Track Stereo. (Source: GIM International, 2004, vol. 18, nr. 7)

4 Dissemination to the Right Persons

The end-users of geo-information in a disaster management context use geo-information commonly occasionally. One will be in general unaware of the special characteristics and limitations of geo-data. The approach of geo-data in terms of quality, which has become a second nature to the surveying and mapping community, is an unknown line of thought for laymen. For this reason, it is not only the delivery of the data set that needs to be at issue but also its proper use by the end-user. As a consequence, part of the product delivery by the analyst should include advisory about its use. A good understanding of the nature and size of errors that may be present in the data set is of essential importance to every user to gain insight into the value of the information derived. Another need involves easily understandable visualizations. Since end-users are operating under superstress the presentation (visualization) of the derived information should be simple and straightforward to understand. End users in an emergency should be able to interpret a map in half a minute. Therefore, analysts should be trained in basic cartographic principles and the most effective visual presentation of quantitative information. Since good communication is crucial, it is proper to state here that presentation is more important than content. Rather than making geo-data available, the analyst should thus take care of an optimal use of the data by the end-user. For that purpose, it is necessary that disaster managers are thoroughly informed on GIS and remote sensing capabilities, without going too deep in technological issues.

5 Discussion

GIS analysts are used to process data in a vector environment. Therefore they underestimate the importance of raster data. They use remote sensing data often just as visual backdrop without using the wealth of information present in the digital reflectance values. This is one of the lessons learned after the Twin Tower disaster (Huck and Adams, 2002). The many uses and analytical potential of raster data thus need to be promoted. Wider use should be made of programs specifically designed for the processing and analysis of grid-based remotely sensed data.

Today risk assessment, mitigation, preparedness, response and recovery are of major concern to governments at all levels: national, state, provincial and municipal. Both authorities and disaster managers increasingly acknowledge that acting without accurate detailed and timely geo-information results in human suffering and huge economic loss. Fortunately, today's geo-information technology is able to provide the fundament upon which disaster managers can found their vital decisions. The quality of geo-data should be adjusted to the aim of the application; it should be fit for use. Unfortunately, during the 1990s an invisible disaster has blown through management land. The geo-data providers of the past were mainly restricted to authorative and respectable governmental agencies, which worked on a non-commercial basis. Since the early 1990s the common consent is that geo-information is a commodity like any other good. As a result, governmental organizations have made a significant step back in providing geo-information to citizens and organizations. Privatization and commercialization has brought much geo-data to a diversity of commercial firms, which are not seldom allied to former governmental agencies. The user may buy the geo-data off-the-shelf and the physical delivery of the data set may go via the internet. One of the consequences of this is that the non-central collected data is stored at distributed locations whilst the quality may be heterogeneous. Today professionals, who are manager by training, run surveying and mapping organizations. Irrespective in which direction they gaze they seem to see just one steering parameter: currency, appearing in three manifestations: return of investment, cost-efficiency and profit. From this business-centered approach, which is in many ways a decent approach of doing business, they will grant the company of which the tender shows the lowest production costs not the company which offers the best quality. However, short-term gains may convert in huge loss on the longer run when the use of collected geo-data results in wrong decisions.

Today, countries under development, which were ruled not that far back in history by foreign nations and mapped by them solely for reasons of exploitation, are becoming increasingly aware of the importance of geo-information for sustainable development purposes. On 27th September 2004 Nigeria launched its first earth observation satellite. The micro-satellite, weighting 100kg, provides every five days images with a spatial resolution of 32m and a spectral resolution of three bands. Notwithstanding this noteworthy achievement, rulers of countries often still do think in military and exploitation terms. During the NigeriaSat-1 workshop, recently held in Abuja, participants recognized that the federal government shows little obligation to provide funding and political support to a national space program (Kofoniyi, 2004). Indeed, it happens too often that governments consider geo-information primarily as a strategic means for consolidation of own power. Consequently, they keep geo-information out of the hands of managers in the private and public sector, frustrating those who have in mind the well-being of the country over a period of time which exceeds greatly own live span. Within this context it is also important to note that Geo-information should be made available free of charge in the framework of disaster management.

6 Conclusions

Without information human is living in blindness. Without knowledge - the ripen outcome of processed information - the sum of all human activities and actions becomes nothing more than null and void. According to the United Nations, the global community is facing a critical challenge with respect to disaster management: "How to better anticipate - and then manage and reduce - disaster risk by integrating the potential threat into its planning and policies?" Geo-information technology has a vital role to play in this challenge. Today, given the abilities of the available technology, a proper approach of disaster management has become mainly a matter of willingness and prioritizing rather than of technology.

References

Harris A et al (2003) Ground-based infrared monitoring provides new tool for remote tracking of volcanic activity, Eos, vol 84, no 40, pp 409-424

Huck CK, Adams BJ (2002) Emergency response in the wake of the World Trade Center Attack: the Remote sensing Perspective, MCEER Special Report Series, Volume 3, MCEER, Buffalo, New York

Kofoniyi O (2004) NigeriaSat-1 Satellite National Workshop, GIM International, vol 18, no 12, pp 55-57

Murrai S, Araki H (2003) Earthquake prediction using GPS: a new method based on GPS network triangles, GIM International, vol 17, no 10, pp 34-37

Stevens D (2005) Emergency response satellite imagery in developing countries: Space-based technologies for disaster management, GIM International, vol 19, no 2

Villegas H (2004) Volcanic disaster seen from space: understanding eruptions using Landsat-TM, GIM International, vol 18, no 5, pp 68-71

The Use of GIS Technologies within the NOAA Climate Prediction Center's FEWS-NET Program

Timothy B. Love

NOAA Climate Prediction Center, 5200 Auth Rd; Rm 811D, Camp Springs, Maryland 20746 USA.
Email: Tim.Love@noaa.gov

Abstract

Geographical Information Systems applications are used daily within the United States Agency for International Development (USAID) Famine Early Warning System Network (FEWS-NET) to provide relevant meteorological and climatic information to support weather-related natural disaster management efforts (Love, 2004). It is through the use of GIS technologies that a more user friendly, value-enhanced, and distinctly directed set of information may be created and disseminated to the end user. However; the data flow cycle does not end there. Users will ingest this information and modify it for their own purposes, resulting in a new product possibly benefiting the original institution. It is the purpose of this paper to inform the disaster management community of our experience with weather and climate related GIS technology in order to gain critical feedback in the direction of improving the data and services we provide and to add to the common pool of GIS knowledge. Within this document, examples of manual and automated Climate Prediction Center's FEWS-NET products will form the base of the discussion. History of GIS use within FEWS-NET will provide background information, while a discussion of our experience determining user requirements will give a view into the task at hand. A summary of possible future directions will culminate the report.

1 Background

The first attempt to incorporate Geographic Information Systems within the NOAA Climate Prediction Center (CPC) was certainly not as fruitful as it could have been; however the initial experience with GIS did provide some useful knowledge of core processes and software capabilities. In the mid to late 1990's, a project began with the goal to convert many of CPC's Joint Agricultural Weather Facility products to GIS format. Without readily available guidance, training, or support, the learning curve was too steep to justify continuation of the project, and the plan was scrapped in favor of a more traditional methodology. The initial attempt at GIS incorporation had failed, partially due to the fact that the group was not enlightened to the entire realm of possibilities that the technology had to offer.

Toward the end of the 1990's, ESRI's ArcView Version 3 was introduced to members of the CPC's Joint Agricultural Weather Facility, whose task was (and continues to be) to monitor global weather and climate conditions with respect to agriculture vulnerability. Though little automation occurred at this time, group members were able to ingest data from internal groups as well as outside organizations into the GIS software for the purpose of locating potential areas of crop damage. For example, regions where South Africa winter wheat crops provide important regional or international food stock were monitored for adverse weather conditions. More specifically, GIS layers of ground-based station temperature data and satellite-based snow cover fields were manually placed over country borders, and given known thresholds for crop damage, meteorologists were able to pinpoint trouble spots and pass this information along to decision makers. By analyzing this information and formulating a conclusion about the negative weather effects on the wheat crop, a potentially disastrous food emergency situation was mitigated, with an early response initiated to transport other food reserves. When examining the South Africa station temperature data, a choice needed to be made whether to import this field as irregularly spaced point data or to use an analysis tool to create a gridded array of values. In this case, original point data were used in areas throughout the country where the concentration of gauges was high, while interpolated fields were used for areas far from any station locations. Though this choice was an important decision to be made in this early time of CPC GIS activities, it remains a question that must be dealt with in many current weather-related parameter datasets to this day and will be discussed later in this paper.

As JAWF members began working closer with the CPC FEWS-NET group, it became apparent that GIS technology was now ready for application within the project (or that CPC FEWS-NET was ready for GIS!), and it was incorporated into an early version of a Southern Africa weather monitoring product. By 2001, ESRI had discontinued production of ArcView Version 3 and had moved to ArcGIS 8. Though transition between the two software versions was fairly straightforward, it did require time to become familiar with intricacies of the latter. Once the FEWS-NET group became familiar with ArcGIS 8, it was used rudimentarily in the operational weekly production of CPC Southern Africa Weather Threats Assessments. Acting more like a mouse-based drawing software such as CorelDraw, GIS was used to outline regions of potentially hazardous weather and place them over a set of standard country borders. ArcGIS was not used in any part of the actual meteorological analyses of weather patterns, but areas of forecast extreme cold temperatures, torrential precipitation, or abnormal dryness were drawn on the map as shapefiles. It was not until a tighter collaboration existed between partners of FEWS-NET outside of NOAA that a clearer vision of the benefits of GIS technology was obtained.

2 Africa Weather Hazards Assessment: Example of a Manually Produced GIS Product

The FEWS-NET Southern Africa Weather Threats product (Figure 1) evolved through the use of GIS, extensive user feedback, and collaborating organizations' participation into the current form that has been renamed the Weekly Africa Weather Hazards Assessment (AWHA). The product, which incorporates many of GIS' most useful applications, is designed to give management personnel interested in Africa food security an early warning of potentially compromising weather related conditions. Precisely, the scope of the AWHA product deals with short and long term drought, regional flooding, excessive temperatures, and other severe weather events such as tropical storms and intense cold fronts. Transformation of the assessment during the past few years was a direct result of user feedback and increased knowledge of GIS technology. Understanding that one of the most important uses of GIS is a spatially precise, simultaneous comparison of information layers, knowledge of this concept was passed on to the user community to gain critical feedback. Almost immediately, it was recognized that while the end product served users an analysis of recent and near-term trends in weather related food security conditions, a spatial representation of actual food producing regions was absent from the map. Working with members of the Food and Agriculture Organization of the United Nations and the United States Geological Survey (USGS), areas of typical, widespread cropping were established. This data was subsequently converted to polygon shapefile format with an identical geographic projection to that of the country border base layer and was imported to form the background map of the product.

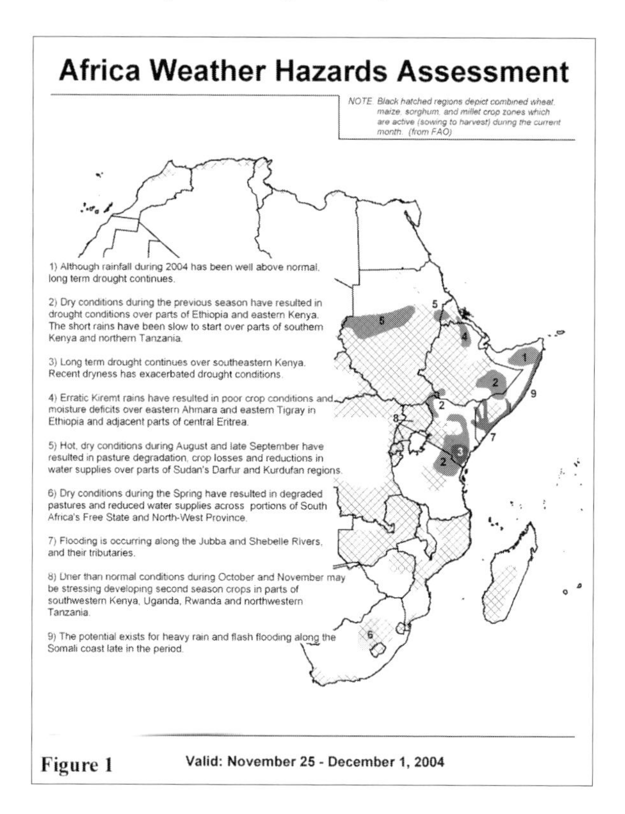

Fig. 1. The CPC FEWS-NET Africa Weather Hazards Assessment is created using manual GIS techniques and automated input data. The product provides an early warning of weather and climate related hazards throughout the Africa continent.

The result of this effort was a layer of known planting areas that could be easily compared to regions of persistent dryness so that a more precise target of problematic conditions could be identified. A similar data identification process yielded the remainder of layers to be used in the GIS Hazards project. Currently, the available layers used in creating each weekly assessment include:

- Country and administration borders (polygon shapefile)
- Monthly agriculture land use (polygon shapefile)
- Lakes (polygon shapefile)
- Streams, rivers, and river basins (polyline shapefile)
- Regional roads and transportation corridors (polyline shapefile)
- City and town locations (point shapefile)
- Topography and elevation (grid)
- Weekly accumulated satellite-derived precipitation (grid)
- Forecast 7-day precipitation from meteorological model (grid)

An example showing the use of hydrological GIS layers by Climate Prediction Center scientists to evaluate an area of potential stream flooding in the southern Somalia region is shown in Figure 2. In Figure 2a, a grid of one kilometer resolution Digital Elevation Model (DEM) data is imported into the desktop version of ArcGIS 9.0. To complement this dataset, USGS HYDRO1k stream vector data is used. Since the overall area of flooding concern is known based on past events, it is determined that three main rivers have the potential for overflowing their banks if rainfall is sufficiently intense over the corresponding catch basin. A simple selection process within the ArcGIS software yields all streams and tributaries which drain into the main southern Somalia rivers (Figure 2b). The addition of a high level river basin dataset enables the creation of spatially represented catch basins for each of the river paths (Figure 2c). A daily precipitation estimate that uses satellite based rainfall sensors to supplement irregularly distributed ground based rain gauge data is produced for the entire Africa continent. This gridded dataset is summed into seven day accumulations, converted to GIS GRID format, and added to the Africa Weather Hazards project. Overlaying the past week of rainfall with relevant river polylines and most recent flood polygon analysis layers yields an image that allows the user to qualitatively determine the latest areas of probable inundation (Figure 2d). By also adding gridded accumulated rainfall forecast datasets from local National Centers for Environmental Prediction model outputs, an overlay with river catch basin layers provides a quick analysis of most likely flooding during the next week (Figure 2e). By synthesizing these GIS products while aggregating critical feedback from partner organizations and taking important field reports, an accurate depiction of short term river flooding is added to the Weather Hazards document (Figure 2f).

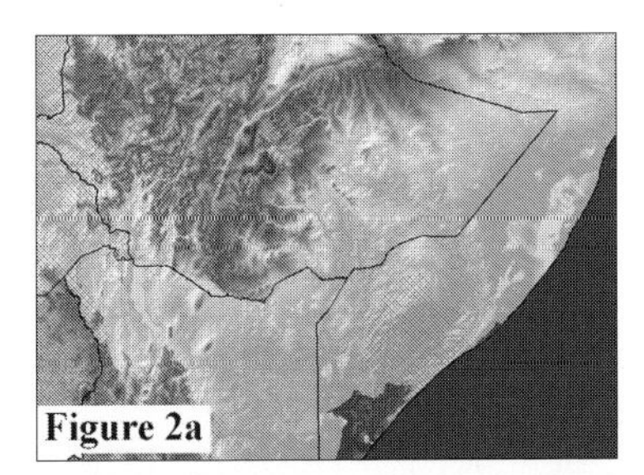
Figure 2a

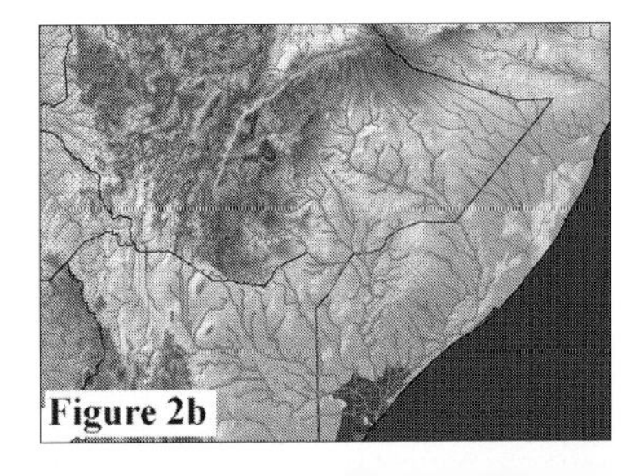
Figure 2b

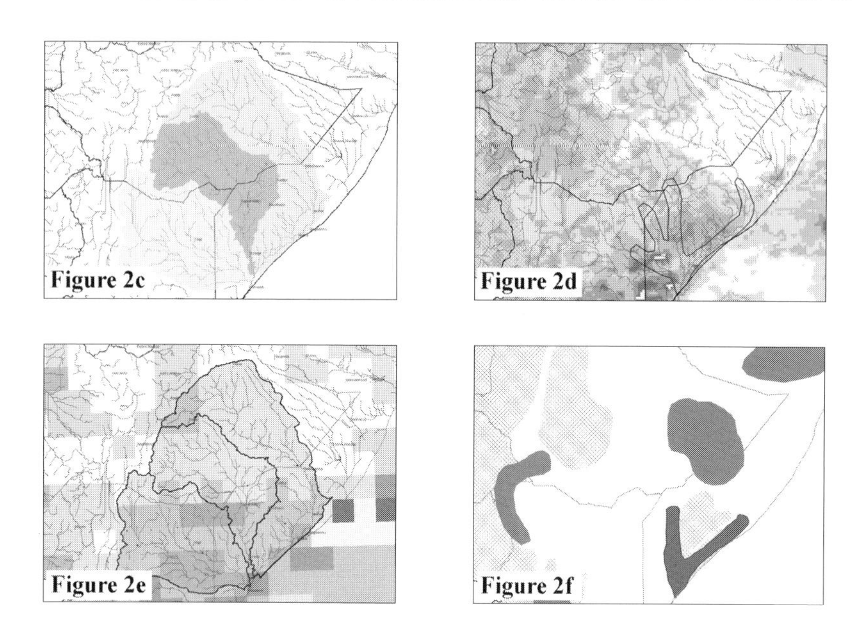

Fig. 2. Example components used in determining flood risk zones for the Africa Weather Hazards product. a: an elevation layer with appropriate hill shading; b: known area rivers prone to flooding after heavy rainfall; c: segmented river basins for determining flow of rainfall; d: recent weekly gridded precipitation with past the most recent flood hazard region; e: forecast rainfall from the GFS model with river basin layer; f: final river hazard flood risk polygon used in the Hazards product

Much of the data used in the analysis of African weather hazards is global in spatial extent, and thus may be applied to other international locations. With this in mind, the expanding data availability is a precursor to a global weather related natural disaster analysis. A process similar to the previous example may be an excellent problem solving technique that could be applied to a vast array of situations. After a statement of the problem that must be solved, a thorough analysis of the most applicable type of input data is one of the most crucial first steps in assessing the situation. Without the proper information, the path leading to a solution will not be as straightforward as possible. Once the input components are determined, a potentially useful set of data layers must be transformed geometrically to an identically similar spatial coordinate system. Only then will an accurate analysis of the situation become feasible.

3 Automated GIS Products

The previous analysis technique used manual interpretation of a variety of GIS layers to determine the spatial extent of river flooding. Though many of the imagery layers (such as the creation of river basin zones) were created using a hands-on approach, other products are produced from automated GIS scripts. For CPC FEWS-NET purposes, all automation is accomplished via ESRI's Arc Macro Language (AML).

3.1 Satellite-Based Precipitation Estimates

Perhaps the most valuable tool that is currently used by scientists at the Climate Prediction Center's FEWS-NET group is the satellite-based rainfall estimate product, which provides a continuous spatial domain of analyzed precipitation over virtually the entire globe. This data type is of particular interest to meteorologists tasked with analyzing international areas of problematic hydrological dryness or wetness, including natural disasters such as drought and floods. The decision to use a satellite-based precipitation estimate rather than a ground-based rain gauge interpolation as a GIS project layer stems from the sparse spatial distribution of rain gauges and the irregular degree of available ground-based data in many countries.

Take for example the country of Sudan in eastern Africa. Due to internal conflict, the structure of their national meteorological agency is not sufficiently structured to maintain a network of countrywide surface-based rain gauges. The station data that is retrieved and transmitted to the global network is not free of errors, and the distribution within the country (roughly one station for every 100,000km^2) is certainly not adequate to capture the regional variability of precipitation. Unless the area of interest coincidently lies within close range of a specific rain gauge point, one must perform an interpolation on this set of point data to yield rainfall amounts elsewhere. Two main interpolation techniques that use irregularly spaced point data are generally used to accomplish this. The first method, an inverse distance weighting (IDW) procedure, performs the interpolation based solely on the distance a particular pixel is away from the point in question (Figure 3a). This procedure generally is appropriate for a somewhat evenly spaced set of point data. From the figure, one can see that each rain gauge point carries a moderate weight outward. In comparison, the second standard of interpolation, kriging (Figure 3b), uses both distance and direction to determine the interpolated value at a point. It can be seen that this technique produces a somewhat more homogeneous grid than that derived using solely inverse distance weighting.

Unfortunately, neither method captures both the natural spatial distribution and intensity of rainfall, with IDW failing to correctly interpolate the area and kriging having the tendency to miss the quantity.

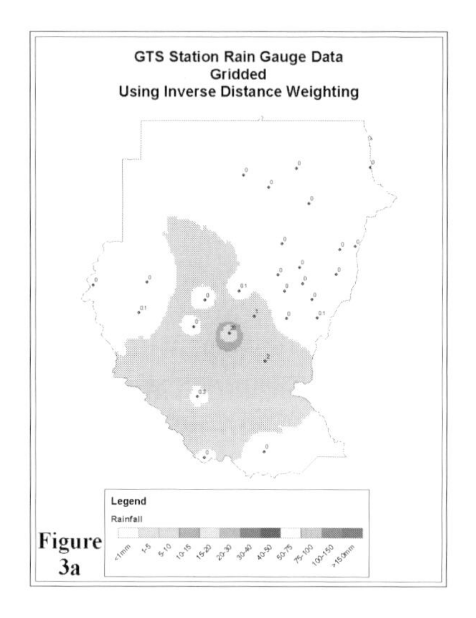

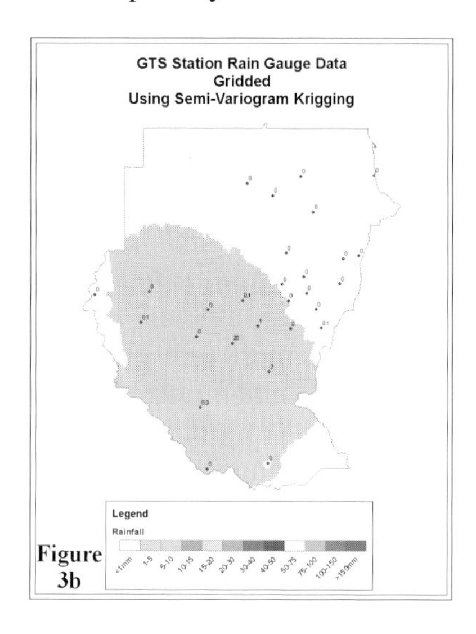

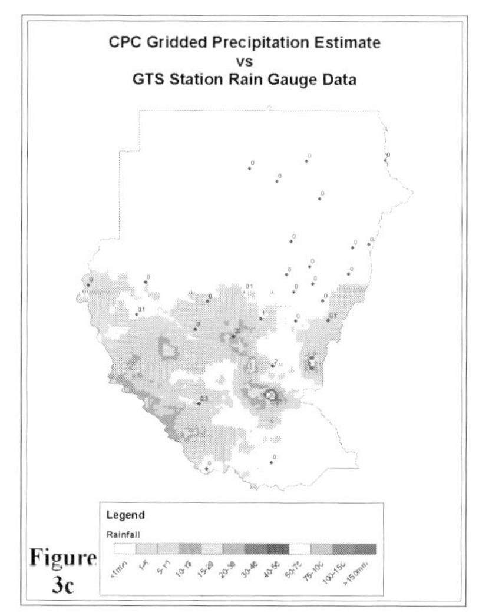

Fig. 3: Techniques of spatially represented rainfall over a selected 24-hour period in Sudan. a: station-based inverse distance weighted rainfall, exhibiting a poor spatial distribution pattern at distance from each rain gauge; b. a station-based kriging interpolation, showing a tendency to over-diffuse rainfall; c. a satellite-enhanced product which uses infrared and microwave sensors to estimate precipitation in areas without sufficient rain gauges.

Using GIS to overlay a faulty layer of information will preclude an erroneous assessment of the situation: Inaccuracies in the rainfall field used in the Weather Hazards Assessment project will make the scientist's analysis virtually worthless in detecting areas of potential river flooding. Thus, for international agencies that do not have access to non-public meteorological data, an enhancement is necessary.

Figure 3c presents the primary rainfall estimate used at the CPC (Xie et. al, 2002) (Xie et. al, 1996). The algorithm uses a combination of ground based rain gauge data and satellite based rainfall estimates to produce a spatially continuous gridded dataset that may be easily ingested into GIS. The fact that this precipitation estimate methodology has the ability to accurately capture rainfall at distance from a rain gauge point makes it a useful layer when attempting an analysis of hydrological phenomena over remote geographic locations. Since the product is disseminated to users in a geographical gridded format, adding this information to any GIS project is as straightforward as is the automated production of the dataset. The proper software (Workstation ArcGIS) and a few lines of fundamental AML code are all that are required to create the precipitation GRIDs on a daily basis. In the end, determining the end use of this data is the main concern that must be addressed before applying the product to an analysis.

3.2 Tropical Cyclone Monitoring

In an effort to enhance the Climate Prediction Center's tropical weather monitoring and produce a more specialized product for the end user, a GIS Southern Indian Ocean cyclone tracking project was recently initiated. The fact is that, though a wide range of similar storm tracking products are available from other international organizations, their characteristics do not meet the needs of the CPC FEWS-NET user. Again, it is apparent that a precise evaluation of the needs of a customer is crucial in the likelihood of project success. For this product, the customer required information that would provide an early warning of an imminent cyclone threat to the continent of Africa while integrating seamlessly into their routine schedule. Past methods of cyclone threat dissemination relied solely on forwarded emails of sometimes cryptic and irregular information from an outside group, and this certainly did not fit the need of the user. More often than not, the data provided via email was more of a burden than anything.

Through a process of information gathering, the requirements from the customer were determined: 1) Near real time cyclone position, intensity, and forecast information, 2) Archived cyclone track and intensity, 3) Graphical format that may be both posted to internet and emailed directly,

4) Available GIS formatted layers of all required information to allow overlaying of relevant information. Creating this product using GIS technology became the obvious, most straightforward direction to proceed. An example of an operational Indian Ocean Basin cyclone monitoring product is given in figure 4a.

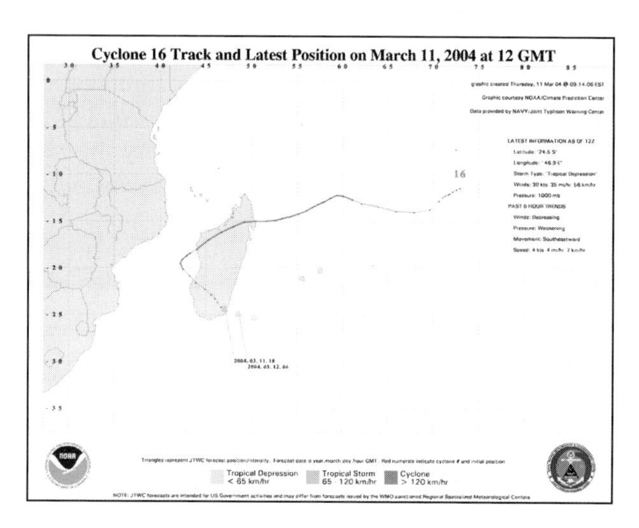

Fig. 4a. The CPC FEWS-NET GIS-based cyclone monitoring project: an example of an automated product using near-real time cyclone tracking data from the Joint Typhoon Warning Center

The graphic is produced solely using automated AML scripting techniques with the workstation ArcGIS software, though some initial data formatting is accomplished via Unix programming. The input information used to produce this cyclone monitoring graphic now covers virtually the entire tropical oceanic region. In fact, similar projects have been implemented, for the northern Indian Ocean and Western Pacific basins, which satisfy the requirements of the CPC FEWS-NET customer.

Figure 4b shows a possible application of the GIS data produced under the cyclone monitoring project. Through operational scripts, a set of GIS formatted cyclone data is produced. Continuing the GIS theme of data sharing and open access, these files are publicly available to any organization that may find the information useful. In Figure 4b, a shapefile of past cyclone position and intensity is overlaid upon an accumulated, gridded precipitation estimate for the Africa region. Due to the public unavailability of ground based rain gauge data, as discussed in section 4a, this satellite estimated rainfall field becomes the most appropriate dataset for the application. Sharing the automatically produced data from the CPC could allow the regional institutions to better coordinate their internal cyclone related disaster management efforts. Again, thc spirit of free and open information sharing has its reward.

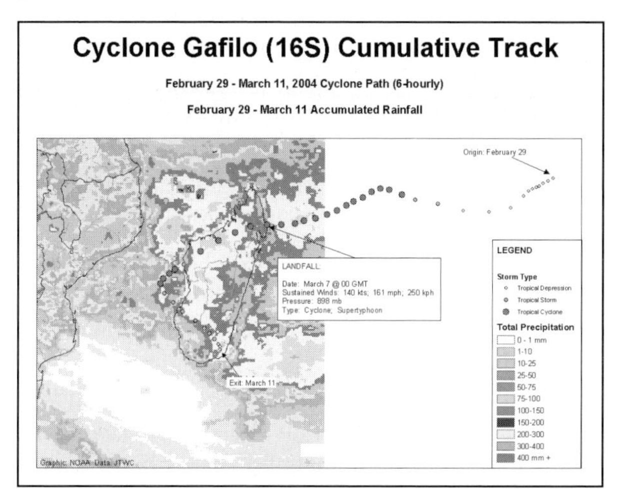

Fig. 4b. The CPC FEWS-NET GIS-based cyclone monitoring project: a combination of automated and manual techniques yields a concise, informative product

4 Determining User Requirements

Determining the user requirements of GIS formatted meteorological data by the disaster management community is the immediate crux of the of the CPC FEWS-NET Geographical Information Systems project. This is not a surprise at all, as this should be one of the fundamental issues for any project, and is critical to its success. The use of international meteorological and climatic data does have its specific problems however, and the fact that many areas requiring assistance from non-domestic sources do not have a sufficient structure in place to maintain a cohesive dataset of information demands the use of non-traditional practices.

Take for example, the renewed humanitarian effort that has taken place over the Afghanistan region since late 2001. Due to a nearly continuous 20 year period of internal livelihood disruptions, the state of affairs of their national meteorological institute has been in disarray and unable to provide sufficient weather related disaster management support. Rain gauges have been neglected, meteorological satellite feeds have not been received, and climatological records have all but been lost to the ages. International agencies have thus been providing assistance toward maintaining a system of meteorological monitoring while at the same time working to build internal capacity in the region. To support humanitarian efforts in the area, a first step required relevant transportation related information to be provided continuously to decision makers throughout participating organizations. Current and forecast weather products were required to be compatible with other information such as roads, towns, and rivers. To determine the best solution to the task at hand, an inventory of available information was warranted.

- Very few near-real time ground based rain gauge reports
- A somewhat larger set of rain gauge data available at delay
- Satellite-enhanced precipitation estimate data from various sources
- Gridded rainfall forecast data from many sources
- Gridded temperature forecast data at various resolution
- Existing GIS layers of roads, rails, rivers, lakes, towns

Due to the fact that near real time rain gauge information was nearly unavailable, the requirement of users that the data be available for quick analyses was unfulfilled. The poor spatial distribution of data did not meet field workers' request for information about road conditions in areas away from major country centers (all rain gauges were in the largest cities). On the other hand, the only available gridded precipitation estimate that covered the entire region continuously had a poor resolution of nearly 25 km. In order to fulfill the user's data request, a new gridded rainfall estimate was created by CPC, with a spatial resolution of 8 km and accuracy near that of the previous coarse resolution product. The selection of the appropriate rainfall forecast data was for the most part guided by resolution. The two possibilities were a 100 km resolution forecast out to seven days (EMC, 2003) and a 15 km resolution product (Grell et. al, 1994) out to two days. While the two day short term forecast was the appropriate dataset for users requiring high resolution forecast information over the local scale, the seven day forecast product was best suited for a more nonspecific hint of weather patterns in the medium time frame. Both products were determined to be useful for the task at hand and were selected to accompany the 8 km rainfall estimate in the package of disseminated data. Easily converted to the GIS GRID format, these three meteorological products were overlaid upon existing transit-related shapefile information, and a continuous rainfall vs. transportation analysis was created.

5 Future Directions

It is apparent, perhaps solely due to the continuously increasing number of user data requests, that GIS formatted meteorological information is becoming widely accepted by organizations throughout the international disaster management community. It is a goal of the CPC FEWS-NET project to convert the majority of its weather and climate related products to GIS format by the end of the year. Continuing the data sharing theme of GIS, these products will be made publicly available, without charge, to any user who may find the data useful. Of particular interest to the disaster management community will likely be a new global weather monitoring project that is currently gaining momentum. When completed, GIS formatted global weather forecast and analysis data will be openly available to the public. It will provide many of the necessary tools so that organizations dealing with disasters may have sufficient meteorological information available as needed to properly and rapidly analyze the situation.

An issue that has been receiving much feedback in recent times is the use of an Internet Map Server to display data created at the CPC. For the FEWS-NET group, this is somewhat problematic due to the nature of IMS in general. Many of the meteorological products created within the group have a dual target audience: a) International organizations' decision makers located away from the geographic area of interest, and b) Field workers on the ground. Due primarily to group b, most weather-related products must be sufficiently small in file size to accommodate slow internet connections. Therefore, converting the existing static products to the somewhat flashy IMS technique may compromise this requirement. This problem will continue to be explored throughout the future.

6 Summary

Geographical Information Systems applications are used daily within the USAID Famine Early Warning System Network to provide relevant meteorological and climatic information to support weather-related natural disaster management efforts. Phenomena such as drought, floods, excessive heat, cyclones, and other severe weather are monitored by the NOAA Climate Prediction Center for all FEWS-NET regions including Africa, Central America, Afghanistan, and Southern Asia. While various levels of humanitarian support are offered by the group, the common thread that holds partner organizations together is uninterrupted and accurate information exchange. The CPC FEWS-NET group has implemented GIS during the past few years, and its importance is clearly evident when performing situational analyses and distributing meteorological datasets. This data must be shared between regional and international partners in such a way to maximize value. A major goal for the CPC FEWS-NET project during the next year is to enhance its GIS capabilities by soliciting feedback from end users and to automate and reformat virtually every product created by other methods into GIS. A first task that must be completed is to determine the needs of partner organizations and investigate other groups' involvement with GIS technologies as they relate to humanitarian aid. Establishing user requirements and tapping into the resources and experiences of the GIS community as a whole is a crucial step in the process of providing weather and climate related humanitarian support using Geographical Information Systems.

References

Environmental Modeling Center (2003) The GFS Atmospheric Model. NCEP Office Note 442, 14 pp

Grell GA, Dudhia J, Stauffer D.R. (1994) A Description of the Fifth-Generation Penn State-NCAR

Mesoscale Model (MM5). NCAR Tech. Note NCAR/TN-398+STR, 122 pp

Love T (2004) GIS Enhancement of Climate Prediction Center's Africa FEWS-NET Products. Preprints, 20th Conf. on IIPS, Seattle, WA, Amer. Meteor. Soc.

Xie P, Yarosh Y, Love T, Janowiak JE, Arkin PA (2002) A real-Time Daily Precipitation Analysis Over South Asia. Preprints, 16th Conf. of Hydro., Orlando, FL, Amer. Meteor. Soc.

Xie P, Arkin PA (1996) Analyses of Global Monthly Precipitation Using Gauge Observations, Satellite Estimates, and Numerical Model Predictions. J. Climate: 9, 840-858.

Geo-Information for Urban Risk Assessment in Developing Countries: The SLARIM project

Cees J. van Westen[1], Birendra Kumar Piya[2] and Jeewan Guragain[3]

[1] International Institute for Geo-Information Science and Earth Observation (ITC), P.O. Box 6, 7500 AA Enschede, the Netherlands.
Email: westen@itc.nl
[2] Department of Mines and Geology, Kathmandu, Nepal.
Email: piya03277@alumni.itc.nl
[3] Department of Local Infrastructure Development, Lalitpur, Nepal.
Email : guragain@alumni.itc.nl

Abstract

The aim of this paper is to present the first results of a research project entitled: Strengthening Local Authorities in Risk Management (SLARIM). The main objective of this project is to develop a methodology for spatial information systems for municipalities, which will allow local authorities to evaluate the risk of natural disasters in their municipality, in order to implement strategies for vulnerability reduction. The project concentrates on medium-sized cities in developing countries, which do not yet utilize Geographic Information Systems in their urban planning, and which are threatened by natural hazards (such as earthquakes, flooding, landslides and volcanoes). The methodology concentrates on the application of methods for hazard assessment, elements at risk mapping, vulnerability assessment, risk assessment, and the development of GIS-based risk scenarios for varying hazard scenarios and vulnerability reduction options, using structural and/or non-structural measures. In the development of elements at risk databases use is made of interpretation of high-resolution satellite imagery, combined with extensive field data collection, using mobile GIS. Although the methodology is primarily designed to assist municipalities in the decision-making regarding vulnerability reduction strategies, the resulting databases are designed in such a way that they can also be utilized for other municipal activities. Within the project a number of case study cities have been identified. Here results are presented on earthquake loss estimation for the city of Lalitpur in Nepal, for buildings and for population losses. Databases have been generated of the buildings of the city, and of the subsurface conditions. Soil response modeling was carried out and vulnerability curves are applied to estimate the losses for different earthquake scenarios.

1 Introduction

The fast-growing world population is concentrating more and more into urban areas. Nowadays, almost half of the world's 6 billion inhabitants already live in cities, and in the next thirty years it is predicted that 90 percent of the expected 2.2 billion newcomers will be born in cities in developing countries (USAID, 2001). Many of these cities are located in areas that are endangered by natural disasters, such as earthquakes, flooding, cyclones/hurricanes, landslides, volcanic eruptions, subsidence etc. It is estimated that over 95 percent of all deaths caused by disasters occur in developing countries and losses due to natural disasters are 20 times greater (as a percent of GDP) in developing countries than in industrial countries (Kreimer et al. 2003).

Local authorities in cities are responsible for the proper management of the area under their jurisdiction, and the well being of the citizens, which includes an optimal protection against disasters. Unfortunately, until recently most of the emphasis has been on the post-disaster phases, and mostly was under the responsibility of national civil defense organizations. Recently, the emphasis is being changed to disaster mitigation, and especially to vulnerability reduction. Since the International Decade for Natural Disaster Reduction in the 1990's many initiatives have been launched worldwide to assess and reduce urban vulnerability. Some example are the program on Risk Assessment Tools for Diagnosis of Urban Areas against Seismic Disasters (RADIUS, 2000), the Earthquakes and Megacities Initiative (EMI, 2002), and the Cities and Critical Infrastructure Project (Cities Project, 2004).

Due to the diversity and large volumes of data needed, and the complexity in the analysis procedures, quantitative risk assessment has only become feasible in the last two decades, due to the developments in the field of Geo-Information science. When dealing with GIS-based hazard assessment, elements at risk mapping, and vulnerability/risk analysis, experts from a wide range of disciplines, such as earth sciences, hydrology, information technology, urban planning, architecture, civil engineering, economy and social sciences need to be involved.

2 The SLARIM Project

In order to contribute to urban risk assessment in developing countries the International Institute for Geo-information Science and Earth Observation (ITC) launched a research project with the acronym SLARIM, which stand for "Strengthening Local Authorities in Risk Management". The main objective of this research project is to develop generic methodologies for GIS-based risk assessment and decision support systems that can be beneficial for local authorities in medium-sized cities in developing countries. For local authorities being able to handle this tool properly implies a lot of attention in this research for user requirements, institutional issues and spatial data infrastructure, connected with the methodologies of hazard and risk assessment on the one hand and the relevant DSS based GIS applications in urban planning and management (what can local authorities actually do with this data) on the other hand. The ultimate objective of this project is to improve the safety of communities, and consequently make them more sustainable and prosperous.

The methodology for the use of GIS in urban risk assessment and management is developed on the basis of a number of case studies. After carefully evaluation and visits to potential case study cities, a number of case study cities have been selected. The willingness of local authorities to participate actively in this project has been considered as one of the main criteria, besides the availability of data, and the types and severity of the hazards in the urban areas. Three case study cities have been selected (See Figure 1): Dehradun (India), Naga (Philippines) and Lalitpur (Nepal). In the following sections an example of the results of a number of the components will be given, namely on seismic loss estimation in Lalitpur (Nepal).

3 Earthquake Loss Estimation in Lalitpur, Nepal.

3.1 Case Study City

The Lalitpur Sub-Metropolitan City is located in the Kathmandu valley, neighboring the capital of the Kingdom of Nepal, Kathmandu. Lalitpur has a population of 163,000, in 35,000 households, according to the 2001 census. The municipality is divided into 22 wards. Lalitpur is one of the oldest cities in Nepal, supposedly founded in 299 A.D., with one of its most important periods during the Malla dynasty from 1200 – 1768.

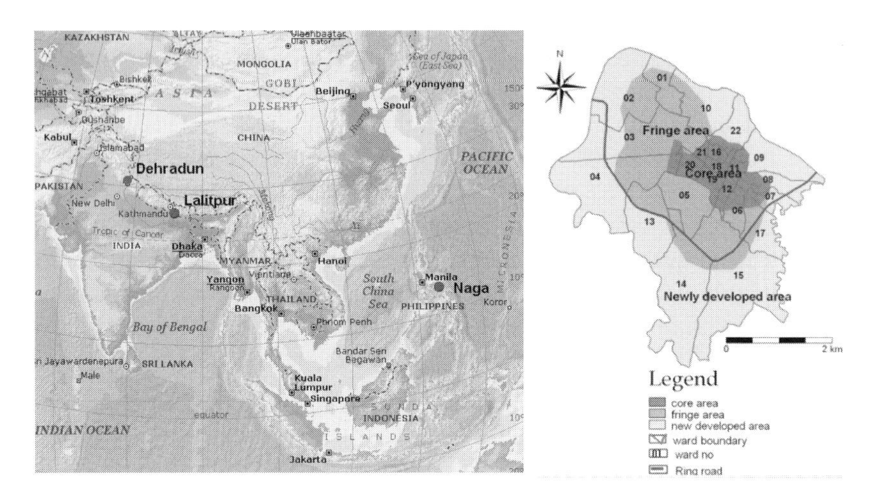

Fig. 1. Left: Location of the three case study cities for the SLARIM project. Right: Structure of Lalitpur

The old core area is famous for its cultural heritage, and has a very dense structure, with a majority of buildings with load-bearing masonry, with mud mortar and adobe. Many houses are built in a courtyard pattern, with very narrow streets. With the increase in population, and the vicinity of the capital, the city started to expand considerably, especially after the construction of the ring road in the 1980s. In the fringe area, which was developed between the core area and the ring road, the majority of buildings are masonry with brick in cement and RCC. In the last year, also rapid construction takes place in the areas, on the outer side of the ring road, where the majority consists of RCC buildings. Until recent, Lalitpur did not have a system for building permit issuing and evaluation. Lalitpur has suffered from damaging earthquakes in the past, such as in 1255, 1408, 1810, 1833, 1934, 1980 and 1988. In the earthquake of 1934, which had a magnitude of 8.4, it was estimated that about 19,000 buildings were heavily damaged within Kathmandu valley, causing the death of more than 3800 people (JICA, 2002).

Lalitpur is located on a former lake in the middle Himalayan mountain range, of which the surface materials are mainly consisting of alluvial terrace deposits on top of a thick sequence of lake sediments, with a thickness up to 400 meters (Fujii et al., 2001).

Various institutions have carried out studies on earthquake hazard and risk in Kathmandu Valley. After the earthquake in 1988, a first study was carried out by the Ministry of Housing and Physical Planning (MHPP), with technical assistance from the United Nations Development Program. In this project a regional scale seismic hazard map for Nepal was produced, and a National Building Code was established, which was unfortunately not implemented (UNDP, 1994). In 1998 this was followed by the Kathmandu Valley Earthquake Risk Management Project (KVERMP), which was implemented by the National Society for Earthquake Technology – Nepal (NSET), with support from the Asian Disaster Preparedness Centre (ADPC), which focused on awareness raising and capacity development at different levels of society, including school reinforcement, mason training, organization of an earthquake safety day, and development of an earthquake risk management plan together with local authorities (Dixit et al., 2002). The project was based on a simple earthquake loss estimation, estimation the losses if the 1934 would occur today. A more detailed study on earthquake loss estimation was made by the Japanese International Cooperation Agency (JICA, 2002). This study divided Kathmandu Valley into large grid cells of 500 by 500 meters, for which the number of damaged buildings were calculated using three scenario earthquakes.

In all of the previous studies, the basis of the loss estimation has always been at a rather general level. The spatial distribution of the earthquake losses is a very important basis for a proper earthquake vulnerability reduction and emergency planning at municipal level. Municipalities would need to have databases at individual building level, in order to be able to carry out proper control over building construction. This study used high-resolution satellite imagery, together with aerial photographs and field survey in the generation of a building database for seismic loss estimation in Lalitpur. An overview of the method is used is given in Figure 2.

3.2 Generation of the Dataset

The Lalitpur Sub-Metropolitan City Office did not yet have a GIS section, nor did they have GIS data on the building stock and other characteristics within their municipality. The Information Department of the neighboring Kathmandu Metropolitan Office, however, was able to provide a series of large-scale topographic maps at scale 1:2,000 in digital form, containing information on drainage, roads, contourlines (1 meter resolution) and building footprints, which was generated in the framework of a European project. These topomaps were in AutoCad format and were converted into a usable GIS database, consisting of separate layers for buildings, roads, contours and drainage.

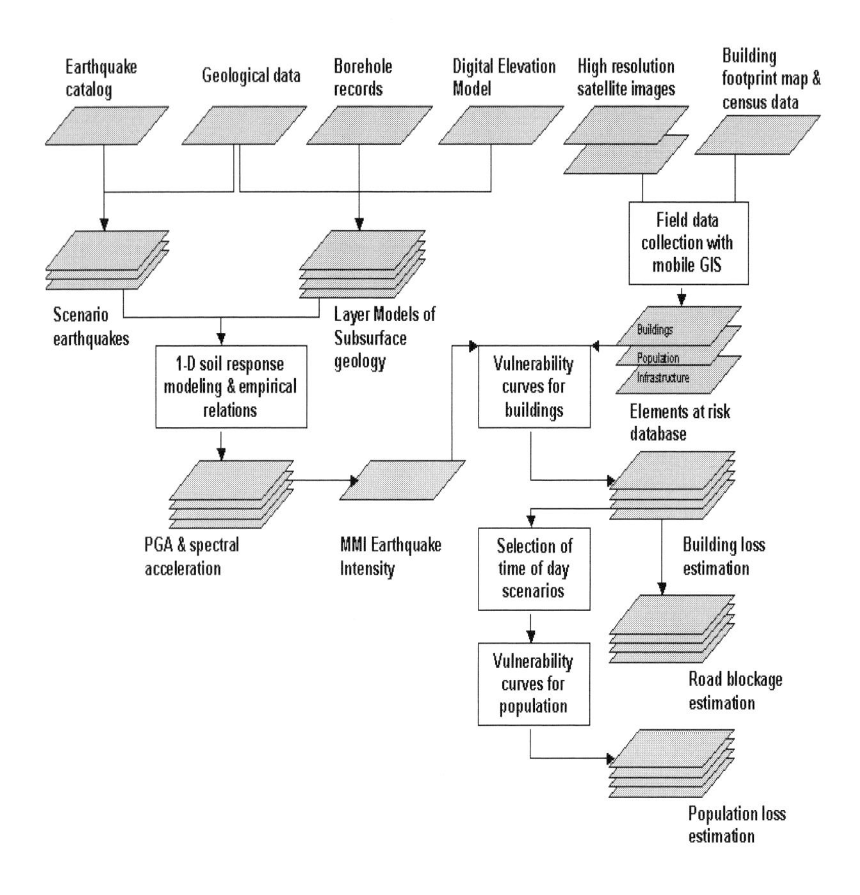

Fig. 2. Flowchart of the procedure for seismic loss estimation in Lalitpur, Nepal. See text for explanation

Especially the generation of building polygons from the segments in the building footprint layer proved to be very cumbersome.

The building footprint map was prepared based on aerial photos of 1981 and 1992 and was updated in 1998. All the buildings constructed after this year as observed in the available IKONOS image from 2001 were digitized on screen to create the building data set for the year 2001. As also a CORONA image was available from 1967, this image was used to delete those buildings that were not yet present in 1967, and generate a building footprint map for that year. In the old center of the city, where most of the buildings are attached to each other and form large complexes around courtyards, the existing building footprint maps did not make a separation between individual buildings, but rather displayed entire complexes of buildings as a single polygon. To calculate the number of buildings the building footprint area of these polygons was divided by the average plinth area of a building, which was taken as 45 m^2 based on samples. A total number of 26,873 buildings are estimated for the year 2001. These buildings were compared to the number of households from the census data (34,996).

The original digital building footprint maps did not contain any attribute information regarding the buildings within the city. Since a complete building survey would require too much time for field data collection, it was decided to use so-called homogeneous units as the basic mapping units within the city. In the field mobile GIS was used to characterize the buildings within each unit according to age (based on procedure outline earlier), occupancy class, land use type and building type, which was a combination of construction material and number of floors.

Population data were available from the latest population census in Nepal, which was held in 2001, published by the Central Bureau of Statistics (CBS) of Nepal. Information was only available at Ward level, according to age and gender. In order to calculate the population distribution per homogeneous unit, which was taken as the basic unit for the loss calculation, wardwise population figures had to be distributed over the various units within the ward. This was done by calculating the percentage of floor space in residential buildings within each homogeneous unit as percentage of the total floor space of residential buildings in the ward. The average population density within different types of buildings (residential, commercial, institutional etc.) was estimated based on 196 detailed samples of buildings carried out by a local NGO, the National Society for Earthquake Technology (NSET) and the Lalitpur Sub-Metropolitan City Office. Based on these samples estimations were made of the population amounts present in different types of buildings during different periods of the day.

3.3 Soil Response Modeling

In order to be able to analyze the seismic hazard in Lalitpur and its surroundings a sub-surface database was generated for the entire Kathmandu valley. A geological database was made for storing the information for 185 deep boreholes, with depths ranging from 35 to 575 meters, of which 36 boreholes actually reached to the bedrock, and 328 shallow boreholes with depths less than 30 meters. Only the shallow borehole records contained both lithological and geotechnical information.

All boreholes were divided into main stratigraphical units, for which the depth was determined and used in GIS for subsequent layer modeling. The horizontal and vertical distribution of the valley fill within the Kathmandu valley is very complex, mainly consisting of intercalations of fluvial and lacustrine deposits. In order to generate layer models for such a heterogeneous environment, a certain degree of generalization had to be accepted. In this case, all sediments of the basin were divided into four layers: Holocene alluvial and anthropogenic deposits, lacustrine deposits formed between 2,500,000 to 29,000 years B.P. (Yoshida and Igarashi, 1984), alluvial deposits below the lacustrine sediments, and the underlying bedrock (See Figure 3). The depth of each of the layer boundaries, including the surface elevation was used in GIS and Digital Elevation Models of each of these surfaces were obtained through point interpolation.

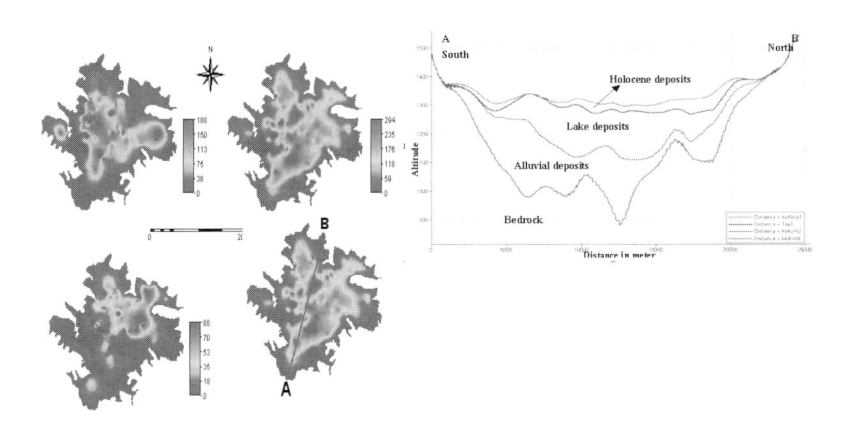

Fig. 3. Left: Thickness maps of the three main material types in the Kathmandu valley. Right: Cross section based on the layer thickness maps

The GIS layer models were used for one-dimensional calculations of the ground response, with the help of SHAKE2000, which is derived from the original SHAKE software, used widely for soil response analysis since 1971 (Ordonez, 2002). For each material type, average values for shear wave velocity, and unit weight, were used, and 5% damping was selected. Unfortunately no strong motion records are available for Kathmandu valley, so comparable records were used from other locations. Three earthquake scenarios were selected in line with the ones used in the study by JICA (2002): one comparable in magnitude and epicentral distance to the 1934 earthquake (called Mid Nepal earthquake), one located North of Kathmandu valley (North Bagmati earthquake) and a local earthquake in the valley itself. The analysis was carried out by sampling the depths of the GIS layers at regular intervals. Each of the sampling points was transformed into a soil profile, which was entered in the SHAKE2000 program, and which was analyzed using the above mentioned scenario earthquakes. The results were calculated as Peak Ground Acceleration (PGA) as well as spectral acceleration for frequencies of 5, 3 2 and 1 Hz. These values were later linked back to the sampling points and maps were obtained through point interpolation.

An analysis of liquefaction potential was made using both qualitative and quantitative methods. In the qualitative analysis the methods of Iwasaki et. al (1982) and Juang and Elton (1991) were used and the quantitative analysis was carried out using simplified methods developed by Iwasaki et al. (1984) and Seed and Idriss (1971). The qualitative methods are based on weights, assigned to a number of factors such as Depth to water table, Grain size distribution, Burial depth, Capping layers, Age of deposition and Liquefiable layer thickness. Following this method, the analysis was carried out for 69 boreholes located at 40 different sites, resulting in 35 boreholes where liquefaction is likely to occur at a particular depth. The final liquefaction susceptibility map was prepared by combining the point information of the boreholes with geomorphological units in a GIS.

3.4 Loss Estimation

For analyzing seismic vulnerability, the buildings in Kathmandu valley have been divided into a number of classes (See table 1). The vulnerability curves used in the GIS analysis were derived by NSET-Nepal and JICA (JICA, 2002). For each MMI class and building type, minimum and maximum values are given of the percentage of buildings that would be heavily damaged (collapsed or un-repairable) or partly damaged (repairable, and available for temporary evacuation).

The earthquake intensity maps were used in combination with the building map and the vulnerability relations indicated in table 1 to calculate the range of completely and partially damaged buildings for each earthquake intensity. In figure 4 some of the results are given. This figure gives the total number of vulnerable buildings in different damage grades and in the four earthquake-intensities used ranging from VI to IX. For example, if an earthquake of intensity IX would occur a number of buildings ranging from 9,192 to 13,710 might be partially damaged and 6,104 to 8,583 might collapse in Lalitpur.

Building type	MMI	VI	VII	VIII	IX
	PGA (% g)	**5-10**	**10-20**	**20-35**	**>35**
Adobe + Fieldstone Masonry Buildings	Total Collapse	2-10	10-35	35-55	55-72
	Partial Damage	5-15	15-35	30	30
Brick in Mud (BM)	Total Collapse	0-6	6-21	21-41	>41
	Partial Damage	3-8	8-25	25-28	<28
Brick in Mud (BMW) and Brick in Cement (BC)	Total Collapse	0-1	1-5	5-18	>18
	Partial Damage	0-11	1-31	31-45	<45
R. C. Framed (≥4 storied)	Total Collapse	0-2	2-8	8-19	19-35
	Partial Damage	0-4	4-16	16-38	38-65
R. C. Framed (≤3 storied)	Total Collapse	0-2	2-7	7-15	15-30
	Partial damage	0-4	4-14	14-30	30-60

Table 1. Damage matrixes for different types of buildings in Kathmandu (The values represent percentages of buildings with the same material type. Source: NSET Nepal)

In a next step specific damage estimations were made for three earthquake scenarios mentioned before. For each of these scenarios the ranges of partially and heavily damaged buildings have been estimated, with and without the effect of liquefaction. In order to take into account the liquefaction effect, the intensities in areas with high liquefaction susceptibility have been increased with 1 on the MMI scale. For the Mid Nepal and Local Earthquakes the amount of partially damaged building ranges from 5,380 to 9,192 and heavily damaged buildings from 2,748 to 6,104. If liquefaction is also included the estimations for partly damaged buildings rise to the range 5,804 – 9,779 and for heavily damaged buildings between 3,034 and 6,412.

The number of casualties was estimated at homogeneous unit level for the three different earthquake scenarios mentioned earlier. The data used for this calculation were the population distribution for different periods of the day and within different occupancy classes, the building loss estimation discussed in the previous section and vulnerability and casualty ratios with respect to building damage. These casualty ratios were derived from the HAZUS methodology, which uses the widely accepted ATC-13 vulnerability curves (FEMA, 2004) and which make a separation into fours severity classes (See table 2). With these relations, the number of casualties was estimated for the three different earthquake scenarios, and for both a daytime and nighttime scenario, with a different distribution of population over the various occupancy classes. Preliminary results are shown in Figure 5 for the Mid Nepal Earthquake scenario.

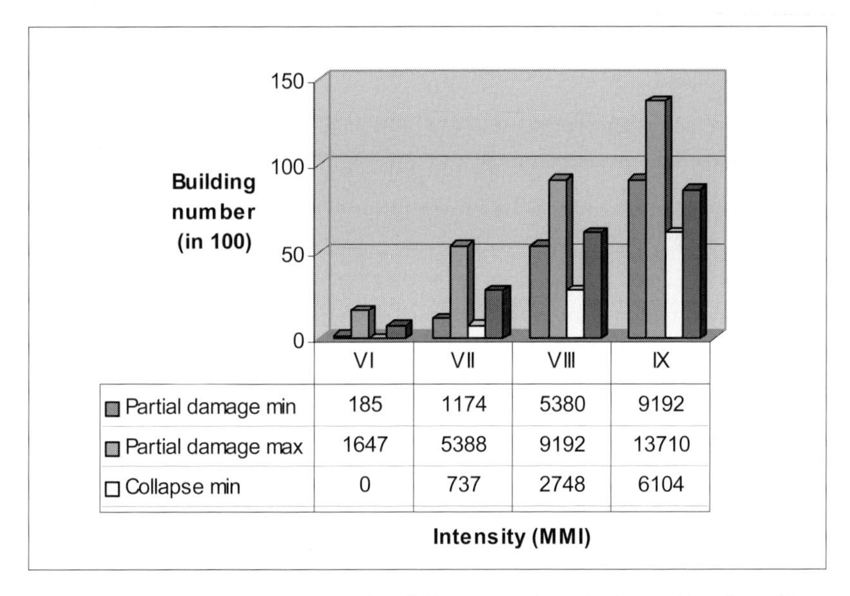

Fig. 4. Building loss estimations for different earthquake intensities in Lalitpur

Building damage level	Injury level (in %)			
	Severity 1 Slight injuries	**Severity 2** Injuries requiring medical attention	**Severity 3** Hospitalization required	**Severity 4** Instant death
Partial damage	1	0.1	0.001	0.001
Complete damage	40	20	5	10

Table 2. Various injury levels according to building damage. Modified from HAZUS

Normally, nighttime scenarios are expected to result in higher casualty numbers. The deviation in this case might be related to the inaccuracy of the population input data, as original data was only available at ward level, and also in the distribution of population over the city in different periods of the day.

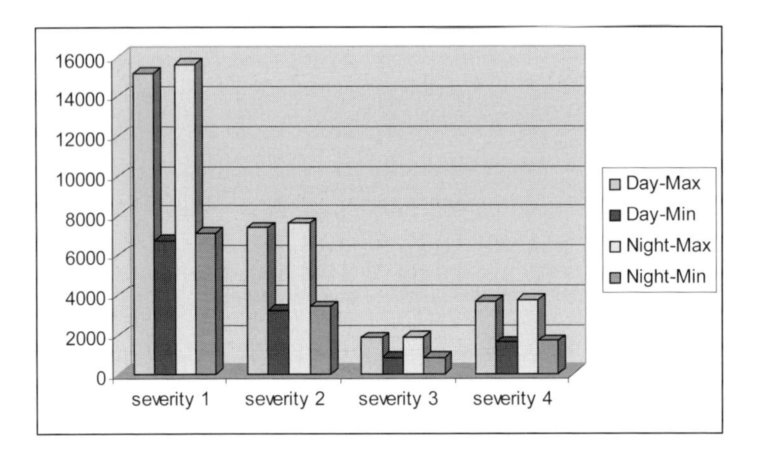

Fig. 5. Casualty estimation for the Mid Nepal Earthquake scenario in Lalitpur

Clearly more detailed information for this should be collected. It might also be caused by the fact that many of the buildings where people are during the daytime, such as schools, shops etc. are equally vulnerable, or sometime more vulnerable than the residential buildings.

4 Conclusions

The results from the earthquake loss estimation for Lalitpur Sub-Metropolitan City in Nepal illustrates the urgent need to support local authorities in developing countries with methods to collect and manage information used for risk assessment in order to be able to implement strategies for vulnerability reduction and disaster preparedness. The collection of basic data is of prime importance, and should be carried out by staff from the municipality in collaboration with local institutions and the local communities. The data collected thus far was mostly in the framework of rather short MSc fielddata collection campaigns, and should be further verified and extended. In the initial period of the project contacts with the Lalitpur Sub-Metropolitan City Office (LSMCO) have been established, and the results of the research was shared with their staff in a workshop. Also a user needs assessment was carried out, leading to the installation of a GIS center within LSMCO and basic GIS training of 12 of their staff. With LSMC a number of phases have been outlined, starting with the collection of base data and the development of a municipal database, leading to the integrated use of this data for various urban planning and management activities, including disaster prevention and preparedness. One of the priority areas for the application of the municipal GIS in the framework of vulnerability reduction is the development of a building permit issuing and control system, that takes into account seismic vulnerability as one of the factors. Some other high priority GIS applications outlined by the LSMCO are the set-up of a proper addressing system for the city, which can be linked to geographic positioning using GPS, and urban heritage management. In a later phase LSMCO plans to apply it to other aspects such as solid waste management, infrastructure management, revenue management, etc. What has become clear in the case study with the Lalitpur Sub-Metropolitan City so far is that specific GIS based Decision Support Systems for Disaster Management at municipal level can only be implemented if a municipality has experience with GIS and has developed a municipal database. Even then, such a system would be less useful for disaster prevention, as vulnerability reduction measures should be an integrated part of all common municipal activities, than for disaster preparedness.

Acknowledgements

We would like to thank the MSc students that have been involved in the studies in Lalitpur: Umut Destegul, Jayaweera Somasekera, Mazharul Islam and Pho Tanh Tung, as well as our colleagues involved in the research in Lalitpur: Luc Boerboom, Mark Brussel, Paul Hofstee, Erik de Man, Lorena Montoya, and Siefko Slob. Also we would like to thank the staff of the Lalitpur Sub-Metropolitan City Office, the Nepalese Society for Earthquake Technology (NSET), the International Centre for Integrated Mountain Development (ICIMOD) and the Asian Disaster Preparedness Center (ADPC) for providing support to the SLARIM research team. More information on the SLARIM research project can be found at the following website: http://www.itc.nl/research/policy/spearhead3/vwesten.asp

References

CITIES Project (2004) Cities and Critical Infrastructure Project. Geoscience Australia, http://www.ga.gov.au/urban/projects/cities.jsp

Dixit AM, Welley RD, Samant L, Nakarmi M, Pradhanang SB, Tucker BE (2000) The Kathmandu Valley Earthquake Risk Management. A paper presented in the 12th World Conference on Earthquake Engineering (30 Jan-4 Feb), Auckland New Zealand

EMI (2002) Earthquake and Megacities Initiative. Website: http://www-megacities.physik.uni-karlsruhe.de/

FEMA (2004) Federal Emergency Management Agency. HAZUS-MH. Website: http://www.fema.gov/hazus/

Fujii R, Kuwahara Y, Saki H (2001) Mineral composition changes recorded in the sediments from a 248-m-long drill-well in central part of the Kathmandu Basin, Nepal. Journal of Nepal Geological Society , vol 25 pp 63-69

Iwasaki T, Tokida K, Tatsuoka F, Watanabe S, Yasuda S, Sato H (1982) Microzonation for soil liquefaction potential using simplified methods, 3rd Intl. Microzonation Conf. proceeding, 1939-1329

Iwasaki T, Tokida K, Arakawa T (1984) Simplified procedures for assessing Soil liquefaction during earthquakes, Soil dynamics and Earthquake Engineering, 1984, vol 3, no 1, pp 49-58

JICA (2002) Japan International Cooperation Agency. The study on Earthquake Disaster Mitigation in the Kathmandu valley Kingdom of Nepal.- Final report, Vol - I,II,III &IV

Juang, CH, Elton DJ (1991) Use of fuzzy sets for liquefaction susceptibility zonation, in Proc. Fourth Intl. Conf. on Seismic Zonation, vol II, Standford Univ., USA. Earthquake Engineering Research Institute, pp 629-636

Kreimer A, Arnold M, Carlin A (eds) (2003) Building safer cities. The future of disaster risk. Disaster risk management series Nr 3, The Worldbank

Ordonez G (2002) A computer program for the 1D analysis of geotechnical earthquake engineering problems, Berkeley

RADIUS (2000) Risk Assessment Tools for Diagnosis of Urban Areas Against Seismic Disasters, http://geohaz.org/radius.html

Seed, HB, Idriss IM (1971) Simplified procedure for evaluating soil liquefaction potential, Jour. of the Soil Mechanics and Foundation division, ASCE, vol 107, pp 1249-1274

UNDP (1994) 1. Seismic Hazard Mapping and Risk Assessment for Nepal, 2. Development of Alternative Building Materials and Technologies, 3. Seismic vulnerability analysis, (Appendix c) His Majesty's Govt. of Nepal, Ministry of Housing and Physical Planning, UNDP/UNCHS Habitat), Subproject NEP/88/054/21.03,1994

USAID (2001) Making Cities Work. Making Cities Work: USAID's Urban Strategy. An Initiative Launched by the Administrator and Prepared by the Urbanization Task Force. February 2001, http://www.makingcitieswork.org/files/docs/MCW/MCWurbanstrategy01.pdf

Yoshida M, Igarashi Y (1984) Neogene to Quaternary lacustrine sediments in the Kathmandu. Valley, Nepal, Jour. Nepal Geol. Soc., vol 4, pp 73-100

Mass Movement Monitoring Using Terrestrial Laser Scanner for Rock Fall Management

Arnold Bauer[1], Gerhard Paar[1] and Alexander Kaltenböck[2]

[1] Institute of Digital Image Processing, JOANNEUM RESEARCH, Wastiangasse 6, A-8010Graz, Austria.
Email: arnold.bauer@joanneum.at; gerhard.paar@joanneum.at
[2] DIBIT Messtechnik GmbH, Gewerbepark 3, A-6068 Mils, Austria.
Email : alexander.kaltenboeck@dibit.at

Abstract

The danger of a rock fall or rockslide event is omnipresent, mainly due to dense settlement, excessive land usage even in alpine regions, and the global warming. In the case of a rock fall event the rapid operational availability of a measurement system is important for disaster management to assess the risk and to take appropriate measures.

The evaluation and classification of instable surfaces need fast and cheap automatic sensing methods with accuracy in the range of a few centimetres. It is shown that a terrestrial laser scanning system is able to successfully perform an efficient change survey.

We report on the sensor and software set-up, the logistics, and the procedure for data evaluation to perform the proposed monitoring task. The system (Laser scanner LPM-2k produced by Riegl Laser Measurement Systems, Austria, combined with software for scanning and data evaluation by JOANNEUM RESEARCH and DIBIT Messtechnik GmbH, Austria) is capable of automatically detecting changes and motion on the surface of an active rockslide area.

1 Introduction

A landslide is the movement of a mass of rock, debris or earth down a slope. A rockslide involves a downward, usually sudden and rapid movement of newly detached segments of bedrock over an inclined surface. A rock fall is the fastest moving landslide. A newly detached segment of bedrock of any size suddenly falls down from a very steep slope.

Once a landslide is triggered along a plane of weakness, material is transported by various mechanisms including sliding, flowing or falling. After falling or precipitously moving, the mass of materials deposits at the base of the slope. The moving mass is greatly deformed and usually breaks up into many smaller slides. Rockslides can vary in size from a single boulder in a rock fall or topple to tens of millions of cubic meters of material in a debris avalanche.

Rockslides can be triggered by natural causes or by human activity. Natural causes include extreme weather conditions like heavy rainfalls, saturation of slope material from rainfall or seepage, vibrations caused by earthquakes, or undercutting of banks by rivers. Also rockslides frequently occur in high mountain areas during spring and autumn when there is repeated freezing and thawing. Another reason is the climate change observed in the last decades. In particular the global warming increases the permafrost level and material formerly frozen gets instable.

Human activities may include the removal of vegetation like the removal of protection forest, interference with or changes to natural drainage, the modification of slopes by construction of trails, roads, railways or buildings, housing sprawl, mining activities, vibrations from heavy traffic or blasting, and excavation or displacement of rocks.

Rockslides are extremely hazardous. They involve a rapid sliding of large masses of fractured rock and regolith. They can move millions of tons of rocks in a short time. Rockslides may endanger or damage buildings, roads, railways, pipelines, protection forest, agricultural land and crops.

The risk assessment of landslides requires an accurate evaluation of the geology, hydrogeology, landform, and interrelated factors such as environmental conditions and human activities. It is of particular importance for engineers and geologists to assess slope stability and landslides in order to take appropriate, effective, and timely measures. Potential measures may vary from road closure, and excavation of buildings within the area of risk, to investments in protection buildings like protective barriers, retaining walls, retention capacity, and rock anchors. Disaster alert plans have to be developed. Above all engineering and geo-technical investigations have to define the landslide hazard and risk.

Potential indicators of active landslides include slope cracks, curved tree trunks, tilted poles, tilted walls, and the presence of wet or seepage areas. Before a rock fall, in most cases a slight shear continuous distortion over a specific period can be observed.

Rock fall models can be useful tools to predict the risk posed by individual falling rocks (Dorren 2003). One essential part of the management strategy for slope instability risk mitigation is a remote monitoring system for a continuous observation of the mass movement, which is often performed over long periods of time (Jaboyedoff et al. 2004). Remote monitoring of slope movement of unstable or potentially unstable slopes normally is a multidisciplinary approach incorporating several sensors. For example, movements and deformation can be measured with inclinometers, tiltmeters, extensometers, time-domain reflectometry, radar, and GPS. Water levels can be observed using vibrating wire piezometers (Kane and Beck 2000).

2 Terrestrial Laser Scanner Technology

In the standard case the area of a rockslide event is not achievable for the application of standard geodetic targets. Therefore remote techniques must be used that are continuously available, which excludes air-based or satellite remote sensing (Kenyi and Kaufmann 2003) from the list of candidate techniques.

Terrestrial scanning laser imaging has turned out to be an essential component of geo-technical disaster monitoring, since it provides high resolution, a wide field of view, medium accuracy and high availability over long periods of time with comparably low cost. The available laser devices have reached a technological fitness for this class of applications just in the recent past. There are quite a few systems on the market with an operating range between near-range (up to 10 m) and 300 m (Lemmens and Heuvel 2001, www.cyra.com, www.mensi.com). However, an integrated scanning device covering a range of more than 1 km for non-reflective targets is a new technology, e.g. offered by Riegl Laser Measurement Systems (www.riegl.co.at).

Laser scanners generate dense clouds of data. Distance measurements are mostly based on the time-of-flight principle, emitting a burst of several hundred laser-pulses for each single measurement. A digital signal processor analyses the reflected return pulses and compiles them to a single distance measurement. Since each single measurement consists of a multitude of laser-pulses, different measurement modes (“first pulse”, “last pulse”, “strongest pulse”) give proper results even on bad weather conditions and surfaces that may otherwise lead to ambiguous measurements like vegetated, moist or roughly structured terrain. For example, the last pulse technique allows detecting the range of the last target even if the measuring beam partially hits or penetrates other targets (like fog) before. Long-range laser scanners can achieve measuring distances up to a few kilometres of range to naturally reflecting targets.

Single time-of-flight measurements with distance accuracy of better than 5 cm are automatically combined to a measurement grid. The distance measurement unit is mounted on a pan and tilt orientation unit motorized by step engines, the exact pan and tilt angles are read out by encoders similarly to a motor theodolite. Each individual measurement point consists of the distance to the surface, the exact angular positions, the reflectance, and an estimated root mean square error (RMSE) of the distance measurement for reliability check.

A monitoring system based on a terrestrial laser scanner is shown with the example of the *DIBIT Geoscanner* (www.dibit-scanner.at), which integrates the long-range laser scanner LPM-2k by Riegl Measurement Systems. The relevant technical specification is shown in Table 1. The device is controlled by an off-the-shelf PC, which handles both the device control and the data transfer and analysis. The control software on the scanning device allows the acquisition of a (almost) rectangular regular grid of measurements in sensor coordinate space, which is stored as one measurement data file on the control PC. The fully automatic monitoring system provides an on-line database both of pre-processed measurements and results accessible for the customer for visualization and interpretation.

Scanner parameter	Value (range)
Measuring range for	
Good diffusely reflective targets	Up to 2500 m
Bad diffusely reflective targets	> 800 m
Minimum distance	10 m
Ranging accuracy	+/- 25 mm
Positioning accuracy	+/- 0.01 gon
Measuring time / point	0.25 s to 1 s
Measuring beam divergence	1.2 mrad
Laser wavelength	0.9 μm
Scanning range horizontal / vertical	400 gon / 180 gon
Laser safety class	3B, EN 60825-1
Power supply	11-18 V DC, 10 VA
Operation temperature range	-10 to +50 °C

Table 1. Technical parameters of LPM-2k by Riegl Laser Measurement Systems

Although the data generated by the measurement devices can in principle be directly used for measurement and further visualization, several methodological, technical, and logistic problems are to be encountered when establishing a fully automatic monitoring system. This includes, among others, the stability of device control software, the automatic sensor orientation, the high number of measurements, the compensation of weather influences, and the selection of reliable measurements. In addition it is of particular importance to consider the highly heterogeneous surface in terms of material (rock, vegetation, and humidity in general) and structure. Rock fall management is not the only application of this sensor setup.

Current experience in this respect include the following scenarios:

- Snow cover monitoring: Feasibility of snow avalanche prognosis using a sensor framework containing a laser scanner as key component (Paar and Bauer 2001)
- Glacier monitoring: Measuring deformation and 3D motion of glaciers and rock glaciers (Bauer et al. 2003)
- First system application for slope monitoring 1999 in Schwaz / Austria (Paar et al. 2000, Poisel et al. 2002, Scheikl et al. 2000)

Fig. 1. Left: Laser scanner LPM-2k by Riegl Laser Measurement Systems. Middle: Laser scanner with weatherproof heated housing mounted on a concrete pile. Right: Reflective reference target for sensor orientation

3 Measurement, Data Processing and Visualization

3.1 Data Acquisition

The laser scanner performs tasks of a predefined measurement schedule to automatically measure regions of interest (ROIs) round the clock. A ROI defines both the measurement raster, and the distance measurement parameters like integration time and mode. Several ROIs are combined to a measurement task. A measurement *task list* defines the measurement strategy in terms of order, priority, and point of time. On the scanner control PC a server is continuously parsing the task list and revaluating the measurement schedule. As soon as the scanner is idle, the next ROI is selected and executed. Unsuccessful measurements (i.e. due to bad environmental conditions, or communication problems) are detected automatically; they are repeated and reassumed into the measurement list.

Different automatically selected strategies, like for example increasing the integration time for a single distance measurement in case of bad weather, or focusing on ROIs with top priority, allow retrieving maximum information dependent on the environmental situation.

3.2 Pre-Processing

An essential pre-processing step is the classification of the measurements to examine the quality and reliability of the distance measures. Uncertain measurements are smaller weighted or even dismissed for the subsequent data evaluation. The classification is based on an analysis of the reflectivity, the RMSE, the structure (e.g. detected artefacts due to fog), and external sensors.

3.3 Sensor Orientation

To guarantee the comparability of measurements both orientation (e.g. due to subtle misalignments of the scanner platform) and distance measurements (due to atmospheric influences) are compensated.

Repeatable sensor orientation is performed continuously using reflective targets fixed on stable surfaces somewhere in the spherical field of view of the sensor. A centroid localization algorithm on the laser reflectance image gains the angular components of the target coordinates. The distance is calculated as weighted average of all individual distance measurements covering the target. Since the current version of LPM-2k does not contain an electronic levelling sensor like standard theodolite it is necessary to determine all unknown position and orientation parameters using the reference targets.

This step is the key to measurement accuracy; it involves several individual tasks that are still subject to improvement:

- Distribution of the orientation targets

- Robust automatic evaluation of the usability of target measurements
- Stable mathematical methods and algorithms for sensor orientation under various restrictions
- Current research aims at using natural such as (non-moving) distinct landmarks (like rock formations) for continuous sensor orientation.

In addition to the determination of sensor orientation and location the reference targets can be used to determine compensation values for atmospheric effects on the distance measurements. The combination of the compensation values (multiplication factor as simplest approach) together with current transformation vector and rotation matrix is called *sensor state*, which is stored together with each measurement.

3.4 Surface Model Generation

To represent each measurement in a reference coordinate system a dense digital surface model (DSM) is generated of the scene to be surveyed. A DSM is a regularly spaced grid in desired resolution on an analytical model of the local surface, in the simplest case a horizontal or vertical plane. It is used to store the elevation as a vertical distance at the grid points. We generalize the DSM to an arbitrary reference surface, to be able to represent the surface data in best resolution, since most of the potentially insecure surfaces are characterized by steep fronts. This data structure well complies with the practical requirements such as difference measuring, volume change evaluation, and various visualization tasks. Neighbourhood relations of measurement data points are directly described in the DSM structure; therefore operating on DSMs allows quick access to the surface heights in a well-defined geometry.

Direct mapping from the sensor spherical system to the DSM Cartesian coordinate space would result in a sparse and non-uniform elevation map, especially at large distances. To avoid interpolation artefacts, the Laser Locus Method (Kweon and Kanade 1992) for DSM (Bauer and Paar, 1999) generation proves to be a robust tool for data acquisition from flat angles, and supports error detection and utilization of additional confidence values provided by the range sensor.

Since the DSMs of (temporally) different surface measurements are geo-referenced, simple differences between the DSMs reflect the changes in elevation. In consequence we can derive a full description of change in volume, spatial distribution of shape, or arbitrary profiles on the surface.

3.5 Measurement database

All measurements are managed in a simple database that contains meta data (Sensor state, reliability, ROI data, etc.) and the measurement itself. The user can access different processing products in various levels of detail such as

- Interactive motion diagrams on a single ROI
- Video sequences of ROI structure
- 3d surface rendering with pseudo – colour overlay of distance change
- Lists of ROI statistics (time dependent) in different formats

4 Motion Detection

In a first step distance changes from the observer to selected points and ROIs can be measured and statistically evaluated immediately. Further analysis quantifies more complex movements, just as changes by debris/rock fall or accumulation. Eventually the objective is to classify the meaning of the event semantically.

The DSM difference describes only the component of the surface change perpendicular to the analytical DSM. In order to examine the complex kinematics of surface deformation the knowledge on surface motion in all three dimensions is required. 3D motion as well as structure changes like rock falls can be calculated by means of optical flow detection on the grey level images using correlation-based matching. A dedicated matching technique (Hierarchical Feature Vector Matching) (Paar and Pölzleitner 1992) can use both the surface structure (distance measurement texture) and/or the radiometric surface texture (RGB sensor or laser beam reflectivity). In such a way a single ROI behaviour can be categorized into simple classes such as *insignificant change, significant change, unusable measurement, material loss,* or *aggregation* (Figure 3 and 4). This enables to quickly focus the attention of experts and further automatic decision steps to potentially hazardous areas and events.

A possible hazard warning system uses the ROI tracking results and sets off an alarm if motion exceeds a critical threshold, or the structure has changed due to a recent landslide or rock-fall. Using higher-level information such as the spatial distribution of deformations on the entire hazardous site, the temporal behaviour of singular or multiple ROI motions, or a knowledge-based expert system could perform the semantic classification of the kind and relevance of the change event.

5 Results

The prototype monitoring system was field-tested at a rock fall near Gries, Austria. In June 2003 pieces of cliff as big as humans had dislodged due to mass movements in terms of a rockslide. Safety measures including the evacuation of several houses and road closures in the affected residential area were quickly initiated.

To assess the risk of succeeding rock falls, a laser-scanner monitoring system was installed which measured potentially instable regions over a period of more than two months.

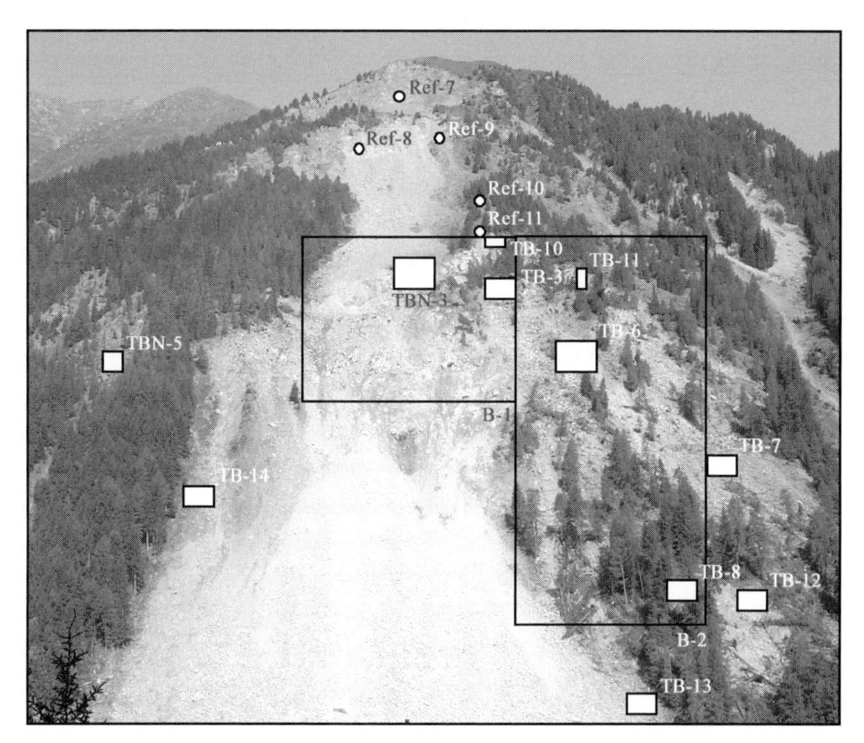

Fig. 2. Overview of the Gries rock fall area, Austria. A set of 13 monitoring ROIs were scanned, together with 5 reflective targets

Figure 2 shows an overview of the Gries rock fall area. In collaboration with geologists, a set of 11 ROIs were identified to be measured in high resolution round the clock. Additionally 2 ROIs, covering almost the whole instable area, were defined to be measured in lower resolution once a day. 5 ROIs were used for the reflective reference targets to compensate atmospheric influences, and another 5 ROIs in the close-up range of the scanner position were used for sensor orientation. Each reference target was measured once using standard geodetic methods, which enabled further geocoding and comparability of the scanner data. Standard resolution on the slope ROIs was 0.05 gon, and the reference targets were scanned with 0.02 gon resolution.

For continuous monitoring the scanner was mounted on a stable console on the opposite slope with an average distance to the target area of 800 m.

The sensor orientation was determined 4 times a day to compensate slight movements and misalignments of the scanner. Before each slope measurement, a reference target was measured to compensate for atmospheric effects that have an influence on the scanner distance measurements.

Fig. 3. Grey-level coded distance images of a ROI (10 x 7 m) in the Gries rock fall area measured over a period of two days (data acquisition every 6 hours). Bright .. 635m, dark .. 620m distance. The motion between frames 55 and 56 was considerably slow, whereas a rock fall occurred between frames 58 and 59 (note that the structure changes completely)

```
078 TB8_20030721_054935 TB8_20030720_234921 0.86 0.13 0.0013 ROI_M_IrrelevantDisplacement
079 TB8_20030720_234921 TB8_20030720_174926 0.10 0.24 0.1860 ROI_M_StructureChanged
080 TB8_20030720_174926 TB8_20030720_114934 0.61 0.19 0.0200 ROI_M_LargeDisplacement
081 TB8_20030720_114934 TB8_20030720_054929 0.64 0.20 0.0023 ROI_M_IrrelevantDisplacement
082 TB8_20030720_054929 TB8_20030719_234942 0.81 0.16 0.0019 ROI_M_IrrelevantDisplacement
083 TB8_20030719_234942 TB8_20030719_174933 0.64 0.20 0.0188 ROI_M_LargeDisplacement
084 TB8_20030719_174933 TB8_20030719_114919 0.61 0.20 0.0194 ROI_M_LargeDisplacement
085 TB8_20030719_114919 TB8_20030718_235057 0.76 0.18 0.0006 ROI_M_IrrelevantDisplacement
086 TB8_20030718_235057 TB8_20030719_054929 0.76 0.17 0.0014 ROI_M_IrrelevantDisplacement
```

Fig. 4. Output of a simple deformation categorization process based on the data depicted in Fig. 3. In the rightmost column the decision of the system is displayed (structure change caused by a rock fall, irrelevant motion of the terrain surface, or considerable side motion of the terrain between subsequent measurements). The decision is based on statistics of matching results between subsequent distance images (some of the statistical parameters – matching area covered, reliability, angular displacements – are displayed in the output list as separate columns)

Figure 3 shows an example of subsequent distance measurements of one ROI over a period of two days. The accuracy obtained is within a range of 5 cm in all three coordinate axes. Involving image processing and classification allows deriving simple deformation categories as shown in Figure 4.

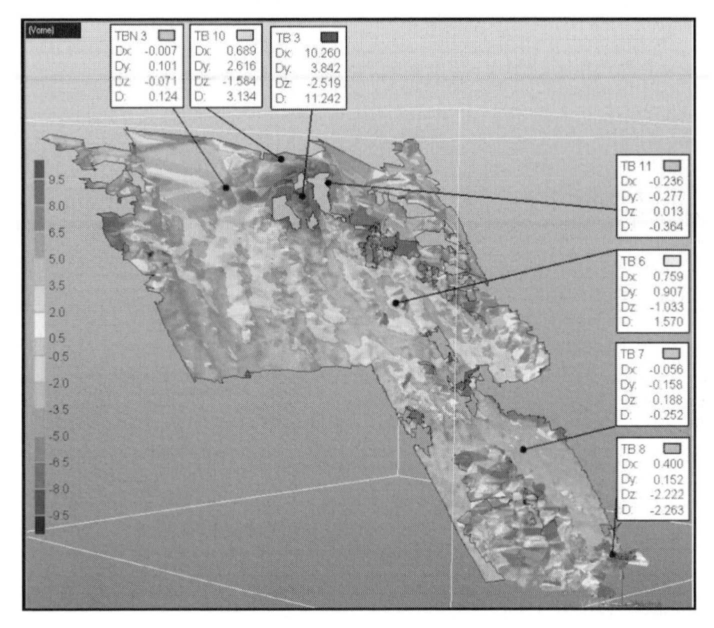

Fig. 5. 3D deformation model of the whole instable area (ROI B-1 and B-2 in Figure 2) over a period of three weeks. The deformation is grey level coded with white colour indicating almost no changes regarding to the reference measurement at the beginning of the monitoring period, and black colour indicating up to 9 m of break-off or accumulation. For significant terrain regions the deformation amount is displayed in all three dimensions

Figure 5 illustrates the 3D deformation of the whole instable area, and Figure 6 shows an example of the deformation of one ROI as a result from region tracking.

Currently the measurement results are verified using geodetic measurements and ground truth information. Future research will emphasize the application dependent knowledge – based systems for ROI tracking, detection and correction of measurement outliers, accuracy investigations, an optimised sensing strategy and additional sensors (Reiterer 2004), as well as the automatic selection of regions of interest.

6 Conclusion

An integrated system consisting of terrestrial laser scanner, control and evaluation / visualization Software and attached logistics for project set-up and handling has been introduced, that can be used for rock fall monitoring and as valuable source for further continuous risk assessment.

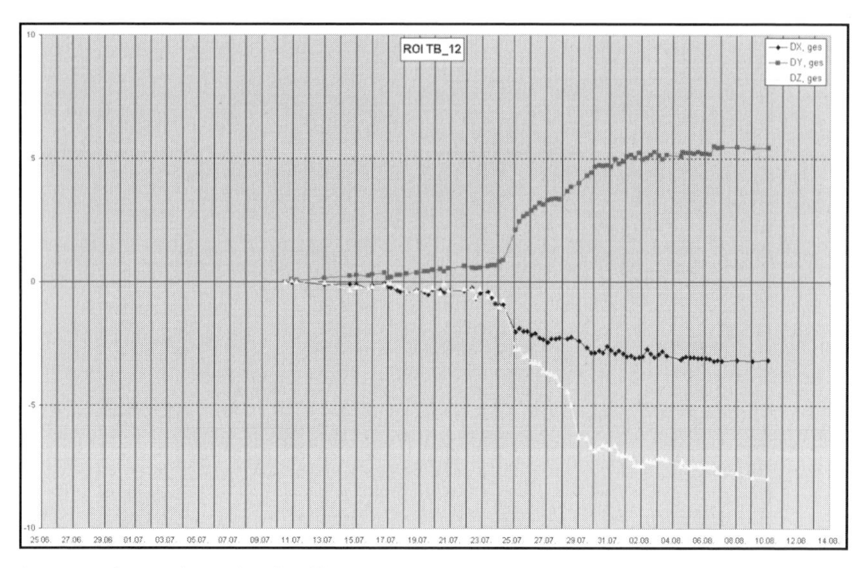

Fig. 6. Deformation of a significant area over a period of one month. Starting from July 11th the average position was tracked. The area moved slightly until July 24th. Mainly due to a heavy rainfall period the region slid up to 5 m until July 31st. Subsequently the region movement stabilized

The system is in operational state, providing the whole processing chain from description and support software for proper sensor and project set-up to the interactive deformation display. Sensor orientation turned out to be the key component in the real application, the combination of methods from computer vision, geodesy and photogrammetry lead to a stable system state in this respect.

Improvement and further development potential exists mainly in the field of sensor and target placement for sensor orientation, the robust detection of false measurements, further reduction of atmospheric effects, a proper measurement and results data base, as well as an integrated alert framework.

The results show the usability of this approach in various related application areas such as landslides, glaciers and the monitoring of large edifices. All the results are available immediately, which makes the terrestrial laser scanner monitoring a valuable tool for risk evaluation and prediction.

Acknowledgements

This work has been carried out in part within the K plus Competence Centre *Advanced Computer Vision.* This work was funded from the K plus Program and FWF Project P14664.

References

Bauer A, Paar G (1999) Elevation Modelling in Real Time Using Active 3D Sensors. Proc 23rd Workshop of the Austrian Association for Pattern Recognition, AAPR, Robust Vision for Industrial Applications, pp 89-97

Bauer A, Paar G, Kaufmann V (2003) Terrestrial laser scanning for rock glacier monitoring. Proc Eight International Conference on Permafrost Zurich vol 1 pp 55-60

Dorren L (2003) A review of rock fall mechanics and modelling approaches. In: Progress in Physical Geography, 27(1):69-87

Jaboyedoff M, Ornstein P, Rouiller JD (2004) Design of a geodetic database and associated tools for monitoring rock-slope movements: the example of the top of Randa rock fall scar. In: Natural Hazards and Earth System Science, Vol. 4, pp 187-196

Kane WF, Beck TJ (2000) Instrumentation Practice for Slope Monitoring. In: Engineering Geology Practice in Northern California. Association of Engineering Geologists Sacramento and San Francisco Sections

Kweon IS, Kanade T (1992) High-resolution terrain map from multiple sensor data. IEEE Transactions on Pattern Analysis and Machine Intelligence 14(2):278-292

Kenyi L W, Kaufmann V (2003) Estimation of Rock Glacier Surface Deformation Using SAR Interferometry Data. IEEE Trans. in Geoscience and Remote, Vol. 41, No 6. June 2003, pp. 1512-1515

Lemmens MJPM, Heuvel FA (2001) 3D Close-range Laser Mapping Systems. In: GIM International, pp 30-33

Paar G, Almer A (1993) Fast Hierarchical Stereo Reconstruction. Proc 2nd Conference on Optical 3-D Measurement Techniques, Zurich, pp 460-466

Paar G, Bauer A (2001) Terrestrial long range laser scanning for high-density snow cover measurement. Proc 5th Conference on Optical 3D Measurement Techniques Vienna pp 33-40

Paar G, Nauschnegg B, Ullrich A (2000) Laser scanner monitoring – technical concepts, possibilities and limits. Proc Workshop on Advances Techniques for the Assessment of Natural Hazards in Mountain Areas, Igls

Paar G, Pölzleitner W (1992) Robust Disparity Estimation in Terrain Modelling for Spacecraft Navigation, in Proc. 11th ICPR, International Association for Pattern Recognition.

Poisel R, Roth W, Preh A, Tentschert E, Angerer H (2002) The Eiblschrofen rock falls - interpretation of monitoring results of a complex rock structure. Proc 1st European conference on landslides, Prague

Reiterer A (2004) Knowledge-Based Decision System for an On-line Video-theodolite-based Multisensor System. PhD thesis, Vienna University of Technology.

Scheikl M, Poscher G, Grafinger H (2000) Application of the new automatic laser remote monitoring system (ALARM) for the continuous observation of the mass movement at the Eiblschrofen rock fall area – Tyrol. Proc Workshop on Advances Techniques for the Assessment of Natural Hazards in Mountain Areas, Igls

Scheikl M, Angerer H, Dölzlmüller J, Poisel R, Poscher G (2000) Multidisciplinary monitoring demonstrated in the case study of the Eiblschrofen rock fall. Felsbau, 18(1):24-29

Findings of the European Platform of New Technologies for Civil Protection: Current Practice and Challenges

Jérôme Béquignon[1] and Pier Luigi Soddu[2]

[1] Direction de la Défense et de la Sécurité Civiles, 87-95, quai du Docteur Dervaux, 92600 Asnières sur Seine, France.
Email : jerome.bequignon@interieur.gouv.fr
[2] Dipartimento della Protezione Civile, via Vitorchiano 2, 00189 Roma, Italy.
Email : pierluigi.soddu@protezionecivile.it

Abstract

A recent study carried out by the French Civil protection with support of Italy and Spain, reviewed the current usage of new technologies for civil protection, in particular for rescue operations (Béquignon and Nardin 2004). The technologies considered here were Geographical Information Systems, GPS and other locating devices, telecommunications, satellite and airborne remote sensing. A number of challenges were identified for further development. The paper will provide some examples and discuss related issues.

1 Current Practice

1.1 GIS and GPS

While the situation in Europe varies considerably from one centre to another, there has been a rapid and widespread development of GIS in the framework of emergency planning, hazard mapping and operation management. The transition to digital maps is complete at national or provincial level, yet many local centers still use printed maps, and use of wall maps still commonplace. The process of digitizing the bulk of hazard and regulatory maps is still ongoing, while recent hazard maps and related information are always stored in GIS systems, often shared among different administrations and sometimes across boundaries, such as the GISMOSEL developed within European INTERREG Rhine-Meuse Activities.

Similarly, GPS is more and more used by firefighters and GPS receivers are now present in nearly every vehicle. Data collected on the field are used to update existing GIS layers such as track viability, location of water tanks (fig.1), field reconnaissance of forest areas damaged by fires or storms, etc. As an example, a GPS device coupled with a GSM equips every vehicle of the Austrian Red Cross of the town of Graz and allows efficient ambulance fleet management.

Fig. 1. Firefighters locate a water tank during a GPS field campaign.
Source: Service Départemental d'Incendie et de Secours du Var, France

Several attempts to use various brands of portable or even wearable computing devices such as tablets, Personal Digital Assistants (PDAs), coupled with GSM or GPRS have been carried out, like in the EGERIS (Rossi and Folino 2003) and PREMFIRE projects, with the participation of French, Italian and Portuguese firefighters. The Swedish rescue service agency investigated the "fireman of the future" fully equipped with sensors and a lightweight camera.

1.2 Satellite Imagery

Over the last decade, many projects of different maturity attempted to exploit satellite imagery for prevention, early warning or mitigation of natural disasters (Béquignon 1999). Very high resolution satellite imagery is being used in lieu of classic aerial photographs for mapping purpose such as in the aforementioned GISMOSEL, while the Italian Civil protection uses burnt area maps derived from LANDSAT images and the Spanish Forest Service gets daily fire risk maps derived form NOAA AVHRR and MODIS.

For emergency operations, the International Charter "space and major disasters" is a remarkable and successful initiative (Bessis et al. 2004) that makes imagery from major satellites such as SPOT, ENVISAT or RADARSAT-1, readily available to civil protections during emergencies. Recent floods of the Danube or Rhone rivers, 2003 summer fires in Europe or the Algiers earthquake are a few examples where information was delivered within hours.

Building upon this experience, future satellite missions (Achache 2004) complemented by tailored services for hazard management at large are being devised in the Global Monitoring for Environment and Security program, as part of the European strategy for space.

1.3 Airborne Imagery

Several watershed were mapped by airborne LIDAR campaigns and digital elevation models suitable to hydraulic modeling were produced.

Other imaging systems, such as an infrared video camera onboard a lightweight aircraft, were successfully tested in the fight against forest fires, through the French-funded PAREFEU project. Such a system has a unique ability to "see" the fire front and its evolution (jumps) through smoke and haze, as well as the effect of water bombing, to follow moves and safeguard field personnel.

1.4 Communications

A wide spectrum of modern telecommunication systems has been deployed in Europe. While Germany and Spain have a robust satellite telecommunication network of command centers, other countries have been deploying, or are in the process of deploying networks based upon the competing TETRA and TETRAPOL standards. Despite its known weaknesses (e.g. saturation in case of emergencies) the GSM network is widely used both for voice and data exchange, for instance the Portuguese water institute uses GSM to collect real information from its nationwide gauge network.

2 Challenges

Yet a series of challenges arise and must be addressed before a complete exploitation of such technologies in operations.

2.1 Cooperative Data Management

An ideal overall emergency management system, ranging from prevention to forecast, operations to lesson learning, would involve a series of actors, sometimes across political boundaries, that would cooperate and exchange a wealth of data through sophisticated processes. Beyond national or proprietary solutions adopted today, ready-to-use and open technologies supporting cooperative work, such as GRID, may provide a solution. Methodologies, such as Italy's AUGUSTUS were designed to describing such processes and a possible implementation investigated in the FORMIDABLE project (Rossi and Folino, 2003). Furthermore, sharing geographically oriented, specialist data or even reference cartography, raises financial, technical and political issues. The INSPIRE initiative of the Commission might ease this situation. Dedicated trans boundary experiments would be required to verify the efficiency of such solutions in a bottom-up approach.

2.2 Field Decision Support Systems

Since the 90's, disaster-specific models complement decision support systems. Some of these models are highly probabilistic and they are used for emergency planning using a priori best case/worst case scenarios. As an example, Italy's SIGE – information system for decision support in case of a seismic event - produces real time tabular reports and statistical scenario maps (Soddu and Martini 2003). Others are based on empirical observations on e.g. fire propagation. With the availability of real time data collection systems, improved forecasts, fast evolving phenomena, such as flash floods, there is a need to move towards real-time and robust algorithms.

Sophisticated algorithms such as earthquake damage forecasts including site effects were developed in Italy, in Portugal and France. They call for a 3-dimension model of a city, at individual building scale, thus requiring the collection and maintenance of an enormous amount of data. Routinely available meter and sub-meter satellite digital imagery, together with 3-dimension simulation software make it possible to develop emergency simulation tools that could be used in near real time situations. For instance, interactive 3-dimensional views of burnt forests or flooded areas were provided in 2003 to French fire brigades using satellite data, thanks to the International Charter. They were found valuable mostly as a communication vehicle to local authorities and citizens, for lessons learning or training. For the time being such information is found too difficult to use during operations.

A final example of complexity, combining time and space dimensions is provided by the continuous infrared video flow of the fire front provided in the PAREFEU project (Chevassus 2003). This was found to be too distracting by incident commanders who asked for 2-dimension fire extension maps updated at regular intervals.

Evaluation of such systems in near-real operation conditions proved to be invaluable and opportunities of large-scale international exercises, such as the Forest fire exercise sponsored by the European commission in 2003, should be exploited by all means.

2.3 Broadband Field Communications

A last issue relates to the communication capabilities available during an emergency and its efficient use. While many civil protections have been acquiring sophisticated tools using e.g. satellite communications, the large communication bandwidth necessary to carry voice, video and data is rarely available on the operation theatre. Technical solutions initially devised for the defense sector are being proposed to the emergency and security sector, and new standards are being proposed, such as the joint European American Project MESA (Ring 2001). As an example the "fireman of the future" was re-used in such a context. Scenarios for an effective use of such technology for rescue operations yet need to be devised.

3 Conclusions

New technologies are widely used in Europe although their deployment varies considerably from one place to another. Several examples suggest that GIS and GPS technologies, modern communication systems, remote sensing are being used on operational basis. Availability of next generation of satellites will increase the suitability of Earth observation-based information systems to emergency management. Availability of new standards and policy shifts towards cooperative work may ease acceptance of such technologies. New challenges arise with the complexity, speed and amount of information to be brought to the field. They call altogether for field broadband communications. Proposed solutions should be assessed in near real conditions during large scale exercises.

Acknowledgements

This study was funded by the European commission in the framework of the action program for civil protection. The authors are grateful to all those colleagues in Europe who provided us with first hand information reflected in this article.

References

Achache J (2004) Les sentinelles de la Terre, Eyrolles, Paris

Béquignon J and Nardin P (2004) The European platform of new technologies for civil protection, European commission, Brussels

Béquignon J (1999) Space technologies for disaster management In: Ingleton J (ed) Natural disaster management, Tudor Rose, Leicester, pp125-126

Bessis JL, Béquignon J, Mahmood A (2003) Three typical examples of activation of the International Charter "Space and Major Disasters". Adv Space Research vol 33 (2004) 244-248, Elsevier

Chevassus F (2003) L'expérimentation PAREFEU , Proc. Rencontres Euroméditerranéenne Feux de Forêts.

CEOS (2002) http://disaster.ceos.org Final report of the disaster management support group, NOAA, Silver Spring.

European Commission (2004) Proposal for a Directive on infrastructure for spatial information in the Community (INSPIRE), COM(2004)516, Brussels

Ring S (2001) Mobility for Emergency and Safety Applications – MESA, Proc. 2nd Conf on Disaster Communications, Tampere

Rossi F, Folino M (2003) Esperienze di ricerca scientifica e tecnologica finalizzata all'innovazione nei sistemi di gestione ambientale, Proc 7th Conf. Nat. Agenzie Ambientali, Milan

Soddu PL and Martini MG (2003) Seismic emergency planning management : SIGE and EGERIS projects Proc. Euro Mediterannean Symposium, ENPC, Madrid.

Geo-Information at the Belgian Federal Crisis Centre

Monique Bernaerts[1] and Philippe Hellemans[2]

[1] Emergency Planning, General Directorate Crisis Centre, Belgian Ministry of Interior Affairs, Rue Ducale 53, 1000 Brussels, Belgium.
Email: monique.bernaerts@ibz.fgov.be

[2] Geographical Information Engineering, Geo-6, Rue Metsysstraat 91, 1030 Brussels, Belgium.
Email: phe.geo6@skynet.be

Abstract

The Federal Crisis Centre in Belgium was created in 1988 in the aftermath of a series of tragic events during the 80's. Its mission is to manage crisis situations at national level: floods, nuclear incidents, chemical pollutions, major strikes, risks for food chains, terrorist threats etc. It consists of 60 professionals, fulfilling their mission 24h a day, offering to partners a "high tech" and ready to use environment for crisis management.

The Crisis Center introduced GIS-technology in the early 90's mainly as a mapping tool to help decision-makers. Today, the aim of Gis is to collect and maintain as many critical data's as possible, normalized and integrated in a system that is continuous, homogenous and regularly updated. This allows us to produce thematic maps as well as case studies showing spatial analyses, disaster consequences, risk distribution for our partners dealing with crisis management. Our strategy regarding Geo-information technologies can be summarized as follow: **autonomy** in management of base maps, setup of **"Crisis" databases** where geographical objects are linked to external databases, easy to use customized mapping functions for our own needs, **partnerships** with other bodies in order to make data exchanges easier especially with local authorities.

1 The Federal Crisis Centre: Basic Missions and Structure of the Organization

The Federal Crisis Centre in Belgium was created in 1988 in the aftermath of a series of tragic events during the 80's. Its mission is to manage crisis situations at national level. The word "crisis" means here an event that by its nature or its consequences becomes a threat against vital interests or needs of the population, requires decision on short notice and coordination between several ministerial departments and organizations. As a consequence, the Crisis Centre plays an active role in organizing responses to many critical situations - floods, nuclear incidents, chemical pollutions, major strikes, risks for food chains, terrorist threats etc. The Crisis Centre became a General Directorate of the Belgian Federal Ministry of Interior Affairs in 2003.

The Crisis Centre consists of a team of more than 60 professionals, fulfilling the missions of 24h a day duty permanence, planning, coordinating and following major events and crisis situations.

The Crisis Centre is also an infrastructure for crisis management offering to its customers and partners a "high tech" and ready to use environment including powerful computers and telecommunication networks, documentation centre, databases, GIS and video-conferencing. Available 24h a day.

Fig. 1. The Crisis management Room : Projection maps and videoconferencing system at work

The Crisis Centre acts as the contact point for International data exchange regarding Disaster Management and Planning, Alert systems, cross border cooperation like ECURIE (nuclear alert), MIC (Monitoring and information Centre for EU) or EADRCC (Euro-Atlantic Disaster Response Coordination Centre of NATO).

2 Crisis Management in Belgium

Belgium is a small country located at the heart of Europe that may be considered as a laboratory for Disaster management by many aspects:

- highly industrialized and densely populated, with many SEVESO industries, nuclear plants within and around the country;
- politic, administrative and linguistic entities that may sometimes seem complex to our foreign partners, with three official languages -Dutch, French, German plus English for international and technical purposes, and competences spread between Federal, Regional and local authorities (provinces, cities and municipalities);
- emergency services having their own competences and territories – Police zones, Fire zones, Civil Protection zones, judicial zones,..;
- seat of international institutions like European Commission or NATO, welcoming regularly foreign personalities and being siege of demonstrations;

Having experienced several major crises during the last years, – the disaster of the gaspipe explosion on 30 July 2004 is still in our memories.

Crisis Management in Belgium is organized in 4 phases and 5 disciplines.

4 phases:

Phase 1. Municipal - Fire Brigade Cpt
Phase 2. Municipal - Mayor
Phase 3. Provincial - Governor
Phase 4. National - Ministry of Interior

5 disciplines:

1. Fire Brigades
2. Healthcare services
3. Local Police
4. Logistic support
5. Information

Fig. 2. Disaster Management in Belgium: phases and disciplines

The first responsible body for population security is the *Mayor*. Security is organized at the municipal level by the *Officer of the Fire Brigade* (phase 1) or by the *Mayor* himself (phase 2) depending on the seriousness of the situation and the importance of the means required. If the situation worsens or requires more means, the *Governor of Province* will take the relay (phase 3).

The law of 31/01/2003 defines the conditions to start up the phase 4 and determines the legal bases for the Crisis Centre actions. The aim of the law is to improve population security against all types of threats and to fill the gap in having a National Emergency plan.

The criteria for initiating a Phase 4 plan are not only geographical (when more than one Province is concerned) but the Phase 4 Plan will also be activated when the threat or the disaster may bring a large number of casualties, represents important risks for the environment or the food chain, when there is a need for complementary resources, coordination, communication means or general information to the population.

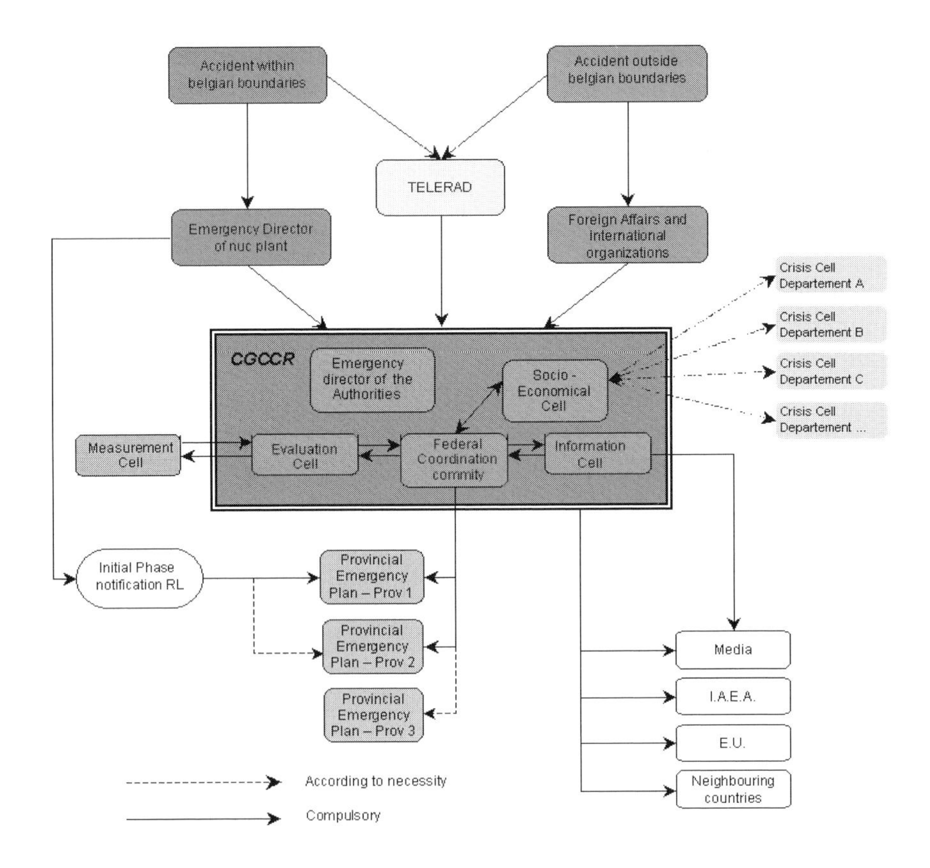

Fig. 3. The workflow of a typical Phase 4 Plan: The nuclear plan

It is also important to further explain our perception of *critical and vital points*.

Critical points are structures and infrastructures presenting a potential danger (chemical industries, nuclear sites) or requiring specific protection measures (a school, an embassy).

Vital points are structures that must be protected in order to ensure the essential needs of the population (security, public health, transport, telecommunication,..).

It is of course our priority to identify and localize those points in order to evaluate potential risks, define planning and intervention zones, avoid "domino effect" and evaluate socio-economic consequences of disasters. Geo-information technology plays here a very important role.

3 Geo-Information for Crisis and Disaster Management at Federal Level

The Crisis Centre started its GIS experiences in the early 90's. At that time, the amount of available data and the complexity of the technology made the system quite confidential, only accessible to a limited number of GIS specialists. Most of the requests could be satisfied with a few colored boundary maps submitted to decision-makers.

Today, things are different and Geographical data are regularly collected and processed in order to meet the following requirements.

3.1 Production of "Reference maps" for Crisis and Disaster Management

With "Reference maps" we understand maps showing territorial competences of local authorities and emergency services -Fire Brigades, Police, Civil protection- but also maps showing planning zones around dangerous sites or structures. "Reference Maps" are continuous, homogenous and regularly updated for the whole country. Typically overview maps of the country will be plotted at a scale of 1/300.000 but source data are accurate enough to produce local maps at the 1/50.000 scale or even better.

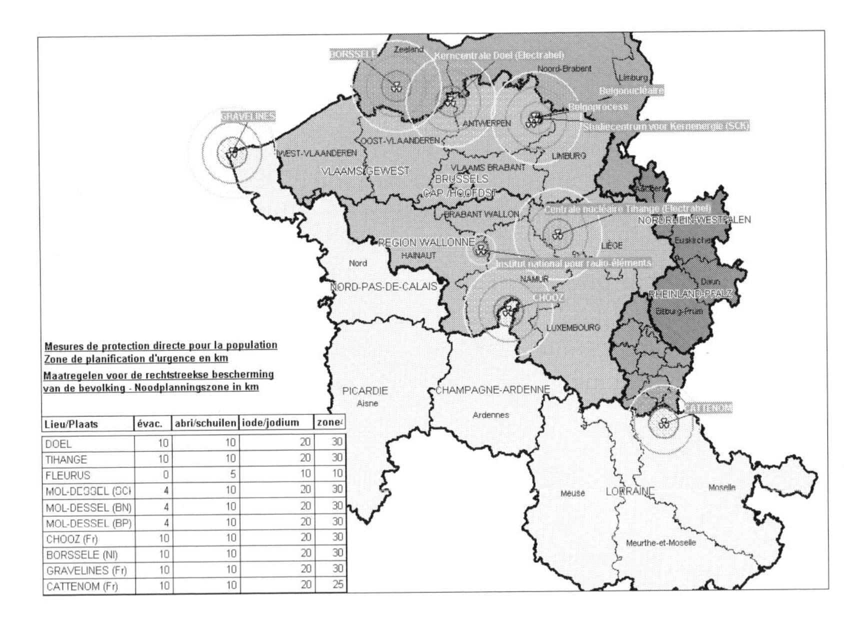

Lieu/Plaats	évac.	abri/schuilen	iode/jodium	zone
DOEL	10	10	20	30
TIHANGE	10	10	20	30
FLEURUS	0	5	10	10
MOL-DESSEL (SCI	4	10	20	30
MOL-DESSEL (BN)	4	10	20	30
MOL-DESSEL (BP)	4	10	20	30
CHOOZ (Fr)	10	10	20	30
BORSSELE (Nl)	10	10	20	30
GRAVELINES (Fr)	10	10	20	30
CATTENOM (Fr)	10	10	20	25

Fig. 4. Production of Reference Maps for Crisis and Disaster Management

3.2 Maintenance of Geocoding Tools and Databases in order to Easily Locate any Critical and Vital Structure

Precise geocoding is a basic requirement for Crisis management but precise data is not always available. Most of the time data will be made available in the form of paper maps, address listings, aerial photography or satellite images without any or with few geographical references. For this reason it is necessary to maintain efficient geocoding procedures translating raw data into usable geo-information.

3.3 Spatial Analysis of Risks or Disaster Consequences

Classical GIS tools or dedicated models must be available to perform analysis on Geographical data. The results of those analyses must be integrated into the system in the form of complementary map layers helping decision makers to take the right decisions. Proximity and accessibility analysis are considered as basic GIS-functions always available "in house" while dedicated models like pollutant dispersion models requiring specific tools and knowledge will typically be produced by partners, making their results available in the form of easy to integrate map layers.

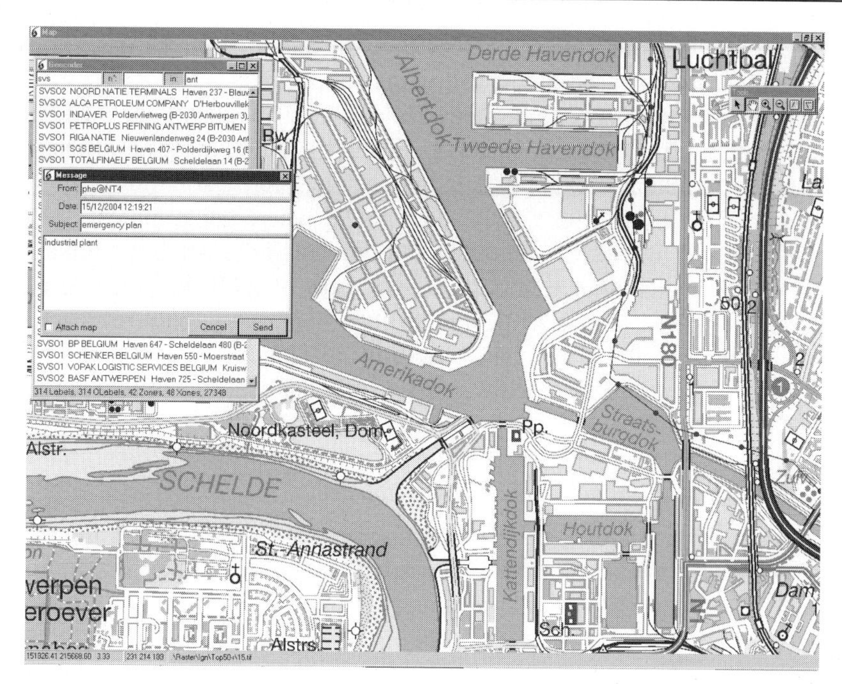

Fig. 5. Geocoding tools

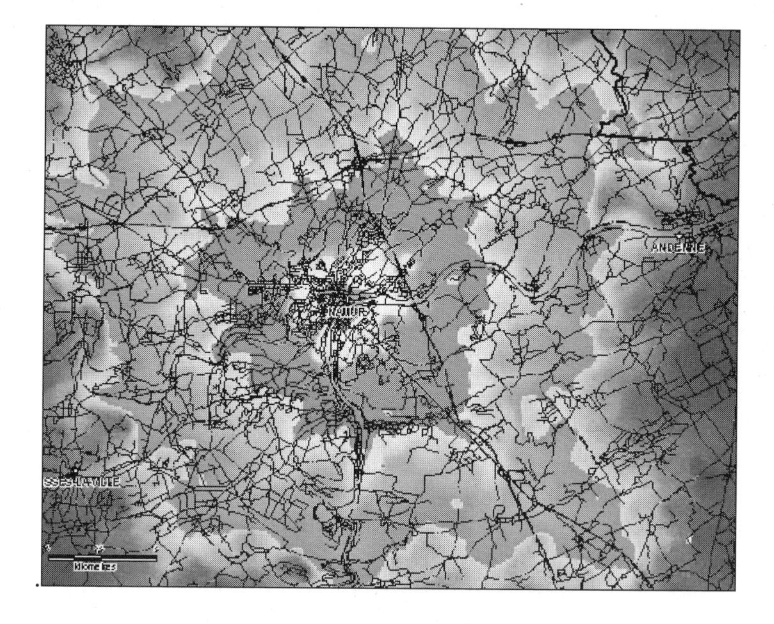

Fig. 6. Spatial data analysis of risks or disaster

Regarding Geo-information technology developments, the strategy of the Crisis Centre is based on the following statements:

- autonomy in the management of base maps by acquiring our own datasets from multiple classical publishers and by maintaining an infrastructure able to translate existing data into usable Geo-information;
- setup of an easy to use "Crisis" data model where geographical objects can be linked to external databases through network or Internet links;
- use of basic mapping functions in the form of well documented dll's (dynamic linked libraries for map display, map query, map registration, geocoding, proximity analysis or routing) that can be customized for our own needs and implemented within existing applications or into customized applications.
- partnership with other bodies dealing with disaster management in order to make data collection and data sharing easier.

This also implies the management of contractual and juridical aspects for data diffusion as well as data security that can be considered as "heavy tasks".

4 Case Study: Cartography of Pipelines

The disaster of Ghislenghien (30/07/04) where a pipeline for gas transport exploded after a leak of pressurised gas has been followed by a deep analysis of data availability concerning pipelines and potential risks associated to it.

It appeared that the transport pipeline networks were administratively monitored through authorizations delivered by Ministry of Economy and by Local and Regional Authorities, and were technically monitored by Organizations and large companies such as the Army (NATO pipelines) or FLUXYS (gas pipelines) all of those having very detailed technical information about their networks. After several meetings, we noticed that most of the pipeline-owners were already involved in long term projects of mapping pipelines in XYZ coordinates but also that there was nowhere to find an exhaustive overview map/database of existing pipelines covering the entire country!

We decided to collect the data and bring them into one general medio-scale map (reference scale 1/50.000) showing the position of transport pipelines and giving the necessary attributes to access more detailed information like contacts, products or risks associated to every single pipeline segment.

The work has been organized under the umbrella of the Crisis Centre but required input from different Federal organizations, like Ministry of Economy, Ministry of Defence and also from Private Companies owning and managing pipelines.

We received hundreds of paper maps, several CD-ROM's and pages of information! And started our scanning, conversion and harmonisation processes.

No doubt that this exercise backed us up in the necessity to maintain our ability to convert various data formats and also to set up simplified data models for Crisis management.

This exercise has also resulted into maps and databases that were presented to operational emergency services like Fire brigades and Civil Protection units. One must know that those operational services receive regularly very detailed information about pipelines but still, they showed great interest in our work, asking copy of the results in digital and paper formats! Putting all those data together and giving access through a simple viewer/browser interface for their own territories was simply new!

5 Priorities for Developments in Geo-Information

If the security of the population remains a Federal mission -and this working together with several administrations that do not consider crisis management a priority - other federal and regional authorities have taken up numerous connected fields like environment, education or transport.

Our challenge is to coordinate the action of those actors in order to offer the highest degree of security to the population.

Therefore, we pursue our efforts of data collection, harmonisation and integration for better planning and response to crisis situations. This implies the set up of effective partnerships for exchanging data with other authorities.

Support to local authorities through adapted regulations and technical assistance is also a primary goal and will be organized by delivering "ready to use" data sets and procedures for crisis management to local partners.

On the International side, the priority goes to operational collaboration with neighbouring countries. An example is the "Disaster Management" task force set up for the Benelux in a European project. On the longer term, the Crisis Centre would be interested in initiating or participating to European projects dealing with standards and norms for data exchanges and communication during crisis.

Further developments of our Geo-information infrastructure will therefore be integrated into a global process of risk and disaster management in order to meet the requirements for planning, prevention, detection, response and recovery process. From the disaster of Ghislenghien we have learnt that **coordination** is the keyword.

6. Conclusion about Geo-Information at Crisis Centre

Our experience shows that you'd better start from rational solutions that satisfy immediately your needs rather than build huge projects far from the reality. Dealing with crisis situations requires first of all coordination and as a consequence a clear overview of the situation, easy to use tools and data models and a clear communication. The technology must be open to "non-specialists" and lead to reliable partnerships. Like an old Chinese proverb says: "You better light a candle than damn darkness".

Real Time 3D Environment Simulation Applied to the Disaster Management Field: Our Experience

Pedro Branco[1], Carlos Escalada[2] and Ricardo Santos[1]

[1] DigiUtopikA, Lda, Rua do Moinho, Lt. 30, Urbanização Casal Labrusque, Areia Branca, 2530-065 Lourinhã, Portugal.
Email : pbranco@utopika.net, rsantos@utopika.net

[2] SAMU, Avda. Américo Vespucio s/n, Edificio Cartuja Blq.E, locales 7-8-9, Isla de la Cartuja, Sevilla 41092, Spain.
Email: carlose@samu.es

Abstract

In this paper we described our experience in the development of real time interactive 3D systems in the form of a framework – HorizoN Sentry – that allows not only the real time simulation of environments based in Geographical Information Systems data but also its fusion with life sensors like traffic, Global Navigation Satellite Systems and weather based sensors. A design to implementation view of the system is shown together with demonstration results based in the European Space Agency framework projects and private developments.

1 Introduction

In this paper we describe our experience in the development of the Horizon framework and it's presently use in the demonstration of several European projects and inside our commercial activities in the field of Civil Protection and Disaster Management.

Horizon is a fully integrated Geographical Information solution that is able to re-use data from past and present independent systems, and fuse them in a single and highly adaptable product capable of various utilizations in civil protection, planning, monitoring and marketing projects. The development of such a system was focused in creating new software architecture to allow an unprecedented Environmental Data Interoperability and Interchange capability.

2 The Vision

Synthetic environment, as used in today's networked, interoperable heterogeneous systems, means more than just the visual scene of the simulated environment. In addition to the visual aspects of the natural environment (terrain, ocean or atmosphere) and objects on the field, the synthetic environment must now encapsulate non-visual information to allow entities under computer control to properly interpret and navigate the environment.

A synthetic environment is created through a costly/time-consuming process resulting in an integrated, fused data set referred to as a synthetic environment database. This database contains sets of objects which define and describe a natural environment. The data objects describe a geographical region and the elements and events expected to occur there. A synthetic environment database also encapsulates the geometric and topological relationships between the data objects. These relationships are critical in ensuring that the run-time databases, derived from the synthetic environment database, will be correlated so that all "views" of the environment are the same.

Fig. 1. HorizoN Runtime Viewer – 3D Bird View – Park Expo Compound (Lisbon, Portugal)

Data interchange, is not just for reuse in building synthetic environment databases but is a central element for achieving interoperability between distributed, heterogeneous training system networks. To successfully interchange environmental data, the interchange mechanism must account for all data types and their relationships used to describe the synthetic environment. The goal is a loss-less, unambiguous transfer of data from one database directly into another. This interchange capability will save Euros through synthetic environment data reuse and will improve training effectiveness through interoperability of training systems. An unambiguous, loss-less data interchange will minimize the potential inter-training system correlation concern [3].

3 What Constitutes a Simulated Environment?

A simulated environment database, for use in M&S applications, is an integrated set of data elements that describes a defined geographic region. It must contain a consistent and correlated description of the environment (terrain, ocean, atmosphere, and space) that is appropriate to the simulation objective.

Additionally, it often includes data describing simulation elements and events expected to take place during simulation execution. For example, data representing trees in a forested region may be found in the database along with data describing the geometry of the vehicles that may drive through the forest and impact trees during a training exercise. There are several general classes of data that pertain to simulated environment databases and the database generation process: surface and volumetric data, 2-D and 3D feature data, 3D models and icons; texture, image and colour data; material attributes; and various animations that describe predetermined effects.

Surface and volumetric data can be derived from both regular or irregularly spaced gridded fields based on measured or remotely sensed observations. Examples include digital elevation models, digital bathymetry, and complex grids that describe various properties of the ocean, atmosphere, and space. From these source matrices and 3D grids, supplemented by value added updates, various simulation database generation systems derive the terrain skin, ocean and river bottom, ocean and river surface, cloud and pressure surfaces in the atmosphere, thermal boundaries in the water column, and various space effects. To create earth surfaces, these source data are further processed into regular or irregular networks of typically triangular shaped polygons.

Feature data represent both manmade and natural objects. Examples include weather fronts, forests, agricultural fields, water bodies, roads and river networks, volumes of atmosphere, ocean or space, and individual objects such as trees and buildings. Some objects may have moving parts (e.g., trees, drawbridges, windmills, and industrial cranes).

Usually these data represent a separate source input to the database generation process and therefore must be correlated with the applicable surface (e.g., ensure rivers flow downhill, roads are driveable, small water body and rice paddy surfaces are at constant elevation). In terrain representation, if feature ground elevations are provided, they are the preferred source for generating the terrain skin in dense feature areas. It is becoming common practice to integrate feature ground elevation data (either provided from direct observation or interpolated from elevation matrices) prior to generating the terrain skin and thus reduce the editing required to ensure correlation.

3D models and icons are also used in simulation. These data reflect detailed 3D surface geometry often derived from Computer-Aided Design (CAD) files, as-built blueprints, or from various image extraction techniques.

In this manner, high-resolution data that describe stationary and movable systems, equipment, and structures can be entered into the simulated environment database. When visualizing surface and feature data, colour and texture are important in conveying added information. The colour and texture data are either extracted from image sources or derived from geographic area or engineering inference techniques.

Rendering data in the non-visual spectrum also requires attributes such as acoustic characteristics, surface materials, or electromagnetic properties. These data include sensor signal reflectance, absorption and transmission attribution on surface materials and volumes, and traffic data that affect sensor system performance and human or vehicle dynamics during simulation execution.

The resulting data are then integrated with surface data to generate the desired appearance (e.g., field and water surface patterns, road and building surfaces, sound reflections, electromagnetic signature, and clouds). Other attribution must also be carried in the simulated environment database to support computational tasks required by the simulation.

A simulated environment database may also contain data describing predetermined events that are captured in the form of short animations. These could include explosion effects, surface drainage patterns, or state changes for specified objects. Other data are included to convey connectivity (topological relationships), system specific data necessary for computations (e.g., precomputed visibility, occulting planes) and to produce electronic or paper maps and charts (e.g., names and labels).

All of these required data and information components must be addressed by a robust interchange specification designed to support both source data input to the database generation system and reuse of its integrated database output products.

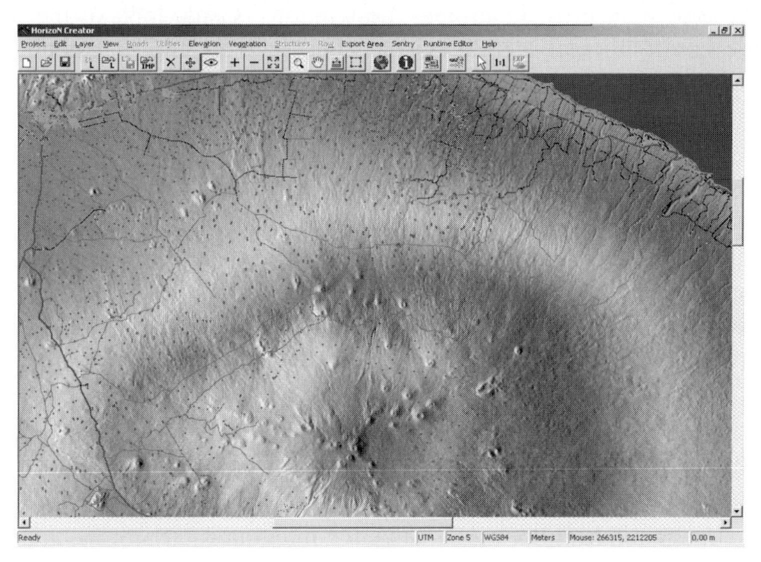

Fig. 2. HorizoN Creator – preparation of GIS Data importation and manipulation for 3D real time simulation display

4 Requirements

With the functional uses defined and the objectives stated for the tool, we developed a series of requirements and assumptions that had to be followed in development. They include:

- A distributed information processing environment in which applications are integrated.
- Applications and data independent of hardware to achieve true integration.
- The tool has to be portable and high-resolution to effectively visualize and simulate the port of choice for operational support and realistic schoolhouse, and shipboard training.
- The tool has to be extensible so that it can be easily modified.

- The tool will use standards and software developed by the M&S community to ensure the third objective is met.
- The tool shall be PC based and shall visualize three-dimensional models of the environment to include air, land and sea.
- The tool will be open-architecture based for ease of extending the functionality.
- The tool shall provide an "out-the-window" visual look of the interactive environmental effects (visual cues such as eddies around buoys, tides, and currents).
- There shall be a multi-sensor, object oriented database (radar, night, day, IR, etc.).
- The tool shall be able to interoperate with other simulations utilizing such technologies as DIS, HLA [8], and SEDRIS [2] [6].
- The tool shall be capable of supporting multiple scenarios such as training and mission rehearsal.
- Develop a PC based system that can be used to visualize 3D models of environments to include air, land, sea and terrain.

5 Past and Present

Truly Integrated Monitoring Systems for Integrated Environmental use are almost inexistent even at the present time. Traffic monitoring, civil protection issues, the real-time monitoring of power lines, underground systems, communication lines are normally controlled and visualized using (independent systems) incompatible tools and limited to using only two dimensional systems.

Complex environments with multiple sensors or 3D requirements are not truly represented in 2D visualization systems. A need for 3D real time interactive systems is increasing due to the growing of monitorization applications requirements.

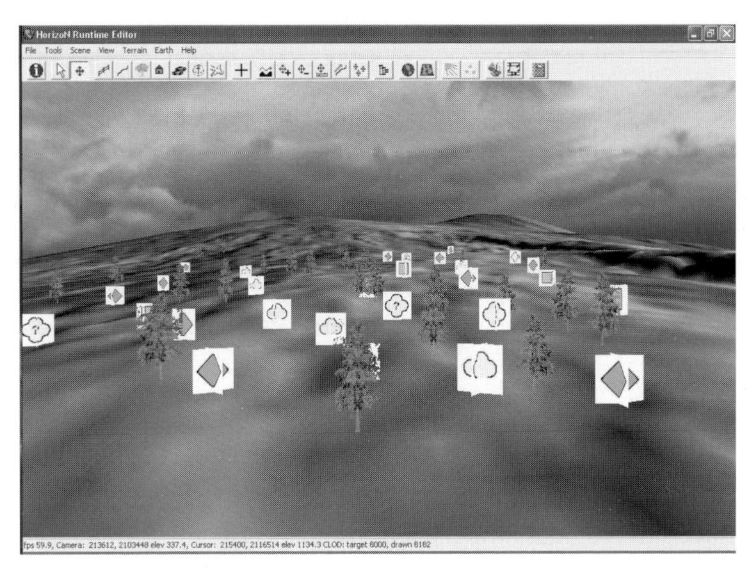

Fig. 3. Military Units displayed thought the NATO standard symbols attached to GNSS input sensors.

6 Functionality

The Sentry system is the last step of the integration of different HorizoN modules (HorizoN Creator, HorizoN Object Manager, Object Runtime Viewer and Editor, HorizoN Vector and Space HorizoN). It was conceived to integrate an enormous range of data formats, from typical Geographical Information System formats to satellite images. Live sensors can also be implemented including Global Positioning System and ground sensors.

Our concept is that in the "perfect information world" behaviour sharing is required. Using this statement the need to create an architecture that can reflect it is obvious. This architecture and related software must establish a communication structure through Live Sensors, GPS Signals, Geographical Information Systems and Satellites and a computer that acts has the "decision maker" and gives the necessary indications to other systems.

It assimilates disparate geographic data into seamless, global databases. For example:

– Include any number of overlapping and adjacent elevation layers,
  - GIS Static Data
  - 3D Models
  - Databases

  - Web Data
  - Live Sensors
  - Satellites (live images and positioning)
- Integrate satellite, radar, aircraft and scanned images from several map projections,
- Drape multiple vector GIS layers simultaneously. We maintain all the development features inside the following worldwide known standards:
  - OpenGL (Open Graphics Library)
  - HLA/DIS (High Level Architecture/Distributed Interactive Simulation)
  - SEDRIS
  - EAX (Environmental Audio Extensions)
  - OpenAL (Open Audio Library)
  - We already give full import support for 60 GIS filters, GPS and 5 three-dimcnsional formats in thc present system.

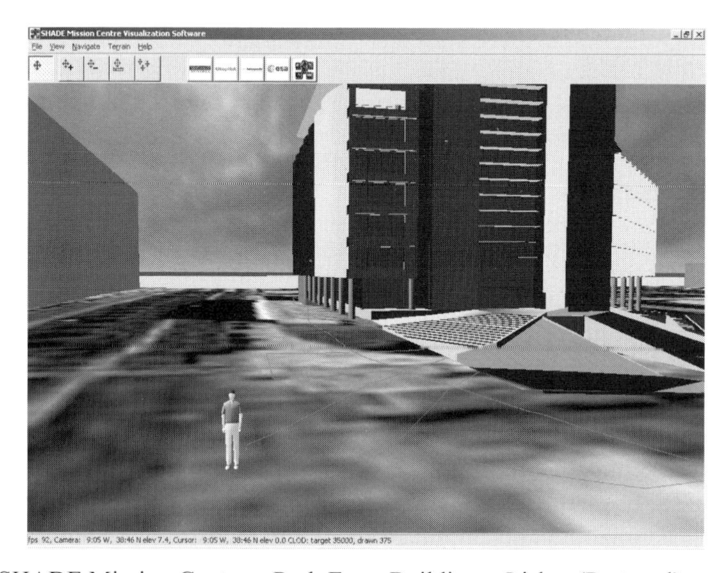

Fig. 4. SHADE Mission Centre – Park Expo Building – Lisbon(Portugal)

The program is installed in a computer connected to the internet and/or intranet and also can have other input devices for gathering GIS and sensor data. The install is going to be only the action of deploy the system in a network and it will start working on itself gathering and discarding data using it's own Artificial Intelligence features. The system makes copies of itself and auto-installs modules in other machines with the required permissions and authorization levels and then it starts working.

The data collection and live inputs are used to provide one or several real-time virtual reality representations of the real world that can be used for navigation, visualization, interpretation or control in/of the world environment.

It is a real environment Sentry that can give you all the information in real time.

7 HorizoN Sentry

HorizoN is an integrated set of tools developed by DigiUtopikA that is used in all our simulation-related projects. HorizoN Sentry is a global solution, joining all software modules that we developed during the last two years in an integrated and fully live solution. An HorizoN Sentry solution is build with static and live data, from satellites to normal maps; inputs can be provided in the form of Geographical Information System (GIS) Static Data, 3D Models, Database, World Wide Web Data, Live Sensors and Satellites (live images and positioning – GNSS inputs). Filters for most of the common civil and military data files already exist (a total of about 60 filters at the present time only for the GIS sector).

HorizoN Sentry is being developed in the form of a Customizable Commercial Of The Shelf (COTS) product. The installation is very fast and simple. After deploying the system in a network, it will start working by itself gathering and discarding data, using its own Artificial Intelligence features.

Full Database Access and Control using multiple sources of data is allowed. The system supports common query and statistical functions.

Amazingly complex visualizations and scenarios can be created with no programming required. The user can even edit the scenario while he is walking or flying inside it. The HorizoN Creator, HorizoN Object Manager and the HorizoN Runtime Editor are some of the tools that provide an easy learning interface for scenario development.

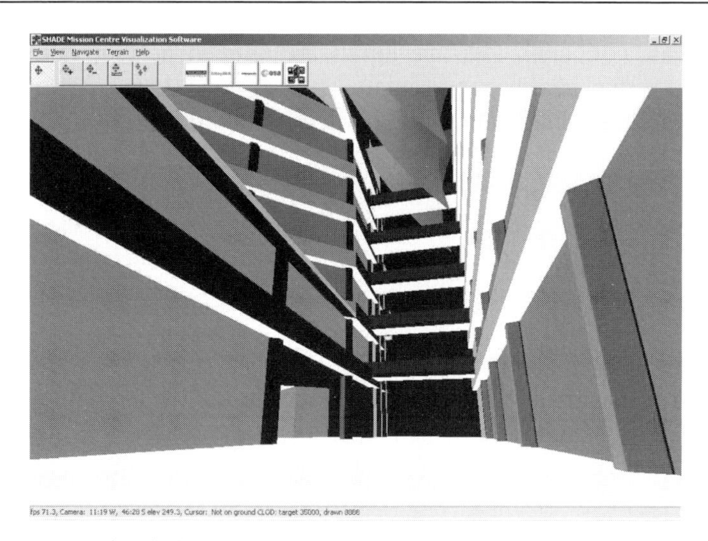

Fig. 5. Indoor 3D simulation – SHADE Mission Centre

Levels of access are achieved by ranks contained in the Sentry Management Database. The Sentry Management module has the functions of control of the user access level to the data visualization and edition system. The traditional client-server architecture is replaced by the High Level Architecture (the NATO standard for distributed simulation). It is a real environment Sentry that can give you all the information in real time.

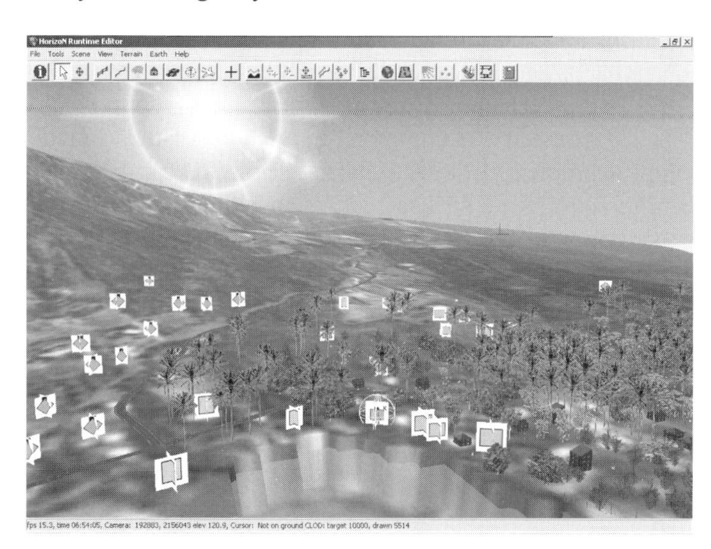

Fig. 6. Military Units displayed through the NATO standard symbols attached to GNSS input sensors

HLA is the North Atlantic Treaty Organization (NATO) Distributed Simulation Worldwide Adopted Standard. The system should provide interfaces for use of HLA in order to be easily adaptable for distributed simulations.

A portable (and emergency) solution was also developed – a one CD setup kit – that contains a fast installation software and the GIS database of the specific location related to the CD function (some of the operational difficulties – related to buildings architectural information and electrical and communications lines exact locations - that became apparent immediately following the 11th of September tragedy, can be eliminated this way). This sub-system can easily be used in a laptop and its use in a handheld is being developed at the present time. An Internet connection via satellite using a laptop can also solve the problems usually found when phone lines and cellular antennas go down during catastrophes [1].

Fig. 7. City scenario generated through object parameterization

8 Demonstration and Use

The software is already working in Windows 95/98/Me/XP/2000, Unix and Linux environments and was developed in C++ using the most advanced standards and methodologies like OpenGL (Open Graphics Library), OpenAL (Open Audio Library), EAX (Environmental Audio Extensions), SEDRIS and High Level Architectures – HLA (the NATO standard for distributed simulation). The portability for Apple Macintosh and other similar systems can easily be achieved thanks to the portability requirements of initial architecture design.

Integration with existing systems was implemented:

- Filters for standard GIS and 3D data are already developed and tested;
- Integration with earth observation projects like Aqua and Envisat is being studied;
- GPS inputs are accepted at the present time (for real time positioning). Applications involving the EGNOS (European Geostationary Navigation Overlay System) were already developed. Applications using the future Galileo system are being proposed;
- Web Integration is already being developed (including interactive real-time 3D navigation over the internet).

The Integrated solution and the independent modules were demonstrated in various companies, military and governmental International institutions during the last months of 2002. A demonstration at the European Space and Research Technology Centre (ESA-ESTEC) was performed in the first week of December 2002. The system or an adapted sub-system is going to be used in an international military exercise during January 2003.

HorizoN was used in several projects during 2004, the most visible ones were:

- SHADE (Special Handheld based Applications in Difficult Environment)

  An ESA Telecom (European Space Agency [9]) funded project leaded by Teleconsult (Austria), where DigiUtopikA, Lda. (Portugal) and Telespazio (Italy) are sub-contractors. The key-objective of the SHADE project is to evaluate and demonstrate the capabilities of different positioning and augmentation sensors for safety critical personal mobility applications (pedestrian users) in difficult environments. Public demonstrations of the SHADE system have been performed in Lisbon, Bolzano, and Rome from May through July 2004. A final public demonstration will be performed on September 29, 2004 at European Space Research & Technology Centre (ESTEC) premises (Noordwijk, The Netherlands).
- WICOR (Wireless Communication Router)

  An ESA Telecom funded project leaded by Telematica (Germany), where DigiUtopikA supplied the software for the visualization interface, to design an intelligent user terminal router for wireless communications has completed its testing phase. The project, known as WICOR for Multifunctional Wireless Communications Router, demonstrated exceptional reliability in routing connections between mobile devices. A 3D

imaging tool is also included in the software giving such potential users the ability to determine the location of their vehicles and visualising them in a three-dimensional easy to view graphic display.
The tests were conducted in Germany, Austria, Italy and Switzerland in areas where connection methods varied considerably.

9 Results

Tests with real data obtained from a wide variety of sources (including sets of GIS data from Lisbon, Hawaiian Islands and New York).

The results were generated in real time interactive mode in a Windows 2000 based system installed in a Pentium IV at 1.2 Ghz, with 256 Mb of Ram and 32 Mb of VRam. The minimum configuration is a Pentium III at 500 Mhz, with 64 Mb of Ram and 8Mb of VRam. Laptop computers are also able to use this software.

HorizoN rendering features have being tested by various military, defence related and civilian organizations. The results of the visualization system were 25 frames per second in the configuration given above. The medium frame rate is about 75 FPS in the same configuration.

The system was tested with data with a higher variety of size magnitude – including extended areas of hundreds of squared kilometres to simple urban scenarios – with results that allow the visualization in real time of the same results in an effective interactive way that can be used with efficiency in information tasks.

10 The Future

We must take some in consideration several issues before talking about the future of HorizoN Sentry:

- Database generation cost is a significant portion of the total cost in any virtual simulation system;
- Creating good databases (content) depends on many parameters, is nontrivial, and quite expensive;
- Access to source data is limited (this will change!);
- Efficient, well-integrated, and inexpensive tools are needed to reduce database creation cost and time;
- Sharing and reuse requires development of standards, and this will add value to any project.

We are currently working in the following future solutions:

- Web Browser (with live 3D data);
- Improved underground and above ground layer functions;
- Off-the-shelf solution deployment;
- Integration with Live Sensors (traffic lights, power and traffic sensors, etc);
- Live Satellite Data Integration(including Earth Observation Data);
- Handheld Integration and Portability;
- Maintain compatibility with latest SEDRIS SDK release;
- Provide more analyses for other geometric interactions;
- Improve the attribution analyses;
- Import data directly from other formats;
- Continue expansion and improvement of the edit and save capability (using the SEDRIS write API);
- Utilize future SEDRIS inter-application interface.

The initial feedbacks gathered from many specialists during the development and testing phases demonstrated clearly that there is a very wide market for this software system and related architecture.

The data and services infrastructure that we envisioned in the future will be based in the SEDRIS specifications. HorizoN Sentry as stated before was designed to be in accordance with the SEDRIS data encoding and information exchange standards.

11 Present Strategy, Partnerships and Future Directions

The concept and partial working demonstration was already presented to several organizations including:

- Portuguese Ministry of Defense;
- Portuguese National Protection Services;
- North Atlantic Treaty Organization (NATO);
- Several GIS, multimedia, aerospace and navigation commercial companies;
- Several partnerships with strategic companies and organizations are being developed regarding software development and optimization and solutions integration in the real world;
- Several proposals to the European Space Agency (ESA) are being and will be posted, especially under the Telecommunications division.

The system and related tools after that first exercise being presented are hopefully going to be used worldwide exercises, real situations and demonstrated in conferences starting next year.

In the multimedia sector Creative Labs [7] was our first and most important partner from the first steps of the project. They are the lead worldwide multimedia hardware supplier. Their worldwide well known sound standards and the recent acquirement of the famous 3D Labs by Creative speaks for the future of this company.

Several development partnerships and advices were and are being established. All the software follows currently accepted standards from NATO and other military organizations worldwide. Technically the already developed software modules are running in Windows and Linux/Unix.

12 Conclusion

Horizon Sentry is a fully integrated and adaptable solution that can be used for many independent or integrated purposes in different market sectors from urban planning to civil protection.

The benefits of such as system are in general the best use of resources (import all the 2D data to create 3D with no extra cost and re-using existent live sensors designed with non-related design purposes), fully customizable (COTS product), works with low-end hardware, Internet and intranet support, integrated systems can be customized for all departments and sectors.

The quality of the results obtained, the effectiveness and esthetical beauty allow HorizoN Sentry to be a product that has conquered the interest from first users, programmers and technicians to various owners of GIS data who seek a tool that makes it possible to extract extra value from all the data.

The methodology used in the solution development allowed that any educated individual without specific programming knowledge will be able to access about 95% of the software functions. These amazing achievements make possible our product's market penetration in a wide range of areas and users. This has been a successful strategy.

Presently DigiUtopikA is building a long term partnership with SAMU (Spain) in order to implement worldwide Disaster Management Centres based in the HorizoN framework.

Talking about results is also talking about numbers in the real life. Tragedies like earthquakes, the 11th of September, unexpected violent storms and out of control large fires have show in recent times the vulnerabilities and the relatively slow response time currently available with traditional systems. The HorizoN Sentry system is not only an amazing achievement in multimedia terms but also a useful contribution for a better and safer world.

Fig. 8. Search and Rescue Real time Simulation in a volcanic eruption scenario (Demonstration for NATO)

References

[1] "Use of Geographic Information Technology in Municipality Civil Protection Services (SMPC)", SAN-PAYO, Margarida; TELHADO, Maria João; PAIS, Isabel

[2] "Using Environmental Information Efficiently: Sharing Data and Knowledge From Heterogeneous Soruces", Ubbo Visser, Heiner Stuckenschmidt, Holger Wache, Thomas Vögele - TZI, Center for Computing Technologies, University of Bremen, Germany

[3] "Moving Toward Higher Fidelity Environments" (2002) http://virtualcities.ida.org, SEDRIS™ Technology Conference, Vancouver, B.C., 23 August 2002, Dana Magusiak, Institute for Defense Analyses

[4] "SHADE: Special Handheld-based Applications in Difficult Environments" (2004) Elmar Wasle, TeleConsult Austria GmbH, Bjoern Ott, TeleConsult Austria GmbH, Austria, Pedro Branco, DigitUtopika, Portugal, Riccardo Nicole, Telespazio, Italy, Guenther Abwerzger, TeleConsult Austria GmbH, Austria, presented at ENC-GNSS2004 and NAVITEC 2004

[5] "Integrated Virtual Solution for Planning, Development, Control and Protection - HorizoN Sentry" (2003) Pedro Branco, DigiUtopikA Ltd., paper presented at the ENC-GNSS 2003

[6] http://www.sedris.org

[7] http://www.europe.creative.com/
[8] https://www.dmso.mil/
[9] http://www.esa.int
[10]http://www.esa.int/esaNA/index.html

Geo-Information for Disaster Management: Lessons from 9/11

Michael J. Kevany

PlanGraphics, Inc., 1300 Spring Street, Suite 210, Silver Spring, Maryland 20910, USA.
Email: mkevany@plangraphics.com

Abstract

This paper will address multiple conference topics but will focus on user needs and technology developments. The paper will describe issues of geo-information for disaster management developed around the experiences prior to, during, and following the World Trade Center event of 9/11/01 in New York in which the author was a participant. The events of 9/11 have had a dramatic impact on the state of geo-information in disaster management. Thus, it is a focal point for the paper. The paper will be presented from the perspective of experience and thus is practical rather than theoretical.

1 Introduction

The purpose of this paper is to provide a specific perspective, that of New York City and the World Trade Center emergency experience, on the topic of geo-information for disaster management. It will substantiate the fact that geo-information and technology are very important to disaster management by describing how they played a key role in the World Trade Center response. The paper will also explore the lessons that were learned from that experience and some of the activities and developments that have occurred since 9/11 both in New York and in the field in general.

The paper consists of five major sections:

- Background of geo-information for disaster management
- The 9/11 event
- Lessons from the World Trade Center

- Solutions developed from the 9/11 lessons
- Concluding recommendations.

2 Background of Geo-Information for Disaster Management

Location is one of the most critical elements of necessary information in disaster management and response. Almost everything in a disaster is related to a location, and often location is the most important attribute of information. Management of location information is the purpose of "geo-information" technologies, the most pervasive of which is the geographic information system (GIS).

There is a growing recognition in the disaster management community of the value of location or geo-information in the form of maps and spatial data. Thus, GIS technology is increasingly being employed in disaster management. Decision-making requires knowledge of the location of disaster impacts, response requirements and response resources, and the location relationships between them.

The management and delivery of location or geo-information and the tools to view and analyze it have proven essential in disaster response operations, as clearly evidenced in the World Trade Center response. This capability is much more than providing maps or an interface to the organization's GIS. It requires imbedding geo-spatial data and capabilities in all aspects of the disaster information system and operating procedures. It means the careful design of tools for displaying and using geo information. That design must be based on an understanding of both the operating procedures of disaster management and the geo-information technology that may be applied to support the procedures. In that manner, the geo-information and tools can be used effectively by disaster managers and responders.

2.1 Location as a Key Element in Disaster Management

The impacts of a disaster occur at locations, and many emergencies are in motion across a series of locations that must be known to managers and responders. The response resources exist at locations, and the relationships between the locations of the disaster impacts and the locations of response resources are critical to effective disaster management. Knowing the availability of an ambulance for a response is useful. Knowing, however, that it is actually located across the river from the need is critical. Location data is an essential element in the recording of information about the disaster and its impacts and for the resources available to respond to the disaster.

Location is recorded in most emergency and general local government operations. A variety of location codes or descriptors are used, including:

- Address
- Street intersection
- Building identifier
- Administrative or statistical area
- Postal code
- Place or landmark name.

Historically, the location capabilities of disaster management have been quite limited, often relying primarily on text descriptions and a reference map of the area posted on the Emergency Operating Center (EOC) wall. Prior to the World Trade Center attack of 9/11, GIS technology was beginning to be used, but often merely as a map production tool operated by a GIS expert and not a disaster manager or responder. This capability was activated by a request from the disaster manager or responder to a GIS expert providing a description of a required map. The GIS expert would interpret the request and produce a paper map of that interpretation. Because of the different perspectives of disaster responders and GIS personnel, miscommunication often occurred. Initial maps didn't actually meet the requirements and proved useless so new maps had to be delivered. This was quite limiting because of the mechanism for communication between emergency and GIS, the time required, the inflexible media in which the product was produced, and the cumbersome logistics of paper product delivery to the end user.

2.2 Early GIS Disaster Management Experience

Numerous examples of GIS use in disaster management exist, though most don't represent the comprehensive support that geo-information can provide. For example, GIS applications and data were used in Osceola County, Florida, in 1998 to analyze the damage caused by a tornado that passed through the area. Using GIS technology, a comparison was made of aerial photography and planimetric maps created prior to the disaster with a comparable set created shortly after the event to determine the location and extent of damage. (Reference Dave Nale)

Extensive use of GIS has been made in California to analyze the impact of earthquakes and to support recovery planning, including the 1989 Loma Prieta and other major earthquakes.

In New York City, prior to the 9/11 emergency, the value of GIS technology was recognized by the Office of Emergency Management (OEM). Rudimentary use was being made of GIS with success, and management wished to expand the capabilities significantly. Thus, a GIS requirements analysis was performed and, based on that analysis, a system design and implementation plan were prepared. This plan defined a system architecture, database, and series of applications for use in emergency planning and operations. The design included an interface with the City's GIS Utility, a central repository of geo-spatial data that would provide a wide range of essential up-to-date data on the physical, legal and administrative conditions in the City. That OEM GIS plan was in the process of being implemented at the time of the emergency.

3 The 9/11 Event

The first plane struck WTC 2, the North Tower, at 8:56 A.M. on September 11. The EOC and its GIS operation were immediately activated in WTC 7 to generate maps to support the response. It was quickly recognized, however, that the building was in grave danger, and the EOC was evacuated an hour after the first strike.

The GIS Utility office was one block from WTC 7 at 75 Park Place. Following the attack, the building was evacuated and was unavailable for occupancy for several months. Therefore, the City's GIS response was dramatically affected.

The next day, the Emergency Mapping and Data Center (EMDC) was established in a temporary EOC at the Police Training Academy. Two days later, a long-term EMDC was established in the EOC on Pier 92 on the Hudson River. The establishment of GIS capability to support the emergency effort was quite challenging—the loss of two key facilities, the establishment of replacement facilities, and the acquisition of backup data copies and of a complete complement of hardware and software. Staffing was less of a challenge with personnel from OEM, the GIS Utility, and its contractors immediately available.

Fig. 1. World Trade Center Location of EOC on 9/11

The primary tool of the EMDC was GIS, but several related geo-information technologies were also used for acquisition, processing, or display of data. These related technologies included:

- GPS
- Photogrammetry
- Digital orthophotography
- Lidar
- Thermal imagery
- Laser ranging
- Digital cameras
- Wireless mobile computing
- 3D modeling

A few of these technologies had been planned for in advance and proved useful. Other technologies whose use had not been planned were less useful, while other technologies were not useful at all for several reasons, including the interference caused by high rise buildings surrounding the site, damage to communications facilities, ineffective design, and lack of prior training.

EMDC provided a wide range of geo-information services to all aspects of the response on an around-the-clock basis for more than two months. EMDC was established as a section of EOC. It was comprised of a complete GIS configuration of hardware, software, database, applications, and staff.

The database included an extensive base of geo-information that was available in the City's GIS Utility prior to the event. The database also included a considerable amount of data generated through the emergency operations and acquired during the response period.

EMDC provided geo-information products and services to a wide range of customers, including responders, search and rescue teams, numerous City departments performing support operations, state and federal agencies, utility companies, volunteer organizations, the Mayor and other elected officials, and the general public.

Customers were served in several ways. All participants in the EOC had direct access to the EMDC database and applications through the EOC LAN. EMDC operated a customer desk at which people could request products and services. In one form of service, EMDC personnel partnered with customers to perform operations and generate products in cooperation with and at the direction of the customers. Information was also published over the City intranet for authorized personnel, and selected information was published through the Internet for the general public. Linkages were also provided to related GIS operations at the FDNY, FEMA, State, and other organizations. The EMDC operated 24 hours a day seven days a week for several months providing these services.

Throughout the operation, the database was constantly updated with data from transactions of changing conditions and the addition of new data sets necessary for emergency operations. These latter data included data from the utility companies; daily updates of orthophotography, thermal data, and other remote sensing products; and digital photography taken from the air above the site by FDNY personnel.

4 Lessons Learned from the World Trade Center

During and after deactivation of the event, numerous critiques and after-action reviews were conducted by various parties to understand the part that geo-information played in the World Trade Center response. From some of these reviews, lessons were identified for future disaster preparation and operations. The following lessons were compiled by the author,

ing of the field situation for those in EOC. This quickly available imagery taken at the scene or from a helicopter frequently proved very useful in the WTC response. This imagery would benefit from improved location registration techniques to relate positions on the imagery to underlying map locations.

Thermal imagery can be useful in disasters involving fires, but scale and timing are critical—Thermal imagery can provide information on the location and intensity of fires for disaster management and operations. To be effective, however, it must be of adequate scale to represent the detail necessary for a particular event. It must also be updated frequently to be useful.

4.6 Customer Lessons

Range of customer groups—Geo-information has found a wide range of customers for its capabilities and its products in a disaster. The customers range from management and command, to operations, to the news media and the general public. The requirements for geo-information range widely across these customer sets.

Range of customer technical skills and needs—Customers of geo-information products and services in a disaster have a potentially broad span of skills ranging from GIS experts to persons who are unfamiliar with technology or even maps. The requirements to support these persons, therefore, will also range widely from access to a GIS workstation or data that the person will process himself to very easily interpreted hard-copy maps.

Accurate timely public information is important—All emergencies generate a need by the public for information. The nature of a terrorist emergency generates a greater emotional impact and thus a greater demand for information. Geo-information is a useful format that is easily understood by the public, though it must be presented differently from that provided to the responders.

Disaster response customers require maps with easily identifiable ground locations and orientation—The impacts of a disaster can change the landscape, and responders unfamiliar with a vicinity may be called in for emergency operations. There are special needs, therefore, for maps that are easily interpreted and provide easily recognized locations when used in the field. Additional landmarks and annotation are often useful.

4.7 Logistical Lessons

Distribution of information is critical—Geo-information is only useful in a disaster if the information produced is available to managers and response personnel. The information must be distributed in multiple forms (digital or hard copy) to all participants, and the electronic or manual logistics for that must be established prior to any emergency.

The logistics must supply customers in many locations—The facilities for distribution must service customers in office and field locations. Distributing hard copy products to field locations is especially challenging.

Logistics are simplified by pre-definition of standard products and production capability—As noted above, standard products and production applications facilitate the distribution of information in a disaster.

Electronic media is the most efficient form for distribution—Electronic media is a more efficient approach for distribution than hardcopy. Intranet, Internet, wireless, and CD distribution allows wide access and flexibility for distribution and use.

Paper distribution is necessary—The current reality is that paper products are still essential and in high demand in a disaster. However, production and distribution of paper products is challenging.

Mobile GIS delivers geo-information to responders in the field—Because much disaster response activity takes place in the field and disaster operations are dispersed, mobile GIS can be very important. Mobile GIS includes not only use of mobile wireless computing devices, but also placement of GIS capabilities at multiple locations or installation of GIS in a vehicle.

Communications capabilities are particularly critical and vulnerable in a disaster—Wire and wireless communications among the various disaster operations and support locations are vital but vulnerable to interruption from the impact of the disaster. Loss of communications, as was the case in the WTC response, can cause severe hardship during emergency operations. Redundancy and back-up are critical for communications.

4.8 Notable Challenges

The author has identified four notable geo-information challenges applicable to numerous types of disaster situations that were faced during the WTC response.

Underground conditions—"Deep Infrastructure" was the term applied to the underground utilities and other infrastructure below the WTC and surrounding area. These elements ranged from small pipes and conduits to very large subway tunnels and underground parking facilities. They were particularly important for several reasons. They posed risks to the rescue workers in terms of potential fire, explosion, electrocution, and collapse, and their damage eliminated services to the undamaged buildings in a large area of lower Manhattan. Prior to the event, no central repository of information about them existed. The information that did exist was retained by several independent organizations, some of them private and sensitive to sharing for liability or competitive reasons. The formats, definitions, scales, datum, and other characteristics were incompatible. Acquisition and integration of these essential data sets proved a formidable task that was never completed.

Precise location—Pinpoint location on the "pile," as the Ground Zero area of rubble came to be known, was a challenge. There were requirements for precise location of various items, including equipment, hazards, and the body parts that were found. The conditions in the area made accurate location very difficult. The very tall buildings surrounding the site interfered with normal GPS operation. The condition of the debris was treacherous, and many of the items to be located lay below overhangs preventing measurement from above. Various approaches to measurement were employed with varying levels of success.

Damage assessment—Throughout the response, it was necessary to conduct damage assessments and inspections to determine the extent of damage, the safety of buildings for placing and releasing restrictions on occupation of buildings, and for reporting damage conditions and losses. The conventional building code enforcement procedures and tools were not appropriate for the requirements of this situation. Automated tools and procedures were tested and used with varying success. A serious impediment to the automated processing of geo-information was the lack of linkage between the mapped building footprints and the Building Identification Number (BIN) used to record and retrieve attribute data. Considerable effort was required to remedy this obstacle.

Public information—The emotional nature of the event created tremendous public interest. EMDC provided a steady stream of products for public distribution via the news media (newspapers, TV). The City's public Web site provided a channel for public information. The Emergency Management Online Locator (EMOLS) Web site application was quickly modified to provide public information on WTC conditions.

5 Solutions Developed from the 9/11 Lessons

5.1 Post 9/11 Geo-Information Activities

The threat raised by the World Trade Center attack generated a great deal of interest in the development of improved methods and techniques for response to disasters. Since 9/11, considerable activity has taken place to develop solutions for enhanced use of geo-information based on the lessons of 9/11. The developments include both technology and procedural solutions.

It became evident to the responding organizations that timely allocation of resources to specific locations is a critical aspect of the response to an emergency. Without sound knowledge of geo-information or "location," including street networks, buildings, utility infrastructure, floor-plan layouts, assets, and personnel, the emergency response can be seriously compromised. Alan Leidner, Assistant Commissioner of the New York City Department of Information Technology & Telecommunications at that time, would later comment, "This stuff saves lives!"

Through the lessons, we have gained a better understanding of the dynamics of the relationships between disaster managers, disaster management processes, disaster response, and geo-information technology. This understanding is allowing technologists to improve methods for design and implementation of technology, as well as disaster personnel to make better use of the technology. Therefore, the solutions discussed here are not just technology but include adaptation of methodology and revision of disaster business processes to achieve full value from technology.

5.2 Methodologies and Business Processes

The experiences of the World Trade Center, as well as those of GIS use in other emergencies, have indicated that geo-spatial information is extremely valuable. They indicate also, however, that delivering this valuable asset to meet the demands of disaster managers and responders poses challenges for IT, and geo-information technologies in particular. These persons can benefit from access to a wide range of information, quick response time, and access to reliably accurate real-time data. Geo-information applications must be flexible and easily operated. People not familiar with sophisticated technology must be able to quickly retrieve needed information.

One of the most significant lessons of the 9/11 experience was the importance of the realistic usefulness of technology, which has implications for both the design of technology tools and the procedures for their use. Design, development, and implementation of effective geo-information capabilities for disaster management now involve more than providing a GIS expert and software in the EOC. It requires careful definition of functional requirements for disaster management that include the perspective of disaster persons. It requires design of a system architecture that will support the demands of disaster management, including a mix of software tools and a comprehensive database and development and implementation of that architecture. It also requires the definition of operating procedures that incorporate geo-information for effective operations and the support of the system and database during emergency operations.

Technical design must include responders for determination of requirements and deployment.

5.3 Interoperability and the Information Access Portal

The primary information management tool in disaster management has been the Incident Management System (IMS). Such systems are used to manage information about the incident, including action reports, resources, and other activities of disaster response. The systems focus on tabular data, though some provide geo-information capabilities, usually in a quite limited way. IMS is used to record and process the data generated by the disaster activities. IMS does not provide much capability for integration or presentation of data from other sources.

An important geo-information development, therefore, has been the use of geo-information data and technology as an integrator of a wide range of information from multiple systems and sources. As noted earlier in this paper, location is a critical element in virtually all aspects of disaster management. Location and location data such as address can be used as an integrator of data. The location attribute of data from damage reports, response resources, response actions, etc., as well as the location attribute of data from systems such as surveillance cameras, document management, automated vehicle locators and other related sources, can be used to relate and integrate information from these sources for use in disaster management.

One of the most significant post-9/11 developments in the area of geo-information has been the concept of the Unified Access Portal. This concept provides a single Web-based portal interface in the EOC to access several disparate systems and databases that contain information useful for disaster management. This interface addresses the need for an integrated information system architecture for disaster response. The key focus is to provide timely access to all data necessary to make informed decisions when responding to an actual disaster or for training and simulation purposes. This interface's design will also provide enterprise data integration for day-to-day operations outside of an emergency situation.

The functionality includes a central Web-based interface capable of integrating many diverse kinds of location-based data. A GIS supplies the base map, and geographic attributes in data tables are also related or joined to base map features. These data tables may reside in many different software package formats. Other data types such as CADD files can be integrated onto the map layers on-the-fly to display building floor plans over building footprints. Selected video traffic or surveillance camera locations may be accessed on a map to view the current images in a live data stream without leaving the portal application. Plume models and other emergency tools may be accessed and integrated through the portal. The portal will provide direct access to information such as pipe valve shutoff locations, valve types, and properties to provide precise information for response personnel in the field. This interface will access data sources in real time to support the real-time decisions required in emergency situations.

Integration and interoperability are particularly important to disaster management and can be provided by the portal concept. A wide range of potential information sources is necessary or of interest and can be made accessible through the portal. Included in the systems of interest for disaster management are:

- Emergency incident management
- Computer Aided Dispatch (CAD)
- Electronic Document Management
- Surveillance and road video
- Emergency alert notification
- Plume modeling
- Weather reporting
- Hazardous materials (HAZMAT) response
- Debris management
- Emergency public information Web site
- Emergency secure Web site
- Mobile computing with GPS or a wireless locator
- Digital camera
- Weather, chemical, and other sensors
- Computer Aided Design and Drawing (CADD)
- 3D visualization

- Thermal data and imagery acquisition systems
- GPS, Automated Vehicle Locator (AVL)
- Work order management systems
- 311/Customer call services
- SCADA process monitoring systems

The portal concept involves accessing both the internal disaster management database and the various external data sources indicated. In addition, it can access conventional GIS and tabular data in participant databases without the need to copy and duplicate data and with the advantage of accessing the most up-to-date version of necessary data.

The portal provides a single, easily used source of information important to disaster managers and responders. Portals for integration and access to multiple information sources for disaster management are now being implemented in various locations.

5.4 Mobile Computing

Most disaster response activity occurs in the field, at the scene of the disaster, at the response resource depots, at the evacuation shelters, and at the command posts. Geo-information is as vital to persons at all of these sites as it is to the managers and coordinators in the EOC. Delivering paper products to the responders in the World Trade Center response was a formidable logistical effort. Therefore, mobile computing offers a vital service in disasters. It may take various forms from conventional laptop computers without wireless communication devices to specially designed, hardened tablets that incorporate wireless communications and GPS locators. Applications that provide displays of geo-information and allow simple query and reporting are required. In addition, data entry applications, especially when wireless communication is available, can be very valuable for collecting information at the scene of the disaster and reporting to the EOC. Integration with digital cameras for recording scenes of the disaster conditions can be a very useful enhancement to mobile computing.

In addition to these general capabilities, versions of the mobile application should include capabilities for specific field functions such as damage assessment, work order management, on-site permit application, HAZMAT response, and emergency management. The mobile application should also include versions for specific day-to-day field activities such as inspections, public safety, and work management.

5.5 Disaster Management Geo-Information Applications

Initially, disaster managers were largely dependent on general-purpose geo-information applications or those developed for other purposes. With the impetus of the post 9/11 period, the requirements of disaster management have moved to the front. Specific applications of geo-information applied to disaster management have been developed recently. Among these are applications for:

- Definition of emergency impact areas to support resource allocation, evacuation planning, and decision-making
- Support for disaster management resource allocation and logistics
- Emergency routing
- Damage assessment
- Transportation network status tracking
- Forecast evolving disaster situation
- Emergency planning
- Debris management
- Plume modeling
- Support for emergency warning notification system
- Secure Web site for the exchange of geo-information among participating units in a disaster response
- Public emergency information Web site

5.6 Disaster Management Geo-Information Data Solutions

From the experiences and lessons of the World Trade Center disaster, geo-information data requirements have been defined in various locales and are now evolving to become disaster management standards. This section addresses the view of data requirements that has evolved from the New York experience.

As stated above, disaster management, operations, and response require a wide range of location-based geo-information. The requirements can be classified as:

- General-purpose GIS and related tabular data
- Disaster-specific data items
- Real-time spatial data generated and acquired during disaster operations

The operations also require information with a spatial component from several disaster management tools and from various other systems and technologies available in the jurisdiction.

5.7 Disaster Management Geo-Information Data Requirements

General-purpose GIS data include items such as:

- Physical conditions—Streets, buildings, water bodies
- Digital terrain model or contours
- Remote sensing/imagery—Orthophotos, recent and post-impact satellite imagery, images of buildings
- Land ownership—Ownership parcel map, selected parcel attributes, rights of way
- Buildings—Footprints as polygons, building attributes, contact persons, floor plans
- Transportation—Streets, centerlines, bridges, as-built drawings, routing network
- Utilities—Selected electric, gas, water, sanitary sewer, storm sewer, telecom, cable infrastructure facilities maps
- Public facilities—Schools, government offices, libraries, parks inventory
- Political and administrative boundaries
- Master address and locations
- Landmarks—Selected notable landmark locations
- Environmental conditions
- CADD engineering and building drawings
- Raster images—Engineering drawings, floor plans, equipment manuals, documents, forms, photos
- Attribute databases—Parcels, census demographic data, business license/tax database, inspection and maintenance, permit tracking databases

Disaster specific data include:

- Critical infrastructure/facilities—Emergency services, water supply systems, government services, telecommunications, electric power systems, gas and oil production, storage and distribution, banking and finance
- Risk sites or areas
- Special facilities
- Emergency/Evacuation routes
- Emergency resources
- Emergency facilities
- Medical facilities
- Hazardous materials
- Pre-fire plans

Real-time data generated through the disaster operations include:
- Emergency call reports
- Street closures and barriers
- Location of emergency impact problem areas
- Flooded areas
- Status of tree and snow removal
- Status and locations of emergency responders
- Location of wires down
- Emergency resource status
- Damage assessment reports
- Utility outage areas

5.8 Disaster Geo-Information Data Management Requirements

In addition to the definition of data components, disaster management generates requirements for the management of geo-information data in addition to the general data management requirements usually applied in an organization. Some of the key requirements for emergency spatial information management include:
- Real-time data management
- Access to data across administrative and legal boundaries
- Access to data and capabilities must be easy to use and require minimal training
- Support for decision-making
- Deliver data in a useful format and not generate information overload
- Highly available
- Provide appropriate security especially in intentional destruction events
- Replicable
- Support mission-critical operations
- Operate in an environment of time and life and death urgency
- Support a high level of public interest in disaster response

6 Concluding Recommendations

The field of geo-information has undergone a tremendous surge of activity and development in the period since 9/11/01. There has been an outpouring of technology, some of it enhancements to existing technology, some entirely new technology, and some adaptation of existing technology from other functional areas. In fact, much more technology is available today than the disaster management and responder communities are able to absorb and make effective use of.

An important lesson that the author would like to highlight in conclusion is the importance of participation by the disaster and responder representatives in the design and deployment of technology. So much of the recent technology is designed by the technologists with little or no input from those who will actually use the technology. A professional field of disaster management exists. The methods and procedures for responding to and managing disaster response are well known. Many constraints and challenges exist to the effective use of technology in this field. The field has not been highly automated in the past, in part, due to these special conditions, so professionals in the field are not particularly technology-oriented. In addition, few of the technologists have any real understanding of disaster response. These factors are critical to the design and effective use of technology. Therefore, methods must be employed to bring the responders and managers into the design, development, and deployment activities along with the technology developers.

Sharing experiences, lessons, and technology advances among organizations can be very important in this time of rapidly changing conditions. Members of the New York City EMDC team have presented their experiences at numerous conferences and meetings and have published several articles in the interest of sharing their experiences with others. Several specialty meetings have been conducted since 9/11 in which the advances in geo-information as applied to disaster management has been the topic. New York has been engaged in negotiating a Memorandum of Understanding for a cooperative agreement with the Netherlands Council for Geo Information (Ravi) for the sharing of information and experiences in the area of disaster management geo-information.

References

Adams Manion K, Dorf W, Havan-Orumieh M Deep Infrastructure Group Provides Critical Data for Disaster Relief

Hall J, Key Activities, Products and Lessons Learned from New York City's Emergency Mapping and Data Center

Kevany MJ (2002) GIS in the World Trade Center Attack—Critique, What was Done, What Can we Learn, Proceedings of the Annual Conference of the Urban and Regional Information Systems Association (URISA)

Kevany MJ (2003) GIS in the World Trade Center Attack—Trial by Fire; Computers, Environment and Urban Systems, 27, pp 571-583, Elsevier, Ltd

Kevany MJ (2003), The Importance of Accessing a Variety of Data and Interoperability in Emergency Operations, Proceedings of the Annual Conference of the Urban and Regional Information Systems Association (URISA)
Langhelm R (2002) Presentation at the Annual Conference of the Urban and Regional Information Systems Association (URISA) WTC GIS Chronology, URISA

Soft Real-Time GIS for Disaster Monitoring

Robert Laurini, Sylvie Servigne and Guillaume Noel

LIRIS, INSA de Lyon, F-69621 Villeurbanne Cedex, France.
Email : Robert.Laurini@insa-lyon.fr; Sylvie.Servigne@insa-lyon.fr;
Noel.Guillaume@insa-lyon.fr

Abstract

The goal of this paper is to underline the importance of real-time systems for managing information during the phase of disaster monitoring. We stress the importance of soft real-time GIS, and we present a list of barriers to overcome in order to get this kind of system. Among the barriers, we present a solution for real-time indexing of spatio-temporal data based on a data structure named PO-Tree.

1 Introduction

The monitoring of natural phenomena is a key element of any situation which can provoke some disasters, such as flooding, volcanic eruptions, landslides and so on. Usually a lot of sensors are used to measure data about the phenomenon under monitoring. Those data are regularly sent to some control center and are stored in a real-time GIS, and displayed in real-time with animated cartography. So decision-makers can follow the phenomenon and be prepared to make relevant decisions. As a direct consequence, computers must always be on.

In addition, real-time characteristics of GIS can also be important during the response phase, for instance to monitor shelter and hospital capacities.

The goal of this paper will be to give a complete list of functionalities that a real-time GIS will offer, to examine them, and to identify the actual existing barriers in order to set a research program. We will continue by providing a new structure for real-time indexing of spatio-temporal data.

2 What is a Real-Time GIS?

Presently are emerging new GIS applications for which the time is a critical factor. For instance during disaster management, information must be collected in real-time, and made immediately available to a lot of potential users. In this kind of application, the characteristics of the database contents are far from the conventional view (for instance, administrative boundaries), for which the information about feature presents none or very slow evolution. In other words, in contrast, in the past, we were dealing with applications for which the date of storing or updating information was not a very critical factor.

Let us take another example in environmental monitoring such as river and flood monitoring. Several sensors are distributed along the river, regularly measuring several parameters, chemical, physical or biological, in addition to water height. After having made the measures, the sensors send the corresponding data to a control center, by using any kind of telecommunication system (could be satellite-based, could be using cellular phones attached to the sensors, etc.). In the control center, a front-end system manages the dialog with the sensors and stores the information into a database. Then, another system visualizes this information to give the decision-maker relevant information about the river. And especially, facing any crucial event, the periodicity of collecting information must be increased, for instance passing from every hour to every minute. Doing so, we are multiplying the number of data to be transferred, so increasing the importance of time: now all transactions must be committed very rapidly, and no system crashes will be permitted.

Several kinds of real-time systems can be considered according to the type of temporal constraints to follow. Usually, we speak about hard real-time when transactions must be fulfilled with hard deadline; for instance for military applications, constraints can be of millisecond magnitude. In disaster management, the phenomena under monitoring are slower, and so we speak about soft real-time systems.

With other words, three aspects are the more critical:

- when a bunch of data arrives, it must be stored very rapidly; that is to say no queue is allowed to install data in their optimal location; differently said no indices must be reorganized; as a consequence conventional indices such as based on quadtrees or any kind of R-trees are totally outside consideration;
- secondly, even if new data arrive, the database must never be saturated or completely full; a special mechanism must be provided in order to

avoid this drawback, per instance by regularly flushing elder data into a datawarehouse;

- thirdly, the computer system must be very robust, that is to say no failures must be accepted; especially the computer must always run, perhaps using electrical batteries.

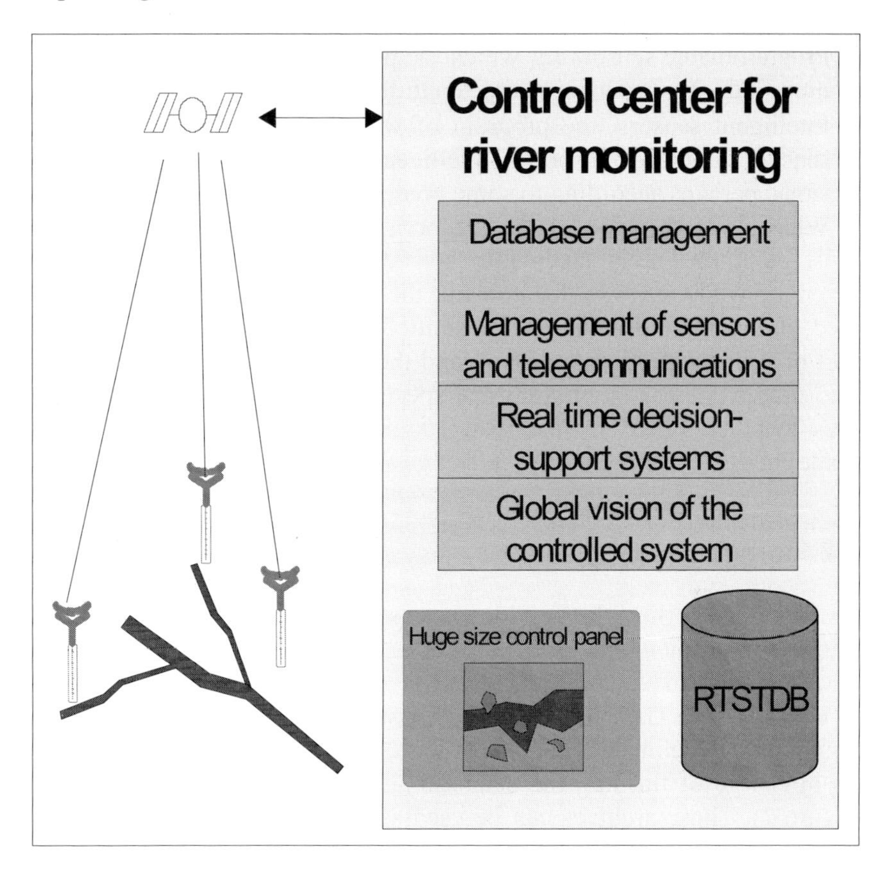

Fig. 1. Architecture for river monitoring

In order to set all the specifications of a real-time GIS, several additional functionalities must be taken into account. Let us examine, or rediscover some of them.

2.1 Input: Role of Sensors

In conventional databases, the main data entry procedure is based on some kind of human-interface dialog. But in our concern, the main procedures will be based on sensors. By sensors, we mean any kind of electronic devices able to make measures on some physical phenomena, and to send this information to a control center by means of any telecommunication system. Different kinds of sensors can be distinguished:

- Passive sensors are built to send regularly their measures, for instance every 10 seconds.
- Programmable sensors for which the periodicity of the measures can be modified by using some remote control statements.
- Intelligent sensors, the program of which can be downloaded. Those sensors can present a sort of "intelligent" behavior by making local decisions, perhaps according to some predefined rules.

Whatsoever the type of the sensor is, the data which are transmitted must be immediately taken into consideration. One crucial problem is when a crisis occurs. In this content, the measuring periodicity is accelerated, and then multiplying the transmitted data: the consequence could be a sort of congestion in data arrival, and the system must be able to manage this sudden increase within critical specified times (for instance all measures must be stored in less than 10 milliseconds after acquisition was made).

2.2 Storage

As explained earlier, the key idea is to store the maximum of information into the main memory, especially more recent data, and to flush ancient data into an archive.

For real-time GIS, this key idea must be accepted: in essence, a first computer can be in charge of the system, whereas a second computer must be in charge of flushing the database and of managing the archives (See Figure 2). Those archives can be used as datawarehouses, in which some OLAP procedures can be launched for analyzing data.

Usually, the phenomena under consideration are continuous, and the sensors only measure a few points. So, some powerful mechanisms for interpolating data, either at temporal, or at spatial, or at both levels must be provided, especially for data of corresponding to spatio-temporal continuous fields such as temperatures, pressures, winds, etc. See (Laurini-Gordillo 2000) for more details.

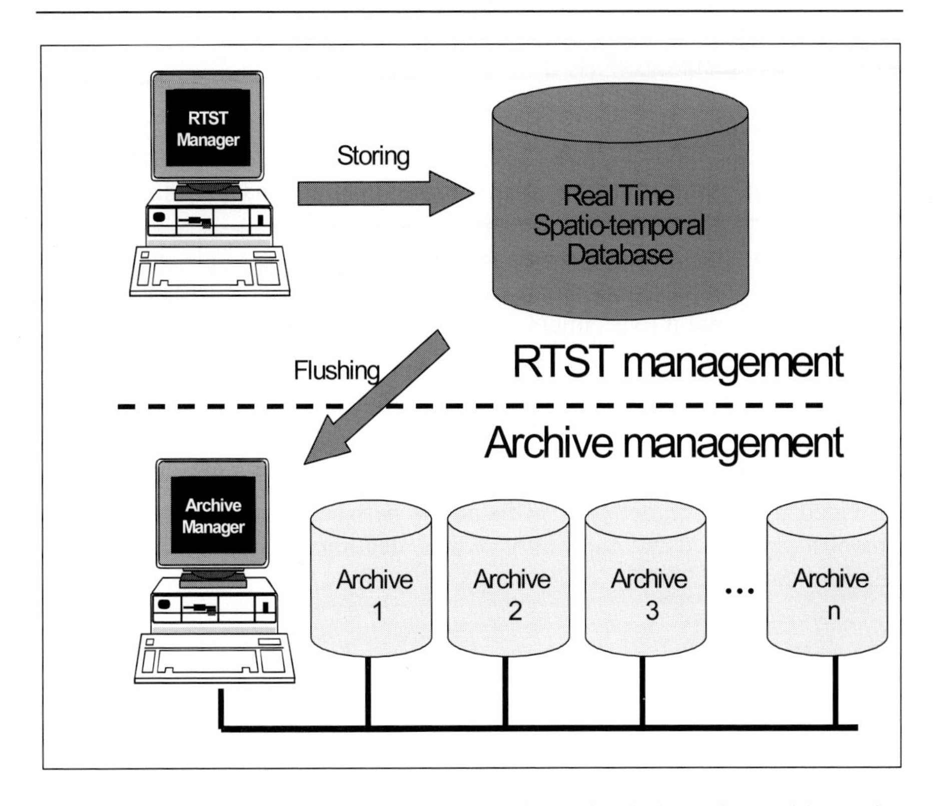

Fig. 2. Structure of a real-time system for geographic information with regular flushing of data into archive (RTST meaning real-time spatio temporal)

2.3 Output: Real-Time Animated Visualization

The privileged output is visualization, and especially real-time visualization. Among visualization systems, animated cartography is a simple way to present the evolution of an object, or a process into a territory. Especially real-time animated cartography in huge format real-time interactive panels must be studied, including:

- cognitive aspects of decision-making in real-time,
- graphic semiology for animated cartography,
- automatic selection of relevant data for synthesizing information for decision-makers.

2.4 Querying and Indexing

Extensions of SQL for real-time (Prichard-Fortier 1997) are now proposed. Those extensions must be harmonized with extension dealing with spatio-temporal issues. This must be done not only for relational DBMS, but also for object-oriented DBMS.

For indications about languages for moving objects, please refer to Güting et al. (2000). Concerning indexing, the Historical R-trees look to be a good starting point (Nascimento et al. 1998).

2.5 Interpolation and Extrapolation

In real-time GIS querying languages, automatic interpolations must be provided to retrieve not only in locations between stored times, but also amongst spatial data, especially when dealing with continuous data (Laurini-Gordillo 2000).

2.6 Integrity Constraints Correction, Sensor Failure

In addition to robustness to sensor- or system-failures, spatial and temporal integrity constraints must be checked and maintained in real-time. It is interesting to note the difference between sensor and system failures. Indeed it is important to discern failures linked to a faulty sensor and failures linked to a system error. Different studies deal with failure detection in sensor networks using redundant data, state prediction, sensor data validation through degrees on confidence in the measurements and fusing of estimated and measured values (Satnam et al. 2001). Real-time studies tend to lead to more flexible constraints so as to maintain a higher level of transaction commits. However this leads to lower the precision of the measured queries as higher flexibility comes from less accurate measurements.

2.7 Interoperability

Even so interoperability is not the goal of any real-time GIS, this requirement is nagging for users. A solution is that the data model must be compliant with standards such as those proposed by OpenGIS consortium (Buelher-McKee, 1996). Spatial models were the first goal of Open GIS, then they tackle spatio-temporal data. As far as we know, there is not yet standards for real-time spatio-temporal data. At best, we can take notice of the preliminary works done on real-time spatio-temporal data queries for the PLACE project, aiming at continuous queries (Mockbel et al., 2004).

2.8 Links to the Internet

The links for Internet must be provided. A URL must be considered as an abstract data type in order to embed hypermap potentialities (Laurini-Milleret-Raffort, 1990) in the real-time GIS.

2.9 Security, Failures

As any real-time system, robustness is a key issue for any real-time GIS. The system must provide security for any user, especially when facing computer failures.

2.10 Conceptual Data Models

For analysis and design purpose, conceptual data models must be invented taking, not only spatio-temporal aspects, which is already done (Price et al. 1999) but also real-time characteristics (Douglas, 1998).

So those tools will facilitate the user for the designing of any new GIS applications integrating real-time components.

3 A Barrier to Overcome: Spatio-Temporal Real-Time Indexing

A state of the art on the different existing methods is essential to better understand the pros and the cons, the main concepts of the situation.

3.1 Soft Real-Time

For the real-time approach, the main idea is to answer queries within time constraints (Noel, Servigne, Laurini, 2004). It is possible to separate three kinds of constraints (Lam, Kuo, 2001). The soft one, used in our case, imply that transactions should be fulfilled within time limits, yet it is understandable that some transaction can not comply with the limits. The firm constraints, more restrictive, allow some transaction not to be fulfilled within the time limits, yet in this case the whole system can be slightly impaired. The hard constraints, finally impose that under no circumstances a transaction should miss a deadline. Otherwise, the system could come to a halt. Priorities are generally used to define which transaction is more important than another, which is not equivalent to define which one should occur before another. Different techniques can be used so as to assign priorities: Earliest Deadline First, Rate Monotonic and other variants (Lam, Kuo, 2001).

Real-time computing is not similar to Fast-computing. Fast-computing does not prevent a low priority transaction to block high priority transactions (priority inversion) because they have already locked the access to some resources, data. Moreover, for databases, the current paradigm is to keep the index, and even the whole base in main memory so as to reduce the number of slow disk access. Index Consistency Control (ICC) methods can then be used to make sure no priority inversion occur while accessing the index (Haritsa, Seshadri, 2001).

3.2 Spatial Approach

The spatial approaches in indexing often tend to linearize the data so as to use known "fast" structures. Such is the case for quadtrees, kd-trees (Ooi, Tan, 1997) or other methods for spatial objects, or more accurately spatial points. Kd-trees are related to binary trees. A reference point is taken, along with a reference dimension. Every other point that falls below the reference point for this dimension shall branch to the left, all points with higher values branch to the right. At the next level, a new reference point is taken in each branch and the next dimension is used as a reference. It is relatively fast, can be updated on-line but the final shape of the tree depends on the insertion order of data, which can lead to unbalanced trees.

Another widely accepted approach is to use rectangles, bounding structures to match the position of objects. The bounding rectangles can then be regrouped within bigger rectangles so as to create a balanced tree. The R-tree, and its sibling R*-tree (Ooi, Tan, 1997) are examples of this. While the R-trees allow to work with complex objects (approximated as rectangles and not points), their higher building and querying time make the use of lighter structures appealing to index points.

3.3 Temporal Approach

For the temporal approach, it is important to note that different notions of time can be used for databases (Ooi, Tan, 1997*). The Transaction Time allows users to perform "rollbacks" so as to find past-values. It does not allow to modify previously entered values, nor to enter future values. One can only append new data issued at the present moment. The Valid Time represents the time when a fact is considered true. It allows users to modify past data, and to enter future data. However, it does not allow rollbacks. The Bi-Temporal Time is a mix between the two others, allowing rollbacks, post-modifications and future updates.

There mainly are two ways of considering temporality. The latter is to consider that time is monotonous (time goes in one direction) and to use B-trees as index structures. An interesting variation of this is to consider that the data flow constantly, therefore it becomes possible to link the root of the tree more closely the last leaf. This leaf containing the most recent data. Such an idea has been developed in AP-trees (Gunahdi, Segev, 1993).

The other way of considering temporality is to consider time just as a spatial dimension and to use R-trees, with on one dimension the timestamps and on the other dimension the validity duration.

3.4 Spatio-Temporal Approach

Spatio-temporal approaches have to face the variety of possible types of data: points, ranges, intervals (Wang, Zhou, Lu, 2000). This leads to a distinction between three families of indexing trees. Those that work with objects in continuous movement, those for discrete changes and finally those for continuous changes of movements. Another way of differentiating the families of index has been brought by (Mockbel, Ghanem, Aref, 2003). They have focused on approaches aiming at indexing past positions, present ones and future ones.

Many trees have been developed to answer specific needs. Some trees tend to consider the temporal aspect as yet another spatial dimension, which has led to 3DR-trees (Theodoridis, Vazirgiannis, Sellis, 1996). However, these trees doesn't take into account the monotonicity of time and usually need to have a previous knowledge of the data to index. Another family of trees make a difference between spatial and temporal dimensions. In HR-tree (Nascimento, 1998), typical of this case, snapshots of spatial R-trees are linked in a time indexed balanced tree. The main problem of this kind of trees is the size of the tree. While nodes that do not change between two snapshots are shared among the R-trees, only minor changes force to duplicate some of the data. Furthermore, they tend not to be optimal for interval queries.

Yet, from these aspects we can take some ideas. First of all, the differentiation of temporal and spatial data can be used to segment the tree. Then, a variation of B+-tree, the AP tree , offers the idea of a direct link between the root of the temporal data and the latest node. All of these ideas have been associated to provide our solution, the Po-tree.

3.5 Specification of a Real-Time Indexing Structure

Our first solution, the Po-tree, is based on the differentiation of temporal and spatial data, with a focus given to the latter. This way the notion of information sources is linked to a specific spatial location. It has been devised so as to deal with volcanic activity monitoring, which involves a number of different sensors with measurement frequencies going as high as 100 Hz. The spatial aspect is indexed through a Kd-tree, while the temporal aspect uses modified B+-trees (see Figure 3). As for now, mobility is not managed by the structure. However the specificities of both of these trees allow on-linc and batches updates: it is possible to update the structure on real-time or using batch files.

The monitoring stations being immobile, this structure does not allow mobile sources of information. This way, every spatial location, akin to spatial object (sensor) is directly linked to a specific temporal tree. Requests shall first determine the spatial nodes concerned and later on deter-

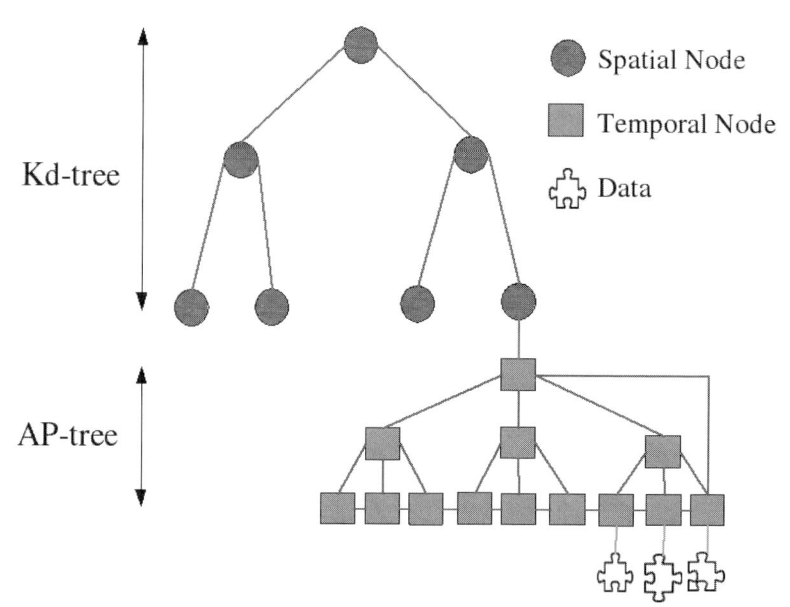

mine the temporal nodes.

Fig. 3. Po-tree structure

3.6 Po-Tree Structure

Kd-trees are simple structures, but they are not perfect structures. One of their main problem being the fact that they rely on the order of inserted data. If the data are entered in different orders, the final trees may have different shapes. Another issue is the fact that they are not perfectly suited for mobility, which is not part of our needs so far. However, ICC methods, originally designed for B-trees can easily be adapted to cover Kd-trees. Different tests have also proven that these trees fared reasonably well compared to R-trees for small number of data (Paspalis, 2003). As each B+-tree is linked to an object, it is possible to develop a secondary structure so as to access directly the temporal data of specific objects, without the need to first determine their position. This can be useful for the notion of hierarchy of information sources.

Furthermore, it has been noticed that the most recent data are considered of higher interest than the older one. It has also been noticed that inserts are generally held at rightmost of the structure, where are found the newest nodes. Therefore, the temporal tree has been modified to add a direct link between the root and the latest node. While maintaining this link requires minimum work for the system, a simple test prevents being forced to traverse the whole tree so as to append or to find the requested data. This direct link is useful to save processing time.

As most, if not all, of the updates take place to the rightmost part of the temporal tree, the fill factor of leaf nodes can be placed higher than usual. Deletes should be somehow rare under normal conditions, and updates that does not concern the newest data should be even rarer, unless the systems experiences lag time due to network problems between the sensors and the database. Therefore the split and merge procedures can be changed so that the nodes can be filled almost at their maximum capacity.

This configuration implies that this tree is more specifically designed for queries on the most recent data. Spatial range / temporal interval requests that does not ends at the present time does not take any advantage of the specificities of the tree.

3.7 Tests

Different tests have been conducted between the Po-tree, Pas-tree and R*-tree structures (thanks to Hadjieleftheriou's implementation, Hadjieleftheriou 2004). Randomly generated data have been generated and sequentially issued to a fixed number of random points acting as information sources. Tests have been conducted changing the total number of data to index (1000-200000), the number of information sources (10-100) used and the portion of the base to scan for interval queries. The tests have been conducted on a 1.6 G Hz, 128 Mo RAM computer, running Linux. The programming language used was Java.

Due to the differentiation of spatial and temporal component, and due to the fact that the data were coming from a finite set of spatial points, the Po-tree built time has been greatly reduced compared to the R*-tree. While 25 000 points stemming from 100 different locations were indexed in less than one second with the Po-tree, it took nearly 45 seconds with a R*-tree. Other tests have shown that the construction time of the Po-tree evolved linearly with the number of stations, the number of different spatial locations...

The different queries have shown interesting properties as well. The interval queries took an advantage of the linking of the temporal nodes of the Po-tree. For point-interval queries, the Po-tree can be up to 8 times faster the the R*-tree. While for interval queries the difference has shown much lighter, it still remains in favour of our solution, as shown in Figure 4. On this figure, the last 10% of the entered data where fetched. The spatial range covered the whole possible locations. It is visible than for a low number of data the R*-tree fares better, yet when the amount of data rises past 6000, it is the Po-tree that gains the advantage.

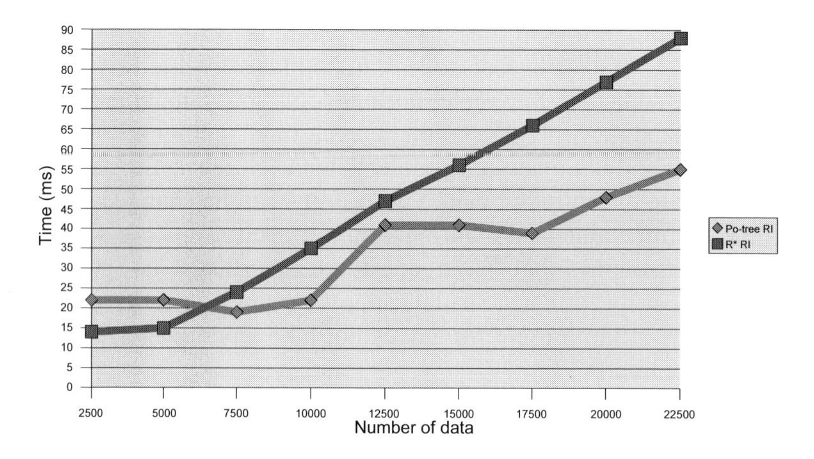

Fig. 4. Range-Interval queries

The results obtained have shown that the Po-tree was compatible with the constraints set by our application case: favoring the newest data, processing of big quantities of data in a given time, fixed set of spatial sources, possibility to use in a real-time system. Even though the mobility is not yet easily managed, the Po-tree meets the initial specifications.

3.8 Managing the Sensors Positions Changes: the Pas-Tree

While the first structure uses two structures to index spatio-temporal data, it does not allow sensor position changes. If a position was to change, a new temporal sub-tree would have to be created. This could be troublesome if a specific location becomes irrelevant, or in order to index some specific data. For example, seismic epicenters are usually determined by specialized processes but can be considered as indexable data. Those epicentres can be considered as data collected from evolving sensors. Thus the need to improve the Po-tree, so as to create the Pas-tree. This tree has to deal with position changes, yet remain focused on prioritizing the newest data and update transactions.

Sensors can move from time to time. The spatial sub-tree should be able to track these modifications. Different approaches exist, yet we shall aim at introducing a multiversion approach. A given node records the presence of past sensors (with an end-time) and of actual sensors. So as to offer other querying options, quadtrees could be used instead of Kd-trees.

The temporal sub-trees should also keep tracks of the position changes. Each of these sub-trees is related to a specific sensor. Keeping track of their position allows to follow the movement of sensors through a time interval without querying the spatial sub-tree. The temporal sub-trees of the Po-tree can be suitably adjusted to differentiate two kinds of entries: measurement data and position changes.

A tertiary structure, based on B+ tree keeps record of the sensor IDs. As a matter of fact, some queries do not need spatio-temporal properties. Scientists have grown used to sensor identifiers, and does not always rely on spatio-temporal properties. Even more so when mobility is scarce. Therefore this structure links directly to the temporal, sensor related sub-trees without using the spatial sub-tree.

The Pas-tree suffers from data duplication, yet it also allows for more request types than the Po-tree. As a matter of fact, queries can be based on the sensor ID as well as on spatio-temporal properties. It allows users to follow a specific sensor through its position changes or to have a look at different sensors passing through a region during a lapse of time. It stills focuses on update transactions and on the newest data.

4 Conclusions

As a conclusion, let us advocate for developing soft real-time GIS for disaster management, or more especially for disaster monitoring. Before reaching this goal, several barriers must be overcome such as robustness and fast storing which implies fast indexing.

References

Barbara D (1999) Mobile Computing and Databases - A Survey. In: IEEE Transactions on Knowledge and Data Engineering, vol. 11 (1), pp 108-117

Behr FJ (1995) Mobile GIS: Contributing to Corporate Benefits. DA/DSM Seminar, 20 November 1995, Rome, Italy http://www.graphservice.de/papers/mobile_g.htm

Buelher K, Mc Kee L (eds) (1996) The OpenGIS Guide, Introduction to Interoperable Geoprocessing. Open GIS Consortium. http://www.opengis.org

Gunadhi H, Segev A (1993) Efficient indexing methods for temporal relation, In IEEE Transactions on knowledge and Data Engineering, 5(3), pp 496-509

Guting RH, Bohlen MH, Erwig M, Jensen CS, Lorentzos NA, Schneider M, Vazirgiannis M (2000) A Foundation for Representing and Querying Moving Objects. ACM Transactions on Database Systems, Vol. 25 (1) pp 1-42

Hadjieleftheriou M, Spatial Index Library [Online], viewable at: http://www.cs.ucr.edu/~marioh/spatialindex/, (last consulted 01/11/04)

Haritsa JR, Seshadri S (2001) Real-time index concurrency control. In real-time Database System – Architecture and Techniques, Kluwer Academic Publishers, Boston, ISBN: 0-7923-7218-2, pp 60-74

Lam KY, Kuo TW (2001) real-time database systems: an overview of systems characteristics and issues. In real-time Database System – Architecture and Techniques, Kluwer Academic Publishers, Boston, ISBN: 0-7923-7218-2, pp 4-16

Laurini R (2001) Information Systems for Urban Planning: A hypermedia Cooperative Approach. Taylor and Francis, Forthcoming February 2001. See http://lisi.insa-lyon.fr/~laurini/isup

Laurini R (2001) real-time Spatio-Temporal Databases. In "Transactions on Geographic Information Systems", Guest Editorial, Vol 5(2), pp.87-98

Laurini R, Gordillo S (2000) Field Orientation for Continuous Spatio-temporal Phenomena. International Workshop on Emerging Technologies for Geo-based Applications, Ascona, Switzerland, May 22-26, 2000. Edited by S. Spaccapietra, published by the Swiss Federal Institute of Technology at Lausanne, pp 77-101

Laurini R, Milleret-Raffort F (1990) Principles of Geomatic Hypermaps. In: 4th International Symposium on Spatial Data Handling. Zurich, 23-27 Juillet 90. Edited by K. BRASSEL, pp 642-651

Mokbel MF, Ghanem TM, Aref WG (2003) “Spatio-temporal Access Methods”, IEEE Data Engineering Bulletin, Vol 26, n°2, pp 40-49

Mockbel MF, Xiong X, Aref WG, Hambrusch S, Prabhakar S, Hammad M (2004) PLACE: a Query Processor for Handling Real-Time Spatio-Temporal Data Streams, In VLDB 2004

Nascimento MA, Silva JRO (1998) Towards Historical R-trees, In: Proceedings of ACM Symposium on Applied Computing (ACM-SAC) Atlanta, USA, pp 235-240

Noel G, Servigne S, Laurini R (2004) "Real-time spatiotemporal data indexing structure", Proceedings of 2004 AGILE 7th conference on Geographic Information Science, Heraklion, 2004, pp. 261-268

Ooi BC, Tan KL (1997) Spatial Databases. In Indexing Techniques for Advanced Database Systems, Kluwer Academic Publishers, Boston, ISBN 0-7923-9985-4, 39-75

Paspalis N Implementation of Range searching Data-Structures and Algorithms [Online], viewable at: http://www.cs.ucsb.edu/~nearchos/cs235/cs235.html, (last consulted 20/11/03)

Prichard J, Fortier P (1997) Real-Time SQL. In: Second International Workshop on Real-Time Databases September 18-19, 1997 Burlington, Vermont, USA, pp. 289-310

Satnam A, Agogino A, Morjaria M (2001), A methodology for intelligent sensor measurement, validation, fusion and fault detection for equipment monitoring and diagnostics, Artificial Intelligence for Engineering Design, Analysis and Manufacturing, Vol 15, No 4

Theodoridis Y, Vazirgiannis M, Sellis, T (1996) Spatio-temporal indexing for large multimedia application, In Proceedings of the 3rd IEEE conference on multimedia computing and systems (ICMCS)

Wang X, Zhou X, Lu S (2000) Spatiotemporal Data Modeling and Management: A Survey, In Proceedings of the 36th International Conference on Technology of Object-Oriented

Wolfson O (1998) Moving Objects Databases: Issues and Solutions. In: the Proceedings of the 10th International Conference on Scientific and Statistical Database Management (SSDBM98), Capri, (Italy), July 1-3, 1998, pp. 111-122.

Step-Wise Improvement of Precursor Services to an Integrated Crisis Information Center for Mountainous Areas

Ulli Leibnitz

VCS Aktiengesellschaft, Borgmannstrasse 2, 44894 Bochum, Germany.
Email: ulli.leibnitz@vcs.de

Abstract

From a technical perspective, the integration of existing information networks and evolution of appropriate precursor services to high reliable operational services for crisis management is one of the main challenges in the context of GMES[1].

Due to the existing budgetary constraints, governmental organizations, decision makers and all users that would like to build up such decision support centers are looking for cost and time efficient ways to build up the required operational (GMES) services.

During the conference, VCS will present a concept for a step-wise evolution of a risk warning and crisis management system, with particular focus on risks typical to mountainous areas e.g. avalanches, landslides, debris flows, floods, etc.

1 The Challenge - Risk Warning and Crisis Management within Mountainous Areas

Some 30% of the EU territory consisting of 30 million inhabitants encompassing numerous mountain ranges or chains, including the Alps, the Sierra Nevada, the Island of Crete, the Pyrenees, the Apennines, the Sierra

[1] Global Monitoring for Environment and Security. The GMES initiative seeks to bring together the needs of society related to the issue of environment and security with the advanced technical and operational capability offered by terrestrial and space borne observation systems.

da Estrela, the Massif Central, the upper Tatra, the Highlands, the Carpathians, etc. Mountain areas represent over 50% of the territory in Italy, Spain, Greece, Austria, Switzerland and Portugal.

Alpine regions are very sensitive ecosystems and the pressure on them is far greater than on other environments. This is due to an aggressive development drive in the past such as damage and uncontrolled felling in forests, the construction of new roads and tracks, the rerouting of rivers and streams, pollution originating in industrial areas which adversely affects alpine vegetation, settlement activities in regions which cannot be considered as safe and increased recreational activities (hiking, biking, climbing, paragliding, canoeing, skiing, snow-shoe tours, etc.)

Natural disasters partly resulting from these developments are problems that occur regularly in alpine regions, posing a major threat to the safety of settlements, tourists and traffic infrastructures. An increasing numbers of rock falls, mud slides, avalanches, floods and windfalls in recent years have shown that natural disasters may even strike areas that have generally been considered as safe.

The following developments will further aggravate the situation in alpine regions:

- The danger of property damage and personal injury caused by natural disasters is rising as a result of increasing settlement pressure and tourism.
- The willingness of people to voluntarily subject themselves to ever-greater risks in their leisure time is also increasing.
- Leisure and industry lead to growing mobility demands resulting in increased vulnerability along traffic routes in the alpine area.
- Due to the recent reduction of mountainous forests and climate changes (increase of extreme conditions) the frequency and intensity of natural disasters has increased.

These developments make it necessary to implement and improve safety measures, such as prevention activities and early warning / event driven systems, to enhance the communications infrastructure for rescue actions in case of a crisis. Simple, safe and fast communications routes and emergency-oriented information structures facilitate the deployment of mountain rescue, fire fighters, civil defense forces and other rescue organizations. Hazard mapping and monitoring and risk classification provide essential decision support for adequate rescue operations in this context.

Although event management services are already using IT tools like GIS there are still several aspects that remain in an unsatisfactory state:

- The systems and platforms used for regular service should be better integrated in the tools for event management.

- EO data for event management is not used at present state. These data could help in decision-making processes and for the documentation of damages. There are neither standardized methods for requesting and using such data nor acceptable time frames for the delivery.
- The usability of the IT tools could be much improved.
- Additional data could enrich the SAR (synthetic aperture radar) tools.
- There are no efficient standards across borders handling communication and regulatory aspects for crisis management.

One important topic is the further collaboration between the Alpine countries to develop common standards which can be applied also to non-Alpine crisis events.

The vision for the future is that space technology-based methods can provide a significant contribution to risk management. Web based information systems including 2D and 3D visualization technologies and mobile systems including Global Navigation Satellite Systems (GNSS), could be used to assist rescue teams to determine their position and navigate in difficult and unfamiliar terrain even under bad weather conditions and support co-ordination with an emergency center. Satellite communications systems can provide an important contribution to efficient and safe communication between the rescue teams in the field and the management center especially in areas with no / limited / damaged communication infrastructure.

Remote sensing methods can be used to assess the conditions and development of alpine regions (deforestation, calamities, erosion damage, path construction, assessment and monitoring of surface deformations, mass movements, etc.). Geographic Information Systems (GIS) in combination with remote sensing data and other spatial information such as geological maps or digital elevation models are powerful tools to assess the hazard potential and vulnerability of a region.

2 The Approach - Alpine Safety, Security & Informational Services and Technologies (ASSIST Project)

The project ASSIST[2] aims at improving the capabilities of risk warning and risk management in the alpine region by implementing an integrated pre-operational service based on existing precursor services and related infrastructure.

[2] ASSIST – selected by the European Commission for negotiation. Successfully evaluated during 2nd Call of the Aeronautics and Space Priority FP6-2003-Space-1, AERO-2003-2.3.2.1c "Risk Management". Specific Targeted Research Project (STReP) activity led by VCS AG (Germany).

ASSIST will focus on risks typical to mountainous areas e.g. avalanches, landslides, debris flows, floods, etc.

The main objectives of the project are:

- To improve the existing methodology for risk assessment, monitoring and management in Alpine regions in order to evolve services delivered by regional risk management centers.
- to implement pre-operational services and realize advanced integrated safety and information services for the Alps. The advanced services will incorporate existing crisis and information centers, and crisis communication systems.
- to specify, design, implement and validate a generic system architecture for the production and exchange of data products used for risk prevention and risk event management.

Considering the actual conditions for risk management in the alpine region (poor data quality at borders of responsibility, limited availability of input data in time, missing regulations for exchange of data among different organizations, etc.) it is expected, that ASSIST will be highly beneficial to improve the means for risk management in mountain areas.

2.1 Step-Wise Service Evolution

For the risk and crisis management center scenario, related technical aspects are system infrastructure, effective use of Earth Observation data, communication, integration of satellite navigation, product generation and seamless integration of Geo-Information Systems (GIS).

Modularity and configurability are mandatory requirements for future-oriented systems. Consequently, the credo is: “Start with a basic service and improve the service according to user needs and budgetary constraints”.

a) Geographically (regional, international, global scale)
b) Service portfolio (core service, advanced, fully integrated service)

The step-wise improvement of existing or even new information services to an operational or even advanced service can be structured in more detail according to the various “technical layers”. Each layer has its own evolution path to follow during the step-wise improvement. A high level example is given in figure 2.

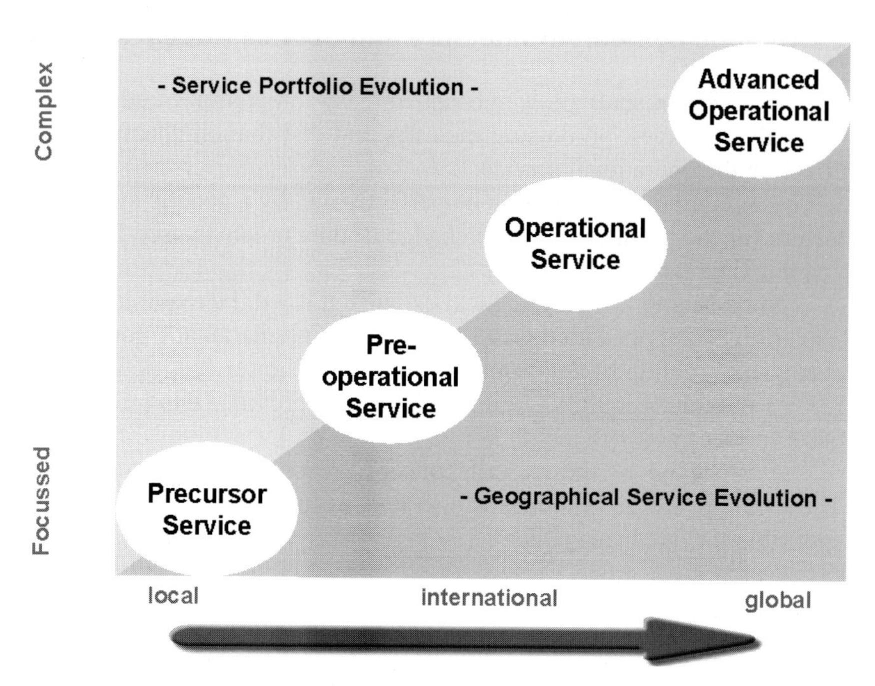

Fig. 1. Step-wise evolution of the Risk and Crisis Management Service Center

	Precursor Service	Pre-operational Service	Operational Service	Advanced Service
Geographical Relevance	Local/Regional Service Defined Test Areas	Regional Service inclusive supports cross-border scenario	International mandate and interface to global services	Fully established international or global service network
Communication/ data distribution	Internet, LAN, GSM, telephone, email, SMS	Increased Communication Flexibility: introducing GSM/GPRS/UMTS, selected WAN/SatCom links	Redundant lines/networks, IP-based Services,secure lines, mobile SatCom (at least as backup) end-to-end M&C of data flows	(Access and contribution to) Global communication networks
Satellite Navigation	Not integrated	Localisation of event	Localisation of events and field staff resources	Fully integrated Location based services (LBS) (integration of EO and Telecom)
Processing/ System	Manual/supervised processing base maps	Processing Framework Increased capabilities for generation of service specific dynamic information layers	Redundancy interfaces to other entities (distributed processing/archiving) also unsupervised processing decision support tools	Advanced Forecasting increase of unsupervised processing Service Intelligence
Data Sources	Static information no or very limited access to EO data lack of actual data (even maps)	Up-to-date data use of airborne data regular use of EO data (single source)	Real-time data access avoidance of single sources for EO data	more sensors = new products
	0-2 years	1-3 years	2 to 5 years	5 years and beyond

Fig. 2. Service Evolution Scenario

Depending on the complexity of a service, each step might need 1 to 5 years for development, engineering, integration and testing.

2.2 Generic System Architecture and Service Nodes

The final system shall allow subsequent easy integration of additional external data sources, processing modules and distribution channels without affecting the operational service.

These requirements can only be perfectly met by a generic system architecture for the production and exchange of data products used for risk prevention and crisis management.

With respect to the geographical expansion it will be required to provide a distributed services architecture to support international information exchange and routing of data sources throughout the services network (e.g. a service provider acquires EO information and provides this information to other services network members).

The backbone of the overall concept are so-called “Service Nodes”. These nodes are autonomously operated by organizations which are responsible for risk management.

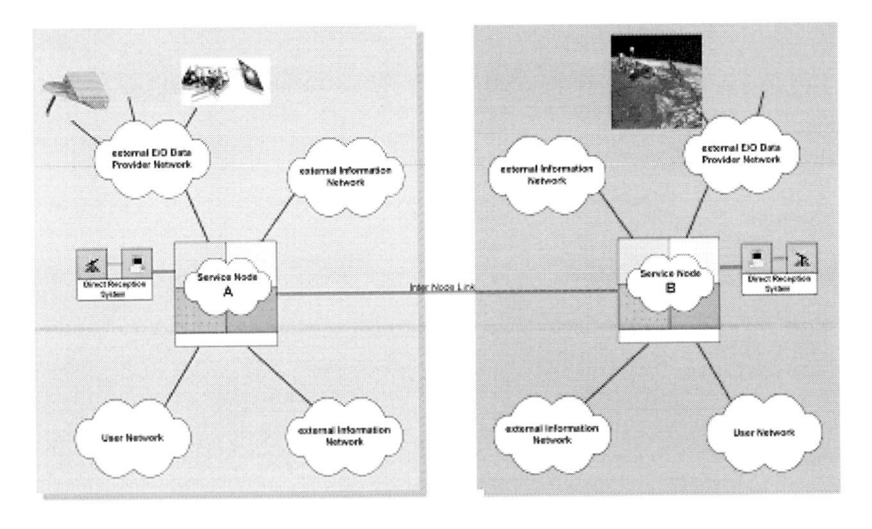

Fig. 3. ASSIST service node interconnection

The individual service center is considered as a node of the overall provider network. The technical implementation shall therefore foresee the definition and provision of a service node that can be instantiated for various service centers and configured as a part of the overall provider network.

The Service Nodes are capable to

- request and ingest raw input data (satellite-borne, air-borne and terrestrial)

- process the input data into products suitable for risk prevention and crisis management
- distribute the products within the "ASSIST user network" (fixed and mobile regional risk management centers and - in case of risk events - up to the mobile staffs in field)
- exchange products with services operated by organizations outside the consortium (e.g. police, hospitals, ambulance services, air rescue, fire fighters, etc.)

These Service Nodes will be laid out to support both

1. day-to-day monitoring and predictions of risk, establishment of mitigation scenarios
2. operation during concrete crisis situations

with different requirements in terms of data products, data timeliness and distribution.

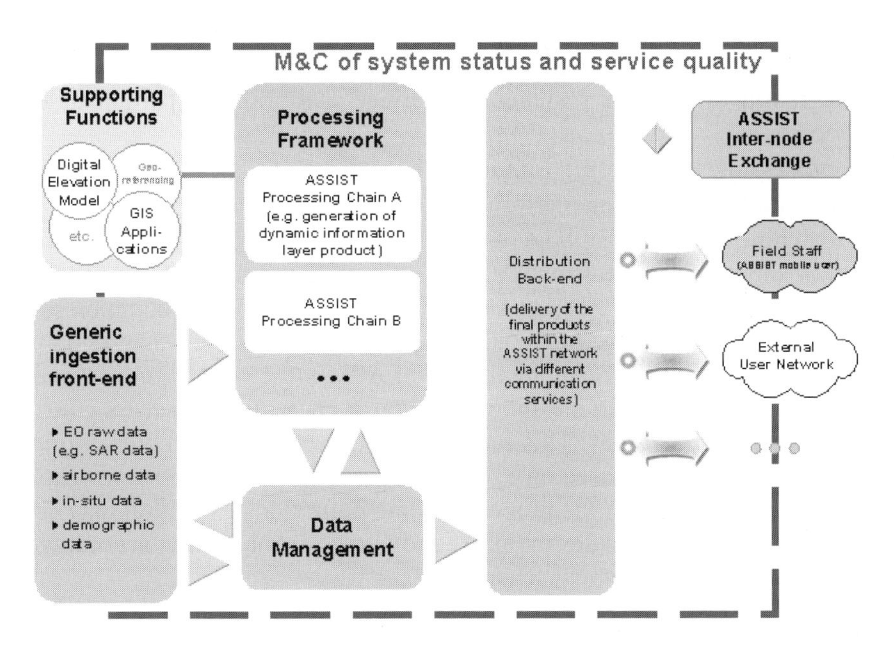

Fig. 4. ASSIST: Generic Service Center Architecture

The related project consortium is composed to reflect both scenarios. It consists of organizations with different scopes (avalanche prediction, mountain rescue, development of complex server infrastructures, generation of enhanced Earth Observation products, integration of mobile field equipment with positioning and communication capabilities, etc.) and nationalities with corresponding regulatory constraints (Austria, Germany,

Italy, Switzerland). The developed architecture will be such generic that it can be transposed to other mountainous areas in the world.

2.3 Key Aspects for a Modular and Open System Approach

This section briefly outlines the key aspects that are relevant for building up state-of-the-art service center architectures. The identified aspects are based on long lasting experience from realizing complex mission critical operational systems with maximum demand for high reliability and flexibility. This experience was mainly gathered throughout numerous space projects.

The implementation of complex and high reliable information and communication solutions is one of the disciplines, where space technology and know-how could perfectly be used and transferred.

The first aspect to be mentioned is the capability to **support heterogeneous system environments**. This is of great importance for the step-wise integration of existing systems and service platforms. In particular, this concerns:

- processing software, algorithms, data handling
- distributed processing platforms / service providers
- computer hardware and the related operating system.

For the service infrastructure itself, a **consistent overall approach** is mandatory. This can be ensured by defining a basic level of common rules and interfaces to allow for monitoring and control as well as scheduling and data handling (enforced by the use of APIs).

Furthermore, the service should rely on a **centralized process control and data management**. This requirement can be met by implementing a kernel infrastructure based on a central process control processor executing process steps based on the process scheme definitions. A central data control processor co-ordinates the input and output data handling according to a defined data management scheme.

In addition, **decentralized process execution and storage** (archive and retrieval) should be foreseen. In our opinion, process management does not neither need a central processing system nor a central data storage. These components can be decentralized in a distributed environment.

The use of **common components** fits with a cost-effective approach. Within a distributed infrastructure it should be avoided to re-invent the wheel for the same category of data processing or management task. The use of common components is very much appropriate for storage mechanisms, access mechanisms and meta data models.

An issue of more theoretical kind is the elaboration of **clear system and interface specifications**. This can be easily assured by the use of a common terminology.

The last aspect to be mentioned within this short excurse is **"Interoperability"**. This interoperability should be mandatory for all service centers that might have to interact during risk prevention and/or crisis event. Interoperability means the capability of seamless product exchange and data integration as well as communication between the involved parties. Otherwise the management of joint actions is not possible. Interoperability can be implemented by a cascaded service center concept. Among others, this might include the establishment of a service center network relying on a common product catalogue.

We are fully aware, that the above explanations can only highlight the relevant key aspects, but do not address the full scope of relevant issues related to the implementation for realizing risk and crisis management centers. However, a first impression on the complexity of such a task and the right approach might have been provided to the reader.

2.4 Use of Earth Observation (EO) Data and Product Strategy

Natural hazards in mountainous regions influence people's living conditions, their work and their recreation activities. The management of these risks has to act on four levels: risk assessment, prevention, crisis management and regeneration, as shown in the figure below:

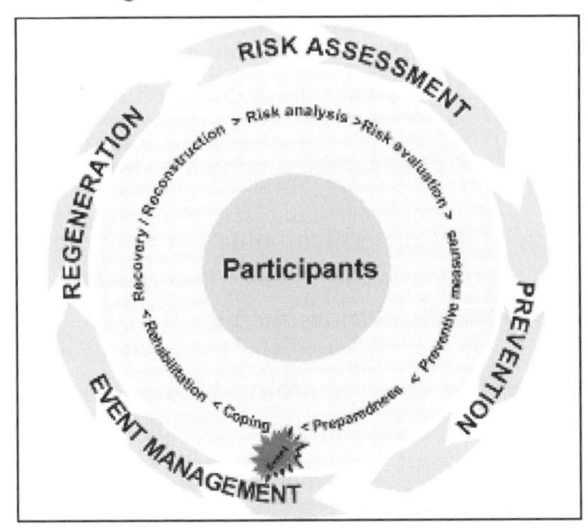

Fig. 5. Risk management level

In the spirit of an integral risk management all components have to be regarded within an equal context. For example, prevention measures on one hand and evacuation and rescue plans on the other hand are not independent and should support each other.

Events of interest are gravitational mass movements such as glacier- and rock-falls, snow avalanches, landslides, debris-, slush- and mud-flows, and floods. Furthermore forest fires, wind- and hail-storms can be of interest.

Weather forecast is important for risk prevention, but in many cases also for crisis management, particularly in search and rescue operations.

Therefore the project ASSIST will be strongly linked with meteorological forecast services.

For risk analysis as well as damage analysis it proved to be appropriate to subdivide endangered values into three main categories:

- buildings and persons inside,
- persons and values on traffic lines (like traffic roads or railways)
- persons in free terrain.

Therefore Alpine settlement structure, traffic networks and mobility plays an important role.

Weather services started to add to their conventional weather forecasts warning products for potential dangerous processes. Good examples can be found at the national weather services of France, Germany or Switzerland.

- MeteoFrance (French meteorological service): "carte de vigilance" http://www.meteo.fr/meteonet/temps/activite/mont/mont.htm
- DWD (German weather service): "Warnsituation" http://www.dwd.de/de/WundK/Warnungen/index.htm
- Meteoswiss (Swiss meteorological service: „aktuelle meteorolische Gefahren" http://www.meteoschweiz.ch/de/Prognosen/Warnungen/IndexWarnungen.shtml

Since recently, in case of an imminent critical situation a 4 level scale warning is published covering stormy weather, heavy precipitation, thunderstorms, icing rains, heavy snow. At present, combined efforts among the Alpine countries are made to further improve the forecasts of extreme meteorological events in the Alps (e.g. Interreg project MeteoRisk).

The situation is entirely different concerning the accompanying natural hazard processes: Detailed forecast and warning systems exist only for snow avalanches, see e.g. http://www.slf.ch/avalanche/avalanche-en.html. The intense collaboration between the European avalanche warning organizations has lead to an European standard in avalanche risk mitigation (e.g. the European danger scale) which we believe can be extended to the management of other natural hazard risks.

It is widely believed that, due to climate change effects, hydrological "summer" natural hazards will gain more and more importance in the near future. We see good chances to extend warning procedures e.g. to mudflows, floods or glacier falls.

Although avalanche warning has reached a high level with respect to technology used as well as quality achieved and public acceptance, several aspects remain in an unsatisfactory state:

- EO data for prevention and prevision are not fully exploited. Satellite data could be used to quantify snow coverage in mountainous regions and, in a later stage, even snow properties. These data are crucial for an improved spatial and temporal accuracy of forecasts.
- Development of forecast and warning system for "summer" hazards needs the recording of further surface and soil properties, and again the detection of events (landslides, mud flows, etc.) These data can, in principle be obtained by earth based observation and measurement system. But they are in general extremely costly and can therefore applied only to a very limited extent. Satellite based information should be used as far as possible.
- A very important aspect for all information is the sensitivity to bad weather conditions. The information particularly needed during bad weather conditions. Thus the priority will necessarily be on radar or infrared type data.
- The recording and communication of a large part of the data is absolutely time critical. The evaluation of the best methods will be very important.
- Detection of events, particularly during bad weather periods, is still a great challenge. The possibility of radar detection of avalanches could strongly support the forecasters as well as security officials. The following chart shows an example of regular avalanche recording in Switzerland. Although this map plays a very important role in the Swiss operational warning, it suffers from the necessarily incomplete information during bad weather periods and at remote locations.

  There are several benefits using EO data within the proposed service:
- EO data can be used standalone to generate new products for the service portfolio that cannot be obtained by other sources.
- EO data can be used as a complementary information source to complement, enhance or validate information from other sources.
- EO data can be used together with models to tune or initialize the model.

- EO data can replace other more costly information sources.

EO data is particularly useful for:

- regional monitoring
- regional mapping (classification)
- local scale monitoring
- rapid mapping
- change detection

Depending on the product type different benefits are offered.

EO data can contribute to basic information on the terrain such as Digital Elevation Models (DEM), land use maps etc. to the service portfolio. These base information layers are static and are processed before the crisis, so that they are at hand when needed.

The forecast and prevention capabilities are improved by the use of EO data through:

- regional monitoring capabilities (such as snow cover, deformation etc.)
- new products that are not available in a similar way from terrestrial methods like terrain stability and permafrost processes

The forecast and prevention information layer, is a dynamic information layer (e.g. snow or vegetation state) that needs to be updated on a regular base so that in case of a crisis up to date information is available for the crisis management.

In case of crisis the availability of processed (near) real-time EO data will support:

- decision-making thanks to improved data/information basis.
- co-ordination of disaster relief/crisis management activities which should contribute to safe live and infrastructure, and to reduce costs.

Depending on the scale of the product and the illumination of the target, different systems are at hand that can be combined. Optical sensors that are restricted to day and cloud-free situations are complemented by airborne and space borne microwave sensors, which operate independently of light and weather conditions.

In the post-crisis phase EO data can contribute information providing spatial information on damage extent and change detection. It helps to understand the disaster and supports post-crisis management.

As already mentioned before, alpine regions are very sensitive ecosystems and the pressure on them is far greater than on other environments. Therefore it is important to implement and improve safety measures. Risk mapping and the analysis of vulnerability can be seen as an essential part in the prevention phase and their results can significantly help to reduce losses in life and goods. During a crisis and for the post-crisis analysis, both historical and recent information data are needed for the management.

To address the needs we have three types of higher level products, base information layers (maps of static parameters), dynamic information layers (maps of dynamic parameters) and derived products such as risk or vulnerability maps. The information layers are obtained from ancillary and/or EO data. Base information layers are height maps, infrastructure maps, land use maps etc. Dynamic information layers cover maps derived from monitoring, such as terrain deformation and snow cover maps, or change detection such as flood maps, avalanche maps etc.

Derived products are higher level products combining information layers with models to risk or vulnerability maps.

SAR (synthetic aperture radar) image processing is also part of the ASSIST service concept. We will use state of the art tools for the product generation (following a defined workflow). As input raw SAR data (level 0), single look complex (SLC) or pre-processed images (PRI) will be used, whereas Level 0 data will be preferred. The data will be received from various data providers.

The processing approach is modular. This helps ensure maximum flexibility for new available SAR data and service node integration. Depending on the application, different modules will be used for data processing. Available are Interferometric Point Target Analysis, Interferometry, Differential Interferometry, Classification and Geocoding and Terrain correction.

Product quality checks will be done during all processing steps to ensure product quality. The final products will be geocoded and terrain corrected and will be ready for GIS integration.

EO Data availability is an important issue for an operational service. EO Data needs to be acquired timely and for the location of interest. The timeliness is affected by three factors:

- Reaction time of the sensor
- Delivery time from EO Data Provider to the Service Node
- Data processing capabilities and performance

The reaction time of the sensor is the time until the system illuminates the target area and acquires data. The reaction time is reduced with the amount of sensors and the flexibility of the EO system (e.g. beam steering). Nowadays it is in the order of days for operational systems.

2.5 Integration of Navigation and Telecommunication

The ASSIST concept includes the interchange of relevant data for risk monitoring and crisis management within the "ASSIST user network". The network consists of fixed or mobile regional risk management centers and

- in case of risk events - mobile equipment used by the staff in the field, too. This information has to become quickly available at the scene of distress. Therefore the mobile working teams will be equipped with robust PDA based user devices combined with positioning sensors (GPS, EGNOS) and hybrid communication technology (GSM/WLAN; optional: SatCom) for facilitating the location of victims, overall coordination and situation awareness.

In detail the mobile devices will assist the rescue teams in terms of navigation by visualizing the current position and a track (position history) on a digital display showing either relevant EO data, thematic maps derived from EO products, or conventional maps. In addition to that the user's positions will be stored for post-evaluation after the rescue operation. For example the actual positions or tracks of all rescue teams could be displayed in the mission center to provide effective support for the decision makers (situation awareness) whereupon immediate commands could be sent via offered mobile communication links to the mobile teams in the field and vice versa. Furthermore such a mobile client will allow accessing the central database of the regional risk management centers via mobile data connections to upload updated spatial information products, photos or reports on the current situation.

Today many applications in the domain of navigation suffer from the limited availability of a single (standalone) communication source. This limitation is well known from Line-of-Sight (LoS) obstructions using Satellite Communication (SatCom), but also from limitations in the coverage of cellular networks (like GSM, UMTS) or of terrestrial communication links (e.g. VHF, UHF) in case of topographic obstructions. New developments in the communication infrastructure like Bluetooth and Wireless Local Area Networks (WLAN) even highlight the vision of a "fragmented communication environment". The users who need to have permanent access to communication links to support safety or security related applications will have to be equipped with a number of communication devices to make sure that one of those might be usable in case a communication link is needed. Further on the intelligent combination of these different communication links might offer the possibility to use the cheapest, the most secure, etc. data link of those available and support new applications and opportunities. Within the ASSIST project a first attempt to increase the flexibility and robustness of communication during crisis situation will be performed.

ASSIST, in its first attempt, should run through an integral risk management not only in a theoretical sense, but also with possible practical effects. A good test area was found in the border triangle of Austria, Italy and Switzerland. This area is dominated by two main alpine passes. The region also has well established tourist resorts, in summer as well as in winter and therefore is generally exposed to natural alpine hazards the whole year round.

3 Concluding Remarks

ASSIST offers the unique chance to combine the existing, mostly ground based data networks for operational warning and rescue systems with the powerful possibilities of EO systems. In stressing these combinations, an increased safety in the sometimes rough Alpine environment, especially in regard to the mobility in these areas will result. With view to the economical, scientific and technological objectives the outcome of ASSIST will be representative and relevant for other mountainous regions in the EU and all over the world.

Alsat-1: First Member of the DMC

Azzeddine Rachedi[1], Benmohamed Mohamed[1], Takarli Bachir[1] and Martin N. Sweeting[2]

[1] Centre National des Techniques Spatiales, BP13, Arzew 31200, Algeria
Email: (rachedia, benmohamedm, takarlib)@cnts.dz
[2] Surrey Satellite Technology Limited, Guildford, Surrey GU2 7XH, UK
Email: m.sweeting@sstl.co.uk

Abstract

Alsat-1 was initially a know how transfer program. It was to allow to 11 young engineers acquiring knowledge by the active participation in all Alsat-1 project phases.

Two options were suggested to us:

- Carry out a traditional micro-satellite of the UoSat family having a proven reliability but with limited characteristics. All the modules were in this case already designed.
- Or, to choose "new generation" micro-satellites that have similar possibilities to those of the commercial satellites and take part of a constellation dedicated to the disaster monitoring with what that supposes like rigorous in the satellite operation.

We chose the second configuration without hesitating and the course of the project gave us later reason because it was carried out starting from a new design (scratch). Several modules were new in Alsat-1 (the imager, the storage unit SSDR) and others derived from demonstratives versions having flown once (redundant storage module SA1100 and the High Rate Transmitter). This allowed the Algerian engineers to be implied since the first phase of the project (Mission Analysis phase).

1 Alsat-1 Mission Constraint

The goal of such important phase of the project is to give answers to the following questions:

- Define GSD, swath with and camera FOV
- How can we obtain 3 spectral bands? What bandwidth for the mission?
- What is the required quantization depth for the imager?
- What are the trades between sensors for this mission?
- What optics has to be used?
- Identify payloads software tasks.

Alsat-1 being an experimental satellite built with Algerian funds. It had to satisfy Algerian needs first then DMC needs for the remaining time (DMC is Disaster Monitoring Constellation - the program to which Algeria adhered): Alsat-1 was to be able to cover both missions.

The answers had, to take into account the mission constraints related to the constellation and in particular the daily revisits which had a major influence on a camera swath width and by consequence on the ground sampling distance - GSD. Therefore we defined two tables: one for Algerian needs and the other for DMC needs. We thus listed all the possible applications for the two missions then from our experience as satellite image users (mainly Spot and Landsat), we defined three parameters:

- Swath width: large=300km, average=200km, reduced=100km
- Ground sampling distance: Low=30m, Medium=25m, High=20m
- Spectral Bands: Visible = [.51-.73 μm], Close Infra-red = [.75-.90 μm]

We formulated these parameters for each application then we sought the best possible compromise which satisfies both missions.

1.1 National Needs

They rise from the geographical nature of the country. The map clearly shows that Algeria consists of two distinct areas:

- The north, where 90% of the population in 95% of the cities are concentrated is characterized by small agricultural areas and a dense and rather dark geography with much of relief.
- The Sahara, desert with only 10% of population and 5% of the cities is characterized by a large homogeneous and luminous extent.

The topics observed in this case are: drought, marine/industrial pollution, geophysics, agricultural activity, mapping, etc.

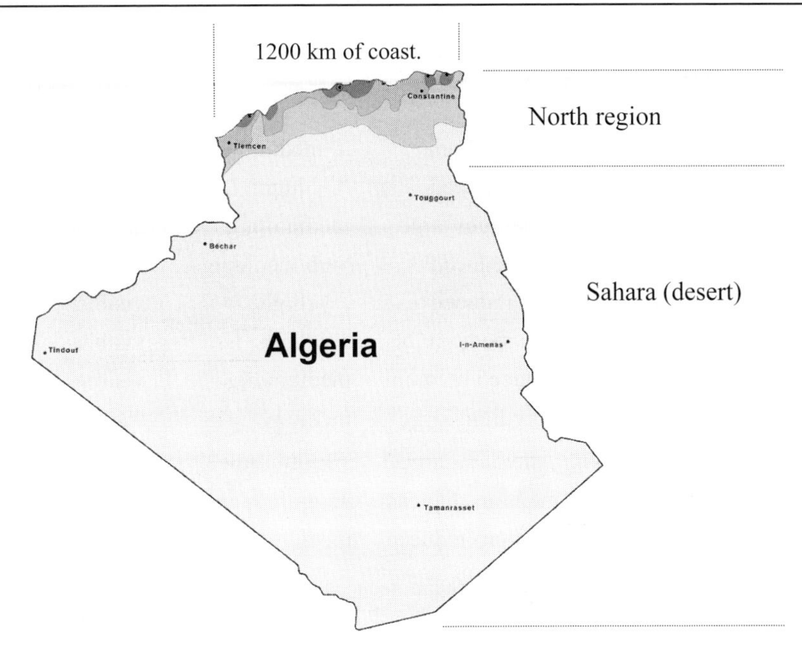

Fig. 1. Geographical configuration of Algeria

Sets of themes observed	Swath width	GSD	Spectral Band
Drought.	large/medium	low / medium	visible
Industrial/marine pollution	Medium	Medium / high	visible / NIR
Geophysics	medium/reduced	medium / high	visible
Agricultural monitoring	medium/reduced	medium / high	visible / NIR
Mapping	medium/reduced	medium / high	visible

Swath width: large=300km, medium =200km, reduced =100km
GSD: low =30m, medium = 25m, high=20
Spectral Band: visible=[.51-.73 μm], NIR=[.75-.90 μm]

Table 1. Topics observed for Algerian needs

1.2 Constellation Needs

The DMC was configured to function with a minimum of five satellites to guarantee a daily image of any point of the world. By joining the DMC, Alsat-1 has to conform to this requirement and thus to cover a band of 600km wide. This implies that the camera of Alsat-1 must have the same field of view as the others.

The topics observed in this case are summarized in Table 2:

Themes observed	Swath width	GSD	Spectral Band
Drought	large	medium	visible
Earthquakes	medium	medium	visible
Volcanoes	reduced	high	Visible/NIR
Fires	medium/large	medium/high	visible/NIR
Floods	reduced	high/medium	visible
Industrial accident	reduced	high	visible
Refugees Movement	medium	high	visible
Camps establishment	reduced/medium	medium/high	visible
Infrastructure	medium/large	medium/high	visible
Agricultural monitoring	medium/large	medium/low	visible/NIR
Hydrological mapping	medium/reduced	medium/high	visible
Hazard mapping	medium/reduced	medium/high	visible

Swath width: large=300km, medium =200km, reduced =100km
GSD: low =30m, medium = 25m, high=20
Spectral Band: visible=[.51-.73 μm], NIR=[.75-.90 μm]

Table 2. Topics observed for Disaster Monitoring Constellation

In conclusion, we can say that Alsat-1 must have a payload able to carry out both missions: Independent imaging for Algeria and joint imaging as a member of the constellation of DMC spacecraft. Thus, the characteristics found for the payload were:

- Large swath width: 600km
- Normal GSD: 32m
- High temporal resolution: daily revisit (with 4 other satellites).
- 3 spectral bands: two in the visible and one in near infra red (identical to those of Landsat-TM for convenience).

2 Satellite description

Alsat-1 is an earth observation satellite which evolves in a sun-synchronous orbit. It is equipped with two banks of cameras giving a total of 600km field of view at 32 meters ground sampling distance in three spectral bands: Red, Green and Near Infra-Red. This field of view allows the constellation to cover the whole earth within 24 hours. In absence of disaster, Alsat-1 is dedicated for Algerian purposes: mainly for remote sensing applications.

The imaging system allows windowing and it is supported by a total storage capacity of two 0.5Gbytes of data which is downloaded to a ground station at 8Mbps (within 10min/SSDR). The downlink and the uplink, both operate in S band at 8 Mbps in normal operation and 38.4kbps in commissioning for the downlink and 9.6kbps for the uplink.

Alsat-1 has an accurate nadir pointing attitude and as part of the constellation, it is equipped with a propulsion system to carry out:

- The circularization of the satellites constellation to correct the launch errors.
- The station acquisition to equally separate the satellites from each other so the daily global coverage is respected.
- The station keeping to compensate the drag effect and to maintain the separation time between the satellites.

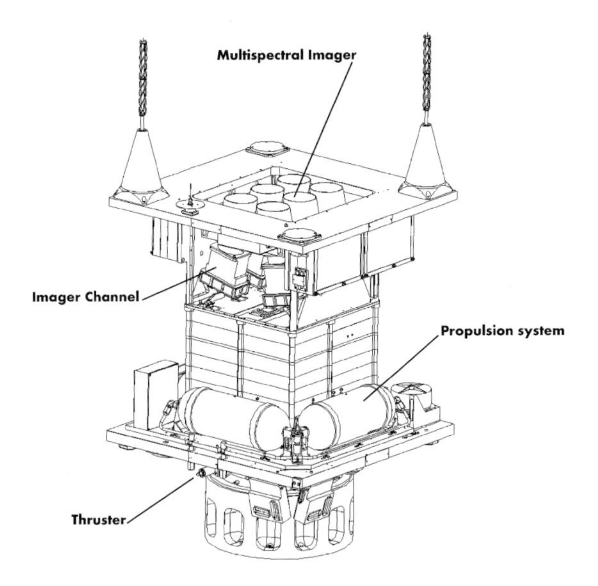

Fig. 2. Alsat-1 structure

2.1 Payload Description

The Alsat-1 payload is a multi-spectral camera which works in a pushbroom mode (forward scan is provided by the spacecraft motion). It is in fact a couple of two imagers (of three channels each) which work separately or both together. This adds flexibility to program images (satellite operations).

For each spectral band (green, red, near-infrared), two channels (from both banks) provide a 600 km swath with (5% overlap between them) from its 686km altitude.

Fig. 3. Alsat-1 Imager

Camera

It is a line scan camera system designed to provide medium resolution (32m GSD), high dynamic range and low noise imagery from orbit [10]. It consists of 6 channels split into two banks and fixed on a V plate. Due to the extremely wide swath, each bank acquires half of the scene (with 5% overlap) and can be operated individually or synchronized together to work in a full 6 channels mode.

Each channel is in fact an independent camera and contains a complete optic-system plus a PCB supporting the sensor & linked to the main PCB board in the stack (one main board per bank) thanks to a flexi-rigid PCB cable.

The storage is achieved thanks to two adding up Solid State Data Recorder (SSDR) based on Power-PC processor and totalizing 1Gbytes data and one redundant unit of 128Mbytes based on StrongARM (SA1100) and acting as a backup unit. Each bank is linked to the three storage units via a synchronous serial input streams at nominal 20Mbit/s allowing a flexible satellite operation.

2.2 Alsat-1 Characteristics

Three major characteristics distinguish Alsat-1 from other micro-satellites.

Wide swath width: The main characteristic of Alsat-1 is to achieve a daily revisit in conjunction with the other satellite of the constellation. Hence, a large FOV was implemented allowing covering 33.6 million hectares in full size image. (ie: whole Algeria with its 2381741km^{2}could be covered by 10 images).

Windowing: This function was added during the design of the satellite to avoid a saturation of the storage units and add more flexibility during the satellite operation. Alsat-1 being an experimental satellite, this function should allow us to image various types of area (around the world).

Short Revisit time: The exceptional swath width of Alsat-1 allows obtaining a particularly short pseudo period (for a satellite not equipped with an Off-pointing capability). This possibility allow to quickly react when a disaster occur.

2.3 Alsat-1 First Results

356 images were taken during the first year and 380 during the second year. Their examination showed qualities close to those of Landsat. The spectral characteristics are almost identical on the other hand the Alsat-1 images showed a lack of sharpness (especially in urban zones) and some distortions more classic like:

- The visible “vignettage” effects (unequal optical transmission factor between the centre and the edge of the optics) due to the exceptional wide swath width.
- The visible "stripping" effects on both bank (more visible on bank0). This effect is created by the CCD sensor. It would result from bad "Reset" of the output stage and unequal sensitivity of photo-sites.

These effects were identified during the pre-flight tests and were easily corrected.

3 Disaster Monitoring Constellation - DMC

The idea of the Disaster Monitoring Constellation was proposed for the first time by Professor Chen F.Y. [3] [4] who said that seven satellites at 772km altitude, on the same sun-synchronous orbit, inclined at 98°, was adequate to observe the same point on the ground every 12 hours with a 400km swath width. It was a good idea but the cost of the large satellites was such that no single owner could afford the cost of conforming to the constraints that such constellation imposed (orbit, inclination, swath, GSD, etc.).

The originality of the DMC established by Surrey Satellite Technology Ltd. (SSTL) is that it was proposed to several countries seeking access to space technologies. The low-cost of the micro-satellites included a know-how transfer and team training program. Hence, the objectives could easily be met:

- Train a team of engineers in space techniques while taking part to the micro-satellite project
- As a constellation member, gain a continual training for a new team in satellite operations in a useful operational field. Experience showed that the majority of the micro-satellites carried out within know-how transfer program, were rarely exploited beyond the commissioning phase for lack of valid operational objectives and for lack of resources also (the micro-satellite launch itself is often misjudged as the main goal for the project).
- By working together in a coordinated orbit the constellation could achieve daily imaging capability for its members and for disaster response.

3.1 Current Configuration of the DMC

The constellation currently functions with four micro-satellites, and the fifth is under development for launch in 2005 [11]. Three of them are equipped with the same multi-spectral camera, identical to that described above. The fourth provided with a narrow angle camera (CCD matrix, GSD=28m, 4 spectral bands) covers the same 600km swath width thanks to off-pointing capability.

Disaster image requests are currently coordinated by SSTL, and are sent by email to each satellite operator. The data downloads are automatically done on the respective ground stations or at SSTL ground station (U.K.) (a software up-grade to install automated downloads is in progress at the Alsat ground station). In addition, the individual ground stations will be linked together through Internet broadband thanks to a Mission System Planning software (MPS) which should give more operationality to the constellation by speeding-up the check of available resources (imager, memory free,...) and the images request while preserving the authority of each owner on his micro-satellite.

After one year in space together, including commissioning period, the four satellites are currently in position (on the same orbit and equally-spaced). The DMC is currently operating as a group of individually operated satellites. The DMC has shown, within this limit, the ability to react quickly to urgent image requests when major disasters occur. Recently in response to the UN request for images covering the Philippines floods, the DMC provided the RESPOND consortium with regular (5, 6, 8, 10, 11, 14 and 17 December) images from the DMC satellites of the area of Manila, of which an AlSat-1 image was used to produce a spatio-map for distribution to aid agencies (see Fig. 4).

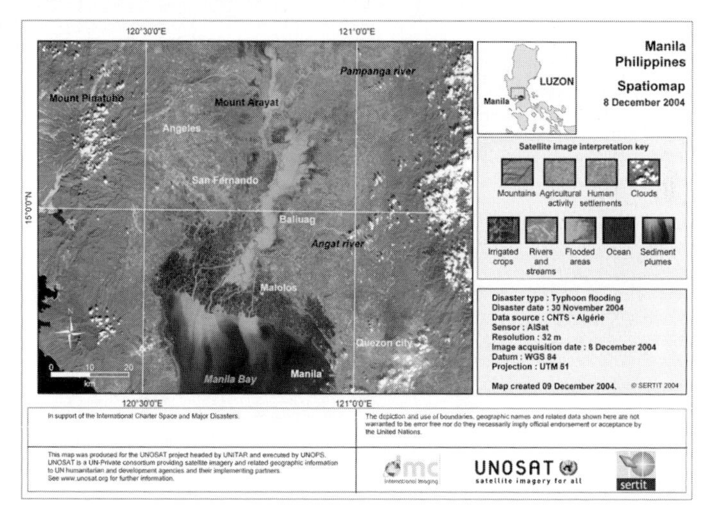

Fig. 4. Spatio-map produced using Alsat-1 image

4 Illustration of Disaster Monitoring Images

The first months were spent for satellite commissioning. The images taken were for demonstration to check modules and operationality of the satellite. The first applicative images were taken on behalf of the Remote Sensing Division (CNTS), whose aims was to re-start studies already done with other satellites images (mainly Landsat and Spot), to compare the results and thus check the quality of Alsat-1 images for traditional remote sensing applications. The results of these studies were presented at the workshop ***Alsat-1 - Users*** organized by the Algerian Space Agency on 14-15 July 2003. The conclusions were that as a whole, for traditional remote sensing applications, Alsat-1 images gave the same results as those of other satellites with similar characteristics. The only reported remark was that the non processed Alsat-1 images, present some radiometric defects (variation of sensitivities between pixels, "vignettage" effect, etc.) which had to be corrected.

4.1 Boumerdes Earthquake

The first disaster image was acquired on May 24th, 2003, three day after the devastating earthquake of Boumerdes (magnitude 6.8° ml) which occurred on May 21th, 2003 at 18h44 UTC.

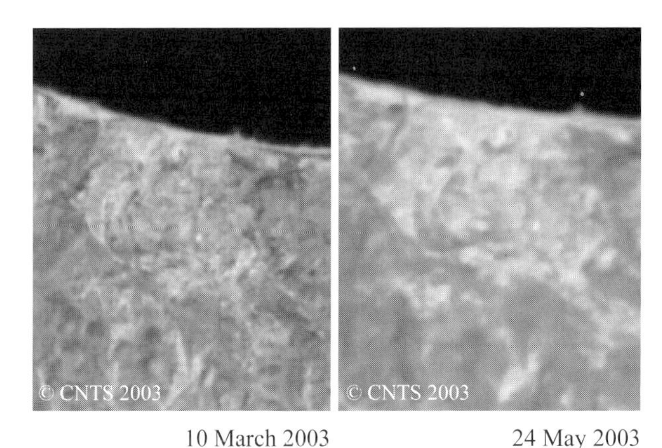

Fig. 5. Two Alsat-1 images of Boumerdes before and after earthquake

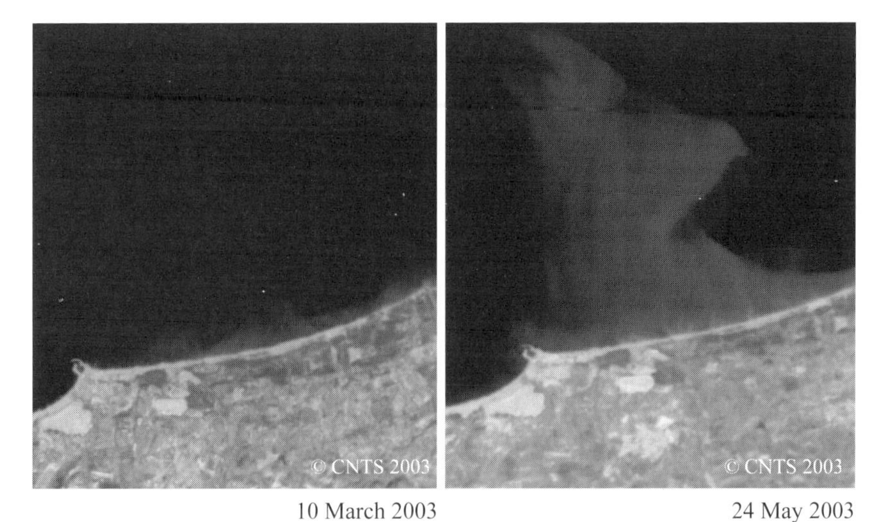

Fig. 6. Sand banks supposed due to Tsunami effect, close to Zemmouri seaport

The pressure was important as it was the first time ever that Alsat-1 had to show its capacities on a real problem. Unfortunately, due to its medium resolution, Alsat-1 images could highlight neither the damage assessment nor camp settlement.

We could highlight only the new distribution of the sand banks close to the Zemmouri seaport (known as the closest city to the epicentre) by comparison with an Alsat-1 image of the same area taken on Mars 7th, 2003. It would be due to the ***Tsunami*** which followed the earthquake (the sea was withdrawn 200m approximately from the coasts before returning to its position 15 minutes later with a shift of 10m compared to its first level).

4.2 Forest Fires Monitoring

In the summer 2003, we launched a discrete monitoring campaign of forests fires to test Alsat-1 under normal conditions of operationnality. The main aim was to show that in the case of forests fires, Alsat-1 images could be used as a basis for a fast to-date evaluation of the burned zones (in middle of season or following an important fire). This related to all the North of Algeria, main forested region of the country (with 3.2 million hectares) which is concerned each summer, by important fires.

Forest Fire Campaign - 2003

Summer 2003, was exceptional as well from the point of view of the frequency of the fires as of the importance of the caused damage (Southern of France, Portugal, Corsica, etc). Algeria didn't escape from this fatality because this period was marked by a clear temperatures raise exceeding the usual averages and increasing by consequences the fire hazards.

The exceptional size of Alsat-1 images (up to 600x600 km) made possible to cover the north of Algeria with only two images full size (67 million hectares approximately). This wide swath width reduces the revisit time up to 2 days in certain cases as for the town of Tlemcen on Table 1. We can see that during the summer 2003, 8 pass of Alsat-1 was predicted during the month of August. This made possible to "shot" any forest fire up.

Fourteen (14) images were taken during the period from June to August 2003. The method adopted for this first monitoring campaign was:
- To take scenes of the target (all the north of the country) by supervising the information sources (TV, newspapers, Internet, etc.) to take the images at the beginning of fire.
- To examine the false color image to locate the burnt zones (grey on the image) and to make a coarse estimate of burned surface. Calculation

was done by pixels counting at nominal size (32x32m). The aim was to make a fast up-to-date inventory of burned surfaces (following each important fire) to show how fast is to have even coarse but useful information for decision makers.

N°	Imaging date and time	Sun angle	Distance from target
1	Sun 03 August 2003 at 09:58:43	53.3°	298.30km W from Tlemcen
2	Tue 05 August 2003 at 09:36:24	52.6°	204.88km ENE from Tlemcen
3	Fri 08 August 2003 at 09:51:36	52.5°	134.38km W from Tlemcen
4	Wed 13 August 2003 at 09:44:29	51.6°	29.23km ENE from Tlemcen
5	Mon 18 August 2003 at 09:37:21	50.6°	192.41km ENE from Tlemcen
6	Thu 21 August 2003 at 09:52:32	50.5°	146.48km W from Tlemcen
7	Tue 26 August 2003 at 09:45:22	49.5°	18.13km ENE from Tlemcen
8	Sun 31 August 2003 at 09:38:13	48.3°	182.09km ENE from Tlemcen

Table 3. Result of the Prediction software from the 01 to 31 August, 2003 for the town of Tlemcen, Algeria [Long. 35.0167°, Lat. -1.4667°]

The results were compared during a technical day, with those of the Forest Conservations and the Civil Protection, obtained by the conventional methods (land surveying, visual estimate...). An important variation appeared at the first reading of the results. It was due to several factors:

- The pixel counting was carried out without processing the image.
- The nominal pixels size was supposed (32x32m), identical on all swath.
- No consideration was given to the relief (ground was supposed flat).
- All the grey zones on the image were assumed as burned whereas for the Forests Conservation, a clear differentiation existed between "burned" and "licked by fire".
- All the zones appearing burned on the image were included in calculation contrary to the Forests Conservation, where only the burnt grounds belonging to the forest field were indexed.

The objective which was to promote the use of Alsat-1 images appeared achieved. A thorough study was ordered with the Division of Remote Sensing (CNTS). It was carried out in collaboration with the same partners (Direction of the forests, Civil Protection and office of meteorology). The objective was to refine the results by taking into account the issued reserves. The method was modified [13] by including the ground verifications.

Thus the image processing taken was done in several stages:

- Identification of the burnt surface on the image (grey color).
- Windows extraction of the image around the burnt surface.

– Images correction with respect to the topographic map by using anchoring points.
– Identification of the burnt surfaces affecting only the forests field.
– Distinction between "burned surfaces" and those "licked" by flames,
– Measurement of burnt surfaces.
– Calculation of the total burned area (with administrative boundary).
– Ground verification of the result.

This new step made possible to bring the results closer to those of the Forests Conservation. Total burnt surface affecting the forests field was brought back to 31859ha compared to 48413 ha of the total burnt surface.

The objective of sensitizing was achieved and the authorities in charge of the forests protection requested the same campaign in 2004.

Forest Fire Campaign - 2004

With the experience gained during the 2003 campaign, a more rigorous methodology was defined:

– Take scenes of the north of the country to evaluate the up-to-date situation before the beginning of the campaign.
– Take images of the targeted area, during all summer.
– Make a final estimation of the surfaces burned (end of the campaign).
– Take into account the forest inventory map (study carried out on behalf of the Forests Conservation). The map gathers twelve (12) forest topics which were combined in one class, for the needs of this study.
– Fieldwork for "*rising of doubt*" and checking at the same time the situation of the vegetable recoveries on the last campaign burned area.

The 2004 summer was characterized by a more moderated average temperature; consequently, less spectacular fires were recorded.

Six scenes taken between May the 17th and June the 25th were used to make the to-date inventory before the beginning of the campaign. It was necessary to note the last campaign burned area to avoid including them in the 2004 one and to check at the same time the state of the vegetable recoveries.

Twelve scenes in total were taken between the beginning of July and mid-October. They were used for evaluating the burnt surfaces. As in 2003, the steps were the same with the difference that the geometrically corrected image was to be superimposed to the forest inventory map to exclude from calculation the fires which wasn't in the national forest field

The result for this campaign was close to the one obtained with traditional methods. The total surface burnt was 20% lower compared to that of 2003.

Finally for the management needs, a forest fires map was dressed [2]. It includes all the 2004 burnt area superimposed to the national forest inventory layer. It was to help the decision maker visualizing in a graphic way all the 2004 assessment.

Fire of the Var Department (France)

During the 2003 campaign, several images were taken including foreign countries. It was the case for the Var department where took place one of the most important fire since 1990. Three (03) images on the whole were taken; it followed the activation of the charter by French Civil Protection [16]. Fortunately, an image of the south of France was taken on June the 13th. The extraction of a window corresponding to the Var department showed a regular Chlorophyllous activity (red colour on the image) mainly on the massif of the Maures, massif of Esterel and the Rouet forest (Fig. 7).

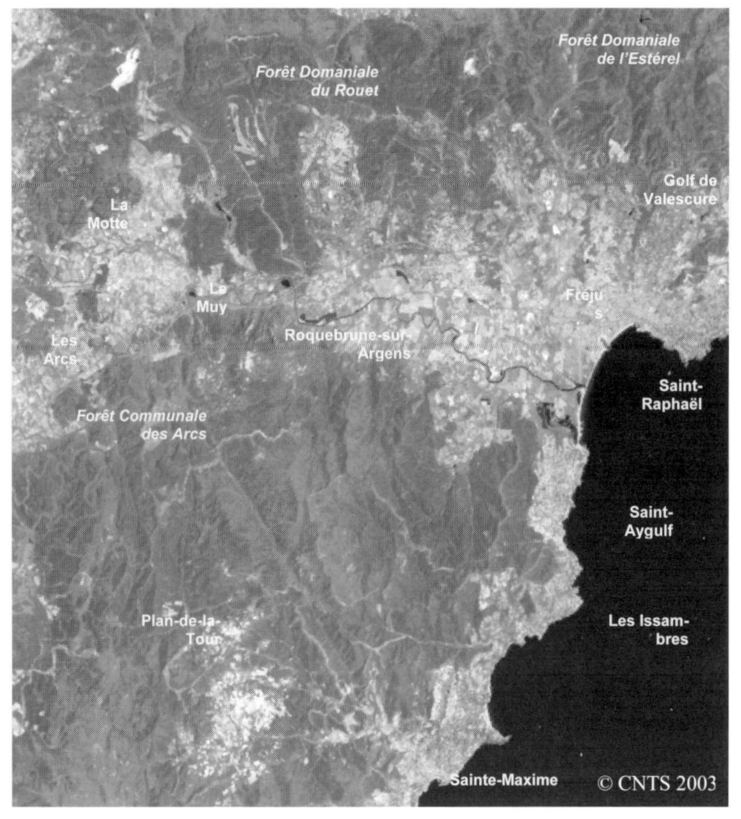

Fig. 7. Region of Var extracted from Alsat-1 image taken on 13 June 2003.

The first disaster image was taken after the 2^{e} fire, probably after the fires resumption in the massif of Esterel, massif of the Maures having burned in the first phase. We followed the same procedures to those of the 2003 campaign for Algeria, explicitly: visualize false color image to locate a burnt area and make a coarse estimation of the burnt surface. Calculation was done by pixel counting at nominal pixel size (32x32m). The result approached those reported by the press at that time.

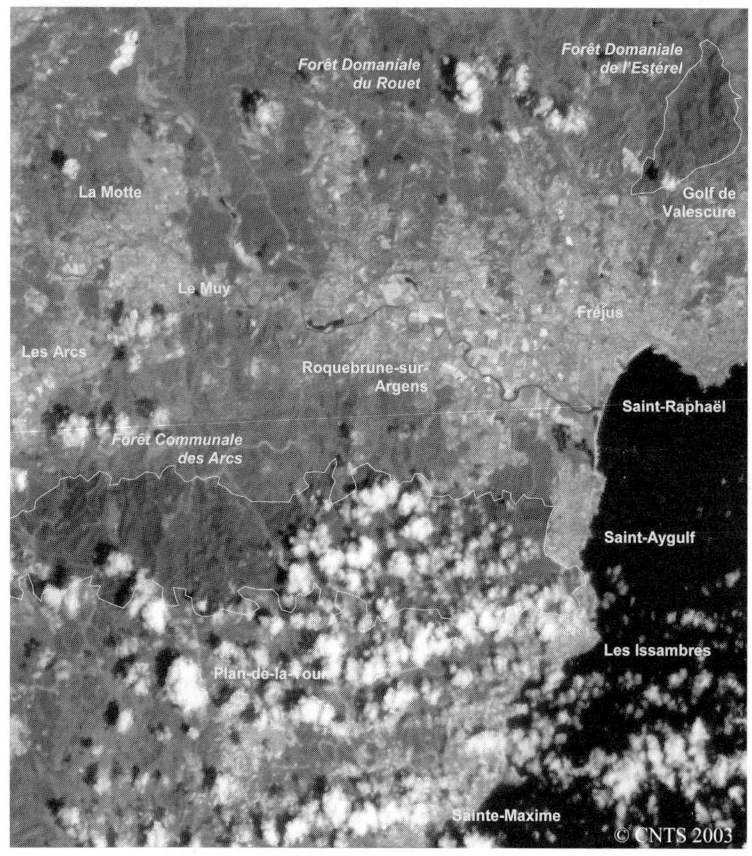

Total estimated surface burned = **7277 ha**
(Esterel forest: 821ha, Arcs forest: 6456ha)

Fig. 8. Region of Var extracted from Alsat-1 image taken on 27 July 2003

A third image was taken one week later on August the 4th, when fire restarted again due to a violent wind. The result was an extension of the fire to reach approximately the double burnt surface.

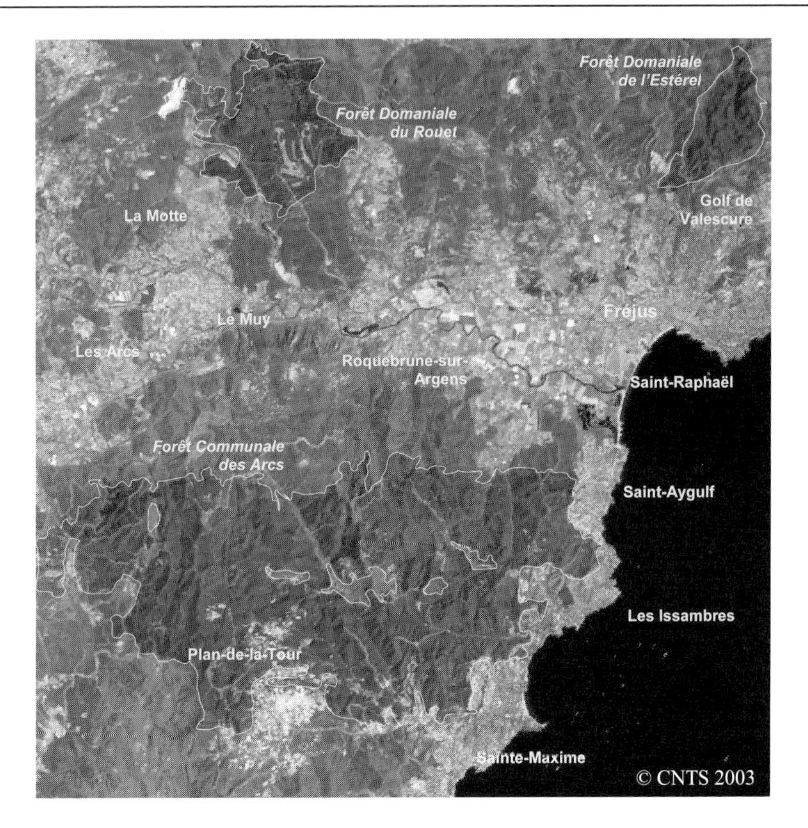

The total estimated surface burned is: **15316ha**
(Esterel Forest: 937ha, Arcs Forest: 12499ha, Rouet Forest: 1880ha)

Fig. 9. Region of Var extracted from Alsat-1 image taken on 04 August 2003

During this period and knowing that the charter was activated, we tried to obtain other satellites images for the same site (Var department) and we could obtain from Spot-Image web site [15], two Spot-5 images (at 10m) taken in July the 19th and 29$^{th.}$ The examination of these images and their comparisons with those of Alsat-1 made it possible to see a spectral similarity of the answers even if the geometrical quality is limited. We can say that at this scale, the Alsat-1 images are as valuable as those of Spot (or other satellites) especially if they are taken during "eclipse period" (when target is not visible from any of the big satellite). It was the case of these images with those of Spot; therefore we could reconstitute the scenario of the Var department Fire by using a combination of corrected Alsat-1 images (rotation applied) and those of Spot-5 (see Fig. 10).

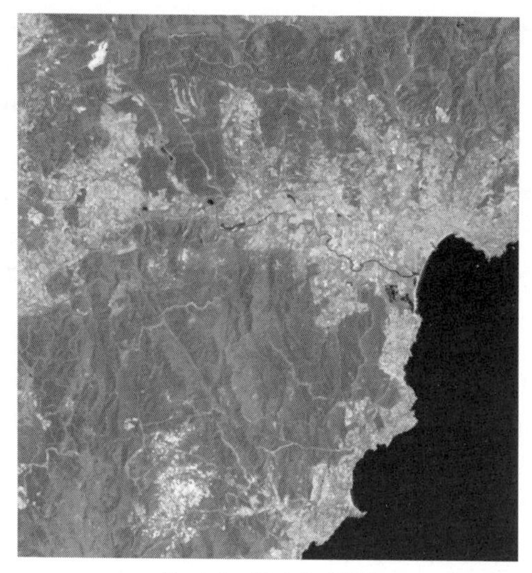

Alsat-1 - 13 Juin 2003

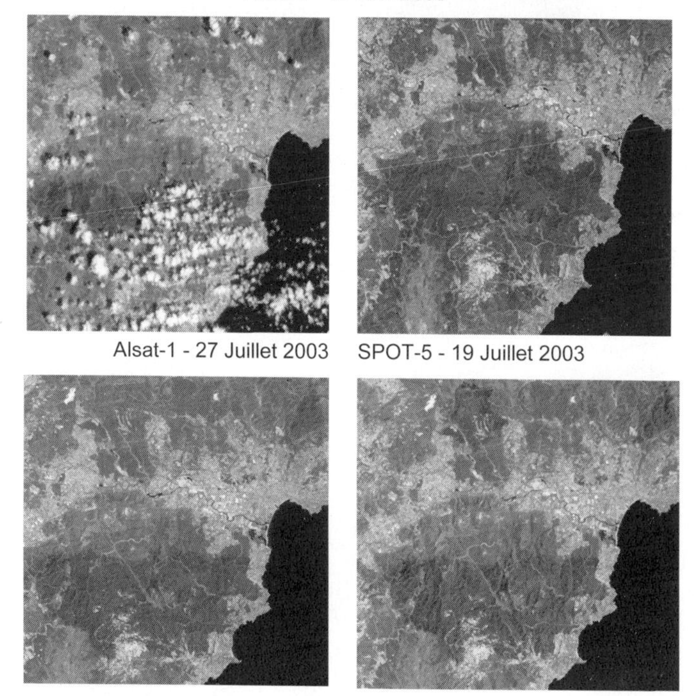

Fig. 10. Reconstitution of the 2003 forest fire of the Var department (France)

Southern of France Floods

In December 2003, an important flood took place in the south of France following strong precipitations (up to 200 mm on 04 December). The flow of the Rhone river increased to reach a record level of 13000m^3/s [6] creating four breaches along the river from where flowed more than 300millions m^3 (quantity estimated for the Camargue only). The Alsat-1 ground station team answering the activation of the charter [17] by the French Agency of Civil Protection started the image scheduling of the satellite for imaging the areas affected by the floods.

The swath width and consequently the short pseudo period made possible to multiply the attempts to find the opportunity of taking clear image as the bad weather conditions (dense cover cloudy) which often follows this sort of disaster preventing the cameras from seeing the target. Four (04) images were thus scheduled between the 07 & 15 December 2003.

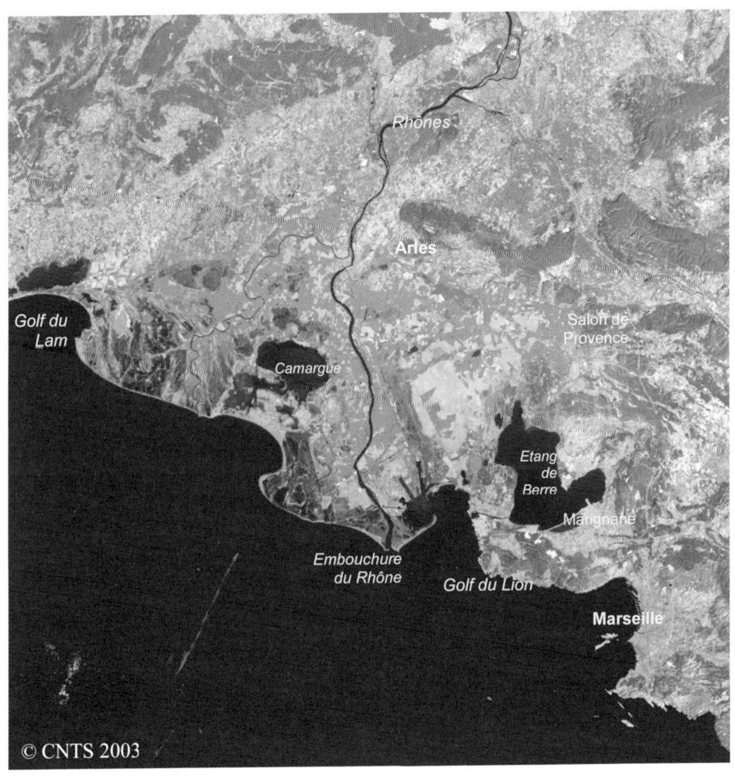

Fig. 11. Region of Marseille extracted from Alsat-1 image taken on 13 June 2003

We see on Fig. 11 extracted from a full size image of the south of France taken on June the 13th, 2003, that the region is characterised by a lot of lakes and ponds. All seems under control and the Rhône river look calm. The bright reddish area inform about rural activities (normal as we were still on the spring). The dark reddish area is more likely to be forest (north-west of Fig. 11).

Fig. 12. Marseille region extracted from Alsat-1 image taken on 07 December 2003

On Fig. 12, extracted from a full size image taken on 07 December 2003, the river seems broader compared to the image taken during summer, on June 13th 2003 (Fig. 11).

We can see the extent of the zones flooded along banks of the river and distinguish at least two of the four broken dykes. Sediment carried by the floods is tinting the seawater a bright blue on the mouth of the Rhône River on the Mediterranean Sea.

Fig. 13. Marseille region extracted from Alsat-1 image taken on 15 December 2003

Ten days later, we succeeded to take another free clouds image from the flooded area (Fig. 13) which showed that the river started to drop down except for Grand Rhône. We can see that the sediment is still carried by this branch of the river making the water brighter.

The two small images show the restored dykes after they were repaired.

5 Conclusion

Alsat-1 was designed by taking into account the disaster specifications. The first tests showed satisfying results. The image quality was good and. the artefacts were easily corrected.

A wide swath width was valuable characteristic as it reduced significantly the delay of having images back by allowing a short revisit time, beneficial when the target is in a bad meteorological area (as we can schedule many images until we succeed).

During the first two year of exploitation, several disaster was imaged and first processing done.

For flooding and forest fire case, Alsat-1 resolution permits a quick evaluation of damage by pixels count. Even if it is approximate, it is often valuable especially for 3rd world country where the road network are inadequate and a decent map hard to find.

However, for earthquake or industrial accident (Alsat-1 image was tested in the Skikda GNL explosion in January 19th, 2004), it seems not adapted as we couldn't clearly see any damaged area even if we knew where it is. We suppose that the limit comes from the medium resolution of the camera and the lack of sharpness (some image pre-processing have to be done).

Acknowledgements

The authors wish to acknowledge all members of the Alsat-1 team for their support.

References

[1] Bekhti M et al (2002) AlSAT-1: The first step into Space for Algeria, Proceedings of the 53rd IAC and World Space Congress, Houston, TX, IAC-02-UN/IAA.3 rd.01

[2] Benhanifia K, Smahi Z (2004) Inventaire des feux de forets du nord de l'Algérie pour l'année 2004, Rapport final, Oct. 2004

[3] Chen FY et al (1992) Earth Environment Observing Satellite System and International Co-operation, 43rd IAF Congress, Aug-Sept. 1992, Washington, USA

[4] Chen, FY et al (1996) Composite Satellite System for Observation of Earth Environment and International Co-operation, IAA-96-IAA.3.3.02, 47th IAF Congress, Oct 1996, Beijing, China

[6] Compagnie Nationale du Rhône (2003) Crue de décembre synthèse météo-hydrologique, Dec. 2003, in http://www.cnr.tm.fr (access 17 Dec. 04)
[7] Curiel A et al (2003) First Steps in the Disaster Monitoring Constellation, 4th IAA Symposium on Small Satellites for Earth Observation, Berlin, Germany
[8] Rachedi A (2001) Study the ALSAT-1 Camera Payload Proposal for an improvement. Internal report, Alsat-1 project, 29 pp
[9] Rachedi A (2002) Utilisation de l'Outil Spatial pour le Suivi des Catastrophes Naturelles - Cas de la DMC, 3e Forum des Assurances, El Aurassi – Algiers
[10] Rachedi A, Hadj Sahraoui N, Brewer A (2004) Alsat-1: First Results of Multispectral Imager, XXth International Congress for Photogrammetry and Remote Sensing, 12-23 July 2004 Istanbul, Turkey
[11] Stephens P (2004) Development of a Commercial Interface for the DMC, IAC-04-B.5.03. 4, Oct. 2004 Vancouver, Canada
[12] Sweeting MN, Chen FY (1996) Network of low cost small satellite for monitoring & mitigation of natural disasters, 47th International Astronautical Congress, Beijing, IAF-96-C.1.09
[13] Yousfi D, Hassani A (2003) Inventaire des Incendies de Forêts dans le Nord de l'Algérie par Utilisation des images Alsat-1, Rapport interne, CNTS, Nov. 2003, 8 pp
[14] Ward J et al (1999) Microsatellite Constellation for Disaster Monitoring, Small Satellite Conference 1999
[15] www.spotimage.fr (access on 11 Dec. 2004)
[16] www.disasterscharter.org/disasters/var_f.html Incendie du var (access on 15 Dec. 2004)
[17] www.disasterscharter.org/disasters/france4_f.html Inondations dans le sud de la France, (access on 16 Dec. 2004)

Experience and Perspective of Providing Satellite Based Crisis Information, Emergency Mapping & Disaster Monitoring Information to Decision Makers and Relief Workers

Stefan Voigt, Torsten Riedlinger, Peter Reinartz, Claudia Künzer, Ralph Kiefl, Thomas Kemper and Harald Mehl

Center for Satellite based Crisis Information (ZKI) of the German Remote Sensing Data Center (DFD), Oberpfaffenhofen - German Aerospace Center (DLR), Germany.
Email: stefan.voigt@dlr.de

Abstract

Recognizing an increasing demand for up-to-date and precise information on disaster and crisis situations the German Remote Sensing Data Center (DFD) of DLR has set up a dedicated interface for linking the available and comprehensive remote sensing and analysis capacities with national and international civil protection, humanitarian relief actors and political decision makers. This so called "Center for Satellite Based Crisis Information" (ZKI) is engaged in the acquisition, analysis and provision of satellite based information products on natural disasters, humanitarian crisis situation, and civil security. Besides response and assessment activities, DFD-ZKI also focuses on the provision of geoinformation for medium term rehabilitation, reconstruction and prevention activities. DFD-ZKI operates in national, European and international contexts, closely networking with public authorities (civil security), non-governmental organizations (humanitarian relief organizations), satellite operators and other space agencies. ZKI supports the "International Charter on Space and Major Disasters", which is a major cooperative activity among international space agencies in the context of natural and man-made disasters.

Different examples on the analysis and fusion of satellite imagery for information extraction and mapping actions in several disaster events are presented. These include examples from earthquakes, floods, forest fires as well as man made disasters and humanitarian relief. The full cycle from emergency call, satellite tasking, data acquisition, pre-processing, interpretation, map generation and provision of the information to the end-user is presented. In all the demonstrated cases the data were generated and provided either through in house satellite acquisitions, the International Charter on Space and Major Disasters or through other national and international data provision networks. It is shown how satellite imagery can be assessed, processed and turned into information products provided to decision makers within hours. Furthermore, gaps between research and real world, near-real-time mapping requirements are discussed.

1 Introduction

During the recent years there has been an increasing demand for up-to-date and precise information on disaster and crisis situations in Europe and world-wide. The reasons are manifold. The vulnerability of societies and infrastructure has increased. The weather pattern most probably has been shifted to more extreme precipitation, wind speed etc. Finally the regional and global cooperation of relief actors has been extended strongly. Satellite imagery serves as a source of information in crisis, emergency as well as during natural disaster situations. Having recognized this need, the German Remote Sensing Data Center (DFD) of the German Aerospace Center (DLR) has set up a dedicated interface called ZKI for linking its comprehensive operational remote sensing data handling and analysis capacities with national and international civil protection and humanitarian relief actors as well as with political decision makers.

In order to provide up-to date and relevant satellite based cartographic information and situation analysis, it is necessary to establish very efficient and operational data flow between satellite operators, receiving stations and distribution networks on the one hand and the decision makers and relief workers on the other. Service lines and feedback loops have to be established to allow best possible data and information provision as well as optimized decision support.

2 Servicing All Pillars of Civil Human Security

The work experience of the past years in the domain of civil disaster and crisis analysis using satellite imagery shows that the separation lines between disaster management, humanitarian relief, development cooperation

and civil security have been vanishing more and more. When for example a camp for internally displaced people in Darfur/Sudan is set up, this is a classical humanitarian relief logistics and support task. Nevertheless, it also involves aspects of disaster management, when heavy rainfalls flood wadis and block land transportation. It may also involve security related questions, when camp inhabitants or relief workers are exposed to threads of attacks or riots. As a result of this, DFD-ZKI serves all involved pillars of civil human security: basic disaster management support, humanitarian relief activities, civil security efforts as well as sustainable development projects.
The following examples on the analysis and fusion of satellite imagery for information extraction and mapping actions in several disaster events illustrate the work in detail.

During the flooding of the river Elbe in Germany in August 2002 the DFD acquired in cooperation with the European space agency (ESA) and other national and international partners all available satellite data (e.g. ERS, IRS-P3-MOS, MODIS, NOAA, IRS 1C/1D, Landsat 7, ENVISAT ASAR) and made them available to external users for situation analysis and flood inventory (Fig. 1). The DFD used its own satellite receivers and prepared the raw data in-house. After the Charter was triggered, additional satellite data from the European ERS-2, the Canadian RADARSAT, the French SPOT, the Indian IRS Satellite as well as DLR's BIRD satellite system were provided to assess the situation and support crisis management. More than one hundred satellite scenes were acquired.

Because of the continuing refugee situation and the onset of the rainy season in western Sudan in august 2002, the humanitarian relief organizations working there were in urgent need of up-to-date, detailed maps (Fig. 2). In consultation with UN-OCHA, Germany's disaster relief organization (THW) and the German Red Cross (DRK), the crisis regions around the cities of Al Fashir and Al Junaynah were mapped. The focus was on ascertaining the road network, its condition, and the traversability of possibly flooded wadis and river valleys. Recording settlements and refugees camps and their current size was also of high interest. Satellite data for the maps were made available through the "International Charter on Space and Major Disasters" and processed and interpreted in the context of the ESA GMES Service Element "RESPOND".

Fig. 1. The Elbe flood between Torgau and Wittenberg mapped with Landsat 7 on August 8, 2002. The flooded areas can be recognized in light blue, in dark-blue the original course of the Elbe River

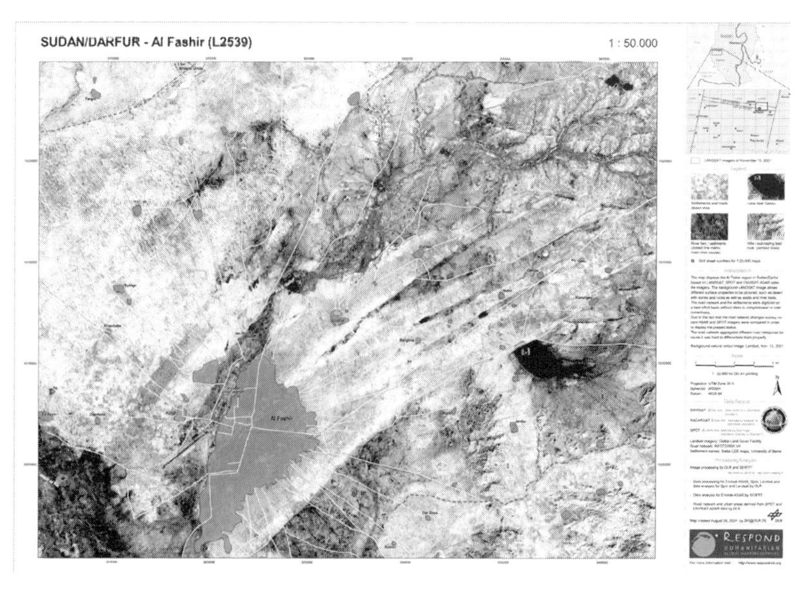

Fig. 2. The Al Fashir region in Sudan/Darfur based on LANDSAT, SPOT and ENVISAT-ASAR satellite imagery. The changing road network and settlements were digitized comparing recent SPOT and ASAR imagery from August 2004

On May 6, 2004 a military arms dump close to the village of Novobogdanovka in southern Ukraine exploded (Fig. 3). According to press statements 10.000 people in the surrounding villages had to be evacuated and a major highway and railway line connecting the cities of Melitopol and Zaporizhzhya had to be blocked. The arms dump was completely destroyed and large amounts of debris were hurled hundreds of meters and even kilometers into the neighboring villages and agricultural land. The satellite imagery shows that some fires were still burning 36 hours after the explosion.

Fig. 3. The explosion site south-east of Novobogdanovka in southern Ukraine in detail, as mapped by satellite on May 8, 2004, approx. 46h after the onset of the disaster

For the "Afghanistan-Conference" in Bonn and the "Geberkonferenz" in Berlin in 2001, the Foreign Ministry needed maps for presentation purposes (Fig. 4). Also the humanitarian relief organizations like the "German Red Cross" (DRK) and the German disaster relief organization "Technisches Hilfswerk" (THW) had an urgent need for information about the status of possible supply routes into Afghanistan from neighboring countries. Therefore DFD analyzed archive data and collected all kind of available geographical information from different sources. In a short time various maps, satellite images and recommendations for the best routes into Afghanistan could be reported to the public authorities.

Fig. 4. Topographic map of the border region Afghanistan-Tadzhikistan with updated information about the infrastructure

3 Earth Observation for All Phases of the Disaster Cycle

Besides response and assessment activities in emergency cases DFD-ZKI provides geoinformation products and services during all phases of the "disaster" or "crisis cycle" (Fig. 5). This includes mapping and analysis for risk assessment and prevention measures before a crisis or disaster occurs. During an event assessment, localization and quantification of a given crisis situation are provided. In the transition phase between relief and reconstruction information products for coordination and planning are developed. Finally, mapping and monitoring services during the reconstruction and preparedness/prevention phases are produced. It is only natural that these different phases require information products at different scales in time and space. The in-crisis information products, such as damage assessments for cities, or flood extent maps may be generated for distinct areas and at given scales. The planning and risk assessment mostly covers much larger areas. For example the localization of potential natural disaster risk areas or humanitarian crisis regions can be extremely difficult beforehand. Furthermore it proves to be difficult to convince funding agencies to provide the required financial basis for such preventive detailed crisis hot spot mapping.

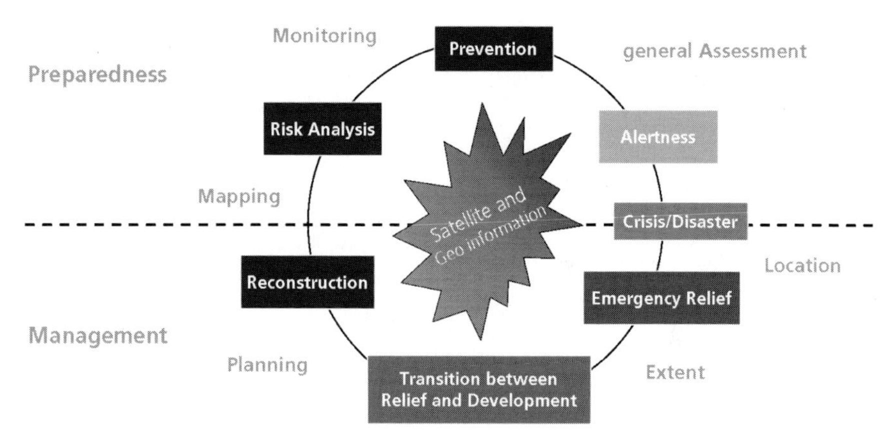

Fig. 5. Satellite and Geoinformation can serve civil human security during all phases of the disaster or crisis cycle. Any information products such as a base maps or a damage assessments can also often be used during several phases of the cycle

The examples given in the section before can be grouped into the disaster cycle as follows: While the activities during the Elbe flood were an in-crisis disaster respond, the satellite image products generated at a later stage were used in many cases for the planning of reconstruction and prevention. The mapping of the area around Al Fashir in Darfur/Sudan can be characterized as a base mapping activity in a humanitarian emergency relief situation. The analysis of the explosion in the Ukrainian arms dump can be considered an emergency relief mapping; however in the civil security domain. The consultancy and disaster reduction work conducted for Afghanistan is primarily dedicated to the reconstruction and prevention phase of the cycle.

4 Interfacing between Space Technology and the Relief Community

The acquisition, handling, analysis and provision of earth observation data, especially when performed in very short time frames, is still a very complex task. However, it is important to support an easy use of this information source, not only for scientists and military experts, who often have the required financial and technological means to access these systems. It is the scope to facilitate and simplify the use of existing space technology and information sources for decision makers in the civil protection and the humanitarian relief sector (Fig. 6). They often do not have access to these technologies, despite the need for topographic base information or situation analysis. In order to fulfill its tasks DFD-ZKI operates in national, European and international contexts and collaborates closely with public authorities (civil security), non-governmental organizations (humanitarian relief organizations), scientists, satellite operators and other space agencies.

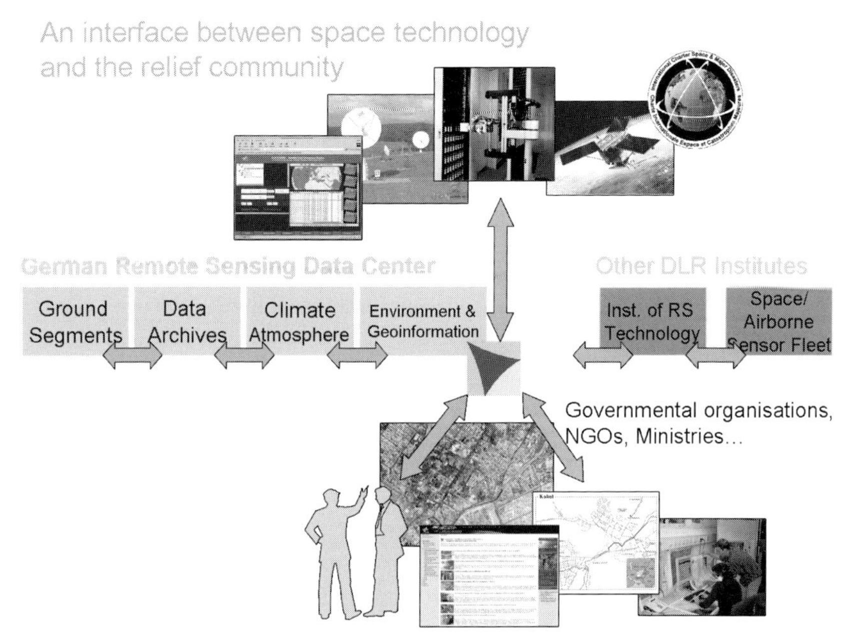

Fig. 6. The DFD-ZKI supports the internal and external coordination between space providers and the relief community; it provides training and support, conducts research and development projects and supports the international charter on space and major disasters

One of the key elements of the process of linking space technology and the relief community is mutual education of needs, requirements and to carefully communicate the potential and the limits of space based technology. It often occurs that the capacities and capabilities of space systems are highly overestimated and it is often difficult to overcome resentment, once expectations could not be met. Thus, it is of primary importance to keep up an open the dialog to ensure that the geospatial sector does not oversell its capabilities and capacities to the relief community.

5 Supporting the "International Charter on Space and Major Disasters"

DLR strongly supports the "International Charter on Space and Major Disasters" through the DFD-ZKI. The Charter is a major cooperative activity among international space agencies in the context of natural and man-made disasters. It is based on a frame work agreement between half a dozen space agencies to provide, in case of major natural or man made disasters, recent or archived satellite imagery in an informal way to authorized users free of charge. This Charter can be triggered through the member states, the UN or the European Union and it operates at 24/7 and best effort basis. After an initial phase, each activation of the Charter is coordinated by a so called project manager, who interfaces between the authorized user and the involved space agencies. DLR has committed itself to support the Charter through such project manager (PM) work and has coordinated and supported several Charter activations in Germany, Europe and world wide. The role of the PM is to translate the user needs and the given disaster situation into appropriate satellite commanding, archive retrieval, image analysis and mapping. Whereas the Charter formally only commits to provide raw satellite imagery it is commonly agreed that the end users can only make use of satellite imagery acquired from optical or radar satellites if they are transferred into satellite maps or thematic interpretations. Thus, the PM is also heavily involved in image interpretation and map production at his premises or mandates the tasks to third parties. DLR has set up a tight in-house network of image analysis experts from all thematic areas of Earth Observation, which can be consulted to support image interpretation at very short notice.

Since the flooding of the river Elbe the DFD-ZKI has supported a number of Charter activations as project manager; two examples will be illustrated in the following. On August 5, the Portuguese fire brigade requested activation of the Charter and DFD-ZKI took over the project management at the request of ESA and coordinated data distribution and analysis. In addition to the satellite data already provided by DFD-ZKI and its partners from NOAA-AVHRR, MODIS and DLR's own BIRD satellite (Fig. 7), other information was made available for evaluating the situation and for crisis management. During the 18 days of the Charter activation 52 satellite images were delivered as georectified data sets. NOAA-AVHRR fire masks were provided daily in the morning, only a few hours after the acquisition.

On December 27, 2003 the area of Bam in the Kerman Province located in south east Iran was stuck by a severe earth quake in the early morning hours. Supporting the international relief activities undertaken by various humanitarian organizations in the area around the city of Bam and supporting the activation of the Charter, the first damage analysis maps based on IKONOS satellite imagery were made available. The satellite image showed extreme damages resulting from the earthquake. In some extremely affected areas, whole quarters of buildings were leveled (Fig. 8).

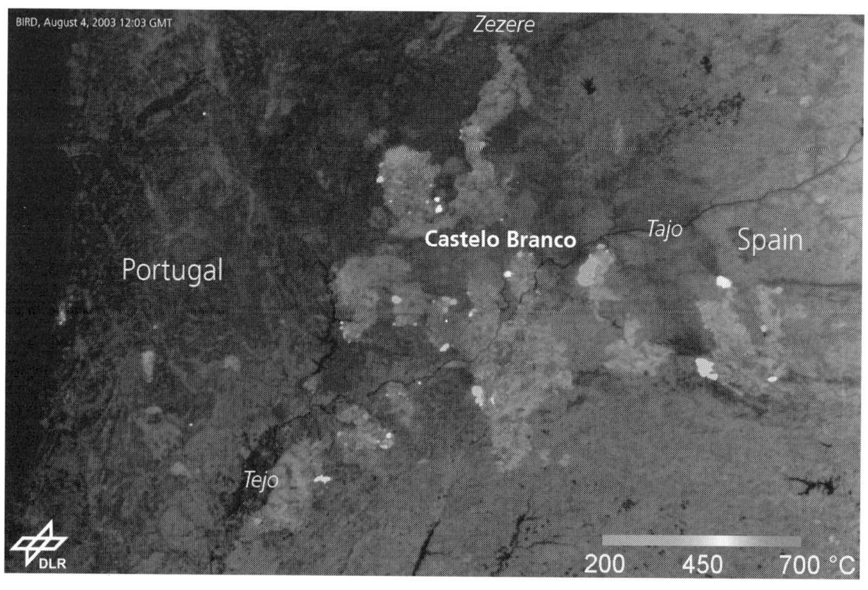

Fig. 7. Forest fires near Castelo Branco, Central Portugal, mapped by BIRD on August 4, 2003 (temperature release indicates recent fires, burned areas are shown in light grey)

Fig. 8. Intact (left) and damaged (right) infrastructure after the Earthquake in Bam, South-Iran, mapped by IKONOS on December 27, 2003

6 The DFD-ZKI Service Cycle

In order to serve the full cycle from an emergency call or request for assistance through satellite tasking, data acquisition, analysis, map provision and interpretation staff at DFD-ZKI has to go through a long chain of steps involving coordination of satellite commanding and data reception tasks as well as data ingestion, pre-processing, correction and analysis (Fig. 9). Later on, the image data have to be interpreted in order to extract the required information and generate an appropriate mapping or reporting product. For this purpose algorithms for hot spot detection, vegetation analysis, generation of digital elevation models or for feature extraction are of great support for the image analyst. In many cases, however, these are not accurate or reliable enough for automated information extraction and thus also still a lot of visual interpretation and image fusion skills are required to generate fast and reliable information products and maps. Sometimes even maps or images are not the proper way of transporting the derived information. In these cases, reports or statistics are used to even better aggregate and communicate the satellite information, such as estimated number of affected people, area of land flooded, most severely hit areas, etc. During the past years it has also been shown, that training and consulting of the decision makers and field workers in the ministries, NGOs and other relief agencies plays a key role in proper understanding and accepting the space based information products as one information source for decision making or mission planning.

In many cases it is not enough to go through this service cycle once. Often, especially in very complex, evolving or unclear crisis situations it is necessary to re-iterate parts of the process. If, for example a large forest fire situation (e.g. Portugal 2003) evolves strongly over time, first coarse satellite observation, e.g. NOAA AVHRR, may be used to identify the hot spots and most severe fire concentrations at a coarse cartographic scale. At a later stage on-command high resolution satellites such as the IKONOS system are tasked for a damage assessment in the most affected municipalities at a scale of 1:10.000 or better.

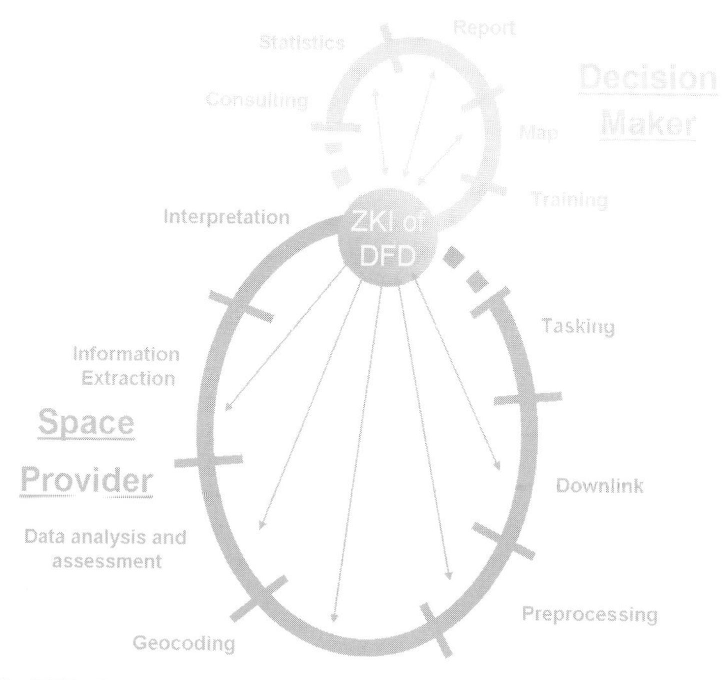

Fig. 9. The DFD-ZKI service cycle

7 Conclusion

With the given examples it could be shown, that earth observation can successfully provide a beneficial support of disaster management, humanitarian relief and civil security. It was shown how satellite imagery can be acquired, assessed, processed and turned into information products for decision makers within hours. It was also discussed that research and operationalization gaps exist between current state of the art automated feature extraction or automated map generation on the one hand side and real world, near real time requirements in actual disasters events on the other. Space technology today is still very complex and the different satellite systems require sophisticated processing techniques, which can not be handled by an individual relief organization for example. As a consequence it is of primary importance that the space technology and geo information sector provides easy to use and ready to access information solutions to the relief community. Care has to be taken at all instances that overselling of capabilities has to be avoided by all means, in order to build reliability and credibility in the suggested high-tech solutions for often low-tech problems.

Due to the complexity not only of space technology but also of the geospatial information sector as a whole, efficient cooperation networks have to be established along the full data acquisition, analysis and provision chains in order to set up meaningful and acceptable geoinformation solutions for the relief sector.

Survey Methodologies for the Preservation of Cultural Heritage Sites

Eros Agosto, Paolo Ardissone and Fulvio Rinaudo

Department of Georesources and Land, Politecnico di Torino,
10129 Turin, Italy.
Email: eros.agosto@polito.it; paolo.ardissone@polito.it;
fulvio.rinaudo@polito.it

Abstract

In the Italian landscape there are a lot of Cultural Heritage sites, thirty seven of which are parts of the Unesco world heritage list. It means that frequently the disaster management has to deal with Cultural Heritage problems, such as preservation, safeguard, reconstruction, etc. Preserving Cultural Heritage for future generations is a duty; the importance of this task is proved by the attention which important international organizations have towards it (UNESCO, ICOMOS, CIPA).

It is possible to conceive two different risk typologies for Cultural Heritage runs. The first one, which could be called "ordinary" is the normal (inevitable?) decay due to age, so a low traumatic but continuous action; the second one refers to isolated impulsive events (earthquakes, fires, floods…). The correct approach to face these two kinds of risk is to get an accurate documentation of objects, buildings and sites. In fact, a whole knowledge of the object is able to lead both the common planned maintenance, and possible extraordinary restorations.

To correctly document an object, it would be important to have a 3D realistic model; its exploration could let the user achieve differently detailed data (both shape and radiometric), depending on the level of inquiry. Nowadays, the preferred way to achieve this goal passes through an integrate use of different survey technologies; in addiction a unique site reference system should be adopted. As a result of such choice, it would be possible to integrate all the surveys made at different epochs; moreover, alphanumeric information coming from different kind of studies could be associated to this precise geometric base in a GIS environment. This kind of tool is important for the management, registration, maintenance, and updating of the data; besides it makes consultation easier and offers the chance to join data for interdisciplinary analysis.

This article focuses on the integrate use of digital photogrammetry and LIDAR, which in recent years were involved in a deep technology progress; an interesting recently conceived tool, that can be easily created by adopting these two techniques, is the so called *solid image*. It lets the user access and manage 3D data through simply viewing a 2D monoscopic image; it adds correct 3D metric information to simple photos, so that information is much easier to be gained also by people who are not survey experts. The results of some test applications are exposed. The tests were carried out on some decoration stones in the Guarini Chapel, in Turin, where the Holy Shroud was held before the tragic fire in 1997. Another set of tests was carried out in cooperation with the Cultural Heritage Safeguard Office of Valle d'Aosta (northern region of Italy) and centred around the creation of a GIS for the management of the archaeological data on the walls of the important and well preserved medieval Castle of Graines.

Fig. 1. Guarini Chapel Fire in 1997

1 Involved Technologies

Photogrammetry and LIDAR applications have a great spread, mainly because of the facilitations brought by their technological progress; furthermore their integrate use, makes it possible to easily create complete and accurate survey products.

Photogrammetry is a well known and consolidated technique; with the development of computer and electronics, it became "digital", both on the on site operations side, and on the office operation one (digital plotters, cameras, etc.).

As a result, photogrammetry became a very quick technique: *in situ,* there is thc chancc to immediately view the taken photos, and see if they are as planned; it is possible to immediately pass to *in office* operations, because there is no waste of time due to development and digitalisation process; radiometry can be adjusted via software, and digital images have no problems of long-term deformations; moreover the internal orientation process is automatic, and complex computations are possible.

LIDAR (terrestrial laser scanner) is a relatively new technology; many efforts from both the University and the both industry side have been made to test its applications in different contexts. A laser scanning machine can be considered as a high automation reflectorless total station; by means of a laser based measurement of distance and accurate angular movement, a target object is sampled in a regular mesh of 3D points.

One of the most interesting applications of such an instrument is the fast and economic way to create DDSM (Dense Digital Surface Model); by using other techniques (e.g. total stations, photogrammetry) it would be an incredibly time-consuming process. Besides, most of laser scanner machines can nowadays have a digital camera mounted on them: in this way it is possible to immediately assign to each point a RGB information. In the near future, then, it will be possible to have these machines equipped with a GPS antenna: this would allow to insert survey results almost immediately into a cartographic system.

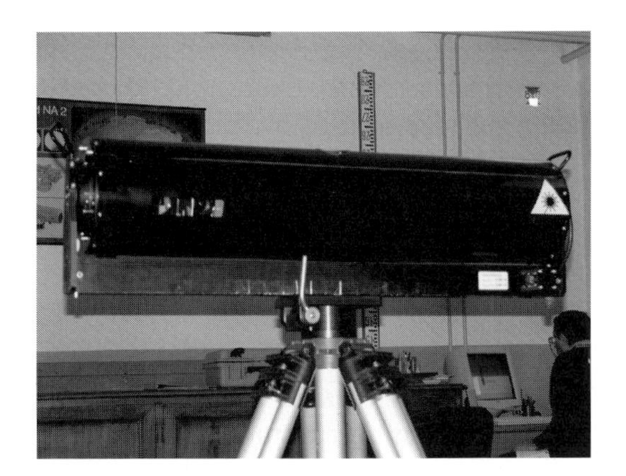
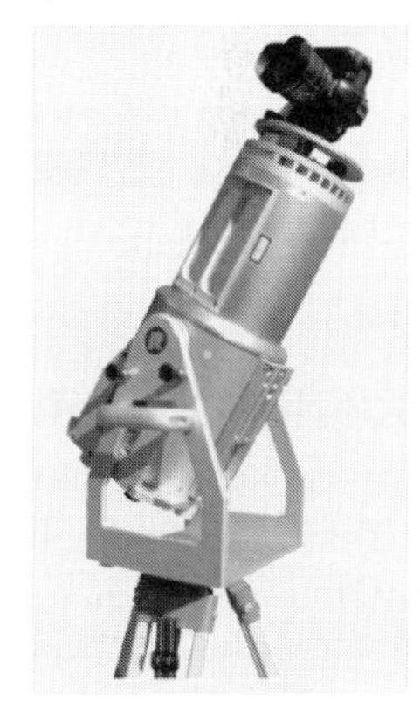

Fig. 2. Mensi S-10 and Riegl LMS Z420i

It is however important to insert all survey results into a GIS environment; this makes it possible to access all the information about a site by a unique tool. The GIS project is essential in integrating and structuring all the information about a site, both geometric, and alphanumeric. From the geometric side, data can be of different types.

Vector elements (e.g. CAD format), can come from a photogrammetric process or be the results of manual or instrumental surveys (architectural or topographic); the main aspect is the fact that this kind of information is punctual, referred to specified elements of interest.

Raster data, can have the same informative contents of the previous ones (e.g. cartography); besides they can be an image of a specific site characteristic: e.g. from a LIDAR survey, it is possible to get an elevation image of the site (DDSM). Moreover, it is possible to have metrically correct realistic views: orthophotos rectify photographic images, so that it is possible to make measurements on them, but their radiometry decays because of the resampling process. In this way the interpretative contents of this product is less than the original photos' one.

Some products, like 3D surface models and solid image, are able to augment the understanding of a site; they can't be directly managed in a GIS environment, but can be easily accessed through it. The first ones are able to give a whole surface understanding (e.g. VRML models), but users are not yet used to them, still preferring traditional 2D supports; they come from 3D point clouds, acquired by LIDAR, which have been elaborated by specific software. The second ones turn the philosophy of orthophotos upside down: they have the original radiometric contents unaltered and they add 3D information to it, by reprojecting the DDSM. In this way it is possible, through a common photo looking 2D interface, to have the 3D coordinates of every points in the image, and to use a set of predefined tools to make analysis and measurements. This product can be created using a specific software called LSR2004® developed by the Politecnico di Torino geomatics research team, whereas it can also be used by means of common commercial software such as Adobe Photoshop®, using a free plug-in.

2 Guarini Chapel Survey Test

The first test was held in the Guarini Chapel. The Politecnico di Torino geomatics research team was involved in integrating the geometry base that had to be the GIS project framework: in fact restorers filed to find a satisfactory documentation, in particular regarding some details of the internal decoration. This meant a not reliable reference model could be recreated. Also because of the fact that the statics of the building were involved, a tool able to let restorers analyse the object in safe place was needed.

The team tried to complete this shape model by an integrate use of digital photogrammetry and very high precision laser scanner.

In reconstructing the shape model of some decoration stones, a high accuracy was necessary; all the fractures of the stone had to be documented. The survey context and the small room available on the scaffolds, practically determined the survey configuration.

Because of the practical impossibility of connecting to the vertexes of the global reference network, local reference systems, located by their position in the scaffolds, were adopted. Furthermore, to speed up the photogrammetric in office operations, the Cyclop® system was used; it is made up of a calibrated bar, where a camera (in the specific case Nikon D1x) can move to different prefixed positions, and a specific software to immediately create the stereoscopic model and start the restitution. During the image acquisition, photographic lamps were used to illuminate the object without altering its radiometry.

Fig. 3. Cyclop bar with D1x camera

The object was then scanned by a high precision triangulator LIDAR, a Mensi S-10, able to measure distances with an accuracy of about 0.1 mm. Different scan resolution were used trying to balance the time needed and the dimension of the scanning grid. A good compromise was found using a step of 1.5 mm at an average distance of 2 meters: a couple of hours was needed to survey an approximate 2*1 m^2 object; some problems due to the low temperature of the environment were encountered at the startof the laser operation. Digital images were oriented on the DDSM created using the LIDAR technique and *solid image* of the objects were generated. Focusing the survey process on the creation of the *solid image*, it made it possible to store a great deal of information for possible future deepening in the study process:

- Point clouds; using them it was possible to get surface models, that could also be used to create prototypes for the restoration;

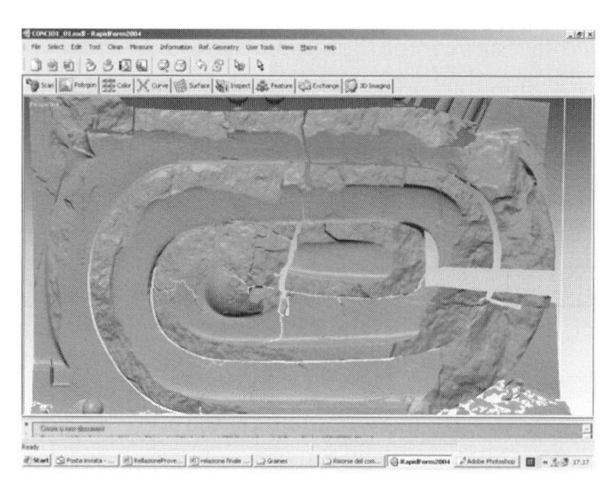

Fig. 4. A 3D surface model of Guarini Chapel decoration stone

- digital photos coming from the use of Cyclop; at any point in time, it is possible to create a stereoscopic model and to return elements of interests, even only in some specific points of the surveyed object.

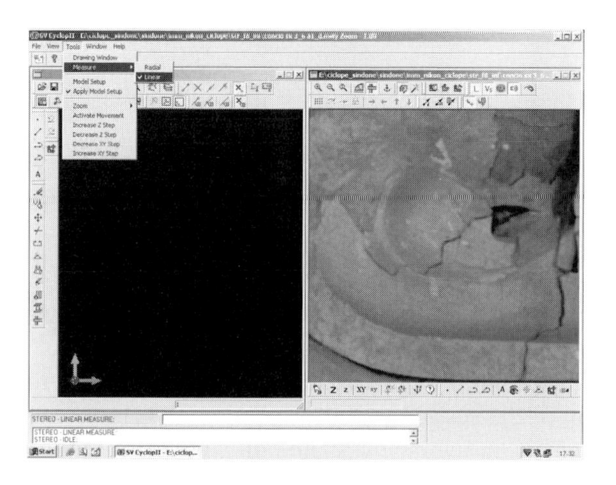

Fig. 5. Cyclop restitution window

Creating surveys products after a disaster happened made more evident the importance of doing such a kind of operation in advance; in this way restorers could have a precise documentation of the object as it was before the event, a model to guide their work.

3 Graines Castle Survey Test

The geomatics research team of Politecnico di Torino designed a GIS for the collection of data coming from many disciplines (geology, history, archaeology), about a territorial census of Valle d'Ayas archaeological sites.

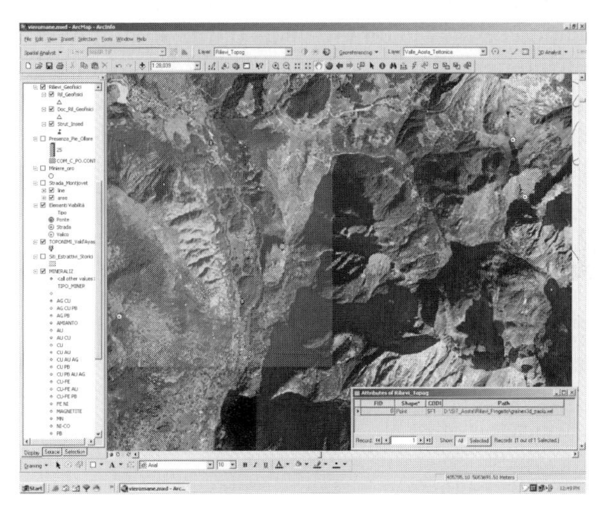

Fig. 6. Graines Caste in the Archeological census GIS tool

A kind of data stored in the tool were high scale surveys: they were located by a medium scale cartography (1:10.000). In this way it is possible to achieve an overview of the whole area, from which users can reach detailed survey data of a specific site using hyperlinks.

A survey was carried out on a well-preserved medieval castle (Graines, in Aosta Valley); this site was not hit by any particular disaster, except the usual decay due to age. The attention was focused on a part of the boundary wall, particularly interesting from an archaeological point of view. In this case there was less need for precision, but a longer acquisition range was required: different kind of instruments were therefore used. The main difference was the use of a Riegl LMS-Z420i "time of flight" laser scanner: this meant the chance to also record information about the reflectivity of the materials. This machine, then, was equipped with a digital camera: besides the photo taken in a normal photogrammetric work, a greater number of images were available. In this test it was interesting to try the use of photos coming from the different cameras: in both case it is possible to get *solid image* but the use of the camera mounted on the laser scanner machine, makes it possible to immediately have the external orientation parameters of the photos. In this way there is no need for a time consuming orientation process; the main disadvantage is to be bound by the chosen scan positions: to sample the object in an accurate way, a proper distance has to be adopted in the taking, and this can limit the field of view of the photos. As 3D supports are still not popular among users, orthophotos were also used as a familiar base for the GIS. This in fact was useful for the stratigraphic analysis of the boundary wall made by archaeological experts: the creation of the vector elements of the stratigraphic units could be done and alphanumeric data could be associated to them.

Solid image can be completely integrated in GIS environments. By an hyperlink, it is possible to open a *solid image* by selecting an object this tool refers to, and vice versa, it is possible to insert vector points in the project by recognizing them in the image (Fig. 7). This was great potential compared to photogrammetry: there is no intermediation in choosing which elements of interest have to be returned.

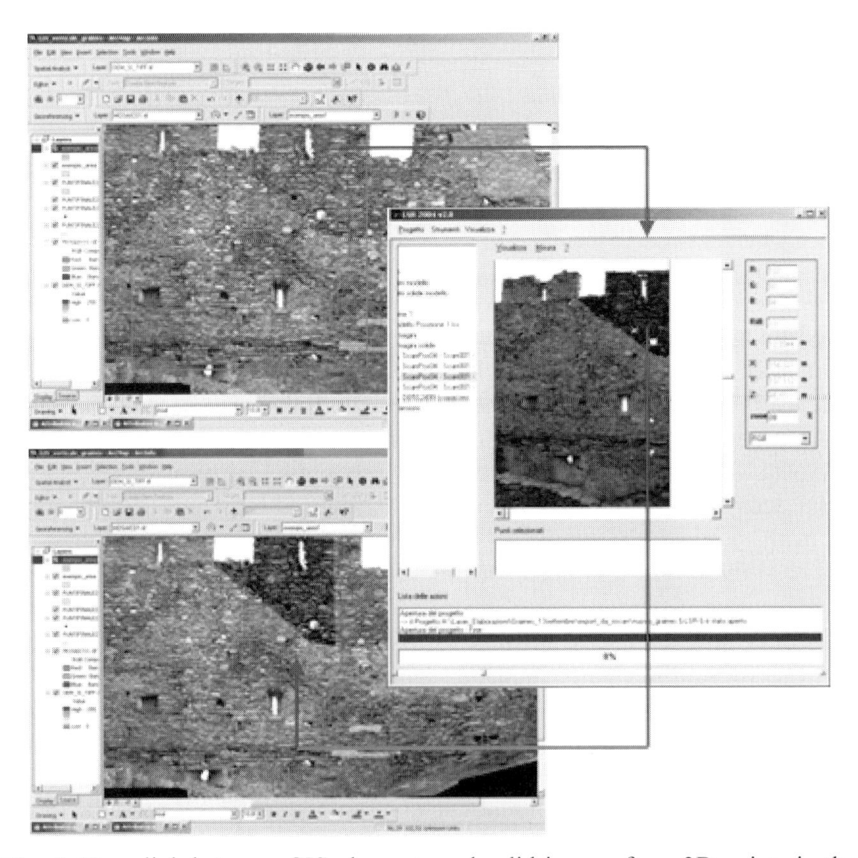

Fig. 7. Hyperlink between GIS elements and solid image; from 3D points in the solid image it is possible to create vector elements in the GIS.

4 Conclusion

Correct documentation is the proper way to deal with Cultural Heritage sites: the building up of a GIS, makes it possible to store all the study results and information, which can lead an ordinary or extraordinary restoration action. The choice of a unique reference system and the integrate use of modern survey techniques makes it possible to have an accurate geometric base, built with the best benefits/cost ratio. The study cases showed the potential of the *solid image:* this product could be used in many different situations in the near future. This is due in particular to three reasons: the automation of the production process, its user friendly approach (in fact there is 3D information in a 2D environment, easier to manage), and the completeness of the 3D information joined to an RGB image.

The creation of this tool is highly automated; it makes use in the best way of the potential of advanced technologies available today, like digital photogrammetry and LIDAR technique. It speeds up the creation of a final result, using which non technician people can select interesting elements and eventually can decide to deepen the study of the object by creating stereoscopic models where necessary, or 3D surface models.

References

Riegl J, Studnicka N, Ullrich A (2003) Merging and processing of laser scan data and high-resolution digital images acquired with a hybrid 3D laser sensor, Proceedings of CIPA XIX international Symposium, Commission V, WG5, Antalya Boehler W, Bordas Vicent M, Hanke K, Marbs A (2003) Documentation of German Emperor Maximilian I's Tomb, Proceedings of CIPA XIX international Symposium, Commission V, WG5, Antalya

Bornaz L, Rinaudo F (2004) Terrestrial laser scanner data processing - XXth ISPRS Congress Istanbul – July

Bornaz L, Dequal S (2004) The solid image: an easy and complete way to describe 3D objects - XXth ISPRS Congress Istanbul – July

Menci L, Rinaudo F (2001) SV Cyclop: a new instrument for close-range photogrammetry, CIPA International Symposium, Postdam

Biasion A, Bornaz L, Rinaudo F (2005) Laser scanning applications on disaster management, The First International Symposium on Geo-information for Disaster Management, Delft

A New Geo-Information Architecture for Risk Management

Christian Alegre, Hugues Sassier, Stephane Pierotti and Pascal Lazaridis

ALCATEL SPACE, 100 Bd. du Midi ,BP 99, 06156 Cannes La Bocca Cedex, France.
Email : win.gmes@space.alcatel.fr

Abstract

1 Introduction

The current document briefly reviews the risk and disaster management issues to identify the major issues to be dealt with, then it defines WIN project contribution to these issues.

2 Problem Statement

Risk and disaster management include various activities performed all along the disaster management cycle:

- risk prevention and risk mitigation activities,
- response phase activities,
- post-disaster and reconstruction activities.

The various activities are performed by a large set of actors at regional, national, and European levels; these actors play different roles and thus have complementary point of views on the problem ; some may perform both real time tasks in the response phase, and non real time tasks for other phases.

Studies like BICEPS have already pointed out the necessity to interconnect the different data and actors.

The definition, the set-up and the deployment of a disaster management environment to support some critical part of disaster management is a challenging action, that faces a problem with complex components:

- the various communities involved in disaster management have their own organization and there is few or no overall links between them;
- every thematic field involves some specific actors and practices,
- the European context implies work performed in various mother-tongue languages,
- there is a extremely wide range of data generated from multiple sources (including Space and/or airborne earth observation and in-situ data) but there is no global catalogues or easy entry-point on which the activities can be based,
- there is an extremely wide range of useful services (like monitoring, modeling, analysis, etc.), but currently few action of integration of these services,
- even if there are on-going standardization actions, these actions are not completed, so there is a wide multiplicity of format and tools involved in the process,
- there is a need for a wide open environment, but there is also a need to protect the data and the access on these data.

The "challenge" to be faced is not limited to provide some useful information but:

- to be in position to provide qualified information or services,
- to provide every actors with global view on relevant data and information, and easy access to these relevant sources of data/information,
- to provide actors with reliable co-operative working environment allowing to minimize usage of operational flows like faxes or phone call and replace them by electronic multimedia exchanges,
- to allow definition by actors themselves of work-flow that it is possible to automate, to act along well-defined process with defined actions.
- to circulate information through computers network with secure exchanges instead of telephones calls and other paper-like exchanges,
- there is also place for combination of Space data with in situ data in order to get strong advantages resulting from both sources.

To perform a qualitative progress with respect to current situation, it is needed to work in parallel on several axes:

- Multi-lingual terminology : to clarify and harmonize the terminology all along the risk management cycle, to enlarge the scope of the terminology tasks to the multilingual context.
- Modeling and Automation of existing processes : need to analyze with relevant risk management actors their processes, to derive generic proc-

ess models and to provide communities with means to set-up their own automated work-flows.

- Data and Information:
  - to help actors to easily access multiple data-source and formats in heterogeneous environment taking into account European geo-information standards (INSPIRE), with a potential benefit on data management and on further value added data processing,
  - to support combination of EO and in situ data , thus breaking through the limitations of each kind of data , with a potential benefit on new value added data processing and optimization of sensors budget,
  - to analyze the needs for information display and provide relevant generic services, with a potential benefit on easier operations for users.
- Networking of actors : there is a need to facilitate co-operative working between actors and circulation of information, but with secure access controls and secure exchanges, with a combination of benefits on efficiency of operations and reliability.

3 Project Summary

Wide Information Network (WIN) Integrated Project has started in September 2004 as part of the GMES FP6 European framework. WIN has the objective of integrating all existing reference results or initiatives to contribute to the design, development, and validation of a "European Risk Management information infrastructure ". WIN will be a major element of the future overall ESDI and as such represents an important innovative component with potential great benefits in the European picture of Operational Disaster Management.

The main issues tackled in WIN relate to:

- The definition of a data/information model valid for several thematic issues of risk management in Europe.
- The architecture of the info-structure optimized in terms of use of state-of-the-art information technologies (Web, Grid,…), and high capability to inter-operate data, services, and risk management actors.
- The coverage of business and organizational aspects through the sub-network and charters concepts.

The overall WIN duration is 36 months. WIN project is first focused on the identification of user requirements and of technical requirements, and on the prototyping of the potential solution ; in the thematic field analysis, performed through use cases, emphasis is put on the identification of needs that are "generic" or transverse to various thematic fields; the goal is to derive a generic services architecture , relevant for several thematic fields; for each thematic field, the info-structure can be reconfigured based on specific field models and some customization, and specific complementary services required for the thematic field are connected to the generic services platform.

Three main milestones allow to validate the progress at the end of the 3 years of the project:

- The Preliminary Design Review performed at T0+12 months allows to check both the user/system requirements and the preliminary design of the information infrastructure ; the preliminary design is checked on the basis of architecture document and through demonstration of mock-ups and prototypes, illustrating the critical points of the architecture.
- The Critical Design Review performed at T0+24 months allows to check the detailed design the progress of the system integration, and the definition of WIN experiments and evaluations.
- The Final Review allows to get lessons learned from WIN experiments and evaluations, to check final version of WIN deployment plan, including practical aspects like ready-to-deploy kits to be delivered to the different classes of WIN users.

Following points considered as factors for WIN success are constantly taken into account as major principles of WIN project:

- Involvement of key stakeholders (national and European agencies, service providers, final users,...) and operational actors to develop a shared vision of organizational model and implementation strategy.
- Synergies and co-ordination with complementary initiatives on operational services development, operational risk management platforms design and development in order to find the best convergence between both projects for the future Risk Management architecture.
- Program strategy taking into account reference GMES initiatives: IST or DG Research Integrated Projects on risk management as well as GMES Services Elements projects of the European Space Agency will be considered in the phase of WIN requirements specification and possibly also as contexts for actual WIN results dissemination.
- It is proposed to have a full synergy between WIN project and the Marine and Coastal Environment Information Services implementation Project. This project will focus, as current ESA ROSES GSE project, on ocean water pollution. The oil spill part of the proposed service portfolio is very much representative of risk management domain.

4 Thematic Analysis

The European Union suffers regularly from major disasters. Examples in previous decades include earthquakes, floods, landslides, forest fires in southern Europe, environmental emergencies etc. The analysis of lessons learned reports such as Prestige disaster, Floods in south of France, activation of the International Charter Space and Major Disasters..., stresses the necessity to improve the exchanges between actors at local, national and European level.

To address this complex issue we needs first to identify and understand the actors segmentation in order to derive as much as possible common requirements.

The main goal is to derive actors requirements through use-cases analysis such as oil spill pollution (with CLS), flood and forest fire (with TELESPAZIO and CNES).

The classification of actors according to their common missions versus risk phases and activities facilitates the requirement analysis and the validation process. The following actor segments have been identified:

- Data/Service Providers,
- Policy/Decision Maker/Support International/EU,
- Policy/Decision Maker/Support National,
- Policy/Decision Maker Local,
- Scientific/technical support.
- On scene Operators

The actors exchanges issue is mainly due to the wide diversity of actors, procedures and existing systems used in the different European countries. First of all WIN will consider the following generic requirements:

- Access for specific geographical area and risk to overall data repositories whatever sources and format.
- Access to common services and tools capable to establish end-to-end information flows in an as much as possible automatic manner (such as electronic data request forms, workflow tool to follow-up the activity progress, delivery services...).
- Set-up a disaster management process capable to deliver value added Earth Observation based information products whatever the type of disaster, the geographical coverage, and ensuring a high quality service (on duty operators, electronic exchanges, data formats compatible with GIS...)
- Communicate between different actors communities using multi-lingual reference systems (risk terminology, common geo-information databases, ...).

- Establish real-time discussions between remote and distant actors to exchange voice, data and video, facilitating the expertise process and information exchanges.
- Archive the overall information exchanged to build-up lesson learned databases and perform user training.

Even though some actors are systematically involved in all types of risks, some requirements are specific when a disaster occurs. To cover such requirements, WIN will address the overall oil spill thematic dealing with trans-border, terrestrial and marine & coastal issues, and requesting for space, in-situ and modeling information networks. The fire and flood thematic will be addressed with a focus on the use of products and services derived from satellite based images.

5 Geo-Information Infrastructure

This information structure will be a major element of the future European Spatial Data Infrastructure (ESDI) and represents an important an innovative step in operational risk management in Europe.

This task is done in co-ordination with other European projects and in line with INSPIRE initiative. Co-ordination with these other activities will assist the convergence towards a common GMES architecture for risk management.

- Some of the WIN goals are:
  To design an open service-oriented architecture for risk management in liaison with other EU and GSE projects.
- To develop an implementation of the architecture.
- To propose an organizational model suitable for risk management deployment.
- To develop generic services that are relevant for various risk management applications.
- To validate the WIN results in real-life scenario. One of the principles to achieve these goals is to obtain the involvement of key stakeholders and operational actors to develop a shared vision of an organizational model and implementation strategy. Periodical work- shops will be held with them to discuss an efficient organizational model.

A Service Oriented Architecture (SOA) is proposed and is based on :

- Open standards for maximum interoperability,
- Contains a set of technical components,
- Web services connector toolkit enabling flexible integration and data exchange between actors,

5.1 Service Description

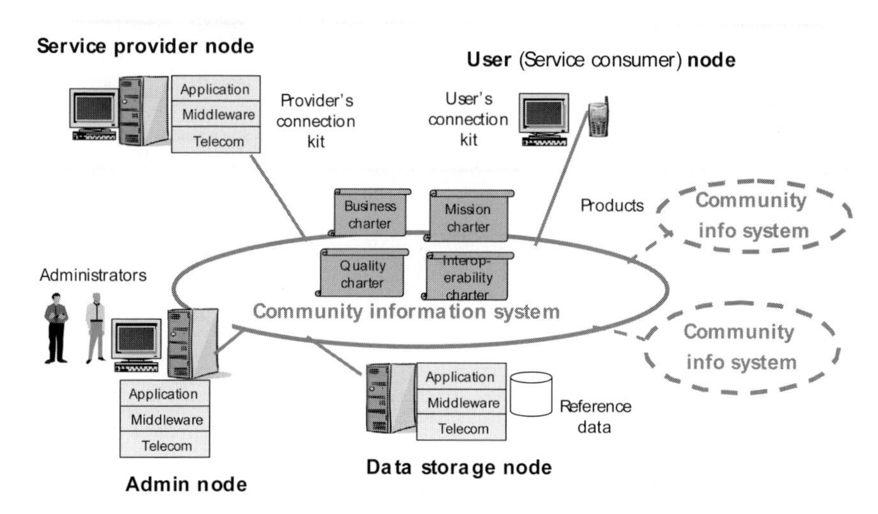

Fig. 1.: Service description

Some of the WIN services characteristics:

- WIN is based on an open architecture allowing to structure the information system on which will be plugged the applications and services deployed for the Risk management. Integrating a new service is facilitated by using of latest technologies as XML, Web services.
- WIN organizes the data and allow all the risk actors involved (end users, operational actors, value added data and services providers) to get the reliable pertinent and certified information in the shortest time, and this, according to the users profiles.
- WIN offers architecture based on sub networks which means in communities of actors dealing with the same kind of data and services which will then benefit from common resources.

These sub networks are ruled by charters: mission, business, quality, and interoperability charters which are specific to each community. This efficient, reliable and flexible organization is the way to limit the complexity and optimize the info structure administration (local ad-ministration for the sub network and global administration for the generic aspects).

- WIN offers generic services: shared GIS, data fusion, workflow management, data mining, multilingual interoperability, collaborative work.
- These generic services are structured in a multi-layer software framework.

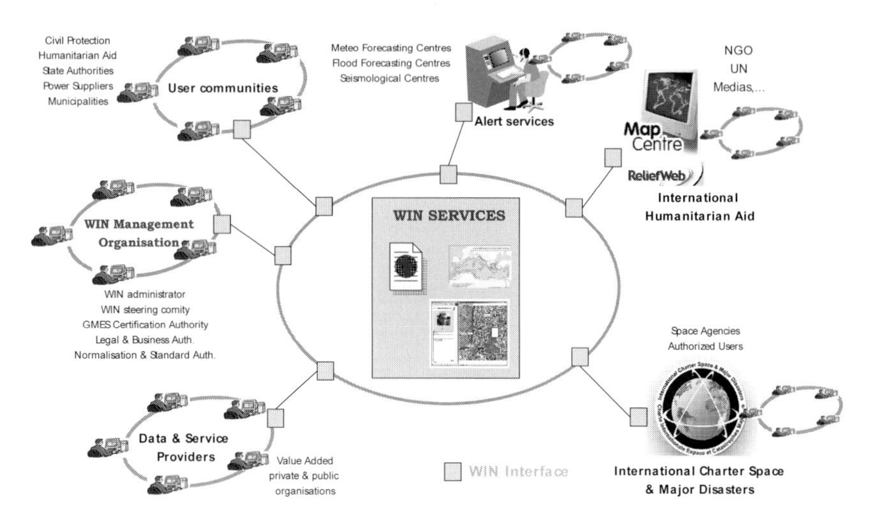

Fig. 2. Example of WIN Community Information System deployment for Risk management

Software to Support a Specialized Bank of Electronic Maps

Rafael Arutyunyan, Alexandr Glushko, Vladimir Egorkin, Vladimir Kiselev, Daniil Tokarchuk and Nikolay Semin

Nuclear Safety Institute of Russian Academy of Sciences (IBRAE RAS), 52, B.Tulskaya, Moscow, 115191, Russia.
Email: kis@ibrae.ac.ru

Abstract

For a number of years, IBRAE RAS has run the activities to form the bank of topographic and subject-matter electronic maps/plans of various scales.

Principal tasks of the IBRAE Bank of Electronic Maps are formulated as follows:

- Development of a cartographic basis to support specialized GIS, information-reference systems, information-modeling and training systems developed at IBRAE;
- Cartographic support of works on analysis of environmental impacts of facilities of FAEA (former Minatom) and the whole Fuel and Power Complex (FPC) on personnel, population and environment;
- On-line cartographic support of IBRAE activities in the areas of ensuring safe operation of radiation and radiation-hazardous facilities in Russia and analysis of implications of potential radiation incidents.

Based on original software and utilization of MAPINFO GIS opportunities, a number of specialized information and prognostic geoinformation technology-based systems has been created Special software modules also are designed for the following - organize an enquiry to retrieve and look through a required map, navigations all over Russia with automatic loading of required electronic maps; access from the map window to information and enquiry data on radiation hazardous objects of FAAE of Russia, provide on electronic maps the results of visualization of radiation monitoring data.

1 Introduction

Since 1990 work has been performed at IBRAE RAS [1] to generate a bank of digital maps. At the first phase the main objective of the bank of digital maps consisted in supporting the IBRAE's activities under the program of analysis of the long-term Chernobyl accident consequences. Later on, as the scope of tasks to be addressed by IBRAE broadened, the range of gazetteer (digital map nomenclature) and their storage formats extended further.

The main actual tasks of the Digital Map Bank (DMB-IBRAE) may be summarized as follows:

- development of a cartographic basis to support specialized Geographical Information Systems (GIS), information-reference systems, information-simulating and training systems generated at IBRAE;
- cartographic support of works performed at IBRAE on analysis of the consequences of radioactive contamination due to accidents at radiation-hazardous facilities; and
- on-line cartographic support of IBRAE RAS's activities in the areas of ensuring safe operation of radiation and radiation-hazardous facilities in Russia and analysis of implications of potential radiation incidents.

Most of vector digital maps and charts stored at DMB-IBRAE were generated at other organizations using different cartographic classifications of layers, objects and parameters that hinders their export to work formats of other GIS (e.g., to Spatial Cartridge BD ORACLE system). Quite often the use of different classifications also results in non-identical graphical representation of similar-type objects that complicates map reading by users. Moreover, to organize storage of digital maps at DMB-IBRAE and their subsequent use in specialized GIS developed at IBRAE, they should comply with some requirements on topology and object composition and be made on basis of a unified classification of layers, objects and parameters. Accordingly, every map layer includes graphic objects of only one geometric type (point, line or polygon).

In compliance with the above-said, work was performed on bringing all IBRAE's digital maps to a unified cartographic classification of objects and layers. When developing the classification, an object-and-layer classification developed at GosGISCenter on basis of object classification of 1:1.000.000 - 1:25.000 scale maps employed by Roscartographiya enterprises was used. To generate and store digital charts of enterprise sites and different thematic layers (e.g. radioactive and chemical contamination of territories, radiological monitoring systems, etc.), the unified classification was enlarged via introducing new objects, cartographic layers and parameters. In parallel with preparing industrial site charts and thematic layers, work on the classification enlargement is performed continuously via adding new specific layers and categories of objects.

2 DMB-IBRAE Structure and Functioning

When developing DMB-IBRAE, the 'MapInfo' system [2] was chosen as a tooling GIS. In our opinion, the 'MapInfo' system fully complies with the requirements of the present-day ideology of open-systems on computational procedures, operational systems, way of writing applications, openness of the format storing cartographic information, operation with external databases and communication potentialities. In addition, as distinct from other widely used commercial GIS, 'MapInfo' is a fully russified system down to application-development tools.

All electronic maps and charts are stored under a single geographical coordinate system 'latitude/longitude Pulkovo-1942' in the following two formats: inner 'MapInfo' format and 'Mif-Mid' exchange text format. Raster maps stored at DMB-IBRAE also have a coordinate reference and thus may be loaded by user in the form of a substrate of the relevant vector map or of an individual layer, if necessary.

To date the DMB-IBRAE comprises the following types of vector and raster maps and charts of different scales:

- a general World's map;
- general maps of Russia and the former USSR's republics;
- separated different-scale gazetteers (nomenclature map lists);
- maps of regions of location of Russian Nuclear Power Plants (NPPs) and other Russian radiation-and nuclear-hazardous enterprises;
- maps of neighborhoods of NPPs and main radiation- and nuclear-hazardous enterprises;
- charts of sites of NPPs and main radiation- and nuclear-hazardous enterprises;
- individual thematic layers (e.g., data on territory radiation and chemical contamination); and
- raster maps (referenced to geographical coordinates) of the regions of foreign NPP locations.

The DMB-IBRAE presently comprises over 1.000 different types of maps with ~10 GB integral volume.

All cartographic information is stored at an individual file-server accessible from local work places within IBRAE's network, or directly via 'MapInfo' system, or through 'MAPVIEW' software (IBRAE's development) installed at local work places. In the latter case users are provided with a friendly interface for searching and viewing the required cartographic information. Work place of the databank administrator is located at the same cartographic file-server and has special software modules for digital map preparing and export to DMB-IBRAE.

For example, a 1:500.000-scale map of the Bryansk region with a ^{137}Cs-contamination layer is demonstrated in Fig. 1.

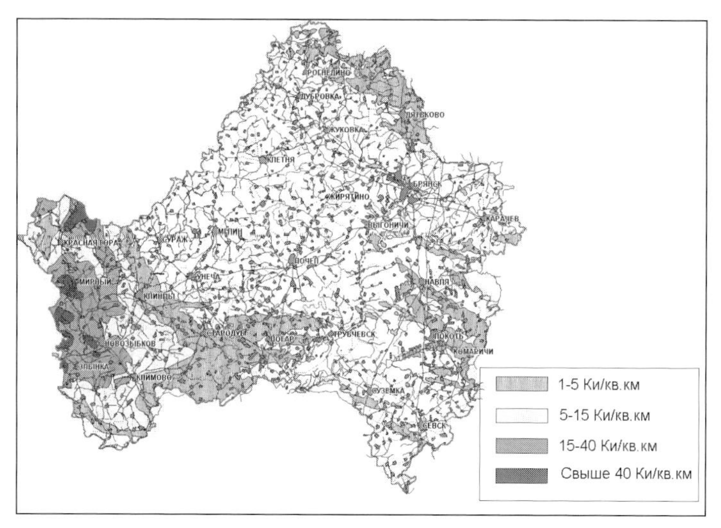

Fig. 1. A map of the Bryansk region contamination by ^{137}Cs

3 MAPVIEW Software to Work with DMB

Various specialized GIS-applications developed at IBRAE interact with DMB-IBRAE in a different way. Some of them operate with the DMB directly: in such a case they are most often integrated into work of 'MAPVIEW' software as one of its options. Other applications use the DMB only as a source of prepared maps to generate a map bank under their own inner formats or under 'Spatial Cartridge BD ORACLE' universal format.

Using the first procedure, software of the mobile complex to support work of radiation safety experts in field conditions (below "the mobile complex") has been developed.

A set of programs developed at IBRAE for the Situation Crisis Center of the Federal Agency for Atomic Energy (Rosatom) is functioning using the second procedure.

Below only the first option of DMB-IBRAE operation will be considered providing for full integration of the DMB into specialized GIS. As an example, let us consider operation of software of the mobile complex.

The whole set of operations related to the mobile complex, such as:

- displaying a required digital map;
- performing *in situ* gamma-background measurements and visualizing the measurement data on the relevant digital map;
- forecasting the radiation situation in the accident area with data displaying on a digital map;
- obtaining information on radiation-hazardous facilities; and
- DMB supporting,

is performed using 'MapView' software module to work with DMB.

'MapView' software module has been developed as a 'MapBasic' application using a large set of modern development facilities. The main active window of the software is demonstrated in Fig. 2.

This is a standard active window of 'MapInfo' GIS wherein a new tool bar - "Bank visualization" - has been added for purposes of work with DMB-IBRAE, the options 'Bank MapInfo', 'Trace' and 'RAMO' being added to the upper functional menu.

The "Bank visualization" tool bar allows rapid finding of the needed map and its displaying either as a standard generalized map, wherein one or another cartographic layers are visualized depending on the scale, or in the form of the so-called "contour maps". Quite often such contour maps are the most easy to use (e.g., when displaying the results of simulating the radiation situation or tracing itineraries). If the required map is lacking, one has the possibility of generating such a map virtually using individual gazetteers (nomenclature lists) that is especially important in case of "transport accidents" with unknown-in-advance location. Moreover, using a contour map one has a possibility of not only finding the needed map in the DMB, but also maintaining the DMB itself. In particular, there is a possibility of arbitrary arrangement of the user's thematic digital maps in the 'MapInfo' GIS format inside DMB rigid structure: note that such thematic maps may not fully comply with the requirements to DMB-IBRAE's maps, the only requirement being their opening as standard 'MapInfo' work sets. Thus one may generate one's own cartographic environment best complying with specific tasks to be addressed.

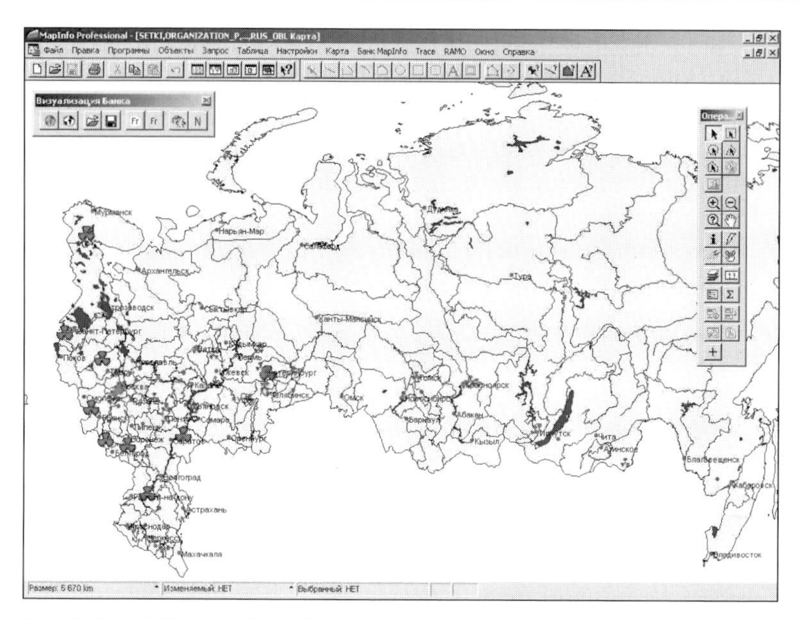

Fig. 2. 'MapView' active window operating with DMB of the mobile complex

However modularized map representation often leads to appreciable time expenditures when performing specific emergency-response activities. This is due to the fact that, depending on the task specificity at a given instant, one has to change many times the map load: for example, add individual layers from the "enterprise site" map to the "enterprise surroundings" map, hide/display individual object inscriptions, etc. Keeping of such map composition versions as "the user's maps" allows saving much time always lacking during emergency response activities.

An example of building "user's maps" of grounds and population evacuation itineraries in "Balakovo NPP surroundings' map" is depicted in Fig. 3.

If measurements of the radiation situation in the emergency area and the result displaying on a map are necessary, one needs (using the "Bank visualization" tool bar) displaying the required map and select 'RAMA' function in the upper functional menu of 'MapView' window. In such a way 'DBMG-200' sensor, connected earlier to computer port, will be activated, and a panel of control over "Gamma background" sensor will appear in the module's active window. The data of measurements will be displayed on line on a digital map and after work termination will be saved as an individual cartographic layer or a DBF-file.

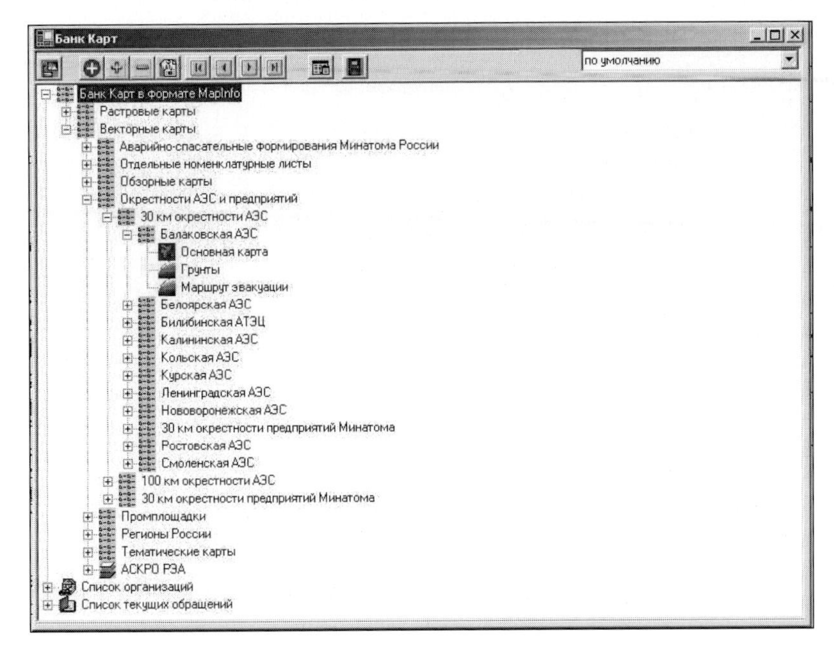

Fig. 3. An example of arranging "users' maps" in the bank structure

There is possibility to transfer a map with displayed data of radiation situation measurements directly to 'Trace_Mi' calculated-forecasting module designed to perform express-analysis of the radiation situation due to radionuclide release to the atmosphere. For this purpose one needs to move from the upper functional menu to the "radiation situation forecast" mode using "Trace" option without leaving the gamma-background measurement mode 'RAMO'. The map with data of measurements will be transferred to active window of the software module for radiation situation forecast, and one will be able to begin both simulation of the radiation situation in the emergency area and displaying of the results on this map. Simultaneously one may continue gamma-background measurements and data displaying on a map with the forecast result. This will make it possible to correct the forecast data in order to increase reliability of the radiation situation progression forecast in the affected area and thus to issue more adequate proposals on minimizing the accident consequences. An example of displaying the results of simulation for release spreading under a hypothetical accident with simultaneous visualization of *in situ* measurement results is demonstrated in Fig. 4.

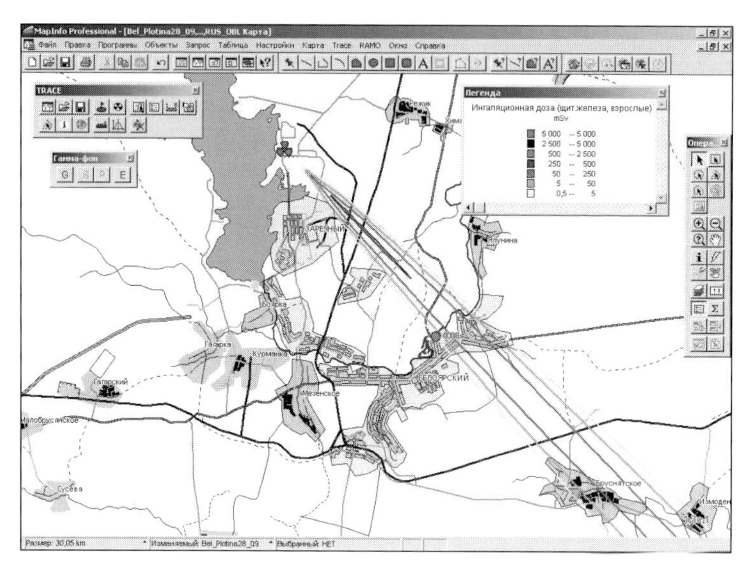

Fig. 4. An example of simultaneous displaying of forecast data and results of 'in situ' radiation measurements on a digital map

4 Conclusions

The developed at IBRAE software to support 'MapView' specialized banks of cartographic information forms a basis for the development of various GIS applications in the emergency-response area, the related financial expenditures and manpower being considerable reduced and the interfaces of generated GIS-applications being unified. In addition, the use of Mif/Mid open format to store cartographic information virtually eliminates the problem of data export into software products developed on basis of widely used commercial GISs.

References

Evdokimova ZA, Zhilina NI, Kiselev VP, Pechenova OI, Semin, NN, Tokarchuk DN, Tokarchuk AD, Yurchenko AS, Dynnik AY, Rogachev AV (2002) Establishing an IBRAE RAS's Bank of Digital Maps, IBRAE's Preprint #2002-04, Moscow, 60 pp (in Russian)

'MapInfo Professional 6.0' User Manual (2000), Esti-Map Publishers, (in Russian)

Project OCTAGON: Special UAVs - Autonomous Airborne Platforms

Simion Dascalu

European Business Innovation & Research Center S.A., Blvd. Ficusului 44A, Sector 1, 013975 Bucharest, Romania.
Email: Simion_Dascalu@Yahoo.com

Abstract

The main objective of this OCTAGON Project is to produce Special Autonomous UAVs to monitor the Low Altitude Electromagnetic Turbulence in the Earth' atmosphere. This kind of UAV will be also used in Special Operations such as the research of the low altitude atmosphere, dangerous atmospheric phenomena, hazard weather behavior, in local disaster aerial surveillance and monitoring, aerial investigations of volcanic activities or over any high contaminated area up to the use in space exploration, such as the exploration of the planet Mars.

The Low Altitude Electromagnetic Turbulence in the atmosphere of the Earth was discovered in 1996, and a complete scientific presentation of it was made since then to several international institutions and authorities, including NTSB in US, the Romanian Space Agency and the United Nations, the Office of Outer Space Affairs during the Workshop for the Disaster Management in Europe, at Brasov, Romania in 20th of May 2003.

1 Introduction

The main problem solved by this Autonomous UAV Project is to provide an unique low cost airborne platform available every where and always able for hovering at high altitude of which there is no aircraft available yet, using all equipment required to detect, monitor or to provide global and local geo-information regarding any disaster situation and very dangerous atmospheric phenomena.

This Special Autonomous UAV has new technical solutions such as the electrical propulsion system the main design is an octagon aerodynamic stealth profile with one wing including the fuselage and no vertical tail with all flight controls for the hovering mode provided only by the thrust vectoring of the 'Fan Power Rings' using a patented airfoil cross-section, or by the aerodynamic forces created during the horizontal flight by the position of the ailerons and flaps on the rear edges of the wings (initially a standard APU with two electric generators will provide the energy required, a model only with electrical propulsion system can be made for small size UAVs).

The UAV's structure has a geodesic design able to be replicated at any size maintaining the same cross-sections (the minimum size is 0.30 x 0.30 x 0.05 meters, with a total weight of 0.5 kg).

The main UAV project dimensions and basic operational characteristics are the followings:

- the wing span over all is 5.00 meters;
- the length over all is 5.00 meters;
- the height over all on wheels is about 2.00 meters;
- maximum take-off weight is estimated at 1200.00 kg;
- payload estimation is 250 kg minimum in addition to the fuel maximum weight of 350 kg;
- cruise speed of around 550 km/h (270 knots) and 2.5 hours a minimum operation mission;
- the operational altitude is unlimited due to the new electrical propulsion system.

The estimated costs of this UAV platform is dependent on its size and on the required equipment such as detectors, recording of video, data monitoring disaster development and communications. The UAV is estimated from 50,000 up to 6,500,000 EUR for the above prototype size, including that special detection and monitoring system for the Low Altitude Electromagnetic Turbulence.

This OCTAGON Project is under development now at the European Business Innovation & Research Center in Bucharest.

2 Scientific and Technical Objectives

The present project is focused to offer and to provide for the air transport industry an unique flying and autonomous platform with all the equipment required to detect and monitor the very dangerous atmospheric phenomena impending threats to air traffic in general, to civil and military aircraft, or used for local disaster aerial surveillance and monitoring.

The Special Autonomous UAV named as the OCTAGON Project is an autonomous flying platform special designated to monitor the Low Altitude Electromagnetic Turbulence in the atmosphere of Earth. It is a new technical solution and development based on the actual level of the aerospace technology that includes inter alia an electric propulsion system assisted named as the Power Ring System which is using the aerodynamic elements of a new Romanian Patent on the airfoils design (the electrical propulsion system is in the due procedure to be protected by a separate patent license).

Fig. 1. The Special Operations UAVs

The minimum special operational requirements are about 20 to 35 minutes hovering over exact designated areas at several altitudes monitoring all weather flight conditions and providing all kind of data including the electromagnetic local field characteristics and the real local data and images. Therefore, this Special Autonomous UAV which is using an electrical propulsion system assisted is able to operate at any altitude with no limitations, which is something that any helicopter platform can not do yet, and also able to take-off and land using both standard and the vertical procedures.

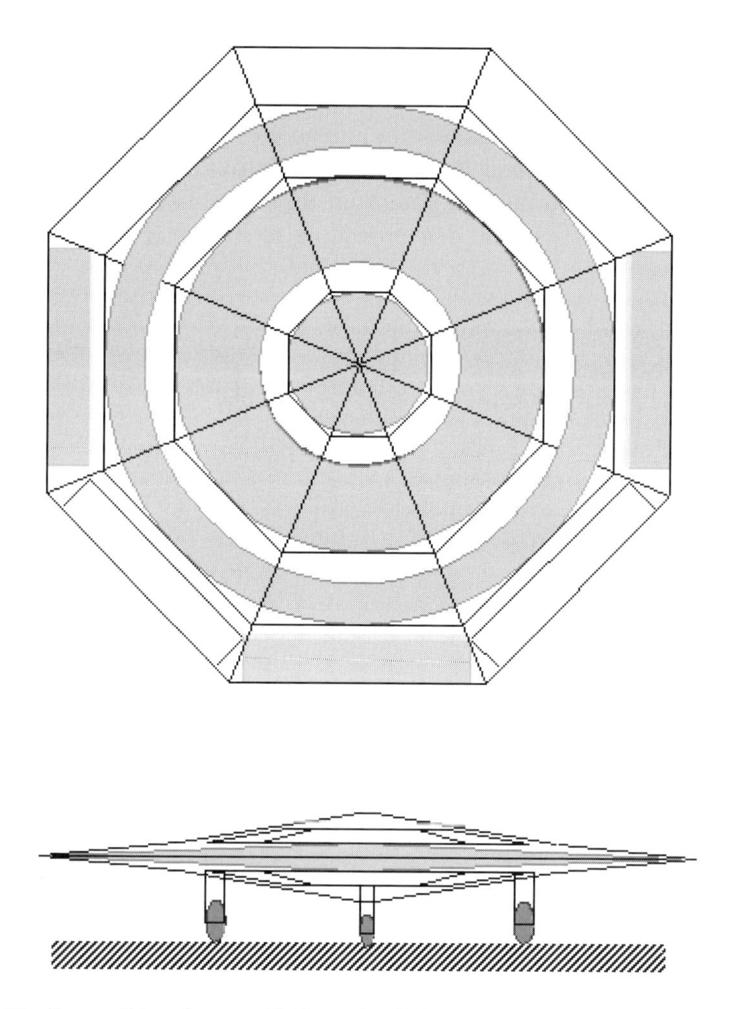

Fig. 2. The Power Ring System (© Dascalu, 2001, 2002, 2003 and 2004)

The Special Operations UAV's structure is a composite laminar octagon geodesic type that is also a single flying wing design with no fuselage or vertical wings. The general UAV's airfoil shape is laminar symmetrical with a classic three wheels landing gear, retractable, with external central wing edges movable as flaps and ailerons providing the full aerodynamic controls during the horizontal flight at high speed. For the low speed flight and hovering operations, the thrust vectoring is the best option in the simplest design mode, but for an advanced design configuration can be used a fixed Fan Power Ring with the air jet speed and fan control system, including the parallel fan-wings pitch control.

The UAV's propulsion system is the Fan Power Ring which can be fixed in the simplest design or that can be inside of a cardan joint to provide the thrust vectoring in an advanced design mode. In all design cases the Fan Power Ring is a double fan system consist of two great fans rotating in the opposite directions to reduce the gyroscopic moments and forces, and each of these fans is also a complex system of parallel short wings between two of each fan's vertical rings.

Each of these two fans has a double blades classic system due to the fact that each fan has a larger diameter and is more suitable for a short parallel wings configuration (see on the biplanes design of some airplanes of the First World War).

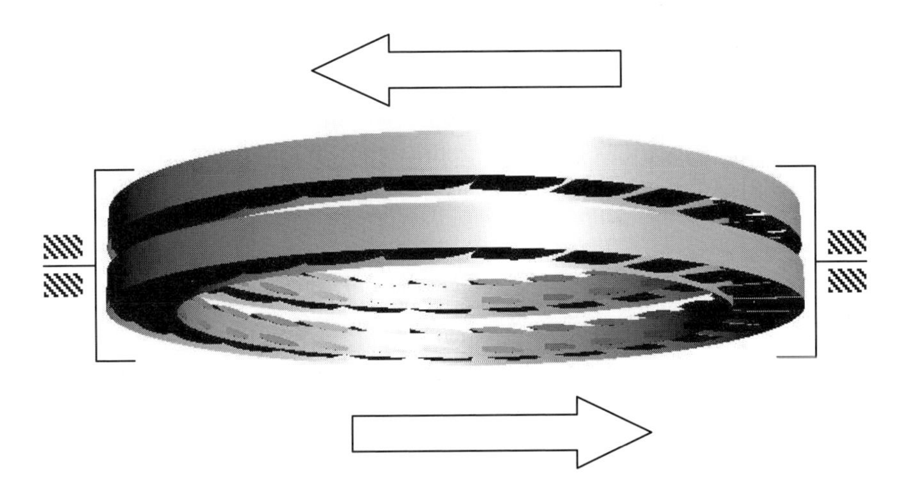

Fig. 3. The Fan Power Rings Propulsion System

This was our main design by the purpose to have similar characteristics with an helicopter rotor and inside of the UAV's structure for the main reasons as: to provide more lift at the same rotor's area and power, to reduce the external noise, to increase the speed of the rotor up to the supersonic limits, and to improve operational safety relative to the hazard impact with birds, high voltage lines or cables, trees, buildings, high terrain, airport equipment, antennas and technicians, etc.

The over all the propulsion efficiency is higher with about of 30% up to 50% relative to the similar and classical helicopter propulsion systems. The Fan Power Ring electric propulsion system assisted is required also due to the fact that in the special flight operations of detection and monitoring the Low Altitude Electromagnetic Turbulence or in any aerial surveillance and monitoring of a disaster development situation and in complex emergencies there are layers of air with high concentration of metal particles, or dust and other unknown chemical components that can be very dangerous and hazardous for the classic propulsion systems using any of the actual thermal driven engines.

This is the main reasons of which the Special Operations Autonomous UAV is more suitable for the aerial investigations of volcanic activities, radioactive high contaminated areas, over any high contaminated areas poluted with the unknown chemical components or even up to the use in space autonomous exploration, such as the exploration of the planet Mars.

The high efficiency of the Fan Power Rings Propulsion System is achieved by the using of a combination of three main technical solutions such as the Fan Rings, the biplane wing effect and the double fan system with the opposite rotations mainly to reduce the well known gyroscopic effect of a single rotor-fan, and using cardan joints for the thrust vectoring.

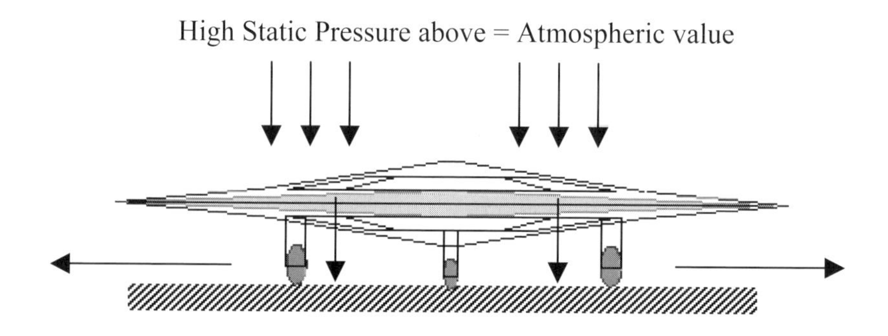

Fig. 4. The static pressure above is higher as the static pressure under body-wing

The main UAV problem is the above aerodynamic paradox specific for such design that was not solved until now that is as follows: as fast the jet current is required to push the aircraft upwards the static pressure under is lower then the static atmospheric pressure above the aircraft which is pushing it downwards, increasing the jet speed under it will generate even more reduction of the static pressure under the aircraft relative to the fixed atmospheric static pressure value above, and due to this important reasons the UAV aircraft will not be able to take-off despite the high power of those engines.

The aerodynamic paradox that was visible during the aviation history in several well known cases is in the present project solved by using the vacuum effect over the main body-wing surface of the UAV with the main purpose to reduce the static pressure above the aircraft faster as is may happened under the aircraft, as you can see that in the next picture.

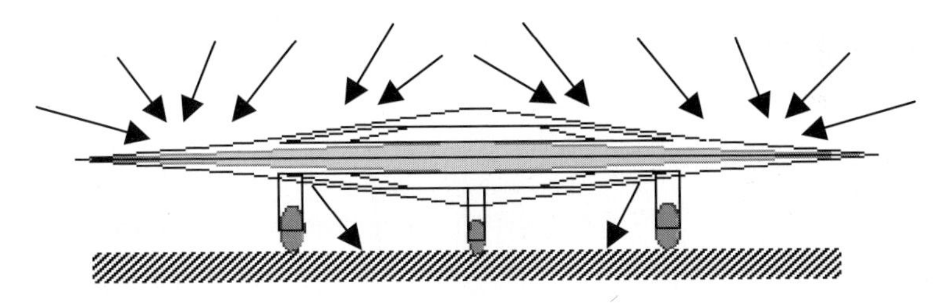

Fig. 5. The aerodynamic paradox is solved using the vacum effect over the body-wing area

The air above the aircraft is diverted to enter in the propulsion fans system using two rings of intake opening systems leading the air direct to the Fan Power Rings Propulsion System and the eject thrust vectoring system is creating the jet current deflected with up to 30^0 from the vertical direction during the take-off and landing procedure, that is able to increase the overall pressure under aircraft as well as using the ground effect, and therefore, the total efficiency of the Fan Power Rings Propulsion System is higher as any classic helicopter rotor system.

The Fan Ring Propulsion is a result of spreading around the circle the classic helicopter rotor blades equally divided into small wings and each part is appropriate adjusted between two ring-walls, and the result is like a fan-ring that is more efficient with about 15% relative to the equivalent helicopter rotor. In addition, each small wing within the fan-ring is a system of two wings one above the other very similar as the classic biplane aircraft of the First World War, that is more efficient at low speeds with 30 up to 50% as any single wing system which also improving the overall power efficiency.

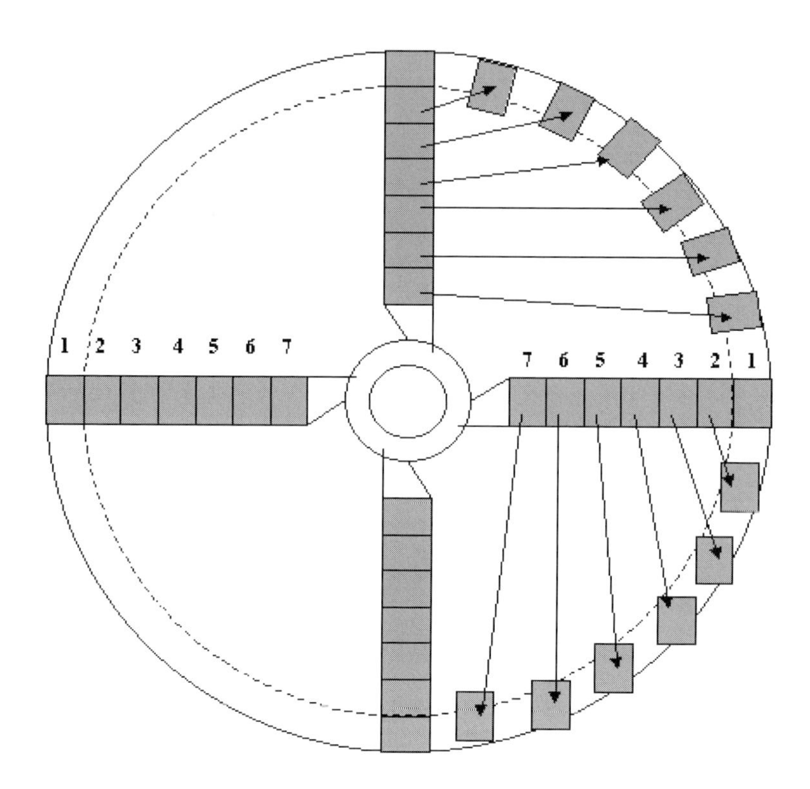

Fig. 6. The classic helicopter rotor blades are divided into wing parts and spread around the circle

The system of dual wings is moving through the air between two ring-walls and therefore, the induce drag is also zero, and the angle of incidence (the pitch control) is able to be adjusted in a more complex design configuration or can be maintained at the optimal levels and then just only the speed of the fan-ring that will increase the total lift force for propulsion system as may be required during the flight operations.

If we take into the account that a ring propulsion is also a dual system required to have the overall reduced gyroscopic momentum and the UAV aircraft to be full controllable, the result in the high efficiency of the power ring propulsion system is even much higher as the equivalent helicopter rotors.

3 Concluding Remarks

The OCTAGON Project – using the electric propulsion system assisted – is an environmental friendly UAV with almost zero noise level, no toxic emissions, with a very low thermal signature, and with the aerodynamic shape and profile this UAV is also stealth with low radar image. The entire surface is protected against the impact of the electromagnetic waves required to have a full protection for all the avionics, flight control system and for the electrical systems against the Low Altitude Electromagnetic Turbulence effects.

Based on all of these the flight operation areas of this autonomous airborne platform are almost unlimited starting with special operations over any kind of disaster, complex emergencies, civil or military aerial surveillance and monitoring.

We are looking to have partners in the development of this Special Autonomous UAV Octagon Project for several other applications, including civil and military operations and also the future commercial use of the Autonomous UAVs.

This Octagon Project is under development now at the European Business Innovation & Research Center in Bucharest, Romania, as new member of European Business Innovation Centers Network.

Disaster Prevention for Alpine Routes

Markus M. Eisl[1], Klaus Granica[2], Thomas Nagler[3], Florian Böhm[1], Helmut Rott[3], Herwig Proske[2] and Mathias Schardt[2]

[1] GEOSPACE Austria, Jakob Haringerstr. 1, A-5020 Salzburg, Austria. Email: Markus.Eisl@geospace.co.at
[2] JOANNEUM Research, Inst. of Digital Image Processing, Wastiang. 6, A-8010 Graz, Austria.
[3] ENVEO, Exlgasse 39, A-6020 Innsbruck, Austria.

Abstract

Natural disasters are an age-old problem that occur regularly in alpine regions, posing a major threat to the safety of settlements and transport routes. Within the project "Safety of Alpine Routes – Application of Earth Observation Combined with GIS (Hannibal)" information has been extracted from satellite remote sensing and integrated into a newly developed GIS based Decision Support System (DSS). Some of the required map information were inferred from ERS- and from SPOT5- and QUICKBIRD satellites. Forest and land cover parameters have been derived from the satellite data. Change detection techniques have been applied to monitor e.g. windfall and clear-cut areas. Methods for interferometric processing of radar data have been optimized to detect landslides and to generate maps of motion fields along the slope surfaces. The results show that this information integrated into the DSS is an efficient tool for focusing the attention to potentially endangered areas.

1 Introduction

Natural hazards such as avalanches, landslides or torrents have been a permanent threat to traffic routes across the Alps as long as man has been traveling them. Within the project "Safety of Alpine Routes – Application of Earth Observation Combined with GIS (Hannibal)" a decision support system (DSS) has been developed which is based on Earth Observation and GIS methodology. The project was funded by the Austrian Federal Ministry for Traffic, Innovation and Technology (BMVIT).

In close co-operation with end users the most relevant indicators for the processes covered within the project were identified, in addition to that the quantitative aspects of the indicators (range of values, thresholds, assessment methods) were investigated. As far as possible, these indicators were derived by methods using Earth Observation systems, others were taken from conventional sources. All data were integrated into the decision support system. The data base for the analysis of Earth observing systems included radar data from ERS satellites and optical data from the satellites SPOT 5 and QUICKBIRD.

Complex analysis algorithms have been applied to extract the required information on the Earth's surface from remote sensing data, which were integrated into a decision support system based on a GIS (Geographical Information System). Methods for interferometric processing of radar data have been optimized and used for the derivation of digital elevation models and the detection of landslides and the generation of maps of motion fields along the slope surfaces. Forest parameters (e.g. age class distribution, tree species distribution, crown closure) and land cover parameters relevant for the assessment of potential threats have been derived by classification of satellite data to integrate them into the decision support system. Specific methods based on multitemporal satellite data allowed to analyze temporal changes within the vegetation cover, e.g. to detect windfall areas.

The decision support system developed within the project uses logical and arithmetic operations for the combination of the basic information layers. By this it is possible to both keep the large area overview and yet focus the awareness of users to areas, where action or at least more detailed analysis is required. An important aspect of this wide area approach is that it allows to assess both direct and indirect threats to traffic routes, e.g. mountain slides directly threatening a road below indirectly by modifying the water runoff in a neighboring catchment basin.

During the assessment of the system by potential users, i.e. the Austrian railway company (ÖBB), Torrent and Avalanche Control Services and Forestry Boards, it has been considered a powerful tool for the envisaged applications.

2 Indicators

The purpose of indicators in the given context of natural hazards is to find a set of parameters which can be used to assess an enhanced probability for given processes such as avalanches or landslides. With respect to threats imposed on alpine routes this question is extremely complex and by far not fully solved. This is a consequence of the complex nature of the involved processes as well as of the interrelations between different factors influencing a process.

On a high level it is necessary to discriminate between indicators for (enhanced) risks on one hand (e.g. steep terrain) and those for the processes themselves (e.g. moving slopes). During the work performed both types of indicators have been used.

Within a series of workshops during the initial phase of the project the most relevant indicators for the processes covered within the project have been identified. The tight integration of experts from end user organizations (national torrent and avalanche services, forestry administrations, Austrian railway company) as well as of Earth observation experts made it possible to construct a set of indicators optimized for the given purpose, regarding both the envisaged processes and the focus on indicators accessible by remote sensing methods.

In addition to identifying the indicators, the quantitative aspects of the indicators (range of values, thresholds) have been investigated. The results of this work have been compiled in an extensive table listing a total of about 20 indicators for 6 different processes (avalanches, superficial landslides, deep reaching landslides, torrents, falling rocks, and direct threats to routes by falling trees). The indicators have been grouped into the categories topography, geology, geomorphology, hydrology and land cover, in addition to that, historic events have been integrated into the considerations.

An important and not unexpected result of the work was that for all analyzed processes the topography of the terrain plays the most important role. In particular, the terrain slope and exposition are the basic parameters which in any case need to be integrated in the analysis, as all processes are gravitationally triggered. This information can be derived from available digital elevation models.

The second important class of indicators is the land cover with a special emphasis on forest parameters. An example for the importance of this type of indicators is the age class of trees – young forests and pole wood are able to stop avalanches, whereas old forests with their larger distances between trees and often sparse ground cover by bushes can leave enough open space for avalanches to proceed. This type of information is well accessible by Earth observation data, whereas the accuracy of the data has to be high both spatially and thematically in order to achieve reliable results.

3 Decision Support Concept

The management and the visualization of information are key tasks of Geographical Information Systems, and in a wide sense they can be viewed as Decision Support Systems (DSS) as they help operators to interprete a given situation. In the given context the term DSS is used in a narrower sense by restricting it to systems allowing further operations on input data, which make it possible to focus the awareness of the operator to areas of special interest.

3.1 Requirements and Concept

In addition to their support in deriving indicators usable for the assessment of threats, the integration of the end users was an important factor in the development of the DSS, both in the initial phase of the identification of requirements and finally for the evaluation of the results.

A high level diagram showing the information flow and the role of the GIS/DSS is given in the Figure below.

The most important result of the requirements assessment was that the end users unequivocally insisted that – in contrast to a black box system – the system should give full control to the operator, with respect both to the visualization of the information and to the set-up of the rules used to apply and combine the extraction of the indicators from the input data. This means that the system should be a tool for experts rather than for the wider public, and that it should help experts to get a large area overview on one hand and on the other hand to focus their attention to areas with an enhanced risk for the disastrous processes listed above. This approach also copes with the general finding in our information society, that not a lack of information but retrieval of relevant information is the biggest challenge (awareness).

As a consequence the DSS was designed in a way that – for a selected process – it provides suggestions to the operator, which input data, indicators and rules should be used (default configuration). Starting with this, the operator introduces his expertise on the area and on the process by modifying the parameters.

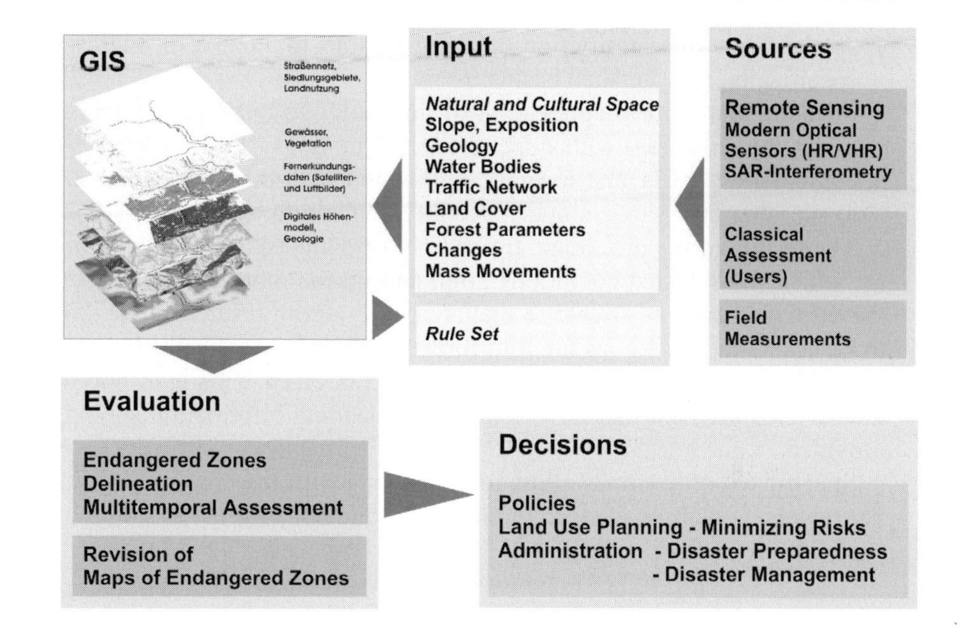

Fig. 1. High Level Concept of the Data Flow around a Decision Support System

3.2 Implementation

In order to meet the requirement of compatibility with existing GIS of the end users, the DSS has been implemented as an ArcView 3.x extension in Avenue. The base data layers comprise, as indicated in the Figure above, a topographic map, infrastructure information – especially the traffic network – and the digital terrain model and layers derived from it (exposition, slope, curvatures).

In a second step information layers derived from Earth observation data are integrated, comprising land cover data with a focus on detailed forest information and data derived from SAR interferometry, such as slope motion fields or alternative terrain models.

Usually the procedure of an analysis is started by opening the input dialogue of the Scenario Analyst. Here it is possible to identify the process of interest, to select a layer to display (e.g. slope), and to apply rules to restrict the displayed layer to an area for which a set of conditions is given. All raster data provided as GRIDs can be integrated into the set of rules. The rules applicable to the input data comprise the following operations:

1. spatial operations: restriction of the operations to an area of interest or, specifically, to watershed basins,
2. thematic restrictions: application of thresholds to displayed layers, e.g. to display only areas with slopes steeper than 40°,
3. combinations of layers: logical combination with the information in up to six other layers (e.g. slope above 40° AND no forest).

The results of the procedure are on one hand a mask indicating all areas for which the selected conditions hold, and on the other hand a subset of the displayed layer restricted to that area. All interim results can be stored to be used later on, e.g. as input to further analyses. Note that no quantitative information on the risk (e.g. probability) is given. This is in line with the user requirement that no potentially misleading information should be provided.

Specific export routines allow to print maps with the prepared information layers as well as to store the layers in one out of a list of commonly used formats.

4 Earth Observation Methodology

To fulfill the above described user requirements a review on the existing satellite systems has been performed and the satellites SPOT5 and QUICKBIRD were selected for the investigations. These images are of high interest because they encompass a range of spatial resolutions from 0.6m over 2.5m to 10m/20m. The investigations should focus on the usability of large area coverage imagery for the derivation of the parameters needed (described above).

Radar is one of the few (quasi) operational active remote sensing techniques. A persuasive reason for using radar is its high degree to penetrate clouds and, to a high degree, rain. The information content of radar is mainly sensitive to the physical and electrical properties of the surface and is, therefore, complementary to the optical sensors. In this project the focus was put on motion aspects for landslide analysis, consequently interferometry techniques have been applied to observe surface movements.

Optical Satellite Data

For the derivation of the input parameters to the GIS system very high resolution QUICKBIRD and SPOT5 bundle scenes have been used. A land cover classification with the focus on detailed forest categories was assigned to yield the needed landscape information. For the different image types different methodologies had to by tested, i.e. visual interpretation, segment-based classification and pixel-based classification. The latter was used for the derivation of the GIS land cover information. The following methodological steps have been applied:

Geocoding

Displacement errors caused by topographic relief must be removed to optimisz the absolute geometric location accuracy of the geocoded image data (Raggam et al., 1991). In the course of geocoding, these errors were removed through the integration of a digital elevation model (DEM), i.e., the consideration of terrain relief information. Geocoding was performed with the RSG software (Remote Sensing software package Graz) of Joanneum Research.

Topographic Normalisation

An ideal slope-aspect correction removes all topographically induced illumination variations so that two objects having the same reflectance properties showing the same digital number despite their different orientation to the sun's position. As a visible consequence, the three-dimensional relief impression of a scene disappears and the image looks flat. In order to achieve this result, several radiometric correction procedures have been developed. Besides empirical approaches, such as image rationing, which do not take into account the physical behavior of scene elements, early correction methods were based on the Lambertian assumption, i.e. the satellite images are normalized according to the cosine of the effective illumination angle (Smith et al., 1980). However, most objects on the Earth's surface show non-Lambertian reflectance characteristics (Meyer et al, 1993). The cosine correction had thus to be extended by introducing parameters simulating the non-Lambertian behavior of the surface (Civco, 1991; Colby, 1991). The estimation of these parameters is generally based on a linear regression between the radiometrically distorted bands and a shaded terrain model. A comparison between four correction methods, including the non-parametric cosine correction, confirms a significant improvement in classification results when applying the parametric models (Schardt et al., 2000). In this mapping project, the parametric Minnaert correction was used for topographic normalization, as this method has been proven to achieve satisfactory results (Schardt et al., 2000; Schardt & Schmitt, 2001).

Supervised Classification

The training areas required for supervised classification were selected on the basis of CIR aerial photographs from different data sources and acquisition times or by field work. After the signature analysis, which is an important step in the supervised classification, the classification was carried out with selected signatures using the maximum likelihood method. A detailed evaluation of several classification runs based on different parameter settings and combinations of training areas resulted in the following "best practice procedure":

1. *Classification according to altitude*: the classification was carried out separately for areas above and below 1500 m, the results being subsequently combined into an overall result to account for the altitude effect.
2. *Weighted classification*: the "probability values" were reduced for uncertain training areas that were difficult to evaluate.
3. *Classification for the derivation of tree species*: the error probability increases for stands below 50% crown closure due to the strong influence of the ground vegetation. Only training areas with a crown closure of more than 50% were therefore selected for tree species classification.
4. Dwarf mountain pine and green alder stands were excluded for determining the *even-aged forest below 1500m altitude.*
5. For age class determination, ideal clusters were generated in the feature space and integrated into the training data set.
6. Computation of a *textural parameter* file out of the 2.5m panchromatic SPOT5 scene, for better discrimination of overlapping classes as settlements vs. fallow land.

Visual Interpretation

For a more detailed derivation of land cover / land use classes the very high resolution Quickbird data were used for visual interpretation. For this purpose a fused image, from the panchromatic and the multispectral channels, has been generated using the adaptive Brovey transform algorithm. The results show high details on air-photo level and were integrated into the GIS.

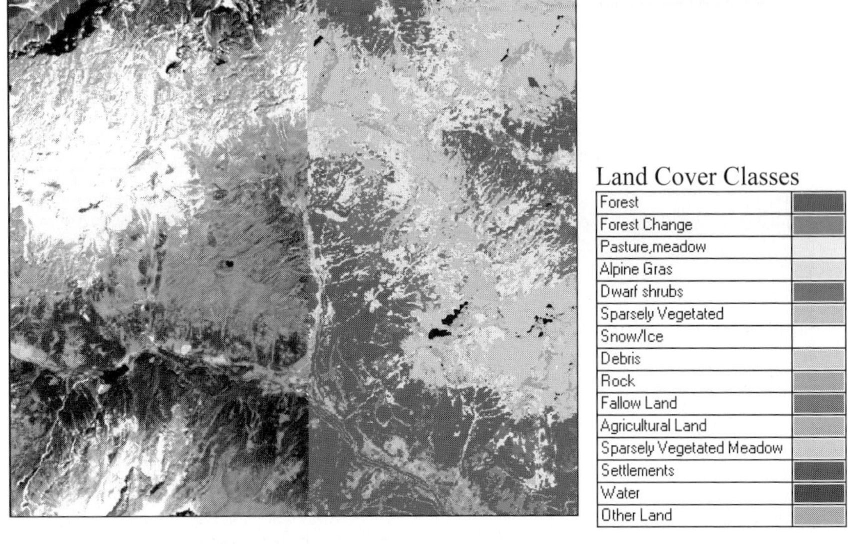

Fig. 2. Fusioned SPOT5 image (left), result of land cover classification (right) (Zederhaus, Tauerntunnelsüdportal)

Co-Registration

For the comparison of two scenes it is a prerequisite to co-register them. In the Remote Sensing Software Package Graz (RSG; Joanneum, 1998) a DEM (digital elevation model) based geocoding was performed, because of the high-mountainous terrain, and scene 1 (t1) was matched to scene 2 (t2).

Topographic Normalisation

In the next phase a topographic normalization was applied on both scenes to compensate for illumination (see description above).

Calibration

The calibration of scenes (t1) and (t2) was achieved applying linear histogram matching. The goal of this processing step should be the exact spectral adaptation of the multi-temporal scenes. To avoid per-pixel errors a "moving-window" approach has been applied. Within the "moving window" of a defined size, e.g. 5x5 pixel, the mean value has been used to compensate for local disturbances. Furthermore, a height effect according to atmospheric effects had to be considered for the calculations.

Change Detection

Based on the above described steps image differencing was applied and as a result the forest changes occurred between 1987 and 2003 could be derived (see Fig. below). Large areas are mainly windfall areas from 2002.

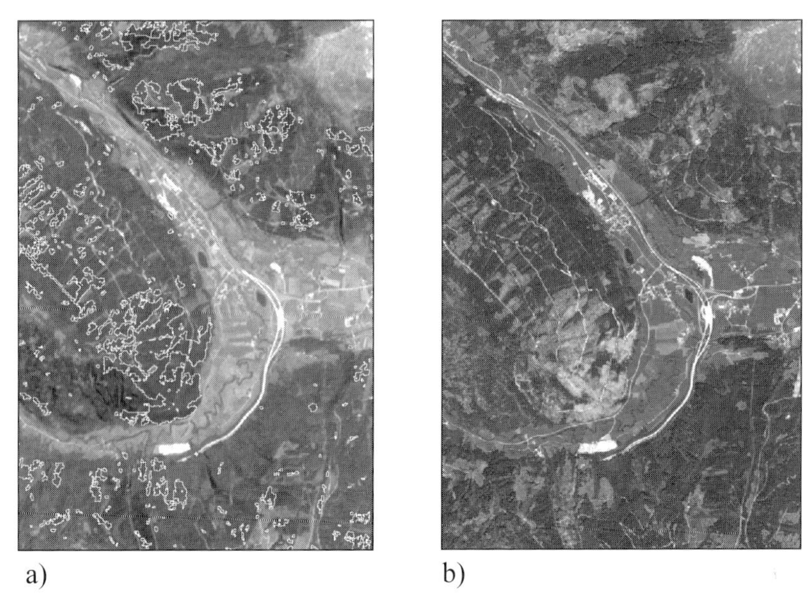

a) b)

Fig. 3. Part of the region were semi-automatic change detection was performed. a) SPOT1 scene from 14.9. 1987; b) SPOT5 scene from 20.7.2003 (red lines indicate the changes between 1987 to 2003)

4.1 Diffential SAR Interferometry

Synthetic aperture radar interferometry (InSAR) offers the possibility to detect and monitor movements at the Earth's surface with high precision over extended areas (Hanssen, 2001). The project focused on detecting and mapping slopes with very slow movements of the order of centimeters per year as indicator of possible instability. The basis for the analysis are SAR images of the European Remote Sensing satellites ERS-1and 2, and from the Advanced SAR (ASAR) of Envisat. ERS SAR operates at the wavelength λ= 5.66 cm (5.3 GHz), with an incidence angle $\theta = 23°$ in the centre of the 100 km-wide swath. The spatial resolution is 9.5 m across track $\times$ 5.5 m along track, the standard orbital repeat period is 35 days.

Across-track SAR interferometry is based on the detection of phase differences between two SAR images acquired from similar orbital positions (Figure 4).

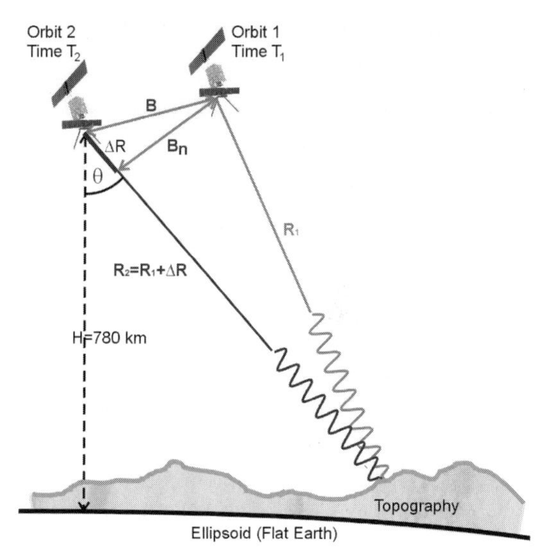

Fig. 4. Imaging geometry of across track SAR interferometry. B - baseline, B_n – perpendicular baseline. θ – radar look angle. R_1, R_2 … slant range distance sensor-target

The interferometric phase is a sensitive measure of the change of the path length of the radar signal in slant range direction. In repeat-pass interferometry the phase difference, $\Delta\phi = \phi_2 - \phi_1$, includes the following contributions, which determine the differences in the propagation path length (R_2-R_1) between the two images:

$$\Delta\phi = \frac{4\pi}{\lambda}\left(R_2 - R_1\right) = \phi_{flat} + \phi_{topo} + \phi_{dis} + \phi_{atm} + \phi_{noise} \qquad (1)$$

where

ϕ_{flat}, ϕ_{topo} … phase differences due to changes of the relative distance satellite-target for flat earth and topography

ϕ_{atm} … the phase difference due to changes in atmospheric propagation

ϕ_{dis} phase difference due to displacement of the target in slant range direction

ϕ_{noise} phase noise (thermal noise, processor noise, etc.)

In order to determine terrain motion (corresponding to ϕ_{dis}) from interferograms, it is necessary to correct for the other phase terms in Eq. 1. $\Delta\phi_{flat}$ can be calculated accurately using the precise ERS orbit data, as e.g. provided by the Technical University of Delft, NL. The topographic phase ($\Delta\phi_{topo}$) depends on the imaging geometry; on the perpendicular baseline (B_n) and the wavelength. $\Delta\phi_{topo}$ was estimated using ERS-1/-2 tandem pairs with 1-day time span, because the motion-related phase of the slowly sliding slopes can be neglected and the coherence is higher than for longer time spans.

Atmospheric phase variations (ϕ_{atm}) are dominated by water vapour (Hansen, 2001). It is not possible to correct for atmospheric effects in a single interferogram without spatially detailed information on the atmospheric state from other sources. Because the typical scale of atmospheric phase variations is much larger than the size of landslides, effects of ϕ_{atm} can be strongly reduced by using tie points on adjoining non-moving surfaces (Rott *et al.*, 1999). Atmospheric effects can also be reduced by stacking interferograms of different dates.

InSAR is sensitive only to motion in direction of the radar beam. The displacement (R_2-R_1) in line of sight can be derived from ϕ_{dis} according to Equation 1. The sensitivity to motion depends on local incidence angle. Assuming only horizontal or vertical motion a phase shift of 2π ($\Delta R = 2.83$ cm) observed by ERS SAR corresponds to a displacement of $\Delta x = 7.24$ or $\Delta z = 3.07$ cm, respectively. In the case of landslides often surface parallel motion is assumed, but especially in the depositional zones of landslides deviations from this assumptions can be significant (Gianni, 1992).

A prerequisite for applying InSAR is interferometric long-term coherence, which describes the phase stability on a pixel basis. In areas with sparse vegetation, rocks, bare soil and urban areas the coherence over time intervals of months to a few years is usually adequate for interferometric analysis. However, in densely vegetated areas (e.g. forests, agricultural fields) the radar signal often decorrelates already within a few days (Rott et al., 2000).

5 Results

5.1 Mapping of Landslides above the Felbertauern-road

Along the north-south traffic routes crossing the Austrian Alps several landslides were detected and mapped by means of InSAR. The analyzed data base includes more than 60 ERS SAR images acquired during ascending and descending passes over the investigation area Hohe Tauern, Salzburg, in the period 1992 to 2002. Most of the detected landslides were located in high alpine areas above the tree line (about 1800 to 2000m) where the coherence is preserved over long periods (several months to a few years).

As an example the landslide in the Ödtal, Hohe Tauern, is shown. The Ödtal is a narrow valley through which a main north-south traffic route, the Felbertauern road, runs. The slopes are very steep, and the road is endangered by rockfall and landslides. The analyzed slope extends from about 1400 to 2600m in elevation and has a mean surface inclination of about 30 degrees. The lower part of the slope is about 35° steep and partly covered by forest up to about 2000 m. Above the tree line rocks and bare soil and low alpine vegetation dominate.

Figure 5 shows a 3-D view of the area, with the InSAR derived motion map overlaid on a pan-sharpened Landsat TM image. The slope parallel assumption was applied for generating the motion map. The existence of the landslide has been known previously, because rock fall from the slopes affected the road from time to time, but no quantitative information has been available. The InSAR analysis revealed a typical velocity of the main landslide body in the Gamskar of about 2.3 cm/a in 1992/93 and 1.8 cm/a in 1997/98. Below the tree line, the coherence was not sufficient for retrieving interferometric information.

The results of the SAR interferometry analysis have been integrated into the DSS. In combination with the slope inclination it was possible to extract areas with both "fast" movement and steep slopes, which is an indicator for an enhanced risk for landslides. During the evaluation phase of the project a rock fall event took place in the Felbertauerntal, which had its origin exactly in the area depicted in the figure below in red – a coincidence, which certainly shows the significance of the results.

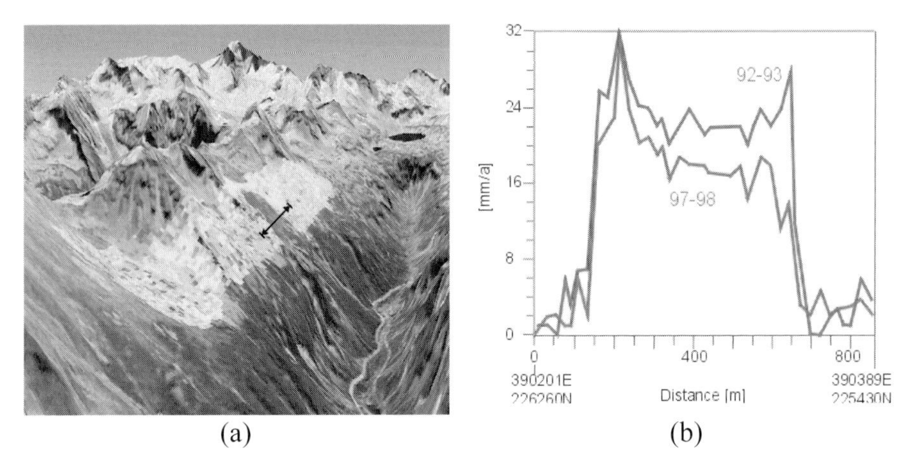

Fig. 5. (a) 3-D view of the Felbertauerntal (viewing direction is south). The surface parallel motion field from InSAR (colour coded) is superimposed to a Pan-sharpened Landsat image; (b) profile of surface parallel motion (black line in (a))

5.2 Test Area Mallnitz

The results of the forest classification proofed that it is possible to derive forest parameters from SPOT5 multi-spectral data over a rather large region. With these parameters the experts have the opportunity to evaluate the actual situation in forests with respect to the defined hazards. For instance, if the crown coverage is below a certain threshold the probability that snow avalanches could be triggered is increased. Thus the derived parameters will be used for protection forest planning. Furthermore, critical areas, i.e. caused by clear cuts or windfall, within forest have been identified by applying change detection.

The results of the land cover classification showed that the derived indicator classes are useful on a river catchment basis to enable quantitative analysis within the DSS. Most of the assigned classes could be derived from satellite imagery. The results could satisfy the requirements in terms of spatial and thematic accuracies.

A more detailed result of land cover / land use classes was obtained by applying a visual interpretation on the very high resolution Quickbird data. The results show high details on air-photo level and were integrated into the DSS. However, as the image covered only a small part of the whole test site primarily SPOT5 data were used in the DSS.

Integrating these data into the DSS it was possible to extract narrow avalanche strips, which are characterized as narrow (width about 10m) treeless strips following the terrain through steep terrain, often along steep torrents. In a further step these strips have been evaluated with respect to their vicinity to infrastructure elements or settlements.

In a similar way the consequences of a disastrous windfall event in 2002 have been analyzed. This was done by combining the change detection results (deforested areas) with the slope inclination and the distance to traffic routes, as these parameters give a good indication of the risks for these routes both from avalanches and from landslides.

6 Conclusions

In this paper we have described the concept of a decision support system for the assessment of risks along alpine routes, followed by a presentation of results obtained for test areas.

For the investigation and prevention of natural hazards in an Alpine environment a GIS based Decision Support Systems has been developed. The need of a wealth of quantitative input data has led to the use of satellite remote sensing data. For this purpose optical as well as radar data have been selected. Based on the most recently developed methodologies the assigned indicators have been derived and incorporated into the DSS. It could be shown that SPOT5 imagery could satisfy many of the requirements in terms of quantitative assessment and spatial resolutions. The ascertainment of detailed forest parameters was possible. ERS-SAR interferometry methods have been applied to derive slope motion fields. This information has been combined with other GIS data to focus the attention to areas with potentially enhanced risk, with the main input for the envisaged gravitational processes coming from the digital terrain model.

During the assessment of the system by potential users, i.e. the Austrian Railway Company, Road Companies, Torrent and Avalanche Control Services and Forestry Boards, it has been considered a powerful tool for the envisaged applications. The project results will be further exploited by transferring the prototype system into a marketable product and by promoting a data evaluation service.

Acknowledgements

The project "Sicherheit von Alpentransversalen unter Verwendung von Erdbeobachtung und GIS - Hannibal" has been supported by the Austrian Federal Ministry for Traffic, Innovation and Technology (BMVIT).

References

Civco DL (1991) Topographic normalisation of Landsat Thematic Mapper digital imagery. Photogrammetric Engineering and Remote Sensing, vol 55, no 9, pp 1303-1309

Colby JD (1991) Topographic normalisation in rugged terrain. Photogrammetric Engineering and Remote Sensing, vol. 57, no 5, pp 531-537

Gallaun H, Schardt M, Granica K, Flaschberger G (2001) Monitoring von Schutzwäldern mit Satelliten-Fernerkundung. VGI - Österr. Zeitschr. für Vermessung & Geoinformation, Heft 2001/3

Gianni P (1992) Rock Slope Stability Analysis. A.A. Balkema, Rotterdam

Hanssen RF (2001) Radar Interferometry. Kluwer Academic Publ., Dordrecht

Meyer P, Itten KI, Kellenberger T, Sandmeier S, Sandmeier R (1993) Radiometric correction of topographically induced effects on Landsat TM data in an alpine environment. ISPRS Journal of Photogrammetry and Remote Sensing, vol 48, no 4, pp 17-28

Raggam H, Almer A, Strobl D, Buchroithner MF (1991) RSG - State-of-the-Art Geometric Treatment of Remote Sensing Data. In Proc. of 11'th EARSeL Symposium: Europe: From Sea Level to Alpine Peaks, from Iceland to the Urals, pp 111-120, Graz, Austria, July 3-5 1991

Rott H, Scheuchl B, Siegel A, Grasemann B (1999) Monitoring very slow slope movements by means of SAR interferometry: a case study from a mass waste above a reservoir in the Ötztal Alps, Austria. Geophysical Res. Letters, 26, pp 1629-1632

Rott H, Mayer C, Siegel A (2000) On the operational potential of SAR interferometry for monitoring mass movements in Alpine areas. Proc. of 3rd European Conference on Synthetic Aperture Radar, Munich, May 2000, pp 43-46

Schardt M, Granica K, Schmitt U, Gallaun H (2000) Monitoring of Protection Forests in alpine Regions. Proceedings of the 4.02.05 Group Session „Remote Sensing and World Forest Monitoring“, IUFRO XXI World Congress, 7-12 August 2000, Kuala Lumpur, Malaysia, 2000 (edited by the Institute of Geodesy and Cartography in Warsaw

Schardt M, chmitt U (2001) Inventory of Alpine Relevant Parameters for an Alpine Monitoring System Using Remote Sensing Data. Proceedings of the International Workshop on „Geo-Spatial Knowledge Processing for Natural Resource Management, University of Insubria, Varese, Italy, June 28-29, 2001

Schardt M, Gallaun H, Häusler T (1998) Monitoring of Environmental Parameters in the Alpine Regions by Means of Satellite Remote Sensing. Proceedings of the International ISPRS-Symposium on "Resource and Environmental Monitoring, Local, Regional, Global, Commission VII, September 7-4, 1998, Budapest, Hungary

Smith JA, Tzeu Lie Lin, Ranson KJ (1980) The lambertian assumption and Landsat data. Photogrammetric Engineering and Remote Sensing, vol 46, no 9, pp 1183-1189

Step by Step Constitution of an Emergency Management Based Object Model and Database System on Linux for the I.T.U. Campus D.I.S.

Himmet Karaman and Muhammed Şahin

Istanbul Technical University, Geodesy & Photogrammetry Engineering Department 34469, Maslak, Istanbul, Turkey.
Email: hkaraman@ins.itu.edu.tr

Abstract

To mitigate the property and life losses at the İ.T.U. Ayazağa Campus, after a probable disaster that may occur in İstanbul, it is strongly necessary to constitute an emergency management based object model and database in order to apply the Turkey Disaster Information System (TABİS) which was developed within the scope of "Constitution of the National RSS-GIS Based Database and Emergency Management Focused Decision Support System Project".

1 Introduction

The first aim of the Disaster Information Systems is to mitigate the loss and hazards after the possible natural disasters like earthquakes, floods, landslides or minimize the possibilities of manmade and technological disasters like fires, traffic accidents and terrorism. The second aim of the Disaster Information Systems is to make the recovery and response phase like, search and rescue, loss detection and restore, as rapidly as and as accurate as possible [1].

In this project, an infrastructure and a data catalog have been constituted for the Campus Disaster Information System, which was established in the İ.T.U. Ayazağa Campus, using the created object model and data base. Furthermore, the object model was connected to a geographic information system platform, and by this way spatially referenced visual queries were constructed.

2 The Aim of the TABiS

The aim of the Turkey Disaster Information System (TABiS) project is to develop the standards for the GIS based information and management system model basically by using the Remote Sensing System (RSS), Global Positioning System (GPS), and other data acquisition techniques, specially for emergency planning, practice and in any disaster situation, disaster management and prediction of losses, and for the applications of central and provincial authorities at the ordinary times as a decision support system [8].

TABiS supply a harmonious work and coordination standards between the provinces about disaster planning and management and constitutes a regional, environmental and managerial information system model.

TABiS is an interdisciplinary project. The reasons for this are [8];

- The ratio of the losses are varies according to the population and construction density at the disaster zone. Natural or manmade disasters are becoming one within the other while the industrialization is increasing.
- In Turkey, the information substructure for the prediction and systematic research of the risks caused by the natural disasters are not enough. The mechanisms that fit the aim are not developed yet.
- Similarly, it can not be analyzed well that how the disaster effects to the community's economic and social life.
- There are no tools to orientate the relevant units to the integrated information.

The basic users of the system are; Civil Defense Units, Administrative Units (governorships, district's head officers, municipalities), Service Sector (banks, insurance companies, construction companies), Research Institutions, Non-governmental Organizations works for the community [8].

TABiS Object Catalog (TABiS-OK)

The base of the Turkey Disaster Information System is Basic Spatial Database. The reference model of the TABiS system comes into existence from two vectoral components. These components are [8];

1. Digital Spatial Model (SMM) and
2. Digital Disaster Model (SAFM)

Both digital models form the space by separating it to its components based on object oriented basis. This process is called as atomizing of the space in the database modeling. The atomized data of the both digital models prepared as an object catalog. These catalogs are [8];

- TABiS-Basic Topographic-Spatial Object Domains Catalog (TABiS-TOK)
- TABiS-Disaster Management Object Domains Catalog (TABiS-AOK)

The aim of the TABiS-TOK is the modeling of the concrete objects which are the characteristic parts of the topography of the region where the system will be constructed. Parallel to this aim, the components of the TABiS-TOK are named as "Basic Topographic-Spatial Object Domains". TABiS-TOK is also has the quality of being a data standard for the country wide public and private institutions who want to set up a detailed spatial information system for their own purposes. Because of the object modeling, object definitions, attribute definitions, data types for the attributes and attribute values can be matched with analog topographic map contents, a disaster management based GIS which is constituted convenient to the TABiS-TOK model can work totally harmoniously with the other GISs of the same region. Even if the aims of the systems are different [8].

A virtual map which was modeled according to the TABiS-TOK is named as "Digital Spatial Model" (SMM), and a virtual map which was modeled according to the TABiS-AOK is named as "Digital Disaster Model" (SAFM).

With the data generalization processes at the Digital Spatial Models which were modeled according to the TABiS-TOK, "Digital Topographic Maps" (STH), and with these STHs, many various "Digital Disaster Management Maps" (SAH) which are constructed by the data aimed in analyzing and managing the disasters are obtained [8].

By using the same method, 1:5000 and 1:25000 scaled "Analog Topographic Maps" (ATH) and similar or different aimed, disaster management based analog thematic maps can be produced.

Modeling approach of the TABiS-OK is given below under the name of TABiS Reference Model [8].

The subjects explained in the Object Catalog are given below as headlines [8]:

1. Concrete Objects with Topographic Characteristics
2. Specialization Objects which are related to the Disaster Management
3. Object Definitions and Attribute Data to define the objects more precisely

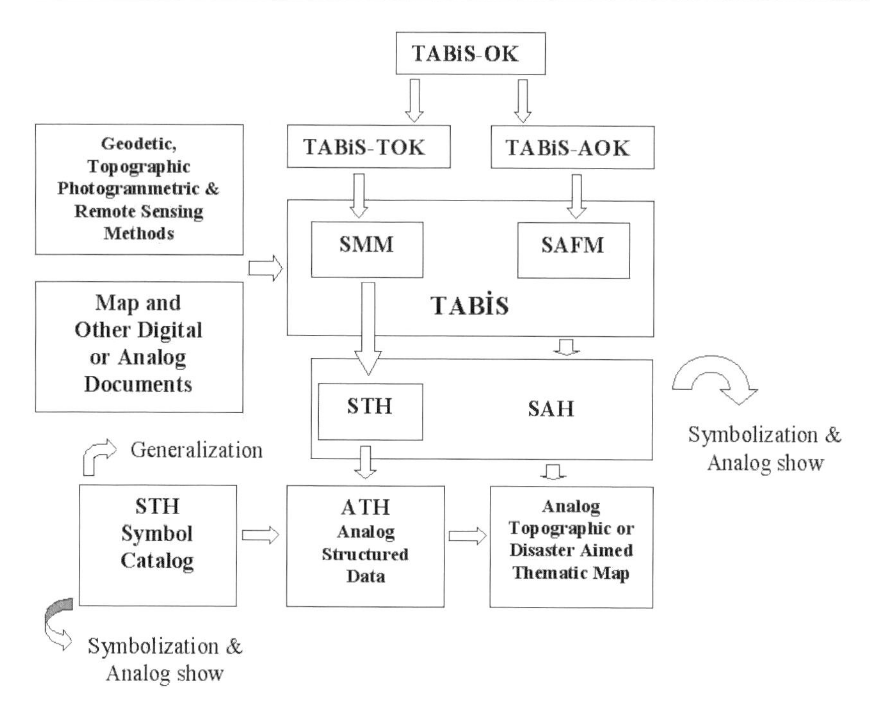

Fig. 1. TABiS Reference Model

Object Concept in TABiS-OK

In TABiS-OK, the object is everything that can be monitored or its existence can be determined with various tools and which has a historicity at the space, existing there, or has a possibility to exist at the space [8].

Existing at the space means that related object has spatial reference or can be geometrically atomized. Atomizing has three types; point, line and area based.

An object with a spatial reference must obtain the conditions listed below [8];

1. Must have a plan metric geometry that the edges with the other objects can be determine
2. Must have topological relations
3. Must have self attributes
4. Must have been defined in any wise

Complex objects are the objects that were formed by combining of concrete object with a new name which were referenced to the space as point, line or area. As an example, a hospital, a veterinary, a clinic and similar objects are combined in the name of "health complex" as a complex object [8].

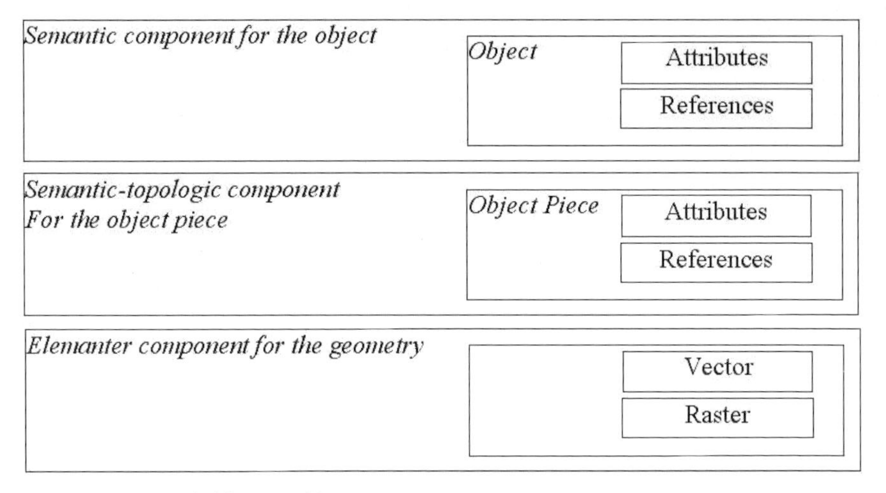

Fig. 2. Data Model in TABiS

An object may have separated piece objects. Any object in the TABiS-OK depended to plan metric-geometric dimensions, can be a point, a line, or an area. One object can not be both or all of them. These characteristics of the objects are named as "Object Type" in the TABiS-OK [8].

Object Group

This concept includes the same kind of objects, like railways, highways. An object group must belong to only one basic topographic-spatial object domain. If an object domain does not have object groups, related object domain is evaluated as an object group, and after the two letters in the code comes a "0". This coding explains that related object domains has only one object group in itself [8].

Object Type

This is the concept that was defined for the same objects as the TABiS-TOK and TABiS-AOK were constituting, like "Deformation Velocity", "Forest Fire", "Religious Buildings", "Monument Tree", etc.. One object type can only belong to one object group [8].

Object Piece (Piece of Object)

Object Piece is the subpart of an object. It has self geometry and self attributes. Object Pieces begin or end on an object where an attribute's or the topologic node point exists. Object Piece absolutely belongs to an object.

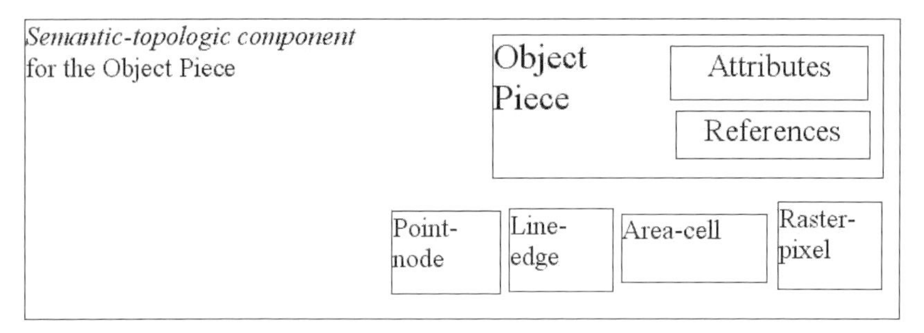

Fig. 3. Topologic Component of the TABiS Object Piece Model

Attribute Data

Attribute values are qualitative or quantitative information about an object or object piece. For example, a buildings year of built, the materials used to built, elevation of the top point, number of the floors and the area of the buildings base [8].

Some of the object type's possible values of the attributes are also given both in the TABiS-TOK and TABiS-AOK. Those attributes must be shown to the user by the system by taking the cardinality degrees into account. Controlling the cardinality degrees by the system has a designating importance on the consistency of the system. The cardinality degree of an object shows the obligation on the appointment of the attribute values to the related object type and single or multiple value situation of the relation.

Disaster Management Object Domains Catalog (TABiS-AOK)

In the concept of TABiS-AOK, individual spatial referenced object domains which were foreseen to be in a GIS for disaster management purposes were taken up. Those domains includes, subject domains that require definite specializations like geology, soil structure, critical establishments, risk areas [8].

Basic object domains, object types and the attribute values for the related object types were organized by constituting the required data for the risk analysis which forms a sound basis for the disaster management. Urban planners, disaster planning experts, geologists, experts for soil structure were applied for data constitution.

As it is widely known, modern disaster management strongly requires disaster planning. It matters too much to use the data which were created special to district by GIS. At the moment of disaster, response and recovery, those data must have been used to decide rapidly and correctly [3].

TABiS-AOK also have "B.G. Plans" object domain which would help on planning or receiving the plans that were created before. The objects created here can be related to the spatial objects and spatial analysis can be made on disaster planning. The "B.F. Environmental Pollution" object domain of the TABiS-AOK has three object groups named as, water pollution, air pollution and noise pollution. The data of these object groups are based on the measured values obtained from one point of an area and used to characterize the related pollution. Those values inserted to the system as an attribute data.

3 Designing the System

In this study, the attributes of the concrete components of object types which are included under the "B.A. Geology" B.B. Zones with a Disaster Risk", "B.C. Critical Establishments", "B.D. Historical and Cultural Zones" and "B.G. Plans" object domains of TABiS-AOK were extended and managed under control of a different database from the Basic Topographic-Spatial Object Domains. The basic reason for this is the spatial reference of the disaster area or on a probable disaster analysis the model area where the disasters may occur is unsteady. The relation of the data managed by the database and the spatially modeled disaster is provided by the object-id of the objects which were created by the GIS software, for various analyses. Disaster management database constituted of a five group of tables.

Conceptual Data Models are known as high level data models. They include some concepts and rules for defining the reality at a high level and independent from any software and hardware. In this study, the Entity-Relationship model used to constitute the conceptual data model. The basic components of the entity-relationship model are entity, attribute, and the relation. With this method, the relations between the entities are explained according to the relational data model. After the data were classified, it is determined that how the related data classes stored in the tables and how these tables related to each other [2].

OBJECT_DOMAINS	
OD_NAME	OD_CODE
GEOLOGY	B.A.
RISK_AREAS	B.B.
CRITICAL_ESTABLISHMENTS	B.C.
HISTORIC&CULTURAL_ZONES	B.D.
SOIL_STRUCTURE &EROSION	B.E.
ENVIRONMENTAL_POLLUTION	B.F.
PLANS	B.G.

OBJECT_GROUPS	
OG_NAME	OG_CODE
NATURAL_DISASTER_RISK	B.B.1
TECHNOLOGIC_DISASTER_RISK	B.B.2
POLITICAL_DISASTER_RISK	B.B.3
ACUTE_DISASTER_ZONE	B.B.4

OBJECT_TYPES	
OT_NAME	OT_CODE
AVALANCHE	B.B.1.01
EARTHQUAKE	B.B.1.02
HAIL	B.B.1.03
COLD_WEATHER/FROST	B.B.1.04
STORM/HURRICANE	B.B.1.05
TORNADO/TWISTER	B.B.1.06
SNOW	B.B.1.07
ROLLING_ROCK	B.B.1.08
DROUGHT	B.B.1.09
FOREST_FIRE	B.B.1.10
FLOOD	B.B.1.11
HOT_WEATHER	B.B.1.12
FOG	B.B.1.13
RAIN	B.B.1.14
LANDSLIP	B.B.1.15
WARMINT_INVASION	B.B.1.16
VOLCANIC_ERUPTION	B.B.1.17

ATTRIBUTES		
ATT_ID	ATT_NAME	OT_CODE
OTR	EVENT_DATE	B.B.1.05
OCK	LOSS_OF_LIFE	B.B.1.05
SSA	DURATION_HOUR	B.B.1.05
MRH	MAXIMUM_WIND_SPEED	B.B.1.05
YON	WIND_DIRECTION	B.B.1.05

ATTRIBUTES_CODES			
ATT_ID	ATT_VALUE	ATT_CODE	OBJECT_ID
YON	SOUTHWESTER	101	10
YON	NORTHEASTER	102	107
YON	NORTHWEST	103	-
YON	SOUTHEASTER	104	1003
YON	DIRECTION_OF_MECCA	105	-
YON	EASTERLY_WIND	106	-
YON	WESTERLY_WIND	107	-
YON	STAR	108	-

To the raster or vector data

Fig. 4. Grouping of the object model

The conceptual model consists of a normalized model that is generally set to third normal form. The third normal form is no data or information which can be acquired from any other row or columns or any other tables must be stored in the database's any field [7].

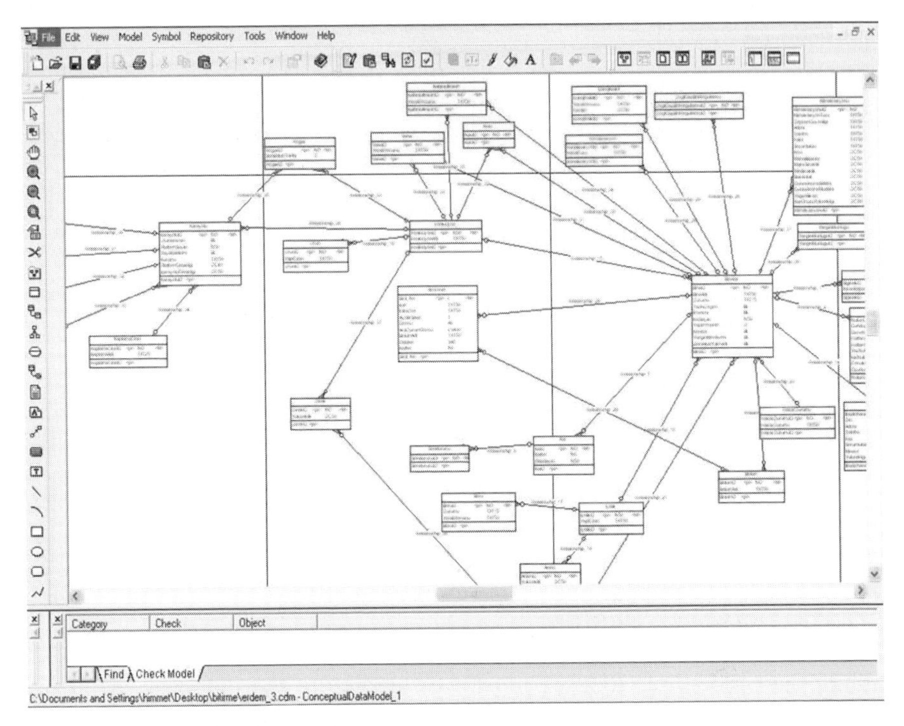

Fig. 5. Conceptual data model of the system

It includes many elements that make up a database, but it is not specific to any software or database implementation. Performance factors are not a major consideration at this point nor are the applications that will be using the database. The main concern is building a model of what the database would look like when capturing the data needed by the users [5]. It is very important to prevent the repetition of the data, while the table structure is designing. The conceptual model can be confronted as logical model in different resources. The physical data model design means the denormalization process. The conceptual model was taken in to consideration and optimized for the queries, specific database implementations, and applications that were planned to talk to the database like Grass GIS Software. Then the physical data model was mapped back to the conceptual data model. The differences found between the two data models and modifications were done according to the physical data model. By this way the physical data model also got some modifications and it continued iteratively until the desired model determined [2].

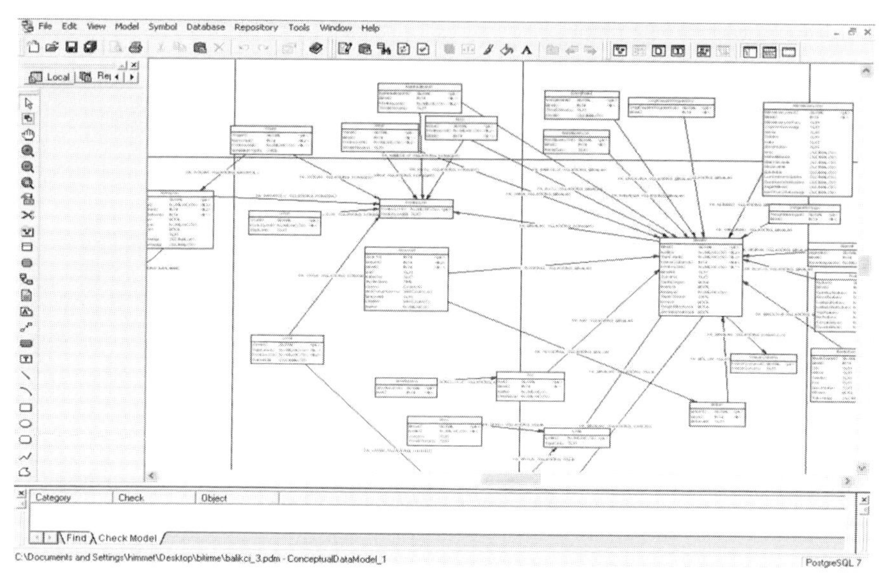

Fig. 6. Physical data model of the system

3.1 The Database Design

While data modeling focuses mostly on depicting the database, database design encompasses the entire process from the creation of requirements, business processes, logical analysis, and physical database constructs to the deployment of the database. For example, in database design, physical data modeling includes the modeling of not only tables and columns but also table spaces, partitions, hardware, and the entire makeup of the database system. Database design includes uncovering and modeling requirements, the business processes (as they are today and where they are going in the future), the activities of the business, the conceptual models, and the physical database models, as well as addressing issues of what information is needed, how the different parts relate, how applications communicate with the data, and how the entire system is implemented [5].

In this study, PostgreSQL is chosen as database management system, because of its high level of security, and the high data storage capacity. The database management system installed to the UNIX based Suse Linux 8.0 operating system via relational structure. PostgreSQL gives good performance and security because of the Linux structure and offers wide alternatives to the user. Almost all UNIX and operating systems derived from UNIX structure like Linux, FreeBSD, etc., can support PostgreSQL. Furthermore all the windows systems that have NT structure can run this database management system. PostgreSQL is free and an Open Source Software [1].

According to the aim of the project, the database should have been accessed and queried from multiple platforms, should have handle high capacity and heavy tasks, should have programming interfaces for multiple programming languages, should have to be run in the multiple processors, and supply fast access on heterogeneous systems which have multiple client. And the most important part is, it must have work on a high level secure structure. The architecture of the PostgreSQL allows simultaneously access from heterogeneous systems. Access from the windows based computers or platforms obtained via Open Database Connectivity (ODBC) support. In addition to this via JDBC, C, C++, PHP, Perl, Tcl, ECPG, Pyton, and Ruby, new interfaces can be added or new and private connection structures can be created. It is also possible to use various ready to use connection and management interfaces [4]. By this way, both user friendly structures, this brings rapid data and work flow and secure data transfer.

In this project, database system server is constituted on Linux operating system and queries made both from Linux and Windows based computer from the network. The query structures and results show no differences according to the platform used.

3.2 Creating the Relation between the Database and the Spatial Model of the Region

The vector formatted campus data imported to the Grass GIS software and point, line and area objects were created according to TABiS-OK to spatially reference the region. This process automatically created object-ids for the each point, line or area object for the GIS software. Because of a shorter way to relate those ids to the database is not known by the system generator this relation process done one by one for the each object.

Geometric objects like lines are used for roads, streams, or utility networks, while areas can represent soil types, land use categories, lakes, or zoning in urban areas. Vector data are stored using their coordinates. In Grass the vector data model includes the description of topology. To assign the attribute information to vector data, a label point is required which links the attribute information to the geometrical data. Grass is capable of managing internally only one attribute per vector, but a quasi infinite number when connecting the system to the PostgreSQL [6].

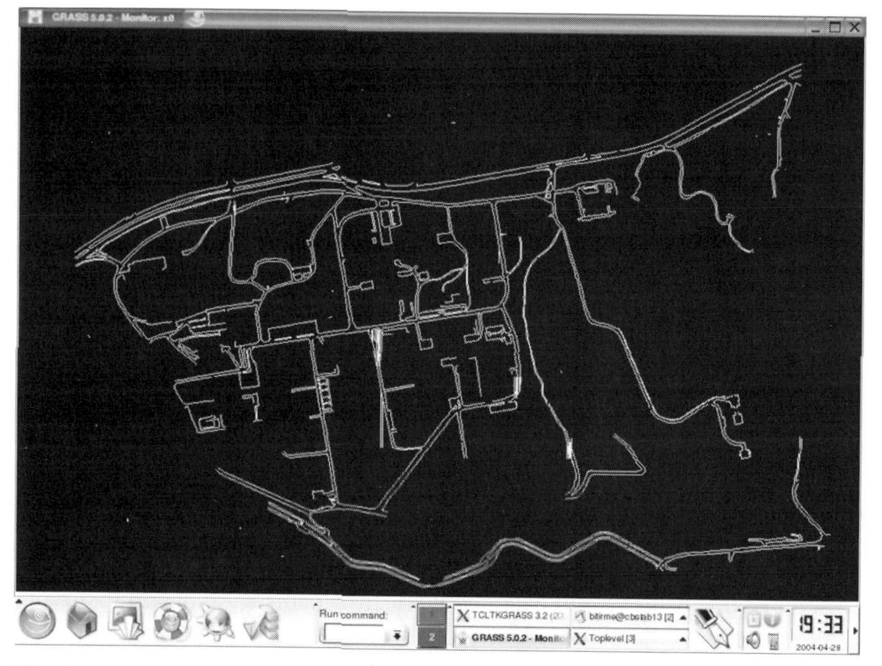

Fig. 7. Query results of the roads of the Campus

4 Conclusions

Using GIS for disaster management is the easiest and the quickest way for these days. While the technology is developing for the information systems it is getting faster to use more complicated systems. However the system mentioned in the paper uses relational model, the object-relational and the object-oriented system modeling are becoming more popular and having more usage areas with the object-oriented programming languages.

The next step for this study must be the updating the system design to the object-oriented model and creating user defined interfaces for both the database and the GIS parts.

References:

[1] Karaman H (2003) I.T.U. Campus Disaster Information System; Constitution of the Emergency Management Based Object Model and Construction of the Related Queries. MSc. Thesis, I.T.U. Institute of Science and Technology, Istanbul

[2] Karaman H, Sahin M (2004) I.T.U. Campus Disaster Information System; Constitution of the Emergency Management Based Object Model and Con-

struction of the Related Queries. 24th Urban Data Management Symposium, Chioggia, Italy, October 27-29, 2004

[3] Karaman H, Sahin M, Uçar D, Baykal O, Türkoğlu H, Tarı E, İpbüker C, Musaoğlu N, Göksel Ç, Coşkun MZ, Kaya Ş, Yiğiter R, Erden T, Yavaşoğlu H, Bilgi S, Üstün B (2002) GIS Standards of Turkey based on Emergency Management. International Symposium on Geographic Information Systems, İstanbul, Turkey, September 23-26, 2002

[4] Momjian B (2001) PostgreSQL: Introduction and Concepts. Boston, MA: Addison-Wesley

[5] Nailburg EJ, Maksimchuk RA (2001) UML for Database Design. Addison Wesley

[6] Neteler M, Mitasova H (2003) Open Source GIS: A GRASS GIS Approach. Kluwer Academic Publishers, Boston Dordrecht London

[7] Perkins J (2001) PostgreSQL: A Better Way to Manage Data with PostgreSQL. Premier Press, Indianapolis, Indiana 46204

[8] Sahin M, Uçar D, Baykal O, Türkoğlu H, Tarı E, İpbüker C, Musaoğlu N, Göksel Ç, Coşkun MZ, Kaya Ş, Yiğiter R, Erden T, Karaman H, Yavaşoğlu H, Bilgi S, Üstün B (2002) TABİS: Development of a National Database Using Geographical Information Systems (GIS) and Remote Sensing System and Standards for a Disaster Management Decision Support System Project, I.T.U., Istanbul

Development of a Web-Based GIS Using SDI for Disaster Management

Ali Mansourian[1], Abbas Rajabifard[2] and Mohammad Javad Valadan Zoej[1]

[1] Faculty of Geodesy and Geomatics Engineering, K.N.Toosi University of Technology, Vali-e-Asr St., Mirdamad Cross, Tehran, Iran.
Email: alimansourian@yahoo.com, valadanzouj@kntu.ac.ir
[2] Department of Geomatics Engineering, The University of Melbourne, Melbourne, Australia.
Email: abbas.r@unimelb.edu.au

Abstract

Spatial data and related technologies, particularly Geographical Information System (GIS) with the capability of display, retrieval, analysis and management of spatial data, have proven crucial for disaster management. However, there are currently different problems with availability, access and usage of spatial data for disaster management. The problems with spatial data become more serious during disaster response in which reliable and up-to-date spatial data describing current emergency situation are required for planning, decision-making, and coordination of activities. It is suggested that by having an effective and efficient spatial data framework and institutional arrangements and through cooperative efforts of involved organizations in disaster response for spatial data production and then sharing these data, it is possible to have required information, always available and accessible for use. In this respect, Spatial Data Infrastructure (SDI) can provide an appropriate environment for influencing participation of organizations in spatial data production process and sharing. Meanwhile, web-based GIS can provide appropriate tool for data entry and sharing as well as data analysis to support decision-making.

This paper aims to describe development of a web-based GIS using SDI framework for disaster response. It is argued that the design and implementation of an SDI model and consideration of SDI development factors and issues, together with development of a web-based GIS, can assist disaster management agencies to improve the quality of their decision-making and increase efficiency and effectiveness in all levels of disaster management activities. The paper is based on an ongoing research project on the development of an SDI conceptual model and a prototype web-based system which can facilitate sharing, access and usage of spatial data in disaster management, particularly disaster response.

1 Introduction

Disasters have long been presented a tragic disruption to the humans' lives, properties, infrastructure, economy, capital investment and development process. In today's world that different nations around the world aim achieving sustainable development, disasters would present a major threat to such an aim, or a sign of its failure. Under such situation, different nations around the world have had particular attention on managing disaster as a part of their activities. Meanwhile, the recent terrorist activities such as World Trade Center and Pentagon attack on September 11, 2001 and October 2002 bombing in Bali, have increased worldwide attention on disaster management.

Disaster management is a cycle of activities (Figure 1) beginning with *mitigating* the vulnerability and negative impacts of disasters; *preparedness* in responding to operations; *responding* and providing relief in emergency situations such as search and rescue, fire fighting, etc.; and aiding in *recovery* which can includes physical reconstruction and the ability to return quality of life to a community after a disaster.

Fig. 1. Disaster Management Activities Cycle

Considering that most of the required information for disaster management has spatial component or location (Cutter et. al 2003), GIS with the capability of spatial data display, analysis and management has proven crucial in detecting, mitigating, preparing for, responding to, and recovering from disasters (Amdahel 2002). However, current studies show that although spatial data and GIS can facilitate disaster management, there are substantial problems with collection, access, dissemination and usage of required spatial data for disaster management (SNDR 2002 and Jain and McLean 2003). Such problems become more serious in the disaster response phase, with its dynamic and time-sensitive nature.

Disaster response is dynamic and decision-makers need to be updated on the latest emergency situation. Disaster response is also time-sensitive with little allowance on delay in decision-making and response operations. Therefore, any problem or delay in data collection, access, usage and dissemination has negative impacts on the quality of decision-making and hence the quality of disaster response. With this in mind, it is necessary to utilize appropriate frameworks and technologies to resolve current spatial data problems for disaster management.

It is suggested that Spatial Data Infrastructure (SDI) as an initiative in spatial data management can be an appropriate framework and a web-based system can be an appropriate tool for resolving current problems with spatial data. With the other word, web-based GIS and SDI as an integrated framework can facilitate and improve disaster management, particularly disaster response, by resolving current problems with collection, dissemination, access and integration of spatial data for use.

This paper aims to describe the development of an SDI conceptual model and a prototype web-based system that facilitate spatial data collection, access, dissemination, and usage for proper disaster management. This is based on an ongoing research and case study in Iran which investigates the role of SDI in disaster management with emphasize on the response phase.

2 Collaboration in Data Collection and Sharing

Different organizations (such as Fire, Medical and police departments; Red Cross Society; and Utility Companies) collaborate in disaster management activities due to diversity of disaster response operations. Inter-organizational coordination of disaster response operations and controlling the emergency situation is generally conducted through Emergency Operation Center (EOC) where the representatives of involved organizations are gathered.

Considering search, relief, rescue, firefighting, medical service, debate removal, sheltering, and repairing utility network as some examples of disaster response activities, a large number of spatial data layers are required for planning and coordinating such operations. In this respect Road network, closed road, hospital, disaster area, damaged building, location of victims, location of emergency workers, available resources, and utility network are some examples of required spatial data layers for disaster response operations.

Due to dynamic nature of emergency situation, required data for disaster response should be collected regularly in order to be *available* for decision-makers. However, due to variety of required data, individual involved organizations in disaster management activities can not handle exclusively the collection and maintenance of all required data layers for disaster response. As a result, collection and maintenance of required spatial data layers for disaster response should be conducted based on collaborative effort of different organizations for spatial data collection and updating. If so, required spatial data layers are always available and accessible for producer. The required datasets should also be *accessible* for decision-makers (involved organizations and EOC) to be utilized for planning and decision-making purposes. This is achieved if collected data by each of the participants in data collection to be shared to wider disaster management community (Mansourian et al 2004 and Rajabifard et al 2004).

One of the challenges of this collaborative effort in data collection and sharing (Figure 5.5) is to choose the collaborator organizations. In this respect, Mansourian *et* al. (2004) highlighted that considering their daily and disaster response businesses, involved organizations in disaster management community are potentially the main producers and maintainers of required spatial data for disaster response, based on and during their normal or disaster response activities. If such potential is turned into act and the results of data production and updating efforts are physically recorded in appropriate databases, the required spatial data for disaster response is always available to the producer. As described earlier, by sharing available data, they will be accessible to other organizations.

In addition, the required datasets need to be easily integratable with each other and interoperable with decision-makers' systems for real-time use. This is achieved by utilization of common and appropriate standards and specifications for data collection and sharing in the mentioned collaborative effort.

Although a collaborative effort for spatial data collection and sharing can resolve the problem with collection, access and dissemination of required spatial data for disaster response, however, different researches on collaborative efforts for data collection and sharing (Rajabifard and Williamson 2003; McDougall *et* al. 2002; Nedovic-Budic and Pinto 1999) show that there are different technical, institutional, political, and social issues that create barriers for such participation to occur. With this in mind, by creating an environment in which such issues are taken into consideration and resolved and consequently the access of decision-makers to spatial data is facilitated, the concept of partnership in data production and sharing can become a reality. In this respect, Spatial Data Infrastructure (SDI), as an initiative in spatial data management with related concepts and models, can be used as a framework for creating such an environment and consequently, facilitating disaster response.

3 Role of Spatial Data Infrastructure in Disaster Management

Spatial Data Infrastructure (SDI) is an initiative intended to create an environment that will enable a wide variety of users to access, retrieve and disseminate spatial data in an easy and secure way. In principle, SDIs allow the sharing of data, which is extremely useful, as it enables users to save resources, time and effort when trying to acquire new datasets by avoiding duplication of expenses associated with generation and maintenance of data and their integration with other datasets. SDI is also an integrated, multi-leveled hierarchy of interconnected SDIs based on collaboration and partnerships among different stakeholders. With this in mind, many countries are developing SDIs to better manage and utilize their spatial data assets. As a result of these activities different models have been suggested for facilitating SDI development.

Recent studies on SDI initiatives (Rajabifard and Williamson 2003) have highlighted that development of SDIs is a matter of different challenges such as social, cultural, political and economical challenges beside technical issues.

With respect to core components, an SDI encompasses the policies, access networks and data handling facilities (based on the available technologies), standards, and human resources necessary for the effective collection, management, access, delivery and utilization of spatial data for a specific jurisdiction or community (Rajabifard et al 2002). Based on these components, Figure 2 illustrates a basic SDI model. According to this model, appropriate accessing network, policies and standards (which are known as technological components) are required for facilitating the relation between people (data providers, value-adders and decision-makers in disaster management community) and data.

By clarifying each of these core components, an SDI conceptual model can be developed which can contribute to facilitating the availability, access and usage of spatial data for disaster management and hence facilitation of disaster management.

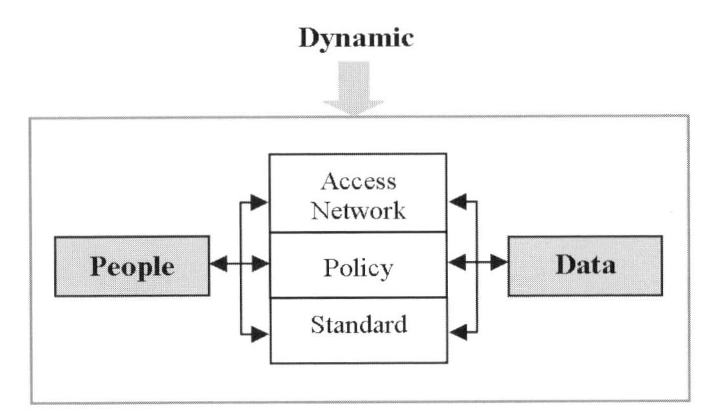

Fig. 2. SDI Components (Rajabifard et. al 2002)

Considering Geographical Information System (GIS) as underpinning technology for SDI and its role in facilitating data collection and storage as well as facilitating decision-making based on spatial data processing and analysis, GIS is a good tool for improving decision-making for disaster management. In this respect, a web-based GIS can be a good tool for facilitating disaster management due to need to high interaction between decision-makers in disaster management community, particularly during disaster response.

With this in mind, a web-based GIS using SDI can facilitate disaster management by providing a better way of spatial data collection, access, management and usage.

4 SDI Conceptual Modeling and Development of Web-Based System for Disaster Management – A Case Study

With respect to above description, a research study has been designed and conducted in Iran with the aim of development of an SDI conceptual model and a prototype web-based system for disaster response. Main steps of this research included:

- Assessing disaster management community from different technical and non-technical perspectives with respect to spatial data,
- Development of an SDI conceptual model, based on the results of the assessment,

- Development of a web-based GIS based on the SDI conceptual model,
- Conduction of a pilot project to test the developed SDI conceptual model and prototype web-based GIS, and
- Refinement of the SDI conceptual model and the developed prototype web-based system.

At the first stage disaster management community was assessed with respect to spatial data and those technical and non-technical factors that affect development of SDI. Results of organizational assessment showed that development of SDI for disaster management in Iran is a matter of *social*, *technical and technological*, *political*, *institutional* and *economical* challenges. Based on the results of organizational assessment, at the second step, the SDI conceptual model was developed by examining and expanding each of the components of SDI within the context of disaster response. This model is a framework that can create an appropriate environment for participation of organizations in collection, sharing and usage of spatial data for disaster management.

At the third step, a prototype web-based system using GIS engine with a user-friendly interface was also developed as a tool for spatial data collection, sharing and analysis. Figure 3 shows the overall structure of this system. As Figure 3 shows the web-based system is based on five core components including user interface for clients to access and analyze data, web server and application server for getting the clients' request and sending it to map server, map server for data analysis and query based on clients' request, data server for retrieving data from a database and serving them to map server for analysis, and database that includes spatial data.

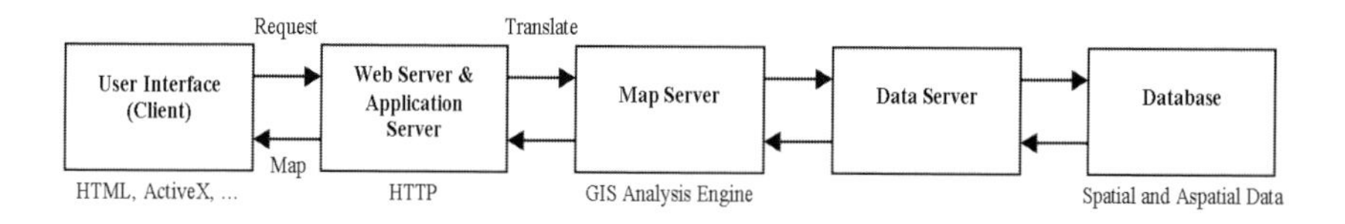

Fig. 3. Core components of web-based System and their relations

At the fourth step, a pilot project was conducted. This pilot was conducted in Tehran, the capital of Iran with collaboration of different organizations from disaster management community in order to test the web-based system and developed SDI conceptual model. Considering the important role of awareness for SDI development, increasing the awareness of disaster management community on advantages of developed system that works using SDI, was another aim of this pilot project. This pilot project was about responding to an assumed earthquake in Tehran.

In this pilot, a maneuver scenario was defined with which involved organizations could experience a coordinated disaster response based on spatial data sharing and analysis. During the maneuver, each organization updated its own spatial datasets within responding operations, and shared them with the disaster response community. Therefore each individual responding organization had access to required spatial datasets to integrate and analyze their datasets using GIS functionalities to support their own decision-making for disaster response.

At the last step, based on the results of the pilot project, the developed prototype web-based system and the developed SDI conceptual model were refined. Figure 4 shows the schematic presentation of the developed SDI conceptual model for disaster response.

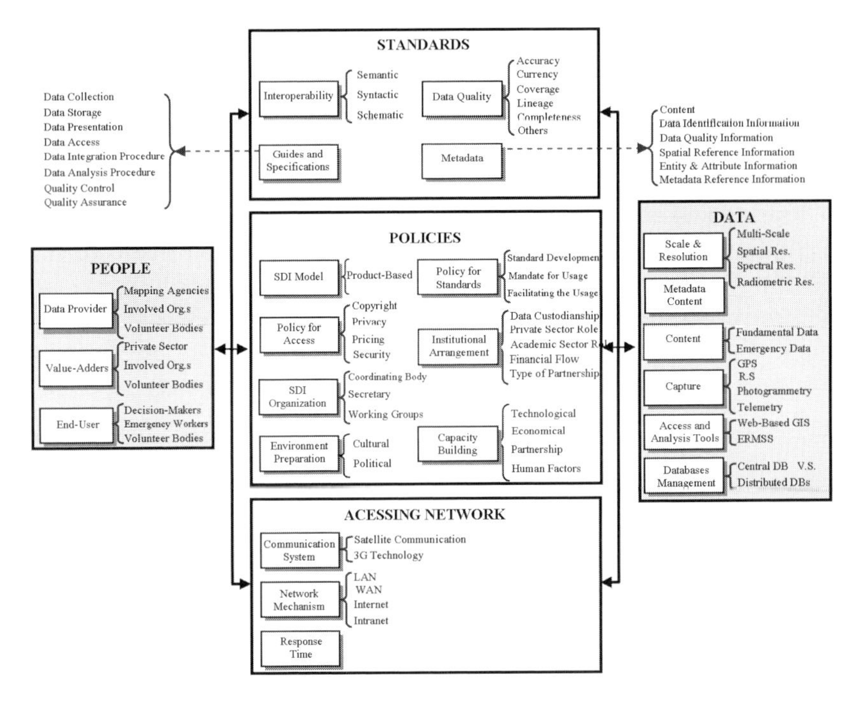

Fig. 4. Schematic presentation of the developed SDI conceptual model for disaster response

5 Conclusion

The results of the case study and its pilot project showed how useful a web-based system that works using SDI can be for effective and efficient disaster response management. Using SDI framework, reliable and up-to-date spatial data for disaster response is always available and accessible for decision-makers. A web-based system is also an appropriate tool which can be used for data analysis and consequently coordinating and controlling emergency situation.

The effectiveness and efficiency of the system can be interpreted by different elements, however, in this research reducing the response time and removing chaos by better management and coordination were considered as two evaluating factors.

It should be noted that such SDI conceptual model and web-based system facilitates and improves not only disaster response, but also other phases of disaster management including mitigation, preparedness and recovery.

References

Amdahl G (2002) Disaster Response: GIS for Public Safety, Published by ESRI, Redlands California, http://www.esri.com/news/arcnews/winter0102articles/gis-homeland.html - visited on October 2002

Cutter SL, Richardson DB and Wilbanks TJ (2003) The Geographic Dimension of Terrorism, New York and London: Toutledge

Jain S and McLean C (2003) A Framework for Modeling and Simulation for Emergency Response, Proceedings of the 2003 Winter Simulation Conference, Fairmont Hotel, New Orleans, Louisiana, USA.

Mansourian A, Rajabifard A, Valadan Zoej MJ, Williamson IP (2004) Facilitating Disaster Management Using SDI, Journal of Geospatial Engineering, vol 6, no. 1, June 2004, Hong Kong

McDougall K, Rajabifard,A, Williamson IP (2002) From little things big things grow: building the SDI from local government up, Joint AURISA and Institution of Surveyors Conference, 25-30 November 2002, Adelaide, South Australia

Nedovic-Budic Z and Pinto JK (1999) Understanding inter-organizational GIS activities: a conceptual framework, Journal of Urban and Regional Information Systems Association, vol. 11, no. 1, pp53-64

Rajabifard A, Mansourian A, Valadan Zoej MJ and Williamson IP (2004) Developing Spatial Data Infrastructure to Facilitate Disaster Management, Proceedings of Geomatics 83, National Cartographic Center, Tehran, Iran

Rajabifard A, Williamson IP (2003) Anticipating the cultural aspects of sharing for SDI development, Spatial Science 2003 Conference, 22-26 September, Canberra, Australia

Rajabifard A, Feeney MEF, Williamson IP (2002) Future Directions for SDI Development, International Journal of Applied Earth Observation and Geoinformation, ITC, the Netherlands, vol 4, no 1, pp 11-22

SNDR (2002) A National Hazards Information Strategy: Reducing Disaster Losses Through Better Information, National Science and Technology Council, Committee on the Environment and Natural Resources, Subcommittee on Natural Disaster Reduction (SNDR), Washington, DC, April 2002

Visual System for Metric 3D Data Gathering and Processing in Real Time

Petr Rapant, Jan Stankovic, Eduard Sojka and Emil Gavlovsky

VSB - Technical University of Ostrava 17. listopadu 15,
708 33 Ostrava – Poruba, Czech Republic
Email: petr.rapant@vsb.cz, jan.stankovic@vsb.cz
eduard.sojka@vsb.cz, emil.gavlovsky@vsb.cz

Abstract

Emergency events can create situations, which exclude presence of people in hazardous places or at least create conditions hostile or dangerous for people. Using special equipment for assessment of a situation can help to do some operations without direct presence of people in place. There are some stationary or mobile means, mostly remotely controlled, which can be directed or navigated to risky place to do some visual investigation. These tools are usually equipped by simple camera system, which permits to do some visual investigations but they provide no metric information about observed scene. Presented paper describes modular system which permits 3D metric measurements in dangerous or inaccessible places.

1 Introduction

Emergency events can create situations, which exclude presence of people in hazardous places or at least create conditions hostile or dangerous for people. Using special equipment for assessment of a situation can help to do some operations without direct presence of people in place. Especially emergency response teams would appreciate such system. But it can be used also under other circumstances like terrorist attacks, police, Special Forces and military operations, etc.

There are some systems used for reconnaissance purposes, for example [1], [2] or [3]. The common feature of these systems is that they use only one camera system, which usually does not permit to gather stereo pairs of images to do 3D metric measurements (like measurements of distances, sizes or profiles).

Nowadays technology like small frame video cameras, small (but very powerful) computers, wireless communication tools on the one side and small and powerful robot systems on the other side gives us tools for creating new instruments to avoid above mentioned disadvantage.

This paper reports present state of development of a such system.

2 Conception

We have decided to develop new mobile system with following features:

- Low cost solution,
- Modularity so we can combine different components which best fit special conditions without losing power,
- Using non-metric small frame digital (video) cameras,
- Using mobile robot as a primary mobile platform,
- Fully digital processing,
- 3D measurements in nearly real time,
- Wireless communication and control (as a final solution),
- Real time processing of 3D data,
- Post-processing on professional digital photogrammetric workstation.

That is why the team is composed from people representing different specialties:

- Digital image processing,
- Phogrammetry,
- Robot technology,
- Radio electronics,
- Computer hardware and communications.

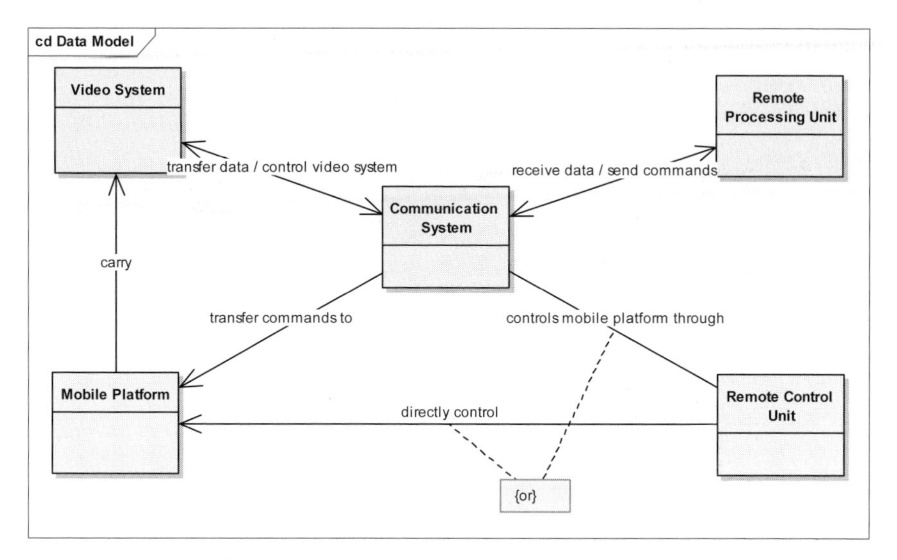

Fig. 1. General structure of system under development

Final conception is presented in the figure 1. The developing system is composed from following modules:

- Mobile platform – robot, helicopter, airplane,
- Visual system – two high resolution non-metric digital cameras or two medium resolution digital video cameras, mounted on rotary arm, of single high resolution non-metric digital camera (in the case of helicopter or airplane),
- Positioning system – GPS, DGPS, EGNOS,
- Communication system – wire connection, wireless connection,
- Remote control – RC system, computer control system, remote desktop,
- Remote processing – our own software, professional digital photogrammetric workstation (for example ERDAS Imagine), and
- On-board processing unit – small PC-based computer for in-place processing of image data gathered by visual system (alternatively).

It is clear that every module has some alternatives.

Fig. 2. Chassis of the robot – development version

The **mobile platform** is equipped by many sensors:

- a pair of cameras for gathering stereo pairs of images, mounted on rotary arm,
- a PC camera for direct on-line visual investigation and navigation,
- an electronic compass to get direction of camera system,
- an GPS receiver to get absolute position if it is possible,
- a pair of inclinometers to be able to put camera system to horizontal position,

which permit, among others, to determine absolute or at least relative position of the camera system in the space. Absolute position is possible to obtain only in the case of open-air measurements. GPS receiver is not usable in buildings. But at this case we are able to measure at least direction of camera system and fix it in horizontal position. Open-air measurements of position will be very exact in most cases. This is ensured by online DGPS corrections and by use of wide area augmentation systems in places without DGPS range. For future stages of the development we are also planning to implement inertial sensors (electronic accelerometers) that will allow us to keep knowledge of absolute position for several minutes without GPS signals.

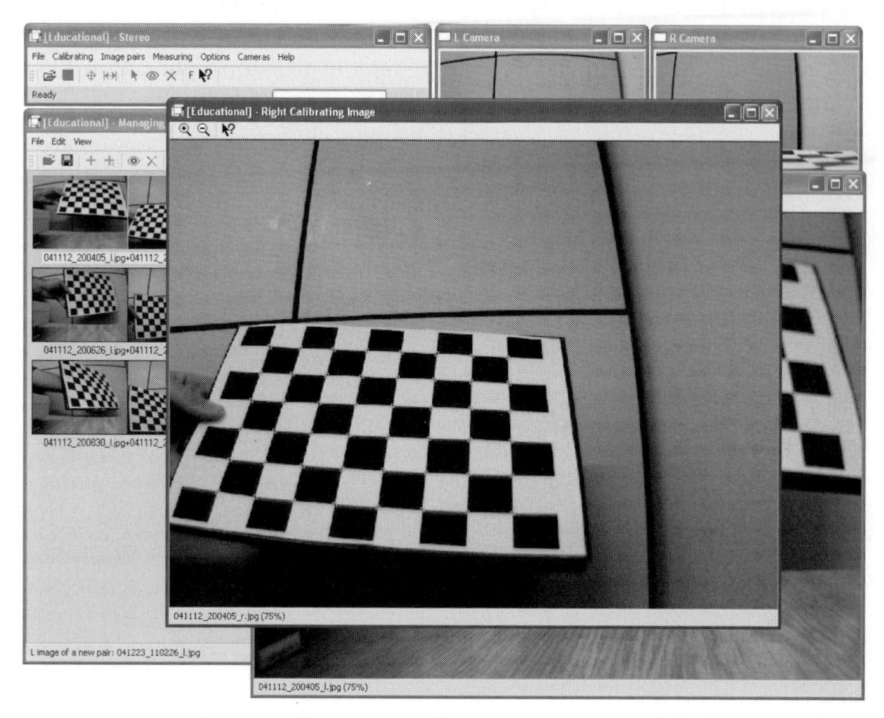

Fig. 3. Example screen shot of software developed

The main chassis of the robotic platform is made of aluminum alloys combined with steel parts and equipped with caterpillar bands. Caterpillar crawler gives the robot excellent mancuvering abilities (it may turn 360 deg. right on the place) and allows it to pass trough rather rough terrain. We will use modified design of such a chassis that was developed by department of robotics at our university [4]. Development version (but fully functional) is shown in the figure 2.

The **optical measurement system** (or video system) is composed of two digital cameras DFK 31F03 (by Imaging Source) with resolution 1024x768 pixels and IEEE 1394 interface (FireWire), mounted on rotary arm (see figure 4). Cameras are equipped with H416 (Pentax) lenses with 4 mm fixed focus. Physical connection between cameras and computer processing unit is done by two separate IEEE 1394 PCI interface cards. The imagery data are transferred by communication system to remote processing unit.

Alternatively, on-board processing unit can be situated at mobile platform and all the necessary processing can be done directly on the mobile platform which can be used as a server for 3D data gathering and process9ing. In such case the remote processing unit plays role of remote client – all processing software is running on mobile platform, remote unit use only remote management console (e.g. remote desktop connection from Windows XP) to control whole the process of 3D measurements.

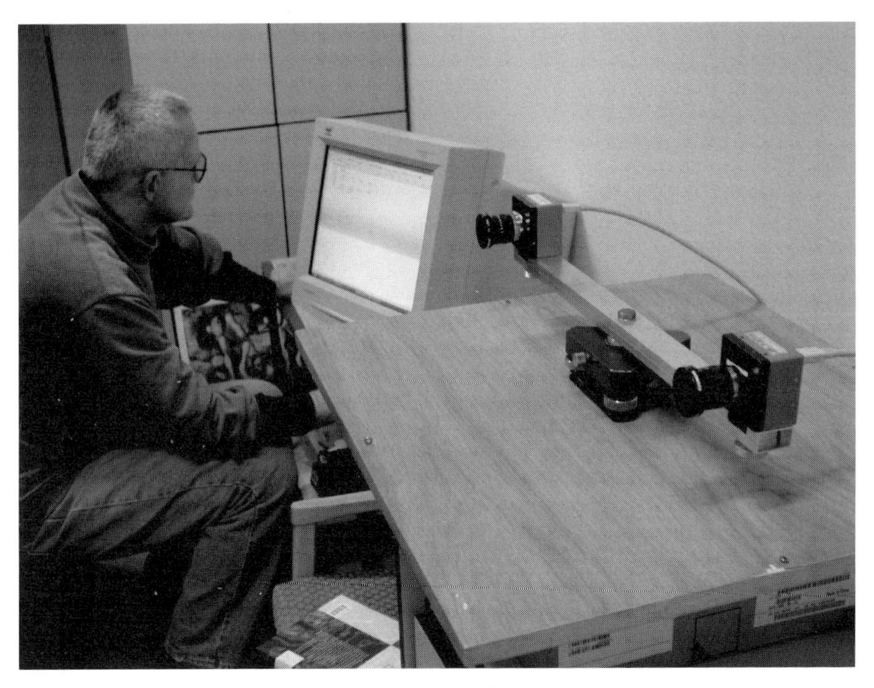

Fig. 4. Example screen shot of software developed

This configuration reduces volume of data transferred from platform through communication system up to remote processing unit and helps to solve potential bottleneck of the system. Due to very wide data streams (over 270 Mbps) from each of FireWire cameras and not existing wireless technology capable to transmit so much data in real time. This is good reason to do the primary data processing directly on-board. For this reason we use powerful computing on-board unit derived from IBM PC platform (modified micro ATX board, 2 GHz Mobile Celeron CPU, 512 RAM, 40 GB HDD).

We have developed special software for 3D data gathering, preprocessing and on site 3D measurements. All the functions are integrated into single program environment that ensures comfort of the user with little need for manual operations (see figure 3).

In case that measuring system is calibrated it is possible to provide measurements. We can determine point coordinates, line lengths (distances) and lines angles in the scene. Polygon measurements (generation) is under development. Measurement is done with assistance of operator – that means it is necessary to mark on the images what should be measured. Measured values may be commented with notes. Software for 3D measurements creates and incrementally updated measurements file. The file can be edited (add more measurements, drop unnecessary values). Basic task that had to be implemented here has been determination of 3 dimensional coordinates of the points in the scene associated to relative coordinate system (in our case coordinate system of the left (L) camera). Computation of coordinates has been solved like optimization task – by minimization of sum of length square distances between theoretical and real projections of reconstructed points. Implementation of remaining mentioned measurements (lengths, angles, polygons) is elementary while knowing coordinates of corresponding points.

Communication and control system is based on several wireless technologies. In fact we are using three separate technologies. For movement control of the platform and of the rotary arm we use R/C interface (manual control panel common in hobby modeling) that gives the operator maximal sensitivity and freedom of move together with long range accessibility even in problematic environments. Visual data from movable navigation camera are transmitted with use of analog radio system. The reason for analog system is again in long range accessibility in problematic environments. All other communication is done by 802.11 b/g (WiFi) technology. Robotic platform is equipped with WiFi access point that ensures multiuser access to the resources (computer unit, sensory systems, etc.) and to data stored on-board. For simplicity we are using standard TCP/IP protocol and standard services over it. The computing unit runs secure ftp server (FTPS), SSH server, http server (web) and two types of remote desktop sharing that allow direct control and configuration of the operating system and system environment. With this solution the robot in fact acts to users as any ordinary network server. Multiple users may download stored imagery, see the view of "robots eyes" and cooperate in this way. Security of data transfers is managed by encryption algorithms implemented in hardware and software layers of wireless networks (WEP) and TCP/IP protocol and services over it (SSL, SSH).

Primary data processing is done directly on-board as mentioned before. Capture of image stereo pairs suitable for precise measuring is done manually by operator using remote desktop feature of the system or automatically in scheduled time intervals. Both ways require previous stabilization of rotary arm to horizontal position. Near real time post processing (measurements) may be done by on-board processing unit with our software as well as using any other computer that has access to the data (images) stored on the robot. Post processing may also be done later, with use of advanced photogrammetric software like ERDAS Imagine. We are planning to use Tablet PC computer (Penbook with WiFi) to communicate with robot while working in the field and to perform near real time measurements. The reason we want to use Tablet PC is great mobility it gives the operator while working outdoors and walking during the work. The other but related reasons are natural control with use of touch screen and voice commands recognition systems.

The modular system can do these tasks in the lab:

- single camera calibration,
- double camera calibration,
- calibration of compass,
- calibration of leveling sensors,
- checking of the whole system.

Modular system can do these tasks in the field:

- on a site camera recalibration/calibration check,
- move mobile platform to necessary position,
- support visual investigation of site using single PC camera,
- direct the double camera system to the target,
- obtain stereo pair of images,
- transfer the images to remote processing unit,
- nearly real time visual 3D investigation of the site,
- nearly real time 3D measurements on the observed scene,
- 3D post processing of the gathered images using ERDAS Imagine.

3 Results

We have developed some of mentioned above modules until now and have performed successful separate testing of particular modules. Photogrammetric and data capture software is finished and functioning well. The software has been successfully tested with hardware it will work on at the time the whole robotic device will be completed. Robotic platform is also done and tested, but we want to rebuild it to make it lighter. Construction design of the rotary arm is prepared as well as electronic microprocessor unit that will take care of vertical stabilization and rotation monitoring of the arm holding cameras. The robot should be completely put together and first on-board tests performed by the end of March. We have also tested wireless remote control of the photogrammetric system – online hi-resolution picture view, data capture, measuring, and other software functionality. For the tests we have successfully used HP TC 1100 Tablet PC with Buffalo WiFi cards at Ad-Hoc mode. The project should be finished up to the end of 2005.

Acknowledgement

This project is supported by Grant Agency of the Czech Republic, contract number 105/03/0719, named "Visual system for gathering and processing of metric 3D information in real time".

References

[1] Terzic O (2004) Robotska izvidnica. Ekonomist on-line.
http://www.ekonomist.co.yu/magazin/em206/nit/nit1htm
[2] --- (2004) Linux bazeti roboti dodas uz Iraku.
http://www.capital.lv/index.php?id=8489
[3] --- (2004) About ROBHAZ. KIST Intelligent Robotics Research Center.
http://www.robhaz.com/about_dt3_main.asp
[4] Turon M (2004) Mobile robotic system on caterpillar bands
http://robot.vsb.cz/uspechpos/turon2003/TURON2003.htm

Risk Assessment Using Spatial Prediction Model for Natural Disaster Preparedness

Chang-Jo F. Chung[1], Andrea G. Fabbri[2,3], Dong-Ho Jang[4] and Henk J. Scholten[2]

[1] Geological Survey of Canada, 601 Booth St., Ottawa, Canada K1A 0E8. Email: chung@gsc.nrcan.gc.ca

[2] SPINlab, Vrije Universiteit, De Boelelaan 1087, 1081 HV Amsterdam, the Netherlands.
Email: andrea.fabbri@ivm.vu.nl; hscholten@feweb.vu.nl

[3] University of Milano-Bicocca, Piazza della Scienza 1, 20126 Milan, Italy.

[4] Gong-Ju National University, Gong-Ju, Korea.
Email: rsgis@gongju.ac.kr

Abstract

The spatial mapping of risk is critical in planning for disaster preparedness. An application from a study area affected by mass movements is used as an example to portray the desirable relations between hazard prediction and disaster management. We have developed a three-stage procedure in spatial data analysis not only to estimate the probability of the occurrence of the natural hazardous events but also to evaluate the uncertainty of the estimators of that probability. The three-stage procedure consists of: (i) construction of a hazard prediction map of "future" hazardous events; (ii) validation/reliability of prediction results and estimation of the probability of occurrence for each predicted hazard level; and (iii) generation of risk maps with the introduction of socio-economic factors representing assumed or established vulnerability levels by combining the prediction map in the first stage and the estimated probabilities in the second stage with socio-economic data. Three-dimensional dynamic display techniques can be used to obtain the contextual setting of the risk space/time/level distribution and to plan measures for risk avoidance or mitigation, or for disaster preparedness and risk management. A software approach provides the analytical structure and modeling power as a fundamental tool for decision making.

1 Introduction

Risk is a condition that we all have to bear in our daily life and that, when known, we should try to avoid or minimize. Natural risk is due to processes that take place independently of human presence, while technological risk has human processes and activities as the main cause. The concept of risk relates to the probability that an event, either natural or human induced, affect human presence and activity and indeed it concerns the geographical distribution of human settlements. An extensive set of risk terms has been defined by the Society of Risk Analysis (www.sra.org/resources_glossary.php). A distinction is made between **risk analysis** that deals with the quantification of the probabilities and expected consequences of identified risks, and **risk evaluation** that deals with the socio-political and moral-ethical component in which judgments are made about the significance and acceptability of risks. In addition, **risk management** is an activity aimed at reducing risks that are found to be unacceptably high. It combines risk analysis and risk evaluation in a decision process to implement economic and technical measures (Plattner, 2004).

This contribution focuses on **spatial risk assessment**, i.e., the methods of generating risk maps in which the different levels of relative risk are represented for analysis, evaluation and management. Industrial risk management is often based on the risk management cycle of identification, analysis, mitigation and follow-up, or simply identification, prioritization, response and control, using risk check lists should risk maps be unavailable. Natural risk management has to be based on the distribution of natural processes whose spatial boundaries and characteristics need to be recognized, understood and represented in order that remedial action can be taken. Examples of important actions with respect to natural risks are pre mitigation strategies in development planning and hazard preparedness. Hazard is the probability estimate that a damaging event will occur in a given area at a given time. For instance, rather than wait for a disaster to occur, society can take action to prevent, minimize or mitigate the impact of a disaster to reduce social vulnerability (Hansson, 2004; www.unisdr.org).

Recent examples of risk maps of landslide hazards (Komac, 2004; Bell and Glade, 2004; Liguori and Mortellaro, 2004) provide some ways in which the territory is zoned into classes of risk levels for different types of risks, e.g., risks to life, to assets, to economy, etc. Clearly, without the spatial expression of the risk levels provided by such risk maps, natural risk management would hardly be feasible. This contribution discusses how, with spatial data models hazard prediction can be empirically validated in order to obtain risk maps to potentially bridge the gap with disaster management. The process of using hazard predictions, vulnerability analyses and risk assessments must be linked to the subsequent stage of disaster preparedness.

An application example to a study area affected by landslide activity allows discussing how risk maps can be visualized and used for disaster preparedness planning. Works by the authors in the last ten years have proposed a variety of predictive models (Chung and Fabbri, 1993, 1998, 1999, 2001 and 2005) and analytical strategies of validating the results of spatial predictions (Chung and Fabbri, 2003; Fabbri *et al*, 2002). More recently cross-validation techniques have been used to estimate the probability of occurrence of hazardous events and, with appropriate scenarios, to generate risk prediction maps (Fabbri *et al*, 2004). This contribution discusses the usefulness and indispensability of such maps for disaster mitigation planning.

To plan a strategy to manage the occurrence of future natural disasters in a given area, one of the critically indispensable preparation steps is to perform the risk assessment for all natural hazardous events in the area such as earthquakes, floods and/or landslides. Risk assessment for natural hazard is linked to several socioeconomic elements such as the distribution of economic activities, infrastructures, human lives, and valuable assets. The risk of any one given socioeconomic element at a location in the assessment area is defined as the multiple of three components: the value of the element, the vulnerability of the element with respect to the natural hazard, and the probability of the future occurrence of the hazardous event at the element location. The priority in the preparedness should follow the level of the estimated risk at every location in the area. For example, if the risk is estimated near the zero value, then the priority should be the lowest within the area. Of the three components of risk, the only one related to the future is the probability of the occurrence of the natural hazardous event.

The analytical strategy discussed in this contribution is based on favorability function models applied to spatial databases and uses the results of empirical validation to construct acceptable scenarios for the introduction of socioeconomic data for vulnerability and risk evaluation. The different levels of risk can also be based on establishing how the various hazard uncertainties (e.g., Chung, 2005), identified via the validation techniques, impact on consequent risk uncertainties.

2 An application to a Landslide Hazard-Risk Study Area near Boeun, South Korea

The spatial mapping of risk is critical in planning for disaster preparedness. An application from a study area affected by mass movements is used here as an example to portray the desirable relations between **hazard prediction** and **disaster management**.

The Boeun area, in South Korea, has been affected by numerous mass movements. It covers approximately 48.6 km2 (8.12 km x 7.22 km, or 1642 x 1444 pixels of 5m x 5m resolution). Before the year 1997, at least 420 surficial debris flow landslides took place in the study area. Data on 44 more landslides that occurred in 1978 were also available so that a spatial database was constructed that consisted of digital maps of (a) the landslide distributions up to 1997 and in 1998, (b) the digital elevation models or DEM (continuous data layers), surficial geology, forest coverage, land use, and drainage maps (thematic data layers), and (c) a variety of additional socioeconomic maps (thematic data layers) with the associated statistics on values and vulnerabilities to surficial debris flow landslides.

About 45,600 people are living in the area in 1,500 households. The 44 landslides that occurred in 1998 caused 3 casualties, damages to properties up to US $ 3.3 millions (of which $ 200,000 to man-made infrastructures and $ 3,300,000 to forests) and occupied 2,000 5m x 5m pixels. The data in (a) and (b) in the database were used to generate landslide hazard prediction maps and later the data in (c) were used to generate three risk maps: risk to population, risk to man-made infrastructures, and risks to forests.

Figure 1 shows a population risk map of the Boeun area. The ratios in the legend indicate the number of expected victims to be affected by the landslides. The histogram below the risk map provides the estimated probability for each 5m x 5m pixel to be affected by the landslides. It was obtained through cross-validation of the hazard prediction map shown in Figure 4. In individual pixels the risk appears low but combining the risks together the total number of casualties is 3.

The risk map for man-made infrastructures is shown in Figure 2. The same probability histogram was used to generate the risk maps in Figures 2 and 3. The legend indicates the estimated property damage in US $ per pixel. The total damage was US $ 200,000: we have less that one $ per pixels in the red class. The risk is low for an individual pixel, however, it is spread over the entire area. Collective risk is high. Indeed at the location of the damaged house (see Figure 6) the risk is high.

333,697.730m N
258,810.487m E

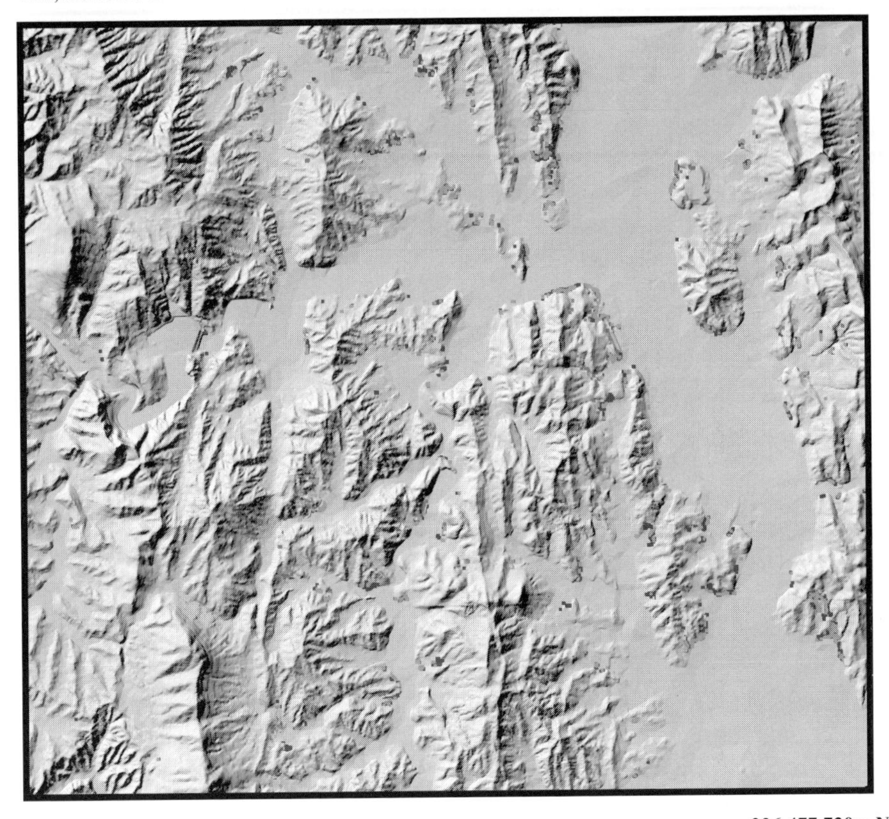

326,477.730m N
266,930.487m E

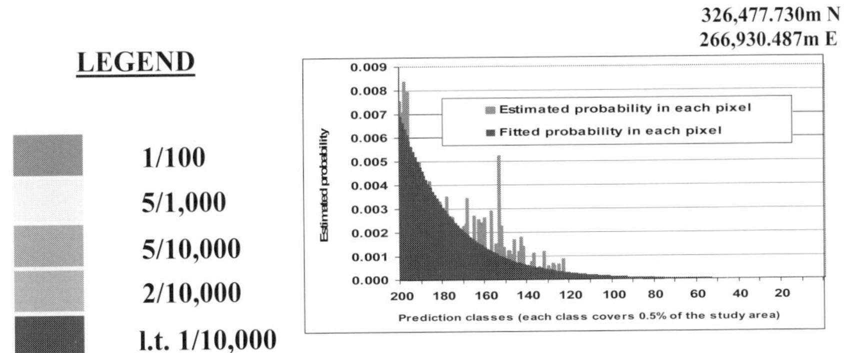

Fig. 1. 5m x 5m population risk map of the Boeun area, South Korea. The total number of casualties is 3. The histogram below provides the estimated probability for each pixel from the hazard cross-validation. The ratios in the legend indicate the number of expected victims to be affected by the landslides. The colored risk values have been draped on an enhanced digital elevation image, a shaded relief

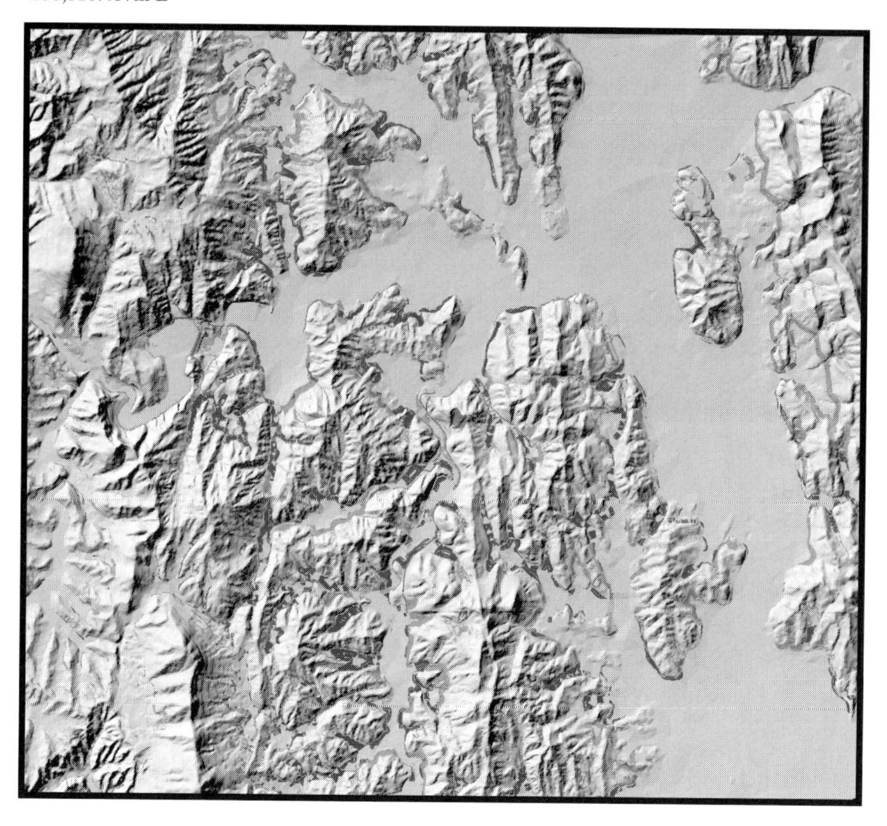

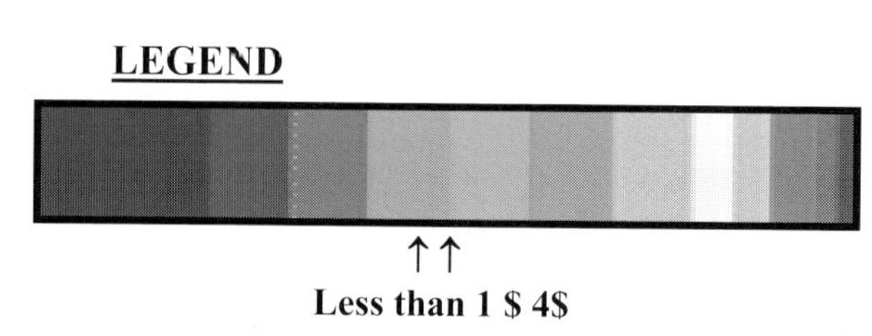

Fig.2. Risk assessment for man-made infrastructures in the Boeun area, South Korea. The estimated property damages were 200,000 $. A three-dimensional view of a particular in the upper right quarter image, looking in the south-southeast direction is shown in Figure 6A. The same shaded relief background used in Figure 1 is used here. The legend indicates the estimated property damage in US $ per pixel

The forest risk map is shown in Figure 3 where the legend indicates the damages expected also in US $. The situation represented in this map is somewhat special in that it was produced because the local land owners in the area had been compensated by the government for the damages to the forest. That situation may be uncommon elsewhere. The red class indicates US $ 7 per pixel.

To generate the three risk maps, the landslide hazard prediction map in Figure 4 was critical in conjunction with the prediction-rate curves obtained from the cross-validation to empirically measure its spatial support by the database of map data layers. These curves are shown in Figure 5 that provides both the prediction rates and the corresponding estimated probabilities of occurrence. The three stage approach used to obtain the risk maps is discussed in Section 4 and in Figure 7. The software that has been developed for predictive modelling and risk analysis is described in Section 5 and in Figure 8.

The socio-economic data layers used for the computation of risk were: (1) the distribution of population density, (2) the distribution of road networks (construction costs and costs of a two-day interruption), (3) the spatial distribution of buildings of different type, and (4) the distribution of drainage patterns and the construction costs of embankments etc. For all these data layers values and vulnerabilities were estimated per pixel. The 1998 landslide data were used to construct a scenario for risk assessment. In practice, because the 44 landslides that occurred in 1998 occupied 2000 pixels, an area of that many pixels was expected to be affected after 1997. It is remarkable that the risk maps provided results in harmony with the damages paid to the local land owners. For instance the total expected costs to man-made infrastructures were estimated as US $ 106,396.80, i.e., about half of the actual damages of $ 200,000 that have been paid by the local government. Our estimates of damages were: $ 31,041.16 to buildings, $ 34,280.19 to road networks, and $ 30,710.91 to traffic. Other estimates of damages were not significant because the locations of those infrastructures were away from hazardous zones. The total expected casualties were estimated as 3.14, rather close to the actual casualty number of 3. It is remarkable that in general, the highest risk areas are in fact in the middle hazard areas predicted by the analysis. A complete discussion of the results of this study, the prediction method used and the database is in Fabbri *et al* (2004).

333,697.730m N
258,810.487m E

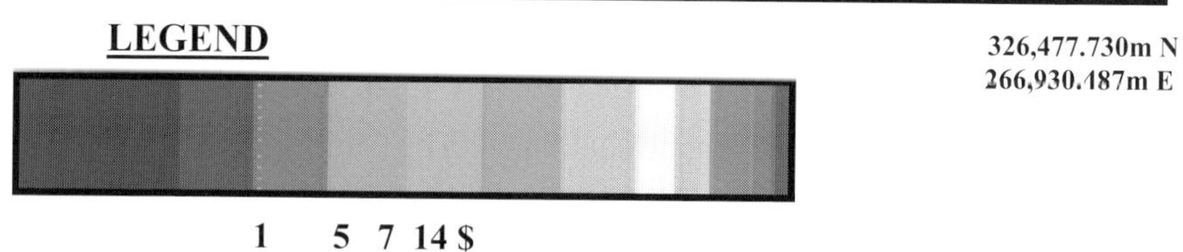

Fig. 3. Forest risk map for the Boeun area, South Korea. Estimated damage was $ 3,300,000. The legend indicates the forest damages expected also in US $. The red color indicates 7$ per pixel. It was obtained knowing the sum that the local government paid the land owners to compensate for the damages caused by the landslides

3 Risk Visualization and Disaster Preparedness Plans: What is so Special about the Risk Maps?

The risk maps generated for the Boeun area have several new aspects that should be underlined.

First of all they were obtained using an empirical cross-validation that represents the degree of support for each hazard prediction class. That means that we know how good the classes are in terms of predicted hazard level. Secondly, the risk maps were based on clear assumptions concerning the data layers (e.g., uniformity principle and reliability of the data base), the prediction models (each model requires assumptions such as conditional independence of the data layers, etc.), and the number/size of the future damaging events (e.g., the 2000 pixels expected to be affected). The latter represents a scenario for an acceptable realistic situation that provides the necessary setting for a decision maker. Other more optimistic or pessimistic scenarios can be used to explore further the risk assessment results.

In addition, even if that was not discussed here, hazard predictions can also be obtained imposing various degrees of uncertainty on the map unit boundaries of the thematic map data layers, thus enabling comparisons of predictions via cross-validations. From those different hazard maps different risk maps can be generated and compared.

Three-dimensional dynamic displays can be easily generated in which a background is added of all important transportation routes and other points of interest for disaster preparation or risk management. Indeed, the various classes or levels of risk represent spatial indicators of when given proportions of future damaging events are likely to occur. This is because no prediction will be able to locate all the future events: a more or less satisfactory proportion will be obtained in practice (say 60-75%). The associated levels of uncertainty are also easily estimated and visualized. The multidimensional renderings with fly through techniques, now commonly available, can provide realistic landscapes for analysis and planning for relief action. They can be used either in proactive mode, before the disasters happen, or in reactive mode, after the damage.

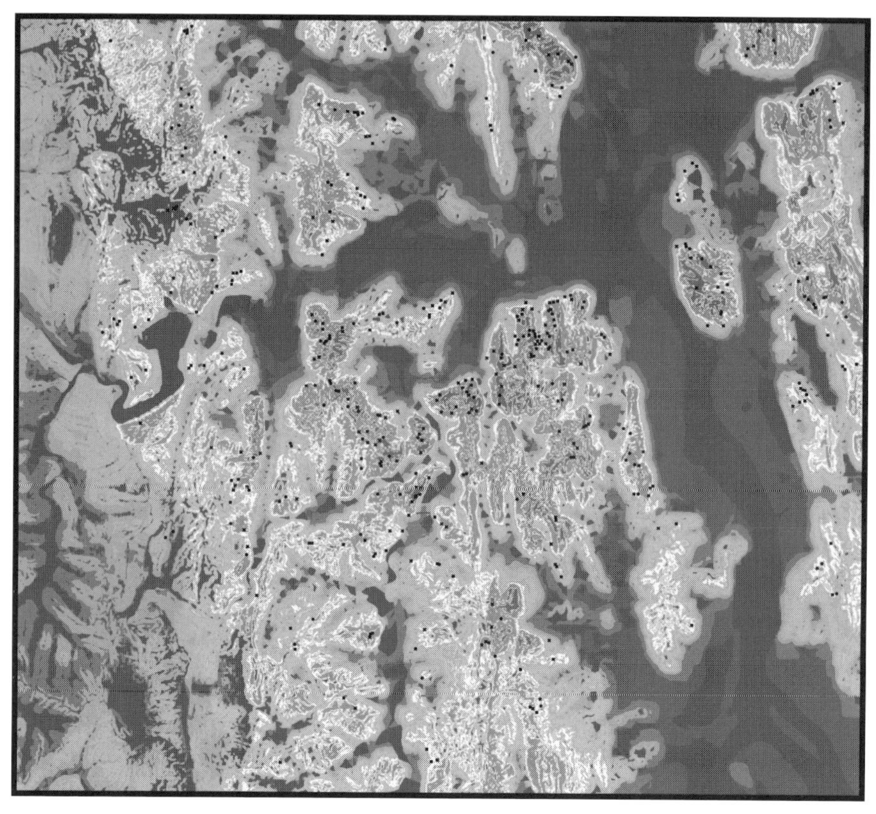

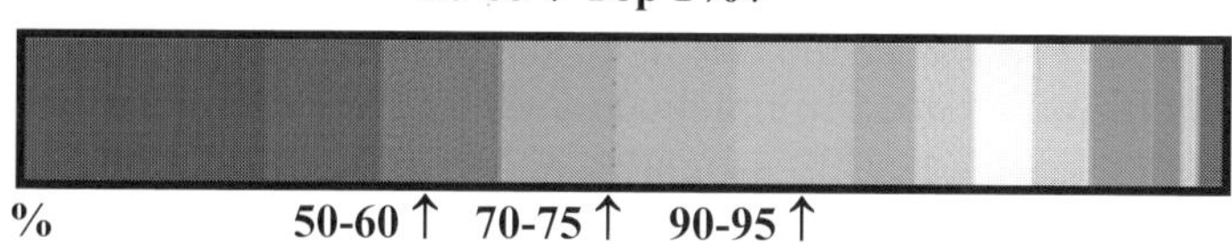

Fig. 4. Landslide hazard prediction map for the year 1977 for the Boeun area, South Korea. It was obtained using the Empirical Likelihood Ratio function model and distribution of the 420 landslides, shown as black dots, that occurred prior to 1977, the DEM, surficial geology, drainage and forest coverage. The prediction was interpreted generating another prediction using a random half of the 420 landslides and validating it with the distribution of the remaining half of the landslides to obtain the prediction-rate curves in Figure 5

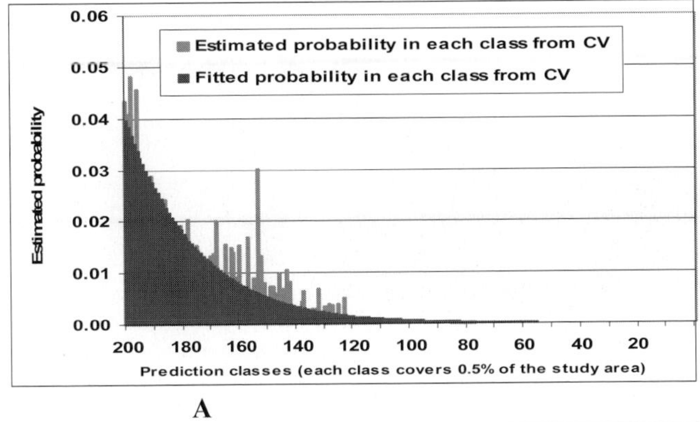

A

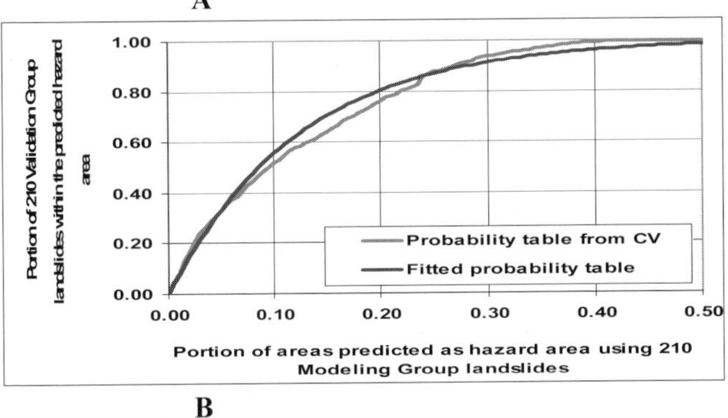

B

Fig. 5. Estimated probability of landslide occurrence for each prediction class shown in Figure 4 and the corresponding prediction-rate curve in the Boeun area, South Korea. The histogram in (A) provides the estimated probability for each of the 200 classes in Figure 4. It was obtained from the prediction-rate table used to generate the curve in (B)

The illustrations in Figure 6 are just an indication of how we could use visualization techniques for applying spatial decision support techniques to risk analysis. For instance, we should be able to decide in what order or priority sequence are we to deal with risk avoidance or risk mitigation or even risk acceptance. If the risk is estimated to have a low value, then the priority should be the low within the area.

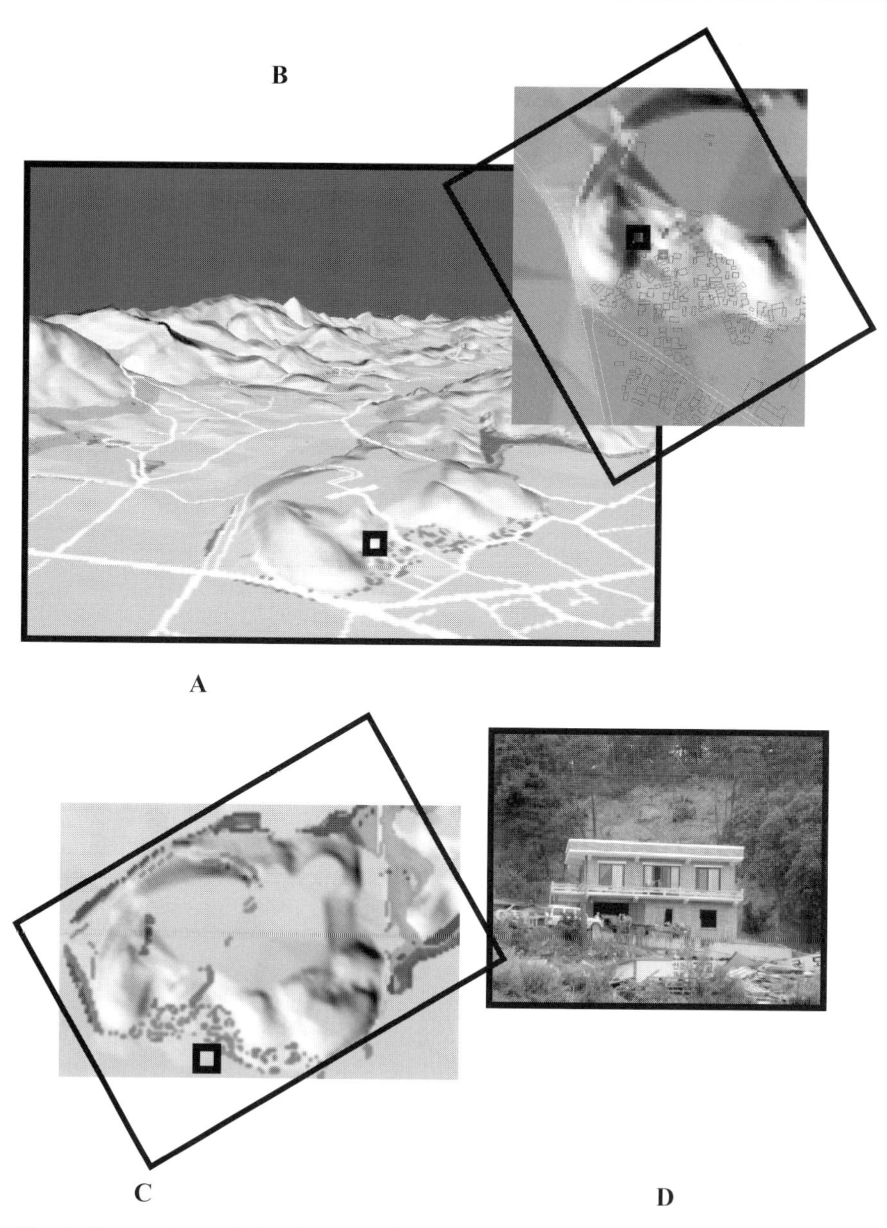

Fig. 6. Some views of the risk map for man-made infrastructures shown in Figure 2. In (A) a 3-D view is shown looking in direction SSE from the top right of the risk image. In (B) a vertical view of the sub-area with the distribution of man-made infrastructures. In (C) a vertical view of the same sub-area with the risk map. In (D) a house that was damaged by a landslide and in which a casualty occurred. The house is located on the other three views

4 An Approach to Spatial Predictions

We have developed a three-stage procedure in spatial data analysis (Fabbri et al, 2004) not only to estimate the probability of occurrence of hazardous events but also to evaluate the uncertainty of the estimators of the probability. The three-stage procedure consists of: (i) construction of a hazard prediction map of "future" hazardous events; (ii) validation/reliability of prediction results and estimation of the probability of occurrence for each predicted hazard level; and (iii) generation of risk maps with the introduction of socio-economic factors representing assumed or established vulnerability levels by combining the prediction map in the first stage and the estimated probabilities in the second stage with socio-economic data. Figure 7 shows a flow diagram of the approach used.

4.1 Hazard Prediction: Stage I

The initial step in the **Hazard Prediction Stage I** is the preparation of the input data layers, i.e., (a) the thematic data layers that express the natural setting of the landslide events (bedrock geology, quaternary cover, land use, vegetation, etc.), (b) the continuous data layers that characterize the topographic landscape (elevation, slope and aspect making up the digital elevation model or DEM), and (c) the occurrence data layers with the distribution and shape of various types of hazardous events possibly including the year of occurrence.

Data preparation requires a uniform resolution, geo-referencing and co-registration, the extraction of map unit boundary images for the thematic data layers, the assignment of unique labels to identify each individual occurrence, and if necessary, the identification of relevant portions of the areas of the occurrences (e.g., the scarps representing the trigger zones of the landslides) and possible derived occurrence maps that partition the occurrences in time periods, or spatial subsets or random groups for the validation experiments to follow in the successive Stage II.

The *1st Prediction Step*, after the data preparation, consists of using all thematic and continuous data layers and all hazardous occurrences of the same type, selecting the appropriate prediction model out of an available set of mathematical models (e.g., Fuzzy Sets, Empirical Likelihood Ratio, etc., see Chung and Fabbri, 1993, 1998, 2001). This generates a prediction map or image (we will be using the terms image or map interchangeably to indicate a digital prediction map) in which each point or pixel is assigned a prediction value that ranges between a minimum and a maximum according to the model specification (e.g., between 0.0 and 1.0).

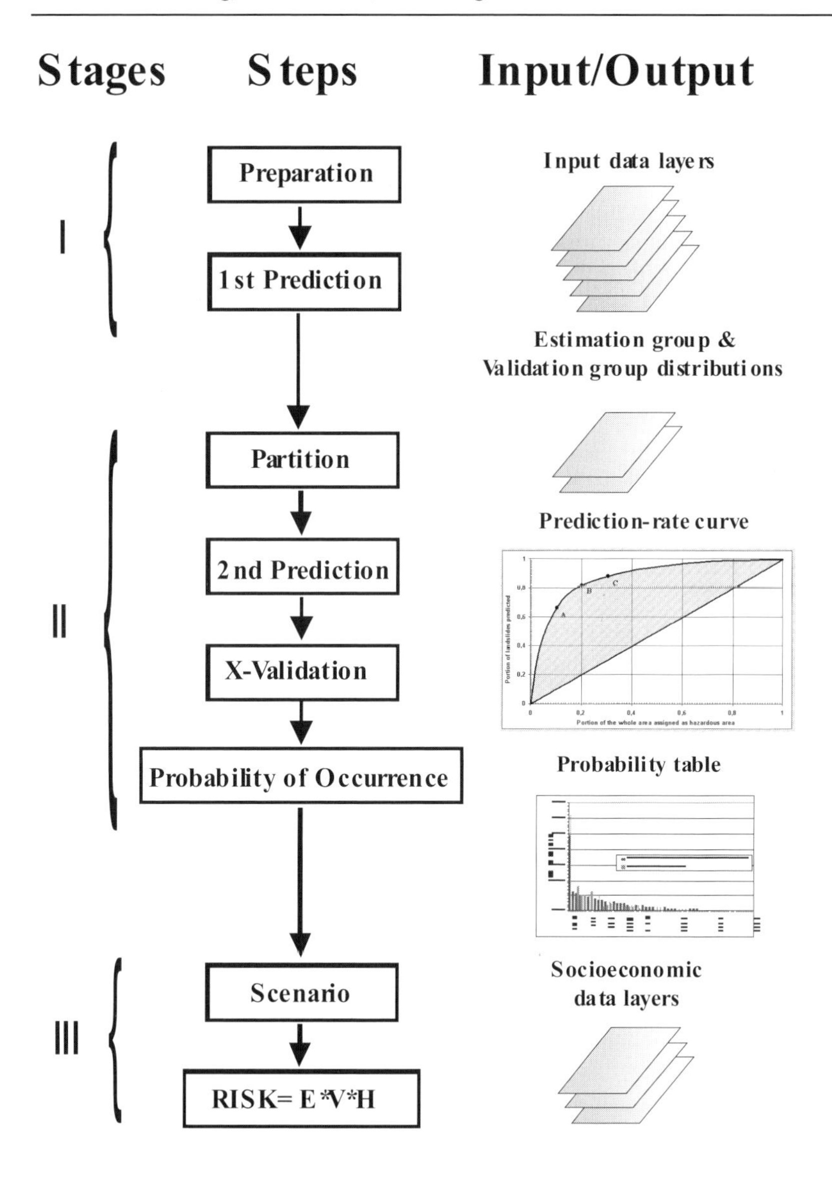

Fig.7. Flow diagram of the approach proposed for spatial predictive modeling of hazard and risk

To observe that prediction image, the prediction values can be sorted in descending order and the ranks of fixed pixel percentages of study areas, say 0.5%, can be assigned 200 successive colors in a pseudo-color look-up table. Such a simple visualization technique completes the **Hazard Prediction Stage I**. An example of a hazard map is shown in Figure 4. The *1*st

prediction uses all occurrences available of a given type and is usually the one with most supporting input data layers (thematic and continuous). It now needs to be interpreted and well understood. As is, we do not know much of its significance in terms of relative hazard representation. How good is it? To answer the question we need another analytical stage.

4.2 Cross-Validation: Stage II

The **Cross-Validation Stage II** requires the partitioning of the distribution of hazardous occurrences in time intervals, or spatial subsets of parts of the study area (say upper and lower or left and right halves) or of random half occurrence subsets for later *cross-validation.* One subset of occurrences (for instance the distribution of the older ones) will be used to obtain a *2nd Prediction* using the same model used in the *1st Prediction*, and the other subset of occurrences will be used to cross-validate the prediction result (the prediction image) by counting the number and measuring the area of the validation occurrences falling within each prediction class, ranked as described in Stage I.

The spatial comparison generates a prediction-rate table or curve describing the effectiveness of the prediction. We can display the *2nd Prediction* image as we have done for the 1st one in Stage I, however, such a display will not be as informative as the prediction-rate table. Obviously, while the *cross-validation* is functional to interpreting the prediction, it can happen that the prediction image is unsatisfactory (e.g., high predicted values might not have good concentrations of validation occurrences), so that various iterations are desirable in which Stage I is repeated modifying the input data layers used (e.g., using a greater or smaller number of data layers, or different sets or combinations of data layers).

If the *2nd Prediction Step* appears satisfactory, the prediction-rate table can be transformed into a *probability of occurrence* table according to some scenario based on assuming some similarity of occurrence distribution in time, for instance. This was done to obtain the diagram in Figure 5 and the histogram at the bottom of Figure 1. At this point we should be ready to consider the analysis of risk if the necessary socioeconomic data layers can be made available.

4.3 Risk Assessment: Stage III

The **Risk Assessment Stage III** necessitates a realistic *scenario* based on the uniformity of occurrence distribution of hazardous events during the first time interval of a same or similar duration in the future. Should a trend towards an increasing or decreasing intensity of occurrences be available, a corresponding modification of the scenario can be applied. The scenario provides input on the average dimension and number of hazardous events expected in the future.

The distribution of such future hazardous events among the predicted hazard classes has been represented in the probability of occurrence table obtained in **Stage II**.

A number of socioeconomic thematic data layers have to be prepared or generated as digital images of the same resolution and co-registration of the remainder of the database used so far for hazard prediction. Such thematic socioeconomic data layers, for instance, represent population density distribution, infrastructures, transportation networks, drainage systems, etc., for which values and vulnerabilities per unit area or length have been calculated. Such values, stored in tables linked with the socioeconomic feature identifications, have to be expressed in the same spatial resolutions of the remainder of the database (e.g., 5m x 5m, or 10m x 10m, or 20m x 20m, etc.). While the values will be expressed in monetary units, the vulnerabilities will be ranging between 0, no damage, and 1, total destruction.

The *risk computation* will generate a risk map using (1) the data from the socioeconomic data layers (values and vulnerabilities), (2) the previously generated 1st hazard prediction map of Stage I, and (3) the probability table for each area resolution unit, each pixel, from **Stage II**. The risk map obtained satisfying the expression $RISK = E*V*H$ (where E is the element at risk, V is its vulnerability and H is the hazard) now needs to be, studied and interpreted via a variety of static and dynamic techniques (e.g., 3-D visualizations with different contexts and enhanced backgrounds).

A spatial modeling software package was developed to satisfy some of the requirements of this approach. The following section describes it.

5 Software Tools for Spatial Prediction Models and Spatial Risk Analysis

The software approach summarized here has been proposed by the authors (e.g., Fabbri *et al*, 2004) to provide an analytical structure and modeling power that are a fundamental tool for decision making. No other computer system, at present, allows the validation of predictions and the estimations of probabilities of occurrence. A number of technical and analytical issues have been tackled by SPM, the Spatial Prediction Modeling system, and by SRA, the Spatial Risk Analysis system (see www.statialmodels.com).

The realization that the common analytical functions in most GIS are still inadequate for prediction modeling has encouraged the programming of special analytical tools, SPM and SRA, complementary to GIS, to satisfy the following tasks:

a. to make available most mathematical methods of predictive modeling under a unified framework such the one of Favorability Functions (Chung and Fabbri, 1993), and the subsequent representation and visualization criteria (e.g., Chung and Fabbri, 1999);
b. to make simultaneous use of thematic and continuous data layers without the transformation of one data type into the other and subsequent loss of information (Chung and Fabbri, 2005);
c. to apply validation procedures to assess the degree of data support to prediction results, to compare different predictions and select, out of many groupings of occurrences, the better one, to perform sensitivity analysis to assess the robustness and stability of the prediction results, and to estimate the probability of occurrence of unwanted events (i.e., the hazardous occurrences) given appropriate scenarios;
d. to exploit the fuzzy boundary concept for thematic data layers enabling the explicit introduction of spatial uncertainty into predictive modeling (Chung, 2005);
e. to provide an operational framework to risk assessment and risk map generation based on values and vulnerabilities of socioeconomic input thematic data layers, on the probability of hazardous occurrences estimated from cross-validation, and the result of predictive modeling of hazard, i.e., the hazard map (Fabbri *et al*, 2004).

The Spatial Prediction Modeling system, SPM, as shown in Figure 8, provides tools for inputting raster images of continuous, thematic and occurrence data layers (DP), to select various mathematical models of prediction image generation (SP) such as Fuzzy Sets, Empirical Likelihood Ratio, Logistic and Linear, and Bayesian Probability models. The interaction can be pipelined into a multilayer simultaneous analysis or on a gradual layer-by-layer process to obtain initial confidence in the processing.

The fuzzy boundary concept is implemented as a spread parameter that is applied to the boundaries of thematic data layers units. It can account for either the integration of different map scales or of the uncertainties associated with the thematic map units.

The cross-validation module (CV) provides several ways to handle different groupings of hazardous occurrences in the interpretation of prediction results.

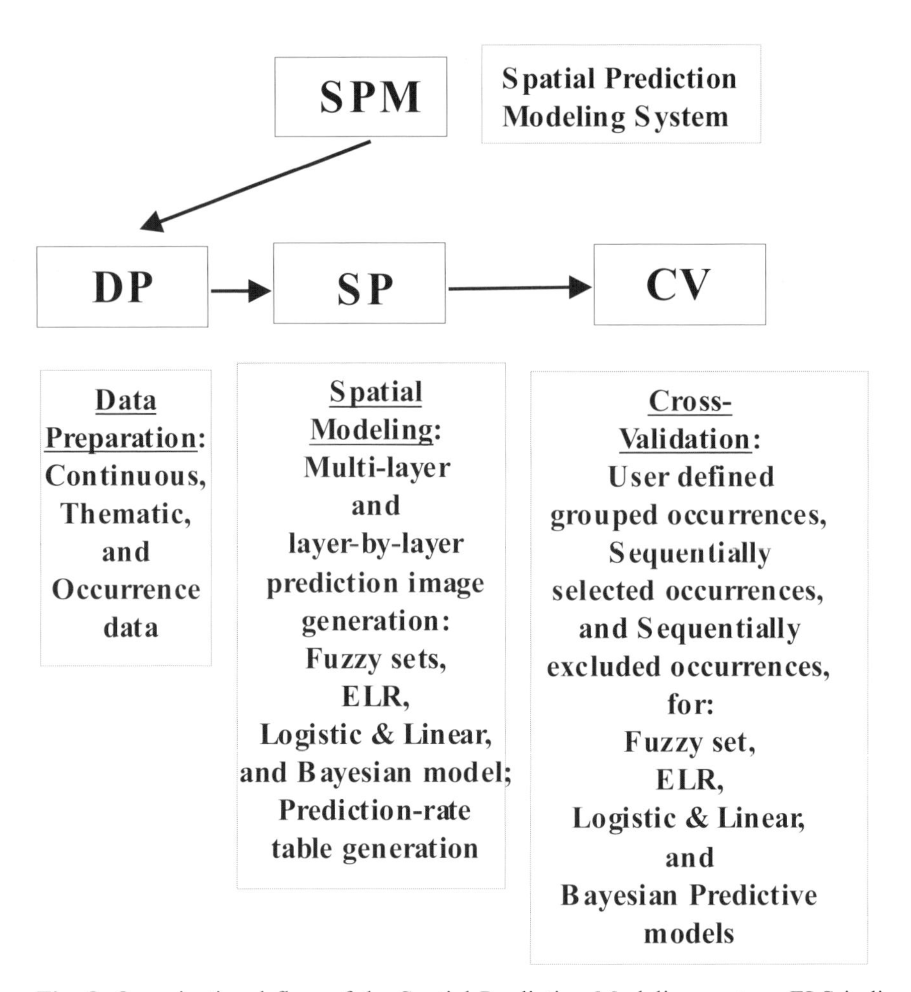

Fig. 8. Organizational flow of the Spatial Prediction Modeling system. ELS indicates the Empirical Likelihood Ratio model

The Spatial Risk Analysis system, SRA, described in Figure 9, shows a sequence of processing steps in which the socioeconomic input thematic data layers and tables input are generated (T), a probability table is computed for each pixel by transforming the prediction-rate table, obtained from SPM, according to a scenario based on the average number and size of hazardous occurrences, that provides the number of pixels expected to be damaged in the future (PrTa). Should such a probability table not have good monotonic non-decreasing slope properties, a further transformation can be obtained (M-Ta). Should still that modification show much irregularity or fluctuations, a smooth curve fitting can be computed (F-Ta).

Finally, using one of the probability tables (i.e., the results of either PrTa, M-Ta or F-Ta) as input, a socioeconomic thematic data layer with the associated value and vulnerability table, and the initial most satisfactory hazard prediction image, a risk map is generated (RISK).

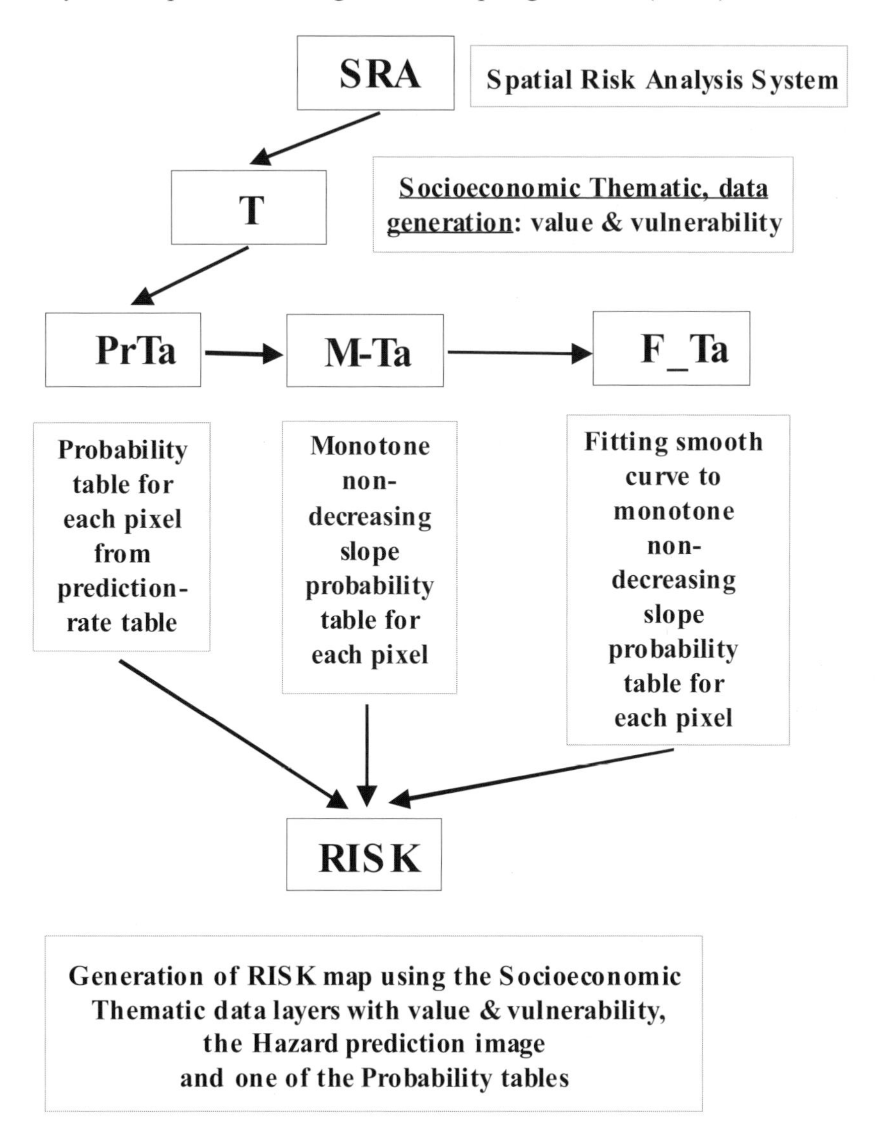

Fig. 9. Organizational flow of the Spatial Risk Analysis system

All SPM and SRA outputs, images and tables, are in formats that can be transferred to or used by other software systems such as GIS, spreadsheets, etc., for further analysis and visualization. SPM and SRA have been programmed to be independent from such systems and complementary to them.

6 Concluding Remarks

What more should be done? To this question we can try to provide a partial answer with the following considerations. In this contribution we have presented three risk maps and described how they have been generated. The prediction process requires spatial data, mathematical models, assumptions, scenarios and strict analytical strategies. We have presented an approach that uses favorability modeling and three analytical stages that we believe has advantages over many risk mapping methods to date and certainly over a variety of GIS techniques. Many extensions or improvements can be further thought of. Depending on the data available, various dynamic visualizations can be designed, for instance providing a set of risk maps for subsequent time intervals thus using trend information. Other types of natural or technological risk representations can be generated, such as earthquake, flooding, forest fires, or aquifer vulnerability to pollution. Essential, in spatial data analysis is to generate the spatial context of causal factors that integrate much of the spatial distribution that leads to estimating the probability of occurrence of damaging events or processes.

The fuzzy boundary concept can also be used to refine the approach to risk mapping so that degrees of uncertainty are analyzed or tested with cross-validation techniques.

Another aspect, on which the authors of this contribution are presently working, is the assessment of multi-risks to generate an aggregated risk map. Indeed, we wonder how little would be possible in risk management without risk maps of the nature of the ones presented here. A proper use of GIS technology for disaster preparedness is to use the risk maps as a basis for planning the distribution of mitigation measures, such as alarm systems, access to affected areas by emergency teams and vehicles, protection barriers, sensors to detect changes in condition of vulnerable elements, escape routes, safe meeting points, information nodes or centers, relief units, disaster foci bypass routes, to mention a few. The thematic and continuous spatial support to a risk map should provide an understandable and reliable expression of the potential damage distribution and of the associated levels of uncertainty with the realistic background and landscape setting of the past events and of the existing communication network. The risk maps can easily become integral part of spatial decision support systems.

Acknowledgements

This research has been partly supported by a research network project on the "Assessment of Landslide Risk and Mitigation in Mountain Areas, ALARM" (Contract EVG1-CT-2001-00038) of the European Commission's Fifth Framework Programme (http://www.spinlab.vu.nl/alarm). This research was also partly supported by the Gong-Ju National University that granted a one-year fellowship to Dr. Dong-Ho Jang as visiting scientist with the Geological Survey of Canada. Additional partial support was provided by the "Sustainable Development Through Knowledge Integration" or SDKI Program of the Pathways Project of Natural Resources Canada's Earth Science Sector.

References

Bell R, Glade T (2004) Multi-hazard analysis in natural risk assessment. In Brebbia CA, ed., Risk Analysis IV, Southampton, Boston, WIT Press, pp 196-206

Chung CF (2005) Using likelihood ratio functions for modeling the conditional probability of occurrence of future landslides for risk assessment. Manuscript submitted for publication to Computer and Geosciences

Chung CF, Fabbri AG (1993) The representation of geoscience information for data integration, Nonrenewable Resources, vol 2, no 2, pp 122-139

Chung CF, Fabbri AG (1998) Three Bayesian prediction models for landslide hazard. Proceedings of the International Association for Mathematical Geology Annual Meeting IAMG 1998, Ischia, Italy, October 1998, pp 204-211

Chung CF, Fabbri AG (1999) Probabilistic prediction models for landslide hazard mapping. Photogrammetric Engineering & Remote Sensing, vol 65, no 12, pp 1389-1399

Chung CF, Fabbri AG (2001) Prediction models for landslide hazard using a fuzzy set approach. In, M. Marchetti and V. Rivas, eds., Geomorphology and Environmental Impact Assessment, Balkema, Rotterdam, pp 31-47

Chung CF, Fabbri AG (2003) Validation of spatial prediction models for landslide hazard mapping. Natural Hazards, vol 30, pp 451-472

Chung CF, Fabbri AG (2005) Systematic procedures of landslide hazard mapping for risk assessment using spatial prediction models. In, Glade T, Anderson MG, Crozier MJ, eds., Landslide Hazard and Risk. New York, John Wiley & Sons, in press

Fabbri AG, Chung CF, Jang DH (2004) A software approach to spatial predictions of natural hazards and consequent risks. In, Brebbia CA, ed., Risk Analysis IV. Southampton, Boston, WIT Press, pp 289-305

Fabbri AG, Chung CF, Napolitano P, Remondo J, Zezere JL (2002) Prediction rate functions of landslide susceptibility applied in the Iberian Peninsula. In, Brebbia CA, ed., Risk Analysis III. Southampton, Boston, WIT Press, pp 703-718.

Hansson K (2004) The importance of pre mitigation strategies in development planning. In Brebbia C. A., ed., Risk Analysis IV, Southampton, Boston, WIT Press, pp 713-723.

Komac M (2004) Statistical landslide prediction map as a basis for a risk map. In Brebbia CA, ed., Risk Analysis IV, Southampton, Boston, WIT Press, pp 318-330.

Liguori V, Mortellaro D (2004) Geomorphic hazard and risk in Platani's basin: landslide risk valuation. In Brebbia CA, ed., Risk Analysis IV, Southampton, Boston, WIT Press, pp 163-175

Plattner T (2004) An integrative model of natural hazard risk evaluation. In Brebbia CA, ed., Risk Analysis IV, Southampton, Boston, WIT Press, pp 649-658

Automatically Extracting Manmade Objects from Pan-Sharpened High-Resolution Satellite Imagery Using a Fuzzy Segmentation Method

Yu Li, Jonathan Li and Michael A. Chapman

Geomatics Engineering Program, Department of Civil Engineering, Ryerson University, 350 Victoria Street, Toronto, Ontario, Canada M5B 2K3T. Email: {y6li, junli, mchapman}@ryerson.ca

Abstract

The paper describes a new method for extracting objects from high resolution color remote sensing images. This method is based on the fuzzy segmentation algorithm which has been developed in our previous works. The proposed object extraction method is following three steps. (1) Segmenting color images, (2) Detecting objects from segmented images, and (3) Post-processing of extracted object. The paper also gives experimental results from using the proposed method to extract centerlines of road networks and roofs of building from QuickBird and Ikonos Images.

1 Introduction

Very high spatial resolution satellite imagery such as 1 m Ikonos and 0.6 m QuickBird, particularly its high temporal resolution (e.g., 1-3 days), has implied that this kind of image data acquired from spaceborne sensors can provide a viable alternative to aerial photography for emergency response planning. Unfortunately, such satellite imagery has not been readily adopted by metropolitan mapping agencies and emergency response personnel for quick detection of emergency change information for planning, monitoring and damage assessment. In the applications of high resolution satellite imagery, object extraction is the most basic and important task. Though many object extraction algorithms for satellite or aerial images have been proposed in the past years, most of them processed grayscale imagery only. The objective of this paper is to present an effective approach to man-made object extraction from the pan-sharpened high-resolution satellite imagery. This method is based on the fuzzy segmentation algorithm which has been developed in our previous works. The proposed object extraction method consists of three steps: segmenting color imagery, detecting objects from the segmented imagery, and post-processing of the extracted objects. Several high-resolution satellite images with different scenes in urban residential areas have been examined and the results presented illustrate the potential of the proposed approach.

The paper is organized as follows. Section 2 introduces notations and algorithms for color segmentation from our previous work. The proposed new method for object extraction is described in Section 3. The results on extracted objects from pan-sharpened QuickBird and Ikonos color images by using the proposed method are illustrated in Section 4. Finally, conclusions are drawn in Section 5.

2 Fuzzy-Based Segmentation

2.1 New Fuzzy C-partition Method

Fuzzy c-partition algorithm has been wildly used method to solve the clustering problems in pattern recognition (Tou and Gonzalez, 1974; Zeng and Starzyk, 2001), image segmentation (Liew and Yan, 2001), unsupervised learning (Langan et al., 1998), and data compression (Zhong et al., 2000).

Consider a vector set $\boldsymbol{V}$ formed by n vectors in L-dimensional real number space R^m, i.e., $\boldsymbol{V} = \{\boldsymbol{V}_1, \boldsymbol{V}_2, \dots, \boldsymbol{V}_n\}$, $\boldsymbol{V}_j = [\boldsymbol{V}_{j1}, \boldsymbol{V}_{j2}, \dots, \boldsymbol{V}_{jL}]$ and $j = 1, 2, \dots, n$, a fuzzy c-partition on $\boldsymbol{V}$ is represented by

$$\boldsymbol{P} = [p_{ij}],\ i = 1, 2, \dots, c \text{ and } j = 1, 2, \dots, n \qquad (1)$$

where $\boldsymbol{P}$ is a fuzzy partition matrix and satisfies

$$\sum_{i=1}^{c} p_{ij} = 1,\ \ for\ \ j = 1,\ 2, \cdots, n \qquad (2)$$

$$0 < \sum_{j=1}^{n} p_{ij} < \mathrm{n},\ \ \text{for}\ \ \mathrm{i} = 1, 2, \cdots, \mathrm{c} \qquad (3)$$

where c is the positive integer to indicate the number of the clusters in the partition, and $p_{ij} \in [0,1]$ is the fuzzy membership value of V_j belonging to ith cluster (George and Bo, 1995).

Before using fuzzy c-partition to design a clustering algorithm, the following two issues should be solved. First one is how to determinate the number of clusters for a clustering. Another issue is how to calculate the fuzzy c-partition matrix.

Unfortunately, in most of situations, the number of clusters is unknown a prior and sometimes it is difficult to specify any desired number of clusters. For example, the situations often happen in the segmentations of remote sensing images, because the ground truth is always not available for these images. Based on our previous work, a histogram-based procedure is used to obtain the number of the clusters (Li, 2004). The second issue is solved by following procedure.

Given a vector set $V = \{V_1, V_2, \ldots, V_n\}$, and the number of the clusters c. Based on this number, a central vectors set can be selected, that is, $V_{CV} = \{V_{CV1}, V_{CV2}, \ldots, V_{CVc}\}$ and $V_{CV} \subset V$ The fuzzy c-partition matrix can be calculated as follows.

$$p_{ij} = \frac{\mu(V_{CVi}, V_j)^{\frac{1}{m-1}}}{\sum_{k=1}^{c} \mu(V_{CVk}, V_j)^{\frac{1}{m-1}}}, \text{ for } i = 1, 2, \ldots, c \text{ and } j = 1, 2, \ldots, n \qquad (4)$$

where $m \in (1, \infty)$ is the weighting exponent on each fuzzy membership. The larger m is the fuzzier the partition is, $\mu(V_i, V_j)$ is a similarity measure between vectors V_i and V_j and can be calculated by

$$\mu(V_{CVi}, V_j) = \exp(-k_1 d(V_{CVi}, V_j)) \cos(k_2 \theta(V_{CVi}, V_j)) \qquad (5)$$

where $d(V_i, V_j)$ and $\theta(V_i, V_j)$ are the distance and the angle between V_i and V_j as follows, and k_1 and k_2 are parameters.

$$d(V_{CVi}, V_j) = \left(\sum_{l=1}^{L} | V_{CVil} - V_{jl} |^2 \right)^{1/2} \qquad (6)$$

$$\theta(\boldsymbol{V}_{CVi}, \boldsymbol{V}_j) = \arccos\left(\frac{\sum_{l=1}^{L} \boldsymbol{V}_{CVil} \boldsymbol{V}_{jl}}{\sqrt{\sum_{l=1}^{L} \boldsymbol{V}_{CVil} \sum_{l=1}^{L} \boldsymbol{V}_{jl}^{2}}} \right) \quad (7)$$

In order to obtain the best fuzzy c-partition, a set of the best central vectors is tried to find. It is modeled as an integer programming (IP) problem as follows.

Given a vector set $\boldsymbol{V} = \{\boldsymbol{V}_1, \boldsymbol{V}_2, \ldots, \boldsymbol{V}_n\}$, and the number of the clusters c,

Max

$$\sum_{i=1}^{c} \sum_{j=1}^{n} p_{ij} \quad (8)$$

Subject to

$$\boldsymbol{V}_{CVi} \in \boldsymbol{V} \quad (9)$$

After finding the best centre vector set $\boldsymbol{V}_{BCV} = \{\boldsymbol{V}_{BCV1}, \boldsymbol{V}_{BCV2}, \ldots, \boldsymbol{V}_{BCVc}\}$, the best fuzzy c-partition matrix is calculated with Equation (4).

2.2 Segmentation by the Proposed Fuzzy C-Partition Algorithm

Based on the above fuzzy c-partition algorithm, the color segmentation approach is developed. The approach consists of three steps: (1) Pre-clustering. This process includes determining of the number of clusters, finding an initial centre vector set V_{CV0}, and indicating the ranges in which the centre vectors are chosen in the following optimal procedure. This procedure is finished by using a histogram-based technique. (2) Searching the best fuzzy c-partition. It is realized by solving an integer programming problem to find a good fuzzy c-partition. (3) Post-processing. It means a defuzzification procedure to convert the fuzzy c-partition matrix to the crisp c-partition matrix.

Pre-clustering. For the given color image ***CI***, the color histogram H (***CI***) can be obtained (Li, 2004). It is obvious that if an image is composed of distinct objects with different colors, its color histogram usually shows different peaks. Each peak corresponds to one object and adjacent peaks are likely to be separated by a valley. The height of a peak implies the number of the pixels falling in the bin corresponding to the location of the peak.

The pre-clustering procedure is carried out by thresholding the color histogram of a color image. For a selected threshold, the peaks having higher magnitudes than the threshold can be detected. The number of all detected peaks is chosen as the number of clusters, and the bins corresponding to the detected peaks determine the ranges in which the centre vectors are investigated for the purpose of the optimization. The initial centre vectors consist of the minimum vectors of all bins. On the other hand, they can also be produced randomly, as long as they are located in the selected bins.

The threshold is determined by either a manual or an automatic way. In the manual case, the number of clusters is determined by observing the color image and the color histogram of the image. In the automatic case, the criterion to determine the threshold should be given first. For example, the mean of all peaks can be used as the criterion. It means that the peaks with the higher magnitudes than the mean are valid.

Optimizing. To solve the optimization model introduced in the previous section, there are many methods, such as a branch-and-bound approach (Winston, 1991), and genetic algorithms (Goldberg, 1989). However, they are very time consuming and not practical in the real world. Therefore, the use of a heuristic, which gives a good but sometimes not optimal or the best solution, is necessary.

Post-processing. In order to obtain the segmented image, it is necessary to transform the fuzzy c-partition matrix to the crisp partition matrix. In this study, the following defuzzification scheme is used.

Let $\boldsymbol{P} = [p_{ij}]$ $i = 1, 2, \ldots, c$ and $j = 1, 2, \ldots, n$ be the fuzzy c-partition matrix, it is well known that p_{ij} presents the membership grade for pixel j belonging to cluster i. A percent partition matrix, $\boldsymbol{P}_p$, is defined as

$$p_{pij} = \frac{p_{ij}}{\sum_{j=1}^{n} p_{ij}} \qquad (10)$$

In terms of the percent partition, the crisp partition matrix, $\boldsymbol{P}_c = [p_{cij}]$, is defined as

$$p_{cij} = \begin{cases} 1, & p_{pij} = \max_{i=1}^{c}(p_{pij}) \\ 0, & otherwise \end{cases} \qquad (11)$$

It is clear that in the crisp-partition matrix each pixel belongs to a certain cluster.

3 Our Object Extraction Method

In this section, the segmentation method based on the fuzzy c-partition algorithm is utilized to extract objects from high-resolution remote sensing images. It consists of three main steps: (1) segmenting color images based on the above segmentation method; (2) detecting objects from segmented images; (3) post-processing of extracted object, for example, delineating road centerlines from the extracted road networks or . The discussions are mainly focused on the Steps 2 and 3.

3.1 Extraction objects

Once the segmented images are obtained by the above segmentation method, the binary object image can be extracted from it by selecting the pseudo-color corresponding to the object regions. In general, the objects in the binary image are corrupted by noise objects, which have the similar colors to objects. In order to make the object regions clear, it is necessary to filter the corrupted object image. To this end, binary morphological operations are used. For example, depending on the shapes of noise objects, the appropriate combinations of binary dilation, erosion, opening, and closing should be chosen.

3.2 Post-Processing of Extracted Objects

Delineation of road centerlines. An important process for representing the structural shape of the detected road regions is to reduce it to a graph. This work can be accomplished by a thinning algorithm. The thinning algorithm developed by Zhang and Suen (1984) for thinning binary regions is utilized in this study. It is assumed that the road pixels in the binary road network images have value 1 (black), and those background (non-road) pixels have the value 0 (white). The method consists of the successive passes of two basic steps applied to the contour pixels of the given images, where a contour pixel is any pixel with value 1 and has at least one 8-neighour value 0. With reference to the 8-neighbourhood definition shown in Figure 1, the first step indicates a contour pixel p for deletion (from black to white) if the following conditions are satisfied:

- $2 \leq N(p) \leq 6$
- $S(p) = 1$
- $p_0 \cdot p_1 \cdot p_3 = 0$
- $p_3 \cdot p_5 \cdot p_7 = 0$

where $N(p)$ is the number of nonzero neighbors of p, i.e.,

$$N(p) = \sum_{i=0}^{7} p_i \qquad (12)$$

and $S(p)$ is the number of 0-1 transitions in the ordered sequence of p_0, $p_1, \ldots, p_6, p_7$.

In the second step, first two conditions remain the same, but the last two conditions are changed to

- $p_0 \cdot p_1 \cdot p_7 = 0$
- $p_0 \cdot p_5 \cdot p_7 = 0$

Extraction Build Profiles. To extract the building regions according to the color features of the buildings and uses an edge extraction algorithm to detect the skeletons of the detected buildings. To this end, a boundary extractor is designed and described in this section.

Following the definition of 8-neighborhood shown in Figure 1, the boundary pixel for building is determined if it is a contour pixel and satisfies the following condition:

- $0 < N(p) < 8$

where $N(p)$ is the number of nonzero neighbors of pixel p.

p_7	p_0	p_1
p_6	p	p_2
p_5	p_4	p_3

Fig. 1. Neighborhood arrangement

4 Experiments and Results

The proposed road extraction algorithm has been tested on two types of high-resolution satellite images, including 0.6 m QuickBird and 1 m Ikonos image data (see Figure 2). All test images have a size of 150 × 150 pixels and cover a sub-scene of a typical urban residential area in Toronto, Ontario.

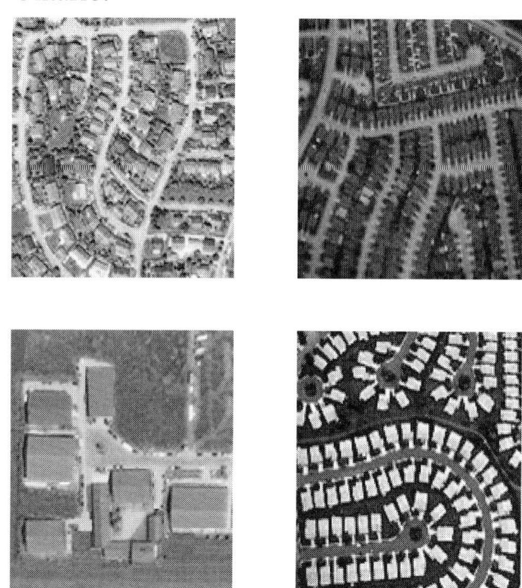

Fig. 2. Tested images: (a) and (c) QuickBird, (b) and (d) Ikonos images

The pseudo-color segmented images generated from the test images shown in Figure 2 are illustrated in Figure 3.

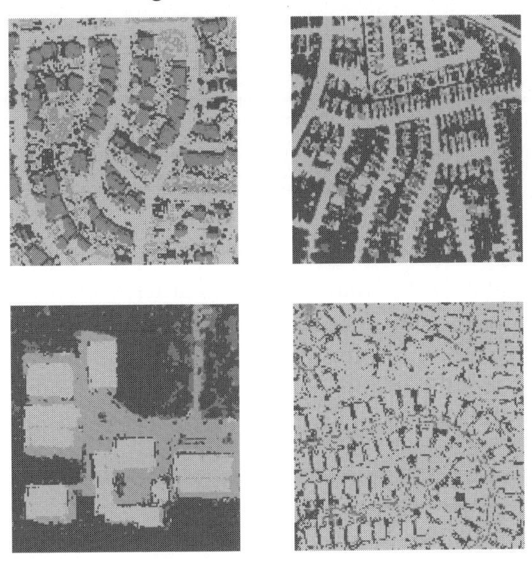

Fig. 3. Segmented images: (a) and (c) QuickBird, (b) and (d) Ikonos images

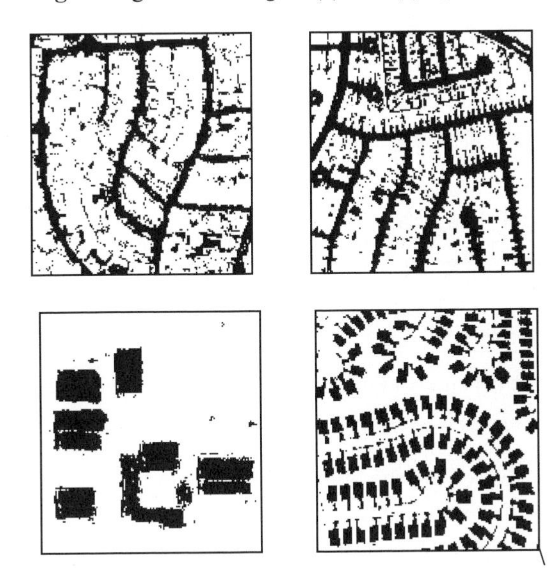

Fig. 4. Binary images of object regions: (a) and (c) QuickBird, (b) and (d) Ikonos images.

Figure 4 shows the binary images of the objects extracting from the segmentation images shown in Figure 3. It can be observed in all four images shown in Figure 4 that the segmented objects are corrupted by other objects with similar colors to objects

Figure 5 shows the object regions obtained after filtering the segmented images depicted in Figure 4 using the binary morphological operators. A visual comparison of the images clearly favors the filtered images (see Figure 5) over the segmented images (see Figure 4). Figure 5a shows the results obtained by filtering Figure 4a using binary dilating with a structuring element of 3 × 3, followed by eroding with a structuring element of 5 × 5. Figure 5b shows the results obtained by dilating Figure 4b with a structuring element of 3 × 3 and eroding with a structuring element of 5 × 5. Figure 5c shows the results obtained by closing Figure 4c with a structuring element of 4×4. Figure 5d shows the results obtained by eroding Figure 4d with a structuring element of 3×3 followed by closing with a structuring element of 4×4.

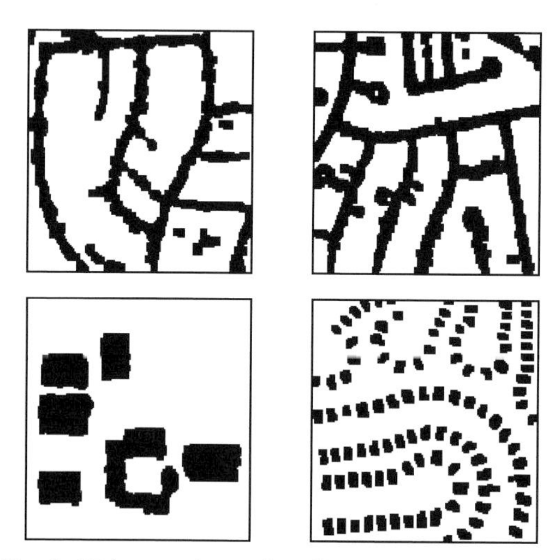

Fig. 5. Object regions after filtering: (a) and (c) QuickBird, (b) and (d) Ikonos images

The road centerlines and the edges of the extracted building roofs are delineated using the thinning algorithm and the proposed boundary extractor discussed above, and the results are shown in Figure 6.

In order to illustrate the accuracy, the extracted road centerlines and the edges of the extracted building roofs are overlaid on the original image, see Figure 7. In the overlay images the thin red lines indicate the road centerlines and the edges of the extracted building roofs. It can be obverted in

Figure 7 that most centerlines and the edges of the extracted building roofs match well the roads and buildings.

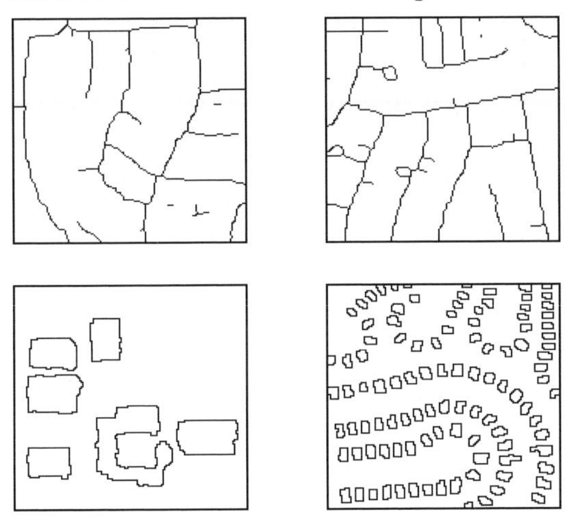

Fig. 6. Road centerlines and building roofs: (a) and (c) QuickBird, (b) and (d) Ikonos images

Fig. 7. Road centerlines and building edges (in red) overlaid on tested images: (a) and (c) QuickBird, (b) and (d) Ikonos images

5 Conclusions

A new method for extracting manmade objects from high-resolution satellite imagery such as QuichBird and Ikonos has been presented in this paper. The method employs a segmentation algorithm proposed in our previous work and works in three steps: (1) segmenting color images, (2) detecting objects from the segmented images, and (3) post-processing of the extracted objects. The proposed method has been examined by extracting road networks and buildings from pan-sharpened QuickBird and Ikonos images. The results demonstrate that the proposed method for object extraction is very effective.

Acknowledgements

This research was partially supported by a Natural Sciences and Engineering Research Council of Canada (NSERC) discovery grant.

References

George JK, Bo Y (1995) Fuzzy Sets and Fuzzy Logic: Theory and Applications, Prentice Hall PTR, Upper Saddle River, New Jersey, USA

Goldberg DE (1989) Genetic Algorithms in Search, Optimization and Machine Learning, Addison-Wesley, Reading, Mass, USA

Langan DA, Modestino JW, Zhang J (1998) Cluster validation for unsupervised stochastic model-based image segmentation. IEEE Transactions on Image Processing, 7(2), pp 180 -195

Li Y (2004) Fuzzy Similarity Measure and Its Application to High Resolution Color Remote Sensing Image Processing, Master Thesis, Ryerson University

Liew AW, Yan H (2001) Adaptive spatially constrained fuzzy clustering for image segmentation. Proceedings of 10th IEEE International Conference on Fuzzy Systems, University of Melbourne, Australia, Dec., 2001, Vol 2, pp 801-804

Tou JT, Gonzalez RC (1974) Pattern Recognition Principles, Addision-Wesley, Reading, Mass, USA

Winston WL (1991) Introduction to Mathematical Programming: Applications and Algorithms, Duxbury Press, Belmont, Ca, USA

Zhang TY, Suen CY (1984) A fats parallel algorithm for thinning digital patterns. Communications of the ACM, 27(3), pp 236-239

Zhong JM, Leung CH, Tang YY (2000). Image compression based on energy clustering and zero-quadtree representation. IEE Proceedings on Vision, Image and Signal Processing, 147(6), pp 564 –570

Extension of NASA's Science and Technology Results, Earth Observations for Decision Support

Stephen D. Ambrose

NASA, Science Mission Directorate, Sun-Earth Systems Division, Applied Sciences Program, Washington, DC 20546, USA.
Email: sambrose@nasa.gov

Abstract

The Office of Science of the National Aeronautics and Space Administration (NASA) is focused using the vantage point of space to improve our knowledge of the Earth system, space systems, and exploration. NASA spaceborne satellites provide measurements that are used in science research associated with the water and energy cycle, the carbon cycle, weather and climate, atmospheric chemistry, and the solid Earth and natural hazards. The NASA science mission has a focus on improving the prediction capacity in the areas of weather, climate, and natural and technological hazards.

The data and knowledge resulting from the Sun-Earth observing systems and science models of the Sun and the Earth are available for assimilation into decision support systems to serve society. Through partnerships with national and international agencies and organizations, NASA contributes to benchmarking practical uses of observations from remote sensing systems and predictions from Sun-Earth science research. This objective is to establish innovative solutions using Sun-Earth science information to provide decision support that can be adapted in applications of national and international priority. A common modeling framework is followed as well as utilization of an enterprise architecture.

Space-based data acquired by NASA contribute to Sun-Earth science models that enable understanding and forecasting of weather, climate, and to disaster management to serve in primary applications including the related applications in wildfire management, food security, aviation safety, homeland security, tropical weather, human health, invasive species management, and water and air quality management. Common systems engineering approaches are employed, including dependence on geospatial standards and interoperability, verification and validation, benchmarking, visualization, and workforce development.

This paper describes the Sun-Earth observing missions, the science models, and the decision support systems in the context of a systems engineering approach to enhancing decision support systems for disaster management relevant to decision makers. The architecture for systematically delivering results from research to operations is described. Recent results and solutions are highlighted specific to disaster management as related to U. S. National and International efforts.

1 Introduction

The objective of NASA's Applied Sciences Program is to expand and accelerate the realization of economic and societal benefits from Earth science, information and technology in fulfillment of the Agency's goal to apply Sun-Earth system science to improve prediction of climate, weather, and natural hazards and advance the vision of earth and space exploration. The NASA Science Mission Directorate (SMD) program accomplishes this objective by using a systems approach to facilitate the assimilation of Sun-Earth observations and predictions into partner organizations' decision support tools through which they provide essential services to society. These services include management of forest fires, water, coastal environments, agriculture, air quality, weather prediction and hazard mitigation, and enhancement of aviation safety. It is through this pathway that NASA's longer-term research programs yield near-term practical benefits to society. SMD is citizen-centered, results-oriented, and market-driven in accord with the President's Management Agenda for federal agencies and vision for space exploration and protection of our home planet.

NASA is well poised to make major contributions in these areas. It's global and solar system observational capability and developments, systems approach to the study and understanding of environmental phenomena, and ability to turn science and data to useful information for decision-making make it uniquely situated to lead major new global initiatives. In the development and application of information, tools, models and predictions for risk and consequence assessments, sustainable mitigation activities, and predictive understanding of natural phenomena, NASA will assist decision-makers and ultimately reduce losses and save lives from weather, climate, and natural disasters.

The purpose of NASA's SMD is to increase our knowledge of the Earth-Sun system, including its response to natural and human-induced changes to enable improved predictions of climate, weather, and natural hazards. SMD and the Applied Sciences Program serve NASA and society by expanding and accelerating the realization of societal and economic benefits from Earth science, information, and technology research and development.

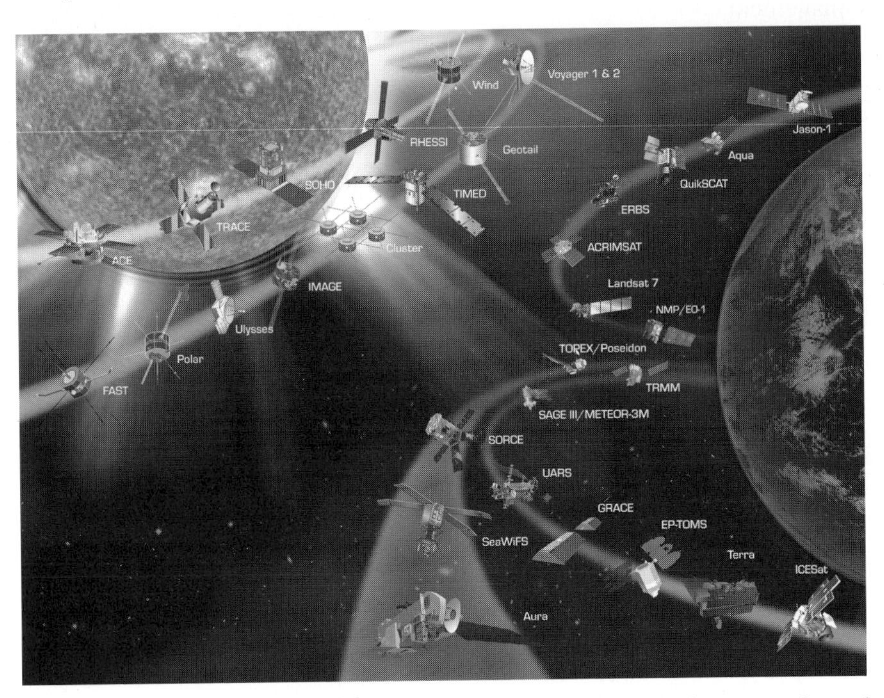

Fig. 1. NASA's cadre of Earth and Sun-Solar System observation research satellites

The overarching goal of the Applied Sciences Program is to bridge the gap between Earth-Sun science research results and the adoption of data and prediction capabilities for reliable and sustained use in decision support. Our strategy is to employ system-engineering approach in support of "completing the information cycle" and to connect the research domain and the operational domain associated with Sun-Earth science and remote sensing technologies.

The SMD works through partnerships with public, academic, and private organizations to develop innovative approaches for using Sun-Earth science information and technology. The Applied Sciences Program enhances the availability, interoperability, and utility of Enterprise and private sector data sets, communications, computing, and modeling capabilities to serve specific national applications.

Overall, the approach is to enable the assimilation of Earth and Sun science data and model outputs to serve as inputs to decision support tools and assessments. The outcomes and impacts are manifest in enhanced decision support and the social and economic benefits gained through improved decision-making, public information, and efficient and effective management.

2 Partnerships

NASA has many partners in the advancement of Sun-Earth system science and its application to societal concerns.

- We consider this a major part of a successful program for NASA. We partner with ten other federal agencies in the U.S. Climate Change Science Program (CCSP) to provide a sound scientific basis for the policy decisions the Nation will make in the years ahead. We participate in the Climate Change Technology Program (CCTP) to develop means for adaptation and mitigation.
- We have extensive partnerships with other U. S. Federal Agencies that use the products of research to improve the essential services they deliver to the nation.
- We are working with the commercial remote sensing industry to enhance the scientific utility of commercial imagery, and with the broader industrial community on the technologies of the future.
- We have over 290 agreements with 60 countries, and make leading contributions to major international research programs and scientific assessments of environmental change. NASA is a leader in the effort by the world's space agencies to design and deploy an integrated global observing system.
- We have nearly 2000 research and technology grants and contracts with universities nationwide for basic research, applications demonstration and technology development.
- We work with the international community on sustainable development and internationally coordinated satellite missions for global monitoring.

The Applied Sciences Program has developed partnerships with a number of agencies and organizations and include: ***Federal***: Department of Defense (DoD), Department of Homeland Security (DHS), Environmental Protection Agency (EPA), U. S. Geological Survey (USGS), National Oceanic and Atmospheric Administration (NOAA) Center for Disease

Control (CDC), Army Corps of Engineers (COE), Department of State (DOS); ***State-Local,*** over 25 state, local,regional, and tribal organizations engaged in a wide variety of applications from emergency management to invasive species control.: ***Independent Association*:** International Society of Photogrammetry and Remote Sensing (ISPRS), American Society of Photogrammetry and Remote Sensing (ASPRS), Hazards Research and Applications Information Center, Pacific Disaster Center, Humanitarian information Unit, American Association of State Geologists, National Emergency Management Association and others.

The Earth-Sun Systems program seeks to enable the practical use of Earth and Sun science, information, and technology in ways that are systematic, scalable and sustainable, thus magnifying the benefit of Sun-Earth system science to the nation. The approach to doing so is to partner with other federal agencies having decision support systems that can be enhanced to assimilate observations from remote sensing systems and predictions from NASA's Sun-Earth system models and data. NASA works with its partner agencies to benchmark (measure) the improvement from use of new observations and predictions. In addition to observing systems and models, NASA has the systems engineering expertise that can enable their effective use by our partner agencies. NASA and its Centers engage the capacity of academia, industry, and others through competitive solicitation for products, tools and techniques employing remote sensing observations and Sun-Earth System models.

NASA is engaged in twelve national applications with partner federal agencies that can be served by NASA's observations and research in Sun-Earth System science: (1) Energy Forecasting; (2) Agriculture Efficiency; (3) Carbon Management; (4) Aviation Safety; (5) Homeland Security; (6) Ecological Forecasting; (7) Disaster Management; (8) Public Health; (9) Coastal Management; (10) Invasive Species; (11) Water Management; (12) Air Quality Management.

Also under the Applied Sciences Program is a Cross-Cutting and Network Solutions Program. The Crosscutting Solutions Program contains a portfolio of tactical and strategic program activities in order to meet the current requirements and prepare for the future challenges of the Science Mission Directorate. Tactical activities include direct support to the National Applications Program in elements such as the Integrated Benchmarked Systems, Community Growth, and some aspects of Solutions Network. Strategic activities include investments in human capital development, standards and interoperability, and Earth Science information management in elements such as the DEVELOP student development project, the Geospatial Interoperability Office, and some aspects of Solutions Network. The desired impact for the resulting networks, standards, prototypes, processes, guidelines, and outreach is for the partner organizations and their customers to benefits from operational use of Earth science in serving their decision-making processes.

3 Societal Benefits

The Applied Sciences Program is producing significant returns on the federal investment in science and technology research. For example, NASA is integrating science and technology research results to produce improved warnings and predictions of hurricanes, tornadoes, and other severe weather events, thus enabling more cost effective damage mitigation, emergency preparation, and subsequent emergency management through a partnership with the FEMA. With regard to agricultural efficiency, NASA is working with the US Department of Agriculture to benchmark predictions of El Nino and La Nina events to support decisions on agriculture planning and practices to assimilate improved observations and predictions of the health and condition of crops and forests around the globe. In Aviation Safety, measurements and predictions from NASA weather and environmental satellites are being integrated with other traditional aviation weather information to improve information delivered to pilots and air traffic controllers for hazard avoidance. In the area of space weather, we are working with our partners in space science to advance applications for energy, homeland security, disaster management, public health, and astronaut safety with respect to fluctuations in the Earth's magnetic field and solar radiation. These are just a few examples of how NASA works through partnerships to extend the benefits of Earth science research to serve society.

The figure below depicts the Applied Sciences Program to systems engineering. NASA remote sensing systems and science models are inputs generating the observations and prediction outputs (which are inputs from the standpoint our partners decision support systems). The outcomes are improved decision support tools generating positive impacts on national policy and management decisions in a range of activities from coastal evacuations due to hurricanes to positioning of fire fighting resources in national forests. The target impacts are improvements to the quality and effectiveness of operations and policy management by enabling decision makers to benefit from decreasing uncertainties associated with complex and dynamic Sun-Earth System processes. This leads to improved policy and management decisions.

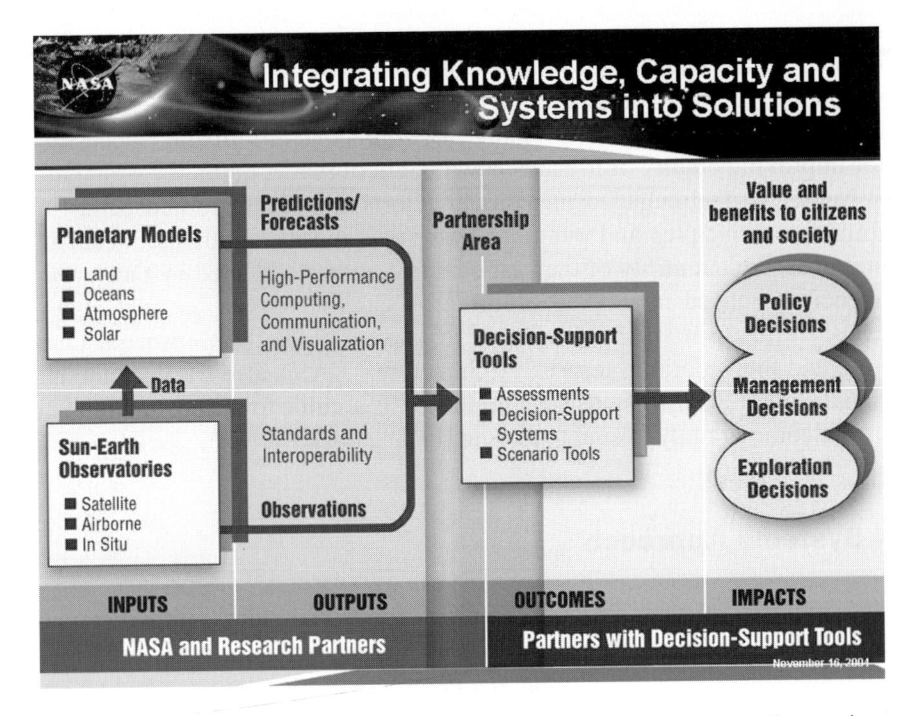

Fig.2. NASA's Integrated engineering architecture towards moving observations to models to improved decision support for society

The source of Applied Sciences Program inputs (scientific knowledge, observations, and predictions) is the Sun-Earth System Science program. NASA research and development of Sun-Earth System Science has yielded significant accomplishments in the form of a flotilla of remote sensing satellites and the Earth Observation System Data Information System (EOSDIS) database of global measurements of key geophysical parameters. The Sun-Earth System Science program has developed sophisticated Sun-Earth System models of complex and dynamic process and their predictions of key types of information.

The Applied Sciences Program purpose, approach, and intended outcomes are consistent with the recognition by both the Executive and Legislative branches of the value of using Earth science, information and technology to improve the many decision support systems employed by the public and private sectors to deliver essential services. The Global Earth Observations (GEO) program has embraced the paradigm of observing systems to models to decision support systems, underpinned by research, as a structured means of delivering science for society. This basic approach is also serving as one basis for the next phase of international collaboration on Earth observation. The beginning of this new phase was marked by the international Earth Observation Summit held in July 2003, which inaugurated planning for a Global Earth Observation System of Systems (GEOSS) aimed at providing environmental information necessary for improving quality of life and stewardship of planet Earth.

NASA is also engaged in national efforts to coordinate geospatial information principles and practices (inter-operability, standards, metadata, etc) to assure the utility of the vast quantity of data collected by the several agencies engaged in *in situ*, airborne, or satellite remote sensing. Interagency efforts, in which NASA participates, include the Geospatial One-Stop and the Federal Enterprise Architecture. NASA also employs the national Commercial Remote Sensing Policy as a guide to the acquisition and use of commercially available remote sensing data.

4 Systems Approach

The Applied Sciences Program employs functional steps of Applications Research and Evaluation, Verification & Validation, and Benchmarking in its systematic approach to bridge the gaps between the research and operational domains.

Applications Research and Evaluation: Evaluate the requirements and technical feasibility of Earth science data and models, information products, and predictions in partners' operational decision support needs.

Verification & Validation: Measure the performance characteristics of data, information, models, and predictions to meet requirements for the applications.

Benchmark: Enable the adaptation and adoption of geospatial information and methods derived from Earth science results to serve decision support.

The desired outcome of the national applications is for the partner organization to use the project results, such as guidelines and prototypes and procedures, as benchmarks for operational use and enhance their management and decision-making capabilities with appropriate Earth science products and tools.

Fig. 3. Example of NASA science and technology inputs to support the Department of Homeland Security Federal Emergency Management Agency's (FEMA) "HAZUS-MH" decision support tool

5 NASA's Role

Clearly, remote sensing science and technology is poised to continue to make a major contribution in the Disaster Management field. Over the last decade the number and capability of remote sensing systems in the government and private sectors has increased to a point where there is broad recognition of their potential. We look at Disaster Management in the entire cycle. Below is an analysis of potential contributions in each of major area of the disaster cycle.

Preparedness - Except for the weather satellites (and the modeling and data integration behind weather forecasting) providing continuous tracking of meteorological hazards, little remote sensing data and science has been utilized in the preparedness phase of a disaster management. There are two areas where NASA science and technology can help in the preparedness phase of a disaster: better time/space monitoring and prediction of natural phenomena as they are developing into extreme events; and better tools to incorporate physical factors about those phenomena into models that predict the impact of the impending disasters. The former is necessarily a longer-term activity because the observational systems that are evolving today due to both private sector (e.g. high resolution sensors) and government-sponsored systems are simply not capable of providing this, but the later is a research activity that is part of the Applied Sciences Program.

Mitigation - Much like the preparedness phase of disaster management, little science and technology has made its way into routine mitigation strategies and activities over the past several years. NASA potential contributions here are enormous, particularly considering the new paradigm in disaster management of trying to be more proactive and less reactive and preparing communities for disaster. This necessarily requires better observation and understanding of natural phenomena as well as better tools to assess, predict, and model natural events and their impact. NASA's philosophy of a "systems" approach to scientific problem solving is particularly relevant to the new paradigm for mitigation since mitigation necessarily targets the intersection of the environment, society, and infrastructure, and needs to include scientists, sociologists, economists, planners, disaster managers, resource managers, as well as others.

Response/Recovery - Remote sensing data are useful for a synoptic view of a damaged area. Although environmental damage of large regions may be identified using imaging systems with tens of meters (and perhaps up to the 100m) of spatial resolution, very high resolution information (1-10m spatial) is usually necessary for the emergency manager to assess damage to infrastructure and buildings. One of the major problems of satellite remote sensing data for response is the timeliness, as the orbits may not allow revisiting of an area for days to weeks, and in extreme cases, months after an event. Bad weather over the affected area is also a factor. The disaster management community continues to see images of disasters or damage areas after the disaster strikes, but the real question that has not been answered is whether or not those images are useful for the decision-making process.

6 Disaster Management Application

Disaster Management is one of 12 National Applications of the Applied Sciences Program. The Disaster Management Application of this program specifically addresses the use of NASA basic and applied science, data, and technology in the decision-making processes of Disaster Management. This National Application area includes two ongoing activities, the Solid

Earth and Natural Hazards Program (SENH) and the Disaster Management Application.

The Disaster Management program is divided into 7 program or theme areas. There may be significant overlap between program activities. Each application is supported by individual applications research initiatives, grants, and research proposals. These theme areas are Earthquakes, Volcanic Eruptions, Wildfires, Floods, Hurricanes (wind/surge), Climate Change, and Drought.

Over 100 projects have been sponsored through NASA Research Announcements (NRA) over the years. The results of these projects are continually being fed to agency partners so the results of this research can be utilized in real-world applications or further research.

The Disaster Management Program of NASA's Applied Sciences Program continues to work with national and international partners to improve disaster response, risk mitigation and build sustainability across all communities that can benefit from NASA's science and technologies.

7 Goals

The goals of the Disaster Management Application are to: 1) develop applications for holistic understanding of Earth processes which lead to natural disasters; 2) improve risk assessment and disaster mitigation capabilities for vulnerable regions; and 3) establish NASA relationships with disaster management practitioners and transfer research, science, and technology to operational use.

The Disaster Management program spawned from the Solid Earth and Natural Hazards program in the year 2000 to do applied research related to natural hazards. The program is closely linked to the other NASA programs such as the Hydrology Program, the Climate Program, Earth Observing System (EOS) and others, as the basis for basic scientific understanding of both geological, oceanic, and atmospheric hazards. The program has grown to include validation of applications in operational settings with sister agencies responsible for operational disaster management.

The current set of imaging systems and technology that are utilized in this program number 18 satellites and 80 sensors along with a variety of commercial and airborne systems, and *in situ* surface and subcrustal observing system, and other data sets. They include: ***Satellite Systems*:** Landsat, Terra, ASTER, Aqua, Aura, GOES/POES, TRMM, QUIKSCAT, GRACE: ***Data Sets*:** Satellite Radar Topography Mission (SRTM) Data for topography; ***Other***; GPS arrays, Information Systems technology, and

GIS. All of this and more data are available from NASA's Distributed Active Archive Centers (DAAC).

The approach includes the conduct of applied/applications research that is fundamentally tied to the science and technology programs. The program works with the end customer to understand information needs and builds partnerships among the science community, the disaster managers and practitioners, state/local government agencies, and commercial data and service providers. In addition, the program focuses on unique NASA capabilities (a "One NASA" philosophy as "Only NASA Can") so as not raise end-user expectations unnecessarily.

The decision support systems enhanced under the Disaster Management Program work with several national centers in the development of applications that can be put into operational use for our partner decision support systems.

As with all the applications, many disaster management applications require the use of geospatial data. The Geospatial Interoperability Office closely collaborates with the Disaster Management Program to assure that geospatial technologies are applied to decision support system solutions.

8 Summary and Grand Challenge

As stated before, a primary challenge and objective is to assess the continuity of Earth science data sets, observations, and predictions to serve the community into the future as part of an integrated Sun-Earth information infrastructure that contributes to economic and environmental security. A primary challenge and objective is to assess the continuity of Sun-Earth science data sets, observations, and predictions to serve the community into the future as part of an integrated Earth information infrastructure that contributes to economic and environmental security.

To do this NASA has adopted a systems approach to solving many of society's problems and disasters. We are doing this as "Only NASA Can". This statement is not one that says only NASA can do it, it means that we will participate in activities that reap the benefits of NASA's research specific to NASA's vision to improve life here, to extend life to there, and to find life beyond in alignment with our space exploration vision.

The National Applications of NASA is well poised to take on this responsibility through our programs that specifically focus on the improvement to the Nation's important Decision Support Systems and specifically focus on utilizing NASA's science and technology for societal benefit. This is the Grand Challenge NASA is taking on for stakeholders.

References

GeoData.Gov, Geospatial One Stop, http://www.geodata.gov/
NASA, Destination Earth, Sun-Earth System Program, http://www.earth.nasa.gov
NASA, Distributed Active Archive Centers, http://nasadaacs.eos.nasa.gov/
NASA, Earth Science Applications Program, http://science.hq.nasa.gov

A Concept of an Intelligent Decision Support for Crisis Management in the OASIS project

Natalia Andrienko and Gennady Andrienko

Fraunhofer Institute AIS, Schloss Birlinghoven, 53754 Sankt Augustin, Germany.
Email: natalia.andrienko@ais.fraunhofer.de

Abstract

Within the OASIS project, intelligent decision support tools shall improve situation analysis and assessment, provide assistance in finding appropriate recovery actions and scenarios, and help in handling multiple decision criteria. This functionality will be developed in two major directions: 1) knowledge-based decision support, and 2) advanced methods for handling decision complexity. Knowledge-based modules include data analysis support tool, event recognition and prediction, intelligent checklist, information prioritization, and knowledge base on relevant norms and regulations. Modules for handling specific decision complexities include data mining, data aggregation, argumentation and collaboration support, and multi-criteria decision analysis tools.

1 Introduction

OASIS is an Integrated Project focused on the Crisis Management part of "improving risk management" of IST (IST-2003-004677, see http://www.oasis-fp6.org/). Conducted over 48 months (starting September 2004), OASIS aims to define a generic crisis management system to support the response and rescue operations in case of large-scale disasters. The project is coordinated by EADS (France).

Taking full advantage of, and leveraging work from, the previous projects in the relevant domains, from the dual-use technologies, and in continuity of the successful developments in the EGERIS project (see http://www.egeris.org/), OASIS will:

1. Analyze the users requirements to extract European generic system requirements,
2. Specify and design a true generic, interoperable and open system architecture, which will allow easy deployment at every level of the action chain (local, regional, national and European). This generic architecture will rely on the integration of mature state-of-the-art technologies.
   The project will provide the definition of:
   - the system backbone (data bases, common operating environment and fully interoperable message handling system), supported by a reliable and secure communication network,
   - the deployable broad-band wireless communication network,
   - the command and control functions,
   - the decision support software modules.
3. Implement these architectural concepts through the development of 2 versions of a pre-operational system, representative of the future European and national target system(s),
4. Validate and evaluate pre-operational systems with users from different EU countries. The evaluation sessions will be performed in the frame of operational scenarios.
   In this paper we present a concept of the decision support in the project.

2 Decision Support in Crisis Management

Decision support modules will be developed within an applied research sub project (SP4) coordinated by the Fraunhofer Institute AIS (FhG AIS), other participants are WG RAS (Working Group of the Russian Academy of Sciences) and BAES (BAE Systems). It will undertake the applied research to develop a set of software tools to support the C3I developments (sub-project SP3 - C3I - Command, Control, Communications and Intelligence) with more advanced decision making methods in the process of crisis management, in particular:

- support for more effective and comprehensive situation analysis and assessment;
- assistance in finding appropriate recovery actions and scenarios;
- help in handling multiple decision criteria and conflicting goals of multiple actors.

SP4 will develop techniques and implement tools intended to facilitate and enhance the work of crisis management personnel, in particular, people working in emergency control centers. The tools of SP4 are suggested as an addition (but not replacement) to the C3I tools provided in SP3. The main intention is to counteract the specific factors that complicate the work of emergency management personnel:

- Time constraints. In an emergency situation, the need in information, maps, and analyses is very urgent. Therefore, it is desirable that intelligent tools could automatically determine what information and analyses are required, run the analyses with minimal or no involvement of human analysts, and provide the results as quickly as possible. This procedure should run iteratively since the situation rapidly changes. Intelligent software can also provide quick access to relevant knowledge sources such as pre-existing action plans, precedent cases, etc.
- Stress factor. Human analysts and decision makers are prone to emotions and distractions; a stress situation may impede their capabilities, and time pressure can lead to some aspects being overlooked. Therefore, emotionless intelligent software tools could provide valuable services. They can reduce the necessity to keep all information, knowledge, plans, etc. in one's mind and the risk of forgetting or overlooking anything important. Besides, utilization of decision support tools can increase the confidence of the users in the validity of the situation assessment and actions undertaken.

There are also other factors that make decision problems complex, such as the necessity to deal with processes developing in both space and time, large data volumes, uncertainties in data and information, multiple conflicting criteria, and multiple actors with conflicting interests involved in the decision-making process. SP4 will also address these issues and broaden the capability envelope of C3I by including more advanced decision making methods and techniques. It will achieve this by researching into a targeted set of technology areas in Information and Knowledge Management, Artificial Intelligence and Cognitive Science, that are relevant to the decision making process.

Current state-of-the-art in the decision support area offers a large number of generic approaches and technologies. However, since effective decision support in any domain crucially depends on domain-specific knowledge, it is impossible to mechanically reuse systems created for some particular problems and activities (e.g. for market analysis) for different problems and activities (e.g. for crisis management). Therefore, in OASIS we apply the generic approaches and methods from multiple disciplines to create tools specifically adapted to crisis management.

3 Technical Concept

SP4 will provide the aids to decision support based on the methods of artificial intelligence (AI), machine learning and data mining, information visualization, multi-criteria decision analysis and optimization, and computer-supported co-operative work. This is to try to address the technical elements of the decision-making process cycle - Intelligence - Design - Choice [1].

SP4 will increase the capability envelope of C3I by adapting and applying methods from a number of fields to aid decision-makers to handle complex decision problems in emergency situations involving stress and time pressure.

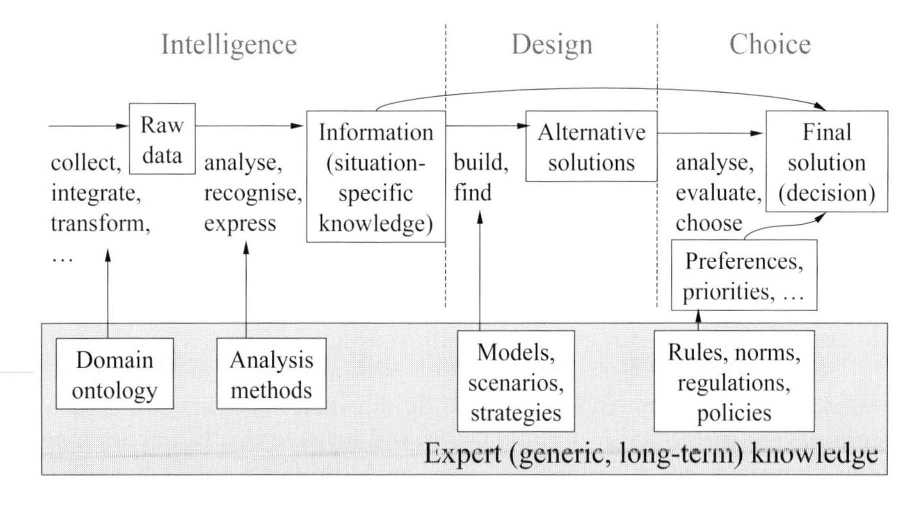

Fig 1. The Intelligence - Design - Choice – cycle of decision-making

SP4 tools relate to the first three stages of the OODA (Observe-Orient-Decide-Act) loop: they will support situation establishment, finding possible approaches to tackling the situation, and deciding which of the approaches to choose. In the research community, it is adopted to divide these activities into 3 phases, Intelligence, Design, and Choice (figure 1):

The tools to be built in SP4 will be integrated in the C3I software framework, which will be developed in SP3. The SP4 modules will employ the services developed in SP1, in particular, the data management services for accessing real-time data as well as archived data and documents. The SP4 tools will also utilize some of the innovations resulting from SP3, specifically, the ontologies and semantics for crisis management and the multi-source information fusion modules.

The decision support tools to be built in SP4 are conceived as optionally used components integrated in the C3I software framework. The users invoke them when some complexities arise that are difficult or impossible to handle using the usual tools. The main requirement is that the decision support tools must not complicate the usual work of emergency personnel but must facilitate it as much as possible. In particular, the user interface must be very intuitive and friendly, e.g. using whenever possible "natural gestures" like the user pointing or drawing on a map, which may activate an appropriate tool to run in a background mode. For releasing the users from many routine operations, it is planned to achieve a high level of automation. However, this does not mean that the decision support tools will automatically make any decisions. They may only suggest some actions to users, but it is users' choice whether to follow these suggestions. Automation takes place mainly on the phase of situation establishment (intelligence). For example, instead of suggesting the user to run some simulation model for predicting flood development, an intelligent tool could start this model automatically. It would also be appropriate to detect automatically what objects that might require particular attention (e.g. chemical plants) are situated on the flooded territory now or may be affected if the flooding develops as predicted.

The R&D activities in SP4 can be divided into 2 categories:

1. Knowledge management and knowledge-based decision support. This involves building of knowledge resources and knowledge-based software to be used on different stages of situation analysis and decision-making. The main idea is to capture knowledge of experts and make it accessible to less skilled and less experienced personnel. However, the knowledge is not just passively stored like in a library. It is utilized in knowledge-based software modules, which recommend appropriate actions and solutions to the users depending on the current situation and remind them about important things to take into account.
2. Advanced technologies for handling decision complexity. The sources of complexity in crisis management are large amounts of data, existence of multiple decision criteria, in particular, specific spatial and temporal criteria, necessity of on-site situation analysis and decision-making, and multiple actors with conflicting interests. The main idea is to cope with these complexities by applying various state-of-the-art techniques and approaches, which have to be adapted to the specifics of crisis management applications.

In the next section, a vision of intelligent decision support in crisis situations is represented. This should be viewed as a long-term goal to aspire to. Most probably, it cannot be fully achieved in the course of the project. The foreseen minimum to be accomplished within the project includes development of generic tools and approaches that will be usable in different types of crises and specific tools for some selected types of crises. On this basis, the feasibility and effectiveness of the proposed solutions will be demonstrated. The ways of extending the system to other emergency cases will be described.

4 A vision of Intelligent Decision Support in Crisis Situations

Let us imagine an intelligent software component, which will be further called Emergency Assistant, or simply Assistant.

In an emergency situation, a decision maker (DM) can activate Assistant (we assume that the decision maker is not necessarily a single person but may be a group of people working cooperatively in an emergency control centre). Assistant has an access to the database with all actual data and watches the display, in particular, the map (or maps) the decision maker uses for the investigation of the situation. Assistant also maintains a lookout for any updates in the current data. Assistant "understands" the meaning of all the data. This means that the semantics of data components is previously described and represented in a machine-readable form.

Upon the activation, the assistant asks DM about the type of the current emergency situation (i.e. whether this is flooding, fire, toxic substance spill, etc.). Then, DM outlines the problem area on the map. Since this moment, Assistant starts its *autonomous* functioning, i.e. tries to do as much as possible by itself, without disturbing the decision maker. However, DM can at any moment inspect the current activities of Assistant, its findings and suggestions. All this information is displayed in a special Assistant's window on the screen. When some information has a primary importance, or Assistant cannot proceed without an input from DM, Assistant attracts DM's attention by a special signal.

Step 1: Situation Establishment and Problem Identification

Assistant begins its work with determining what data are relevant for understanding the current situation. This depends on the type of emergency. For example, in the situation of flooding, relevant data are water levels in the river(s), current weather and weather forecast, as well as such persistent geographical data as elevation, locations of roads, inhabited sites, industrial plants, water supplies, and so on. The information about data relevance is specified in Assistant's knowledge base and can be automatically interpreted.

Then, Assistant checks whether it has sufficient data for a quick assessment of the situation and detecting the major dangers and problems. When some essential data is missing and cannot be derived from the existing data, Assistant signals about data absence to the decision maker. Whenever possible, Assistant not only indicates what data is missing but also recommends appropriate ways of getting necessary information. For example, Assistant may recommend reconnaissance flights for determining the geographical extent of the disaster area.

When it is possible to operate with the currently available data, Assistant automatically starts appropriate analytical computations. This may include:

- data interpolation from sample measurements;
- identifying what is affected by the disaster;
- predicting situation development, for example, by running simulation models.

In OASIS, Assistant may run the tools for situation assessment provided by SP3.

For determining what analysis methods are appropriate in the given situation, Assistant consults its knowledge base on data analysis methods. The knowledge base contains information what input is required for each method and what is its output. Method sequences (possibly, with conditional branches) may be also specified. The knowledge base may be viewed by users and extended with new analytical procedures.

When some analysis results are obtained, Assistant immediately represents them to DM in an appropriate way. Thus, if the results have geographical reference, they are displayed on the map. Assistant may also put some messages in its window. Messages of primary importance are specially marked. This concerns first of all problems Assistant has detected, for example, inhabited sites isolated due to flooding, risk of toxic spills from a chemical enterprise situated in the flooded area, endangered historical monuments, etc. Assistant may assign various "degrees of criticality" to its messages.

Step 2: Generating and Evaluating Solution Options

Assistant may have access to existing emergency plans, which are represented in a computer-interpretable form. In this case, Assistant may retrieve the relevant emergency plan and identify which items of the plan apply to the current situation. Then, Assistant displays the appropriate parts of the plan to DM, so that DM can further act according to the plan. As the situation changes, Assistant checks whether any other parts of the plan need to be activated and, if so, notifies the decision maker.

If no previously prepared plans are available, or there is no adequate plan for the current emergency situation, Assistant tries to find precedents (similar cases) in its historical database. If such cases are found, Assistant determines what actions and action sequences (scenarios) were used earlier to solve similar problem(s) and checks whether these scenarios are applicable to the current situation. If yes, Assistant recommends these scenarios to DM with references to the corresponding precedents. For each scenario, it is indicated who (e.g. which service or authority) should fulfill what and by what time moment. For example, "Local authorities must notify the citizens about the necessity of evacuation. The evacuation must be finished by 16:30".

It may happen that neither appropriate emergency plan nor precedent exists for a specific problem. In this case, Assistant tries to apply its general knowledge in order to find appropriate variants of solution. Thus, if Assistant detects that the expected level of water rising makes usual flood fighting measures (e.g. sandbagging) ineffective or there is an indication of possible dam failure, it may conclude that the evacuation of inhabitants from the inundation-endangered area is indispensable.

On the basis of available data and knowledge, Assistant makes an evaluation of each scenario it suggests: resources and time required, existing constraints (technical, economical, legal, etc.), expected results, to what extent this solves the problem, contingencies, possible side effects. Whenever necessary, Assistant may run appropriate simulation models to determine the possible outcomes of the scenarios ("what if" modeling). There may be several alternative solutions to the same problem. All the information thus found is provided to DM in an appropriate form. Thus, initial representation of a scenario may be rather compact due to high aggregation and focusing on the highest priority information, but all details are provided on demand.

DM may choose any of the suggestions provided by Assistant or add his/her own variant of solution to the list of options. If this variant is familiar to Assistant (i.e. such scenario is present in its knowledge base), Assistant can also evaluate the new option on the basis of available data.

When DM chooses a particular scenario, Assistant further elaborates it into a more detailed action plan. For example, for an evacuation scenario, Assistant may propose the roads and directions of movement from each location and where to organize shelters for the displaced people.

Step 3: Choosing a Particular Decision Option

For the most critical decisions, there are typically not many options. Thus, in a flooding situation, there may be just two alternatives: continue the regular flood fighting measures or start evacuating people. Which of these alternatives to choose, depends on the current situation and its expected development. If there is a high probability that the preventive measures will not withstand the expected water rise, it is clear that people must be evacuated. The only source of doubts for the decision maker is whether the prediction can be trusted. Anyway, whatever decision will be chosen, DM must appropriately justify it, and the decision along with its justification must be logged and stored for further inspection. Assistant may help DM by displaying a special decision form, which is already filled with the relevant information, including maps and charts representing the current situation and the prognosis. DM may only need to fill a few fields that cannot be filled automatically. All the information concerning the decision made will be stored in a structured form that makes it easily searched. The process of substantiating decisions is served by the argumentation support tool, which is automatically started by Assistant.

In other decision situations, DM needs to choose from several *comparable* options, for example, where to allocate shelters for displaced people or assistance centers on a disaster-affected territory, or which fire-fighting strategy to use. In such situations, the issue is decision optimization, i.e. finding the best suitable option(s) from those available, so that the problem is solved with the least possible use of time and resources and maximum possible safety and conveniences for the population and the emergency personnel. This is where Assistant may summon the tool for multi-criteria decision analysis, which will help DM to make an optimal choice according to the existing criteria and their relative importance. Again, the argumentation support tool will help DM to substantiate the decision made by referring to relevant information and background knowledge.

Where different SP4 tools fit in this scenario

Emergency Assistant is a collective name for several knowledge-based software components (to be developed in SP4.2):

- Data analysis support tool, which watches the current situation (i.e. its representation in the database), determines what data analysis methods are relevant, and automatically runs these methods;
- Event recognition and prediction tools are among the data analysis methods that may be invoked by the data analysis support tool;
- "Intelligent checklist" finds appropriate emergency plans or relevant precedents and adapts them to the current situation;
- Information prioritization tool is used for the most effective representation of incoming data and analysis results to DM;

- Knowledge base on relevant norms, regulations, and policies is used for evaluating decision options from the perspective of existing legal constraints.

Whenever necessary, Emergency Assistant invokes one of the tools for handling specific decision complexities, which will be developed in SP4.3:

- Data mining tools are particular tools for data analysis, which may be useful when the amounts of data are very large while the situation permits spending some time for comprehensive analysis. An example situation where data mining could be appropriate is an epidemic of a disease. Data mining tools may be used to reveal the spatio-temporal patterns of disease occurrence and thereby may help to find appropriate strategies for fighting the epidemic.
- Data aggregation tools are used for an effective representation of information to DM, i.e. in a compact but still comprehensive way, with an appropriate level of detail depending on the importance of each part of the information. The same tool can be used for representing information to public, mass media, various authorities, as well as for generating instructions for emergency personnel.
- The use of the argumentation and collaboration support tool for substantiating and documenting the decisions made was already mentioned in the scenario description. Besides this, the tool may be used for the preparation of documents to be communicated to the public or authorities and instructions for the emergency personnel. Another important use of the tool is in cooperative decision making, when multiple participants suggest their variants of problem solutions, express their opinions regarding the importance of various criteria, and present their expert judgment concerning different options. Any contribution can be built with the use of the tool, with including appropriate maps, tables, and/or charts, which help to substantiate the presented opinion. The whole process is automatically documented and logged, which allows an effective access and thorough analysis afterwards.
- Multi-criteria decision analysis tools (optimization tools) were also mentioned in the scenario description. They are used for an optimum choice from several comparable variants of problem solution on the basis of multiple conflicting criteria.

5 Sub-Project Overall Description

On the basis of a study of user requirements, SP4 will actively involve various advanced technologies in developing tools tailored to particular needs and conditions of crisis management.

The core of the work in SP4 is applying state-of-the-art technologies and approaches from various research disciplines to provide situation-related assistance to decision makers in crisis management. The work is divided into 3 work packages:

– SP4.1. Management and technical coordination;
– SP4.2. Knowledge management and knowledge-based decision support;
– SP4.3. Advanced technologies for handling decision complexity.

As it was pointed out, domain-specific knowledge is critical for effective decision support. For this reason, any generic approach needs to be specialized and fed with relevant knowledge, and any technique approbated in a different domain needs to be adapted to the specifics of the domain of crisis management.

SP4.2 focuses on knowledge-based technologies resulting from the research in Artificial Intelligence (AI). Any knowledge-based system needs a particular knowledge resource, i.e. certain knowledge typically used by human experts, which has to be captured and represented in a machine-readable form. Therefore, the process of building knowledge-based software consists of two parts: building of knowledge resources (knowledge acquisition, structuring, formalization, and representation) and development of program modules capable of using these resources or adaptation of existing software to the new knowledge resources. It is important to note that knowledge resources (i.e. externalized and structured knowledge of experts) are extremely valuable even without the software tools utilizing them. They can be made accessible to rescue personnel and used for reference, education and training, as well as for supporting mutual understanding in co-operative problem solving and decision-making.

It is not supposed to develop in SP4 any new methods or algorithms in AI but rather to use existing methods and approaches. Of course, some software implementation of these methods will be needed. Whether some existing software will be used or new modules developed, depends on whether it will be possible to find existing software that satisfies the specific requirements of using in OASIS (in particular, platform independence, speed, external interfaces). It may happen that it will be easier to implement such software anew than to try to fit something pre-existing into the overall architecture. However, this will not be a major problem for the project, since the effort for implementing a program module capable to read a knowledge base and do inferences on this basis is typically significantly smaller that the effort for building the knowledge base itself.

SP4.3 centers on techniques and approaches from the disciplines other than AI: data mining, statistics, information visualization, multi-criteria decision analysis, and computer-supported co-operative work. These tools will address different aspects of decision complexity in crisis management.

In SP4.3, it is supposed to intensively use existing software, in particular, tools for data mining and for multi-criteria optimization. These software modules will be appropriately interfaced and thereby integrated into the overall software framework.

The partners involved in SP4 will contribute with their experience and earlier developed software. Thus, FhG AIS has developed a prototype knowledge-based component for guiding users in analysis of spatial data [2]. This component will be extended and adapted to the needs of OASIS. FhG AIS will also contribute with their tools for exploratory analysis of spatio-temporal data [3], visual data mining [4], and multi-criteria site selection [5]. WG RAS has created the NOSTRADAMUS computer package designed to support decision making on mitigation of the impact of accidents on the environment and population at the initial and "acute" stages. NOSTRADAMUS incorporates modules that can predict initial parameters of release in the case of fire and explosion. BAES will contribute with advanced display and visualization technologies from the military environment, which will be adapted into the civil environment. These and other assets will be used in the course of the project.

An important activity in the sub-project is an intensive study of problems and requirements of personnel involved in crisis management. The main difficulty is that such people usually have no experience in using advanced decision-support tools and, hence, cannot consciously formulate their requirements to such tools. Moreover, potential users may even feel afraid of introducing any new techniques and approaches to their work practice. Therefore, prior to collecting requirements, it is necessary to properly inform representative users about the purposes of the new tools, the ideas and technologies standing behind them, and the benefits the users can expect from them. It would be advisable to demonstrate some of currently existing tools and prototypes (probably, with examples from different domains) to the users to raise their understanding of what is suggested to them. It is expected that the study will define which of the suggested decision support tools are actually needed in crisis management applications, how they must be integrated into the working environment, and what requirements to the user interface, response time, and presentation of the outputs they must satisfy.

The implementation will include following stages:

1. User problems and requirements are studied. The set of tools to be developed is defined.
2. Stand-alone prototypes of the tools are developed.
3. The prototypes are evaluated from the perspectives of their functionality, performance, and usability.
4. On the basis of the evaluation results, the tools are refined and improved.
5. The decision support tools are integrated with the other software tools (in particular, with software built in SP3) to enable the enhancements offered.
6. The functioning of the tools within the integrated environment is tested and the tools are further refined to ensure their smooth integration.

The development of the decision support tools must be done with ensuring the possibility of their integration into the overall framework. Hence, the definition of data formats, software interfaces, performance constraints, and requirements to the user interface must precede the work on software implementation. The tasks on defining software specifications and architectural requirements as well as software integration and validation activities are included in SP4.1.

Here is a summary of the tools that we preliminary plan to build in SP4:

- Data analysis support tool involving a knowledge base on data analysis methods.
- Tools for event recognition and event prediction.
- "Intelligent checklist": a case-based reasoning system on problem solving actions and scenarios.
- Information prioritization tool.
- Knowledge base on relevant norms, regulations, and policies.
- Data mining tool.
- Data aggregation tool.
- Multi-criteria decision analysis tool.
- Argumentation and collaboration support tool.
- Helmet mounted display tool.

However, we anticipate that this list may need to be amended according to outcomes of the user requirement study.

References

Simon H (1960) The New Science of Management Decision. Harper & Row: New York

Andrienko N, Andrienko G (2001) Intelligent Support for Geographic Data Analysis and Decision Making in the Web, Journal of Geographic Information and Decision Analysis,vol 5 (2), pp115-128

Andrienko N, Andrienko G, Gatalsky P (2003) Exploratory Spatio-Temporal Visualization: an Analytical Review, Journal of Visual Languages and Computing, v.14 (6), pp 503-541

Andrienko N, Andrienko G, Savinov A, Voss H, Wettschereck D (2001) Exploratory Analysis of Spatial Data Using Interactive Maps and Data Mining, Cartography and Geographic Information Science, vol 28 (3), pp 151-165

Andrienko N, Andrienko G (2003) Informed Spatial Decisions through Coordinated Views, Information Visualization, 2003, vol 2 (4), pp 270-285

Mapping World Events

Clive Best, Erik van der Goot, Ken Blackler, Teofilo Garcia, David Horby, Ralf Steinberger and Bruno Pouliquen

Institute for the Protection and Security of Citizens (PSC), Joint Research Centre, Via Enrico Fermi, 1, 21020 Ispra (VA), Italy.
Email: clive.best@jrc.it

Abstract

This paper describes new methods used for mapping news events gathered from around the world. Web based graphical map displays are used to monitor both the real time situation, and longer term historical trends. The results are derived from a synthesis of world events based on 20,000 news reports collected from the Internet each day by the Europe Media Monitor (EMM) (http://emm.jrc.org). EMM has invented a powerful method of rapidly classifying multilingual articles by matching weighted combinations of phrase and word patterns. The articles are cross-classified according to the countries mentioned and to general themes like "Conflict" "FoodSecurity" "Natural Disaster" "Ecology" and so on. The Alert system runs 24 hours per day keeping continuous hourly statistics on article populations. This then allows to generate real-time graphical presentations of "news maps", time series of Indicators, and animations of crisis developments.

In a second step, a more detailed geolocation of news articles based on town names identification in text and a reference gazetteer is performed. This has been coupled with a daily "Top News" clustering algorithm and results in 24 hourly News distributions for each EMM alert topic. This technique can also link clusters through different language versions and track stories in time. Geospatial representation of news developments enables decision makers to better assess a complex situation in a fast and objective fashion.

1 Introduction

The European Commission like other governmental bodies needs to continuously monitor press reporting of political developments and world events. Press reports from member states and across the world must be monitored in multiple languages. With the expansion of the Union, a strategic decision was made to move to electronic monitoring rather than manually selected reviews. In 2002 a software system to automate the gathering and sorting of news reports was introduced. This system is known as the Europe Media Monitor (EMM) [1]. EMM currently monitors nearly 700 on-line news sources and 15 news agencies in 30 languages. The automatic system has another major advantage, namely the ability to track news in real time. EMM processes and classifies about 25,000 articles a day. An automatic live newsletter (http://emm.jrc.org) summarizes top stories and topic alerts in each language.

This paper studies the spatial representation of news. The EU external affairs directorate is concerned with developments in any country in the world. Desk officers need to keep continuously updated on developments. Crisis monitoring and long term situation monitoring are both suited to a geospatial representation. These studies rely both on subjective and objective input. The objective measure is derived from statistical indicators derived from world news reports. News Maps are intended to represent current affairs analogous to weather maps. Human affairs evolve rapidly just like the weather.

The work is divided into two separate parts. Firstly, thematic live news have been derived from statistical measures of classified incoming articles. Secondly, a daily summary based on a news clustering algorithm and place name recognition is performed once per day and gives detailed long term spatial maps.

2 EMM Technical Overview

EMM exploits an XML/XSLT [2] technique for headline extraction on news web-sites to create RSS feeds [3][1], but additionally it uses some new "web intelligence" to categorise and filter the identified articles. A major innovation of EMM is an "on the fly" filtering of the textual content of articles to create topic specific real-time alerts. This fits well the require-

[1] XML standard format containing lists of hyperlinks, with some metadata.

ment for immediate and efficient "breaking news" detection with the less urgent requirement for a-posteriori search and retrieval.

The EMM software is built on a JAVA/XSLT [2] framework using Apache/Tomcat and public domain software. This provides a platform independent architecture, which allows for a scalable expansion in the future. Crucially, EMM fully supports multiple languages across all software components. All alerts, search criteria and web interfaces use UTF-8 character encoding and therefore support all world languages. This is an important requirement for the European Commission.

3 EMM Alert System

The EMM Alert system filters articles gathered from the Internet and/or from News Agency wires according to topic specific criteria. An alert definition consists of groups of multilingual keywords which can either be weighted (positive and negative) or combined in Boolean combinations. The alert system extracts the pure article text from the web page and then processes each word against all keywords defined for all alerts, in a single pass. An article triggers an alert if either the sum of detected keyword weights passes a threshold, or if a Boolean keyword combination is valid, or both. An alert can consist of one or more combinations and/or a weighted list of keywords. The alert system thus provides a very flexible and tunable 'topic definition' for automatically classifying articles in different languages. In practice the quality of the alert detection depends on the fine-tuning of the alert definitions.

News alerts detect and sort articles as they appear in on-line media. Alerts can be thought of as rather like "fishing" for articles that have yet to be published. Each alert definition consists of a list of multilingual keywords (the bait) designed to catch future articles (the fish). When caught, the article is placed into the appropriate alert, which contains up to some maximum number of most recent catches. Alerts are intended to cover a single topic area. Each on-line source is checked as frequently as every 15 minutes for new articles. The alert scan on a new article is done in a fraction of a second.

Running in parallel with the Alert system, is a breaking news systems which detects sudden increases in similar stories by keeping track of entity keywords appearing in news headlines. It keeps a time rolling average of expected occurrences. If a keyword appears many times more often than expected, and appears from several independent sources, then a breaking news item is created. The strength of the breaking news story is measured by the number of articles per hour. The breaking news system thus detects the new unexpected topics which are not already predefined in the alert system.

4 Thematic Statistical Indicators for World Countries

The objective is to derive statistical indicators which reflect the current political state within a country with regard to world news reporting. Advantages of this approach are that it is fully automatic, multilingual and can cover all countries in the world simultaneously. The method builds on the Alert detection system developed within EMM as follows.

A “World News” section was added to EMM with the aim of geolocating news stories by content. It consists of 218 alerts, one for each country in the world. An individual country alert simply consists of variations of that country name in all European Languages plus usually the capitol city. Capitol cities are avoided where false triggers occur due to person name clashes. An individual article will therefore trigger each country to which it explicitly refers. Note that we are not concerned here with which country published the article, but rather which country is being talked about. Therefore articles on any subject which refer to a given country will be flagged for that country by the alert system. In addition a number of global subject “themes” were defined such as “conflict”, “food aid” etc (see table 1). Likewise these theme alerts consist of groups of characteristic keywords, which occur frequently in articles about the particular theme. These are standard EMM alerts and clearly the quality of classification depends on fine tuning these multilingual keywords.

A feature of the EMM alert system is the recording of which (multiple) alerts an article has triggered. This information is stored in XML files held for each alert. At the same time the ALERT system also maintains statistics on the number of articles triggering for each ALERT per hour and per day. The results are coded in XML as shown below.


```
<alert id="Israel">
        <count>1085</count>
    <stats>64,18,11,19,22,25,21,37,55,69,4
    1,47,84,135,98,76,34,33,36,38,35,37,28
    ,22</stats>
</alert>
```


Finally for this new work on indicators a statistical combination measurement was added to the alert system specifically to cross-correlate Countries and Themes. Combinations occur when a single article triggers BOTH alerts referenced in the combination. A combination statistic coded in XML automatically by the EMM Alert processor is shown below.

```
<combination>
      <alertRef id="Conflict" />
      <alertRef id="Israel" />
   <count>112</count>
   <stats>5,2,1,8,2,8,4,7,0,4,3,4,14,16,8,5,3,2,3,5,3,2,3,0
   </stats>
</combination>
```

For both XML snippets, the <count> tag gives the total number of articles published in a single 24 hour period and the <stats> tag gives the hourly statistics. One full XML file containing statistics for all EMM alerts and combinations between countries and themes is generated for each day. These files therefore record time series of theme and country statistics, and are then available for trend studies.

5 Definitions of Socio-Political Indicators

An indicator represents a numerical measurement of the relative importance of a given theme for a given country, as reported by the world's media. To be useful the indicator must be normalized to allow cross-country comparisons and trend studies.

Two different normalized Indicators for a given theme and a given country can be defined as follows :

Definition 1:

$$I_{cj} = \frac{N_{cj}}{N_c}$$

where N_{cj} = article <count> for Combination of country C and theme j and N_c = total <count> for country C.

Icj is therefore a measure of the fraction of all articles written about that country which also refer to the particular theme. In some sense it measures the relevance of a given theme to a particular country as sampled from the world's media and varies between 0 (no relevance) and 1 (100% relevance). This standard definition becomes particularly relevant for trend studies. In practice we renormalize the indicator to be a percentage by multiplying by 100.

Definition 2:

$$I_{jc} = \frac{N_{cj}}{N_j}$$

; where N_{cj} = article <count> for Combination of country C and theme j and N_i = total <count> for theme j.

I_{jc} thus measures the fraction of articles written on a given theme which also refer to a given country. There is a subtle difference between I_{jc} and I_{cj}. I_{jc} measures the focus of the world's media attention when writing about a given subject. I_{jc} is useful for finding forgotten crises. Some countries are rarely in the news, but when they are the media report mainly problems. Therefore I_{cj} can have a very high value, but this does not mean that the world's attention is focused on that countries problems. I_{jc} will identify this because it will have a low value, reflecting the fact that the crisis is rarely reported in the media priorities. The signature for a forgotten crisis therefore is a country with a high I_{cj} and a low I_{jc}

Both indicators are measured automatically on a daily basis by EMM, and all values are archived. Note that EMM actually measures them on an hourly basis, but in practice this is usually too stochastic. For similar reasons it is sometimes necessary to aggregate some indicator measurement over longer time periods than one day, such as a week or n-days. This is usually the case for countries which are rarely in the news but where there is still the need to measure the indicator.

The results are visualized using Scalable Vector Graphics (SVG) [4]. The country polygons are defined as vectors in the SVG file. The country data values are then transformed to a proportional fill colour for each polygon using an XSLT. In this case the data is aggregated as numeric values per country. Clicking on a country gives direct access to all the articles for that country and the chosen theme. SVG animations are also an effective way to see time changes. This is achieved by stepping through daily EMM indicator values for a month or more. This gives an effective visual impression of crises growing and decaying. The results are illustrated in figures 1 and 2 which show both conflict indicators for one day – 9th December 2004.

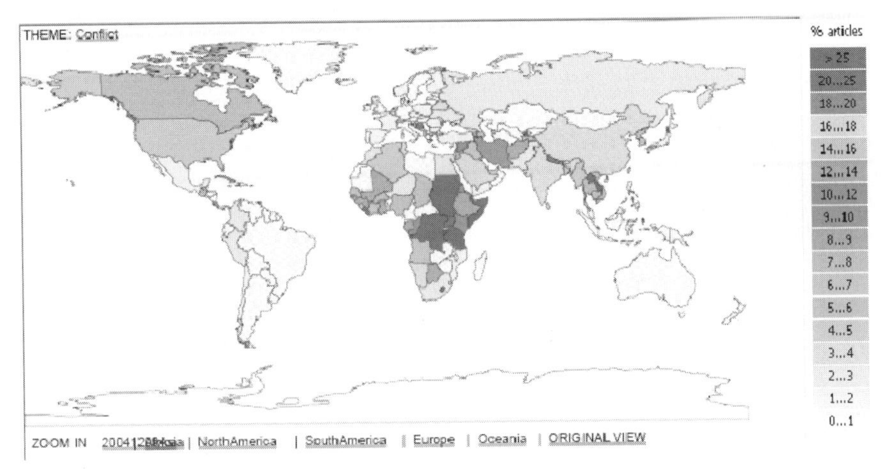

Fig. 1. Conflict indicator I_{cj}. This reflects the percentage of articles concerning a country which also refer to conflict. Data is for one day – 9th December 2004

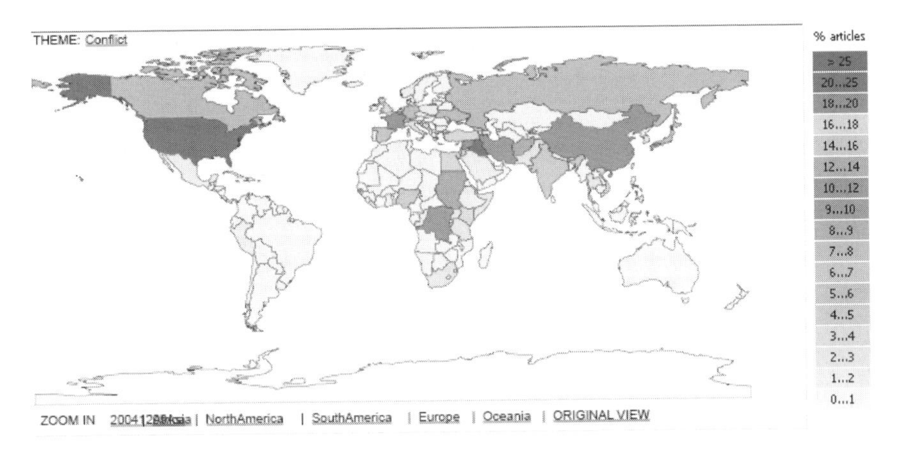

Fig. 2. Conflict Indicator I_{jc} . This measures the percentage of all conflict articles referring to a country. Data is for one day – 9th December 2004. Note here that USA and Iraq have high values reflecting the world's media attention, whereas figure 1 highlights those countries in conflict with relatively low media coverage

Similar maps to Figures 1 and 2 can be generated for each EMM theme. The current list of available themes are given in table 1. Time series animations and graphs track changes with time. Figures 3 shows the conflict indicator normalized by theme (I_{jc}) for Sudan and the Ivory coast over a 3 month period.

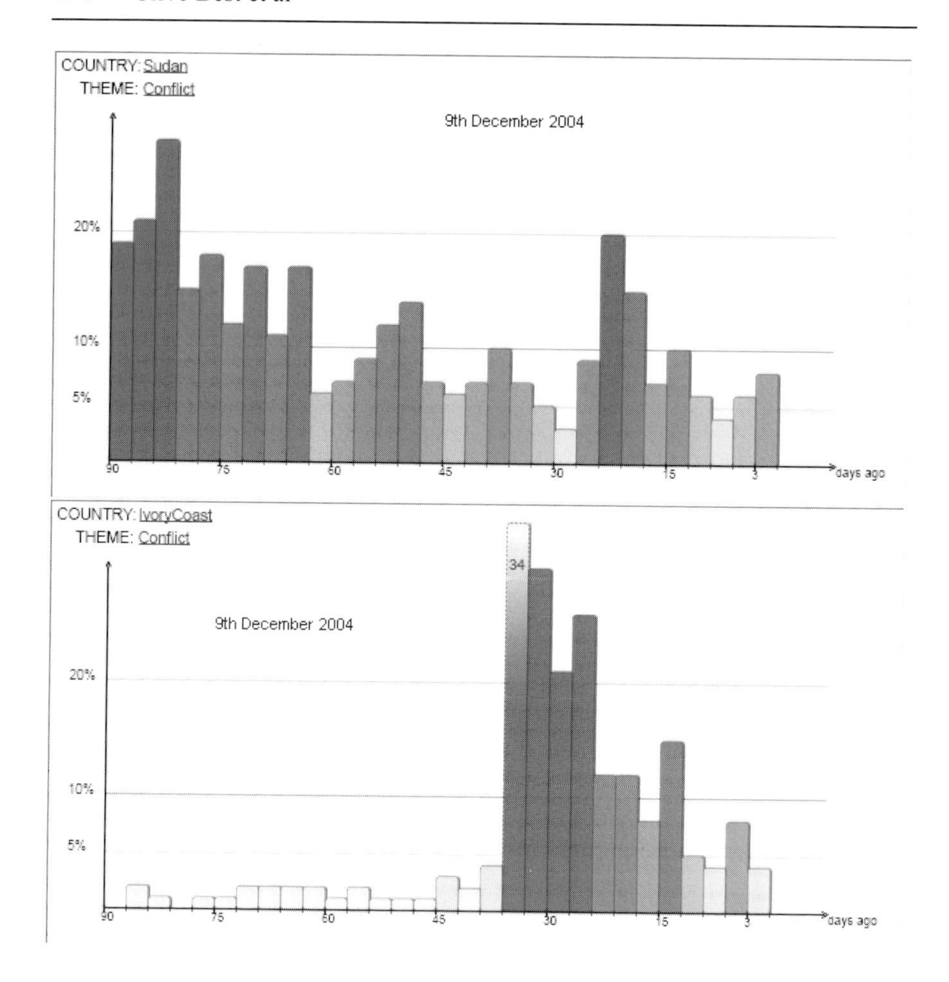

Fig. 3. Conflict indicator Ijc for Sudan and Ivory Coast aggregated in 3 day increments over a 3 month period. These plots are taken directly from the web displays in SVG.

Conflict	Terrorist Attack	Humanitarian Aid	Security	Development
Ecology	Food Security	Man Made Disasters	Natural Disasters	Political Unrest
Human Rights	Drugs	Financial Crime	Human Trafficking	Militancy

Table 1. Current EMM themes. These are subject alerts which are cross-correlated with the 218 country alerts. This allows the software to generate automatic thematic indicator maps as described in the text

6 Daily Cluster Analysis

At the end of each 24 hour period a clustering analysis is performed to identify the top stories in each language. The first step is to identify the major keywords in each article. This is done by comparing each word against a word frequency reference based on a large corpus of articles in the given language. In this way a keyness value is defined based on the uniqueness of the word within the corpus. Next an iterative clustering algorithm groups articles according to keyword overlapping. The details are described elsewhere [5]. For each day the 10 or so largest clusters represent the top stories of the day.

7 Geographic Place Name Recognition

Geolocating articles involves identifying the latitude and longitude of recognized places names mentioned in the text. Once candidate place names have been identified their coordinates can then be looked up in a gazetteer. This is complicated for EMM because place names have different spellings in different languages, and duplicate place names exist in different countries. Disambiguation and language variants are a major challenge for this work. Currently the 'Global Discovery gazetteer [6] is being used, but work is underway to merge it with the various sources compiled by the KNAB project [7] of the Institute of the Estonian Language.

Four important concepts must be in the database: the place name, its spelling variants for various languages, its relative 'importance' (a size information value from "1" = capital of a country, "2"= major city, up to "6" small village/place), its geographical co-ordinates (latitude/longitude) and the country it belongs to. Such gazetteers are becoming more easily accessible as many countries are providing free place name lists in order to normalize the denomination of their cities. Significantly, the United Nations has a Group of Experts on Geographical Names. This organization aims at helping the normalization of geographical place names in different countries spelled in different languages[7]. In addition to the place name list, we added lists of country ISO codes, of currency names, of adjectives pointing to the country and of names of the people of the country. This means that a hit for the country is generated even if only its currency or its people (e.g. 'Iraqi') are mentioned.

The recognition of geographical information in text can be divided into the two sub processes *geo-parsing* and *geo-coding* [5], where the former refers to the recognition of place names and the latter to the disambiguation and marking-up process.

7.1 Geo-Parser

The first tool aims at analyzing a natural language text and recognize 'potential' place names. As we are analyzing various languages, we must know which language a text is written in. For this purpose, we use an n-gram-based language guesser. The parsing must take into account the language. 'Monaco', for example, is non-ambiguous in English, but in Italian it can also be used to refer to 'München' (Munich) in Germany. In our system, each place name will be recognized if written in either the text language or in the local language of the place name, but not in any other language. A German text will refer to the city of Brussels either by using the German name 'Brüssel' or one of the two local names 'Bruxelles' (French) or 'Brussel' (Flemish), but not by using other variants such as 'Brussels' (English). A full description of this work can be found elsewhere [5].

7.2 Geo-Coding

After the recognition of potential place names during the geoparsing step, these potential place names need to be disambiguated and linked to the relevant data base information. Resolving ambiguities is not a trivial task. The main difficulties are linked to:

(a) Place names that are also words in one or more languages, such as 'And' (Iran) and 'Split' (Croatia);
(b) Place names that are homonymic with people's names, such as 'Victoria' (Hong Kong, Canada and others) and 'Annan' (UK);
(c) Places that have varying names in different or even in the same language ('Saint Petersburg', 'Saint Pétersbourg',' ' [Sankt-Peterburg], 'Leningrad', Petrograd', etc.).
(d) Multiple places that share the same name, such as the fourteen cities and villages in the world called 'Paris';

The software relies on a simple dictionary lookup in the text, which means that issue (a) can only be solved using extensive language-dependent geo-stop-lists (namely a list of location names that our system should never recognize. Such lists do not necessarily have to be generated manually. Instead, they can be created by first extracting all potential place names from a corpus and by then looking at those place names that appear more often than expected.

For problem (b), it is possible to put the most frequent ambiguous person names (e.g. 'Bush', 'Chirac', 'Annan') into our geo-stop-list, but common first names like ‘Victoria’, which can refer to important geographical places, should not be discarded automatically as this would lead us to miss some major locations. The best way to avoid this problem is to previously recognize person names.

Problem (c), caused by the fact that place names have many different translations, can only be solved by completing the database of place names. We have recently merged our database with the Estonian KNAB database and hope to incorporate more sources in the future.

The first attempt to solve problem (d), i.e. the ambiguity between places that share the same name, is to reduce the size of the database. As we were mainly interested in European place names, we reduced the half million place names in our database to about 85,000 by taking out the smaller places outside Europe. The potentially lower recall is acceptable for us because the majority of our text sources is from Europe and the US and tends to mention the country name when talking about smaller places outside Europe so that the reference to the country is not lost. This reduction improved the computational efficiency and reduced noise. As a result, we then have a number of clearly unambiguous place names such as ‘Vladivostok’, but we also have some ambiguous ones, such as ‘Victoria’, which could belong to Canada, to the Seychelles, to Hong-Kong or to over one hundred other places around the world. We try to further disambiguate and to identify one single location for each of these ambiguous place names, by looking at the combination of two parameters:

The relative ‘importance’ of the place, according to the size information in our database (the name ‘Paris’ will preferably be associated to the capital of France);

The other place names mentioned in the same text: an ambiguous place name will be identified as belonging to a certain country if this country is already being referred to in other parts of the text. In the short text ‘Victoria is the business and cultural centre of the Seychelles’, ‘Victoria’ will be geo-coded as the capital of the Seychelles.

After the disambiguation process each article is then marked up with the latitude and longitude coordinates for each place name found in the text. These same articles may also have triggered one or more EMM alerts and these are also marked up with the article. The XML format we use is RSS 2.0 [3] and the geocoding is based on the GeoURL [9] proposal.

Once the articles have been geocoded distributions of articles can then be plotted on maps. These maps are delivered over the web to end users who then interact with them to investigate further the results. This is achieved by linking html display of relevant articles to each location plotted on the map. For the point data we do not use SVG. Instead we have used the Worldkit [10] software to superimpose location data onto a world map. Worldkit is a Flash [11] application which takes as input an RSS file with each item encoded with GeoURLs. The title and description of the item are displayed together with the data points. A magnitude value can also be applied to alter the size of the point icons. The Worldkit display can be used for live updates which refresh the data from the remote RSS file. Each location is an active link which when clicked on gives access to all the relevant articles which refer to that location.

Any of EMM's 400 topic alerts can be displayed as a place distribution map for a given time period. Currently one day of data is displayed aggregating the number of articles referring to a single place name. Figure 3 shows such a distribution for the "Terrorism" alert for one day following the Beslan siege.

Fig. 4. Map of geocoded articles concerning terrorism for 6th September 2004. This was just after the Beslan tragedy. Beslan is the largest concentration of articles, and Moscow is the second largest

The second example of a map distribution is based on the daily clustering analysis. This identifies the 6-7 most important news stories for that day in each language. The most referred to place within the stories of each cluster is used to locate the overall story. In this way a daily news map is generated for each top story associated with a cluster. The user can navigate to a given date and follow the same story through different languages and different days. A live breaking news map is also generated showing what is happening now. This does not use the clustering algorithm since it is currently too slow for live results. Instead it tracks the most mentioned country within the current top breaking news stories for that language. The stories are then geolocated at the capital city for each country.

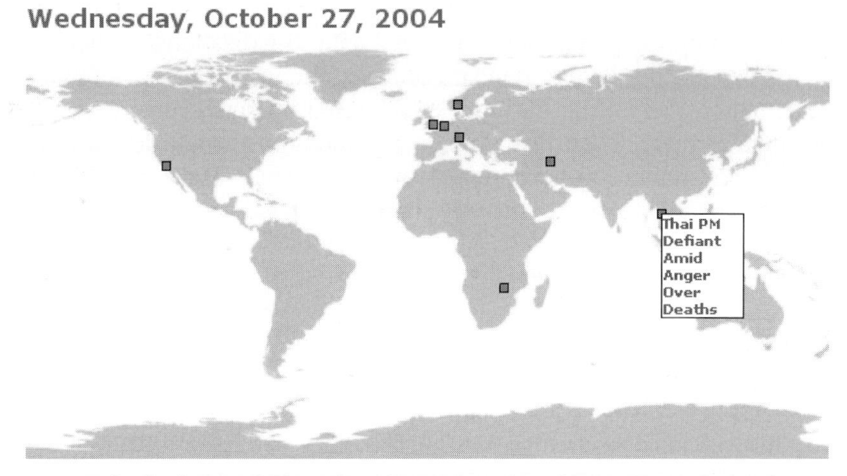
Wednesday, October 27, 2004
Thai PM Defiant Amid Anger Over Deaths
Barroso backs down over EU vote fr de it nl
A vote to approve the new European Commission has been delayed after the incoming president withdrew his proposed line-up of commissioners. | Jose Manuel Barroso said more time was needed to choose a ... (photo: EC)

Fig. 5. Map of top stories in English for 27th October. These correspond to the 7 largest clusters of articles found for that day, which are located at the place names most mentioned in each cluster. Similar maps exist for 6 languages and for any date

8 Conclusions and Future Work

News maps have proven to be an effective method to visualize world events for a given time. Users find this type of presentation clear and informative, especially when reviewing many thousands of reports. Two types of news map derived from EMM gathered news articles have been presented. The first map is used to present country based thematic indicators. These maps measure the media attention for individual countries for a particular theme. The values are derived directly from EMMs statistics system. The second news map relies on accurate geocoding of place names in each article text. These places can then be displayed directly as point data over world maps. The maps themselves reflect groups of articles. These can be any EMM alert for a given day or form one of the top news stories for the day. The top stories are identified live using the breaking news detection system and off-line using a more sophisticated clustering algorithm.

On-going improvements are being made to the geocoding process, both by expansion to new languages and improving the disambiguation process. Work is progressing on more advanced data visualizations, particularly to avoid dependence on non standard browser plugins.

Situation monitoring during a crisis based on small regions is another area under study. These "situation maps" focus on a single area such as the Israeli Palestinian conflict and track events on a map of the area with time.

References

[1] Best C, Van der Goot E, De Paola M, Garcia T, Horby D (2002) Europe Media Monitor – EMM. JRC Technical Note No. I.02.88. Ispra, Italy
[2] XML/XSLT, http://www.w3.org/TR/xslt
[3] Really Simple Syndication RSS 2.0, http://blogs.law.harvard.edu/tech/rss
[4] Scalable Vector Graphics, http://www.w3.org/TR/SVG/
[5] Pouliquen B, Steinberger R, Ignat C, De Groeve T (2004) Geographical Information Recognition and Visualisation in Texts Written in Various Languages. In: Proceedings of the 19th Annual ACM Symposium on Applied Computing (SAC'2004), Special Track on Information Access and Retrieval (SAC-IAR), vol 2, pp 1051-1058. Nicosia, Cyprus, 14 - 17 March 2004
[6] Global Discovery Gazetteer, From Europa Technologies Ltd, http://europa-tech.com/
[7] KNAB project http://www.eki.ee/knab/knab.htm
[8] Pouliquen B, Steinberger R, Ignat C, Käsper E, Temnikova I (2004) Multilingual and Cross-lingual News Topic Tracking. In: Proceedings of the 20th International Conference on Computational Linguistics (CoLing'2004). Geneva, Switzerland, 23-27 August 2004
[9] GeoURL, http://geourl.org
[10] Worldkit, http://www.brainoff.com/worldkit/
[11] Macromedia Flash, http://www.macromedia.com/

Allocation of Functional Behavior to Geo-Information for Improved Disaster Planning and Management

Timothy J. Eveleigh, Thomas A. Mazzuchi and Shahram Sarkani

School of Engineering and Applied Science, The George Washington University, 1505 Oakview Dr. McLean, Washington DC, VA 221010 USA. Email: eveleigh@gwu.edu

Abstract

This paper describes a novel modeling strategy that couples a systems engineering design approach borrowed from industry that allocates spatial features in a GIS to a design-driven hierarchical functional model. We show that establishing a design goal such as 'evacuation' can be decomposed into a hierarchy of related functions that enable evacuation (e.g., provide shelter, provide mobility) and that these functions can be mapped to a requirements perspective and used to develop a functional feature attribution scheme that is more risk-relevant than traditionally-attributed geo-information features in the disaster modeling context. We present a riverine flooding example of this methodology applied to assessing the vulnerability of lifeline systems, and describe how this modeling approach and the systems perspective may bring efficiencies to geo-information management and provide new insights to the modeling of risk in complex interdependent systems.

1 Introduction

Geographic Information Systems (GIS) have revolutionized disaster planning, and analysis. Notable is the success of GIS technology in using geo-information to assess where unusual or extreme environmental events place human activities and systems at risk. GIS-based models, once an expensive novelty, are now available to the general public and require only modest investment in computational resources. By way of example, the U.S. Federal Emergency Management Agency (FEMA) now provides public access to its "HAZUS" bundle of GIS-based natural hazard models. Recent successes of GIS technology applied to various phases of natural and man-made disasters are well documented. Recent examples include the GIS response to the September 11th 2001 terrorist attacks on New York City in Greene (2002) and western U.S. forest fires and the 1994 Northridge Earthquake in Amdahl (2001). Underpinning these successful applications of geospatial technology are ever-expanding geo-information holdings. Local and state agency awareness that foundational geospatial information is a critical resource that supports governance, commerce, and national security has led to a large-scale digitization of the landscape and markedly improved access to geo-information datasets. Hundreds of terabytes of geo-information of varying provenance and form can be found online, purchased, or can be ordered from local and regional planning agencies. Yet geo-information presents unique management challenges. It also has properties that may limit its full application in the disaster context. This paper explores some of these challenges, and presents a possible methodology to overcome them.

2 Challenges with Geo-information Modeling in the Hazard Assessment Context

Geo-information supports disaster impact analysis by modeling the earth's surface and the manmade and natural features on it. The intrinsic spatiality of geo-information can be used to identify hazards like gas lines crossing faults and residential areas with increased landslide risk. Subjected to physics-based models of disaster phenomenology one can further model where natural hazards subject representative human activities and the function of manmade infrastructural elements to risk. This is a very effective technique and hazard cases like identifying hospitals located in areas that will likely flood, bridge spans that may be damaged by certain earthquakes, or estimating the probabilistic hurricane damage to structures in a census tract are typical.

An example of this classical approach is shown in Figure 1. In this example, we see the modeled depths of a 50 year riverine flood generated by FEMA's HAZUS-MH flood model and the relationship of the floodwater inundation to attributed feature data. Whether this were a hypothetical flood and its potential effects or an actual flood in progress, the disaster manager will quickly notice such potential hazards as the school in the flood zone and elements of a road network being inundated by flood waters. While these discoveries may in of themselves be useful, they represent isolated findings which will likely have to be fused with many other such findings before a synoptic picture of vulnerability or disaster impact emerges. Before this aggregation occurs, however, it is necessary to determine the relevance of these findings.

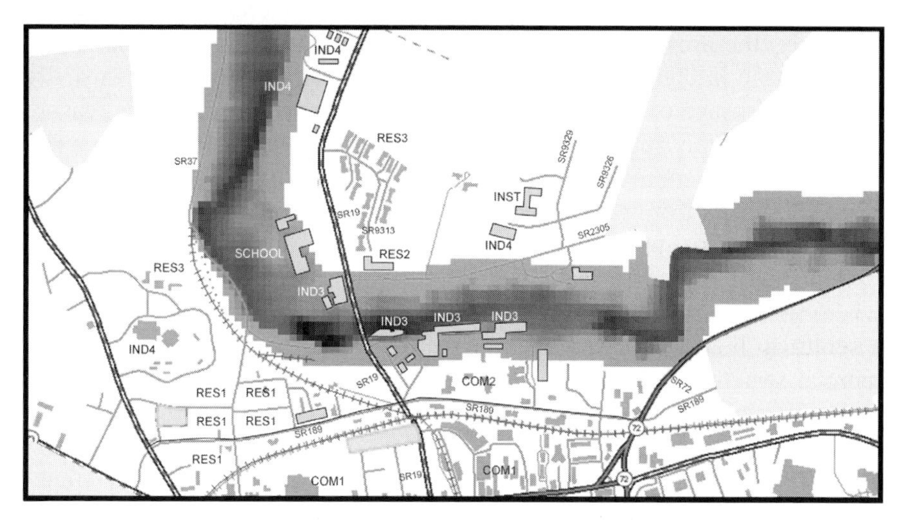

Fig. 1. Fifty-year flood generated by HAZUS inundates attributed features

It is difficult to judge the significance of these example findings without knowledge of the context of the disaster. For example, if the flooding shown in Figure 1 were the result of a powerful hurricane in progress, how significant would the school in a flood zone finding be if schools were closed due to regional evacuation? Then again, if the school also functioned as temporary shelter for evacuees, a key command and control node, or as a field hospital during regional disasters, the findings would be very significant.

2.1 Modeling the Context-Dependent Functional Behavior of Features

To suggest that the school feature in the above example might be a school in one context and a shelter in another suggests that this feature exhibits context-based functional behavior. It turns out that many of the features in this simple example also manifest this behavior. The inundated road segments, for example, may in fact be key segments of evacuation routes, or may play key roles in the transport of relief supplies. To effectively model the impact of the hazard (or potential impact) we need to be able to bring the awareness of this conditional functional behavior into the disaster modeling context.

The function of objects in geospatial databases is typically inferred from object attributes or the attributes of a thematic layer that contains objects. For example, an object with the attribute 'School' is inferred to function as a school and a 'parking lot' polygon is inferred to function as a parking lot. Similarly, the arcs contained in a 'Roads' layer are inferred to be roads. Indeed these inferences may hold true in the general sense as it is usually the general usage of features that drives their attribution. For this reason, attribution usually records the dominant design intent of features in a default, non-crisis context. A further inference occurs when feature attribute is itself inferred from remotely sensed imagery by some combination of its literal and spectral properties.

Functional inference from feature attribute is reasonable for features where function is not strongly context-driven (e.g., a radio antenna, a lake, a suburban house); the approach is synoptic (e.g., the population of commercial structures in this census tract, the average home in this community); or where the attribution is purposefully contextual (e.g., tornado shelter, storm drainage culvert). In many cases, however, functional inference is strongly context-dependent and nowhere is inferred object functional behavior more tenuous than in the context of disaster. A disaster can redefine a school as a shelter, a football pitch as an evacuation helicopter landing zone, and a two lane road bridge as a lethal evacuation chokepoint.

The tradition approach to capturing extended or variant functional behavior is to either create a new thematic layer to collect features relevant in a particular context (e.g., buildings that serve as evacuation shelters, evacuation routes) or to add additional attribution to features. These approaches, however, can be problematic. Generating additional thematic layers can add redundancy and additional data management burdens when features are copied between thematic layers and thus exist as duplicate features on two or more layers. This problem is exacerbated when the same process is repeated for multiple possible contexts. Furthermore, the task of capturing all crisis-dependent attributes on the entirety of geospatial elements in the many databases that contribute to a disaster management effort is daunting and according to Moore and Abraham (1994) may break down the decision process or complicate it greatly as in Greene's (2002) World Trade Center disaster example. Finally, there is little guarantee that such a quest would be exhaustive unless a firm grasp of the relationship of the piece parts to the systems they describe were possible.

2.2 Relating Geo-information Features to Aggregate Systems

For many analysis tasks, an aggregation of piecewise effects is desired. The value of the FEMA HAZUS software, for example, is in its ability to assess the probabilistic environmental effects on thousands of attributed structures and compile statistical summaries that quickly allow generalized estimates for the loss of aggregate function (e.g., percent of hospital beds available, percent of damage to police stations) and economic damage (estimated cost to repair or replace, indirect costs to labor force, etc.). HAZUS also generates aggregate estimates of shelter burden and debris type and density. Aggregate damage estimation information is essential in the planning process to estimate community vulnerability to disasters, in the mitigation stage to assess the efficacy of potential mitigation actions, and during the response and recovery phases to direct services and establish the regional requirement for disaster aid and assistance. Aggregate effect estimation derives its power from the averaging properties of accumulating damage estimates on a very large sample population; its answers then are neatly probabilistic and synoptic in nature.

The value to the disaster manager of aggregate estimation is that the answers it provides are high-level and general in nature, they tend to be systems-level answers. Institutional consumers of disaster information ultimately need synoptic information about the effects of disasters such as: "How is my transportation grid affected?" and "How is my ability to evacuate, shelter, and protect my population affected?" In the time-compressed domain of a burgeoning disaster and its immediate aftermath, it is these top-level systems-of-systems assessments that are the most crucial to prioritize resource allocation and to guide strategic decision-making. There is simply not enough time to assimilate and summarize the tide of feature-level information that pours in; it must instead be related to macro systems and core community functions to become manageable.

2.3 Modeling Lifeline Infrastructure Functional Interdependencies

Recent experiences with cascading infrastructure failures resulting from the loss of large regional electric power transmission capability in the United States during the summer of 2003 and elevated concerns for the possible widespread effects of terrorist attacks on civil infrastructure have brought new emphasis on the need to understand both the behavior of infrastructure layers and key dependent relationships between the layers. While GIS technology is certainly a key contributor to this understanding, the sheer complexity of the problem demands new approaches to modeling and the use of geo-information. The 2003 U.S. National Strategy for the Physical Protection of Critical Infrastructures and Key Assets suggests the investigation of systems-of-systems approaches (Rinaldi 2004). Such approaches will require the transformation of traditional geo-information into systems frameworks.

3 The Need for a Systems Perspective

The challenges described in the preceding paragraphs prescribe capabilities that transcend the traditional approach to GIS-based natural hazard modeling. We believe that the present approach to developing and managing geo-information in this context should be augmented to provide the following capabilities:

- The capture and representation of the context-derived functional behavior of objects
- The aggregation of piece parts into the systems they compose
- The capture and representation of functional interrelationships between systems
- A means to make the process systematic and purposeful

The systems approach and systems engineering, its process manifestation, offer these capabilities.

3.1 The Systems Approach and the Systems Engineering Process

The International Council on Systems Engineering's Systems Engineering Handbook (Sage and Rouse 2000) defines a system as "an integrated set of elements that accomplish a defined objective." The handbook also states that the "elements include hardware, software, firmware, people, information, techniques, facilities, services, and other support elements." Sage and Rouse further define systems engineering as "an interdisciplinary approach and means to enable the realization of successful systems." The systems approach to engineering is characterized by its systematic and repeatable processes, full consideration of alternatives, emphasis on interoperability and harmony, refinement and convergence, and satisfaction of stakeholder requirements (Eisner 1997).

Systems engineering is driven to evolve designs that satisfy stakeholder objectives codified as requirements. The systems engineering design process as described by Buede (2000) includes "the decomposition and definition of both the requirements, or statement of the design problem, and the architectures, functional and physical representations of the system." Systems engineering design approaches typically include a decomposing process that exposes and records in a series of linked "views" of the hierarchical detail of system requirements, system functions, and possible physical manifestations of elements that might perform these functions. As we will show, this thought process and the modeled views it can generate have a direct utility in the disaster management context.

3.2 Systems Engineering Applied

Translated into the disaster modeling context, the systems engineering process provides a means to identify and describe the systems in the community that provide lifeline functions and an integrated context to assess the vulnerability of these and interrelated functions. Systems engineering also provides integrated views necessary to relate the physical vulnerability of features to the functional vulnerability of systems and ultimately to the design vulnerabilities of community disaster objectives.

Combining the systems engineering design approach with the traditional geo-information approach to natural hazard and vulnerability modeling suggests the following steps:

1. Establish community disaster design objectives
2. Model the functional behavior of systems that seek to achieve these objectives
3. Allocate this understanding to features in the physical world as represented by the GIS
4. Model the possible effects of key hazards on the geo-information feature population
5. Assess the resulting vulnerability of or impact to the functional systems
6. Assess the vulnerability of the community disaster objectives in the disaster context

We will now review each of these steps and present an example that illustrates a realization of this methodology.

4 An Application Example

We will revisit the scenario that generated Figure 1, a HAZUS-generated inundation of a river valley and its potential impact on lifeline systems.

4.1 Establishing Community Disaster Design Objectives

Establishing community disaster design objectives provides a context for the development of policies, systems, and procedures to achieve these objectives. Much like in the traditional engineering design context, this amounts to defining the stakeholder requirements for a specific desired capability. In the disaster management context, examples of design objectives might include the requirements to provide for the evacuation of 19,000 people from a region or resilience to the loss of the community power grid for seven days. Typically, in systems engineering approaches, top-level requirements are further decomposed into a hierarchy of requirements. A good designer will use this tree both to understand requirements traceability (to other requirements, and source documents) and as a checklist to be sure that solution designs accommodate all the design requirements (and thus satisfy the stakeholders who levied the requirements). For our example, we used Vitech's popular CORE systems engineering software to generate such a requirements tree for our riverine natural disaster context (see Figure 2). CORE allows multiple views of traceability so that the dependency of requirements on policy (as instanced by documentation), the interdependence of requirements, and prevailing issues and concerns can all be viewed. Each of the boxes in Figure 2 is linked to detailed holdings in CORE's requirements database.

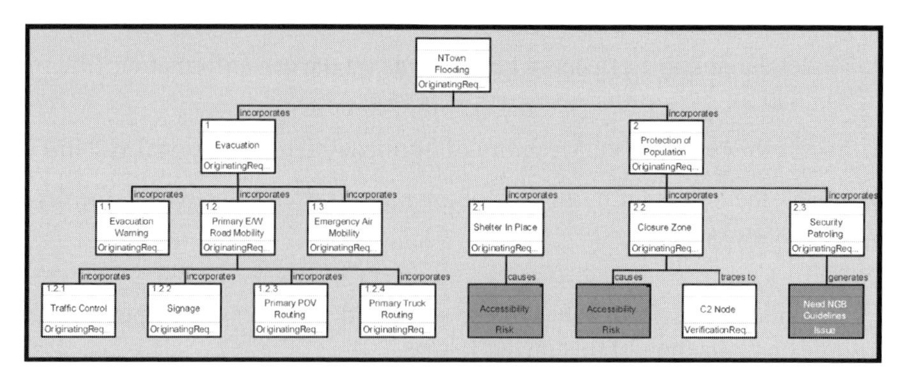

Fig. 2. Top four levels of an example requirements traceability model

This step in the modeling approach provides us with a top-down view of our expectations for community lifeline performance, its traceability to policy, and a means to know that we have studied the whole problem.

4.2 Modeling the Functional Behavior of Lifeline Systems

The next step is to generate a functional understanding of the community with respect to the requirements tree. This step of the modeling codifies precisely what functions must exist to meet our design goals. Llinas (2002) refers to this as modeling the 'operational elements' of the disaster environment. If done correctly, functional modeling is 'solution agnostic' in the sense that it does not attempt to prescribe how we are to satisfy the requirements tree, simply what functions must be performed. Functional modeling is hierarchical in nature, and allows for a structured decomposition of specific sub-functions that when linked perform the top-level functions. Key advantages of this approach are the formal separation of the *what* from the *how*, and a means to span the generic to the specific. These advantages make the approach highly adaptable and transferable by allowing for generic community system functional models to be developed and then mapped to specific community instances and powered by specific community requirements trees (and ultimately to the specific physical entities that provide those functions). The analogy is very similar to the design of an aircraft. From a top-level functional view, most aircraft are quite similar. Accordingly, a top-level functional model for aircraft is a good foundation to design a particular aircraft through detailed decomposition of its top-level functions into detailed sub-functions that provide capabilities (such as launching missiles) suited to specific operational design objectives.

Figure 3 presents a section of a functional model we built for this example. Here we see the top-level function "Provide Evacuation" decomposed into a series of lower-level functions. Notice how the modeling shows parallel and serial functional relationships and interdependencies between these functions. Figure 3 also suggests information flows between functional elements. At this point, however, we still do not have a sense for what entities (objects in the real world such as road segments, traffic lights, and people) actually realize these functions.

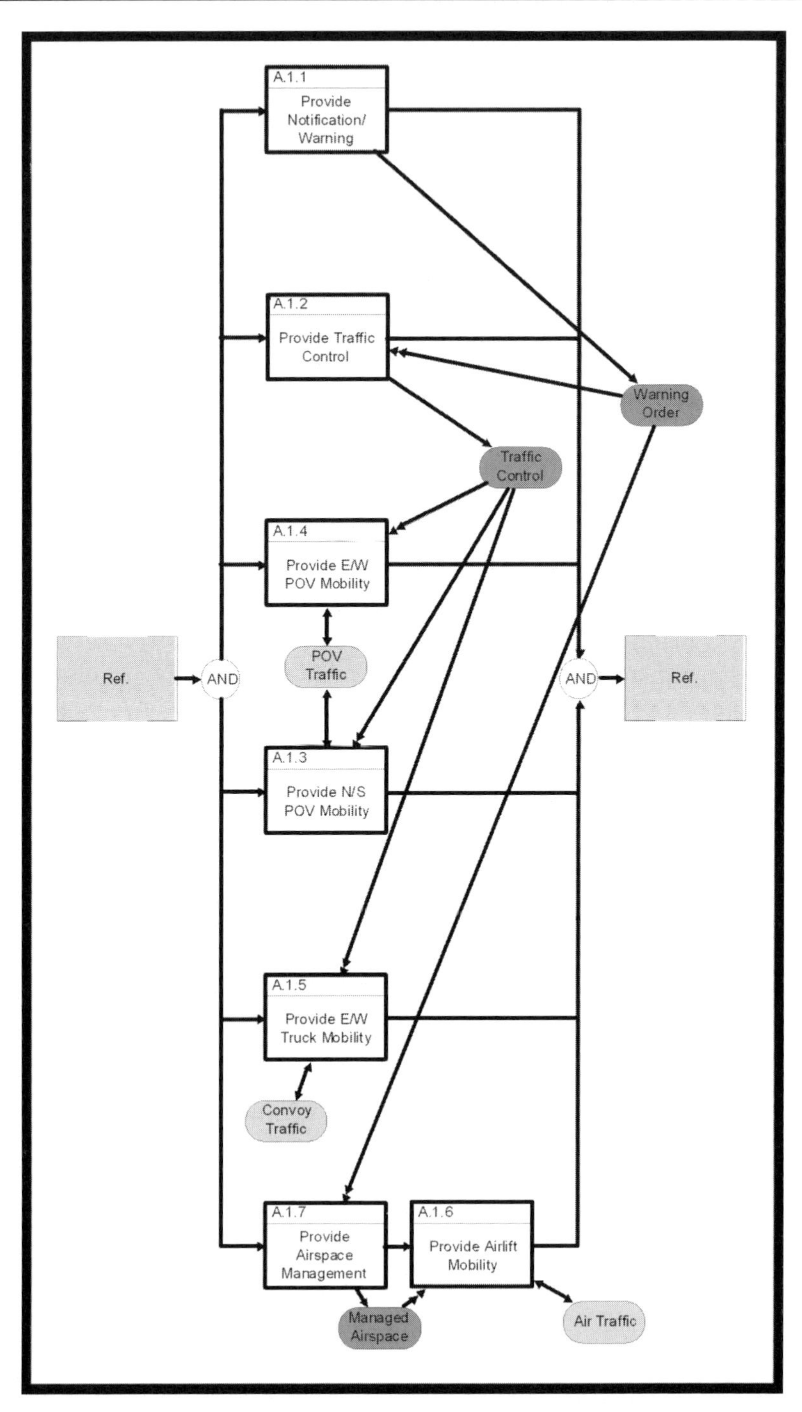

Fig. 3. "Provide Evacuation" Functional Flow Model

4.3 Allocation of Functional Behavior to Physical Systems

Now that we have modeled the structure and behavior of functions that we believe will satisfy the evacuation design objectives, we must allocate this behavior to the actual entities that will perform them. In the top-down design context that drives much engineering, at this point we would typically prepare several possible solution alternatives for consideration. To use the aircraft example again, we might propose a rotary wing and several fixed wing designs that provide the functions described in the functional model. The engineering process would then continue by trading cost and performance requirements against each other to determine a preferred design. In our natural hazard example, however, the physical instance of our lifelines already exists as the result of other previous engineering and design activities. For example, our road network exists as the result of many years of progress toward satisfying a flux of regional transportation requirements. Similarly, commercial and residential structures exist as a result of accumulated personal, commercial, and civil objectives.

In most systems engineering modeling tools, functional behavior is allocated to a description of, or reference to, objects in the physical world. To our knowledge however, none have extended this into a geospatial modeling space where full topologic rigor can be used to provide a rich spatial view. Geo-information modeling through a GIS provides just such a geospatial view. Brimicombe (2003) suggests that the GIS spatial model is an ideal couple to engineering models. The focus of our research has been to explore the nexus of these two model spaces and the value each brings to the other.

There are several useful new perspectives possible when the systems engineering views and the spatial view are brought together - particularly when the relevant object base is rigorously traceable. These new perspectives include the:

- Geospatial view of objective requirements
- Geospatial view of functional behavior
- Functional view of geospatial objects
- Requirements view of geospatial objects

Perhaps the perspective most relevant to natural disaster management modeling and the simplest to realize is the geospatial view of functional behavior. A relatively simple way to realize this is to allocate functional behavior to features by attributing the features with functional purpose. Figure 4 is just such an example.

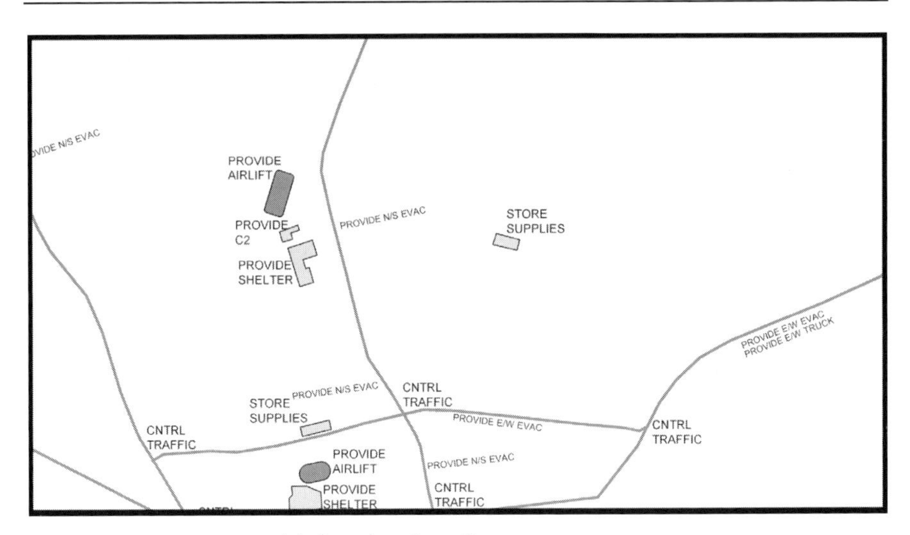

Fig. 4. Features recast with functional attributes

In Figure 4, which is a re-attributed version of key features from the database that generated Figure 1, we now see the features described by their functional role within the evacuation design context. The athletic field polygon next to the school is now cast by its functional behavior "provide air mobility" (a helicopter landing zone), the school is now cast as "provide shelter" and "provide command and control," and elements of the road network are cast with various mobility functions. Preserving the upward links back to the functional and requirements views retains our systems perspective of the functional roles of these elements and their contribution toward satisfying the overall design objectives of evacuation.

4.4 Model the Possible Effects of Key Hazards on the Feature Population

With the next step, we return to our traditional GIS-based hazard modeling approach which pits a physics-based natural hazard model against the geo-information held in the GIS. For example, we again generate 50 year flood depth predictions using HAZUS-MH's riverine flood model. As before, by spatial intersection and query we determine which geospatial features are impacted by the predicted inundation. Figure 5 shows the functionally-attributed features in the context of the predicted hazard. One could also model the effects of other hazards such as earthquakes, wind forces, fire propagation, landslides etc. in this manner.

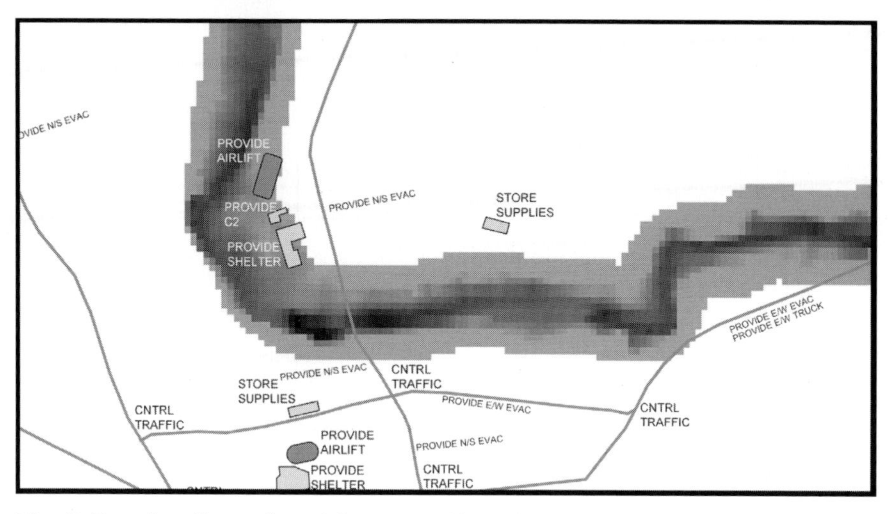

Fig. 5. Functionally-attributed features subjected to inundation

4.5 Assess the Resulting Vulnerability of or Impact to the Functional Systems

Regarding Figure 5, we can now begin to assess how this hazard may have a functional impact. We can directly infer this in the geospatial view by assessing, for example, where key evacuation function-providing elements are inundated by riverine flooding. What was formerly an inundated school now becomes an inundated "provide shelter" and "provide command and control" feature. Similarly, the important "provide air mobility" and "provide road mobility" function-providing features are also impacted by this hazard.

At this point, it is useful to link back into the functional view of the geospatial to assess the greater impact of this inundation on the top-level functions. This view is shown in Figure 6. Here, we can assess the impact of the modeled hazard in the purely functional perspective now the geospatial view has mapped it to function-providing features. In this view, we see that the inundation of the road segments in Figure 1 was not as significant in functional space as it first appears in physical space because other non-impacted features provide parallel functions.

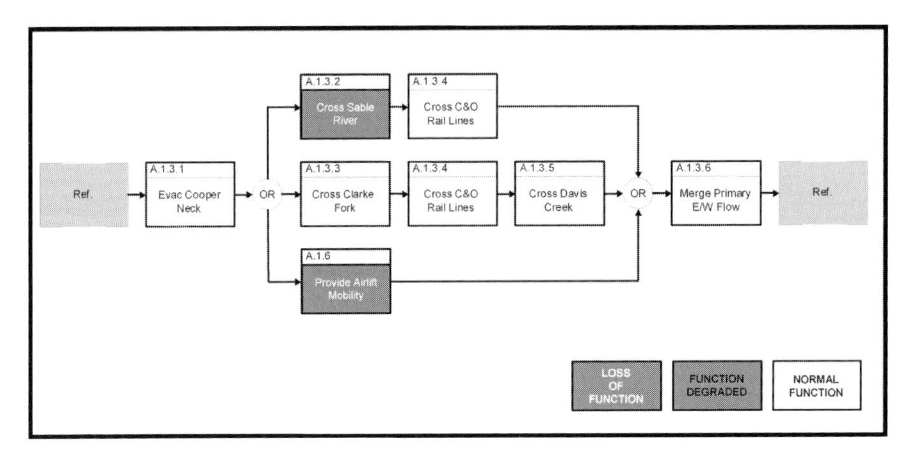

Fig. 6. A decomposed mobility function with functional impact of inundation

4.6 Assess the Vulnerability of the Community Disaster Objectives in the Disaster Context

The final step of the approach is to reflect the impact of the hazard back into the originating requirements space and in doing so, determine the impact the hazard will likely have on the stakeholder objectives. In our toy example, for instance we see that the 50 year flood as modeled by HAZUS would adversely impact the requirements to provide shelter, command and control, and air mobility given the features assigned to provide those functions.

This final perspective is an ideal vantage from which to fully assess mitigation measures. In this particular example, that might mean selecting other air mobility, shelter, and command and control facilities or building a levee (change to the physical dimension). Mitigation might also entail providing another means to achieve evacuation such as to provide an amphibious means (a change to the functional dimension), or rethinking evacuation objectives (a change to the requirements dimension). Each or some combination of these changes may be assessed independently or in combination. The key is that with this modeling approach, the three views are systematically linked and any change to one will propagate to the others.

5 How this Approach Would Improve Geo-Information Management

The approach we have just described imparts a systems understanding on geospatial features. While it does not necessarily reduce the amount and type of geo-information feeding the assessment process, it does afford several management advantages. These include:

- A means to focus the collection of geo-information toward those that provide functions that are relevant to key disaster management objectives
- A means to make the collection of geo-information systematic and complete
- A means to abstract complex non-spatial behaviors and requirements attribution outside the GIS instead of in multiple partially-attributed feature layers

When performing a hazard vulnerability assessment, there is natural inclination to try to collect as much geo-information detail as is available on the off chance that features within it may prove significant to either the understanding or portrayal of the impact of the hazard. While this in-of-itself is not a bad philosophy, it does tend to result in a far greater data management burden. Furthermore, when time is of the essence, data are scarce, or a summary analysis is sufficient, the top-down systems perspective may make more sense. In this approach, the data collection effort becomes focused on collecting only the physical features that provide the functions that meet community disaster objectives and that are necessary to model the hazard. For example, we may find that in order to estimate the gross functional effect of an earthquake in a region with little feature coverage, it makes more sense to try to collect information on particular features that represent unique lifeline functional vulnerability versus collecting as much data as we can and then picking out salient features. These key features may be found across numerous traditional thematic layers (e.g., transportation, utilities, and structures) that previously may have to have been exhaustively collected at great expense and management burden. Reduction in data collection and management complexity might help reduce the cost of assisting geo-information poor regions like those described in Stage and von Meyer (2004).

Accomplishing the top-down systems engineering process approach has helped engineers ensure both that alternate physical designs are investigated and that designs under consideration address the entirety of the stakeholder requirements. Unsatisfied objectives and unallocated functions become readily visible and this visibility systematically drives the design process toward satisfying the outstanding entities. The same concept would hold true of the structured approach when applied to the collection of geo-information for disaster management modeling. Changes to any of the linked requirement, functional, or physical views would bring changes to each of the other views and geo-information collection would be driven to fully-realize the new functional view which in turn would try to fully realize the new requirements view.

By abstracting the complex non-spatial behaviors of features into a separate data model we obviate the need to develop duplicative and perhaps complex feature layers to record this information in the spatial data model. This should reduce spatial feature complexity and the need to transform features into other thematic layers to support the modeling approach. In turn, reducing feature complexity should reduce feature management burden.

6 Applicability to the Study of Complex Interdependent Systems

Rinaldi (2004) describes the daunting challenges of modeling and simulating critical infrastructures and the importance of this task to national security and economic prosperity. To be effective, Rinaldi indicates that infrastructure modeling and simulation will need to include modeling the dependencies between the infrastructural elements. Thissen and Harder (2003) indicate that systems approaches should be brought to this problem and that the socio-technical character of infrastructures should also be considered.

An advantage of the functional view in the systems engineering approach is that it does not prescribe the means that instantiate the particular component functions. Functional modeling also makes explicit the relationships between functions. Taken together these two properties suggest that a functional perspective might have strong applicability to the modeling of complex interdependent systems where entities from numerous physical systems may be related in the functional perspective. Indeed Modarres (1999) uses the functional modeling approach to investigate the reliability of the interdependencies between physical components within complex systems, and Larssen (2000) applies the same approach to assess the reliability of nuclear power stations. We believe that the modeling approach described in this paper may bring new insight into the investigation of the effect of natural hazards on complex interdependent systems and may even provide the risk framework called for by Grabowski et al (2000) in their study of risk modeling for distributed, large-scale systems.

7 Work in Progress

To date we have only applied our methodology to small demonstration cases and have not applied it to the study of a major large-scale disaster planning, mitigation, or impact analysis effort. To prepare for this, we are presently investigating the human interface considerations required to make the approach easier to apply in the disaster management context. The structured analysis and design methodology familiar to many systems engineers may at first appear foreign to geo-information managers. Similarly, the capabilities, interfaces, and performance of a GIS are likely new territory for many engineers. Of key interest is the human machine interface that connects the functional and requirements view with the spatial view providing the four new perspectives presented in Section 4.3. This interface is requiring experimentation and may ultimately necessitate the design of several new display methods to make the cross-view allocation and traceability easier to visualize and follow. We hope to apply some of the findings and advice of Medieros et al (1996), Andrienko and Andrienko (2003), and Brimicombe (2003) toward developing coordinated and manageable manifestations of these views.

The second focus of our current research is investigating the stability and portability of disaster management functional models. We feel that these attributes are key to the rapid application of the approach to multiple scenarios and reaping the disaster management benefits propounded in this paper.

References

Amdahl G (2001) Disaster response: GIS for public safety. ESRI Press, Redlands CA USA

Andrienko N, G Andrienko (2003) Coordinated views for informed spatial decision making. Proc of IEEE Coordinated & Multiple Views in Exploratory Visualization, IEEE Computer Society Press, Washington

Brimicombe A (2003) GIS, environmental modeling and engineering. Taylor & Francis, London New York

Buede DM (2000) The engineering design of systems: models and methods. John Wiley & Sons, New York

Eisner H (1997) Essentials of project and systems engineering management. John Wiley & Sons, New York

Grabowski M, Merrick JRW, Harrald JR, Mazzuchi TA, Van Dorp JR (2000) Risk modeling in distributed, large-scale systems. IEEE Transactions on Systems, Man, and Cybernetics – Part A: Systems and Humans, vol 30, no 6 pp 651-660

Greene RW (2002) Confronting catastrophe: a GIS handbook. ESRI Press, Redlands CA USA

Larssen JE (2000) Knowledge engineering using multilevel flow models. Proc 2nd International Symp on Engineering of Intelligent Systems, Natural & Artificial Intelligence Systems Organization New York

Llinas J (2002) Information fusion for natural and man-made disasters. Proc of 5th IEEE International Conf on Info Fusion, pp 570-576

Medeiros CB, Bellosta MJ, Jomier G (1996) Managing multiple representations of georeferenced elements. Proc of the 7th International Workshop on Databases and Expert Systems Applications, pp 364-370

Modarres M (1999) Functional modeling of complex systems with applications. Proc 1999 Annual Reliability and Maintainability Symposium, pp 418-425

Moore KL and JK Abraham (1994) An architecture for intelligent decision support with applications to emergency management. Proc International Conf on Systems, Man, and Cybernetics, vol 2, pp 1571-1576

Rinaldi SM (2004) Modeling and simulating critical infrastructures and their interdependencies. Proc of the 37th Hawaii International Conference on System Sciences, IEEE Computer Society, Washington

Sage AP and Rouse WB eds. (1999) Handbook of systems engineering and management. John Wiley & Sons New York

Stage D and N Von Meyer (2004) Parcel data for emergency response. GeoIntelligence, Sep/Oct 2004, pp 38-43, Advanstar Communications Cleveland

Thissen WA and PM Herder (2003) Critical infrastructures: challenges for systems engineering. Proc International Conf on Systems, Man, and Cybernetics, vol 6 pp 2042-2047

Towards an Integrated Concept for Geographical Information Systems in Disaster Management

Richard Göbel[1], Alexander Almer[2], Thomas Blaschke[3], Guido Lemoine[4] and Andreas Wimmer[2]

[1] University of Applied Sciences Hof, Alfons-Goppel-Platz 1, D-95028 Hof, Germany.
Email: richard.goebel@fh-hof.de

[2] Joanneum Research, Wastiangasse 6, A-8010 Graz, Austria.
Email: alexander.almer@joanneum.at

[3] University Salzburg, Hellbrunnerstraße 34, A-5020 Salzburg, Austria.
Email: thomas.blaschke@sbg.ac.at

[4] Joint Research Centre, Via E Fermi, 21020 Ispra (VA), Italy.
Email : guido.lemoine@jrc.it

Abstract

Disaster management takes current information technology to its limits due to the very large amount of relevant data and tight response times. This is in particular true for satellite data which needs to be retrieved from different archives, processed an interpreted. Typical issues are efficiency of database access, integration of multiple heterogeneous sites and quality of image interpretation. Unfortunately experiences from many different projects and systems have shown that these issues cannot be considered independently from each other. Therefore this paper proposes a concept addressing the mentioned issues by providing an integrated solution. For this purpose the paper focuses on content based searches for satellite images in a distributed environment.

1 Introduction

Geographical Information Systems are essential tools for rapid decision making in disaster management. This application context has specific requirements ranging from fast response times over data fusion from multiple sources up to the handling of very large amount of data for some applications. In many cases these requirements take current information technology to its limits:

- The access to very large databases of pre-disaster data is often slow. This problem comes from the fact that existing database indexing technology has its limitations in supporting complex search requests.
- In many cases recent information about disasters needs to be derived from raster data (e.g. aerial photographs, satellite images). Although manual interpretation of raster data usually provides reasonable results in this context, automatic extraction of information is still difficult. A semi-automatic interpretation is however essential if tight response times have to be met.
- Existing technologies, as for example web services, provide a stable platform for interconnecting heterogeneous information systems if information needs to be retrieved from multiple sources. But even in this context there are still issues which need to be resolved, as for examples defining precise semantic attributes and combining incomplete and/or unreliable information.

These issues are clearly interrelated. As an example the query concept of a distributed system needs to consider limitations of database indexing technology to ensure reasonable response times. Also data processing methods (e.g. image interpretation) need to be harmonised with the query concept, if queries may refer to derived information.

This paper describes a first tentative concept integrating spatial databases, image processing and web services. The concept supports queries to distributed database systems via criteria referring to the content of images. As an example the query may ask for all images containing forest areas above a certain size.

A query of this type may be processed by applying an appropriate algorithm to every image in the database. This approach is not feasible for a very large number of images. A more efficient approach requires preprocessing of images before storing them in a database. For instance, an algorithm could identify forest areas and stores their sizes in special fields. The query would need to check only these fields to identify all relevant images. A database index (index structure) speeds up the search even further by supporting the navigation to relevant images.

The pre-processing in our example supports only queries referring to sizes of forest areas. This approach is feasible if all possible queries are known in advance. In a distributed environment however it is difficult to predict all potential queries since application contexts at other sites are usually not known by the site operating the image archive.

In this paper we suggest a more generic approach by deriving general parameters from an image and store them in a database. Then the application will generate secondary conditions for these parameters which need to be satisfied by images containing objects specified by the initial query condition.

The secondary conditions will usually return also false results. Therefore the object recognition algorithm needs to be applied to all images from the intermediate search result. This approach works efficiently if secondary conditions ensure a small number of false results.

Our concept in this paper is based on a generic segmentation procedure where key parameters of segments are stored together with their geographical position. Search requests for this database will be supported by an optimized R-Tree as the index structure. Customised WebServices facilitate remote access to the database. Figure 1 summarizes the data flow for this concept.

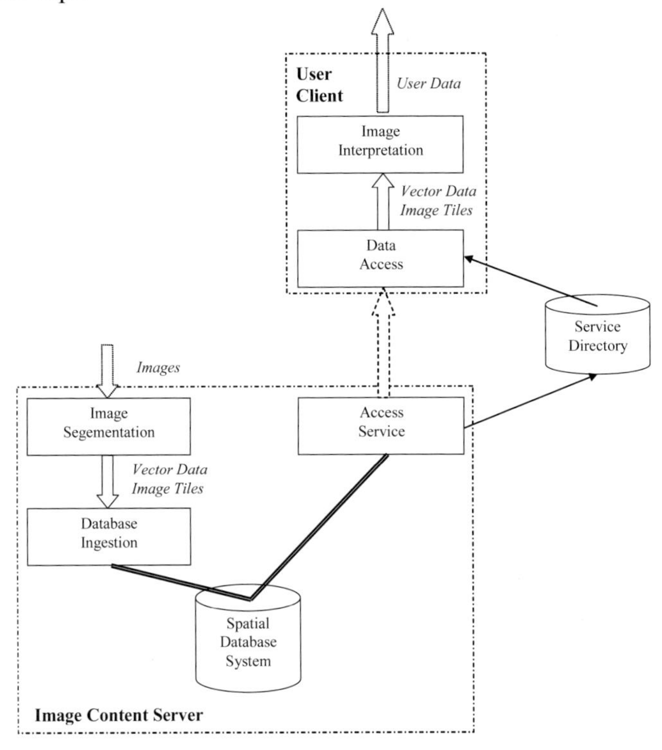

Fig. 1. Data Flow Diagram for proposed concept

This paper starts with a summary of challenges from image segmentation and database indexing technology as they occur in this application context. Then the description of the integrated concept follows. Finally the concept is analyzed in the context of pre-fire risk analysis as a potential application.

2 Challenges

2.1 Image Segmentation

Image segmentation is often used as an intermediate step in object recognition procedures. After initial pre-processing steps, as for example restoration and image enhancement, the segmentation builds connected segments of pixels. These segments are either parts of objects or already candidate objects. As such these segments are the input for further steps like feature extraction and object labeling.

Basic segmentation approaches focus on pixel colors or reflection values, respectively: pixels with similar values (in all relevant image bands) are grouped into the same segment. The similarity is defined by a distance function (either explicit or implicit) or through the size and shape of a kernel. Several sophisticated image segmentation algorithms make use of the context of a pixel. As an example a texture could be used for grouping pixels into a region (Chaudhuri, Sarkar 1995, Laine, Fan 1996, Hofmann et al. 1998). Several options exist for describing textures (e.g. distribution functions, fractals, probability functions for neighborhood definition etc.) and even more methods exist to recognize these textures.

In general no unique assignments of pixels to segments exist. In fact all criteria provide only plausibility values for grouping pixels into segments. Newer methods try to optimize the segmentation by minimizing segment heterogeneity. The ambiguity of grouping pixels can be significantly reduced if the segmentation algorithm is optimized for one or few objects which shall be identified in an image (Baatz, Schäpe 2000). Although image segmentation is definitely not new (Fu, Mui 1981) recently more algorithms have been developed for creating segments. The homogeneity is derived through geostatistical analysis (Hofmann, Boehner 1999), unsupervised texture recognition by extracting local histograms, Gabor wavelet scale-space representation with frequency (Hofmann et al. 1998), image segmentation by Markov random fields and simulated annealing, or Markov Random Field (MRF) using a Maximum a posteriori (MAP) probability approach. The MRF method generally classifies a particular image into a number of regions or classes (Tso, Mather 2000). The image is modeled as a MRF and the MAP probability is used to classify. The problem is posed as an objective function optimization, which in this case is the *a posteriori* probability of the classified image given the raw data which constitutes the likelihood term, and the prior probability term, which due to the MRF assumption is given by the Gibb's distribution. MRF was already exploited for an unsupervised classification by Manjunath and Chellappa (1991).

Dubuisson-Jolly & Gupta (2000) developed an algorithm for combining color and texture information for the segmentation of color images. The algorithm uses maximum likelihood classification combined with a certainty based fusion criterion. Hofmann et al. (1998) developed a promising approach based on a Gabor wavelet scale-space representation with frequency-tuned filters as a natural image representation. Locally extracted histograms provide a good representation of the local feature distribution, which captures substantially more information than the usually used mean feature values. Homogeneity between pairs of texture patches or similarity between textured images in general can be measured by a non-parametric statistical test applied to the empirical feature distribution functions of locally sampled Gabor coefficients. This algorithm systematically derives a family of pair wise clustering objective functions based on sparse data to formalize the segmentation problem.

Region growing algorithms cluster pixels starting with seed points and growing into regions until a certain threshold is reached. This threshold is normally a homogeneity criterion or a combination of size and homogeneity. A region grows until no more pixels can be attributed to any of the segments and new seeds are placed and the process is repeated. This continues until the whole image is segmented. These algorithms depend on a set of given seed points, but sometimes suffering from lacking control over the break-off criterion for the growth of a region.

Additionally, various texture segmentation algorithms exist. They typically obey a two-stage scheme (Mao, Jain 1992, Hofmann et al. 1998). In the modeling stage characteristic features are extracted from the textured input image which includes spatial frequencies (Jain, Farrokhnia 1991, Hofmann et al. 1998), Markov Random Field models (Mao, Jain 1992; Tso, Mather, 2000), co-occurrence matrices (Haralick et al. 1973), wavelet coefficients (Salari, Ling 1995), wave packets (Laine, Fan 1996) or fractal indices (Chaudhuri, Sarkar 1995). In the optimisation stage features are grouped into homogeneous segments by minimising an appropriate quality measure. Other groups of algorithms include watershed transformation or 'split-and-merge' algorithms (Cross et al. 1988). They start by subdividing the image into squares of a fixed size, usually corresponding to the resolution of a certain level in a quad tree. These leaves are then tested for homogeneity and heterogeneous leaves are subdivided into four levels while homogeneous leaves may be combined with three neighbors into one leaf on a higher level etc.

In our integrated concept the segmentation procedure needs to be able to derive general parameters from images for supporting efficient searching. These parameters are used to decide whether a given object is contained in an image or not. The segmentation procedure needs to satisfy the following requirements:

- The segmentation procedure needs to generate segments which allow to distinguishing different types of objects. As such the procedure does not necessarily identify complete objects. Instead sufficiently unique parts of objects may be sufficient.
- The segmentation procedure has to be flexible and generic. Object sizes or neighborhood relationships are difficult to be used.

2.2 Database Indexing Technology

The key challenge for the management of very large amounts of data is the efficient access to (small) parts of this data. In general users will access this data via a search request defining constraints for attributes of objects. As an example a user may ask for all images containing forest areas greater than 10 km^2. Other search conditions may define upper and/or lower bounds for one or more attributes (e.g. build up areas smaller than 0.1 km^2 and altitude above 500 m).

A database could process such a search condition by checking the search condition for every stored object (sequential scan). This sequential scan may however not be acceptable for large numbers of stored objects. Therefore most database systems provide index structures facilitating efficient navigation to relevant objects.

The most commonly used index structure is the so called B-Tree (Bayer, McCreight 1972). A B-Tree supports searches which define a specific value for a single attribute. With a small extension B-Trees support also searches where lower and/or upper bounds are defined for a single attribute. In this context a B-Tree guarantees response times which grow logarithmic with the total number n of objects in the database and linear with the number m of objects which satisfy the search condition $(O(log(n) + m)$.

The efficiency of a B-Tree however degenerates if lower and/or upper bounds are specified for more than one attribute. In the worst case the search time grows linearly with the number of entries in the database.

Search conditions with upper and/or lower bounds for multiple attributes (multidimensional range searches) are increasingly used in modern database application. Also Geographical Information Systems make use of multidimensional range searches as for example in the case of geographical search conditions.

Since the late nineties major vendors of commercial databases have recognized the need for supporting multidimensional search conditions and integrated new index structures (multidimensional index structures). The most commonly used index structure is the R-Tree (Gutmann 1984).

A multidimensional index structure is build for *d* attributes generating a *d*-dimensional value space. Most multidimensional index structures, including the R-Tree, are based on a tree structure where every node represents a region in the *d*-dimensional space. As a general rule the region of a parent node needs to contain the regions of its child nodes. The leaf nodes of the tree either contain entries or references to entries.

In the case of the R-Tree the nodes represent hyper cuboids. In addition an R-Tree has to be balanced (length of every path from root to leaf is equal). An example of an R-Tree for two dimensions is given in figure 2 where rectangles represent the regions of nodes from the tree.

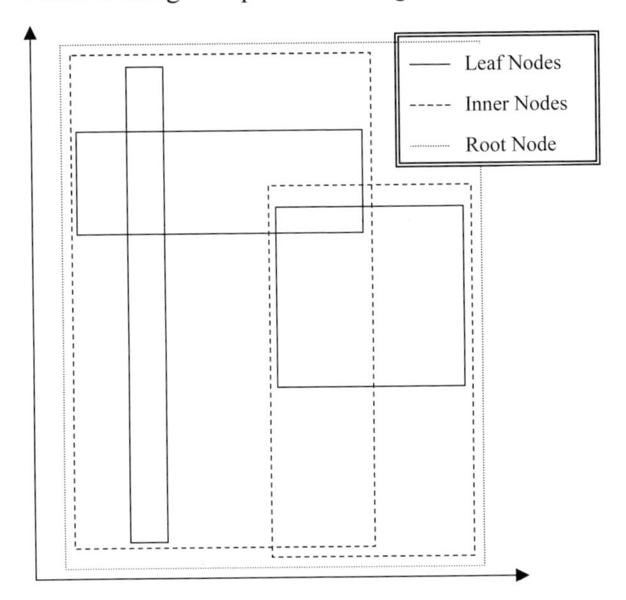

Fig. 2. Example for a 2-dimensional R-Tree

In comparison to B-Tree applications the use of R-Trees shows mixed results. In some contexts the use of R-Trees does not seem to help at all since the search time grows almost linearly with the number of entries in a database.

Significantly better time complexity can be achieved by so called replicating index structures, as for example the range tree (Bentley 1980) and the k-range structure (Bentley, Maurer 1980). Unfortunately the space requirements of these structures make it difficult to use them in most applications. Therefore research has concentrated on so called non-replicating index structures as for example the R-Tree.

Theoretical analysis of non-replicating index structure has identified a lower bound of $O(n^{(d-1)/d}+m)$ for the search time complexity (see for example Mehlhorn 1984 and Ravi Kanth, Singh 1999) These theoretical results show that search time for non-replicating index structure grows at least linearly for a high number of dimensions. This lower bound is however a result of a worst case analysis. Therefore response times may be better in some situations.

In fact a later result (Göbel, Hornsteiner 2002) derives a better complexity of $O(log(n)*m)$ if data is sufficiently equally distributed and only "larger search ranges" are considered. The situation for unequally distributed data is however not yet clear.

Real world examples show even more problems. As an example the incremental generation of R-Trees often results in suboptimal tree structures (e.g. overlaps of leaf nodes). As a consequence significant research has been performed and different methods were proposed to generate optimized R-Trees (see for example Sellis et al. 1987, Beckmann et al. 1990, Berchtold et al. 1996, Ang, Tan 1997) sometimes at the cost of a more complex insertion method and the use of more memory resources.

Although these solutions optimize some criteria they do not really ensure efficient search times in some situations. Even worse a theoretical analysis of these methods is difficult since most mentioned methods do not guarantee certain properties. The reason is the incremental generation of the R-Tree which makes global optimizations according to considered criteria difficult.

A global optimization of an R-Tree is possible if a static tree is generated from a given set of entries. This approach is called a tree packing algorithm. Using such an approach it is possible to generate a tree structure where no leaf nodes overlap if the tree contains only point data.

Several of these tree packing algorithms have been proposed (see for example Roussopoulos, Leifker 1985, Kamel, Faloutsos 1993). So far the best published method seems to be the Sort Tile Recursive method (STR) of (Leutenegger et al. 1997). The authors of (Leutenegger et al. 1997) showed by extensive tests that this method generates better tree structures than previous methods.

3 Concept of an Integrated Information System

3.1 Image Segmentation

Based on the discussion of image segmentation above we searched for a procedure for the integrated concept based on the following ideas:

- The procedure should be centered on spectral information in order to avoid loss of generality of more sophisticated algorithms.
- The procedure should cater for several different segmentation levels with different sizes of the segments.
- The procedure shall optimize homogeneity of regions and accepts even non-assigned pixels for this purpose.

Technically, we apply the fractal net evolution approach developed by Baatz and Schäpe (2000). Successful applications demonstrated the applicability and the high potential of transferability (Blaschke et al. 2000, Blaschke et al. 2001, Bauer, Steinnocher 2001, Schiewe, Tufte 2002, Neubert, Meinel 2003, Blaschke 2003, Koch et al. 2003, Collins et al. 2004, Tiede et al. 2004, Langanke et al. 2004, van der Sande et al. 2004). The merging technique starts with 1-pixel image objects. Image objects are pair wise merged one by one to form bigger objects. In this conceptualization the procedure becomes a special instance of an assignment problem, known as pair wise data clustering. In contrast to global criteria, such as threshold procedures, decisions are based on local criteria, especially on the relations of adjacent regions concerning a given homogeneity criterion.

This segmentation is incorporated in an object-based GIS/remote sensing methodology. Burnett and Blaschke (2003) developed a multiscale segmentation/object relationship modeling, or MSS/ORM. It is based on GIS objects and/or objects derived from image analysis. Based on the delineation of image objects at several levels, a semantic network is built. This is done using a commercially available, object-oriented, GIS/remote sensing software environment called *eCognition.* The segmentations generate objects at several user-defined levels and allow for the generation of image objects on an arbitrary number of scale-levels, taking into account criteria of homogeneity in color (reflectance values in a remotely sensed image) and shape. Thereby, a hierarchical network of image objects is generated, in which each object knows its neighboring objects in the horizontal and vertical direction.

Applying this methodology to our workflow some pre-processing of the original search condition is required. In our context the original search will specify an object which should be contained in images of the result set.

The pre-processing step will derive key properties from this object or a significant part of it relating to these parameters:

- lower and/or upper limits for relevant bands.
- lower and/or upper limits for the size of object (in m^2)
- lower and or upper limits for form parameters.

In addition to standard information derived from image metadata such as for example sensor and acquisition time the segmentation procedure will store the following information for a segment in the database:

- Spectral information for every band used including average values per segment, minimum and maximum value, median, standard deviation etc.
- Segment size in square meters and number of pixels.
- Descriptive measures such as perimeter, number of bordering segments, direction of the longest axis.
- Constructed parameters describing the form of a segment such as the ratio between perimeter and size. This helps to distinguish long and narrow objects from objects which are more like a square or a circle.
- Geographical location of segment on the ground given by the circumscribing rectangle and the centroid.

The database will use these properties for efficiently identifying candidate images. Afterwards an application specific image analysis could be performed which may also use a more appropriate segmentation procedure.

3.2 Database Indexing Technology

According to our previous analysis the database needs to support queries for segments which contain constraints for spectral ranges of some channels, segment sizes and forms. In addition the query may contain spatial, temporal or sensor constraints. These types of search criteria require the support by a multidimensional index structure, as for example the R-Tree.

An issue in this context is that the set of channel relevant for a certain application is not clear. Detecting different types of objects may also require different combinations of channels.

A solution for this issue might be an index structure which includes all channels. For those channels, which are not relevant, the value range is set to the minimal value for the lower bound and the maximal value for the upper bound. Theoretical analysis however shows that this is probably not a good idea, since the efficiency of the index structure degrades with the number of dimensions. Even worth the wide ranges for one attribute might require that large parts of the index structure need to searched even though these parts do not contribute to the search result.

A better approach is the definition of different index structures for different sets of channels. The definition of an index structure for every combination of channels would however result in a very high number of index structures. As an example 7 channels would require 2^7-1 index structures. In fact the other criteria and different combinations of them would be considered as well. This makes such a generic approach not feasible.

Since generating indexes for every combination of criteria is not feasible the set of potential search conditions needs to be analyzed. Even though we loose generality, this limitation is unavoidable. It however helps to use an index structure containing a subset of attributes from the search condition. In this case the search would identify all candidates by the index structure using the subset of attributes and then apply the full search condition to the hopefully small set of intermediate results.

Considering these constraints it is necessary to distinguish mandatory attributes and optional attributes in a search and identify frequent combinations of optional attributes.

Mandatory attributes should be all attributes related to object identification to reduce the intermediate result set as much as possible. All other criteria are probably optional. Therefore the index structures should be generating according the following rules:

- Every index structure contains the size and the form parameter.
- An index is generated for every channel. Further indexes might be generated for combinations of channels if this is required by applications.
- Indexes should exist for every combination of geographical position, time and sensor (8 possible combinations).

This means that we get (c+d)*8 index structures if *c* is the number of channels and *d* are application specific index structures. This number is already high and leaves little room for further indexes. The number of indexes could be reduced by half if data from different sensors is stored in separate tables. In this case the indexes need not consider sensor attributes.

As an index structure R-Trees generated by tree packing methods might be the right solution for our concept. Most applications in this context deal with a large amount of data changing only slightly over a time period. In this scenario it makes sense to generate a static R-Tree on a regular base (e.g. every week) and store the changes in a separate location. All operations will then be applied to the static database and the separate data store. Since this separate data store contains only a small amount of data the processing time of an operation is dominated by the processing time of the static database. This approach will exclude certain difficult situations for which R-Trees cannot guarantee acceptable response time.

3.3 Integration of Heterogeneous Information Systems

Access to geo-information is often hindered by a number of factors, ranging from costs to a lack of knowledge on availability of appropriate data sets and software tools to integrate this data in the geo-application scenario. Development coordinated by the Open Geospatial Consortium address a number of standards for geospatial and location based services. The aim of the OGC (www.opengeospatial.org) is to create open and extensible software application programming interfaces for geographic information systems and other mainstream technologies. Historically, the efforts of the OGC have been strongly focused on the GI backgrounds of the respective consortium members. One of the most visible results of the OGC standardization process is the increasing availability of OGC compliant web mapping and web feature servers (see http://www.w3.org/TR/wsa-reqs/).

Web services address integration of heterogeneous information systems at a more generic level, for instance, by not defining a priori what types of data, or indeed, functionalities are to be distributed between services. GI specific exchange of data and functionality can be regarded as a thematic subset of web services. In fact, one of the most recent OGC developments is the implementation of OGC standards directly as web services.

A Web service is a software system identified by a URI, whose public interfaces and bindings are defined and described using XML. Its definition can be discovered by other software systems. These systems may then interact with the Web service in a manner prescribed by its definition, using XML based messages conveyed by Internet protocols. The key factors in this definition are the standardization of the interfaces, the possibility to discover unknown services, usually via a registry and the (implicit) potential to chain and combine different web services to generate a new service. The actual implementation details of both the services themselves and the mechanisms for publication and discovery of services is beyond the scope of this paper. It is sufficient to say that mature standards such as SOAP, WSDL and UDDI are already widely used in many software APIs, including Open Source, and implemented in a wide range of business scenarios, include e-commerce. Open Source implementations of web services is of considerable importance to our application context, as it allows the implementation of the required functionalities without restriction and at very low cost, for instance, in emergency situations in developing countries.

In the context of this paper, the use of web services for (1) the supply of vector data, for instance cadastral maps from municipalities in the risk area, (2) for the near real time delivery of high resolution georeferenced satellite and airborne imagery and (3) as an interface to a configurable segmentation algorithms. The latter may interface with a spatial data store, which is searchable via its own dedicated web service. To illustrate the concept of chainable web serviccs, a service can be envisioned that would combine output of the risk analysis output with outputs from an in-situ sensor network and weather forecast to define an alert system.

In figure 3 a simulated web service request is given for the SOAP implementation of the web service that interrogates the spatial data store for segments that have been derived from recent SPOT image data for which the segment size is above 1 hectare, segment width is at least 100 m and for which the red and infrared channel values are in a range that is typical for a forest coverage. The output format is requested as GML, an open XML based format for GI data that can be integrated in user inter-faces that combine outputs from the web service with other map layers.

Datei Bearbeiten Ansicht Favoriten Extras ?


```
- <!--
    POST /examples HTTP/1.1
    User-Agent: Axis SOAP/1.1 (Linux)
    Host: gmoss.emergency.org:81
    Content-Type: text/xml; charset=utf-8
    Content-length: 899
    SOAPAction: "/segment_lookup"
  -->
  <?xml version="1.0" ?>
- <SOAP-ENV:Envelope SOAP-
    ENV:encodingStyle="http://schemas.xmlsoap.org/soap/encoding/"
    xmlns:SOAP-ENC="http://schemas.xmlsoap.org/soap/encoding/"
    xmlns:SOAP-ENV="http://schemas.xmlsoap.org/soap/envelope/"
    xmlns:xsd="http://www.w3.org/1999/XMLSchema"
    xmlns:xsi="http://www.w3.org/1999/XMLSchema-instance">
  - <SOAP-ENV:Body>
    - <m:getSegments xmlns:m="http://www.gmoss.org/">
        <imageid
          xsi:type="xsd:string">SP4_200409011073423_11_56</imageid>
      - <bandconstraints SOAP-ENC:arrayType="xsd:ur-type[6]"
          xsi:type="SOAP-ENC:Array">
          <channel_id xsi:type="xsd:string">XS1</channel_id>
          <channel_minimum xsi:type="xsd:byte">20</channel_minimum>
          <channel_maximum xsi:type="xsd:byte">35</channel_maximum>
          <channel_id xsi:type="xsd:string">XS4</channel_id>
          <channel_minimum xsi:type="xsd:byte">40</channel_minimum>
          <channel_maximum xsi:type="xsd:byte">75</channel_maximum>
        </bandconstraints>
        <segmentsize xsi:type="xsd:int">10000</segmentsize>
        <segmentwidth xsi:type="xsd:int">100</segmentwidth>
        <outputformat>GML</outputformat>
      </m:getSegments>
    </SOAP-ENV:Body>
  </SOAP-ENV:Envelope>
```


Fig. 3. Web service request for segment information

4 Application Scenario

The project Fireguard (http://dib.joanneum.at/fireguard) is a good candidate to apply the above concept in a real applications scenario. The aim of Fireguard is to assess the risk of forest fires in the Mediterranean area by the integrated use of very high spatial resolution satellite imagery (QuickBird, Ikonos), GIS, mobile wireless devices, Internet technology, and broadband terrestrial and satellite telecommunication networks. The parameters derived form these data and information sources will build the input for accurate fire hazard, risk and behavior models to be used by local forest and fire-fighting services in a pro-active approach to fuel management and fire pre-suppression activities.

For the development of methods which are capable of such an analysis, seven test-sites in the Mediterranean area, specifically in Greece and Portugal have been selected and will be fully analyzed within the project.

The core information extraction method is making use of Quickbird and Ikonos imagery in order to be able to provide information on a very detailed level. After geometric correction these images undergo various image processing, segmentation and classification steps to extract the relevant base parameters like species type, cover, distribution, crown base diameter, and height. These parameters are generated by a segmentation based approach which is necessary to properly analyze the detailed datasource. Also a segment based technique makes it easier to handle the large amount of raster data.

The current status of the project produced first results for a robust extraction of basic parameters like tree species, crown base diameter and tree height. The tree species parameter is obtained via a segment based classification method, whereas the crown base diameter makes use of a specialized segmentation method which will be applied in forested areas within open forest stands. The tree height parameter is derived from an Ikonos stereo image pair which is used to create a DSM. Hence in areas with relatively open stands it is possible to create a digital terrain model from the DSM by including information from the classification results. In general, these processing chains a relatively complex and need a considerable amount of CPU time. Consequently this part will have to be included in the data ingestion part of the database system (see figure 1).

Once this information has been obtained, the corresponding parameters are ready to be stored in a GIS database system, where they could be accessed by the local forest- and fire-fighting services. These local service centres will then use this information to build-up their fire-hazard and risk models as guidance for their fire fighting activities. Additional user applications will also use these basic parameters to create a virtual simulation of the area of interest by simulating the trees and the other classified land cover objects from a set of prototype objects parameterized by the height and diameter features.

Without an optimized access to the information in the database the calculation of their modules will be severely hindered when frequent updates are needed or when near real time information is required.

At this point an integrated GIS concept like the one described above will start to give a big advantage over conventional systems. The GIS system will decouple the basic information from the specific application used by the local authorities. The basic information can be easily updated by new image acquisitions and applying the processing chain for the basic parameters. On the other hand the applications in the local service centers have easy access to the data they need. The querying capabilities of such a database system would both, greatly reduce the complexity and simultaneously increase the performance and response time of the fire hazard, risk, and behavior models of the local services. Typical access patterns like looking for all trees of “Pinus pinea” with a crown base diameter over 2m in a specific area of interest will greatly benefit from such a GIS system.

Since the primary goal of the Fireguard project is the development of processing chains which are capable of robustly derive the mentioned basic parameters from very high resolution satellite imagery, an elaborate GIS system has not yet been defined or implemented. However, if the fireguard concept will be set into an operational scenario an integrated GIS concept like the one described in this paper must be considered since it provides clear advantages over conventional GIS systems.

5 Summary

This paper proposes an integrated concept for content based image searches in the context of disaster management. For this purpose the paper clearly identifies critical issues in this context and proposes an integrated concept as a solution dealing with these issues. The core of the concept is a generic segmentation procedure combined with a multidimensional index structure. These methods are key parts of nodes in a distributed information system connected via web services.

The concept does not only propose a first solution for the mentioned issues but also provides a framework for future research in this area. In fact other types of segmentation procedures or even completely different methods could be used to derive parameters about images which are stored in a database. In the case of index structures further research is needed to identify potential and limitations of multidimensional index structures. This research would facilitate a better optimization of index structures for a given application domain.

The next step for the proposed concept is a prototype implementation. The purpose of this implementation is the demonstration of this concept and its optimization with respect to different application scenarios.

References

Ang C, Tan T (1997) New linear node splitting algorithm for R-Trees. In Scholl M, Voisard A (eds.), Advances in Spatial Databases, LNCS, Springer-Verlag 1997

Baatz M, Schäpe AA (2000) Multiresolution Segmentation – an optimization approach for high quality multi-scale image segmentation. In: Strobl J, Blaschke T, Griesebner G (eds): Angewandte Geographische Informationsverarbeitung XII, Wichmann-Verlag, Heidelberg, pp 12-23

Bauer T, Steinnocher K (2001): Per-Parcel land use classification in urban areas applying a rule-based technique. GIS – Zeitschrift für Geoinformationssysteme 6/2001, pp 24-27

Bayer R, McCreight C (1972) Organization and Maintenance of Large Ordered Indexes. Acta Informatica 1(3), pp 173-189

Beckmann N, Kriegel HP, Schneider R, Seeger B (1990) "The R*-tree: An Efficient and Robust Access Method for Points and Rectangles. Proc. ACM-SIGMOD International Conference on Management of Data, Atlantic City (NY), pp 322-331

Bentley JL (1980) Multidimensional divide and conquer. Communication of the ACM, 23(6), pp 214-229

Bentley JL, Maurer HA (1980) Efficient worst-case data structure for range searching, Acta Informatica

Berchtold S, Keim DA, Kriegel HP (1996) The X-Tree: An Index Structure for High-Dimensional Data. Proc. International Conference on Very Large Data Bases, Mumbai (Bombay), India, pp 28-39

Blaschke T, Lang S, Lorup E, Strobl J, Zeil P (2000) Object-oriented image processing in an integrated GIS/remote sensing environment and perspectives for environmental applications. In: Cremers, A. und Greve, K. (Hrsg.): Environmental Information for planning, politics and the public. Metropolis Verlag, Marburg, pp 555-570

Blaschke T, Conradi M, Lang S (2001) Multi-scale image analysis for ecological monitoring of heterogeneous, small structured landscapes. Proceedings of SPIE, Toulouse, pp 35-44

Blaschke T, (2003) Continuity, complexity and change: A hierarchical Geoinformation-based approach to exploring patterns of change in a cultural landscape.

In: Mander Ü, Antrop M (eds.): Multifunctional Landscapes Vol III: Continuity and Change. Advances in Ecological Sciences 16, WIT press, Southampton, Boston, pp 33-54

Burnett C, Blaschke T (2003) A multi-scale segmentation / object relationship modelling methodology for landscape analysis. Ecological Modelling 168(3), pp 233-249

Chaudhuri B, Sarkar N (1995) Texture segmentation using fractal dimension. IEEE Transactions on Pattern Analysis and Machine Intelligence. Vol. 17, Nr. 1, pp 72-77

Collins CA, Parker RC, Evan DL (2004): Using multispectral imagery and multi-return lidar to estimate tree and stand attributes in a southern bottomland hardwood forest. Proceedings ASPRS 2004 Annual Conference, Denver, USA

Cross A, Mason D, Dury S (1988) Segmentation of remotely-sensed images by a split-and-merge process. Intern. Journal of Remote Sensing 9 (8), pp 1329-1345

Dubuisson-Jolly MP, Gupta A (2000) Color and texture fusion: application to aerial image segmentation and GIS updating. Image and Vision Computing (18): pp 823-832

Fu KS, Mui JK (1981) A survey on image segmentation. Pattern Recognition, vol. 13, pp 3-16

Göbel R, Hornsteiner G (2002) A Non-Replicating Index Structure Supporting Efficient Range Searching"Interner Bericht, Fachhochschule Hof

Guttman A (1984) R-Trees: A Dynamic Index Structure for Spatial Searching. Proc. ACM SIGMOD Conference, Boston, pp 47 - 57

Haralick R, Shanmugan K, Dinstein I (1973) Textural features for image classification. In: IEEE Transactions on Systems, Man and Cybernetics. Vol. 3, No. 1, pp 610-621

Hofmann T, Puzicha J, Buhmann J (1998) Unsupervised texture segmentation in a deterministic annealing framework. In: IEEE Transactions on Pattern Analysis and Machine Intelligence. vol 20, no 8, pp 803-818

Hofmann T, Boehner J (1999) Spatial pattern recognition by means of representativeness measures. In: IEEE 6/99

Jain A, Farrokhnia F (1991) Unsupervised texture segmentation using Gabor filters. In: Pattern Recognition vol 24, no 12, pp 1167-1186

Kamel I, Faloutsos G (1993) On Packing R-Trees"Proc. 2nd International Conference on Information and Knowledge Management (CKIM-93), Arlington, pp 490-499

Koch B, Jochum M, Ivits E, Dees M (2003) Pixelbasierte Klassifizierung im Vergleich und zur Ergänzung zum objektbasierten Verfahren. In: Photogrammetrie Fernerkundung Geoinformation 3/2003, pp 195-204

Laine A, Fan J (1996) Frame representations for texture segmentation. In: IEEE Transactions on Image Processing. vol 5, no 5, pp 771-779

Langanke T, Blaschke T, Lang S (2004) An object-based GIS / remote sensing approach supporting monitoring tasks in European-wide nature conservation. Proceed. Mediterranean conference on Earth Observation. First Mediterranean Conference on Earth Observation, April 21-23, 2004, Belgrade, pp 245-252

Leutenegger ST, Lopez MA, Edgington J (1997) STR: A Simple and Efficient Algorithm for R-Tree Packing. Proc. 12th International Conference on Data Engineering, pp 497-506
Manjunath B, Chellappa R (1991) Unsupervised texture segmentation using Markov random field models. In: IEEE Transactions on Pattern Analysis and Machine Intelligence, vol 13, pp 478-482
Mao J, Jain A (1992) Texture classification and segmentation using multiresolution simultaneous autoregressive models. In: Pattern Recognition, vol 25, pp 173-188
Mehlhorn K (1984) Data Structures and Algorithms 3: Multidimensional Searching and Computational Geometry, Springer Verlag
Neubert M, Meinel G (2003) Evaluation of segmentation programs for high resolution remote sensing applications. In: Proceedings High Resolution Mapping from Space, Hannover (CD-ROM)
Ravi Kanth KV, Singh A (1999) Optimal Dynamic Range Searching in Non-Replicating Index Structures"Proc. International Conference on Database Theory, Springer Verlag Berlin Heidelberg, pp 257-276
Roussopoulos N, Leifker D (1985) Direct Spatial Search on Pictorial Databases Using Packed R-Trees. Proc. ACM SIGMOD
Salari, E, Ling Z (1995) Texture Segmentation using hierarchical Wavelet Decomposition. In: Pattern Recognition, vol 28, no.12, pp 1819-1824
Schiewe J, Tufte L (2002) Potential und Probleme multiskalarer Segmentierungsmethoden der Fernerkundung. In: Blaschke T (ed.) Fernerkundung und GIS: Neue Sensoren – Innovative Methoden. Wichmann-Verlag, Heidelberg, pp 42-51
Sellis T, Roussopoulos N, Faloutsos C (1987) The R+-Tree: A Dynamic Index for Multi-Dimensional Objects. Proc. International Conference on Very Large Data Bases, pp 507-518
Tiede D, Burnett C, Heurich M (2004) Objekt-basierte Analyse von Laserscanner- und Multispektraldaten zur Einzelbaumdelinierung im Nationalpark Bayerischer Wald. In: Strobl J, Blaschke T, Griesebner G (eds): Angewandte Geoinformatik 2004. Wichmann Verlag, Heidelberg, pp 690-695
Tso B, Mather P (2000) Classification of Remotely Sensed Imagery Using Markov Random Fields. ARCS 2000 Proceedings, CD ROM
Van der Sande C, De Jong SM, De Rooc, AP (2004) A segmentation and classification approach of IKONOS-2 imagery for land cover mapping to assist flood risk and flood damage assessment. In: ISPRS Journal of Photogrammetry & Remote Sensing 58 (2004), pp 217-229

A Distributed Spatial Data Library for Emergency Management

Tony Hunter

Risk Research Group, Geoscience Australia, GPO Box 378, Canberra, ACT 2601, Australia.
Email: Tony.Hunter@ga.gov.au

Abstract

Australia has a three-tiered hierarchal model of government. A single Federal government, eight State/Territory governments and approximately seven hundred municipal councils make up the three tiers. Each of these tiers, and the separate jurisdictions within the tiers, can have their own standards and arrangements for managing information useful for Emergency Management (EM). Other information resources are held by private organizations. The business drivers for a coordinated national approach to 'data collection, research and analysis…' was identified by the Council of Australian Governments (COAG) review and documented in their report 'Natural Disasters in Australia – Reforming mitigation, relief, and recovery arrangements' in 2001 and released in August 2002. Representatives of all tiers of governments were signatories to this report. Later in 2001 the events in New York on September 11 reinforced the business drivers for access to data that transcends jurisdictional boundaries, as did the 2003 bushfires in Canberra. Against this backdrop there are several projects that are addressing the infrastructure and data requirements at the state/territory level. The 'LIST' in Tasmania. 'VicMap' in Victoria, the 'EICU' project in NSW, the 'SIS' project in Queensland, the 'SLIP' project in Western Australia and the ESA CAD system in the ACT are examples of spatial information Infrastructure initiatives that partially support EM at the jurisdictional level.

At the national level the Australian & New Zealand Land Information Council (ANZLIC) proposed a national Distributed Spatial Data Library in

2003. Previous attempts to create centralized repositories have failed but maturing web services and the ability to produce hard-copy maps on-demand have moved this concept to a practical reality. Underpinning the distributed library is the development of a community 'All Hazards' Data Taxonomy/Model for the EM community. The majority of the state jurisdictions provided input to the taxonomy, while additional expertise in the modeling and socio-economic domains were provided by Geoscience Australia (GA). The data identified by the taxonomy is sourced from varied and complex sources and formatted into a simplified, coherent form suitable for Emergency Management. The benefits of sharing data through a standardized framework are being progressively demonstrated to organizations through the ability to provide early warning of threats to their assets and services, while ensuring they maintain control of their data. There are still many hurdles to overcome before an infrastructure to support a Distributed Spatial Data Library can be realized. These hurdles can be broadly categorized as technological and cultural. The technological hurdles are no longer a significant barrier as bandwidth steadily increases, and major GIS systems support web service based data integration. It is arguably the cultural hurdles that are the most difficult. The process of consultation and review used in creating the 'All Hazards' taxonomy has created a realization among the jurisdictions of the benefits of closer ties and co-operation in data sharing and delivery arrangements. There is still some distance to travel but the implementation of an Australian Distributed Spatial Data Library for Emergency Management is moving closer to reality.

1 Introduction - Emergency Management in Australia

Australia was formed as a federation of states in 1901, although the idea was first raised in the 1840's it took sixty years to reach fruition. It has a three-tiered hierarchal model of government. The single Federal government has a virtual monopoly over taxation, immigration, security, navigation and defense. Eight State/Territory governments have responsibility for education, Emergency Management (EM) and most other domestic issues. Responsibility for health is shared between the state and federal governments. Approximately six hundred and eighty municipal councils make up the third tier, although they exist at the discretion of the state governments they play an important role in service provision and influence the workings of society through their close interaction with people's day-to-day lives. Each of these tiers, and the separate jurisdictions within the tiers, can have their own standards and arrangements for managing information useful for Emergency Management.

The Federal governments ability to influence the arrangements for EM in Australia are predicated on its capacity to provide specific funds that the states would otherwise struggle to find from their limited ability to raise revenue. This is a very powerful mechanism for influencing policy and administrative arrangement at all levels of government. The federal government also has the ability to invoke various extraordinary legislation that can be used during times of national emergencies; these are not generally used unless the situation exceeds the capacity of local resources and governance arrangements.

Many agencies are involved in the field of emergency management in Australia. These range from those who have a strong reliance on volunteer labor, such as State Emergency Services (SES) and Volunteer Bushfire Services to fully professional services such as Police, Fire Brigades and Ambulance Services. There are also many private sector organizations, such as water, electrical, gas, liquid fuels and mining companies who have highly trained staff that would be drawn upon if a disaster occurred that affected their company's assets.

Volunteer community organizations provide the extended human resources that underpin most major disasters in Australia, where volunteering has a long tradition. The Armed forces can also play a role in major disasters, and former distinguished military commanders continue to play significant senior management roles in of disaster organizations.

The realization of limited funding and the catastrophic nature of some recent disasters have triggered a more strategic approach to prioritization of EM resources. The exemplary work of response and recovery is now being augmented by a shift toward disaster mitigation; through risk assessment, risk reduction and readiness as governments seek to target their limited funds in the most vulnerable areas. This was highlighted by a report commissioned by the Council of Australian Governments (COAG) and released in August 2002 [1] 'Natural Disasters in Australia – Reforming mitigation, relief, and recovery arrangements'. Representatives of all tiers of governments were signatories to this report.

COAG identified the need for some major reform commitments in the Australian EM sector. In the context of this paper two are of notable significance:

1. Reform Commitment one - Develop and implement a five-year national program of -systematic and rigorous disaster risk assessments.
2. Reform Commitment two - Establish a nationally consistent system of data collection, research and analysis to ensure a sound knowledge base on natural disasters and disaster mitigation.

Later in 2001 the events in New York on September 11 reinforced the business drivers for access to data that transcends jurisdictional boundaries. The recent report by Dawes et al [2] 'Information, Technology and Coordination: Lessons from the World Trade Center Response' has reinforced the usefulness of Geographic Information Systems (GIS) delivery and highlighted how the response was hampered by inadequate data/infrastructure resources as described by the following quotes from the executive summary:-

'Information was crucial to every aspect of the World Trade Centre Crisis. It's existence, availability, quality and distribution clearly affected, sometimes dramatically, the effectiveness and timeliness of the response and recovery efforts.'

'Of the many kinds of data put to use, geographic data and information systems (GIS) emerged as the most versatile analytical resource associated with the response'.

At the national level, the Australian & New Zealand Land Information Council (ANZLIC) proposed a national Distributed Spatial Data Library in 2003. Previous attempts to create centralized repositories have failed due to the resources required to co-ordinate and maintain a centralized system.

The Disaster Mitigation Australia Program (DMAP) administered by the Department of Transport and Regional Services (DoTARS) distributes federal government funds to the state and territory governments for mitigation and relief of disasters in Australia. Geoscience Australia's (GA) Risk Research Group (RRG) in the Geohazards Division act as technical advisors to DoTARS for the DMAP program. This program is aligning its work with the COAG [1] reports recommendations to 'move to evidence based mitigation programs for disaster management' in Australia based on a coordinated national framework.

2 A Community Data Taxonomy Model

The ability for a coordinated response to a major disaster from cross-jurisdictional agencies is enhanced when a common data model has been adopted. Visiting agencies can adapt to the local conditions and optimize their efforts when they can quickly understand the Geo-information that is supplied. This extended capability is extremely useful during a large-scale disaster that requires resources from the wider EM community. This message is being increasingly accepted at the state/ territory level. The next step is a coordinated framework at the national level.

The path that will lead to implementing a common vision for disaster management information is not completely clear. Should it be a top down approach with conformance legislated by a management elite? Or will the traditional methods applied at the 'coal face' of disaster management evolve toward a common system, the bottom up approach? Most likely a combination of these two approaches will be the most successful, as either method is unlikely to work on its own. There are several methodologies in use for implementing computer & software projects. These commonly map the lifecycle of a project through the initiation, requirement analysis, design, implementation and maintenance phases. Another popular approach is more agile and involves creating prototypes and using them to gather feedback from the users. The development of a major **distributed** spatial data library is a significant project that requires the co-operation and consensus building from a large range of diverse organizations. It is probable that a distributed development cycle may be required, with co-ordination augmented by some centralized agencies. This collaborative approach will see elements of the lifecycle methodologies integrated toward a positive outcome.

The benefits of a national emergency management sector that can respond to catastrophic disasters will be feasible when the capability to operate in a unified manner supported by a common data taxonomy/model can be demonstrated. Another advantage of adopting a national framework is that gaps in data availability and access can be better conveyed to relevant governments by the EM organizations. This is the environment that has facilitated the development of an 'All Hazards' data taxonomy in Australia. ANZLIC recognized the need for a national common framework and the Risk Research Group of GA took a leadership role in creating a prototype or 'straw-man' 'All-Hazards' data taxonomy/model to stimulate discussion.

The rationale behind an 'all-hazards' approach is quite simple. The data needs of disaster scenarios are remarkably similar. Whether it is a fire, flood, road accident or explosion the same environment and people are effected, dependent of course on location. The cause of these events, either naturally or technologically based, may require additional data for effective planning, response and relief management, but the core data sets remain constant. Figure 1 shows the inter-relationships between the various phases of EM and the commonality of the data that support them.

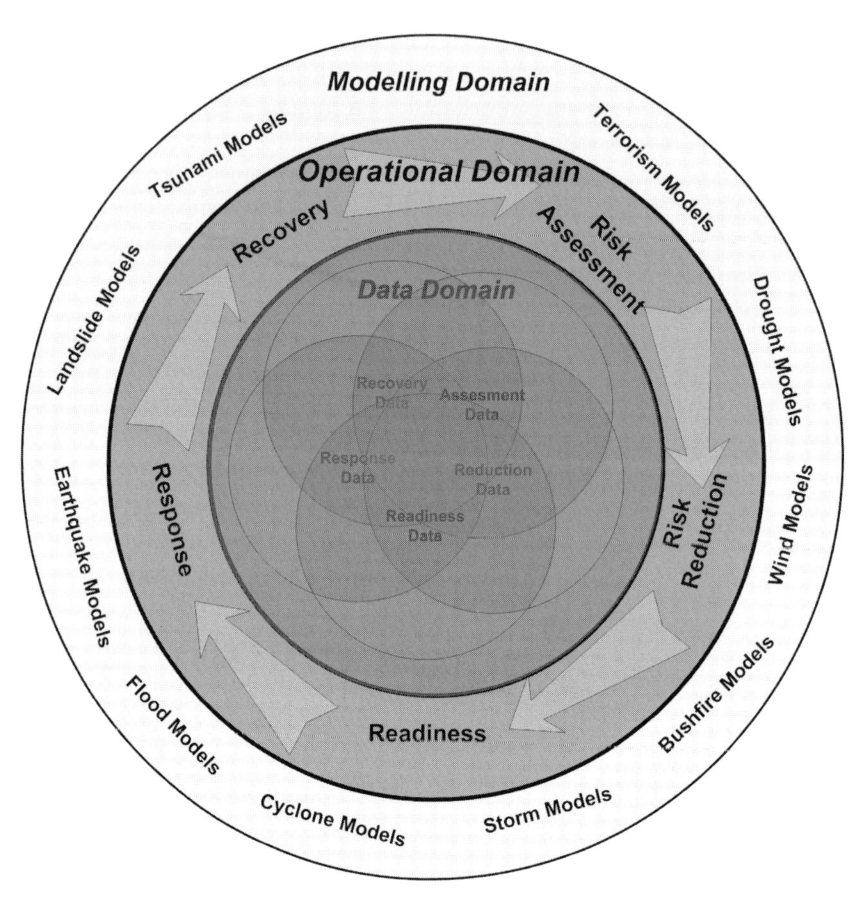

Fig. 1. Conceptual model of the modeling, operational and data domains and their interrelationships

Similarly the organizations that respond to disasters overlap to a large degree. Police, State Emergency Services, Ambulance Services and Fire Services are often involved. All these organizations will benefit from access to consistent, accurate and up-to-date geo-information. The Intelligence community, and the military, will also benefit from a national infrastructure, as they make greater use of these assets in their fight against terrorism. Their own specific data/information needs can be stored in even more secure systems and used to augment the more widely available distributed spatial data library (figure 2).

Developing the Model

The majority of the state jurisdictions, Tasmania, Victoria, ACT, NSW, Queensland, NT and WA and GA (federal), provided input to the taxonomy in the form of unsorted data layers, sourced predominately from response agencies. The initial task was to sort the data and remove duplicate data types. In doing this a list of which jurisdiction identified the data and any priority they supplied was recorded.

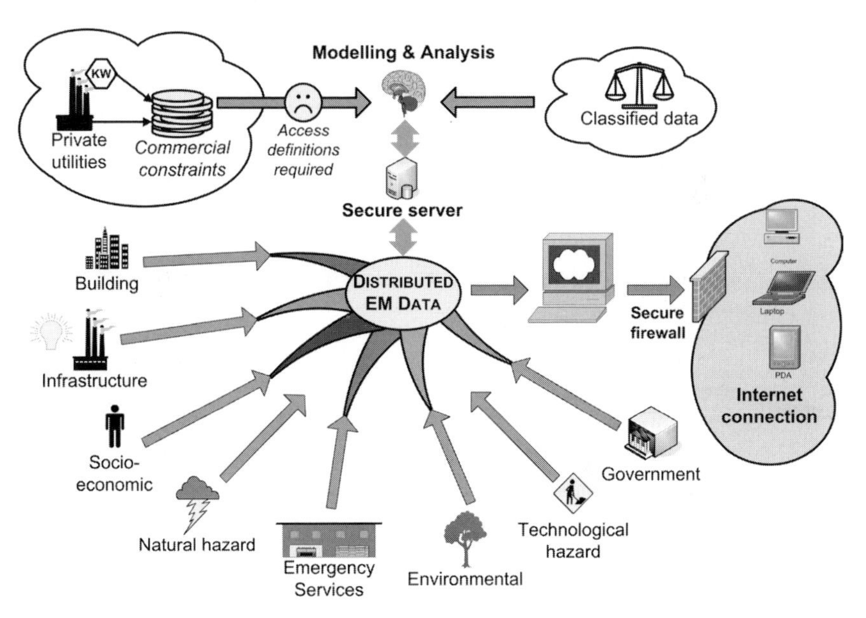

Fig. 2. A simplified conceptual mind-map of the infrastructure required to support an emergency management distributed spatial data library implemented using Web Services.

The next phase in creating the model involved classifying the data into fundamental data and derived data. This was necessary to avoid multiple listings of essentially the same data. Creating a classification system inevitably leads to making decisions on data that could fit into more than one category. An example of this is digital elevation models; they are usually derived from surveyed spot heights, contours and satellite-derived heights. This could see them classified as derived data. However their widespread use and acceptance, once they are created has lead to them being included as fundamental data.

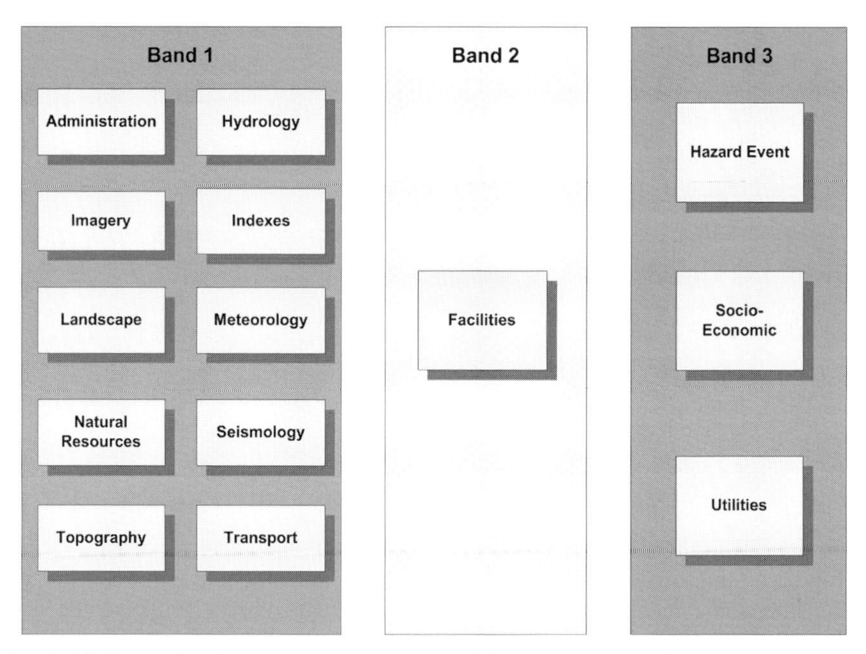

Fig. 3. All-Hazards Taxonomy themes grouped by resources needed to fully acquire the data

3 Feedback and Review of the Taxonomy

The Emergency Management Information Development Plan (EM IDP) Working Group is working along side the ANZLIC Working Group reviewing and aligning business drivers and data gaps for the EM sector. This group, EM IDP, is focusing on the processes involved in all phases of EM and identifying the gaps in information resources. It contains experts from a broad range of EM agencies and is taking a holistic view of the sector. The taxonomy has been distributed to the members of this group for reference and information will be shared across the two complimentary projects.

Theme	Description
Administration	Political definitions of Australia and environs. i.e. borders, reserves, administration areas.
Facilities	Man made features and structures
Hazard Event	Events that have significant destructive potential. (includes Hazard Event Impact)
Hydrology	Water related features, Onshore and Offshore i.e. Rivers, lakes, reefs.
Imagery	Remote sensed and photographic resources
Indexes	Map sheet and map tile indexes
Landscape	Geomorphologic measurements of the earth i.e. Geology & natural physical structures
Meteorology	Weather events
Natural Resources	Biota, vegetation and agriculture
Seismology	Seismic events.
Socio-Economic	Social and economic data sets.
Topography	The measurement of surface terrain i.e. spot heights, contours, isobaths, Digital Elevation Models
Transport	Road, rail, water and air transport features. i.e. roads, railways, airports
Utilities	Man made networks and structures - Related to facilities but specific to Utility

Fig. 4. Definitions of the themes in the ANZLIC 'All-Hazards' data taxonomy version 1.0

The Emergency Management Spatial Information Network Australia (EMSINA) has also previously canvassed their members for availability of and access to band 1 data (figure 3). EMSINA decided to share the results of their survey with the ANZLIC taxonomy and the EM IDP Working Group. The Emergency Information Co-ordination Unit (EICU) is a specialized unit within the NSW government. They are developing a state based infrastructure and host a Database Working Group. The taxonomy has been distributed to this group and two-way feedback is currently being incorporated into the models being developed. There is a large degree of conformance between the two models.

The taxonomy is a work in progress and was created to stimulate discussion and analysis from the EM community. All avenues available are being used to distribute the taxonomy and elicit feedback from the EM community. The state/ territory jurisdictions are viewed as being vital participants in a national framework. Wherever possible they are used to forward this important work of building and enhancing the national capacity to respond to disasters.

Creating a National EM Infrastructure

The business drivers for creating a coherent infrastructure of Geo-information can be now readily identified and have established their bona-fides as described previously [1], [2], [3]. The capability for GIS to provide spatial modeling and analysis was also demonstrated around Australia in twenty-four national workshops by the 'GeoInsight' program in 2002. This highly successful program highlighted the value of GIS delivery of information in the risk assessment, risk reduction, readiness, response and recovery phases of emergencies. The role of GIS in the response and recovery phases of 9/11, the Canberra Bushfires and other recent disasters have emphasized the point.

The impetus for change appears to be strongest in jurisdictions that have been touched by these large-scale disasters. This can be seen in the (ACT) were the extensive 2003 bushfires in Canberra claimed four lives, destroyed over four hundred houses and burnt 75% of the territory. An ad-hoc GIS facility that was set up by volunteers during the disaster made a significant contribution to the response and recovery effort. They produced custom on demand hard copy maps for specific purposes. This ability to provide pre-formatted but customized printed maps that can be tailored to the changing circumstances of a disaster augments the computer based delivery and analysis capability of GIS. It can also help surmount the cultural barriers and gain the support of more traditional disaster practitioners. The ad-hoc nature of this map production facility highlighted its inadequacies. There was a realization that the information made available could be far more extensive and used for all phases of disaster management.

The experience of this catastrophe has lead to the formation of the ACT Emergency Services Authority (ESA). They have implemented a cutting edge GIS based dispatch system in their control room, and access to an extensive range of geo-information.

The ACT is a small territory surrounded by the much larger state of New South Wales. Some of the fires started in the neighboring state jurisdiction of NSW, and that the response involved services from those areas. The inability for the services from the neighboring jurisdictions to be easily coordinated highlighted the need for better co-operation across political boundaries. This has lead to closer co-ordination of services between the ACT and NSW.

Many practitioners of GIS havc been aware of its ability to provide information in a timely and targeted form [5] for some time. This awareness is rapidly spreading as both managers and field workers gain more experience and confidence with the technology. Systems that utilize the power of GIS are emerging at the state/territory jurisdictional level. The 'LIST' in Tasmania, 'VicMap' and the 'ANGORA' project in Victoria, the 'EICU' project in NSW, the 'SIS' project in Queensland, the 'SLIP' project in Western Australia, and the ESA CAD system and ACT online in the Australian Capital Territory (ACT) are examples of spatial information Infrastructure initiatives that can be utilized to support EM. These projects have all addressed, to differing geographical and informational levels, the need for a cohesive information infrastructure that could be accessed in disasters. They are all based around a state/territory or lower jurisdiction and are at different stages of development and implementation.

Many of these projects utilize components of land information management scenarios. This has been one of the domains where GIS has widespread use. Other sectors such as Fire Services and State Emergency Services have also utilized GIS to varying degrees. These practical examples of effective dissemination of information over distributed systems are a useful way of demonstrating capability to the EM sector.

4 Legal Barriers

Navigating through the maze of jurisdictional related Intellectual Property (IP) constraints can be problematic. There are numerous organizations that are currently the custodians of the data needed for disaster management. Any system that utilizes this data must negotiate and implement legal access. A list of organizations that must be approached includes:

rail infrastructure, airports, marine authorities, road traffic authorities, energy distribution and production, municipal councils, water authorities, shopping centre management, defense, education, emergency service, national parks, environmental, fire services (rural and urban), health and weather services (Figure 2.). This list is by no means exhaustive and each type of organization may have numerous instances within any jurisdiction. This is especially true of municipal councils; there are approximately 680 of these in Australia. If the system is based on a state/territory jurisdiction, or lower, then any IP agreements may also need to be reached with organizations from the neighboring jurisdictions.

Within any state jurisdiction there is a variability of arrangements in place. The jurisdictions with smaller geographical footprints have been able to create more unified GIS environment for their needs. This can be seen in the excellent work done by the 'LIST' in Tasmania, and the Emergency Services Authority (ESA) and Urban Services in the ACT. Both of these jurisdictions have relatively small populations and fewer agencies and municipal governments and therefore less complex negotiations. No matter what the size of the jurisdiction consensus building among the stakeholder community is an important step in any successful systems implementation. In the case of the ACT legislation was implemented specifically to create a single agency that would have the authority to co-ordinate and manage the operations of the fire, ambulance and SES.

In the larger jurisdictions the number of organizations that a lead state agency must deal with is even greater. These can include commercial or government organizations that run on a cost recovery model. This effectively makes them a commercial concern and there are cases where the EM organizations are denied access to government data because they can't afford the purchase price. In these situations it is necessary for the jurisdictional government to have the political will to facilitate the supply of data. This may be in the form of funding arrangements or, in extreme cases, legislative inducements or letters from the state premier's office. Experience shows that it is preferable to avoid the use of heavy-handed approaches to force organizations to supply data, wherever possible. There may be some advantage in the judicious use of legislative requirements or letters from senior government offices to elicit the supply of data. Using such a process once may be enough to demonstrate the resolve of the government. The down side is that the damage done to the relationships between the supplier and collector agencies can take a long time to repair. It is preferable to be able to demonstrate some mutual benefit to the agencies that hold the data sets. This can be in the form of transformation and data scrubbing services for those whose data is not currently spatially coherent.

It is important to develop relationships with agencies and re-assure them that they will not loose operational control of their core business by sharing their information resources. After integrating the component data into a comprehensive GIS the ability to more effectively plan and respond to disasters can be demonstrated. The ability for early notification to a responsible agency when an impending disaster such as fire or flood will threaten their operations is a persuasive argument for contributing to the infrastructure. Often it is better to engender co-operation 'one step at a time'. Getting a first cut of the data, integrating it into a GIS system and demonstrating the power and capability for planning and timely response should be the first step. Other advantages for the submitting organizations may be that the state body can co-ordinate and manage license requests and extraction of data clips from third party organizations from a centralized repository. This has the effect of freeing up GIS resources in the original supplier organizations. When the agencies see that this will increase their control of a situation, rather than erode it, the next step of ongoing data supply, negotiating appropriate data sharing with other organizations and related IP issues can be addressed. It has been found that agencies then start to volunteer more of their data sets so that they can leverage the power of the system for their increased benefit, and that of the wider community.

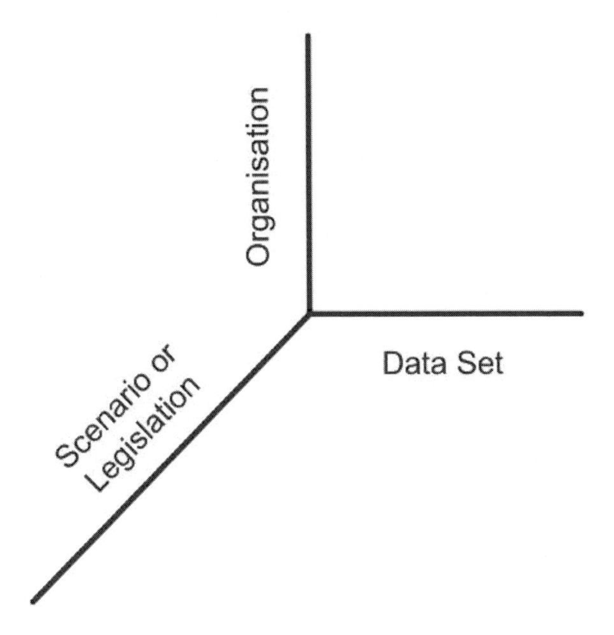

Fig. 5. Conceptual matrix of data availability to Emergency Management Community during disaster scenarios

Access to inter-agency data must also be contained within the theatre of operation and disaster scenario that requires the restricted data. Making the data too freely available may compromise national security. Restricting access to the data may unnecessarily endanger lives and property during a disaster. Establishing the appropriate balance continues to be a major challenge for all those involved. The agencies that receive the data must demonstrate that they have appropriate governance in place.

The realization of these security related issues has lead the Intergovernmental Committee for Surveying and Mapping (ICSM) has to recommending a security attribute for inclusion in the 'Harmonized Framework' structure.

It may be that some sort of 3, or 4, dimensional matrix needs to be implemented to control access to restricted data in a web delivery environment. A less formal access control method may suffice if web access is not required. Any access arrangements must be predicated on documented data access agreements, in all but the most catastrophic circumstances.

5 Technological Barriers

A national infrastructure for delivery of spatial information will most likely be implemented in a staged manner.

1. Establish business case for the system
2. Develop community data taxonomy/model
3. One time load of data into a jurisdictional based system for use in EM control rooms
4. Develop model for managing IP and data access to third party organizations.
5. Develop ongoing data sharing and updating arrangements with contributing organizations.
6. Develop web based delivery mechanisms
   a. Short term based around specific incidents
   b. Ongoing delivery and access – ability to plan and asses readiness

It is possible to establish a distributed data library using web services now. Such a system would be notionally constrained to specific organizations that were pre-granted access. There is still considerable work required on automating the management of IP and on-line web access.

The lack of a functioning robust Web Services Catalogue is an impediment to web delivery of a distributed spatial data library. Dissemination of, less sensitive, data will be impaired until the Open Geospatial Consortium (OGC) Web Services Catalogue functionality is built and implemented.

Another major unresolved issue hampering the implementation of a national multi-agency Geo-information infrastructure for EM is the lack of an agreed symbology set. It is important that a core set of scalable symbols can be defined for use across the sector. Any symbol set must be able to be read after the image is degraded by photocopying and faxing of the material. The symbols also need to address the fact that an element of spatial information needs to be expressed cartographically as a point, line or polygon.

6 Cultural Barriers

There are still many hurdles to clear before any national infrastructure can be implemented. The range of political and operational organizations involved is large. Each has its own cultures and practices. There are still areas where it is difficult to gain acceptance for the worth of GIS. Funding for 'boots and suits' is seen as the highest priority among more traditional EM practitioners. Operational control rooms of the police services are staffed by hard headed people who have to deal with the grim consequences of disasters, and make life and death decision under considerable duress. These people need to see overwhelming evidence of the benefits of GIS before they support the effort involved in implementing systems, especially if additional resources are not made available. It appears that the fire, SES and ambulance organizations are more readily adopting the Geo-information paradigm among the first responder communities.

The cultural barriers that have hampered a coordinated national framework for disaster management in Australia are being steadily eroded. Peak bodies are getting the message and using their influence among the community. The Australian Emergency Management Committee (AEMC), during it's March 2004 meeting, have endorsed the major recommendations of the COAG report [1] on Reforming mitigation, relief and recovery arrangements for natural disasters in Australia. They have recognized that 'significant limitations exist in national capability to deal with the consequences of a catastrophic event, whether as a result of natural, technological or human-caused hazards. Catastrophic disasters are defined as extreme hazard events which impact on a community or communities, resulting in widespread, devastating economic, social or environmental consequences' [4].

This approach also recognizes that the existing arrangements state and Territory arrangements, while working for small-scale specific hazards, can't necessarily be extended to cope with catastrophic disasters. There is an uneven level of planning between the jurisdictions for coping with very large disasters and currently no national framework in place.

This situation is changing and the analysis and planning for improved capability has begun. This involves identifying the vital elements of the systems that are needed for managing disasters in Australia. Any improvements in contingency planning for natural disasters will extend the capacity to manage man-made and technological disasters and their consequences.

By creating a community taxonomy and data model of EM information it is possible to asses the capability of the jurisdictions against a common set of criteria. This is only one of the elements needed in a national infrastructure for EM in Australia, but the process of creating the taxonomy and the dialogue that it has involved between the jurisdictions and organizations has been productive. By gaining a better understanding of each others information and business requirements the stakeholders in emergency management in Australia are strengthening their relationships and enhancing their ability to work collaboratively for a safer Australian community.

7 The Road Ahead

There are still many hurdles to overcome before an infrastructure to support a Distributed Spatial Data Library can be realized. These hurdles can be broadly categorized as technological and cultural. The technological hurdles are no longer a significant barrier as bandwidth steadily increases, and major GIS systems support web service based data integration. The lingering technological issues include a robust web services catalogue, and management of IP and security in a web delivery environment. Another unresolved issue is the standardization of scalable symbology. It is arguably the cultural hurdles that are the most difficult. The cultural barriers of data sharing and making use of GIS data still persist in pockets of resistance. These are becoming less entrenched as projects that demonstrate the power of Geo-information are utilized. As more and more agencies sign on to a coordinated framework and leverage its information sharing capabilities, their neighbors will find they are at a disadvantage if exist in isolation. It is quite likely that this EM taxonomy may be able to be used as a data exchange model between the various jurisdictions. This will allow for the phased implementation of a national infrastructure as the various jurisdictional components reach maturity.

The process of consultation and review used in creating the 'All Hazards' taxonomy has expanded the realization among the jurisdictions of the benefits of closer ties and co-operation in data sharing and delivery arrangements. The federation of Australia took sixty years from inception in the 1840's to realization in 1901. The federation of Geo-information into a distributed spatial infrastructure available for emergency managers is mover closer to a reality. The rate of technological development in the twenty first century exceeds that of the nineteenth century. This should lead to the implementation of a national infrastructure in the foreseeable future.

List of Acronyms

ABS	Australian Bureau of Statistics
ACT	Australian Capital Territory
AEMC	Australian Emergency Management Committee
ANZLIC	Australia and New Zealand Land Information Council
CAD	Computer Aided Dispatch
COAG	Committee of Australian Governments
DoTARS	Department of Transport and Regional Services
DMAP	Disaster Mitigation Australia Program
EICU	Emergency Information Co-ordination Unit
EM	Emergency Management
EM IDP	Emergency Management Information Development Plan
EMSINA	Emergency Management Spatial Information Network Australia
ESA	Emergencies Services Authority (of the ACT)
EMA	Emergency Management Australia
GA	Geoscience Australia
GIS	Geographic Information System
ICSM	Intergovernmental Committee for Surveying and Mapping
IP	Intellectual Property
LIST	Land Information System Tasmania
OGC	Open Geospatial Consortium
NSW	New South Wales
NT	Northern Territory
RRG	Risk Research Group (of Geohazards Division, GA)
SLIP	Spatial Land Information Platform (of WA)
SES	State Emergency Service
SIS	State Information System (of Queensland)
SA	South Australia
WA	Western Australia
WG	Working Group

References

Natural Disasters in Australia – Reforming mitigation, relief, and recovery arrangements' (2002) A report commissioned by the Council of Australian Governments (COAG) and released in August 2002

Dawes et al (2004) 'Information, Technology and Coordination: Lessons from the World Trade Center Response' New York 2004

Report of the ANZLIC Counter-Terrorism Project (2002) Using Australia's Spatial Information Infrastructure for Counter-Terrorism, Conybeare C, Australian & New Zealand Land Information Council (ANZLIC) 2002

Catastrophic Disaster Working Group (2004) scoping document, Australia Emergency Management Committee (AEMC), August 2004

Scott G (2004) Counter Terrorism & Critical Infrastructure Protection: Managing Them with GIS Technologies

On Quality-Aware Composition of Geographic Information Services for Disaster Management

Richard Onchaga

International Institute for Geo-Information Science and Earth Observation (ITC), Postbox 6, 7500 AA Enschede, the Netherlands.
Email:onchaga@itc.nl

Abstract

Dynamic chaining of geographic information services (geo-services) is emerging as a viable framework for evolving flexible geo-information systems, integrating heterogeneous geographically dispersed geo-information systems, and for providing on-demand access to geographic information in many application domains and location-based services. Alongside functionality, quality of service (QoS) is basic to successful chaining of dispaprate geo-services. This paper explores QoS provisioning in the context of geo-service chaining for disaster management. The paper presents a QoS model for disaster management and illustrates how user-level QoS requirements can be supported in a QoS-aware geo-service architecture.

1 Introduction

Developments in geo-information systems over recent years have centered on geographic information services (geo-services). Primarily, focus has been on defining open interface specifications and standards for dynamic chaining of disparate geo-services (OGC 2003, ISO/TC211 2002) in accordance with the service oriented architecture (SOA). SOA is a conceptual architecture that specifies interoperable, modular, loosely-coupled, self-contained and self-describing applications, systems or services that interact only at well-defined interfaces (McGovern et al. 2003). Accordingly, geo-services are loosely-coupled, interoperable and modular network addressable geo-spatial application modules which can be invoked across heterogeneous networks to access and process geographic data.

Loose-coupling means that atomic geo-services can be deployed, deprecated, or reconfigured independently, for example to adapt to changes in technology or to achieve performance targets, without need to modify other geo-services or infrastructural components. Interoperability and modularity enable dynamic chaining of disparate geo-services to create more elaborate composite geo-services that are capable of value-added services (Alameh 2003).

Clearly, geo-services are set to change the way geo-processing functionality is composed and delivered and the way the geographic information value chain is defined and executed. First, geo-services offer a framework for modular composition of evolvable geo-information systems. Geo-services chained at run-time facilitate dynamic composition of ad-hoc applications customized to requirements. Secondly, geo-services present a framework for integrating otherwise heterogeneous geo-information systems. Legacy geo-information systems are inherently heterogeneous and not interoperable but encapsulated as geo-services, the systems can be integrated to enable seamless sharing of information and processing resources. Moreover, as geo-services, legacy geo-information systems can also be integrated with mainstream enterprise information systems thus allowing them to play a greater and more strategic role in the enterprise. Lastly, geo-services facilitate automation of the geographic information value-chain therefore making on-demand access to value-added geographic information increasingly feasible (Di 2004).

Implicitly however, the dynamically composed ad-hoc application, the integrated legacy geo-information systems and the automated geographic information value-chain should be compliant with user requirements – both in terms of functionality and quality of service (QoS). QoS concerns the non-functional characteristics of a service that determine its utility and usability in an application context. Alongside functionality, QoS is therefore essential for commercial exploitation of geo-services. In our previous work we defined a framework for QoS provisioning in geo-service architectures (Onchaga 2004). In this paper, we employ the framework we previously defined in the context of dynamic chaining of geo-services for disaster management. The paper presents a QoS model for disaster management and illustrates how user-level QoS requirements can be supported in a QoS-aware geo-service architecture.

2 Defining Quality of Service

Quality of service (QoS) is a term that is widely used in many domains with equally many interpretations. In the context of service-oriented geo-processing, QoS is a set of qualities related to the collective behavior of one or more collaborating geo-services (ISO/IEC 1998). More precisely, QoS is the totality of characteristics of a service delivered by collaborating geo-services that bear on its ability to satisfy stated or implied needs in an application context.

In service-oriented geo-processing we identify two broad components of QoS;

- Quality characteristics of geographic information delivered, and
- Non-functional (operational) characteristics of the service.

We call the components QoS dimensions (Onchaga 2004) i.e. *information* and *operational* dimensions. Figure 1 is a UML class diagram showing the relationships among various QoS components in service-oriented geo-processing. The figure shows that deliverable QoS *(QualityOfService)* comprises quality aspects of information *(InformationQuality)* and qualities related to the operational aspects of the ad-hoc application realized by a geo-service chain *(QualityofApplication)*.

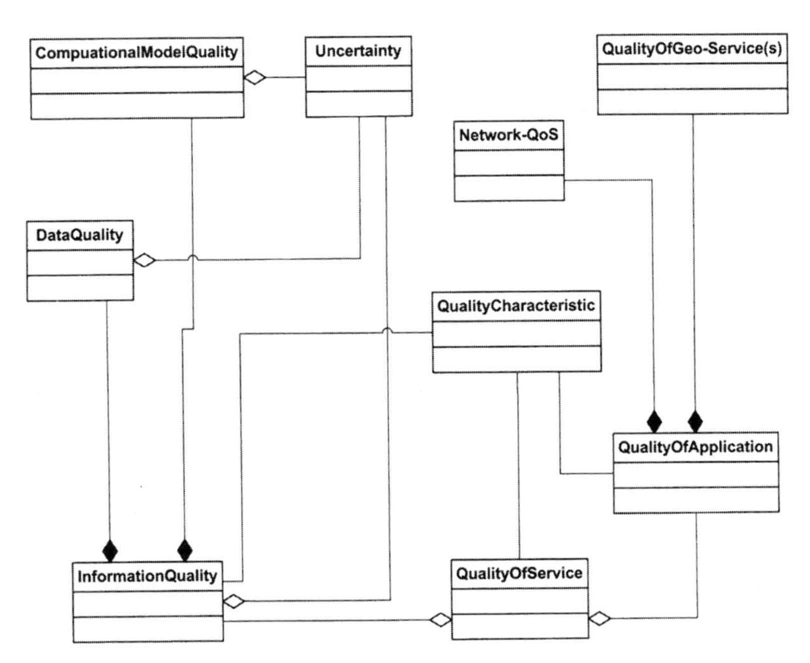

Fig. 1. QoS meta-model

Figure 1 further refines *InformationQuality)* into data and computational model qualities (*DataQuality* and *ComputationalModelQuality* respectively) and *Uncertainty*. Similarly, *QualityofApplication* is refined into quality of geo-service(s) *(QualityOfGeo-Service(s))* and network-QoS *(Network-QoS)*.

Further, in distributed service-oriented geo-processing, different architectural levels are identified and QoS abstractions at the different architectural levels are different. Consequently, QoS can also be specified according to the architectural (QoS) level. Figure 2 shows the different architectural levels and the interdependencies between them.

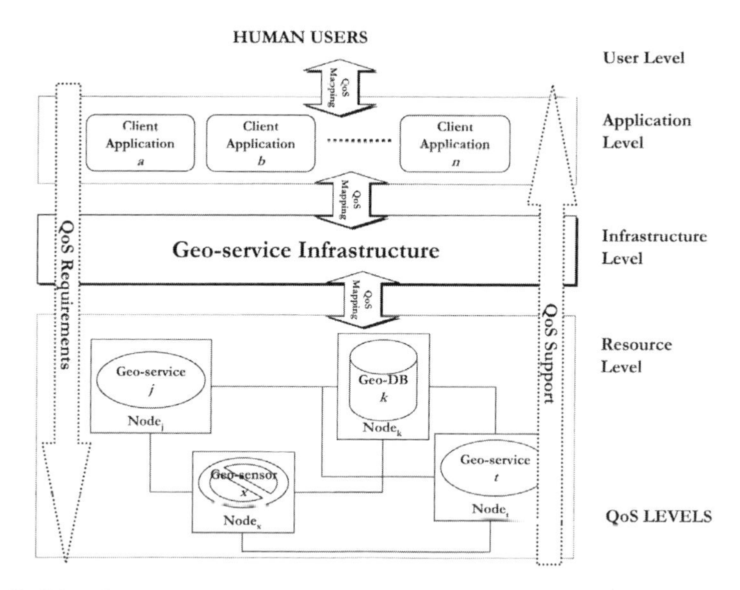

Fig. 2. QoS levels

At the top of the hierarchy are human end-users within a specific problem domain. Inherently, user requirements are domain-dependent. Because application level QoS requirements are geared towards meeting higher level user requirements, they are similarly domain-dependent. However, QoS at the infrastructure level is domain independent i.e. the notions of QoS supported at the infrastructure level are independent of any particular problem domain. Thus, infrastructure level QoS comprises generic notions of QoS that apply across the entire geographic information processing domain. QoS specified at the user, application and middleware architectural levels apply to the entire geo-service chain that realizes a desired service. This is in contrast with QoS at the resource level which typically applies to

atomic resources and is also technology and domain dependent. QoS at the resource level is with respect to:
– Geographic data (information)
– Autonomous geo-services
– Transport and communication networks

3 An Overview of Disaster Management

A disaster is an unforeseen and often sudden event that causes great damage, destruction and human suffering[1]. Disasters are normally categorized according to the cause. For example natural disasters are caused by naturally occurring phenomena e.g. earthquakes, landslides, etc. Similarly, technological disasters are caused by design and management failures in technological artifacts. Other categories of disasters can similarly be defined. Nonetheless, disasters share a number of common features:

- Disasters are a threat to life, property and livelihoods
- Disasters are rapid onset events i.e. the time between the moment it becomes apparent that a disaster event is eminent and the onset of the event is rather short
- Disasters occur with intensities that demand emergency response and external intervention
- A greater proportion of the direct loss occasioned by a disaster is suffered within a relatively short time after onset of the disaster event.

Disaster management concerns the organized efforts focused on eliminating or reducing the risk of a disaster and minimizing the impact of the disaster when it happens. The process of disaster management consists of four broad phases; disaster mitigation, disaster preparedness, disaster response, and disaster recovery (Montoya 2002). These phases can generally be grouped into pre-disaster and post-disaster phases. Pre-disaster phases are disaster mitigation and preparedness and generally concern activities that take place before a disaster event happens. In contrast, post-disaster phases concern activities that take place after a disaster event. The post-disaster phases are disaster response and disaster recovery.

[1] http://www.disasterrelief.org/Library/Dictionary/

4 QoS Model for Disaster Management

Disaster management is a multi-faceted, multi-disciplinary process that demands collaboration among disparate agencies and organizations. The various partners in the disaster management process need to share and integrate multi-disciplinary information seamlessly and continuously to execute disaster management programs. Geographic information forms a major component of the information requirements for disaster management. The role of geo-services in the context of disaster management will therefore include:

- Integration of disparate information systems in collaborating agencies and organizations therefore enabling seamless access and integration of information
- Provision of on-demand access to customized information and services to aid disaster response and other disaster management activities.

Information system integration is essential for seamless sharing of information and geo-processing resources across different administrative, technology and management domains. An integrated system allows partners to seamlessly share and process desired information. A useful integrated system however is one that is secure, resilient, available every when it is required and that offers adequate performance to support tasks in a dynamic and heterogeneous environment. Furthermore, the quality and integrity of the information shared must be assured.

On-demand access to information and services is particularly relevant to disaster response situations but it has more stringent QoS requirements. Typically, in disaster response, the people at risk need to be evacuated quickly and the destruction of property because of secondary damage checked as far as possible. In typical disaster response situations therefore search and rescue teams need quick answers to basic questions that are typically of spatial nature e.g. *shortest* route to and from the disaster area for evacuation purposes, locations of vulnerable populations, optimal locations for emergence medical services, etc.

It is significant that the answers sought are expected in near real-time. Moreover the information should be accurate otherwise it will lead to sub-optimal decisions and, as a consequence unnecessary loss of life, property and livelihoods. Moreover typical in disaster response situations, communication has to be maintained with mobile and wireless devices that have significant limitations in their capabilities.

In general, the rather unpredictable and rapid nature of disaster events, have important implications on the architecture of geo-services for disaster management. Services should exhibit high levels of availability and resilience, and the ability to adapt to varying environments.

Table 1 summarizes QoS characteristics relevant to disaster management. The characteristics are organized according to quality dimensions and architectural levels. While more architectural levels are possible, Table 1 shows only the *user* and *application* levels because these two levels abstract from the technology specific concerns that are outside the scope of this paper.

We use the term *user* rather loosely to imply notions of human end-users in a specific domain. A more rigorous meaning of the term *user* refers to a role played by any entity that requires or depends on the functions or services of another entity for its successful operation. For example, the various architectural levels of Figure 2 facilitate dynamic chaining of geo-services through nested *user-provider* interrelationships.

	Information dimension	Operational dimension
User-level	• Accuracy • Integrity • Fidelity	• Availability • Interactivity • Price • Security • Reputation (dependability)
Application-level	• Data quality elements • Computational model quality • Uncertainty	• Reliability • Performance • Cost • Security

Table 1. QoS model for disaster management

In the following paragraphs we elaborate on the QoS characteristics shown in Table 1. Under the information dimension two prime user-level QoS characteristics are identified; accuracy and integrity (fidelity).

- Accuracy – the notion of accuracy used here is synonymous with fitness-for-use as applied to data quality. Accurate information in an application context implies that the information is appropriate for that specific application. As such, user-level accuracy subsumes the notion of spatial data quality which is shown in Table 1 as an application level information QoS characteristic. Other accuracy-related application level QoS characteristics include computational model quality and uncertainty (with respect to both the data and the computational model).
- Integrity – integrity defines the extent to which information is presented appropriately, clearly, without distortion and consistent with user-expectations e.g. using well known symbols and other visualization

tools. Integrity can be interpreted as fidelity in cases where audio content is delivered. Fidelity is the requirement that the audio content is intelligible.

User-level operational QoS characteristics include availability, interactivity, price, and dependability (reputation). The corresponding QoS characteristics at the application-level include reliability, performance, cost and security. We briefly elaborate each of these in the following sequel.

- Availability – is the extent to which a service is available and ready for use when needed
- Interactivity – is the extent to which interactions with the service are responsive i.e. the swiftness with which a response is delivered. In an interactive service, interactions e.g. zoom, database queries, etc. receive swift responses
- Price – is the perceived value of a service as compared with the cost of using a service
- Dependability – is the extent to which a service can successfully execute required tasks. Dependability confers a sense of trustworthiness on a service.

The application-level QoS requirements are:

- Performance – concerns timing aspects and may refer to the time it takes to execute a task at a given service node, the time to transmit data packets from a source-node to a destination-node, or both
- Reliability – is the ability to maintain a service without violating service level agreements (SLAs)
- Cost – is the money charged for using a service as prescribed by the pricing policies of a service
- Security – defines security-related characteristics supported by a service e.g. protection, access control, authentication, etc.

5 Geo-Service Infrastructure

Geo-services are autonomous entities that are discovered and chained on-demand to deliver desired functionality. To a user, the resultant ad-hoc system should, as far as is practicable, appear homogenous and should ideally support desirable QoS characteristics. Stated differently, the user should be masked from the complexities of the chaining process and QoS provisioning. In (Onchaga 2004) we introduced the concept of a *geo-service infrastructure*. The *geo-service infrastructure* is a distributed computing infrastructure that provides the necessary distribution and QoS transparencies. It shields users from the complexities of QoS-aware geo-service discovery, chaining and execution. The infrastructure facilitates QoS-aware discovery, chaining and execution in which disparate geo-services are discovered and chained on the basis of their quality capabilities and subsequently executed so as to comply with user requirements.

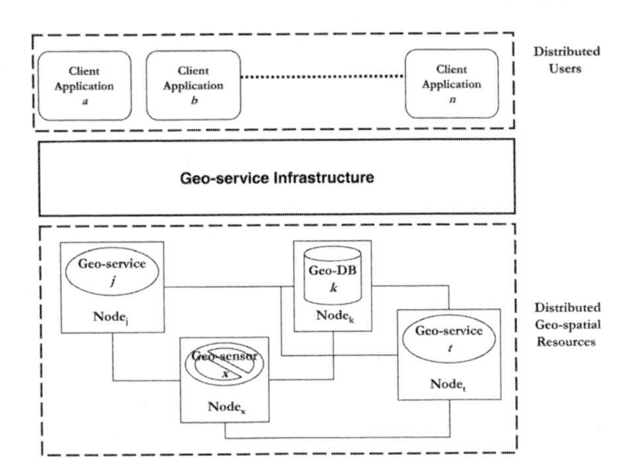

Fig. 3. Geo-service Infrastructure

Figure 3 shows that the geo-service infrastructure sits between client applications and geo-spatial resources (geo-services, geo-databases and geo-sensors) in distributed nodes that are interconnected by communication networks. The geo-spatial resources together with their underlying communication networks constitute a distributed resource platform for service-oriented geo-processing.

Client applications are user-side applications that enable specification of user-level requirements. Typically, users employ client applications to interact with infrastructure components and exploit distributed geo-spatial resources. The client application offers an (user) interface through which users can express their requirements. In essence both functional and QoS user requirements are specified but we focus on QoS requirements.

QoS requirements at the user level need to be propagated through the interceding layers down to the resource layer to realize a service that is compliant with requirements. Accordingly, user requirements must be translated into corresponding QoS requirements at the interceding levels and the resource level. The process of specifying QoS requirements for a given application is called QoS specification and the corresponding process of translating requirements across different architectural levels is called QoS mapping. In the next sections we elaborate on QoS specification and QoS mapping respectively.

5.1 QoS Specification

QoS specification concerns QoS requirements and the QoS policies that govern control and management of requested QoS. The QoS concepts we use here are drawn from (ISO/IEC 1998). A QoS requirement is QoS information that specifies all or part of the requirements to manage one or more QoS characteristics. A QoS requirement therefore expresses a set of constraints that are applied to one or more QoS characteristics e.g. a minimum value, maximum value, average value, range, etc., that specifies a desired level of service.

A QoS policy is a set of rules that determine the QoS management functions that are applied to control and manage a specified QoS characteristic. QoS policies define how QoS is managed including how QoS is monitored and reported. QoS policies are typically defined by providers depending on the QoS management functions available and technology applied. Before delivery of a service, users and providers of a service negotiate on required QoS and offered QoS to establish an agreed QoS. The agreed QoS together with the relevant QoS policies constitute a service level agreement (SLA).

In the preceding sections, QoS deliverable in the context of service-oriented geo-processing was defined as comprising two dimensions – the information and operational dimensions. It follows then that a QoS specification will generally comprise:

- Information quality requirements
- Operational quality requirements
- QoS policies
- Price of service and pricing policy

	Gold	Silver	Bronze
Accuracy	:high	:high	:high
Fidelity	:Good	:Medium	:Good
Availability	:≥ 99.99%	:≥ 99%	:≥ 95%
Interactivity	:≤ 2sec	:≤ 4sec	:≤ 8sec

Table 2. QoS classes for disaster management

To illustrate QoS specification, consider the example of disaster management. In disaster management it is apparent that disaster response demands relatively higher levels of QoS when compared to the other three phases. Therefore disaster response requires a different class of QoS that has more stringent requirements than the other three. Further, different users can be distinguished in a disaster response situation; those on wired devices and those on wireless devices. The two users will typically have different QoS requirements even when used in disaster response situations. Therefore disaster response can further be assigned two refined QoS classes depending on the kind of networked device from which a service is accessed. Logically, the wireless devices will have inferior QoS compared to wired devices but still much superior to QoS required for the phases disaster mitigation, preparedness and recovery.

Using the QoS model of Table 1, we define three different user-level classes of QoS service for disaster management. In our classification scheme, disaster response has two distinct QoS classes – the Gold and Silver classes corresponding to wired and wireless devices. The other three disaster management phases are assigned the Bronze QoS class. The QoS classes are illustrated in Table 2. For the sake of simplicity, Table 2 only shows a subset of the user level QoS requirements shown in Table 1. The values used in the table are hypothetical and are used to illustrate the difference between the QoS classes. Knowing the phase of disaster management and the type of end-user device a user, the infrastructure can assign an appropriate QoS level to the end-user. An appropriate QoS policy will be for example to warn end users of eminent degradation of QoS. This could be helpful for instance when critical decisions have to be made based on information delivered in which case other local sources of information can be used e.g. intuition and experience.

5.2 QoS Mapping

QoS mapping concerns translating QoS requirements across the different architectural levels of a service-oriented geo-processing architecture. For example *interactivity* as a user-level requirement can be mapped onto: *performance* at the application level; *delay* at the infrastructure level; and, *reliability, availability, latency* and *performance* at the resource level. In general every user level QoS requirement needs to be successively translated into lower level QoS requirements. However, QoS requirements at the different levels typically exhibit many-to-many type of relationship. To achieve reliable and effective QoS mapping, the many-to-many relationships among QoS requirements across different levels should ideally be resolved into one-to-one relationships

6 Implementing the QoS Framework

Geo-services that are compliant with the principles of SOA are loosely coupled, interoperable, modular and can therefore in principle be deployed on any network computing infrastructure that enables service discovery and collaboration. Such a computing infrastructure can be extended with QoS provisioning mechanisms, if necessary, to enable QoS aware discovery, chaining and execution of geo-services.

The World Wide Web (the Web) and the Grid are two evolving network technologies on which geo-services can be deployed. Developments on the Web are coordinated by the World Wide Web consortium (W3C), and the Internet Engineering Task Force (IETF) among other parties. For the Grid, the coordinating agency is the Global Grid Forum (GGF).

While the Grid and the Web are both key paradigms driving distributed systems research, they are different in the underlying philosophy. The philosophy behind the Web is that any user on the Web should be able to access and exploit Web resources. In contrast, resources on the Grid are only accessible to known partners in what is called a virtual organization (VO) (Foster et al.2003). Moreover, the Web is primarily engineered for information sharing whereas the Grid is biased towards providing access to high-end computing resources, data stores, visualization tools and instruments. Nonetheless, the technologies are converging as is evidenced by the Open Grid Services Architecture (OGSA) that extensively uses core Web technologies. We briefly outline the two technologies.

6.1 Web Service Technologies

The thrust of research on geographic information services has been focused on the Web and web services. Web services are application modules that can be discovered and accessed over the Web. Web services are based on standards developed by the W3C and other standardization bodies. A number of standards have been developed to enable interoperability, description, discovery, and chaining of web services i.e. XML[2], SOAP[3], WSDL[4], UDDI[5] and BPEL4WS[6].

The Open Geo-spatial Consortium (OGC) is the geo-spatial industry consortium that leads efforts towards open specifications for geo-services. In its OWS-2 program the OGC aims at migrating its interface specifications to be compliant with the industry standards for web services. The goal of OWS-2 is to adapt OGC specification to be compliant with the de facto SOAP, WSDL, UDDI and BPEL4WS suite of standards for web ser-

[2] eXtensible Markup Language.

[3] Simple Object Access Protol.

[4] Web Service Description Language.

[5] Universal Discovery Description Integration.

[6] Business Process Execution Language for Web Services.

vices. Accordingly, the OpenGIS[7] services architecture fronted by the OGC follows the Publish-Find-Bind interaction pattern employed by the web service architecture. In this interaction pattern, Providers of services create descriptions of their service offerings and *publish* then in a Registry (also called the Broker). The de facto standard for the Registry is UDDI. The service descriptions are done using WSDL. Users (Consumers) access the Registry to *find* services to which they can then *bind.* Interaction among collaborating services is achieved by SOAP messages over HTTP (FTP and SMTP are also possible) and chaining is achieved through BPEL4WS.

SOAP, UDDI, and the WSDL descriptions (in UDDI registry) and BPEL4WS realize the basic distributed computing infrastructure for web-based geo-services. OpenGIS services based on SOAP, UDDI, WSDL and BPEL4WS standards are interoperable and can be discovered on a web environment and chained to deliver value added services. This infrastructure for OpenGIS services can be extended with QoS provisioning services and mechanisms to enable QoS-aware chaining of geo-services. Towards QoS-aware composition of geo-services in a web environment, we propose the following enhancements:

- Service discovery in Web service architecture is achieved through query and search of service descriptions maintained in a service registry. The service descriptions constitute service metadata that are used to discover, evaluate and access suitable web services. Accordingly, the first step towards QoS-aware composition of web-based geo-services is to include QoS metadata in geo-service descriptions. The "*features*" and "*properties*" components of WSDL (W3C 2003) can be used to achieve this. With QoS properties as part of the standard service descriptions, the filters and search mechanism used to query and search service registries need to be similarly enhanced to enable query and search capabilityies that incorporate QoS properties
- QoS requirements at the user and application levels apply to an entire service chain and not to individual geo-services. To achieve meaningful QoS-aware service composition, these aggregate QoS requirements must be appropriately apportioned onto lower level QoS of individual resources i.e data, geo-services and network QoS. The main challenge here is to be able to discover and select appropriate resources that when chained will realize desirable application and user-level QoS. We call this process QoS composition (Onchaga 2004). Successful QoS composition entails 1) explicit knowledge of how various QoS properties of interest propagate in a service chain comprised of heterogeneous re-

[7] OpenGIS services are geo-services that comply with OGC specifications.

sources, and 2) appropriate algorithms that take application level QoS requirements and available resources (considering their respective qualities) as operands to define an optimal instance of a service chain with respect to the higher level user QoS requirements. QoS composition is particularly critical given the potentially large number of resources with which a geo-service chain can be instantiated

- QoS mapping across different architectural levels requires appropriate QoS mapping functions to effectively resolve the many-to-many relationship types that are typical among QoS requirements at different architectural levels into one-to-one relationship types. Appropriate QoS mapping functions need to be designed and deployed in OpenGIS service architectures.

6.2 Grid Technologies

The Grid is a distributed computing technology that is geared towards open sharing of distributed resources in a virtual organization (Foster et al. 2003). The Grid initially targeted applications with high-end performance demands but is increasingly positioned as technology for enterprise computing, business to business (B2B) collaboration, etc. The scope of the Grid is much broader and encompasses access to CPU cycles, data stores, powerful visualization devices and instruments. At the core of the Grid architectures is coordinated resource sharing that ensures that resource owners in the VO have exclusive control of local resources while allowing remote access and usage by members of the VO. Resource scheduling and QoS are therefore basic concerns of the Grid architecture.

The Grid architecture comprises basic protocols, services and toolkits that are a base for interoperability and establishing complex resource sharing relationships in the VO. The architecture is structured in a set of horizontal layers, with each upper layer using services of the lower layer to access and exploit distributed resources. We briefly outline the layers:

- Fabric Layer – is at the low end of the stack. Fabric components implement local resource specific protocols and services necessary to service the higher levels. The Fabric layer at the minimum should implement enquiry mechanisms to allow enquiry on the state of a resource and management mechanisms to provide some control level QoS control.
- Connectivity layer – is atop the Fabric layer and is essentially for reliable and secure communication among collaborating partners in the VO.
- Resource layer – resource layer services and protocols are for controlling and managing the sharing of a single resource.

- Collective layer – the protocols and services on the collective layer manage and control sharing of collections of resources. Services on this layer include; directory, co-allocation, scheduling, and brokering services, etc.
- Application layer – is the last layer in the stack and concerns the applications that operate in the Grid.

QoS-aware driven composition of geo-services in the Grid environment is intrinsically much simpler because QoS is an underlying concern of Grid computing that pervades all the architectural layers Therefore, advantage can be taken of existing QoS functions and services on the Grid architecture for QoS-aware composition of geo-services. In particular, mechanisms and functions on the collective layer are core to QoS-aware composition, discovery and execution of geo-services chains because the mechanisms and functions apply to collections of resources.

7 Summary

In this paper we defined quality of service and presented a QoS model for disaster management in the context of service oriented geo-processing. We defined QoS as comprising of qualities related to the operational behavior of a service and the quality of geographic information that the service delivers. We introduced the notion of a geo-service infrastructure that is a middleware platform that enables QoS-driven composition of geo-services. The geo-service infrastructure enables discovery, chaining and execution of geo-services in compliance with user requirements. It provides the distribution and QoS transparencies necessary for QoS-driven composition of geo-services. The paper illustrated how desired QoS in different phases of disaster management can be supported.

References

Alameh NS (2003) Chaining of geographic information web services. In: IEEE Internet Computing 7 (5) pp 22-29

Campbell A, Aurrecoechea C, Hauw (1998) A review of QoS Architectures. Center for Telecommunications Research, New York

Di L (2004) GeoBrain – A web services based geospatial knowledge building system. In: Proceedings of NASA Earth Science Technology Conference. Palo Alto USA

Foster I, Kesselman C, Tuecke S (2003) The Anatomy of the Grid. In: Berman F, Fox CG, Hey GJA (eds) Grid Computing making the global infrastructure a reality. John Wiley & Sons, West Sussex, pp 171-198

Halteren van A (2003) Towards an adaptable QoS aware middleware for distributed objects. Center for telematics and information technology, Enschede

ISO/IEC (1998) Information Technology – Quality of Service: Framework (ISO/IEC 13236). ISO/IEC, Geneva

ISO/TC211 (2002) Geographic information – services. ISO, Geneva

ITU/ISO (1996) Open distributed processing – reference model Part 1 Overview. ITU-T Recommendation X.901

McGovern J, Tyagi S, Stevens EM, Mathew S (2003) Java service oriented architecture. Elservier Science, USA

Montoya LA (2002) Urban disaster management A case study of earthquake risk assessment in Cartago, Costarica. Utrecht University, Utrecht

OGC (2003) Opengis reference model. OGC, Wayland

Onchaga R (2004) Quality of service provisioning in geographic information service architectures. Submitted: International journal of Applied Earth Observation and Geo-Information.

W3C (2004) Web Services Description Language (WSDL) Version 2.0 Part 1: Core Language. http://www.w3.org/TR/wsdl20/ (Accessed 2004)

Web-Based Assessment and Decision Support Technology

Nicole Ostländer[1] and Lars Bernard[2]

[1] Institute for Geoinformatics (IFGI), University of Münster,
Robert-Koch Str. 26-28, 48149 Münster, Germany.
Email: ostland@uni-muenster.de

[2] Institute for Environment and Sustainability, European Commission,
Joint Research Centre (JRC), TP262, 21020 Ispra (VA), Italy.
Email: lars.bernard@jrc.it

Abstract

This paper presents and discusses an approach for web-based assessment and decision support using multi-criteria evaluation methodology to be integrated into an interoperable service infrastructure. The research is imbedded in the EU-funded BALANCE project, which tries to assess possible impact and vulnerability to climate change. The results will be incorporated into an Assessment and Decision Support System (ADSS) for the arctic, which shall raise awareness among stakeholders in the BALANCE study area about possible climate change impacts on their environment and way of live and support decisions concerning possible adaptation and mitigation strategies. The paper elucidates the concept of geoprocessing and service chaining with the help of a use case in the field of reindeer herding.

1 Introduction: What Decision Support and Risk Assessment Have in Common

In modern risk assessment the use of spatial information is essential. It is often needed for the development of scenarios in precautionary disaster and emergency management as well as for ad hoc decisions in emergency situations. While in the first case knowledge discovery plays an important role (i.e. assessing new threats), the latter case is normally dependent on up-to-date spatial data and the rapid extraction of significant information. The trend to store information where it is produced has been initiated through the internet and nowadays information needed for a certain application can be stored at machines all over the world. The standardisation of web services for displaying, accessing and processing distributed spatial data offers a series of new opportunities and challenges, and the use of service infrastructures for disaster and emergency management is covered within several research projects (see e.g. ACE GIS (Probst and Lutz 2004) (http://www.acegis.net) and *crossborder* (Riecken, Bernard et al. 2003) (http://www.gdi-nl-nrw.info)).

In this paper, we present and discuss an approach to knowledge discovery using spatial decision support techniques through an interoperable spatial data infrastructure (SDI) based on actual specifications as well as recommendations of the Open Geospatial Consortium (OGC). The overall design of the approach and the planned specifications are to support an ad-hoc, transparent and flexible decision making process through the internet. Thus, the approach can be beneficial for both cases named above: precautionary disaster and emergency management and actual emergencies. The research presented is conceptual, focusing on the possibilities of web enabled spatial decision support using multi-criteria evaluation (MCE) (Malczewski 1999) and the required operations, specifications and workflows (Bernard, Ostländer et al. 2003; Ostländer 2004) (with a prototypical proof-of-concept) rather than on operational systems.

In the following paragraphs, the paper introduces the BALANCE project, and the strategy of impact assessment and vulnerability to climate change in combination with the used multi-criteria evaluation. With the help of a use case the developed SDI and the to-be-specified services and the actual workflow is explained.

2 Impacts of Climate Change in the Arctic: The BALANCE Project

The research described in this paper is undertaken by the Institute for Geoinformatics (IFGI) as a member of the BALANCE consortium. BALANCE (*Global Change Vulnerabilities in the **Ba**rents Region: **L**inking **A**rctic **N**atural Resources, **C**limate Change and **E**conomies*) (www.balance-eu.info) is an EU-funded project whose aim is to assess impacts on the Barent Sea Region's environment and society caused by climate change. A major goal is to assess the vulnerability of the Barents Sea Region to climate change based on a common modelling framework for major environmental and societal components (BALANCE 2002). The area under consideration is presented in figure 1. Although the BALANCE investigations are generally limited to the northernmost parts of Fennoscandia and Russia, the BALANCE modelling approach itself and the integration of the ocean model requires taking the enlarged area into account.

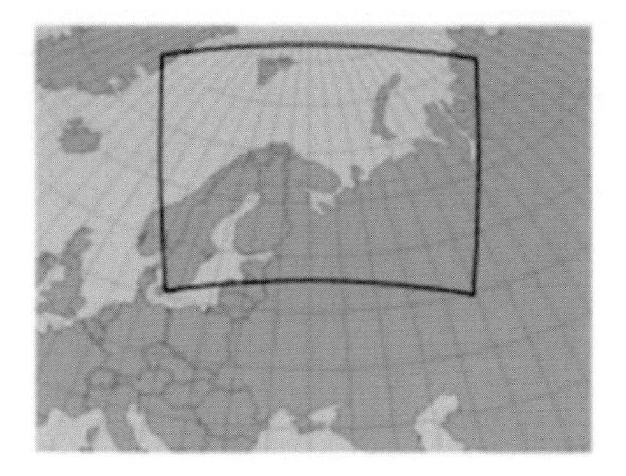

Fig. 1. BALANCE study area

A regional climate model developed by the Max-Planck Institute for Meteorology in Hamburg, with a 6-hour temporal resolution for a period of 140 years and a spatial resolution of ½°, is the driving force of all other terrestrial, marine and socio-economic models. In combination the modelling framework will provide data about the changes that the Barents region will undergo in the next 100 years and identify areas and systems that will be especially vulnerable to climate change.

3 Web-Based Impact and Vulnerability Assessment Using Multi-Criteria Evaluation: The BALANCE ADSS

The Institute for Geoinformatics develops an online Assessment and Decision Support System (ADSS) grounded on the research undertaken and Coverage data produced by the project partners to offer information to the stakeholders of the Barents region about future climate trends and expected changes. The ADSS places emphasis on impact and vulnerability assessment for people active in the local renewable resource industries forestry, fishery and reindeer herding and corresponding policy makers in the different countries. It shall raise awareness about possible climate change impacts on their environment and their way of live and support decisions concerning possible adaptation and mitigation strategies. To achieve this, the ADSS will provide means for a quantified, spatially distributed and interactive impact and vulnerability assessment using multi-criteria evaluation methods and map algebra for use cases defined for the different stakeholder groups.

3.1 What Is Impact Assessment of Climate Change and How to Quantify It?

Impact can be seen as a function of the sensitivity for a certain change and the actual exposure (or risk). The sensitivity is herein seen as the "[..] degree to which a system will respond to a given change in climate, including beneficial and harmful effects" (McCarthy, Canziani et al. 2001), e.g. if a system is exposed to a certain change but is not sensitive to it, then there is no impact, and vice versa. Possible impacts (PI) are e.g. changes in temperature, water regime, and vegetation cover as well as changes in the socio-economic situation like the rate of employment. To be able to show possible impacts of climate change within the ADSS, it is necessary to quantify these factors, which is rather unproblematic when having values on a metric scale like mentioned above, but complications can occur when using nominal or ordinal values.

3.2 What Is Vulnerability to Climate Change and How to Quantify It?

The vulnerability assessment extends the described above static impact assessment. As defined by the *Intergovernmental Panel of Climate Change (IPCC)* it combines the sensitivity of a system with systems' possibility to cope with the resulting change. It is a function of the stated above possible impact and a systems adaptive capacity. Adaptive capacity is the degree ".. to which adjustments in practices, processes, or structures can moderate or offset the potential for damage or take advantage of opportunities created by a given change in climate.." (McCarthy, Canziani et al. 2001). Therefore in the BALANCE ADSS, the quantification of vulnerability (V) will be done by subtracting a systems quantified adaptive capacity (AC) from the quantified possible impact (PI):

$$V = PI - AC$$

This will be done only for plain systems like described in the following use case, which can be parameterised by the user rather than overall vulnerability (see e.g. (Metzger, Leemans et al. 2004)).

3.3 The Use of Multi-Criteria Evaluation

Multi-criteria evaluation (MCE) is a methodology from the decision support domain, where decision alternatives are ranked based on a number of aggregated criteria (attributes) that can influence a decision. Within the ADSS, possible impacts (PI) are used as criteria for the evaluation of given areas (grid cells) concerning climate change impacts and their vulnerability to climate change.

To be able to compare and combine criteria, an initial standardisation of variables is required. Standardisation can be done using common operations such as 'linear scale transformation' and 'score range transformation'. These methods transform the current values of a criterion (e.g. °Kelvin) to a nondimensional scale with values from 0 to 1. The importance of a criterion (generally in comparison with others) can be defined by a criterion weight (Malczewski 1999).

For an impact assessment, the standardised and weighted PI can be combined. A typical MCE method to combine criteria is weighted linear combination (WLC), where the weighted criteria are aggregated into a single map (Malczewski 1999). For vulnerability assessment, the standardised possible impacts and their corresponding adaptive capacities are combined in the formula named above by subtracting the adaptive capacity from the possible impact. Additionally a user defined factor weighting can be applied for the adaptive capacity.

3.4 An Example Use Case: Summer Warming

The ADSS is based on a number of different use cases for the groups named above. As the conditions for the inhabitants of the four selected countries (Sweden, Norway, Finland and north-western part of Russia) can differ widely in terms of economic wealth and technical devices, separate use cases and information pages are specified per country.

The following (simplified) use case has been created for the sectoral interest group "reindeer herders". This profession is mainly taken by indigenous people in the different countries under consideration, e.g. the *Saami* in Sweden and the *Nenet* in north-west Russia. The use case is build from the preliminary results of the BALANCE project partners. The research is still ongoing, for further information see (Keskitalo 2004).

"Vulnerability to rising summer temperature of reindeer herders in Sweden"

Reindeer herders in Sweden are sensitive for summers that are too hot, as they make reindeer suffer and might cause death among the young offspring. The mean July temperature is the chosen indicator for the rising summer temperature, as it appears to be the warmest month in northern Sweden where the reindeer herders have their summer pastures. While a July mean temperature below 20°C (threshold: minimum, PI = 0) is considered as positive, a mean temperature above 26° is considered as intolerable (threshold: maximum, PI = 1). A full grown forest where the reindeer can hide from the heat is beneficial and can counterbalance the problem of the rising summer temperature to a certain extent. Therefore, the availability of forest (which in the short term is dependent on country-specific policies rather than climate change) is considered as an adaptive capacity.

Figure 2 shows the calculations that have to be carried out for each grid cell, when executing this use case performing scale range transformation and vulnerability calculation for the following two Coverages

1. mean July temperature in common grid with attribute values (between 20 and 26) in °C and
2. forest fraction per common temperature grid cell with attribute values from 0 (= no forest cover) to (1 = 100% forest cover)

and with a weighting of 50% for the adaptive capacity.

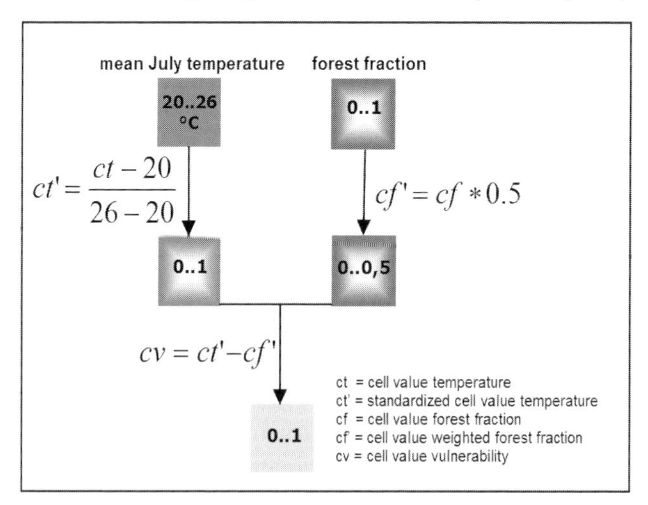

Fig. 2. Calculations performed for use case “summer warming”

4 Service Architecture and Service Chaining

For the BALANCE ADSS a service infrastructure is being developed, that includes information management services and processing services (ISO/TC-211&OGC 2002) and allows MCE as described above.

4.1 Components of Service Architecture

The following web services are part of the ADSS architecture:

- The terrestrial, marine, hydrologic and climatologic modelling results needed for the vulnerability and impact assessment are offered through Web Coverage Service (WCS) interfaces (OGC 2003c).
- The data portrayal is done through the Web Map Service (WMS) interface, which might not necessarily belong to a WMS, but could also be part of a Coverage Portrayal Service (CPS) that fully implements the WMS interface (OGC 2002).
- A Web Map Algebra Service (WMAS) is currently being developed that offers nested algebraic operations on 1..n Coverages requested from 1..n Web Coverage Servers through the WCS interface. The algebraic operations are based on the Map Algebra proposed by (Tomlin 1990). The functions shall include automatic resampling and interpolation.
- A Statistical Information Service will be developed. It shall produce a statistic report describing the result of a specific getCoverage request. Examples are the minimum and maximum values, standard deviation, mean and weighted mean of an attribute. This information is important for attribute standardization and user information and can not be requested through the current WCS interface.
- Finally a thick client will be developed for service chaining and user interaction. The client can request the statistical calculation service, interpret the result and generate a corresponding request for the WMAS and a request for a WMS. Furthermore it will guide the user through the service chain encoded in the client.

The two processing services will implement SOAP (Simple Object Access Protocol) messaging and have WSDL (Web Service Description Language) descriptions to allow a subsequent integration into executable service chains and UDDI (Universal Description Discovery and Integration) registration. The WMAS additionally implements a WCS interface for the following reason: In the current conceptualisation, the WMAS can store the result Coverage and return a reference (ID) instead. This reference will be used by the thick client to request and display the result Coverage using a Coverage Portrayal Service or a Web Map Service. This procedure allows service chaining in a loosely coupled web service infrastructure and might furthermore be beneficial in case of long processing times.

4.2 Service chaining

To explain the service chaining, the use case "summer warming" is run and chained exemplary in the following 3 steps (see also figure 3).

- Step 1 (statistical information about area of interest): The user starts the ADSS client and chooses the predefined use case "summer warming". For this use case two Coverages are available, the mean July temperature and the forest Coverage. The user defines an area of interest with a Bounding Box (BBOX) and the client requests statistical information for the two Coverages. The client interprets and presents the service's response to the user.
- Step 2 (geoprocessing): According to the presented information, the user accepts or changes the predefined temperature threshold values (min 20° and max 26°) and sets the adaptive capacity of the forest criterion to 50%. The client sends a request to the WMAS, which defines an algebraic expression to be solved using the two getCoverage requests. The WMAS requests the mean July temperature with the range 20 to 26°C and the forest Coverage and executes the algebraic expression. The service returns a reference ID to the client that can be used to request the processed image with a getCoverage request.
- Step 3 (portrayal): The client requests the Coverage through the WMS interface using a CPS (OGC 2002). The user can navigate through the resulting vulnerability map and restart the geoprocessing with different settings if needed. For an overview over the whole service architecture and chain, see figure 3.

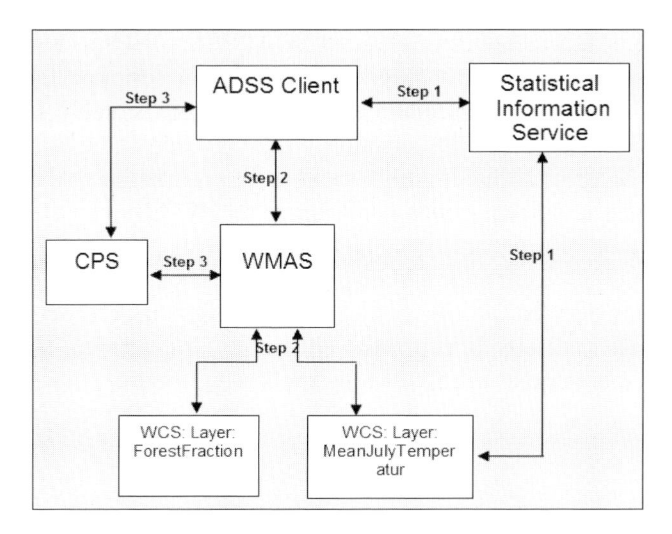

Fig. 3. ADSS architecture and use case service chain

5 Conclusions and Future Work

It is expected that the findings of the GI research undertaken in Balance can be transferred to the field of emergency and disaster management, as the concept of MCE and map algebra can be used for e.g. assessing the risk for earthquakes (Rashed and Weeks 2003) or landslides, and help fire fighters to detect hot spots. The proposed concepts can therefore be seen as beneficial for emergency and disaster management. Additionally the BALANCE results may serve as input to the upcoming INSPIRE implementing rules for network services (EC 2004). Especially the drafting of implementing rules for the mentioned *INSPIRE transformation services* and *INSPIRE invoke "spatial services" services* may benefit from the BALANCE work.

To allow service chaining, that is independent from specific (thick) client applications and not restricted to user-defined transparent chaining, future work will consider the description of service chains (e.g. by using the Business Process Execution Language for Web Services, BPEL4WS) that can be executed with an appropriate workflow management service (e.g. the Oracle BPEL Process Manager[1]). This requires a WSDL description of all involved web services. Approaches towards descriptions of existing OGC Web Services undertaken taken by OGC (see e.g. (OGC 2003a; OGC 2003b)) may serve as a starting point. Moreover, future work needs to consider the portrayal of field information encoded as Coverages provided by a WCS and the preparation of legends. The planned prototypical implementation of the described geoprocessing services and the testing of an appropriate service chain for the described use case will help to prove the defined concepts as well as it will help to identify the future research needs.

Acknowledgements

This work is carried out with support of the European Union in the BALANCE project [EVK2-CT-2002-00169]. Further thanks go to Carina Keskitalo from University of Lapland and Florian Stammler und Fiona Danks from Scott Polar Research Institute, University of Cambridge for professional advice on the use case.

---

[1] Available from http://www.oracle.com/technology/products/ias/bpel.

References

BALANCE (2002) Global Change Vulnerabilities in the Barents Region: Linking Arctic Natural Resources, Climate Change and Economies. Description of Work. EVK2-CT-2002-00169

Bernard L, Ostländer N et al (2003) Impact Assessment for the Barent Sea Region: A Geodata Infrastructure Approach. 6th AGILE Conference April 24-26 in Lyon, France, pp 653-661

EC (2004) Proposal for a DIRECTIVE of the European Parliament and of the Council: establishing an infrastructure for spatial information in the Community (INSPIRE). Brussels, Council of the European Union, 23.7.2004

ISO/TC-211&OGC (2002) Geographic information Services Draft ISO/DIS 19119, International Organization for Standardization & OpenGIS Consortium

Keskitalo ECH (2004) Vulnerability and Adaptive Capacity in Forestry, Fishing and Reindeer-Herding Systems in Northern Europe. The ACIA International Scientific Symposium on Climate Change in the Arctic: Extended Abstracts, Reykjavik, Iceland, 9 - 12 November 2004: Oral Session 8: Paper 4 (Pages 1 to 5)

Malczewski J (1999) GIS and Multicriteria Decision Analysis. New York, John Wiley & Sons

McCarthy JJ, Canziani OF et al (eds.) (2001) Climate Change 2001: Impacts, Adaptations, and Vulnerability. Contribution of Working Group II to the Third Assessment Report of the Intergovernmental Panel on Climate Change. IPCC Third Assessment Report - Climate Change 2001. Cambridge, Cambridge University Press

Metzger MJ, Leemans R et al (2004) A multidisciplinary multi-scale framework for assessing vulnerability to global change. Millennium Ecosystem Assessment conference: Bridging Scales and Epistemologies, 17-20 March 2004, Alexandria, Egypt

OGC (2002) OWS1 Coverage Portrayal Service, OGC (Open Geospatial Consortium Inc.)

OGC (2003a) OWS 1.2 SOAP Experiment Report, OGC (Open GIS Consortium Inc.)

OGC (2003b) OWS 1.2 UDDI Experiment, OGC (Open GIS Consortium Inc.)

OGC (2003c) Web Coverage Service, Version 1.0.0, OGC (Open GIS Consortium Inc.)

Ostländer N (2004) Interoperable services for web-based spatial decision support. 7th AGILE Conference, Crete (Greece)

Probst F, Lutz M (2004) Giving Meaning to GI Web Service Descriptions. 2nd International Workshop on Web Services: Modeling, Architecture and Infrastructure (WSMAI-2004), Porto, Portugal, INSTICC Press

Rashed T, Weeks J (2003) "Assessing vulnerability to earthquake hazards through spatial multicriteria analysis of urban areas." International Journal of Geographical Information Science 17(6), pp 547-576

Riecken J, Bernard L et al (2003) North-Rhine Westphalia: Building a Regional SDI in a Cross-Border Environment / Ad-Hoc Integration of SDIs: Lessons learnt. 9th EC-GI \& GIS Workshop ESDI

Tomlin CD (1990) Geographic information systems and cartographic modelling. New Jersey, Prentice Hall.

Evaluating the Relevance of Spatial Data in Time Critical Situations

Hardy Pundt

University of Applied Studies and Research, Friedrichstr. 57-59, 38855 Wernigerode, Germany.
Email: hpundt@hs-harz.de

Abstract

Accuracy and relevance are properties that are often mentioned in context with usability and quality of data. The use of GIS and mobile GI services in spatial planning and management requires *accurate* and *relevant* data. For usage of spatial data in time critical situations, e. g. disaster management, this is especially true. Decisions must be made in short terms, within hours or minutes. Such situations do not leave any time to evaluate whether the data used for decision making are "accurate" or "relevant". This evaluation has to be done before. The use of inaccurate or irrelevant data, however, can lead to decisions that are inadequate or in the worst case harmful.

Increasingly, spatial data are provided via the Internet, and through wireless connections. This is potentially a great chance for GI services in general, but especially for time critical, spatial decision making. An important question that arises is how accuracy and relevance of data can be evaluated, a difficult task taking into account that data come from various sources, are distributed over several servers, and provided by different data producers. Such an evaluation is important because users must have confidence in the data. They must be sure that they use information that is reliable and adequate in a given situation. This requires not only a syntactical control, but especially an understanding of the semantic content of information.

The paper pursues these thoughts by investigating methods and techniques to search for specific spatial data sets and simultaneously evaluating the *relevance* of data. "Fitness for use" can be examined - to a certain extent - using metadata. But metadata have shortcomings as they describe a fixed number of properties not taking into account adequately data semantics. Ontologies - an approach propagated currently within the framework of the semantic web (Fensel 2004) - are aimed at the identification of *relevant* data sets.

1 Introduction

It is a special property of geographic information models that these differ, even if they represent the same objects. Users of geographic information tend to act in contradiction to other scientists. The latter try to develop "common" models of the reality that are unique and accepted in general. Providers of geographic data specify fairly different models for same objects depending on their notion and with regards to their specific application, point of view and understanding of the reality (Giger and Najar 2003). Taking into account this fact, accuracy and relevance are terms that are not assessable in general: the same object, represented in different ways, is possibly relevant for one application, but not for the other. Accuracy and relevance, however, depend on a specific view of the reality that is consistent *within one* geospatial information community. This kind of "subjectivity" has to be considered properly when evaluating accuracy and relevance of spatial data sets.

Accuracy has been chosen to be one of the data quality parameters that are, for example, part of the metadata standard defined by ISO 19115. The ISO standard mentions that consistent methods of reporting data quality will not be enough to assure consistent evaluation of data set quality, because the quality information reported for a geographic data set will also depend on a consistent application of standardized methods for measuring the quality of geographic information (ISO TC211 2004). This means that such measuring methods are required - but which method to measure "accuracy" or "relevance" of data is available and adequate? Another aspect is mentioned by ISO: The results of one method of measuring quality may not be readily comparable to another although each is valid (ISO TC 211 2004). This points to the fact that data can be relevant for one application, but lack relevance for another. Concerning the measurability of relevance exists obviously a gap. This paper aims not at proposing such a measuring method, but to discuss some steps for a pathway to deal with the problem.

Accuracy shows different facets, such as spatial, thematic, and semantic accuracy (Guptill and Morisson 1995). Relevance of data concerns especially semantic accuracy. Many metadata standards define specific properties of data in a manner that enable users to evaluate fitness for use. Quality parameters are important, but are useless if they cannot be interpreted correctly, which means in direct relationship to semantics.

For disaster management this means that data must meet conditions that are specific for a (*one*) disaster situation. The crux is that in time critical disaster situations no time is left for users to look on data quality properties, or metadata. When spatial information is needed within a few minutes to save lives or to avoid large damages, the data must be *available in time,* and users must be sure that they can *trust* the data.

The research question that arises is how an *automatic evaluation of relevance* of data can help in time critical situations to increase the reliability of data. This makes formal ontologies an interesting approach to support data and object identification. The usefulness of domain ontologies for mobile GIS has been shown in (Pundt and Bishr 2002), and the aspect of automatically controlling relevance is an add on that is important for disaster management in particular.

2 Heterogeneity and Relevance

From an etymological point of view, relevance is a term used to describe the correctness of a theory or even the ability of a theory to explain specific things or processes. If there is information that supports the explanation of things or processes, this information is considered as *relevant.* This means that relevance underlies objective, but also subjective criteria. If something is relevant, or not, is in many cases dependent from the person that explains and interprets a thing or process. This makes apparent that relevance is a term that only makes sense within a specific context. The same information might be relevant in one context, but useless, or *irrelevant*, in another. To conclude, the evaluation of an information concerning its *relevance* requires

1. to define the context in which the information is used, and
2. to evaluate the relevance of information (only) within this context

Spatial data sets are collected by geospatial information communities (GICs). They collect data for specific purposes, specific to the tasks that are carried out by such GICs. The GICs define real world objects using classes, entities, attributes, properties, relationships, uses, etc.. They describe these aspects using their specific terminologies. This means that the database objects reflect specific semantics, not necessarily clear and understandable to users outside that information community. We face the well known problem of semantic heterogeneity that occurs when there is a disagreement about the meaning, interpretation or intended use of data (Xu and Lee 2002). Semantic heterogeneity has been classified in previous years into schematic, syntactic, and semantic heterogeneity. Schematic heterogeneity refers to the categorization of real world features in a certain context; here, features are represented as classes which have attribute structure and are arranged in hierarchies, in other terms schemata. Features are captured and represented as objects with thematic and geometric descriptors, which is the syntactic aspect. When such features are used, they attempt to relate database objects to the context world view of a specific geospatial information community, which is the semantic component (Bishr 1997, Pundt and Bishr 2002). The semantic heterogeneity has various facets that are summarized in figure 1.

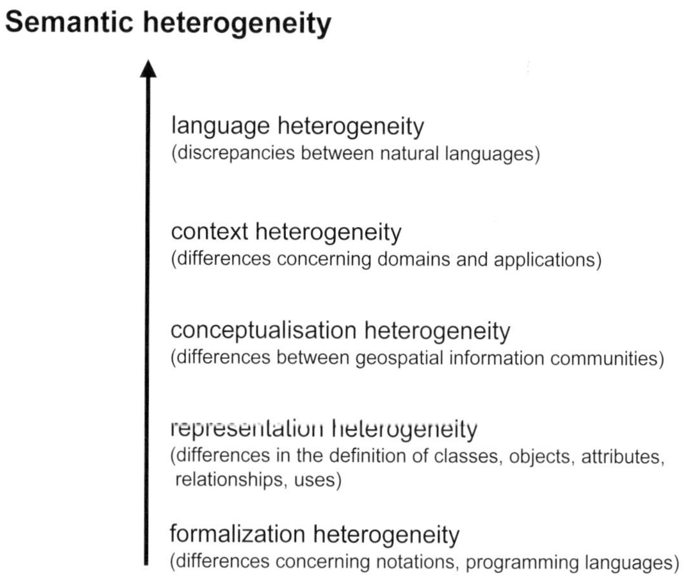

Semantic heterogeneity

language heterogeneity
(discrepancies between natural languages)

context heterogeneity
(differences concerning domains and applications)

conceptualisation heterogeneity
(differences between geospatial information communities)

representation heterogeneity
(differences in the definition of classes, objects, attributes, relationships, uses)

formalization heterogeneity
(differences concerning notations, programming languages)

Fig. 1. Facets of semantic heterogeneity

Figure 2 shows a model of relevance for geospatial data. It enlightens the fact that the same objects can be represented differently. Information about the differences must be included in the data models to make them interpretable.

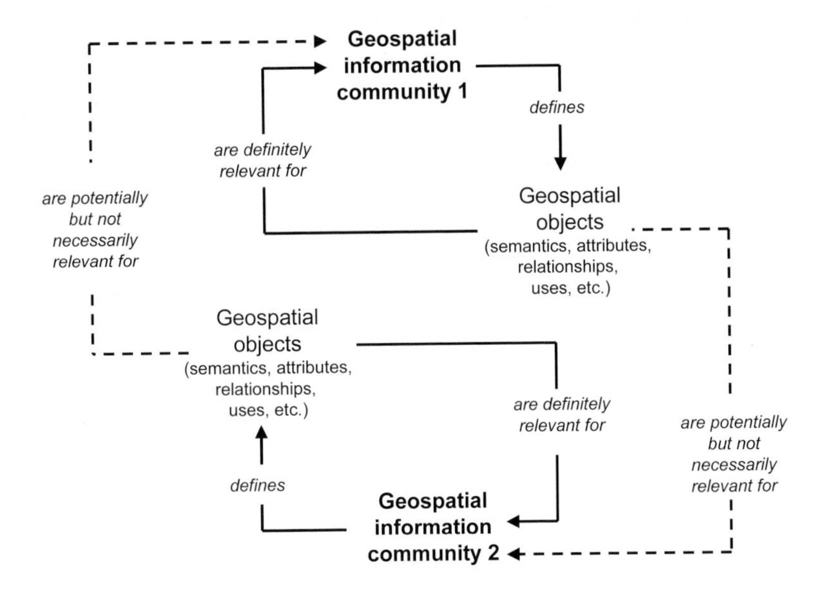

Fig. 2. A model of *relevance* for geospatial information

Interpretation should be carried out by the computer, which is due to the two following aspects:

1. Disasters afford decision making in very narrow time slots. Human interpretation of metadata costs time that is not available in disaster situations.
2. Semantics of data are often not considered adequately within metadata sets because metadata are limited concerning the model that underlies them. But using data without considering semantics adequately can lead to wrong or even harmful decisions.

The conclusion so far is that, if spatial data are used in disaster management, an *automatic support* to identify *relevant* data sets is needed. This requires the inclusion of metadata that are readable and interpretable automatically. Additionally, the consideration of data semantics is required. Ontologies - developed in artificial intelligence to facilitate knowledge sharing and reuse - provide a machine-processable semantics of information sources that can be communicated between different agents, software and humans (Fensel 2004). Both, metadata and ontologies, can be formalized using XML grounded languages to make them machine-processable.

3 Formal Ontologies

Utilizing geospatial data held in different commercial and public sectors is still a hard problem. Experts estimate that there is a growing market potential for the use of geographic data, while at the same time, adequate information about their content and availability is still missing (Greve 2002, cited in Vögele and Spittel 2004). On the one hand, this concerns the availability of metadata for spatial data sets. On the other hand this concerns the increasingly growing interest in ontologies. Being on the way toward the semantic web, ontologies are seen in a very pragmatic manner: converting machine-readable into machine-understandable information by providing well-defined meaning for the content distributed within the WWW, which is the main goal of an ontology (Vögele and Spittel 2004). If computers become able to *understand* information, they should also be able to *evaluate*, to a certain extent, the relevance of the data in a given disaster situation.

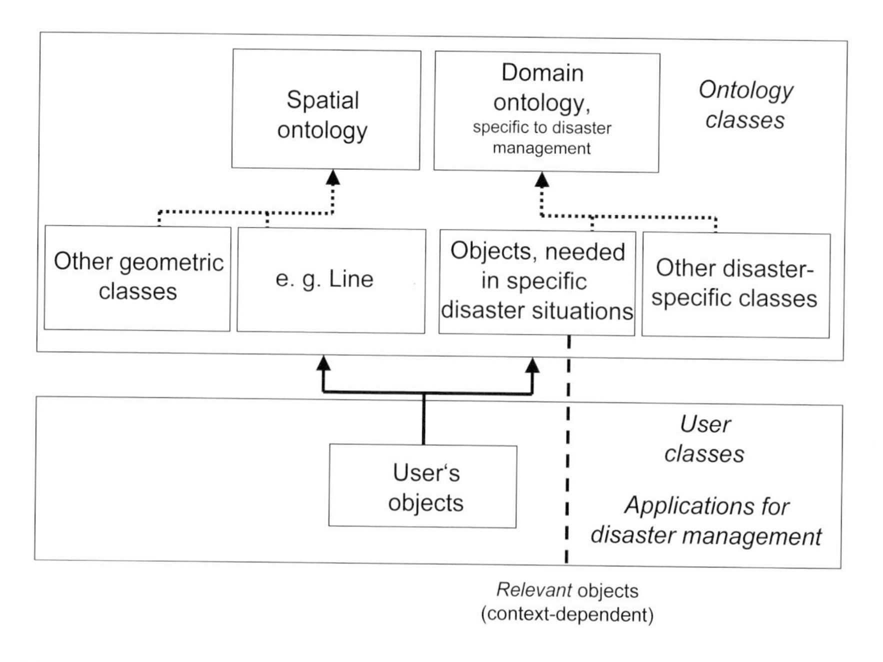

Fig. 3. Categorization of ontologies and *relevant* objects (Basic model from Fonseca, Egenhofer et al (2002), modified)

Taking into account semantic heterogeneity as described briefly before, ontologies act as an interface between different domains (Visser et al

2002). This requires a communication process and a resulting consensus between different actors. These must develop common, in this case disaster management specific, ontologies. They can be based on existing ontologies of certain information communities, or existing object catalogues. In numerous cases completely new ontologies must be designed. In all such cases, ontologies represent a shared and common understanding of *some domain* that can be communicated between people and applications (Broekstra et al 2000, Fensel 2004). Such a common understanding requires semantically enriched data models that enable users to assess data's fitness for use in a *comprehensive* manner (Pundt 2002). They help to identify only those spatial data sets that are *usable* in a specific context. Such ontologies help to avoid situations that we face at daily work again and again: Search engines, as powerful as they are, return too often too large or inadequate lists of hits. What is needed is machine-processable information that can point a search engine to the relevant pages, or data sets, and can thus improve both, precision and recall (Berendt et al 2002). For such purposes modeling techniques are required that are superior to those that are used in general, such as entity-relationship- (Chen 1976) or object-oriented modeling (Rumbaugh et al 1998). These considerations must necessarily lead to the hypothesis that ontologies are *the* means that support the access to and sharing of spatial information, including the evaluation of relevance.

To make ontologies usable within a GIS or modeling framework, a proper design and formalization of ontologies is required. The design phase includes a decision to which category an ontology belongs. Fonseca, Egenhofer et al (2002) provide a categorization of ontologies that has been adapted for the case of disaster management. Figure 3 shows that there is not only one general ontology, but various, and that the ontology belongs to a category that describes its universality or degree of relation to a specific domain. A general ontology of geometric objects describes the objects concerning their spatial extent and geometric form. A specific domain ontology describes the objects including their semantics, the latter defined in a language of a geospatial information community. It is such a domain ontology that gives the objects a meaning within a specific context.

Currently, XML-based languages are used in many different application areas to design ontologies. The World Wide Web Consortium (W3C) proposed these languages to provide formalized specifications of conceptual models (it was Gruber (1993) who defined an ontology as a "specification of a conceptualization"). The RDF (Resource Description Framework) and OWL (Web Ontology Language) are under development at the W3C. Both, RDF and OWL (as well as DAML + OIL which is, in a certain sense, a stage between the two others) were already used for ontology generation in the spatial domain (see, for example, Pundt and Bishr 2002, Hart et al 2004, or Redbrake and Raubal 2004).

At this stage it is clear that it is not intended to argue for the development of "an (*one*) ontology for disaster management". As mentioned within the discussion of figures 2 and 3, this would not represent adequately the various *different* possibilities to model one and the same spatial object, or class of objects. In other terms: no ontology will eliminate semantic heterogeneity.

But ontologies can help to overcome the problems that occur due to semantic heterogeneity. The only way to support information access and sharing is to make data sets understandable for humans, as well as computers. This goal is supported via formal ontologies. In future an increasing number of ontologies will appear, especially *domain ontologies* that capture the knowledge within a particular domain (e. g. electronic, medical, mechanic, traffic, urban and landscape planning, or disaster management).

4 Conclusions and Next Steps

The aim of the paper is to focus the attention on the relevance and therefore the reliability of spatial data used in disaster management. It discusses briefly a currently popular approach to model data for usage via networks, formal ontologies. Such ontologies support intelligent data discovery, together with metadata, and enable data providers to model spatial data, their properties, quality parameters, relationships, and potential uses (Wilde and Pundt 2004). XML grounded and W3C conformant standard languages such as the Resource Description framework (RDF) and the Web Ontology Language (OWL) enable data providers (specific GICs) to formalize ontologies for their domains.

The next step to be done must bridge theory and practice. This would include the development of concrete domain ontologies for disaster management. Domain ontology, however, would mean a specific domain within disaster management. Floods, earthquakes, hurricanes, and other disasters are examples, for which specific ontologies are required. Up to now it is not clear, if such disaster categories are specific enough. Perhaps, more concrete and detailed ontologies are needed to represent different flooding or earthquake types, for instance. An unanswered question is, whether such ontologies are usable in general, or if regional and local spatial properties must be included. This would even underline that not one, but various ontologies are required for the numerous and very different potential disaster situations. The availability of semantic reference systems will play an important role within this framework (Kuhn and Raubal

2003). This requires much work on ontologies by data providers, but this would enhance extremely the reliability of data used. It would improve qualitatively the decisions that are based on such data, and support the transparency of decisions for members of different information communities. The use of standardized languages would contribute to interoperability and would mean that not only data, but also ontologies can be "reused". The current activities to realize the semantic web within the W3C and the computer science community (Fensel 2004, Mc Guiness and Van Harmelen 2004, W3C 2004) along with results from this first international symposium on Geoinformation for Disaster Management (Gi4DM) should motivate to deal more concretely with ontology based data modeling. Tests must focus on ontology-based data set identification, simultaneously accompanied by the evaluation of relevance of data, in disaster situations. A test scenario could include two or more spatial data sets that contain data potentially usable within a disaster simulation framework. GIS-based simulation models require data that must be available in time when a decision in a disaster situation has to be made. Under defined test conditions the simulation results will give significance whether the underlying data were usable, or not. If the results are transparent and support significantly a fast and reliable decision, the ontology-based data identification was successful, because only relevant data were delivered for the specific task. Such scenario-based tests haven't been carried out until recently, however, the methods and techniques are there to start with them.

References

Berendt B, Hotho A, Stumme G (2002) Towards Semantic Web Mining. In: Horrocks, I., Hendler, J. (eds), The semantic web - ISWC 2002. Springer, Berlin, New York, pp 264-278

Bishr Y (1997) Semantic aspects of interoperable GIS. ITC Publication no56, International Institute for Aerospace Survey and Earth Sciences, Enschede, the Netherlands

Broekstra J, Klein M, Decker S, Fensel D, Van Harmelen F, Horrocks I (2000) Enabling Knowledge Representation on the Web by extending RDF Schema, AIDministrator, The Netherlands, 20 p

Chen PPS (1976) The Entity Relationship Model – Toward a Unified View of Data. Association of Computing Machinery Transactions on Database Systems 12 (1), pp 9-36

Fensel D (2004) Ontologies: A Silver Bullet for Knowledge Management and Electronic Commerce. 2nd edition, Springer, Berlin, New York

Fonseca F, Egenhofer M, Clodoveu A, Borges K (2002) Ontologies and Knowledge Sharing in Urban GIS. Computer, Environment and Urban Systems

Giger Ch, Najar Ch (2003) Ontology-based integration of data and metadata. In: Gould M, Laurini R, Coulondre St (eds) Proceedings of the 6th AGILE conference on Geographic Information Science. Presses polytechniques et universitaires romandes, Lausanne, pp 586-594

Gruber Th (1993) A translation approach to portable ontologies. Knowledge Acquisition 5 (2), pp 199-220

Guptill SC, Morrison JL (eds) (1995) Elements of spatial data quality, Elsevier Science Ltd., BPC, Wheatons Ltd, Exeter, United Kingdom

Hart G, Temple S, Mizen H (2004) Tales of the River Bank, First Thoughts in the Development of a Topographic Ontology. In: Toppen, F., Prastacos, P. (eds), Proceedings of the 7th Conference on Geographic Information Science, Crete University Press, pp 169-178

ISO TC211 (2004) http://www.isotc211.org (page accessed on 2004-12-01)

Kuhn W, Raubal M (2003) Implementing Semantic Reference Systems. In: Gould, M, Laurini R, Coulondre S (eds) Proceedings of the 6th AGILE Conference on Geographic Information Science, Presses Polytechniques et Universtitaires Romandes, Lausannc, pp 63-72

McGuiness DL, Van Harmelen F (2004) OWL Web Ontology Language Overview. W3C Recommendation February 2004 http://www.w3.org/TR/2004/REC-owl-features-20040210/

Pundt H, Bishr Y (2002) Domain ontologies for data sharing - an example from environmental monitoring using field GIS. Computer & Geosciences, vol. 28, no 1, pp 95-102

Pundt H (2002) Field Data Acquisition with Mobile GIS: Dependencies Between Data Quality and Semantics. GeoInformatica, vol 6, no 4, pp 363-380

Redbrake D, Raubal M (2004) Ontology-Driven Wrappers for Navigation Services. In: Toppen F, Prastacos P (eds), Proceedings of the 7th AGILE Conference on Geographic Information Science, Crete University Press, pp 195-205

Rumbaugh J, Blaha M, Premerlani W, Eddy F, Lorensen W (1991) Object-Oriented Modeling and Design. Prentice Hall, Englewood Cliffs, NJ

Visscr U, Stuckenschmidt H, Schuster G, Vögele Th. (2002) Ontologies for geographic information processing. Computer & Geosciences, vol. 28, no. 1, pp 103-117

Vögele Th, Spittel R. (2004) Enhancing Spatial Data Infrastructures with Semantic Web Technologies. In: Toppen F, Prastacos P (eds) Proceedings of the 7th AGILE Conference on Geographic Information Science. Crete University Press, Heraklion, Greece, pp 105-111

Xu Z, Lee YC (2002) Semantic heterogeneity of geodata. Proceedings of the symposium on geospatial theory, processing and applications, Ottawa, Canada, 2002. http://www.isprs.org/commission4/proceedings/ paper.html#3 (page accessed 2004-11-28)

Wilde M, Pundt H (2004) Development of an ISO-compliant, internet-based metadata editor for the EU project MEDIS. In: Strobl J, Blaschke T, Griesebner G (eds) Angewandte Geographische Informationsverarbeitung 2004. Wichmann Verlag, Heidelberg (Germany), pp 782-787

W3C (2004) http://www.w3c.org/ (page accessed 2004-12-13)

Dealing with Uncertainty in the Real-Time Knowledge Discovery Process

Monica Wachowicz[1] and Gary J. Hunter[2]

[1] Centre for Geo-Information, Wageningen UR, PO Box 339, 6700 AH Wageningen, the Netherlands.
Email: monica.wachowicz@wur.nl

[2] Department of Geomatics, University of Melbourne, Parkville VIC 3010, Australia.
Email: garyh@unimelb.edu.au

Abstract

This paper will examine where uncertainty may lie in the knowledge discovery process through the use of case studies in disaster management, in turn leading to a discussion of what future action is required to address the uncertainty that may lie within knowledge obtained through these techniques. We describe our approach to address three types of issues: accuracy, efficiency, and usability. Typically, data mining techniques have higher false positive rates than traditional data exploratory methods, making them unusable in real-time systems. Also, these techniques tend to be inefficient (that is, computationally expensive) during the steps of a knowledge discovery process, particularly during training and evaluation. This prevents them from being able to process data and detect anomalies, hot-spots, or patterns in real-time applications. Finally, disaster management applications require large amount of training data and are significantly more complex than traditional GIS applications. These problems are inherent in developing and deploying any real-time data mining based system, and although there are trade-offs between these three groups of issues, each can generally be handled separately. The paper concludes by presenting the key design elements for supporting a real-time knowledge discovery process and group them into which general issues they address.

1 Introduction

There is now a growing trend towards the application of knowledge discovery processes (KDD) to scientific data, including real-time data mining of large volumes of data sources. While these processes have traditionally been used for intrusion detection applications, in the past few years this situation has started to radically change. For instance, current examples of their use in disaster management include:

- Spatial data (find spatial features such as eddies and fronts to detect cyclones and fires);
- Temporal data (find temporal anomalies to detect *a priori* unknown events such as road collisions); and
- Spatio-temporal data (the most complex area in KDD - examples would be disease correlation between SST/AVHRR data, and association rules in similar catastrophes).

At the same time, however, there is increasing concern about the uncertainties that may be inherent in these techniques and their effects upon the ultimate knowledge gained. There are in fact several steps associated with knowledge discovery, ranging from assembling the raw data, preprocessing it, transforming the data, recognizing anomalies or patterns, and finally interpreting the results. Clearly, uncertainty has the potential to be present at any of these stages, just as it can reside in any of the various stages of applying spatial data for decision-making in a disaster management situation.

While there is a well-established research agenda for the treatment of uncertainty in spatial data, there does not appear to be a similar, comprehensive approach to dealing with uncertainty in all stages of the knowledge discovery process. Certainly, some researchers are studying uncertainty in individual stages, such as in data mining (that is, in pattern recognition), but there does not appear to be a unified approach to dealing with uncertainty from start to finish (that is, from turning raw data into information and then to knowledge).

These difficulties can be grouped into three general categories: accuracy, efficiency, and usability. Typically, data mining techniques have higher false positive rates than traditional data exploratory methods, making them unusable in real-time systems. Also, these techniques tend to be inefficient (that is, computationally expensive) during the steps of a knowledge discovery process, particularly during training and evaluation. This prevents them from being able to process data and detect anomalies, hot-spots, or patterns in real-time applications. Finally, disaster management applications require large amount of training data and are significantly more complex than traditional GIS applications. These issues must be addressed in order to be able to deal with uncertainty in the a real-time knowledge discovery process.

In this paper we discuss these issues inherent in dealing with uncertainty when developing real-time data mining systems, and present an overview of our research, which addresses these issues. These issues are independent of the actual data mining techniques and models used, and must be overcome in order to implement a knowledge discovery process for disaster management applications, which require large amount of training data and are significantly more complex than traditional GIS applications.

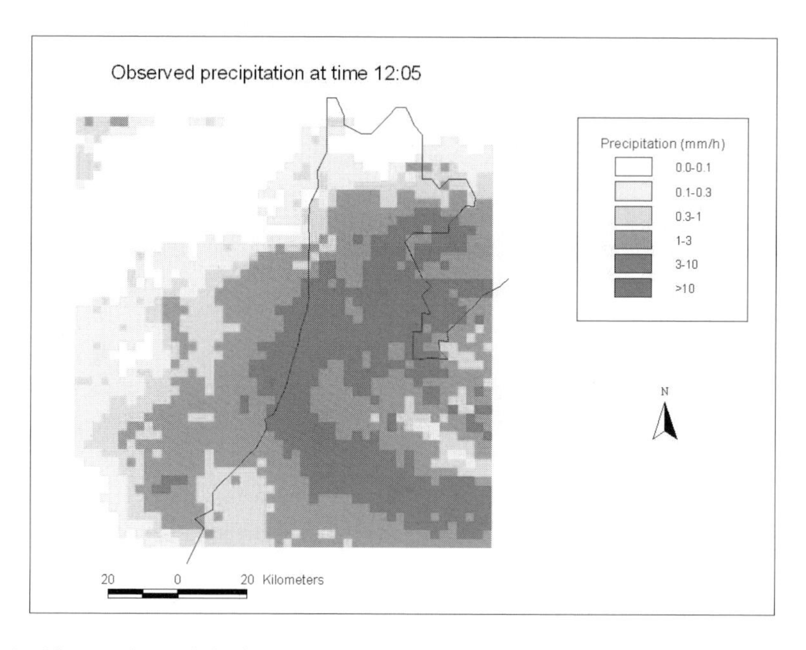

Fig. 1. Observed precipitation at 12:05 on 19 September 2001 (source: KNMI, 2004)

Supervised inductive learning method (Rao and Raao 1993) has been used for both anomaly detection and misuse detection at the same time. A single model was used for training a set of data that contains both normal records and records corresponding to floods. The learned model can detect anomalies and misused rates concurrently. In order to evaluate this combined approach, observations have been gathered into a number of small clusters. This was carried out to simulate the real world process of developing and discovering new anomalies and incorporating them into the training set. The results contain the misuse rules for the observations that are known in the training data, anomaly detection rules for unknown observations in left-out clusters, and rules that characterize normal behaviour. Figure 2 illustrates some of the results.

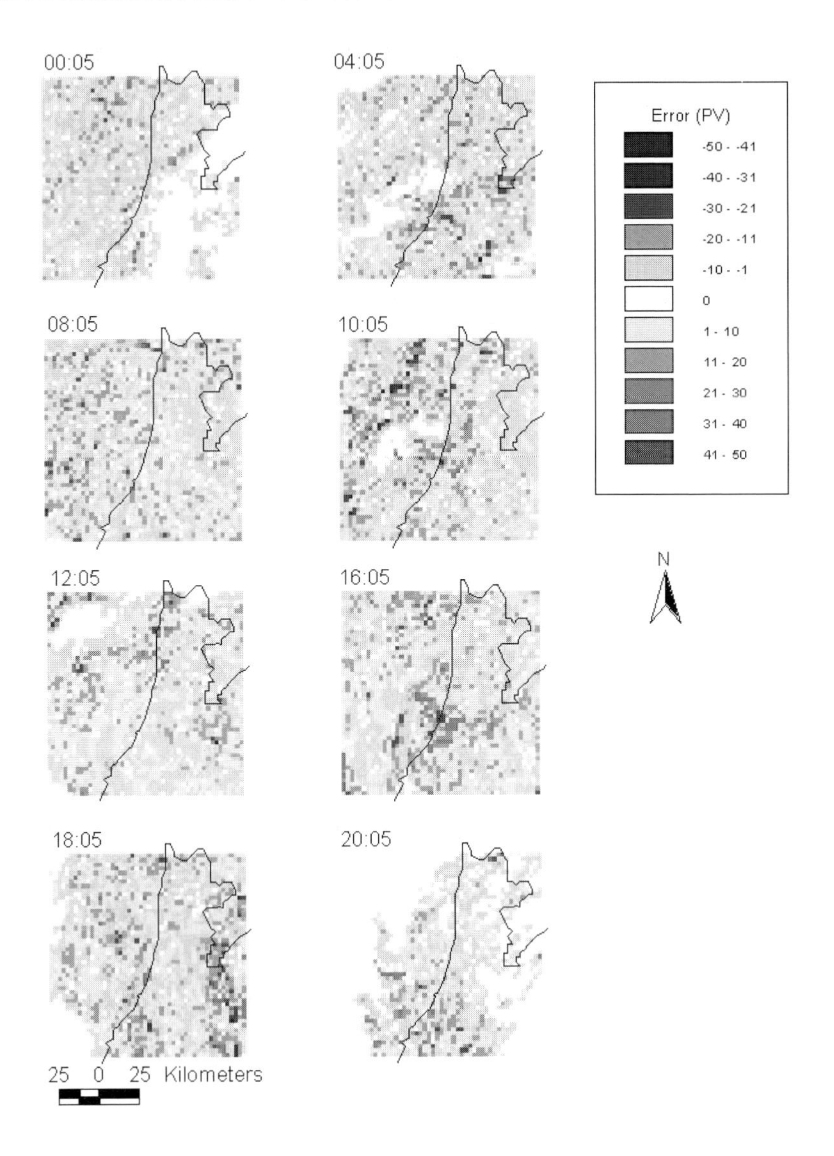

Fig. 2. Error maps artificial neural networks trained 00:00h-00:15h

2 Efficiency

In current applications of data mining to disaster management applications, techniques are performed off-line because the difficulties of learning algorithms in processing tremendous amounts of environmental data. There is a clear need to respond to new data analysis challenges posed by the overwhelming volume and high resolution data sets generated today, with remotely sensed data from Earth Observation Systems alone projected to yield 1 terabyte per day, and therefore, more than can be analysed by conventional means. Besides, the environmental data sets are showing a high variability in data formats, scale and content. They are also getting more complex, due to, in part, to the high dimensionality of these data. Effective KDD processes should happen in real-time, as anomalies take place, to improve policy decision, strategic planning, and research on global climate change, natural hazards, and land degradation, among others.

In contrast to off-line KDD processes, a key objective of real-time KDD processes is to detect anomalies, hot-spots, or patterns as early as possible. Therefore the efficiency of the detection model is very important to be taken into account. Data mining techniques are performed using off-line data, and they implicitly assume that when an event is being discovered all observations related to the event have been gathered in a database. As a consequence, if we use these models in real time without any modification, the event will be only detected once complete information about this event has been recorded and been processed, and all features (e.g. the temporal and spatial statistics features) are computed. Moreover, the volume of an event string is very high, making the detection of patterns severely delayed.

From the perspective of cost analysis, the efficiency of a KDD process is its *computational cost*, which is the sum of the time delay of the features detected by system. Based on the feature detection approaches discussed in the previous section we can categorize features used within a KDD process into 4 cost levels:

- Level 1 features can be computed from the first observation (e.g. *source*);
- Level 2 features can be computed at any point during the capture of the observations (e.g. *state*);
- Level 3 features can be computed at the end of the capture of the observations, using only information about the connection being examined (e.g. the *total number of bytes sent from source to destination);*

- Level 4 features can be computed at the end of the data capture, but require access to data of potentially many other prior observations. These are the temporal and spatial features and are the most costly to compute.

In practice to conveniently estimate the cost of a rule, a cost can be assigned as being 1 to the level 1 features, 5 to the level 2 features, 10 to level 3, and 100 to level 4. These cost assignments are very close to the actual measurements obtained via extensive real-time experiments. However, we have found that the cost of computing level 4 features is linearly dependent on the amount of observations being gathered by a KDD process within the time window used for computation, as they require iteration of the complete set of recent observations.

In the domain of disaster management applications, these four different levels of costs can be used to compute features, as discussed in the previous section. Features of costs 1, 5, and 10 can be computed individually and features of costs 100 can be computed in a single lookup of all the observations in the past. With the above costs and goals in mind, the following multiple rule set approach can be used to generate:

- multiple training sets $T_{1\text{-}4}$ using different feature subsets. T_1 uses only cost 1 features. T_2 uses features of costs 1 and 5, and so forth, up to T_4, which uses all available features;
- rule sets $R_{1\text{-}4}$ that are learned using their respective training sets;
- a precision measurement p_i that is computed for every rule, except for the rules in R_4;
- a threshold value that is obtained for every single class which determines the tolerable precision required in order for a classification to be made by any rule set, except for R_4.

In real-time execution, the feature computation and rule evaluation is proceed by computing all cost 1 features used in R_1 for the observation being examined. After that R_1 is evaluated and a prediction p_i is made. This computation continues for all rules, until a final prediction is made.

In our experiments, the same data sets from the previous section were used for the evaluation. In summary, this approach can reduce the computational cost by as much as 80% without compromising predictive accuracy, where the cost for inspecting an observation is the total computational cost of all unique features used before a prediction is made. If multiple features of cost 100 are used, the cost is counted only once since they can all be calculated in a single iteration through the table of recent observations.

3 Usability

A number of possible definitions of usability are available in the literature, and the needs of usability have been compared and contrasted with broader data-related activities of providers and users of geoinformation. For example, one official definition of usability is given by the ISO 9241-11 standard on Display Screen (VDU) Regulations, Use of Ergonomics for Procurement and Design. In this definition, usability comprises " the extent to which a product can be used by specified users to achieve specified goals with effectiveness, efficiency, and satisfaction in a specified context of use, where:

- Effectiveness measures the accuracy and completeness with which users achieve specified goals;
- Efficiency measures the resources expended in relation to the accuracy and completeness with which users achieve goals;
- Satisfaction measures the freedom from discomfort, and positive attitudes towards the use of the product."

Usability elements outline the features and characteristics of the product that influence the learnability, effectiveness, efficiency and satisfaction with which users can achieve specified goals in a particular environment (Hunter et al. 2002). Therefore, the context of use of real-time KDD processes will determine the types of users, tasks, equipment, and the physical and social environments in which a product will be used.

A data mining-based task is significantly more complex than a traditional GIS. The main cause for this is that data mining systems require large sets of data from which to train. The first step of a real-time KDD process will be about find out relevant data that can be mined for particular purposes. Environmental data sets are usually from different sources and they are collected for multi-purpose use, having different spatial and temporal scales, accuracy, and map thematic classes.

Country name	Address Web-page	Phase description[1)]	Year, first Implementation version	Number of data sets[2)]	Number of visitors per month[3)]	Standard	Language[5)]
Europe							
Belgium	http://www.vlm.be/OC/welcome1.htm (Flanders)	Prototype	1998	150	*n.f.*	CEN TC/287	Dutch (English)
Denmark	http://www.daisi.dk/ http://www.geodata-info.dk/	Developing	1997	180	104	CEN TC/287	Danish/ English
Germany	http://www.ddgi.de/ http://www.atkis.de/	Developing	1997	2 657	*n.f.*	CEN TC/287	German/ English
Finland	http://www.nls.fi/ptk/infrastructure/index.html	Initial	-	-	*n.f.*	-	Finnish
France	http://www.cnig.fr/	Developing	1995	105	1 082	NF52000	French
Hungary	http://www.fomi.hu/hunagi/	Initial	-	-	52	-	English/ Hungarian
Ireland	http://www.tcd.ie/Geography/GIS/Geoid/	Developing	1999	237	152	other	English
Italy	http://195.110.158.111/index.html	Initial	-	-	492	-	Italian/ English
Luxembourg	http://www.etat.lu/ACT/acceuil.html	Initial	-	-	*n.f.*	-	French
The Netherlands	http://www.ncgi.nl/	Developing	1995	1 533	1 070	CEN TC/287	Dutch
Austria	http://www.ageo.at/	Initial	-	-	615	-	German
Poland	http://www.wloc.ids.pl/wodgik/sieci/gispol/edzia_g.html	Initial	-	-	69	-	English/ Polish
Portugal	http://www.cnig.pt/	Developing/ Mature	1994	4 263	1 725	CEN TC/287	Portuguese (English)
Russia	http://www.fccland.ru/	Initial	-	-	249	-	Russian/ English
Slovenia	http://www.sigov.si:81/index-1.html	Developing	1997	407	535	CEN TC/287	Slovenian
Spain	http://mercator.org/aesig/	Initial	-	-	*n.f.*	-	Spanish
Czech Republic	http://labgis.natur.cuni.cz/cagi/	Prototype	1998	120	*n.f.*	CEN TC/287	Czech
United Kingdom	http://www.ngdf.org.uk/	Developing	1999	2 103	2 250	ISO TC/211	English
Iceland	http://www.hi.is/pub/gis/	Initial	-	-	269	-	English
Sweden	http://www.uli.se/	Developing	1998	2 398	550	other	Swedish
Switzerland	http://www.sogi.ch/	Initial	-	-	*n.f.*	-	French/German/English

n.f. not found - not applicable

[1] There are four different levels: a) 'Initial' (not built an actual 'internet'-clearinghouse); b) 'Prototype' (a built 'Internet-application', however not completely operational); c) 'Developing' (clearinghouse with only access to metadata files), d) 'Mature' (clearinghouse with access to the 'real' data).

[2] Number of data sets: based on the information given by the 'webmaster' of the clearinghouse or it has been counted.

[3] Number of visitors: based on the information given by the 'webmaster' of the clearinghouse or read from the 'counter' available at web page.

Table 1. Overview of the National Clearinghouses for Geo-Information (17 March 2000) (Crompvoets 2000)

4 Conclusions

Having the right environmental data set is not sufficient for helping to frame and monitor policies required for improving the state of the environment. Neither for responding to a variety of complex issues and their interrelations concerning the support for sustainable development and processes such as those related to global concerns, regional disparities, and local implications. We need to go beyond the delivery of data to the delivery of information and knowledge derived from these data. Therefore, data mining methods and tools are the fundamental importance for disaster management applications. Without a systematic effort to generate real-time data mining solutions, the environmental databases being created today will be greatly under-exploited, and our efforts to develop a data-information-knowledge-decision strategy will be considerably diminished.

In this paper, we have outlined the breadth of our research efforts to address important and challenging issues of accuracy, efficiency, and usability of real-time knowledge discovery processes. We have applied feature extraction and construction algorithms for environmental data (i.e., when both normal and anomaly data sets are given) Further research work is needed to discuss approaches to reduce computational cost and improve the efficiency and usability of the real-time data mining techniques.

References

Behnke J, Dobbinson E, Graves S, Hinke T, Nichols D, Stolorz P (1999) NASA Workshop on Issues in the Application of Data Mining to Scientific Data. Final Report, Goddard Space Flight Center, USA

Crompvoets JWHC (2000) Methodology to measure effects of the use of Geospatial Data Infrastructure for rural planning. Presented at 3rd AGILE Conference on Geographic Information Science. Helsinki/Espoo, May 25-27, 2000

Ester M, Kriegel H-P, Sander J and Xu X (1996) A Density Based Algorithm for Discovering Clusters in Large Spatial Databases with Noise. Proc. 2nd International Conference on Knowledge Discovery and Data Mining (KDD-96), pp 226-231

Ester, M, Kriegel, H.-P, Sander J (1998) Algorithms for characterization and trend detection in spatial databases. Proc. 4th International Conference on Knowledge Discovery and Data Mining (KDD'98), New York, USA, pp 44-50

Hunter GJ, Wachowicz M, Bregt A (2002) Understanding Spatial Data Usability. Special Section on Spatial Data Usability, Data Science Journal, 2, pp 79-89

Rao VB, Raao HV, (1993) C++ Neural network and fuzzy logic, management information, Inc., New York, NY, pp. 408

Wachowicz M (2001) GeoInsight: an approach for developing a knowledge construction process based on the integration of GVis and KDD methods. In: Geographic data

Experience in Applying Information Technologies to Ensure Safe Operation of Russian Nuclear Industry Facilities

Alexandr Agapov[1], Boris Antonov[2]†, Igor Gorelov[2], Rafael Arutyunyan[3], Igor Linge[3], Vladimir Kiselev[3], Igor Osipiants[3] and Daniil Tokarchuk[3]

[1] Federal Agency for Atomic Energy of Russia (Rosatom), 24/26, B.Ordynka, 101100, Moscow, Russia. Email: agapov@minatom.ru

[2] Rosenergoatom Concern, 25, Ferganskaya,109507,Moscow, Russia. Email: gorelov@rosenergoatom.ru

[3] Nuclear Safety Institute of Russian Academy of Sciences (IBRAE RAS), 52, B.Tulskaya, Moscow, 115191, Russia. Email: kis@ibrae.ac.ru

Abstract

In Russia, a Unified System of Emergency Response (RSER) one of the parts of which is Russia's FAAE Branch System of Emergency Response has been created.

To provide scientific and technical support for the tasks on evaluation of consequences for the environment and the population in case of crisis situation at NPP, in 1996 IBRAE RAS has established the Technical Support Centre (TSC) for the Crisis Centre (CC) of REA Concern. IBRAE RAS TSC is part of the system to render assistance to nuclear power plants in radiation hazardous situations.

Since 1999 a united Technical Crisis Center (TCC) has been operating at IBRAE RAS on a twenty-four hour basis to support the Situation Crisis Center of FAEA (former Minatom) and crisis centers of the Russian Ministry for Emergency Situations (EMERCOM) and Rosenergoatom Concern.

In recent years TCC experts have worked through a number of requests related to real incidents at nuclear-hazardous facilities throughout the world. Within the scope of everyday activity, the check-up of TSC workability is under way through exercises and trainings.

The report represents the examples of practical approbation for the activity of FAAE and CC of REA Concern in their interaction with TSC during the exercises and actual incidents.

1 IBRAE RAS's Emergency Response Activities

The issue of managing man-caused accidents or emergency situations is of equal importance to Russia and other industrial countries. To manage different-type emergencies in Russia, a unified Emergency Response System (RERS) has been established, the Agency-level Emergency Response System (AERS) of the Federal Agency for Atomic Energy (Rosatom) being one of its components. The AERS ensures efficient emergency response on federal, branch and facility levels. Since 1993, IBRAE RAS has become an active participant of various activities within the AERS framework, the attached-to-IBRAE Technical Crisis Center (TCC) [1] representing an important element of the emergency response system in case of emergencies at nuclear- and radiation-hazardous facilities.

Estimates and forecasts of accident progression, effects on personnel, population and environment as well as elaboration of recommendations on anti-damage and protective measures are important components of the whole emergency response system. For this purpose the activities aimed at developing information-simulation and expert systems for decision-making and scientific and technical support of measures on population and territory protection in case of radiation and/or chemical emergency have been rapidly developed at IBRAE. Such work is based on wide practical experience of leading IBRAE's specialists in nuclear physics, physics of reactors, radiation safety, radiation protection and radioecology collected when assessing and eliminating emergency implications at the Chernobyl NPP (1986), Siberian Chemical Complex (1993), PA "Mayak" (1957), nuclear weapons testing grounds and nuclear submarine accident in Chazhma Bay Primorskiy Region (1985).

To date the hardware, software and procedure complex developed at IBRAE comprises the following components:

- Databases describing radiation-hazardous facilities and regions of their location including population, environmental components and infrastructure;
- Databases on radiological scenarios of potential accidents;
- Computer systems to simulate radionuclide spreading in the atmosphere;
- Computer models for radionuclide migrations in soils and water systems;

- Computer models for radionuclide migrations via the chain: soil – plant – animal – human being;
- Software to estimate internal and external exposure doses; and
- Cartographic databank and Geographical Information System (GIS) for sites of radiation-hazardous facilities and regions of their location.

The major part of the above computer models has been developed using GIS-technologies providing large opportunities for visualization and on-line displaying of simulation results Fig.1.

All computer systems and databases operate within a network. TCC is equipped with up-to-date communication facilities providing for rapid and effective interfaces with other participants of the emergency-response procedure including data acceptance and transfer and holding audio- and videoconferences.

The main activity of IBRAE's TCC consists in rapid response on inquiries on estimating potential implications of incidents capable of producing radioactive contamination of the environment or/and exposure of population. Every year TCC's experts process about 30 inquiries coming from the Emercom's Crisis Situation Management Center, Rosatom's Situation & Crisis Center and Rosenergoatom Emergency Crisis Center, some examples being listed below: an estimate of potential consequences of destruction of radiation-hazardous facilities due to an earthquake in Turkey; estimate of radionuclide release and population exposure doses caused by the accident at nuclear fuel fabrication plant in Tokaimura (Japan); a forecast of radiological implications of incident at 'Volsung-3' NPP (South Korea); a forecast of potential radiological consequences of destruction of some NPPs and research reactors during bombing in Yugoslavia, and so forth.

High-level of IBRAE's TCC expertise is maintained via participation of the TCC experts in a variety of trainings and exercises. Since 1996, over 10 large-scale national and international exercises with TCC participation have been performed including those at Russian nuclear power plants (Smolensk NPP, Novovoronezh NPP, Kalinin NPP, Balakovo NPP, Kursk NPP and Bilibino NPP), 2 national-level exercises at Saclay Nuclear Research Center (France) in 1996 and in 2000, international exercise at Armenian NPP, and others. During those exercises TCC experts performed real-time estimates of the radiological situation on basis of incoming in formation and elaborated recommendations on protection of population and environment.

When developing scenarios and performing exercises with IBRAE TCC's participation, new approaches and up-to-date technologies are

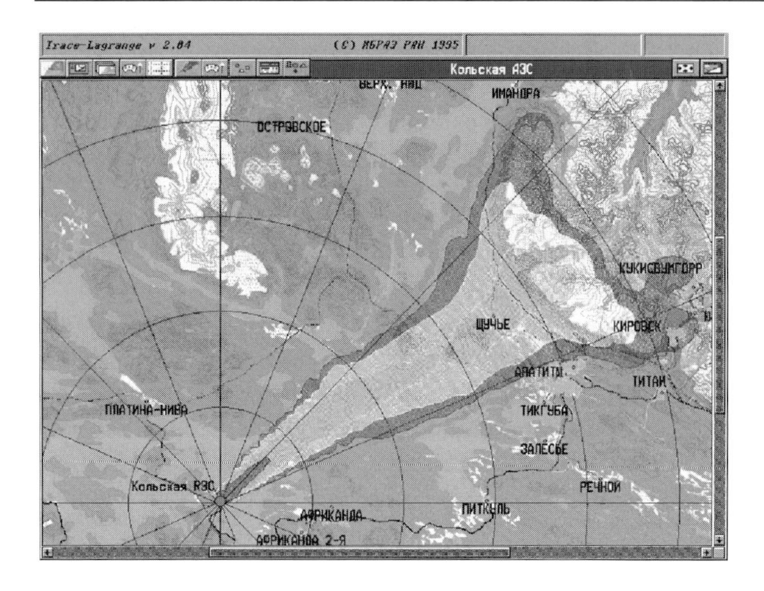

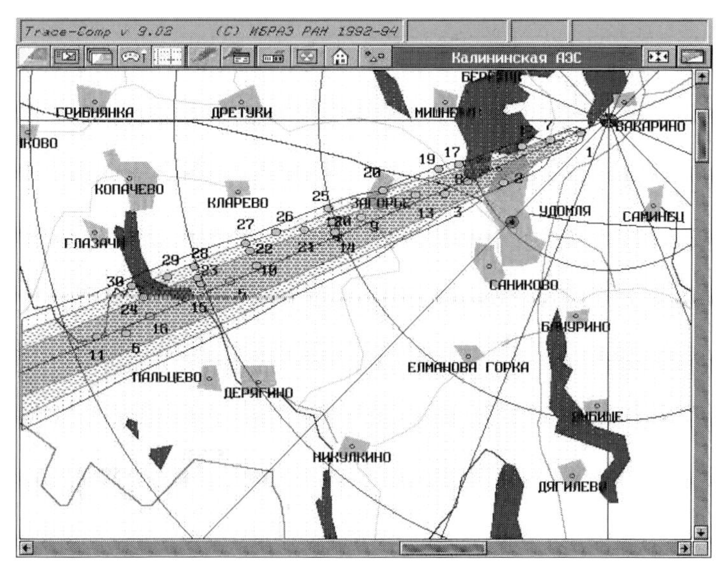

Fig. 1. An example of a computer system for decision-making support

widely used. For example, a new procedure of full-scale simulation of radiological consequences of accidents has been recently developed using GIS-technologies.

2 Programs of Full-scale Simulation of Accidents

When running trainings and exercises focused on population and environment protection in a case of radiation accident, one faces a problem of conformity of the developed scenario of accident and its consequences to those of real accidents. An increasing-over-time information flow comprising of different type of data represents a characteristic feature of expected real situations. Such information includes: the results of measurements in the environment and doses to population, source term estimate data, results of calculations using different models, characteristics of affected territories, population, agriculture, etc. Simulation of a variety of data by participants of trainings and exercises is of crucial importance for scientific support of the decision-making process. However in today's real practice only a simplified approach is applicable in most cases providing for the use of a limited data amount under the relevant scenarios generated using a single deterministic model in order to avoid data contradictions.

Under a full-scale simulation of the radiological consequences, integrity is assumed for methods and procedures on preparing for exercise participants all types of data on emergency consequences for environment and population, namely: dose rates, data of spectrometric and radiometric measurements of air, soil and surface water contamination, aero-gamma-survey data, contamination of agricultural production, foodstuff, wild flora and fauna, data on doses dosimetric examination of population and personnel (radioactivity in the thyroid gland, whole body counting and individual external exposure dosimetry). The following data are also prepared: information on the source term, characteristics of the affected territories, meteorological data and their forecasts, agriculture-related data and some economic characteristics of the region. It is only in such a way that one would be able to create as-real-as-possible conditions to analyze and estimate radioactive contamination of the environment components and population dose commitment with due regard for dynamic, spatial and individual variability, statistical peculiarities and possible protective measures.

The procedure of full-scale imitation of the radiological consequences of accidents was realized in the form of computer systems tested during many trainings and exercises [2]. A tabletop exercise in Saint Petersburg (1993) was one of the first steps of using full-scale simulation technologies. It was for the first time that for the exercise purposes huge data arrays were prepared simulating the results of measurements in the environment and dosimetry examinations of population at the intermediate and remote phases of large-scale terrestrial contamination.

The exercise's participants were supplied with that information in the form of "dynamic databases" in compliance with the developed scenario: the experts directly used such data when preparing recommendations on population protection and territory remediation. Already the early exercise's results confirmed efficiency of the used approach for trainings of experts. Further development of the procedure of measurement result simulation made it possible to create the "PARIS" Code designed to generate measurement databases; see Fig. 2. In the "PARIS" Code 8 types of spectrometric, radiometric and dosimetric measurements are modeled over any time interval after radioactive contamination of the territory depending on contamination level and countermeasures to be implemented [3].

Comparison of simulation results using contamination parameters similar to the Chernobyl-origin contamination in Bryansk region with the results of "practical" measurements showed their satisfactory agreement. That system was also used when preparing "Poliarnye Zori-95" exercise at Kola NPP. Integrated use of databases simulating real measurements, GISs and up-to-date information-visualization facilities made it possible to enhance considerably the interfaces between experts, decision-makers, pressmen and population.

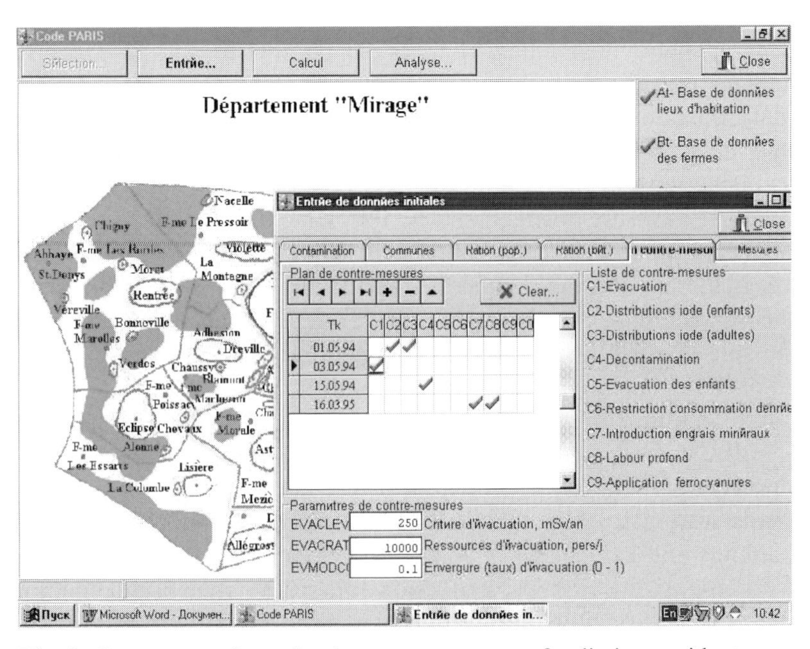

Fig. 2. Computer code to simulate consequences of radiation accidents

The full-scale simulation procedure was developed further during preparation of French national exercise "Becquerel" (1996) [4]. It is under that exercise that the “Envelope” computer code was tested for the first time; see Fig. 3. The “Envelope” Code installed at a laptop computer includes a GIS of the exercise area supplemented with pictures of individual areas, a database on equipment and radiation situation characteristics depending on time and location. The user may move over the region and query in real-time conditions performing at any area measurements of: dose rate and radionuclide composition of air and soil contamination. The results of measurements, depending on gage capabilities, time, area and local peculiarities of contamination, are displayed on monitor or are saved to an individual file. Thus it was for the first time that radiation survey teams faced a close-to-reality situation during the exercise. Obtaining radiation situation-related information *in situ* depended directly on on-line data analysis and rapidity of response of the involved services. The “Becquerel” exercise results have demonstrated that, even under a high level of communication-facility development, the problem of opportune transfer and interpretation of radiation situation data still remains topical.

Within the frames of “Aragats-99” exercise at Armenian NPP held by IAEA, in addition to the mobile radiation survey simulation system, the “ARMS-ARAGATS” system was developed to simulate readings of sensors of a real Automated Radiation Control System (ARCS). The system emulates readings of 8 sensors in real-time conditions.

During the exercise the system simulated ARCS sensor readings that provided a possibility for the experts to estimate changes in the radiation situation in real-time conditions and work through the processes of data transfer to decision-makers authorized to take protective measures.

Thus full-scale simulation of radiological implications of radiation accidents represents an indispensable element of trainings and exercises, the computer systems developed on this basis being an efficient tool to train specialists on radiation situation identification and estimation of emergency implications.

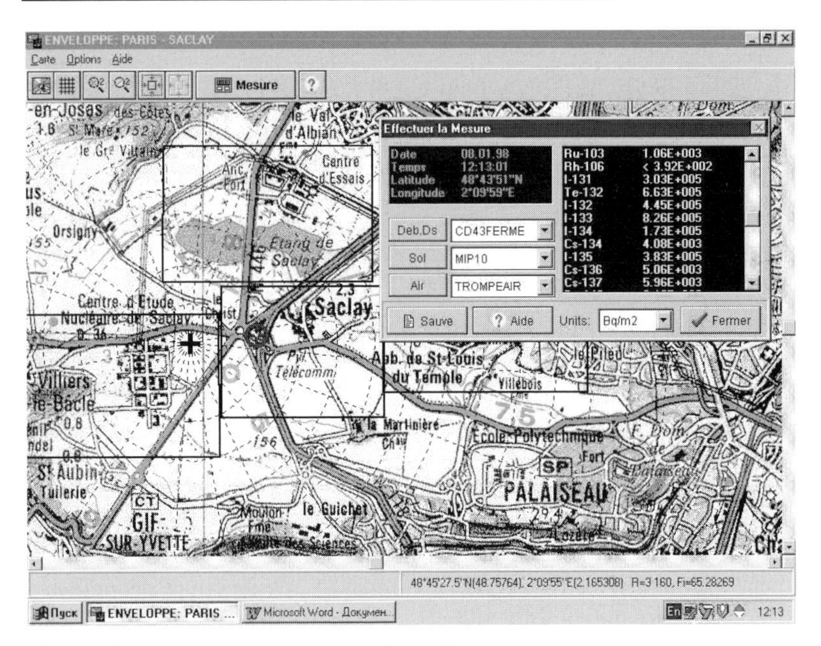

Fig. 3. Training computer system for radiation survey teams

3 Integration of Different Software Modules

Developed at IBRAE, forecasting software complexes - simulating the radiation situation at both the acute accident phase with radionuclide release to atmosphere and the late post-accident phase with migration of radionuclides via water, soil and food chains - operate on basis of a unified bank of digital maps (DMB). The DMB uses a specially developed classification of layers, objects and parameters. Information is stored within inner format of 'MapInfo' GIS, being also duplicated in 'Mif-Mid' open exchange text format of the system.

Forecast results are issued in the form of various reports but also as individual cartographic layers in 'Mif-Mid' exchange format of the 'MapInfo' GIS. The latter allows visualizing forecast results on digital maps and transferring them from one forecasting system to the input of another system. For example, forecast data for surface contamination obtained using 'Trace_Mi' calculated module may be transferred to the input of 'Kassandra' forecasting module dealing with radionuclide migrations via water media. An example of visualizing on digital map of forecast data on contamination of a lake system due to radioactive cloud deposition is demonstrated in Fig.4.

Using such data ‘Inter’ forecasting module allows predicting dose commitment and radiation risk for nearby population at late phases of a radiation accident involving, for example, consumption of water and fish.

Thus, though every of the software has been developed by different designer teams, the use of integrated classifiers and data storage and exchange formats allows their easy integration into a unified software-forecasting complex.

The above approach has been fully implemented when developing software of the mobile complex supporting work of radiation safety experts. Here under a single shell not only forecasting modules, but also measuring and information-reference modules have been integrated forming thereby a unified calculated-information medium for managing emergency implications. When developing multifunctional software complexes, such approach of using unified databases and databanks along with standard data presentation formats allows saving much time and money, avoid duplication and, ultimately, discrepancy between input data typical for the cases when every individual software module operates using its own database.

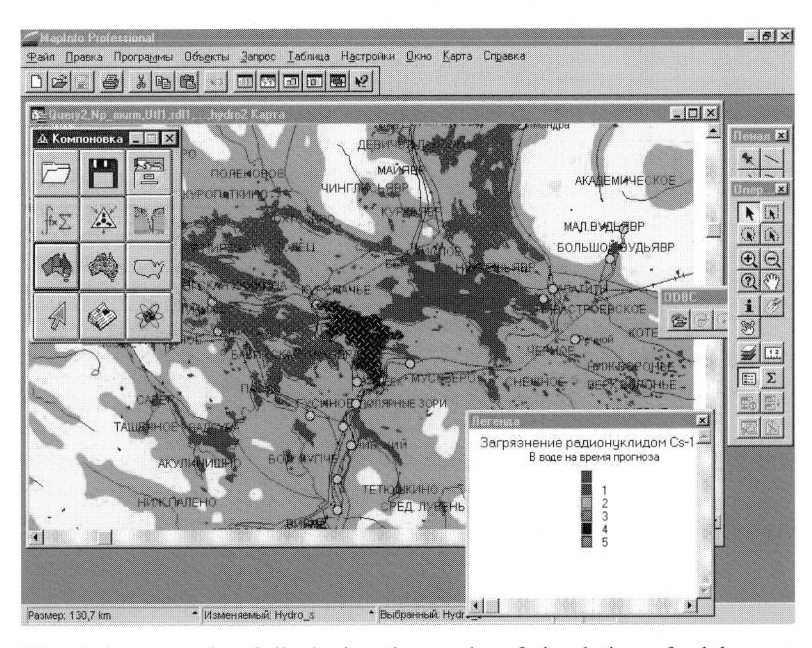

Fig. 4. An example of displaying the results of simulation of a lake-system contamination by ^{137}Cs

4 Conclusions

Within the scope of everyday activity, the check-up of TSC workability is under way through exercises and trainings. The items of interaction with FAAE of Russia, REA Concern and Russia's EMERCOM CDCS in case of emergency occurrence at NPP, research reactors and other ionizing radiation sources are developed during trainings.

The report represents the examples of practical approbation for the activity of FAAE and CC of REA Concern in their interaction with TSC during the exercises and actual incidents. Apart from the work in stationary conditions, the modes to support the FAAE operative group being at the place of exercises have been developed, including those to transfer the data via different communication links in field environment.

References

Bezrukov BA, Gorelov II, Eremin AF et al (1999) Experience of Establishing at IBRAE RAS a Technical Support Center of Rosenergoatom Concern's Crisis Center, Preprint of BRAE RAS, Moscow (in Russian)

Arutyunyan RV, Linge II, Ossipiants IA et al (1998) New technologies in off-site emergency training, In proc. of "The European Conference on safety and reliability ESREL-98, 16-19 June 1998", Trondheim, Norway, pp 147-153

Renaud P, Maubert H, Robeau D, Linge II., Pavlovsky O, Ossipiants IA (1998) Using of PARIS software for the preparation of post-accidental situations. In "OECD Proceedings of the International Workshop «Emergency Data Management»", 13-14 September 1995 Zurich, Switzerland, OECD NEA Publication

Arutyunyan RV, Linge II, Kiselev VP et al (1999) Experience and preparation of exercises and practical games on emergency preparedness and response in case of radiation accidents. Report on 7-th Topical Meeting of American Nuclear Society on Emergency Preparedness and Response. Santa Fe, New Mexico, USA, 14-17 September. (CD-version)

Building Disaster Anticipation Information into the Ghana Development and Poverty Mapping and Monitoring System

Emmanuel Amamoo-Otchere and Benjamin Akuetteh

Centre for Remote Sensing and Geographic Information Services (CERSGIS), University of Ghana, Legon, Ghana.
Email: eamamoo@ug.edu.gh; akuetteh@ug.edu.gh

Abstract

The Centre for Remote Sensing and Geographic Information Services (CERSGIS), a self-supporting, no-profit organization within the University of Ghana is an implementing agency an European Union supported project for "Establishing Mapping and Monitoring System for Development Activities in Ghana (EMMSDAG)". The current EDF-funding cycle of three years covers the "Rural Phase" of the EMMSDAG". The goal of the Rural Phase of the EMMSDAG is an on-line multi-functional/multi-application ArcGIS-driven government-owned data sets on social infrastructure facilities for rural community development projects and impact monitoring and evaluation. Five frontline government institutions are the primary stakeholders of the EMMSDAG – Ministries of Finance and Economic Planning, Local Government and Rural Development, Project Monitoring and Evaluation Unit of the Office of the President, National Development Planning Commission and House of Parliament. These are to be connected on-line through wide area network for direct use of the data resources of the EMMSDAG Laboratory at the CERSGIS.

The project should facilitate widespread use (by government and non-governmental organizations, donors, investors and the private sector in general) of the GIS-driven socio-economic databases to be sustainably maintained by CERSGIS.

It will be the primary task of CERSGIS to assist potential and real users in defining profitable/practical uses of the data for problem-searching/defining and problem-solving. The National Disaster Management Organization (NADMO) will be one of the expected users of the EMMSDAG's data resources.

1 Potential and Real Disasters in Ghana

Seismic-Based Disaster

Historical records of earth tremors in Ghana indicate that the national capital (Accra) in particular and the country in general are vulnerable to earthquakes. The epicenter of the tremors experienced so far is in the Atlantic Ocean. Intensities of the tremors have however not reached catastrophic level, but the recent Indian Ocean experience has awakened the country as a whole to a potential seismic-based disaster. There is an on-going TV-program (Build it Right) which is awakening the public to the need to anticipate seismic-based disaster and to prepare for it henceforth.

Urban Flash Floods

The greater part of the Ghanaian urban landscape is of informal housing development. About tow-thirds of the city of Accra is of informal development. The worse part is the slum landscape. Flash floods are serious in the lowland built-up areas, and have become a rainy season dilemma to National Disaster Management Organization (NADMO).

Fires (Bush and Domestic/Indudtrial)

Domestic and industrial (including market) fires are becoming frequent. Loss of properties is heavy. It is mostly associated with markets, slum and near-slum housing areas. The Fire Service usually arrives late either because of inaccessibility or late communication. In several cases water for quenching the fire will be in short supply.

Though bushfires are dry-season occurrences mostly in the savanna, the forest zone is not spared in places where grass and other herbaceous fire-prone vegetation occur. The victims to bush fires include small farming settlements, wildlife parks and commercial farms. The introduction of forestation with single specie fire-prone plants has brought with it fire-caused disasters. The teak plantations particularly are serious victims to the dry season bush fires

Deforestation

This disaster falls into two types – biodiversity depletion and stream drought. The Forest and Savanna Ecosystems are under severe threat from deforestation. Depletion of the fresh waters of Ghana is now a recognized disaster associated with deforestation.

Among the causes of deforestation surface mining is very serious, and its disastrous side is the geographic spread of the small scale concessions throughout the country associated largely with watershed areas of the country's fresh-water streams. The most fragile yet precious watersheds plant and animal communities are coming under threat. A countrywide degradation lies in waiting as mining concessions are being issued to investors under the country's poverty reduction program.

There attention directed to large-scale commercial farming for export diversification is another disaster trigger development activity. The flip side of this sub-sector development of the country's agriculture is where soil loss will accelerate in the uplands with the corresponding sedimentation of the water bodies. The Volta Lake, which generates about four-fifth of the country's electricity power is loosing its storage capacity to siltation, which is basically being caused by the bush fires and large-scale semi-mechanized agriculture. The country now suffers from dry-season hydropower generation capacity reduction.

Epidemiological Risks

Malaria, other widespread insect-borne and water-borne diseases are the main cause of the high mortality rates, but the HIV/AIDS, Bruli Ulcer are adding to the disastrous situation in the Health Sector. In some localities water blindness and guinea worm have reduced the labor force to near-zero agricultural productivity.

Potable Water Shortages

Potable water shortage in towns is becoming chronic and overshadows even more chronic dry-season rural water shortages. The disaster associated with this have different dimensions. Some large boarding schools of educational towns (Cape Coast in particular) are compelled by the water shortages to close down until storage dams return enough water from the early rains. Substantial part of the low-income urban household budget can go into water buying from tankers, which themselves are not free from contamination during the delivery process. The health risk side is the prevalence of cholera, etc. the high medical bills of which exert very heavy burden on the national budget.

Rural water shortage is a disaster in many ways including which are maternal and child health problems and low productivity of the women. It entrenches the poverty cycle.

Food Security / Food Safety

Thanks to CNN and other global information media to the stereotypic pictures on the TV-screen of hunger-associated sub-Saharan Africa. The situation is very real in communities even few kilometers from the African cities and within the cities. Far away from the cities in the most deprived communities seasonal food shortages may occur, associated with crop failure through drought, fire, floods. In Ghana this type of disaster may be localized because there will be at the same time some food-surplus localities because of favorable conditions. Intra- and inter-country food availability disparities occur and are even more real in the conflict countries.

Food safety is another area of potential economic as well as health disaster. The producing and exporting country can suffer economic losses if safety measures are compromised. The health hazard from fruit and vegetable contamination can be catastrophic. To safeguard such possible occurrences there is now a paradigm for fruit and vegetable products safety regulations under which product origin and route traceability is becoming a subculture in international trade.

Oil and Gas Spillage

Ghana is not an oil and gas producing country yet even though sub-marine prospecting is seriously going on within its Atlantic front and part of the Volta and Tano River Basins. Of recent development however is the West African Gas Pipe Line from Nigeria across the eastern coastline of Ghana, which has in-shore wetlands including Ramsar sites of international importance. There must be anticipatory measures to cope with any possible spillage, when it occurs.

Transportation (Road and Volta Lake) Accidents

Ghana is rated as one of the highest road accident countries; currently there is an initiative for the development of Road Accident Information System by the Ministry of Road and Transport with the Police and Fire Services as stakeholders.

Volta Lake transport is very important for the lake-shoreline communities, yet periodic accident occurs with several casualties.

Political Refugees

Refugee is becoming a menace in the sub-region for several reasons including social and environmental problems associated with cultural practices of the affected population. Ghana has been centripetal to refugee movement from Liberia and Côte d'Ivoire and Togo. The host localities that embrace the refugee settlements have mixed stories to tell. Of particular concern as a potential disaster, but not yet researched into, is the possibilities for cross-border small arms transfer into the host countries.

Plastic Waste Disposal

This plastic technology has come with it widespread, difficult-to-control garbage management which is a real threat to health and environment as a whole. The "take-away" sub-culture and shopping into polythene bags are today's major environmental enemy. The numerous plastic waste dumps appearing in the urban and rural landscapes are serious danger to urban drains and the fresh water wetlands. Among the causes of urban flash floods is the choking of the drainage channels by plastic garbage.

2 Information for Disaster Management

The Information and Communication Technology Resources

The Information and Communication Technology can be utilized as part of the disaster management kits. In nearly all the cases the following are to recognized as part of the information management outfits:

- Media facility – FM-Radio, Television
- Global Positioning System for point and line measure of the sources and coverage of the events and mapping for monitoring and interventions.
- Mobile phones which can allow rural communities otherwise inaccessible to reach the outside world within and outside the borders of the country
- Web-based map management for informing world-wide the event occurrence and progress.

- Spatial data infrastructure which supports dissemination of adequately geo-referenced information for pre-event, during- event and post event management.

All these help in information management of the most important question concerning *who* and *where* are affected by the *event*? But long term disaster management planning will require demographic, housing and infrastructure maps in relation to the risk zones.

Ghana's telecommunication space is fairly adequately resourced by the private sector space phone operators. Most parts of Ghana are now being served with mobile telephones. This has facilitated communication of events to nearby FM-Radio stations spread throughout the country. What exists at the moment needs improvement but it has facilitated timely intervention of some flood, fire and motor accident cases.

Evidence of Use of Spatial Information

Over the past three to four years CERSGIS has been responding to requests from government and governmental organizations, the public sector in general, some maps and associated databases which give indication of the public awareness about the need to build up some databases for the management of natural and human disasters. Among the various demand for maps the following are significant:

- Spacefone Ltd: Countrywide Mobile phone Service Development maps at district level for the ten administrative regions
- Ministry of Local Government: Accra Hydrological basins map for flood control planning.
- EU/ACP Pesticide Initiative Project: Traceability Information System for Pineapple Export to the EU market
- IITA-STCP: Kokoo Kuapa Union Members' cocoa crop production information system for ethical and environmental conditions consideration under the Fair Trade
- UN-HABITAT- Water Resources Commission: Sustainable water supply to Accra from the Weija Lake of the Densu Basin – GIS capacity building for multiple uses and protection of the Lake and the Densu River system.
- Water Resources Commission: Coastal wetlands/Ramsar sites degradation monitoring GIS.
- Ministry of Science and Environment: Space image-based maps for Weija Lake shoreline vegetative buffer planning to stem the pollution of the storage dam.

- Highway GIS-driven base map for accident information system development.
- Noguchi Memorial Medical Research Institute: Malaria Surveillance system – Development of the map system; GIS-Capacity building for Parasitic disease control management.
- Volta River Authority/Department of Fisheries: Volta Lake Fisheries communities monitoring of bad fishing practices; Lake transport accident reduction planning.
- IUCN-World Wildlife Conservation: National Parks database for biodiversity conservation of Kyabobo Park.
- UNESCO/Government of Netherlands: Mole National Park conservation planning database.
- The Ghana Aids Commission: Demographic Map system for monitoring HIV/AIDS in selected townships.

3 The EMMSDAG Structure

The Framework

The structure of the project for "Establishing Mapping and Monitoring System for Development Activities in Ghana (EMMSDAG)," is schematized in the diagram at the next page. It has been explained already that it is currently the "Rural Phase," with the map compilation scale at 1:50 000, and on the projection of the 1;50 000 Topographic Map series of the Survey Department of Ghana.

Progress with the Present Phase of EMMSDAG

The Project commenced officially in May 2004 with the digital conversion of the Ghana Statistical Service (GSS) Population and Housing Census Enumeration Area Map (EAM) of the 2000 Census. The conversion will be completed in April. Parallel to the map conversion is the activity for the conversion of Census Locality data in ACCESS to be linked to the map.
The district assemblies are involved in defining the administrative boundaries with the collaboration of CERSGIS. Space image maps are being used as part of the materials for the assembly-consulted jurisdictional boundary mapping. The output is a geocoded settlement map showing the administrative and electoral boundaries of the district assemblies.

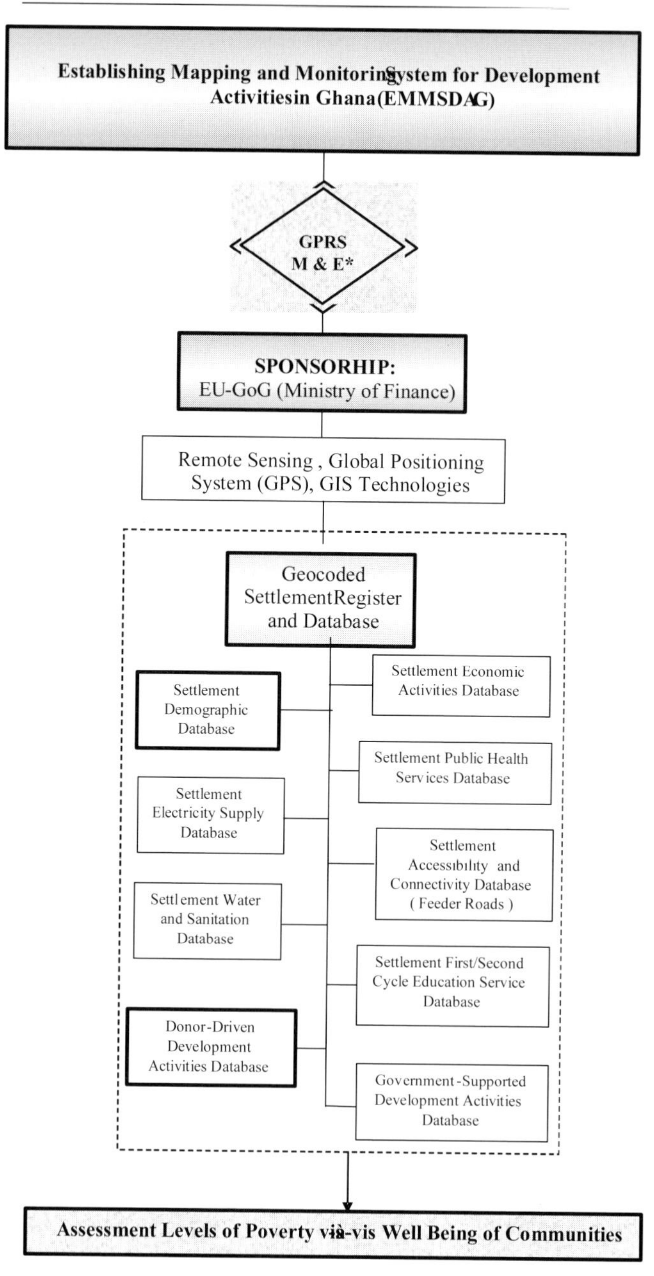

* Ghana Poverty Reduction Strategy Monitoring and Evaluation

Fig. 1. CERSGIS conceptual framework for the EMMSDAG operations

All the other data sets will be linked to this settlement base map. The present phase is at a compilation scale of 1/50 000. It is taken as the Rural Phase.

Add-on for Disaster Management in the Urban Sphere

Talking about disaster the major population centers are of major concern because of numbers. The Urban Phase of the EMMSDAG will be suitable for urban disaster management information system development. The map compilation will be at township scale at 1: 2,500. The same range of data sets being developed for the present EMMSDAG will needed for the Urban Phase. The Local Government administrative structure map system will have to be compiled, which implies that the Metropolitan and Municipal structure will be recognized. Demographic map system based on the Census Locality data will be developed as a primary data layer for spatial analysis at the neighborhood level (street level, area committee/electoral area level, town council/area council level and sub-metro/municipal/district assembly level).

On-line Data Availability

The envisaged on-line accessibility of the EMMSDAG database will pose data management challenges to the CERSGIS. It requires a team of highly skilled spatial information management technicians and professionals for the database management, network management, system designing for specific problem solution, demand-driven map conceptualization and production, web-based map posting for mass education, etc.

There is no doubt that CERSGIS has daunting task ahead, which it is developing itself for. Its independence and self-supporting status allows it to collaborate/partner for knowledge and skill sharing, and this will give it the opportunity to experiment too with methodologies and techniques as and when positive partnership will allow.

7 The EMMSDAG Infrastructure

The information flow from the EMMSDAG system at CERSGIS offices in Legon, and the clients analyzing the data at the focal institutions.

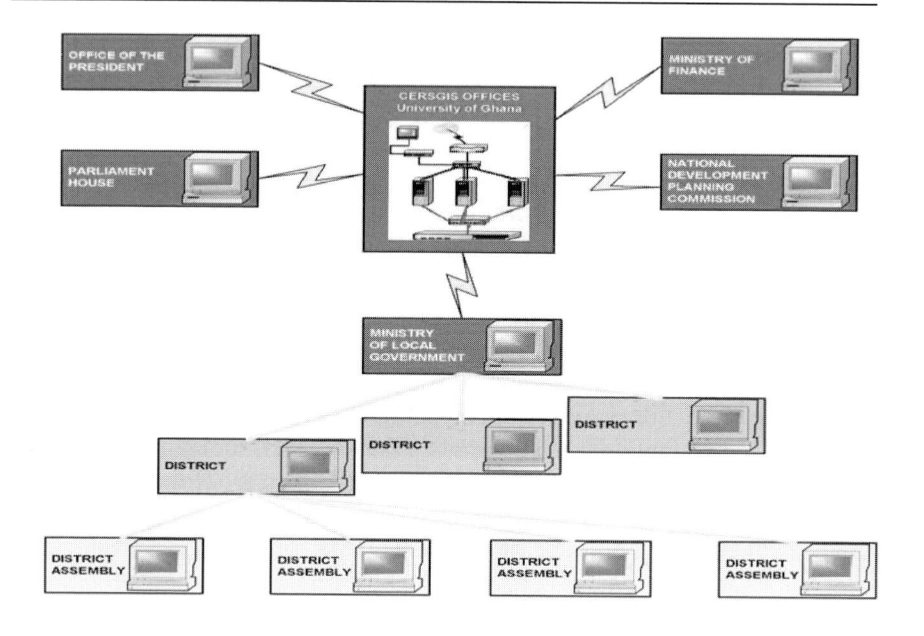

Fig. 2. Data Exchange Traffic

8 Conclusion

The EMMSDAG initiative will solve part of the data shortage problem and make the core spatial data sets in the database available for widespread use by the public including the stakeholders/institutions of the National Disaster Management Organization. However, it is the rural phase of EMMSDAG which is currently being put in place. The urban phase is still waiting for funding.

The media situation in the country is very active. Countrywide mobile phone service is available. Internet services are available also. But there must be good and reliable geographic database especially on population distribution to which all the disaster types and scenarios can be analyzed. The EMMSDAG is likely to assist other mapping initiatives to reduce the current data shortage disaster. At that stage all the resources of the information and communication technology now available will be adequate for building disaster anticipation information system in Ghana.

Vulnerability Assessment for Food Crisis Management in the Sahel Region

Maurizio Bacci[1], Tiziana De Filippis[1], Andrea Di Vecchia[1], Bakary Djaby[2], Francesca Incerti[1], Moussa Labo[3], Leandro Rocchi[1], Fabio Straccali[1] and Patrizio Vignaroli[1]

[1] CNR, Institute of Biometeorology , via G. Caproni 8, 50145 Florence, Italy.
Email: f.incerti@ibimet.cnr.it
[2] AGRHYMET Regional Centre, B.P. 11011, Niamey, Niger.
Email: b.djaby@agrhymet.ne
[3] Direction de la Météorologie Nationale du Niger, BP 218,Niamey, Niger.
Email: dmn@intnet.ne

Abstract

In the Sahel region the unfavourable climatic conditions and the natural resources degradation are recognised to be the main constraints affecting the agricultural productivity; nevertheless more complex and less evident causes of food insecurity of population groups need to be investigated.

The analysis of biophysics and socio-economic dynamics acting in disadvantage areas is considered a basic element to put in action effective programs for the prevention and management of food crisis. The availability of effective and timely information during the development of agricultural season is strategic for the evaluation of famine warning level: it constitutes a support to the planning activity of decision makers.

In this context a schedule for food crisis prediction has been designed as an operational framework enabling to integrate different tools to produce appropriate information to define and monitoring the risk zones during the whole agro-pastoral season. The spatial extension and the importance of the population touched can be timely estimated in order to define the action plans to minimize the crisis damage.

An analysis module, a plugin of the free Geographic Information System VisualCarte, has been developed aiming at increasing the Sahelian National Services' responsibility in emergencies management and at encouraging their involvement in the long-term resources planning for crisis prevention. The plugin PRVS (Structural Vulnerability Mapping Procedures) enables the elaboration of thematic maps through indicators of vulnerability and the creation of scenarios about the current vulnerability.

A structural frame of reference for the identification of vulnerable zones and populations at regional and national level is made available: homogeneous areas and administrative units are classified according to theirs productive capacity related to the economic structure of productive system. As consequence the monitoring of negative events can include their evaluation according to the incidence in the context of structural vulnerability.

The plugin PRVS can process different data type. The use of real or estimated data allows to assess the impact of unfavourable conditions on the primary productivity during the rain fed season. The estimation of the capability of the primary sector to sustain a given population leads to the identification of self-sufficient, surplus or deficit areas.

By means of integrated analyses tools and seasonal forecast, the early assessment of current vulnerability may occur. The continuous availability of reliable information for the mapping of vulnerability satisfy the needs of the various institutions involved in crisis prevention and natural resource management.

The work described in this paper has been conducted by Ibimet - CNR in collaboration with Agrhymet Regional Center within the SVS (Suivi de la Vulnérabilité au Sahel) project funded by Italian Cooperation.

The program's objective is the strengthening of regional and national capacities of CILSS countries (Permanent Interstate Committee for Drought Control in the Sahel) in the crisis prevention and in the poverty reduction. Methodologies for vulnerability analyses have been pointed out to support different users with appropriate information. Furthermore environmental monitoring for the assessment of zoning associated with climate change is also included in the objective of the program.

1 Introduction

The collaboration between Ibimet – CNR (Institute of Biometeorology of the National Research Council - Italy) and the ARC (Agrhymet Regional Center - Niger) started ten years ago pursuing the development of an Early Warning System for the identification of zones under food risk.

This partnership was born to meet the needs of the various institutions concerned with crisis prevention and natural resources management in nine CILSS (Permanent Interstate Committee for Drought Control in the Sahel) countries of the Sahelian Region.

During this period a three years project, called SVS (Vulnerability Monitoring in the Sahel) funded by Italian Cooperation, has been designed in order to consolidate the experiences acquired by the former project AP3A (Early Warning and Agricultural Productions Forecast Project).

The SVS project aims at supplying Institutions operating in the domain of food security with updated information, as well as providing them with instruments allowing them to increase their capacities in the analysis process of food vulnerability.

The activities that were carried out led to the dissemination of a set of tools and information to be integrated with the different methodologies utilized in single country's policy formulation.

The main end users are the Technical National Services and new products and methodologies have been designed to improve their skills in the production of information for decision makers and in order to increase their responsibilities in defining and operating food security programmes.

A set of tools for the monitoring of the agro-pastoral season has been set up together with procedures for the structural vulnerability representation. These tools can be combined in a food vulnerability analysis, so called schedule for food crisis prediction: a process in which high importance is given to the sequence through which each tool is coherently used in the crisis monitoring according to the scale analysis (national, regional or local).

In the context of food security, the schedule for food crisis prediction finds a new operational framework: the phase foreseeing its transfer to the National services guarantees its sustainability. Moreover it finds a possibility of combinations with other methodologies actually adopted at national and regional level.

Considering the difference of needs of each country, it is necessary to run a process of validation: a first testing occurred in Niger, Burkina Faso and Mali during the 2004 season in order to define its constrains or to suggest possible improvements. The validation implies a cross check of scenarios produced by the analysis with the data collected by National Early Warning Services.

2 Methods

Following the numerous food crises that occurred from the seventies and the famine which took place in 1973 and 1984, food security became a main topic in Sahelian Africa.

In the region the unfavorable climatic conditions and the natural resources degradation are recognized to be the main constraints affecting the agricultural productivity; nevertheless more complex and less evident causes of food insecurity of population groups need to be investigated.

The analysis of biophysics and socio-economic dynamics acting in study areas is considered a basic element for the prediction of food crisis.

In this context the software Visualcarte was implemented for the displaying of thematic cartography and also for the management and the analysis of georeferenced data. The main goal Main of the tool is to provide policy-makers with useful information for timely emergency interventions and long-time development plans.

The software Visualcarte has been implemented through the design of PRVS (Procedures for the Representation of Structural Vulnerability), a plug-in to analyze and to elaborate the main issues of food structural vulnerability. Its functions also allow to generate and compare different current scenarios according to variations concerning the food availability and accessibility.

2.1 The Study Area

The study area includes the following countries of the Sahelian region belonging to CILSS committee: Mauritania, Senegal, the Gambia, Guinea Bissau, Mali, Burkina Faso, Niger and Chad. Geographically they are located between 28° and 4° North latitude and -20° and 25° East longitude.

One of the main characteristics of the region is the strong discrepancy between the population growing rate and the natural resources that brings about a high pressure on the environment with heavy and fast land degradation. The agricultural production is mainly consists of cereals (sorghum and millet) used for household' feeding purpose. Cash crops are diffuse in areas characterized by high water availability due to the favourable rainfall regime or to the presence of great rivers (Senegal, Niger).

2.2 The Indicators

The analysis that aims at the identification of vulnerable zones is organized in two complementary phases.

The structural evaluation aims at defining a frame of reference through statistical analysis over a defined period of observation (in this study 1985-2002). The current study considers the progressive development of a phenomenon, in other words its time dimension. The significance of the current analysis depends on the structural frame of reference.

According to the convergence of evidences principle, meaningful analysis about the food vulnerability can be processed by identification of different factors as signals of food risk. The combination of these parameters can give a more accurate description of the context of food insecurity affecting areas and population.

In order to identify and characterize the determining factors involved in the manifestation of crisis, many indicators have been chosen for the definition of the Sahelian structural frame. Homogeneous areas have been identified according to the productive system and the coping strategies adopted by the local populations. In the proposed approach this aspects are considered fundamental for the evaluation of the regional capability to face food crisis.

The knowledge of main phenomena acting on a land system allows the investigation of the effects of negative events on the food self sufficiency of the population.

The description of main indicators applied in the structural study is reported.

The *productive system* identifies the kind of agriculture or pastoral exploitation of the land. This indicator points out the productive constraints due to the limited natural resources or productive factors.

The cereals balance quantifies the capacity of the productive system to sustain the population living in a given area. The productions of cereals, cash crops and livestock are converted on the basis of their exchange possibility with a reference cereal at market prices. The food needs are estimated according to the rural population and to the national official consumption rates. The difference between the exchange value of the whole food production and the cereals demand allows the classification of surplus or deficit administrative units. The index representing the percentage value of food satisfaction is defined Virtual Ratio of Cereals Needs Coverage (*VRCN*).

The *agriculture pressure* is the indicator that compares the agriculture sustainability with the exploitation induced by farmers' activities. The productive attitude is based on climatic and pedologic factors determining respectively the rain fed crops' cycles and the productive capability of the soil. The land's exploitation is evaluated according to the degree of occupation of soils unapt to agriculture practices. The location of villages, as well as theirs demographic distribution, allows the estimation of the area needed for rural activity: the comparison between the effective cultivated area and the cultivable area is carried out for the indicator assessment. The *agriculture dynamics* represent the trend of crop yield related to the evolution of cropped areas. Together with the agriculture pressure, this indicator can characterize the *vulnerability dynamics*: they give a description of the nature and intensity of the exploitation of natural resources.

2.3 Data Processing Project

Within the project activities of AP3A a system for the management of thematic maps was designed. It was funded by the Italian Cooperation. Main objective of the software Visualcarte was the dissemination of statistical and geographic data contained in the databases SAT (Structural Analysis System) and SAC (Current Analysis System).

To complete its functions the software Visualcarte was extended with a new plug-in. PRVS is a tool developed in the Visual Basic® environment with the use of the ActiveX library MapObjects2© (ESRI). The new plug-in displays the maps of all vulnerability indicators, that define the structural context of the Sahelian regions. Moreover it allows to process the current analysis for the VRCN indicator.

Once installed the PRVS plug-in gives access to the thematic maps of structural analysis and for all this data it allows different choices: the display, reclassification and querying. It is also possible to create new indicators on a geographic basis. Regarding the current analysis, two kinds of vulnerability representations are available.

The *Scenarios making* tool allows the computation of VRCN indicator at administrative level. The use of demographic data and the population growth rates enables the predictions for upcoming years. With this analysis it is possible to perform simulations of geographical areas that will be vulnerable and make assumptions on main vulnerability causes.

The *Sensitivity analysis* aims at calculating the VRCN resulting from the variation of one or more parameters in term of availability and access to food in relation to the structural characteristics of administrative units.

Based on actual or assumed changes in production and prices of cereals, cash crops and livestock, the analysis enables to determine the primary production sector that will impact most on the food balance. It allows identification of vulnerable areas according to variations of food accessibility for all the administrative units of a chosen country.

The first analysis produces data concerning the capability of each administrative unit to sustain the resident population: in other words it describes the effects of the manifestation of a risk on the food security level of each unit. The second analysis may highlight for each unit the most important risk factor between relative variations of productive potential of crops or livestock and variations of agricultural and livestock prices.

The analysis procedures are designed to guide the users through the different steps for the choice of the required parameters. The designed wizard allows the insertion of statistical, demographic and economic data. If update data are not inserted, the analysis process will run with values referred to the historical dataset produced by the National Services.

Below, the kinds of interface representing the final request of the two tools are described.

In the *Scenarios making* interface only data concerning the simulation year can be updated: the system supplies two options for compiling data. In fact the scenario making command provides tradable production values for average, below average or above average production conditions (average values and 25 or 75 percentile). Prices (expressed in the most spread money: FCFA) and productions values can be further decreased utilising the reduction factors usually applied for the computation of production costs and of losses production (determined for example by pests). In the computation the animal husbandry are considered according to the livestock exploitation rate.

In the *Sensitivity analysis* tool, the interface gives the possibility to insert the percentage values representing the variation of production or the price that user wants apply to the VRCN estimation.

The final products of the described analysis are tabular and graphic outputs in which the computed VRCN is expressed in a desegregated way for each kind of agricultural production. In order to facilitate the comparison with the frame of reference, the system also reports the values concerning the structural VRCN.

When performing the *Scenarios making* process, the updated VRCN values and the structural VRCN are reported for the administrative units chosen through the wizard. The *Sensitivity analysis* shows the result for all the administrative units of the chosen country: moreover it reports the differences of applied variations relative to the structural VRCN.

For both the analysis the system allows the saving of processed results by creating a theme in *.shp format. In this way the user is able to geographically display the vulnerability assessment.

Therefore the current analysis can be represented by maps that can easily highlight dynamics of vulnerability. These maps constitute an important tool for the different actors involved in the decision-making process.

3 Conclusions

The aim of the schedule for food crisis prediction is the enhancement of information contained in the databases produced by the Statistic National services and the strengthening of the National Service's capacities to produce information useful in the interventions aimed at the food security. The importance of this methodology is function of its potential use within planning strategies and analysis at national and sub-national level.

In the schedule, PRVS plug-in constitutes a fundamental tool: it enables the evaluation of impacts determined on the productive systems by negative events through an analysis summarizing the results obtained during the entire schedule process.

The schedule for food crisis prediction found a real application in the National Services' activities during the 2004. In Mali, Niger and Burkina Faso PRVS was run in order to supply information to be checked with usual monitoring methodologies. The vulnerable areas that were determined by this process were confirmed to be the same geographic entities estimated sensible to food crisis by the National Early Warning Systems.

Considering that first data about the yield estimation are usually processed at the end of September and according to the prices dynamics evaluation, PRVS can supply current vulnerability analysis starting this month.

The free distribution of the PRVS tool via CD-Rom to the interested Technical Services makes it an instrument of easy and spread diffusion and utilization.

Since it was designed for a large number of users, its has a multilingual user interface in order to guarantee its dissemination in different countries.

The distribution of the PRVS plug-in implies a technical support in all the steps, methodologies and tools foreseen by the schedule for food crisis prediction, guaranteeing its operational use.

References

AP3A (Project) (2000) Cadre de Référence Préliminaire de la Vulnérabilité Structurelle dans les Zones Agricoles du Burkina Faso, Mali, Niger et Sénégal, AGRHYMET-OMM-Coopération Italienne, mars 2000

AP3A (Project) (2000) Potentialité Agricole en fonction des sols et de la climatologie – Résultats des analyses au Burkina Faso, Mali, Niger et Sénégal, AGRHYMET-OMM-Coopération Italienne, mars 2000

AP3A (Project) (2000) Approche méthodologique pour l'intégration de l'élevage dans la caractérisation des zones d'insécurité alimentaire structurelle au Burkina Faso, Mali, Niger et Sénégal, AGRHYMET-OMM-Coopération Italienne, juin 2000

AP3A (Project) (2001) Le contexte de la Vulnérabilité Structurelle par Système de Production dans les pays du CILLS, AGRHYMET-OMM-Coopération Italienne, novembre 2001

Bacci L, Maracchi G, Senni B (1992) Les stratégies agro-météorologiques pour les pays Sahéliens, IATA-CeSIA, juillet 1992

Chambers R (1989) Editorial Introduction: Vulnerability, Copying and Policy, IDS Bulletin, no 2, vol. 20 University of Sussex, Brighton

Di Vecchia A, Vignaroli P, Djaby B (2002) Les crises alimentaires et les systèmes de prévision au Sahel, Actes de la Réunion annuelle du Réseau de Prévention des Crises Alimentaire au Sahel, Bruxelles décembre 2002

Using Remote Sensing Data for Earthquake Damage Assessment in Afghanistan: The Role of the International Charter

Joseph Maada Korsu Kandeh[1], Abdul Wali Ahadi[1] and Lalit Kumar[2]

[1] UNDP AIMS, P.O. Box 5, Foreign Ministry Rd., Kabul, Afghanistan.
Email: joseph.kandeh@undp.org

[2] Department of Ecosystem Management, School of Environmental Sciences and Natural Resources Management, University of New England, Armidale NSW 2351 Australia.
Email: lkumar@une.edu.au

ABSTRACT

Afghanistan is located in a zone of high-seismic activity. Given the rugged and mountainous nature of the country and the location of villages, towns and cities, there is propensity for widespread death and destruction due to landslides whenever an earthquake occurs.

Use of satellite imagery by humanitarian agencies in Afghanistan in preparation for and response to natural and man-made disasters has been very limited, mostly to International organizations such as the United Nations. Earth Observation Satellites (EOS) due to their vantage position have demonstrated their ability to rapidly provide vital information and services in a disaster situation. EOS has been used in emergency situations where the ground resources are often lacking.

The perception amongst humanitarian agencies and civil protection authorities in most developing countries is that the cost of satellite imagery is not cheap. With limited budgets available for purchasing satellite data, they tend to opt for less expensive solutions such as interagency survey teams to assess damages. The rugged and mountainous nature of Afghanistan and the lack of roads in most parts of the country, survey teams are most often hampered, leading to delays in delivery of information from the field to the decision makers.

Recent earthquake in the Hindu Kush of the country in April 2004 witnessed the triggering of the International Charter for free delivery of satellite imagery.

Image analysis and interpretation of both pre and crisis data did not show observable features of damages. The damage assessment maps were used by the humanitarian community for decision-making.

Availability and access to space technology in addressing natural disasters have been the main obstacles facing developing countries particularly those poor countries without their own space programs. This problem has been solved through the introduction of The International Charter for major disasters. However, knowledge about the Charter is not common knowledge in most developing countries; Disaster Management Authorities, the Academic Institutions, humanitarian agencies and the affected communities have very little idea about the availability and access to free satellite imagery. There is need for a massive awareness campaign to educate decision makers about the International Charter and the potentials of using space technology in addressing problems relating to disaster management and the environment. The skills to process satellite imagery and integrate it with other GIS layers are lacking in most developing countries; there is need to embark on a massive capacity building exercise to ensure optimization of the benefits of the technology. The Charter needs to find innovative ways of quickly sending value added information products to disaster management authorities instead of relying on in-country skills in image processing.

This paper elaborates on the experiences gained working with images received from the International Charter, and the immense pressures from the humanitarian community for rapid delivery of information.

1 Introduction

Natural disasters are extreme events within the earth's system that result in death or injury to humans, and damage or loss of valuable goods, such as buildings, communication systems, agricultural land, forest, natural environment etc (Van Westen, 2002). Alexander (1993) distinguished between a disaster from a hazard when the disaster occurs in a populated area, and brings damage, loss or destruction to the socio-economic system.

Afghanistan is located in a zone of high-seismic activity. Given the rugged and mountainous nature of the country and the location of villages, towns and cities, there is always a high propensity for widespread death and destruction whenever an earthquake, landslide, mudslide, avalanche, or flooding occurs.

According to the EM-DAT International Disaster Database (http://www.cred.be/), about 1,919 people have been killed, and a total of

76,550 made homeless due to yearly flooding since 1972. Landslides and avalanches have also made their mark on the lives and properties of Afghans; 1,373 people have been affected with 799 killed. The Red Cross estimates that since the early 1980s, natural disasters in Afghanistan have killed an estimated 19,000 people and displaced 7.5 million people (IFRC/RC, 2002).

Decades of War and civil conflict, as well as environmental degradation, have all contributed to increasing vulnerability of the Afghan people to natural disasters. Several assessments by the humanitarian agencies in Afghanistan have revealed significant shortcomings in the areas of water, sanitation, health, security and natural resource management.

Furthermore, the high level of poverty, lack of livelihood and income generating opportunities, chronic health problems, and poor state of the infrastructure all add to the burden of natural disasters on the people of Afghanistan. The Government of Afghanistan is so weak it relies heavily on the humanitarian community particularly the UN in responding and managing major natural disasters. There is lack of coordination amongst government departments, and clear-cut strategy for disaster management in the country.

The success of disaster management depends largely on availability, dissemination, and effective use of information for decision-making. The information needs will vary depending on the users, and the scale, but generally will include information on weather and climate, all elements at risk, socio-economic and demographic data, and past information on previous disasters depicting their location, characteristics, extent, and impact. Since the overthrow of the Taliban Government in 2001 which, witnessed the return of the United Nations and International Non Governmental organization into the country, disaster management data are being generated by multiple users, stored in different formats and media, making it extremely difficult to bring the data together to support disaster management activities.

Lack of critical and timely information for disaster reduction is typical of developing countries particularly those emerging from decades of war and civil conflict; Afghanistan is not an exception. Reliable information in a timely manner saves lives and properties as actionable decisions are made that lead to fore warning of populations, provision of relief and making communities resilient and knowledge based.

Space technologies have proved to contribute unique and significant solutions in disaster management: disaster mitigation, disaster preparedness, disaster relief and also disaster rehabilitation. Space technology based solutions have become an integral part of disaster management activities in many developed and some developing countries (http://www.oosa.unvienna.org/SAP/stdm/index.html).

In Afghanistan, use of satellite imagery by humanitarian agencies in the country in preparation for and response to natural and manmade disasters has been very limited. The humanitarian community generally assumes that the cost of satellite imagery is not cheap. With limited budgets available for purchasing satellite data, they tend to opt for less expensive solutions such as interagency survey teams to assess damages. The rugged and mountainous nature of the country and the lack of roads in most parts of the country, survey teams are most often hampered, leading to delays in delivery of information from the field to the decision makers.

In March 2002, an earthquake measuring 6 on the Richter scale occurred in the Northeast part of Afghanistan killing about 1000 people (http://earthquake.usgs.gov/activity/past.html). For the first time in Afghanistan, satellite imagery was used to assess damages as over 35,000 people were affected.

Recent earthquake in the Hindu Kush of the country (5th April 2004) witnessed the triggering of the International Charter[1] for free delivery of satellite imagery. AIMS[2] requested a set of satellite imagery through UNOSAT covering both pre and post earthquake periods. Pre disaster imagery was immediately released, but it took over a week before crisis data could be taken and delivered. Problems of availability of cloud free images and too much snow cover delayed delivery of crisis data.

The objectives of the satellite imagery received from the International Charter for Major Disasters were as follows:

1. To provide information on damage assessment as a result of the 6th April 2004 earthquake through the use of satellite imagery,
2. To show the extent and impact of the earthquake on lives and properties,
3. To inform the key players of disaster management in Afghanistan, the Government of the Transitional Islamic Republic of Afghanistan and the humanitarian community of the role of the International Charter,
4. To educate key players of disaster management in Afghanistan, the Government of the Transitional Islamic Republic of Afghanistan and the humanitarian community of the potentials of satellite imagery in damage assessment,
5. Provide a decision support system for Government Authorities and the Humanitarian Agencies in addressing rescue, relief, rehabilitation, reconstruction and mitigation.

---

[1] The International Charter for Major Disasters

[2] Afghanistan Information Management Service (AIMS) is a United Nations Development Program (UNDP) project building information management capacity in Government Institutions in Afghanistan.

The methodology used is shown in figure 1.

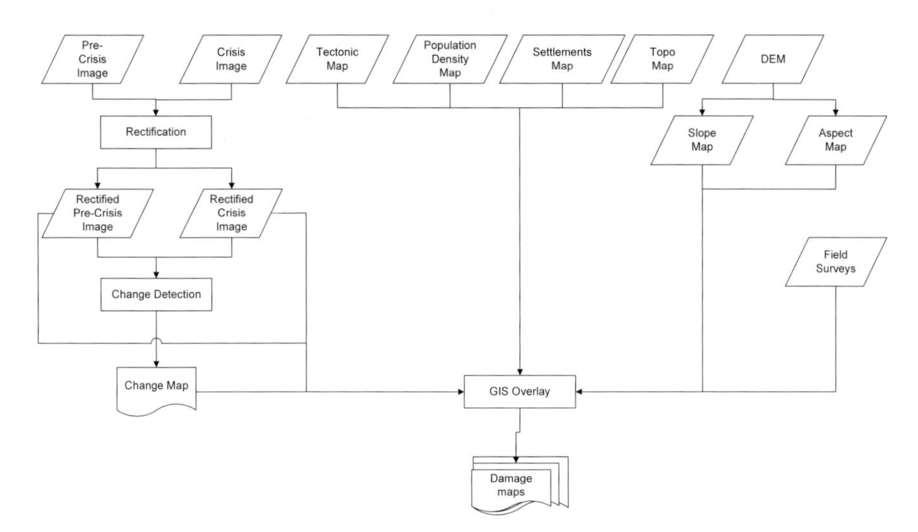

Fig. 1. Schematic representation of the methodology

Using ERDAS IMAGINE 8.4, both pre and post earthquake images (crisis data) were pre-processed, rectified and a change detection map derived. In ArcView GIS 3.2a, the change detection map, and the pre, and post earthquake images were each overlaid with the following GIS layers; major fault lines, population density, settlements, Topographic datasets (1:50K), Slope and Aspect maps to produce damage assessment maps.

2 Results

Table 1 shows that since 2000, about 9 major earthquakes have occurred in Afghanistan, of which 8 occurred in the northeast of the country. The most disastrous earthquake was in March 2002 where about 1000 people were killed and over 35,000 affected.

Table 2 shows the turn around time in the delivery of satellite image following the request for data. Table 2 shows a considerable delay in the delivery of satellite imagery particularly the crisis data after the occurrence of the earthquake (05/04/04) and the request of the imagery from the International Charter. The delay in delivery imagery consequently resulted in the late delivery of processed value added disaster damage maps for decision-making.

Image analysis and interpretation of both pre and crisis data did not show observable features of damages. Change detection performed also did not reveal any structural change. The change maps overlaid with the other GIS layers such as Major Fault Lines (1:6,000,000), Slope and Aspect Maps generated from DEM (90 x 90m), Topographic dataset (1:50,000), Settlements, Population Density, and ground surveys produced a number of value added products.

The damage assessment maps were widely distributed and used by the humanitarian community for decision-making relating to planning of relief assistance.

No.	Date	Scale (Richter)	Killed	Affected	Total Affected	Lat.	Long.
1	19/01/00	6.3				36.2	70.4
2	25/02/01	6.1				36.4	70.9
3	01/06/01	4.9	4	0	270	35.1	69.38
4	03/03/02	7.2	150	3,500	3,500	36.543	70.424
5	25/03/02	6	1,000	35,000	35,200	36.011	69.371
6	12/04/02	5.8	65	6,000	6,150	35.88	69.25
7	29/03/03	5.9	1			35.976	70.585
8	05/04/04	6.6	3			36.512	71.029
9	10/08/04	6				36.456	70.775
	TOTAL		**1,223**	**44,500**	**45,120**		

Table 1. Earthquake Statistics, 2000-2004
Source: (http://earthquake.usgs.gov/activity/past.html)

Disaster damage maps overlaid with major fault lines (Figure 2) and slope maps proved valuable inputs to the development of disaster management plans ranging from national plans to sub-national levels particularly in an earthquake prone country. 50 Km Buffer analyses (Figure 3) showed less densely populated settlements, which according to reports from the limited ground surveys were barely affected by the earthquake. Lack of baseline data makes comparability of damage assessment maps to pre-disaster phase extremely difficult except if local knowledge and ground truthing is used for validation. Change detection analysis (Figures 4 and 5) of images taken in 2003 and those taken few days after the earthquake of 6 April 2004 showed little or no structural changes.

The mountainous and rugged terrain makes most areas inaccessible except by Donkeys, which might take weeks to get there. Due to its vantage position, EOS provided crucial data, which would have taken quite a while due to the inaccessibility of the areas to vehicles.

Data Type	**Imagery**	**Specification**	**Date Image Taken**	**Receipt Date by AIMS**
Pre-crisis	SPOT	HRG 2 (12000 by 12000) 5m, level 1 A processing	18/10/03	9-Apr-04
Pre-crisis	SAC	MS Image – 120 m Resolution	6/10/01	15/04/04
Pre-crisis	SAC	Pan Image – 35 m Resolution	6/10/01	15/04/04
Pre-crisis	IRS-P6	LISS-3 Medium Resolution MS 24m	26/03/04	15/04/04
Crisis	IRS-P6	LISS-4 High Resolution 5m, PAN	14/04/04	16/04/04
Crisis	IRS-P6	LISS-4 High Resolution 5m, PAN	14/04/04	16/04/04
Crisis	SPOT	HRG 1 PAN (24000 by 24000) 2.5m, level 1 A processing	17/04/04	21/04/04
Crisis	SPOT	HRG 1 MS (Image Corrupt)	17/04/04	21/04/04

Table 2. Data turn around time

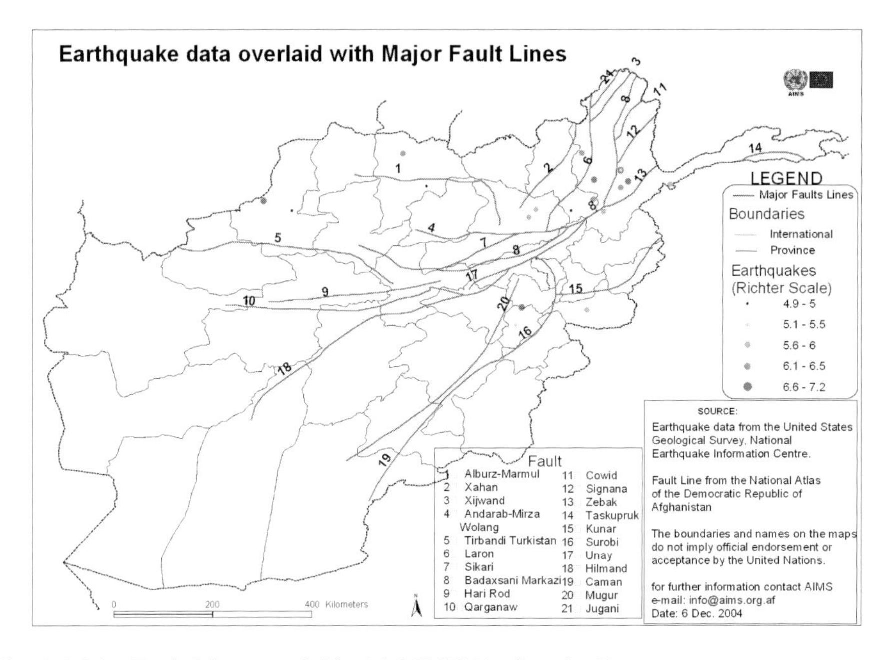

Fig. 2. Major Fault Lines overlaid with USGS Earthquake Data

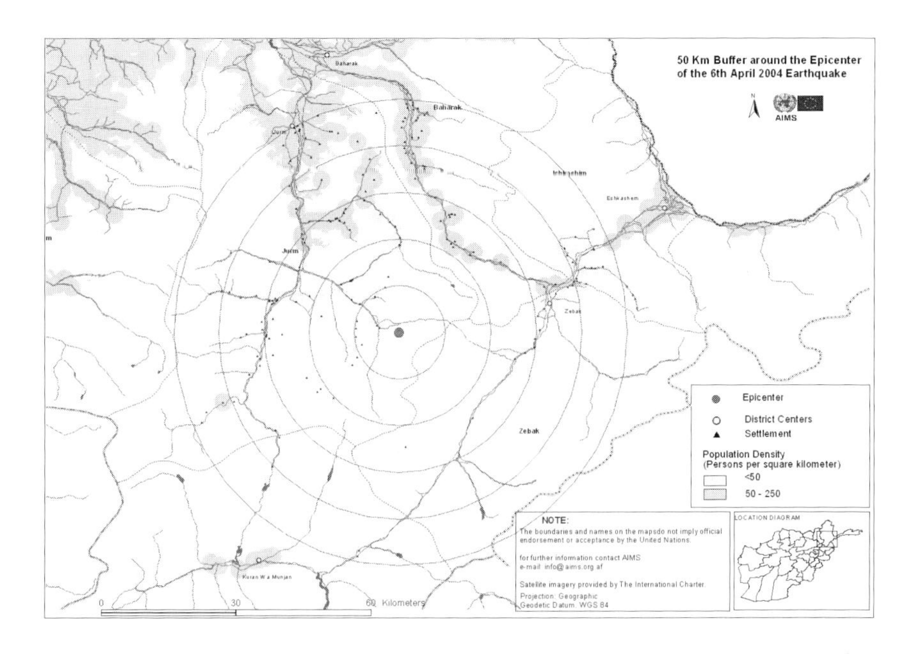

Fig. 3. Settlements within 50 Km Buffer around the Epicenter of the 6 April 2004 Earthquake

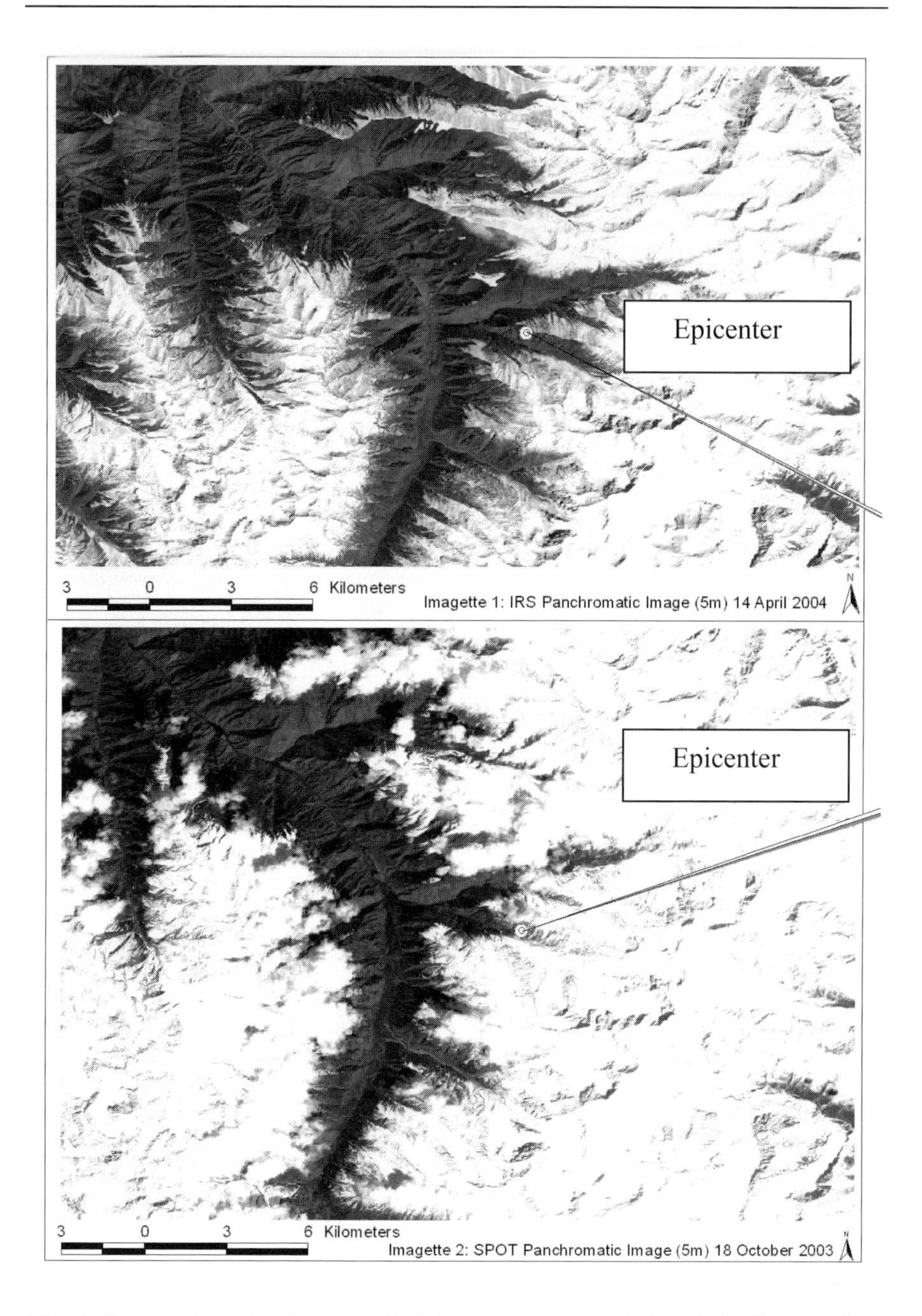

Fig. 4. Change detection between SPOT image taken in 2003 and IRS image taken on 14 April 2004

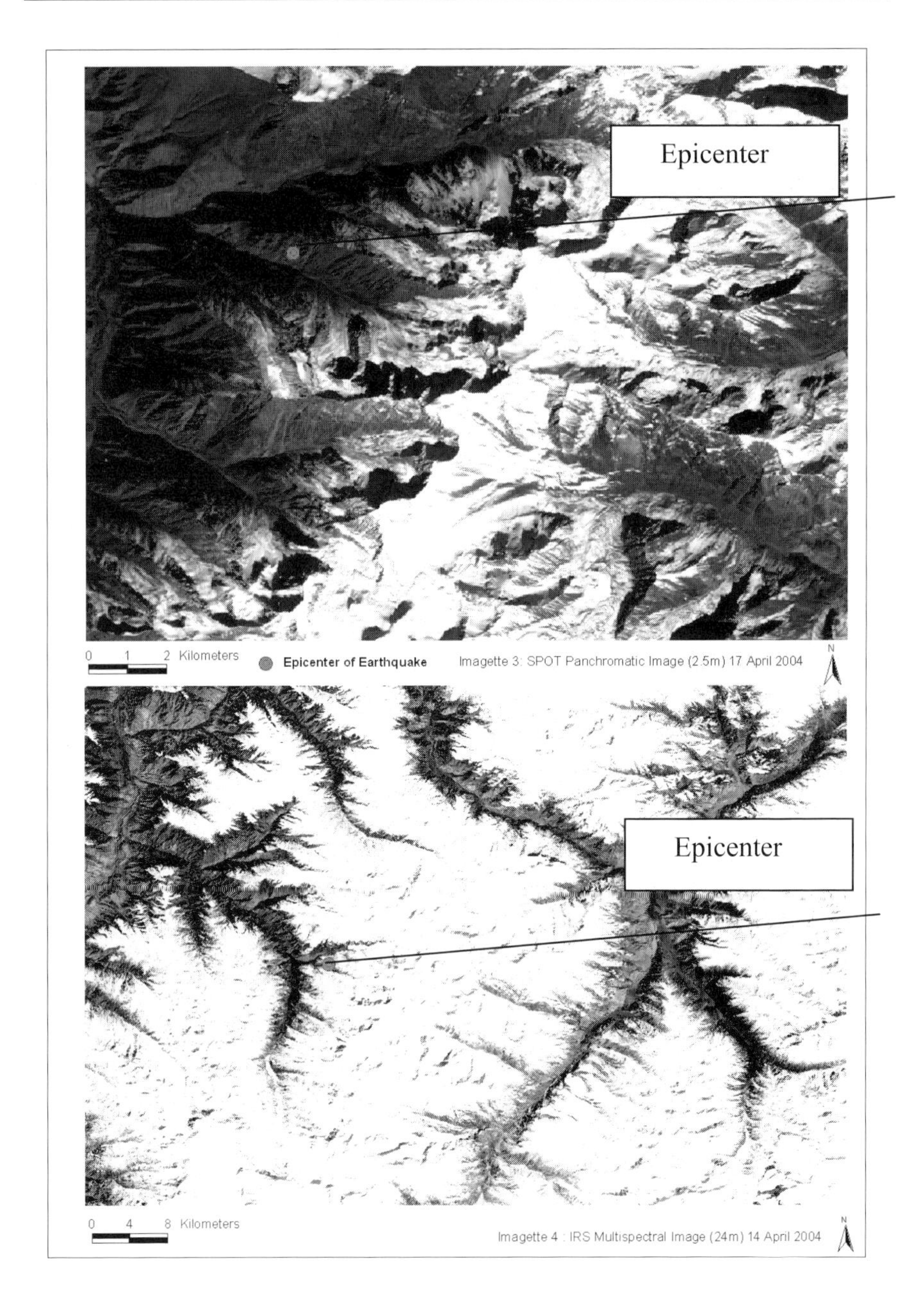

Fig. 5. Spatial variability between SPOT Pan and IRS Multispectral

3 Conclusion

Though there has been a remarkable improvement in the turn around time in delivery of space data, there are still notable delays in image data acquisition and delivery particularly due to problems of availability of cloud free images and too much snow cover. Most disaster management authorities in developing countries lack the infrastructure to access, process and deliver operational products. Lack of near real time data to monitor, assess and map the extent and impact of natural disasters has also hampered disaster management authorities in responding in a timely manner. The spatial and temporal resolution of the imagery provided by the Charter makes it difficult to monitor and assess the damage if the earthquake occurred in a built-up area with high population density.

Availability and access to space technology in addressing natural disasters has been the main obstacles facing developing countries particularly those poor countries without their own space programs. This problem has been solved through the introduction of The International Charter. Knowledge about the Charter is not common knowledge in most developing countries; there is need for a massive awareness campaign to educate decision makers about the Charter and the potentials of using space technology in addressing problems relating to disaster management. The skills to process satellite imagery and integrate it with GIS are lacking in most developing countries; there is need to embark on a massive capacity building exercise to ensure optimization of the benefits of the technology. The needs of the users are varied, and most often the users themselves do not know what they want. Data and information providers, and disseminators need to be fully aware of user requirements when space technology based applications are developed.

It is often assumed that disaster reduction information once acquired would be disseminated to the affected communities and the operational decision makers. A simple mechanism should be designed to ensure that information based on the needs of the users at the various levels is disseminated in a timely manner.

References

Alexander (1993) Natural disasters, UCL Press Ltd. London, pp 57-191

Gupta A (2000) Information Technology and Natural Disaster Management in India, The 21st Asian Conference on Remote Sensing, December 4-8, 2000 in Taipei, Taiwan

http://earthquake.usgs.gov/activity/past.html

http://www.cred.be – The International Disaster Database
http://www.oosa.unvienna.org/SAP/stdm/index.html.
IFRC/RC (2002) World disasters report 2002 : focus on reducing risk International Federation of Red Cross and Red Crescent Societies, 2002. 239 p
Van Westen CJ (2002) Remote sensing and geographic information systems for natural disaster management. In: Environmental modelling with GIS and remote sensing / A. Skidmore (ed.). London etc.: Taylor & Francis, 2002. pp 200-226

3D Buffering: A Visualization Tool for Disaster Management

Chen Tet Khuan and Alias Abdul Rahman

Department of Geoinformatics and Institute for Geospatial Science and Technology, Universiti Teknologi Malaysia, 81310 Skudai, Johor, Malaysia.
Email: {kenchen,alias}@fksg.utm.my

Abstract

Nowadays, 2D GISs are common and their related theories, concepts and models like for geometrical modeling and spatial relationships of objects are also well addressed and investigated. Most of the tasks related to 2D GIS applications are quite straightforward and relatively easy to handle. 2D GIS spatial analysis such as proximity analysis or proximity computation, network analysis, overlay function, neighborhood function, metric measurement and other analytical operations are also well understood and well researched by the GIS community. However, problem started to surface once we move towards 3D domain, i.e. to add an additional dimension to the current 2D GIS situations such as in spatial data modeling, analysis and application. Manipulating and handling spatial objects become more complicated as we move toward 3D. This paper attempts to address one of the problems in 3D analytical operation, i.e. 3D proximity analysis.

As we know that vertical component of spatial data directly interacts with the X and Y from the planimetric plane and makes the description of an object even harder to define. Data structure that supports object generation, preserves and maintains relationship with the neighboring objects is important in the 3D geometrical modeling. Several researchers in this problem domain have stated that 3D conceptual model, topological relationships, data collection, and spatial analysis might comprise a wide spectrum of questions and needs a lot of efforts to realize the solution. Although advancement in computer graphics have benefited to community in terms of 3D visualization and display but some other critical aspects like

3D spatial modeling together with the semantics information and spatial operators are hardly addressed, defined, and implemented commercially.

The commercial GIS packages that able to handle 3D datasets are rather limited to surface analysis and visualizing them in 3D. GIS accepts the fact that 2.5D GIS involves a single height attached to the planimetric positions (*X, Y*) whereas real or "true" 3D GIS should able to handle data like planimetric data with multiple heights, e.g. solid objects. Some advanced 3D tasks such as 3D overlay functions, and network functions are not available in some commercial GIS software, for examples, ERDAS's Imagine VirtualGIS, Intergraph Inc's GeoMedia, PCIGeomatics's Geomatica where they provide excellent tools for 3D visualization and 3D texture models. The systems also provide some operations like surface generation, volume computation, image draping, and terrain inter-visibility can be carried. However, "true" 3D operations are hardly available. Inevitably, many issues need to be investigated.

To move on to the 3D GIS, the third dimension must not be constrained by the single XY plane only. Considering the 3D analysis is the core component of the 3D GIS, therefore, an investigation that involves data input and 3D analytical operation will be addressed in this paper. Other aspect such as databasing is out of the scope. The developed analytical operations have been tested using real datasets that cover Universiti Teknologi Malaysia (UTM) main campus.

In general, managing disaster scene is quite demanding and needs rapid spatial information on the spot where some of the required information is in the form of 3D display of a spatial query and analysis. This paper discusses the development of proximity analysis that is the 3D buffering. The corresponding algorithms that work for most of the spatial primitives, i.e. point, line, and polygon in 3D will be discussed. We tested our buffering approach by using photogrammetrically captured datasets. Finally, the paper provides outlook to the proposed work towards the development of advanced 3D analytical solutions in 3D GIS domain.

1 Introduction

At the moment, commercial GIS packages that able to handle 3D datasets are rather limited to surface analysis, and 3D visualization only. Example of such analyses like triangulation based surface analysis (or TINs) by Abdelguerfi, *et al* (1998) and surface visualization (Batty, 2000; Berry, *et al.* 2001; Comer, 2000). All these analyses are 2.5D in nature and it does not involve true 3D volumetric datasets. Topological relationships of the 2.5D datasets is basically has no major different compared with the 2D counterpart. 3D GIS needs 3D datasets and should be able to analyze the data within the 3D environment (e.g. 3D spatial analysis). That is the main reason why 2.5D GIS fails to produce real 3D spatial information. Some advanced tasks such as 3D overlay functions, and network functions are not available in some commercial GIS packages like ERDAS's Imagine VirtualGIS, Intergraph Inc's GeoMedia, PCIGeomatics's Geomatica. These systems are only able to provide excellent tools for 3D visualization and 3D texture models (Zlatanova, *et al.* 2002). Such systems also able to perform certain surface analysis using like surface generation, volume computation, image draping, and terrain inter-visibility. However, true 3D analytical operations are hardly available from these commercial packages.

The use of 2.5D representation of surfaces has some disadvantages for modeling true 3D datasets - lack of volumetric capability (Bernhardsen, 1999) as well as solid modeling (Bajaj, *et al.* 1996; Boissonnat, 1984). On the other hand, simple and efficient surface generation can be useful for tasks that do not require extensive 3D capability or functions. For examples, slope and terrain analysis (De Floriani, *et al.* 1998), line of sight, topographic feature classification and drainage networks (Yu, *et al.* 1996).

Some applications in meteorology, geology, geoscience, and urban planning require three-dimensional modeling and representation. Several existing systems attempt to represent and model the phenomena by using existing 2D tools and obviously some of the objects could not be analyzed fully. This paper describes the possible 3D analytical operation, i.e. 3D buffering for simple primitives objects like points, lines, and polygon surfaces. The discussion on the algorithms forms major part of this paper. To map the 3D dataset in a true three-dimensional space would not be difficult nowadays due to the advancement in computer graphics. However, to describe its spatial model and relationship is another great challenge because the 3D topological model for 3D objects is still insufficiently studied. The topological aspect of the constructed buffering objects is not part of the paper. The proposed method has been tested with real datasets of the Universiti Teknologi Malaysia (UTM) campus.

2 3D Buffering: A Vizualization Tool of Geospatial Primitives

3D buffering zone could be defined as an operation to generate proximity information of a spatial object, for instance, phenomenon along linear features or polygon features in a three-dimensional space. The operation could be considered as one of the 3D analytical functions for 3D GIS. In GIS, such analytical tool is useful for analyzing spatial objects. User may want to know the shortest route from one point to the other, the distance between two places, or within an area of a housing estate, etc. These kinds of demands need of an analysis tool. This project deals with:

1. Develop algorithms for 3D buffering tool as a framework for 3D analytical solution for 3D GIS. The module consists of three main components respectively to the 3D geospatial primitives, i.e. point, line, and polygon, and
2. Test the developed tool using the real datasets (UTM campus).

Other aspect of 3D analytical operations such as topology and database technology are out of the scope.

In the next section, each of the buffering objects will be studied in detailed (i.e. how to construct them geometrically) based on the geospatial primitives (point, line, and polygon). To initiate the object model for buffering zone, the approach of the buffering object for each of the primitives will be mentioned. Later, all the mathematics that follows and the expected buffering result will be highlighted. On the other hand, the structures of each of the geospatial primitives need to be identified. This is due to the construction of the geospatial primitives affects the development of a buffering object model. Fig. 1 shows the structural model of the 3D geospatial objects.

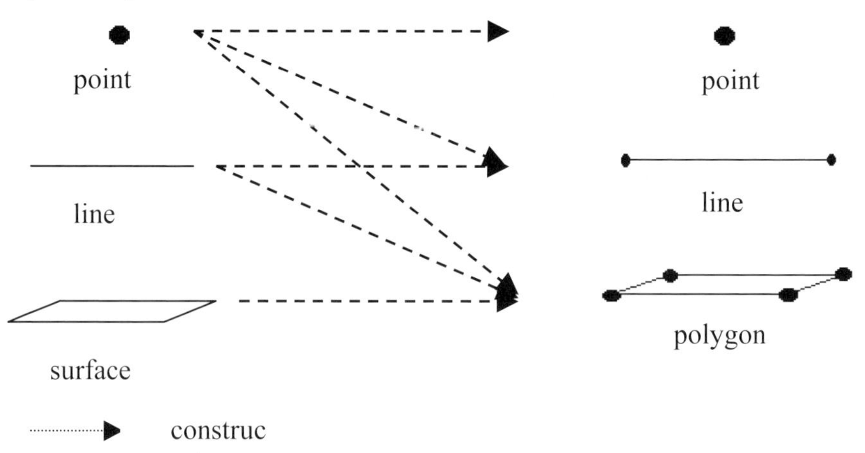

Fig. 1. Geospatial object

Visualization is crucial in any decision making process including for managing the chaotic situation like huge gas pipeline burst, etc. Here, the buffering tool could be utilized to estimate the extent of the damage based on the simple object primitives.

From Fig. 1, points, lines, and a surface form a polygon, whereas a line consists of points. However, a point remains the same structure. After identifying the formation of each of the geospatial primitives, the buffering object modeling are easier to be performed since the scope of the modeling are narrowed into three buffering model. There are point (for point, line, and polygon), line (for line and polygon), and surface (for polygon) buffer. Yet, the final buffering model for the geospatial primitives is not that superficial as mentioned before. Therefore, the detailed studies about the solid buffer model for point, line, and polygon will be discussed in the following sections.

2.1 3D Point Buffering

Point is defined as single coordinate triplets (*x, y, z*) in 3D. Points are used to represent objects that are best described as shape- and shapeless, single-locality features (Rolf, 2000). The point only consists of node feature. Therefore, its buffering zone generated by a fixed distance produces a sphere in three-dimensional space. Consequently, the point itself appears in the center of a buffering zone (see Fig. 2).

The method to generate a sphere begins with the creation of polygon surface. The same approach adapted by ESRI (1998) where the implementation of the *PolygonZ* to create a solid buffering object (sphere). The *PolygonZ* is a one of the spatial features appears within the ESRI's shapefile library. The *Polygon* represents the plane surface, whereas the *Z* ensures the *Polygon* occupied in the three-dimensional space. The research implements the *PolygonZ* to create surfaces due to the lack of curve surface appears within that shapefile library. Therefore, joining all related polygon surfaces creates a sphere.

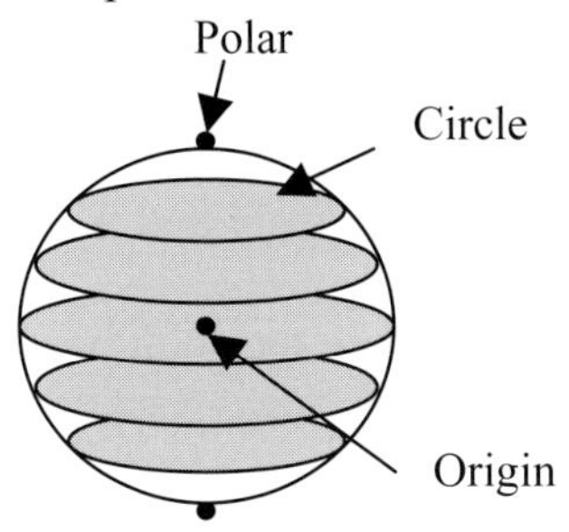

Fig. 2. Sphere created from a point

Again Fig. 3 shows the construction of circles for 3D point buffering, whereas Fig. 4 shows the method to create a sphere.

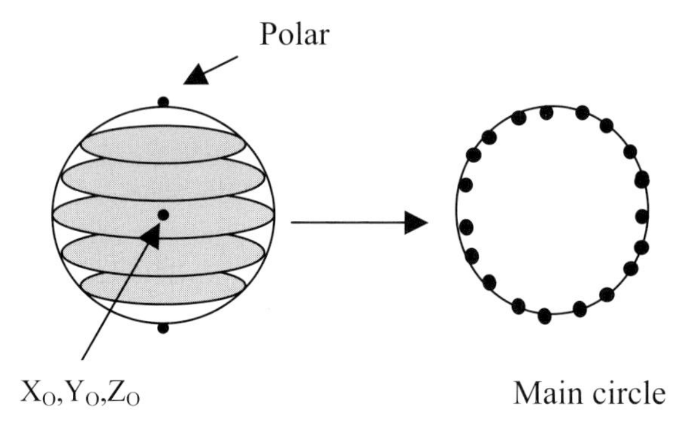

Fig. 3. Method to create circles

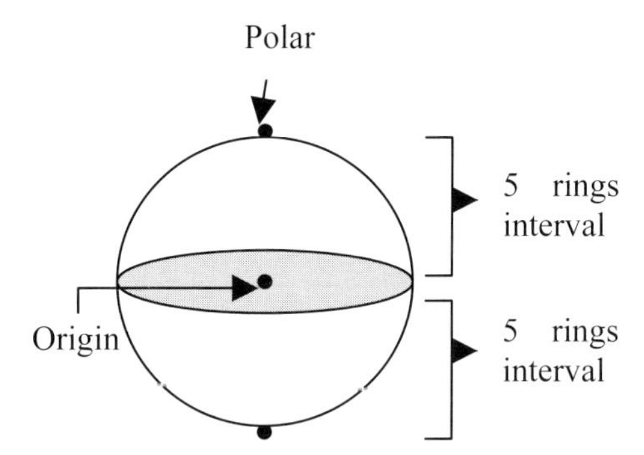

Fig. 4. Method to create sphere

The construction of the main circle appears in the first step. Later, there are five circles are created on upper and lower side of the main circle, respectively. The *PolygonZ* are built using four points (two at upper points and two at lower points) from the two consecutive circles. This process is repeated until the surfaces are patched completely and the whole sphere is created.

The method to generate the buffering zone for point has been described and their corresponding algorithm is in the following section.

2.1.1 The Mathematics for Point Buffering

This section gives a detailed description of the point-buffering algorithm. The problem definition is as follow:

Consider a coordinate point $O(x, y, z)$ as an origin for the buffering zone, then calculate the array of the input coordinates, say H, of heights $H(x, y)$ for each interval ring from a sphere. The purpose of calculating $H(x, y)$ is to divide the radius into 5 equal parts. This will maintain the same height distance between two successive circles. Later, the array $H(x, y)$ will be used to combine with the 2D point sets of circle, (x_i, y_i). From that combination, a complete point sets for each circle is prepared to perform further geometrical construction in three-dimensional. The method to calculate the H is given in the Fig. 5.

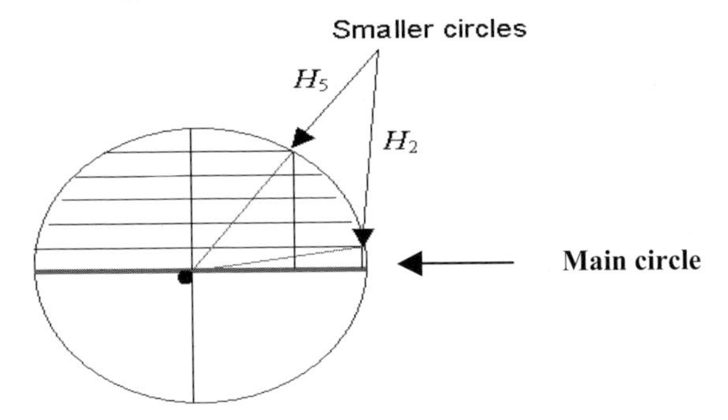

Fig. 5. Calculate the height **array, H_i**

$$H_i = z \pm \left(r \times \frac{i}{5} \right) \tag{2.1}$$

where H_i denotes the array of height respective to the ring from the sphere (see Fig 5); z is the coordinate value from origin; r is the buffer length; i is the array number; $\pm$ are for the upper ring and lower ring, respectively.

The height of each ring changes simultaneously with the buffer length. After calculated the height, we need to define precisely the buffer length corresponding to the H_i. Therefore, the buffer length for each ring, respective to the H_i is:

Refer to the *Phytagoras Teorem*,

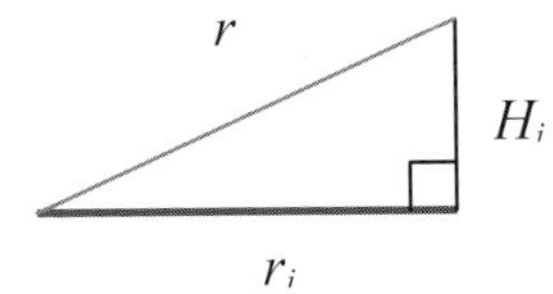

Fig. 6. Calculate the buffer length array, r_i

$$H_i = \frac{r}{5} \times i \qquad (2.2)$$

$$r_i = \sqrt{r^2 - \left(\left(\frac{H_i}{5}\right) \times (i)\right)^2} \qquad (2.3)$$

where r_i denotes the array of buffer length corresponding to the H_i (see Fig. 6). From the equation 2.3, r_i is calculated. This array determines the size of a circle. Therefore, to perform calculation for generating point sets for a circle is denoted below:

$$x_j = r_i \times \cos(\theta_j) \qquad (2.4)$$

$$y_j = r_i \times \sin(\theta_j) \qquad (2.5)$$

Upper ring :

$$z_i = z + \left(\left(\frac{r}{5}\right) \times i\right) \qquad (2.6)$$

Lower ring:

$$z_i = z - \left(\left(\frac{r}{5}\right) \times i\right) \qquad (2.7)$$

where r is the buffer length from origin; i is the number of ring from 1 to 5 for both upper and lower ring; j is the angle from 0^0 to 360^0; r_i is the buffering length corresponding to the upper and lower ring k; x_j, y_j, z_i are the coordinates for ring i.

For the polar point,

$$x_p = x_o \qquad (2.8)$$

$$x_p = y_o \qquad (2.9)$$

Upper polar point:

$$z_{p(upper)} = z_o + r \qquad (2.10)$$

Lower polar point:

$$Z_{p(lower)} = z_o - r \qquad (2.11)$$

where x_o, y_o, z_o are the coordinates for the origin; x_p, y_p, $z_{p(upper)/(lower)}$ are the processed coordinates for each ring.

The first idea of creating the point sets for the rings is to prepare the input array to construct the point buffering zone. Thus, it strongly depends on the insertion order of the point sets. Starting from the main circle (see Fig. 4), either moving up or down to the successive circle (optional), both of them need to be joined to form a surface of point buffering surface. Fig. 7 shows the formation of the surface starts from the main circle to the successive upper circle.

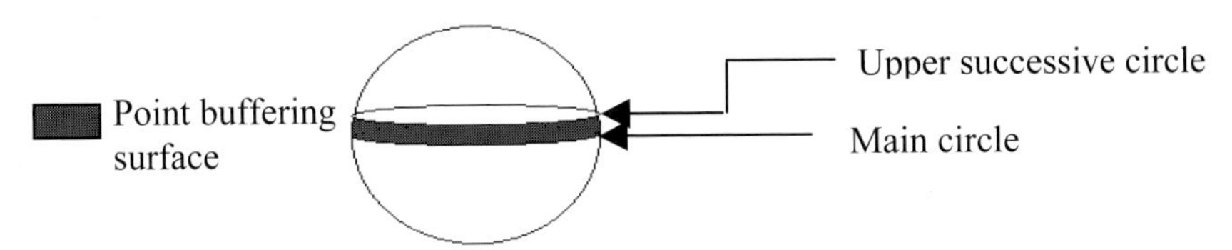

Fig. 7. Point buffering surface

Suppose that we assume the processed datasets for each circle is denoted to x_i, y_i, z_i, where i is the incremental number starts from 0° to 360°, representing the moving angle of a arc.

The array for the joint is:

$$A = \lfloor (x_i \quad y_i \quad z_j), (x_{i+1} \quad y_{i+1} \quad z_j), (x_{i+1} \quad y_{i+1} \quad z_{j+1}), (x_i \quad y_i \quad z_{j+1}) \rfloor \quad (2.12)$$

where A denotes the point buffering surface (inline to the ESRI's ArcView shapefile - PolygonZ) joined by 4 points from 2 successive circles. However, there is a slightly different in the both of the polar section. Since both of the upper and lower polar consist of only a single point, each surface are formed by three points.

The array for the joint at polar section is:

$$A_{Polar} = \lfloor (x_i \quad y_i \quad z_5), (x_{i+1} \quad y_{i+1} \quad z_5), (x_{Polar} \quad y_{Polar} \quad z_{polar}) \rfloor \quad (2.13)$$

Joining all the surfaces from the processed datasets completes the point buffer zone. This section is a vital section because the implementation will be re-used in the line buffer modeling.

2.2 3D Line Buffering

Two end nodes together with zero or more internal nodes define a line. Another phrase like polyline, arc or edges are also being used in GIS. Rolf (2000) had mentioned line data are used to represent one-dimensional objects such roads, railroads, canal, rivers and power lines. Refer to the Fig.

1, line is a combination of both nodes and edge. The straight parts of a line between two successive vertices (internal nodes) or end nodes are called line segments. Therefore, in order to line buffer in 3D space, the joined components of a single line need to be identified (i.e the nodes and edge(s)), see Fig. 1.

As mentioned in the section 2.1, point buffering output is a sphere, whereas the line buffering output is a cylinder. Since nodes appear in the line segment, adding the point buffering zone into the line buffer model becomes a necessity. Fig. 8 shows the method to create a line buffering output.

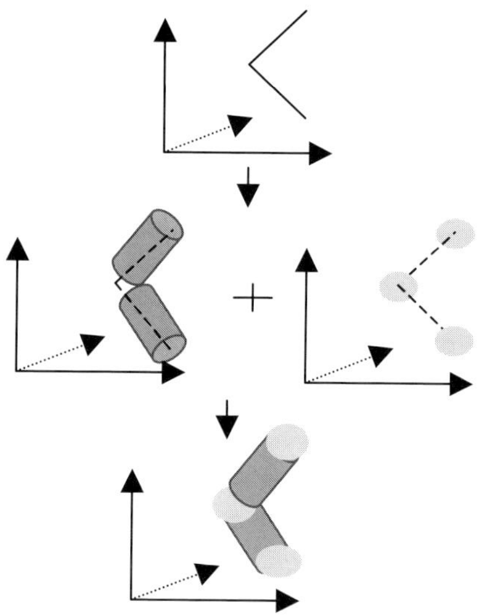

Fig. 8. Method to create line buffer

2.2.1 The Mathematics for Line Buffering

This section gives a detailed description of the line-buffering algorithm. The problem definition is as follows.

A series of input coordinate point, P_i forms a center line as the origin of a buffering zone. At the initial step, our goal is to create a cylinder for each line segment. Two rotated circle need to be created by using two consecutive nodes from a line segment.

The mathematics of the rotation follows:

$$\theta_{YZ} = Tan^{-1}\left(\frac{\Delta z}{\Delta y}\right) \tag{2.14}$$

$$\theta_{XZ} = Tan^{-1}\left(\frac{\Delta z}{\Delta x}\right) \tag{2.15}$$

$$\theta_{XY} = \mathrm{Tan}^{-1}\left(\frac{\Delta y}{\Delta x}\right) \tag{2.16}$$

where θ_{XY}, θ_{XZ}, θ_{YZ} are the rotated angle for XY, XZ, and YZ plane respectively.

The rotated circle should be 90° (perpendicular) toward the line. Fig. 9 shows the result after having three steps of rotation.

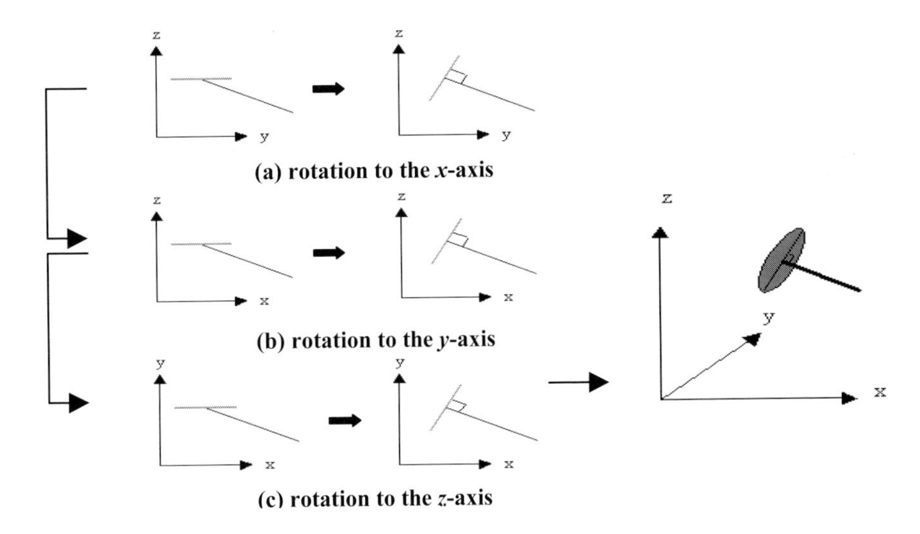

Fig. 9. Final rotated circle

After calculating the angle (θ) for each plane, the plane circle will be rotated using that processed angle. Firstly, we rotate the plane circle toward the x-axis. It is followed by the y-axis and z-axis rotation, subsequently.

$$x_{(YZ)_i} = r \times \cos(i) \tag{2.17}$$

$$y_{(YZ)_i} = r \times \sin(i) \times \cos(\theta_{YZ}) \tag{2.18}$$

$$z_{(YZ)_i} = r \times \sin(i) \times \sin(\theta_{YZ}) \tag{2.19}$$

where $x_{(YZ)_i}$ i, $y_{(YZ)_i}$, $z_{(YZ)_i}$ are rotated coordinates for YZ plane (see Fig. 9(a)).

Rotation towards the x-axis will maintain the location of the x-coordinate value for each point. Therefore, the x-coordinate value applies the formula to generate a circle (see Equation 2.17 and Fig 10), whereas the rest applies the rotation angle (θ_{YZ}) (the first phase of rotation towards the x-axis).

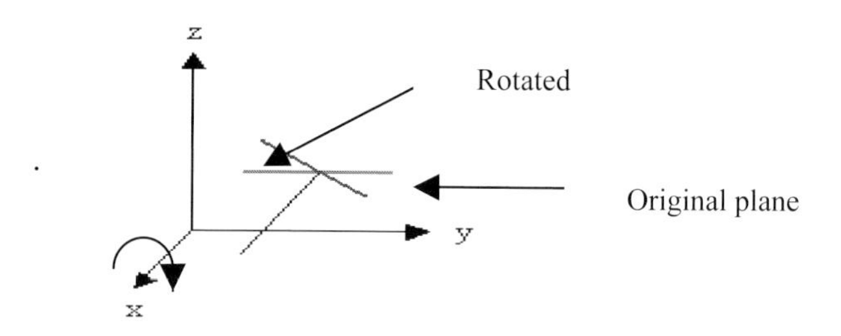

Fig. 10. Rotated circle towards x-axis

Moving on, the results will be manipulated for the 2nd rotation, which towards the y-axis. Before entering this step, the latitude and longitude of the points from the circle need to calculated (see Equation 2.21 & 2.22).

$$B = \theta\{x_i, z_i\} \quad (2.20)$$

$$LongB = B + (\theta_{XZ}) \quad (2.21)$$

$$LatB - \cos^{-1} (Y_i / r) \quad (2.22)$$

$$x_{(XZ)_i} = r \times \cos(LongB) \times \sin(LatB) \quad (2.23)$$

$$y_{(XZ)_i} = r \times \cos(LatB) \quad (2.24)$$

$$z_{(XZ)_i} = r \times \sin(LongB) \times \sin(LatB) \quad (2.25)$$

where B is the angle for each x_i, z_i; LongB is the longitude of x_i, z_i; LatB is the latitude of x_i, z_i; $x_{(XZ)_i}$, $y_{(XZ)_i}$, $z_{(XZ)_i}$ are the coordinates after the rotation of the y-axis (see Fig. 9(b)).

Rotation towards the y-axis maintains the location of the y-coordinate value for each point. This is because the y-coordinate value applies formula to re-generate a circle (see Equation 2.17 and Fig 11) after the latitude was calculated, whereas the rest applies the rotation angle (θ_{XZ}) to re-define the circle (the 2nd phase of rotation).

Finally, it is followed by the third phase of rotation (for the z-axis). Refer to the equation 2.16, it gives the angle for the XY plane (z-axis rotation). However, that is not the final rotated angle. The second phase of rotation already served a partial portion of rotation (θ'_{XY}) towards the z-axis. It needs to be substrated by the rotation angle of z-axis (θ_{XY}) (after applying the second phase of rotation) in order to get the actual rotation angle (θ''_{XY}). Fig. 11 shows the reason that equation 2.16 needs to be finalized in further equation (see equation 2.26).

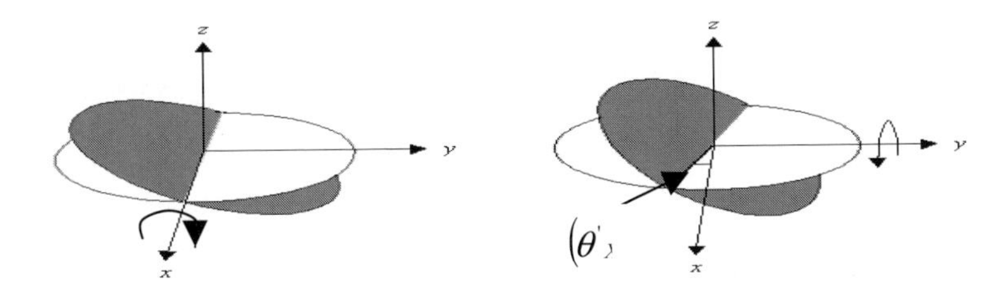

(a) First phase of rotation (x-axis)

(b) Second phase of rotation

Fig. 11. Two phase of rotation (x-axis and y-axis)

$$(\theta''_{XY}) = (\theta_{XY}) - (\theta'_{XY}) \tag{2.26}$$

Figure 12 devotes the result after the subtraction of (θ'_{XY}) by (θ_{XY}).

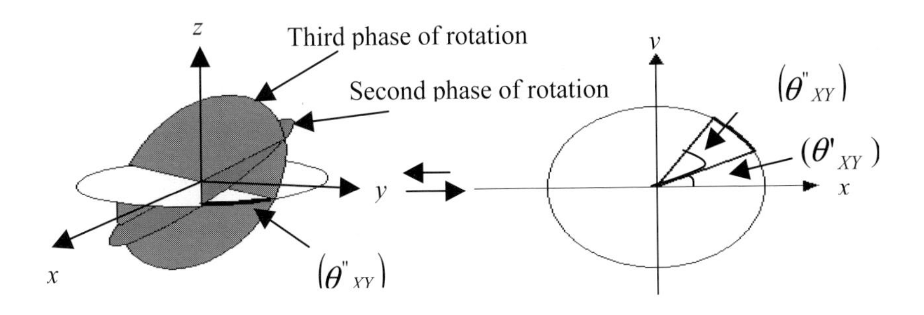

Fig. 12. Third phase of rotation (z-axis)

The latitude and longitude of the points from the circle need to compute (see Equation 2.28 & 2.29) after (θ''_{XY}) is calculated. The purpose of calculating the longitude is to locate the point and perform the third phase rotation toward the z-axis, whereas the use of latitude is to create circle before rotation. The final mathematics of the line buffering follows:

$$A = \theta\{x_i, y_i\} \tag{2.27}$$

$$LongA = A + (\theta''_{XY}) \tag{2.28}$$

$$LatA = \cos^{-1}(Z_i / r) \tag{2.29}$$

$$x_{(XY)_i} = r \times \cos(LongA) \times \sin(LatA) \tag{2.30}$$

$$y_{(XY)_i} = r \times \sin(LongA) \times \sin(LatA) \tag{2.31}$$

$$z_{(XY)_i} = r \times \cos(LatA) \tag{2.32}$$

where A is the angle for each x_i, y_i; LongA is the longitude of x_i, y_i; LatA is the latitude of x_i, y_i; $x_{(XY)_i}$, $y_{(XY)_i}$, $z_{(XY)_i}$ are the coordinates after the rotation of the y-axis (see Fig. 9(c)).

Two rotated circles are the core components of a cylinder. The cylinder is created by joining that two successive circles that will cover the origin (line segment) inside. Recall the properties of a line, which it consists of two nodes and an edge itself for a single line segment. Those two nodes are used to generate point buffering. Once the cylinder is created, the spheres (points buffer zone) need to be connected with the cylinder to become a complete 3D line buffering. As what had mentioned in the section 2.1, the method for sphere was given. Therefore, in the following stage, we need to remove the unnecessary internal sphere's surface covered inside the cylinder. Fig. 13 shows the result after both point and line buffering zone is connected.

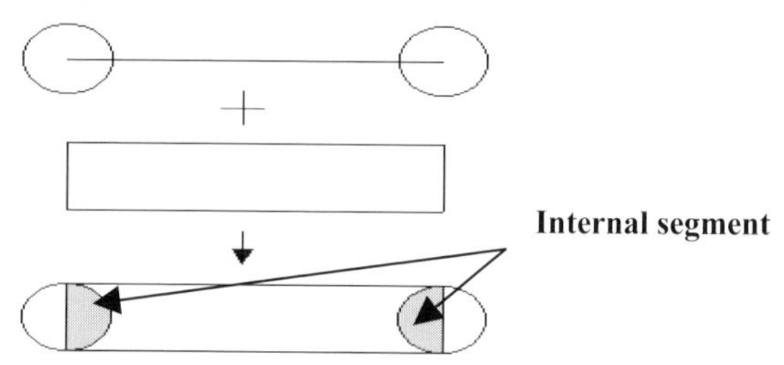

Fig. 13. Side view of line buffering

Fig. 14 denotes the method to remove internal segment of line buffering zone.

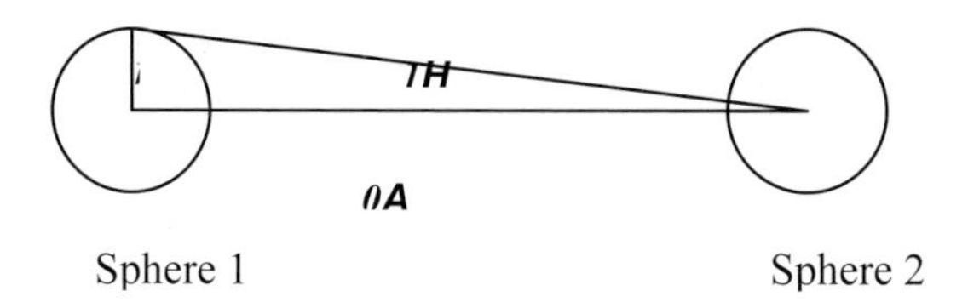

Fig. 14. Method to remove internal segment

This method implements the Teorem Phytagoras as mentioned in Fig. 6. Suppose that 2 buffering zone for both point and line were created. We need to calculate the distance between the origin of sphere 2 and the tangent of sphere 1, which is H. r denotes the buffering length, whereas As denotes the distance between the origin of both sphere 1 and 2. Therefore, the mathematics to derive the H follows:

$$As = \sqrt{\left((x_1 - x_2)^2 + (y_1 - y_2)^2 + (z_1 - z_2)^2\right)} \quad (2.33)$$

$$H = \sqrt{\left((r)^2 + (As)^2\right)} \quad (2.34)$$

The algorithm implements the H as the limit for distance comparison. The distance between two points from sphere 1 and the origin of sphere 2 that exceed H, it will not be removed. The same situation applies for the distance between points from sphere 2 and the origin of sphere 1. Fig.15 gives the final result of line buffering.

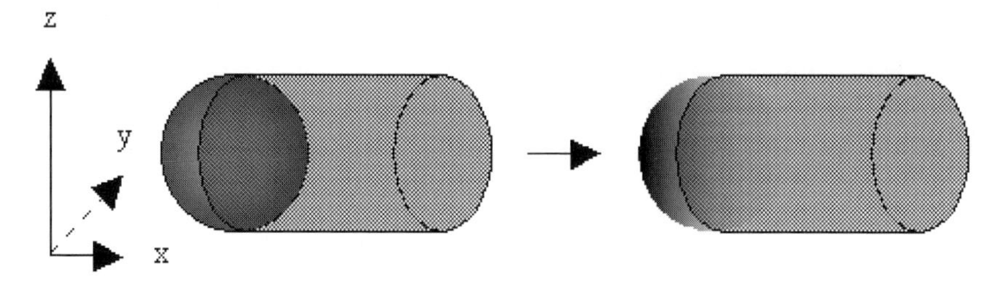

Fig. 15. Method to remove internal segment

With the combinations of the spheres and cylinders that produce a line buffering zone is the best approach model. Comparing to the model mentioned by Kim et al. (1998), the absent of sphere in the line buffering model make the approach incomplete. This is due to start node and end node does not produce the buffer zone Moreover, the interval nodes that create the buffer zone exceed the buffer length (see Fig. 16).

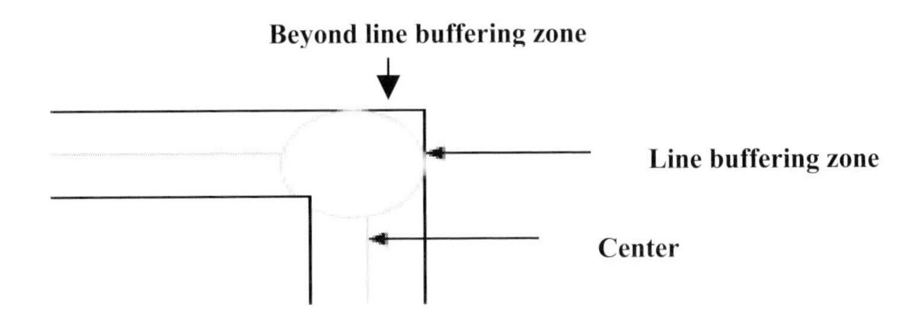

Fig. 16. Weakness of Kim, et al's model

2.3 3D Polygon Buffering

A polygon feature is represented by some arc/node structure that determines an area's boundary. With at least three or more nodes joining together consecutively and return to the starting node define a polygon. To perform geometrically a buffering operation towards a polygon in three-dimensional space, the joined components of a single polygon need to be identified, which are the nodes, edge(s), and a surface, see Fig. 1.

2.3.1 The Mathematics for Polygon Buffering

To generate the surface buffer works in the first place. Thus, the buffering surfaces consist of an upper surface and a lower surface as shown in Fig. 17.

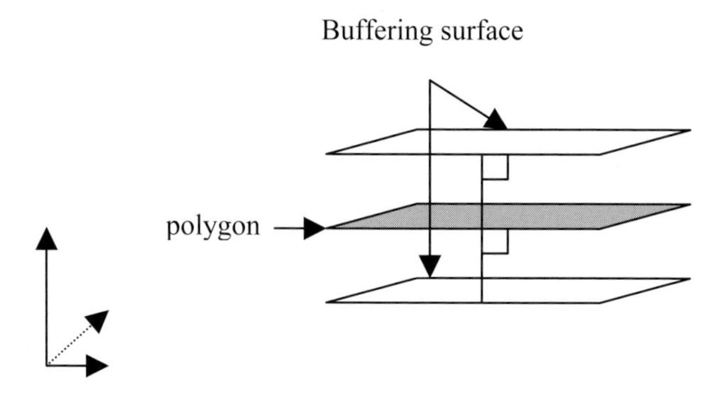

Fig. 17. Buffering surface

The buffering layers are expanded upper and lower starting from the center of polygon, respective to the buffer length. Both of them should be parallel to the main polygon. To define these polygons buffer, the vector's cross product is used. This vector produces a result that is perpendicular to the main polygon (v and u). For vectors u and v, the cross product is defined by (see Fig. 18):

$$\begin{aligned} u \times v &= \hat{x}(u_y v_z - u_z v_y) - \hat{y}(u_x v_z - u_z v_x) + \hat{z}(u_x v_y - u_y v_x) \\ &= \hat{x}(u_y v_z - u_z v_y) + \hat{y}(u_z v_x - u_x v_z) + \hat{z}(u_x v_y - u_y v_x) \end{aligned} \quad (2.35)$$

This can be written in a shorthand notation that takes the form of a determinant (see equation 2.36).

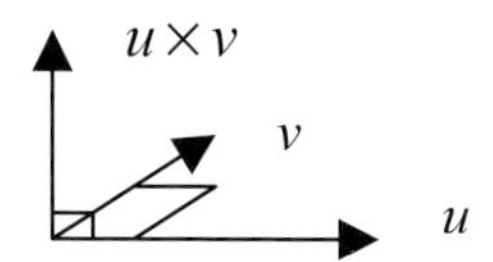

Fig. 18. Vector (u x v)

Here, the (u x v) is always perpendicular to both u and v, with the orientation determined by the right-hand rule.

$$u \times v = \begin{vmatrix} \hat{x} & \hat{y} & \hat{z} \\ u_x & u_y & u_z \\ v_x & v_y & v_z \end{vmatrix} \tag{2.36}$$

Some of the mathematics are illustrated below (refer to Fig. 18). The vector u (A) and v (B) need to be compute in the early step (see Fig. 19 and Equation 2.37 to 2.40). The purpose of computing those vectors is to find out the vector scale [$|N_X|$, $|N_Y|$, $|N_Z|$]. Later on the vector scale are used to compute the point sets that form the polygon.

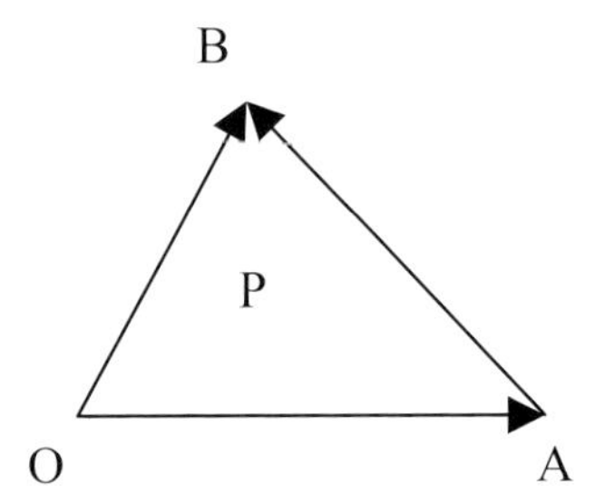

Fig. 19. Vector for polygon P

$$\overrightarrow{OA} \text{ x } \overrightarrow{OB} = \left[|n_1| \quad |n_2| \quad |n_3| \right] \tag{2.37}$$

$$|n_1| = \begin{vmatrix} Y_{oa} & Z_{oa} \\ Y_{ob} & Z_{ob} \end{vmatrix} \tag{2.38}$$

$$|n_2| = \begin{vmatrix} Z_{oa} & X_{oa} \\ Z_{ob} & X_{ob} \end{vmatrix} \tag{2.39}$$

$$|n_3| = \begin{vmatrix} X_{oa} & Y_{oa} \\ X_{ob} & Y_{ob} \end{vmatrix} \tag{2.40}$$

$$\vec{AB} = \sqrt{\left(|n_1|^2 \quad |n_2|^2 \quad |n_3|^2 \right)} \tag{2.41}$$

$$\hat{AB} = \frac{\left[|n_1| \quad |n_2| \quad |n_3| \right]}{\vec{AB}}$$

$$= \left(|N_x| \quad |N_y| \quad |N_z| \right) \tag{2.42}$$

$$x_c = x_o + |N_x| \times r \tag{2.43}$$

$$y_c = y_o + |N_y| \times r \tag{2.44}$$

$$z_c = z_o + |N_z| \times r \tag{2.45}$$

where x_c, y_c, z_c are the processed coordinates for the point C.

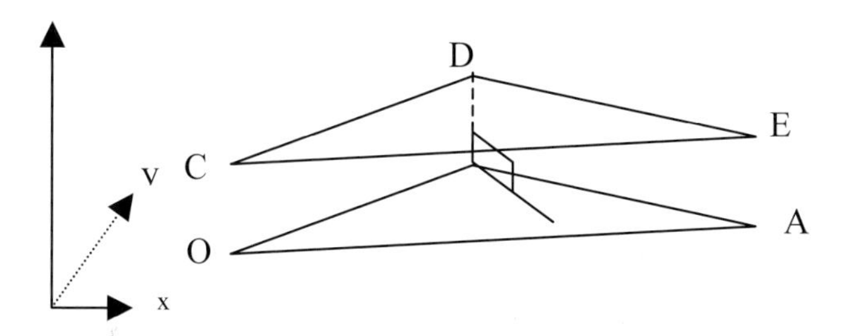

Fig. 20. Upper polygon buffer CDE

Next, point D and E will be calculated (see Fig. 20) by using the same equations start from 2.37 to 2.45. From the Fig. 20, three points (C, D, and E) create a polygon. The polygon becomes an upper polygon of a buffer model. However, the complete polygon buffer shouldn't appear as Fig. 20 only. The lower polygon surface needs to be computed (see Fig 17). The only difference between the upper and lower polygon buffer is the direction determine by the right hand rule. Both of them are directed to the opposite direction based on the main polygon. Therefore, identity that involves the cross product includes:

$$u \times v = -v \times u \tag{2.46}$$

Refer to the right-hand rule, -v x u refer to the lower surface buffer, which is opposite to the Fig. 21.

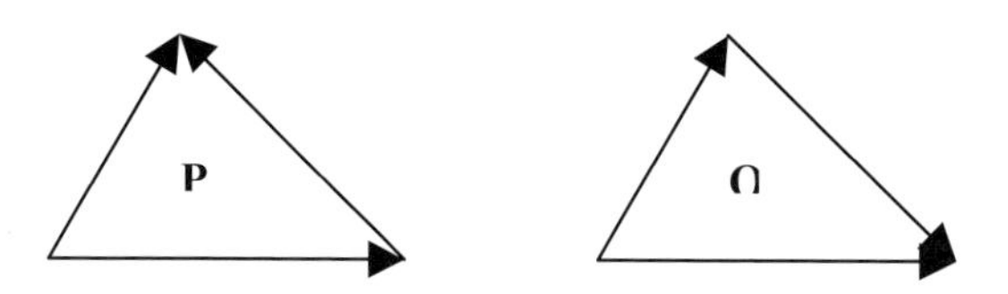

Fig. 21.: Vector for polygon P(upper), and Q(lower)

Refer to the identity, some minor changes must be done starts from the equation 2.42. The mathematics for the lower buffer surface follows:

$$-\hat{AB} = \hat{AB} \quad (2.47)$$

$$x_{c'} = x_o - |N_x| \times r \quad (2.48)$$

$$y_{c'} = y_o - |N_y| \times r \quad (2.49)$$

$$z_{c'} = z_o - |N_z| \times r \quad (2.50)$$

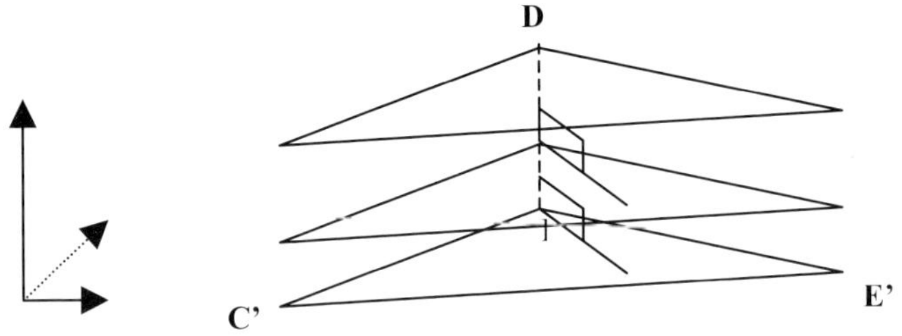

Fig. 22. Upper surface buffer CDE & lower surface buffer C'D'E'

Recall that a polygon is a combination of point line and surface. The polygon buffering zone should implement point and line buffer model. After the upper and lower polygons buffer are defined, the line (cylinders) and point (sphere) buffering objects need to be combined. However, the internal segment between line buffer and the polygon buffer are still appears within the model. The unnecessary internal segment needs to be removed. The same approach (as shown in section 2.2.1) is used for removing them.

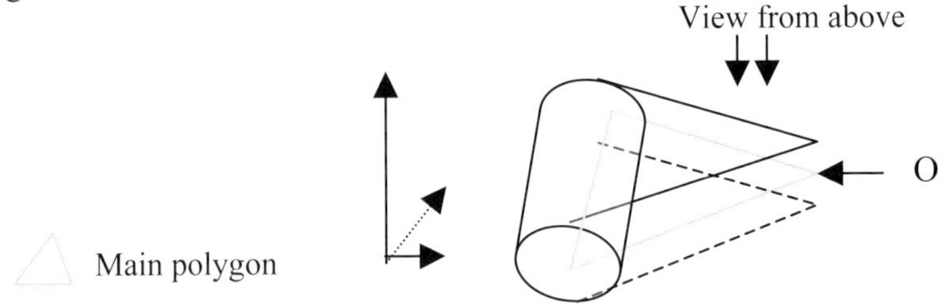

Fig. 23: Combination of cylinder and surface buffer

After the internal segment of sphere was removed (mentioned in the section 2.2.1), there is no intersection between the sphere (point buffer) and the box (polygon buffer). Thus, attention is given into the internal segment between cylinder (line buffer) and box (polygon). To remove all the internal segment of line buffer, we just need to focus on the intersection between both models. First, we need to concentrate on the cylinder and the point O (refer to the Fig. 24), which is the point opposite to the cylinder and it is being a part of the main polygon.

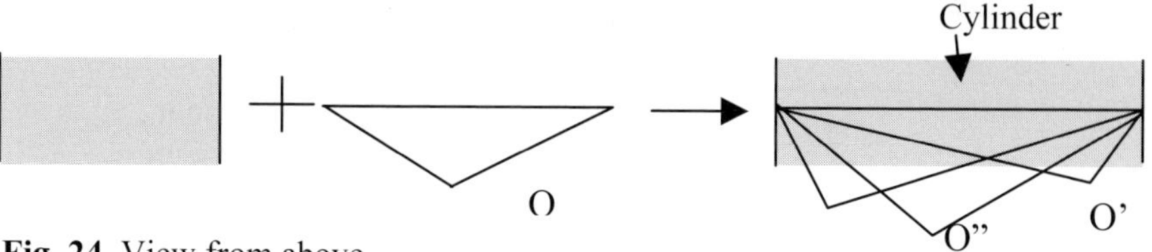

Fig. 24. View from above

Either the third point that creates a polygon is O, O', or even O" from the Fig. 24, the representation remains the same as in Fig. 25.

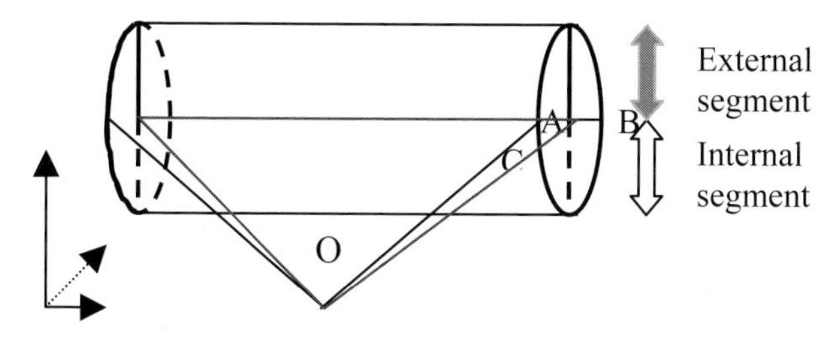

Fig. 25. Method to remove internal segment

The cylinder is divided into two segments, which are the internal and external part. The main problem is to determine where the internal is and the external segment. Refer to Fig. 26, d is the distant limit. The distance between cylinder segment and O (from the polygon) that exceeds d are identified as external segment.

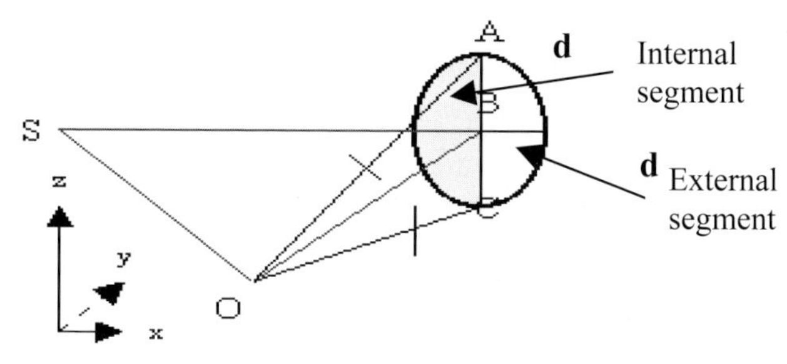

Fig. 26. Calculate distance between OA, OB, and OC

The distance of OA, and OC will be computed. Both of them are in the same situation because the surface of circle is perpendicular to the line AC. Points A and C are the upper and lower point of the circle (see Fig. 26).

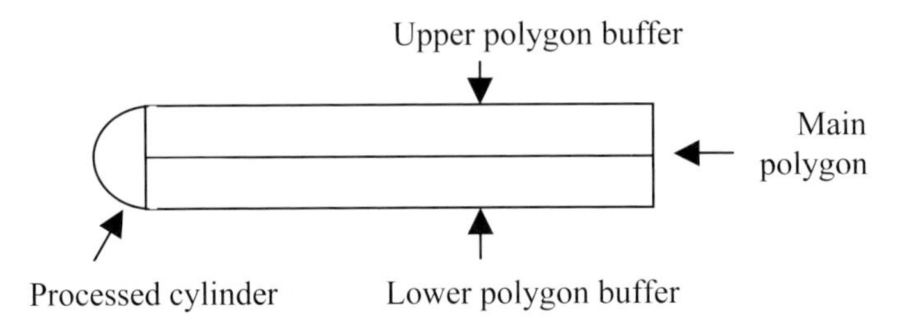

Fig. 27. View from above

Therefore, there is a condition that needs to be fulfilled in order to remove the internal segment of the cylinder. That is the distance of either OA or OC is the benchmark of that condition. If any distance from the circle to point O is less that OA or OC, then it should be the internal part of the cylinder and need to be removed. Finally, Fig. 27 shows the processed result.

3 Experiments and Discussions

In this project, we use the C++ language to create a software module called *3D Buffering Tools* and test it with the real dataset (UTM campus). The data was captured using digital photogrammetric system (Leica-Helava). Our 3D buffering module / software works with ArcView. The input is in ASCII format and the result is in the shapefile (*.shp). Fig.28, Fig. 29, and Fig. 30 show the snap shots of the interface and output.

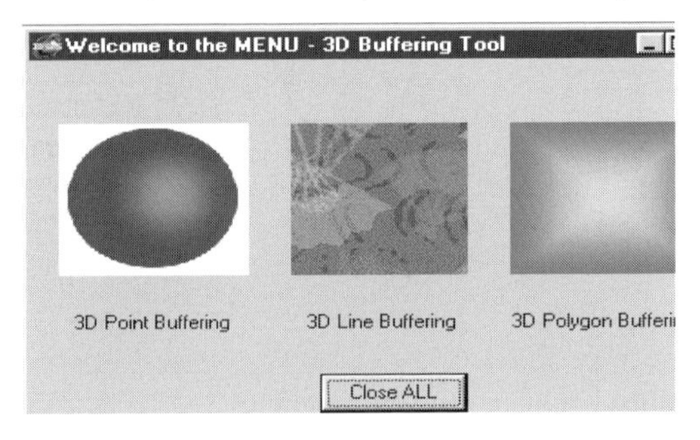

Fig. 28. 3D Buffering Tool's menu interface

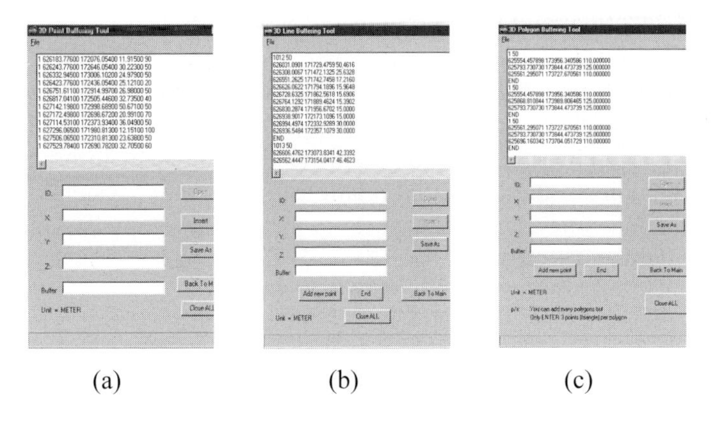

(a) (b) (c)

Fig. 29: (a) Point, (b) Line & (c) Polygon buffering tools' data input interface

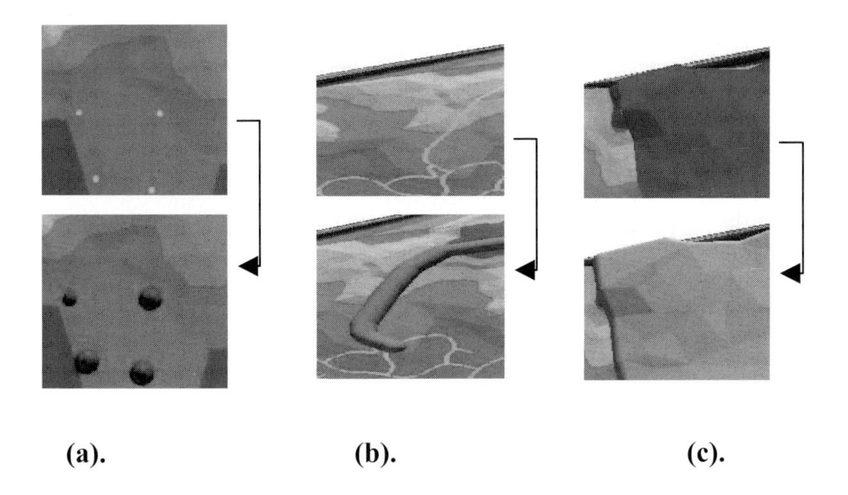

(a). **(b).** **(c).**

Fig. 30. (a) Point-, (b) Line-, and (c) Polygon-buffering output

4 Conclusions

The 3D buffering methods were discussed. The entire algorithm relates to the geometrical modeling, the mathematics, the expected outputs are presented. From this research, the new tools, procedures and methodologies related to the 3D buffering have been developed and could be applied for visualizing some forms of disastrous and chaotic situations. Dynamic display of the situation could form one of the aspects that need further attention for real disaster application. However, more focuses on the development of 3D analytical operation for 3D GIS are needed as well in the future. Any geometrical modeling involving the 3D analysis needs to be investigated. Moreover, current 3D GIS need topological information to represent the objects and its spatial relationships. Finally, system developers need to adopt true 3D in all aspects of system development unlike the current available GIS software.

References

Abdelguerfi M, Wynne C, Cooper E, Roy L (1998) "Representation of 3-D Elevation In Terrain Databases Using Hierarchical Triangulated Irregular Networks: A Comparative Analysis." International Journal of Geographic Information Science (IJGIS), vol 12, no 8, pp 853-873

Bajaj CL, Bernardini F, Chen J, Schikore DR (1996) "Automatic Reconstruction of 3D CAD Models." In Theory and Practice of Geometric Modeling, Springer-Verlag

Batty M (2000) "Virtual Cities: Representations, Models, Designs, Futures."
http://okabe.t.u-tokyo.ac.jp/okabelab/atsu/COEpaper-japan.pdf

Bernhardsen Y (1999) "Choosing a GIS." In Longley, P.A. et al.(eds.), Geographical Information Systems : Principles, Technical, Management and Applications(2nd edition), John Wiley & Sons, Inc: pp 580-600

Berry JK, Buckley DJ, Ulbricht C (2001) "3-D GIS: Visualize Realistic Landscapes"
http://www.geoplace.com/gw/1998/0898/898vis.asp

Boissonnat JD (1984) "Geometric Structures For Three-Dimensional Shape Representation." ACM Transactions on Graphics, 3(4), pp 266-286

Comer DC (2000) "Three-Dimensional Mapping of Environmental and Prehistoric Cultural Features in the Beidha Region with Spaceborne and Airborne Imagery"
http://www.esri.com/library/userconf/proc00/professional/papers/PAP124/p124.htm

ESRI Shapefile Technical Description (1998).
http://www.esri.com/library/whitepapers/pdfs/shapefile.pdf

De By RA (2000) "Principles of Geographic Information Systems." ITC, the Netherlands, 230 p

De Floriani L, Magillo P, Puppo E (1998), "Applications of Computational Geometry to Geographic Information Systems"
http://citeseer.nj.nec.com/rd/39975883%2C63576%2C1%2C0.25%2CDownload/http://citeseer.nj.nec.com/cache/papers/cs/1813/http:zSzzSzwww.disi.unige.itzSzpersonzSzPuppoEzSzPSzSzhandbook.pdf/applications-of-computational-geometry.pdf

Yu S, Van Snoeyink MJ (1996) "Drainage Queries in TINs: from local to global and back again". Proceedings Symposium on Spatial Data Handling (SDH'96), pp 13A.1-13A.14

Zlatanova S, Abdul-Rahman A, Pilouk M (2002) "Present Status of 3D GIS." G.I.M. International, pp 41-43

A GIS-Based Spatial Decision Support System for Emergency Services: London's King's Cross St. Pancras Underground Station

Christian J.E. Castle and Paul A. Longley

Centre for Advanced Spatial Analysis
University College London, 1-19 Torrington Place, London, WC1E 7HB, England.
Email: c.castle@ucl.ac.uk, plongley@geog.ucl.ac.uk

1 Introduction

The fire at London's King's Cross St. Pancras underground station on November 18, 1987, resulted in 31 fatalities, more than 60 severe injuries, and major structural damage to the station. Disasters such as this not only affect lives, as well as physical structures, but they also impact upon the immediate street level environment in ways that considerably reduce the speed of emergency response. The complex internal structures of the buildings and the restricted number of access points at the street level also render speedy escape and rescue particularly difficult in any emergency. When disasters occur within complex multi-level structures, a short period of time (e.g. 5 minutes) may lead to significant changes in the disaster environment within which trapped people need to escape and emergency services personnel have to operate (Kwan and Lee, in press). In this respect, the efficiency and effectiveness of the emergency services is critical, and directly related to their disaster and emergency management.

Preparedness is a key component of disaster and emergency management, and can play a significant factor in the event that emergency response efforts become necessary. Unfortunately, information from comparable incidents is usually unavailable and the way that some emergency scenarios evolve is unknown. Cova (1999) describes how Spatial Decision Support Systems (SDSS) can provide an effective tool for emergency services to monitor, identify, and mitigate vulnerabilities, ultimately improving and enhancing an organisation's level of preparedness for incidents.

London's King's Cross is going through a period of major change that will help to transform the area into a distinctive part of a world city. By 2015, approximately £4 billion will be invested into the area, developing it into the largest integrated transport hub in Europe. In addition, the area will benefit from the creation of 1,800 new homes, 30,000 jobs, and major business opportunities. This paper introduces a prototype SDSS designed to evaluate, revise, and contribute to emergency services preparedness of a major disaster within the London King's Cross redevelopment. Specifically, the framework of an evacuation model incorporating current social systems, and network and infrastructure systems, as well as initial evacuation scenarios will be discussed. These will be coupled via an integration link interface to a GIS, controlled and implemented through a Graphical User Interface (GUI) which displays the current state of the evacuation process, allowing the user to request information or run the simulation. Results from this study will facilitate the coordination, implementation, and allocation of health and emergency resources during future emergencies or disasters. It is anticipated that the SDSS will be of use to emergency medical service administrators, medical and public health professionals, and other community policy makers and planners who must prepare for future mass trauma events.

2 Disaster and Emergency Management

Disaster and emergency management has historically focused on the immediate and urgent aspects of an incident i.e. response and post-disaster recovery. However, there is a growing awareness that disaster and emergency management is much more complex and comprehensive than traditionally perceived (Gunes and Kolel, 2000). Although the primary function of emergency services is to protect life and property, this involves more than just crisis-reactive responses to emergencies. It also incorporates methods to avoid problems in the first place and preparing for those that will undoubtedly occur. Disaster and emergency management can be defined as the discipline and profession of applying science, technology, planning and management to deal with extreme events, that can injure or kill large numbers of people, cause extensive damage to property, and widespread disruption to society (Kreps, 1991). Disaster and emergency management processes can therefore be separated into four distinct components, each critical to ensuring the success of emergency response. These component activities are mitigation, preparedness, response and recovery, which can be perceived as a cyclic process (Fig. 1). According to

Quarantelli (1997); Cova (1999); ESRI (2002); Nakanishi *et al.* (2003); and Gerrad 2004, these are described as:

1. Mitigation involves activities such as risk assessment to accomplish steps that will limit or, in some cases, eliminate the effects of an emergency. Preventing an incident or minimizing its effects involves planning and prevention activities at many levels. These activities can be as systemic as community health monitoring or as targeted as identifying and limiting access to buildings or information. Mitigation measures are carried out through changing policy or operational procedures (i.e. altering the normal routine), or by physical actions (i.e. reinforcing or relocating structures, posting security guards).
2. Preparedness measures are crucial for risks that can not be sufficiently mitigated. These measures limit the loss of life and property and enhance response. Advanced planning is used to identify and evaluate risks, develop emergency procedures, ensure coordinated interagency response and inter / intra-agency communications, define a clear chain of command, conduct training, etc.
3. Response activities are those conducted immediately after an event to assist victims, stabilise the situation, and limit secondary damage.
4. Recovery starts after an emergency has ended, and continues until all systems return to normal or better. Typically this is a two step process. Short-term recovery returns vital life-support systems to minimum operating standards. Long-term recovery may continue for a number of years after a disaster.

3 Successful Emergency Response Depends on Preparedness

The National Institute for Occupational Health and Safety (NIOHS, 2004) describes why the emotionally charged, chaotic environment in the immediate aftermath of a major disaster should not be the test bed for evaluating emergency response resources or procedures. Strategic planning and management well before the event, along with standardised systems and procedures, are key to efficient emergency response. Preparedness is the crux of effectiveness, and emergency management provides a classic example of a situation where modelling and simulation can be used to anticipate unexpected outcomes.

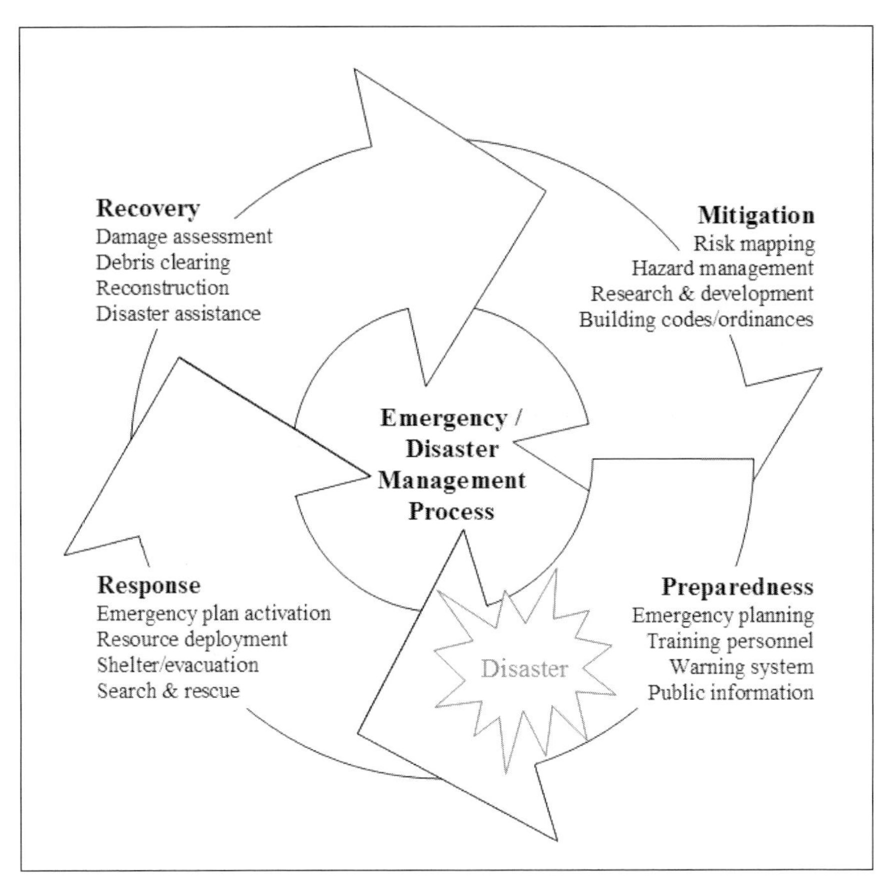

Fig. 1. Emergency and disaster management process, with examples during each phase where GIS plays a role (adapted from Cova, 1999)

The distinctive characteristics of major disasters make the case for scenario analysis of emergencies especially strong. The multiple hazards inherent in major emergency situations requires a flexibility from the response community that can only come through anticipation; especially as major disasters can take so many different forms and evolve in so many different ways. The response community may be called upon to carry out a very wide range of different activities as crises arise and develop. Effective emergency response necessitates the capabilities and resources to be put in place in advance to deal with the widest range of possible outcomes.

The NIOHS (2004) continues to emphasise that, because major disasters are rare and the scenarios medical care responders face are without precedent, emergency service organisations get little or no 'real world' practice in managing them. In this context, scenario based planning assumes added value. Moreover, it is also important to develop a response strategy that can meet the needs of an emergency or disaster, as well as to evaluate organisational operating procedures where possible. Although using an emergency response strategy during smaller-scale events, limited in scale and complexity, will never be directly comparable to a major disaster, the experience gained makes it more likely they can be applied effectively when they are most needed. Whilst no disaster situation is entirely predictable, the more prepared emergency response services are to deal with expected hazards, the more attention and energy they will be able to devote to handling unanticipated issues as they arise.

Taken together, this argues that emergency planners might benefit greatly from a computer based decision support tool which can be used to examine the effects of different assumptions or contingencies.

4 The Role of a SDSS

SDSS are explicitly designed to provide the user with a decision-making environment that enables the analysis of geographical information to be carried out in a robust, yet flexible manner (Densham, 1991). Fundamentally, SDSS involve the coupling of GIS and analytical / decision models to produce systems especially able to cope with spatial problems (Batty and Densham, 1996). They are designed to aid in the exploration, structuring, and solution of complex spatial problems such as the evacuation process (de Silva, 2000). The aim is to support decision making by employing quantitative approaches with the use of geographic information that is stored in a manipulable form within the GIS. A primary advantage is the consequent ability to display the critical information related to an incident on maps, satellite images, or digital terrains in relation to time (Mondschein, 1994).

Densham (1991), describes a typical SDSS as having four components:

1. **Analytical tools** enabling data investigation;
2. **Decision models** enabling scenario based investigations; a
3. **Geographic / Spatial database** enabling storage and analysis of geographic information; and a
4. **User interface** providing easy access to components 1, 2, and 3, as well as an attractive and comprehensive display of the output.

Computer simulation offers an efficient way of modelling evacuations within the SDSS because it allows the model to incorporate what is likely to be realistic behaviour as opposed to making concessionary assumptions for the sake of easy computation; necessary when using many analytical models (de Silva, 2000). It is crucial to ensure that evacuation plans are based on realistic assumptions.

Work in the design and application of tools formulated on automata methodologies has been at the forefront of computer simulation research (Benenson and Torrens, 2004). Two classes of automata tools, cellular automata and multi-agent systems, have been particularly popular; their use has dominated research literature. Both classes of tool offer significant advantages for general simulation, and pedestrian simulation specifically. Of particular interest to our research is the agent-based approach, which seeks to represent individual actors (or groups) in a given system. Agents may interact with each other and / or with an environment. The simulation of escape panic presented by Helbing *et al.* (2000) treats every member of a crowd as thinking and reacting individuals rather than an identical particle, which generates some interesting observations.

5 A Prototype SDSS for Kings' Cross Emergency Services

In the context of the previous discussion, the aim of research is to create a prototype KXSDSSES (King's Cross Spatial Decision Support System for Emergency Services) to evaluate, revise, and contribute to emergency services preparedness of a major disaster within the London King's Cross redevelopment. KXSDSSES is designed for contingency planning before an emergency or disaster evacuation, rather than for real-time emergency management use. Similar to the general constituents of a SDSS outlined by Densham (1991), and the Configurable Evacuation Management and Planning Simulator (CEMPS) discussed by de Silva and Eglese (2000), it is anticipated that the KXSDSSES will consist of the following four main components:

1. A multi-scalar evacuation model which incorporates current social systems (pedestrian simulations) and network and infrastructure systems (route optimisation) consisting of the required dynamic analysis and decision modelling components;
2. A GIS component which includes the spatial database and geographical analytical tools;

3. An integration link interface which consists of mechanisms developed for dynamic communication and data and information exchange between the GIS and evacuation model; and
4. A user interface which displays the current state of the evacuation process and allows the user to request information or run the simulation.

Some of the deliverables that the completed KXSDSSES will produce include, but are not limited to (NB: the time frame for this study is the immediate post-event i.e. less than 24 hours):

1. What quantity of patients with event-related injuries / conditions could be expected from varying degrees of emergency?
2. How will patients with event-related injuries / conditions arrive at on-site medical care delivery?
3. Depending on the type and degree of incident, where should the on-site medical care delivery be located?
4. How many ambulances are required, and where should they be stationed?
5. What mode of alternative transportation (i.e. other than ambulances) might be appropriate for varying incident types / scales (e.g. mini buses, coaches, etc)?
6. What are the shortest / quickest transport routes to non-on-site medical care delivery depending on the type and severity of injuries e.g. burns, toxic inhalation, neurological, etc?
7. What is the estimated time between dispatch and patients' arrival at points of non-on-site medical care delivery?

6 The Multi-Scalar Evacuation Model

Models exist for various purposes, but even models for the same purpose employ different theoretical concepts of physical or social processes; different numerical algorithms for solving applications of the theoretical concepts; and different programming languages, file types, data sources, etc, to execute the model (Johnston, 2001). A wide range of models are currently accessible, some better than others because of their superior resolution (encapsulating finer detail of the phenomena), greater efficiency, etc. However, from a user's perspective there remain considerable limitations of implementing a pre-existing model, even if it is for the same location or purpose. Indeed, models of the King's Cross area already exist, and in some instances it might be possible to combine components to construct a model that addresses a particular scenario spanning several processes. Nevertheless, this is not an approach that will be adopted by this study.

Primarily because of the assumptions and purpose these models were created for, but also because of the current advancements in computer simulations that these models lack, let alone the difficulties associated with incorporating second-hand models.

The evacuation model proposed will allow for a broad spectrum of scenario outputs to be generated by integrating several pedestrian simulation programs. Specifically, current pedestrian simulation programs such as buildingEXODUS, Simulation of Transient Evacuation and Pedestrian movementS (STEPS), and Myraid, will be coupled with the network analysis and route optimisation of ArcLogistics™ Route within ArcGIS 9 (Dangermond, 2002). ArcGIS 9 provides extensive built-in capabilities of analysis tools in addition to standard data storage, manipulation and representation tools required for a SDSS. In short, the SDSS will link together the topographical support and analysis provided by ArcLogistic™ Route, to determine the distribution of patients to medical care, whilst the simulation programs will imitate the dynamics of a pedestrian evacuation process in detail. ArcGIS 9 will be used for storing information and providing the display facilities required by the simulator. A link interface will be designed to integrate the evacuation model with the GIS, while a user interface will encompass the various decision support and scenario generation functions that will aid the evacuation planning process. It is anticipated that the SDSS will evaluate the outcomes of scenarios ranging from a fire, to an explosion, or CBRN (Chemical, Biological, Radioactive, or Nuclear) terrorist attack, arising from interactions between individuals and the environments in which they find themselves.

6.1 Social System: Pedestrian Simulations

In terms of the social system incorporated into the evacuation model, there is much to consider. Developing theories of crowd movement complexities can be used to both drive and evaluate emergency scenarios and represent an important analytical cutting edge of current urban and transport science. In academic terms, this is part of a shift away from aggregate, static conceptions of how cities are structured to a concern with the more detailed and varied micro-scale dynamics that characterise cities (Batty *et al.*, 2003). Such models are based on the premise that unusual and important behaviours, and outcomes emerge when interactions between the individuals involved with an event accumulate to a degree where distinct changes occur in how people react (Batten, 2000). Crowding is a phenomenon that generates panic, flight, sometimes mass hysteria, all of which are very relevant to highly concentrated spatial events requiring rapid exit or entrance from or to high-capacity facilities (Canetti, 1962). Mathematical models of these kinds of event are in their infancy. Most are based on fine-scale traffic models with the focus shifted to pedestrians, often called 'agents'. Batty *et al.*, (2003) explain that even within this emerging paradigm, very different approaches to simulating pedestrian movement have been adopted at different geographical scales. In confined spaces of tens of square metres, models based on social force and fluid flow (Still, 2001; Helbing, 1991; Henderson, 1971) have been used to predict the onset of panic situations at confined events such as football matches and music concerts (Helbing *et al.*, 2000). For events related to elongated areas associated with flow order, queuing theories have been adopted (Lovas, 1994). For buildings and urban spaces such as shopping malls, event simulation based on task scheduling, often using cellular automata, has been applied (Baer, 1974; Dijkstra *et al.*, 2002; Burstedde *et al.*, 2001; Kirchner and Schadschneider, 2002). For areas measured in square kilometres, accessibility models which simulate decisions between competing attractions have been developed (Borgers and Timmermans, 1986). Recently, methods which embody properties of self-organisation characterising how crowds form and disperse have become significant (Vicsek *et al.*, 1995; Helbing *et al.*, 1997). Within these examples the density of flows to measure crowding and vulnerability to accidents is emphasised.

Clearly, the accuracy of predicting evacuee behaviour and the detail required for modelling this behaviour depends on whether the simulator uses a micro, meso or a macro modelling approach. Micro simulations access individual parameters e.g. aggression, agility, size, frustration, etc. However, this method involves a much higher degree of validation. One of the main criticism for microscopic techniques is the problems of analytical intractability which is best described by the "little old lady" effect (see Crowd Dynamics, 2004). This kind of model building is also very expensive, in terms of both time and expertise to interpret the results. Conversely, macro simulations use flow equations to average evacuee or entity behaviour, for which there is a wide body of science associated. The bulk of pedestrian planning and design, evacuation and contingency planning, and building codes of practice are based on macroscopic models. According to Crowd Dynamics (2004) they are applied because, by and large, they are proven to work. Meso simulations are a compromise between the two approaches by averaging behaviour among groups of evacuees. The choice among the three approaches very much depends on the trade-offs that must be made in order to maintain realistic computing power when processing large amounts of data. However, flexibility for scenario generation increases as the micro approach becomes more and more detailed (de

Silva, 2000). The social system incorporated within the evacuation model of KXSDSSES will include simulations at different scales to produce a spectrum of evacuation outputs for evaluation. For example, Myriad is a hybrid of a number of different modelling systems. It includes elements of microscopic and macroscopic (e.g. Intelligent Space/Space Syntax) techniques (Crowd Dynamics, 2004).

6.2 Network and Infrastructure System: Route Optimisation

At present the review of network analysis and route optimisation programs has been curtailed. Traditional route optimisation software depends on Euclidean or as-the-crow-flies distance measurements to determine ideal route sequencing (Shivaram, 1999). This can be extremely inaccurate and often results in gross underestimates of time and travel distances. ArcLogistics™ Route has been identified for use as the network and infrastructure systems within the SDSS because it applies street logical topology-based networks that include variables such as underpasses, one-way streets, speed limits and turn restriction conditions (Dangermond, 2002). The ability to visualise the interrelation between logistical and topographic factors provides better interpretation of data and improves the decision-making process with optimisation techniques, this results in more practical efficient route planning.

6.3 Initial Scenario Assumptions

The initial scenarios modelled within KXSDSSES will be based on the London Underground Limited (LUL, 2000) engineering standard for station planning. These are: 1) Train on fire in station; and, 2) Fire within the station structure. Over time, scenarios involving an explosion or CBRN terrorist attack will be incorporated into the SDDS.

Although there may be scope for passengers to be evacuated by train, according to LUL (2000) this should not be relied upon as a means of escape. Consequently, the train on fire in station scenario will be defined as a train on fire in the busiest platform being considered, where the busiest platform is that with the greatest total of passengers on the train plus passengers on the platform. Evacuation capacities are calculated for evacuation routes from each platform in turn, until the point in the station at which evacuation routes from different platforms merge. At this point, it is assumed that the train on fire is at the busiest platform feeding into that evacuation route. It is assumed that 'down' escalators are stationary, and can be used as stairs in the 'up' direction. 'Up' escalators are assumed to be working, however, one up escalator for each change in level is assumed to be unavailable, even as a fixed staircase. Lifts will obviously not be used in emergency evacuations. The evacuation model will be calculated for the busiest period in the traffic week, and all passengers within the station are assumed to be on platforms at the start of the evacuation (as this is the worst case).

The number of people to be evacuated from the busiest platform being considered is given by the number of passengers on the train after a gap in the service of one cancelled train, plus the number of passengers waiting for that train. The maximum number of passengers on the train will be practically crush capacity (individual load data has been provided by LUL). The normal train service (with no cancellations) is assumed on all other platforms, and the first train to arrive at any platform after notification of the incident will not have time to prevent the train doors being opened.

The fire in station scenario is defined as a fire in the most capacious route in the station, were the most capacious route is therefore assumed to be unavailable for emergency evacuation. Similar to scenario 1, it is assumed that 'down' escalators are stationary, and can be used as fixed stairs in the up direction. 'Up' escalators are assumed to be working, however one up escalator for each change in level is assumed to be unavailable, even as a fixed staircase. Lifts shall not be used in emergency evacuations. Evacuation capacities will also be calculated for the busiest period in the traffic week, and a normal train service shall be assumed at each platform. All passengers within the station are assumed to be on platforms at the start of the evacuation (as this is the worst case). Finally, the number of people to be evacuated is the normal boarding and alighting loads on each platform, supplied by LUL.

7 The Current State of KXSDSSES

At the time of writing, KXSDSSES is in development. At the moment the authors are currently evaluating the various pedestrian simulation programs available in relation to their robustness, scalability, and integration possibilities with ArcGIS 9. Incorporating the SDSS into ArcGIS 9 will provide an opportunity to build upon our knowledge and expertise gained during the development of a demonstrator produced with the Environmental Systems Research Institute (ESRI) Inc., for their 2004 user conference plenary. Additionally, potential contributions of expert and local knowledge / advice, as well as data contributions are being sought. Specifically developers and contractors involved with King's Cross redevelopment are being approached for AutoCAD or Microstation floor plans, and any previous pedestrian movement analysis and research conducted for the associated facilities and current redevelopment. At present, LUL has supplied all 2 & 3-Dimensional architect drawings of their new and renovated facilities, which will provide the preliminary framework for the evacuation model (Fig. 2). Furthermore, local emergency medical service administrators, medical and public health professionals, and other community policy makers and planners who must prepare for future mass trauma events have been approached to participate in the KXSDSSES Advisory Panel. The advisory panel is actively supported by Inspector Michael Burnham of the British Transport Police who states that one of the main benefits of participating would be to assist all the statutory responders in meeting their future obligations under the UK Civil Contingencies Act 2004.

Fig. 2. 3D view of the King's Cross St. Pancras underground stations renovated main, and new western and northern ticket halls (maroon areas, Allies and Morrison, 2004)

8 Conclusion

Emergency and disaster preparedness is a significant factor in the outcome of emergency services response efforts. Unfortunately, information from past incidents is usually unavailable and the way that some emergency scenarios evolve is unknown. Thus, a SDSS can be used by emergency services to monitor emergency preparedness by identifying and mitigating vulnerabilities, ultimately improving and enhancing an organisation's level of emergency preparedness.

The KXSDSSES will couple current pedestrian egress simulation programs with the network analysis and route optimisation of ArcLogistics™ Route to evaluate, revise, and contribute to emergency services preparedness of a major disaster within the London King's Cross redevelopment. It is anticipated that results will be used by emergency medical service administrators, medical and public health professionals, and other community policy makers and planners who must prepare for future mass trauma events.

Acknowledgements

We gratefully acknowledge the financial support of the Economic and Social Research Council (ESRC, CASE award PTA-033-2004-00034, http://www.esrc.ac.uk), and Camden Primary Care Trust (PCT, http://www.camdenpct.nhs.uk/). A special note of gratitude is extended to Dr. Muki Haklay. I would also like to thank David Murray the Assistant Director of public health intelligence at Camden PCT for his guidance, and John Calkins of ESRI Inc. for his work on the Bloomsbury demonstrator using ArcGlobe.

References

Allies and Morrison (2004) King's Cross Underground Station. Available at http://www.alliesandmorrison.co.uk [December 14th 2004]

Baer AE (1974) A Simulation Model of Multidirectional Pedestrian Movement within Physically Bounded Environments, Report 47. Institute of Physical Planning, Carnegie–Mellon University, Pittsburgh

Batten D (2000) Complex Landscapes of Spatial Interaction. In: Reggiani A (ed) Spatial Economic Science: New Frontiers in Theory and Methodology. Springer Verlag, Berlin, pp 51–74

Batty M, and Densham P (1996) Decision Support, GIS, and Urban Planning. Available at http://www.geog.ucl.ac.uk/~pdensham/SDSS/s_t_paper.html, [November 20th 2004]

Batty M, Desyllas J, Duxbury E (2003) Safety in numbers? Modelling Crowds and Designing Control for the Notting Hill Carnival. Urban Studies, 40(8), pp 1573-1590

Benenson I, Torrens PM (2004) Geosimulation: Automata-Based Modelling of Urban Phenomena. John Wiley and Sons Ltd, London

Borgers A, Timmermans HA (1986) A Model of Pedestrian Route Choice and Demand for Retail Facilities within Inner-City Shopping Areas. Geographical Analysis, 18, pp 115–128

Burrstedde C, Klauck K, Schadschneider A, Zittarz J (2001) Simulation of Pedestrian Dynamics Using a Two-Dimensional Cellular Automaton. Physica A, 295, pp 507–525

Canetti E (1962) Crowds and Power. Victor Gollanc, London.

Cova TJ (1999) GIS in Emergency Management. In: Longley, PA, Goodchild, MF, Maguire, DJ, Rhind, DW (eds) Geographical Information Systems: Principles, Techniques, Applications, and Management, (vol 1). John Wiley & Sons, New York, pp 845-858

Crowd Dynamics (2004) Micro/Macro Simulations. Available at http://www.crowddynamics.com/Myriad/micro%20v%20macro.htm, [December 14th 2004]

Dangermond J (2002) More Efficient Route Planning for Cleaner Air, Available at http://www.esri.com library/reprints/pdfs/fleetowner_reprint.pdf, [December 14th 2004]

De Silva NF (2000) Challenges in Designing Spatial Decision Support Systems for Evacuation Planning. Available at http://www.colorado.edu/hazards/wp/wp105 /wp105.html, [December 15th 2004]

De Silva NF, Eglese RW (2000) Integrating Simulation Modelling and GIS: Spatial Decision Support Systems for Evacuation Planning. Journal of Operational Research Society, 51 pp 423-430

Densham P (1991) Spatial Decision Support Systems. In: Maguire DJ, Goodchild MF, Rhind DW (eds) Geographical Information Systems: Principles and Applications (vol 1). Longman, UK

Dijkstra J, Jessurun J, Timmermans HJP (2002) A Multi-Agent Cellular Automata Model of Pedestrian Movement. In: Schreckenberg M, Sharna SD (eds) Pedestrian and Evacuation Dynamics. Springer Verlag, Berlin, pp 173–180

ESRI (2002) Supersizing Emergency Management with GIS. Available at http://www.esri.com/news/arcuser/0102/steps.html, [November 23rd 2004]

Gerrad, RM (2004) Emergency Management. Available at http://www.pmkgroup.com/Emergency%20Management.htm, [November 17th 2004]

Gunes AE, Kolel JP (2000) Using GIS in Emergency Management Operations. Journal of Urban Planning and Development, 126(3), pp 136-149

Helbing D (1991) A Mathematical Model for the Behaviour of Pedestrians. Behavioral Science, 36, pp 298-310

Helbing D, Farkas I, Vicsek T (2000) Simulating Dynamical Features of Escape Panic. Nature, 407, pp 487-490

Helbing D, Schweitzer F, Keltsch J, Molnar P (1997) Active Walker Model for the Formation of Animal and Trail Systems. Physical Review E, 56, pp 2527-2539

Henderson LF (1971) The Statistics of Crowd Fluids. Nature, 229, pp 381–383

Johnston DM (2001) Computation, Communication, and Data Storage. In: Case MP, Goran WD, Gunther TA, Holland JP, Johnston DM, Lessard G, Schmidt WJ (eds) Decision Support Capabilities for Future Technology Requirements. US Army Corps of Engineers, USA

Kirchner A Schadscheider A (2002) Simulation of Evacuation Processes Using a Bionics-Inspired Cellular Automaton Model for Pedestrian Dynamics (Mimeograph). Available at http://arxiv.org/abs/cond-mat/0203461 [December 14th 2004]

Kreps GA (1991) Organizing for Emergency Management. In: Drabek TE, Hoetmer GJ (eds) Emergency Management: Principles and Practice for Local Governments. International City Management Association, Washington, pp 30–54

Kwan M-P, Lee J (in press) Emergency Response After 9/11: The Potential of Real-Time 3D GIS for Quick Emergency Response in Micro-Spatial Environments. Computers, Environment, and Urban Systems

Lovas GG (1994) Modeling and Simulation of Pedestrian Flow Traffic. Transportation Research, 28(B), pp 429–443

LUL (London Underground Limited) (2000) Engineering Standard: Station Planning (document E1024 A2). London Underground Limited, London

Mondschein LG (1994) The Role of Spatial Information Systems in Environmental Emergency Management. Journal of the American Society for Information Science, 45(9), pp 678-685

Nakanishi Y, Kim K, Ulusoy, Y, Bata, A (2003) Assessing Emergency Preparedness of Transit Agencies: A Focus on Performance Indicators. Annual Transportation Research Board, USA

NIOHS (National Institute for Occupational Health and Safety) (2004), Protecting Emergency Responders: Safety Management in Disaster and Terrorism Response. RAND Science and Technology, USA

Quarantelli EL (1997) Ten Criteria for Evaluating the Management of Community Disasters. Disasters, 21(1), pp 39-56

Shivaram S (1999) Optimization Models and Analysis of Routing, Location, Distribution, and Design Problems on Networks, PhD Thesis, Virginia Polytechnic and State University. Available at http://scholar.lib.vt.edu/theses/available /etd-042499-225537 [December 14th 2004]

Sill GK (2001) Crowd Dynamics, PhD thesis, University of Warwick. Available at http://www.crowddynamics.com/, [December 14th 2004]

Vicsek T, Czirok A, Ben-Jacob E, Molnar P (1995) Novel Type of Phase Transition in a System of Self-Driven Particles. Physical Review Letters, 75, pp 1226–1229

CityGML: Interoperable Access to 3D City Models

Thomas H. Kolbe, Gerhard Gröger and Lutz Plümer

Institute for Cartography and Geoinformation, University of Bonn,
Meckenheimer Allee 172, 53115 Bonn, Germany.
Email: {kolbe|groeger|pluemer}@ikg.uni-bonn.de

Abstract

Virtual 3D city models provide important information for different aspects of disaster management. In this context, up-to-dateness of and flexible access to 3D city models are of utmost importance. Spatial Data Infrastructures (SDI) provide the appropriate framework to cover both aspects, integrating distributed data sources on demand. In this paper we present CityGML, a multi-purpose and multi-scale representation for the storage of and interoperable access to 3D city models in SDIs. CityGML is based on the standard GML3 of the Open Geospatial Consortium and covers the geometrical, topological, and semantic aspects of 3D city models. The class taxonomy distinguishes between buildings and other man-made artifacts, vegetation objects, waterbodies, and transportation facilities like streets and railways. Spatial as well as semantic properties are structured in five consecutive levels of detail. Throughout the paper, special focus is on the utilization of model concepts with respect to different tasks in disaster management.

1 Introduction

Virtual 3D city models provide important information for different aspects of disaster management. First, they memorize the shape and configuration of a city. In case of severe destruction of infrastructure e.g. caused by earthquakes, immediate access to this reference data allows to quickly assess the extent of the damage, to guide helpers and last but not least to rebuild the damaged sites. Second, 3D city models enable 3D visualizations and facilitate localization in indoor and outdoor navigation. Augmented reality systems provide helpers with information that is visually overlaid with their view of the real world. Such systems need 3D city models in order to compute the positions and occlusions of the overlay graphics. Third, 3D escape routes inside and outside of buildings can be determined with an appropriate city model. Fourth, in flooding scenarios 3D city models allow to identify even affected building storeys.

In the context of disaster management, up-to-dateness of and flexible access to 3D city models are of utmost importance (Zlatanova and Holweg 2004). Spatial Data Infrastructures provide the appropriate framework to cover both aspects, integrating distributed data sources on demand (Groth and McLaughlin 2000). However, the prerequisite is syntactic and semantic interoperability of the participating GIS components (Bishr 1998).

Syntactic interoperability can be achieved by using the XML-based Geography Markup Language (GML3, see Cox et al. 2004) of the Open Geospatial Consortium (OGC). GML3 is an XML-based abstract format for the concrete specification of application specific spatial data formats. It is open, vendor-independent, and based on ISO standards; it can be extended and specialized to a specific application domain; and it explicitly supports simple and complex 3D geometry and topology. Furthermore, GML is the native data format of OGC's Web Feature Service (WFS), a standardized web service that implements methods to access and manipulate geodata within a spatial data infrastructure (Vretanos 2002).

Semantic interoperability presumes common definitions of objects, attributes, and their interrelationships with respect to a specific domain. However, no common semantic model for 3D city models has been established yet. In the following we present CityGML, a multi-purpose and multi-scale representation for the storage of and access to 3D city models. It has been developed during the last two years by the Special Interest Group 3D of the initiative Geodata Infrastructure North-Rhine Westphalia (GDI NRW). The data model behind CityGML is based on the ISO standard family 191*xx*. The implementation is realized as an application schema for GML3.

CityGML covers the geometrical, topological, and semantic aspects of 3D city models. The class taxonomy distinguishes between buildings and other man-made artifacts, vegetation objects, waterbodies, and transportation facilities like streets and railways. Spatial as well as semantic properties are structured in five consecutive levels of detail (LoD), where LoD0 defines a coarse regional model and the most detailed LoD4 comprises building interiors resp. indoor features. Included thematic objects, which are especially relevant for disaster management, are different types of digital elevation models, building features like rooms, doors, windows, balconies, and subsurface constructions.

The paper concludes with first results of the ongoing evaluation project "Pilot 3D", in which six groups from Germany including the municipalities of Berlin, Hamburg, Cologne, Düsseldorf, and Leverkusen implement the CityGML application schema and exchange 3D city models.

2 Related Work

3D city modeling is an active research topic in distinct application areas. Different modeling paradigms are employed in 3D geographical information systems (3D GIS, Köninger and Bartel 1998), computer graphics (Foley et al. 1995), and architecture, engineering, construction, and facility management (AEC/FM; Eastman 1999). Whereas in 3D GIS the focus lies on the management of multi-scale, large area, and geo-referenced 3D models, the AEC/FM domain addresses more detailed 3D models with respect to construction processes (Kolbe and Plümer 2004). Computer graphics concentrates rather on the visualization of 3D models.

The representation of geometry and topology of 3D objects has been investigated in detail by Molenaar (1992), Zlatanova (2000), Herring (2001), Oosterom et al. (2002), Pfund (2002), and Kolbe and Gröger (2003; 2004). The management of multi-scale models was discussed (among others) by Coors and Flick (1998), Guthe and Klein (2003), Gröger et al. (2004).

The ISO standard ISO/PAS 16739 'Industry Foundation Classes' (IFC, Adachi et al. 2003) is a semantic model for buildings and terrain which has been developed in the AEC/FM domain. It defines an exchange format and contains object classes for storeys, roofs, walls, stairs, etc.. Nevertheless, since IFC is lacking concepts for spatial objects like streets, vegetation objects or water bodies, it is not appropriate for the representation of complex city models. Similar problems arise with respect to 'green building XML' (gbXML 2003), an AEC/FM standard for building energy and environmental performance analysis, and 'Building-construction XML' (van Rees et al. 2002), a standard for the mapping of construction taxonomies.

LandXML/LandGML is a standard for land management, surveying and cadastre, providing a semantic model for parcels, land use, transportation and pipe networks (LandXML 2001). Although LandXML supports 3D coordinates, it does not comprise volumetric geometries. Buildings are only represented by their footprints. Further concepts for 3D man-made objects are missing.

Computer graphics (CG) standards like VRML97 (1997) and its successor X3D model only the appearance of 3D objects. They do not provide concepts for the representation of thematic aspects, attributes, and interrelationships of the graphical objects.

Since thematic information are crucial for disaster management, CG standards are not sufficient. AEC/FM standards concentrate on man-made constructions and are lacking concepts for the representation of natural objects. Furthermore, none of the discussed AEC/FM and GIS standards supports multi-scale models resp. multiple levels-of-detail (LoD).

3 CityGML: Unified 3D City Modeling

CityGML is a profile of GML3, which implements an interoperable, multifunctional, multi-scale and semantic 3D city model. This section presents the highlights of CityGML, starting with general concepts in the first section. CityGML covers the thematic objects which are relevant for city models, including transportation objects like streets or traffic lights, or vegetation objects. However, in this section the focus is on the most important components of city models, on the building model and on the Digital Terrain Model. Both are discussed in detail.

Initially, CityGML is specified using the graphical *Unified Modeling Language (UML)* (see Booch et al. 1997). From UML diagrams, the XML schemas are derived by applying the transformation rules given in Cox et al. (2004). Thus, CityGML may be processed by standard GML3 readers.

3.1 General Concepts

CityGML implements several novel concepts to support interoperability, consistency and functionality.

3.1.1 Levels-of-Detail

CityGML supports different *Levels-of-Detail (LoD),* which may arise from independent data collection processes and are used for efficient visualization and efficient data analysis. In one CityGML data set, the same object may be represented in different LoD simultaneously, enabling the analysis and visualization of the same object with regard to different degrees of resolution. Furthermore, two CityGML data sets containing the same object in different LoD may be combined and integrated.

CityGML provides five different LoD, which are illustrated in **Fig. 1**. The coarsest level *LoD0* is essentially a two and a half dimensional Digital Terrain Model, over which an aerial image or a map may be draped. *LoD1* is the well-known blocks model, without any roof structures or textures. In contrast, a building in *LoD2* has differentiated roof structures and textures. Vegetation objects may also be represented. *LoD3* denotes architectural models with detailed wall and roof structures, balconies, bays and projections. High-resolution textures can be mapped onto these structures. In addition, detailed vegetation and transportation objects are components of a LoD3 model. *LoD4* completes a LoD3 model by adding interior structures like rooms, interior doors, stairs, and furniture.

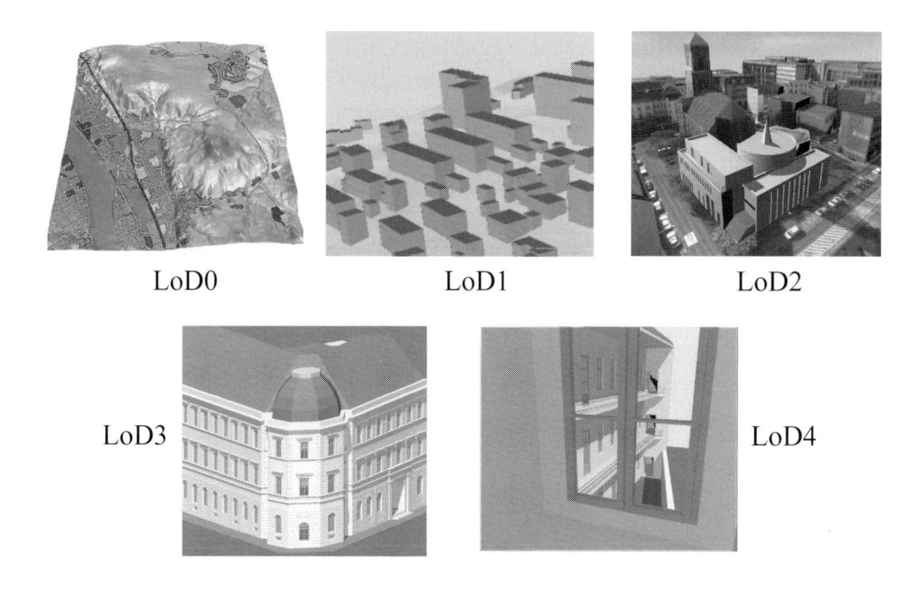

Fig. 1. The five Levels-of-Detail (LoD) defined by CityGML

The different LoD are also characterized by accuracies and minimal dimensions of objects. In LoD1, the positional and height accuracy of points may be 5m or less, while all objects with a footprint of at least 6m by 6m have to be considered. The positional accuracy of LoD2 is 2m, while the height accuracy is 1m. In this LoD, all objects with a footprint of at least 4m by 4m have to be considered. Both types of accuracies in LoD3 are 0.5m, and the minimal footprint is 2m by 2m. Finally, the positional and height accuracy of LoD4 must be 0.2m or less. By means of these figures, the classification in five LoD may be used to assess the quality of a 3D city model data set. Furthermore, the LoD category makes data sets comparable and thus supports the integration process of those sets.

3.1.2 Geometric-Topological Modeling

Spatial properties of thematic objects in CityGML are represented by a geometrical-topological model, according to the well-known *Boundary Representation* (Foley 1995). For each dimension, there is a geometric-topological primitive: a zero-dimensional object is a *node*, a one-dimensional an *edge*, a two-dimensional a *face*, and a three-dimensional a *solid*. A solid is bounded by faces, a face by edges, and an edge by nodes. An edge is restricted to be a straight line, and a face must be planar, i.e. its boundary and all interior points are forced to be located in one plane. Edges, faces and solids may be aggregated to *CurveGeometries*, *Surface-Geometries* and *SolidGeometries*, respectively (Kolbe and Gröger 2003). These geometries are used to define the spatial properties of application objects. The primitives and the aggregates are implemented using the GML3 geometry and topology classes, similar to the two-dimensional 'Simple Topology' profile of the standard ISO 19107 (Herring 2001).

The primitives node, edge, face, solid, and the aggregates must satisfy a number of integrity constraints, which guarantee consistency of the model. The interiors of the primitives must be disjoint, and if two primitives touch, the common boundary must be a primitive of lower dimension (Herring 2001). These constraints assure a clean topology without any redundancy. Since solids must be disjoint, the computation of volumes of solids does not yield erroneous results, which would be the case if solids overlap.

3.1.3 Coherent Semantic-Geometrical Modeling

Another feature of CityGML is the coherent modeling of semantics and geometrical/topological properties. On the semantic level, real-world entities are represented by application objects, for example buildings, including attributes, relations and aggregation hierarchies (part-whole-relations) between objects. On the spatial level, geometric-topological objects are assigned to semantic objects, which represent their spatial properties. Thus, the model consists of two hierarchies, the semantic and the geometric-topological, were the corresponding objects are linked by relations. The advantage of this approach is, that it can be navigated in both hierarchies and between both hierarchies.

3.1.4 Closure Surfaces and Subsurface Objects

A novel concept in CityGML is the *ClosureSurface*, which is employed to seal objects, which are in fact open, but must be considered as closed to compute its volume. An airplane hangar is an example for such an object.

ClosureSurfaces are special surfaces which are taken into consideration when needed to compute volumes and are neglected, when they are irrelevant or not appropriate, for example in visualizations.

The concept of *ClosureSurfaces* also is employed to model the entrances of *subsurface objects*. Those objects like tunnels or pedestrian underpasses have to be modeled as closed solids in order to compute their volume, for example in flood simulations. The entrances to subsurface objects also have to be sealed to avoid holes in the digital terrain mode (see **Fig. 2**). However, in close-range visualizations the entrance must be treated as open. Thus *ClosureSurfaces* are an adequate way to model those entrances.

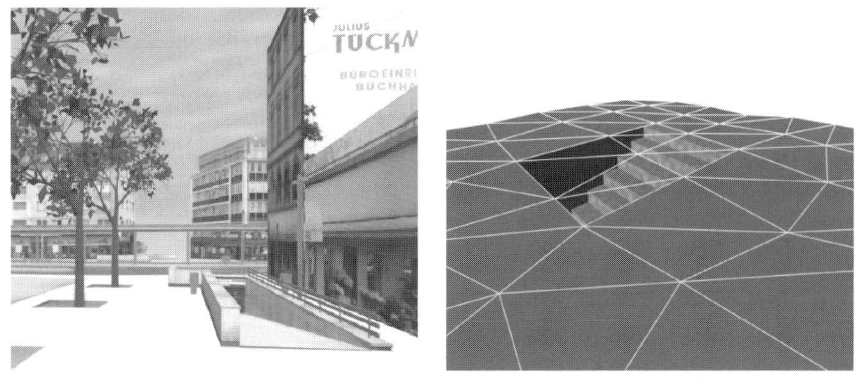

Fig. 2. Passages are subsurface objects (left). The entrance is sealed by a virtual *ClosureSurface,* which is both part of the DTM and the subsurface object (right)

3.1.5 References to Objects in External Data Sets

CityGML objects often are derived from or have relations to objects in other databases or data sets. For example, a building may have been constructed from a two-dimensional footprint in a cadastre, or may be derived from an architectural model. The reference of a 3D object to its corresponding object in an external data set is essential, if an update must be propagated or if additional data, for example the name and address of a building's owner, is required. In order to supply such information, each CityGML thematic object may have *External References* to corresponding objects in external data sets (see **Fig. 3**, upper right part of diagram). Such a reference denotes the external information system and the unique identifier of the object in this system. Both are specified as *Uniform Resource Identifier (URI)*, which is a generic format for any kind of resources in the internet.

3.1.6 Dictionaries and Code Lists for Attributes

Attributes, which classify objects often have values which are restricted to a number of discrete values. An example is the attribute *roof type*, whose attribute values are saddle back roof, hip roof, semi-hip roof, flat roof, pent roof, or tent roof. If such an attribute is typed as string, misspellings or different names for the same notion obstruct interoperability. In CityGML, such classifying attributes are specified as *Code Lists* or *Dictionaries*. Such a structure enumerates all possible values of the attribute, assuring that the same name is used for the same notion. In addition, the translation of attribute values into other languages is facilitated.

3.2 The Building Model

The building model is the core of CityGML. It allows the representation of thematic and spatial aspects of buildings, building parts and accessories in four levels-of-detail, LoD1 to LoD4. The UML diagram of the building model is depicted in **Fig. 3**. The pivotal class of the model is *AbstractBuilding*, which is specialized either to a *Building* or to a *BuildingPart*. Since an *AbstractBuilding* consists of *BuildingParts*, which again are *AbstractBuildings*, an aggregation hierarchy of arbitrary depth may be realized. Since *AbstractBuilding* is a subclass of the root class *CityObject*, the relation to the *ExternalReference* (see section 3.1) is inherited. In LoD1, the spatial extent of an *AbstractBuilding* is given by a *SolidGeometry*, which in this case is a simple block. A Building may be part of a *BuildingComplex*, which has at least one main building.

A crucial issue in city modeling is the integration of buildings and the terrain. Problems arise if buildings float over or sink into the terrain, which is particularly the case if terrains and buildings in different LoD are considered. To overcome this problem, the *TerrainIntersection* curve of a building is introduced. This curve denotes the exact position where the terrain touches the building, and is represented by a closed ring surrounding the building (see **Fig. 4** for an example). If the building has a courtyard, the *TerrainIntersection* curve consists of two closed rings. This information can be used to integrate the building and a terrain by 'pulling up' the surrounding terrain to fit the *TerrainIntersection* curve. By this means, the curve also ensures the correct positioning of textures. Since the intersection with the terrain may differ depending on the LoD, a building may have different *TerrainIntersection* curves for each LoD.

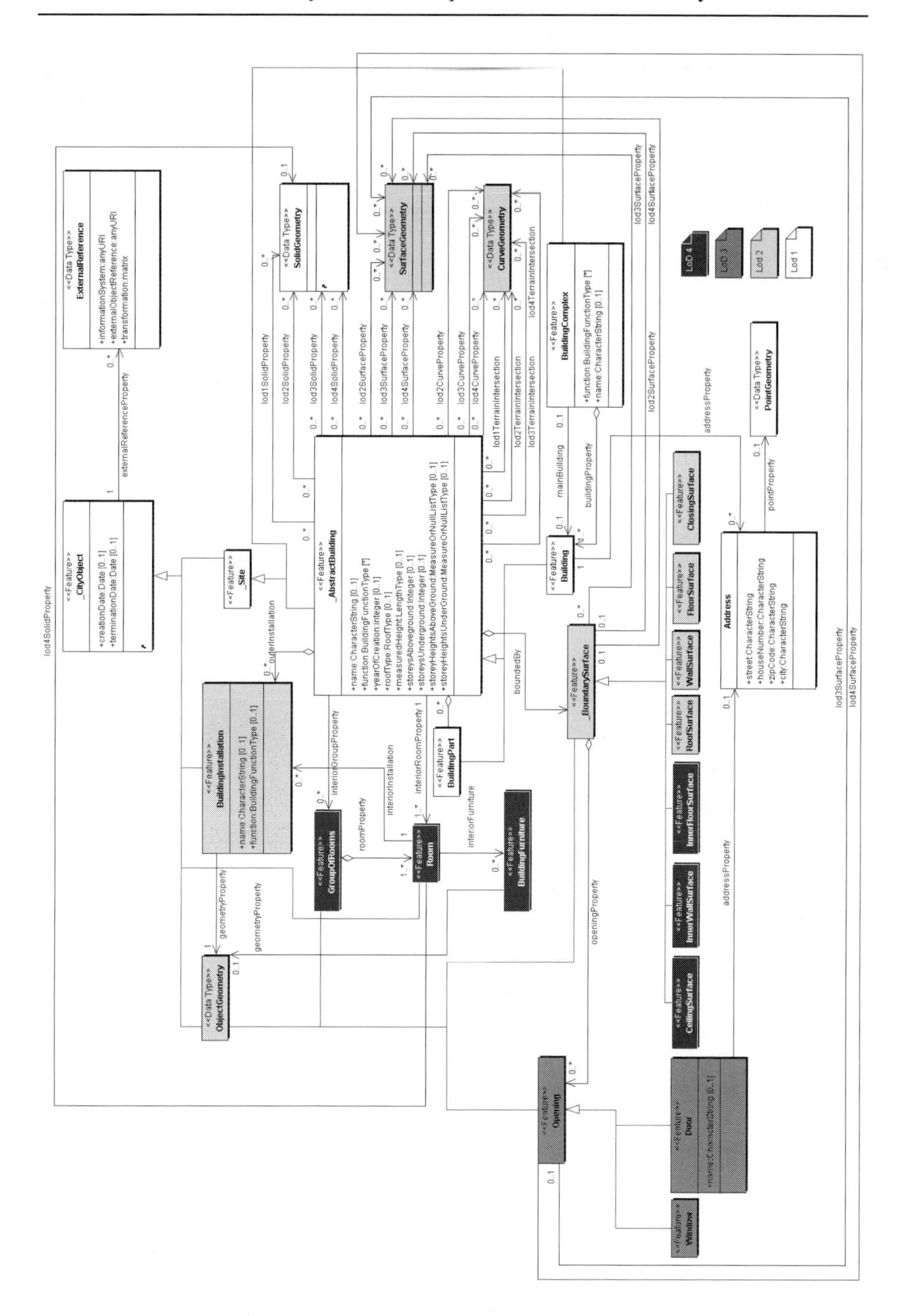

Fig. 3. UML-Diagram (Booch et al. 1997) of CityGML's building model

In a LoD2 building, it is possible to distinguish the bounding surfaces as own thematic objects. These surfaces may be classified as *Roof*, *Wall* or *Floor Surfaces*. The geometry of these surfaces, however, is shared with the *SolidGeometry* that defines the whole building. An opening in a building is modeled by a *ClosureSurface*; this concept was already discussed in sec. 3.1. The geometry of a LoD2 building is given by *SolidGeometries*, and additionally by *SurfaceGeometries*, which represent surfaces that are part of the building, but do not bound the solids of the building. The overhanging part of a roof is an example for such a surface. A LoD2 building also may have *BuildingInstallations*, for example chimneys, balconies or outer stairs. The geometry type of a *BuildingInstallation* is not restricted. It is specified by a *ObjectGeometry*, which is the super class of the aggregates *CurveGeometries*, *SurfaceGeometries* and *SolidGeometries*. In contrast to *BuildingParts*, *BuildingInstallations* are smaller and only accessories, but not a constituent part of the building.

Fig. 4. *TerrainIntersection* curve for a building (left, black) and a tunnel object (right, white). The tunnel's hollow space is sealed by a triangulated *ClosureSurface*

In LoD3, buildings additionally may have *Openings* such as *Windows* and *Doors*. The class *Opening* is a sub class of *CityObject*, thus the reference to objects in external data sets is inherited. As discussed in section 3.1, the accuracy requirements of LoD3 are much higher than those of LoD2.

LoD4 complements LoD3 by adding interior structures of buildings such as *Rooms*, which are bounded by *Ceiling-*, *InnerWall-* and *InnerFloorSurfaces*. Rooms may be aggregated to a *GroupOfRooms*, and may have *BuildingFurnitures* and interior *BuildingInstallations*. A *BuildingFurniture* is a movable part of a room, such as a chair or furniture, while a *BuildingInstallation* is permanently connected to the room. Examples are stairs or pillars. Doors are used in LoD4 to connect rooms topologically:

the surface that represents the door geometrically is part of the boundarics of the solids of both rooms.

Note that all these objects inherit the references to objects in external data sets. Important data sources for LoD4 models are IFC data sets (c.f. section 2), which can be converted accordingly (Benner et al. 2004).

As discussed in section 3.1, the different accuracy requirements of LoD1 to LoD4 have to be applied to the building model as well.

3.3 The Digital Terrain Model

An essential part of a city model is the terrain. In CityGML, the terrain may be specified as a regular raster or grid, as a TIN (Triangulated Irregular Network), by break lines or skeleton lines, or by mass points. These four types are implemented by using standard GML3 elements. A TIN may either be represented as a collection of triangles, or implicitly by a set of 3D points, where the triangulation may be reconstructed by standard methods (Okabe et al. 1992). A break line is a discontinuity of the terrain, while skeleton lines are either ridges or valleys. Both are represented by 3D curves. Mass points are simply a set of 3D points.

In a CityGML data set, these four terrain types may be combined in different ways, yielding a high flexibility. First, each type may be represented in different levels-of-detail, reflecting different accuracies or resolutions. Second, a part of the terrain can be described by the combination of multiple types, for example by a raster and break lines, or by a TIN and break lines and skeleton lines. In this case, the break and skeleton lines must share the geometry with the triangles. Third, neighboring regions may be represented by different types of terrain models. To facilitate this combination, each terrain object is provided with a spatial attribute denoting its *extent of validity*. This extent is represented by a 2D footprint polygon, which may have holes. This concept enables, for example, the modeling of a terrain by a coarse grid, where some distinguished regions are represented by a detailed, high-accuracy TIN. The boundaries between both types are given by the extend attributes of the corresponding terrain objects. This approach is very similar to the concept of *TerrainIntersection* curves introduced in section 3.1.

4 Application of CityGML for Disaster Management

CityGML was designed as a data model and exchange format for the multifunctional utilization of 3D city models. In the following, we discuss the specific use of key concepts for disaster management tasks.

4.1 General Modeling Aspects

The coherent semantic modeling of the spatial and thematic properties of 3D objects and their aggregations is one of the most important features of CityGML. Object classes have thematically rich attributes which allow for specific queries like 'What are the buildings with more than 10 storeys above ground?' or 'Where are buildings with flat roofs which are large enough that a helicopter could land on them?'. Since the terrain model may consist of neighbored or nested patches having different resolutions, high resolution DTMs for e.g. regions with high flood risk may be embedded into large area DTMs at low resolution.

Since the geometry of all 3D objects must be represented by at least one closed solid, the computability of volumes and masses is always facilitated. For example, in flooding resp. fire scenarios it could be estimated how much water resp. smoke or gas will flow into a tunnel, pedestrian passage, or a building. The estimation of masses from volumes is also interesting for planning the removal of debris after an incident.

The possibility to provide external references can be used to associate any CityGML object and its parts with data sets of other applications like facility management systems or the cadastre, which is important to determine the owner of a building. By using external references it is also possible to relate *BuildingInstallations* (c.f. section 3.2), which are relevant for disaster management like hydrants or fire protection doors with databases that hold the technical data about these.

4.2 Building model

In flood situations resp. simulations, the representation of storey heights above and under ground allows to determine to which degree buildings are affected. This information is especially useful for planning evacuations and for damage assessment by aid organizations and insurance companies.

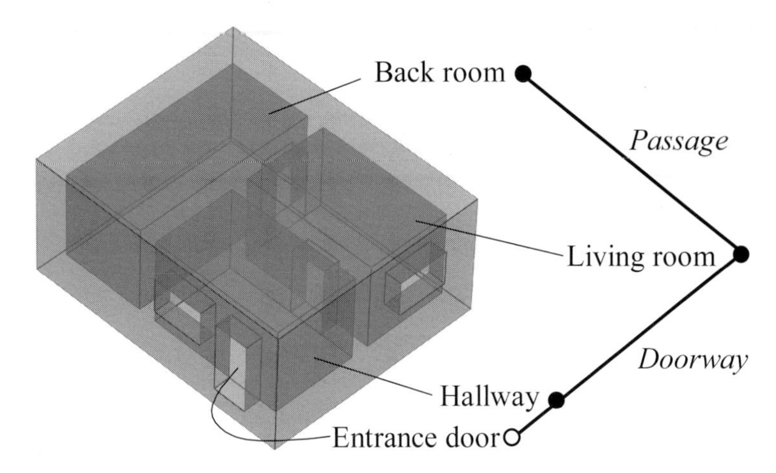

Fig. 5. Building interior (left) and accessibility graph (right) derived from topological adjacencies of room surfaces for the determination of escape routes

Building interiors are modeled by rooms. Their solids are topologically connected by the surfaces representing doors or closure surfaces that seal open doorways. This adjacency implies an accessibility graph, which can be employed to determine the spread of e.g. water, smoke, gas, and air, but which can also be used to compute escape routes using classical shortest path algorithms. The edges of the accessibility graph can be marked by the corresponding distances and types of connection like normal door, fire protection door, open doorway etc. (see fig. 5).

5 Pilot 3D: First Implementation and Evaluation

The 'Pilot 3D' is a testbed of the GDI NRW in which CityGML is currently evaluated by five project teams. The teams consist of participants from academia, software manufacturers, and the cities Berlin, Hamburg, Düsseldorf, Cologne, Leverkusen, and Recklinghausen. In order to keep the implementation simple, only building models and DTMs in levels-of-detail 1 and 2 are included. Furthermore, only object geometries are exported – topology will be added in the future. Fig. 6 gives an impression of the contents and the structure of CityGML data files.

```
<?xml version="1.0" encoding="UTF-8"?>
<CityModel>  <!-- declaration of namespaces omitted here due to limited space -->
  <gml:boundedBy>  <!-- extent of this dataset -->
    <gml:Envelope srsName="EPSG:31466">  <!-- Gauss-Krüger 2nd meridian -->
      <gml:pos srsDimension="3">2571000.0  5677000.0  41.8</gml:pos>
      <gml:pos srsDimension="3">2573000.0  5679000.0  213.7</gml:pos>
    </gml:Envelope>
  </gml:boundedBy>
  <siteMember>
    <Building gml:id="Building0815">
      <externalReference>  <!-- This is a reference to the German cadastre ALKIS -->
        <informationSystem>http://www.adv-online.de</informationSystem>
        <externalObject>
          <uri>urn:adv:oid:DEHE123400007001</uri>
        </externalObject>
      </externalReference>
      <function>31001_1010</function>   <!-- this code means residential house -->
      <yearOfConstruction>1985</yearOfConstruction>
      <roofType>3100</roofType>          <!-- this code means saddleback roof -->
      <measuredHeight uom="#m">8.0</measuredHeight>
      <lod2SolidProperty>                <!-- building geometry for Level-of-Detail 2 -->
        <gml:Solid srsName="EPSG:31466">    <!-- Gauss-Krüger 2nd meridian -->
          <gml:exterior>
            <gml:CompositeSurface>
              <gml:surfaceMember> <!-- first roof face -->
                <gml:OrientableSurface orientation="+">
                  <gml:baseSurface>
                    <gml:Polygon>
                      <gml:exterior>
                        <gml:LinearRing>
                          <gml:pos >2571013.0  5678371.0  64.2</gml:pos>
                          <gml:pos >2571018.4  5678375.5  64.2</gml:pos>
                          .........................
                        </gml:LinearRing>
                  ...... end tags of the elements above
              <gml:surfaceMember>
                <!-- second roof face --> .............
              </gml:surfaceMember>
                  ...... further faces (walls, floor)
      </lod2SolidProperty>
    </Building>
  </siteMember>
    ...... further Buildings
  <loD2TerrainMember>
    ...... here comes the digital terrain model for Level-of-Detail 2
  </loD2TerrainMember>
</CityModel>
```

Fig. 6. Excerpt from a CityGML dataset containing buildings and a digital terrain model in level-of-detail 2 (LoD2)

6 Conclusions and Further Work

CityGML provides substantial information for urban disaster management tasks. Spatial objects and terrain models are represented by their geometry, topology, appearance, and semantic resp. thematic properties. The ability of maintaining different levels of detail makes it suitable for small to large area utilization. Since GML was designed by the OGC to serve as the standard exchange format for spatial data infrastructures, processing of CityGML is immediately supported by corresponding web services like the Web Feature Service (WFS), Web Catalog Service (WCAS), and Web Coordinate Transformation Service (WCTS).

The availability of data is steadily increasing as more and more municipalities decide to build up virtual 3D city models. In the SIG 3D of the GDI NRW we are currently working on the required methods and workflows in municipalities to generate and maintain LoD1 and LoD2 models in conjunction with cadastre. In the long-term, 3D cadastres could provide the main elements for 3D city models (Stoter and Salzmann 2003). LoD3 and LoD4 models currently are only available for selected areas resp. building complexes. Their automatic acquisition is still a topic of current research. However, the semantic properties of buildings and their parts as supplied by CityGML play an important role for e.g. the development of procedures for the automatic reconstruction of 3D building models from aerial images (Fischer et al. 1998).

Future work will focus on the development of more detailed models for roads, waterbodies, and vegetation. Above, concepts for the representation of history in the sense of a timeline and concurrent (planning) versions shall be integrated. On the international level, CityGML will be further discussed within the European Spatial Data Research organization (EuroSDR) and the Open Geospatial Consortium in the next months. It is intended to evaluate CityGML regarding practical issues with respect to big city models, interoperability to AEC/FM, and possible extensions to GML in the future. Finally, it would be interesting to discuss possible extensions of CityGML for the domain of disaster management.

Acknowledgements

We wish to thank the members of the modeling working group of the SIG 3D of the GDI NRW: Joachim Benner, Frank Bildstein, Martin Degen,

Dirk Dörschlag, Rüdiger Drees, Heinrich Geerling, Ulrich Gruber, Frank Knospe, Andreas Kohlhaas, Kai-Uwe Krause, Ulrich Krause, Klaus Leinemann, Marc-Oliver Löwner, Hardo Müller, Hanns-Florian Schuster and Frank Thiemann. Furthermore, we like to thank our colleagues at the Institute for Cartography and Geoinformation, in particular Viktor Stroh. Thanks go to Michael Haas for assistance in preparing the illustrations.

References

Adachi Y, Forester J, Hyvarinen J, Karstila, K, Liebich T, Wix J (2003) Industry Foundation Classes IFC2x Edition 2, International Alliance for Interoperability, http://www.iai-international.org

Benner J, Leinemann K, Ludwig A (2004) Übertragung von Geometrie und Semantik aus IFC-Gebäudemodellen in 3D-Stadtmodelle. In: Schrenk, M (ed) Proc. CORP 2004 & Geomultimedia04 (in German)

Bishr Y (1998) Overcoming the semantic and other barriers to GIS interoperability. Int. Journal on Geogr. Information Science, vol 12, no 4

Booch G, Rumbaugh J, Jacobson I (1997) Unified Modeling Language User Guide. Addison-Wesley

Coors V, Flick S (1998) Integrating Levels of Detail in a Web-based 3D-GIS, Proc. 6th ACM Symposium on Geographic Information Systems (ACM GIS 98), Washington D.C., USA

Cox S, Daisy P, Lake R, Portele C, Whiteside, A (2004) OpenGIS Geography Markup Language (GML3.1), Implementation Specification Version 3.1.0, Recommendation Paper, OGC Document no 03-105r1

Eastman, CM (1999) Building Product Models: Computer Environments Supporting Design and Construction, CRC Press

Fischer A, Kolbe TH, Lang F, Cremers AB, Förstner W, Plümer L, Steinhage V (1998) Extracting Buildings from Aerial Images using Hierarchical Aggregation in 2D and 3D. Computer Vision & Image Understanding, vol 72, no 2, Academic Press

Foley J, Van Dam A, Feiner S, Hughes J (1995) Computer Graphics: Principles and Practice. Addison Wesley, 2nd Ed

gbXML (2003) Green Building XML Schema. http://www.gbxml.org

Gröger G, Kolbe TH, Plümer L (2004) Mehrskalige, multifunktionale 3D-Stadt- und Regionalmodelle. Photogrammetrie, Fernerkundung, Geoinformation (PFG) 2/2004 (in German)

Groot R, McLaughlin JD (2000) Geospatial Data Infrastructure - Concepts, Cases, and Good Practice. Oxford University Press

Herring J (2001) The OpenGIS Abstract Specification, Topic 1: Feature Geometry (ISO 19107 Spatial Schema), Version 5. OGC Document no 01-101

LandXML (2001) LandXML Schema 1.0. http://www.landxml.org

Köninger A, Bartel S (1998) 3D-GIS for Urban Purposes, Geoinformatica, 2(1), March 1998

Kolbe TH, Gröger G (2003) Towards unified 3D city models. In: Schiewe, J., Hahn, M, Madden, M, Sester, M (eds): Challenges in Geospatial Analysis, Integration and Visualization II. Proc. of Joint ISPRS Workshop, Stuttgart

Kolbe TH, Gröger G (2004) Unified Representation of 3D City Models. Geoinformation Science Journal, vol 4, no 1

Molenaar M (1992) A topology for 3D vector maps. ITC Journal 1992-1

Okabe, A, Boots, B, Sugihara, K (1992) Spatial Tessellations: Concepts and Applications of Voronoi Diagrams. John Wiley & Sons

Oosterom P, Stoter J, Quak W, Zlatanova S (2002) The balance between geometry and topology. In Richardson D, Oosterom P (eds): Advances in Spatial Data Handling. Proc. of 10th Int. Symp. SDH 2002, Springer, Berlin

Pfund, M (2002) 3D GIS Architecture. GIM International 2/2002

Stoter J, Salzmann M (2003) Towards a 3D cadastre: where do cadastral needs and technical possibilities meet? Computers, Environment and Urban Systems, Theme Issue: 3D Cadastres, vol 27, no 4, July 2003

Van Rees R, Tolman F, Beheshti R (2002) How BcXML Handles Construction Semantics. In: Proc. of the Int. Council for Research and Innovation in Building and Construction, CIB w78 conference 2002, Aarhus, 12-14 June

Vretanos PA (2002) Web Feature Service Implementation Specification Version 1.0.0, OGC Document no 02-058

VRML97 (1997) Information technology – Computer graphics and image processing – The Virtual Reality Modeling Language (VRML) – Part 1: Functional specification and UTF-8 encoding. Part 1 of ISO/IEC Standard 14772-1:1997

Zlatanova S (2000) 3D GIS for Urban Development. PhD Thesis, ITC Dissertation Series no 69, The International Institute for Aerospace Survey and Earth Sciences, the Netherlands

Zlatanova S, Holweg D (2004) 3D Geo-Information in Emergency Response: A Framework. In: Proceedings of the 4th International Symposium on Mobile Mapping Technology (MMT 2004), March 29-31, Kunming, China

Population Density Estimations for Disaster Management: Case Study Rural Zimbabwe

Stefan Schneiderbauer and Daniele Ehrlich

Joint Research Centre, European Commission
Institute for the Protection and Security of the Citizen (IPSC)
Via E. Fermi, 1 - TP 267, Ispra 21020 (VA), Italy.
Email: stefan.schneiderbauer@jrc.it

Abstract

This paper tackles the need of enhanced population data for disaster management and aid delivery studies in developing countries. It analyses the usefulness of a set of spatial data layers, including medium resolution satellite imagery, for population density estimations in rural Zimbabwe. The exercise conducted on a 185 x 185km area at a grid cell size of 150m allowed us to develop a methodology that can be extended to the whole of Zimbabwe.

The surface modelling of population density was implemented by integrating 4 main variables: land use, settlements, road network, and slopes. During the modelling procedure, pixel weighting values were allocated according to pre-defined decision rules. In a final step the district population counts of the recent Zimbabwean census were distributed among all pixels of the relevant district according to the pixel weighting values. The resulting land use information and population data can be linked to vulnerability and food insecurity.

In order to be transferred to other countries, the modelling procedure needs to be adapted to case specific characteristics, the determination of which requires a certain level of local / expert knowledge. In addition, passive sensors might not provide sufficient cloud free satellite data for regions lying within the moist tropics.

1 Introduction

Population data is crucial information for disaster management, early warning systems and emergency actions. Population density data for developing countries seem to be an easily accessible asset to the unexposed researcher but are in fact a scarce resource. In developing countries population censuses have started to be conducted only in the second part of the 20th century (Vallin 1992). Censuses are typically carried out every 10 years and in less prosperous countries often supported by donor institutions. They are usually elaborated by national statistical offices and made available to the public in aggregated form as national statistical yearbooks. Aggregated statistical information at country level, often referred to as secondary data, is also available through international organisation and commercial yearbook atlases.

Population data in developing countries are also recorded in the frame of development aid projects and in the aftermath of disasters for needs assessment estimation. These surveys are limited in extent by the geographical coverage of the disaster or the development project. The information generated usually targets certain aspects of population but lack the spatial and temporal continuity and coverage of traditional censuses. Due to the surveys' specific application there is no strategy to (1) update the information or (2) make it available to wider user groups. Hence, these data tend to remain with the institutions that carried out the work.

To address geographically based processes such as those related to agriculture, environment, disasters or communities' vulnerability/poverty estimations, population data need to be available at finer scale units than those available from censuses. In addition they should be in a format more amenable for modelling. The research community has developed techniques to disaggregate population information to finer spatial units, typically the raster grid cell. The raster format containing population counts is also referred to as population density representation and is increasingly used for disaster management, early warning and rapid response to crises.

Research is currently being conducted to improve population distribution accuracy and develop spatial units in an attempt to model population attributes. This paper focuses on improvement of the spatial resolution of population datasets available at district levels over Zimbabwe. It (1) uses country specific information layers, (2) evaluates the use of medium resolution satellite imagery and map based data for a population density estimation made available at a grid size of 150 meters. The methodology is developed over a region in Zimbabwe that includes the capital Harare.

2 Background

Population data from censuses are commonly made available per political unit. The datum includes the graphical representation of the administrative unit that constitutes the spatial reference system and the associated attribute. The attribute reports the value that can be expressed as total population or urban / rural population of that unit. Countries are subdivided in a hierarchical system of administrative units up to 4 levels deep: country, regions / provinces, counties / districts, and municipalities / communes that make up the administrative hierarchy.

Administrative borderlines are not usually designed to represent geographical phenomena. Therefore, on choropleth population maps, the populations seem to be homogenously distributed over the area of the administrative unit, notwithstanding possibly significant variations in real population densities. The dasymetric mapping approach aims to delimit regions with similar population densities by applying ancillary data (Mennis 2003; Langford and Unwin 1994). Dasymetric mapping of populations at large and medium scale has been successfully applied in developed countries by using land use / land cover information stemming from earth observation data (Langford et al. 1991; Holloway 1999; Chen 2002; Mennis 2003; Lo 2003; Liu 2004).

Population data associated with vector boundaries are difficult to process. Therefore, researchers often disaggregate these data into grid cells of a finer spatial unit for the following reasons. Grid data format provides uniform scale, even if the original administrative units cover different areas. Grid cells are easy to perform computation on, which is of particular importance for analysis within a GIS framework. Grid cells are particularly useful in a number of decision support applications implemented when population estimations have to be computed for areas that intersect the administrative units for which population data are made available. In fact, raster data can be re-aggregated to accommodate any aerial arrangement, making population counts particularly easy to carry out.

The first global population density estimation in raster format was initially developed under requests from international agricultural research institutes (Deichmann 1996). Since then other raster based global population densities have been produced (Sutton et al. 2003), of which Landscan (Dobson 2000) has received the most attention because it is available and has been updated up until 2002. Unfortunately it has no qualitative measures attached to it and the model used in disaggregating the data is not documented. Potential errors within Landscan include the integration of the Nighttime lights dataset. These data tend to underestimate the population density of urban centres and to overestimate the population density of suburban areas (Sutton et al. 1997). Exclusion areas are calculated by digital slope and by the global land cover map of the world; two datasets, for which no accuracy measures are attached. Also, Landscan uses a more limited measure of administrative boundaries and population counts than the Global Population of the World, an alternative global population database in gridded format.

The global population density datasets are invaluable. They are used to provide preliminary estimates of casualties in the aftermath of natural disasters such as earthquakes, tropical storms or large floods. Also, in the absence of better population figures, these datasets are used in the assessment of slow onset disasters, namely droughts or epidemics, that are typically geographic specific. The demand for more precise population density data at a finer scale remains unmatched.

Converting point values or polygon values into grid format is referred to as surface modelling. A commonly used technique in population modelling is smart interpolation. Smart interpolation for population density estimation is based on two major steps. (1) The calculation of a grid based population potential and (2) the allocation of population numbers, usually available at a certain administrative level, to the gridded population potential. The population potential estimation relies on a series of variables including the location and size of urban settlements, and thematic data layers such as those depicting national parks, protected areas and inaccessible areas that are supposedly not populated. Smart interpolation has been used to produce the global population database Landscan. Landscan was developed by allocating census counts at 30" X 30" cells through a "smart" interpolation based on the relative likelihood of population occurrence. The allocation to cell is based on weighting computed from slope categories, distance from major roads, and land cover. The weighting is undertaken by type, with exclusions for certain types. The Nighttime lights of the World are weighted by frequency.

Improvement of the smart interpolation techniques relies on the availability of fine scale data that are country specific, and Earth Observation data. Aerial photography and earth observation imagery were already tested for population density mapping in the 1980's. Remote sensing techniques have been applied in order to improve or overcome census shortcomings in developing countries' censuses (Adeniyi 1983; Olorunfemi 1984). Coarse resolution imagery has been used in improving global population density assessments (Eldvige et al. 1997), and medium resolution for zone based estimations (Lo 2003). Harvey (2002) specifically tested estimations of population numbers from satellite imagery. With the availability of very high resolution imagery, formal settlements such as cities and informal settlements such as refugee camps can be assessed for population density (Giada et al. 2003a; Giada et al. 2003b). A recent overview of techniques used for counting population in cities is provided by Mesev (2003).

This study addresses the use of medium resolution imagery for zone based estimates of population densities in developing countries. It aims at further developing methodologies that allow the relatively rapid generation of information on population distribution at a fine scale that is required for improved disaster management.

The study area extends over 185 x 185 km, coinciding with the coverage Landsat TM scene path 170 and row 72 (Fig. 1).

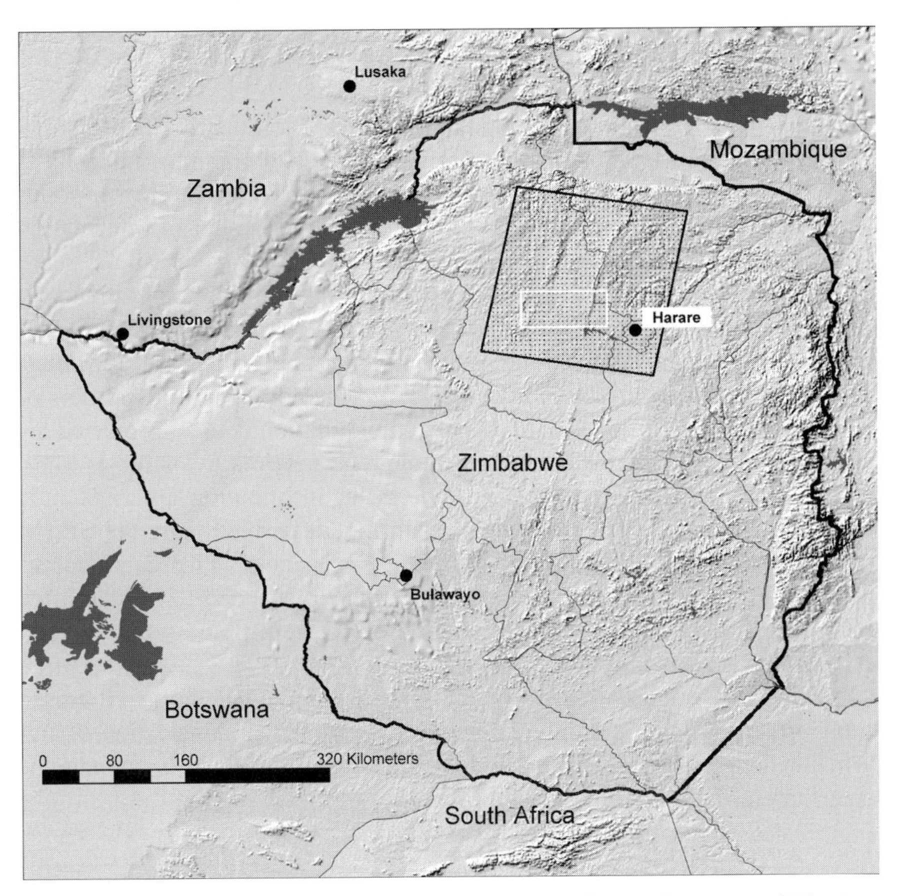

Fig. 1. Study area within Zimbabwe, the white box indicates the extent of Fig. 5

It includes the economic centre and capital Harare, and makes up part of the most productive agri-ecological zone of intensive farming in Zimbabwean's Highveld and Middleveld (Vincent and Thomas 1960). The study area covers commercial private farmland as well as government owned land, referred to as communal land[1]. The study is therefore representative of the land cover and land use diversity in Zimbabwe. Within the study area are three million people, accounting for 25% of the overall Zimbabwean population, living on an area of 33206 km2, that is, 12% of the total land cover. The methodology devised and tested within the study area will eventually be applied to the whole of all Zimbabwe.

3 Methodology

We model the distribution of population within each district of our study area in Zimbabwe by creating a continuous surface of population density at a grid cell size of 150 m. The methodology for surface modelling of population data relies on combining information from a number of information layers listed in Table 1.

The input data were available as explicit digital information, with a spatial and attribute component, and data layers that required information extraction to provide delineation of spatial units and associated attributes. The ready to use information layers included the administrative borders of Zimbabwe at district level with associated population counts, protected areas, urban areas available as polygons, the road network available as lines, and towns, available as point data. The raster information layers include the 1 km land cover and digital elevation model derived from the SRTM (Shuttle Radar Topography Mission) elevation dataset. Two data sources required information extraction. Satellite imagery needed to be classified into land cover / land use classes, while maps were processed to extract specific features such as village location.

All data layers required pre-processing that included the standardised transformation into a common geographical projection. A selection of significant data layers was processed in order to allow their input into the smart interpolation model.

[1] In southern Africa 'communal land' generally refers to an area of land owned by the State, which confers certain use rights (for cultivation, livestock grazing, timber harvesting, settlement, etc.) to rural populations that do not hold individual proprietary deeds. Many such areas in Zimbabwe are used for traditional agro-pastoral farming activities.

Dataset	Format / Resolution / Scale	Date	Coverage / Size	Theme extracted	Source
Population Census data	District level	2002	Countrywide	Population per district	CSO Zimbabwe
Administrative subdivision of the country	Polygons	2002	60 Districts	District boundaries	Survey-General Zimbabwe
Urban areas	Polygons	NA	Areas of 8 main cities	City areas	HIC
Protected areas	Polygons	NA	Countrywide	Protected areas	Survey-General Zimbabwe
Road network	Lines	2002	Countrywide		HIC
Settlements	Point	NA	Countrywide	Towns	DCW
Elevation	Raster / approx. 90 m (3 arc seconds)	2002	SRTM	Slope	USGS
GLC 2000	Raster / 1 km	2000	Countrywide	Selected land cover classes	JRC
Landsat ETM	Raster / 15 m & 30 m	2002 /2003	24 images	Land cover / Land use	LANDSAT
Maps	1:250,000	1973-1995	34 maps	Villages	Survey-General Zimbabwe
Woody cover map	Paper map 1:1 000,000	1998	1 sheet	Woodland (Photointerpreation)	Zimbabwe Forestry Commission

CSO – Central Statistical Office
HIC – Humanitarian Information Centre
USGS – United States Geological Survey

Table 1. Data layers used for modelling population densities within the study area

The modelling procedure produced a weighting value for each pixel in the study area according to the relative probability of population numbers living within the pixel area. The weighting values permitted the redistribution of one population value per district to a number of grid cells within this district.

3.1 Pre-Processing

Pre-processing consisted of scanning maps, geo-coding all information layers in a common geographic projection, and then conversion of the data into raster format.

Zimbabwe is covered by two UTM zones making standard UTM difficult to use. The national geographic projection and the most common projection for mapping Zimbabwe as a whole country, is the Transverse Mercator projection based on the Modified Clarke 1880 spheroid (Mugnier 2003). Therefore all the data were re-projected into this coordinate system. The re-projection of the satellite images was carried out after the classification in order to avoid information loss. After the re-projection all vector layers were transferred into raster layers of 15 m pixel size, fitting with the panchromatic band of Landsat ETM.

3.2 Information Extraction

Information extraction was performed on scanned maps in order to extract the location of villages, and on satellite imagery in order to produce a land cover map for the area.

3.2.1 Processing Map Data

The objective of processing the scanned topographic maps at a scale of 1:250,000 was to extract the location of small settlements. Villages and buildings indicating the centres of settlements, such as schools and hospitals, are represented within these maps as black dots and rectangles of slightly different size (Fig. 2, top left).

A combination of feature recognition and image processing algorithms allowed for the extraction of these objects. First we identified black features of a defined size within the scanned maps by applying the mathematical morphological methods of erosion and dilation. This resulted in the production of a binary file (Fig. 2 top right). Following this, we separated erroneously selected objects such as fragments of black labelling based on their shape and shape / size combinations (Fig. 2 bottom left). Fig. 2 shows the working step results for a cut-out of the study area, where mining symbols printed in the same colour as the village dots, hampered the extraction procedure. The extraction process resulted in the identification of 6823 settlements within the study area.

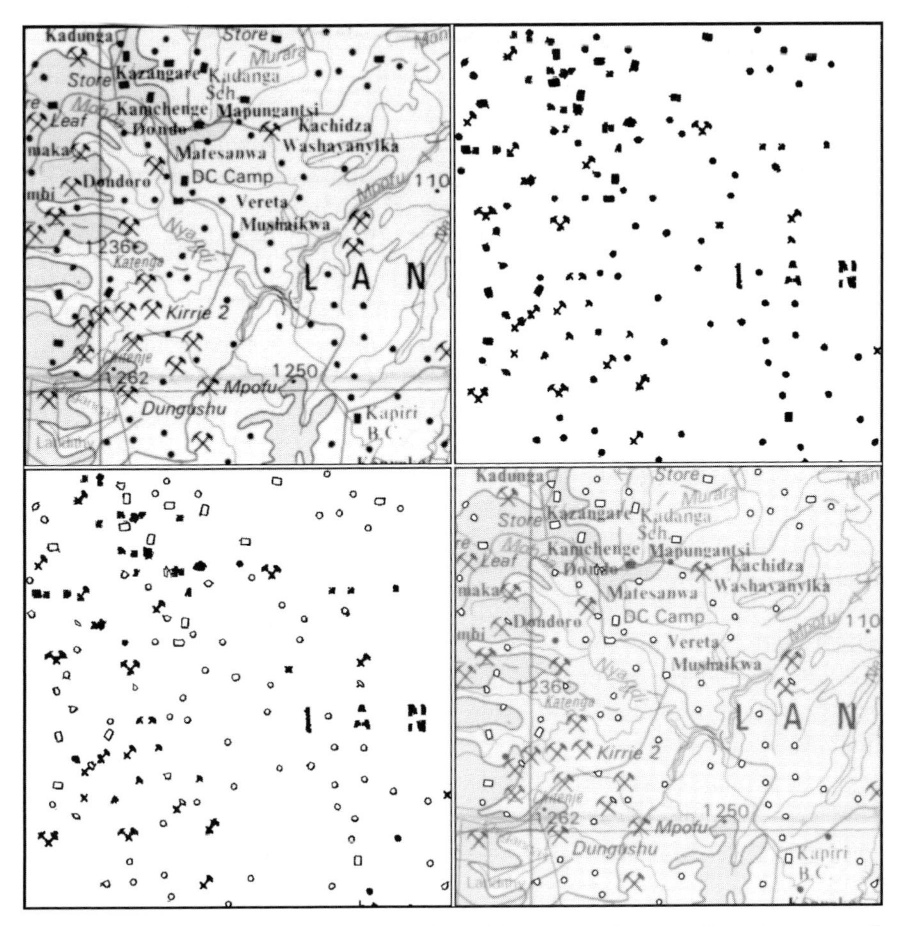

Fig. 2. Processing stages of the extraction of human settlements from the scanned maps. The figure shows the original data (top left), the features extracted from the map (top right), the extracted features classified - for which villages are coded as small circles (bottom left) and the village location superimposed on the original map in order to check for validity (bottom right)

3.2.2 Processing Satellite Imagery

Two Landsat ETM scenes were used in the exercise. The dry season image (Aug 03, 2002) was used for the classification process, whilst the scene recorded at the end of the wet season (May 02, 2003) was used for validation purposes. 4 major land use classes were produced that included intensive small scale farming on communal land, large scale commercial farming, woodland, and land unsuitable for population. The classification process used object oriented image analysis tools available in the software package e-cognition Definiens. The classification process was based on data from the Landsat ETM multi spectral band with 30m and the panchromatic band 8 at 15m resolution (Fig. 3a). The NDVI calculated from the TM band 4 and TM band 3 was used as an additional channel. The image was first segmented at two different levels. The first level was used to identify relatively small objects of similar spectral characteristics such as large fields in commercial farm areas (Fig. 3b). Typical feature characteristics for object differentiation are shape, mean reflection values in a certain band or their standard deviations. The second segmentation level was used to identify larger regions with similar general land cover patterns (Fig. 3c). Each object at this level included a number of sub-objects stemming from the first segmentation level. The differentiation process at level 2 is predominantly based on the type and characteristics of the sub-objects included combined with the sum of the area covered by sub-objects of the same class. The classification procedure required frequent verification and supervision supported by the Landsat image of the wet season and the woody classification map.

We identified 23 land cover classes based on segmentation level 1, which were converted into 8 land use classes derived from segmentation level 2. For integration into the model these 8 classes needed to be aggregated resulting in 4 final land use types (Fig. 3d). The division into these final classes relied on local and expert knowledge relating to the main variances in population densities that exist between intensive small scale subsistence agriculture, large scale commercial farming, bush and woodland, and non-populated areas.

3.2.3 Geo-Processing

Geo-processing consisted of processing and sometimes merging the input layers into information layers to be used in the smart interpolation modelling exercise. 4 final input layers were derived comprising land use, human settlements, road network, and slope classes.

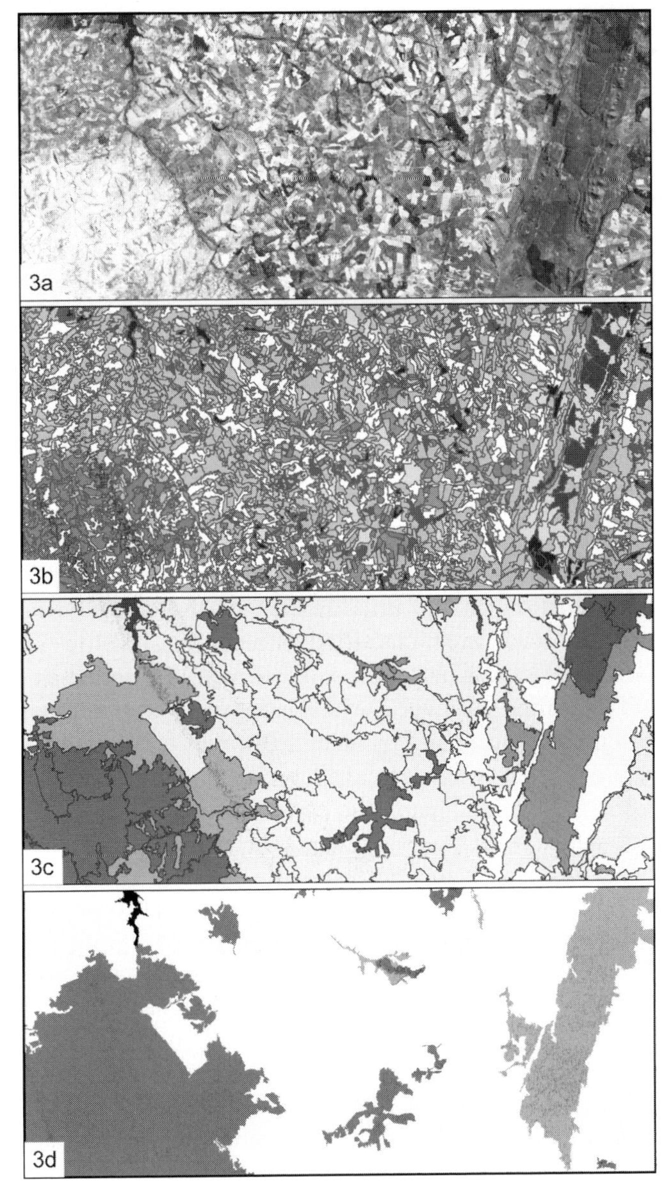

Fig. 3. Landsat imagery classification, study area extract including commercial farms and communal land [scale approx. 1:500,000]. 3a: Original Landsat 3b: Level 1 segmentation resulting in 23 classes, 3c: Level 2 segmentation and aggregation of level 1 into 8 classes, 3d: Aggregation of classification results into 4 classes for integration into the model

3.2.4 Land Use

The classification of different land use types is based on medium resolution satellite imagery (Landsat ETM) of the dry season in 2002, enhanced by ancillary information stemming from medium satellite imagery from the end of the wet season in 2003 (Landsat ETM), a woody cover map from 1998 and the GLC2000 [2]. The classification of the study area resulted in 4 main rural land use classes, each accounting for a specific average population density: (1) intensive subsistence small scale farming, which is the predominant use of communal land, (2) large scale commercial farming, (3) bush and woodland not (recently) used for agricultural purposes and (4) areas not suitable for human settlements.

3.2.5 Human Settlements

The human settlements layer merged information from three data sources that have been allocated to 4 hierarchy levels. (1) A polygon vector file - that includes the 8 most important Zimbabwean urban areas. This layer provided information on the urban extent of the cities Harare and Chitungwiza for the study area, classified as hierarchy level 1. (2) The settlements from the Digital Chart of the World (DCW), coded as point information and classified as level 2 when large or district capitals (only two present in the study area) and as level 3 if towns. (3) Smaller settlements, typically villages that where not available from DCW, were derived from the 1:250,000 topographic maps and coded as level 4.

The settlements available as single pixel (level 2, 3 and 4) were buffered to produce an associated area corresponding in size to the relevant hierarchy level. Towns of level 2 where buffered with a 1500 m radius, those of level 3 with 600 meters and the villages extracted from the map with 100 meters.

3.2.6 Road Network

Information on the road network is based on a dataset from the Humanitarian Information Centre of the UN in Zimbabwe, last updated in 2002. The roads were classified into two groups, primary and secondary roads. The class allocation was carried out by considering tarmac quality, which is included in the dataset, and level of road traffic. All roads missing at-

---

[2] Global Land Cover 2000 database. European Commission, Joint Research Centre, 2003, http://www.gvm.jrc.it/glc2000 based on SPOT VEGETATION satellite images.

tribute information were classified as secondary roads. The primary and secondary roads were buffered by 300m and 150m respectively, based on the assumption that population density close to roads increases and the concerned area is correlated with the road importance.

3.2.7 Slope

Slope information was calculated from the SRTM elevation dataset. This dataset is produced at 90 meter resolution. We use the CGIAR-CSI (Consortium for Spatial Information) SRTM data product on which a number of processing steps have been applied in order to represent the elevation values as continuous surfaces. The SRTM dataset served as the foundation for computing the slope, based on a 3 by 3 neighbourhood around each pixel. The results were divided into three classes of slopes with < 10 degree, 10 – 20 degree and > 20 degree.

The resulting land use classification was available at 15m grid cells, corresponding to the resolution of 15m of the panchromatic band 8 of the Landsat ETM scene. All other vector and raster layers were also converted into raster grid files of 15m resolution.

3.3 Smart Interpolation

Modelling of the population density was carried out by allocating weighting factors to the pixels of each input layer, according to a number of decision rules. The process of weight allocation is summarised in the decision tree diagram of Fig. 4. The first rule relies on the land use class input layers. The weight zero was allocated to all pixels within the land cover class representing areas unsuitable for hosting populations. All other pixels, unless they are in the vicinity of the buffered areas of roads or human settlements, were allocated weight 1 for bush and woodland, weight 2 for large scale farming and weight 8 for small scale intensive farming.

In the second decision step, weighting values for all pixels lying within the defined vicinity of roads and human settlements were assigned. If the pixel was close to transport infrastructure, then it obtained the weight 5 or 20 respectively, according to whether the transport line crossed a wooded area or any type of farm land. The weighting values allocated to pixels within or close to settlements depend on the hierarchy class of the settlement. Pixels associated with class 2 (small cities) and class 3 (towns) received a weighting of 60, while those related to class 1 (villages) received a weighting of 20[3]. All weights allocated to pixels in consideration to their vicinity to a road network and human settlements were overlaid and summed up.

The last decision rule took into account the interrelation between population density and terrain. It has applied to all pixels lying within areas of a certain degree of slope. If the slope value is between 10 and 20 degrees the pixel weighting value was halved. If the slope was steeper than 20 degrees then it was halved again. After giving consideration to the degree of slope, each pixel received its final weighting value. For example, a pixel lying within the land use class 'large scale farming', in the vicinity of a road and close to a village located on a slope of 13 degrees would have received the value: (20 + 20) / 2 = 20.

Following this allocation procedure, the population counts at district level available from the Zimbabwean Census of 2002 were distributed according to the weighting value pw of each pixel according to Eq. 1.

$$pop\ (i,j)\ [D] = a_D * pw\ (i,j) \quad (1)$$

In Eq. 1 pop(i,j) [D] is the number of estimated people living within the area of pixel (i,j) that is lying within district D, a_D is a constant computed for district D, and pw (i,j) is the weighting factor of pixel (i,j) received from the model (Fig. 4). The constant a_D is computed by dividing the overall population number of district D by the sum of the weighting values of all pixels within district D (Eq. 2).

$$a_D = \frac{pop\ [D]}{\sum \{pw(i,j)\ \forall\ (i,j) \in [D]\}} \quad (2)$$

In Eq. 2 pop [D] is the census population of district D and pw (i,j) is the weighting value of pixel (i,j). Hence, the sum of the estimated population values for all pixels within a district area is equal to the census value of the same district. For those districts of which only part of the area is overlapping with the study area the number of population considered in the model (pop [D]) was computed according to the proportion of the area lying within the study area, assuming a homogeneous population distribution.

---

[3] Large cities corresponding to human settlements class 1 obtain a unique population density value that is based on the average population of the entire city. Population density estimations within cities require a different model and will be addressed in future work.

Following the modelling process, the pixels were aggregated to a raster layer with 150 m resolution. A lowpass filter was applied to smooth differences between adjacent pixel values.

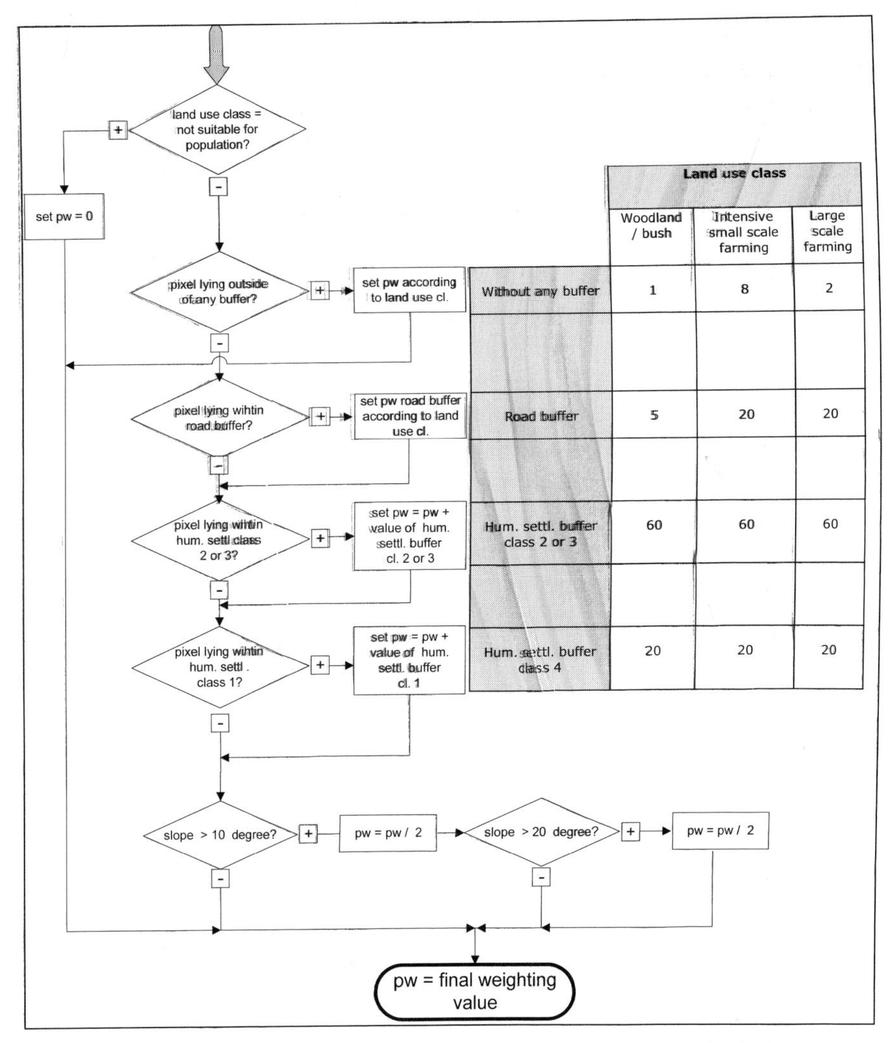

Fig. 4. Decision tree diagram used to assign weighting values to pixels (pw)

4 Results and Discussion

The resulting population density estimation for a part of the study area is shown in Fig. 5. The figure also shows the equivalent area as a choropleth map representing population counts at district level and as the Landscan 2002 population dataset. Light grey values represent low, and dark grey values high population densities. The figure highlights the advantage of having population densities in grid format and at finer resolution. The choropleth map provides single counts for large areas, thus averaging areas with high and low population densities. Landscan inevitably provides coarse results. The outcome of the population modelling exercise at the fine resolution of 150 m reflects typical population density patterns in Zimbabwe, for example communal land with high densities and commercial farm land with low population densities.

Combining different sources of input data to provide weighting for population assessment is an empirical and error prone exercise. The input datasets are of different quality and the errors accumulate when merging the dataset. Errors introduced into the model applied for this study can be linked with a number of sources. First, the location accuracy of human settlements within the DCW is probably low. The DCW are in fact derived from 1:1,000,000 maps. The quality of the topographic maps and the density of mapped villages vary with different years that they have been updated. The automatic dot detection allows for false hits and inevitably misses some villages. The location of some villages may also have changed since the production of the map.

Medium resolution Landsat imagery provides valuable information to identify land cover. The pixel size is in general sufficiently small for distinction to be made between fragmented communal land used intensively by subsistence agriculture and large scale commercial farm areas. The combination of satellite images recorded during dry and wet seasons assists in the identification of irrigation schemes and regions with little vegetation accounting for areas with the likelihood of high population pressure.

The object oriented classification procedure proved effective and efficient in processing large datasets, such as a full Landsat ETM scene. The image segmentation at two different levels as a base for a land use classification of 8 classes proved to be most appropriate and allowed a clear discrimination of the main Zimbabwean land use systems.

Additionally the ETM imagery with a 15 meter panchromatic band can be used to identify the spatial outline of larger towns. Densities within urban areas will have to be addressed with very high resolution imagery.

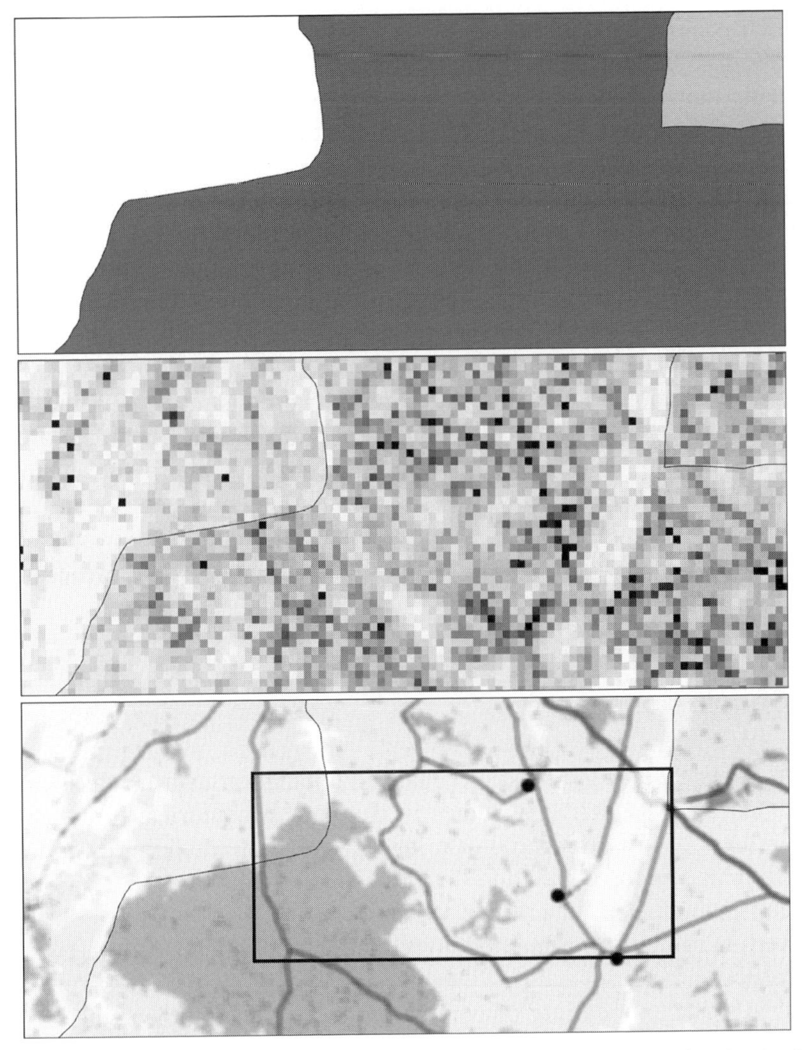

Fig. 5. Population distribution as choropleth map (top), population density in the Landscan 1km dataset (centre), and in the resulting 150m dataset of this study (bottom) [scale approx. 1:800,000]. The black box in the bottom map indicates the extent of Fig. 3

The global availability of the Landsat imagery makes it a unique dataset for use in population density studies in developing countries. In moist tropics however, the use of active sensors has to be considered in order to solve the problem of permanent cloud coverage.

The availability of a 90 meter resolution digital elevation model is critical for at least two reasons. It allows computation of the slope gradient, which is strongly correlated with population densities, and thus identifies areas that are not suitable for hosting any population. Also, population density may vary with the elevation. For example, in Zimbabwe, the highlands are favoured areas of settlement due to climate and geological conditions.

At last, the selection of weighting factors allocated in the modelling process is based on subjective local and expert knowledge. The modelling result, the final population density estimation, remains very difficult to evaluate. Research findings related to the creation of continuous population surfaces in developed countries are often cross-checked with small enumeration units (Langford et al. 1991; Eicher and Brewer 2001). In Zimbabwe, population data for the smallest administrative unit, the 'ward', are available but they are often incorrect and the ward boarders are continuously being changed for political reasons. An alternative method of evaluation could be to get the census' population counts before the data are aggregated into administrative units. This would require the cooperation of the relevant statistical institution and their willingness to provide the data for research purposes.

5 Conclusions

This paper tackles the need of enhanced population data for disaster management and aid delivery studies in developing countries. It analyses the usefulness of a set of spatial data layers, including medium resolution satellite imagery, for population density estimations in rural Zimbabwe. The exercise conducted on a 185 x 185km area at a grid cell size of 150m allowed us to develop a methodology that can be extended to the whole of Zimbabwe.

The surface modelling of population density was implemented by integrating 4 main variables: land use, settlements, road network, and slopes. A central determinant for estimating population numbers was the type of land use, which was successfully obtained from Landsat imagery by applying object oriented classification techniques. The location of small human settlements was only available on paper maps requiring the automatic extraction of printed symbols. Feature recognition algorithms based on mathematical morphology and size / shape relationship proved to be useful tools for solving this problem. During the modelling procedure, pixel weighting values were allocated according to pre-defined decision rules and based on the characteristics of the 4 input layers. In a final step the district population counts of the recent Zimbabwean census were distributed among all pixels of the relevant district according to the pixel weighting values.

Due to the lack of any calibration data, the results of the exercise are difficult to evaluate. The error range is determined by the quality of the input layers, unavoidable faults in the feature extraction or classification procedure and by the choice of the weighting values, which are based on local and expert knowledge. However, a comparison with the Landscan dataset from 2002 clearly demonstrates the advantages of the applied methodology.

In most developing countries recent data about land use, infrastructure and population at fine scale are poor. The method applied for this study is a relatively quick way to extract information from and add value to datasets that are normally available, also for developing countries. The methodology elaborated and the tools implemented are therefore of general importance for population density estimations at fine resolution in those countries. The results can feed into early warning or alert systems and support decision making in order to enhance disaster management and humanitarian aid activities. Finally, land use information and population data can also be linked to vulnerability and food insecurity. Therefore, mapping land use and population data is essential in order to come up with needs assessments in the case of drought or other disasters.

However, population distributions are determined by physical, socioeconomic and historical factors that are country specific. Hence, in order to be transferred to other countries, the modelling procedure needs to be adapted to case specific characteristics, the determination of which requires a certain level of local / expert knowledge. In addition, passive sensors might not provide sufficient cloud free satellite data for regions lying within the moist tropics.

Acknowledgments

We would like to acknowledge the support of the HIC in Zimbabwe, namely the expert knowledge provided by Tinago Chikoto.

References

Adeniyi PO (1983) An Aerial Photographic Method for Estimating Urban Population. Photogrammetric Engineering and Remote Sensing 49, pp 545-560

Chen K (2002) An approach to linking remotely sensed data and areal census data. International Journal of Remote Sensing 23, pp 37-48
Deichmann U (1996) A review of spatial population database design and modeling. Technical Report 96-3. National Center for Geographic Information and Analysis, Santa Barbara, USA
Eicher CL, Brewer CA (2001) Dasymetric Mapping and Areal Interpolation: Implementation and Evaluation. Cartography and Geographic Information Science 28, pp 125-138
Elvidge CK, Baugh E, Kihn H, Kroehl ED, Davis C (1997) Relation between satellite observed visible-near infrared emissions, population, economic activity and electric power consumption. International Journal of Remote Sensing 18, pp 1373-1379
Giada S, De Groeve T, Ehrlich D, Soille P (2003a) Information extraction from very high resolution satellite imagery over the Lukole refugee camp, Tanzania. International Journal of Remote Sensing 24, pp 4251-4266
Giada S, De Groeve T, Ehrlich D, Soille P (2003b) Can satellite images provide useful information on refugee camps? International Journal of Remote Sensing 24, pp 4249-4250.
Harvey JT (2002) Estimating census district populations from satellite imagery: some approaches and limitations. International Journal of Remote Sensing 23, pp 2071-2095
Holloway SR, Schumacher J, Redmond RL (1999) People and Place: Dasymetric Mapping Using ARC/INFO. In: Morain S (ed) GIS Solutions in Natural Resource Management, Onword Press, Santa Fe, New Mexico, pp 283-291
Langford M, Maguire D, Unwin DJ (1991) The areal interpolation problem: estimating population using remote sensing in a GIS framework. In: Masser I, Blakemore M. (eds) Handling Geographical Information: Methodology and Potential Applications, Longman, New York, pp 55-77
Langford M, Unwin DJ (1994) Generating and mapping population density surfaces within a geographical information system. Cartographic Journal 31 (1), pp 21-6
Lo CP (2003) Zone-based estimation of population and housing units from satellite-generated land use/land cover maps. In: Mesev V (ed) Remotely Sensed Cities, Taylor & Francis, London, pp 157-180
Mennis J (2003) Generating Surface Models of Population Using Dasymetric Mapping. The Professional Geographer 55, pp 31-42
Mesev V (2003) Remotely Sensed Cities (ed), Taylor & Francis, London, pp 372
Mugnier C (2003): Grids and Datums, Republic of Zimbabwe. In: Photogrammetric Engineering and Remote Sensing 69, pp 1206-1207
Olorunfemi JF (1984) Land Use and Population: A Linking Model. Photogrammetric Engineering and Remote Sensing 50, pp 221-227
Sutton P, Roberts D, Elvidge C, Meij H (1997) A Comparison of Nighttime Satellite Imagery and Population Density for the Continental United States. Photogrammetric Engineering and Remote Sensing 63, pp 1303-1313

Sutton P, Elvidge C, Obremski T (2003) Building and evaluating models to estimate ambient population density. Photogrammetric Engineering and Remote Sensing 69, pp 545-553

Tobler W, Deichmann U, Gottsegen J, Maloy K (1995) The global demography project, Technical Report TR-95-6, National Center for Geographic Information and Analysis, Santa Barbara

Vallin, J (1992) La population mondiale. La Decouverte, Paris

Vincent V, Thomas RG (1960) An agricultural survey of Southern Rhodesia: Part I: agro-ecological survey. Government Printer, Salisbury

Xiaohang L (2004) Dasymetric mapping with image texture. In: Proceedings of the ASP R S 2004 Annual Conference, Denver, USA, pp 23-28

The Fourth Column in Action: Dutch Municipalities Organizing Geo-Information for Disaster Management

Margo de Groot

DataLand, P.O. Box 246, 2800 AE Gouda, the Netherlands.
Email: m.degroot@dataland.nl

Abstract

Disaster management depends on large volumes of accurate, relevant, on-time geo-information that various organizations systematically create and maintain. In The Netherlands municipalities are an important party in the safety management chain when accurate geo-information is concerned . As opposed to municipalities in many other countries Dutch municipalities not only carry out central government policies but also have several policy areas in which they operate autonomously. Hence Dutch municipalities have a wealth of geo-information. This is why in the safety management chain Dutch municipalities are often referred to as the fourth column besides the police, the fire department and medical care. However, due to the municipal autonomy the geo-information is structured and described in as many ways as there are municipalities. This results in a situation in which the semantics of geo-information is not even always clear to the producer and whereby formal semantics are almost never available. On the other hand following several large scale disasters such as the fire in a pub in Volendam and the exploded fireworks factory in Enschede Dutch government has decided to reorganise disaster management in The Netherlands in so called safety regions. These regions consist of several municipalities. Thus this development increases the demand for easily accessible and readily available relevant, standardized and accurate geo-information from municipalities on a regional level.

Working from the same municipal autonomy that has created the described difficulties Dutch municipalities have organised themselves to found DataLand. DataLand is the non-profit on-stop-shop for municipal geo-information. The mission of this initiative is to make the municipal geo-information widely accessible. In fulfilling its mission DataLand together with its participating municipalities standardizes the registrations and formalizes the semantics so that third parties know what information they get. Also DataLand actively monitors the data quality.

This paper will focus primarily on how DataLand as a one stop shop offers a solution to the problem of data management between the different columns involved in disaster management. The paper will demonstrate how challenges for data management, data collection, translation, integration, classification and attributes schemes, temporal aspects (up-to-dateness, history, predictions of the future) are tackled by using simple, low cost IT solutions. Also the paper will describe how open communication is of overriding importance in creating co-operation between different institutions involved with disaster management. Secondly how this solution works in practice will be demonstrated by discussing the results of a real-life, real-time test that DataLand recently conducted together with the Regional Rescue Service Rotterdam-Rijnmond (RHRR) over the course of a year. This experiment not only showed the improved response time of this rescue service body but also for the first time gave clear insight in the user needs of disaster management bodies with regard to municipal geo-information. In closing the paper will reflect on the lessons learned in developing a one-stop-shop for municipal geo-information. This is important as the concept of a one-stop-shop to facilitate smooth data flows between different partners involved in disaster management is also feasible for other parties involved in disaster management.

1 Introduction

Disaster management depends on large volumes of accurate, relevant, on-time geo-information that various organizations systematically create and maintain. In The Netherlands municipalities are an important party in the safety management chain when accurate geo-information is concerned . As opposed to municipalities in many other countries Dutch municipalities not only carry out central government policies but also have several policy areas in which they operate autonomously. Hence Dutch municipalities have a wealth of geo-information that is not available on other levels or in other institutions. Dutch municipalities are the sole source for much of The Netherlands administrative geo-information such as addresses, floor space, and dimensions of the house. In other words: data that objectively describe a building. Besides the fact that municipalities have important disaster management and prevention tasks, this wealth of information is why in the safety management chain

Dutch municipalities are often referred to as the fourth column besides the police, the fire department and medical care.

There is however a price to be paid for municipal autonomy: the geo-information gathered and registered by the municipalities is structured and described in as many ways as there are municipalities. This results in a situation in which the semantics of geo-information is not even always clear to the producer and whereby formal semantics are almost never available. This situation renders the information very inaccessible and difficult to use for parties outside the municipality.

The difficulties that external parties that need the information to carry out their tasks already experience when one municipality is concerned take are brought to a new level when the latest development in disaster management in The Netherlands is concerned. Following several large scale disasters such as a devastating fire in a pub in Volendam and an exploded fireworks factory in Enschede, Dutch government has decided to reorganize disaster management in The Netherlands. The former organization on a local level is being replaced by a regional disaster management organization. These so called safety regions encompass a multitude of municipalities. Naturally this development increases the demand for readily accessible, standardized and accurate geo-information from municipalities on a super municipal level.

Working from the same municipal autonomy that has created the described difficulties Dutch municipalities have organized themselves to found DataLand. DataLand is the non-profit on-stop-shop for municipal geo-information. The mission of this initiative is to make the municipal geo-information widely accessible. As such the initiative is proving to provide a welcome solution for parties involved in disaster management when it comes to accessing municipal geo-information. In fulfilling its mission DataLand together with its participating municipalities standardizes the registrations and formalizes the semantics so that third parties know what information they get. Also DataLand actively monitors the data quality.

This paper will focus primarily on how DataLand as a one-stop shop offers both a technical and an organizational solution to the problem of data management between the different columns involved in disaster management. In the second section the history of this municipal initiative will be briefly described. Thirdly a closer look will be taken at the organizational challenges that DataLand faced – and still faces - in disclosing municipal information for external parties. The fourth section will focus on the technical issues involved in realizing disclosure and interoperability of the data. This section will also expand on the relevance of metadata in the process. The fifth section will zoom in on how DataLand has demonstrated its clear added value for the disaster management chain over the course of a one-year real life test. In closing the epilogue will reflect on which lessons can be learned from the DataLand initiative, as the concept of a central point of data distribution in disaster management is widely applicable.

2 DataLand: A Brief History

Municipalities in The Netherlands come in various sizes, the largest municipalities being the home over 100.000 citizens each. Among these municipalities are of course Amsterdam, Rotterdam, Den Haag and Utrecht but also municipalities such as Apeldoorn, Emmen and Groningen. Altogether this group consists of close to 30 municipalities. Because of their size these municipalities see themselves confronted with similar challenges. Hence they meet at a regular basis.

Already in the late 1990's the problem of accessibility, standardization and interoperability – or rather: lack thereof – was addressed in these meetings as an increased demand for municipal administrative geo-information was noted on the one hand, whereas on the other hand central government started several initiatives to realize a central geo-information infrastructure in The Netherlands.

At the time already many municipalities, small and large, were busy trying to get their own geo-information service in order digitally. The Netherlands Association of Municipalities, the VNG, made in inventory in spring 1997 which showed that local governments with up to 70,000 citizens spend increasingly more on informatisation. Compared to 1996 there was an 18% increase in 1997 in investment in information, communication and technological (ICT-) aids. Of the total score of investments in local government informatisation, approximately 30% are invested in geo-applications such as taxes, charges and Cadastre[1].

Taking into account the political importance of the concept of municipal autonomy against the various developments in the field, the municipalities were faced with the choice to either provide a solution for the accessibility and interoperability issues from their own autonomy or to await further actions of central government. The municipalities decided to take up the challenge themselves rather than to wait for central government. In taking up the challenge themselves they voiced a clear ambition to hit two birds

[1] Jellema, M. (1999). Geo-information chains present local governments a new challenge, 21st UDMS Symposium.

with one stone. In the first place the municipalities wanted to created a central point of distribution for municipal administrative geo-information data of all municipalities in The Netherlands thus answering to the market demand. Secondly this central distribution point should support municipalities in organizing their data for accessibility and interoperability. On the basis of these two ambitions the initiative could easily evolve into a bottom up contribution to the national geo-infrastructure planned at the central governmental level.

Over the course of a year the different terms and conditions such as privacy restrictions and other legal pitfalls were mapped and a business plan was developed. In April 2001 DataLand was founded. Over the course of three years DataLand has managed to become the one stop shop for municipal geo-information, distributing data over more than 4.000.000 objects. This is 75% of the available data. This steady growth has mainly been due to pro-active account management, no frills management and a solid business plan.

3 Organizational Challenges

In realizing its ambitions, DataLand faced a number of organizational challenges as well as technical challenges. This section will briefly discuss the three most important organizational challenges that played a role in the development of this central distribution point:

- municipal autonomy
- history of failed initiatives
- ownership issues

3.1 Municipal Autonomy

As was discussed previously Dutch municipalities enjoy a relatively large measure of autonomy in forming local policies and in running the municipal administration. This has lead to a situation in which geo-information data are registered in as many different forms using almost as many different standards as there are municipalities. In itself standardization does not conflict with the concept of municipal autonomy. However, historically this situation has grown and in shaping up DataLand it has proven a challenge to demonstrate that co-operation and common standardization does not interfere with municipal autonomy. On the contrary: co-operation in this field strengthens the position of municipalities in the field of governmental information management.

The fact that the DataLand initiative is firmly rooted in the concept of municipal autonomy has proven a crucial success factor and has been a prime unique selling point in the marketing strategy. The municipal autonomy is further more reflected in the self-sustaining business model of DataLand. While it is beyond the scope of this paper to discuss the business model in detail, it is worth noting that the business model of DataLand ensures that municipalities not only own DataLand and thus govern its course, but also they share in the revenue that is generated.

3.2 History of Failed Initiatives

Although DataLand is the first initiative in its sort in the field of geo-information, the idea to create central distribution points for municipal administrative information is far from new. Over the years many initiatives have seen the light of day, and despite the fact that there are several very successful initiatives there are of course also the failures that cost money. "Once bitten, twice shy" as the saying goes. The concern for a potential financial loss is aggravated by the economic recession in The Netherlands that of course also takes its toll on the budgets of municipalities.

One way of facing this challenge is to prove success. Now that DataLand has proven its added value for example in the disaster management chain and now that the first strategic successes are a fact municipalities are decidedly more enthusiastic about participation and realize their participation on more than solely economic grounds. Another way in which DataLand has faced this challenge is to make participating municipalities part of the success by sharing the – albeit modest – revenues. The revenues are modest because following European directives and Dutch law it is not possible to make a profit with governmental information as the taxpayer has already paid for this information. Thus the selling price DataLand uses for the information does not reflect the actual value of the information. It only reflects the cost of distribution.

3.3 Ownership Issues

Within municipalities geo-information has more than any other field grown into a field that has many collectors of data. The municipal tax department for example gathers administrative geo-information for tax purposes, while the Geo-information department gathers similar information to support e.g. spatial planning. The information gathered at different departments within the municipal organization is similar, but not the same. From a technical point of view it would be an easy statement to simply match the different databases, do quality checks etc. and so construct one database with the correct information that is accessible for those who need the information. The municipal reality however is slightly more complex as the exercise of matching databases and centralizing information immediately brings ownership issues into the equation. If there is one database, who owns the information? Who is responsible for the information? Who will manage the distribution of the information? Who will guard the quality of the information? Of course behind these questions there are financial motivations and clear political and power related issues. Precisely these ownership issues have proven hard to tackle.

While DataLand does not provide consultancy in this field, in terms of a yearly quality monitor DataLand does give insight to municipalities in how their peers organize their data. Also because of the active account management DataLand is able to direct questions of municipalities to other municipalities that are further advanced. In this way DataLand supports the allocation of ownership within municipalities without actually getting involved in individual municipalities.

The municipal autonomy, the history of failed initiatives and the ownership issue are three of the most important organizational issues that are fundamental to understanding the complexity of the Dutch geo-information situation and thus to the difficulties that disaster management bodies face when they want to access this information. In describing how DataLand faced these challenges to build a central distribution point that works, some insight has been given to how DataLand is a solution at an organizational level to facilitate accessible and available geo-information as a backbone for sound disaster management. The organizational level however is only one side. The other side is technology.

4 Technical Challenges

In the third section the main organizational challenges have been outlined. Besides several organizational challenges it will be clear that with a multitude of standards the technical challenges to realize a central distribution point with accessible and useable data are potentially enormous. Especially when once again the autonomy of Dutch municipalities are taken into account. This section will take a closer look at the technical issues involved in realizing disclosure and interoperability of the data. Special attention will be given to the importance of metadata in this respect.

Developing a system for the distribution of data over millions of buildings proved to be relatively simple. Getting the data in the system proved the hard challenge, because there is a clear need for standardization in order for a database to function. Also the users of the information distributed by municipalities via DataLand need to know and understand the information they get.

The two most important closely related instruments with which DataLand faced the challenge of creating uniformity of and insight in data are:
a pre-programmed query
metadata

As described previously different departments gather similar but slightly different administrative geo-information in Dutch municipalities. Due to its tasks inevitably in nearly every municipality the tax department is the department that has the best geo-information data in terms of accuracy and actuality. The reason for this is that as this is the department that sends out the invoices for real estate taxes the control of the data is actually done every year by the citizens. Because if a tax invoice is incorrect a complaint will be made, thus providing the accurate data to the department.

Virtually all Dutch municipalities use a standard exchange format for their taxation process. This exchange format contains nearly all data that DataLand distributes centrally. Hence this exchange format provided a major opportunity for DataLand to facilitate simple distribution of data from municipalities to DataLand.

A query was pre-programmed that extracts the data for DataLand from the multitude of data that the exchange format contains. The query is distributed free of charge to all municipalities. It is also available as a free download on the website www.dataland.nl. The data extracted by the query can be burned on a CD or zipped in an e-mail and be send to DataLand. The frequency of distribution from the municipalities to DataLand is six-monthly.

While this pre-programmed query does provide a good instrument to facilitate the distribution process from municipalities to DataLand, it is not in itself an aid towards the much-needed insight in the data that the municipalities distribute.

That is why the query contains an obligatory field for metadata. Metadata are information about information, data about data. For DataLand metadata are the instrument to create accessible information. When a municipality distributes its information to DataLand it is obliged to fill out the metadata field in the software, thus giving DataLand insight in which registration was the source for the data, which standard was used etc.

It needs to be noted here that whereas metadata are the instrument for DataLand to gain insight and to provide the users of information with the same insight, metadata do not realize a de facto standardization of the data. It does however allow for interoperability.

The quality monitor fuels a further move towards bottom up standardization that DataLand conducts every year. This monitor shows municipalities which standards are used by most municipalities for which data thus allowing for municipalities to make an informed choice when reorganizing their data. Over the course of several quality monitors a clear move towards voluntary, bottom-up standardization is demonstrated.

A pre-programmed query containing an obligatory field for metadata are the instruments that DataLand employed to meet the challenge of different standards and different administrations. As this pre-programmed query and the metadata in themselves are instruments that establish clear insight in the data, they do not provide de facto standardization. However, the annual quality monitor that DataLand conducts supports de facto standardization. The uniform insight in the data that is created by using metadata and the central accessibility that is established through

DataLand have proven to contribute to the quality of disaster management.

5 DataLand and Disaster Management

Reasoning from the argument that disaster management bodies need a wealth of geo-information to carry out their tasks effectively DataLand assumed that a central distribution point for municipal administrative geo-information should have a clear added value. To see if this assumption was correct DataLand decided to conduct a real life experiment.

Over the course of 2003 DataLand cooperated with several municipalities in the province of South Holland and the Regional Rescue Service Rotterdam-Rijnmond (RHRR). InAxis - the national committee for the stimulation of governmental innovation, subsidized the project. The goal of the project was fivefold:

1. to create a real life pilot environment in which distribution of municipal administrative geo-information to the RHRR takes place via DataLand;
2. to test the use of these data in real life situations with the RHRR;
3. to realize an improved service of the RHRR by facilitating their access to high quality data;
4. to gain insight in the qualitative and quantitative advantages of central distribution of data for disaster management;
5. to demonstrate the usability of this form of cooperation between municipalities and disaster management bodies for other municipalities and other regionally and nationally operating disaster management institutions.

The project started by making an inventory of the wishes of the RHRR in relation to administrative geo-information. The most important wishes were:

1. accuracy
2. actuality
3. accessibility
4. interoperability
5. availability

It should be noted that in terms of accuracy the demands of disaster management bodies are clearly higher than the quality that is necessary for the tax department. The margin of failure that is acceptable for disaster management bodies is much smaller, seeing that an inaccuracy can ultimately cost human life.

Having said that, the central availability of a wealth of data to the RHRR was a glass that was half full rather than half empty. In the present situation the RHRR only had some information of 4 out of the 40 municipalities in its safety region. To obtain the necessary data the RHRR had to actively seek contact with all municipalities in their safety region. Besides the small amount of data of very few municipalities that the contacts produced, another problem of course again was that the information of the responding municipalities was delivered to the RHRR in various standards. Hence the data had to be processed before being useable. Through cooperation with DataLand the RHRR not only had more information of nearly all the municipalities in its safety region at its disposal but also the data came standardized and in a format that was directly useable. This already was a clear added value of this new form of cooperation.

Concrete added value was furthermore shown in a real life accident when over the course of the project a gas explosion on the border of two municipalities took place in the safety region of the RHRR. As this accident took place exactly on the border of two municipalities prior to cooperation with DataLand the data with regard to the location, surface etc. would have been very hard to access – if they would have been available at all. The different standards, lack of uniformity and hence lack of interoperability would have meant virtual unavailability of the data and thus a less efficient and less effective disaster management service. Due to the cooperation with DataLand the RHRR had a significantly quicker response time and was better prepared both in terms of manpower and material for the situation at hand.

The exchange of data between DataLand and the RHRR was relatively easy to establish by using a standardized exchange format, which naturally included the metadata thus allowing for swift implementation of the data in the current systems of the RHRR.

The concrete results of this project have been widely disseminated, as the project was one of the first to clearly demonstrate the added value of close cooperation between parties involved in the disaster management chain. Also the RHRR is now a client of DataLand, thus underlining the importance of a central distribution point to its peer institutions.

The experiment has shown that cooperation between municipalities and disaster management bodies through one central distribution point has clear added value in terms of availability, accessibility, accuracy and interoperability of the data. Moreover the experiment has shown that this form of cooperation leads to a shorter response time and more efficient planning process in the event of a disaster.

6 Epilogue

Over the past two decades it has become increasingly clear that local governments maintain a great number of external business relations with other organizations, as part of their own geo-information service. This paper has demonstrated that in this respect local governments are providers as well as users of information. In this respect we mainly focused on the role of the municipal government as a provider of information for disaster management bodies. It has become clear how the Dutch situation of relatively autonomous municipalities has given rise to a situation in which the geo-information that is of vital importance for the functioning of disaster management bodies is barely accessible. A clear need has appeared for closer cooperation between municipalities on the one hand and between municipalities and disaster management bodies on the other hand.

We have shown how by organizing central accessibility in the form of DataLand Dutch municipalities have hit two birds with one stone. On the one hand the market demand and the demand of disaster management bodies is met through the central organization of the data. The experiment that was conducted with the RHRR has clearly shown the added value of central and direct accessibility of high quality data for the performance of this disaster management body. On the other hand the annual quality monitor that DataLand conducts supports the standardization processes within municipalities. The paper has described how a pre-programmed query containing an obligatory field for metadata are the instruments that DataLand employed to meet the challenge of different standards and different administrations. As this pre-programmed query and the metadata in themselves are instruments that establish clear insight in the data, they do not provide de facto standardization. However, the annual quality monitor that

DataLand conducts supports de facto standardization. The uniform insight in the data that is created by using metadata and the central accessibility that is established through DataLand has proven to contribute to the quality of disaster management.

Taking into account the different disasters that have taken place in The Netherlands over the past years there is a clear need for closer cooperation between municipalities and disaster management bodies. Firmly rooted in the autonomy of the Dutch municipalities DataLand provides municipalities with the possibility to take up their responsibility of disaster management. As this paper has clearly shown solely technical solutions do not do justice to the complexity of the organizational situation. The DataLand initiative has achieved its success because it addresses both the organizational and the technical issues underlying the organization of geo-information for high quality disaster management.

From the development of DataLand we feel that at least three important lessons can be learned on a more general level for cooperation in the safety management chain.

Firstly there is the lesson of the value of close cooperation between parties involved in disaster management in which there is a clear understanding of and recognition of the role and position of every party in the safety management chain. Over the years the role of municipalities in The Netherlands in disaster management has sometimes been underestimated. However, following several large-scale disasters it has become clear that municipalities are a link in the safety management chain that is equally important as the fire department or medical care. The safety management chain is no exception to the rule that a chain is only as strong as its weakest link.

The second lesson to be learnt from the DataLand initiative is that when one of the columns in the safety management chain decides to make information available to the other columns via a central distribution point, there is a clear need for active account management, firm project management and clear communication about the ambitions of the innovation. In other words: one should not buy into the myth of a scenario in which a central distribution point on the basis of its added value is self-selling and will grow organically. An innovation – even a relatively simple one such as the creation of a central distribution point to enable easy re-use of already existing information – implies a change in current patterns of work and cooperation. Change in times of economic recession and high work pressure is rarely welcome. Thus (pro-) active marketing and active account management are necessary tools for the implementation of an innovation, no matter how obvious the innovation or its added value are. Innovation needs to be sold in its early stages until the time that the parties involved consider what once was an innovation as a normality.

This statement leads to the third lesson that can be learned from the DataLand story. An additional link in the safety management chain must be able to continuously demonstrate its added value to the parties involved. Without clear added value there is no ground to build on.

These are just a few of the lessons that can be learnt from the development of the DataLand initiative. Through this paper we have tried to share these lessons and our experiences as we feel that the concept of a central distribution point for the distribution of information that is needed by many parties in the safety management chain is widely applicable. It can be concluded that in The Netherlands this central distribution point delivers a clear and positive contribution to the improvement of the quality of disaster prevention and management.

References

Dierendonck JH (2004) Rood-wit-blauw bij ongelükken. Cicero, 16 april 2004, nr 5

Groot, M de (2003) Waar rook is……; Gebouwgegevens in de OOV sector. B&G, jaargang 30/9, pp 18-19

Groot M de (2004) De vierde kolom in beweging. Incident, 2004

Jellema M (1999). Geo-information chains present local governments a new challenge, 21st UDMS Symposium

Koops LJ (2004) Rood, wit, blauw en grijs kunnen en willen samenwerken; iedereen aanwezig op Nationaal Congres Rampencoördinatie. GIS Magazine, okt./nov. 2004

GRIFINOR: Integrated Object-Oriented Solution for Navigating Real-Time 3D Virtual Environments

Lars Bodum, Erik Kjems, Jan Kolar, Peer M. Ilsøe and Jens Overby

Centre for 3D GeoInformation, Aalborg University, Niels Jernes Vej 14, DK-9220 Aalborg, Denmark.
E-mail: lbo@3dgi.dk

Abstract

The ability to navigate a 3d virtual environment in real-time has a high priority in connection with different applications for disaster management. This paper presents GRIFINOR – a platform for applications within this area. As a part of GRIFINOR, three new innovations have been promoted. They are presented in this paper as well. In many situations it is important that geoinformation can be accessible for queries within a very short time-frame (minutes). Questions about spatial reasoning and volumetric calculations in connection with different types of simulation has been very dependent on a priori models and very fast computer graphics hardware and software. It is also essential that the features of the model become real objects with attributes etc. Urban 3d models have traditionally been built as wire frame models. This makes it very difficult and in some situations impossible to attach geoinformation to the spatial structure in a way so that it is useful for real-time navigation in 3d. These wire frame models are not suitable for either a connection to a spatial database or for spatial queries. Centre for 3D GeoInformation at Aalborg University, is developing a system that handles 3D data structures as objects with facilities to support real-time geovisualization. The need for a new concept in this area has been one of the major motivations for the development of GRIFINOR. Instead of dealing with simple geometry in a CAD-based environment, GRIFINOR is developed to support object-oriented technology.

1 Introduction

Access to information through the use of a geographic interface such as a map, has been promoted and developed for many years as Geographic Information System or GIS. This technology has influenced the way people work, understand and communicate about matters that have a spatial relation. The concepts of GIS was first coined in the 1960′s by geographers that were able to store, manipulate, query and visualize information on a computer and in a realistic geographical context [1-3]. This meant building a computational representation of a map (in 2d) and making relations to tabular data. The representation is based on the model of 1 to n layers of geographical features or raster. To be able to work with these layers interactively, it is necessary that they refer to the same geographic datum and projection, and that the geographic coordinates refer to the same coordinate system. This puts a lot of constrains on the whole concept of GIS. Therefore it is not possible to represent real (global) 3d space within the original concept of GIS. There has been many suggestions to the solution of this conceptual problem, but they all build on projected and local systems [4-7].

Another important issue is the fact that access and query for real-time 3d visualization rarely is possible in commercial GIS products. In real life situations such as disaster management, where time is crucial, the issue becomes even more essential. There is a need for solutions that can minimize the amount of time from question is asked and to answer is visually displayed for the user. This type of communicative challenge can be dealt with in many different ways. There have been a few very interesting initiatives for new concepts within communication of information through multimodal and multidimensional systems with geographical capabilities [8-10]. These initiatives focus mainly on user interface and less on real-time communication as an issue for system optimization. A new concept should focus on the implications for object generation, object storing and object query and visualization.

To facilitate these fundamental system needs and raise the general level of comprehension for spatial information of various kinds, we suggest a technology that allows access to geographically referenced information through three-dimensional graphics. This is done by linking information to a location in a virtual model of the real world. Any type of information can be linked this way.

In order to actually implement such a technology, it is crucial to define a general platform for a solution. The solution suggested here minimizes the number of conceptual and technological problems, while keeping the technology general for adding new features to the system. The vision of a virtual representation of our planet that would enable a person to explore and interact with any natural and cultural information gathered about the Earth addresses an abundance of problems, which would be too overwhelming to cope with. Efforts have been carried out for years attempting to find a suitable solution for such a technology, referred to as Digital Earth[11-13]. In contrast to this approach, which attempts to define the way different information can be visually presented, we define a fundamental technology that provides a base for an arbitrary number of smaller applications focused on visualization of specific information. This solution provides a modularized approach towards Digital Earth.

2 GRIFINOR

The griffins are legendary creatures – half eagle and half lion. Griffins are powerful - and so is GRIFINOR. They can fly across vast areas, but can also walk on the ground, if necessary. Griffins belonged to the Gods. Their appearance inspired respect, and even though they looked malevolent, they were actually here to do good. Their main task was to guard or protect, and they had sufficient skills to carry out this mission autonomously.

Griffins are still here with us. Leaning out from buildings and from the top of roofs, they watch us. Even as protectors of Gods house – the churches – we can see them as part of the ornaments, providing a message from above. As such, Griffins were used to express visual messages to people – as spatial models – as sculptures. GRIFINOR has inherited this feature of expressing itself visually, and when used in geovisualization, the link between the past and the future is established. GRIFINOR is therefore a digital griffin. It is here to foster and navigate our digital world and further to provide a visual perspective of our environment. GRIFINOR can fly even higher than its legendary predecessors and has the whole globe within its field of vision.

Fig. 1. The logo for GRIFINOR system

2.1 System Definition for GRIFINOR

GRIFINOR is a platform for different sorts of applications. A system for 3D geovisualization is, in contrast to an ordinary GIS system, which at most handles surfaces and 2D objects with height information, a system that can store, retrieve, analyze, simplify, generate, and visualize spatial data that are generic 3 dimensional. Furthermore it allows user interaction with these data. GRIFINOR will be able to handle "soft" real-time demands as well as being application and device adaptable - that is the system will be module based and object oriented so it can be adapted to PDA's, PC's, mobile units and so on, without requiring alterations to the code of the applications. GRIFINOR is collaborative so that more than one user per session can experience and interact in the same virtual world. It is build around one or more database technologies, used in a scalable and distributable system, in which large amounts of data will be present (magnitudes of about one TB), powerful server hardware and fast 3D graphic hardware. GRIFINOR is part of a research project and for that reason the users are not specified ahead of time. The user group is potentially vast from system- and application programmers and administrators to users of applications in GRIFINOR.

3 System Architecture

GRIFINOR has four main structures that can be described individually. This is the GeoDB (2D geographic relational database) that supports the construction of 3D objects, the object database (ODB), the viewer (with 3 different viewing platforms) and finally the applications of GRIFINOR. This system architecture can be observed in figure 2.

3.1 GeoDB

The purpose of geographical database (GeoDB) is to store and serve geographic data as a source for generating content of GRIFINOR system. This should facilitate development of generators that can be reused with a minimal effort for any corresponding data that will be later loaded into GeoDB. Another reason for GeoDB is to keep data readily available as opposite to having them stored on multiple media (DVD-ROM, CD-ROM, tapes) archived on different places, in different folders, in various formats and coordinate systems. An instant and uniform access encourages performing small tests with data in much shorter period of time whenever necessary. This increases efficiency of a development based on the data. By GeoDB is understood mechanism for storing and retrieving data in contrast to a mechanism for data processing (e.g., enhancing, generating, interpolating, transforming etc.). Although certain processing capabilities are needed for querying these should be general and proved solutions, which in turn leads to a decision on technology that can be utilized by GeoDB. Geodb should be based on established relational database management system (RDBMS), in order to provide standard set of tools for data manipulation. Re-use of a standard technology is expected to be easier for others when accessing and loading data for GRIFINOR, as well as when GRIFINOR should be deployed at new sites. This expectation is based on the fact that a broad community has experienced RDBMS in practice for decades now. RDBMS has evolved into mature and robust data management solution; backed with SQL standard and with several commercial and open products intensively used in production level of many businesses.

This would provide GRIFINOR with a promising foundation for communication with others.

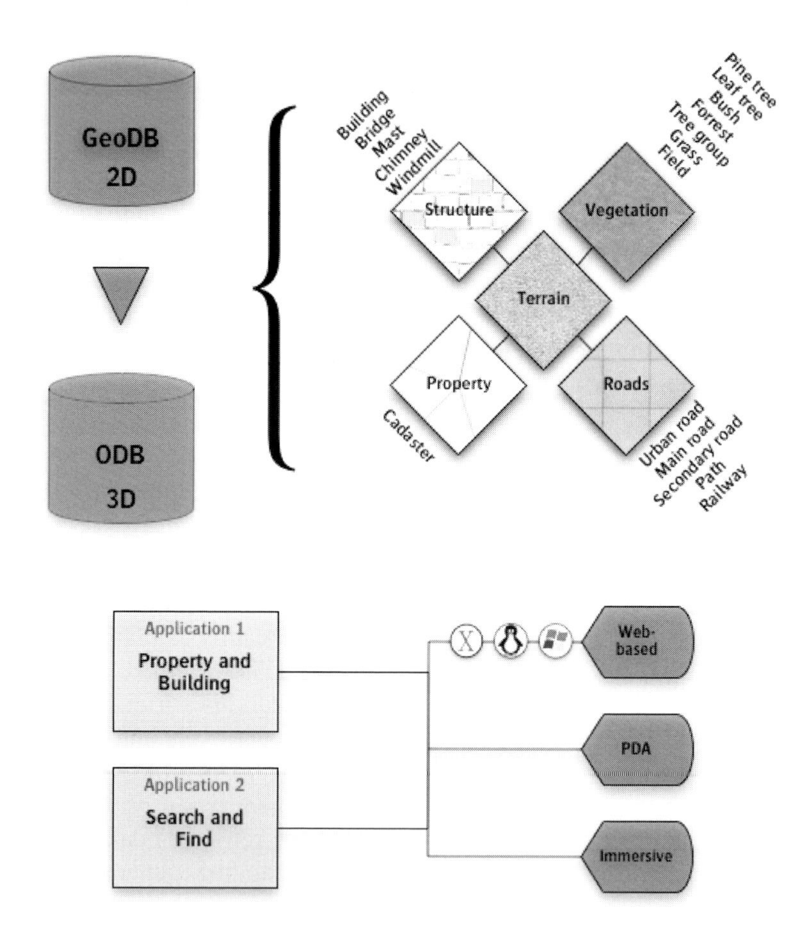

Fig. 2. The system architecture of GRIFINOR

3.1.1 Object Construction

The different geodata will be used as source for the construction of objects. This process is one of the essential parts of the concept behind GRIFINOR. Even though there are several data in the GeoDB that have a z-value, there are generally no spatial representation in 3 dimensions for the data. This representation has to be created for the system. In this specific project, there will only be created 3d objects for a few specific object classes.

3.1.2 Objects

The system of objects for GRIFINOR can be described as follows. The central object is Terrain, which is generated with information from different digital height models. These height models are based on interpolated GRID's produced either from photogrammetric methods or from airborne laserscannings. The terrain becomes the basic element of the geographical model. All other objects must relate and connect to this representation. On top of the terrain four different groups of objects are generated. They are: Structures (such as: buildings, bridges, masts, chimneys or windmills), Vegetation (such as: trees, bushes, grass, fields and tree groups or forests), Roads (such as: urban roads, main roads, secondary roads, paths and railways) and finally Property (cadastre). These object classes will be the main focus and are at the same time the main objects in the two applications we will develop within GRIFINOR.

A lot of effort has been put into the semi-automatic construction of buildings [14]. Through this process, it has been possible to reconstruct more than 85% of the buildings in a test area. Work on the sub-division of the buildings into property-units is still going on [15].

3.2 Object Database (ODB)

The ODB handles the persistent low-level storage of dynamic objects in the GRIFINOR system. An actual system is composed of a network of 1-N ODB servers distributed on the Internet. The system features unique distributed spatial indexing algorithms and persistent data structures optimized for efficient 3d level-of-detail queries. The ODB is unique for GRIFINOR due to the fact that there are no commercial solutions that can perform the needed tasks. Further documentation of the ODB will soon be made available.

3.3 Viewer

The GRIFINOR Viewer is based on Java technology, which provides the flexibility of running on various different platforms such as: MacOS X, Linux, and Windows, directly from a web browser. Future PDAs and cell phones with sufficient hardware should be able to run the Viewer as well with minor modifications. Beside the traditional platforms, such as desktop computers and PDA's, we will be able to port the viewer to more immersive display types, such as panorama. This will also be a part of the development of GRIFINOR.

3.4 Applications

As the first application to use the GRIFINOR system, we will develop specific tools to query and visualize information from the Building and Dwelling Register and Danish National Cadastre in a viewer. As test area for the development will be used an average sized Danish municipality with a substantial urban area in Northern Jutland. The second application will be more focused on the spatial search and find functionalities. Here we will use the intended navigational tools and combine these with very simple spatial queries, so it becomes possible to do a virtual travel from one position to another in the model.

4 Specific New Technologies in GRIFINOR

During the development of GRIFINOR, some specific new innovations have been made. In this paper we will describe three of these. The first one is the indexing method, which is called *Index for Distributed Spatial Object (DSO)*. The second new technology featured here is *Topographic surface representation indexed using global grid* and the third is *Geo-embedded visual navigation.*

4.1 Index for Distributed Spatial Object (DSO-Index)

The DSO-index is a distributable data structure suitable for indexing 3d objects on world scale, suitable for progressive LOD queries for visualization. DSO-index is based on an octree-like data structure. The physical extends of the top node are somewhat bigger than the Earth or the planet it is applied to. In each octree node it is possible to assign a set of objects through distributed object pointers. Child octree nodes are optional and also marked by explicit distributed pointers. The use of distributed object pointers for both the index tree and the objects grants the possibility of distributing data on as many servers as desired. Since all nodes and objects are directly or indirectly referenced from the top node, this node acts as a reference to the world, and multiple versions of the world can be created in the same system by making multiple top nodes.

As an integral part of the concept, it is necessary to define a method for calculating a LOD-measure for 3d objects – e.g. the average triangle size or the precision of the object. A method for relating the LOD-measure of the 3d objects (or specific detail of 3d objects) to the level in the tree must also be defined in such a way that it follows the subdivisions of octree (a length measure is halved for each subdivision, an area measure is square-rooted, and a cubic measure is cubic-rooted). In effect this provides a mapping between the LOD objects and the relevant node in the global octree structure.

Identification of the exact node, into which a 3d object should be inserted, can therefore be calculated without the presence of the data that constitutes the index data structure. The actual storing and insertion is performed according to the distributed object storage system, which can either traverse the tree explicitly to locate the insertion node or it can adhere to a preset (possibly also hierarchical) distribution policy.

When used for visualization, the index allows for querying the visually most important objects at a given viewpoint and dynamically altering the fidelity of the visualization according to the capabilities of the hardware and the network as well as user preferences. This is an important feature in a real-time system running on limited bandwidth connections since it is possible after a relatively short time to show a coarse representation which is progressively being refined over a possibly much longer time into the view of the final fidelity. At anytime the user is free to move to other places where new information is determined to be more important and thus gets higher priority for download.

Technically, since the tree structure is given implicitly, it is possible to identify which nodes are relevant to the visualization and which are not, without actually loading them. However, the depth of the tree at various places is not known in advance since it is data driven, so if only the client is capable of executing queries, only one level in the tree can be expanded for each query. If more levels are desired, a server capable of executing partial queries is needed as well as an octree node distribution policy, which increases the likelihood of a node and its child nodes being stored on the same server. To optimize communication efficiency with servers, queries to the same server can be grouped into multi-object queries and multi-sub-query queries. Additionally, objects can be cached locally and clients can form a peer-to-peer network to relieve object servers. For optimization of the visualization, the octree structure can be utilized for view culling.

4.2 Topographic Surface Representation Indexed Using Global Grid

This provides a solution for a spatial indexing around the globe. The core concept is to have an algorithm, which transforms more complex three-dimensional indexing into a simpler and established linear (one dimensional) index. At the same time it exploits 2.5 dimensional indexing concepts, which are used in currently planar geographic solutions, and relate them to a spherical surface in order to facilitate global geographic applications. This provides a general approach that can be used with any existing DBMS technology with indexing capabilities.

The indexing transformation uses a method that divides three-dimensional space through a tessellation marked off on the unit sphere using a geometric approach known as global grid. An original global grid, based on concepts of Voronoi tessellation, has been devised and implemented. This facilitates proximity spatial queries around planet's surface. A set of algorithms for processing basic spatial queries has been also elaborated.

The result is an unambiguous tessellation, which means that no data describing the geometry of the tessellation scheme are necessary. This feature is similar to the regular property of raster data. Such property enables to identify a linear order among the units of the spatial tessellation. This can be used directly with existing DBMS indices. Moreover the unambiguous tessellation exists for an arbitrary number of levels with different resolution of the tessellation. It is convenient for applications with multiple level of detail (LOD), which is critical for an interactive visualization.

A terrain data representation that exploits the LOD capabilities of the global indexing algorithm is coupled with the solution. The representation of the terrain surface does not store topology of the surface explicitly but uses a surface reconstruction algorithm at run-time. This provides flexibility in storing terrain data. Flexibility means; possibility to distribute storage on many places which constitute the same terrain model; elimination of dealing with topology neither across borders of individual tessellation cells (tiles) nor between LOD; and providing a generic data representation that is suitable for an analytical application in contrast to a data representation devoted to visualization purposes only.[16, 17]

4.3 Geo-Embedded Visual Navigation

In the subject of global visual navigation, works presenting concrete solutions convenient for an implementation are lacking. By providing a simple mathematical solution, we aim at facilitating adoption of similar navigations in other global systems for interactive geographic visualization. In traditional navigation, up is related to models based on flat approximation of the surface. The globally applicable navigation algorithm induces a practical value by estimating of the gravitational up vector at any given position of the user in relation to the globe and aligning the view accordingly. This means that on a local scale, navigation can appear to work like a traditional flat navigation algorithm, which the user might already be familiar with, but this navigation just works on the whole globe instead of a restricted flat space. On a global scale it is more apparent that the navigation works differently, simply because it is more obvious that the world is spherically represented. We try to address this improvement by explaining a possible use of the proposed navigation.

Similar to navigation of planar maps, the user can, for example, look over a 3D area, which is actually part of a spherical model; looking down one could zoom-in and zoom-out and pan to any side. The proposed navigation also allows to raise the view up, look to the sides and survey in a three dimensional context with streets and buildings surpassing the terrain. Heading to a particular direction one can move forth and back following certain elevation level. The user can also move further from the surface and if more global issues are at interest. At this global point of view it becomes apparent that moving forth or back, the navigation actually has to follow spherical shape of our planet in order to preserve the user's elevation level. This provides a natural experience for humans and offers a visual navigation and perception at many scales [18].

5 Conclusions

The presented technologies for GRIFINOR will make it possible to develop applications within the area of disaster management that can utilize both distributed servers in real-time 3d geovisualization and combine these new functions with analytical tools, as we know them from more traditional GIS. This result is reached by using true geographic and global modeling. Further research and development within these matters will be done before the solutions are fully applicable.

Acknowledgements

This initiative under the Centre for 3D GeoInformation is funded by:
- European Regional Development Fund (ERDF)
- Aalborg University, Denmark
- Kort & Matrikelstyrelsen (Danish National Survey and Cadastre)
- COWI A/S, Denmark (formerly known as Kampsax)
- Informi GIS, Denmark

References

[1] Chrisman N (1988) The Risk of Software Innovation: A Case Study of the Harvard Lab, The American Cartographer, vol 15, pp 291-300
[2] Dangermond J, Smith LK (198) Geographic Information Systems and the Revolution in Cartography: The Nature of the Role Played by a Commercial Organization, The American Cartographer, vol 15, pp 301-310
[3] Tomlinson RF (1988) The Impact of the Transition from Analogue to digital Cartographic Representation, The American Cartographer, vol 15, pp 249-261
[4] Pilouk M (1996) Integrated Modelling for 3D GIS, Wageningen Agricultural University and ITC, the Netherlands, Ph.D
[5] Pfund M, (2002) 3D GIS Architecture, GIM International, pp 35-37
[6] Stoter J (2004) 3D Cadastre, Technical University Delft, Delft, Ph.D
[7] Zlatanova S (2002) Present Status of 3D GIS, GIM International, pp 41-43
[8] Kwan M-P, Lee J (2005) Emergency response after 9/11: the potential of real-time 3D GIS for quick emergency response in micro-spatial environments," Computers, Environment and Urban Systems, vol 29, pp 93-113,
[9] Sharma R, Yeasin M, Krahnstoever N, Rauschert I, Cai G, Brewer I, MacEachren A, Sengupta K (2003) Speech–Gesture Driven Multimodal Interfaces for Crisis Management," Proceedings of the IEEE, vol 91, pp 1327-1354
[10]Coors V (2003) 3D-GIS in networking environments, Computers, Environment and Urban Systems, vol 27, pp 345-357
[11]Alexandria Digital Library and University of California Santa Barbara (1998) Alexandria Digital Library Project, http://www.alexandria.ucsb.edu/
[12]Y. Leclerc G, Reddy M, Eriksen M, Brecht J, Colleen D (2002) SRI's Digital Earth Project, SRI International, Menlo Park, Technical Report 560, 1. August 2002
[13] Gore A (1998) The Digital Earth: Understanding our planet in the 21st Century California Science Center, Los Angeles
[14]Overby J, Bodum L, Kjems E, Ilsøe PM (2004) Automatic 3D building reconstruction from airbornelaser scanning and cadastral data using Hough transformation presented at XXth ISPRS Congress, Istanbul

[15]Stoter J, Sørensen EM, Bodum L (2004) 3D Registration of Real Property in Denmark presented at FIG Working Week 2004, Athens
[16]Kolar J (2004) Global indexing of 3d vector geographic features presented at XXth ISPRS Congress, Istanbul
[17]Kolar J (2004) Representation of geographic terrain surface using global indexing presented at Geoinformatics 2004, Gävle, Sweden
[18]Ilsøe PM, Kolar J (2005) Geo Embedded Navigation," in Geoinformation for Disaster Management, vol #, Lecture Notes in Computer Science, NN, Ed. Heidelberg: Springer Verlag

An Intelligent Hybrid Agent for Medical Emergency Vehicles Navigation in Urban Spaces

Dino Borri[1] and Michele Cera[2]

[1] Politechnic of Bari,Department of Architecture and Town Planning, Via Orabona 4, Bari 70125, Italy.
Email: borri@poliba.it

[2] Politechnic of Bari ,Department of Highways and Transportation, Via Orabona 4, Bari 70125, Italy.
Email: michcera@libero.it

Abstract

This paper presents a decision support system integrated in a GIS with the aim of managing optimal navigation of emergency medical vehicles in urban areas. The focus is particularly on individuation of the shortest route for ambulance facilities. We do not simply propose a solution to the calculation of the shortest route in the traditional terms of distance to be traveled but rather the individuation of a route bearing in mind the set of factors that can cause delayed ambulance response. For this purpose, an "expert system" is integrated with the traditional algorithms for calculating the shortest route (such as Dijkstra's algorithm). To build the expert system, a set of rules is deduced from observation of the behavior of ambulance drivers, and a model of the urban road network is built in order to apply the algorithm calculating the shortest route. We thus aim to provide a decision support tool that can maximize emergency service vehicle response, an evidently critical factor with a life or death impact. Finally, we stress the theoretical and practical difficulties of interaction among two such different tools as the "expert system" and a mathematical method calculating the "shortest route".

Key words: expert systems, shortest route, medical emergencies

1 Introduction

Although the concept of risk in urban environments is primarily associated with factors of an environmental nature (land shifts, insufficient coastal protection and erosion, seismic risk, etc.) or the presence of industrial plants and their incumbent risks (pollution, transport of hazardous materials, etc.), inefficiency of the emergency services is also an important potential risk factor that can often entail the loss of human lives.

Medical emergencies are not in themselves a purely urban risk. The need to come to the aid of a person in need of medical care can arise anywhere regardless of location. Nevertheless, the large concentration of people in a city means on the one hand that such needs will arise more frequently and on the other, that in the urban environment events may occur that could hamper the timeliness of the response. All this applies even to the ordinary activity of a medical emergency service, let alone to such circumstances as calamities or exceptional events (earthquakes or terrorist attacks are just two of the myriad possibilities), that again have a greater impact in urban environments and are more likely to cause injury to people and damage to property.

An ambulance service must set up a network of centers to cover such emergencies, generally subdivided into several hierarchical levels in order to be able to assure different levels of assistance. It must also be able to rely on the availability of suitable vehicles, as well as on a network of operative central exchanges, answering to a single national telephone number (118 in Italy, 911 in the U.S.A.), that can coordinate the available resources and the strategy. A failure at any of these levels can impair the efficiency of the service, sometimes with dramatic consequences.

In this paper, we focus on what happens from the time when the ambulance departs from the station up to when the medical attendants come into actual contact with the patient.

The time factor plays an essential role in this phase, indeed it can appropriately be taken as an indicator of the standard of the service. The regional Apulian law, for example, specifies 8 minutes as the time limit for urban calls (time from when the central exchange receives the call to when the ambulance arrives at destination). However, in many cases this time limit seems far from satisfactory.

It can be concluded from the survival probability rates individuated for the various "links" in the "chain of survival" (Cummins, Ornato, Thies, Pepe, 1991) formulated for cardiac arrest, that the third link in the chain, in other words early defibrillation, is the one that has the greatest impact on reducing the mortality rate. The time between the cardiac arrest and the early defibrillation (and hence the time between recognition of the event and the arrival of the medical attendants), should be less than four minutes (Petrino, Aprà,Sardoni 1998). In some Italian areas the mean time for intervention is down to 6-7 minutes but in most cases it takes much longer. In Bologna, a city with a venerable tradition in the medical emergency field, the mean time for intervention is currently 8 minutes and 35 seconds. Bearing in mind that it takes an average of one and a half minutes for the call to be processed and another minute and a half to reach the patient after the ambulance has arrived at destination, this means that the time rises from 8 to 11 minutes. This shows how critical the time factor is and that the time established by the Apulian law, like the standard achieved in many other Italian areas, is insufficient.

The time factor is not the only indicator providing a measure of the efficiency of an emergency medical service. The ratio between the number of deaths and the number of interventions is another possibility, or else the percentage of interventions per unit (Adenso-Diaz, Rodriguez 1997). However, apart from its objective importance, the time factor is easy to understand and is the one which is generally brought to the attention of public opinion when assessing the efficiency of the service (Zaki,Cheng 1979).

Essentially, it is possible to minimize intervention times in two ways:

- by means of appropriate location of the emergency vehicles over the territory;
- by individuating the route enabling the ambulance to reach the site of the event in the shortest possible time.

In the literature, the greatest attention has been devoted to the first of these aspects, largely in the form of searching for models allowing individuation of the optimal location, in the given surroundings, of the ambulance stations to ensure response to all calls within a given time (Revelle 1991).

On the contrary, we are going to concentrate on the second of the two aspects, by describing the building of a Decision Support System based on a GIS.

2 The Problem of the Shortest Route and the Proposal of the DSS

Calculation of the shortest route is highly important in many different applications pertaining to network analysis, especially those having to do with transport. Over the years, the work done by researchers in various different disciplines such as operative research, management sciences, geography, transport engineering and computer science, has led to the formulation of various algorithms for calculating the shortest route (Dijkstra 1959; Goldberg-Radzik 1993).

Given two nodes on a network, an initial and a final one, it is possible to individuate the best route along the points of the same network, in order to reach the second node starting from the first, while minimizing the value of a variable associated with each element of the network. The latter variable is typically "distance", but the other possibilities include "cost" (time or monetary cost) associated with the points on the network. The variable can be expressed as a function of given attributes of the network (generally size) and the flow through each point on the network. In this case the term is the "cost function".

In the transport systems engineering field, by implementing these algorithms on a graph representing a real transport network, it is possible to individuate the minimum cost route for traveling from one point to another along the network. To take into account the phenomenon of traffic congestion of the roads the flows thorough each point in the network need to be estimated. This estimation can be made using d/o interaction models.

However, an approach of this kind entails the need to know a large quantity of data about the physical characteristics of the network and the transport demand, data which are not always available. Moreover, it seems to be difficult to take into account the great fluctuations of the traffic throughout the 24 hours, on the one hand, and a large series of events and circumstances that can cause delay, and that are not included in the cost function calculations, on the other.

For all the above reasons, we have preferred to adopt a cognitive approach to solving the problems. In other words, we have made recourse to "expert knowledge", in the sense of the knowledge possessed by those who have to face the problems of managing emergency services on a daily basis, and have based our system on their experience. In particular, we consider that the approach to the problem of the shortest route should involve acquisition of the knowledge possessed by ambulance drivers. This knowledge is poorly structured and of declarative type. A typical way to manage such knowledge is by using an "expert system" based on rules of the "if-then" type.

It is then necessary to deal with the problem of interaction between a typical method for solving problems based on structured knowledge (the algorithm for calculating the shortest route) and a typical method for representing non structured (or poorly structured) knowledge, like the expert system.

In our experimentation, carried out in the territory belonging to the Bari Municipal area, the knowledge was acquired by means of semi-structured interviews, during which a sample of ambulance drivers was asked what criteria they adopted when choosing the route, and to indicate the route they would take to reach given points and why. Analysis of the answers given allowed a list of the elements conditioning the choice of route to be made. The study sample consisted of seven drivers covering seven different areas.

The aim of the interviews was to explore the methods for choosing the route adopted by the ambulance drivers. These were heuristic, in other words their methods were based on their own experience and practice. They were used as a basis for formulating a hypothesis for building an expert system aiming to simulate human reasoning, by means of a knowledge base expressed in terms of "production rules".

First of all, the interviews allowed us to individuate the elements involved in the choice of the route. These are:

- the presence of heavy traffic flows;
- the width of the roads;
- the presence of priority lanes;
- the presence of traffic lights;
- the presence of level crossings;
- the presence of street markets;
- the presence of bus stops along the routes;
- the presence of the double parking phenomenon;
- the presence of schools;
- the local control exerted by traffic wardens;
- the conditions of the road surface.

Regarding the first element, it is clear that the presence of a high number of vehicles traveling on the road can hinder the passage of the ambulance. This phenomenon fluctuates during the day, congestion being generally more severe during peak hours on working days and less of a problem at night and during the weekend and bank holidays.

The width of the roads is important for ambulance drivers not only because it is presumed that the wider the road the lower the saturation point, but also because it is easier for the driver to maneuver on a wider road. The ambulance driver sample we questioned judged the possibility of invading the opposite lane to be particularly important, because the traffic is frequently heavy only in one direction. For this reason, the absence of a hard shoulder between the carriageways is considered a very good thing.

The presence of a priority lane is another reason why one road may be preferred to another. All the drivers in our sample hoped the number of roads with priority lanes would be increased, although they complained that they are also used by buses and that other drivers do not leave the priority lane clear as they should.

Ambulances are theoretically required to respect the highway code and therefore stop at the traffic lights. Still, this does not always happen and traffic wardens normally turn a blind eye. However, crossing when the lights are red is in any case a problem because the driver is obliged to slow down greatly for safety reasons and above all because queues of traffic form up at the traffic lights that can pose an immovable obstacle to the upcoming ambulance.

Level crossings are frequently an element inducing drivers to avoid routes where they are present, because an ambulance driver cannot run the risk of finding the crossing closed and having to turn back. Nevertheless, in some cases it may be possible to see whether the crossing is closed from a distance and select the route accordingly. Not all level crossings are attributed the same degree of "danger", which depends on the quantity of railway traffic conveyed. On most minor railways, for instance, the level crossings are generally open on holidays and during the night hours.

Street markets obviously make the roads they occupy unusable when they are open, and also hamper circulation in the nearby roads. They are considered as one of the worst problems of road circulation.

The double parking phenomenon is important above all because it narrows the roads. The ambulance drivers pointed out the case of the Borgo Murattiano in the city of Bari, theoretically a two-lane carriageway, which is in fact reduced to a single lane because of this phenomenon, with all the difficulties this entails (the ambulance cannot overtake the cars in front queuing at the lights).

The presence of a school creates difficulties during the hours of entry and exit from school, due to the large number of people (children, adolescents, adults) and vehicles (especially those of parents taking or collecting their children) milling around in front of the school.

The presence of traffic wardens is seen as a positive factor, as they can regulate the traffic and assist the passage of ambulances by directing the traffic to draw aside. Finally, a poor road surface obliges the ambulance to proceed more slowly.

On the basis of the results of our interviews, we were able to hypothesize a method for individuating the best route by simulating the ambulance drivers' reasoning. In the first place it was assumed that the drivers had a mental map of the city, in which the city was subdivided into zones, and that for each of these zones they had in mind one or more "approach" routes of a more or less fixed type. These were chosen *a priori* following the line of reasoning delineated above. As a general rule, we can say that ambulance drivers essentially follow the main route where possible. The choice of one approach route rather than another is made according to a set of rules, that generally take into account the day of the week and the time of day. The rules and approach routes we extracted from the interviews were used to implement a preliminary model of the intelligent module by means of an expert system shell, "Exsys", that serves to individuate the best approach route. Figure 1 shows an excerpt of the decision tree used.

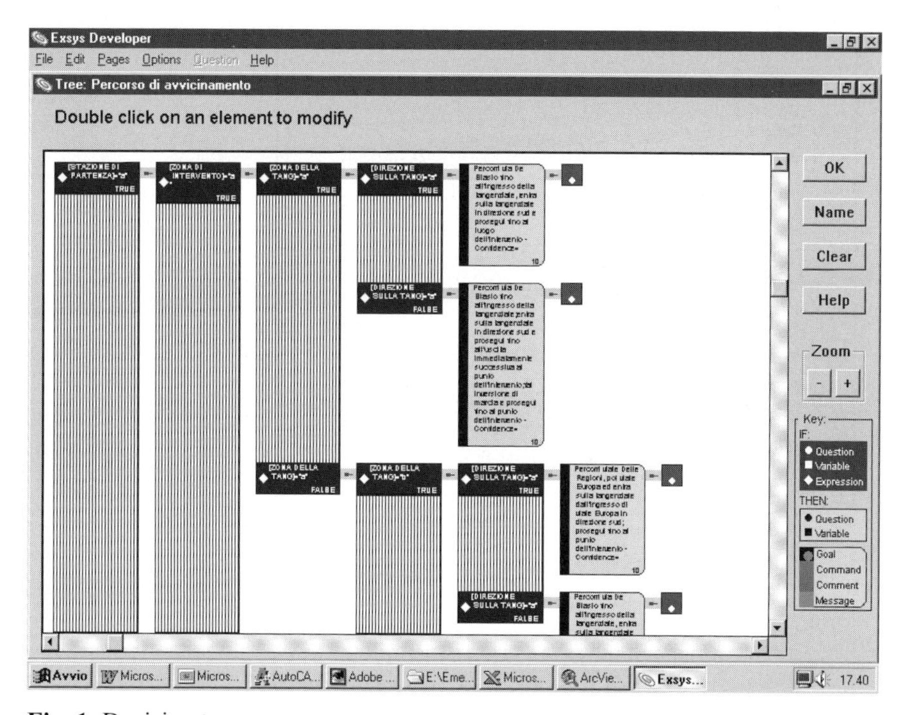

Fig. 1. Decision tree

Once the zone where the call originated has been reached, the problem arises as to how to reach the precise point of need. Herein it was assumed that the shortest route in terms of distance to be covered would be chosen, according to an assessment of its advisability or suitability. If the new route has no "inadvisable" branches it is selected, otherwise the search continues. In the case of several "inadvisable" routes the least inadvisable is chosen.

The calculation of the shortest route must be made using one of the common algorithms, such as Dijkstra's algorithm, used in the "Network Analyst" extension of the "Arcview" software by the ESRI. To assess the "advisability" of the branch, a series of rules extracted from the interviews are used (see figure 2).

1) If:
-there's a priority lane
Then:
-The road is advisable (μ_i=0.9)

2) If:
-the road is broad,
Then:
- The road is advisable [μ_i =(L-2.5)/4]

3) If:
-the road is middle-sized
Then:
-the road is on the average advisable [μ_i =(L-2.5)/4]

4) If:
-the road is narrow
Then:
- The road is unadvisable [μ_i =(L-2.5)/4]

5) If:
-at th end of the road there are traffic lights
Then:
-the road is on the average unadvisable (μ_i=0.6)

6) If:
-along the road there is a FS level crossing
Then:
- The road is advisable (μ_i=0.2)

7) If:
-along the road there is a no-FS level crossing
-it's a weekday,
-it's the daytime
Then:
- The road is advisable (μ_i=0.3)

Fig. 2 Examples of rules

Essentially, if one of the conditions individuated is present then a degree of membership μ_i is assigned to the fuzzy set "advisability". An analysis is then made of the μ_i according to a series of fuzzy multicriteria, so that the final degree of membership of the branch to the set of advisable branches is equal to μ=Min(μ_i).

If 0<= μ <=0.3 the branch is considered inadvisable.

This system can interact with a GIS for management of an emergency medical service.

Below, we describe the system architecture (see figure 3).

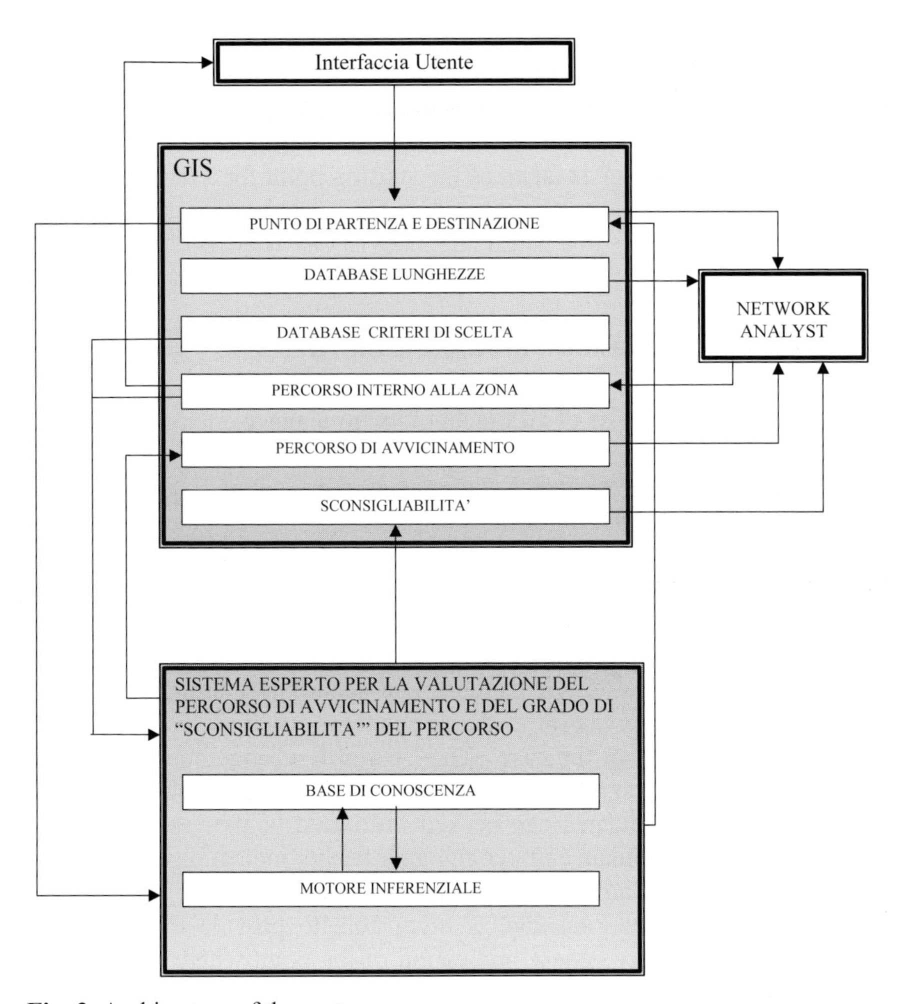

Fig. 3 Architecture of the system

It consists essentially of a GIS and an Expert System. The user, in this case an operator of the 118 exchange, dialogs with the system via the GIS. The Expert System serves to identify the approach route and to assess the degree of advisability of the branches.

The GIS consists of the road network of the city associated with a database with the length and name of each road (length Database), a database with the attributes according to the criteria of choice used by the drivers (criteria of choice Database). The road network and associated database enable individuation by the user of the point where the call originated. At this stage the shortest route is calculated with the "Network Analyst". Using the data extracted from the "criteria of choice Database", associated with the arches belonging to the route calculated with the "Network Analyst", the expert system assesses the degree of advisability of each branch. If none of the branches results inadvisable, the route is considered suitable and highlighted on the GIS map. If not, then the first node of the branch considered "inadvisable" is taken as the starting point for a new route with the same destination as the previous one. The cycle is repeated iteratively until solution but comes to a halt if an "inadvisable" route is chosen.

3 Use of GIS to Manage Medical Emergencies

Below, we outline the use of GIS technology to manage emergency medical services.

The user of such a system is defined as the Central Exchange of the emergency number 118. All requests for intervention reach these central exchanges, organized by province, that have the task of translating the requests into operative action, coordinating the resources available to the Emergency System. From the operative standpoint, the operator of the Central Exchange answers the call, eliciting a series of answers to pre-set questions, assesses the severity of the case and accordingly assigns the case a code. The nearest ambulance to the site is then called up, communicating the degree of severity of the case. The Central Exchange operator remains in constant contact both with the caller and with the ambulance. In this way the medical attendants can be kept informed of the evolution of the situation, and the ambulance driver can gain further indications as to the site of the event while the ambulance is in motion.

In fact, the Central Exchange is often able to provide the ambulance driver only with a very general indication of the point where the patient is located, and more detailed, precise information only becomes available as the call continues (reference points for operators unacquainted with the zone, or details serving to individuate the exact point where the person to be assisted is to be found).

The operator has a digital map on the computer monitor, which shows where the ambulance stations are located and the current availability of vehicles. As soon as the ambulance leaves, the station warns all operators that that vehicle is no longer available, or that no vehicle is free at the given station.

The importance of the Exchange as the centre of coordination of the actions of assistance is obvious. The operators are continually aware of the availability of ambulances and can respond to a request for assistance in the best way, within the limits of the availability of resources.

This is the scenario in which our proposal of a GIS for management of this crucial service is inserted. Construction of such a GIS requires a model of the city transport network to be built. We have used the model built by Ernesto Ciani (Binetti,Ciani 2002). The network model alone allows calculation of the shortest route, in terms of the distance to be traveled, using software such as "Arcview" with the "Network Analyst" extension. Interaction with a module such as the one we describe would enable an ambulance driver to select the best route, when receiving an emergency call. Clearly, this would provide an indication not an imperative, although we hope it would prove to be reliable. The driver might nevertheless be aware of supervening elements making the route inadvisable (road works, unforeseen events). Contingent circumstances may always arise preventing the user from being able to use the indications provided by the System.

The implementation of a demand model and a demand-supply interaction model could make it possible to individuate the accessible areas for each ambulance station, thus providing a tool for better location coverage by the stations.

The GIS requires a database to be created detailing the characteristics of all the hospital facilities and one detailing the features of the road network, both of which are essential for correct functioning of the expert module.

Finally, the GIS can manage an archive of the interventions made, detailing the characteristics of each (time, duration, place, type).

4 Conclusions

In this study we propose the hypothesis of a DSS based on a GIS for the management of an emergency medical service. In this context, the problem of the individuation of the shortest route has been dealt with in a non traditional manner, by creating an Expert System that can simulate human behavior. In this way we aim to create a flexible tool that can provide reliable results in real time. The method proposed has been judged by operators in the sector to be a valid approach. Future experimentation in field will validate its true efficacy and efficiency.

References

Adenso-Diaz B, Rodriguez F (1997) *A simple heuristic for the MCLP: Application to the location of ambulance bases in rural region*, Omega, International Journal of management science, vol 25, no 2, pp 181-187

Benedict JM (1983) Three hierarchical objective models which incorporate the concept of excess coverage to locate EMS vehicles or hospitals, MSc thesis, Department of Civil Engineering, Northwestern University, Evanstone, IL

Berlin GN (1972) Facility Location and vehicle allocation for provision of an Emergency Service, PhD dissertation, The John Hopkins University

Binetti MG, Ciani E (2002) Effects of traffic analysis zones design on transportation models, EWG on transportation, Bari 10-13 June 2002

Cascetta E (1998) Teoria e metodi dell'ingegneria dei trasporti, UTET, Torino

Chapman S., White J., Probabilistic formulations of emergency service facilities location problems, Paper presented at ORSA/TIMS Conference, San Juan, Puerto Rico

Church R, ReVelle C, (1974) The Maximal Covering Location Problem, Papers of Regional Sciences Association 32

Cummins RO, Ornato JP, Thies W, Pepe PE (1991). Improving survival from sudden cardiac arrest: the "chain of survival" concept. A statement for health professionals from the Advanced Life Support Subcommittee and the Emergency Cardiac Care Committee, America Heart Association. Circulation 1991, 83, pp 1832-1847

Daskin M, Stern E (1981) A multi-objective set covering problem for EMS Vehicle Deployment, Transportation Sciences 15

Dijkstra EW (1959) A Note on Two Problems in Connection with Graphs, Numeriche Mathematik 1, pp 269–271

Eaton D, Hector M, Sanchez U, Lantigua R, Morgan J (1986) Determining ambulance deployment in Santo Domingo, Dominican Republic, Journal of the operational research society 37

Felder S, Brinkmann H (2002) Spatial allocation of emergency medical services: minimising the death rate or providing equal access?, Regional Science and Urban Economics 32, pp 27-45

Gendrau M, Laporte G, Semet F (2001) A dynamic model and parallel tabu search heuristic for real-time ambulance relocation, Parallel Computing 27, pp 1641-1653

Goldberg,AV, Radzik T (1993) Heuristic Improvement of the Bellman–Ford Algorithm, Appl. Math. Lett. 6, pp 3-6

Hogan K., ReVelle C. (1986), Concepts and application of backup coverage, Managment Science 32

Leung Y (1997) Intelligent Spatial Decision Support Systems, Springer

Peters J, Hall GB (1999) Assessment of ambulance response performance using a geographic information system, Social Science & Medicine 49, pp 1551-1566

Petrino R, Aprà F, Sandroni C (1998) La catena della sopravvivenza,

ReVelle C (1991) Siting ambulances and fire companies. New tools for planners, APA Journal, vol 57, no 4, Autumn 1991, pp 471-484
ReVelle C, Hogan K, (1989) The maximal availability location problem, Transportation Science 23
Timmermans H (ed) (1997) Decision Support Systems in Urban Planning, E&FN Spon
Toregas C, Swain R, ReVelle C, Bergman L (1971) The location of emergency service facilities, Operations Research 19
White J, Case K (1974) On covering problems and the central facility location problem, Geographical Analysis 6
Zaki AS, Cheng HK (1997) A simulation model for the analysis and management of an emergency service system, Socio-Economic Planning Science, vol 31, no 3, pp 173-189

GIS Solutions in Public Safety: A Case Study of the Broward County Sheriff

Scott Burton[1], Patricia Behn[2] and David C. Prosperi[2]

[1] GIS Administrator, Broward County Sheriff, 2601 W. Broward Blvd, Fort Lauderdale, FL 33312, USA.
Email: scott_burton@sheriff.org

[2] Department of Urban and Regional Planning, Florida Atlantic University, 111 E. Las Olas Blvd, Fort Lauderdale, FL 33301, USA.
Email: pbehn@fau.edu
Email: prosperi@fau.edu

Abstract

This paper discusses GIS and disaster preparedness and management by a public safety agency at the scale of a small metropolitan area or jurisdiction within a larger metropolitan area. After briefly reviewing the local, South Florida, context, the paper focuses first on the agencies experiences with mostly Federal GIS programs – ALOHA, CATS/JACE and HAZUS-MH – geared to managing natural and man-made disasters. The next portion of the paper focuses on an evolving enterprise GIS within the public safety agency that has included projects related to: use of GIS technology for critical incident response; homeland security; and incorporating and anticipating wireless technology in data collection, and distribution. The paper concludes with lessons learned and recommendations in terms of both technology and organizational imperatives.

1 Introduction

One of government's most important jobs is to protect its citizens from crime, terrorism, and natural disasters. Furthermore, it is expected to do so in an increasingly effective manner, to reduce costs, and to improve intergovernmental cooperation though professional collaborations, joint data sharing, and obtaining economies of scale by combining efforts. This paper describes the emergence of the Broward County, USA Sheriff's Office (BSO) as a "full service" public safety agency, their experience with geo-informational technology, and its current and planned GIS operational agenda.

Broward County, Florida, USA is located on the southeast coast of Florida. It lies at the center of the Miami-Fort Lauderdale-West Palm Beach metroplex, an area encompassing some 4.5 million permanent residents (which often swells to over 7.0 million people seasonally). The county is bordered on the east by the Atlantic Ocean and elsewhere by Palm Beach, Hendry, Collier, and Miami-Dade Counties. There are approximately 1.7 million permanent residents located on 409.8 square miles of developable land (the remainder of the total 1,205 square miles is primarily the Everglades Preserve). There are 28 municipalities in the County.

BSO is responsible for providing public safety to all of Broward's unincorporated areas as well as contracted services to 14 municipalities, the Ft. Lauderdale/Hollywood International Airport, Port Everglades (the 12th busiest container port in the U.S.), and the Broward County Mass Transit system. Moreover, in 2003, fire rescue, emergency services, and E911 communications responsibilities were added to the Sheriff's office as the result of a merger. Finally, the Broward County Sheriff is the co-chair to the Florida Regional (Broward, Miami-Dade, and Monroe counties) Domestic Security Task Force.

The GIS Unit was established in 2002 with the end goal of providing efficient and effective integration of disparate crime and other disaster and emergency management databases and operational models. These include: providing Emergency 911 dispatch, law enforcement, detention, probation, fire rescue, emergency medical services and Homeland Security. The range of uses creates a disparate set of requirements leading to concerns over the timeliness and integrity of the data. Additionally, the variety of potential end-users ranges from GIS specialists to disaster-relief volunteers with no technical background. Since that time, the GIS Unit has explored new uses for GIS technology that supports public safety responses to both man-made and natural disasters.

This paper is structured as follows. First, the U.S. GIS and public safety context is briefly described. Second, the GIS software tools currently employed by BSO are reviewed in terms of both capability and experiences. Third, existing practices and evolving technologies within the BSO GIS Unit are described. Lessons learned and recommendations for further implementation of the Enterprise GIS within the GIS Unit of BSO conclude this paper.

2 Literature Review

The majority of the GIS and public safety literature often takes the form of speculative inquiry into the *potential* use and advantages of GIS. The following section outlines these statements in an effort to provide a framework for discussion of BSO's GIS Unit which follows.

The Department of Homeland Security's Information Analysis and Infrastructure Protection's budget request acknowledges the importance of the use of GIS in emergency response. This budget line is stated for purposes of "mapping threat information against our current vulnerabilities, and the development and maintenance of a complete and accurate mapping of the Nation's critical infrastructure and key assets."

The Federal Geographic Data Committee (FGDC) is an interagency committee composed of representatives from the Executive Office of the President, Cabinet-level and independent agencies. Currently, the FGDC is developing the National Spatial Data Infrastructure (NSDI) in cooperation with organizations from State, local and tribal governments, the academic community, and the private sector. The NSDI encompasses policies, standards, and procedures for organizations to cooperatively produce and share geographic data.

In one of FGDC's publications, an emphasis is placed on the "importance of the implementation of a comprehensive national spatial data infrastructure, interoperability of the systems that process this information, and commonality of the processes that collect, manage and disseminate geospatial information." However, there are methodological concerns that need to be addressed to achieve assurance of data and technology accessibility and interoperability. These include: having national data standards to set a framework for providing data that is immediately useful for Homeland Security; consistent and standardized road data for Emergency 911 (E911) capabilities; and having current and accurate data on the Nation's critical infrastructure accessible to relevant agencies.

Cutter, Richardson and Wilbanks (2003) argues that one of the top areas of concentration for the national policy officials and geographic researchers on Homeland Security is the geospatial data and technologies infrastructure. Priority action items to promote this endeavor include establishing "a distributed national geospatial infrastructure as a foundation for homeland security" to simultaneously serve multiple needs, including "local government, planning, environmental protection, and economic development."

The World Trade Center attacks demonstrated the importance of having GIS as a foundation in emergency response. Langhelm (2002), the GIS

Coordinator for the Federal Emergency Management Agency (FEMA), addressed the supporting role GIS had as "invaluable in the response and recovery efforts in disasters in recent years." Much of the initial work focused on search and rescue with an operation staffed by 25 people working 24-hour operations over two GIS nodes. After the initial support, Langhelm continues, "GIS products supported decision-making" and this allowed officials to "see the site in a completely different perspective."

A variety of difficulties arose during the post-9/11 recovery effort that, perhaps still, adequately describes the problem context for the use of GIS in public safety applications. These included: the availability of office space, power, web connectivity, data access and inadequate staffing; data, particularly building addressing issues, since there were many occurrences of multiple street addresses; and data unavailability about subsurface infrastructure.

3 Technology - Software

This section examines the three software programs that BSO has implemented to analyze and respond to emergency situations. The programs are able to examine both man-made and natural events. Each software program has capabilities to both gauge the effects of an event as they occur as well as the ability to create scenario event analysis.

3.1 ALOHA

The U.S. Environmental Protection Agency (EPA) and National Oceanic and Atmospheric Administration (NOAA) use a plume-dispersion model called Aerial Locations of Hazardous Atmospheres (ALOHA) (e.g., EPA, 2002). ALOHA's database includes a chemical library about the physical properties of approximately 1,000 common hazardous chemicals. The air-modeling program is used to predict the spread of chemical vapors, creating a footprint that can be overlaid on other pertinent GIS layers, allowing users to track the chemical's gas cloud or "plume" in order to determine its potential impact on nearby population centers.

ALOHA program is designed to be easy to use. Inputs required are: city, time, and date of the liquid spill; the selection of the chemical from the program library; current weather conditions, including the direction and speed of wind and other variables that would affect the dispersion of the plume; and a description of how the chemical is escaping from containment.

The results may be displayed graphically using an ESRI-based extension, called ALOHA Conversion Analysis and Summary (ACAS), designed to integrate the plume-modeling results directly into ArcGIS. This allows the user to analyze the plume model data automatically within ArcMap to diagram potential scenarios. Data output may also include graphs showing predicted chemical concentrations at any location of concern downwind of a release and the dose of chemical to which people at that location may be exposed to.

ALOHA Implementation by the BSO

ALOHA initially was offered to BSO in 1992. The impetus was the need for a tool in the event of a chemical leak at Port Everglades as well as its close proximity to the Fort Lauderdale International Airport. The program was a stand-alone application that allowed prediction of the footprint - shown on a scaled grid – of a gas cloud after an accidental chemical release. It was intended to be easy to use by those unfamiliar with chemicals, emergency responses, or GIS.

The software was utilized to create several scenario events. During the scenario events, several limitations were identified. Despite improvements to ALOHA between 1995-2003 such as improvements in dispersion and source, (1995 - version 5.2.1), updated information on chemicals (2002 - ALOHA 5.2.3), and inclusion of flammable explosives (2003 - ALOHA 5.3), the product had limited use. Although BSO could generate footprints depicting a chemical dispersion, it does not include any direct connection to GIS. The footprint must be exported from ALOHA into ArcView 3.x utilizing an ALOHA extension in this GIS product.

Since the intention was to provide BSO personnel with a user-friendly tool to measure chemical dispersions with GIS capabilities, the program proved to be difficult since BSO personnel lacked GIS training and at that time there was no GIS Unit established at BSO. Another limitation with ALOHA was its ability to assess and predict 'man-made' explosions. Based upon the lack of interoperability between GIS and the inability to assess and predict the consequences of a 'man-made' explosion, BSO has turned to utilizing the CATS/JACE program.

3.2 CATS/JACE

Developed under the guidance of FEMA and The Defense Threat Reduction Agency (DTRA), the Consequences Assessment Tool Set (CATS)/Joint Assessment of Catastrophic Events (JACE) software provides powerful disaster analysis in real time with a rich set of information integrated from a variety of sources (e.g., DTRA, 2004). The software is deployable for actual emergencies with capabilities including contingency and logistical planning as well as consequences management.

The CATS program integrates hazard prediction, consequence assessment and emergency management tools with critical population and infrastructure data. It uses tools and data to both: (1) predict the hazard areas caused by natural phenomenon, inadvertent human actions and intentional hostile events, including chemical, biological, radiological, nuclear and explosive incidents, earthquakes, and hurricanes; and (2) help estimate collateral damage to facilities, resources, and infrastructure and creates mitigation strategies for responders.

The CATS system provides for a multitude of events. Information is included within the software for nuclear, biological and chemical hazards as well as for hurricanes, storm surges, and earthquake events. The software provides a user-friendly graphic interface and predefined event scenarios. A variety of skill levels are anticipated. The GIS interface enables the user to combine and manipulate multiple layers of information to assess affected persons, property and infrastructure.

3.2.1 CATS/JACE Implementation by the BSO

BSO acquired CATS/JACE software in 2002. The impetus was the preparation for natural or man-made disasters. The GIS Unit has aggressively tested this software utilizing ArcView 3.x as the operating systems. CATS/JACE has enabled BSO personnel to combine multiple layers of information such as aerial photography, census block demographics, location of first responders (i.e., fire stations), hazards, and casualty probabilities to determine total number of persons affected. Of particular interest is the spatial component, which helps to answer the geographic-related questions: (1) what geographic areas will be impacted based upon time, place, chemical/explosion, weather and release, (2) what populations will be impacted and may need to be notified, (3) what adult living facilities and/or schools may need to be evacuated, and, (4) what hazardous sites are nearby.

The implementation of CATS/JACE has provided the BSO with the program it needs to effectively manage emergency situations. The software allows for the organization of resources in response to a broad range of events. As the software is fully integrated into the department, its effectiveness will be fully realized.

Despite the usefulness of the CATS/JACE program, it is not able to closely analyze natural hazards. The common occurrence of major weather events in South Florida has generated the need for additional software that would allow for an intensive analysis of these events. The HAZUS program was chosen for this task.

3.3 HAZUS-MH

In the early 1990's, FEMA sponsored a study by the National Institute of Building Sciences (NIBS) for the express purpose of considering how earthquakes might affect the nation and what methods of mitigation might be useful. These studies resulted in the development of the Hazards U.S. software (HAZUS). The early versions of HAZUS were intended to provide accurate loss estimations in the event of seismic activity.

The program has evolved into Multi-Hazards HAZUS (HAZUS-MH) (e.g., FEMA, 2004). This software acts as a loss estimation and risk assessment program covering earthquakes, hurricane winds, and flooding. By modeling the physical world of buildings and structures and then subjecting it to the complex consequences of a hazard event, users can implement this tool to prepare for a natural disaster, respond to the threat, and analyze the potential loss of life, injuries, and property damage.

In the HAZUS-MH program three levels of intensity or detail can be employed, with each level based upon the quality and detail of the initial data input. A user choosing the level of implementation would likely base the decision on the funds available for data gathering and input, as well as the level of detail required in the output.

Level 1 Analysis: HAZUS software was created with a default database of information for each region of the U.S. These data are highly generalized and their accuracy is only provisional. The data within the level 1 database include the number of buildings in the area and their value, basic population characteristics, costs of building repair, and basic economic data.

Level 2 Analysis: This analysis is the standard level of implementation of the HAZUS software. While the input requirements expected from the user are greater, a far more extensive and accurate set of analyses and outputs can be provided. Costs can still be limited by the amount of information that is collected, but – as before – the greater the input, the greater the output. Local data inputs would include details of local building conditions and construction, local soil conditions, flood areas and local economic data.

Level 3 Analysis: This level showcases the flexibility of the HAZUS program by permitting customization. For example, customized analysis may be applied to examine threats to high-potential loss facilities such as dams, nuclear power plants or military installations. Specialized data structures and methodologies can be deployed.

3.3.1 HAZUS-MH Implementation by the BSO

HAZUS-MH served a practical use for BSO during Hurricane Jeanne, which struck South Florida on September 26, 2004. Utilizing wind swaths generated from FEMA's HURREVAC 2000 software program, BSO overlaid the wind swaths within HAZUS to determine what essential facilities (fire-rescue stations, police stations, and shelters) may be impacted. BSO updated the essential facilities information as it relates to public safety facilities to better assess damage to these facilities as well as to determine how many people would be utilizing designated shelters.

However, the use of HAZUS was limited since the data in the program has not been significantly updated with the local data. Importing information on building stock, essential facilities, land use, and economic data would greatly improve the usefulness of the output. If South Florida emergency responders are to maximize the use of HAZUS, it will require extensive coordination and cooperation among other entities. Working with Broward, Miami-Dade, and Palm Beach counties as well as the South Florida Water Management District (SFWMD), BSO is looking to develop a South Florida Regional HAZUS Work Group to provide the data needed to customize HAZUS for Level 2 and Level 3 Analyses.

4. Evolving Enterprise GIS

This section examines three evolving projects within BSO's GIS Unit. These are: the supply of GIS data to officers in the field for critical incident response situations; planning for Homeland Security initiatives; and implementing the Wireless Technology - Federal Communication Commission (FCC) Phase II.

4.1 Use of GIS technology for Critical Incident Response

In 2000, the Sheriff challenged his agency to develop a portable visualization tool to combat school violence, crimes and other emergencies. BSO subsequently equipped 1,600 road patrol deputies, supervisors, dispatchers, SWAT teams, and school resource deputies with a portable compact disc (CD) that contained, among other things, street/parcel maps, detailed floor plans, aerial photos, and interior pictures of 125 public and private elementary, middle, and high schools in the agency's jurisdiction.

GIS was used to map all strategic locations around schools, utilizing the best and most currently available map layers, including aerials. The data are a combination of computerized addresses that are geo-coded to a specific building, office, or space on a campus. In cooperation with the Broward County School Board, floor plans were obtained and photos were taken on campus and from the sky, via a BSO helicopter. Every piece of information that schools store in their databases can be digitized and applied to a GIS system to provide public safety personnel with a visual depiction of what is facing them prior to their arrival at a critical incident.

In any critical incident, knowing whom to call, where to respond, and how to gain access is crucial. Even if the deputy is not familiar with a given school campus, he/she will be able to access street/parcel level maps, detailed floor plans, aerial photographs and interior photos. Primary and secondary perimeter points and staging areas are pre-designated so that supervisors and dispatchers will instantly know the best places to send responding emergency units. The program includes predetermined command post locations, the closest spot to land a helicopter and a place for parents to gather. A list of all other police agencies and hospitals is also included, as well as hazardous materials information.

Response to the CD has been favorable. Tactical response experts say the CD has proven invaluable in the event of a violent incident on campus, but it will inevitably be useful in other situations, as well. For example, if a school's security system detects midnight intruders, deputies will know how to best approach and apprehend the burglars. If a young student is missing, deputies will have emergency after-hours contact numbers for school officials.

However, significant hurdles remain. Updating the information as it evolves currently requires the creation and redistribution of the CD to the officers. This method is time consuming and limits the frequency of providing updates. As part of the evolving Enterprise GIS, the BSO is implementing ESRI's ArcIMS (ESRI, 2004) platform. This platform, referred to as a 'thin-client' solution, provides Internet capabilities to the Department's GIS Unit. The web-based mapping interface will allow BSO to collect, store and update the information in a timely manner. The system will also allow users to edit features such as setting up primary command posts or helicopter pad landing zones. The Internet functionality is being combined with wireless access for the first responders' laptops. The officers' computers are equipped with Verizon Wireless CDMA 1xEVDO cards. With a live connection, the users can access the most recent data available to the department.

4.2 Homeland Security

The GIS Unit is advocating the development of a Strategic Technology Plan for Homeland Security. This plan is intended to address the implementation factors needed to fully utilize the GIS potential. In its early stages, a 'Technical Homeland Security Work Group' would address the following factors.

Assess the security implication of sharing critical infrastructure data. Of particular concern are the federal and local sources of geospatial information. Geospatial data and information are useful for identifying various geographical features of U.S. locations and facilities, as well as characterizing their important attributes. Although these agencies produce and publicly disseminate such information for a wide range of beneficial purposes, the risk also exists that some types of geospatial information could be exploited by terrorists. One of the issues needed to be addressed is developing security measures to provide this information solely to first responders.

Assess the usefulness of geospatial information for the purpose of Homeland Security. This would require the development of matrix indices depicting which GIS layers are mission critical and to establish procedures regarding data coordination, data updating, and data sharing between internal/external entities.

Increase interoperability of resources to better regionalize responses to emergencies and homeland security – this will require the establishment of data/mapping standards, utilizing best practices in regards to technology and coordination efforts.

The GIS Unit is also working to develop a 'thin-client' mapping application that utilizes CATS/JACE within a web environment (ESRI's Arc IMS platform). The intent is to allow first responders to access the information in the field rather than be relayed the information indirectly by radio dispatch. Also, first responders will have the visual information needed to deploy resources quickly and effectively.

4.3 Next Generation: Wireless Technology - FCC Phase II

The GIS Unit is also examining how to best utilize GIS for Emergency 911 response, particularly in the face of new federal legislation requiring the ability to determine the exact latitude and longitude location of wireless E911 calls.

The public safety community, embodied by several national level professional organizations –National Emergency Number Association (NENA), Association of Public-Safety Communications Officials (APCO), and the National Association of State 911 Administrators (NASNA) -- united in 1994 to officially lobby the FCC for service parity between existing wire line E911 systems and wireless services. They requested wireless subscribers have the same level of service currently provided to wire line subscribers. The result of their efforts was the FCC's "Notice of Proposed Rule Making" (NPRM), or FCC Docket # 94-102.

The magnitude of the technical challenge became evident to the communications industry, as well as the 9-1-1 specialists, who were not previously involved, as soon as the NPRM was released for comment. The result of these comments led the FCC to release a "Report and Order" that identified several phases of implementation, occurring over a specified time, to allow appropriate technological adjustments to bring wireless service up to par with wire line service.

The FCC's wireless E911 rules require wireless carriers to begin transmission of enhanced location information in two phases. Phase I requires carriers to transmit a caller's phone number and general location to a Public Safety Answering Point (PSAP). Phase II requires more precise location information to be provided to the PSAP.

To make sense of FCC's wireless Automatic Location Identification (ALI), the call location must be located and plotted on a map. Plotting the location, along with the existing streets and addresses, electronic serial number (ESN) boundaries, and similar "background" information, will allow the call taker to quickly determine the location of the call. The background information should include the street centerlines, railroads, water features, ESN areas, city boundaries, county boundaries, emergency service agency locations, and other information.

A recent National Emergency Number Association (NENA) Critical Issues Forum identified key concerns of GIS technology in the PSAP as being data quality, integration, and data maintenance.

At BSO, the GIS Unit has developed a GIS infrastructure that will support Phase II. Working in cooperation with Motorola, BSO is currently researching potential mapping applications associated with Phase II. For example, the ability to transmit Phase I wireless 911 cell tower and sector coverage areas, as well as Phase II wireless E911 call locations on a deputy's laptop which also could also show map layers such as streets, parcels, waterways, and aerial photography.

The next component to add to Phase II is the potential of integrating tools like ALOHA or CATS within an E911 Wireless Dispatch Mapping program that can be utilized by BSO's first responders. The goal is to locate the E911 caller's location on screen and obtain the required information where these tools can be deployed such as overlaying a chemical plume dispersion layer in relationship to the E911 call and the area of impact.

5 Lessons Learned and Recommendations

The overall mission of the BSO is to develop an interoperable enterprise GIS that efficiently and effectively uses geographic technology to protect and serve the people of Broward County. To that end, the goal of the GIS Unit is to continue in enhancing mapping technologies that support BSO's ability to effectively address natural disasters, accidents, and other types of major emergencies, including terrorist incidents. These best practices help in creating an enterprise solution in designing a geospatial infrastructure needed to prepare, respond, mitigate, and recover from both man-made and natural disasters. The success lies in terms of integrating various geospatial technologies, disparate databases into the business process to create a 'one-stop' interface to query, map, and report information efficiently and effectively.

To date, BSO's GIS infrastructure includes the development of a comprehensive Enterprise GIS Data Warehouse containing nearly 8.5 million records from disparate databases. Integrating the GIS Data Warehouse with ArcSDE, ArcIMS, and ArcObjects, the agency has a 'one-stop' shop for querying, analyzing, and mapping public safety information quickly and efficiently.

Clearly, BSO has had a long history with computerized programs and software solutions to deal with both man-made and natural disasters. They have emerged from an organization that relied on stand-along programs (such as ALOHA) to an organization that seeks an integrated GIS capable of responding anywhere or anytime.

To optimize the Enterprise GIS in the near future, the organization is considering a number of important issues.

Develop an Infrastructure that is Interoperable - Policy to Promote Integration

Many government agencies and other organizations have been creating and maintaining spatial data for decades, making the United States the most geo-data rich country in the world. However, much of this data is trapped in information silos isolated within departments and organizations. To implement homeland security without costly replication of data will require the type of data integration provided by GIS as well as agreements that inventory and allow access to data by many jurisdictions in a controlled manner.

Manage Access to Data - Policy to Promote Data Quality

The creation of accurate and consistent data standards must be kept to insure the veracity of the data provided. Effective response by safety personnel can only be assured when they have accurate information to act upon.

Develop Standard Operating Procedures to Better Regionalize Responses to Emergencies and Homeland Security - Policy to Promote Data Sharing/Updating

The ability to provide data to everyone who needs it is of utmost importance. At the same time, the ability to secure and protect sensitive data cannot be overlooked.

Increase Awareness and Capacity to Use these Tools through Training and Technical Assistance Workshops

The potential benefits of GIS can only be realized if the personnel involved are properly trained to use it. Understanding the benefits of the system will increase the level of interest among potential users.

Build Relationships

Data coordination and emergency response planning efforts have an added benefit. These activities build relationships between people in agencies, departments, and organizations. An old bromide of emergency management is that "people at the scene shouldn't be exchanging business cards." The process of gathering data develops relationships between people who would not necessarily have any reason to interact except in an emergency.

Develop Thin-Client Solutions (Technology)

The implementation of Internet accessibility coupled with wireless access will greatly enhance the potential of the system. The advantage of this system includes connectivity to the latest data, capacity to receive a broader range of information, ability to collaborate with others, and the capability of providing data updates to the GIS Unit.

An accurate analysis of the current BSO systems for disaster management is difficult, due to the fact that the majority of the work completed thus far has been "scenario" analysis, and not an actual event. Lacking the ability to compare the results with real quantified losses or impacts indicates that the system is operating in a speculative mode. The system must continue to put its faith in the algorithms developed by the software creators. Refinement of the systems to reflect local conditions will need to occur as event analysis occurs.

The lessons to be gleaned from the analysis of this system are many. On purely technological grounds, the analysis of the available tools and methods is critical. The integration of these departments and the ability to incorporate GIS into a multitude of uses effectively and accurately is equally valuable. As the opportunity to respond to large-scale emergencies arises, we will be better equipped to evaluate and transform our technology.

References

Cutter SL, Richardson DB, Wilbanks TJ (2003) The Geo-graphical Dimensions of Terrorism. New York, NY: Routledge

Defense Threat Reduction Agency (2004 CATS. Retrieved December 1, 2004, from http://www.dtra.mil/Toolbox/Directorates/td/programs/acec/cats.cfm

Environmental Protection Agency (2002). ALOHA, May 24. Retrieved December 1, 2004, from http://www.epa.gov/ceppo/cameo/what.htm

Environmental Systems Research Institute, Inc. (2004) ArcIMS, November 30. Retrieved December 1, 2004, from http://www.esri.com/software/arcgis/arcims/index.html

Federal Emergency Management Agency (2004) HAZUS, November 1. Retrieved December 1, 2004, from http://www.fema.gov/hazus/

Langhelm R J (2002) The Role of GIS in Response to WTC – Supporting the first 30 Days [Electronic version]. Accessed 14 December 2004. < http://gis.esri.com/library/userconf/proc02/pap1348/p1348.htm

Information Management Boosts Command & Control

Eric van Capelleveen

Twynstra Gudde Management Consultants, P.O. Box 907,
3800 AX Amersfoort, the Netherlands.
Email: eca@tg.nl

Abstract

The POIRE project focused on making the information demand of crisis managers and disaster fighters more explicit by a stepwise refinement of their need for relevant information. A five step analysis was used to define the basis questions for each working process defined to be activated during crisis management and disaster fighting. Maps appeared a valuable method for communicating the answers on the raised questions. Maps that look more like PowerPoint sheets than GIS based presentations. The presentation of the core information implies focus on what to show and what to omit. The huge amount of data to be effectively searched in moments of crisis, requires a smart information retrieval instrument. An instrument that opens data by location, object, actor, process and organization. The classic item and pull down menu driven GIS tools need extensions for non GIS users. Ontology based menus can probably fulfill this requirement. Managing the sudden tsunami of data requires support of information management and intelligent post processing of logged data. Distinction between 'must know' and 'might need to know' needs implementation. Information should ideally be available within three mouse clicks. GIS technology can also contribute on managing the huge amount of distribution and monitoring of workload. By providing a common operational picture (SITPLOT) and features for having both overview and drill down functionality, users receive a powerful set of information processing tools enabling them to execute their jobs effectively. The merging of SCADA and GIS mechanism promise this functionality. Last but not least, clickable maps that open position located documents and vice versa (G-linked documents) provide the information process tools required. One might think that ICT is the "Haarlemmer Oil" lubricant that solves all information processing problems. One should however bear in mind, that 'ease of use' and not a tsunami of functionality is the factor of success.

1 Orientation regarding Information Retrieval

1.1 Stepwise Refinement towards Using GIS

Information is defined as data with a meaning for its users. A meaning that is most probably derived from the working process the information user is involved in. Therefore, one could state that if information demand derived from working processes is appropriately met, shown data is most likely meaningful for its users. In the POIRE[1] project we have adopted this paradigm and defined information demand by using a five-step mechanism ©Twynstra Gudde . These steps are: 1) Information Demand Definition 2) Identification Information Objects 3) Defining Information Flow including Sources, Messages and Recipients 4) Rules of information Engagement and 5) Preferred presentation.

First of all we modelled the working processes and defined the basic questions raised on discussing the question "What would you primarily want to know when executing this process?" We listed the seven or eight most important questions, all beginning with "*What, Where, How, Who, When*"

1.2 Information Demand Rules

The following example shows how we handled this process. The working process is "Public Communication". The main questions to be answered and part of the press release and internal information distribution are listed below.

Main questions presented in relation to public communication are:

- What is confirmed information?
- Which information can be communicated?
- Do we have reliable information about the situation and effects for the population?
- Where is detailed information available?

[1] POIRE Project Operational Information Regional Firebrigades.

- Which information is already known (media; streets)?
- How are you to be informed on developments?
- Which effects are most likely to be dealt with?
- Which behavior advice ought to be communicated?

Most organizations involved in disaster management are used to include the answers to these questions in a text-based press release. Only occasionally we see the inclusion of a photo, map or a video. This is surprising, when we consider the ICT possibilities and the power of maps and images at transferring a message. It is even more surprising when we consider the power of maps for overcoming language problems in a multi-cultural society. The participants of the POIRE meetings decided to utilize maps as an instrument and designed a basic communication map to be included in the press release.

1.3 Geo-Information Based Communication

The map shown below indicates major information to be communicated parallel to the press release. The evacuated area along with the barriers set by the police is indicated. Here a set of basic, well-known and explicit signs should be used, because in a disaster and crisis situation ambiguous communication is not allowed. Furthermore, the temporary accommodation is indicated on the map. The text of the press release is added to complete the message. The used map is a simple sample. For reference and location purposes of the public, a standard town map with streets, houses and green areas is used. Important here is the grey-lining of the neighbouring district. In this manner the relevant district X is highlighted expressing the message "only meant for district X". In most GIS applications this is a standard feature using 'transparency'

1.4 Modern ICT Integrates

The fast increase of ICT in our homes and the integration of TV and PC/Internet applications might show us how to communicate about disaster in the near future. Using SMS and MMS techniques in combination with dynamic geo-referencing techniques, the persons located in the district and those living there, can be alerted by sending them SMS/MMS messages with maps and/or messages like the one shown below.

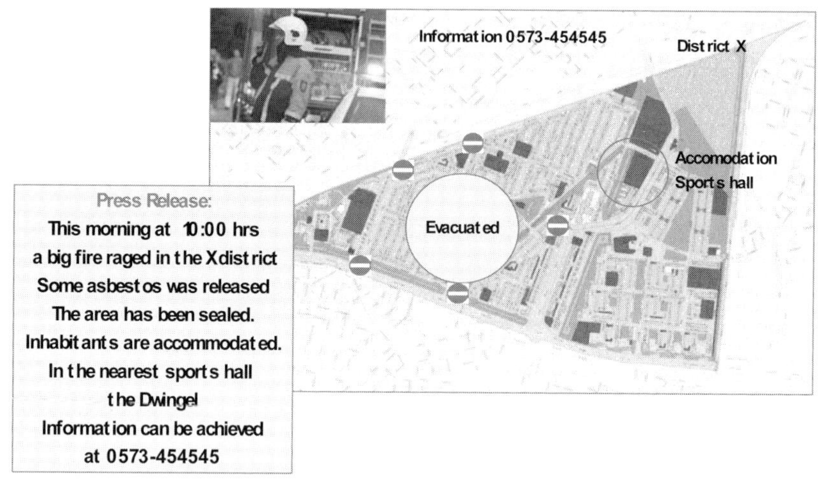

Fig. 1. Communicating the disaster and evacuation

1.5 Scale & Local Knowledge

The example shows how process oriented information is composed and illustrated. The working process was the main trigger for retrieving information, but other triggers are also quite likely. Think about the object oriented approach, indicating an object and trying to find the relevant information about it. In the example shown this might have been the contact information of the sport hall manager, the hall's capacity and the catering firm to be hired, due to a pilot agreement available. Often data is also searched and opened by location. Just indicating an area and requesting the number of known inhabitants, known risk locations etc. is a common occurrence. Due to scale enlargement the following question appears to be more frequently applicable. "Which organizations can be approached for providing mental care, damage survey, environmental control etc.", becomes more and more a relevant question.

Relevant because emergency room operators cover a large area; Characteristics of an area which they do not automatically know by heart, due to being born and/or living there.

1.6 Opening Data

Basically, we see four ways of opening data sources in disaster management information systems:

- by location/object,
- by process,
- by task/actor,
- and by organization.

The question presents itself whether this means that GIS can extend these information retrieval capabilities or not. One could say that GIS systems are perfectly able to register and retrieve data by organization, task/actor and by process if such data is geo-referenced. By geo-referencing we mean giving indications about the persons taking care of a certain task, about the organization to be approached in a certain area. Processes regarding "who does what" are not easily linked to an area. Here the interfacing with process oriented information systems like a WFM (Work Flow Management) system is more likely.

1.7 Predefined & Toolbox GIS

The second issue is coping with flexibility versus the pace of information retrieval and finding relevant matches on search queries. In the POIRE project the participants concluded that both functionalities should be supported. Flexibility, as currently provided by the standard GIS tools, is needed for high-end users with thorough know-ledge of GIS, DIS and WFM systems. Such systems require thorough skills of its users and knowledge of data sources. Information retrieval in disaster situations, however, must not only depend on the availability of the GIS professionals. Disaster applications should provide functionality for "pre-baked" information products on-demand such as maps, checklists etc. Firemen, medics and police officers as well as trained information professionals should be able to use a button based control board for retrieving information as provided in a SCADA (Supervisory Control and Data Acquisition) system. This gives its operator direct access to all relevant data in the preferred format. Data acquired from several sources, combined in one overview, with detail information, three mouse clicks away. It is quite probable that this is both location-based, text-based, and/or process-oriented data. Most likely, these predefined information views also make up the basis for more situation specific views to be built by the information professional.

1.8 Onthology Based Menus

Finally we should point out that finding focus and balance, building and equipping these tools and button board-based applications are among the biggest challenges for the ICT/GIS industry and information managers. Application environments like MS Office and GIS tools show that enlarging the number of possible features results in a slowdown on ease-of-use and performance, which in the end leads to decreased use. Dynamic menus that are different and adapt to the described situation and corresponding use, are considered to be the appropriate tools for finding focus and balance. Onthology techniques might help us out regarding this challenging topic.

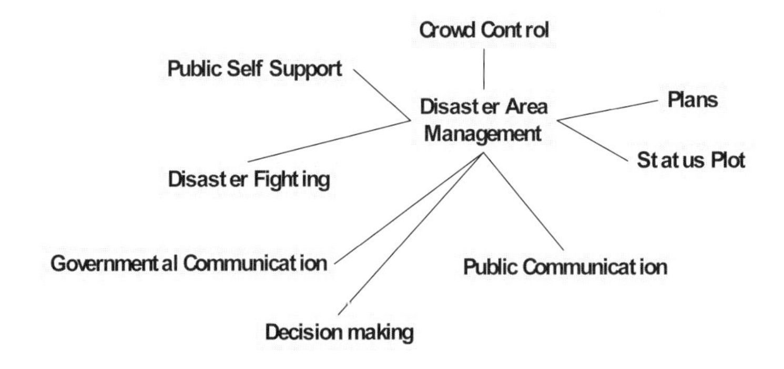

Fig. 2. Example of an onthology-based menu

Using these menus users are encouraged to think about other relevant issues to be dealt with. These menus, however, should be compact and specifically user role-oriented.

2 Information Management Mechanisms to be Supported

2.1 Effective User Interface

A series of information management mechanisms must be supported by Crisis Management Information Systems (CMIS), mechanisms like validation, interpretation, filing, distribution and opening large amounts of data, the relevancy of which is yet to be determined. Historic data, knowledge how to handle situations, accessing knowledge networks in search for relevant aspects to keep in mind, all these information processing activities must be supported in a CMIS. And the system requires a simple and effective user interface.

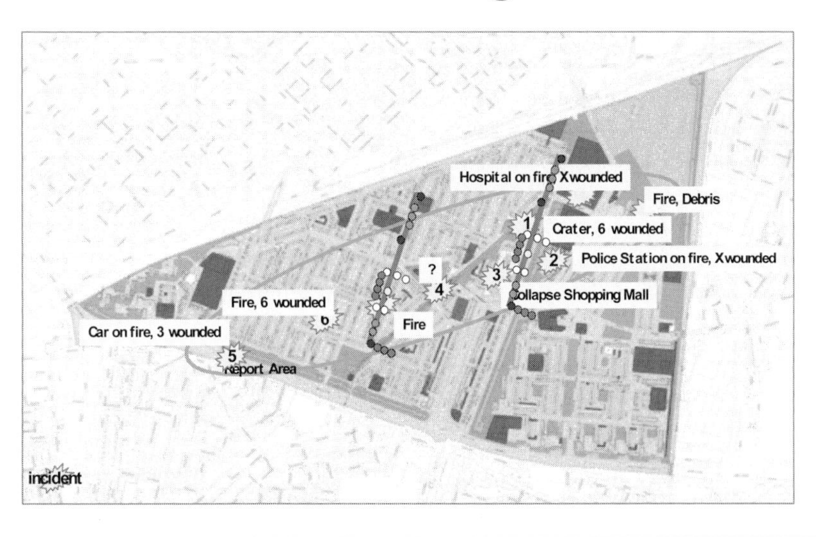

Fig. 3. Debris hits neighborhood, overview by SITPLOT

Most CMIS systems used in the Netherlands focus on data processing and presentation of the large amount of interrelated data using the presentation capabilities of a GIS. Yet crisis and disaster fighters are no GIS professionals. Maps should fit the users need to present overview and should provide possibilities to show detail on some relevant issues. This requires the possibility to focus, omitting irrelevant data or grey-lining them. And at the same time, the perspective of various users and their specific focus should not result in inconsistent presentation and, even worse, inconsistent or ambiguous interpretation. The example above shows us a map meant to give an overview of a disaster of some importance. In a village neighborhood a series of fires and damage due to debris have been reported.

2.2 Information Management Issues

For reasons of information management a series of information management tasks should be supported. Basically, a logging register should be implemented, logging text, drawing data, geographical data, voice, and video. This data should be tagged intelligently, so that quick and intelligent retrieval becomes possible. Time must be included in all registered data in such a way that the log enables a movie presentation of the changes in information. Preferably, the geographical position should also be included thus enabling geographical access. This requires intelligent linking to the data put in and recorded, or direct intelligent post processing with an extreme high rate of correct interpretation. Data logged from certain sources or referring to persons and/or organizations about from which is known where they are situated, can be geo-referenced quite easily. But the question remains of course whether to geo-reference the logged data to the location/area it refers to, the location where the reporter is and/or, for instance, the distribution area of the organization involved. Anyway, disaster experts told us that they want to validate the machine-intelligent linking anyway, without raising a workload that distracts them from the initial target of crisis and disaster fighting.

Intelligent distribution is a third issue to be supported. When a calamity occurs in a certain spot, the system should be able to analyze all the people involved and to be notified.

2.3 Three Mouse Clicks Away

During disaster and crisis situations quick and effective retrieval of data is a valuable asset. Obtained data should be stored quickly in such a manner that retrieval in several ways is possible. Information that requires distribution after/before registering should be distributed either in total to those who "*must know*" and to those who might *"need to know"*. Stored information should, from a user perspective, be easily found. Basically, the user should be able to address the searched data set by just defining a series of search words by clicking on three menu-bars or three onthology-based click points. Search engines, both geographically, theme-based and smart process/theme-oriented menus should support this requirement. This is a major challenge for both information managers and analists, as well as the software builders.

3 Presentation of Core Information on the Basis of GIS

3.1 Alerting

The process of alerting the public presents a series of questions to be answered. Questions like: "In which area do we have and want to alert the public?", "Which target groups require which way of alerting?", "What confidential information of the police force extends or changes the evaluation of disaster picture?" But also questions as: "Which communication facilities are available, working and will possibly gain effect?"

And of course the definition of the message to be sent in terms of:

- what is going on (situation),
- what is the public supposed to do and not to do,
- what are the possible consequences of the disaster,
- which activities are carried out by the emergency services,
- where to obtain more information,
- how the public is informed constantly.

The figure below shows a GIS based plot in which the coverage area of the alert sirens is plotted. This way emergency workers know the likely penetration of their alerting sirens. On the map special attention groups like elderly and disabled persons in the clinic are shown. The capacity numbers give a first impression of the amount of possible aid to be organized. The also plotted school with 200 schoolchildren and the quarter with a lot of native speakers (Arabic) indicates extra activity to be deployed realizing full attention of those being there. This map shows indicative and basic data plotted on a town map level. Mouse-over functionality provides instant extra information for the operator.

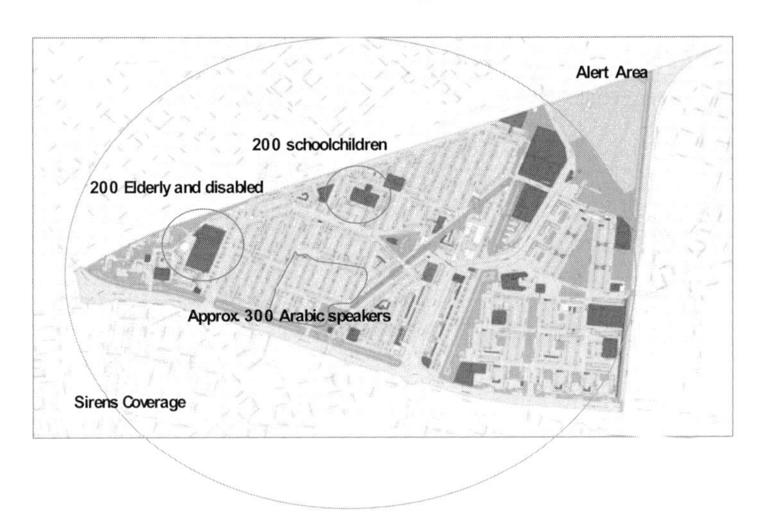

Fig. 4. Sample situation Alerting the public

3.2 Contamination

The process of contamination handling raises a series of questions to be answered. Questions like: "In which area do we have contamination and which emergency vehicles/staff are present in that area and thus contaminated?" But also questions like:

- what is the contamination character,
- what is the contamination volume and intensity,
- who knows how to handle, to treat those contaminated,
- which precautions can be taken,
- what are the possible effects of the contamination,
- what can be done to prevent further outbreak and effects/casualties,
- which might be the political and publicity effects,
- which capacity for decontamination do we have.

The answers to these questions can partially be shown on a map. Contamination area, vehicles and staff present as well as their location is plotted above. A direct link to a contamination knowledge database or network should also be facilitated. A network of known lessons, measures, effects, treatments etc. should be shown in an onthology network, thus supporting the operators/coordinators in carrying out a quick and effective response to the situation. This can be considered as a predefined overlay of applicable geo-information and knowledge structures to be launched if necessary.

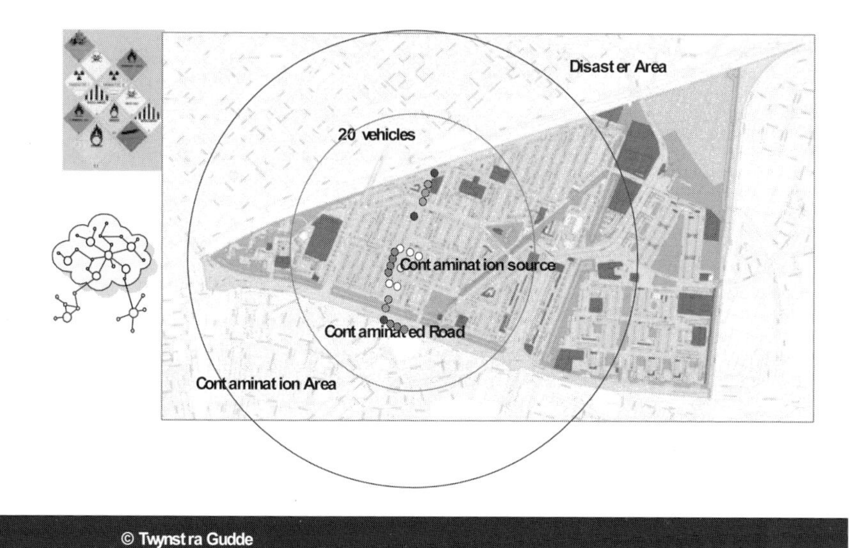

Fig. 5. Sample situation Contamination

3.3 Hazard chemical spill

In this situation a series of questions have to answered. Questions like:

- which spill and chemical material is involved,
- where can/have we gauges or sniffers present,
- which concentrations and volumes are registered,
- which danger might hit the public,
- which areas are most likely contaminated, in which order and intensity,
- what are the chances of escalation,
- which chain effects might occur (water system, sewer),
- which possible spread might have occurred before the sealing of the source.

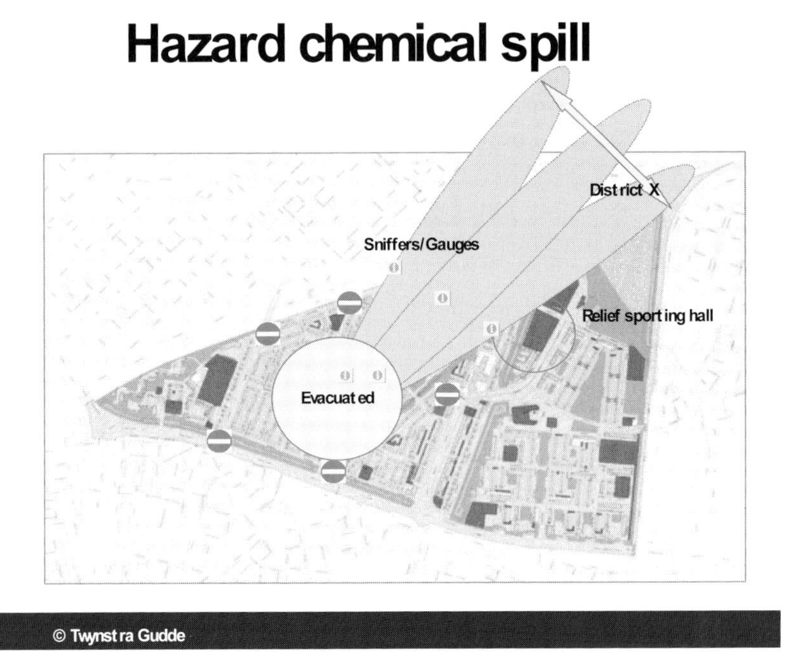

Fig. 6. Sample situation Hazard Chemical Spill

There are a series of questions to be addressed and answered. A series of answers can easily be plotted on a map, thus being a powerful communication instrument. In the map above we chose for indication of the evacuation area, the sealing of this area by roadblocks. The planned locations for the gauges/sniffers in a first and second ring are plotted as well. The possible variation of the hazard, due to winds, is plotted too. Combination of this information with the projected relief location shows that the standard disaster relief location for this area cannot be used.

4 Problem/Workload Oriented Visualization

4.1 Common Operational Picture

Within a disaster situation two major topics are to be dealt with. First of all consistency of information (no different content of the same object/subject) and overview of the situation and ongoing/planned activities.

Provision of this Common Operation Picture (COP) is a major task for geo-information. Standard GIS systems can provide the functionality to draw the COP. Dynamic functionality enables its user to log the changes in the COP and to rehearse the flow of events or to simulate a planned operation. If the contingency plans are predefined in the same formats as the CMIS event and workload monitor, planned and actual situation, as well as planned and actual operations can be presented and monitored. Somehow this might be applicable for those activities that are standardized and predictable. But functionality supporting improvisation based upon training and skills should also be present.

ACTUAL & TARGET

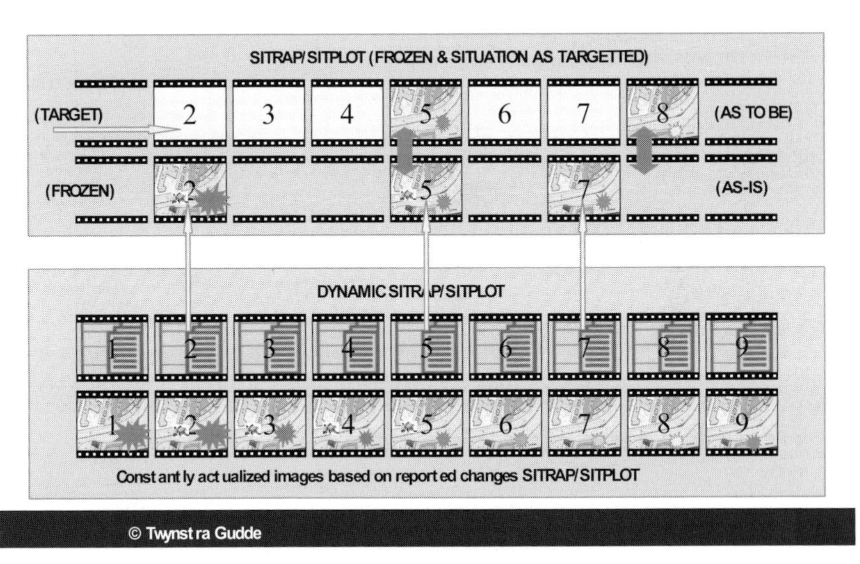

Fig. 7. Mechanism regarding facilitating planned/actual

4.2 Overview and Drill Down Functionality in One Package

At first one would tend to show all operations and all details of the current situation. After some time, users require layers to be switched on and off depending on their needs. This requires however constant navigation. As in SCADA systems, (SCADA stands for Supervisory Control and Data Acquisition) users really want two principles to be joined into one application. Overview on the whole situation, events and activities based on exceptions to the planned situation as well as possibilities to drill down on certain issues and topics based upon located/observed deviation. Discipline-oriented layers can of course contribute to this way of working. Basically, only exceptions to the planned/expected situation and new events are interesting. An example of this principle is shown below.

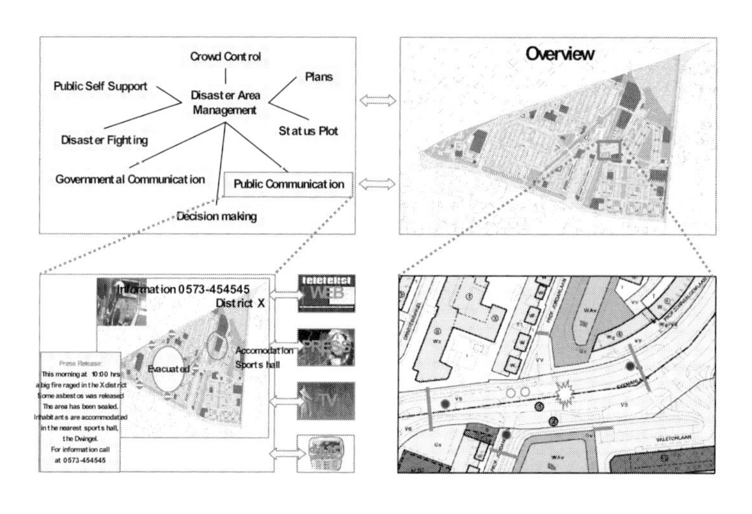

Fig. 8.Mechanism on facilitating planned/actual

The figure shows a two columns approach of data presentation. On the left side the onthology structure of the issues relevant. On the right side the COP. Both views are detailed in a specific part. The COP zoomed in on the incident spot area, providing more detail in which the objects can be consulted by mouse over, popping up identifying information. The 'Public communication' aspect is drilled down, unfolding the composed communication message and proposed media channels (to be) activated. Clicking on the media buttons activates for instance an online video in a special window.

4.3 Signaling

By supporting the coordinating emergency services staff GIS/WFM systems could easily help them to monitor the tidy execution of tasks. One could do this by just making to-do/job lists and check marking and monitoring them regarding status as planned, issued, executed. A geographical positioning would enable analysis and presentation on possible collision of job execution in one spot. The analysis can easily be performed by the operators, although smart analysis could simplify the job for them as in air traffic control. Possible overload of certain emergency services can be spotted as well if allocation and dispatching are monitored. Here modern ICT technology provides tracking and tracing of emergency vehicles on worker level following their GSM or GPS/Gallileo transponders. Basically, areas too crowded with emergency units or black spots with no coverage must be observed at first level.

5 Approach and Operational Procedures Managing Disaster Fighting

5.1 Intelligent Document Access

Most crisis and disaster management information systems include a digitized version of the so-called disaster generic contingency plans, specific object fighting and accessibility plans as well as the more generic disaster management plans. In this first generation CMIS systems these plans are digitized straight forward. A Word or PDF version of the document is attached and in most cases also accessible by pull-down menus. Menus that require its users to know where data is stored and how to address it. One could consider this as one-to-one digitalization. More and more CMIS systems are upgraded and the content of the disaster management and fighting plans are I-mapped. With I-mapping we mean the meaningful apportion of the text in compact parts that represent a certain topic and that cover preferably no more than the viewable part of the window in which they are presented to the user. Therefore, the meaningful compact data can be accessed quickly and header and content are congruous with each other.

5.2 G-Linked

The next generation CMIS systems should contain G-linked data. The data is not only I-mapped, but also geographically positioned. This means that this document, and especially its I-mapped fragments, should be able to show the service/administrative area or object they refer to by just clicking a map symbol. Vice versa indication or selection of an area or object should expose the relevant documents and their relevant content. The answers to questions about authority, responsibilities and (dynamic, resource depended) assigned service areas are included in the data structure and its I- and G-links.

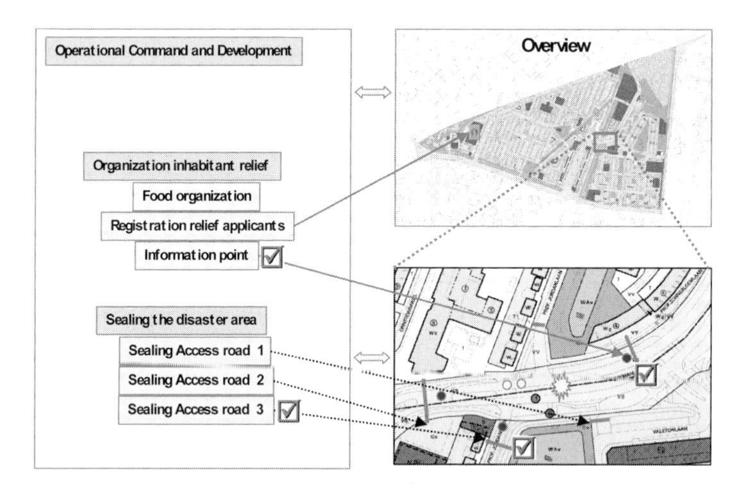

Fig. 9. Example of G-linking

This way of thinking also applies to task management and task execution. Fighting a disaster implies constant issuing of tasks and monitoring their execution. If a disaster is enormous, the principle of division of tasks into geographical units and disciplinary aspects is necessary. This is called "*hakken in vakken , zagen in lagen*" and can be translated into "*chopping in compartments, sewing in layers*". This implies that a series of activities is possibly executed at the same moment in the same geographical spot. Coordination, therefore, is required both regarding the aspects and the location of which geographical information systems can provide a quick overview of possible jostling. What is more, activities should preferably be positioned and linked to objects, subjects and/or areas. In this manner analysis of conflicts can be performed quite easily.

6 Wrap-Up

6.1 Lessons Learned

The POIRE project taught us that not system possibilities should determine how CMIS systems should be designed and built, but that investing in explication of information demand and information support on the job pays off. Using the geographical option contributes enormously to the quick use and effectiveness of information usage. GIS and SCADA-alike functionality should be integrated into the CMIS systems. Demand for frequently used information should be predefined and be retrieved by just pushing a button. Symbol sets form a powerful interface in a meaningful presentation of information. Dual GIS functionality linked with text and data can provide a powerful tool in effective retrieval and linking information. Ease of use and intelligent geo-linking are two mechanisms to be developed in CMIS systems.

6.2 To keep in Mind

If the development of the described functionality results in more buttons, more menus, more navigation, the use of such functionality would require more skills and know-ledge. All resources that are quite likely scarce whilst coordinating disaster fighting. The challenge remains to include the functionality on a logical, easy to use basis. All superfluous mouse clicks and/or keyboard input should be avoided. Ease of use rules.

Task-Centred Adaptation of Geographic Information to Support Disaster Management

Nosakhare Erharuyi and David Fairbairn

School of Civil Engineering and Geosciences, University of Newcastle upon Tyne NE1 7RU, United Kingdom.
Email: dave.fairbairn@newcastle.ac.uk; nosakhare.erharuyi@ncl.ac.uk

Abstract

The ability of an agency or group of agencies to manage any disaster rather than just react to crises is critically dependent on the immediate availability and flow of geographic information to responders in the field for decision support. Such information can include location of emergency management facilities, state of transportation routes, presence of population centres, susceptibility of environmental and habitat zones, scenario and simulation models etc.

However, in order that such information fits the needs of the responders we need to move beyond location-aware computing in which primarily the location of the user is considered. In particular, individual information needs optimized on the basis of location, user, goals and tasks allow for the determination of situation-adapted information solutions. This chapter discusses the potential of task-centred adaptation of geographic information for effective disaster response. It examines issues such as content adaptation, concepts of domain, utility and task-related optimization. The ultimate goal is to achieve task-specific delivery of appropriate geographic information to those in the field who are responding to an emergency situation

1 Introduction

Human and environmental security (together defined as "sustainability") have emerged to become a foremost concern of the 21st Century. The essence of this approach to development is a stable relationship among human activities, environmental change and human vulnerability. Unfortunately, there has been a rapid rise in the number of disasters and severity of their impacts over the last few decades (UNDP 2004), an indication of a failure of human and environmental security. Therefore, if we are to build and maintain liveable communities, a major challenge is the search for effective and efficient ways and methods of planning for, responding to and recovering from disaster. This make the understanding and handling of appropriate geographic information critical.

Geographic Information (GI), easily accessible in real time and capable of being shared amongst users through different emerging technologies (particularly mobile technologies), is essential for a wide range of Federal, State, Local and privately-owned agencies responsible for disaster management tasks. The ability of an agency or group of agencies to manage an emergency rather than just react to crises depends critically on the immediate availability and flow of task-centred geographic information.

No longer is the question of whether or not geographic information can be used in disaster management relevant: its value has been shown in disaster preparedness activities such as risk identification, risk assessment, awareness-raising and warning, preparing risk maps, and carrying out 'what-if' scenario analyses (Erharuyi & Fairbairn 2003; Gui-lian et al. 2000; Gunnes & Kovel 2000; Hall 2002). However, a significantly important requirement is that field responders, engaged in emergency response, benefit significantly from geographic information to support response decisions.

How do we enable access and optimum use of GI by field personnel, so they can handle information about response decisions, resource allocation and other conditions at their location? We propose that this challenge can be facilitated through advances in task-centred adaptation of geographic information i.e. the use of knowledge about user tasks and situation to package content with reference to contexts meaningful for specific users. We examine issues such as content adaptation, concepts of domain and utility, and task-related optimization.

2 Geographic Information and Disaster Response

The identification and adoption of spatial data infrastructures (SDIs) by key strategic bodies around the globe, as a mechanism for effective spatial data service delivery, has potential to benefit disaster management. This benefit is enhanced at a personalised level if we can use the infrastructure to access and interact with task relevant digital geographic information in a user-specific and adaptive way.

Although the underlying cause of disaster may vary, the approach to its management is the same. Disaster Management encompasses a wide range of activities that can be grouped into pre-, during, and post-disaster phases and each of the phases can make use of GI from a variety of sources., User requirement analysis studies for emergency management, and recommendations from various response exercises have shown the importance of GI in disaster management (Baldegger & Giger 2003; Hall 2002; Meissner et al. 2002). However, an important requirement needing further attention is how GI can be optimised to accommodate situational contexts and put in the palm of field workers engaged in disaster response during the incident.

It has been proposed that response activities can be facilitated and benefit significantly from advances in mobile and adaptation technologies (Erharuyi et al. 2003; Fairbairn & Erharuyi 2004). Within the information science community, content adaptation measures have been developed (Brusilovksy 1996; Chalmers et al. 2001; Franti et al. 2002; Reichenbacher 2003; Wai & Francis 2002). However, some of the measures are limited and are applied to geographic information solely in order to provide it in a particular format, limited scale or for adjusting the bit rate of compressed video streams and reducing resolution. These approaches, apart from being techno-centric one-size-fits-all, mainly concentrate on the aspect of the data at the expense of the model of activities implicit in the application and the requirements of the purpose for which the data will be used.

We believe there is a need to refocus geographic information adaptation to a more problem solving process, asking questions such as; what are the activities we use it for? what are the tasks that constitute an activity or phase in disaster management? what actions do we need to perform within a task? Geographic information is produced and used by people to support better informed and faster decision making, but this potential can only be exploited fully if it accommodates user expectations/work flow and decreases cognitive overhead. The ability of geographic information to accommodate user-defined objectives/goals and activities translates as its *utility*. Therefore an adaptation approach with a focus on user's tasks and actions will facilitate effective and unambiguous communication between geographic information providers and seekers.

3 Concepts of Adaptation

According to Holland (1992), the first attempts at technical description and definition of adaptation came from biology. In that context, adaptation defines any process whereby a structure is progressively modified to give better performance within its environment. It can involve the fitting, changing, modifying or selecting of some structure to suit a different purpose or new environment. The concept of adaptation has a critical role in fields as diverse as psychology, economics/optimal planning, machine control, artificial intelligence and computational mathematics.

The adaptation process is largely characterised by a mixture of operators acting on structures. So adaptation, whatever its context, involves a progressive modification of some structure by operators to perform well in the environment confronting it. While the structure is largely determined by the field of study, each field of study is typified as much by its performance measure as by its operators. Table 1 outlines some fields and their corresponding structures, operators and performance measures.

It is important to state that among the specific difficulties resulting from situations that require an adaptation process are;

- The largeness and complexity of the structure, such that there are many feasible sub-structures to be tested and as such it is difficult to determine which sub-structures are responsible for good performance.
- The performance measure is a complicated function (for example it is likely to be non-linear) with many inter-dependent parameters that may vary over time and space.
- The variable characteristics of the environment.

FIELDS	STRUCTURES	OPERATORS	PERFORMANCE MEASURE
Geoinformatics	spatial objects	attributes, constraints, affordances, task	Utility
Genetics	chromosomes	mutation, recombination	Fitness
Control	policies	Bayes' rule, successive approximation	Error function
Physiological psychology	cell assemblies	synapse modification	Performance rate
Game theory	strategies	rule for iterative approximation of optimal strategy	Payoff
Artificial Intelligence	programs	learning rules	Comparative efficiency

Table 1. Outline of fields, structures, operators and performance measures

3.1 Adaptation in a GI Context

Adaptation of geographic information can be seen as an optimization process that enables the provision of objects of high utility that satisfy a user's current situational context. Adaptation of GI can be carried out at data level, at the communications level, at the task-specific level, at the platform level etc. For example, GI can be adapted to a special format, adapted for transmitting over wireless network, adapted to specific device etc. GI is produced and used by people to support better informed and faster decision making: this potential can only be exploited adequately if the purpose (tasks) for which the user needs the data is taken as an important intervening variable (operator) for the optimization process.

Geographic information adaptation methods have been proposed by different authors. The dominant model (paradigm) still concentrates on the aspects of data, although there has been recognition of the importance of user's tasks. Chalmers et al. (2001), in their proposed approach for map adaptation, highlighted user's current task as an important intervening variable, but the focus of their approach was more on the semantics of the data. Reichenbacher (2003) discussing his adaptive concepts for mobile cartography (with an emphasis on cartographic presentation), was among the first to suggest that a good starting point to turn the mobile cartography debate away from being technology-centred is attention to the tasks users perform while being mobile.

This paper agrees with the concept of a task-centred approach to GI adaptation. We assert that adaptation of GI should be seen more in the sense of supporting users in achieving their objectives/tasks as this view of adaptation is much closer to what the user requires. So this approach relates task process to the actionable properties of a geographic dataset, through the concept of affordances, giving effective and unambiguous communication between GI providers and seekers.

3.2 Task-Centred Approach to Adaptation

A great part of our effort to develop a task-centred approach to GI adaptation is derived from the principles of computing science that allow the matching of objects with a set of properties of the desired functionality. The approach can be used in many resource allocation-related applications e.g. the specification of the group of objects (domain) that are of interest to a user, the description of the importance (utility) of each of the items in the domain, and the prioritizing of the data delivery.

3.2.1 Mathematical Framework

We define a function (a mathematical relation such that each element of one set is associated with at least one element of another set) that relates the universe of data objects to some representation of context (in our case the user tasks). For example if Ω be the Universe of data objects (e.g. Ω might consist of all objects retrievable from a database); S be the Domain Set Expression (which results in the domain – the data of interest); U be Utility Equation (the relative value of objects within a particular domain); and UT be the Utility Expression (the method of defining the utility). Therefore,

any domain set expression, s, that is formed according to S, is a predicate,

$$s : \Omega \rightarrow Bool$$

that returns truc of any object in Ω that belongs to the set denoted by s

any equation u formed according to U, is a function,

$$u : \Omega \rightarrow UT$$

that maps objects in Ω to other types of objects in UT

if P be any task-centred profile expressible as $P(S,\ U)$, and D = $\{O_1, \dots\dots, O_n\}$, the set of object it defines, such that for any $O_i \in D$, the utility value of objects O_i is specified by the Utility equation,

$$U(O_i) = W_{i.}$$

Then P can be defined as a function,

$$P: 2^{\Omega} \rightarrow Int$$

that maps any objects to the value of that object within the set of all objects, as specified by the utility equation.

3.2.2 A Syntactic Framework

In order to specify these mathematical relationships effectively, we need to define a syntax which allows us to handle objects, their utility measures, domains, and tasks. Such a generic syntax is language-independent and allows for effective code generation. It was adapted from the initial framework of Cherniack (2003) to include task expression as the main focus.

A Generic Syntax for Specifying the Task-Based Language

APPLICATION : TASKPROFILE IDENT DomainSec UtilitySec END

DomainSec : DOMAIN DomainSet ';' .... ';' DomainSet

UtilitySec : *UTILITY UtEqation ';' .... ';' UtEqation*

DomainSet : *IDENT '=' (ObjectExp)*

UTEqation : *U '(' IDENT ')' '=' UtExp*

UtExp : *INT*
IF Conds THEN UtExp ELSE UtExp
UPTO '(' INT ',' UtExp ',' UtExp ')'

Conds : *SCond*
Conds AND SConds
Conds OR SConds
NOT '(' Conds ')'

SConds : *(TaskExp) Op INT*
(TaskExp) Op (TaskExp)
(ObjectExp) Op (TaskExp)

Op : $=, \neq, <, >, \leq, \geq$

4 Case Study

In evaluating the task-centred approach described above, we used a scenario of adaptation of an environmental sensitivity index (ESI) dataset for oil spill emergency response. Such datasets are widely used in the management of oil spill incidents (Gundlach 2001). To date the datasets have been designed and targeted at emergency managers running static, desktop client systems, and are often used for long-term planning. But during an emergency, the situation is often characterised by activities and locations not amenable to such a desktop information system. In order for on-site workers enabled with mobile devices to benefit significantly from ESI information, to support response decisions at their location, there is a need for optimization (adaptation) of the ESI data content.

Applying the concepts of domain and utility as contained in our framework to this scenario, the first step was to pre-adapt the dataset into different information domains based on the spill management cycle/workflow. Therefore, the classification of domains was influenced by the user's workflow: geo-phenomena come to attention through human activities and salience in the social and physical environment.

Figure 1 shows the relationship between the temporal characterization of activities and a broad classification of users that have need of the ESI dataset and shows the contribution of ESI data to oil spill management. Three parties may have need of the ESI dataset: planners such as emergency managers, the actual frontline responders, and researchers. The result is a matrix where each cell represents a given party's data requirement at each stage of the spill management cycle. This matrix was used to examine the information needs of various parties at various stage of the cycle.

Spill Cycle →

Users	*Pre-*	*During*	*Post*
Planners			
Responders			
Researchers			

Fig. 1. The relationship between activities and users

For example, the shaded cell in the centre of the matrix, which is the focus of our scenario (response by fieldworkers during the oil spill incident), represents the ESI information needed during the incident. It is important to note that there is a significant amount of overlap in the information needed for the different parties during the spill cycle: the matrix highlights that the different information needs of the various parties and the consequent pre-adaptation based on characterisation of their activities. Table 2 and Figure 2 reflect such possible differences.

	PLANNERS	RESPONDERS
Activity characteristics		
Importance of future	High	Low
Extension	Long	Short
Rational	Creative	Disordered
Flow	Slow	Very fast
Density	Many	

Table 2.

Sample of 'pre incident' relevant data *Sample of 'during incident' relevant data*

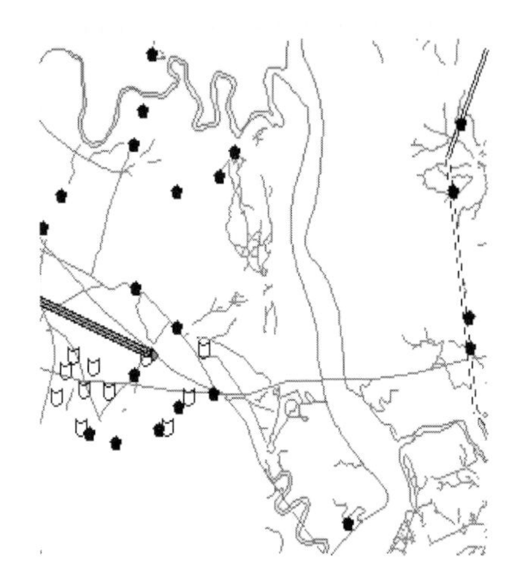

Fig. 2.

To describe the utility of objects within the relevant information space (domain of interest) we had to develop an ESI response task-typology. Developing the task-typology required assessment and understanding of response related tasks, for example what are the objectives underlying the use of the data? what specific tasks are expected to be performed using the data? which are the actions we use to achieve a task? The idea is to include knowledge about common users' objectives, tasks, actions and constraints to handle and enable flexibility, adaptability, and re-use of the content.

Our concept of utility of objects that are goal directed (i.e. those in the relevant domain using which there is the possibility of attaining a defined goal), lies in the correlation between the objects (with their description) and a representation of the user's need for the object. The ESI datasets consists of structured data objects which are geographical features. Features are the basic unit of geographic information storage and exchange within most models of geographic information. One of the products of the Open Geospatial Consortium is the Geographic Markup Language (GML) (OGC 2002), an XML encoding for the transport and storage of geographic information. GML provides support for building feature collections and it describes features as a list of properties. Affordance is one such property and can be defined as what the feature can afford/be used for, and by which action it can participate. It can be modelled as an attribute of the feature class using UML (Booch et al. 1999), and as a feature member property of the feature collection, or in a structured way as a UML association. A system was developed to relate the task characteristics of the task-typology to the affordances possibilities of the relevant features.

<element name="ESIModel"
type="ex:ESIModelType"
substitution-
Group="gml:_FeatureCollection"/>
<element name="River"
type="ex:RiverType"
substitutionGroup="gml:_Feature"/>
<element name="Affordance"
type="ex:AffordanceType"
substitution-
Group="gml:_AssociationAttribute"/>
<complexType
name="ESIModelType">
<complexContent>
<extension base=
"gml:AbstractFeatureCollectionType">
<sequence> ..... </sequence>
</extension>
</complexContent>
</complexType>
<complexType name="RiverType">
<complexContent>
<extension
base="gml:AbstractFeatureType">
<sequence> ..... </sequence>
</extension>
</complexContent>
</complexType>
<complexType
name="AffordanceType">
<complexContent>
<restriction
base="gml:FeatureAssociationType">
<sequence> ..... </sequence>
</restriction>
</complexContent>
</complexType>

Fig. 3. A sample shoreline type from an ESI dataset and its GML schema representation

5. Conclusions

A major challenge in using GI for field based applications, such as disaster response, has been identified. The concepts of domain and the utility adaptation approach were used to show how task-centred GI optimization can be achieved and its potential for facilitating the delivery of relevant content that satisfies user's interests and context.

It is important to note that although most attempts at mobile solutions have focused more on the latest gadgets, communication protocols and semantics of the data, advances in task-centred geographic information optimization will play a significant role in the development of mobile geographic information handling for applications such as disaster management.

Information designers must begin to consider fast-changing working environments and future work must further investigate possible ways of facilitating effective and unambiguous communication between information providers and information users.

References

Baldegger J, Giger C (2003) Wearable GIS: A Smart Assistant in Disaster Management. Paper presented at the 6th AGILE Conference on Geographic Information Science, Lyon, France

Booch G, Rumbaugh J, Jacobson I (1999) The Unified Modeling Language Reference Manual. Addison Wesley, Reading

Brusilovksy P (1996) Methods and Techniques of Adaptive Hypermedia. User Modelling and User Adaptive Interaction 6:87-129

Chalmers D, Sloman M, Dulay N (2001) Map Adaptation for Users of Mobile Systems. Paper presented in the Proceedings of the 10th International Conference on World Wide Web, Hong Kong

Erharuyi N, Fairbairn D (2003) Mobile Geographic Information Handling Technologies to Support Disaster Management. Geography 88: 312-318

Erharuyi N, Fairbairn D, Lakan T (2003) Mobile Handling of Environmental Sensitivity Index (ESI) Dataset. Paper presented in the Proceedings of the 5th International Symposium for Coastal Zone Management, Genova, Italy

Fairbairn D, Erharuyi N (2004) Adaptive techniques for delivery of spatial data to mobile devices. Geowissenschaftliche Mitteilungen 66:11-15

Franti P, Ageenko E, Kopylov P, Gröhn S (2002) Map image compression for real-time applications. Paper presented in the Proceedings of the 10th International Symposium Spatial Data Handling, Ottawa

Gui-lian W, Youg-long L, Jian X (2000) Application of GIS Technology in Chemical Emergency Response. Journal of Environmental Sciences 12:172-178

Gundlach E, Imevbore VO, Witherspoon B, Ainodion J (2001) Incorporating biodiversity into sensitivity maps of the Niger River delta. International Oil Spill Conference, American Petroleum Institute, Washington DC

Gunnes AE, Kovel JP (2000) Using GIS in Emergency Management Operations. Journal of Urban Planning and Development 126:136-149

Hall LS (2002) GIS Aids Oil Spill Response. Maine IS Technology 9

Holland JH (1992) Adaptation in Natural and Artificial Systems. The MIT Press, Cambridge, Massachusetts

Meissner A, Luckenbach T, Kirschner H (2002) Design Challenges for an Integrated Disaster Management Communication and Information System. Paper presented at the 1st IEEE Workshop on Disaster Recovery Networks, New York

OGC (2002) OpenGIS Geographic Markup Language (GML) Implementation Specification, version 2.12 (02-069). Open GIS Consortium Inc, Wayland, MA, USA

Reichenbacher T (2003) Adaptive Methods for Mobile Cartography. Paper presented at the 21st International Cartography Conference (ICC), Durban, South Africa

UNDP (2004) Reducing Disaster Risk: a Challenge for Development. United Nations Development Program, Bureau for Crises Prevention and Recovery, New York

Wai YL, Francis CML (2002) A Context-Aware Decision Engine for Content Adaptation. IEEE Pervasive Computing 1:41-49

The Adoption of Geo-information and Geographic Information Systems for Natural Disaster Risk Management by Local Authorities

Gabrielle Iglesias

National College of Public Administration and Governance, University of the Philippines, Diliman, Quezon City, 1101 PHILIPPINES
Email: gabrielle.iglesias@up.edu.ph

Abstract

The timely availability of relevant information is vital for the operations of local authorities. In the case of dealing with natural disasters like earthquakes and floods, geo-information and geographic information systems (GIS) can be used to improve and organize response and consequently minimize their impact. The paper explores the adoption and use of geo-information and GIS applications within an organisational context bound by legal mandates and official procedures. The paper relies on a socio-technical approach in exploring and understanding this process. This interaction is captured in the organizational routine concept that is a template for understanding mandates and reflects the agreement on the role of the geo-information in organizational activities. Empirically, the paper is based on two local authorities adopting a GIS for natural disaster risk management: Naga City in the Philippines having a flood hazard concern, and Lalitpur Sub Metropolitan City in Nepal having an earthquake concern. The case studies show that the socio-technical interaction of GIS technology for natural disaster management with other organizational needs leads to a continuous re-design cannot be possibly based on pre-given and final users' needs.

1 Introduction

Geo-information needs of local authorities have been expanding with their growing disaster management roles from emergency management towards managing the risk of natural disasters (Cristoplos, 2003; GDIN, 1997; Hoch, Dalton, & So, 2000; Schroeder, Wamsley, & Ward, 2001; UN, 2002; UNDP, 2004). The gradual shifts reflect a recognition of vulnerability as a combination of social, economic and political factors that determine the degree to which someone's life and livelihood is put at risk by a discrete and identifiable event in nature or in society (Blaikie, Canon, & Davis, 1994; Cristoplos, 2003; DMTP, 1994; Durham, 2003; Few, 2003; White, Kates, & Burton, 2001). Local disaster management is now related to planning for sustainable development (DMTP, 1994; Hoch et al., 2000; Pelling, 2004); coordinating disaster management activities with local legislators and community leaders (Cristoplos, 2003; Few, 2003; Sapat, 2001); monitoring the social and economic processes related to poverty, urbanization and underdevelopment that may increase the risk of disaster (Blaikie et al., 1994; Fothergill & Peek, 2004; Kakhandiki & Shah, 1998; Vogel, 2001; White et al., 2001); and dealing with the sometimes conflicting expectations of their citizens on what local authorities should be doing to manage disasters (Few, 2003; Homan, 2003; Mauro, 2004; Morris, 2003; Sjöberg, 2001; Slovic, 2003).

Geo-information is defined for this paper as digital descriptions of geographic locations and characteristics of features, boundaries and phenomena for a given time period. Geographic information (or geo-information) can be obtained from GIS applications for assessing natural hazards have been developed, and specific applications are available for earthquake risk assessment and flooding risk assessment (Carrara & Guzzetti, 1995; Lazzari & Salvaneschi, 1999; Montoya & Masser, *in press;* Todini, 2004; Valpreda, 2004; Zerger & Smith, 2003; Zhang, Zhou, Xu, & Watanabe, 2002). GIS application refers to a computer application for the storage, management, analysis, modeling and mapping of digital spatial data. For the purposes of this study, routine refers to interlocking, reciprocally-triggered sequences of actions of actors who are linked by relations of communication and/or authority.

The important role of GIS was seen in the management of large data sets, the development of models of natural hazards, the assessment of building risk, forecasting casualties, and generating hazard and risk maps.

Organizational factors can affect the adoption of GIS applications and of geo-information within a local authority's planning and decision-making. These factors are part of an implementation process, or the process responsible for transforming the unproven potential of a new GIS application into a taken-for-granted component of the local authority's daily activities (Campbell & Masser, 1995).

Studies of local authorities adopting GIS applications found indications that technical problems tended to reinforce and be reinforced by existing organizational problems (Campbell, 1992; Campbell & Masser, 1995); for example, data consistency, data ownership and control of geographic information are related to issues of scope of activities between bureaus within a local authority. Success of the adoption of GIS by individuals applications was found tied to perceptions of the social and political meaning of the adopted information systems rather than as a particular configuration of equipment, and was therefore improved by individuals' commitment to and participation in the implementation of the system (Campbell, 1992; Campbell & Masser, 1995); by the presence of champions (Campbell & Masser, 1995) and of political support (Budic, 1994); by the presence of a full-time GIS specialist (Budic, 1994); and relevant user training or high trainability (Nedovic-Budic & Godschalk, 1996).

Nedovic-Budic (1998) evaluated studies of the use of GIS by planning agencies and local governments and noted instances when the functionality of GIS for land use planning did not correspond fully to the requirements of planning processes and methods, that the information produced by adopted GIT applications had low impact on planning decisions, and the predominant use of GIT applications was still data processing (storage, retrieval, dissemination, communication). In a later study on the impact of national spatial data infrastructures[1] for local planning, the issue of limited relevance of dataset contents to the mission and needs of local planning practice and limited effects on the decision-making process were again raised (Nedovic-Budic et al., 2004).

If disaster risk reduction components are to be integrated within the mainstream planning and development activities of local authorities, then the geo-information on disaster risk reduction have to be deliberately infused into their targeted development activities, activities that are in fact existing organizational routines. This is not as easy because of the characteristics of organizational routines as emergent from experience rather than explicit decision making, and as having underlying knowledge that is par-

---

[1] An SDI is the set of policies, fundamental data sets, technical standards, physical data access network, and human resources necessary for the effective collection, management, access, delivery and utilization of spatial data at different political/administrative levels (Nedovic-Budic, Feeney, Rajabifard, & Williamson, 2004).

tially inarticulate (the people who execute their roles in routines often do not know the reason for their actions) (Cohen & Bacdayan, 1994).

The problem of low use of geo-information for local disaster risk management therefore refers to how the organization structures its related disaster risk management activities to process geo-information, and if this structuring has an aspect that reinforces geo-information use. Specifically, there is a need for the articulation of the process of redesigning disaster risk management activities, and a formalization of a sub-process for defining geo-information use in terms of procedures, monitoring reports and other reports. Reinforcement of geo-information use through social incentives and disincentives should also be formalized within the process.

Some of these organizational factors that affect GIS adoption may recur for disaster risk management GIS applications. Preliminary visits were made to the two case study sites for this research, and interviews with some local officials have helped to elaborate what organizational factor(s) may be important to the adoption of the GIS applications. The two sites were selected because they signified their intention to have a new GIS application, signified disaster management as one of the reasons for obtaining the GIS application, and represent different levels of experience with GIS. The two sites also come from different political contexts, have different degrees of power and financial independence from their national government, will be adopting different GIS applications, and will represent different natural disaster risks.

2 Organizational Routines

In this paper, capacity for natural disaster risk management is studied in terms of the official routines that the local authority attempted to use with the technology for disaster management, whether or not the attempt was successful. Levitt & March (1996) gave reasons why routine is a good focal point for studying whether an organization is learning:

- Behavior in an organization is based on routines.
- Organizational actions are history dependent. Routines are based on interpretations of the past, respond to feedback, and are changed only incrementally.
- Organizations are oriented to targets. Organizations behave according to relations between observed outcomes and their aspirations for those outcomes.

Routine is a generic term referring to structures, rules, procedures, strategies and technologies that an organization operates to perform certain functions. It also refers to the culture, paradigms and knowledge that support them. A routine requires a minimum level of one or more skills in order to perform them. A routine is considered an organizational resource because from the perspective of evolutionary economics, the organization's knowledge assets are embedded in routines (Szulanski & Jensen, 2004).

More than a resource, it is the official means for a local authority to convert its skills, knowledge and resources into an action. For this research, routine refers to a collective set of official actions whose purpose is to directly serve a mandate (such as the procedure of a safety inspection), not the small, intermediate steps not directly linked to a mandate (such as sending out notifications of safety inspections). Routine serve as the indicator that capacity is being developed. Whether it is the skills, knowledge or resources of the local authority that are targeted for improvement, they have to be part of a routine. There is no point in giving training on how to use a GIS or DSS for natural disaster risk management if there are no related official procedures that reinforce its use. Specifically, the local authority needs a routine for embedding its GIS knowledge and experience, ready to be pulled out and used in response to situations that call for it, limiting its choices, reduces complexity, and saves time and effort (Feldman & Pentland, 2003; Levitt & March, 1996; Pentland & Rueter, 1994; Scott, 1990). It is a mechanism for learning GIS routines of other organizations (Levitt & March, 1996), for creating new routines (Argyris, Putnam, & Smith, 1985; Feldman & Pentland, 2003; Szulanski & Jensen, 2004), and possibly for creating knowledge out of separate knowledge fields (Hendriks, 2000; Scott, 1990; Sproull & Goodman, 1990; Weick, 1990).

The routine is modeled as having two aspects of pattern and of action (Feldman & Pentland, 2003; Pentland & Rueter, 1994). The pattern is the abstract or schematic form of a routine. Official procedures are a good example of the pattern aspect, and represent the compromise among actors on how a routine should be structured. The action is the observable aspect, and involves a greater number of actors, and this is where the conventional concept of "end-user" of an information system can have an influence on the use of the adopted GIS or DSS.

3 Case Studies

Field visits were made to the two case study sites in Nepal and Philippines (Latitpur SMC and Naga City) for this research, and interviews with some local officials have helped to elaborate what organizational factor(s) may be important to the adoption of the GIS applications. The two sites were selected because they signified their intention to have a new GIS application, signified disaster management as one of the reasons for obtaining the GIS application, and represent different levels of experience with GIS. The two sites also come from different political contexts, have different degrees of power and financial independence from their national government, will be adopting different GIS applications, and will represent different natural disaster risks. The interviewed officials were identified as actors with roles in the adoption of the applications. During the interviews, the organizational routine emerged as the concept by which actors framed their needs for the GIS application being adopted.

Lalitpur in Nepal intends to set-up their first GIS Laboratory to support urban development, infrastructure development, city administration, disaster management and preparedness. Lalitpur described the environment for the GIS as: lack of data and information, existing data is disorganized/unmanaged, no digital data sets, lack of awareness of GIS among decision makers, lack of adequately trained staff, and lack of equipment. Lalitpur originally intended to adopt GIS for earthquake safety, but they do not have an earthquake safety regulatory activity other than inspecting blueprints of proposed buildings. They face a risk of an earthquake strong enough to damage much of their buildings, but it does not occur frequently. From the point of view of the local authority, using the GIS for urban planning therefore had a stronger basis than for earthquake safety. Lalitpur is still new to GIS, and they see its usefulness in planning routines for different capacities, and less with earthquake safety regulation, and they can see the value of resources such as digital data sets and trained personnel.

Naga City local authority in the Philippines adopted its GIS for urban flooding mitigation in 1994. It has an annual flooding problem that affects its poorest communities and 90% of its investments. It has used its basic GIS to map floods, identify communities at risk, select sites that need elevated footpaths and buildings, and to develop their disaster mitigation plan, but there appears to be difficulty in connecting the use of geo-information with other flood management activities. One point is that the GIS application is used to generate thematic maps when the database is updated, and then to convert the most commonly requested maps as image files (pdf format) that are printed using other applications. Second, the city planning and city engineering offices, whose functions encompass the development planning and infrastructure planning related to urban flooding mitigation, do not develop additional applications, possibly due to unwillingness of current employees even when the heads of office are supportive of the GIS. Third, in a SWOT analysis of the floodwaters drainage, the use of GIS to address the drainage problem did not arise from the key members of the local authority, possibly indicating an inability to recognize when a GIS application can be made. Fourth, the GIS is not yet developed for important disaster risk management activities such as the construction of drains, the maintenance of the drainage system, and monitoring land use change that can affect flooding. Finally, the mayor would like to use the GIS application for political lobbying, to gain cooperation from communities and other local authorities upstream of the river systems whose activities exacerbate the flooding in Naga City, and as leverage against national agencies whose infrastructure plans may worsen the flooding in the city.

Naga City has ten years of experience, but there is an uneven use of GIS among its offices. The Planning Office does not use GIS directly, have not used maps other than those they used before the GIS was established, and the City Planner sees it as a generator of maps useful for planning and monitoring. The Naga City Engineering Office had some skills at using GIS, and the City Engineer sees GIS as an engineering tool, and wants to develop more applications relevant to their work if they had the staff with appropriate training. The head of the Naga EDP had staff with the most computer skills, had been operating a GIS application since 1994, have difficulty developing GIS useful applications, and would rather have the end users develop their applications. In this case, skill level is not the reason for the uneven use of GIS, because the EDP who develops applications has no use for it in their work.

Both Planning and Engineering offices in Naga City have planning routines but do not have application development routines or a planning GIS. On the other hand, the EDP has an application development routine but not a planning routine. While it was perhaps "easier" for the GIS to remain in EDP, it does not have a planning routine, and as a consequence the EDP staff does not have the urban flooding mitigation knowledge or planning skills required developing useful applications.

The Mayor's actions to ease pressures of in-migration or deforestation, addressing underlying causes like uneven development is consistent with the Pressure-and-Release model's approach of reducing the social production of vulnerability. These approaches are not explicitly part of the disaster management cycle, but are creative and pragmatic responses to a problem that requires multi-actor cooperation to solve.

The Mayor would probably benefit from a system that can help him lobby, convince, and cooperate with other agencies of national government, other local governments, and even with an NGO. It is reasonable that he would like a system that is not limited, either by geography to the city boundaries, or by administrative extent to the city's planning functions alone, but facilitates his interaction with co-actors in disaster mitigation.

The two sites are dissimilar in amount of experience with GIS and access to resources such as digital data and trained staff. Lalitpur SMC is just beginning to use GIS. Naga City has had ten years of working with its urban flooding mitigation GIS application, but is not yet able to expand the use of their GIS application to other flood mitigation activities related to development planning, and its flooding mitigation activities continue without the use of GIS. The low level of use can be related to the level of technical training available among the supporting staff, but a local authority such as Naga City must manage with whatever existing resources they have and with whomever in the staff are willing to learn how to use the application. What both local authorities did have are officials with knowledge of how their organization operates, experience with the recurring flooding problem, and varying degrees of appreciation of the usefulness of GIS. From each preceding case, the local authorities always view GIS technology in terms of its potential for doing a routine better.

4 Conclusions

Organizational concerns are acknowledged to have a potential effect on the quality of a disaster risk management information system's design, architecture and implementation. What is difficult to answer is why information from a GIS application is not used even when the GIS design; system components and user training are adequate. It is hypothesized that a local authority needs direct experience in converting map outputs into disaster risk management information.

In answering this question, the organizational routine is a proposed mechanism for identifying organizational-level user needs and for developing use for geo-information. It is also a mechanism for learning GIS routines of other organizations (Levitt & March, 1996), for creating new routines (Argyris et al., 1985; Feldman & Pentland, 2003; Szulanski & Jensen, 2004), and possibly for creating knowledge out of separate knowledge fields (Hendriks, 2000; Nyerges, 1995; Scott, 1990; Sproull & Goodman, 1990; Weick, 1990) to include disaster risk management. The adopted GIS or DSS can spark new routines and consequently new user needs, as is the case with the new lobbying needs from the mayor of Naga City. This indicates that the GIS can be in a continuous and continual process of development as the usability of the system grows with the users' appreciation of its geo-information output or decision support.

GIS routine formation is a good focusing point because it involves the creation of agreement about what disaster management task has to be done, what is the legal basis for the task, how to do it, how the organization recognizes the spatial aspect or spatial problem within a routine for natural disaster risk management, and how geo-information will be used in this planning and decision context.

References

Argyris C, Putnam R, Smith DM (1985) Action Science: Concepts, Methods, and Skills of Research Intervention. San Francisco: Jossey-Bass, Inc., Publishers

Blaikie P, Canon T, Davis I (1994) At risk : natural hazards, people's vulnerability and disasters. London: Routledge

Budic ZD (1994) Effectiveness of Geographic Information Systems in Local Planning. Journal of the American Planning Association, 60(2), pp 244-263

Campbell H (1992) Organisational issues and the implementation of GIS in Massachusetts and Vermont: some lessons for the United Kingdom. Environment and Planning B: Planning and Design, 19, pp 85-95

Campbell H, Masser I (1995) GIS and Organizations. London: Taylor and Francis Ltd.

Carrara A, Guzzetti F (eds.) (1995) Geographical Information Systems in Assessing Natural Hazards. Dordrecht: Kluwer Academic Publishers

Cohen MD, Bacdayan P (1994) Organizational Routines are Stored as Procedural Memory: Evidence from a Laboratory Study. Organization Science, 5(4), pp 54-568

Cristoplos I (2003) Actors in Risk. In Pelling M (ed.), Natural Disasters and Development in a Globalizing World (pp. 95 to 109). London: Routledge

DMTP, U. (1994). Disaster Mitigation (2nd ed., pp 68)

Durham K(2003) Treating the Risk in Cairns. Natural Hazards, 30, pp 251-261

Feldman MS, Pentland BT (2003) Reconceptualizing Organizational Routines as a Source of Flexibility and Change. Administrative Science Quarterly, 48, pp 94-118

Few, R. (2003). Flooding, vulnerability and coping strategies: local responses to a global threat. Progress in Development Studies, 3(1), pp 43-58

Fothergill A, Peek LA (2004) Poverty and Disasters in the United States: A Review of Recent Sociological Findings. Natural Hazards, 32, pp 89-110

GDIN (1997). Harnessing Information and Technology for Disaster Management: Global Disaster Information Network

Hendriks PHJ (2000) An organizational learning perspective on GIS. International Journal of Geographical Information Science, 14(3), pp 373-396

Hoch CJ, Dalton LC, So FS (eds.) (2000) The Practice of Local Government Planning (3rd ed.). Washington D.C.: ICMA

Homan J (2003) The social construction of natural disaster: Egypt and UK. In M. Pelling (Ed.), Natural Disasters and Development in a Globalizing World, pp. 141-156, London: Routledge

Kakhandiki A, Shah H (1998) Understanding time variation of risk: Crucial implications for megacities worldwide. Applied Geography, 18(1), pp 47-53

Lazzari M, Salvaneschi P (1999) Embedding a Geographic Information System in a Decision Support System for Landslide Hazard Monitoring. Natural Hazards, 20(2-3), pp 185-195

Levitt, B, March JG (1996) Organizational Learning. In Cohen MD, Sproull LS (eds.), Organizational Learning. Thousand Oaks: Sage Publications.

Mauro A (2004) Disaster, Communication and Public Information. In Casale R, Margottini C (eds.), Natural Disasters and Sustainable Development (pp. 239 to 248). Berlin: Springer-Verlag

Montoya L, Masser I (in press). Management of natural hazard risk in Cartago, Costa Rica. Habitat International, In Press, Corrected Proof

Morris A (2003) Understandings of Catastrophe: The landslide at La Josefina, Ecuador. In Pelling M (ed.), Natural Disasters and Development in a Globalizing World, pp. 157-169, London: Routledge

Nedovic-Budic Z (1998) The impact of GIS technology. Environment and Planning B: Planning and Design, 25(5), pp 681-692

Nedovic-Budic Z, Feeney M-EF, Rajabifard A, Williamson IP (2004). Are SDIs serving the needs of local planning? Case study of Victoria, Australia and Illinois, USA. Computers, Environment and Urban Systems, 28, pp 329 to 351

Nedovic-Budic Z, Godschalk DR (1996) Human Factors in Adoption of Geographic Information Systems: A Local Government Case Study. Public Administration Review, 56(6), pp 554 to 567

Nyerges TL (1995) Cognitive Issues in the Evolution of GIS User Knowledge. In . Nyerges TL, Mark DM, Laurini R, Egenhofer MJ (eds.), Cognitive Aspects of Human-Computer Interaction for Geographic Information Systems, pp 61-74 Dordrecht: Kluwer Academic Publishers

Pelling M (2004) Paradigms of Risk. In M. Pelling (Ed.), Natural disasters and development in a globalizing world, pp 3-16, London: Routledge

Pentland BT, Rueter HH (1994. Organizational Routines as Grammars of Action. Administrative Science Quarterly, 39(3), pp 484-510

Sapat A (2001) The Intergovernmental Dimensions of Natural Disaster and Crisis Management in the United States. In Farazmand A (ed.), Handbook of Crisis and Emergency Management, pp 339-356, New York: Marcel Dekker, Inc.

Schroeder, A., Wamsley, G., & Ward, R. (2001). The Evolution of Emergency Management in America: From a Painful Past to a Promising but Uncertain Future. In A. Farazmand (Ed.), Handbook of Crisis and Emergency Management (pp. 357 to 418). New York: Marcel Dekker, Inc.

Scott, W. R. (1990). Technology and Structure: An Organizational-Level Perspective. In P. S. Goodman, L. S. Sproull & associates (Eds.), Technology and Organizations (pp. 109 to 143). San Francisco: Jossey-Bass.

Sjöberg, L. (2001). Political decisions and public risk perception. Reliability Engineering and System Safety, 72, 115 to 123.

Slovic, P. (2003). Going Beyond the Red Book: The Sociopolitics of Risk. Human and Ecological Risk Assessment, 9, 1 to 10.

Sproull, L. S., & Goodman, P. S. (1990). Technology and Organizations: Integration and Opportunities. In P. S. Goodman, L. S. Sproull & associates (Eds.), Technology and Organizations (pp. 254 to 265). San Francisco: Jossey-Bass.

Szulanski, G., & Jensen, R. J. (2004). Overcoming Stickiness: An Empirical Investigation of the Role of the Template in the Replication of Organizational Routines. Managerial and Decision Economics, 25, 347 to 363.

Todini, E. (2004). FLOODSS: A Flood Operational Decision Support System. In R. Casale & C. Margottini (Eds.), Natural Disasters and Sustainable Development (pp. 53 to 64). Berlin etc.: Springer.

UN. (2002). Living With Risk, A global review of disaster reduction initiatives (preliminary version ed., pp. 387).

UNDP. (2004). Reducing Disaster Risk: A Challenge for Development (No. 92-1-126160-0). New York: UNDP.

Valpreda, E. (2004). GIS and Natural Hazards. In M. Pelling (Ed.), Natural disasters and development in a globalizing world (pp. 373 to 385). London: Routledge.

Vogel, R. M. (2001). Disaster Impact upon Urban Economic Structure: Linkage Distruption and Economic Recovery. In A. Farazmand (Ed.), Handbook of Crisis and Emergency Management (pp. 69 to 90). New York: Marcel Dekker, Inc.

Weick, K. E. (1990). Technology as Equivoque. In P. S. Goodman, L. S. Sproull & a. associates (Eds.), Technology and Organizations (pp. 1 to 44). San Francisco: Jossey-Bass.

White, G. F., Kates, R. W., & Burton, I. (2001). Knowing better and losing even more: the use of knowledge in hazards management. Global Environmental Change Part B: Environmental Hazards, 3(3-4), 81-92.

Zerger, A., & Smith, D. I. (2003). Impediments to using GIS for real-time disaster decision support. Computers, Environment and Urban Systems, 27, 123 to 141.

Zhang, J., Zhou, C., Xu, K., & Watanabe, M. (2002). Flood disaster monitoring and evaluation in China. Global Environmental Change Part B: Environmental Hazards, 4(2-3), 33-43.

Web-Based 3D Visual User Interface to a Flood Forecasting System

Mikael Jern

ITN, Campus Norrköping, Linköping University, SE – 601 74 Norrköping, Sweden.
Email: mikael.jern@telia.com

Abstract

Linkoping University (LiU) has developed a Visual User Interface and Web-enabled advanced 3D visualization to a flood forecasting system in an EC funded project named MUSIC. The project develops state-of-the-art precipitation estimation algorithms, assess their uncertainty and use an innovative combination of the output data of the three independent data sources radar, satellite and rain gauges. The basic role of any real-time quantitative precipitation and flood forecasting system lies in its capability, within the forecasting horizon, of assessing and reducing the uncertainty in forecasts of future events in order to allow improved warnings and operational decisions for the reduction of flood risk. In line with this requirement, the MUSIC project is to develop an innovative technique for improving the weather radar, weather satellite and rain gauge derived precipitation data, taken as independent measurement source. Another key objective is seeking to improve the communication and the dissemination of results to the authorities involved in real-time flood forecasting and management. The project tests the system on real world case studies on two test catchments the Reno and Arno rivers in Italy. LiU is developing the innovative 3D Visual User Interface that will enable the users to take a more active role in the process of visualizing and investigating flood forecasting, allowing them to better understand the data and uncertainty behind the forecasting system. LiU also provides easy-to-use collaborative visualization tools enabling the meteorologists, hydrologists, operations managers and the civil defense manager to view and discuss the forecasting results in real time across the network before finally interacting with the media, the police, other officials and the public. The system will considerably improve the flash flood forecasting reliability and precision and will shorten the time required to detect events that lead to catastrophic flood events.

1 Introduction

The problem of flooding is as old as time. However, while natural flooding of large areas did not create situations more dangerous than others in a prehistoric world, the expansion of human activity and cities has made preventing damage caused by floods or harnessing over-bank flows for one's own purposes as in ancient Egypt a necessity that remains vital to this day. Given the large number of high risk situations as well as the extremely high costs in terms of casualties and damages involved, it can be reasonably estimated that a flood and related phenomena DSS, which combines Visualization and GIS capabilities with modeling for the planning and the preparation of risk maps, emergency plans and the real time analysis of possible interventions, has a large potential market.
Linkoping University (LiU) has developed a Visual User Interface and Web-enabled advanced 3D visualization to a flood forecasting system called the "FloodViewer". This is a close collaboration with in an EC funded project named MUSIC with Italian partners University of Bologna, environmental technology supplier ET&P, institute CNR, University of Newcastle and Gematronik Germany. LiU provide the innovative visualization expertise while Bologna focuses on the precipitation estimation al gorithms. MUSIC was completed September 2004.

The basic role of any real-time quantitative precipitation and flood forecasting system lies in its capability, within the forecasting horizon, of assessing and reducing the uncertainty in forecasts of future events in order to allow improved warnings and operational decisions for the reduction of flood risk. A key objective here is seeking to improve the communication and the dissemination of results to the authorities involved in real-time flood forecasting and management. The system is tested on data from the Arno river in Toscana, a river well known from its many flooding.

The FloodViewer will enable the users to take a more active role in the process of visualizing and investigating flood forecasting, allowing them to better understand the data and uncertainty behind the forecasting system. FloodViewer also provides easy-to-use collaborative visualization tools enabling the meteorologists, hydrologists, operations managers and the civil defense manager to view and discuss the forecasting results in real time across the network before finally interacting with the media, the police, other officials and the public through dynamic documents.

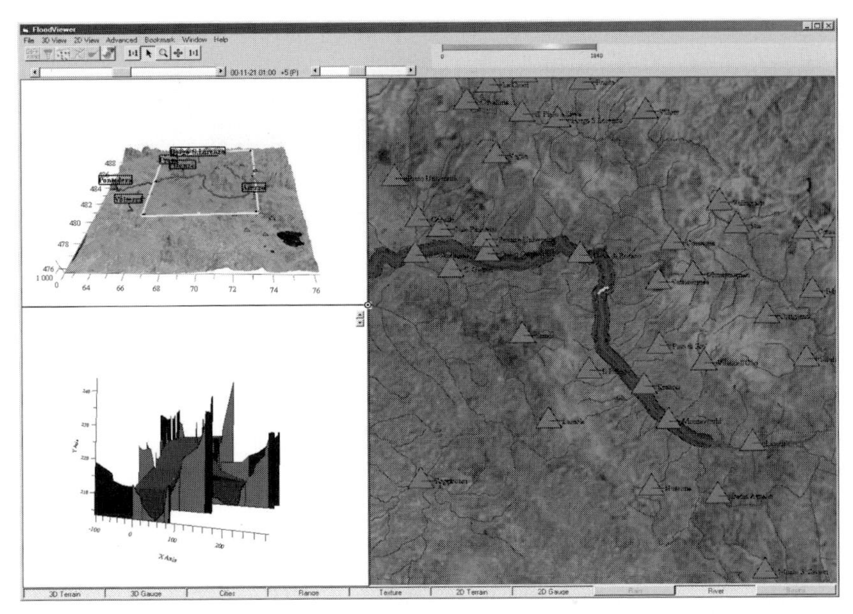

Fig. 1. Visual flood forecasting system - FloodViewer. Data from the Arno river region (Toscana, Italy) is showed in linked and coordinated 2D and 3D windows. The 3D terrain display provides full viewing control, light control, and "focus-and-context" selection. The 2D display shows animated rainfall, gauge stations and water level. With a mouse-move, the user can easily zoom in-and-out or scroll the 2D or 3D display

These are, in turn, made available to the *operations managers*. In order for the operations managers to meet both their own needs and other requests interrogate the data, models and decision support tools. Relying on this, the operations manager may issue warnings to the *civil defense manager*. The civil defense manager interacts with the media, the police, other officials and the public, both directly and indirectly. This forms the *social subsystem* of the overall total warning system that will imply the need for the acknowledging of the following lessons from the sociology of hazards, which are important for the design and operational success of the technological subsystem of the database.

The flood forecasting decision management system can be summarized: The *technological subsystem* is run by meteorologists and hydrologists and consists of precipitation measurement systems, river gauges, and meteorological forecasting systems, which together provide data for hydrological and flood-forecasting models, and decision support tools. The visual-user-interface presents clear, unambiguous predictions and courses of action.

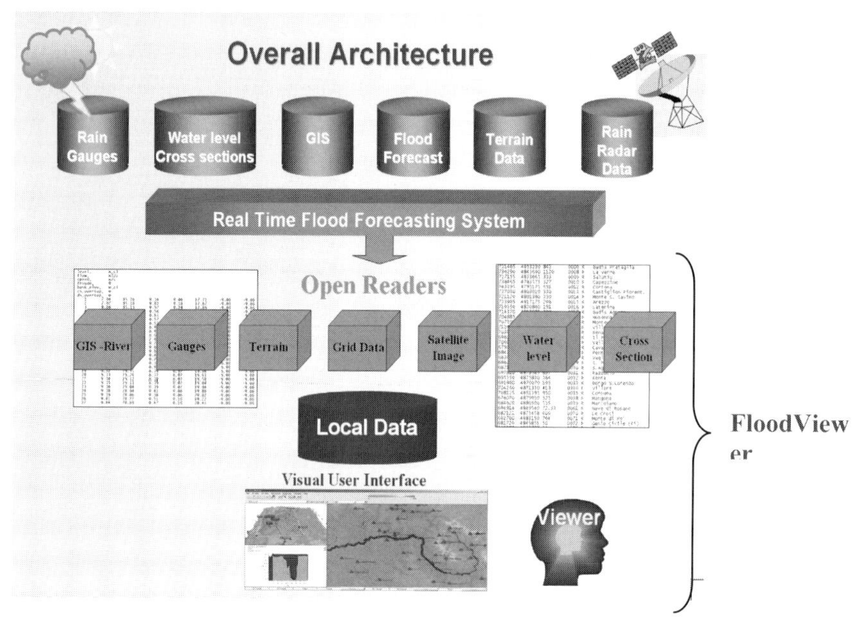

Fig. 2. The overall architecture of FloodViewer. The MUSIC flood forecasting data is transferred to a local data file through AVS Open Reader components. This feature gives maximum flexibility to use FloodViewer in other Case Studies and provide real-time visualization

2 A Visual Approach to Better Forecasting Decisions

Instead of a large general-purpose GIS application that can do everything, our application is based on individual functional components. A set of components, each one performing a specific task, can easily be put together in order to create efficient tailor-made flood forecasting applications. The FloodViewer provides interactive, geo-visualization examination of flood forecasting data, DTED, radar, satellite and GIS data. Visualization methods supported include: 3D surface, 3D isoline, 2D image, 2D contour, 2D isolines, 4D contour - superimposed raster data, vector data (river, basin), bar chart, axis, direct zooming and panning, data-correlated windows.

Our use of Web-based layered atomic component architecture (figure 3) offers the ability to "re-use" low-level algorithmic functions to easily produce application specific components for integration into custom user interfaces. The user interface and low-level component integration is done with Visual Basic development tools. Importantly, the chosen Web architecture will offer full collaborative interaction and the ability to use low-cost PC platforms.

Layered Component Architecture

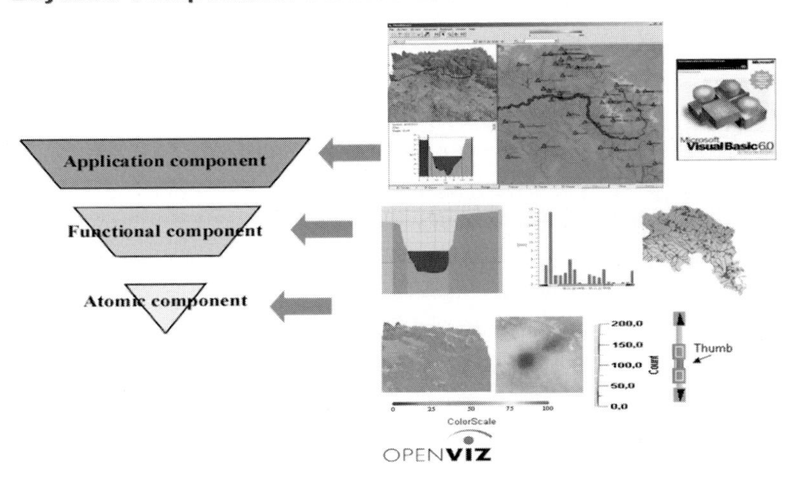

Fig. 3. Web-based layered component architecture is the foundation for a client-based application and allows integration with geo-computational methods such as spatial statistics, classification or even data mining

This approach to system architecture could, for the first time provide the infrastructure for affordable climate and flood forecasting system interpretation facilities. The open component architecture allows integration with geo-computational methods such as spatial statistics, classification or even data mining.

FloodViewer addresses the following three application scenarios:

(1) *Analyze and Explore* – Standalone single user version
(2) *Collaborative Analysis* – Network version
(3) *Presentation and Journaling* – Dynamic document - ActiveX/COM components

In (1), the user analyses and explores the data and prepares the journal (3) for review by other users. This step includes setting bookmarks and annotations of interesting discoveries for embedding into documents and later viewing. During the analyze process the user can also start a collaborative session (2) to share insights with other users. The collaboratory network version includes a collaborative environment where two or more users can meet together in cyber space displaying and interacting with the same dataset remotely. Bookmarks and annotations can also be used in the collaborative session. In the final step, an ActiveX component (3) is integrated into an electronic document such as Microsoft WORD with bookmarks and annotation data.

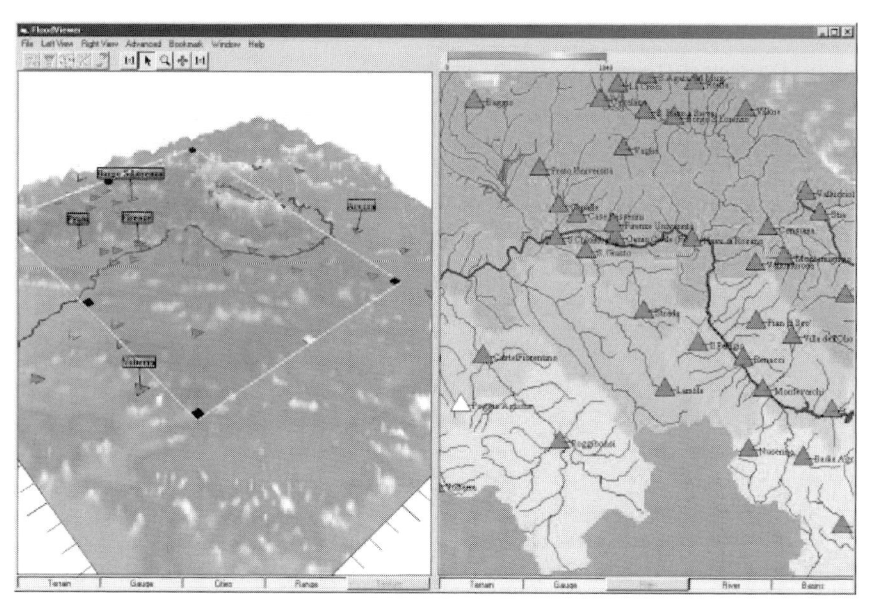

Fig. 4. Example of the Visual User Interface showing cropping of landscape supported in the 3D view with interactive movable handles. The cropped area is shown in the 2D view (right)

2.1 Visual User Interface

Of central importance to the MUSIC project is the Visual User Interface (VUI) that enables the user to directly manipulate graphical objects, which respond interactively and immediately to the user's input actions. The user interacts with both 2D and 3D objects with an immediate graphical feedback response, without the need for moving to a secondary menu window. Example of direct manipulation features: "area-of-interest selection" in 3D landscape to be viewed in the 2D display (figure 1), direct zooming and panning with mouse movement, click on gauge station or sub basin, full-featured scroll bars (see figure 4 left), drill down through forecasting data sets, dynamic control of 2D opacity planes "Z-thru" (figure 4 right), and interactive animation of rain data.

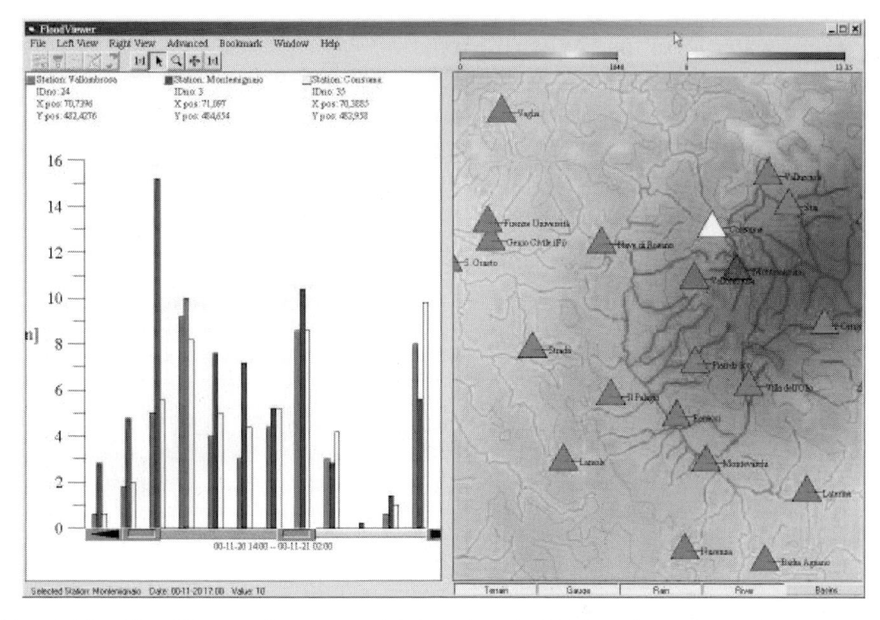

Fig. 5. The FloodViewer interacts with hourly rainfall data from Arno river (period between 20 November 2000 14:00 and the 23 November 2000 14:00). The animated rain data is superimposed on the terrain using opacity blue coloring. The user has selected the Montemignaio Gauge Station by picking on the "red triangle" in the 2D Contour (right window). A 2D bar Chart appears in the second window (left window) and the selected Gauge Station and bar are highlighted. Additional gauge stations can be viewed simultaneously

Our components consistently implement a Model-View-Controller architecture. This implies a separation between the data and the views and analysis of those data. This simple and design enables users achieve a high degree of interactivity with multiple graphs that visualize the same dynamic data source in multiple 2D or 3D views. The FloodViewer components allow you to create multiple views on multidimensional data sets. Since each of these view shows the same data, but is otherwise independent form the other views, you can change the setting for each individual view to highlight a different aspect of your data.

2.2 Water Levels

The FloodViewer has interactive support for water level visualization.

- 2D and 3D cross section charts – move along river
- Select river segment to be viewed
- Alerts for high water levels in red
- Set water level for a What-if Scenario
- Range of time includes both history and forecasts

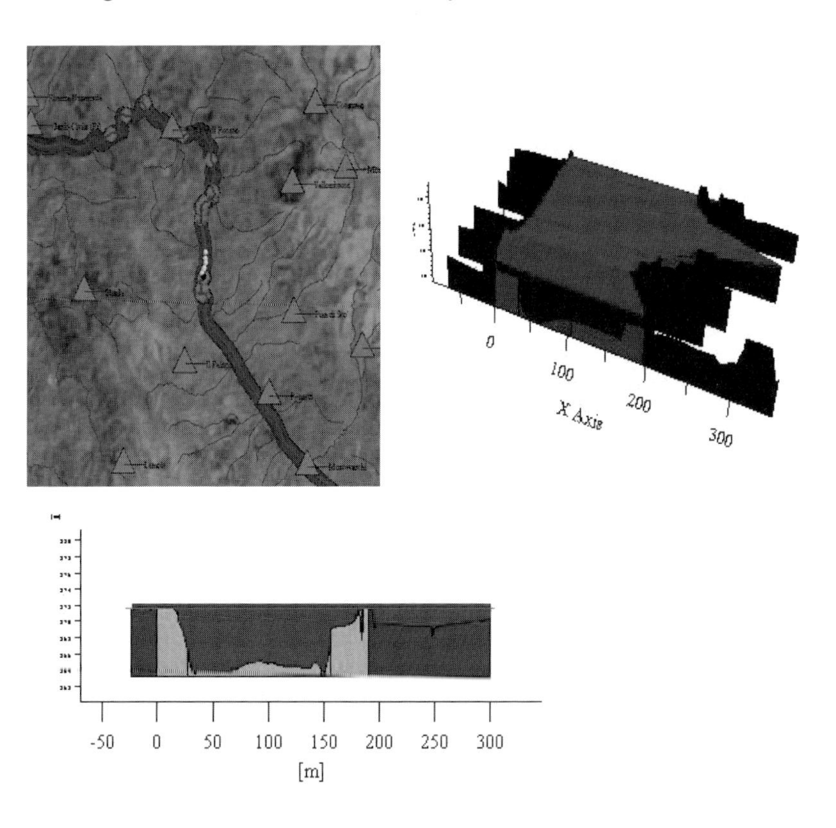

Fig 6. Example of Water level VUI showing "alerts for high water level" in red color along the river segment. The water section can be viewed as 2D images or in a 3D cross section chart. The white dots in the left 2D view show the location of the corresponding 10 3D cross sections (right picture)

2.3 Bookmarks

A "bookmark" can help the analyst to highlight data views of particular interest and guide other users to certain discoveries of a case study. Colleagues can use these descriptive bookmarks to quickly locate key information by simply selecting the view they need. The *Bookmark Manager* remembers and records the status of a data navigation experience. The analyst has selected suitable data dimensions, display properties, filtered data with the slide rangers focusing on the data-of-interest and finally highlighted the "discovery" from a certain angle (viewing properties) and can now save this status "bookmark" in an external file. A bookmark includes visualization attributes such as the time step, viewing matrix, color scale, annotation, etc. When a new user initiates the FloodViewer, it will then start the VUI based on the analyst's bookmarks. The visualization will revert to exactly the same status as defined by the analyst. Users can, step-by-step, follow the analyst's way of work and understand how the results were achieved by.

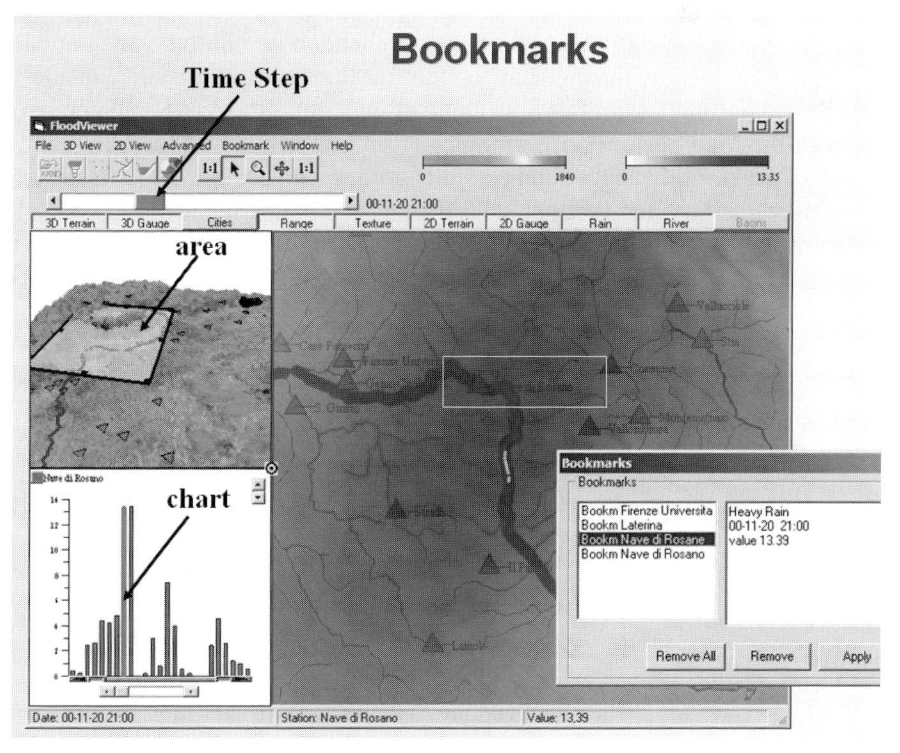

Fig. 7. A "bookmark" can help the analyst to highlight data views of particular interest and guide other users to this discovery. The interactive image above highlights the rainfall data at a selected gauge station "Nave di Rosane" for a certain time step

The analyst can also use the bookmarks to create an interactive and dynamic "story telling" path. This non-linearity of an interactive discovery path leads to a higher responsibility for the analyst to provide powerful and efficient navigation functionality. Due to the possibility to generate dynamic "story telling" paths, the analyst can offer appropriate semantic information for information retrieval. Several associated bookmark files can be provided.

3 Plug-In Viewer

The FloodViewer is sharing a plug-in similar to Adobe PDF documents and Adobe Reader. Our plug in "Viewer" is responsible for visualization, interaction and rendering at the client-side (figure 8). The Viewer is a central part of the collaborative process and is a "freeware" to allow exchange of visualization and VUI components embedded in a FloodViewer application. The Viewer manages all of the 2D and 3D visualization inside the view including taking full advantage of managing the complex interactions with the high-performance graphics layer in OpenGL. The Viewer architecture will allow lightweight applications such as the FloodViewer to be deployed across the Internet. The Viewer that is downloaded only once has a size of about 8Mb, while the embedded visualization components have a small footprint 200-600Kb.

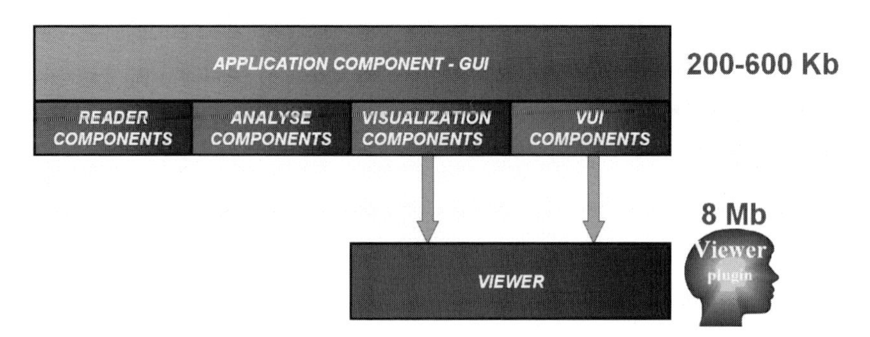

Fig. 8. The FloodViewer and application components share a single Viewer plug-in responsible for client-side rendering of 2D and 3D visualization and VUI interaction. The Viewer architecture will allow lightweight visualization components to be deployed across the Web

4 Multiple Visualization Components

Flood forecasting data are not best analyzed through the use of a single type of graph. In order to detect complex patterns within the data it is necessary to view it through a number of different visualization methods, each of which is best suited to highlight different patterns and features. It's also necessary to achieve balance between 2D and 3D functionality. Some problems are best solved with one or the other, however many require the services of both technologies. In the FloodViewer, each 3D visualization tool can have a 2D counterpart and the object-oriented nature of the technology ensures that most functionality is shared between the two. By using the two together, the power of each is amplified (figure 1, 4 and 5).

The simplest manner in which to employ one or more graphs simultaneously is to view them side-by-side. However, the *context* of each point is lost in the process. One point may be salient in one graph, but may not be identifiable in another. Only through interaction may points like these be located and investigated. The user may select a gauge station that appears interesting in the 2D image chart to see detailed rainfall data in the bar chart. The converse may occur as well. This point may then be deselected from the bar chart and the impact upon analysis performed on the data set as a whole visualized in an entirely different graph. Therefore, the data, analysis and visualization ' flow' together in a seamless process of discovery. For example, figure 5 demonstrates correlation between rainfall data visualized as both 2D animated contouring (right picture) and bar chart with detailed hourly rainfall that identifies how they relate to the each other.

5 FloodViewer Dynamic Document

Projects in flood forecasting require permanent exchange of information between all involved experts, normally working at different locations and having different specializations, e.g. are meteorologists and hydrologists, operations managers, government officials, media, the police, other officials and the public. Traditional reports are characterized by the paper medium, presented in hierarchical chapter structures, passive for the reader and normally restricted to static items such as text, imagery and sometimes animations. Integrating geo-visualization methods with innovative documentation technology can increase the quality and provide better understanding. Dynamic and interactive documents incorporate not only text and images but also the entire interactive data visualization and navigation process (see figure 9, 10 and 11).

A "FloodViewer Dynamic Document" allows an author to produce electronic documents that collaborate and share flood forecasting data, analysis, visualization parameters and insight while distributed over Internet, using intuitive visual navigation techniques. It also brings into focus the question of how to present and evaluate the various forms of uncertainty in forecasts. A dynamic document includes tools enabling scientists to discern, visualize and concretize already discovered patterns and/or new, earlier overlooked ones to provide better, more accurate interpretations of flood forecasting issues.

For example, a flood-forecasting document reports an interactive rainfall scenario for a certain time period. Target groups for example are the project engineers and the decision-makers of flooding operations. The engineers normally require detailed information, e.g. single flow states of physical calculations. The operation management needs condensed information to support their decisions and may issue warnings to public. So both target groups need different information, which either cannot be documented in one single traditional document, or which restricts the readability due to the linearity in documents containing information for more than one target group (figure 10).

The author sets "bookmarks" that highlight data discoveries of particular interest to different readers, who use these descriptive bookmarks to quickly locate key information. The author has selected suitable data dimensions, display properties, zoomed data focusing on the risk area and highlighted the "discovery" from a certain view. The bookmarks are stored as part of the document. When the reader opens the document, it will start the 3D interactive visualization process based on the author's bookmarks. The visualization will revert to exactly the same status as defined by the author. A special "link component" allows the author to link text to defined bookmarks. The author can also use the bookmarks to create a dynamic reading path.

- Dynamic documents use the advances in modern information and communication technology to realize dynamic visualization and interactive editing possibilities even for the reader. The quantity of presented information, visualization, analysis and data structure components are structured in an interconnected document (figure 8). A dynamic document uses Microsoft's Visual Basic, Office2000 Word and ActiveX components as the foundation for a dynamic exploration. These components can be distributed over the Internet embedded in an electronic document, which offers possibilities for efficient and free usage.

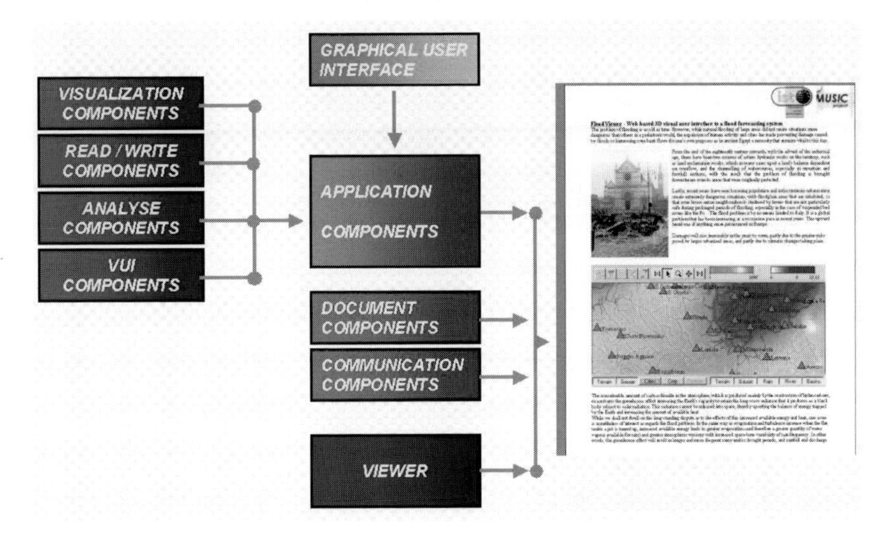

Fig. 9. A set of task-specific functional components (green) is assembled into application components together with the GUI (orange). The Viewer plug-in (blue) allows a client to view 2D and 3D visualization. Document components (red) provide the foundation and tools for integrating the application component, data, hyperlinks and bookmarks into a Dynamic Document or "SmartDoc"

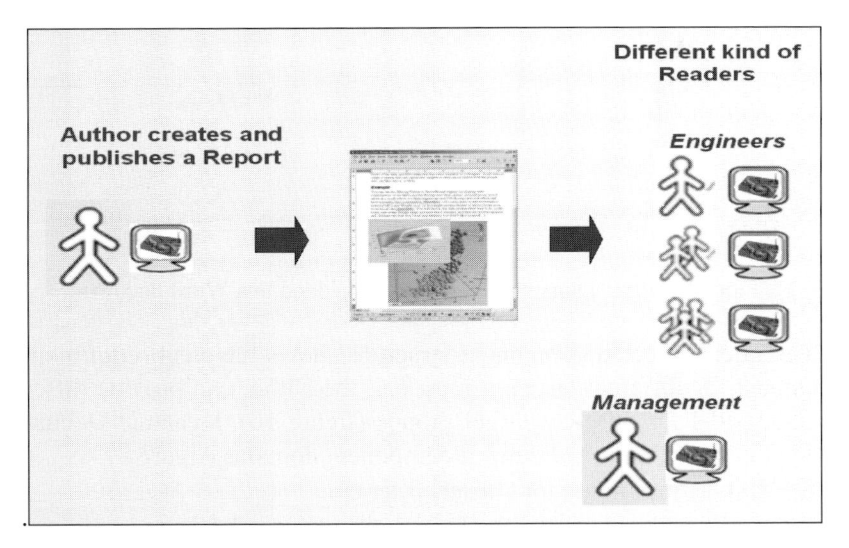

Fig. 10. The dynamics and linking structure in a Dynamic Document allows the creation of a single document for different target groups, e.g. the author can provide different reading paths for different target groups

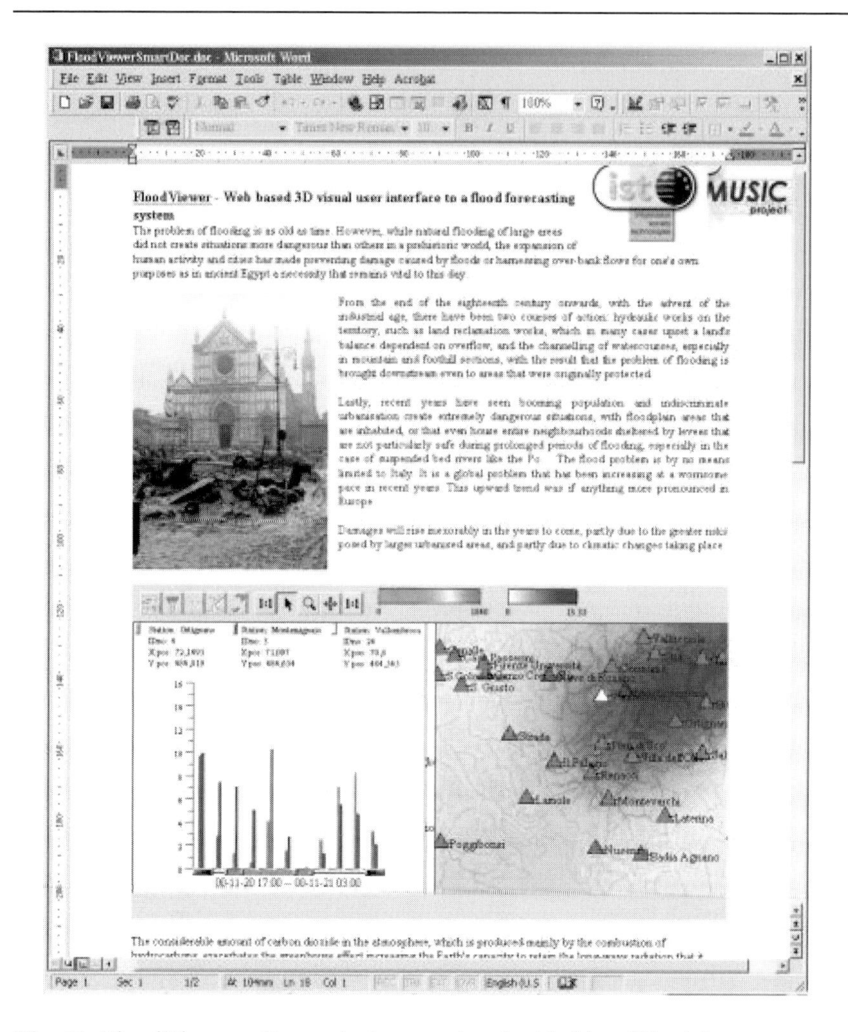

Fig. 11. FloodViewer – Dynamic document embedded in a Word document

The interconnected component structure allows the creation of a single document for different target groups, e.g. the author can provide different reading paths for different target groups (figure 10). Dynamic Document content is classified in specialized component domains (figure 8):

- **Analysis components** for cluster, filter etc
- **Interactive visual user interface components (VUI)** for proving insight
- **Visualization components** for viewing results
- **Document components** for authoring documents, e.g. embedded data, bookmarks, links, reading paths

6 Collaborative Flood Forecasting System

LiU also provides a framework for real time collaborative data visualization and discovery among geographically distributed remote users, linking people and their desktop in a worldwide "Virtual Data Environment". We have defined, developed and validated a conceptual distributed computing infrastructure required to support the development of collaborative visualization applications "Collaboratories". The FloodViewer extends the Web-based collaborative paradigm to the domain of interactive data navigation in an integrated 2D and 3D worlds and VUI technology supporting large climate, flood forecasting, terrain and satellite image data sets. The project has learned from other research projects that have been dedicated to collaborative games.

The network service provider architecture in the FloodViewer application component is based on Microsoft's DirectX API. The FloodViewer integrates component-based 2D and 3D geo-visualization tools and VUI methods with connectivity software based on emerging industry-standard network protocol for collaborative games. The FloodViewer supports most generalized communication capabilities shielding users from the underlying complexities of diverse connectivity implementations, freeing them to concentrate on the real-time navigation scenario. The integration of a network abstraction system, "Application Sharing" and a "Data Navigation Protocol" provide the foundation for real-time collaborative geo-visualization.

Our implementation of a "Collaboratory" provides a layer that largely isolates the user from the problems of an underlying network. With a multi-user application session, each user's VUI is synchronized with that of the other user(s) in the session. Continual stream of messages flow to and from each user. For example, every time a user rotates a 3D object or changes an attribute, a message is sent to update that user's position on the other application client in the session. The FloodViewer supports efficient and flexible messaging between all the computers in a session.

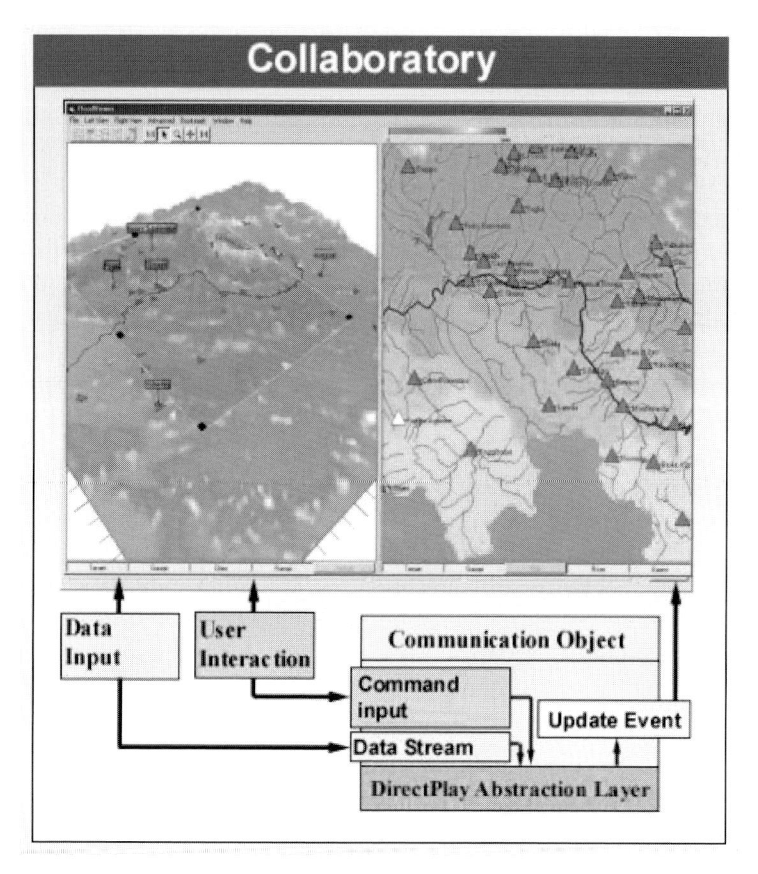

Fig. 12. FloodViewer Collaboratory provides collaborative flood forecasting over the Internet based on Microsoft's DirectX

Collaborative flood forecasting is important to improve decision support by better communication between meteorologists and hydrologists, operational manager and the civil defense manager. Experts and politicians normally don't work in the same offices and an integration of videoconferences and collaborative visualization will allow faster decisions and safe valuable time in a critical situation.

7 Conclusion

Given the large number of high risk situations as well as the extremely high costs in terms of casualties and damages involved, it can be reasonably estimated that a flood and related phenomena DSS, which combines visualization, GIS and advanced visual user interface capabilities with modeling for the planning and the preparation of risk maps, emergency plans and the real time analysis of possible interventions, has a large potential market.

The increased numbers of users with very different backgrounds, who will be using water resource data to make important decisions, elevates the importance of finding reliable and easy-to-use methodologies for flood forecasting. Analyzing, visualizing, and visual user interfaces are important for spatial data in general, but it is especially important for water resource data where a small local change may have a dramatic impact.

We promote the use of a component-based approach to the development and engineering of software systems, applications and services. Customizable and scalable high-level "application" components are designed and developed from low-level "atomic" components. Our application components are based on Advanced Visual Systems' OpenViz, a low-level visualization component framework. Atomic components from several other sources, including data interactors, data filters, analysis, and data access were also integrated. We believe that using lower-level atomic components for developing application components would provide better scalability and more customizable visual data navigator components. Atomic COTS (Components Of The Shelf) components from different vendors (or developed when necessary) were used in assembling the FloodViewer.

Another key objective is seeking to improve the communication and the dissemination of results to the authorities involved in real-time flood forecasting and management. We have shown the possibility of deploying interactive and dynamic visualization in electronic documents. The main barrier in using such innovative technology can be determined as an acceptance problem. Because it is not easy to convince people to change their way of thinking and working strategies, new *technologies* often fail to be introduced to a wide number of users (which should not be misunderstood as acceptance problems in introducing *tools*). For real profit in using our "Dynamic Documents", one has to understand the documentation processes as *integral* part of the project work, or better to embed the project work in the documentation process.

Another overall goal of our research is to make people more effective in their information or communication tasks by reducing learning times, speeding performance, lowering error rates, facilitating retention and increasing subjective satisfaction. We believe that customizable and scalable Visual User Interface (VUI) components in collaborative work can increase effectiveness for users who range from novices to experts and who are in diverse cultures with varying educational backgrounds.

Based on our experience, we have drawn some tentative conclusions regarding 2D versus 3D data visualization. We can conclude that 2D data visualization methods are more easily accessible to the user. The 3D data visualization allow the user to combine more information into a single scene, but these methods are not yet accepted as instruments for decision making among the local agencies.

The collaboratory version of FloodViewer have been tested and validated in Europe with local water quality agencies, geographically distributed users, linking experts and their desktops in a "Virtual Flood-forecasting System". As a final remark, it is the belief of the author that the availability of a system allowing for the hydrological and hydraulic analyses; the evaluation and the preparation of the flood risk maps; the planning of structural and non structural measures; the real time flood forecasting with the possibility of analyzing the advantages of the different intervention scenarios, will constitute a major breakthrough in the four areas related to flooding: planning, real time management, dissemination and training of personnel.

Acknowledgements

The MUSIC project is collaboration between Linköping University, AVS, University of Bologna, ET&P and institute CNR Italy, University of New-castle, Gematronik Germany and 3 end user agency partners (see Web site). The project was partly funded by the EC Commission EVK1-CT-2000-00058. The MUSIC web site can be found at:

http://www.geomin.unibo.it/orgv/hydro/music/index.htm and Flood-Viewer can be downloaded at:http://servus.itn.liu.se/projects/music/

References

Hindmarsh J, Fraser M, Heath C, Benford S, Greenhalgh C (2000) Object-focused interaction in collaborative virtual environments. ACM Transactions on Computer-Human Interaction, 3(4), pp 477-509

Wood J, Wright H, Brodlie K (1997) Collaborative Visualisation, Proc., IEEE information Visualisation '97. IEEEComputer Society, Pheoniz, Oct. 19-24, 1997, pp 253-259

Todini E (1996) The ARNO Rainfall-Runoff model, Journal of Hydrology, 175, pp 339-382

Todini E, Marsigli M, Pani G, Vignoli R, (1997a) Operational Real-Time flood forecasting systems based on EFFORTS. Proc. RIBAMOD Workshop

Todini E, Bottarelli M (1997b) ODESSEI: Open architecture Decision Support System for Environmental Impact: Assessment, planning and management, Operational Water Management, Refsgaard, Karalis (eds.), Balkema, Rotterdam, pp 229-235

Collaborative Visualisation (1998) Jason Wood; Leeds University PhD Thesis, February, www.scs.leeds.ac.uk/kwb/publications95.htm#Love:98

Watson VR (2001) Supporting Scientific Analysis within Collaborative Problem Solving Environments, HICSS-34, Maui, Hawaii, Jan 3-6

Roussev A, Reference Architecture for Distributed Collaborative Applications; http://www.cs.unc.edu/~munson/DARPA/

Chabert A, Grossman E, Jackson L, Pietrovicz S, NCSA Habanero – Synchronous collaborative framework and environment

Fox G, Jin J, Overview of Collaborative Computing and Some NPAC Experience. Syracuse University

Jern M(2000) Collaborative Visual Data Navigation on the Web. Invited Keynote Lecture to INFVIZ 2000, IEEE International Conference on Information Visualisation, London, IEEE Computer Science Press

Jern M (2001) Visual Data Navigators "Collaboratories - True Interactive Visualisation for the Web. Invited Speaker, Mobile and Virtual Media International Conference

Brown JR, Dam A van, Earnshaw R, Encarnação J, Guedj R, Jern M, Preece J, Scheiderman B, Vince J (1999) Human-centered computing: Online communities and virtual environments. Special report to the First Join European Commission/National Science Foundation Advanced Research Workshop, June 1-4, 1999, Chateau de Bonas, France. Computer Graphics 33(3), pp 42-62

AVS OpenViz toolkit web site: http://www.openviz.com

MUSIC web site: http://servus.itn.liu.se/projects/music

http://www.geomin.unibo.it/orgv/hydro/music/index.htm

A Web Application for Landslide Inventory Using Data-Driven SVG

Maurizio Latini[1] and Barend Köbben[2]

[1] Centro di Geotecnologie - Universita' di Siena, Via dei vetri vecchi 34, 52027 San Giovanni Valdarno, Italy.
Email: latini@unisi.it
[2] International Institute for Geo-information Science and Earth Observation (ITC), PO Box 6, 7500AA Enschede, the Netherlands.
Email: kobben@itc.nl

Abstract

The landslide map in the Serchio basin (Central Tuscany, Italy) is an official document that represent the actual state of the landslides in the region. At present, the updating of the map is carried out by different municipalities using paper sketch maps, sent by post to the Autorità di Bacino del Fiume Serchio (AdB). The objective of the work presented here was to significantly speed up and simplify this updating process, while taking into account the severe constraints of the municipalities. This was to be achieved by providing them with a lightweight Web based map application that allows inventory of new landslides and submitting them via the WWW directly to the AdB databases. Open Source technology and Open Standards were employed to build a database-driven application. The data is stored in a spatial database backend (MySQL), following OpenGIS Simple Features specifications. Server applications extract data from the database and deliver it as a client-side application in SVG, the Scalable Vector Graphics format of the W3C. This paper presents the technical background and setup of this application and plans for future development.

1 The Official Landslide Map of the Serchio Basin

The process of updating geological and geomorphological databases is not always a straightforward operation. In Italy for example, landslide databases are in some cases the responsibility of central authorities, such as Regions, in others of local authorities, the so-called "Autorita' di Bacino". This article is focused on one of these, the *Autorità di Bacino del Fiume Serchio* (AdB), responsible for the Serchio basin, located in a zone close to the Apuane Alps in the North West part of Tuscany, Italy.

The landslide map of the Serchio basin is an official document that should represent the actual state of landslides in the region. Nowadays this map is updated every year, but this particular region is regularly affected by landslide phenomena, and a yearly updating sometimes is not sufficient.

The Current Updating Process

The current updating system, shown in figure 1, is depending much on the local municipalities in the region. They usually receive advice of new landslide phenomena from citizens. The location of these phenomena is sketched on a printed map, which is submitted to the GIS department of the AdB via surface mail. At the AdB offices, an operator has to draw new polygons representing the new landslide phenomena on a temporary layer of the landslide GIS. Before updating the official landslide map with the occurrence of the new phenomena, a field check is needed, for which the sketch layer is printed and team consisting of a geologist and a geomorphologist is sent to the field.

Their objective is to decide if the landslide on the sketch map represents a possible damage for the population. If the phenomena can really affect either infrastructures or private properties the official landslide map is updated with the appropriate polygons.

There are many factors that slow down the updating process and make it less efficient, especially the cumbersome sketching and posting of these sketches to AdB where they subsequently have to be digitized.

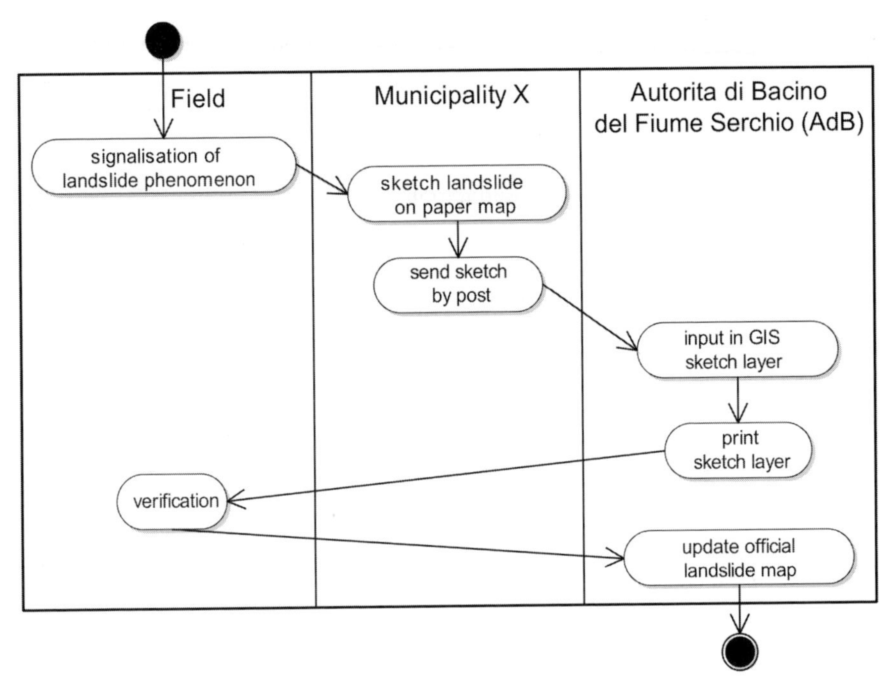

Fig. 1. Current landslide map updating process

Towards a more efficient updating process

The objective of the work presented here was to significantly speed up and simplify the updating process, while taking into account the limited possibilities of the municipalities. Many of these are small and located in inaccessible sites. They also lack the staff and funds necessary to implement and maintain complicated ICT systems. Another problem due to their location is the lack of a fast internet connection. If available, the connection to the web is usually made via telephone line and a 56K modem. All these circumstances dictate that they cannot be expected to use a full-blown Geographical Information Systems.

Making the process more efficient, while at the same time coping with the restraints, had to be achieved by providing the municipalities with a lightweight mapping application running in a Web-browser, that allows the recording of new landslides directly to the AdB database. The map application had to have some basic navigation functionalities, such as zooming, panning and layer and legend control. Beside that, it should offer also a "digitizing" tool that allows the user to draw new landslide features which, once digitized, are submitted via the WWW directly to the AdB GIS. As an added bonus, the same system could be used by the AdB fieldwork teams to digitally finalize the official landslide map, as shown in figure 2.

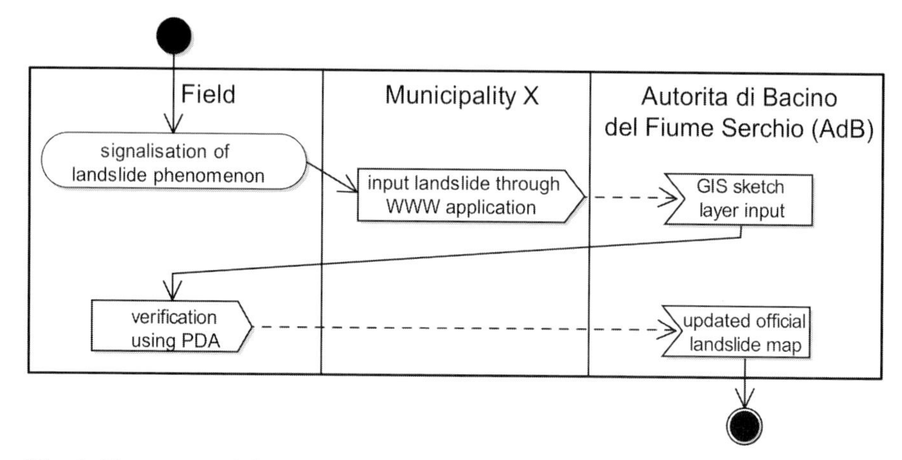

Fig. 2. The proposed future setup of the updating process

In order to achieve the goals described above, Open Source technology and Open Standards were employed to build a database-driven application suitable for use on many platforms. The setup of this system was based upon ideas on a "GDI-light" architecture, being developed at the International Institute for Geo-Information Science and Earth Observation (ITC).

2 GDI^{LIGHT}: Lightweight Geo-Data Infrastructures Based on Open Standards and Open Source Software

The term GDI might be usually connected with (very) large regional or national spatial data warehouses, but it is defined more generally as "the networked geospatial databases and data handling facilities, the complex of institutional, organizational, technological, human and economic resources (...) facilitating the sharing, access to, and responsible use of geospatial data at an affordable cost for a specific application domain or enterprise" (Groot 2000). In many cases GDI data and application infrastructures are being developed using high–end geospatial software solutions and large corporate databases, needing substantial investments in financial and human resources. But the principles of GDIs can be applied in simpler and more cost-effective ways just as well, which is of particular interest for the students and partners of the International Institute for Geo-Information Science and Earth Observation (ITC). ITC is an institute that aims at capacity building and institutional development specifically in developing countries (URL 1).

GDI^{LIGHT} (formerly called GDI-EMERGE) is an internal ITC project to employ web services and a data back end to build light-weight, low-cost Geo-Data Infrastructures, using Open Standards and Open Source software. It serves as a general purpose testbed for applied as well as fundamental research activities, and should provide researchers and students alike with a proof-of-concept platform for relatively simple, low-cost, yet powerful ways of sharing data amongst various distributed offices and institutions as well as the general public. It is not the intention that it should grow out to be a fully working, coherent system, but should be seen as a *testbed* in the broad sense of "equipment for testing". It is the place where we can show fellow researchers, consultants and students as well as possible users (such as GIS users from developing countries) that the things we teach can be made to work quite quickly, in a relatively simple and low-cost setup. The main building blocks of the system are:

- A spatial database back end that stores the geometry and the attribute data; Spatial data should be stored using the OpenGIS (OGIS) Simple Features specifications. In GDI^{LIGHT}, MySQL is used at present.
- A set of interoperable web applications that interface with the database and with each other, and fulfill tasks such as delivering data in SVG for visualization purposes and other XML formats (such as GML) for data exchange, serve data in OGC formats to and from the database using web services, etcetera. In GDI^{LIGHT}, these are developed using Java server technology, and at present deployed using the Open Source Apache Tomcat server.
- Simple Web-based interfaces enabling access to the maps and data for both desktop browsers and mobile platforms, as well as more sophisticated interfaces, for example providing data through an OpenGIS Web Feature Server to GIS clients. At present, we concentrate on we browser clients, using SVG for graphics and interaction.

In figure 3, the conceptual setup of the system, such as it is currently being implemented, is shown. We do intent this to be a flexible setup which is expected to change over time. The current work ongoing is focused on the processes represented by the darker arrows.

A first application of the GDI^{LIGHT} testbed called *RIMapper* (Köbben 2004) was developed in early 2004 to look into the possibilities of generating light-weight, versatile Risk Indicator Maps (RIMs) from online databases. RIMapper uses a JSP Tomcat server application to extract OpenGIS Simple Features stored in MySQL Spatial Extension and delivers these to mobile web clients as interactive SVG maps, based on XML configuration files. These maps are to be part of an urban risk management system, and therefore needed to fit a multitude of use cases, ranging from providing the general public with information about risks to providing local authorities an interface to the underlying risk assessment databases and models. Furthermore, the maps needed to be usable on a wide range of platforms, from the office systems of the local authorities to hand-held devices providing location based services to field personnel. Many of the techniques used in RIMapper could be deployed in the landslide inventory application that is the subject of this paper.

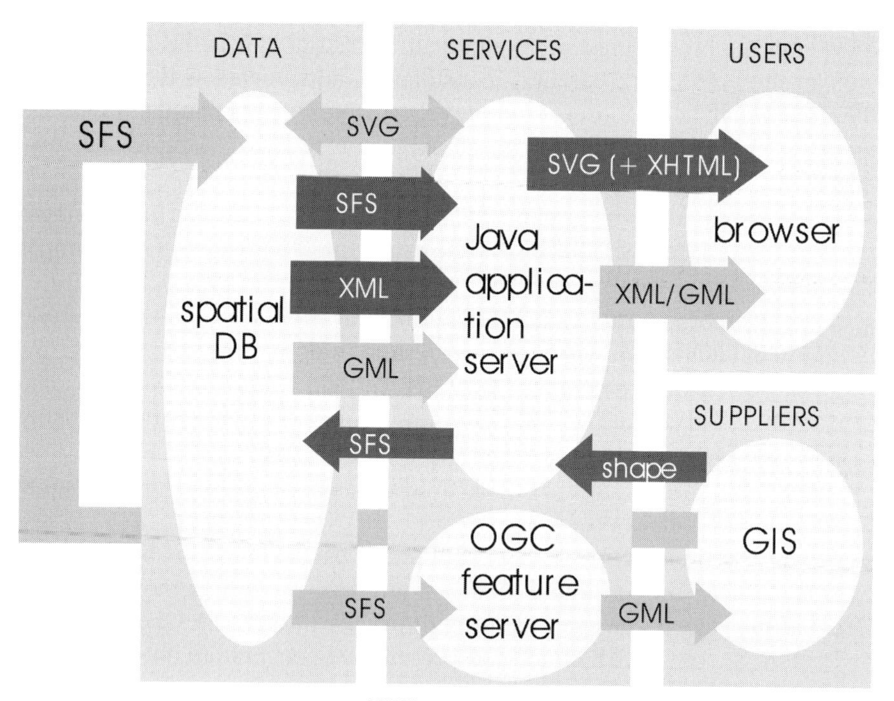

Fig. 3. Conceptual setup of GDI^{LIGHT}. Darkest arrows show current focus (SFS stands for OGC's Simple Features Standard)

The Spatial Database

As the project set out to comply to open standards, there was the need to use a database that supported the standards of the OpenGeospatial Consortium (OGC, see URL 2). Among other things, the OGC has set the Simple Features SQL Specification that provides for publishing, storage, access, and simple operations on spatial features (point, line, polygon, multi-point, etcetera) through an SQL interface.

There are several databases with OGC-compliant spatial extensions, of which Oracle is probably the most prominent, but as another goal was to use open source software, two databases were under consideration: *PostgreSQL* (URL 3) a database system with *PostGIS* as an spatial extension and *MySQL* (URL 4). MySQL's recent versions include OGC-compliant spatial extensions, although not implementing the full set of OGC specifications. The reasons for the choice of MySQL were the native Windows support and the simple 'lightweight' character of the software, as compared to the complicated though more fully-featured PostGIS. By adhering strictly to the OGC standards it should be straightforward to change or even mix database platforms in the future.

Geometric features are stored as OGC Simple Features geometry and transformed to other formats at run-time, by the application tier. The non-spatial attributes of the features are stored per layer in specific tables ('homes', 'roads', etcetera), related with the spatial features through their ID. A 'layers' table is provided as a per layer link to a 'styles' table, for layers that should be styled uniformly, e.g. all roads sharing the same visualization. Whenever the visualization should depend on some data attribute per feature, e.g. for a chorochromatic map of homes viewed by vulnerability type, the link is made from the data specific attribute table directly to the 'styles' table. The choice for the visualization type mentioned above is directed by the XML map configuration file (see next paragraph). A further two tables, 'fragments' and 'actions' are also for use by the application tier, storing SVG code fragments and ECMAscript event listeners, respectively.

The Application Tier

The heart of the RIMapper system are the services provided by a set of Java servlets. In our case, the application tier runs on Tomcat, a well-known open source servlet container from the Apache Software Foundation (URL 5), but the applications should run on any standards-compliant servlet container. The RIMapper system uses a set of generic Java classes to do recurring tasks like extracting OGC features and attribute data from the database, translating these into fragments of SVG and ECMAscript, collecting and structuring these fragments into valid output and delivering this output to the clients.

The glue provided to make all these parts act together are *XML map configurations*. They are parsed to get a description of the map needed and all its component parts. Below is a sample of an XML configuration for a very simple map of roads, which when clicked will show their 'type' attribute in a pop-up message.

```
<?xml version="1.0" encoding="iso-8859-1"?>
<!DOCTYPE RIM PUBLIC "" "/RIMapper/RIM.dtd">
<RIM TYPE="SVG_STANDALONE" DB="rimapper" UN="X" PW="X">
    <TITLE>Clickable Data...</TITLE>
    <AUTHOR>ITC</AUTHOR>
    <HEADER>
       <FRAGMENT DBID="default" NAME="defSVGRoot"
             TYPE="SVG_ROOT"/>
       <STYLES>
          <STYLE DBID="default" NAME="defLine" TYPE="CSS"/>
       </STYLES>
       <FRAGMENT DBID="default" NAME="defInitPlusRIMmessage"
             TYPE="ECMASCRIPT"/>
       <FRAGMENT DBID="default" NAME="showRIMData"
             TYPE="ECMASCRIPT"/>
    </HEADER>
    <LAYERS>
          <LAYER DBID="default" NAME="roads" STYLETYPE="single"
                STYLE="defArea" ATTRIBS="type">
             <ACTION TYPE="simple" NAME="showRIMData"
                   SCOPE="feature" EVENT="onclick"
                   PARAMS="evt, 'rim', 'type'"/>
             </LAYER>
    </LAYERS>
    <FOOTER/>
</RIM>
```

In the header section of the XML map configuration, several fragments of SVG code are loaded to define gradients, symbols, filters and ECMA-script fragments to provide interactivity and other functionality. The last part, the footer section, simply declares the closing part of the output. In between is the body section that lists the actual layers of information to be mapped, used to retrieve the attributes and the geometry from the database. Layers have *types* that determine the way they are visualized cartographically, can have one or more *actions* that set the interactivity and *events* that set which event will trigger the action.

When all data needed has been collected by the system, the SVG output is composed and handed over to the web server for delivery to the client. The SVG generated will adhere to the SVG-Basic profile and will be suited for a broad range of clients, including PDA's.

Using this system, one can very flexibly offer database-driven maps on the web that are generated on the fly from the most recent data, and that can incorporate all the functionality, scalability and graphics quality that the SVG standard offers.

3 Implementation of the Landslide Inventory Tool

Based on the needs of the municipalities described in the first paragraphs, the landslide inventory tool has to respond to some pre-specified characteristics. The steps needed to implement such a tool are:

- Creation of an SVG web map application using open source technology (MySQL, SVG) – based on the existing RIMapper system mentioned above.
- This SVG application should have basic functionalities necessary for the sue as a landslide inventory tool (i.e. legend that can turn on/off the layers, identify features)
- Creation of an edit tool that will allow drawing a polygon in a sketch layer directly from the SVG application.

The map layout has been developed as shown in figure 4. The main map is located in the centre of the web browser/screen and is the only part of the screen that can be zoomed and panned.

The other elements of the map application are a legend, a reference map for the navigation through the main map and a navigation tool where all the buttons and tools are located. At the time of writing the following functionalities have been developed:

Layers Control via Legend

A layer control is necessary in order to manipulate the main map. The map layers can be turned on and off like in a traditional GIS systems. When the map is loaded all the layers are set to visible, but the legend is invisible. To turn the layers off it is necessary to make the legend visible by clicking the button 'Show Legend' in the navigation tool. After this it is possible to set the visibility of the layers by simply using the check boxes next to the layer names. When the legend is turned off and then turned on again, the visibility state of the layers will be remembered.

Panning and Zooming

Due to the restrictions of the screen dimensions, map navigation tools are one of the most important functionalities for a screen map. As suggested by its name any SVG-viewer is 'pan and zoomable' by using key shortcuts. In this application however this intrinsic functionality has been turned off by setting the attribute 'ZoomAndPan' of the <svg> tag element to disable. Panning and zooming functionalities in this case have been obtained via calculations on the viewBox attribute of the svg file and by nesting <svg> elements. This approach was used, so that only the map is scalable and the buttons and the legend remain fixed on the side.

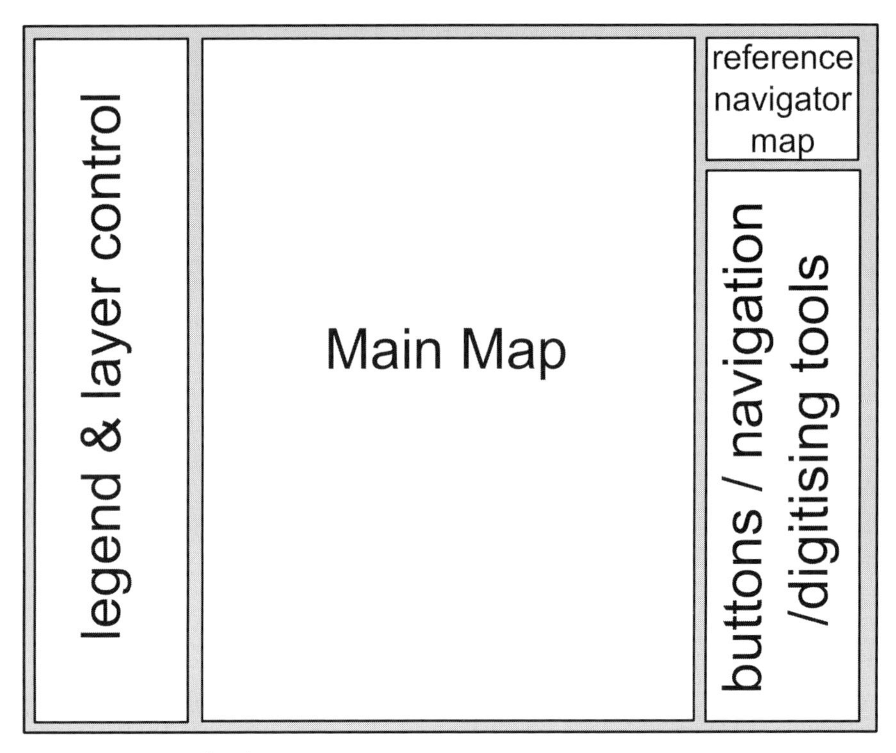

Fig. 4. Map Application Layout

A series of buttons with arrows have been created in the Navigation Tool, by clicking on one of the buttons the map will scroll in the direction of the arrow. The central button is the Full Extent one, by clicking this button the map will come back to its original extent. Clicking on one of the zoom buttons will enlarge or decrease the details of the map view.

Interactive Overview Map

The interactive overview map is a good solution in order to navigate through a map. The overview map shows the whole map extent and it is also useful for an overall orientation. The overview map implemented contains a semi transparent rectangle which represents the corresponding view in the main map. The interactivity is achieved by adding a dragging functionality to the above mentioned rectangle. This will become a control by adding mouse events, if the user holds down the button and moves the rectangle around, a mouse move event is triggered. When the mouse up event is called the rectangle stays and the main map is centered to its new extent.

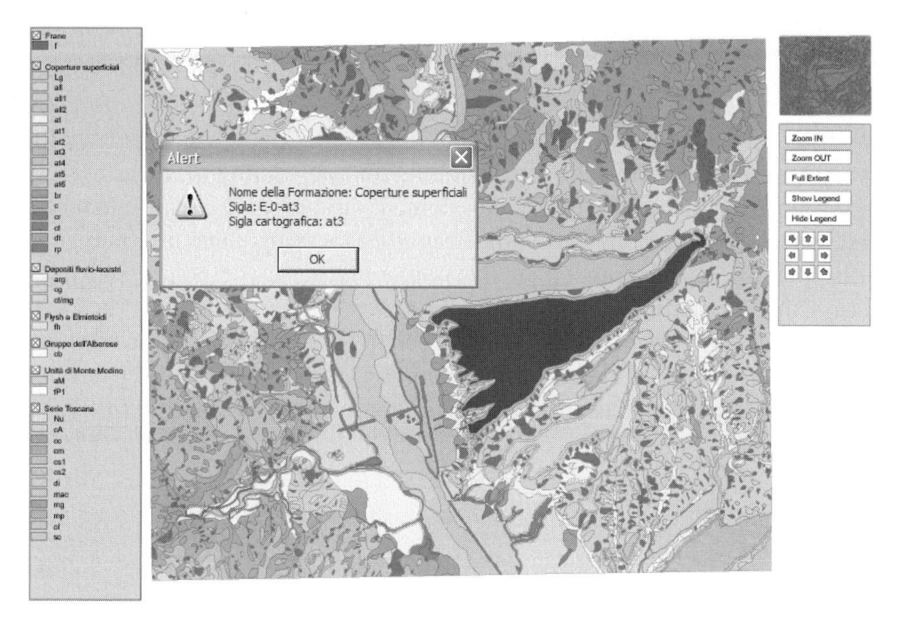

Fig. 5. Screen dump of the application, showing an information message box

Map Layer Information

When the mouse cursor is over the map the layers are highlighted, by clicking the layer of interest, a message box will pop up, as can be seen in figure 5. This message box contains the information regarding the clicked feature like the name of the geological formation, the cartographic reference used and the rendering employed.

Showing Coordinates

When the mouse cursor is over the map, in the bottom part of the control panel the map coordinates are shown. The calculation of the actual coordinates is really important, because the digitizing tool is based on this coordinates. It is also important because in a future development of the application it should be possible drawing the new polygons by entering its coordinates collected directly in the field with a GPS system.

Digitizing Function

Once the map coordinates are calculated, implementing a digitizing functionality for the map was quite straightforward, by using the DOM-functionalities of the SVG map. In SVG, being an XML language, all elements are represented in a Document Object Model or DOM. This *DOM-tree* is exposed through API's to the viewer application. Therefore by scripting one can use dynamic arrays that store the X and Y coordinates for every mouse click. After the dimensioning and populating of the arrays the next step is to create a DOM node element (in our case a path element), that has to store the drawing coordinates and represents them as a polygon.

Before the actual drawing of the polygon a series of points and lines shows the proceeding of the digitizing process. The digitizing process is strictly related to the coordinate system of the map, whenever the map extent changes (because of zooming or panning) the map coordinates have to be recalculated. The steps of the digitizing process are (see figure 6):

- When the *Start Digitizing* button is pushed a message pops up
- During the digitizing the points are added together with the lines
- When the *Stop Digitizing* button is pushed the polygon is closed and displayed in blue color.

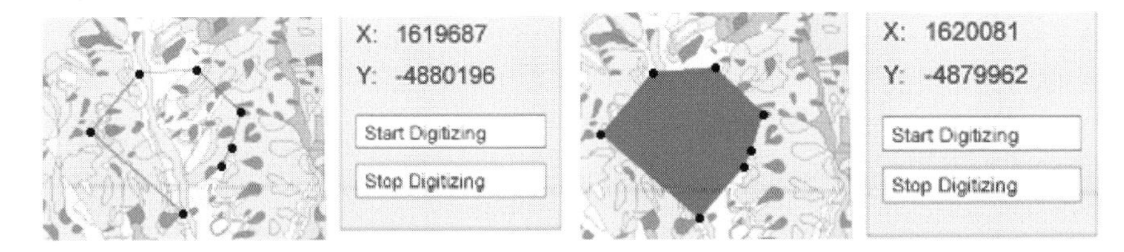

Fig. 6. Screenshots of the digitizing process

Some refinement in the digitizing process should be made, by allowing vertices to be edited. At the moment, the newly digitized polygon is not yet uploaded to the AdB database. The client-side functionality needed for this (the *postURL* function) is already well-established in the current viewers, and will become an official standard soon. The only extra functionality needed is converting back the SVG to OpenGIS Simple Features, this could best be achieved server-side. However, the AdB still will need to rework the server-side to actually receive the inputs and store them in a safe, transactionally sound way.

An overview of the currently implemented and planned features can be seen in table 1.

Feature	Implemented	Planned
Navigation:		
Interactive Overview Map	X	
Panning	X	
Zooming	X	
Show coordinates on mouse move	X	
Map controls:		
Layer control	X	
Identify features	X	
Digitizing operations:		
Digitizing	X	
Move vertex		X
Delete vertex		X
Client/Server Communications:		
Update sketch layer with digitized polygon(s)		X

Table 1. Implemented and planned functionalities for the landslide SVG Map

4 Conclusion and Future Work

The application described in this paper will be used by the municipalities that do not have a GIS system but at the same time need to update a sketch layer with new landslide phenomena. Via the WWW this authorities can access the official landslide map and, based on the new landslide phenomena occurrence draw a sketch polygon on the map. This sketch polygon can be later on submitted (always via WWW) to the central authority responsible for the updating. Based on these sketch drawings, experts will perform a field check and if the landslide phenomena results potentially dangerous for the population, the official landslide map will be updated.

The first step in the application developing has been the creation of a SVG map from an online database, based on the data-driven GDI^{LIGHT} system. The choice of SVG has been made because it is an open source standard and through the use of ECMAscript highly customizable. Furthermore it permits to generate a light weight map that represents what these authorities needs.

The work described here is only the starting point of the Landslide Web Map. Future developments for this project will need to obtain a lighter SVG application, in fact the actual dimension of the test project are too big. The problem is that when the file is loaded all the geological formations are loaded in it; this is the main cause of the dimension of the file. In the future, a file loader will be developed. With it the user will be able to load only the data needed directly from the server. This will speed up the process because only a skeleton map will be loaded at the beginning.

Future developments will include also a more appropriate rendering for the geological formations, using the cartographic standards available for geological mapping. This rendering should include the representation of oriented symbols.

The digitizing tool also needs to be implemented with additional functionalities like the possibility to edit polygons and individual vertices. Another planned functionality will be the possibility to add attributes to the digitized polygon; this will be obtained via a popup window that the user will fill with the appropriate information.

With these functionalities added, a well-designed and useful digitizing tool should finally be achieved.

URLs

[1] ITC web site: see http://www.itc.nl /
[2] OGC web site: http://www.opengis.org/
[3] PostGIS site: http://postgis.refractions.net/
[4] MySQL AB site: http://www.mysql.com/
[5] Apache Foundation Tomcat pages: http://jakarta.apache.org/

References

Groot R, McLaughlin J (2000) Geospatial data infrastructure – concepts, cases and good practice. Oxford University Press, Oxford etc.

Köbben B (2004) RIMapper - a test bed for online Risk Indicator Maps using data-driven SVG visualisation. In: Gartner G (ed) Proceedings of Location Based Services and TeleCartography (Geowissenschaftliche Mitteilungen, Heft 66). Wien: Institute of Cartography and Geo-Media Techniques & ICA, pp 189-195. http://kartoweb.itc.nl/RIMapper/

High-Resolution Satellite Image Sources for Disaster Management in Urban Areas

Jonathan Li, Yu Li and Michael A. Chapman

Geomatics Engineering Program, Department of Civil Engineering, Ryerson University, 350 Victoria Street, Toronto, Ontario, Canada M5B 2K3. E-mail: {junli, y6li, mchapman}@ryerson.ca

Abstract

With the rapid development of Earth observation technology and geospatial information technology, disaster managers (in the broadest sense) now have power tools capable of collecting and integrating data from various sources in an efficient and cost-effective manner. These properties make it particularly attractive to disaster management support activities. This paper examines the problems of geospatial data acquisition for disaster management with a focus, in particular, on urban environments from two perspectives: geospatial data requirements and the role which high-resolution satellite imagery (0.6 – 5 m) can play in satisfying these geospatial information requirements, and effective image exploitation methods. We focus on the potential of available very high-resolution commercial satellite image data for rapid urban mapping and discuss the example of automated building and road extraction from pan-sharpened IKONOS and QuickBird images.

1 Introduction

The summer flood of 1998 in Yangtze River region, China, the 9-11 terrorist attack of 2001 in New York, USA, the disastrous earthquake of 2003 in the historic city of Bam, Iran, have tragically demonstrated that the whole disaster management sector is under pressure for better, more sophisticated and appropriate means for facing natural and man-made disasters. One of the key issues in disaster management is to effectively monitor and analyze hazards and risks, which is a very complex and challenging task, as many factors can play important role in the occurrence of the disastrous event. Therefore, analysis requires a large number of input parameters, and techniques of monitoring and analysis may be very costly and time consuming. The increase availability of high-resolution satellite image data an innovative development of geographic information systems (GIS) have created opportunities for a more detailed and rapid monitoring and analysis of disasters (Li and Chapman, 2004; Johnson, 2000; Lu et al., 2004; Banger, 2004; Cova, 1999). The proper structure of an information system for disaster management should be presented to tackle the disaster and to manage it. Geospatial information technology can be used to create an elaborate and effective disaster management information system. An integrated approach using scientific and technological advances should be adopted to mitigate and to manage disasters. Moreover there should be a national policy for disaster management.

The application of geospatial information technology begins with the acquisition of data about disasters and culminates in the effective communication of information to those concerned with the outcomes of decisions, which here are disaster managers. In between, there are technologies relating to information manipulation, modeling, analysis, and management. Our work on disaster management concentrates on three levels of this process: information extraction from remote sensing data using digital photogrammetric and image analysis techniques, information management and exploitation using GIS, and information communication using environmental visualization tools. We are seeking to put these into a large context of information management for interactive spatial decision support. In considering both information extraction and visualization, we go beyond the concepts of two-dimensional (2D) mapping and consider the importance of three-dimensional (3D) modeling for disaster management.

For technologies to be effective in disaster management support activities they must be fast but reliable, cost effective, and simple to use by disaster managers, and as far as possible based on the off-the-shelf software components, such as desktop/laptop/pocket GIS. Among these requirements, fast and effective imaging sources and mapping techniques are most critical. This paper principally explores the role of both satellite-based geospatial data collection and touches upon the exploitation of these data in a desktop environment for disaster management. The needs for geospatial data cover a broad range of spatial scales, from regional to local, and thus encompass different imaging sources from Earth observation satellites with different spatial and temporal resolutions. Pan-sharpened imagery is invaluable in land use and land coverr classification for generating map products for disaster managers. This imagery can also provide cues (e.g., color) in the process of feature extraction for automated mapping. A significant goal in this regard is the extraction of buildings and roads to support change detection between pre- and post-disaster for damage assessment.

Initially, we discuss the geospatial information requirements and roles for high-resolution satellite imagery in disaster management. We then consider present sources and future options for the provision of image-based geospatial data products. Finally, the example of building and road extraction is discussed to illustrate the potential of automated spatial information collection.

2 Geospatial Information Requirements

The term "disaster management" encompasses a wide range of activities which can be grouped into five phases that are related by time and function to all types of emergencies and disasters. These phases are planning, mitigation, preparedness, response, and recovery. All phases of disaster management are related to each other and depend on data from a variety of sources (Johnson, 2000). During an actual emergency it is critical to have the right data, at the right time, displayed logically, to respond and take appropriate action. Each phase has different requirements in terms of data types, and specifications (positional accuracy, completeness, currency). Different scales of disaster management can be distinguished, for example, the regional and the local level, mach differing with respect to the granularity of the geospatial data required.

2.1 Mapping Scales

Natural hazard information should be included routinely in developmental planning and investment projects preparation. Development and investment projects should include a cost/benefit analysis of investing in hazard mitigation measures, and weigh them against the losses that are likely to occur if these measures are not taken. According to Banger (2004), satellite remote sensing can play a role at the following levels:

At national level, the objective is to give an inventory of disasters and the areas affected or threatened for an entire country and create disaster awareness with politicians and the public. Mapping scales will be in order of 1:1,000,000 or smaller. The following types of information should be included:

- Hazard free regions for development.

- Regions with severe hazards where most development should be avoided.
- Hazardous regions where development already has taken place and where measures are needed to reduce the venerability.
- Regions where more hazards investigations are required.
- National scale information is as required for these disaster that affect and entire country (drought, major hurricanes, floods, etc.)

At regional level (typical mapping scales between 1:10,000 and 1:1,000,000) and at **all inter-municipal or district level** (mapping scales ranging from 1:25,000 to 1:100,000), the objective is to investigate where hazards can be a constraint and to provide the pre-feasibility study of the development of urban or infrastructural projects. The areas to be investigated range from several thousands to a few hundreds of square kilometers, and the required details of the input data are sometimes up to considerable higher. Slope information at this scale may be sufficiently detailed to generate Digital Elevation Models (DEMs), and derivative products such as slope maps. Spatial analysis capabilities for hazard zonation could be utilized extensively.

At local level, the objective is to generate hazard and risk map for existing settlements and cities, and in the planning of disaster preparedness and disaster relief activities, which are typically that of a municipality. Typical mapping scales are 1:5,000 - 1:25,000. The details of information will be high, including for example cadastral information. The hazard assessment techniques will be more quantitative and based on deterministic/probabilistic models. The size of area under study is in the order of several tenths of square kilometers and the hazards classes on such maps should be absolute, indicating the probability of occurrence for mapping units, with areas down to one hectare or less. At site investigation scale GIS is used in the planning and design of engineering structure and in detail engineering measures to mitigate natural hazards. Typical mapping scale is 1:2,000 or larger. Nearly all of the data is of a quantitative nature. GIS is basically used for the data management, and not for data analysis, since mostly external deterministic models are used for that. A 3D GIS can be of great use at this level.

2.2 Roles of Satellite Imagery

The impacts of floods, forest fires, tornado, and tropical storms, earthquakes, and other natural disasters around the globe in recent years are forcing us to seek new prevention and mitigation methods. This task involves improving predictive models and monitoring tools, drawing up regulatory requirements and planning emergency response. For this purpose, we need to acquire regularly updated, reliable and objective spatially referenced information in a timely fashion.

The roles of satellite image and image analysis in supporting disaster management are many and here are only two examples: (1) Preventing floods. The only way to prevent flooding is through effective land-use planning and a detailed knowledge of land occupancy and the natural phenomena likely to affect a region. In this respect, high- and very-high resolution Earth observation data are a valuable aid for producing and maintaining maps to provide information about flood-prone areas. Such imagery has the potential to improve our understanding of land use, land cover, and flood extents. It also can be combined with cadastral maps for flood risk prevention planning. (2) Evaluating earthquake damage. Optical and radar satellite imagery has already shown its potential for detecting damage caused by natural disasters. High- and very-high resolution satellite imagery makes it possible to interpret earthquake damage zones in sufficient detail and map earthquake damage quickly. By comparing two satellite images (one acquired before and one immediately after the event) within less than 48 hours of image reception could may urban zones affected by the earthquake. Such imagery is able to detect changes to large buildings and determine the probability that they have been damaged.

3 Image Sources

A number of new spaceborne imaging satellites present interesting and viable options for the acquisition of spatial data in disaster management support activities. We will now present a brief summary of the primary characteristics of these high-resolution spaceborne imaging sensors as they relate to disaster management in urban areas.

Investigations are conducted in this section to demonstrate how high-resolution satellite imagery data can assist emergency management and disaster response teams in urban areas. High-resolution satellite images with spatial resolution ranging from 0.6 m to 5 m have a number of advantages over and provide additional applications to aerial photography. These include:

- Normally having four spectral bands from visible blue to near infrared, providing the equivalent of both color and color infrared photography.
- Their digital radiometric characteristics providing the ability to undertake spectral classification and semantic modeling of the data.
- Allowing multi-date analysis of radiometrically calibrated data.

- Capability of radiometric calibration and allowing the compilation of large area mosaics.
- Being captured as 11 bits (e.g., IKONOS and QuickBird) rather than 8 bits giving better dynamic range than aerial photography and airborne scanners.
- With no variable brightness within a single image that is usually associated with aerial photographs.
- Normally already having data in archive over your area or can commence capture within 7 days of placing an order.

Here we group the satellite images with spatial resolution ranging from 2.5 m to 5 m into high-resolution satellite imagery, while those with spatial resolutions between 0.6 m and 1 m are grouped into very-high-resolution satellite imagery.

3.1 High-Resolution Satellite Imagery

High-resolution Earth observation satellites with ground resolution of 2.5 m to 5 m currently in operation are French SPOT 5 and Indian IRS-1D. SPOT 5 carries a High Resolution Geometric (HRG) instrument and has a multi-resolution, wide-swath imaging capability offering the finer resolution of 2.5 m and 5 m (instead of 10 m by SPOT 4) in black-and-white mode, and 10 m (instead of 20 m by SPOT 4) in color mode for studying vegetation. Like all their predecessors, each SPOT instrument covers a 60-km swath, making it possible to image large conurbations in a single pass. In addition, SPOT 5 also carries a (High Resolution Stereoscopic (HRS) instrument, which provides stereoscopic imaging capability along the satellite track, designed specially to acquire wide-area DEMs.

The Indian Remote Sensing (IRS) satellite system collecting 5-m resolution panchromatic imagery, with a pushbroom configuration (much akin to SPOT system), is ideal for urban planning, disaster management, mapping and other applications requiring the unique combination of high-resolution imagery, high revisit frequency (5 days), and broad area coverage (70 km by 70 km). These satellites have stereo imaging capability, adjustable gain and cross-track imaging capability. The elevation accuracy of 5 to 10 m (relative) and 10 to 15 m (absolute) can be expected for the DEMs generated from HRS images. This would support urban mapping up to 1:10,000 scale. It is also seen feasible for mapping to 1:15,000 scale from IRS-1D data, at least in terms of planimetric accuracy (Ravichandran et al. 2002; Raju et al., 2002).

From the disaster management perspective, however, it is not so much the metric quality of the data from satellite platforms which is of prime importance, but the semantic content of the imagery since both the national and the regional level mapping focus upon land use. So-called pan-sharpened imagery "colors" the high-resolution panchromatic image with fused multispectral image at lower resolution to achieve the impression of a high resolution color image. This imagery provides a very useful tool for both land use and land coverr mapping and the collection of basic road networks and other cultural data. In cases where optimal metric quality is sought, both ortho rectified imagery and even stereoscopic analysis can be employed.

Nevertheless, because the spatial resolution of the multispectral bands is relatively coarse and those bands are broad spectrally and only cover the visible to near infrared range of the electromagnetic spectrum, 5 m satellite imagery is unlikely to be a complete source of data for disaster management at the local level mapping, mainly of municipality.

3.2 Very-High-Resolution Satellite Imagery

While the semantic content of SPOT-5 and IRS-1D imagery by and large limits its stand alone application in the generation of a 1:10,000 scale urban mapping, the new generation of commercial 1 m resolution Earth observation satellites will be capable of fulfilling this demand. Satellite imaging systems such as IKONOS from Space Imaging, QuickBird from DigitalGlobe, and OrbView-3 from ORBIMAGE are providing resolution equal to that of about 1:20,000 scale aerial photography, as well as supplying supplementary multispectral image data. The range of applications envisaged for such imagery goes well beyond the provision of topographic map data to 2 – 3 m accuracy, to include infrastructure planning, land and natural resource management, environmental monitoring and disaster management. The 0.6 – 1 m resolution satellite imagery will be extremely valuable as means of providing the local level and the on-site scale mapping capability in disaster management support activity in urban areas. The moderate size of several tenths of square kilometers, means that a single IKONOS (11 km 11 km), QuickBird (16 km x 16 km), or OrbView-3 (8 km x 8 km) image can provide sufficient coverage to meet most needs for managing disasters at the local level, with a potential of virtually immediate delivery of georeferenced, radiometrically and geometrically corrected digital ortho image map products. Thus, the familiar problems of data currency are overcome and the imagery can support temporal studies of dynamic conditions within urban environments, albeit probably at a scale lower than that required to detect the geometric changes of a single house caused by disaster.

The use of 1m satellite imagery for local level disaster management also becomes viable, at least as far as the planning of utilities and services and the general demarcation of land plots under risk is concerned. Similar to the case of 5 m satellite imagery, however, 1 m satellite imagery is not expected to fully suffice as a source for on-site investigation scale disaster management. For example, it remains to be seen whether water points and power lines can be reliably extracted from this imagery. Under such circumstance, 1 m satellite imagery may be supplemented by higher resolution airborne imagery using a principle of multiresolution coverage.

3.3 Multiresolution Coverage

It is readily apparent from the descriptions of the different satellite imaging options presented that no one sensor system offers stand alone information solution to data acquisition for disaster management. Moreover, the capabilities for semantic information extraction from these satellite images do not fully overlap the information requirements of the different spatial scales employed in disaster management. For example, we can observe that 5 m satellite imagery may fulfill the requirements of a regional level disaster mapping at scales between 1: 25,000 and 1: 100,000 and 2.5 m satellite imagery may provide sufficient cartographic integrity for inter-municipal or district scale mapping to scale as large as 1: 10,000. However, at the local mapping level, which is not smaller than 1:2,000, useful satellite imagery will no double be limited to that with 1 m or higher resolution. Even this satellite imagery will probably insufficient and the semantic content cannot be expected to be sufficient to provide both comprehensive thematic information extraction and automated feature extraction functions such as detecting changes in houses. Here we may have to rely on aerial imagery. The use of conventional aerial photography is by no means precluded, but its role will probably be restricted, due to its high cost, to the creation of initial regional- or local-level disaster mapping. For example, first epoch capture of DTMs, and the mapping of major infrastructure and fixed features. Map or GIS database updating with aerial photography is simply not economically viable option for disaster management. It is also noteworthy that, for regional or municipal scales, provision of all data via 1 m satellite imagery, while certainly conceivable, may not be warranted due to cost factors equivalent to those which limit large-format aerial photograph coverage to 5-10 year cycles. A practical solution is to consider image coverage at multiresolution for each of the levels of spatial data required in disaster management.

4 Example of Automatic Image Exploitation: Building and Road Extraction

One of the predominant data requirements in local level disaster management is a spatial inventory of structures. Building data are required for many applications ranging from house counts for residential density analysis to precise building footprint measurement for in situ damage assessment Given the size of many residential areas (many are composed of hundreds of buildings) and the need for quick pre- and post-disaster inventory updating, there is a strong need for automated information extraction tools. Ideally, these tools should be implemented on a desktop/laptop computer and be simple enough to be used by trained emergency managers to exploit their local knowledge in the extraction process. This would also have additional benefits of direct community involvement in disaster management. In the following discussion, we examine development towards automated building and road extraction from high-resolution satellite imagery. We begin by introducing a new fuzzy C-partition method following by presenting a building and road extraction strategy with some promising results.

The section describes a new method for extracting objects from pan-sharpened high-resolution satellite images. This method is based on the fuzzy segmentation algorithm which has been developed in our previous works. The proposed object extraction method consists of three steps: color image segmentation, object detection, and post-processing. The paper also gives practical results from using the proposed method to extract centerlines of road networks and roof outlines of buildings from pan-sharpened QuickBird and IKONOS images.

4.1 Image Segmentation by Fuzzy C-partition Algorithm

Fuzzy c-partition algorithm has been wildly used to solve the clustering problems in pattern recognition (Tou and Gonzalez, 1974; Zeng and Starzyk, 2001), image segmentation (Liew and Yan, 2001), unsupervised learning (Langan et al., 1998), and data compression (Zhong et al., 2000). We are currently developing a man-made object extraction strategy using a fuzzy c-partition-based color segmentation approach (Li et al., 2004). The approach takes full advantages of color cue into the automated feature extraction process. It consists of three steps: segmenting the color image, detecting objects from the segmented image, and post-processing of the extracted objects, for example, delineating road centerlines from the extracted road networks. Here we concentrate our discussion mainly on Steps 2 and 3.

After the color image is segmented by the proposed approach, the binary object image is generated by selecting the pseudo-color corresponding to the object regions. In general, the objects in the binary image are corrupted by noise objects, which have the similar colors to the objects of interest. Then binary morphological operations are applied for filtering the corrupted object image. The appropriate combinations of binary dilation, erosion, opening, and closing are chosen depending on the shapes of noise objects.

In the case of road extraction, an important process for representing the structural shape of the detected road regions is to reduce it to a graph. This work can be accomplished by a thinning algorithm developed by Zhang and Suen (1984) for thinning binary road regions. It is assumed that the road pixels in the binary road network images have value 1 (black), and those background (non-road) pixels have the value 0 (white). The method consists of the successive passes of two basic steps applied to the contour pixels of the given images, where a contour pixel is any pixel with value 1 and has at least one 8-neighour value 0. In the case of building extraction, we develop am.

In the case of building extraction, the building regions are first detected according to the color features of the buildings followed by extracting roof outlines of the detected buildings using the developed boundary extractor (Li, 2004).

Our object extraction method has been tested with pan-sharpened 0.6 m QuickBird and 1 m IKONOS images (see Figure 1). All test images have a size of 150 × 150 pixels, a subset of a typical Toronto residential scene which highlights the residential aspects of an area, including trees, lawns, houses, schools, roads, rivers and parks. This type of imagery could be used to assess and measure damage to buildings, facilities, roads and highways, utility networks and other structures. It could also be used for disaster preparedness, insurance and risk management and disaster mitigation efforts.

4.2 Experimental Results

The segmented images generated from the test images shown in Figure 1 are presented in pseudo-colors and given in Figure 2. The binary images of the objects extracted the segmented images are shown Figure 3. It can be observed in all the four test images shown in Figure 3 that the segmented objects are corrupted by other objects with similar colors to the objects of interest.

Figure 4 illustrates the object regions obtained after filtering the segmented images depicted in Figure 3 by using the binary morphological operators. A visual comparison of the images clearly favors the filtered images (see Figure 4) over the segmented images (see Figure 3). Figure 4a shows the results obtained by filtering Figure 3a using binary dilating with a structuring element of 3×3, followed by eroding with a structuring element of 5×5. Figure 4b shows the results obtained by dilating Figure 3b with a structuring element of 3×3 and eroding with a structuring element of 5×5. Figure 4c shows the results obtained by closing Figure 3c with a structuring element of 4×4. Figure 4d shows the results obtained by eroding Figure 3d with a structuring element of 3×3 followed by closing with a structuring element of 4×4.

Fig.1. Test images: (a) and (c) QuickBird, (b) and (d) IKONOS images

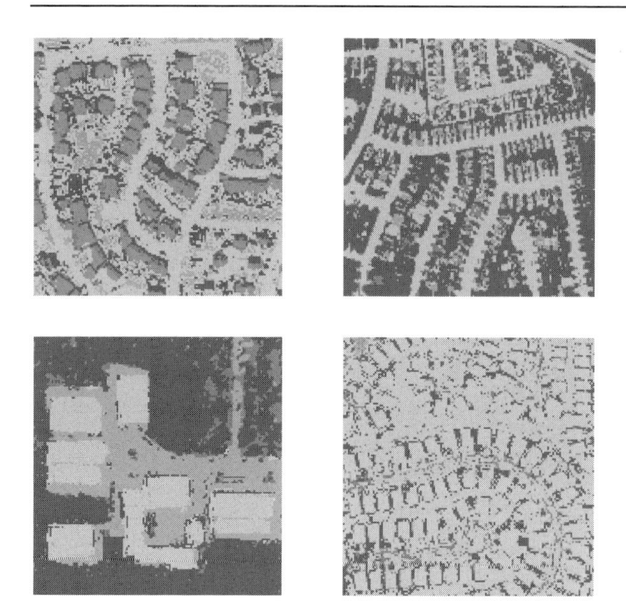

Fig.2. Segmented images: (a) and (c) QuickBird, (b) and (d) IKONOS images

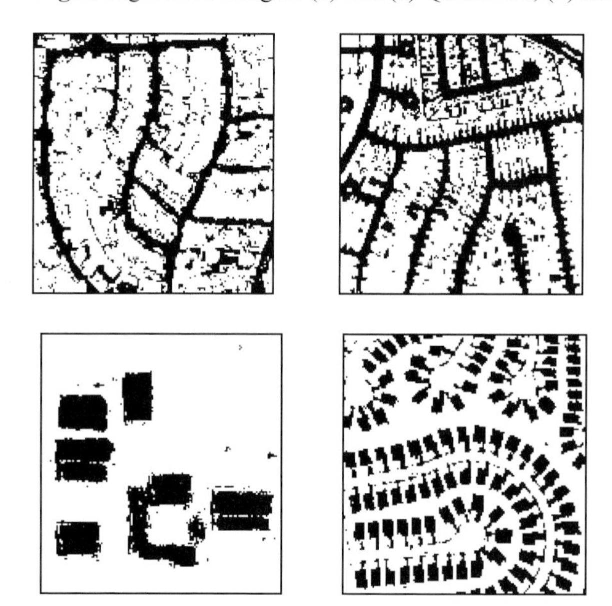

Figure 3. Binary images of object regions: (a) and (c) QuickBird, (b) and (d) IKONOS images

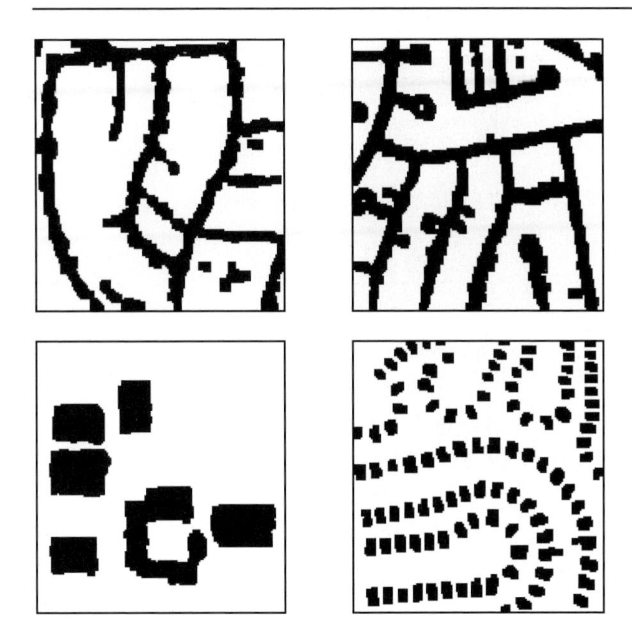

Fig. 4. Object regions after filtering: (a) and (c) QuickBird, (b) and (d) IKONOS images

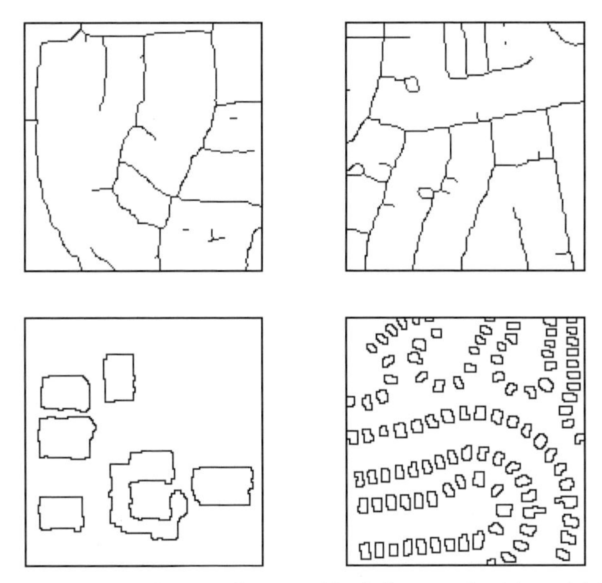

Figure 5. Road centerlines and building roofs: (a) and (c) QuickBird, (b) and (d) IKONOS images.

Fig. 6. Extracted road centerlines and building roof outlines (in red) overlaid on test images: (a) and (c) QuickBird, (b) and (d) IKONOS images

The road centerlines and the outlines of the extracted building roofs are delineated using the thinning algorithm and the proposed boundary extractor (The detailed description about it can be found in Li et al., 2005). In order to illustrate the accuracy, the extracted road centerlines and the outlines of the extracted building roofs (both are presented by red lines) are overlaid on the original test image, see Figure 6. The results presented illustrate the potential of the proposed approach.

5 Concluding Remarks

The spatial data requirements have been presented in the context of disaster management at the regional and local scale. In this paper, we have discussed a multiresolution approach whereby information extracted from high-resolution satellite imagery with different spatial and temporal resolutions is complemented from one to another to cover disaster mapping need from the regional to the local scale. The potential of pan-sharpened IKONOS and QuickBird images has been demonstrated through practical applications of our fuzzy segmentation approach to building and road extraction. In the future, our research will continue on a number of fronts, with work on the imagery aspect emphasizing further development of the principle of multiresolution coverage, additional investigation of multi-spectral classification, fusion of optical and synthetic aperture radar (SAR) satellites (e.g., Canadian 3 m Radarsat-2 and German 1 m TerraSAR-X, to be launched in 2005 and 2006, respectively) data and continuation of automated information extraction tools.

Acknowledgements

This research was partially supported by a Natural Sciences and Engineering Research Council of Canada (NSERC) discovery grant.

References

Banger SK (2004) Remote sensing and geographical information system for natural disaster management, GIS Development, 2 pp

Cova TJ (1999) GIS in emergency management, In: Geographical Information Systems, Principles, Techniques, Applications, and Management, P.A. Longley, M.F. Goodchild, D.J. Maguire, D.W. Rhind (eds.), John Wiley & Sons, New York, pp 845-858

Langan DA, Modestino JW, Zhang J (1998) Cluster validation for unsupervised stochastic model-based image segmentation. IEEE Transactions on Image Processing, 7(2), pp 180-195

Li J, Chapman MA (2004) Remote Sensing, In Telegeoinformatics: Location-based Computing and Services, edited by Karimi H. and A. Hammad, CRC Press, New York, pp 27-68

Li Y (2004) Fuzzy Similarity Measure and Its Application to High Resolution Color Remote Sensing Image Processing, Master Thesis, Ryerson University, 180 pp

Li Y, Li J, Chapman MA (2005) Automated Extraction of Manmade Objects from High-Resolution Satellite Images by a Fuzzy Segmentation Method, paper submitted to the First International Symposium on Geo-information for Disaster Management (Gi4DM), Delft, The Netherlands, March 21-23, 12 pp

Liew AW, H Yan (2001) Adaptive spatially constrained fuzzy clustering for image segmentation. Proceedings of 10th IEEE International Conference on Fuzzy Systems, University of Melbourne, Australia, vol 2, pp 801-804

Raju P, Ghosh S, Saibaba J, Ramachandran R (2002) Large scale mapping versus high resolution imagery, Indian Cartographer, LSTM-03 pp 127-134

Ravichandran V, Srivastava PK, Singh D, Bhatti AH, Krishna BG, Padmanaban D (2002) Large scale mapping from IRS 1D, Indian Cartographer, LSTM-05, pp 144-146

Lu W, Mannen S, Sakamoto M, Uchida O, Doihara T (2004) Integration of imageries in GIS for disaster prevention support system, Proceedings of ISPRS Commission VI, WG II/5, 5 pp

Tou JT, Gonzalez RC (1974) Pattern Recognition Principles, Addision-Wesley, Reading, Mass, USA

Zeng Y, Starzyk JA (2001) Statistical Approach for Clustering in Pattern Recognition, Proceedings of the Southeastern Symposium on System Theory, Athens, OH. 5 pp

Zhang TY, Suen CY (1984) A fats parallel algorithm for thinning digital patterns. Communications of the ACM, 27(3), pp 236-239

Zhong JM, Leung CH, Tang YY (2000) Image compression based on energy clustering and zero-quadtree representation. IEE Proceedings on Vision, Image and Signal Processing, 147(6), pp 564 -570

Geo-Information as an Integral Component of the National Disaster Hazard and Vulnerability "ATLAS"

Dusan Sakulski

United Nations University, Institute for Environment and Human Security (UNU-EHS), Görrestr. 15, 53113 Bonn, Germany.
Email: sakulski@ehs.unu.edu

Abstract

The increase in the frequency of disasters and their associated damages globally is part of a worldwide trend, which results from growing vulnerability and may reflect changing climate patterns. Global risks seem to be increasing.

These trends have significantly initiated the development and implementation of the National Disaster Hazard and Vulnerability Atlas. The main idea was to design and develop database-driven, web-enabled interactive "virtual book" (Atlas). It consists of various "chapters", such as drought, flood, cyclones, storms, severe weather, and fires.

Web-enabled GIS is used as the most important user communication interface for various hazards. User is able to submit input through maps. Results of various calculations, if spatially distributed, are returned back to user in form of GIS.

1 South African National Disaster Management Centre

Intensive implementation of the integrated information technology in disaster management of the South Africa started in the year 1998. It was an El Nino year and the weather pattern was unusual for the average season in this part of the world. Ministry of the Department of Water Affairs and Forestry has instructed experts from the Strategic Planning Directorate to initiate implementation of the information technology as a tool for the rainfall, riverflow, and flood related information management.

In October 1999 author was instructed to join the National Disaster Management Centre (NDMC) of South Africa (at that time Y2K Centre), to continue implementation of the integrated information technology in disaster management. The main idea was to monitor and register, as early as possible, potential hazardous events and to increase lead warning time. It is one of the main activities at the NDMC (http://sandmc.pwv.gov.za) main observation room (Figure 1). It contains 9 workplaces, continuously monitoring various hazardous events worldwide, by utilising integrated database-driven web-enabled technology.

Fig. 1. NDMC observation room

In January 2003 the President of South Africa has signed South African Disaster Management Act, a backbone of the national disaster management legislation. The main attention is on the prevention (pro-active activities), which is a "180 degree turn" from the historically inherited post-disaster activities. Act highlights the role of the information (information flow) as the most significant driver.

2 National Disaster Hazard and Vulnerability "ATLAS"

The increase in the frequency of disasters and their associated damages during the last decades is a global trend. It results both from growing environmental and social vulnerabilities and it reflects changing climate patterns. This trend has significantly emphasized the need for the development of an appropriate tools to support various disaster management activities: National Disaster Hazard and Vulnerability Atlas. The main idea was to design and develop database-driven, web-enabled interactive "virtual book" (Atlas). The Atlas was implemented for South Africa. It consists of various "chapters", such as drought, flood, cyclones, storms, severe weather, and fires.

It enables users, using just web browser, to search and select various data, images, maps, graphs, to perform different calculations, to run certain model on-the-fly, and copy-paste results to the local computer and to print "their own page of the ATLAS".

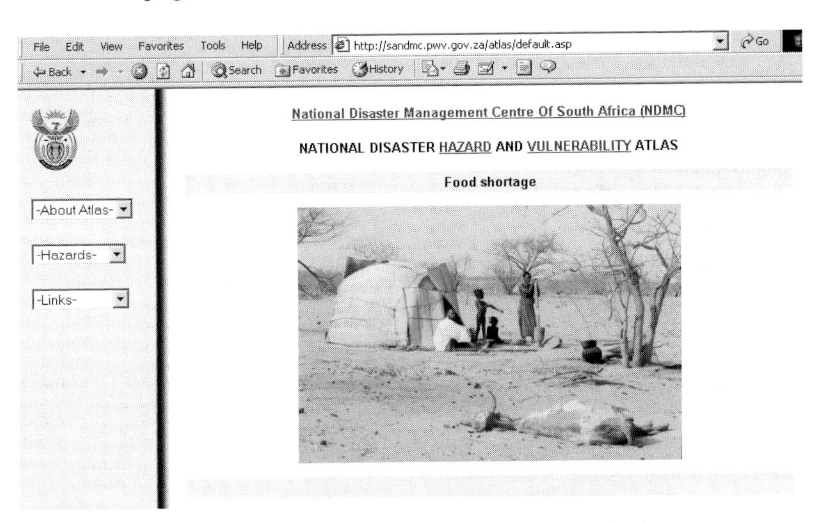

Fig. 2. ATLAS home page (http://sandmc.pwv.gov.za/atlas/)

Key ATLAS elements includes:

- Development of a national disaster related hazard, vulnerability and risk assessment tools, to be able to report periodically on the national exposure to natural hazards, patterns and trends or changes in the exposure and to guide priorities in natural disaster vulnerability reduction efforts.
- Development of an integrated national disaster hazard and vulnerability information network to provide the tools needed by national, provincial and local governments, the private sector, and the general public. The

network will also facilitate much-needed augmentation of education and training.
- Augmentation of comprehensive, hazard specific programs.

In general, here are some of the main elements of the ATLAS:
- Developing a comprehensive database to identify hazard, vulnerability and risk-prone areas.
- Risk understanding and addressing.
- Information assimilation and dissemination.

In particular, ATLAS attempts to:
- Carry out research on factors contributing to disaster hazard and vulnerability and measures to alleviate this vulnerability.
- Develop methodologies for the analysis of disaster related hazard and vulnerability indicators, and to improve disaster management.
- Disseminate the research results and methodologies through national, regional as well as global disaster information network and other channels to promote increased awareness and preparedness to natural and man-made hazards.
- Act as a core of the national early warning system.

3 GIS Setup at NDMC

Geographic Information Systems (GIS) plays very important role in the setup of the database-driven web-enabled integrated hazard, vulnerability and risk information system at the South African national Disaster Management Centre (NDMC). Structure and setup of such system is based on ten years experience from the South African Department of Water Affairs and Forestry (DWAF) Water Management Information System. Realising the importance of the visual information, GIS was considered as the one of the most significant components for the interactive communication between user and system.

NDMC's server room contains four interconnected servers: (1) main web server; (2) database server; (3) GIS server; (4) mathematical server

GIS server serves interactive visual information (maps) using ESRI ArcIMS version 4 and Arc SDE.

4 Hazards, Vulnerability and Risk Assessment and Coordination

At the national, as same as provincial levels, the main disaster management activity is **coordination**. As a complex activity in its nature, coordination is multidimensional. It has, in general, two equally important components: **temporal** and **spatial**.

Temporal component assumes coordination as a continuous (sustainable) activity. Ad hock coordination usually results in confusion. Improvement in coordination of hazards, vulnerabilities and risks related activities requires better understanding of complex mechanisms and interaction between “mother nature”, human society and technology. It takes time (in continuous sense).

Spatial component has three main coordination directions: **horizontally**, **vertically** and **thematically**.

- Horizontally, coordination, as part of the national framework and activities, is already in place bringing together various national, provincial and district institutions and organizations, who carried out active programs in support of disaster reductions: Department of Water Affairs and Forestry, Department of Agriculture, Department of Environmental Affairs and Tourism, South African Weather Services, South African Police Service, National Defense Force, nongovernmental organizations (NGOs), public and private institutions, businesses as well as educational institutions.
- Vertically, coordination between provinces, districts, metros and municipalities.

  Thematically as **hazards**, **vulnerability** and **risks** related coordination.

5 GIS and Hazard Related Activities at NDMC

Spatial and temporal distribution of various hazards differs around the globe. At the ATLAS web site 12 hazards are listed. But, not all of them are equally present over South African territory. The most dominant natural hazards are floods and droughts. For both of them, common input variable is rainfall. That is the main reason why ATLAS puts special attention towards continuous integrated spatial and temporal rainfall monitoring and evaluation.

The remaining part of this paper will highlight implementation of GIS and remote sensing technologies for some hazards, vulnerability and risk related activities at the NDMC.

5.1 Rainfall Dynamics

Since mid 1980s frequency of tropical cyclones occurrences above the Indian Ocean, between Australia and Madagascar, has been increasing. In average during rainfall season, between October and April, 8 to 12 tropical cyclones are born and heading towards African continent (east-to-west). Figure 3 shows two tropical cyclones approaching Mauritius and Madagascar, in February 2003.

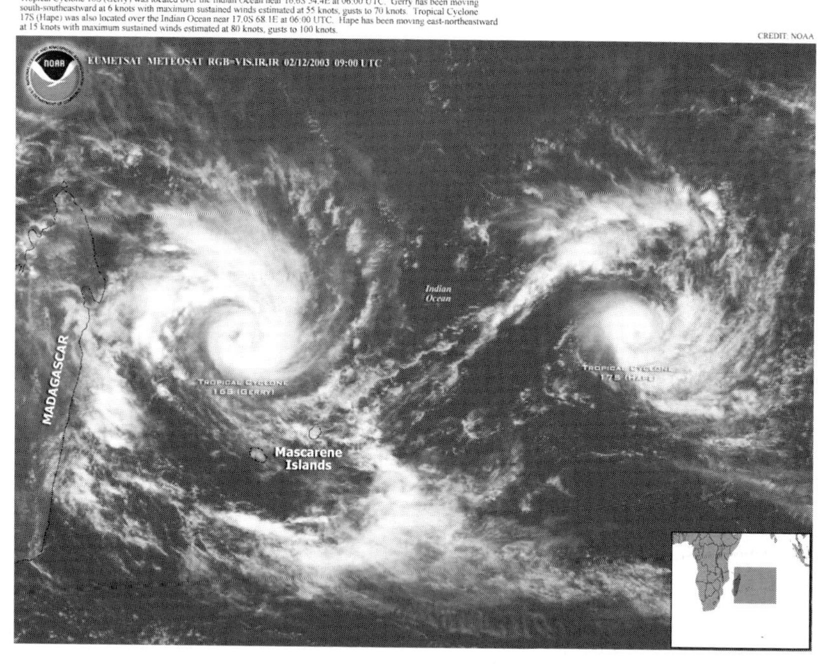

Fig. 3. Two tropical cyclones approaching Madagascar (source NOAA)

One of the workstations at NDMC's observation room monitors continuously the web site from the University of Hawaii (http://www.solar.ifa.hawaii.edu/Tropical/), which on daily basis refreshes the map of Indian Ocean (Figure 4).

Still in our memory is the year 2000. 16 tropical cyclones were born in that region (Figure 5).

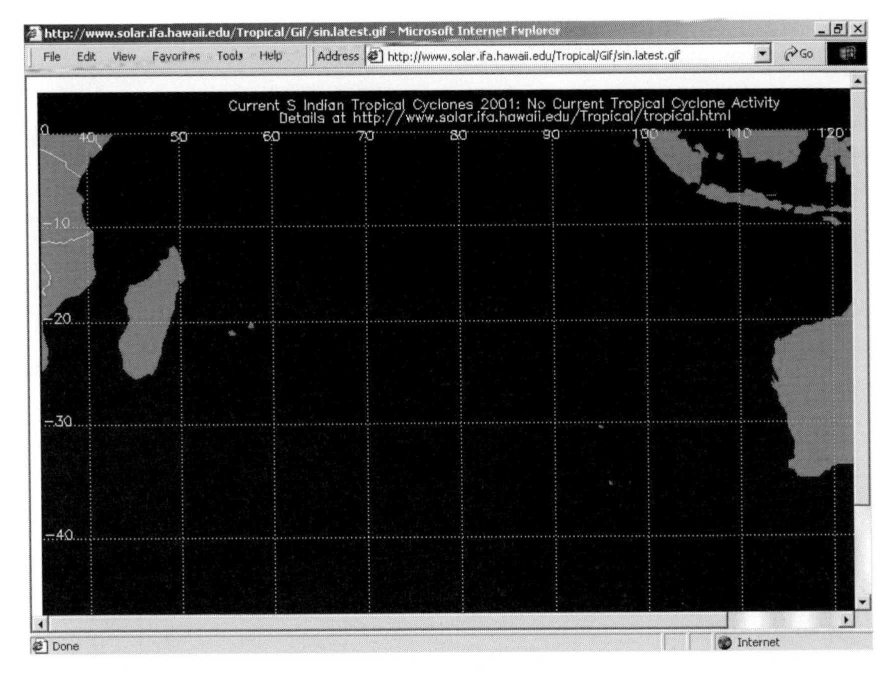

Fig. 4. University of Hawaii tropical cyclones web site

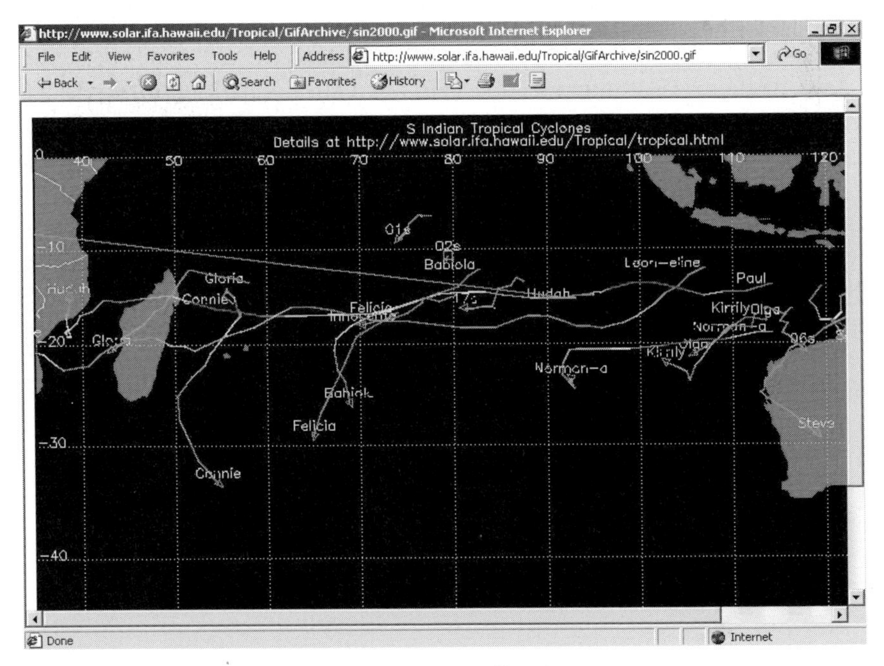

Fig. 5. Year 2000 tropical cyclones above Indian Ocean

Fig. 6. Tropical cyclone approaching Mozambique (source NOAA)

Two of them, Gloria and Elaine, had been strong enough to "walk" all over the African continent (Figure 6). Output is known: heavy rainfall and floods in this region (Mozambique, South Africa).

Sub-Saharan Africa is very well known as a region of sudden heavy rain. It is nothing unusual to have 70 – 150 mm of rain in 24-hour period. Disastrous impact of such rain is on the area of the informal settlements (informal housing, rural communities). People at the NDMC are maximizing effort to have that kind of information as early as possible. Very useful web site is NOAA satellite 24-hours rainfall estimation for 4 days in advance (Figure 7).

South African National Weather Services (http://www.weathersa.co.za) is the biggest national rainfall data collector. Apart from the ground based gauging rainfall measurement, they operate the network of 11 ground radars (Figure 8a).

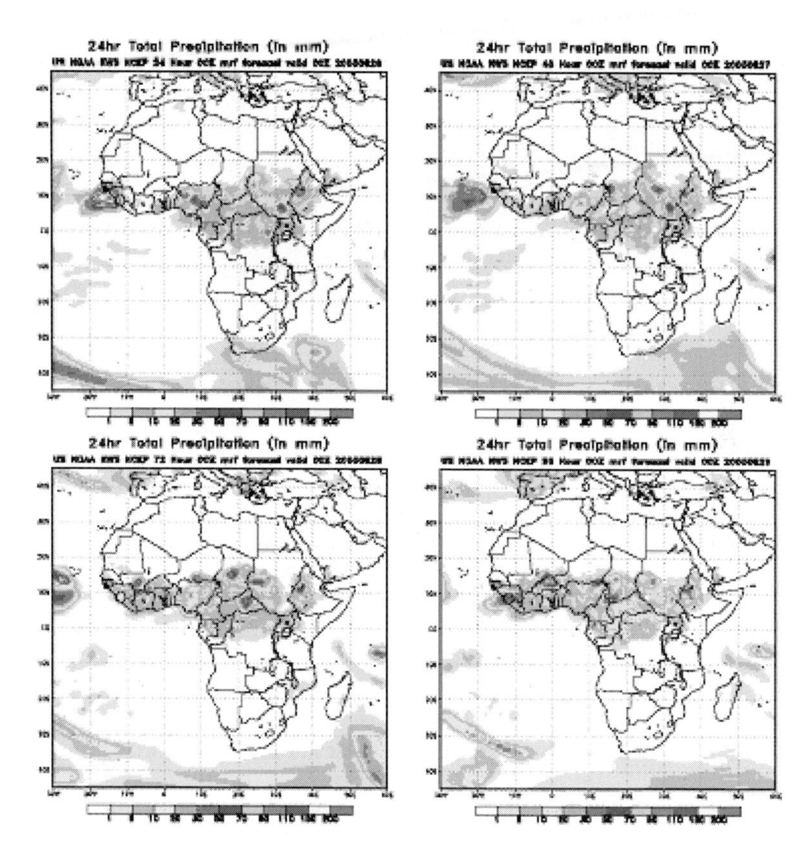

Fig. 7. NOAA 4 days satellite rainfall estimation

From the disaster management point of view, two kind of data/information are extremely important:

- Real time storm development and 30 min estimation (Figure 8b)
- Accumulated hourly rainfall estimation

In the year 2000 NDMC and WITS University Johannesburg have started jointly relationship with NASA and University of Virginia on project SAFARI 2000 (http://www.safari2000.org). One of outcomes was distribution of the historical global monthly rainfall data, available on SAFARI 2000 CDs, Volume 1 and Volume 2. The author has made additional effort to develop a web-enabled version based on the same data set, and included it into ATLAS (http://sandmc.pwv.gov.za/atlass/rain/). User can select country and station via GIS (ArcIMS) enabled browser interface (Figures 9a, 9b).

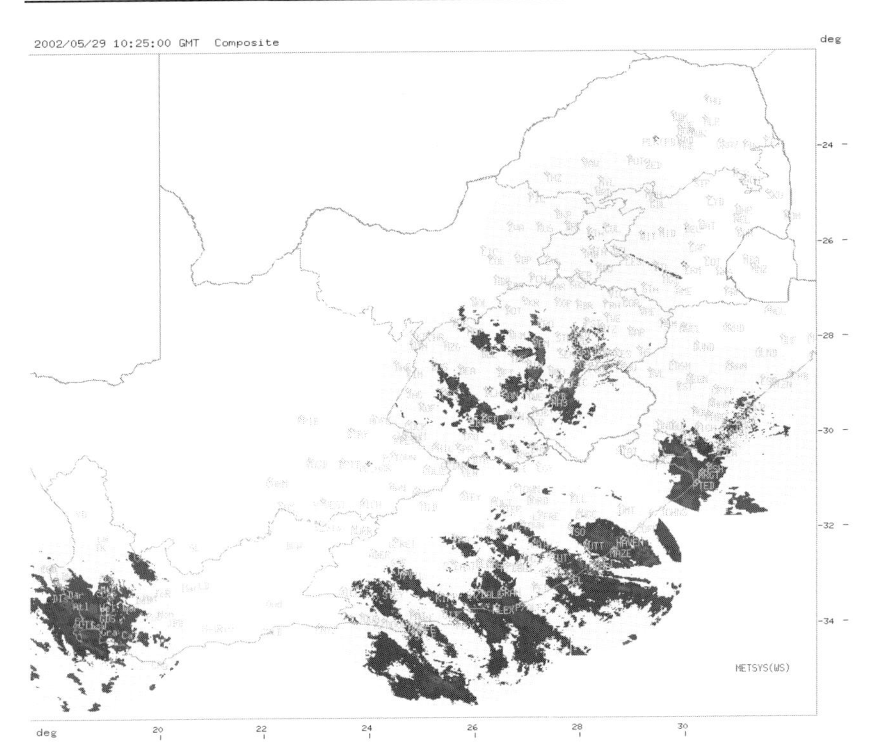

Fig. 8a.:South African Weather Services ground radar network (GRASS)

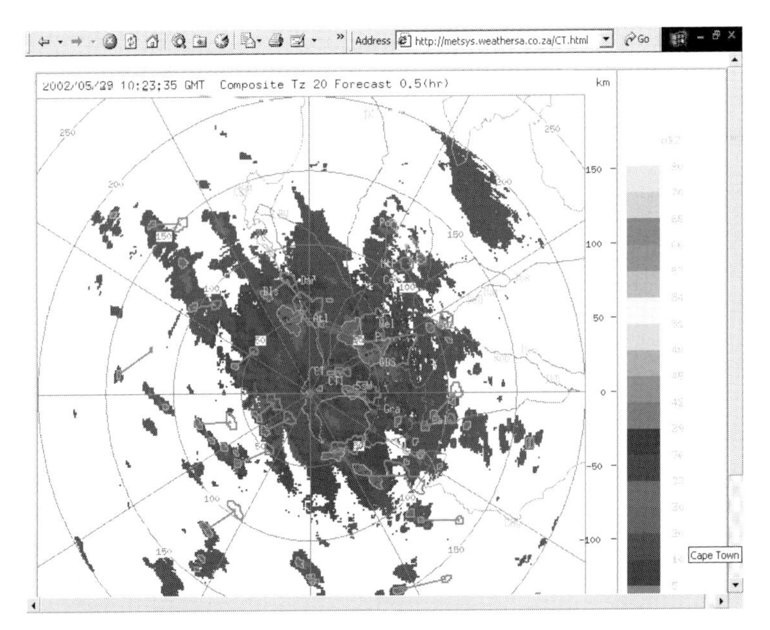

Fig. 8b. Storm development around Cape Town (GRASS)

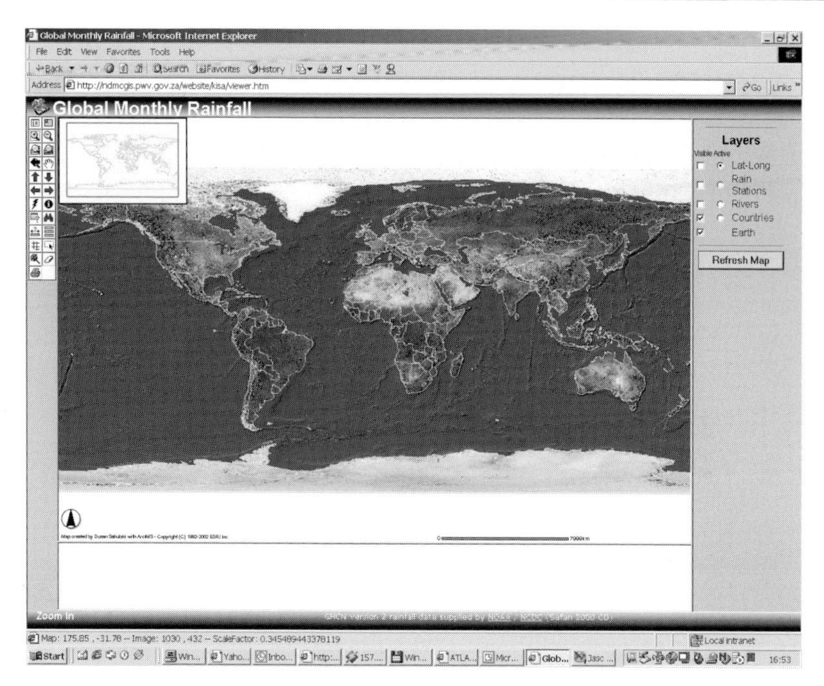

Fig. 9a .Global monthly rainfall GIS user interface (ArcIMS)

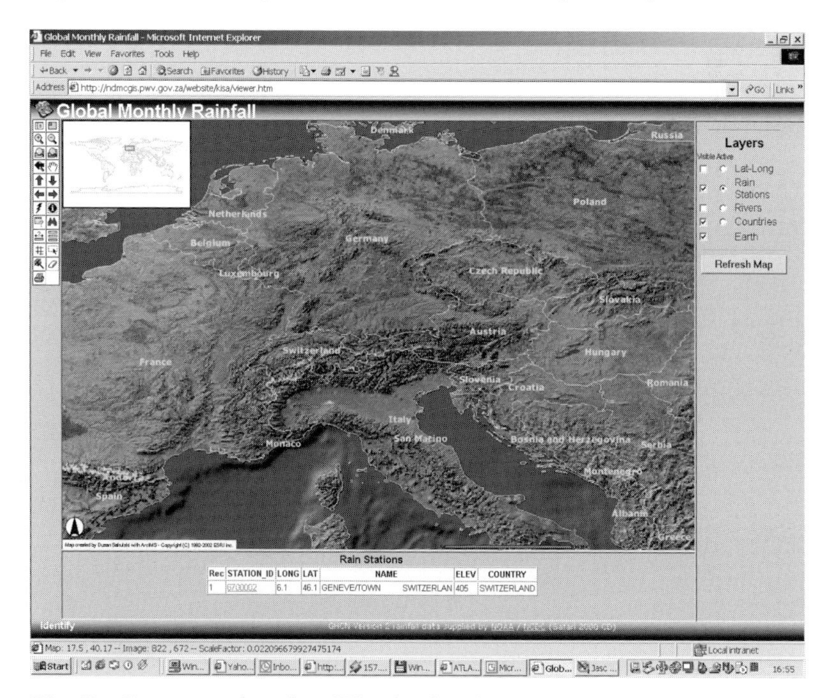

Fig. 9b. Country and station GIS selection front end (ArcIMS)

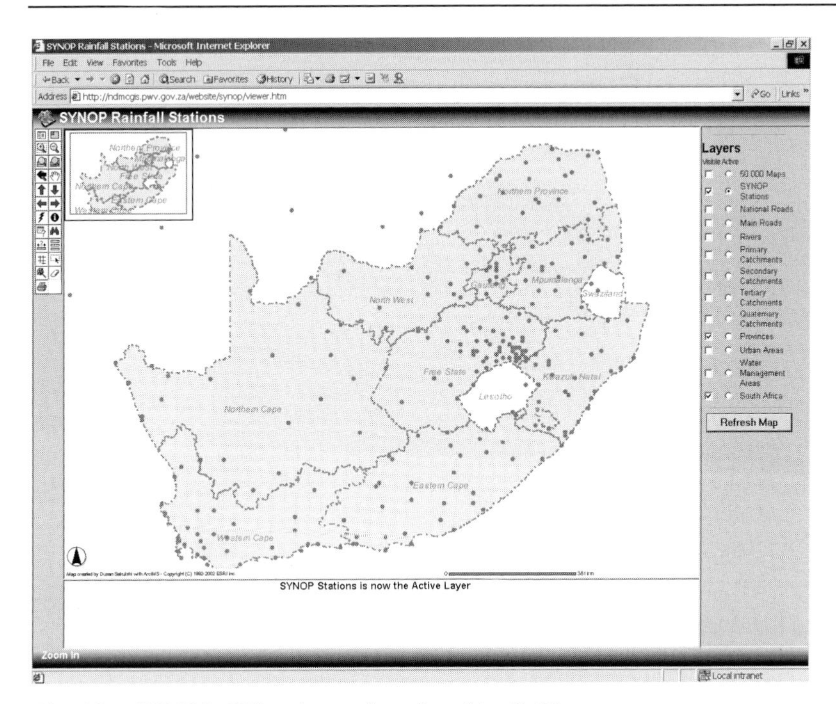

Fig. 10a. SYNOP GIS web user interface (ArcIMS)

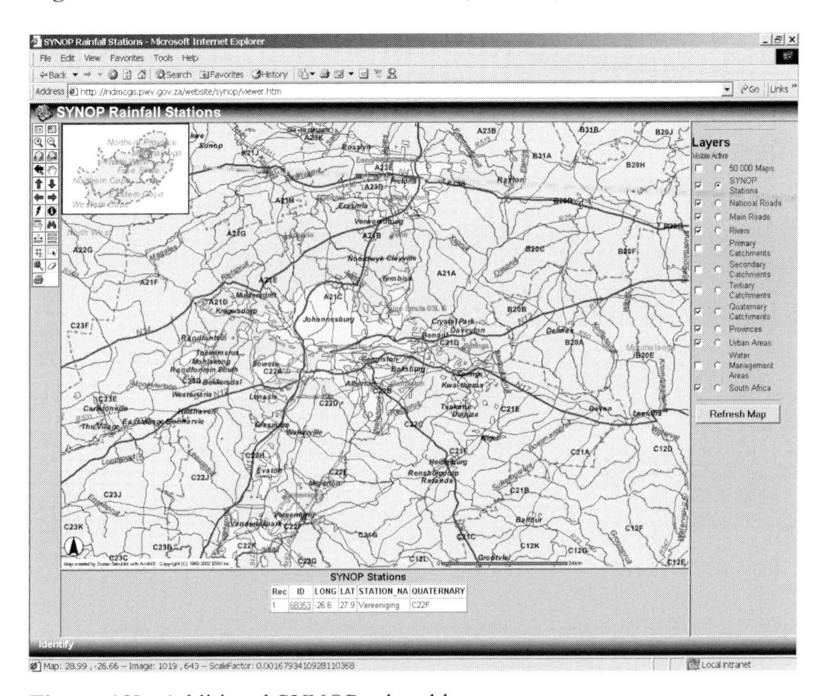

Figure 10b. Additional SYNOP related layers

South African Weather Services provides, every 6 hours, SYNOP file according to the WMO standards. NDMC receives a copy of that file and, automatically, extracts relevant rainfall data and stores it into relational database. Internet and Intranet user (http://sandmc.pwv.gov.za/atlass/rain/) can easily manipulate SYNOP data, in the form of accumulated daily rainfall, usig ArcIMS-based user interface (Figure 10a and 10b).

Apart from the SYNOP stations, additional layers (Figure 10b), such as national and provincial boundaries, primary road network, river network, catchment boundaries, water management areas, urban areas are also available.

5.2 Drought

Drought should not be viewed as merely a physical phenomenon or natural event. Its impacts on society result from the interplay between a natural event (less precipitation than expected resulting from natural climatic variability) and the demand people place on water supply. Human beings often exacerbate the impact of drought.

Recent droughts in both developing and developed countries and the resulting economic and environmental impacts and personal hardships have underscored the vulnerability of all societies to this "natural" hazard

Basis for drought analysis is an index called the Standardized Precipitation Index (SPI). Index was developed by McKee (McKee et al., 1993), with the intention to give a better representation of wetness and dryness. At the NDMC SPI is calculated on monthly basis. Historical spatial and temporal SPI distribution is calculated on-the-fly, utilising integrated database-driven web-enabled technology.

Spatial units are quaternary catchments (polygon entity) as well as gauging stations (point entity). Again, GIS ArcIMS browser user interface has been developed to enable catchment or gauging station selection (Figure 11).

Spatial SPI distribution, as a result of the complex time series analysis, is also GIS enabled, in the form of simple “semaphore” like colored drought levels (Figure 12).

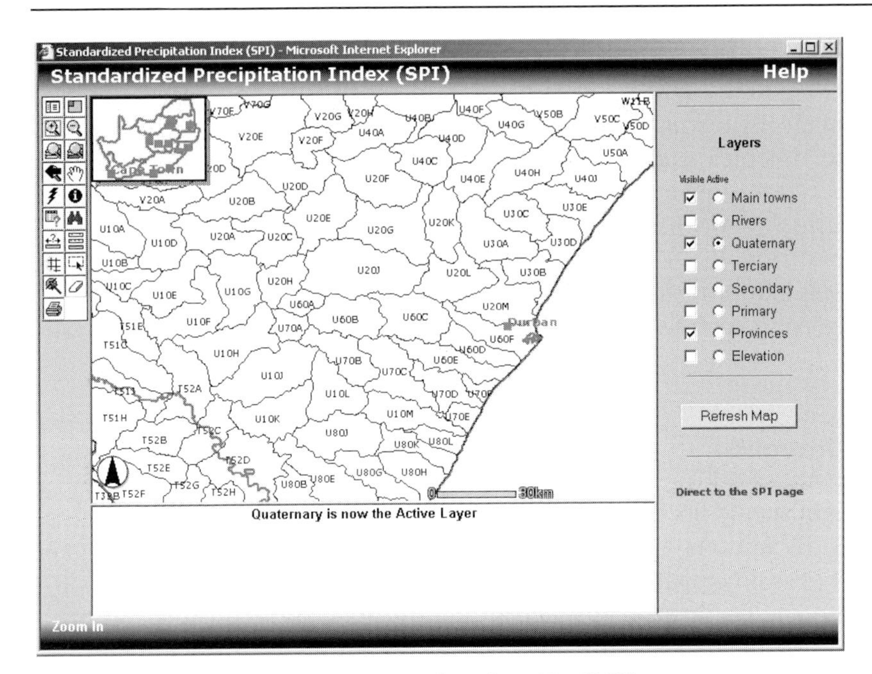

Fig. 11. Catchment selection GIS user interface (ArcIMS)

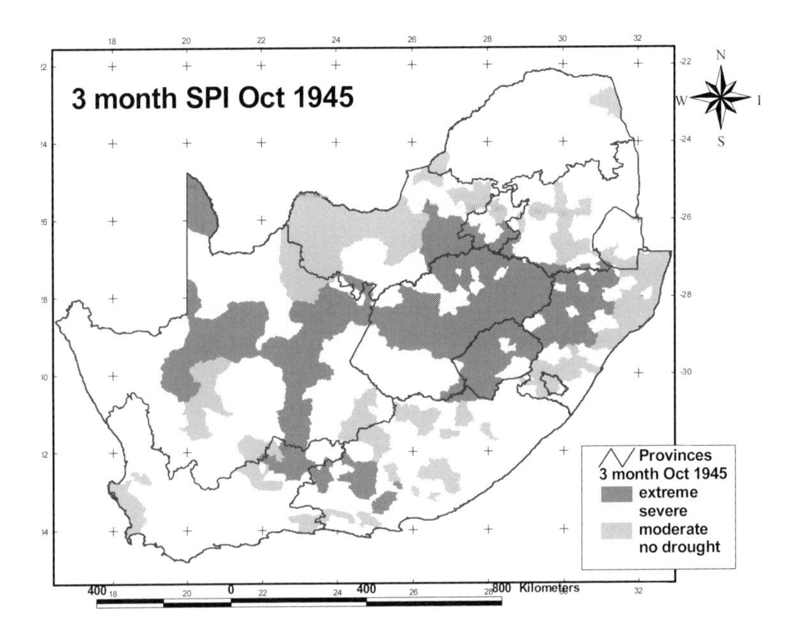

Fig. 12. SPI spatial distribution

5.3 Cholera Outbreak Monitoring

At the beginning of year 2001 NDMC was approached by the National Department of Health to help develop a database-driven web-enabled application for cholera outbreak monitoring and management.

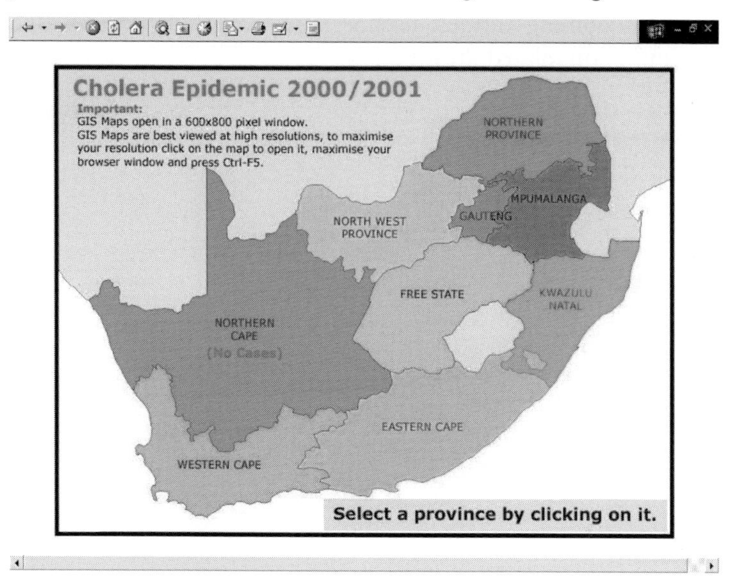

Fig. 13a. GIS enabled area selection screen (ArcIMS)

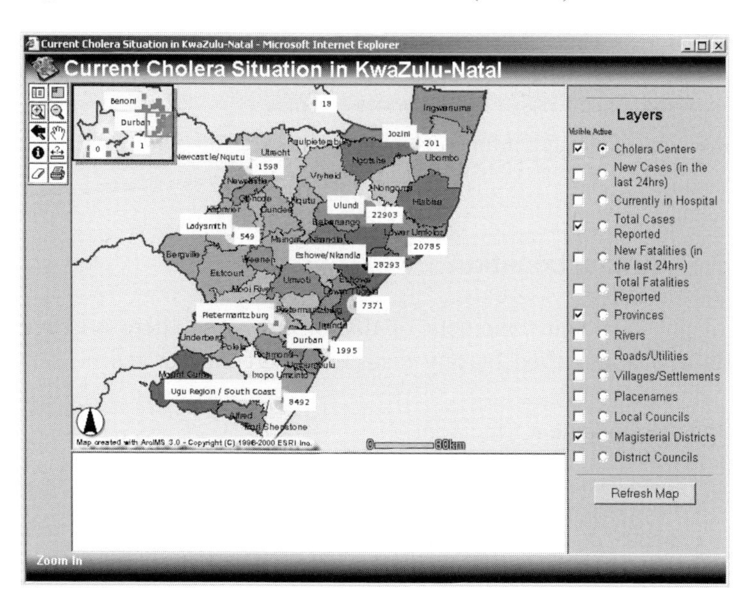

Fig. 13b. Additional layers for selected province (ArcIMS)

Cholera data have been transferred, via ftp, to NDMC database server, and structurally stored into a relational database. Again, GIS technology has been utilized to enable interactive and flexible user interface, using ESRI's ArcIMS technology (Figures 13a – 13c). After selecting a province, user is able to include other spatial information, such as local councils, hospitals, etc. Zooming further, user is able to include schools and their water and sanitation related information.

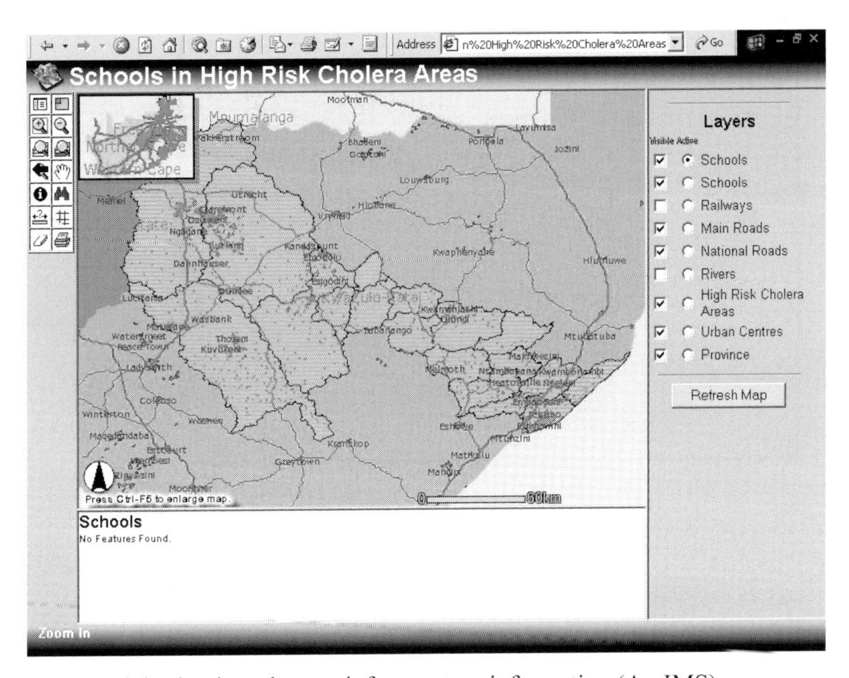

Fig. 13c. Arial schools and water infrastructure information (ArcIMS)

5.4 "Hot Spot" Spatial Location

In the case of any accident / incident or disastrous event, there are five most common questions asked by any disaster, emergency or rescue service:

- Where is "place of accident / incident"?
- Are people there?
- How many people are there?
- What is the capacity of the nearest airport (if any)?
- What is the status of the road network around?

To be able to answer those questions, we at the NDMC use combination of the database-driven, web-enabled Geographic Information Systems (GIS) and alphanumeric tools. Figures 14a-14c are visual answers of those questions.

Since the beginning of this year, this application has been improved enabling users to utilize GIS web version of 1:250000 topographic maps (Figures 15a – 15c). Using web browser user can zoom in, select appropriate section and use selected map.

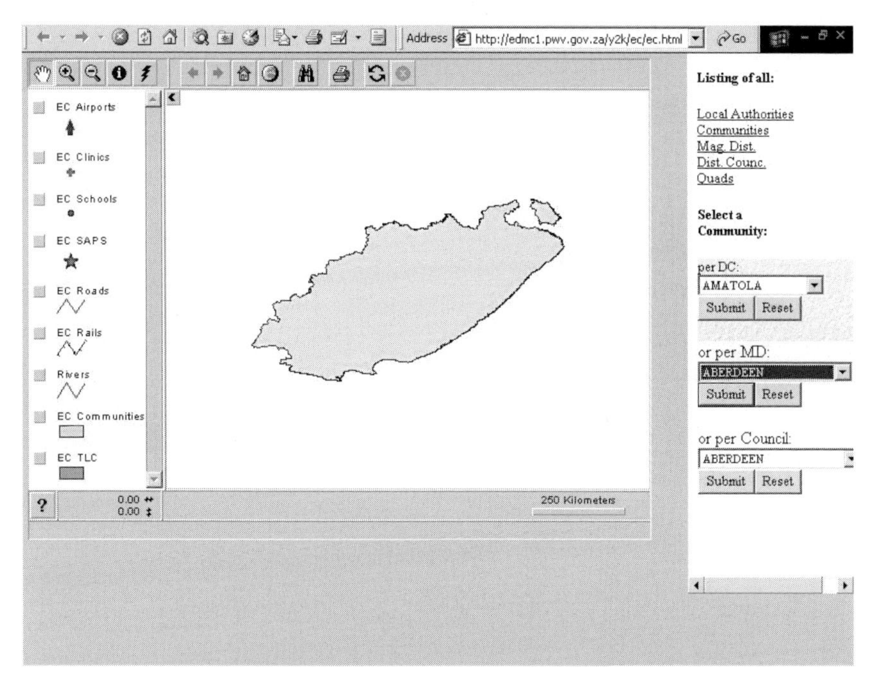

Fig. 14a. Question 1 (Eastern Cape Province, South Africa)

Community Selection - Microsoft Internet Explorer

Address: http://edmc1.pwv.gov.za/y2k/ec/SelectEC.asp?where=EC&what=MAGDI

Province: EC ---> Total of 6 Communities

No. of people: 10327

No	District	Magisterial	Council	Community	Population
1	WESTERN REGION	ALEXANDRIA	ALEXANDRIA	ALEXANDRIA	1750
2	WESTERN REGION	ALEXANDRIA	CANNON ROCKS	CANNON ROCKS	2950
3	WESTERN REGION	ALEXANDRIA	ALEXANDRIA	KWANONQUBELA	713
4	WESTERN REGION	ALEXANDRIA	BUSHMANSRIVERMOUTH	MARSELLE	2552
5	WESTERN REGION	ALEXANDRIA	PATERSON	PATERSON	1439
6	WESTERN REGION	ALEXANDRIA	CANNON ROCKS	RICHMOND	923

Fig. 14b. Questions 2 and 3 (population info)

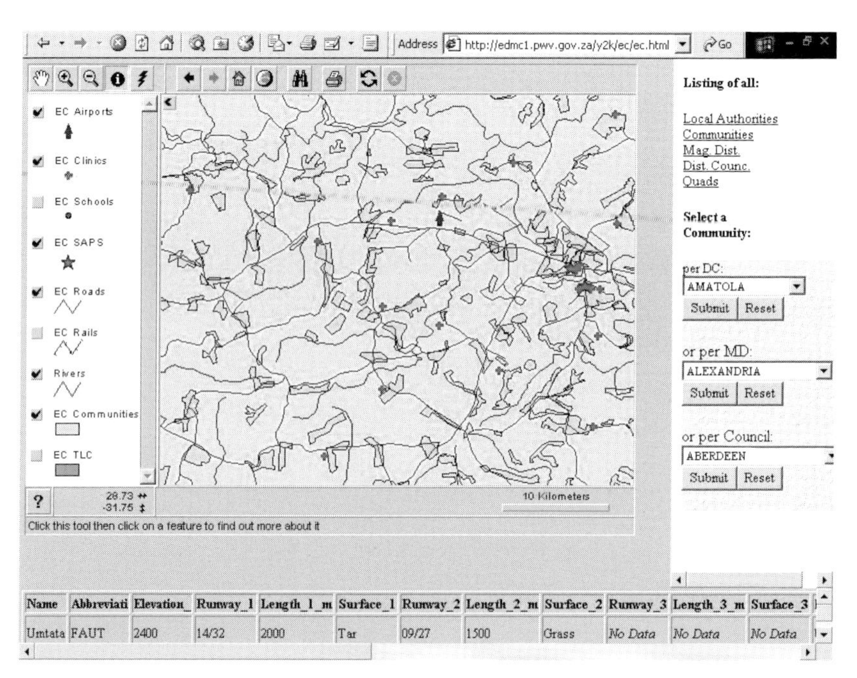

Fig. 14c. Questions 4 and 5 (communication info)

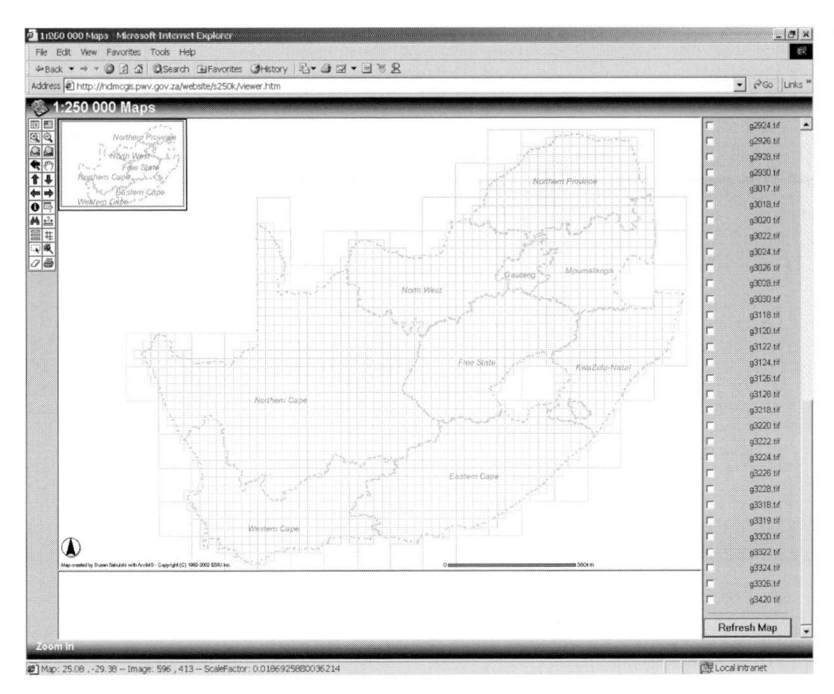

Fig. 15a. 1:250000 map grid (ArcIMS)

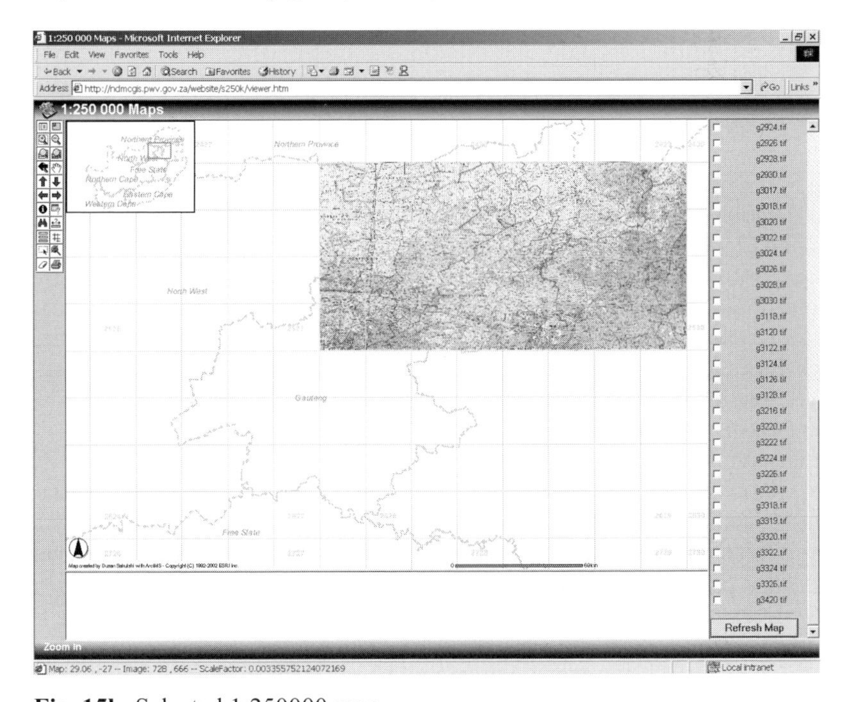

Fig. 15b. Selected 1:250000 map

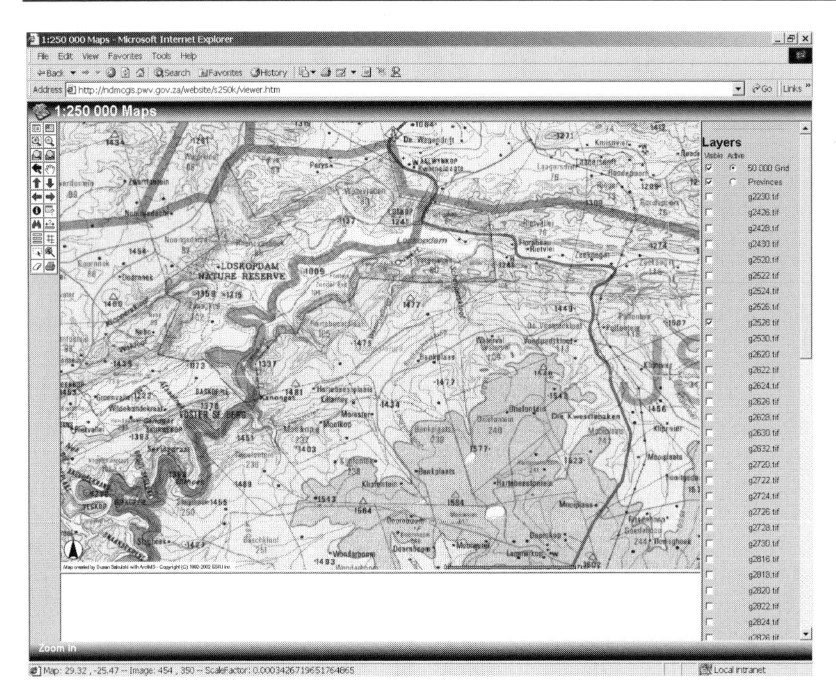

Fig. 15c. Zoom in selected 1:250000 map

References

McKee, TB, Doesken NJ, Kleist J (1993). The relationship of drought frequency and duration to time scales. Preprints, 8th Conference on Applied Climatology, 17-22 January, Anaheim, CA, pp 179-184

Sakulski, (2000). Implementation of the multi-software solution for the on-the-fly calculation of the Standardized Precipitation Index (SPI) as a drought indicator for the South African environment. 8th International Conference Computer Techniques to Environmental Management, ENVIROSOFT 2000. WIT Press.

METSYS, South African Weather Service, (http://metsys.weathersa.co.za)

National Disaster Management Centre (NDMC), South Africa. (http://sandmc.pwv.gov.za)

National Oceanographic and Atmospheric Administration (NOAA), USA. Geostationary Satellites Monitoring Server (GOES) (http://www.goes.noaa.gov)

National Oceanographic and Atmospheric Administration (NOAA), National Weather Service, Climate Prediction Center, African Desk, USA. (http://www.cpc.ncep.noaa.gov/products/african_desk/mrf_fcst/mrf_QPF24_96.gif)

University of Hawaii, Institute for Astronomy, USA. Tropical Storms Monitoring Worldwide (http://www.solar.ifa.hawaii.edu/Tropical/)

Seismic Emergency Management: Technologies at Work

Pierluigi Soddu and Maria Giovanna Martini

Dipartimento della Protezione Civile via Vitorchiano, 2, 00189 Roma, Italy.
Email: pierluigi.soddu@protezionecivile.it;
giovanna.martini@protezionecivile.it

Abstract

This job shows the experience of the office "Servizio Sismico Nazionale" (OSSN - Italian Department of Civil Protection), started in the 1992, to build up the complex information system: SIGE. SIGE, that utilizes a lot of integrated technologies to support the seismic emergency management: G.I.S., G.P.S., G.P.R.S., P.D.A. and W.E.B., is activated in case of an earthquake of magnitude 5 or more that hits the Italian Territory. The article describes the different phases of a seismic crisis: the activities start from the scenario production in the OSSN centre, continue in the involved area with the data collection and close with the production of synthesis maps for the emergency management. SIGE utilizes the positive experience of two IST European project: FORMIDABLE and EGERIS.

1 Introduction

The Italian law 225/ 92 affirms that the followings are Civil Protection activities:

- Forecasting - activity devoted to the analysis of causes of disaster events, to the identification of risks and to detection of risks areas;
- Prevention activities whose aim is to reduce damages due to the disaster events, also taking into account the knowledge gained during the forecasting activities;
- Assistance for all interventions whose aim is to ensure the elementary assistance to the population;

- Emergency management and overcoming - all the needed interventions for pursuing the achievement of acceptable *quality* of life conditions.

The definition, elaboration and management of the emergency plan for the risks on the territory have enacted by the law (225/92), while the main characteristics for the emergency planning have described in the Augustus methodology. Augustus methodology consists in the definition, maintenance and execution of an emergency plan to face any kind of emergency in local areas. The 14 Support Functions foreseen in Augustus represent the basic organization required in order to make any kind of emergency plan effective.

Function	Description
1.	Technical and Scientific Support for the scientific analysis and physical interpretation of the event and related data
2.	Health, Social Assistance and Veterinary Services for the management of all people and means involved in the sanitary area
3.	Mass-media and Information for the distribution of specific information through mass-media and to citizens
4.	Volunteers for the coordination and training of specific volunteers organizations
5.	Material and Resources for estimation of all resources necessary during emergency, their location and availability
6.	Transport and Mobility for transfer of material and people, to optimize evacuation paths and regulate intervention flows
7.	Telecommunications for the provision of TLC networks, as back up and support to operations
8.	Utilities for the coordination and maintenance of all necessary services (e.g. water, electricity, gas)
9.	Damage Assessment for the estimation/evaluation of damages (e.g. to people, buildings, agriculture)
10.	Search and Rescue Operational Organizations for the co-ordination of all entities involved in S&R operations
11.	Local Authorities for the coordination at aids at local level in terms of resources, necessary services, accessible areas
12.	Dangerous Materials for the management of storage location and material census with respect to the impact on affected areas
13.	Population Logistics for coordination of assistance to population, in terms of identification and set-up of suitable areas to provide assistance and necessary services
14.	Operational Coordination for the management of the support functions and the rational interventions of means and people

Table1. Example of support function

The Support Functions, within an Emergency Plan, represent the answers provided to the different needs arising during an emergency management. The relevance of each function may change accordingly to the effects produced by the event, in order to keep any emergency plan flexible and efficient (see FORMIDABLE project).

2 SIGE

Since 1992 the Italian National Seismic Survey (OSSN), actually it is an Office of the Italian Civil Protection Department (DPC), has been developing an information and decision support system named SIGE (contraction of the Italian words: **S**istema **I**nformativo per la **G**estione dell'**E**mergenza), in order to support the activities for:

- The Italian seismic risk assessment and reduction;
- The National emergency management.

In Table 2 there is a road map of SIGE system.

Year	Topics
1992	Developing of the data bank of Servizio Sismico Nazionale (SSN) of whole Italian territory
1994	Developing the first SIGE system (using Unix – Arc Info Oracle). Produce a territorial description
1995	Introducing in SIGE the first model to produce an earthquake consequences data base (see seismic model)
1997	Introducing QUATER (friendly arc view module that use SIGE products.)
1999	Developing SIGE on LINE (Unix cluster system, early warning system, sending the scenario to the local CPA and produce a dedicated files to the operative teams.)
2000	Introducing the Augustus methodology: SIGE produces dedicated data base for different function: Health, Social Assistance, Transport and Mobility and Operational Co-ordination (Activities from FORMIDABLE EU Project)
2002	Developing the PAD, GPS, GPRS technologies (data directly from the involved area- Activities from EGERIS EU Project). Developing new scenario model.
2004	Developing new system to collected data. Introducing project of Earth Observation

Table 2. Time road of SIGE evolution

The main topics of SIGE system can be summarized in the following points:

- A central client server architecture;
- Connection, in emergency, between OSSN and the different operative civil protection units displaced over the national territory (CCS- Provincial level, COM – Municipal level: Operational Centers created immediately after the event in the areas hit by earthquake).
- Selective dispatching for the remote Civil Protection stations.
- Mobile Computing: Hardware and software for in-situ data collection (users: OSSN teams for technical surveys) and store on OSSN server. Dynamic scenarios production:
- Dedicated WEB site (intranet and internet) in emergency.

For the architecture of SIGE system see Figure 1.

The central units of SIGE system consist of:

- A client server architecture based on UNIX platform, Oracle RDBMS and ARC/INFO G.I.S., in an integrated environment;
- A data bank for seismic risk evaluation including both cartographic and thematic information. Nowadays, the databank includes about 80 different databases (more than 600 cartographic layers and 500 alpha numeric attributes). Every database describes the whole national territory with an homogeneous level of accuracy from the North to the South; have the same format (coverage arc/info) and the same projection system (UTM - Zone 32); The SIGE data bank (demography, buildings, services, historical seismology, roads, dams, risk factories, etc) have these focal points:
- Each single database is official.
- It is relative to the administrative municipality areas;
- A metadata structure builds according to European standard.
- A decision support system tool for the seismic emergency management.

To carry out the emergency activities are needs to characterize the scenario.

The scenario in an *emergency planning* phase means to identify and describe the reference event to be used in order to set up the emergency plan and, using that, make a loss evaluation (analysis of the consequences), with the purpose of setting the civil protection answers in the specific emergency plan (i.e. number of needed resources in case of crisis, needed means…).

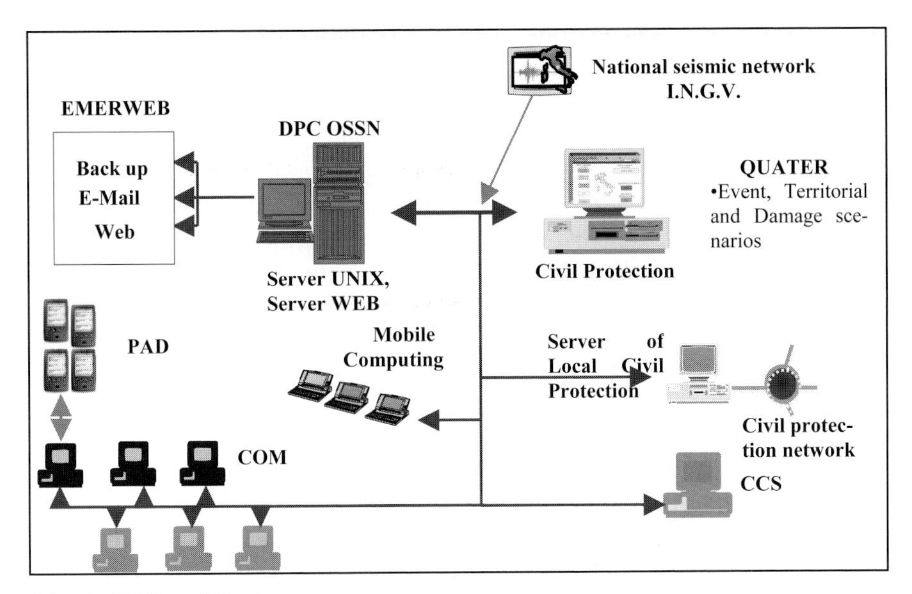

Fig. 1. SIGE architecture

In an *emergency control time,* the scenario means to describe the event, the affected area and the impact that the event has had on the territory, with the aim to support the operative civil protection units in their activities during the emergency (identification of the affected area and recognition of the severity of the event passing from an esteem, given few minutes after the event, to a description, given updating the scenario with the data that arrive in real time from the in-situ observations). In post-emergency phase the scenario means to give a final and detailed description of the event, of the affected areas and of the consequences that it has caused, in order to support the reconstruction planning activities.

In order to support seismic emergency activities at national level, a specific sub-system (SCEN – Scenario Management) has been designed inside SIGE system for seismic scenario analysis.

During an emergency, the seismic scenarios required three analyses:

The event evaluation that means to describe the seismic input and the shaking in all the involved territory;

- A description of the involved territory (lifeline, facilities, population and so on,)
- A loss assessment that means the consequences analysis for the damage evaluation.

In the first emergency with few data available, the system uses a probabilistic model to esteem the event and damage parameters, so the scenario output provides an esteemed value and its relative uncertainty (see seismic model).

The data coming from the field are a control point and a feedback for every further model, and runs so the value uncertainty decreases with the time.

Nowadays, by this system, immediately after the earthquake of magnitude 5 or more, the Italian Civil Protection department, through a direct call to the SIGE system, inserting only the earthquake co-ordinates and magnitude as input data, can run a specific program for the consequences analysis of the interested scenario.

This programs after five - ten minutes, produce up to 20 cartographic databases resuming the crisis scenarios components. The output data describes the territory for a radius of 50 Km from the epicenter (infrastructures, public and private service, population, geological information, ...), and provides a first estimation of the event and of the damage (population, residential real estate and monetary losses). In order to query the first seismic scenarios in SIGE system, there are 3 operative environments available for different user's typology.

For a technical operators SIGE system provides an operative friendly interface named QUARTER. It is used to generate query, report, analysis and display for all the above-mentioned data produced (Figure 2 – example in Modena province).

So, after few minutes the technicians know, esteem and print many topics:

- Ground motion parameters: intensity, PGA, PGV, spectral values;
- The expected structural damage (collapsed, uninhabitable and damaged dwellings, damaged surface, monetary loss);
- Expected number of casualties (fatalities and homeless);
- Territorial scenarios: many maps and data about population, residential real estate, facilities, infrastructure and lifeline, geological and seismological information;
- Possibility to have the selected scenario for different Function: for example the Health scenario is organized follow the statistic of disaster medicine and follow the Triage code (Figure 3 example in Sicily area).

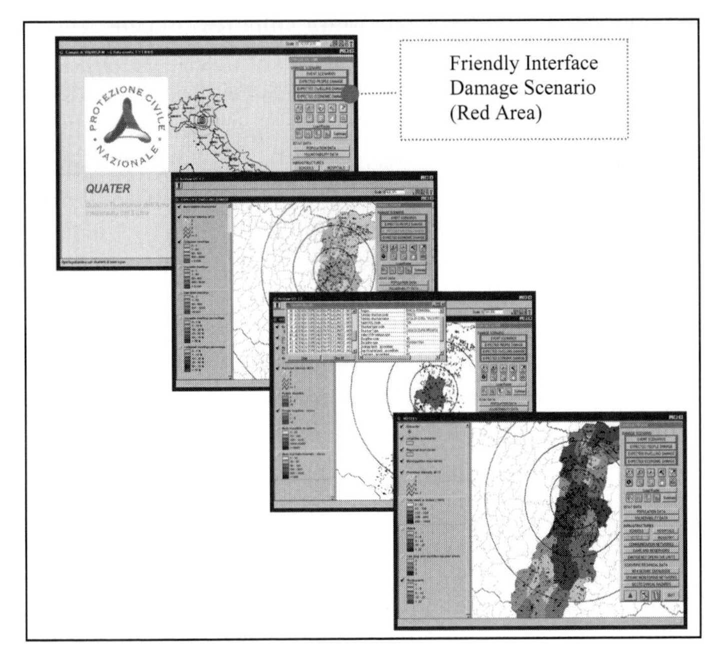

Fig. 2. QUATER

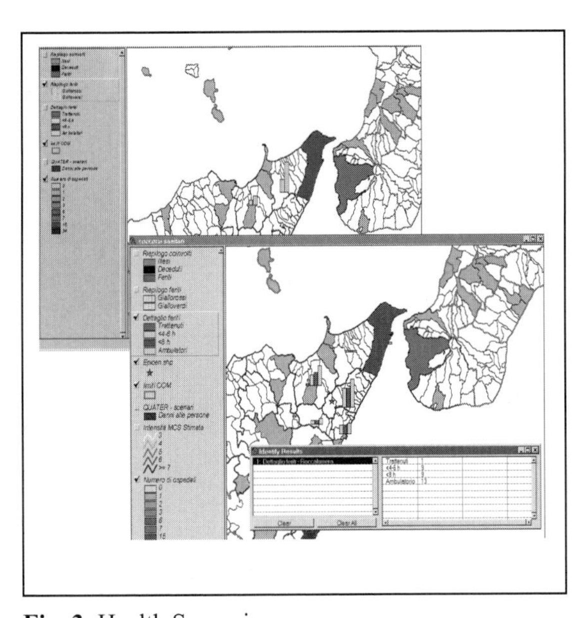

Fig. 3. Health Scenario

A dedicated component produces after 30-45 minutes, depend of the earthquake's magnitude, also an instant book scenario (paper report), consisting of more than 50 pages resuming the crisis scenarios components, for disaster declaration at the Italian government. Report contents cover the following areas: demography, building, services, networks, historical seismology, damage, scenarios (property and people). In intranet there is the whole report (example of the instant book: index in Table 3, maps and tables in Figure 4, earthquake in Sicily Area).

The report (without the damage scenario) is available in Internet at the following address: www.serviziosismico.it.

For public information (population, media and press news, etc) a specific application in Internet environment is now under development.

And after ten minutes? How does SIGE support the emergency?

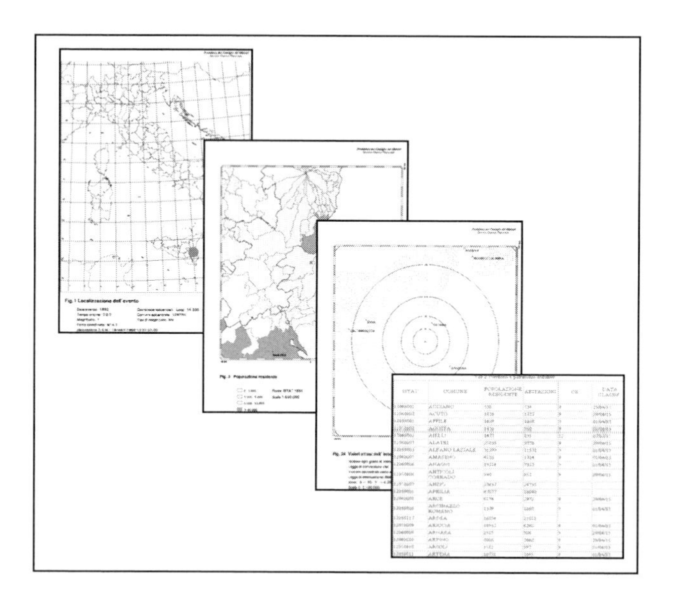

Fig. 4. Instant book

Fig	Description	Fig.	Description
1	Earthquake area		
2	Municipalities around 50 Km of the epicentre	31	Expected total collapsed habitations
3	Resident population	32	Expected unsafe habitations
4	Official seismic classification (municipality)	33	Expected percentage of collapsed habitations
5	Population density (inhabitant/Kmsq.)	34	Expected percentage of unsafe habitations
6	Percentage of habitations for municipality	35	Expected value of people involved
7	Percentage of habitations in class A (MSK scale)	36	Expected value of homelessness
8	Percentage of habitations build before the (municipality) seismic classification	**Table**	**Description**
9	Number of hospitals (public and private)	1	Territory and Population
10	Number of school rooms	2	Vulnerability MSK
11 12 13	Number of potential risk industries (class A B C)	3	Risk factories
14	Vulnerability for landslide	4	Schools and Hotels
15	Dams	5 6	Hospitals (Public and private, beds)
16 17	Roads, Railways, Airports (around 150 and 25 Km of the epicenter area)	7	Dams
18	Earthquake database: NT4 (Intensity MCS >5)	8	Earthquake Database: NT4
19 20 21 22 23 24	Macro seismic field of historical earthquakes: 1349,1695, 1799, 1898, 1904, 1943	9	Seismic and acceleration networks
25	Expected Intensity (MCS scale)	10 10a 11	Building forecast loss (value and percentage)
26 27 28 29	Expected value of PGA (g), PGV (cm/sec), PSA (T =0.2 sec, Hz.=5.00), PSA (T=0.5 sec, Hz=2.00 Hz)	12	Earthquake casualty estimates
30	Seismic and acceleration networks	13	Forecast monetary losses

Table 3. Example of SIGE instant book (index)

3 Local Activities

The OSSN is developing tools for updating both the event and the damage scenario with the data directly coming from the technical teams in the involved area (results of EGERIS project).

The EGERIS aim was to provide the Civil Protection organizations and national or local authorities concerned with emergency management with the most recent ICT developments to support them in their risk-based on-field emergency operations. During the emergency phase the informative system has to support the activity concerning the organization of information coming from technicians teams operating in areas involved by earthquake.

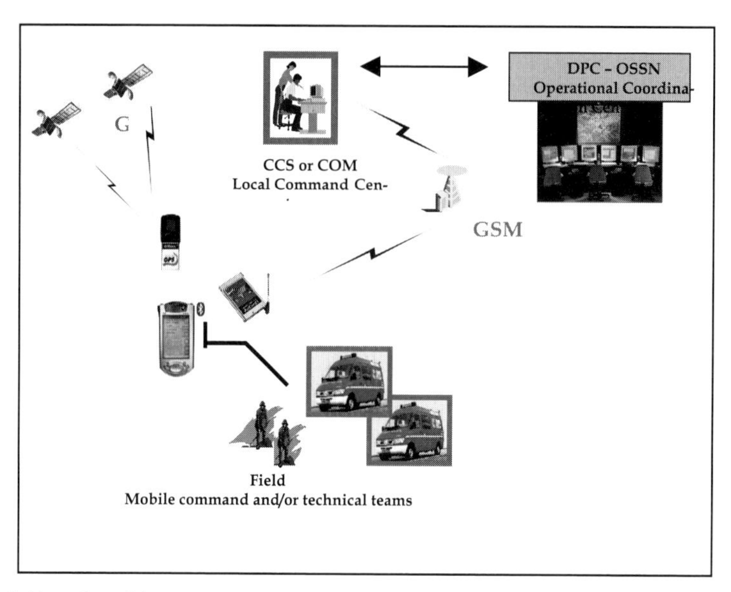

Fig. 5. Local architecture

The architecture (in local area and into the COM or CCS) includes three levels (Figure 5):

- Crews on-the-field equipped with portable devices (hand held terminals e.g. PDA with a GPS receiver and GPRS communication system);
- Mobile command centre installed in vehicles, close to the "hot spots" and coordinating several crews on-the-field; equipped with on-board devices like lap-top computers and GPS receivers;
- Function Specific in COM, hosting the local management and interfaces with Auxiliary functions (Technical & Scientific Support, Viability, Health Assistance, Damage Assessment, etc.), including reporting to upper level of co-ordination.

Function Specific Management	**Mobile Command Centre**	**Portable Tools**
GIS Database Directives management Reports management Centralized squad monitoring and control Planning management Communication support (data, voice, faxes …)	Communication support (data, voice, faxes …)	GIS Database Reports management Priority checklists management Positioning navigation Planning tools Communication support (data, voice, faxes…)

Table 4: EGERIS focal points

The points, in Table 4, are tested in EGERIS and now we are planning to integrate in the SIGE system. Immediately after the earthquake, the scenario is send to the local command centre and here the activities in order to organize and coordinate the different survey start.

The trial (middle of September 2003), that tested methodologies and technologies, has been focalized on these Functions: Transport and Mobility (road condition), Technical and Scientific support (macroseismic activities), Damage assessment (building damage).

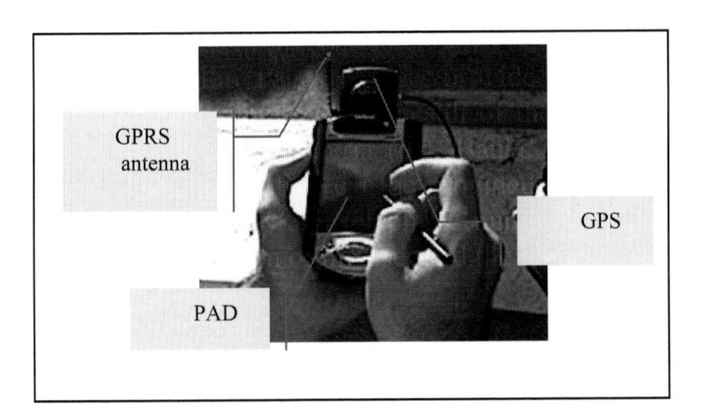

Fig. 6. PAD, GPS, GPRS integrated station

3.1 Road Condition

The road condition is one of the first activities to be carried out. It is important because it aims to producing a real description of the road network status. The data are required for better planning of all the other activities in the area (Figure 7).

Moreover, the teams have, the possibility to send an informal report about critical points (landslides, damaged buildings, etc.) to other functions too.

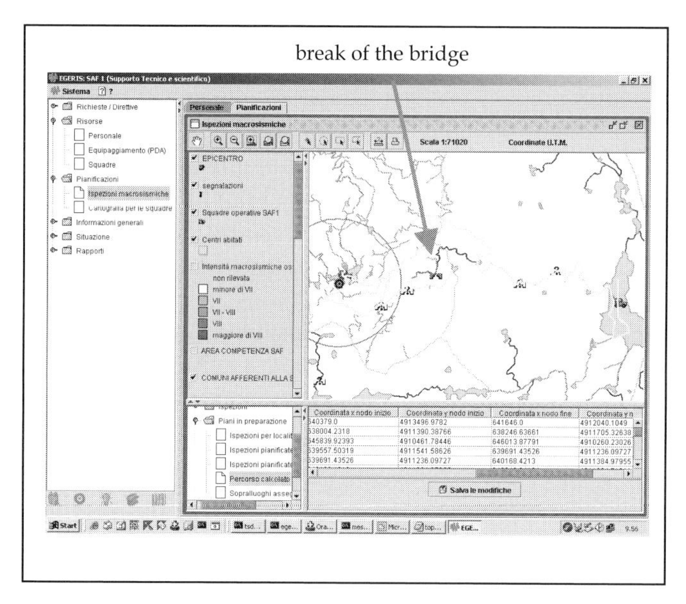

Fig.7. Update Road network and new path

3.2 Macro Seismic Survey

Goal of the activity: produce in real-time the updated macro seismic intensity map. Intensity maps (plans) represent the summary of the impact produced from the earthquake on the land; in fact, the teams quantify the real damage occurred in every locality after earthquake.

According to the representation scale, maps have been produced by associating the MCS intensity value carried out by technicians (by using different parameters) to the polygon of the built-up area or to its centre (cartographical database: ISTAT 1991 Centers and nucleus. ISTAT: National Statistical Institute).

By this representation (called "quoted plan") it has been pointed out:

- The more damaged areas: damages to property have generally checked (verified) starting from a MCS 5-6 degree;
- Earthquake impact on landing (people feeling after earthquake). For all the assessed localities where impact has been greater or equal to 4 degree MCS scale, a detailed map-making has been done;
- The possible presence of local effects producing amplification;
- and lastly a conclusive processing pointing out a damage estimation on the whole municipal territory.

The operative teams have in the PAD a simple cartography and a (update) road network where are selected specific point (locality or municipality) to check. If there are particular points that define better the damage, there is the possibility to take the coordinate and send the macro seismic value to the COM.

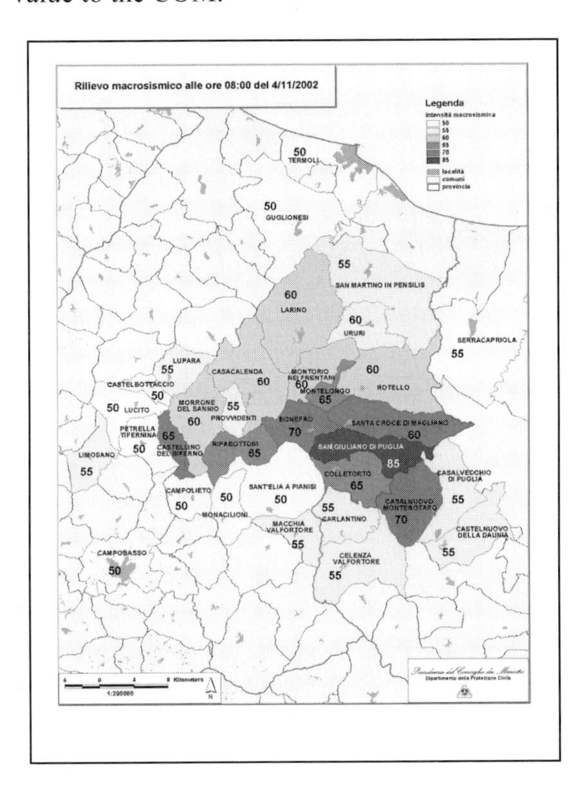

Fig.8. Macro seismic survey: Molise Earthquake

In the COM there is the possibility to look where the teams are. All the collected data (the information coming from different teams in different COM) are sending in the central unit of SIGE that harmonizes the information and produces a summarized map. In Figure 8, there is an example of macro seismic map referred to the Molise earthquake (October 2002).

3.3 Building Damage Assessment

The last activity, tested in the trial, is the Building damage assessment: to manage the safety analysis (buildings private and public inspection).

Normally, the activity can last months: as an example, last earthquake in 2002 in Molise (centre of Italy) required 3 months and 80 squads, it allowed to know immediately which buildings had to be evacuated; it is based upon an official citizen request after the event (only for private building) and it allows a step-by-step refinement of the damage scenario. In Figure 9 there is an example of work planning. The work planning is sent to the teams (different work planning for different teams).

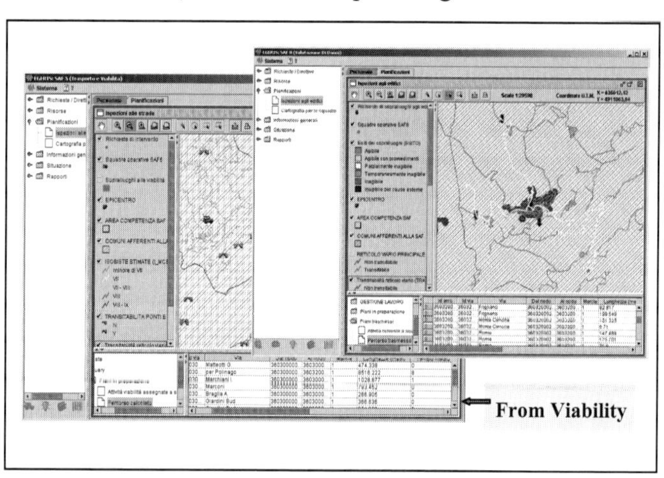

Fig. 9. Work planning of Building Survey and update Road

Each team has an integrated station PAD ready to collect the building data (Figure 10) by a dedicated interface. The collected information is sending to a Local Command Centre (COM) to validate the operation and to store the chronological history of the team's activities. Finally, the daily database is sending form the COM to the SIGE central unit (sub system) where it is re-tested, organized and sent back.

The sub-system allows the data collection from different COM, the data integration and database update. By those data SIGE runs a specific program and produces the thematic map for the national crisis rooms, showing the updated seismic scenarios.

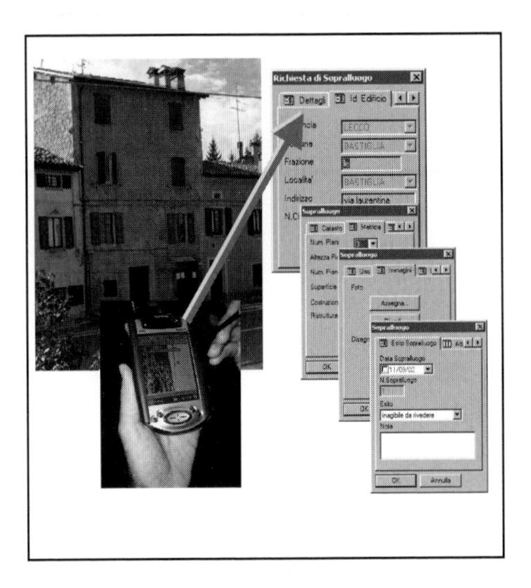

Fig.10. PAD, Interface DB for Building survey

3.4 Emergency Cartography

The last activities of SIGE are those of collecting the information that daily cam from the COM into the central crisis room. In the last earthquake SIGE produce daily-dedicated maps to show the evolution of the emergency.

The maps are organized for the Augustus functions (dedicated tema) and more of 100 maps in one month of observation have been produced (example of theme: helped population in Figure11).

The seismic Model code (since 1995) is running on the SIGE and is used after any event in the national territory capable of producing significant damage. Earthquake consequences such as the expected number of damaged, unusable and collapsed dwellings; the direct economic losses; and the number of homeless individuals, casualties and fatalities in each municipality, are estimated in nearly real time. The performance in terms of global losses is generally satisfactory. However, the code has undergone various modifications and extensions in order to remove limitations and improve the accuracy of the original model.

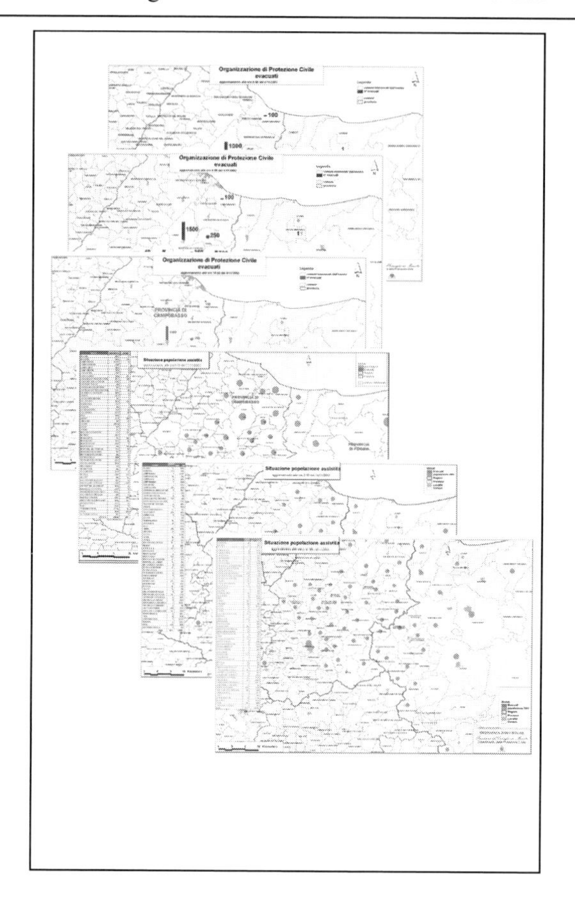

Fig. 11. Molise Earthquake. Time observation October- November 2002. Tema: Helped population

3.5 Seismic Model

In the code implemented in the SIGE, developed for emergency management purposes, the event severity is measured both in terms of macro seismic intensity and in terms of expected ground motion parameters, but only the intensity is used for loss estimation. Furthermore the municipality is considered as an isoseismic unit, i.e. the hazard is defined by a unique value of the macro seismic intensity (I) that has to be considered as a mean value over the municipality territory. The attenuation of the intensity from the epicentre is circular with a unique law for the entire nation.

The inventory of the exposed structures is limited to the dwellings collected by the ISTAT (National Statistical Institute) during the census of the Italian families performed in 1991. About 25 millions of dwellings and 57 millions of occupants are considered. Other strategic elements have been inventoried (transportation networks, hospitals, industries, airports and railway stations, dams), but their vulnerability is not evaluated yet. The code doesn't use the geology information.

The dwellings vulnerability is measured by means of correlation between the description of simple parameters available in the ISTAT census (type of construction and age) and the vulnerability classes described in the MSK - 76 macro seismic scale (Medvedev, 1977).

The damage in each municipality is estimated by means of relationship between damage and intensity expressed by the DPM 'damage probability matrices'.

At last, the consequences of damage are estimated in terms of economic losses, human lives, and injured population.

The first updated version of the program is the FACES code, - Fault Controlled Earthquake Scenario - that allows considering linear sources, directivity effects, and influence of the hypocenter depth.

A second refinement has increased the spatial resolution of the analysis, from the municipality level (8,100 municipalities for the whole of Italy) to "localities" or census tracts (300,000).

At the end of 2001, this project aims at the improvement of the earthquake loss scenario modulus, currently implemented in SIGE, has been activate by the Office National Seismic Survey.

The main goal was to develop a second-generation code to evaluate in real time the damages and losses in post-earthquake scenarios. The code should include several improvements over the existing OSSN capabilities and should be able to better utilize currently available information on seismic sources, regional attenuation relationships, the inventory of physical infrastructure and population at risk, etc.

From a conceptual point of view, the main innovations are two:

- The generation of damage and loss estimated in a probabilistic format (as opposed to single best estimates);
- The capability to modify such probabilistic estimation as relevant information becomes available after the seismic event.

These two innovations are expected to be the key points in order to produce a scenario code conceptually more robust, and so more flexible.

4 Conclusion

The job has underlined how an informative structure using GIS as an operative instrument may be employed to support Civil Protection and help to manage an emergency and to analyze the results coming from damaged areas.

Such operation has allowed to point out both the positive site of the approach and the limit between theoretical employment and operativity whenever the works are under pressure due to the emergency. It is obvious therefore that the operative instruments have to be set up and tested in peaceful time.

Field experience (EGERIS trial) has shown how new instruments have to be arranged: more reliable communication systems, GPS, the PDA and remote sensing as an informative aid etc.

Summary of Technologies utilized in SIGE System					
GIS	GPS	GPRS	PAD	WEB	E/O
Produce (first esteem and update) of the scenario during the seismic emergency. Facility to read the data. Update the scenario with data coming from the involved area. Mapping	Give in real time (by GPS and PAD) the coordinate of the teams in the operative area. Give the coordinates of the checked objects: bridges, landslides, building, etc.	Data Transmission between the technical teams and the COM.	The PAD is fundamental with GPS and GPRS to send and receive the data and work planned between the technical teams and the COM. Use Arc Pad to visualize the cartography and GPS to have the coordinate	For public information	First study to utilize the Earth Observation in Seismic emergency management.

Table 5. The follow table synthesizes technologies and applications that are used in SIGE system

Same of those technologies are ready to use others will be subject to new studies (for example Earth Observation) and the resulting instruments or the products will be integrated in SIGE system.

5 European Project

5.1 FORMIDABLE (Start Project 2000- End Project 2002)

The Friendly Operational Risk Management through Interoperable Decision Aid Based on Local Events (FORMIDABLE) System represents the standardized access to data and knowledge required to manage natural risks, with all means to access, maintain and exchange information for natural hazards, in particular those with a fast evolution time typical of the Mediterranean countries.

The FORMIDABLE Project has the main objective to: contribute to the definition of a European Standard Methodology for Emergency Management based on the consensus of major Mediterranean operational actors, and develop an interoperable support system prototype which integrates the resulting guidelines with data and tools needed to operate during an emergency, in line with Council Decision 98/22/EC of 19 December 1997

A critical analysis of where emergency management operations encountered major problems, causing even partial failure or delay in recovery action, highlights a generic lack of co-ordination, and a poorly unified and homogeneous approach. Often actions have been planned and executed with a limited field of view about the real size of the events and the involved responsibilities for emergency management. Moreover, this "local management" has produced conflicts between the Central Civil Protection authorities and the local administrators, thus hindering the efficient execution of interventions.

Generally the emergency scenarios are quite complex, although in some cases the cyclical occurrence of natural events might help to improve emergency management. Although the repetition of an event can be regarded as a constant factor, the damage extent and the type of intervention required are variable aspects, also due to the changing features (geomorphologic, administrative) of the affected areas.

For this reason the "Emergency Management" of Natural Hazards should include, as main features, flexibility and simplicity to ensure efficient intervention and immediate relief to affected citizens.

WEB site : www.formidable-project.org

5.2 EGERIS (Start Project 2001- End Project 2003)

The ***EGERIS*** objective is to provide Civil Protection organisations and national or regional authorities concerned with emergency management with the most recent ICT developments to support them in their risk-based on-field emergency operations. The proposed Emergency Management tools are intended to be used for the *Response* phase (ACTION) and, to a lesser extent, for the *Preparedness* phase (PLANNING).

The major goals and objectives of the ***EGERIS*** architecture are to ensure:

- Alignment of the requirements for the ***EGERIS*** information system with the operational procedures that support emergency workers missions;
- Broader and federating approach to enable scalability/replicability and interworking at network and application levels;
- Adequate performance, reliability and security;
- Application of a collection of standards warranting operational interoperability (voice and data), investments durability and enabling integration with legacy systems (interface with current applications and public/private networks).
- The target ***EGERIS*** architecture includes three levels:
  1. Crews on-the-field equipped with portable devices (hand held terminals e.g. PDA with a GPS receiver);
  2. Mobile command centres installed in vehicles, close to the "hot spots" and co-ordinating several crews on-the-field; equipped with on-board devices like lap-top computers and GPS receivers;
  3. Function Specific Control Room, hosting the local management and interfaces with Auxiliary functions (Technical & Scientific Support, Health Assistance, Damage Assessment, etc.), including reporting to upper level of co-ordination.

A further level, the operational co-ordination of the emergency that is carried out in the operational co-ordination centre, is out of the scope of ***EGERIS*** as already addressed in the already funded IST Project FORMIDABLE. ***EGERIS*** will help in extending the activity, both in terms of methodology, and in terms of supporting systems, with respect to on-field emergency operations.

An important requirement is that ***EGERIS*** should be **flexible** and **scalable** enough to allow its function deployment to fit the particularities of encountered organisations and of course the technical possibilities of legacy systems to interface with (e.g. band-width, analogue or digital radio technology).

Web site: www.EGERIS.org

Acronyms

COM- CCS	Operational Centers for management of emergency (Municipality and Province level)
DPC	Italian Department of Civil Protection
FACES	Fault Controlled Earthquakes Scenario
FORMIDABLE	European project: Friendly operational Risk Management through interoperable decision aid based on local events
EGERIS	European project: European Generic Emergency Response Information System
INGV	National Institute of Geophysical and Vulcanology
ISTAT	National Statistical Institute
MCS	Macro seismic scale
MKS	Vulnerability classes
OSSN	Office National Seismic Survey
PGA	Peak ground acceleration
PGV	Peak ground velocity
QUARTER	Friendly interface of the SIGE system
SIGE	Information system to management of seismic emergency

References

Braga F, Dolce M, Liberatore,D (1982) "A Statistical Study on damaged buildings and an ensuing review of the M.S.K. - 76 scale", Atti del 7 ECEE - Atene

Coburn AW, Spence RJS, Pomonis A, (1992) "Factors determining human casualty levels in earthquakes: mortality prediction in building collapse", 10th World Conference on Earthquake Earthquake", U.S. - Italy Workshop on Seismic Hazard and Risk Analysis, Varenna,Italy

De Marco R, Colozza R Ferlito C Mercuri G, Orsini F, Papa A, Pizza G (2002) "The scenarios quick assesment methodology for seismic emergency planning at municipal scale", R50th Anniversari of the European Seismological Commission (ESC) – XXVIII General Assembly. Genova

Di Pasquale G, Orsini G, Pugliese A, Romeo RW (1998) "Damage scenario for future earthquakes" presented at the Eleventh European Conference of Earthquake Engineering, Paris

Di Pasquale R, Ferlito G, Orsini F, Papa A, Pizza G, Van Dyck J, Veneziano D (2004) "Seismic scenario tools for emergency planning and management", ESC Assembly, Potsdam, Germany - subsession SCF-2B

European Seismological Commission (1993) W.G. Macroseismic scales "European Macroseismic Scale 1992", Grunthal G (ed), Luxembourg

Medvedev SV (1977) "Seismic Intensity Scale M.S.K.-76", Publication Inst. Geophys. Pol. Ac. Sc., Varsavia

Martini MG, Soddu P (1995) 15th ESRI International User Conference; Palm Spring, CA- USA Study and design of an information system for civil defence interventions in seismic events

Martini MG, Soddu P, Ursino S (1995) 10th ESRI European Conference", Project for a system for the management of emergencies due to natural accidents, employing GIS", Praga

Orsini (1999) "A model for buildings vulnerability assessment using the Parameterless Scale of Seismic Intensity (PSI)", Earthquake Spectra, Agosto, vol 15, no 3

Soddu P et al (1993) 13th ESRI International User Conference; Palm Spring CA-USA, Study, design and implementation of a multimedia data base dedicated to a seismic risk evaluation

Soddu et a+ (1997) 18th ESRI International User Conference;1998 ; Information System and seismic Events. Central Italy Earthquake of 26 September

Soddu P. (2001) 'A European Perspective on disaster management'- Frascati – Esa-Esrin ; "S.o.L.: emergency management information system on line"

Soddu P, Martini MG, (2003) Seismic Emergency Planning Management: SIGE System and EGERIS project. Proc. Foro Euromediterraneo; Madrid

Tiedemann (1992) "Earthquake and Volcanic Eruptions - A Handbook on Risk Assessment", Swiss Reinsurance Company, Zurich

Pedestrian Navigation in Difficult Environments: Results of the ESA Project SHADE

Björn Ott[1], Elmar Wasle[1], Franz Weimann[1], Pedro Branco[2] and Riccardo Nicole[3]

[1] TeleConsult Austria GmbH, Schwarzbauerweg 3, 8043 Graz, Austria.
Email: bott@teleconsult-austria.at
[2] DigitUtopika, Ltd., Rua do Moinho, Lt. 30, Urbanização, Casal Labrusque, Areia Branca, 2530-065 Lourinhã, Portugal.
Email : pbranco@utopika.net
[3] Telespazio S.p.A., 965, Via Tiburtina, 00156 Rome, Italy.
Email: Riccardo_Nicole@telespazio.it

Abstract

Satellite navigation has become an important positioning source for a wide range of applications, many of which going much beyond the traditional transport sector. One example is personal mobility including dense urban, indoor, and outdoor applications. Practically all of the current applications rely on the GPS signals, sometimes also exploiting regional or local augmentations for better accuracy. As applications move into safety-critical and other areas where service reliability is of concern, users and service providers alike are becoming aware of the importance of service qualities and guarantees. Disaster management is one example of such applications. As a first step, an integrity signal is already provided with the SBAS services WAAS and EGNOS. From the year 2008 onwards, full-scale service guarantees will be available on certain signals of the GALILEO system.

In pedestrian user environments like dense urban canyons or indoors, the performance of GNSS (including conventional terrestrial and/or satellite based means of augmentation) for position and integrity reaches well-known technological limits. These limitations can be overcome by adding additional information sources to the system.

The ESA funded project SHADE addresses special handheld-based navigation applications in difficult environments. The main focus of the project lies in the navigation non-transport applications like rescue services, VIP tracking or lone worker protection. A highly mobile demonstration system for pedestrian use has been developed that targets sensor augmented navigation, enhanced through visual representation, in security related fields of operation (e.g. rescue operations). The system architecture is based on the combination of navigation, communication and geoinformation. Geoinformation in this context means the creation of a visualization component for a better 3D orientation at the service Center and for the mobile user.

The modular system architecture of SHADE is built for information exchange between several mobile units and one or more service centers, which regulate and coordinate information and position exchange. The position information is sent from the Mobile Unit via the mobile communication link to the Service Center, where all position information of different users are managed and provided to the SHADE mission Center over fixed internet services. The mission Center accesses its database and the position updates to render images of the surrounding, based on 3D city and/or terrain models. This can be a bird's view, over the shoulder view, or any other view defined by a mission Center operator. The image is continuously updated and can be accessed by the Mobile Unit.

The system that has been designed, built and tested for SHADE uses three different technologies alternatively. Each of the so-called Pilots provides position and integrity data over the mobile communication link.

The first Pilot applies the assisted GPS principle. It provides position information even in dense urban environments where weak navigation satellite signals are still detectable. The EGNOS-TRAN service Center provides rapid acquisition assistance information, such as satellite almanacs and precise local time, to the mobile unit. Raw pseudo range measurements are sent back to the service Center, where the user position is calculated. Additionally, the server uses EGNOS information to differentially correct GPS measurements and to process integrity.

The second Pilot implements innovative dead-reckoning technologies. GPS positioning is aided during periods of poor reception or bridged during complete signal outages, e.g. indoors or in tunnels. A custom furnished Multi-Sensor Box (MSB) is equipped with a GPS/EGNOS receiver and a number of digital sensors (accelerometers, gyros, barometer, magnetometer). In dead reckoning mode, a step detection algorithm uses accelerometer measurements to calculate displacement vectors. Adding direction measurements of the magnetometer and relative altitude data of the barometer in a Kalman filter, a three-dimensional relative coordinate update is computed.

The third Pilot integrates Loran-C terrestrial radio navigation with GPS and EGNOS satellite navigation in the pseudo range domain. The integration of the Loran-C time-of-arrival (TOA) measurements provides position information in dense urban and even in light indoor environments. The performance depends on the coverage of the terrestrial system and the presence of interference.

Besides a continuous position update, reliability information is of main interest in the context of disaster management. Satellite-Based Augmentation Systems and their integrity provisions have been primarily designed for aviation applications. The potential use of such augmentation systems for land-mobile and maritime applications is subject to a number of ongoing research and development projects. EGNOS and WAAS offer two types of information to the user: Differential range corrections to improve the accuracy of GPS pseudorange measurements, and integrity information consisting of differential signal-in-space accuracy input for protection level computation. The protection levels are finally compared to thresholds called the alert limits. Although the differential pseudo range corrections can be applied for non-aviation applications without modifications, the integrity concept needs re-assessment and a number of modifications in order to be useable for land-mobile. Different integrity requirement figures, necessitating changed probability multipliers, and a slightly different probability equation according to other operational procedures and user dynamics, can be handled rather easily. The magnitude and bounding of local error contributions to the integrity equation are more challenging. Atop of this, the applied sensor fusion, as in the SHADE MSB, poses new challenges on the protection level algorithms when additional types of measurement are introduced to the position computation. Consequently, each of the three SHADE Pilots implements specific adaptations to the integrity processing.

Field demonstrations of the SHADE system encompass nine campaigns with different key application objectives. They have been performed over a scheduled duration of six months in spring and summer 2004. The environmental scenarios include the Expo'98 Park at Lisbon (Portugal), a scenic hotel and an office building at Bolzano (Italy) and a road tunnel in the Center of Rome (Italy). Especially for the sensor-based second Pilot, the tests showed promising results even in difficult environments. GPS outages occurring in tunnels, indoors or in urban areas just slightly degraded the navigation performance. The involvement of targeted user groups (public safety authorities, fire brigades, etc.) in the demonstrations showed user group acceptance and provided detailed feedback on the concept and the architecture.

The paper presents in particular the system architecture of SHADE, the design of the three different integrated navigation technologies and the modified integrity processing approaches. Field test results and application domain feed back complement the presented technical information.

1 System Architecture

The architecture of SHADE serves the location sensitive information exchange. The architecture relies on the integration of four elements: navigation, communication, GIS, and multimedia. Especially the navigation element has been under investigation to analyze the system performance in dense urban and indoor environments. Another key-role plays the multimedia component. Communication and the GIS functionality are based on existing systems and have not been under investigation; however some interfaces have been specified and developed. The SHADE project features three different navigation technologies, that are the technological basis of the three Pilots.

Analyzing the requirements of the named applications revealed that the general architecture must include one or more service centers and an unlimited number of Mobile Units (MU). The service centers, are responsible for information dissemination and decision taking. The MU represents a sort of "executive element" of the application.

The system architecture identified five different sub-systems: the SHADE Mission Center (SMC), the EGNOS TRAN Service Center (ESC), and three different navigation units, each including the Mobile Unit user terminal. The subsystems have been designed and implemented by different partners during phase 2 of the SHADE project.

The Mobile Unit either in autonomous operation or in cooperation with the ESC determines the MU position. The position, if not already derived within the ESC, is transmitted via the GPRS network to the ESC, where it is stored in a database. The position information is attributed by different parameters (user ID, velocity, integrity, etc.) and is made accessible to the SHADE Mission Center (SMC). The SMC uses the position information to render images of the surrounding based on a 3D city and/or terrain model. The image is sent to the Mobile Unit on user request.

1.2 SHADE Mission Center

The task of the SHADE Mission Center (SMC) is to implement the GIS functionality and the image rendering component. The Mission Center was designed to provide situational awareness to the decision makers, by displaying the position of the field users and providing additional visual and descriptive information. The SHADE Mission Center is subdivided into a SHADE Server and a Web Server. The tasks of the SHADE Server are to analyze location sensitive information, to visualize the situation around the actual position, and to generate images for display in the SMC. The tasks of the web server are to access the position database of the EGNOS-TRAN Service Center, to route the data to SHADE Server and external agents (police, ambulance, fire brigade, …), to generate images for the Mobile Units, and to provide the MU access to the images. The SMC software uses existing programs and tools of the HorizoN framework, which have partially been adapted for the SHADE software requirements.

1.2 EGNOS-TRAN Service Center

The EGNOS-TRAN Service Center (ESC) is an adaptation of the service Center as it was developed for the EGNOS-TRAN project. Within SHADE, the tasks of the ESC are different for each Pilot. MU1, which applies A-GPS technology, receives assistance data (almanac, ephemeris, ionosphere) from the ESC. In this context the ESC does also compute the EGNOS augmented position of the MU in case that the A-GPS transmits the measurement data to the ESC. For the Mobile Units 2 and 3 the ESC works basically as a router that re-distributes the position information from the Mobile Unit to the SMC. In this context it would be possible that the ESC routes the position information to more than one SMC, e.g. in case of a mission involving several security forces. The ESC has to cover two different scenarios:

The Pilot 1 scenario applies the assistance data procedure, the wake-up procedure (MU logs on to the ESC), and the tracking procedure that allows an external agent (i.e., the SMC) to request periodic positions of a selected Mobile Unit. The Pilot 2 and 3 scenario applies only the wake-up procedure and the tracking procedure.

The A-GPS requires a rough user position (100km), a rough time (100ms), GPS almanac, GPS ephemeris and ionospheric information for commencing operation. The user position and time are provided by the PDA user terminal, the other information is acquired from the ESC.

Before the ESC routes position information to an external agent (e.g. SMC), the external agent has to log on to the ESC first (done by identifying its IP), and then sending a “start tracking” command. Both activities are done by http commands, which are sent via a web-based security portal to the ESC.

1.3 Pilot 1 MU - A-GPS Module

The A-GPS module integrates modern GPS receiver technology with communication infrastructure in mobile networks for receiving assistance data. The assistance data consists of the GPS satellite almanac / ephemeris, ionospheric information, the current GPS time and initial position data of the receiver. The knowledge of the GPS satellite locations at the beginning of the receiver's search process for satellites allows the receiver to predict the GPS measurements, especially the Doppler shift of the signal and the approximate current sequence of the C/A code. Thus the search space for the Doppler shift and the PRN code sequence is reduced, which significantly reduces the time to lock on a satellite and to provide a first position fix. Applying A-GPS technique, the required strength of incoming satellite signals is much lower compared to a conventional search process. Therefore assisted GPS receivers are able to receive satellite signals, decode them and to compute a position even within buildings or at other GPS hostile environments.

1.4 Pilot 2 MU - Multi-Sensor Box

The second Pilot implements a concept, which is especially tailored for the requirements of pedestrian navigation: A GPS/EGNOS receiver is complemented by a number of different autonomous sensors. In case of poor GPS reception conditions, or even total GPS outages, those sensors enable aiding and continuation of the position determination in "dead reckoning" mode: Digital acceleration sensors in combination with a step detection algorithm provide information on the covered distance. A magnetometer provides direction information, and a barometric pressure sensor is used for providing relative altitude data. Thus, three-dimensional relative coordinate changes can be computed.

For the purpose of SHADE, a prototype Multi-Sensor Box that was developed for pedestrian navigation application was adapted to the project requirements.

1.5 Pilot 3 MU - GPS/EGNOS/Loran-C

The third Pilot integrates the terrestrial radio navigation system Loran-C with GPS/EGNOS satellite navigation. During periods of good GPS availability and accuracy, this system is used to calibrate Loran-C measurements. If GPS suffers from poor reception conditions or total outages, the previously calibrated Loran-C can be used to aid the position solution or even to continue positioning in stand-alone mode, but with GPS-like accuracy. Therefore, the integration of these sensors promises to provide position information in dense urban and even in light indoor environments, depending on the coverage of the terrestrial Loran-C system. In the course of SHADE, a previously developed integrated navigation algorithm was tailored to meet the needs of the project.

2 Integrity Determination

The integrity determination for the three Pilots bases on the integrity concept specified in [2], which was originally developed for aviation. This concept had to be modified for the needs of SHADE with respect to the integrity multipliers (κ-values). The approach for the determination of Protection Levels (XPL) follows Appendix J of [2].

2.1 Determination of κ-values

Transferring the aviation approach to the SHADE pedestrian application, the precision approach mode is most applicable due to the vertical accuracy and integrity requirements in SHADE. However, for the horizontal protection level (HPL), the argument for using a one-dimensional distribution does not anymore hold – unlike aircrafts during precision approach, pedestrians have no favored direction in their movements. A one-dimensional (Gaussian) Normal distribution can be used for the pedestrian vertical protection level (VPL), whereas the pedestrian HPL follows a two-dimensional Rayleigh distribution, comparable to the non-precision approach case described in [2].

According to the user requirements identified for SHADE, the applicable integrity risk for the pedestrian user is $1 \cdot 10^{-3}$ per 60 seconds. According to [2], half of the total integrity risk is allocated to protection level bounding within the signal-in-space (SIS) and user equipment domain. This requirement is allocated equally to the horizontal and vertical protection level risk. This results in equal integrity risk requirements of $2.5 \cdot 10^{-4}$ per 60 seconds for HPL and for VPL. The other half of the total integrity risk is allocated to the "system-internal" component. This part comprises, e.g., undetectable ionospheric blunders, XPL formula inaccuracies, UDRE tail effects and others (see also [1]).

For deriving κ-values, it first must be ensured that the probability of missed detection of a faulty position solution associated to the XPL algorithm is expressed per sample (i.e., per each XPL computation). To establish the link between this probability and the given integrity requirement, it is necessary to make an assumption on the number of independent samples per time unit.

For the three SHADE pilots, different intervals of statistically independent samples are applicable. In pilot 2 and 3, the sampling interval of 6 seconds for independent samples is dominated by the update interval of EGNOS fast corrections (carrier phase smoothing is not applied). However, in case of total GPS outages, the position solution implemented in Pilot 2 solely relies on dead reckoning and barometric height determination. In this case, the variance of subsequent samples will become a function of time. First investigations of suitable XPL computation algorithms have been carried out for this special case. The LORAN-C measurements of Pilot 3 have a sampling interval of 5 seconds, thus being lower than the EGNOS update. The A-EGNOS system of Pilot 1 has a sampling interval of 6 seconds, which is equal to the fast corrections update interval, and, hence, does not change the κ-factors.

3 System Demonstrations

Field demonstrations of the SHADE system encompassed nine campaigns with different key application objectives. They have been performed over a scheduled duration of six months in spring and summer 2004. The environmental scenarios include the Expo'98 Park at Lisbon (Portugal), a hotel and an office building at Bolzano (Italy), a road tunnel in the Center of Rome (Italy), and part of ESA-ESTEC premises at Noordwijk (Netherlands). The involvement of user groups (public safety authorities, fire brigades, etc.) into the demonstrations showed application domain acceptance and provided detailed feedback on the current system and future enhancements. This chapter provides details on the demonstration locations and selected results.

3.1 Demonstration Locations

3.1.1 Lisbon/Portugal

In Lisbon (Portugal) the main focus lay on the EXPO 98 administration building within the Parque das Nações and headquarters of the 2004 European Football Cup. The area is characterized by dense urban environment with high office and apartment buildings, but in the area are also places of good GPS visibility. The administration building itself has a front side (southward direction) made of glass. Additionally it is a typical example for office buildings.

3.1.2 Bolzano/Italy

The region of Bolzano is characterized by mountainous terrain. For the demonstrations in Bolzano two scenes have been selected. The first one is the Business Innovation Center (BIC) office building in the industrial area of Bolzano. The environment of the building is characterized by sub-urban to dense urban environment. The building itself is a 10 storey office building with a basement garage. The second scene is about 10 km away in hilly to mountainous terrain. The Schloss Korb is a typical tourist attraction in combination with a hotel. In both areas indoor, dense urban and sub-urban environments could be tested.

3.1.3 Rome/Italy

The Center of Rome is a typical dense urban area. Additionally, the Via del Traforo goes through a tunnel under the "Presidenza della Repubblica" in the Center of Rome. The tunnel and the dense urban area compose a demanding environment for a navigation system. Pavements run on both sides through the tunnel. Within the 3D model of the SMC the size of the pavement (1.5 m) has been exaggerated for better identification of the user in the tunnel.

3.1.4 Noordwijk/Netherlands

The focus of the last demonstration lay on the ESA-ESTEC premises, which are characterized by sub-urban to dense urban to indoor environment. A 3D model was provided to UT for implementation into the SMC software. The model contains the ground floor of the buildings "Bf", "Ba", "Ca" and "Da", where the demonstration was performed.

3.2 Test Results

3.2.1 Pilot 1 MU: A-GPS Module Test Results

The navigation unit of Pilot 1 relies on A-GPS. The A-GPS showed capabilities for dense urban navigation but in general the device could not convince indoors. However, during the acceptance tests it could be shown that under some circumstances even indoors the device provides acceptable position information. Position accuracy gets better with a growing number of solutions that are averaged or filtered.

After initializing the A-GPS at the starting point (urban area) the system provided a position with an offset of about 30 m. As soon as the user walked along the parking lot (good GPS visibility) the position solution converged to the true trajectory. In this area the across-track offset of the system is about 5-10 meters. The system delivered position solutions even when the user walked into a building corridor for several meters. Unknown is the along-track error here, but the across track error is still between 5-10 meters. After about 10 position solutions the system could not solve for a position anymore. Back in the presentation room the GPS antenna was laid onto the window sill. The position computed out of these measurements shows an offset of 100 m, which was attributed to multipath, and limited satellite geometry.

3.2.1 Pilot 2 MU: Multi-Sensor Box Test Results

The first demonstration of Pilot 2 in Lisbon already showed the high potential of the navigation unit. The autonomous sensors bridge any GPS gaps, even for several seconds. Even when using the sensor-only solution the user is able to track back to the starting point after 40 minutes of walking (~1 km in distance) in urban environment, going up by elevator to the third floor and going back. The closed loop error of this track is about 10 meters. Considering, pure "calculative", distance and position error, the error of the system would correspond to 1% (10m / 1000 m) in sensor-only operation.

The system performed as expected under open sky conditions. Even indoors the system delivered good position results. For example, in the corridors of the ESA ESTEC building the system suffered from magnetic fields which lead to a position offset. At the end point of the demonstration the computed position is about 25 m off – this corresponds to about 10% error considering the distance walked indoors.

3.2.3 Pilot 3 MU: GPS/EGNOS/Loran-C Test Results

Pilot 3 combines the strengths of satellite navigation system GPS with the terrestrial radio navigation system Loran-C. The integration of both systems works well in outdoor environments. Despite of GPS outages, the system still outputs valid position solutions. However, accuracy is degraded. In certain dense urban areas and indoors, where local effects influence Loran-C signals and reception, validity of the position solution is strongly impaired.

During the demonstration at ESA-ESTEC, the measurements have been heavily influenced by multipath and geometry effects near the presentation room, where the system was initialised. The accuracy of the position solutions under open sky was as expected. As soon as the user walked into the building, the position solution started to deviate very soon; although the Loran-C receiver still could track Loran-C stations. The offset between computed position and true trajectory finally amounts to several hundreds of meters. Although Loran-C stations could be tracked indoors, the position accuracy was not as expected.

4 User Feedback

Starting with the live demonstrations, discussions of possible evolution of the SHADE system took place between invited guests, ESA representatives, and the SHADE partners. Numerous proposals for future enhancements of the system have been offered. The invited guests also made suggestions on changing the system concept; i.e. the Mobile Units should be equipped with robust, small, and lightweight black boxes without PDA (e.g. considering a number of fire fighters in action, coordinated by local controllers operating within a vehicle). The user representatives showed interest especially in the potential of the 3D visualization software.

Proposals and ideas for future enhancements of the system or certain elements were collected during and after the public demonstrations. Some of the proposed system enhancements have been definitely out of scope of the SHADE project, but have been noted as input for future work.

In total 27 proposals for enhancements of the SHADE visualization and SMC monitoring and control functions have been received during the demonstrations. Out of them 4 have already been implemented during the demonstration phase of the project.

In total 5 proposals for accuracy and acquisition improvements of the A-GPS module have been received during the demonstrations. Out of them 2 have already been implemented during the demonstration phase of the project.

In total 12 proposals for future developments and enhancements of the Multi-Sensor Box have been received during the demonstrations. Out of them 4 have already been analyzed during the demonstration phase of the project.

One proposal for improving the Loran-C positioning by performance has been received during the demonstrations.

In total 5 proposals for enhancements of the communication functions and usage of other communication technologies have been received during the demonstrations.

In total 10 proposals for enhancements of the SHADE user scenarios, refinements of application requirements and market penetration strategies have been received during the demonstrations.

5 Project Results

The primary incentive of SHADE was to overcome satellite navigation line-of-sight limitations in difficult environments, like urban & natural canyons, narrow streets & tunnels, and inside buildings & parking garages.

The targeted applications have been in the area of safety critical personal mobility, like civil protection, police and security, fire fighting inside buildings and mountainous forests, rescue operations inside buildings and mountainous forests, and dangerous/valuable goods & VIP tracking and monitoring.

The SHADE key features are to provide/enhance system integrity by using EGNOS signals and EGNOS based algorithms in the Mobile Units (MU), and to improve system availability by Assisted GPS/EGNOS (MU 1), external sensors as augmentation to GPS/EGNOS (MU 2), and by the terrestrial system (Loran-C) as augmentation to GPS/EGNOS (MU 3).

The complete system with all subsystems and the required communication links was specified, designed, implemented and validated within the project, and reached the expected performance.

Nine demonstration campaigns were successfully performed to test the SHADE system with its three Pilots, and to present it to the interested public. Four campaigns in three different European cities were repeated twice with identical application focus. A ninth demonstration at ESTEC, the Netherlands, concluded the demonstration phase.

The EGNOS-TRAN Service Center showed good performance during all demonstrations. The ESC routed the position solutions to the SHADE Mission Center. Further, the ESC provided assistance data to the A-GPS, and in ESC-mode solved for the EGNOS-augmented position computation.

The SMC software provided a 3D model of the four different scenarios. The user communities welcomed the concept of visualisation of the scenario in the service Center. They also agreed to exchange information with the Mobile Units, but depending on the application, different information shall be exchanged. Several enhancements for the SMC software have been identified.

The A-GPS showed good results under favourable GPS conditions – the GPS specifications could be met. As soon as the receiver operates in dense urban environments or indoors the position solutions show higher scattering. Nevertheless the receiver is still able to track several satellites even in these difficult environments. During the final demonstration at ESTEC the Pilot 1 met all expectations.

The navigation performance of Pilot 2 met the expectations and the demonstration requirements. The Multi-Sensor Box, which integrates GPS/EGNOS position solutions with autonomous measurements, provides continuously reliable position information outdoors and indoors. The position accuracy is satisfying, although there are a number of effects which will be taken into account in future developments. The results of the protection level computation turned out to be slightly too optimistic.

The principle of integrating the complementary navigation concepts GPS, EGNOS and Loran-C could be proven. It was also possible to show that Loran-C signals are available even indoors. However, although the Loran-C receiver could track some Loran-C stations even indoors, the calibrated Loran-C stand alone solutions could not meet the expectations. Further developments in antenna and receiver technology may enhance the position performance in future.

Except during the third demonstration in Bolzano, the EGNOS / ESTB signal could not be integrated into any position solution of the demonstrations. Either no signal was visible, or the signal could not be decoded, or visibility was frequently interrupted by indoor measurements, measurements in dense urban area, etc. Different means of EGNOS information acquisition should be taken into consideration (e.g. EGNOS-TRAN concept) for future developments.

The SHADE architecture proved to be feasible for security relevant applications. The concept of visualisation in the service Center was followed by the user community with interest and the capabilities of pedestrian navigation acknowledged.

5.1 Future Developments

Future developments are currently being carried out individually for all subsystems of the SHADE project. The 3D visualization is introduced into new application areas by DigiUtopika. The Assisted GPS applications are used by Telespazio in new projects. TeleConsult Austria will concentrate on downsizing and enhancing the most promising Pilot 2 navigation component, while partner companies of TCA push forward the implementation of combined terrestrial and satellite data communication and enhanced mission control Center concepts. The currently performed development tasks of the Multi-Sensor Box encompass downsizing, magnetometer error exclusion and enhanced integrity determination.

References

[1] Flament D (2004) EGNOS System and Performance, Presentation to CNIG/PSD: 25/03/04, available at http://www.esgt.cnam.fr/sites/CNIG/cnig.psd/CIAG/CNIG.PSD/reunions/25mars2004/EGNOS_DFlament.PDF (July 2004)

[2] RTCA (1999) Minimum Operational Performance Standards for Global Positioning System/Wide Area Augmentation System Airborne Equipment, RTCA-DO 229 B, October 6, 1999

Location Interoperability Services for Medical Emergency Operations during Disasters

Remko van der Togt[1], Euro Beinat[1], Sisi Zlatanova[2] and Henk J. Scholten[1]

[1] SPINlab, Vrije Universiteit Amsterdam, De Boelelaan 1087, 1081 HV Amsterdam, the Netherlands.
Email: remko.van.der.togt@geodan.nl; euro.beinat@ivm.vu.nl; hscholten@feweb.vu.nl

[2] Section GIS Technology, OTB Research Institute for Housing, Urban and Mobility Studies, TU Delft, P.O. Box 5030, 2600 GA Delft, the Netherlands.
Email: s.zlatanova@otb.tudelft.nl

Abstract

The organizational structure that deals with the Response phase in disaster and risk management is based on a strong co-operation between several organizations, such as the police, fire departments, the local government and the health services[1] The size of the organization depends largely upon the scale of the disaster itself. Van Dijke 2003 identifies 31 processes, that concern information flows and coordination of forces, that are relevant in these cases.

This paper concentrates on the information process at the first aid in hospitals, which is part of 'somatic health care'. Research has been conducted on the information problems during emergency operations at first aid departments in Italy and The Netherlands. The results identify location information (location of patients, equipment, physicians and/or relatives, and so on) as a critical factor for improve quality and coordination of health services.

In most cases the location has to be determined indoors, where the most common global 3D positioning (based on GPS) is not available. It is still a challenge to obtain accurate positions indoors.

In general terms, one can distinguish between two broad classes of location technology: global (telecommunications) and local (WiFi, Bluetooth)

---

[1] Often called "Medical Aid during Accidents and Disasters". In the Netherlands there is a specific health organization involved in disaster management, called 'GHOR'.

network approaches, based on absolute (providing coordinates) or relative (providing speed and direction of movement) positioning. Currently, the most commonly used approaches for indoor positioning are based on WiFi and RFID.

This paper presents a system for indoor positioning and LBS to support hospital teams in emergency management. The paper discusses current information problems, investigates the required functionality of a system for hospital services, and the added value of indoor location technology.

1 Healthcare: Challenges for ICT

The health care industry is in the middle of a process that is often described as 'going from supply-based to demand-based services' Corrigan, 2004. In the supplied-based services, patient needs are 'matched' to the available services and the patient is required to adapt to the quality and level of services available. The demand-based services, in contracts, concentrates on the needs of the patient and adapts the services provided by a hospital to every particular case. The provision of demand-based services would imply higher requirements to the hospital organization; this is to say much more flexible and adaptive service organization. Such a demand-based organization can only be achieved with better coordination between providers of health services, better resource use, better personnel management and coordination, and an integrated information flow. Information Communication Technology (ICT), as in many other areas, plays a critical role in improving flexibility, quality and productivity in health industry and facilitates the shift towards demand-driven services.

Patient needs and quality of services increase in emergency situations. Large disaster, such as fire in a public building, a traffic accident, a plane or train crash, or an industry accident (explosion, leakage of heath-threatening chemicals, etc.), are characterized by:

- Large number of patients, which require mobilization of more hospital personal and equipment than usual;
- A large number of the same type of injuries (e.g. skin damages in fire, or breathing problems in gas leakage) that may require large amounts of the same medicines and specialists;
- Injuries that requite immediate and simultaneous high qualified treatment (e.g. surgery), which has usually limited capacity and is provided based on a long term schedule;
- Ambulances have to deliver several patients;

- Stress and panic situations, and very often bad estimates of the number of injured people and that need treatment;
- The critical need for real-time coordination and communication with police, fire brigade and the local authorities;
- The need for real-time communication and information to several media and the public.

These characteristics underline the need for a much better communication between medical institutions and other organizations involved in disaster management. At the same time, there is a need for a better organization within the hospitals concerning the deployment of specialists, availability of medical supplies, transportation, rooms and equipment. The services to be provided to the patients are necessarily demand-driven.

In this paper we argue that location technology through a process-oriented approach will increase the quality of the somatic health care process during disasters, by supporting the provision of relevant information and by helping coordination within health care and between organizations. LBS can play a very important task in order to facilitate a demand-driven approach to deliver the right information to the right person in order to deliver good care. The Dutch GHOR-organization GHOR, 2004 states that the integration of geographic information systems is essential to support health care services (see also Verhoef, 2001).

2 LBS Background

Location-based services (LBS) revolve around the ability to locate at any point in time the position of resources, people, vehicles and objects. People or objects can be located by a variety of means such as GPS location, telecom based location, indoor location systems such as RFID, Wi-Fi, bluetooth or radio tags. This location can be requested by a user, for instance to find a piece of equipment in the vicinity, or by a control facility, for instance to locate a nurse or a doctor within the hospital facilities.

All mobile applications are based on the ability to provide remote access to data sources from mobile devices. What distinguishes Location-based Services from a pure extensibility service (such as email access from a handset) is how critical location information is to the added value to the user. Services that add value by using the location component are called location-based services.

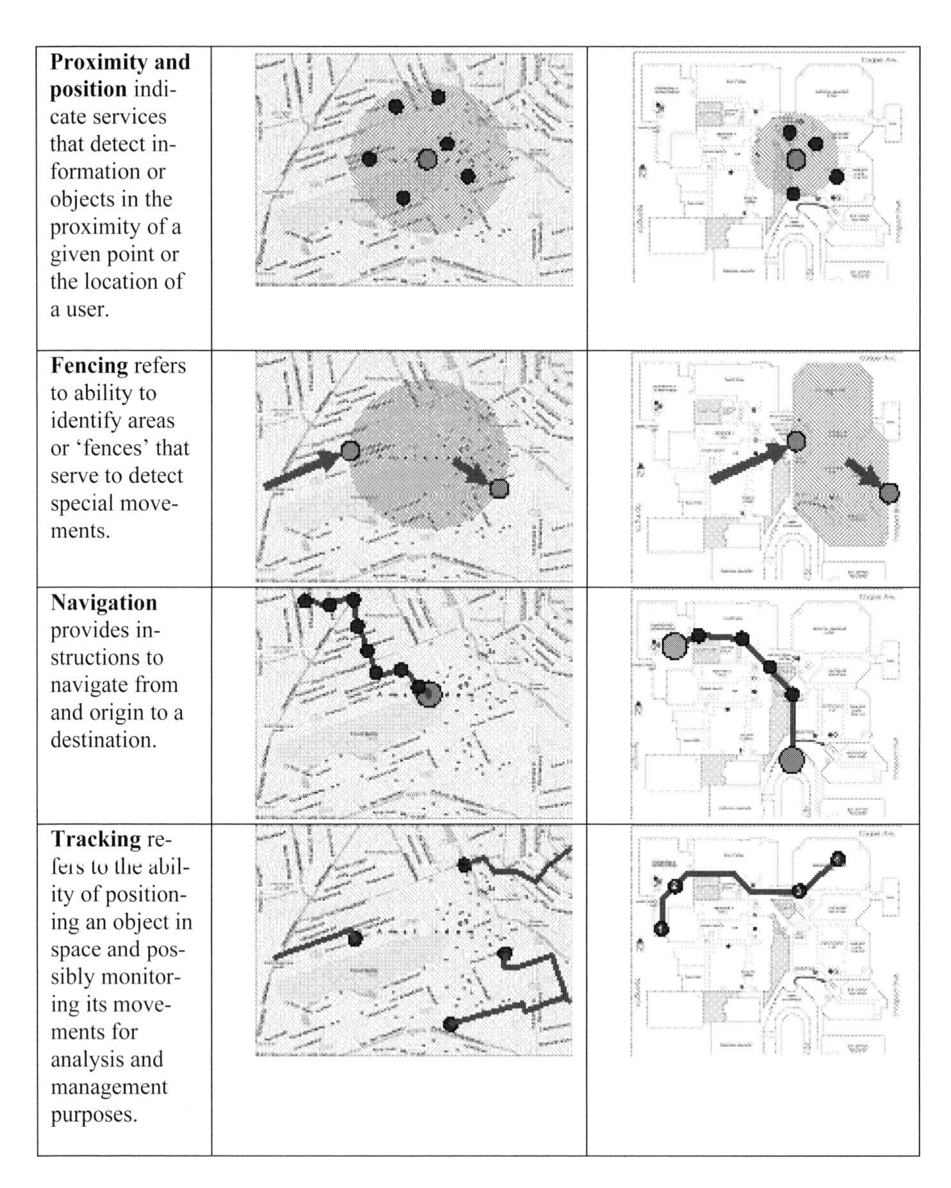

Proximity and position indicate services that detect information or objects in the proximity of a given point or the location of a user.		
Fencing refers to ability to identify areas or 'fences' that serve to detect special movements.		
Navigation provides instructions to navigate from and origin to a destination.		
Tracking refers to the ability of positioning an object in space and possibly monitoring its movements for analysis and management purposes.		

Table 1. Depending on **position** of the user and/or the position of the Point of interest, the LBS can be subdivided into four main types of location services

LBS are both horizontal (services suitable for consumers and business, such as route information) and vertical services (specific for type of business, such as field service engineer management). The importance of LBS is recognized by Open Geospatial Consortium (the formal Open GIS Consortium) and several standards regarding LBS are already either available (OpenGIS Location services) or under development.

OpenLS consider compulsory six core services, i.e. Gateway (related to obtaining position), Directory (access to user-specific information and or Points of Interest), Geocode (conversion from address to co-ordinates), Reverse Geocode (ability to convert co-ordinates to address), Route (commutation of route) and Presentation (presenting the requested information to the user, e.g. a map with the route). The services are intended for all classes of mobile devices. The request should contain the type of the user device (according to a list with well-known devices) and a number of parameters specifying the range and type of requested information.

Depending on the **type** of the information to be provided to the user, the services can be subdivided into the following groups:

1. **Information services**, which provide information about objects close to the user (in terms of distance, travel time or other). Examples are: locate my position, locate an address, check traffic conditions on the highway, etc.
2. **Interaction services**, which are based on the interaction between mobile users/objects and do not require a 'mobile internet' component or a content sources. Examples are: Where is my nearest colleague? Where is the specific device? Where is the closest emergency car to an accident?
3. **Mobility services that** support smart mobility and revolve around navigation capabilities. Examples are: How do I get from A to B? What is the quickest reroute to avoid this traffic jam?

3 Applications using LBS

A variety of applications utilizing location and location-based information are already operational.

In the US, emergency location services prescribe the possibility to locate a 911 call from a fixed and mobile phone, with the request to operators to disclose this information to emergency services. The accuracy requirements are the following: (1) for network-based solutions: 100 meters for 67% of calls, 300 meters for 95 percent of calls; (2) for handset-based solutions: 50 meters for 67% of calls, 150 meters for 95 percent of calls.

Alert Services may be enabled to notify wireless subscribers within a specific geographic location of emergency alerts. This may include such alerts as avalanche warnings, pending floods or accidents that interrupt cir-

culation. No legal or administrative requirements currently exist for Emergency Alert Services anywhere in the world.
Fleet and Asset Management services allow the tracking of location and status of specific service group users. Examples may include a supervisor of a delivery service who needs to know the location and status of employees, parents who need to know where their children are, or a natural park administrator that wants to know where the visitors to his park are. The service may be invoked by the managing entity, or the entity being managed, depending on the service being provided. Fleet Management may enable an enterprise or a public organization to track the location of vehicles (cars, trucks, etc.) and use location information to optimize services.

Asset management services may range from asset visualization (general reporting of position) to stolen vehicle location and geofencing (reporting of location when an asset leaves or enters a defined zone). The range of attributes for these services is wide.

For Fleet and Asset Management services, a distinction may be made between the manager of the fleet/assets in charge of tracking, and the entities being tracked (service group users, etc). The tracking service may make use of handsets with possible specialized functions (Web browsers, etc) to allow for tracking and specific methods for communicating with the managing entity. A managing entity would be able to access one or several managed entities' location and status information through a specified communication interface (Internet, Interactive Voice Response, Data service, etc). The managing entity would be able to access both real-time and recent location and status results of managed entities.

Mobiles in automobiles on freeways anonymously sampled to determine average velocity of vehicles. Congestion, average flow rates, vehicle occupancy and related traffic information can be gathered from a variety of sources including roadside telematic sensors, roadside assistance organizations and ad-hoc reports from individual drivers. In addition average link speeds can be computed through anonymous random sampling of MS locations. Depending on the capabilities of the location method, traffic behavior can only be determined if a vehicle location is sampled at least twice within a finite predetermined period. Traffic monitoring technology is at the basis of innovative road taxes schemes such as the proposed road pricing scheme currently under evaluation in the Netherlands. Location technology and on-board wireless connections are being considered as a means to exercise road taxes based on type of road, time of travel etc.

Location-Based Information services allow subscribers to access information for which the information is filtered and tailored based on the location of the requesting user. Service requests may be initiated on demand by subscribers, or automatically when triggering conditions are met. The purpose of the navigation application is to guide the handset user to his/her destination. The destination can be input to the terminal, which gives guidance how to reach the destination. The guidance information can be e.g. plain text, symbols with text information (e.g. turn + distance) or symbols on the map display. The instructions may also be given verbally to the users by using a voice call. This can be accomplished through carrying a GSM mobile phone that has location technology capabilities down to a few meters. Less granularity impedes the applicability of this functionality.

City Guides would enable the delivery of location specific information to city visitors or residents. Such information might consist of combinations of various services including historical site location, providing navigation directions between sites, facilitate finding the nearest restaurant, bank, airport, bus terminal, restroom facility, etc.

The main characteristic of this service category is that the network automatically broadcasts information to terminals in a certain geographical area. The information may be broadcast to all terminals in a given area, or only to members of specific group (such as members of a specific organization). The user may disable the functionality totally from the terminal or select only the information categories that the user is interested in.

An example of such a service may be localized advertising. For example, merchants could broadcast advertisements to passers-by based on location / demographic / psychographic information (for example 'today only, 30% off on blue jeans'). Similar services would include weather and traffic alerts Steenbruggen, 2004.

Mobile Yellow Pages services provide the user with the location of the nearest point of interest. The result of the query may be a list of service points fulfilling the criteria (e.g. Banks with ATM within walking distance). The information can be provided to the users in text format (e.g. bank name, address and telephone number) or in graphical format (map showing the location of the user and the bank).

4 Technology for LBS

The next step is to choose the right technology to perform the job. Figure 1 shows an overview of different location technologies and their related accuracy. Table 2. Location Systems and summery of their features (from Beinat, 200illustrates the major location systems currently available. Most of them find their application in health care, in particular the indoor systems.

Telecom based technology like Cell-ID and triangulation methods are useful for services that require an accuracy of 50 to 100 meters and are rather cheap and can be used both out- and indoor. For more accuracy outside however the Global Positioning System (GPS) can perform the job as it comes to outside services. For a more accurate solution location services based on wireless local area networks are suitable. These however are suitable for rather small areas. Radio Frequency Identification both active and passive (RFID) are suitable for accurate positioning within buildings. Ultra Wide Band (UWB) seems the right technology for all the services. Unfortunately UWB is not available yet. This will mainly depend on future technological developments, standardization, jurisdiction Steenbruggen et al, 2004 and it's acceptance by users.

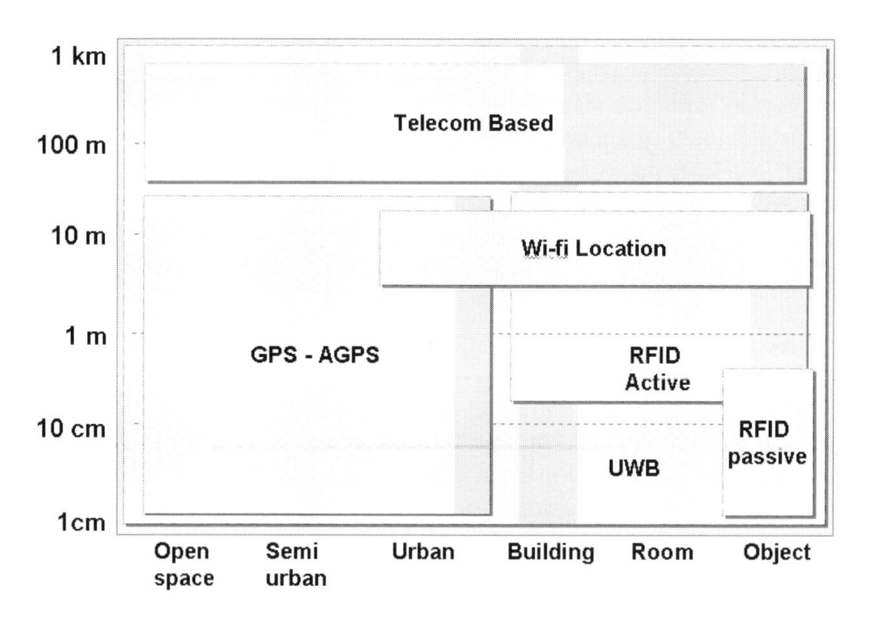

Fig. 1. Accuracy of different location technologies on a logarithmic scale

In order to facilitate processes that cover in- and outdoor activities, location technology needs to adapt to information systems that already exist or are being developed at this moment that will cover these processes. This will finally fulfill the need for, in our case, disaster management processes.

5 LBS for Health Care in Disaster Management

The services for health care in disaster management are much more demanding than most of the application described above. What is the added value of the use of location technology in solving problems during disasters?

Verhoef, 2001 describes the integration of GIS with systems used in control rooms. In this context he describes that integration of these systems will lead to information about the exact location of ambulances, travel times and distances between ambulances and incident locations, and the use of dynamic information like road blocks and such. The national Dutch GHOR organization describes the need for a system which is able to support the entire chain of health care during disasters. Although a lot of initiatives are taken at this moment in order to deploy this, a chain wide registration system is not developed and accepted yet GHOR, 2004. At this moment the Dutch Red Cross has taken the first steps in achieving this goal by developing a system where patients and relatives can be registered and followed trough the whole process of somatic health care Luyendijk and Schoof, 2004.

To Mennecke et al., 2004 it appeared that GIS can be used to extend the range of problems that can be solved using technology by allowing users to more efficiently complete problems that are more complex. A direct outcome of our interviews was the fact that information systems which will be used with the purpose to support processes during disasters have to resemble information systems which are used in the daily normal situations. Users said it is very hard to switch to other systems, which they do not use often and were they have to work under high pressure.

On the basis of interviews with people concerned with medical treatment, management and policy making during disasters the following requirements to a system can be formulated. The system has to be able to provide at any time:

- Patient position and health status. Registration systems can benefit form the use of location technology to know exactly which patient is where in the logistic process. On the basis of the collected and known medical information the nearest suitable hospital with enough resources can be found. When the location of the hospital is known, travel times can be calculated between the scene and the hospital or other health care facilities. This is important because some patients need specific care which can not be provided by every hospital or health care facility. Matching of location and medical information is very important in supporting these decisions.

- Position of health care personal. A clear overview of the entire scene during a disaster is important for management and logistics, and especially rescuers' safety. For health care purposes location of certain objects in the field is very important, for example: location of paramedics and other health care personnel, vehicles, areas of patients and specific areas of risks (chemical substances). Especially managers at the scene can benefit form this in order to have a clear overview of the whole situation at the scene.
- Routing for the first aid team. It is not usually hard to predict at what time and where exactly a critical situation will appear. An appropriate and fast routing to the accident, taking into account the actual situation in the surrounding areas, will save time and greatly support the work of the first aid teams. In some cases (e.g. accidents in metro, tunnels, large shopping centers), indoor routing will be required to avoid all kinds of obstacles and arrive quickly to victims.
- Information about the area affected. Another important requirement is information about the incident area for planning purposes. For managers it can be important to take decision on the basis of geographic information of the scene compounded with dynamic information collected in the field, such as the location of certain objects. With the help of maps displaying this information, managers can for instance decide where to deploy emergency teams or vehicles.
- Relevant information from different sources and information systems. To facilitate management processes it is very important to facilitate an infrastructure that will not only support health care managers but also managers from, for instance, the fire departments, the police and the public authorities. In the management team at the scene the combined information provided on a map can support the process of multidisciplinary decision making on the basis of geographic information combined with dynamic information collected by the resources on the field.
- Decision making. The system has to be fast, flexible and adaptable to any new coming information, suggesting partial or complete solutions for dispatching patients, delivering medicaments and managing emergency teams.

All these requirements show the need and added value of location information at the incident site. The next phase of a rescue operation takes place at the hospital or health care facilities. The case study below describes this next step and the use of location information within the first aid of a hospital.

6 Indoor Location for Emergency Services

Patient logistics play a very important role within health care facilities during disasters. Health care organizations, and the emergency departments of hospitals, are usually subject to an enormous level of stress that takes facilities and resources to the limit of their capacities. In these circumstances it is not always known were patient exactly are within the facility, if certain patient already left certain departments (such as research, intensive care or operating facilities) and have arrived at a ward for less intensive treatment. This is very important for management purposes, as a prerequisite to allocate resources efficiently. With this information, managers can monitor the status of their entire logistic chain all the time, anticipate and manage bottlenecks.

An example of this type of location system can be found in the figure below. This picture below (FIGURE 2) illustrates the emergency floor of a medium-large hospital in northern Italy (700 beds) that has implemented location information system within the emergency department.

In this implementation, tags are associated to important and scarce pieces of equipment. Nurses and doctors can locate a missing piece simply by locating the associated tag, which in this case is located with less than 2 meters accuracy. The same technology can be used to locate patients that are provided with a small tag, such as a bracelet. By connecting this information to the existing hospital information system, users can associate location and information related to the patient. Furthermore, since the location information becomes an element of the underlying information asset, a variety of analysis can be carried out based on this information and on the integration with ERP or CRM systems.

The system registers active users (such as a physician that carries a handheld system) and passive users (e.g. a piece of equipment). Active users can specify the degree to which their location can be disclosed, to who and when in respect of their privacy.

The service can also find a great number of uses in health services not directly linked to disaster management, from operations logistics to theft prevention, to warehouse maintenance and optimization.

In general, these systems are meant to provide services such as:

- Locate patients, doctors, specialists, nurses etc.
- Provide all kinds of routing: 1) to a patient, 2) with a patient to a particular first aid nest and 3) to other health care personnel in the area avoiding blocked stairs and inaccessible exits
- Locate other teams (or specialist) that are already in the building but in another section and establish communication with them

- Search for information on the internet (or particular server) e.g. how many and what type of people (old, young, children etc.) are in the building,
- Communication with the hospital or other managing the disaster centers.

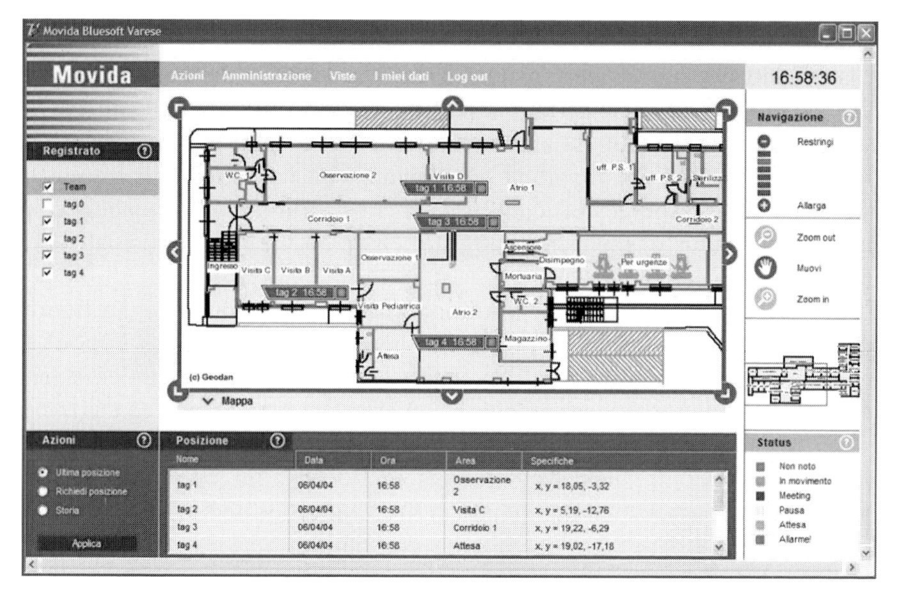

Fig. 2. Screenshot of an application showing the first aid-department of a medium-large hospital in northern Italy

7 System Architecture

The example above is developed on an Java platform[2]. The functionality revolves around three mayor layers:

- The application logic, built on Java and XML/GML, that contains all workflows of the application, and the integration interfaces towards other systems, such as CRM or ERP.
- The application toolbox, that contains service modules (such as web services) that perform specific task and provide basic services. They include:
  - Connector interfaces between legacy databases and the Java/XML platform;

[2] See Geodan Movida: www.geodan,com/movida.

- Messaging, such as SMS, email, voice, etc;
- Maps, portrayal, and spatial functions.
- Routing and Geocoding
- Push mechanism

- The location server. A broker system that connects to position information from indoor location systems while accounting for authentication, authorization and privacy. This component of the platform, in particular, ensures that any location system is made available and hides the complexity of location fixing from the user.

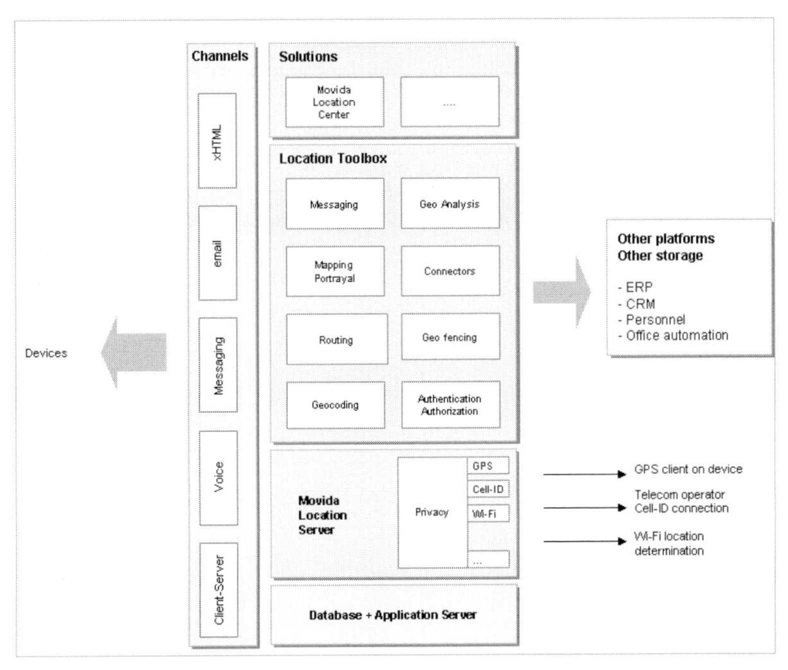

Fig. 3. Components of the indoor location solutions, and main functional blocks.

8 Discussion and Conclusions

An issue which is not addressed so far is the acceptance of users who will finally work with a system which can deliver both in- and outdoor location-based services. Because this is not the objective of this paper we will not describe this in to too much detail. The idea is to make use of the technology acceptance model as first described by Davis, 2004. The model will be applied to obtain a general understanding of how users will complete their tasks more efficiently and if it is to be expected that they will keep on using the system in the future. On the basis of the results policy makers can decide to implement and adapt the system.

Location system	**Object located**	**Operations**	**Coverage**	**Accuracy**	**Limitations**
GPS	GPS receiver	24 satellites broadcast their position. A receiver interpolates x,y,z	Outdoors, poor indoor availability	<1 m .. 20 m	Battery consumption, warm-up time, indoor coverage
A-GPS	A-GPS receiver	GPS signal is processed by receiver and support network - such as UMTS - for higher sensitivity and lower consumption	Outdoors and some indoor environments	1 m .. 20 m	Battery consumption, indoor coverage
Cell-ID, various versions	Mobile phone or SIM based tag	The location is approximated by the position of the connected base station. Sector shape and time advantage can be used to increase accuracy	Network coverage	50 m .. > 1 Km	Accuracy
Telecom based triangulation (e.g. UTOA)	Mobile phone or SIM based tag	Triangulates distance from base stations to estimate position	Network coverage	20 m… 200 m	Accuracy, devices
Wi-Fi passive devices	Wi-Fi device (laptop, handheld) or Wi-Fi tag	Distance from several hot spots is used to triangulate a precise position based on time difference	Coverage of location receivers	2-4 m	Infrastructure calibration, environmental sensitivity
Wi-Fi active devices	Wi-fi device with computation capabilities	The signal intensity model of the environment is used to estimate the position of the device	Wi-Fi network coverage	2-4 m	System calibration, environmental sensitivity
RFID passive tags	Passive tags	The tag is activated by a transmitter/reader field and sends an ID back	Proximity of the receiver, from a few centimeters to a few meters	Na	Proximity detection only
RFID active tags	Active tags	The position of a broadcasting tag is detected by entering in the range of a receiver	Receivers coverage	Rooms, corridors, etc.	Calibration, accuracy

UWB systems	Active tags	Detects distance through travel time on UWB and filters out multipath	Receivers coverage	15 cm	Price
Image systems	Object Pattern	Detects object pattern from camera image	Visible area for the camera	Small areas	Image noise
Hybrid RFID-Infrared	Active tags	The position of a broadcasting tag is detected by entering in the range of a receiver. Infrared signals – blocked by walls or obstacles - are used to increase location resolution	Receivers coverage	Rooms, corridors, etc.	Infrastructure costs, accuracy
Ultrasound	Active tags	The position of a broadcasting tag is detected by entering in the range of a receiver	Receivers coverage. Blocked by walls.	Rooms, corridors, etc.	Infrastructure costs, accuracy, tags

Table 2. Location Systems and summery of their features (from Beinat, 200)

References

Adviesbureau Van Dijke en Ingenieurs, Adviesbureau SAVE (2003) Leidraad Maatramp, 1.3.2003, Directie Brandweer en Rampenbestrijding van het Ministerie van Binnenlandse Zaken en Koniklijkrelaties

Corrigan JM (2004) Institute of Medicine Committee on the Quality of Health Care in America

GHOR Visiedocument (2004) 2002-2006

Verhoef J (2001) Inrichting Ambulancezorg bij Grootschalige Incidenten en Rampen, Zwolle, The Netherlands, Nederlands Ambulance Instituut (NAI)

Steenbruggen JGM (2004) Cell Broadcasting, een nieuwe locatiegebonden informatiedienst. GeoNieuws 3, pp 10-12

Steenbruggen JGM, Vree WG, Dijkstra F, Westerhuijs P (2004) UWB: een nieuwe mobiele communicatietechnologie. GeoNieuws

Luyendijk W, Schoof R (2004) Systeem voor hulp slachtoffers bij rampen. NRC Handelsblad

Mennecke BE, Crossland MD, Killingsworth BL (2004) Is a map more than a picture? The role of SDSS technology, subject characteristics, and problem complexity on map reading and problem solving, MIS Quarterly 24, pp 601-629

Davis FD (2004) Perceived usefulness, perceived ease of use, and user acceptance. MIS Quarterly, 13, pp 319-340

Evacuation Route Calculation of Inner Buildings

Shi Pu and Sisi Zlatanova

Delft University of Technology, OTB Research Institute for Housing, Urban and Mobility Studies, Jaffalaan 9, 2628 BX Delft, the Netherlands.
Email: s.pu@ewi.tudelft.nl; s.zlatanova@otb.tudelft.nl

Abstract

Disastrous accidents (fire, chemical releases, earthquake, terrorist attacks, etc) in large public and residential buildings (discotheques, cafes, trade and industrial buildings) usually result in tragic consequences for people and environments. Such accidents have clearly showed that need for reliable systems supporting rescue operations is urgently appealing. Amongst all, giving appropriate information to the ordinary people in/around the affected area considering the disaster developments (available exists, assessable corridors, etc.) and the human factors (age, gender, disability) are of critical importance for the success of the rescue operation.

This paper promotes a new approach (based on 3D models) for giving evacuation instruction to people. The paper is organized in three general parts. The first part discusses briefly current approaches for alarming people in buildings showing their disadvantages. The second part presents the 3D system architecture and elaborates on the needed components. Discussion on the required developments concludes the paper.

1 Introduction

Since the 'birth' of buildings, the damage from both natural and unnatural disasters never stops. Current buildings are designed higher and more complex than ever before, therefore the potential disasters are also various. Generally, these disasters can be grouped into the following types (William H. Stringfield, 1996):

- Primary disasters such as: fire, power outage, terrorism (bombing incidents, bomb threat, taking of hostages, etc.), chemical releases (radioactive materials, toxic gases, etc.), earthquake, flood, hurricanes, etc.

- Secondary disasters. For example, an earthquake could cause a structural fire, which may in turn burn out circuits resulting in a power failure.

All of the above disasters require people inside buildings to be evacuated as soon as possible. Considering the complexity of modern buildings and the great numbers of people that can be inside buildings, it is rather difficult to organize a quick evacuation. Very often serious problems such as huddle, trample, inaccessibility of exits, etc are observed. Typical examples of such tragic circumstances are the fires in a cafe in Volendam, the Netherlands (2001) and discotheque in Brazil (2004).

In order to make minimal losses from these disasters, many evacuation strategies have been researched and some of them are widely accepted or even integrated into architecture design. This paper will focus on the importance of 3D geo-information. Geo-information has been widely used in all the Disaster Management Phases, e.g. Mitigation, Preparedness, Recovery, but hasn't been really applied to the Response phase (Zlatanova and Holweg 2004) and especially inside buildings. This paper gives a new concept of inner buildings evacuation strategy, which aims at developing of a knowledge-based system able to provide dynamic, specific and accurate evacuation guidance based on indoor geo-information, and sends these instructions to people with interactive instructions.

Next section introduces the evacuation strategies used in current buildings and discusses their drawbacks. After that the overall architecture of our concept model is presented with detailed explanations of the three main technical parts. This paper is closed with conclusion and recommendations for future research.

2 Current Evacuation Systems

A lot of developments and research are aiming at providing more efficient means for alarming and guiding people. Good examples are fire-alarming systems. Many buildings are currently equipped with modern fire detection systems and it is possible to alert people to a fire. However, this gives no clues as to how to escape. Directional Sound Evacuation (DSE) beacons are also available and they can eventually give clear audible navigation to nearest exit (http://www.soundalert.com/dse_buildings.htm). Several large public buildings (Business Design Centre, London, Munich International Airport, etc.) are equipped already with DSE. These systems can be combined with sophisticated analogue addressable Fire Alarm Control Panels (FACP) (e.g. http://www.adt.co.uk/fire_panels.html). These are systems that can locate seats of fires and decide which are preferred evacuation routes. DSE can then be activated only along these routes. It is also possible to teach people what is the meaning of sounds. The problem is still that these kinds of systems react only on the seat of the fire and are 'blind' about situation after the fire alarm is triggered.

In principle, current alarming systems can be subdivided into three groups (Galea et al, 1999):

Automatic detection system involve a sensor network plus associated control and indicating equipment. Sensors may detect heat, smoke or radiation and it is usual for the control and indicating equipment to operate an alarm system. It may also perform other signaling or control functions, such as the operation of an automatic smoke control system.

Alarm system alert people at the early stages of a disaster and give them maximum escape time. Normally any of following devices needs to be incorporated in the building:

- Manually operated sounders;
- Simple manual call points combined with bell, battery and charger;
- Internal speed communication system (telephone, intercom, etc.) should be provided so that conversation is possible between every floor and the control centre.

Emergency lighting is designed to allow occupant to continue to occupy, although they may not operate as efficiently as under normal lighting.

Additionally, building construction requirements to resist potential natural and industrial disasters exist (that may vary from country to country). Evacuation exits should satisfy certain criteria so that people can pass through the exits safely and quickly. Different standards give different requirement for exits' width, numbers, capacities, etc. Generally, there should be:

- 1 exit for up to 60 persons;
- 2 exits for 61-600 persons;
- 3 exits for over 600 persons.

Despite the variety of systems, current emergency alarming systems are only able to provide simple mostly constant evacuation instructions (evacuation plans, green lights, sounds, voice) which is not sufficient because:

- Evacuation plans lack **flexibility**. Evacuation notifications usually directly follow pre-defined evacuation plans, regardless whatever has happened or is happening in buildings. This may lead people into dead ends (collapsed ceilings, destroyed stairs, blocked exits, etc.) or bring more serious problems (lead to a spaces with gas leakage and possible explosions).

- Evacuation plans is not **intelligent** enough. If not controlled, too many people rush to same exits, which results in the 'traffic jam' in inner buildings, and would greatly delay the speed of evacuation, may results in blockage of exits and many injured people.
- Evacuation instructions give **insufficient information**. For people who are not familiar with the building, current evacuation instructions may not be helpful of even of no use. Some serious disasters (power outage, fire, etc.) that reduce visibility can make their situations even worse.

The problems mentioned above are result of three serious deficiencies:

- Lack of indoor geo-information (about the structure of the building),
- Lack of dynamic information (about the current situation),
- Lack of flexible means for evacuation instructions.

A more elaborated systems are needed that can offer intelligent, knowledge-based evacuation navigation. How such a system should react? Let's assume that a fire accident happened on the 10th floor of a 21-floor building, and expanded to the 9th and 11th floor in a very short period. People in this building should be evacuated immediately! In a sequence of scenes we will give a possible development of the situation:

- **Scene 1**. Evacuation lights were turned on, indicating the direction of evacuation exits.
- **Scene 2**. Voice instructions sounded from the speakers all over the building, asking people below 9th floor to evacuate from Exit A, people on 9th, 10th, 11th, 12th, 13th, 14th, 15th floor to evacuate from Exit B, people on 15th, 16th, 17th floor to evacuate from Exit C, and people above 17th floor to evacuate from Exit D. Never use elevators!
- **Scene 3**. Joanne was visiting this building when the accident happened. She was unfamiliar with the building, and she easily found herself totally lost in the building because the power system went down. But she had her PDA and suddenly she noticed that it is blinking. She opened it and realized that she receives instructions hoe to get out of the building. She followed the interactive graphical instructions and finally evacuated safely from the building.
- **Scene 4**. Andy and his colleagues were in the 11th floor when the accident happened. Following the voice instructions, they went to Exit B immediately. But shortly after that they found the stairs of Exit B, 9th floor was totally in fire! 'Where shall we go?' they asked themselves. Some of the group prefer to go to Exit A because it was the nearest. But Andy said 'NO!' because according to his PDA, most parts of Exit A had lost power. If they go to Exit A, they might be lost in total darkness!

According to the instructions on the PDA, the best route was to go back to the 10th floor, go to Exit D, and then go down to the ground floor. And this is what exactly they did.

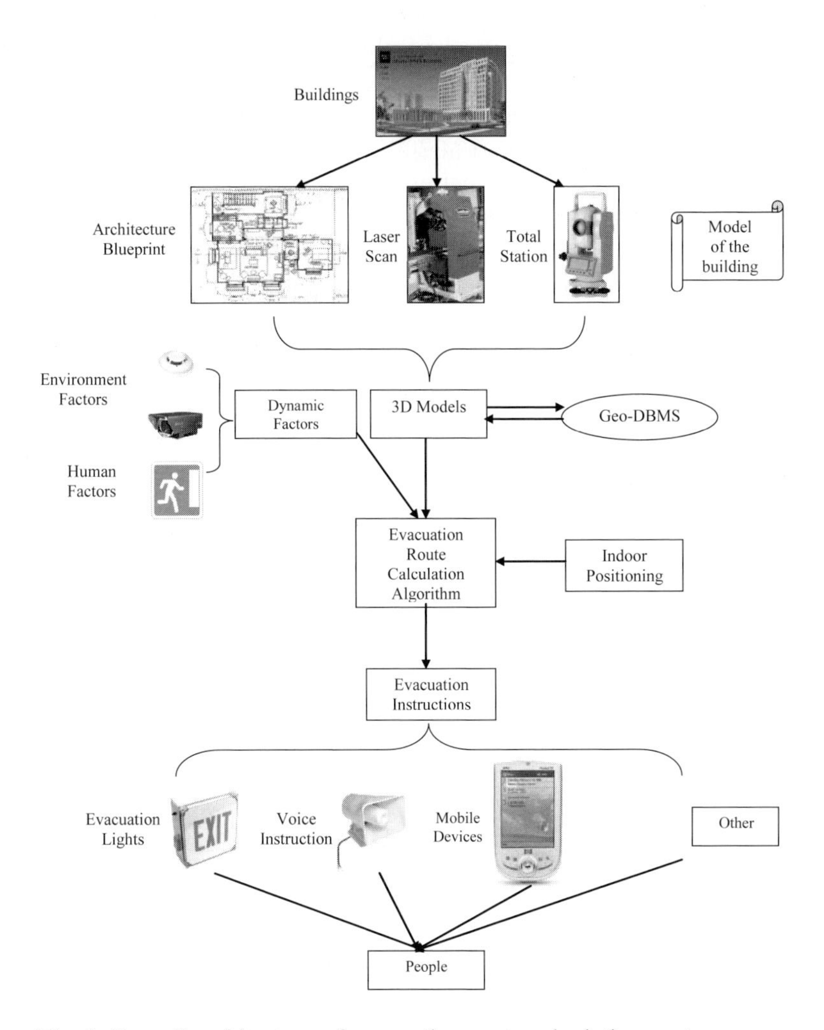

Fig. 1. Overall architecture of evacuation route calculation system

3 A System for an Evacuation Route Calculation

To be able to provide such a dynamic navigation four important components are necessary: system for indoor positioning, communication connection, inner model of the building, information about the people in the building, real-time information about the disaster, engine for route calculation (Figure 1) and engine for adapting the routing presentation.

Presently there're many techniques for positioning (passive and active), for example, the Global Positioning System (GPS), digital compass, and etc. However, most of these positioning techniques are only for outdoor positioning. In the recent couple of years, many indoor positioning techniques have been suggested (Gillieron and Mrminod, 2004, Kaemarungsi and Krishnamurthy, 2004, Fritsch et al 2001) using WLAN, location fingerprinting, vision system or combinations of them. For example, using location fingerprinting, one first collects the local fingerprints by performing a site-survey of the Received Signal Strength (RSS) from multiple Access Points (APs). Then the APs sends the RSS measurements to a central server, so that certain algorithms can be used to determine the estimated position of the signal, or in other words, the position of the individual.

For simplicity, we will consider that 3D indoor positioning (passive or active) and a telecommunication network (WLAN, GPRS or UTMS) are available (see Zlatanova and Verbree, 2003) and will concentrate only on the 3D indoor model, the algorithms for route calculation and the presentation of the information.

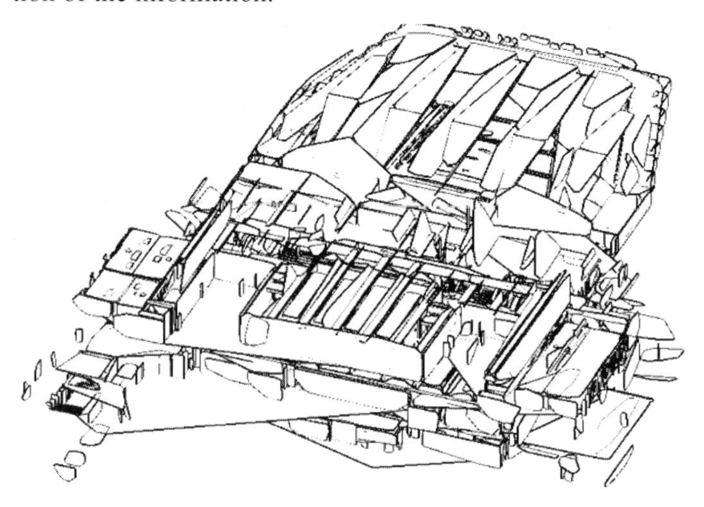

Fig. 2. Geometry of the Aula

3.1 3D Model of Buildings

Modeling buildings will be understood here as the process of converting a building's architecture structure to a model that can be recognized by computer. With the help of this model, fastest and safest evacuation routes can be calculated both globally (for crowd) and locally (for a single person). The model of a building can be either a geometry/topology model, which means, an abstraction of a building is represented with polygons in 3D space (see also Zlatanova et al, 2004); or a logical model (see Lee, 2001), which represents the connections between the rooms. The rooms and important crossings are represented with nodes; the paths are represented as links between nodes.

Geometry model. Building a geometry model of interiors of buildings is more complex compared to outside model. The interior structure has to be seen as an aggregation of several different types of objects (rooms, stairs, etc.) with different shape. Geometry model (e.g., Figure 2) of a building can be generated in many different ways:

- Architecture blueprint of this building
- From point clouds obtained from terrestrial laser scanning
- Close-range images
- Surveying measurements

Each of the methods has advantages and disadvantages. Architectural blue prints are used very often for 3D models. Given the height of the floors, the 3D model can be manually 're-designed' in appropriate modeling software, e.g. 3D StudioMax, AutoCAD, MicroStation, etc. The common bottleneck is that the plans are either not available or differ from the real situation.

Laser scanning is relatively easy but results in a large amount of data (many of them even not necessary for the application, e.g. curtains, tables, etc) that has to be further filtered and modeled to obtain the 3D polygons. In general the human eye can easily understand geometry model and get an overview of a building out of point clouds. However the computer does not distinguish between the points. Therefore object recognition is needed. 3D building reconstruction from point clouds requires determination of rooms, corridors, stairs, etc. Different commercial software (e.g. Cyclone, CloudWorx) exists that allows a primitive (plane, cone, cylinder, etc.) to be fit to a group of points. The fitted element can be further aggregated into objects. This procedure is mostly manual and rather time consuming. Research towards automatic fitting procedures is currently going0n (e.g. Rabani and Van den Heuvel 2004), but still complete automation is not possible.

3D reconstruction using close-range images (taken with conventional camera) allows the operator to select only the points that are needed for the model (e.g. using PhotoModeler). The reconstruction is again in large amount manual (pointing the same points in different images and connecting selected points into triangles and polygons), but thanks to the images the operator orients better (compared to point clouds) in the model. Close-range images are wisely used for 3D reconstruction of historical buildings.

Surveying, in contrast to laser scanning and similar to close-range photogrammetry provides the points only needed for the modeling. To be able to reconstruct for example a simple room of four walls, one would need to measure only four points. Since the measurements are outside the office (in the building), a very good organization of the registered points is needed. Missed important points or unclear identification of points may lead to repeating of measurement. The actual 3D reconstruction is again mostly manual and rather time consuming.

There is not an automatic approach for 3D reconstructing of the interior of buildings and other closed constructions, and it is difficult to extract paths from geometry model. However, the 3D geometry is needed for presentation of the navigation route.

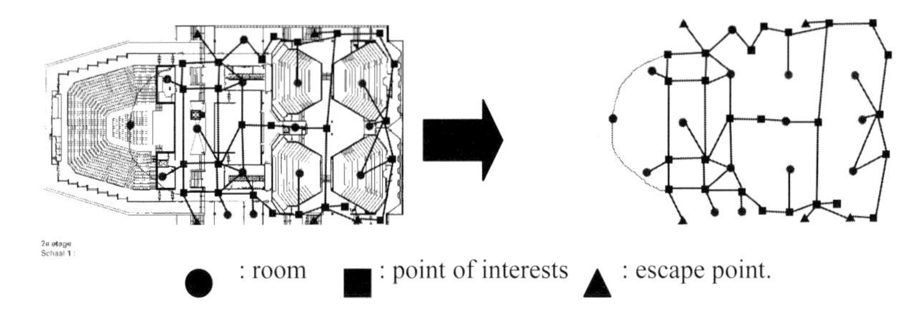

Fig. 3. Simplified Logical model of Aula, 2nd floor

Logical model. Logical model represents each room/crossing/exit in buildings with a node, and represents paths with the links between nodes. Using logical model, the inner structures of buildings instead of geometric information can be analyzed, and routes from node to node can be calculated easily, which is the most significant advantage over the geometry model (Gillieron and Merminod, 2004).

However, it is very difficult to generate the logical model (Figure 3) of a building automatically from its geometry model. We have to manually or half manually (with computer aided applications) create the important nodes and make their links. Especially, the links should be assigned with cost values, to represent the travel distance or travel time for each links. Search algorithms will consider the links' cost values and non-spatial data of nodes (will be discussed later) together, and then determine the most appropriate evacuation routes.

A lot of attention and experience are required to analyze the building and determine the nodes for logical model. This looks a tedious job, but once the logical model is completed, path search within the logical model can be easily done because this have been researched many years and there're already a number of well-developed search algorithms now which can be used directly to calculate the evacuate route. However, logical model can't be visualized directly by people as the geometry model does.

Hybrid model. A combination of geometry model and logical model can become a powerful tool for evacuation route calculation. The logical model will be used to compute the shores available path and the geometry model will be used for visualization. With links between these two models, information can be shared so that any route in logical mode can be reflected in geometry model.

Such combination has been implemented by Oracle and published as the Network Data Model in Oracle Database 10g. There're three kinds of Network Data Models: a logical network (no geometry), a geometry network (has geometry) and a LRS geometry network (has geometry with measure value). All of these networks can be hierarchical. They are all made up of four tables:

- Node table (Table 1). Stores the nodes, columns of this table include id, name, type, partition, cost, and etc.
- Link table (Table 2). Stores links between nodes, columns of this table include link id, name, type, start node, end node, cost (can be used to represent length), and etc.
- Path table
- Path link table

Name	Type	Example
NODE_ID	NUMBER	3
NODE_NAME	VARCHAR2(200)	'Meeting Room A'
NODE_TYPE	VARCHAR2(200)	'Room'
ACTIVE	VARCHAR2(1)	'Y'
PARTITION_ID	NUMBER	2
GEOMETRY	MDSYS.SDO_GEOMETRY	MDSYS.SDO_GEOMETRY (2003, NULL, NULL, SDO_ELEM_INFO_ARRAY (1, 1003, 3), SDO_ORDINATE_ARRAY (100,200, 500,700))

Table 1. Columns of Node table

Path table and path link table are optional, but calculated shortest path can be stored here. Networks can be created by either PL/SQL or Java API, but can only be analyzed by Java API. A shortest path function is already available in the Java API, but only the costs (length) of links are considered. Based on the existing well-structured network model, we can easily implement user-defined java functions that take other problems that affect the evacuation route calculation into consideration.

Name	Type	Example
LINK_ID	NUMBER	2
LINK_NAME	VARCHAR2(200)	'Corridor 2'
START_NODE_ID	NUMBER	3
END_NODE_ID	NUMBER	5
LINK_TYPE	VARCHAR2(200)	'Corridor'
ACTIVE	VARCHAR2(1)	'Y'
LINK_LEVEL	NUMBER	1
GEOMETRY	MDSYS.SDO_GEOMETRY	MDSYS.SDO_GEOMETRY (2003, NULL, NULL, SDO_ELEM_INFO_ARRAY (1, 1003, 3), SDO_ORDINATE_ARRAY (500,2100, 550,3500))
COST	NUMBER	6

Table 2. Columns of Link table

Non-spatial information. All the three models above organize the spatial information of inner buildings well. However, the model of a building with only spatial information is just an abstraction, which is not able to give the whole pictures for disasters in buildings. A complete evacuation model should also contain **non-spatial information** that influences the evacuation route. Some kinds of important non-spatial information are:

- **Type of the spaces.** Types for nodes (logical model) can be room, crossing, exit points and etc; types for links (logical model) can be corridor, stairs, and etc. Different types of element/node have different purposes and attributes. For example, the exit nodes are the target nodes for search algorithms; people have different moving speed in rooms, corridors, stairs, and etc.
- **Population density.** There're always more people in certain areas and less in other areas. For examples depending on the type of the building, in one room can be more than 40 and up to 200 people (theatres, congress centres) or less than two (offices). The system has to have information about the 'daily' distribution of people in the building
- **Construction type.** Different types of buildings have different requirement for the evacuation routes. For example, 'safe' is more important than 'fast' for evacuation routes in hospitals; evacuation route calcula-

tion should bc morc carcful about capacitics of rooms in congress centres and theatres. Importance of different factors should also be adjusted a bit in order to be more suitable for different types of constructions.

- **Type of utilities/networks.** There might be potential dangerous or helpful utilities in certain areas. For example, gas that can explode, high voltage devices, telecommunication networks, medical utilities, and etc. These kinds of non-spatial information should not be ignored because they also have rather large influences on the evacuation routes calculation.

3.2 Dynamic Factors

Emergency situations are not static, and route calculation cannot only consider travel distance. There're many other factors that would affect the optimal evacuation routes. Theses factors can be grouped into two kinds:

Environmental factors. Unknown problems can happen to a building at any time during a disaster. The evacuation routes calculated from fixed search tree can be of no use at all, or even give wrong evacuation instructions. In such cases, an **adaptive search tree**, which reflects the real time status, can be of great help to make valid and accurate evacuation routes. Modern buildings are equipped with detection system that is able to detect some kinds of disasters such as fire/smoke/etc. The detection system can be connected to the buildings' logical models to form the adaptive search trees. Once a disaster is detected by sensors or showed from monitors, the nodes within dangerous areas should be marked automatically or manually as 'unsafe' or 'inaccessible' in the non-spatial information of a building's logical model. Then an evacuation route, which avoids the dangerous areas, should be calculated from current search tree. If more events are detected later, then the evacuation routes should be recalculated and the modified evacuation instructions should be sent to the affected people, if the new evacuation routes are different from the previous ones.

Different types of dynamic factors may be defined. An initial classification follows:

- **Damage status.** Areas that are already damaged should be avoided in an evacuation route. The damage status of an element (geometry model) or a node (logical model) can be True or False.
- **Toxicity status.** Areas that are already full of toxic gases/smokes should be avoided in an evacuation route. The toxicity status of an element (geometry model) or a node (logical model) can be True or False.

- **Power status.** Places that lost power should not be in an evacuation route. The power status of an element (geometry model) or a node (logical model) can be True or False.
- **Capacity of the routes.** If people are given an evacuation route that did not consider the capacities of routes, they would probably get stuck in some low-capacity parts. This is quite dangerous, and should be avoided with no doubt. Evacuating from multi routes will efficiently avoid evacuation jams. Calculation of multi route evacuation will be discussed later in this chapter.

Human factors. There has been a lot of research on human behavior under different situations. From these researches, people's speed of movement varies with:

- **Population density**. If the density is higher than a certain value (crowd-density value), conditions would become very uncomfortable and movement is difficult or impossible. Nelson and Maclennan's research (quoted in Galea et al 1999) on human behaviour suggests that:
  - If population density < 0.54 p/m^2, individuals maintain movement speed.
  - If population density > 3.8 p/m^2, there is no or very little movement.
  - Between these limits, speed is hindered and is a function of the density given by:

    $$\mathbf{S = k - 0.266*k*D}$$

    where 'S' is the practical movement speed (m/s); 'D' is the population density (p/m^2); 'k' is the standard movement speed. 'k' is dependent on type of terrain, for example, k for corridors, doorways and ramps are 1.4.
- **Age and gender**. Ando's research on crowd movement can be summarized as: (quoted in Galea et al 1999):
  - Walk speed peaked at about 20 years of age.
  - Males outpaced females at all ages.
  - Max walk speed for males is about 1.6 m/s.
  - Max walk speed for females is about 1.4 m/s.
- **Level of disability.** Problems for disabled people are not only in slowed down movement speed. People with specific movement aids, such as wheelchairs, cannot use stairs unaided. An evacuation flow with disabled people could have lower movement speed because normal occupants would help them along the way.
- **Terrain effects**. People move slower in certain areas, e.g. stairs. Fruin's research (quoted in Galea et al 1999) on people's behavior on stairs shows that the default stair travel speed are:

Gender	Age	Down average (m/s)	Up average (m/s)
Male	<30	1.01	0.67
Female		0.755	0.653
Male	30-50	0.86	0.63
Female		0.665	0.59
Male	>50	0.67	0.51
Female		0.595	0.485

Table 3. Fruin's research result about movement rates on stairs

The environment and human factors above will be both considered for the evacuation route calculation. Their importance may vary in different kinds of buildings. For example, population density should be carefully considered in theaters, level of disability should be carefully considered in hospitals, and etc. Values for these factors can be collected from investigation over the buildings. Many commonly used investigation methods are mentioned in Galea et al 1999. All the factor values should be stored along with the logical model as non-spatial information in the database.

3.3 Route Calculation Algorithm

Below we give the algorithm for evacuation calculation. It takes a building's model and dynamic factors as parameter, and aims at giving correct and safe evacuation routes, which will be sent to people with different evacuation instructions.

3.3.1 Basic Route Calculation

With the logical model of a building, routes from a certain node to another node can be calculated using a lot of existing search algorithms, if only the travel distances between nodes are taken into consideration.

A search algorithm is generally evaluated from four criteria (Russel and Norvig, 1995):

- Completeness: is the algorithm guaranteed to find a solution when there is one?
- Time complexity: how long does it take to find a solution?
- Space complexity: how much memory does it need to perform the search?
- Optimality: Does the strategy find the highest-quality solution when there are several different solutions?

Among these four criteria, completeness is not critical because we only need the best route (optimality) not every route, with acceptable computation time and memory requirement. Below we just give the definitions and evaluation criteria of four popular search algorithms. More detailed information and proof of the evaluation criteria can be found in the literature (see Russel and Norvig, 1995).

Breadth-first search In breadth-first search, the root node is expanded first, then all the nodes generated by the root node are expanded next, and then their successors, and so on, till find the goal node.

Depth-first search Depth-first search always expands one of the nodes at the deepest level of the tree. Only when the search hits a dead end does the search go back and expand nodes at shallower levels. The drawback of depth-first search is that it can get stuck going down the wrong path.

Depth-limited search Depth-limited search avoids the pitfalls of depth-first search by imposing a cut off on the maximum depth of a path.

Iterative deepening search Iterative deepening search is a strategy that sidesteps the issue of choosing the best depth limit by trying all possible depth limits.

Bidirectional search Bidirectional search algorithm searches both forward from the initial state and backward from the goal, and stop when the two searches meet in the middle.

Dijkstra's algorithm Dijkstra's algorithm differs from the first five in that it is an informed search algorithm, which means the 'goodness' of a node can be estimated. This algorithm keeps two sets of vertices Stringfield, 1996:

- **bbS**: the set of vertices whose shortest paths from the source have already been determined
- **bbV-S:** the remaining vertices.

The other data structures needed are:

- **d:** array of best estimates of shortest path to each vertex.
- **pi**: an array of predecessors for each vertex.

The main steps of Dijkstra's algorithm are:

- Initialise **d** and **pi**,
- Set **S** to empty,
- While there are still vertices in **V-S**,
- Sort the vertices in **V-S** according to the current best estimate of their distance from the source,
- Add **u**, the closest vertex in **V-S**, to **S**,
- Relax all the vertices still in **V-S** connected to **u**. The relaxation process updates the costs of all the vertices, **v**, connected to a vertex, **u**, if we

could improve the best estimate of the shortest path to **v** by including **(u,v)** in the path to **v**.

Table 4 compares the first five algorithms in term of the four criteria.

Criteria	Breadth-First	Depth-First	Depth-limited	Iterative Deepening	Bidirectional
Time	b^d	b^m	b^l	b^d	$b^{d/2}$
Space	b^d	bm	bl	bd	$b^{d/2}$
Optimal	Yes	No	No	Yes	Yes
Complete	Yes	No	Y, if $l \geq d$	Yes	Yes

Table 4. Evaluation of search algorithms. b is the branching factor[1]; d is the depth of solution, m is the maximum depth of the search tree; l is the depth limit (Russel and Norvig, 1995)

Normally there're several hundreds of nodes in logical models of modern buildings. Experiments show that all of the above search algorithms take approximately 1 second to find all the solutions (if any) for trees with hundreds of nodes. Therefore here it doesn't matters too much about which algorithms to use.

3.3.2 Advanced (Evacuation) Route Calculation

Basic search algorithms can only be used to calculate the routes, but not evacuation routes. During a disaster in a building, the route network may change at any time. Therefore we need an advanced algorithm that also considers dynamic factors as discussed in Section 3.2. If not only the cost of links, but also the dynamic environment factors and human factors are taken consideration, then the earlier basic route calculation algorithms should be improved in several aspects:

- Try to avoid traffic jams. Traffic jams are the result of either too high population density or small capacity of routes. People should be evacuated to exits from different routes, which have approximately same population densities.
- Adaptive to real time status. Once the status of some nodes are changed, for example, from 'normal' to 'dangerous', then the calculation procedure should be repeated using the current model.
- Give special evacuation routes for disabled people. If applicable, evacuation elevators are the first choice for disabled people. If not, dis-

---

[1] If in a hypothetical tree where every nodes is connected to b nodes, we say the **branching factor** of this search tree is b.

abled people and their aid workers should be assigned to less used routes because their movement can significantly decrease the speed of exodus evacuation.

Here we propose an algorithm for evacuation route calculation, which aims at improving the basic route calculation algorithm in above three aspects.

- Build a search tree from the logical model. Specify each node with two estimated population density value: for disabled people, and for non-disabled people. Specify the global movement speed values for disabled people and non-disabled people. Specify the crowd-density value.
- Scan the current searching tree, remove the unsafe nodes, and build a new tree.
- Run the basic route calculation algorithm and find the evacuation route for people within each node. This will result in a time line of the evacuation.
- Split the time line into several time segments Ti.
- For a segment Ti in time order, calculate the current population densities for each node. The current population density D is:

  $$D = D_n + D_d \times m$$

  where D_n is the population density of non-disabled people; D_d is the population density of disabled people; m is a constant number which is larger than 1.

  If the current population density D is higher than the crowd-density value or the route capacity of this node, go to 6, else go to 8.
- Go back to the previous time segment (i-1), reallocate the flows which 'will' join in the jam nodes to other routes. Try different allocation till no traffic jam will happen. Notice that people from same initial nodes should be allocated to the same route.
- Step into the next time segment, and repeat 5-7, until finish the last time segment.
- Evacuation route calculation completed.

This algorithm is to be implemented in a hybrid model as described in Section 3.1. Tests to be performed on the model of the Aula (Figure 3) will valid the algorithm and will supply knowledge for further improvements. Expectations for improvement are given in the later in this paper.

3.4 Evacuation Instructions

Once the route is calculated, a very important question is how to represent it to the user. There're many possibilities (depending on the available technology) to give evacuation instructions from calculated evacuation routes (Figure 4).

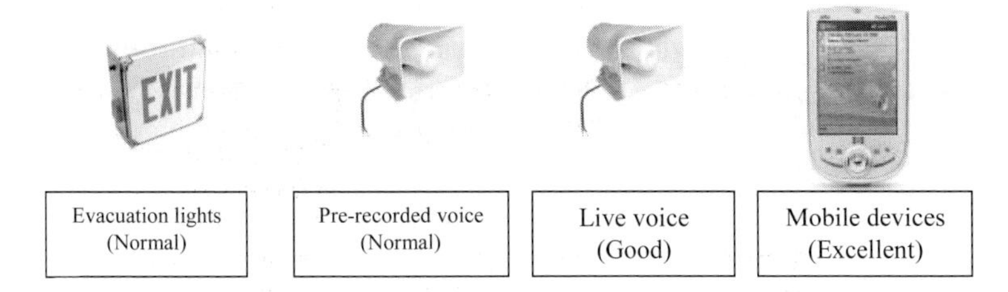

Fig. 4. Effectiveness of different evacuation routes

Evacuation lights and speakers can be used to give evacuation instructions for all the people within a building. Voice usually gives helpful information, but it is still difficult to generate voice instructions automatically from calculated evacuation route. A possible solution is to first generate the path figures that can be easily understood by administrators, so that they can give use microphones to live evacuation instructions.

Fig. 5. 3D models with Nokia GL on the Communicator 9210 as one result of the TellMaris project (IST-2000-28249, http://www.tellmaris.com). © Nokia Research

The most advances possibility is to use mobile devices like cell phone, PDA, etc. Mobile devices can supply people with interactive and graphical evacuation instructions. Using 3D location-based services, the received evacuation routes can be always adapted to the real time position, so that people don't need to remember all the evacuation details. Another advantage of using mobile devices is the graphical instruction. A large number of mobile devices are able to show pictures or videos, some even embedded with 3D rendering engines (Figure 5). Evacuation instructions can be sent from central computer to mobile terminals with either video streams or evacuation directions, or combination of both. Video streams are already rendered in the server from both the geometry model and evacuation directions (from logical model). This method has low requirement for the mobile terminals because they only need to display the videos, but the amount of transferred data is rather considerable. If only evacuation directions are sent via wireless network, the transferred data is not heavy, however, mobile devices need to render the visualization locally, which might be tedious task for slow mobile devices. Balanced visualization between sending video frames and evacuation directions can lead to satisfactory results.

4 Concluding Remarks

This paper suggested a new concept of evacuation route calculation for inner buildings using 3D models and considering human and environment factors. The 3D model (hybrid model) will be organised in DBMS. The evacuation algorithm is based on the basic search algorithms, but it also takes the timeline of an evacuation into consideration. The calculated routes could be sent presented in various ways so that they could reach as many people as possible. Mobile devices like PDA can be very helpful to receive evacuation instructions.

Based on the concepts given in this paper, implementation of the evacuation calculation algorithm and further research will be carried on. **Accuracy** and **speed** are certainly the two most important evaluation parameters for the system. It might be also necessary to give **multiple** evacuation routes to users so that they can have more possibilities to choose. We consider focusing on the following aspects:

- Determine exactly which kinds of environment factors and human factors should be considered, and how can they be organized as non-spatial information in the Geo-DBMS.
- Create an automatic procedure for linking logical model with geometry model.
- How to make more accurate timeline for the evacuation process?
- How can this system be technically fast and robust enough to really be useful in a disaster? And how to validate the calculated evacuation routes?
- What is the balance between bandwidth and rendering capability for evacuation instructions via mobile devices?
- How can the system be modularized and gradually evolved? Since this algorithm aims at giving route instructions for inner buildings in disaster situations, it is also possible to give route guidance for normal situa-

tions. It would be interesting to know how to evolve this algorithm to a guide system for inner buildings.

References

Buchana AH (2001) Structural design for fire safety, Wiley

Galea ER, Owen M, Gwynne S (1999) Principles and practice of evacuation modeling second edition, CMS Press

Gillieron P, Merminod B (2004) Personal navigation system for indoor applications, 11th IAIN World Congress

Fritsch D, Klinec D, Volz S (2001) NEXUS---positioning and data management concepts for location-aware applications, Pergamon, Computers, Environment and Urban Systems 25, pp 279-291

Kaemarungsi K, Krishnamurthy P (2004) Modeling of Indoor Positioning Systems Based on Location Fingerprinting, IEEE INFOCOM

Lee J (2001) 3D Data Model for Representing Topological Relations of Urban Features', Proceedings of the 21st Annual ESRI International User Conference, San Diego, CA, USA

Morris J (1998) Data Structures and Algorithms. Available at http://ciips.ee.uwa.edu.au/~morris/Year2/PLDS210/dijkstra.html

Oracle (2003) The Network Data Model manual of Oracle Database 10g, Oracle

Rabbani T, Van den Heuvel, F (2004) 3D industrial reconstruction by fitting CSG models to a combination of images and point clouds, IAPRS vol XXXV part B5, Istanbul, pp 7-15

Russel, Norvig (1995) Artificial intelligence a modern approach first edition, Prentice Hall

Stringfield WH (1996) Emergency planning and management, Rockville: Government Institutes

Zlatanova S, Holweg D, Coors V (2004) Geometrical and topological models for real-time GIS, Proceedings of UDMS 2004, 27-29 October, Chioggia, Italy, CDROM, 10 p

Zlatanova S, Holweg D (2004) 3D Geo-information in emergency response: a framework, Proceedings of the 4th International Symposium on Mobile Mapping Technology (MMT'2004), March 29-31, Kunming, China 6 p

Zlatanova S, Verbree E (2003) Technological developments within 3D location-based services, International Symposium and Exhibition on Geoinformation 2003 (invited paper), 13-14 October, Shah Alam, Malaysia, pp 153-160

Geo Embedded Navigation

Peer M Ilsøe and Jan Kolar

Centre for 3D GeoInformation, Aalborg University, Niels Jernes Vej 14, DK-9220 Aalborg, Denmark.
Email: {ilsoe, kolda}@3dgi.dk.

Abstract

Current software for estimating potential losses from hazards uses GIS software in order to display results of damage analysis on a map. In parallel, within the fields of geographic visualization and GISc systems capable of visualizing data for entire planets are starting to appear. One of the new challenges in this context is to develop a simple and intuitive general purpose navigation mode that will work well for a single planet from outer space to a street level. Such works are missing today. Although the need for global navigation in disaster management applications is rather conceptual than practical, introducing a conceptually better navigation mode has positive practical consequences also for applications in this field. By better concept for a navigation mode we mean: The navigation become generally usable around the whole globe regardless location of the viewpoint or level of detail used for the scene rendering; the navigation is intuitive for humans; and the navigation is simple by utilizing a single set of straightforward mathematical relations. In this text we introduce such global navigation mode, which has been implemented in Grifinor system.

1 Introduction

Subject of global visual navigation lacks works presenting concrete solutions convenient for an implementation. The GeoVRML specification [4] is the only work we have found, which addresses, but does not solve, this issue. By providing a simple mathematical solution implemented in the Grifinor system [1], we aim facilitating adoption of similar navigations in other global systems for interactive geographic visualization. A globally applicable navigation induces a practical value by improving an estimation of gravitational up vector. In traditional navigations, up is related to models based on flat approximation of the surface. We try to address this improvement by explaining a possible use of the proposed navigation.

Similar to navigation of planar maps, the user can, for example, look over a ground acceleration analysis along Newport-Inglewood fault in Los Angeles, with surface colored according to the acceleration level of the ground; looking down one could zoom-in and zoom-out and pan to any side. The proposed navigation also allows to raise the view up, look to the sides and survey the situation in a three dimensional context with streets and buildings surpassing the terrain. This provides a perspective when volumes of constructions at places with various level of ground acceleration are apparent, which might be useful. Heading to a particular direction one can move forth and back following certain elevation level. The user can also move further from the surface and survey the haze and pollution in the area, which it in turn might become necessary to see from the even broader context of atmospheric circulation. At this scale it might be possible to identify hazard vulnerability near areas forming tropical cyclones, areas with high horizontal temperature contrasts forming strong wind or areas with an exceptional rainfall. At this rather global point of view it becomes apparent that moving forth or back, the navigation actually has to follow spherical shape of our planet in order to preserve the user's elevation level. This provides a natural experience for humans and offers a visual navigation and perception at many scales. At one session the user can approach the surface a little, explore cyclone's vertical structure by looking up and down, and return to the street level underneath, which is in a possibly endangered urban area. If such system exists, the user can go through all of this using a single navigation mode.

In the following section we introduce a theory briefly and concept for the proposed navigation mode. In Section 3 we provide an algorithm that formalizes the navigation. We discuss results at the end.

2 Types of Navigation

We classify navigation modes with at least four degrees of freedom as either *world orientation independent* (independent mode) or *world orientation dependent* (dependent mode).

The independent modes implement navigation in the local coordinate system of the camera. In general this allows translation along the left-right, up-down, and forward-backward axes as well as roll, pitch, and yaw rotation all in the local coordinate system of the camera. Since there is no constrains with regard to position and orientation of the camera, it can be used for navigating every kind of 3D environment. However, this navigation mode may not be the easiest to use in special cases. For example if there is a sense of gravitational up and down, it is up to the user to keep track of those directions as well as aligning the horizon to be horizontal, which we assume the majority of humans prefer. In this case orientation dependent modes can be much easier to use.

In contrast, dependent modes of navigation partially or fully depend on the orientation of the world coordinate system. A well known example of this is used in most first or third person perspective computer games. In this mode the up axis of the player is always parallel to the up axis of the world. The user can move forward and sideways in his local coordinate system, rotate around his local up axis (yaw), and look up and down (pitch) but no more than 90 degrees in either direction from horizontal. It is usually impossible to modify roll permanently so aligning the horizon is unnecessary. This navigation usually requires a flat world to function and thus limits the possible geographical extent. This is not a problem for most games since they deal with a relatively small world.

Dealing with spherical worlds, we have identified two problems that we would like to address: First, we would like to have the advantages associated with a dependent navigation mode since we believe this would make it easier for the user to navigate the world in our system. Second, the huge size of the world makes it convenient to have a mechanism that more or less automatically guesses the speed at which you want to travel. This guess could simply be based on elevation above the spherical or ellipsoid approximation of the earth, or more advanced above the terrain. This is addressed in [4]. In this paper, we focus on the first issue only.

2.1 Desired Navigation

In sake of proposing the desired navigation mode we will distinguish between three coordinate systems:

- The world coordinate system, with origin at the center of the planet.
- The local coordinate system of the user, where the up direction is dependent on the position of the user in the world coordinate system, the forward direction is given by the horizontal orientation of the user, and the right direction is implicitly given by the two others.
- The camera coordinate system is the local coordinate system rotated pitch degrees to allow the user to look up or down.

We have found it easy and intuitive to navigate, at local and global scale, when the up vector of the user is continually kept aligned with the gravitational up vector of the world in the position of the user. Navigation with regard to rotations is performed in the local coordinate system, and movement is performed in the camera coordinate system. In effect this means we have a dependent mode where the position in the virtual world gives the direction of the local up vector, and movement is relative to the camera orientation. Moving along a spherical approximation of the terrain surface with a pitch angle of zero, the user should stay at the same altitude. We have chosen to name this family of navigation modes *Geo Embedded Navigation*.

3 The Algorithm

Our intention is that it should be possible to design the navigation code as if navigation took place in a flat world on a frame to frame basis. For this reason we expect a movement vector in the camera coordinate system as well as changes in the pitch and yaw angles of the camera.

Technically, navigation in our system is performed by continually setting a transform matrix for the camera which positions and orients the camera in the virtual world. We assume a right handed world coordinate system and a graphics system, which furthermore has the x axis to the right, y axis up, and z axis towards the observer.

We keep track of the orientation of the local and the camera coordinate systems by keeping four variables: A position vector in world coordinates, $\overrightarrow{pos}$, a forward vector in world coordinates, $\overrightarrow{localFwd}$, an up vector, $\overrightarrow{localUp}$, also in world coordinates, and a local camera pitch angle, *pitch*. The local coordinate system is illustrated in Figure 1.

Between each frame we receive incremental updates of the *yaw* and *pitch* angles, Δyaw and $\Delta pitch$ as well as a move vector, $\overrightarrow{move}$ in camera local coordinates.

The major tasks of the algorithm are the following, which will be described in more detail below:

- Perform changes to orientation requested by the user
- Make an orientation transform matrix
- Transform the camera local move vector into global coordinates
- Find new position candidate
- Remove vertical position error
- Find new orientation
- Set new orientation and position in the resulting matrix

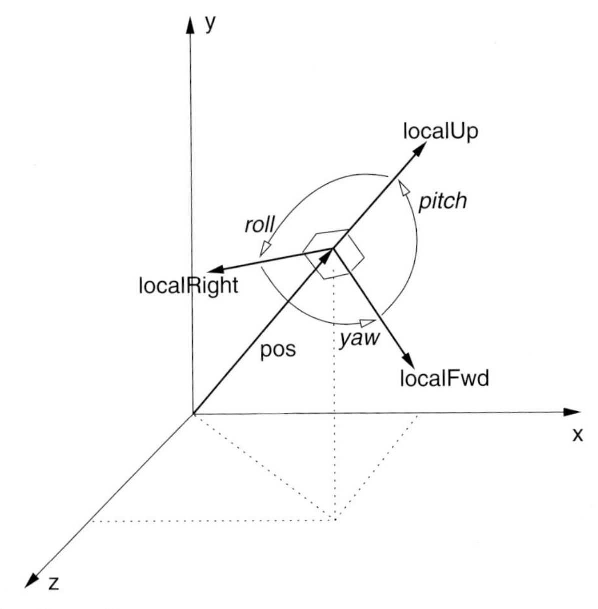

Fig. 1. The local coordinate system

To perform the changes in orientation requested by the user, the variable *pitch* is incremented with $\Delta pitch$ and clamped to range $[-\pi/2; \pi/2]$, while the yaw update is handled by rotating $\overrightarrow{localFwd}$ Δyaw around $\overrightarrow{localUp}$.

Based on $\overrightarrow{localFwd}$, $\overrightarrow{localUp}$, and their cross product, $\overrightarrow{localRight}$, we form a transformation matrix from local to global coordinates. To have a transformation from camera coordinates to world coordinates, the matrix is multiplied by another matrix which rotates *pitch* around the *x* axis. $\overrightarrow{move}$ is then transformed into world coordinates.

A new position candidate, $\overrightarrow{pos1}$ is found by adding $\overrightarrow{move}$ and $\overrightarrow{pos}$.

To take the curvature of the earth into account, the length of $\overrightarrow{pos1}$ is reduced by the length of *error* as depicted in Figure 2. If this operation was skipped, moving horizontally would slowly send the user into outer space.

The orientation also needs adjustment due to $\overrightarrow{pos}$ and $\overrightarrow{pos1}$ not being parallel. A new $\overrightarrow{localUp}$ vector is created based on $\overrightarrow{pos1}$ and the rotation of the old $\overrightarrow{localUp}$ into the new $\overrightarrow{localUp}$ is used for updating $\overrightarrow{localFwd}$.

Finally, the transformation matrix is created based on the new position and orientation.

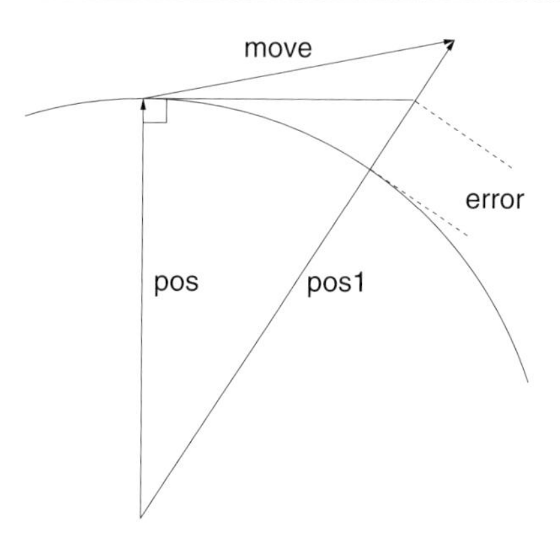

Fig. 2. The error to be removed from $\overrightarrow{pos1}$

3.1 Pseudo Code

In the following we present the pseudo code for the implementation of updating the world camera position and orientation. Checks for numeric precision problems are not included in order to preserve clarity.

The function `updateNavigation()` that implements the navigation, is intended to be called each time navigation parameters have changed and a new frame needs to be drawn.

The `move` vector parameter is in camera local coordinates. The `deltaHAngle` and `deltaVAngle` are changes in local (not camera) horizontal and vertical angles. These parameters have to be supplied by the "flat-navigation" system implemented to meet the specific requirements for the application and user preferences.

The function uses four variables (`pos`, `localUp`, `localFwd`, and `pitch`) to store state, and these have to be initialized to something reasonable. `pos`, `localFwd`, and `localUp` are all in the global coordinate system, and furthermore `localFwd` and `localUp` have to be unit length and orthogonal. `pos` must never become *(0, 0, 0)* as the up vector is undefined in that case. `pitch` is restricted to lie in the interval $[-\pi /2; \pi /2]$ using the function `clampToRange()`.

The following functions have to be available (for resources see [2, 3]).

`function vector rotateVector(vector v, real angle, vector axis)` returns a rotated version of `v`, the angle, `angle`, right handed around `axis`.

`function vector rotateVector(vector v, vector v1, vector v2)` returns a rotated version of `v` given by the rotation from `v1` to `v2`.

`function matrix createXAxisRotationMatrix(real angle)` returns a matrix that rotates `angle` around the *x* axis.

`function matrix getRotationMatrix(vector i, vector j, vector k)` returns an identity matrix but with the upper left 3x3 part set to `i`, `j`, and `k` as the first, second, and third column vectors respectively.

`function matrix getRotationAndTranslationMatrix(vector i, vector j, vector k, vector p)` is like `getRotationMatrix()` but also sets the position `p`.

Additionally, we assume operators + on vectors and * on matrices. The rest of the functions should be self explanatory.

```
vector pos, localUp, localFwd
real pitch

function matrix updateNavigation (vector move,
                                  real deltaHAngle,
                                  real deltaVAngle) {
  # Update local orientation
  pitch = clampToRange(pitch + deltaVAngle, -PI/2, PI/2)
  matrix pitchMatrix = createXAxisRotationMatrix(pitch)
  localFwd = rotateVector(localFwd, deltaHAngle, localUp)

  # Make orientation transform matrix
  vector localUp = normalize(pos)
  vector localRight = cross(localFwd, localUp)
  matrix m = getRotationMatrix(localRight, localUp,-localFwd)
  m = m * pitchMatrix

  # Calculate new position
  move = transform(m, move) # local to world coordinates
  pos1 = pos + move

  # Eliminate vertical position error
  real posLen = length(pos)
  real pos1Len = length(pos1)
  real cosangle = dot(pos, pos1)/(posLen*pos1Len)
  real error = posLen/cosangle - posLen
  real scaleFactor = (pos1Len - error)/pos1Len
  pos1 = scale(pos1, scaleFactor)

  # Update orientation and position
```

```
    vector pos1Norm = normalize(pos1)
    localFwd = rotateVector(localFwd, localUp, pos1Norm)
    localUp = pos1Norm
    localRight = cross(localFwd, localUp)
    pos = pos1

    # Make final matrix
    m = getRotationAndTranslationMatrix(
          localRight, localUp, -localFwd, pos1)
    m = m * pitchMatrix
    return m
}
```

4 Discussion

We have introduced a general purpose navigation mode suitable for global visual navigation in virtual environments representing an entire planet. The article attempts to relate the subject of global navigation to systems for disaster management. The presented theory and algorithm are based on a real implementation in the Grifinor system using Java. We categorize navigation modes as either world orientation independent or world orientation dependent. In order to provide intuitive behavior to most users, we found the dependent mode of navigation most desirable for general purpose navigation on virtual planets.

The navigation can be used at any scale around the globe with the same natural feel, which provides a sense of down in the direction to the center of the planet and movement forward along a current spherical level. There are no poles where the navigation behaves unpredictable, except for the center of the planet where down is undefined.

The presented algorithm addresses a few issues that are worth mentioning. An error in elevation occurs at very high speeds. If the horizontal movement between consecutive frames approaches infinity, the actual movement approaches only a quarter of a turn around the planet (independently of the given height). However, even with a relatively slow frame rate of 12 frames per second it is still possible to go almost three times around the globe in one second without changing elevation.

Another issue is that mathematical approximation of planets by sphere is not exact. In case of Earth, traveling from the pole at zero level along the meridian brings user approximately 20 kilometers below Atlantic Ocean at the equator. This is due to the fact that planets tend to be ellipsoids of revolution rather than a sphere. This is a subject for improvement although in practice this imprecision is not constraining since user can correct the movement direction slightly if there is a drift up or down. Moreover this becomes an issue when navigating across long distances, which are usually performed from elevations above 50 kilometers.

Based on the implementation we can subjectively conclude that the new navigation mode is an improvement over a general world independent mode when browsing globe-like worlds.

References

Bodum L (2005) GRIFINOR – Integrated Object-Oriented Solution for Navigating Realtime 3DVirtual Envirnoments. IBID

Foley J, Van Dam A, Feiner S, Hughes J (2000) Computer Graphics Principles and Practice 2nd edition in C. Addision-Wesley, ISBN 0-201-84840-6

Möller T, Haines E (2002), Real-Time Rendering 2nd edition, Peters AK, Natick, Massachusetts, ISBN 156881-182-9

ISO/IEC 14772-1/Amd. 1:2002 (2002) The Virtual Reality Modeling Language, Part 1- Functional specification and UTF-8 encoding

LoBI-X: Location-Based, Bi-Directional, Information Exchange, over Wireless Networks

Antonis Miliarakis[1], Hamed Al Raweshidy[1] and Manolis Stratakis[2]

[1] BRUNEL University, Uxbridge, Middlesex UB8 3PH, United Kingdom
Email: amil@forthnet.gr; hamed.al-raweshidy@brunel.ac.uk
[2] FORTHnet SA, P.O. Box 2219, Science and Technology Park of Crete, Vassilika Vouton, 71003 Heraklion, Crete, Greece.
Email: stratakis@forthnet.gr

Abstract

Effective warning services enable people to take actions in order to reduce life losses and fear, speeds up recovery and prevent oncoming hazards caused by various disasters. The proposed platform integrates different IT technologies with mobile devices in order to achieve smooth and undistracted collaboration among all rescue units and citizens involved in emergency situations.

1 Introduction

With several kinds of terrorism at the forefront and the continuously increasing natural disasters worldwide due to the greenhouse effect, the need for effective emergency and warning communication systems and platforms has never been more evident than it is today. Fortunately, the evolution of mobile communication systems and wireless networks is paving the way towards the development of new added value information services. The purpose of this project is to combine and integrate different state-of-the-art technologies in order to develop the Location-based, Bi-directional, Information eXchange (LoBI-X) communication platform for urgent risk management.

Effective warnings should reach on time, every person at risk, who needs or wants to be warned, no matter what they are doing or where they are located. A basic concern with current, limited public warning systems like for example sirens and loudspeakers is that they do not reach enough of the people who actually are at risk. Furthermore, warning information is often related to certain geographic areas and so warning messages have to be delivered targeting civilians who live or who are temporarily located in these specific territories.

The technology and specifications [8] to develop warning platforms exists today. The challenge however, is to implement standards and procedures that provide all stakeholders involved in public warning services, the ability to receive warning information and distribute it in a timely and reliable manner to a wide variety of receivers such as the affected civilians, rescue units etc.

The main aim of LoBI-X is therefore to provide all parties involved in emergency situations with the capability to communicate with risk management headquarters using mobile devices such as PDAs, laptops or common mobile phones.

3 Platform Description

The LoBI-X platform consists of three different applications in order to achieve collaboration and communication among headquarters (desktop application), rescue units (mobile application) and citizens (warning notification through the use of SMS messages). All information and data received by these units are displayed on an electronic map that is available in the desktop application. The electronic map receives live positioning feedback from mobile units, (rescue squads, firemen) and project their stigma on the screen.

Each rescue unit is equipped with a PDA and a GPS receiver. The PDA is continuously reading data from the GPS receiver and transmits its geographical co-ordinates to the main server. Immediately after the co-ordinates have arrived to the server, a spot representing the unit is displayed on the electronic map. The position of each unit on the map is updated automatically every 11/2 sec. This way, the main server is always updated with mobile unit's position and therefore, the civil protection headquarter is constantly aware about each rescue unit's location.

Apart from the location, mobile units are able to upload information about emergency situations or remarkable incidents. Once on site, the member of the rescue team is able to send a report about an incident and receive immediate directions from the headquarters. Each uploaded event consists of a segment that describes the type of the event (e.g. car accident, damaged building, blocked road etc), the exact geographic co-ordinates, the priority, a short description about the incident, and the time / date of the submission. These events are displayed on the electronic map as different coloured spots according to the priority of each incident (yellow for normal priority, orange for high, or red for highest). The e-map is reloaded with newly reported events every 2 seconds.

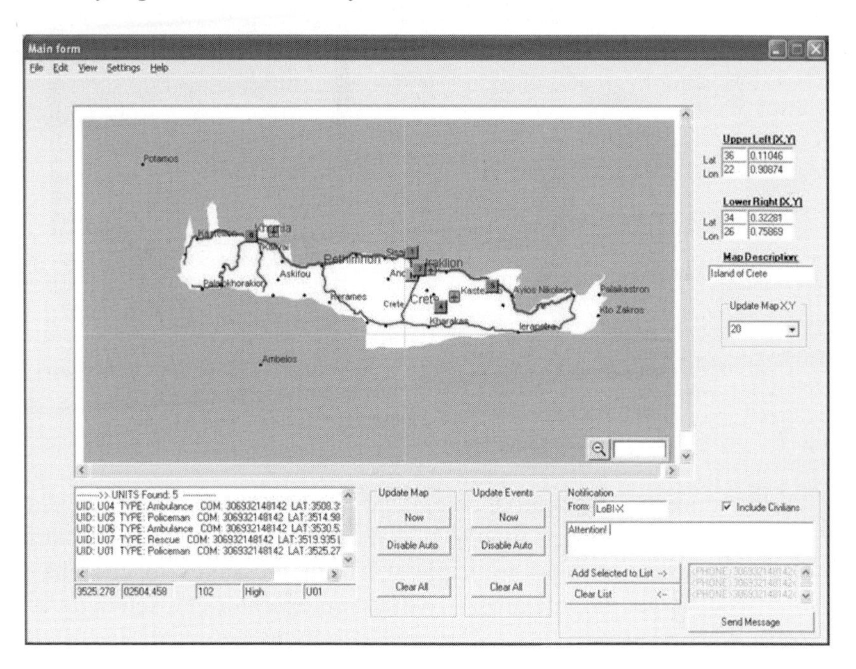

Fig. 1. LoBI-X main application with eMAP

As far as all submitted data (user's location and events) have been stored in the main server's data base, they are displayed on the electronic map. Thus, the chief has a clear view of the rescue unit's position, blocked roads or emergency incidents. The chief is able to select different maps, zoom in and out on a map, or even draw one and obtain a clear view of each area. Therefore he is able to co-ordinate rescue missions and even guide rescue units straight to the points of interest.

A system dedicated in disaster management should be able to provide immediate information to co-coordinators, rescue units and citizens. Thus, an announcement board has been developed, where rescue units and headquarters are able to add comments and exchange opinions and advices about uploaded events.

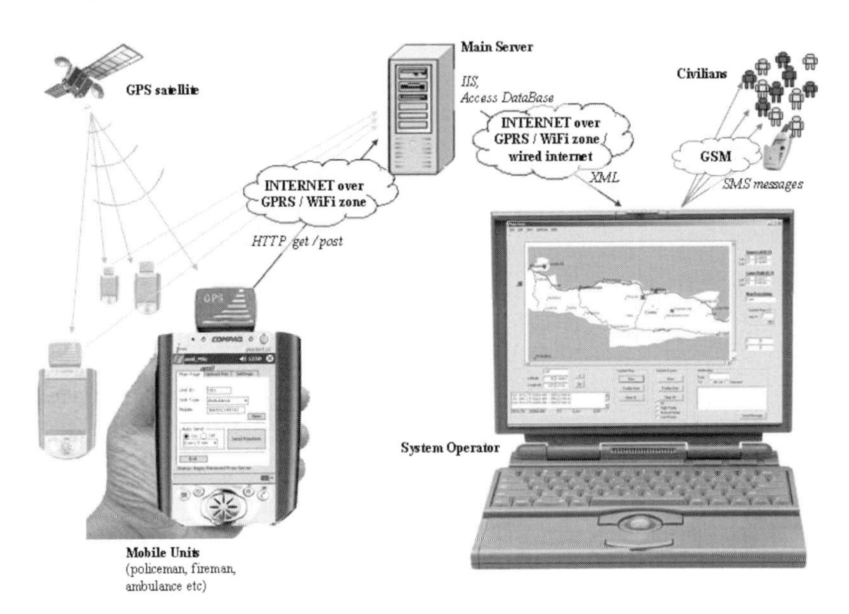

Fig. 2. LoBI-X architecture diagram

Furthermore, an innovative integrated system has been developed for message delivery through mobile phone networks. The LoBI-X platform has the capability to **send multiple warning messages to citizens** who live in specific geographic areas. The system operator can select units and citizens who must be informed or notified about an event. The operator isolates these units and citizens using *"drag and select"* on a specific geographic area on the e-map. Then a notification message is prepared and distributed to all pre-selected recipients though the predefined *groupSMS® gateway* developed by *FORTHnet SA* [3].

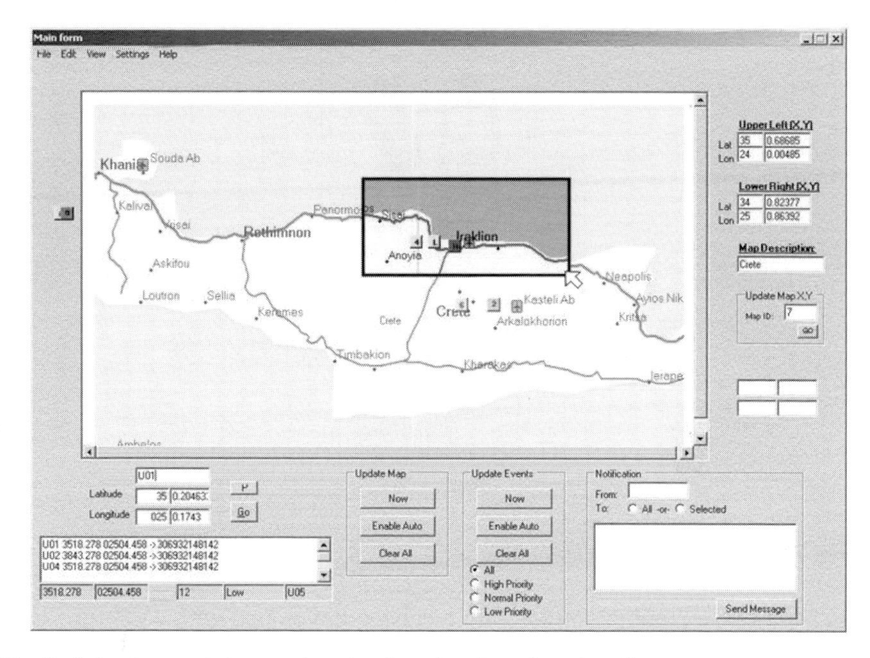

Fig. 3. Selecting recipients using the drag & select functionality

4 Technical Development

The platform was developed based on the latest internet and network technologies and protocols such as ASP.NET, ASP, XML, eVB, GSM, GPRS and 802.11b/g [1,2,4,5]. The challenge of this project was to achieve a sound connection of mobile clients with the LoBI-X data base over wireless networks.

The development of the platform was basically divided into three parts: (a) the mobile application for mobile clients, (b) the server side implementation, and (c) the main application enriched with alerting system for citizens.

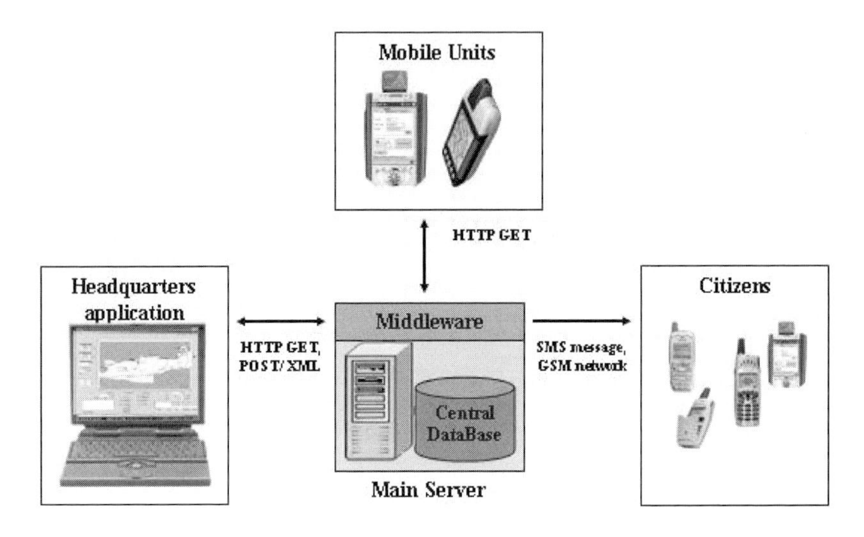

Fig. 4. LoBI-X communication diagram

The LoBI-X mobile application has been developed with eVB for PDAs while communication with the main server was implemented using HTTP 1.1 GET and POST protocols [5]. The PDA should be equipped and configured properly in order to communicate with the main server over WiFi zone, over latest mobile networks such as GPRS or by any other available network. The communication between the unit's mobile terminal devices and the main server (LoBI-X server) achieved through HTTP calls over Wireless network (eg GPRS, UMTS, 802.16x etc) [4].

The server is responsible to receive, and present to client devices and citizens while it is also responsible to store and serve electronic maps and their co-ordinates to the main application. Each map is a jpeg image 700 pixels width by 400 pixels high, which is defined by the upper left and the lower right corner's GPS co-ordinates in degrees format (dd.dddddo) [6], a name and a unique id (mapid) number. All map details are stored in an XML formatted document (map_xml.xml) which is always downloaded from the server to the main application's terminal device.

To allow for the discussion among the different decision makers and rescue units, a discussion board was implemented using ASP. The announcement board and information about uploaded events and units location is accessible from all units using a typical web browser (PocketPC browser for PDA or IE for laptop).

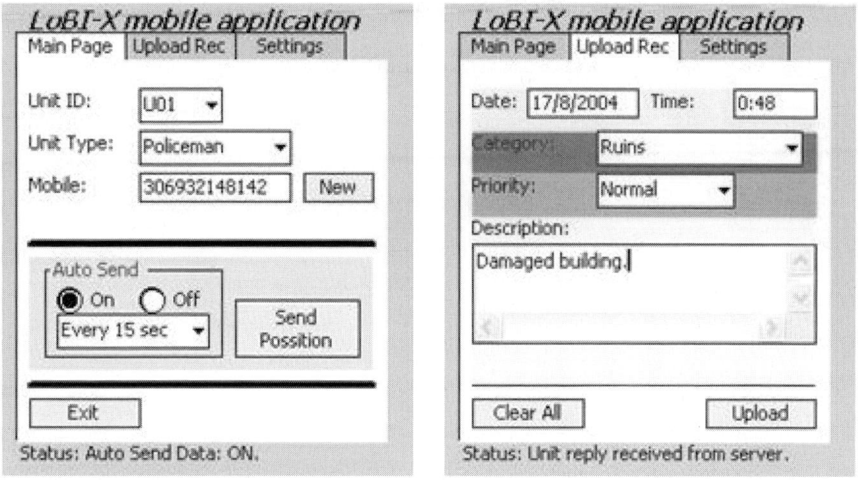

Fig. 5. LoBI-X PocketPC application

The main application has been developed using the VB.NET framework. The .NET framework is an ideal solution for communications based on the XML protocol. Information that comprises the delivered content can be transformed and transmitted automatically to the mark-up language. Information content was described, and transmitted as XML elements. Thus, the main application exchanges XML files with the data base.

The direct notification method has been implemented by sending targeted text messages to citizen's personal mobile phones. This alerting system for citizens is based on short message service (SMS) due to its immediacy and distinguished behaviour even on loaded telephone networks. SMS messages are ideal for such cases as -in opposition to voice calls- it is transmitted using the GSM's data -and not the voice- channel [7]. Practically that means that an SMS message has greater possibilities to be successfully delivered on- or near- time, even over a loaded network that is observed during emergency situations.

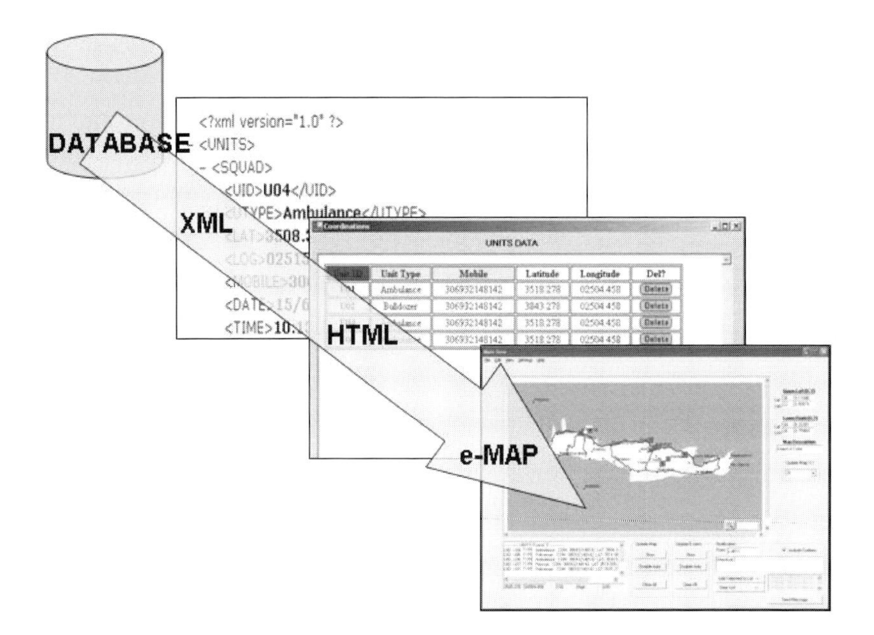

Fig. 6. Data transformation from DB to XML, HTML and eMAP

5 Conclusions

The LoBI-X platform incorporates a number of the latest technologies in data transmission, wireless networks and IT in order to provide civil protection services with a platform, which is able to distribute effective, targeted warning messages to civilians. Furthermore, the LoBI-X platform aims to achieve collaboration between the civil protection headquarters and rescue units during emergency situations by implementing a communication platform among all units involved with disaster management such as ambulances, fire brigade, police department, rescue units etc.

References

[1] Andersson C (2000) GPRS and 3G wireless applications, ISBN:0471414050
[2] Conard, Dengler C, Glynn F, Hollins H, Ramachandran, Schenken, Short, Ullman (2003) Introducing .NET, ISBN:1861004893
[3] FORTHnet SA, groupSMS® service http://www.groupsms.com
[4] Held G (2001) Data over Wireless Networks, ISBN:0072126213

[5] Grattan (2001) Pocket PC, Handheld PC Developer's guide, ISBN: 0130650773
[6] The Navstar Global Positioning System Joint Program
[7] Agarwal N, Chandran-Wadia L, Apte V (2000) Capacity analysis of the GSM short message service, Indian Institute of Technology Bombay, India
[8] OASIS (2003) Common Alerting Protocol, v1.0, Committee Specification, August 12

Integrated Distributed GIS Approach for Earthquake Disaster Modeling and Visualization

Rifaat Abdalla and Vincent Tao

GeoICT Lab, Center for Research in Earth and Space Science,
York University, Toronto, Ontario, Canada, M3J 1L1.
E-mail: (tao)abdalla@yorku.ca

Abstract

In November 2002 a simulated earthquake damage assessment scenario for the Greater Vancouver Region was visualized using GeoServNet. GeoServNet is web-based GIS software developed by York University GeoICT Lab with unique functionality of online 3D visualization and 3D fly. We tested our software in an earthquake simulation exercise that included a simulated Shakemap. Shakemaps are representations of ground motions recorded and extrapolated from knowledge of surface soil conditions. The Geological Survey of Canada is considering applying this technique to Canadian cities at risk. GeoServNet has been used for the demonstration of the utility of Shakemaps and how it could be used in an emergency response scenario. The constructed visual scenarios and information databases is crucial for the purpose of Disaster Management and Emergency Response. Most of currently available tools that are used for disaster management are focusing on the temporal component of the four phases of disaster management leaving an obvious gap in dealing with the spatial element particularly in visualizing disaster and emergency information. This study was conducted as a part of a federal project funded by the former Canadian Office of Critical Infrastructure Protection and Emergency Preparedness (OCIPEP), recently known as the Ministry of Public Safety and Emergency Preparedness (PSEPC). In this study, federal, regional and local authorities along with industry and academic research institutions together coordinated information exchange in a collaborative manner. Results obtained from this project showed that visualization of earthquake disaster scenario was effective and near real time using GeoServNet.

1 Introduction

PSEPC represents the federal government along with British Colombia Emergency Communications Center (ECOMM) representing the provincial level, York GeoICT Lab, representing the academia and two companies represent the private sector, all collaborated together in developing a pilot information visualization system under the Natural Hazards Action Plan. The information structure developed emphasized on the improvement of the efficiency and effectiveness of managing the four phases of natural disasters. i.e. mitigation, preparedness, response and recovery.

The primary purpose of his collaborative effort was to further enhance visualization techniques of Geospatial data for emergency managers. As well as to allow researchers of the collaborating institutions to identify issues associate with the management of data and the infrastructure required. The focus was on a 2D/3D visualization tool that incorporates open standards, wide area access and interoperability for integrating multi-scale Geospatial data.

As the global concern of natural disasters is growing, many researchers have tried to come up with variety of solutions for the question of providing efficient and advanced information system that can handle multiple events with the aid of geographic information systems. The ease of using web based GIS for desktop applications was clearly examined by (G. BERZ, LOSTER et al. 2001), they were able to produce multi-layer GIS database that has incorporated multiple hazards. GIS applications in Disaster Management have always tried to support the concept of all hazards modeling. Hazard is often understood as a quantity that relates the occurrence/frequency and intensity of an event to a specific time interval and so is usually expressed in terms of a probability. The hazard information has three essential components – intensity, frequency and reference period.(G. BERZ, LOSTER et al. 2001).

Based on hazards and disasters widely accepted definitions, Geospatial data are important for managing all the four phases of disasters and emergencies. Geospatial data include images, vector files and location information. Due to the advance and diversity in technology, Geospatial data consists of many formats and representations. To combine these data and form meaningful information products can be a labour intensive and complicated operation. The presentation of the information products to the emergency responders in the way that the embedded information can be easily digested and understood is still a research topic. If emergency responders were given easy access to basic registers of population, building, tenement and property information, and state-of-the-art tools for risk assessment in form of GIS data models, this will certainly ease the processes of disaster management. Many basic infrastructure layers were integrated by (Heino and Kakko, 1998) for producing visual risk models.

The application of GIS for hazards and disasters mapping and visualizations is well known and understood. Beginning in the mid-twentieth century, when the use of aerial photography became widespread, triggered by many more earthquakes in a wider variety of environments. Since the 1980s many researchers used combination of GIS and for seismic microzonation studies to assess the likely effects of earthquakes on urban built on unconsolidated sediments. (Jensen 2000).

Results obtained from different exercises, including those obtained by Keefer, 2002 have demonstrated that Geospatial data are important for managing earthquake disasters in all the four phases. The interoperability of Web-GIS is of great benefit to decision makers in the filed of emergency management, since it will allow them to access and visualize the same data at the same time.

2 Methodology

To provide a relative context to 2D/3D visualization for disasters and emergency situations scenario based enhanced disaster management and risk assessment modeling was applied. With the help of supporting datasets and a mocked storyline, three presentations were conducted to gather more definitive understanding of the user needs in visualization of emergency information. The adopted approach was as following:

2.1 Scenario and Data Preparation

To stimulate discussion on the effectiveness of the visualization tool within the context of emergency response, scenarios were developed for the presentations. The first presentation was during the City of Vancouver emergency exercise at the E-COMM Center and the next presentation was at the Center for Research in Earth and Space Technology (CRESTech) in Toronto. Contents from the last presentation were used for the OCIPEP demonstration.

The scenario for the E-COMM center focused on utilizing vector spatial data and remote sensing imagery to associate with a fictitious subduction earthquake occurring in the Strait of Georgia. Included in the scenarios were demonstrations of an Earth Shake Map to visualize the intensity of the earthquake.

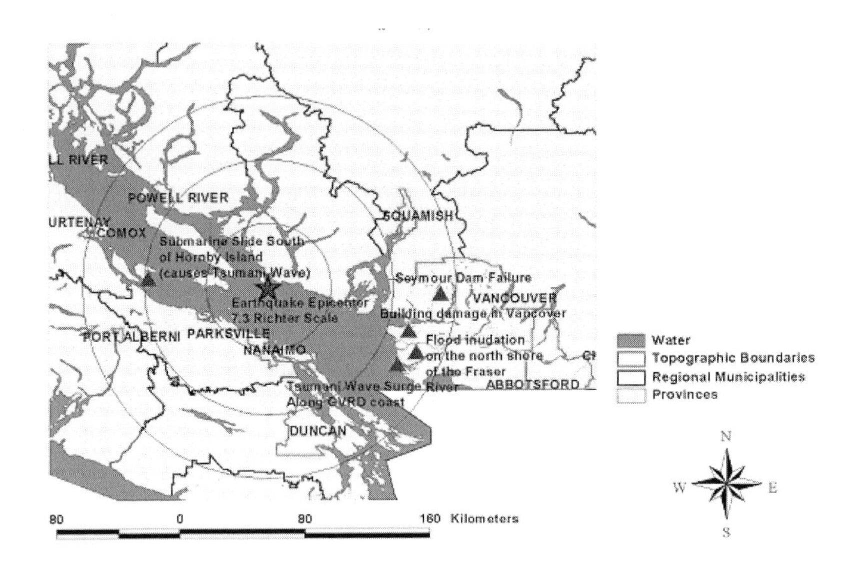

Fig. 1. Scenario event locations

2.2 Storyline

The mock storyline for the earthquakes used focused on a seduction earthquake occurs in the Strait of Georgia, an earth Shakemap containing MMI and PGA was used. Shakemaps are representations of ground motions recorded and extrapolated from knowledge of surface soil conditions. Shakemaps show the 2D map of the event location, color coded based on the magnitude of the earthquake. The area around the epicenter which has high magnitude is colored in Red with gradual color dilution away from this region. The location of the study area is shown in Figure [1].

Referencing information such as population, roads and other infrastructure are then overlaid in 2D. 2D reference information such as roads, water, and population are critical in damage assessment studies, from both critical infrastructure perspective as well as humans concentration.

Manual animation of the topography is done to show the intensity correlating to slope areas.

The advantage of 3D visualization lies in the way we see the information and have our own perception towards the fact of what we are exactly looking at, and whether it is symbolic conceptual or semi real world as in

3D perspective display. Many researchers have recognized that the advantage of 3D lies in the way we see the information. 3D display simulates and enhances realizing the spatial content of the real world. We live in a 3D world; naturally, we perceive and visualize information in 3D much better. Additional point is that 3D visualization provides additional dimension for decision makers. Decision makers are not always heavily involved with mapping, particularly in Disaster Management. 3D GIS is unique in providing them with close to reality models, thus allowing them to more quickly recognize and understand changes in elevation, pattern and features. GeoServNet we demo here is just a beginning, the research and development in the filed of 3D Internet GIS is rapidly evolving.

Constructing visual damage assessment model involved many process from desktop (1) data preprocessing and setting visualization parameters to (2) building web-based visualization model using GeoServNet. The first part involved matching various data sets coordinate systems, data conversion, since all vector layers provided were in 3D format (either polyline z or point z) it was crucial to convert those data sets to 2D shapefiles as initial stage for using GeoServNet. The second stage was setting visualization parameters in terms of layers sequence and colors. This project shows how well to link an earthquake scenario to a web-based 3D GIS together, to come up with integrated damage assessment model. Figure [2] highlight the visual impact that 3D GIS can provide in modeling and visualization. 3D distributed GIS gives a new angle to show the model in both 2D and 3D way, and provides a new protocol to deliver the processed model with the most effective way – Internet. It is just for now. The technologies evolve so fast, and there is a lot of exciting Disaster Management applications needs to be explored.

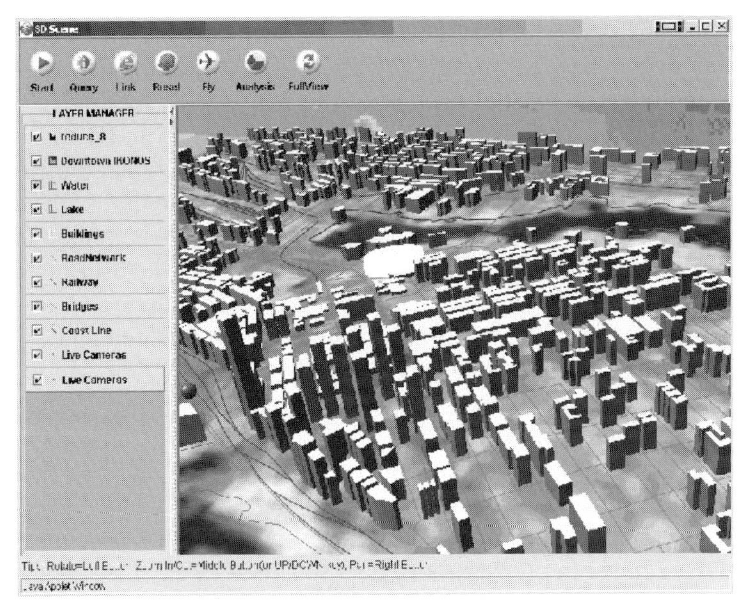

Fig. 2. GeoServNet 3D Model of Downtown Vancouver

3 GeoServNet Architecture

GeoServNet is a distributed web-based 3D GIS which is being developed by GeoICT Lab in York University. GeoServNet provide an easy and accessible data publishing facility. Through the three modules of GeoServNet, i.e. GSNBulider, GSNAdministrator and GSNPublisher, it was possible to make the visual scenario of Santa Barbara Airport available online. GSNBuilder task is to build data i.e. shapefiles, JPEG Raster and ASCII DEM and configure them to make them ready for using GSNAdministrator. GSNAdministrator job is to register the configured datasets to the server and make them assessable to the publishing model. GSNPublisher mainly perform setting visualization parameters in terms of visual effects i.e. color, line thickness and transparency. The other function of the GSNPublisher is generating the application file that would be linked to the web to make the project available online.

GeoServNet is fast. It uses progressive streaming technology and intelligent data transmission strategy, which squeeze each drop of the bandwidth for good use. GeoServNet is platform independent. It is designed and implemented using java and java 3D technology, and can be deployed easily in any machine. GeoServNet is interoperable. GeoServNet follow

OGC standards and has been proved compatible with OGC Web Services. In this project, we use this powerful web-based 3D GIS platform to present a whole new angle to visualize the mock earthquake damage assessment scenario in both 2D and 3D way over the Internet. Figure [3] shows the system architecture of GeoServNet.

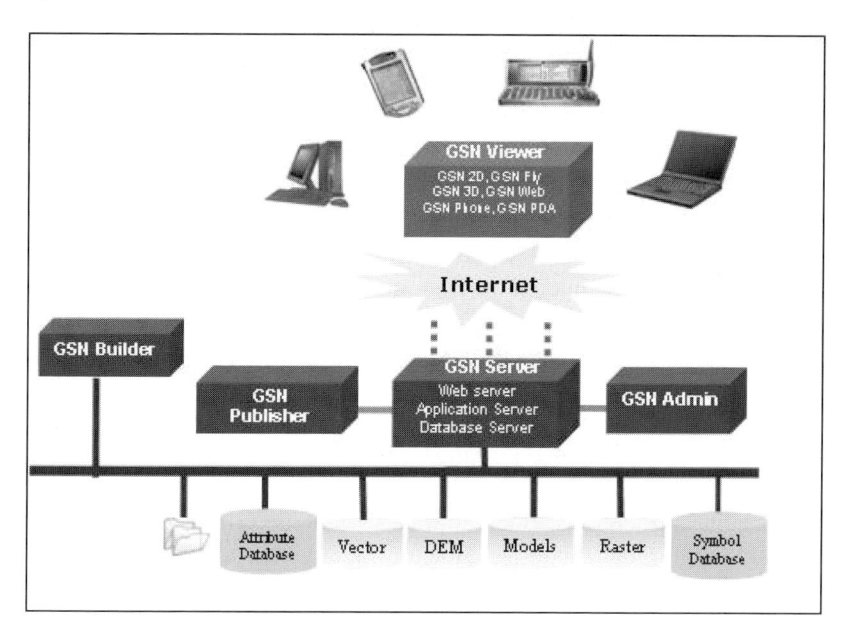

Fig. 3. GeoServNet structure

Geographical information always has a location-based element to it: "the where of information". And the information is usually presented as a map. Not just a picture of a map, mind you, but information as a map.

4 Importance of Distributed GIS

The importance of distributed GIS comes from its accessibility by many users. In disaster management applications, there are many authorities that are involved in planning, decision-making and implementation. Desktop GIS doesn't provide instant and effective multi-user platform for the same project. This capability is mainly one of the most significant advantages of distributed GIS.

Distributed or network based computing is extremely required for processing large volume of data. Certainly, disasters may extend to large space, thus, simulating natural disasters may require advanced and great computing capabilities. This is where the strength of distributed computing and systems interoperability came from. Distributed or network-based computing share the computing power of all systems within the network. This distribution is crucial in functionality execution, it also share the different network resources to achieve high performance computing through multi-user interfaces. However, such strength is useless without clear interoperability standards for systems and data. Such standards would make the use of advanced functionalities and computing power a useful tool in disaster management applications, particularly in resources sharing and coordination. The clear application in disaster coordination could be emphasized by analyzing the earthquake model of this study. The volume of geospatial data, municipal database, and provincial data, may exceed the processing capacity of a desktop computer. In the same time, users from different decision-making levels would not be able to access information at the same time. In such case and without GIS interoperability, it wouldn't be a straightforward task to secure simultaneous data access and processing, this make removing system and data heterogeneity is very useful for disaster management applications. Such applications made GIS interoperability a highly demanded mechanism for disaster management coordination. Another advantage that internet-based GIS doesn't' require deep technical background, which make it user friendly for decision makers to use GIS.

5 Results

Obtained results from this visualization model have clearly demonstrated that GeoServNet was very efficient in handling multiple scenarios. The flexibility and the ease of using this web-based GIS tool were appreciated by all Disaster Management personnel with limited or no GIS background. Web GIS has provided an easy and accessible mean of communication between the various decision-making levels i.e. local, provincial and federal. GeoServNet customizable interface, shown in Figure [4] below allowed for various basic visualization functions and capabilities. GeoServNet 3D visualization functions enabled more sophisticated analysis operations, including, 3D fly, 3D rendering, critical surface functions, and surface profile analysis

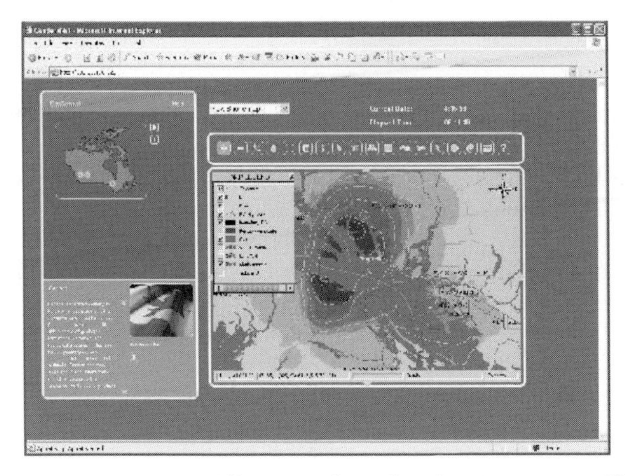

Fig. 4. GeoServNet Client Interface showing Vancouver Shakemap

6 Conclusions

The results obtained from this project have indicated that it is very feasible to integrate various data sources for producing Disaster Management scenarios and models. In particular, data obtained from Canada Centre for Remote Sensing, Statistics Canada, Vancouver City, DMTI Inc., Geological Survey of Canada and the Province of British Columbia was very valuable in this study.

The process of compiling and rectifying the entire vector data used for the project into a common projection and geographic extent by was crucial for having effective visual model. GeoServNet functionality in particular the 3D visualization and analysis functions were of great use for visualizing earthquake simulation and damage assessment models.

Acknowledgements

Special thanks to Dr. Ko Fung and Patricia Pollock from Earth Science Sector, Natural Resources Canada, Steven McArdle from 4DM Inc. and to Michael Morrow from EMIS Technologies for their for providing valuable suggestions and huge input to this study. Many thanks to Dr. Mauricio Artigilo from Geological Survey of Canada for providing study area Shakemap.

References

Ehler GB, Cowen DJ, Mackey Jr. HE (1996) Development for shape fitting tool for site evaluation. In: Kraak MJ, Molenaar M, Proceedings (ed) Advances in GIS Research II. 7th International Symposium on Spatial Data Handling. Vol. I Delft University of Technology, Delft., pp 1-14A12

Bertz G, Loster WKT et al (2001) World Map of Natural Hazards – A Global View of the Distribution and Intensity of Significant Exposures. Natural Hazards 23, pp 443-465

Heino P, Kakko R (1998) Risk assessment modeling and visualization. Safety Science 30, pp: 71-77

Jensen VH (2000) Seismic microzonation in Australia. Journal of Asian Earth Sciences 18, pp 3-15

Keefer DK (2002) Investigating Landslides Caused By Earthquakes – A Historical review. Surveys in Geophysics 23, pp 473-510

Molnár DK, Julien PY (1998) Estimation of upland erosion using GIS. Computers & Geosciences 24(2), pp 183-192

Peek-Asa C, Ramirez MR (2000) GIS Mapping of Earthquake-Related Deaths and Hospital Admissions from the 1994 Northridge, California, Earthquake." AEP 10(1), pp 5-13

Worboys MF (1995) GIS: A Computing Perspective. Taylor & Francis, London

M3Flood: An Integrated System for Flood Forecasting and Emergency Management

Rosanna Foraci[1], Massimo Bider[1], Jonathan Peter Cox[2], Francesca Lugli[3], Rita Nicolini[3], Giuseppe Simoni[4], Leonardo Tirelli[5], Stefania Mazzeo[5] and Michele Marsigli[6]

[1] ARPA-SIM, Regione Emilia-Romagna, Italy.
[2] CHS – SICE S.A. - INDRA S.A - Dragados S.A., Spain.
[3] U.O. Protezione Civile e Difesa del Suolo, Provincia di Modena, Italy.
[4] Servizio Tecnico Bacino Reno, Italy.
[5] Datamat S.p.A., Via Laurentina, 760, 0143 Rome, Italy.
Email: Leonardo.tirelli@datamat.it
[6] ProGeA s.r.l., Italy.

Abstract

M3Flood (Monitoring, Managing, Mitigating) is born from the European project MUSHROOM. The aim of the project was to develop a software product and relative services which can provide a complete, integrated, modular and scalable support, interoperable with external tools and data, to the organizations responsible for managing flood emergencies. The two case studies (in Italy and Spain) have shown the strength of the systems in two directions: 1) applicability of the system in both phases risk forecasting and management and 2) flexibility of the system with respect to various components.

1 Introduction

M3Flood (Monitoring, Managing, Mitigating) is born from the European project MUSHROOM (Multiple Users Service for Hydro-geological Risk Open & Operational Management), as part of the e-TEN (Trans European Network) Program.

The aim of the project was to validate on the market the M3Flood solution: a software product and relative services which can provide a complete, integrated, modular and scalable support, interoperable with external tools and data, to the organizations responsible for managing flood emergencies, in terms of:

- forecasting flood events, using a real-time meteorological-hydrological model chain applied to hydrographical basins;
- real-time monitoring of hydro-meteorological measures recorded and forecast on the basins of reference, and the relative analysis and assessment of the events underway;
- alerting civil defense bodies and providing them with information on the expected and observed evolution of the phenomena underway;
- management of flood emergencies.

The project, run by a consortium including Italian, Spanish and English private companies and public bodies, was coordinated by Datamat S.p.A. (from now on "DATAMAT", Italy), with the active collaboration of ProGeA s.r.l. ("PROGEA", Italy) and Geosys S.L. ("GEOSYS", Spain) as industrial partners, the Province of Modena (Provincia di Modena, "PDM", Italy), Emilia-Romagna Regional Agency for the Environment Protection–Hydrological and Meteorological Service (Agenzia Regionale per la Protezione e l'Ambiente - Servizio Idro-Meteorologico Regionale, "ARPA-SIM", Italy), the Rhine Basin Technical Service (Servizio Tecnico Bacino Reno, "STBR", Italy), the Segura Hydrographical Confederation (Confederación Hidrográfica del Segura, "CHS", Spain) and Murcian Institute for Agrarian and Alimentary Research and Development (Instituto Murciano de Investigación y Desarrollo Agrario y Alimentario, "IMIDA", Spain), in the role of end users for the assessment of the solution, and NuWater Limited ("NUWATER", United Kingdom) for the assessment of the whole project activities.

The system underwent to two specific tests, in Spain and in Italy respectively. For the Spanish case study, a sub-basin of the Segura river, the Mula basin, was selected, while for the Italian case study the basin of the Secchia river, a right-hand tributary of the river Po, which crosses the provinces of Modena and Reggio Emilia in the Region of Emilia Romagna, was chosen.

2 The Italian Operational Context: the Functional Centre and Civil Defense

With regard to the Italian case study, M3Flood is proposed as a support tool for the analysis, assessment and management of flood emergencies in a very complex setting which includes national, regional and local bodies having precise competencies in terms of civil defense.

The recent Decree of the Prime Minister's office "Operating guidelines for the management of the national and regional flood alert system for the purposes of civil defense" (OJ n. 59 del 11/3/2004 Ord. Suppl. n.39), has identified the subjects involved in this kind of risk prevention and management, as well as associated roles and responsibilities in the alert, monitoring, territorial protection and ready intervention operational chain run at a regional level with the co-ordination of the National Civil Defense.

The system was studied and tested in order to enable the bodies responsible for emergency management to perform their institutional tasks which, in the case of potentially adverse weather events, the "Operating Guidelines" distinguish in two successive phases:

- forecasting phase;
- emergency phase.

In the case at hand, the functionality of M3Flood was tested starting from the management of alerts at a regional level, up to the operationalityveness connected to the management of the various emergency phases for which specific hydraulic protection structures of local bodies are responsible. In particular the test phases, on the basis of the tasks assigned by the above-mentioned decree and by the Provincial Emergency Plan, enabled the verification of M3Flood's functionality, to three type of actors of the civil defence system:

- the Regional Functional Centre, whose functions in Emilia Romagna are assigned to ARPA-SIM;
- the provincial civil defense structure, represented by PDM Civil Defense and Soil Conservation Operating Unit;
- the Basin Technical Services, represented by STBR.

The Functional Centre is assigned a fundamental role in the whole forecasting phase with the issue of adverse weather warnings beforehand, containing the description of the forecast meteorological event, and hydrogeological and hydrologic risk warnings further on, containing the assessments of the possible event scenarios forecast and the probable effects of meteorological events on the ground, subsequent to the meteorological expected situation, estimated in concert with the local protection structures.

With the reception of the adverse meteorological conditions warning in the Functional Centre, the Basin Technical Service organizes its own structure to face the forecasted situation and plan the spot check of watercourses conditions. From the flood risk warning on the part of the Functional Centre, the mentioned Technical Service activates the flood service, i.e. the direct and continuous surveillance services for the watercourse, so as to activate the hydraulic maneuvers to guarantee water outflow and the possible emergency interventions with the reaching of crisis situation.

As the forecast event draws closer according to the issued warnings, the Civil Defense structures activate the instrumental monitoring of the event until the passage from the forecasting stage to a state of emergency (organized in various phases) is declared, during which they provide to adopt all the necessary measures to safeguard the population, assets and the territory, as set out in the Provincial Emergency Plan.

During the emergency phase, the Functional Centre provides information support to Civil Defense structures and Basin Technical Services, by continuous monitoring the meteorological-rainfall-hydrometric network and the evolution in the meteorological and hydrological situation forecast for the following hours.

The tools the Functional Centre requires in order to fulfill its task are:

- weather forecast models;
- hydrological forecast models (inflows-outflows and propagation) in real time, fed both by observed and forecast rainfall data;
- the telemetric network (thermometers, rain gauges, hydrometers, radars, etc.);
- a system of forecasting rainfall thresholds enabling to identify potentially critical weather conditions.
- In parallel, the tools required by the Basin Technical Services are:
- real-time data from the observation network, accessible by operators in remote;
- information regarding the evolution of the meteorological situation;
- the comparison between the current flood event and the most significant past events for each watercourse of own cognizance;
- the operational plan of flood service, i.e. an internal procedures and actions protocol to be activated in expectation or at the reaching of specific hydrometric thresholds.

Finally, the tools used by the Civil Defense O.U. of the Province of Modena, during the forecasting and emergency phase are:

- the telemetric observation network (rain gauges, hydrometers, etc.);
- a system of observation thresholds (rainfall and hydrometric) for the identification of risk situations;
- the Emergency Plan, which associates specific actions to the various risk levels;
- territorial information on the state of vulnerability and on elements potentially at risk;
- automatic alert system, which enables the activation of a series of warnings when given thresholds of reference are exceeded, in order to overcome the limits of structures not manned 24Hrs.

Fig. 1 shows a scheme of how the installations were performed and how the information circulated between the various bodies involved in the pilot.

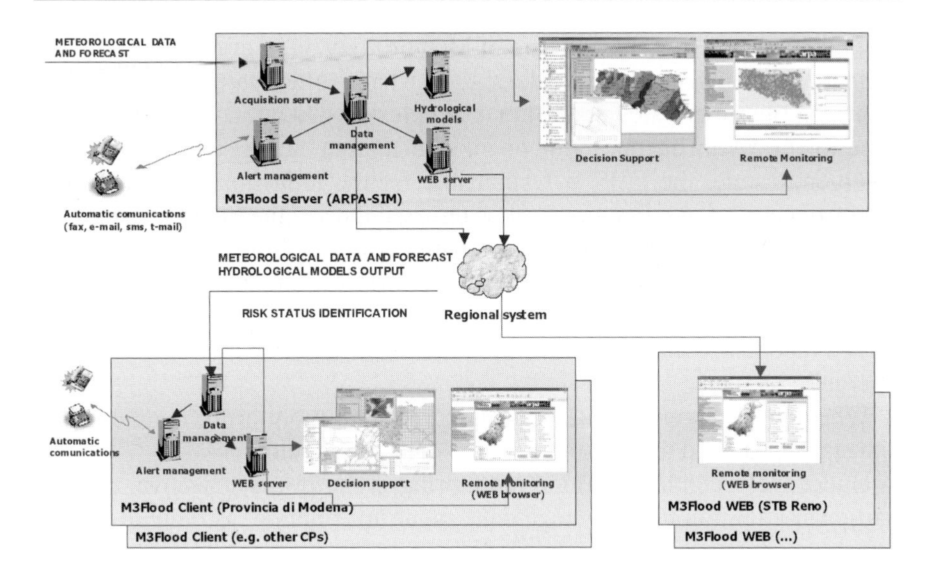

Fig. 1. Bodies involved in the test in Italy

3 The Spanish Operational Context

M3Flood will occupy a niche in the Spanish market as a decision support system aiding the efficient coordination of flood management activities by public bodies in critical circumstances. One of the prime advantages of the system is the logical interface, which can be easily configured in function of the end users requirements, permitting the participation and coordination of the personnel. In Spain, it is envisaged that the end users will be Regional Water Authorities, or Confederations, Civil Protection Authorities, both regional and national, and the National Institute of Meteorology.

At the time of redaction, the M3Flood system is running in two installations in the Segura basin: at IMIDA premises and the headquarters of the CHS in Murcia.

3.1 CHS Role

The role of the CHS under adverse hydro-meteorological episodes is detailed below.

The first stage is the declaration of a Pre-emergency state. This is normally carried out by regional Civil Protection Authorities based on information received from the Institute of Meteorology. However, it is also possible that the CHS facilitates information which could warrant the declaration of a pre-emergency state. For example the detection of precipitation superior to 30mm in one hour or 50 mm in four hours. During the Pre-emergency state, the CHS is responsible for the redaction of detailed reports concerning precipitation and water level data in the Segura basin.

This information, is by default, prepared at three preset intervals: 08:00, 16:00 00:00, however, when circumstances dictate, it can be provided on demand.

The reports are normally transmitted externally via telephone or fax and internally on paper.

When an Emergency state is declared, system monitoring is intensified and staff mobilized to accommodate 24 hour supervision.

Under these conditions flood monitoring is the prime objective, paying particular attention to rain-gauge data, radar information, satellite information, the evolution of river level gauges, hydrographs, and surface discharge measurements.

In addition, the information provided by the system aids decision support management on behalf of a Permanent Committee (Comité Permanente), which is formed under these conditions. The committee duly informs the Civil Protection Authorities (Murcia, Albacete, Alicante, Jaen and Almería as required) of possible dangers and takes the final decisions with regard to resource management.

3.2 M3Flood Segura

The Segura installation is based on the existing network of hydro-meteorological sensors installed in the Automatic Hydrological Information System S.A.I.H. and the IMIDA network of agro-climatological stations, a total of more than 100 installations basin wide. The SAIH information is transmitted via radio and satellite to a control centre in Murcia every 60 minutes. However, when necessity demands, information can be provided every 5 minutes. Data from the IMIDA system is transmitted via GSM once every 24 hours. Under adverse conditions this periodicity of course can be modified to the end users requirements.

The data from the SAIH network is processed by the SAIH server and rapidly transferred via ftp to the M3Flood server which also receives data from IMIDA.

Clients can access the data in near-real time. M3Flood offers several main advantages over the present SAIH Segura system:

- The incorporation of data from distinct sources:

  - o Meteorological forecasts
  - o Precipitation model forecasts
  - o Observed data from other networks
  - o Video/photographic information
- The possibility to manage internal and external communications in an efficient manner:
  - o Fax, SMS, e-mail, etc.
- The association of hydrological models both within the system and as peripheral modules:
  - o TOPKAPI
- The possibility to access information via the web thereby facilitating out of office supervision when required without the need for dedicated software.

Up to the present the Segura tests have been carried out on a flashy basin up river from Murcia capital (Fig. 2) with the following morphology:

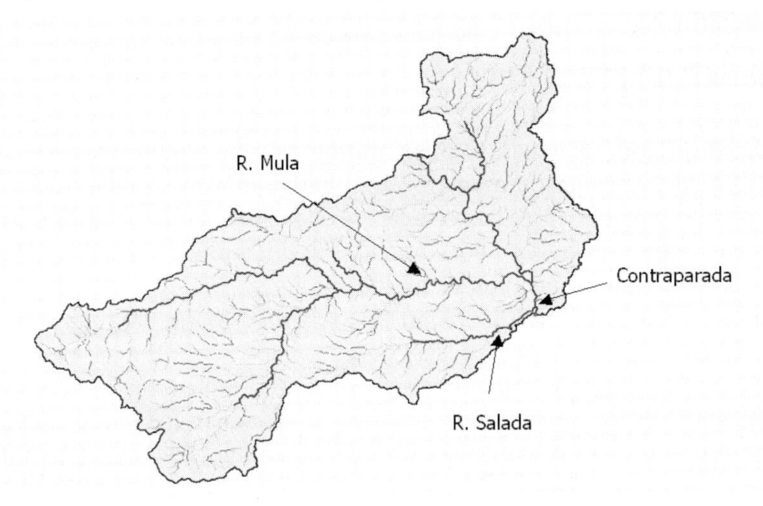

Fig. 2. Drainage network in the study basin

- Area: 1189 km2;
- Max Altitude: 1109 m;
- Min Altitude: 60 m;
- Perimeter: 284 km:
- Longest flow path: 79.3 km;
- Mean precipitation: 330 mm/year.

This basin presents great challenges in terms of modeling above all due to its flashy nature. This can be clearly demonstrated by the limnigraphs presented by the M3Flood system for a particularly extreme episode from

October 2003. It can be appreciated that within a period inferior to fifteen minutes, the water level rose from 0 cm to 388 cm at the Rambla Salada (Figure 3 and 4). It is estimated that celerity exceeded 5m/s.

The underlying problem in the Mediterranean basins of the South Eastern Iberian Peninsula is the lack of sufficient homogeneous episodes to attain a robust model calibration. With a specific calibration, the model associated to the M3Flood system performed reasonable well. However, in favor of M3Flood what must be reinforced is the that optimizing efficiency using M3Flood can save time, which, when dealing with more or less vertical limnigraphs as demonstrated in the above case, mitigates the effects of flooding.

In the near future it is planned to increase the area covered by M3Flood to incorporate the entire Segura basin (19000 km2), and thereby, logically, increase the number of sensors and stations available in the system and determine parameters for wider model application and calibration.

4 Operation in Real-Time

In order to provide adequate support during an emergency, the M3Flood acquires data from the thermometer, rain gauge and hydrometer network together with weather forecasts in numerical format, as maps of the meteorological models and radar observations (Fig. 5), and alphanumeric, in the form of a bulletin as issued by the competent body.

On the basis of the input (forecast and recorded) data, the hydrological and hydraulic models return forecast outputs at 24 hours for the river sections of specific interest. Each time new data are input in the system, they are automatically compared with a system of reference thresholds defined by the user, and identify the "state of risk", enabling the competent structures to be alerted.

From this moment onwards the hydrological and meteorological centre (Centro Funzionale in Italy and Confederación Hidrográfica in Spain), the Civil Defense structures and the Basin Technical Services can monitor the state of the basin and the evolution of an event through the "Risk Analysis and Assessment" interface, by displaying the rainfall trend (Fig. 6) and hydrometric data recorded, even in numerical form, as well as forecasts of levels in the sections at 24 hours, provided in real time by the hydrological and hydraulic models (Fig. 7).

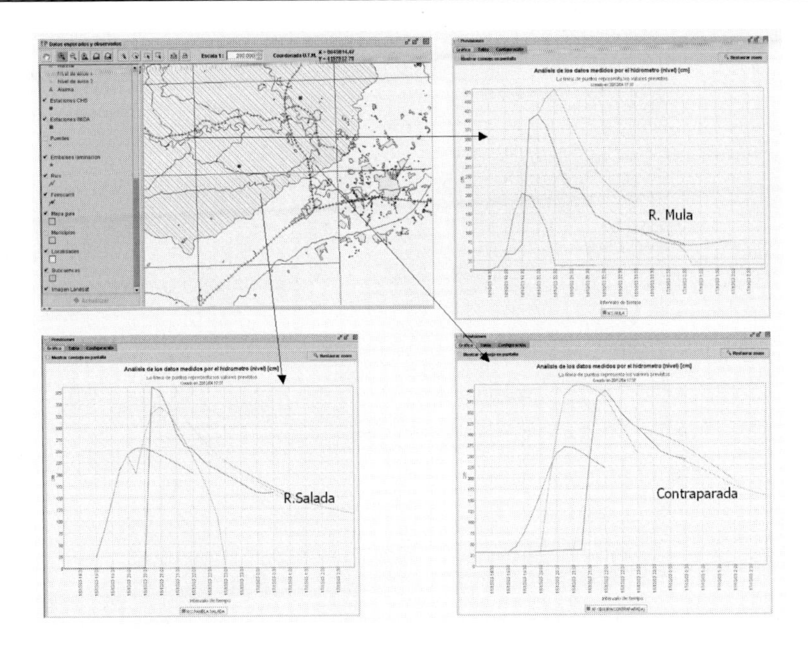

Fig. 3. Limnigraphs detailing observed and predicted levels

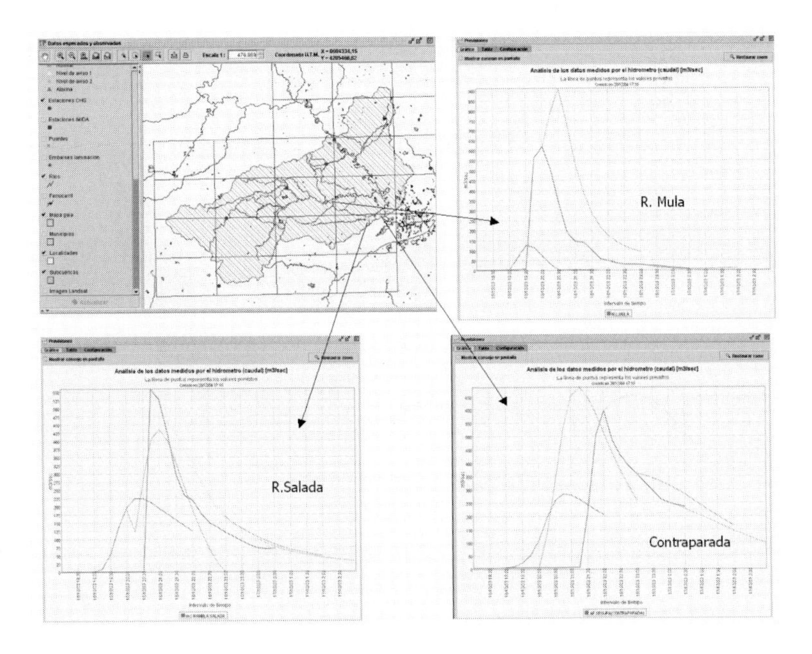

Fig. 4. Hydrographs detailing observed and predicted discharge

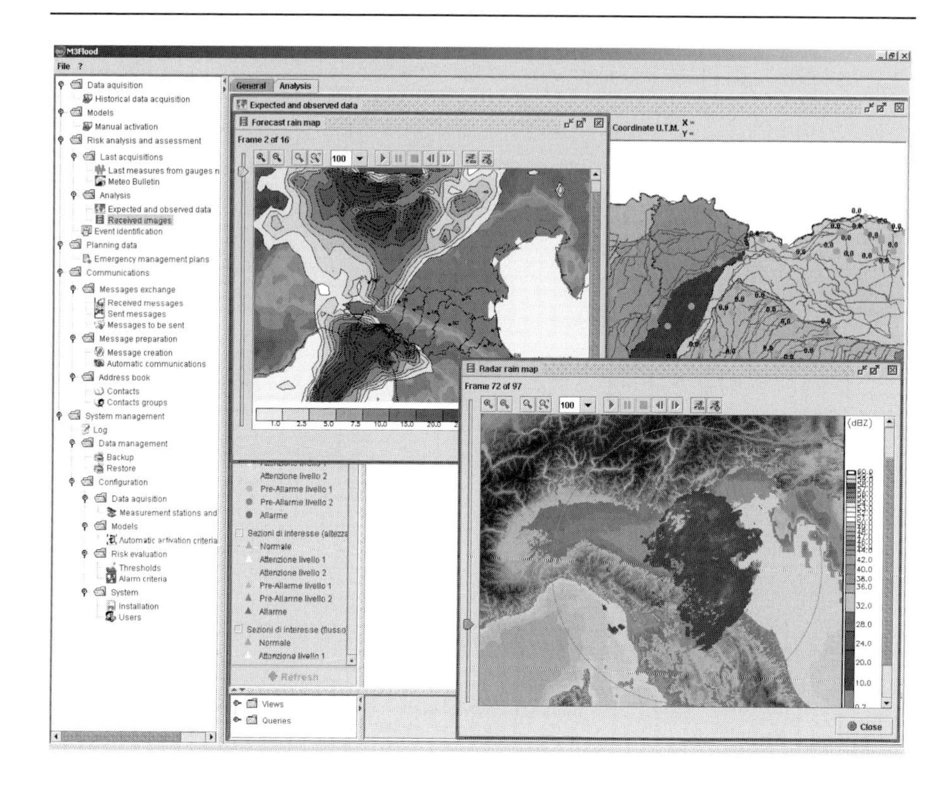

Fig. 5. Radar and weather forecast maps

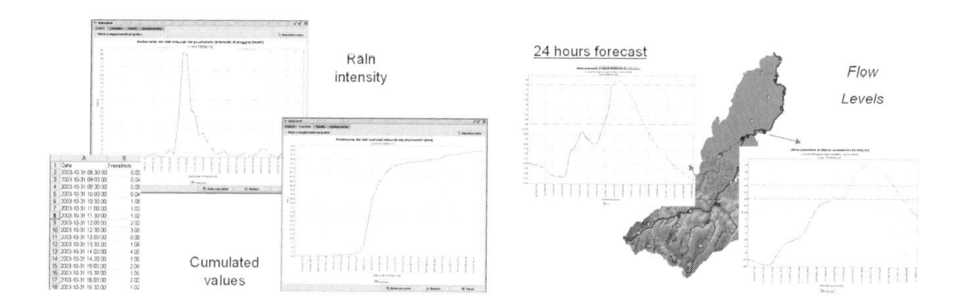

Fig. 6-7. Measured data and forecast flows

The interface also enables the comparison of the phenomenon underway with historical events of reference (Fig. 8), or to perform analyses more sophisticated than forecasting, by viewing various model runs at the same time, or comparing the data recorded and forecast in the past.

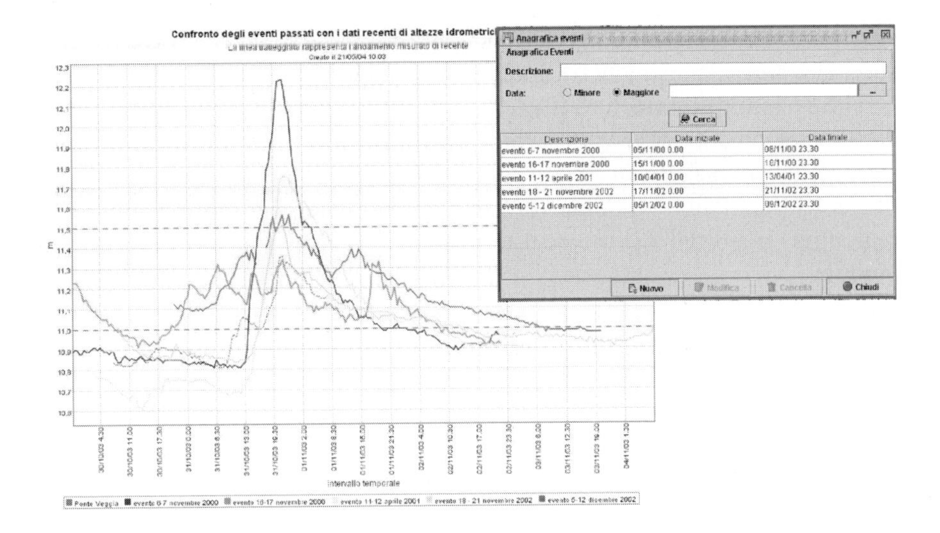

Fig. 8. Comparison of measured data with historical data

All data can then be displayed in a special cartographic application. The system enables the preparation and management of special predefined views which allow rapid access to map databases of specific interest for each of the forecasting and emergency management phases.

In particular, all the hydro-meteorological measuring stations, whose data are updated in real time, can be viewed in the cartography of interest. In function of the analysis being performed, it is possible to display regional-scale views (view of the alert macro-areas for weather forecasts), or those specific towards assessing the event at the basin scale (consultation of the cartography of the Provincial Forecasting and Prevention Plan containing the analysis of the risk areas on the territory, from the hydromorphological map to the localization of critical stretches of the hydrographical grid), down to those specific to civil defence activities at a municipal scale, whose map databases are comprised by the Technical Regional Map in raster format, at 1:5,000 or 1:25,000 scale, which are superimposed by the elements exposed to risk and the areas useful for rescue operations (gathering areas, reception areas for the population, etc.).

5 System Features

M3Flood's main objective is to provide, in one system, all the tools required by the hydrological and meteorological centre, the Civil Defense structures and the Basin Technical Services for all phases of the alerting system, from forecasting and assessment of the event, to the phase of observation and management of flood emergencies.

The system's architecture (Fig. 9) includes a central core comprising the data management component which relies on an Oracle database for the storage and management of all data, both static data, configured through the "GIS population" module for cartographic data, and those acquired in real time, imported by the real-time data "Acquisition and processing" module.

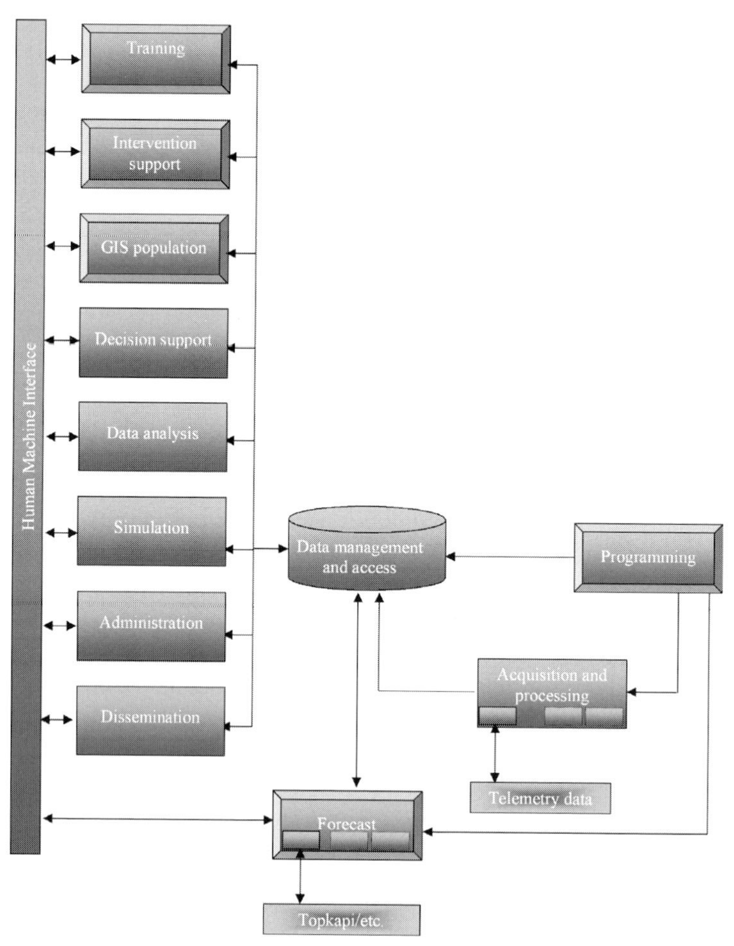

Fig. 9. M3Flood modules

The information contained in the Database is presented to the user through two independent interfaces (given the generic definition of "Dissemination" and "Decision support" in the diagram in Fig. 9): the first of these is more schematic, as it is destined for general WEB users (Fig. 10), and the other, designed for operating centre operators, in GIS format (Fig. 11), through the risk analysis and assessment interface. A third interface, again integrated with the first two but designed for the system administrator and described in the next paragraph, allows the configuration of the system's characteristics.

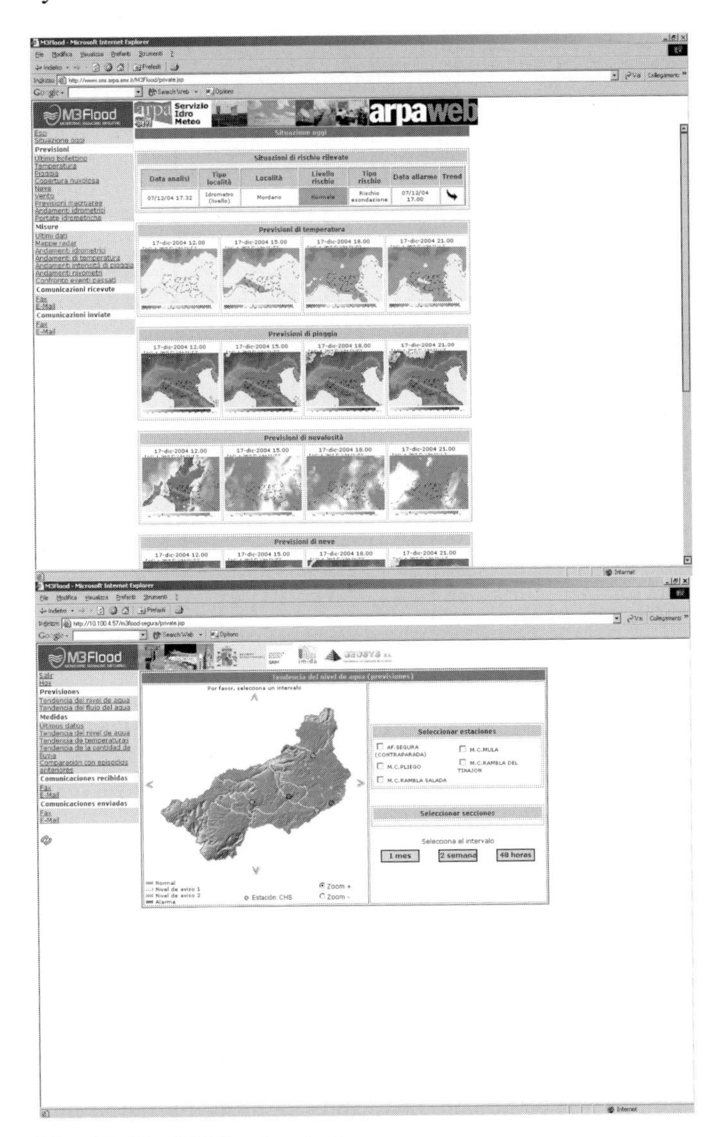

Fig. 10. The M3Flood web site

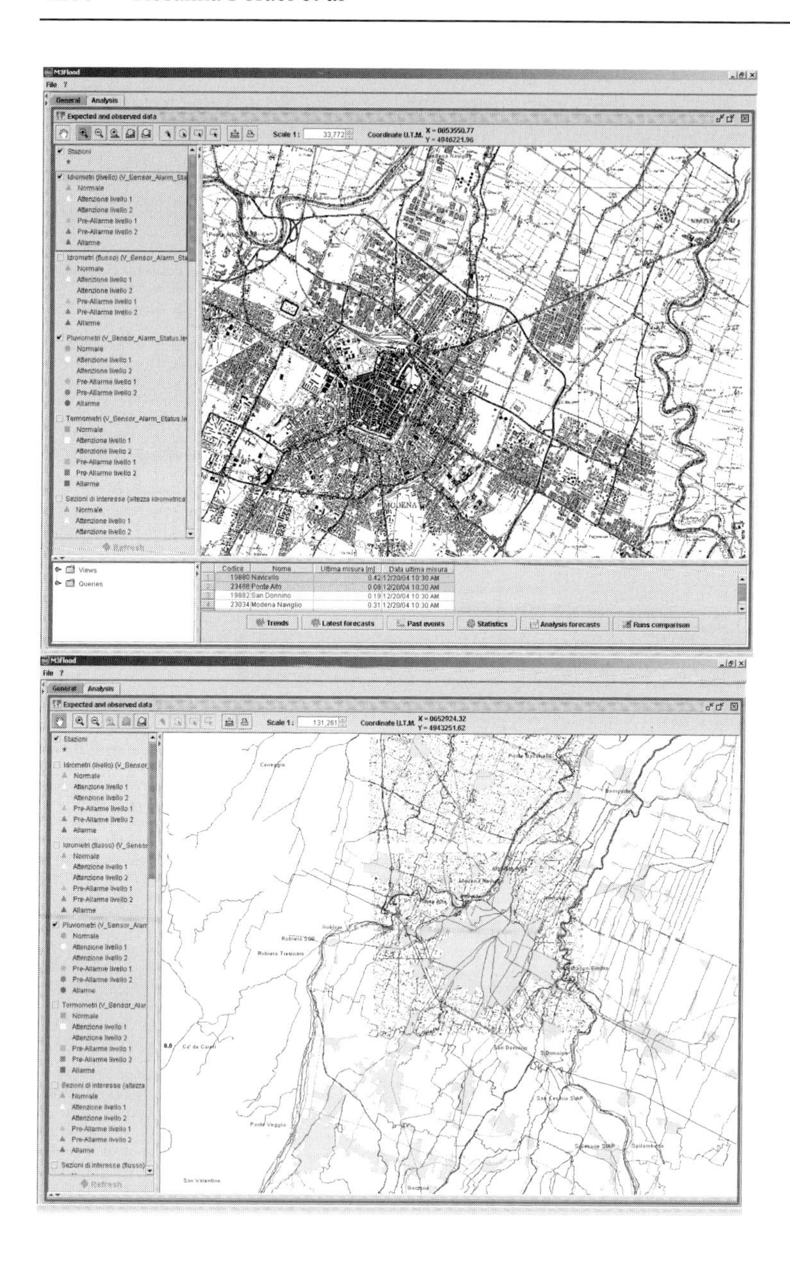

Fig. 11. GIS operating centre interface

The system is entirely developed in Java, thus assuring its portability across different platforms. For the GIS environment, the ESRI MapObjects Java Edition was integrated, while the communication support component runs on David, specifically customized for the purpose. For the web component JSP technology and an Apache Tomcat web server were chosen.

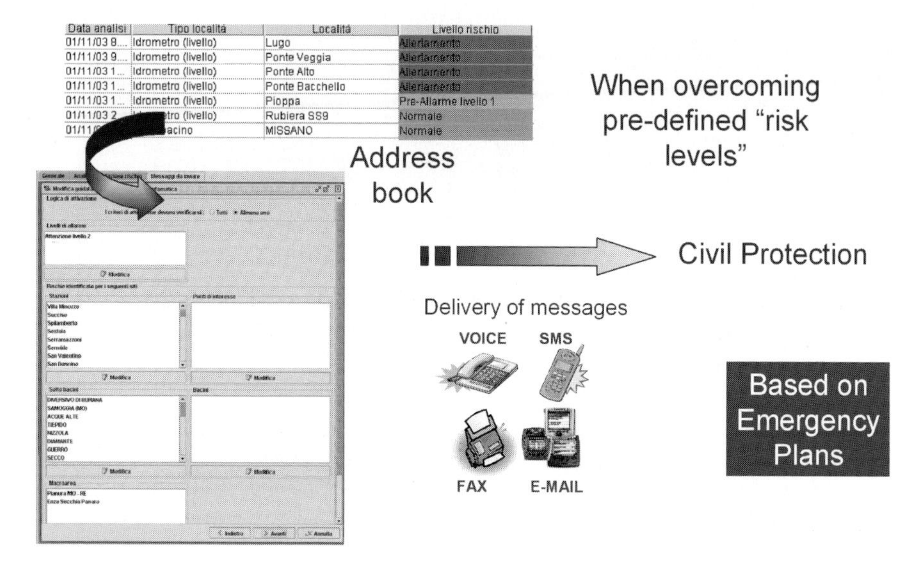

Fig. 12. Alert

6 System Configurability

One of the main strengths of M3Flood is its high degree of configurability on the part of the user, including for the Database (including mapping data), viewing modalities, models, and the operation of the alert system.

The Database and the risk analysis and assessment interface are programmable by the user on the one hand through the GIS population module – a graphic interface developed in a MapObjects environment - and on the other through the system administrator's interface.

The first of these enables the definition in detail of the mapping project of reference, specifying its contents (shp files and tables) with the relative metadata, and performing the population of the Database; in addition, by defining a series of "views" (Fig. 13) and "queries", it is also possible to configure the display modalities for the stored information, be it numerical (sensor data) or cartographic (road network maps, flooded areas, etc. with relative legends), without the need for programming in a specific language.

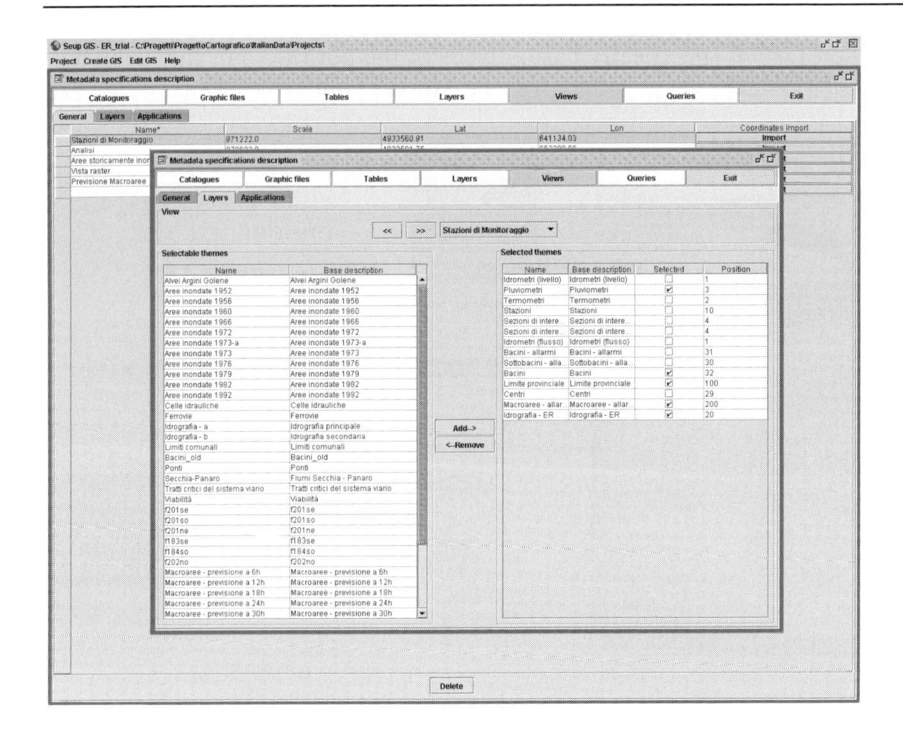

Fig. 13. GIS population component

The second provides a series of simple window interfaces which, in order to make the operation of the system more adapted to the need of the various users, enable the configuration of information such as thresholds, risk levels, sensors, stations, data acquisition parameters, model activation, the characteristics of the messages to be sent, the address book, the logs, the backup-restore, the users, etc.

In particular, the threshold values forecast and recorded can be configured on different objects, for examples alert macro-areas, sub-basins, sensors, etc.

The instances where the set forecast and recorded thresholds are exceeded contribute to the identification of risk levels, on the basis of criteria again definable by the user. For example, the criterion for identification of the “Attention 1” risk level, can correspond to the exceeding of the rainfall forecast threshold 24 hours earlier than estimated, together with the exceeding of the event threshold on the sub-basins.

The system of notification to the interested operators which follows on the identification of risk levels can also be configured by the user, a fact which allows M3Flood to be adapted to the needs of the various bodies involved depending on their function and the various tasks of their personnel. The system enables the definition of an address book for those to whom the messages are to be sent, and keeps an archive of sent messages.

7 Conclusions

M3Flood is an innovative product with respect to the flood risk forecasting and management systems currently available.

Its main strengths are:

- the synthesis in a single system of the tools necessary for both the risk forecasting and the analysis and management phases;
- the configurability of the system's various components.

The advantages it provides are those of having a single program which is able to handle weather forecasts, the reception of data in real time, risk analysis, mapping application and alerts system at once, resulting in time savings with respect to the use of different applications. The structure of the system, in addition, structured in Client applications which acquire data from a same server, guarantees the univocality of monitoring and forecasting data available to the central and peripheral structures, which are thus able to operate in a shared fashion, reducing to the minimum any complications linked to diverging interpretations of the events underway. The configurability of the various models, in addition, means that M3Flood can in any case be adapted for the various requirements of the technical structures operating during emergencies, each having their own competencies in terms of forecasting, event analysis and alerting, and management of operations on the field; the configurability of the various models in addition allows users to rapidly adapt the system at any time when their operating requirements change. Finally, the possibility of publishing on the web and placing at the disposal of the various operators involved the main information relating to the events expected or underway, as organized in the M3Flood system, facilitates the operation of structures which do are not manned 24 hrs and reduces the waste of human resources and materials to a minimum.

For this reason M3Flood is particularly useful for the Functional Centres, the Confederaciónes Hidrográficas and other civil defense organizations in the performance of the tasks assigned to them by the recent legislation. M3Flood will also ease the organization of flood events in Spain. One of the clear benefits being that the terminology used by diverse organisms involved in flood management will be standardized. This should eliminate the uncertainty generated by the incorrect interpretation of observed data, trends and predictions thereby improving the efficiency of flood mitigation.

Mobile Hardware and Software Complex to Support Work of Radiation Safety Experts in Field Conditions

Rafael Arutyunyan[1], Ravil Bakin[1], Sergei Bogatov[1], Leonid Bolshov[1], Sergei Gavrilov[1], Alexandr Glushko[1], Vladimir Kiselev[1], Igor Linge[1], Igor Osipiants[1], Daniil Tokarchuk[1], Alexandr Agapov[2], Andrey Fedorov[3] and Evgeny Galkin[3]

[1] Nuclear Safety Institute of Russian Academy of Sciences (IBRAE RAS), 52, B.Tulskaya, Moscow, 115191, Russia.
Email: kis@ibrae.ac.ru

[2] Federal Agency for Atomic Energy of Russia (Rosatom), 24/26, B.Ordynka, 101100, Moscow, Russia.
Email: agapov@minatom.ru

[3] Inform-Atom Association, 2, Dmitrovskoe s, 127434, Moscow, Russia.
Email: fedorov@minatom.ru

Abstract

The developed software and technical complex represents a mobile working place designed to support the work of radiation safety experts in case of emergencies, when leaving for the place of accident. It comprises required databases on normative documents in the radiation safety area; enquiry databases on objects, the personnel and equipment of emergency-rescue teams (ERT) from FAAE of Russia; the bank of electronic maps; computer systems for operative forecast and radiation environment measurement; the system to define geographical coordinates as well as a wide set of communication links to accept and transfer the data.

The software and technical complex is implemented on the base of a portable IBM-compatible computer. The technical complex is also equipped by a dose rate sensor to run field radiation measurements. Geographical reference to the dose rate measurements is performed via GPS-receiver.

Basic objectives solved by tools of the present mobile complex are as follows - information and analytical support for radiation safety experts, express simulation of radiation environment in the place of accident accompanying by radionuclide release, on-site dose rate measurement and retrieval of data on electronic maps, preparation of on-line required documents, forecasts and recommendations, operative communication and data transfer provision for RF FAAE crisis centers.

1 Introduction

When eliminating the consequences of nuclear and radiation accidents, successful fulfillment by radiation safety experts of their functions requires on-line solution *in situ* of a number of tasks, such as:

- measurements of the radiation situation parameters in the affected area,
- rapid making of a forecast of the radiation situation development, and
- elaboration of appropriate recommendations on protection of personnel, nearby population and environment.

Adequate solution of the above tasks is based on the use of a variety of information-reference and simulating systems and databases. During elimination of consequences of a radiation accident the role of stationary control posts could decrease whereas that of mobile teams and laboratories, especially during early and intermediate post-accident phases, could increase. Moreover, during early phases of transport accidents, thefts/losses of radioactive sources, accidents with mobile radiation sources or potential terrorist attacks no devices for situation control and data transfer could be available at all. Forecasting of radiological implications of a radiation accident and the development of appropriate recommendations on population and environment protection requires permanent refining of the available forecast on basis of experimental confirmations and corrections of previous estimates.

This paper is dealing with description of a mobile hardware and software complex developed at IBRAE RAS to support *in situ* activities of experts on solution of the above tasks when addressing accidents involving the radiation factor. As a matter of fact, the mobile hardware and software complex under consideration represents a diminished version of a stationary information-analytical crisis center being aimed at fulfilling basic functions of such center but in the close-to-emergency area. In addition to information and analytical data typical for information-analytical crisis centers, the mobile complex is supplemented with special devices for on-line measurements of individual radiation situation parameters (firstly, to verify and correct the radiation situation forecasts), a satellite navigator and a variety of voice communication facilities providing for control and on-line data exchange with crisis centers involved into elimination of the accident consequences.

Hardware and software of the mobile complex includes: -a database on radiation protection standard & regulatory documents; -reference databases on radiation-hazardous facilities, personnel and equipment of Emergency-Rescue Teams (ERTs) of the Federal Agency for Atomic Energy (Rosatom); -a digital-map databank; -computer systems of on-line forecast and measurement of the radiation situation parameters; and -different data-exchange communication channels.

The main functions to be fulfilled using the mobile complex facilities may be summarized as follows:

1. Information and analytical support for radiation safety experts;
2. Express simulation (forecasting) of the radiation situation in the affected area involving radionuclide release to the atmosphere;
3. On-line communications with crisis center including data transfer;
4. On-site measurements (including on-line measurements) aimed, firstly, at verifying and correcting earlier forecasts;
5. Rapid preparing of necessary working and reference documents, forecasts and recommendations and their on-line transfer to the higher echelon.

2 Hardware of the Mobile Complex

The mobile complex developed on basis of 'Panasonic CF-29' industrial laptop computer represents a "fully-protected" device complying on its properties with the MIL-STD 810F standard (Military Standard). The laptop computer structure ensures stable operation of the complex under severe weather conditions as well as during transportation by a motorcar via poor roads.

A functional diagram of the mobile complex is demonstrated in Fig.1.

To run field radiation measurements, the mobile hardware and software complex is equipped with 'BDMG-200' dose rate sensor manufactured by "Doza" RPE providing for dose rate measurements within range of 0.1 μSv/h – 3.0 μSv/h for gamma radiation energies of 0.05 – 3.0 MeV. 'BDMG –200' sensor has a built-in processor and 'RS-485' interface allowing its direct connection to personal computer via 'USB' port. The sensor does not require any additional power source, electric power being supplied via 'USB' port immediately from computer.

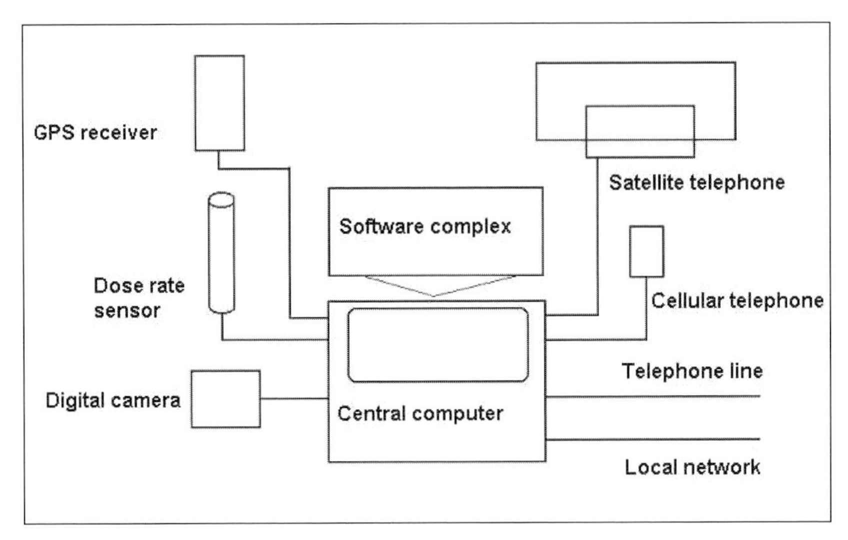

Fig. 1. Functional diagram of the mobile complex

Geographical reference of dose rate measurements is performed using 'Garmin 76 MAP' GPS-receiver, the absolute accuracy being about 15 meters. GPS–receiver is connected to laptop computer via 'RS-232' serial port. The GPS-receiver is supplied form batteries of ten-hour-continuous-work service life.

To be connected to the Internet, the mobile complex under consideration has the following communication channels:

- accessible data-exchange network channel (local network);
- telephone network and telephone modem;
- cellular telephone;
- satellite telephone.

The choice of a specific communication channel will depend on the specificity of conditions in the emergency area.

To ensure data transfer via wire telephone channel, a standard built-in modem (providing for 56 Kbit/sec data transfer rate at a minimum) and the relevant standard software are used.

To connect to the Internet via a cellular communication line, a standard cell-phone with SIM-carte and activated GPRS service is used. The connection between cell-phone and computer is realized via USB–port also performing the recharge function when connecting cell-phone to computer. Slow data transfer rate giving virtually no realistic option for graphic information transfer represents a considerable shortcoming of this procedure (in practice today only text information up to 8 Kbite is transferable). So far 'GPRS' protocol allowing a considerable increase in data transfer rate has been of a limited use because of minor coverage of the Russian territory yet. Enlargement of GPRS-protocol coverage area is expected in the near future; after that communication via cellular phones could become the principal way during work in “field” conditions in Russia.

It is satellite telephony that is presently the most universally used communication channel. The mobile complex under consideration uses ‘Nera World Communicator’ satellite terminal operating under the ‘Inmarsat-M4’ system and providing for data transfer rate up to 64 Kbit/sec. To date ‘Inmarsat’ satellites cover about 98% of the Earth’s surface ensuring “guaranteed” coverage of the area from the latitude 70° North to the latitude 70° South, i.e. include the whole territory of Russia. IBRAE’s specialists successfully tested the procedure during “Rosenergoatom” emergency exercises at the Bilibino NPP (August 2002) and the Beloyarsk NPP (September 2004).

Outward view of fully deployed mobile hardware and software complex is demonstrated in Figure.2.

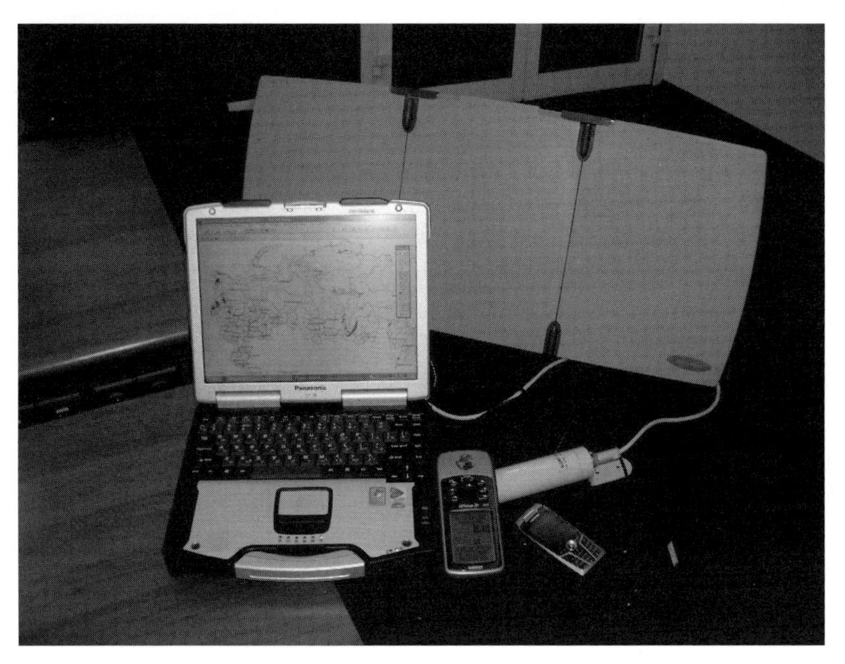

Fig. 2. Mobile hardware and software complex

3 Software of the Mobile Complex

Windows XP Professional' is used in the actual version of the mobile hardware and software complex as operational environment ensuring reliable operation of the whole system and of its individual subsystems. The operating system is supplemented with a full set of office and communication applications supplied by Microsoft Co.

Developed at IBRAE RAS, the mobile complex application software comprises:

- an information-reference system for principal radiation- and nuclear-hazardous facilities of Rosatom (OBJ-MINATOM);
- a system of operation with digital map databank (MAPVIEW);
- a forecast system for analysis of radiation emergency implications for population and environment (TRACE-MI);
- a program for radiation monitoring data acquisition and visualization (RAMO);
- reference information on Rosatom's Emergency-Rescue Teams (RISK-ATOM);
- a reference system on radiation safety standard and regulatory documents (NORMA).

The 'MAPVIEW', 'RISK-ATOM' and 'NORMA' software modules have been developed on basis of widely used MapInfo Geographical Information System (GIS).

For convenience sake, all above-listed application programmes have been integrated into a single software shell - ARM-EXPERT, Fig.3. Depending on the task to be addressed at the moment, user chooses the relevant menu item (RISK-ATOM, NORMA, OBJ-MINATOMA, MAPVIEW, RAMO, TRACE-MI) and thereby initiates the required application program. It should be emphasized that all above programs are functioning independently of each other allowing for the user to work simultaneously with different applications.

A brief description of application programs used by the mobile complex is given below.

To obtain necessary information on Rosatom's ERTs forces and facilities, one should use 'RISK-ATOM' information-reference system developed and maintained by InformAtom Association. This system developed on basis of 'CACHE' database management system is oriented to work in the Internet.

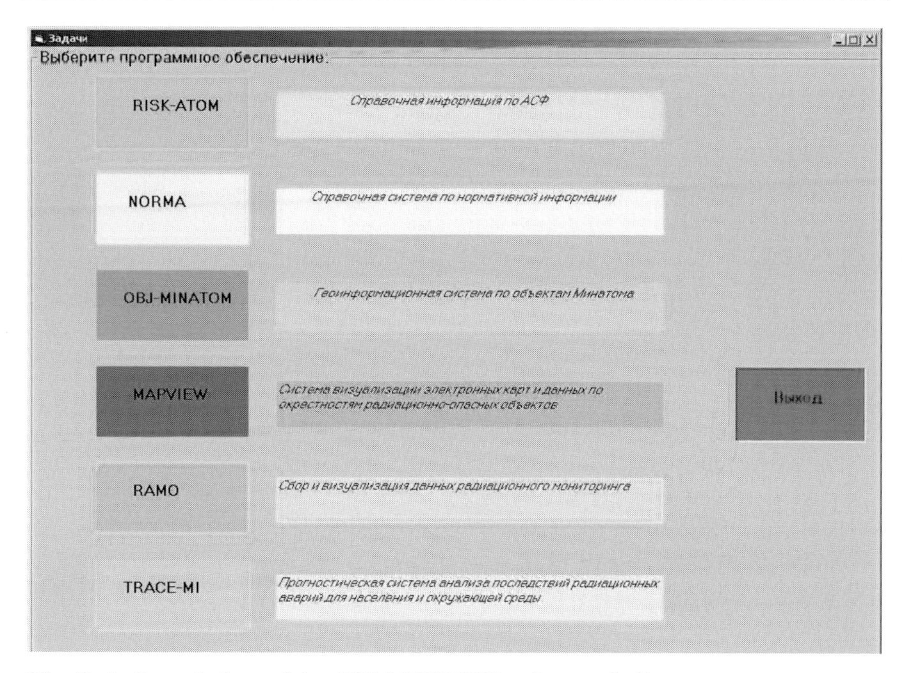

Fig. 3. Active window of the ARM-EXPERT software shell

All information of the 'RISK-ATOM' information-reference system related to work of ERTs is subdivided into logic sections.

Depending on the type of information the user is interested in, the relevant section is selected. The system interface provides for rapid context switch between the sections. To accelerate and enhance the efficiency of search for necessary information, the user has a possibility of specifying the search conditions in the system filter panel. An example of information output by responsibility areas of different Rosatom's Emergency-Technical Centers (ETCs) is given in Fig.4.

The reference system 'NORMA' for radiation safety-related standard and regulatory documentation has been developed on basis of 'Macromedia Flash' technology with the built-in 'Action Script' programming language and comprises the main Russia's and the IAEA's documents in force on regulation of the radiation safety-related activities.

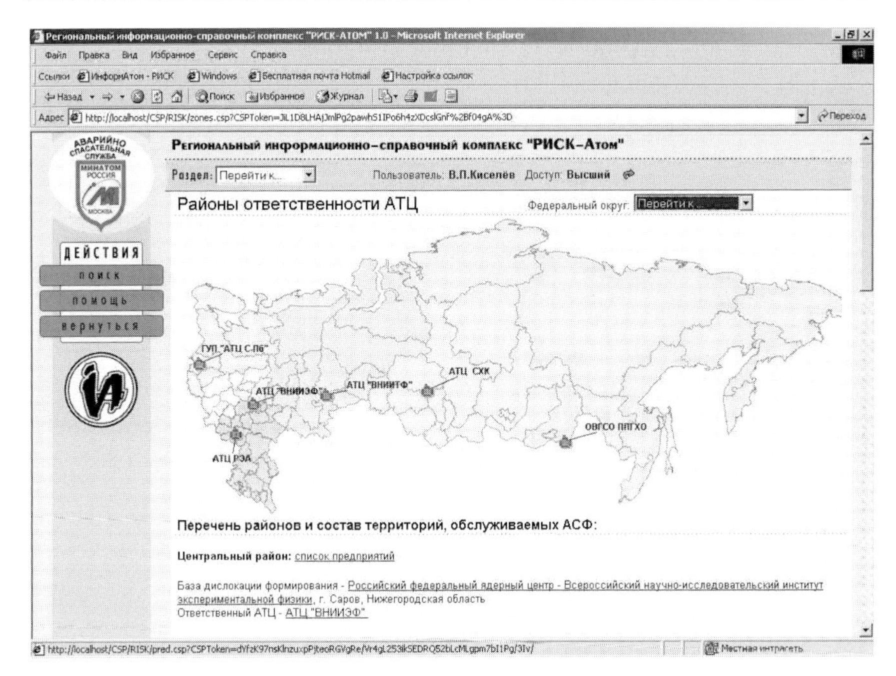

Fig. 4. A page of subsection "Distribution of Enterprises and ERTs over ETC Responsibility Areas"

'OBJ-MINATOM' subsystem is a general information-reference system on principal radiation- and nuclear-hazardous facilities of Rosatom. The system allows obtaining a variety of data on radiation-hazardous facilities including description of the sources of hazard at such facilities, geographical description of the region and its infrastructure, facility-level emergency-response plans, different-type charts, maps, and so forth.

The mobile complex under consideration comprises a large bank of different-scale digital maps available in a single geographical coordinate system. The digital-map bank includes both vector maps in 'MapInfo' GIS format and raster maps of regions non-covered by vector maps yet.

To date the digital map bank of the mobile complex includes the following maps and charts:

- general map of Russia at a 1:8.000.000 scale;
- maps of Russian regions with main radiation-hazardous facilities of Rosatom at a 1:1.000.000 scale;
- maps of Russian Nuclear Power Plant (NPP) surroundings and of main Rosatom's radiation-hazardous facilities at a 1:200.000 scale;
- gazetteers (nomenclature map lists) covering the whole Russia at a 1:1.000.000 scale;

- gazetteer at a 1:200.000 scale covering 30-km areas around every of 10 Russian NPPs, principal Rosatom's nuclear & fuel cycle facilities and the entire territories of Moscow region, Leningrad region, Bryansk region, Kaluga region, Orel region and Tula region;
- maps of sites of Russian NPPs and of principal Rosatom's radiation-hazardous facilities; and
- raster maps of the surroundings of foreign NPPs at a 1:500.000 scale.

Work with the digital-map bank of the actual mobile complex is performed using 'MAPVIEW' software module generated as a 'MapBasic' application of 'MapInfo' GIS.

'MAPVIEW' software module makes it possible to perform:

- search and view of individual maps of Russian regions, 30-km surrounding areas and sites of NPPs and principal radiation-hazardous enterprises of Rosatom;
- generation of a map of any Russian settlement surroundings using individual gazetteer at a 1:1.000.000 and a 1:200.000 scales;
- navigation over the whole map of Russia with automatic loading of necessary gazetteers; when so doing, one has a possibility for a smooth automatic transfer when zooming from one map scale to another;
- obtaining information-reference data for radiation-hazardous facilities; and
- saving the needed map fragment for further work.

The software module under consideration is also used as a software component to prepare cartographic information and a database on settlements for both 'TRACE-MI' forecast software and 'RAMO' forecast module on acquisition and visualization of radiation monitoring data, the latter forming a basis of application software developed at IBRAE RAS for the mobile complex.

An example of active window of the module for work with the digital-map bank with a loaded "surroundings' map" at a 1:200000 scale is demonstrated in Fig. 5.

The main objective of 'RAMO' software module on acquisition and visualization of radiation monitoring data consists in visualizing dose rate measurement data on a map and their saving in a tabular form comprising: dose rates, coordinates of points of measurement, time, date and comments. From 'Atlant-1' sensor the current data of dose rate measurements are transferred to a program of the measurement result visualization 'RAMO'of the mobile complex. Simultaneously the 'RAMO's visualizer program inquires GPS-receiver and obtains the current coordinates of measurement point. In such a way the table is generated simultaneously including both dose rate measurement and measurement point locations data. The table is generated in the inner format of 'MapInfo' GIS allowing its on-line representation on top of the background map selected from the digital map bank. After completion of a session the table can be saved in the inner 'MapInfo' format or in 'DBF' format using standard options of the 'MapInfo' system.

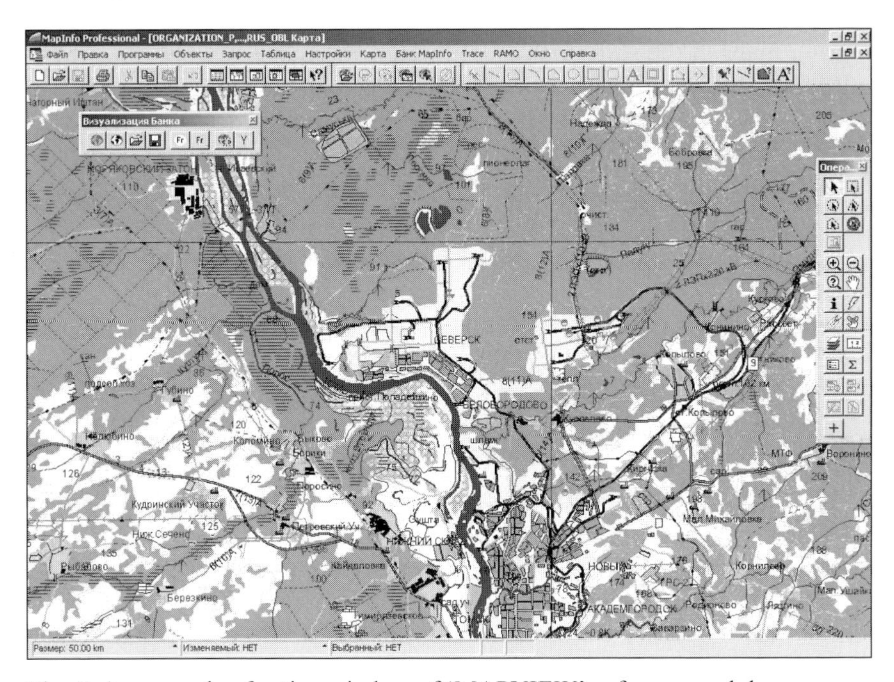

Fig. 5. An example of active window of 'MAPVIEW' software module

An example of outputting gamma-background measurement data on a map is demonstrated in Fig.6.

'TRACE-MI' forecast software module is designed to: -perform express-analysis of the radiation situation as a result of radionuclide release to the atmosphere in case of hypothetical emergencies at radiation-hazardous facilities; -estimate potential doses for personnel and nearby population; -generate thematic maps to support decision-making in emergency situations; and –prepare reports.

'TRACE-MI' simulating unit of the mobile hardware and software complex is based on the TRACE software [1] developed at IBRAE not long ago and simulating radionuclide transfer in the atmosphere in a case of potential emergency.

This software module includes the Gauss atmospheric transfer model for calculations of radioactive release spreading in the atmosphere and estimates of dose effects of such releases on human beings. The calculation procedure is based on recommendations of the actual standard and technical documents [2, 3].

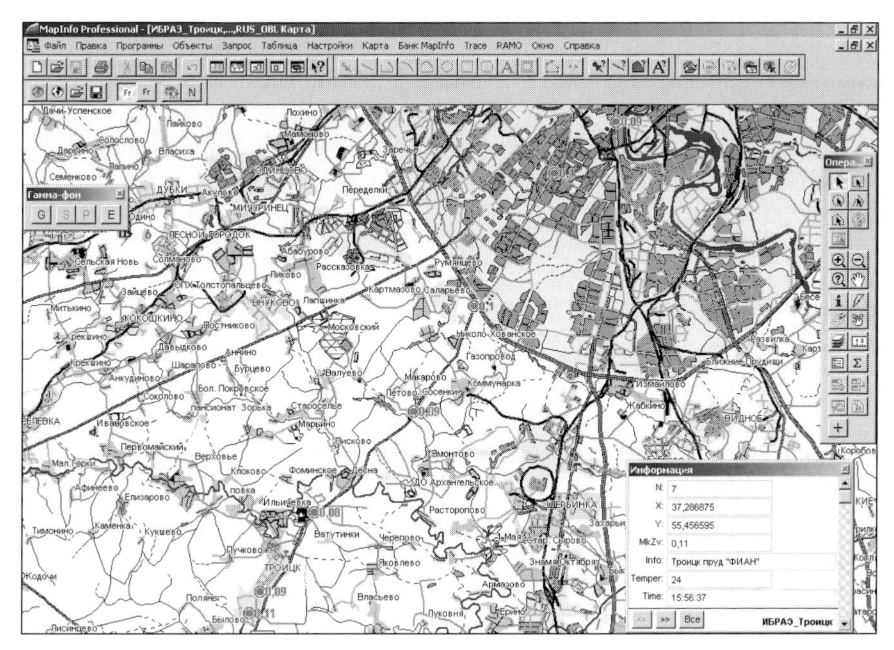

Fig. 6. An example of gamma dose rate measurement data representation on a digital map

The user is provided with the following possibilities:

- Setting radioactive release parameters and performing calculations using the atmospheric transfer model with visualization of simulation results on a digital map;
- Analyzing potential consequences of releases including report generation in the form of a text file;
- Saving the current status of the system in a special configuration file for purposes of its reproducing during subsequent analysis;
- Recording the results of simulation to a standard-format file for subsequent preparing of hard copies.

In the course of simulation the following objective functions are calculated:

- surface contamination;
- external (effective) dose from radioactive cloud;
- external (effective) dose from depositions;
- inhalation (effective) dose;
- inhalation (lung) dose;

- inhalation (thyroid gland, children) dose;
- inhalation (thyroid gland, adults) dose; and
- dose rate.

In most cases the above set of objective functions is quite sufficient for experts to estimate potential radiological consequences for population and environment under the majority of hypothetical radiation accident types.

To analyze the results of calculations, the following set of special information forms useful for work of radiation safety experts is provided in the system:

- 'Input data' –list of objective functions, specified weather conditions, description of the calculation area and list of nuclides selected for simulation;
- 'Values along the trace axis' –table of the results of simulation of all objective functions depending on the distance from the source term epicenter;
- 'Plot along the axis' – depending on the type of selected objective function, distribution of its value along the trace axis is given;
- 'Valuc along the section' – table of the results of simulation of all objective function values across the trace axis;
- 'Section plot' – plots across the trace axis are given with the results of calculations depending on the type of selected objective function.

There is a possibility of highlighting information on individual settlements covered by the trace area and thus affected by the radioactive contamination.

An example of output of the results of radiation situation simulation in the emergency area on a digital map in the form of level lines is demonstrated in Fig.7.

4 Experience of the Mobile Complex Use

Full-scale and real-time testing of all units of the mobile complex was performed during the Roenergoatom's exercises at Smolensk NPP in 2003 and at Beloyarsk NPP in 2004.

On agreement with the leaders of the Rosenergoatom Emergency Crisis Center, complex testing of the mobile complex during the Beloyarsk NPP's exercise was performed on basis of a mobile communications center of the Emergency Crisis Center under active participation of the Concern's specialists. That allowed the complex testing under both the off-line mode in "field" conditions using only its inner power sources and when transporting by a 'KAMAZ' truck with substantial vibrations and mechanical shocks. The complex demonstrated reliable operation in different conditions including proper functioning in the truck during movement via a country road.

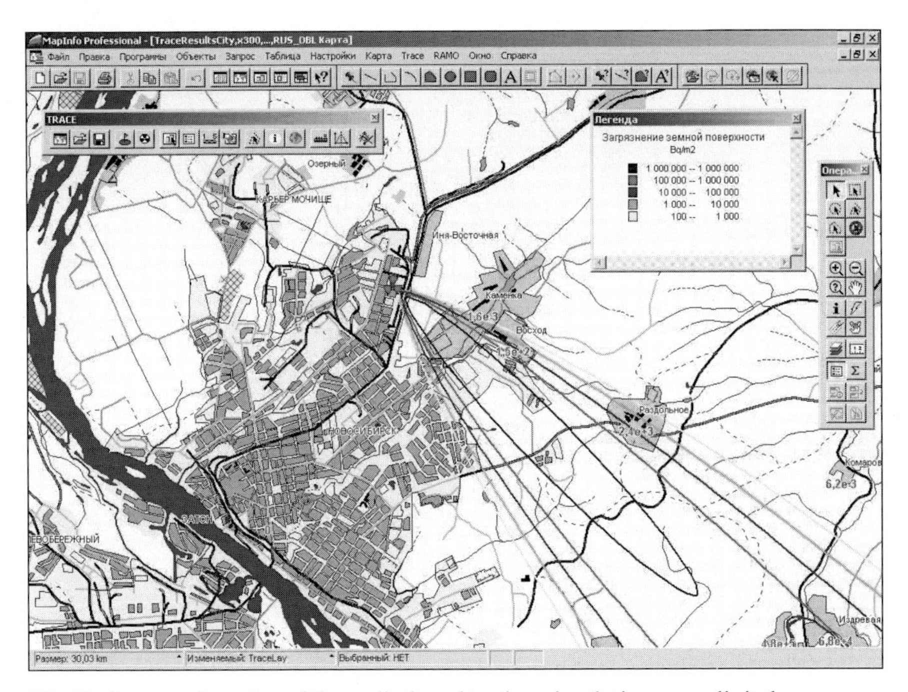

Fig. 7. Output of results of the radiation situation simulation on a digital map

Using the mobile complex gamma-background was measured within all main exercise's locations (save for NPP site itself), the measurement results being illustrated in Fig.8. Then those data were transferred to the IBRAE's Technical Crisis Center via a satellite channel.

In the course of the exercise a similar-type of work on simulation of the radiation situation in the hypothetical accident area was performed using the 'Trace_MI' forecast module of the mobile complex. The results were compared with the relevant forecast data generated at the IBRAE's Technical Support Center and other expert groups involved into the exercise. Their comparison showed a good agreement with the data of other expert groups, which also used the Gauss atmospheric transfer model in their calculations.

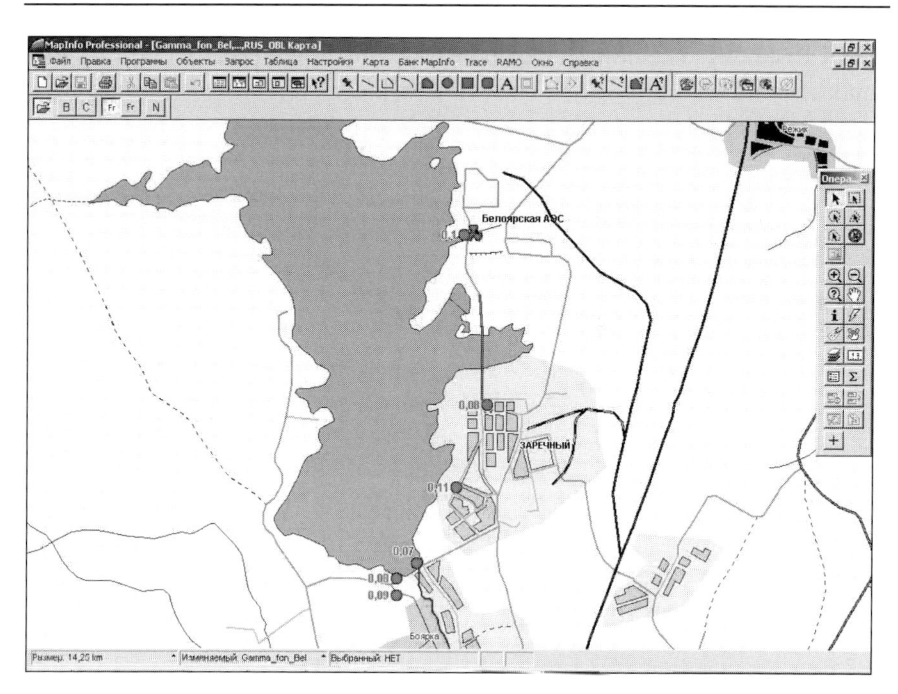

Fig. 8. Gamma-background measurements in the exercise area

References

Arutyunyan RV, Belikov VV, Goloviznin VM et al (1999) Models of radioactive contamination spreading in the environment, Proceedings of the Russian Academy of Sciences, Energy-Series, 1, pp. 61-76 (in Russian)

Methods of calculating spreading of radioactive substances from NPP site and exposure of the nearby population (1984), in: General Provisions of Nuclear Power Plant Safety, Energoatomizdat Publishers, Moscow (in Russian)

Guide Book for Establishing Permissible Releases of Radioactive Substances to the Atmosphere (DV-98) (1999), Goskomekologiya of Russia and Minatom of Russia, Moscow, P. 329 (in Russian)

The New Zoning Approach for Earthquake Risk Assessment

Fatmagül Batuk[1], Betül Sengezer[2] and Ozan Emem[1]

[1] Yildiz Technical University, Geodesy and Photogrammetric Engineering Department, Besiktas-Istanbul, Turkey.
Email: (batuk, oemem)@yildiz.edu.tr
[2] Yildiz Technical University, Department of City and Regional Planning Istanbul, Turkey.
Email: sengezer@yildiz.edu.tr

Abstract

In Turkey, mitigation works have been given more importance since the 1999 Gulf earthquake. The first drafted projects put forth for Istanbul are the seismic microzonation and earthquake risk assessment projects. However, the fundamental plan which provided a ground for thorough assessment deducted from the mentioned projects as well as a road-map for mitigation is the Earthquake Master Plan. YTU (Yildiz Technical University) settlement and urban planning team have been faced some difficulties for risk assessment and priorities, during the IEMP (Istanbul Earthquake Master Plan) project. Therefore, some processes related to spatial decisions such as definition of priority areas were applied just for highly macro level. Risk priority areas could not be determined due to the lack of necessary information. The team has developed a new zoning approach for risk zoning based on density, pattern and other properties according to the requirements of previous projects with the contribution of BU (Boğaziçi University) project team.

In this paper, existing GIS contents of risk maps and other information as the most essential component of the mitigation phase of disaster management, new zoning approach developed by YTU-BU teams during the IEMP project and possible results are presented.

1 Introduction

Earthquakes are the most devastating type of disaster in terms of loss of both lives and properties in Turkey, which seems primary to stem from factors such as faulty land use planning and construction, inadequate infrastructure and services and environmental degradation (BU-KOERI, 2002). Istanbul is the largest, most crowded and beautiful city of Turkey. According to the research conducted by scientists a large-scaled earthquake is almost certain to take place in Istanbul. When the amount of buildings is taken into account, today it seems inevitable to avoid a great deal loss of lives. Scenario A described in the JICA-IMM report foresees that 7 percent of all buildings in Istanbul (51,000 buildings) will either collapse or suffer major damage, causing some 73,000 deaths (JICA-IMM, 2002).

Generally, the legislations and the processes connected with disaster management in Turkey tend to cope with the last two phases of disaster management. Thus, most applications carried out so far have been limited only with response and recovery phases. In fact, in the event of estimating the lives and properties at risk and their effects are spatially determined along with risk reduction activities; there is no need to exert further action under normal circumstances in the following phases.

Mitigation is defined as "sustained action that reduces or eliminates long-term risk to people and property from natural hazards and their effects". Mitigation begins with local communities assessing their risks and repetitive problems and making a plan for creating solutions to these problems and reducing the vulnerability of its citizens and property to risk.

Naturally the first step of mitigation is to determine the risk if the risk in question is spatially unclear. When it is assessed in terms of earthquakes; current disasters and their effects; topography; underground facts and their possible outer effects; artificial elements such as buildings, roads, bridges, technical infrastructure facilities etc. are necessary basic data. With the modern scientific methods, which have been developed up until now, it is possible to put forth the relative spatial distribution of risks, by estimating earthquakes and their effects. The other phase comprises the determination of mitigation alternatives, thus devising plans by carrying out cost-benefit analyses. At this phase it is important to obtain necessary special facts regarding the priority areas. The application and continuation of plans consist the last phase of mitigation process. In mitigation plans, there exist scales in three categories: macro, mezzo and micro levels. Nowadays due to hazard assessment, mitigation planning has become easier and GIS support to multi hazard possibilities are taken into consideration.

2 Earthquake and Loss Estimation

Catastrophe loss modeling for natural hazards comprises four major steps (Chen et al., 2004):

- Hazard analysis: quantifies the physical characteristics of a hazard, including probability of occurrence, magnitude, intensity, location, influence of geological or meteorological factors.
- Exposure analysis: identifies and maps underlying elements at risk or exposures, including the built environment and socioeconomic factors such as population and economic activity.
- Vulnerability analysis: assesses the degree of susceptibility to which elements at risk are exposed to the hazard. A community with strong capacities would decrease susceptibility. A common form of vulnerability analysis uses historical damage records to prescribe relationships between damage to dwellings and hazard intensity. Different building construction classes and occupancy types will have distinct vulnerability curves. Vulnerability curves can also be established for socioeconomic exposures, such as population age groups, although for the time being such relationships are not well developed.
- Risk analysis: synthesizes the above three components and determines the resulting losses as a function of return period or as an exceedance probability.

Each of these four stages is fraught with uncertainty. Limited scientific information in defining the hazard, lower quality data in defining the exposure inventory characteristics, and limited engineering information in estimating the inventory damage result in error bounds in estimating expected losses (Grossi et al., 1998).

Catastrophe loss estimation for natural hazards combines both hazard and exposure data. While hazard attributes such as intensity distributions are usually represented at a spatially explicit raster (or pixel) level, exposure data such as population, dwellings and insurance portfolios are usually only available at spatially lumped census tracts. In current loss estimation studies, this spatial incompatibility is often inadequately addressed and a uniform distribution of exposure data within an areal unit assumed. As a result, loss estimation models overlook a great deal of spatial disparity (Chen et al., 2004).

At earthquake loss estimation applications, generally, parameters are determined as 500 m x 500 m cells. For example, at USA and HAZUS applications, USGS's 10 m gridded-data is frequently used for geological parameters (Lowe 2004). If there is no existing data within the demanded interval, through the interpolation method, data is transferred to grids with no data. As it was stated above, if the inventory data does not exist in the mentioned cell type, either again with interpolation or with loss estimation for small scale spatial units. In the USA, census tract or zip code are used for building inventory and social, demographic data. In this kind of applications, various assumptions during estimation, unreliability of geological, geophysical and geo-technical data, and spatial boundaries utilized in obtaining information etc. paves the way for uncertainty and results in the highly macro form.

3 Istanbul Earthquake Master Plan

In order to mitigate heavy losses in the event of a large-scaled earthquake, Istanbul Metropolitan Municipality (IMM), the main body responsible for all local administrations of Istanbul, has undertaken a variety of precautions since 1999, coordinating with research institutions in many projects covering the land within its responsibility. The first of these projects is the Seismic Microzonation Project (JICA-IMM 2002). Another project which is necessitated on the face of the present threat was launched by the Boğaziçi University (BU) (BU-ARC 2002). Risks determined to pose threat to the existence of the city are all assessed also within a comprehensive project: the Istanbul Earthquake Master Plan Project (IEMP). Members of academic staff and students from the four universities in Turkey have taken part in the project for the purpose of defining maps that could reduce risks dramatically.

The scope of IEMP comprised of activities to be done for assessment of current situation, seismic assessment and rehabilitation of existing buildings, issues about urban planning, legal, financial, educational, social, risk and disaster management and, aimed at planning of the activities in these fields, preparation of implementation programs, and identification of the responsibilities and responsible authorities for earthquake disaster mitigation works to be carried out in Istanbul (BU et al., 2003).

An important aspect to be covered by IEMP is decided to be the assessment of seismic vulnerability of existing building stock in Istanbul, the development of seismic retrofitting methods and the determination of technical, social, administrative, legal and financial measures to be taken in order to be able to implement such methods. In the Plan the work to be done in these fields are examined and the recommendations about the measures to be taken are given (BU et al., 2003).

3.1 Istanbul and Earthquake Risk

In recent years, two main projects have been carried out according to the earthquake scenarios in order to determine the risk, which has been discussed seriously. IEMP has been prepared according to previous projects data motivated to the risk reduction, prior action area definition, disaster management etc. These projects are:

- The Study on a Disaster Prevention / Mitigation Basic Plan in Istanbul including Seismic Microzonation in the Republic of Turkey (JICA-IMM 2002)
- Earthquake Risk Assessment for Istanbul Metropolitan Area (BU-ARC 2002)

Both in projects, earthquake risk assessment in Istanbul are based on the scenario earthquake governed deterministic methodology. A scenario earthquake was determined to take place on the Main Marmara Fault (MMF) at BU-ARC project. Four scenario earthquake models were determined with MMF nearly at JICA-IMM Project. Model A is the same as the BU project. Ground motion and loss estimations based on deterministic assessments are mean values. In other words the reality has a 50% probability of being both below and above those levels (BU et al 2003).

The BU – ARC study assumed that, based on the segmentation model developed, the scenario earthquake would occur on the unruptured segments of the Main Marmara Fault producing an Mw = 7.5 event. The same seismo-tectonic structure is used in the JICA – IMM study and the scenario earthquake model A is equivalent the one used in the BU – ARC study. Compilation and interpretation of topographic, geologic and geotechnical data and the selection of the appropriate attenuation and site response models constitute the remaining main inputs of the earthquake hazard assessment. Following, the transfer of the ground motion values to the 0.005 x 0.005° (400 m x 600 m) cells (polygonal), the spectral accelerations obtained for 20 and 50% probabilities of exceedance and from the deterministic studies have been normalized with respect to their mean value (BU et al 2003). In the JICA-IMM project, however, seismic motion has been determined with 500 m x 500 m cell.

In the JICA-IMM project, the number of buildings in 500 m grid cell using 1/1000 CAD map data has been determined. Other building data has been taken from the building census 2000 on district basis. For surface geology, an improved geology map has been used at 1/5000 (original scale has 1/50000). Population data were used from population census 2000 with sub-district base. Another important point is that the census data has not been collected as geocoded in Turkey. In the BU-ARC project, data collected more comprehensively.

Heavily damaged and total collapse constitute are the most critical building damage classes. Within the context of the BU – ARC study, building damages are computed based on both intensity and spectral displacements. For the computations base done spectral displacements HAZUS99 methodology is used. However, for various building types, the story drift amounts are modified to obtain damage grades compatibles with those defined in EMS-98 (BU et al 2003).

3.1.2 Determining Priority Areas in Risk Reduction

Social and physical losses have been put forth in sub-district basis in the JICA-IMM project and in cell basis in the BU-ARC project. At the IEMP project, some problems were encountered in determining priority areas by using loss estimation data of previous projects.

The main problems in the JICA-IMM project may be placed in two groups. One of these problems arises from the difficulty in comparing various sub-districts with different land-sizes. The other one stems from omitting the third dimension (density) during loss estimation.

In the JICA-IMM project, social and physical losses have been determined on the sub-district bases. In defining risks on sub-district level there emerged two fundamental problems. One of them is the difference in land-sizes of sub-districts in Istanbul – towards the city center sub-districts cover smaller lands (5 Hectares) while in the periphery they occupy larger spaces (3148 Hectares). The reason why they occupy large spaces on the periphery is the fact that following the expansion of the city center towards periphery, former villages were transformed into sub-districts. In these kinds of sub-districts, former village centers are the focal point of housing while the sub-district lands are mostly empty spaces. In JICA-IMM studies, risk definitions were provided as numerical losses varying according to sub-districts. However, the vast difference in lands of sub-districts has made the relative comparison of risk impossible.

For example, in the historical peninsula, sub-districts cover small lands, thus, they have smaller number of buildings. The buildings to be torn down exhibit a low figure but a high rate. On the contrary, Bahcelievler sub-district leaves an impression as to have a larger damage due to covering a larger land.

Secondarily, in sub-districts with up to 3148 hectare lands there may be intensities at various levels. When an intensity value is given on sub-district basis, the level of uncertainty increases due to the increased value of standard deviation.

The solution to this problem is the transformation of heavy and beyond losses on sub-district basis into unit. In other words, the mentioned losses should be divided into the sub-district area in order to calculate losses per hectare. To increase the reliability of the result it would be quite useful to rearrange the densities of sub-district areas. That is, in some sub-district areas there are urban centers like universities, large woods and airports which increase the level of population dramatically. Consequently, these larger urban function areas should be deducted from the total of sub-district areas, rearranging the land of sub-district areas.

The other point of problem in the JICA-IMM project is the statement of losses only on the building count level. It is not possible to reach reasonable and clear conclusions from loss estimations obtained only on the basis of building count. When the damage occurs in terms of building count, then density is omitted. High-story buildings are dense in some sub-districts while low-story buildings are dense in others. Due to not taking the density into consideration, the real level of damage, loss of lives and properties is neglected. Because of omitting the third dimension in the JICA-IMM project, loss of lives statistics are provided only on district basis, not on sub-district one. The main problem seems to stem from the fact that loss estimation is made on the basis of building count, thus, preventing the real level of damage and that priority areas cannot be determined.

JICA-IMM data has been transformed into building losses on unit basis by the YTU project team. The number of buildings likely to be heavily or above damaged has been divided into the area of the sub-district. Because there exist large public centers (public institutions, industrial sites etc.) in the sub-district, these areas are excluded, evaluating other relatively large entities necessary for every sub-district area such as schools, roads, and greenery and density blocks. Density of heavily damaged buildings has been calculated in the mentioned process (Figure 1). In this case, high-density sub-districts with relatively small lands seem to suffer a heavier volume of damage.

In the BU-ARC project, there are comparisons in terms of the number of damaged buildings and construction area. Thus, this project proves to be more useful in order to reach priority areas. Vulnerability on cell basis include loss of lives and properties, construction areas.

The BU-ARC project adapted the HAZUS99 methodology to Turkey and the developed KOERILoss software, social and physical losses were determined. The data entered in the software includes a great amount of detailed inventory data. For example, each cell was entered data regarding the construction type of buildings, number of floors, and the year of construction. Data obtained over a variety of sources have been applied at the project. For this reason, it is hard to say anything clear about the reliability of data and information. The solution to data collecting and storage in order to achieve reasonable loss estimation necessitates the immediate design of NSDI components in Turkey and that core data should include the mentioned data.

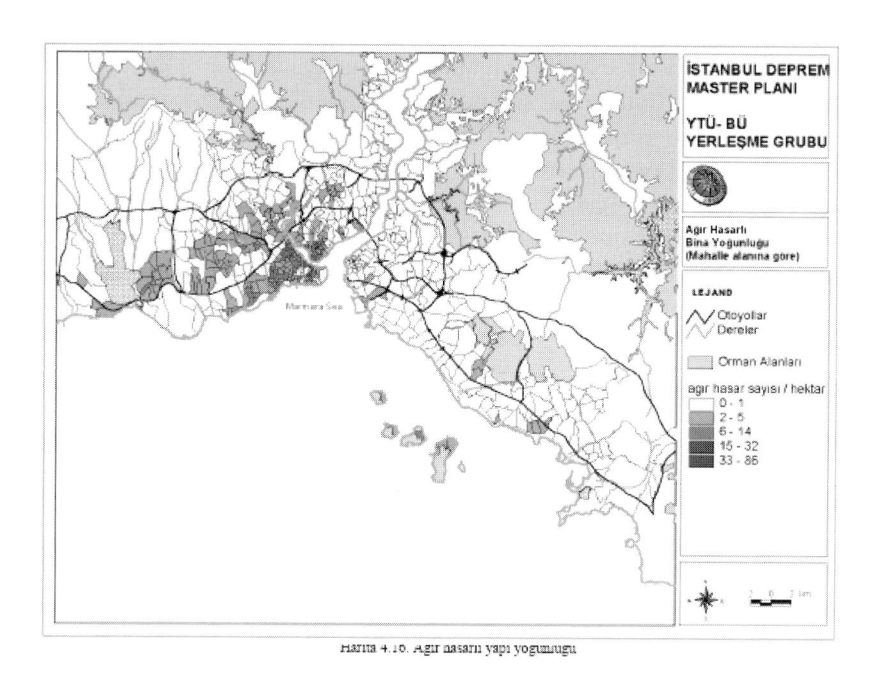

Fig. 1. Heavily damaged building density

Loss estimation for insurance operations is the objective BU-ARC project. Although there are detailed studies in this regard, they still remain insufficient to determine priority regions. There are two main insufficiencies in this kind of approaches in determining the problematic priority areas. The first of these, data include geologic formation are based on middle scaled maps. Density, structural variations different microzonation areas are not taken into consideration due to featuring the grid base. In other words, formations, variations and densities, which affect the loss from separate directions, are all present within a cell.

3.2 Proposed Zoning Design

By utilizing the macro level defining indicators, the YTU team has strived hard to overcome the above-mentioned difficulties by also assessing risk indicators, in search for priority areas. Due to time limitations, as a result of applying some data obtained from the IMM GIS project to the data obtained from this peculiar study, risk areas have been indicated in a general framework, and in this respect strategies have been developed in order to mitigate the problems to be arisen from a probable earthquake. Following a prior study, an approach aiming at describing risks at detail level has been devised. With this paper it is intended to open the aim in question for discussion.

During the project, some difficulties have appeared for risk assessment. Therefore, some processes related to spatial decisions such as definition of prior areas were applied just for macro level. Furthermore, it was required to define sub-areas, which were changed in macro level, estimate the loss in respect of possible damage models, which was found according to different structure characteristics and computational methods, and prepare cost-benefit analyses. Priority areas in terms of risk could not be defined due to lack of proper information. The team has developed a new zoning approach for risk zoning based on density, pattern and other properties according to the requirements of previous projects with the contribution of BU project team.

In the Istanbul metropolitan area, different fabrics of society, income groups, functions and different building groups are at some point in close proximity and at other they are interwoven. An indication to be benefited from in order to classify those groups is density and the other is pattern. The determiners of density are land use and heights at the third dimension. Other characteristics to be classified are geological differences (ground structure) and slope. In addition, earthquake characteristics have been determined in classifications according to the velocity reaction values obtained over the probability method. It is possible to make correct loss estimations by determining sub-district areas featuring different characteristics on macro level and by applying a variety of damage estimation models prepared in line with different structure characteristics. Thus, by assessing the loss estimations of differing sub-district areas on unit basis, it will be possible to determine the risk priority areas.

In the IEMP, based on the current risk analyses, strategies have been devised at metropolitan city level or macro level, by putting forth a general risk-zoning process. However, through the support of studies based on cost-benefit analysis, it will be possible to develop programs to reduce risks in the Istanbul Metropolitan Area.

3.2.1 Creating New Zones

There are digital photogrammetrical maps of Istanbul the first of which dates back to 1987 and the last 2000. These maps are scaled at 1/1000 with UTM 3 degree coordinate system. In 2004 the administrative authority of Istanbul Metropolitan Municipality has been enlarged up to provincial frontiers. However, because the project was given a start before and the lately added areas do not have a digital data, the scope of the project today is much smaller.

During the project, orthophotos scaled at 1/5000 and 1/5000 scaled CAD data deducted from basic maps scaled at 1/1000 were used. On the other hand the data used as sources were land use, building inventory, 5 m x 5 m elevation data and administrative boundaries which were produced from the CAD data during the GIS project of Municipality in 2002.

Since the city is very large and there is no consistent planning concept, the different building patterns have appeared dispersedly. These patterns can be grouped and digitized in 5 main and 22 sub categories:

1. Regular (individual rare/ mixed, adjacent rare/ mixed/ highly congested, rare/ congested/ mixed),
2. organic rare/congested,
3. social housing,
4. develop along highway, and
5. large urban areas (public spaces/+ commercial/ industry, industry/+ rare/ congested/ commercial, commercial, recreation area, sport).

However; the floor numbers were designed to be classified as 1-4, 5-9 and 9+ before digitization process, the classification appeared as 1-4+5-8, 5-8+9+, complicated and no-data during the digitization.

In order to define the areas with the 20% or more slope, TIN data has prepared from the 5 m x 5 m grid data. Then slope values have been calculated and results have been converted to vector from grid.

Soil classification, probabilistic PGA for 50% pass over in 50 years PGA, spectral acceleration for 1 and 0.2 seconds data were obtained from BU-KOERI in digital vector format. However the accuracies of data is not clear enough, it is estimated to be around the accuracy of detailed 1/50.000 scaled map. It is believed that using seismic microzonation base maps with larger scales will be more appropriate.

After all data is collected in the GIS environment, areas were reviewed to avoid edges longer than 500 meters, since it wasn't wanted to have for seismic aspects.

In the IMM GIS project, in addition to the attributes mentioned above, the point building data was arranged as building count according to the construction type (reinforced concrete, masonry, wooden, pre-fabricated, steel construction); according to construction type and floor number; according daily and night population, main function etc. Although it was quite necessary, building construction dates could not be obtained due to being not reliable enough. If they were added to the study, then it would be necessary to state the buildings as those constructed in 1979 and before, those constructed between 1980-1998 and those constructed in 1998 and after, along with the number of floors and construction type. In the mentioned years some alterations were made at the Earthquake Regulations, which, of course, affected the construction of buildings. However, in the damage analyses after the earthquake, it was observed that no significant damage occurred between buildings constructed before 1975 and after the year (Sengezer 2003).

The new zones will be assessed with KOERILoss software of the BU, and with MSK98 scale for each area, the number of collapsed, heavily-medium-lightly damaged or non-damaged buildings and construction site areas will be stated as well as the number of death and injury.

With the obtained results, a more accurate and realistic estimation regarding the risk reduction in priority areas could be made, and thanks to the cost-benefit analyses a variety of solutions could be brought about. To this end, first of all, the below mentioned path should be followed.

The number of damaged buildings = the number of heavily and more damaged buildings / zone area (3.1)

Risk = the number of heavily and more damaged buildings / total number of buildings = A (3.2)

3.2.2 Spatial Analysis Scenarios

According to the results, which are to be obtained in line with the characteristics of zones, the following steps have to be taken:

- If A > 50% (approximate rate, a certain value is not clear for our country), it could be a priority area. The other factors need to be studied closer.
- If A > 50% and TAKS > 0.75, KAKS > 2.5, then the rate of density is high with no good environmental conditions and they have the priority for studies. Detailed microzoning and urban transformation researchers must be done as well as cost-benefit studies.
- In the end, it may be decided to retrofit or reconstruct the building in question. Instead of individually retrofiting buildings, it may be pre-

ferred to increase the total quality of the environment by tearing down all buildings and reconstructing them. Instead of increasing the level of security one by one with a new plan, many benefits may be obtained over the total increase of environmental quality.
- If the risk is high and the income and education levels are low, some incentives may be provided with the help of government. In order to achieve urban transformation, it is possible to elevate the amount of synergy by combining the constructional aims with social development projects.
- For regions emerging with peculiar patterns and function characteristics, different project methods and reinforcement techniques, for example not to spoil their historical pattern, should be thought of.

4 Conclusion

The rapid growth rate in metropolitan cities like Istanbul in addition to the problems regarding city planning as well as problems with construction techniques, resulted in dramatically increased risk of vulnerability in the event of a major earthquake. Projects launched with the pioneering local administrations in cooperation with universities have proved to be initial steps taken in the direction of eliminating the factors that may cost many lives, and disaster management mitigation efforts have been put into action. Within the framework of the IEMP, a variety of roadmaps have been drawn so far.

Within the scope of the IEMP, the YTU settlement and urban planning team has applied indicators defining the problems of Istanbul on macro level, harmonizing them with risk indicators, thus, obtaining general priority areas. As a result, according to their typologies, offers and strategies for solution have been brought about. The level of risk is quite high in Istanbul. Consequently, with the help of aforementioned approaches, the priority areas have been determined but the amount of land equals almost one-third of the city, which necessitates the urgency to narrow and specify the risky areas in order to immediately start to apply solutions. In order to use the time in the best way possible until a probable earthquake, a method to be developed by using the present data provided by JICA, BU-ARC and IMM has been suggested, and the fundamentals that may operate the risk model have been prepared at GIS environment.

Zones have been produced by overlapping vector data that includes administrative borders and other data regarding the pattern, floor count, slope and earthquake. Then building inventory, functions, and population data to each zone transferred to each zone. Areas will be assessed with the BU KOERILoss software and on new areas basic physical, economic and social losses may be calculated. Results will be assessed and risk priority areas will be determined, rendering the strategies clearer for more detailed studies.

In researched studies and current projects, it is not possible to check the correctness of the information obtained as a result of earthquake loss estimation and vulnerability methodology. According to our point of view, the underlying reason of the above situation stems from various data collected from different sources each of which have different scales, reliability and belong to various time periods, at some points the lack of proper and necessary information and the insufficiency of information produced at subdistrict, census tract, zip code or various cell levels. The only solution seems to develop NDSI applications which include multi hazard analysis and loss estimation data, and the subject should also be given more priority in e-government applications as is the case in developed countries.

References

BU-ARC (2002) Earthquake Risk Assessment for İstanbul Metropolitan Area, Project Report prepared by Erdik M, Aydınoğlu MN, Barka A,Yüzügüllü Ö, Siyahi B, Durukal E, Fahjan Y, Akman H, Birgören G, Biro Y

JICA-IMM (2002) The Study on a Disaster prevention/Mitigation Basic Plan in İstanbul including Seismic Microzonation in the Republic of Turkey, draft final report, Main report

BU, ITU, ODTU, YTU, MMI (2003) Earthquake Master Plan For Istanbul, Final Report

Chen K, McAneney J, Blong R, Leigh R, Hunter L, Magill C (2004) Defining area at risk and its effect in catastrophe loss estimation: a dasymetric mapping approach, Applied Geography 24, pp 97–117

Grossi PA, Kleindorfer P, Kunreuther H (1998) The Impact of Uncertainty in Managing Seismic Risk: The Case of Earthquake Frequency and Structural Vulnerability, Project Report

Lowe JW (2004) Spatial Trends and Drivers in Insurance Loss Modeling, Geospatial Solutions

Sengezer B (2003) Seismic Vulnerability of Buildings: The case of Turkey, UIA Summer School Urban Settlements and Natural or Other Disasters Work Programme, Izmir

Application of Remote Sensing and GIS Technology in Forest Fire Risk Modeling and Management of Forest Fires: A Case Study in the Garhwal Himalayan Region

Sunil Chandra

Forest Survey of India, Kaulagarh Road, P.O.-IPE, Dehra Dun.
Uttaranchal, India.
Email: scdangwal@yahoo.co.uk

Abstract

Natural disasters are inevitable and it is impossible to fully recoup the damage caused by the disasters. But to some extent it is possible to minimize the potential risk by developing early warning strategies for disasters, prepare and implement developmental plans to provide resilience to such disasters and to help in rehabilitation and post disaster reduction. Uncontrolled forest fires have adversely affected the local landscape and economy. Climatic, phenology variations and topography, apart from local factors are some of the main causes of frequent occurrence of wild forest fires in Garhwal Himalayas. Understanding the important of forest in the national economy (12% of global plant wealth), conservation of environment and biodiversity, Forest Survey of India(FSI) as a central monitoring agency is assessing and estimating the forest resources in a two years cycle. India is one of the few countries in the world to carry out the forest cover assessment and mapping using satellite data in a two years cycle period. Keeping in view the role of forest in national development, a Central Sector scheme has been implemented that includes- development of Early Warning system for forest fires, mapping of forest fire affected areas, development of a fire danger rating system, monitoring the impact of the scheme and its evaluation, identification and mapping of all fire prone areas, compilation and analysis of data-base on forest fire damage, development and installation of 'Fire Danger Rating System' and 'Fire Forecasting System'. The other measures include building up a strong communication network between the monitoring station and fire suppression teams, effective transportation, watch towers, Fire line creation and maintenance, creation of water harvesting structures, fire management plans, any other technological innovation, assistance to JFM(Joint Forest Management Committees),awareness, training and research. Remote sensing and GIS technology could be effectively used in fire risk zonation .The technology has proved to be a valuable tool in identifying different fire risk zones based on appropriate parameters such as fuel load, slope, aspect, altitude, drainage, distance from roads and settlements. The approach followed for broad based forest type classification in the study was helpful in identifying different forest types available in the area. Fuel load, slope degree, aspect, elevation, drainage, roads and settlement layers were assigned different weight ages depending upon their impact, in identification of fire risk zones. This was followed by ground verification of the generated fire risk zone maps and their comparison with incidences of forest fire in previous years. The response time to disaster relief was calculated based on the friction offered by slope, altitude and other factors. Thus, high to low fire risk zones can be identified and suitable management strategy for controlling the disaster can be prioritized in this region.

1 Introduction

Forest play an important role in conservation of environment and flora. Forest fire cause severe damage in terms of loss of timber, human property, wildlife , loss of certain rare plant species and soil erosion.

Forest cover in recent times have been subjected to pressures not only from human beings, but also from natural calamities such as fire, floods and cyclone. Frequent occurrences of uncontrolled forest fires have caused adverse impacts. As per the studies, 90% forest fires are estimated to be man made. About 3.73 million hectare forest area is affected by fires leading to a loss of Rs. 440 crores annually. About 54.7% of India's forest are fire prone . Of this 9.2% forest area is affected by frequent forest fires whereas 45.5% by occasional fires(FSI, 1997). There has been frequent occurrences of forest fire in Garhwal Himalayas in recent years.

Most of the wild forest fires in recent times are caused by man deliberately or unintentionally. High temperature and dry biomass created as a result of humus, dry herbs ,shrubs & fallen twigs accelerate the fire. Fires could be creeping fires, ground fires , surface fires and crown fires. Certain tree species are more vulnerable to fire than others. Forest density is largely responsible for the weight and compactness of fuel load, which determines the amount of biomass available for burning.

Forest fires are strongly linked to weather and climate (Faannigan and Harrington, 1988). Slope, aspect and elevation have a direct affect on the severity and extent of fire. While elevation and aspect are responsible for the quality of fuel biomass, slope has an important role in deciding the rate of fire spread. On a steep slope, flames are closer to the fuel, wind currents are normally uphill and convection heat rising along the slopes increases the rate of spread. The vast differences in the quantity of solar heat delivered to southern exposures compared to northern, causes differences in soil moisture and vegetation types.

Dependence of forest fire on such spatial parameters has made the application of GIS techniques feasible for classifying a geographical area into different degree of fire risk. Such occurrences have resulted in a need to develop a system that can identify fire risk area based on certain criteria and parameters. GIS has come up as a wonderful tool in analyzing different layers to come up at a conclusion

An effort has been made in the present study to highlight those factors in the Garhwal Himalayas which are primarily responsible for initiating wild fires . A study was taken up in part of Uttarkashi district in Garhwal Himalayas to create fire risk zones. The fire risk index developed as a result of risk zonation model can be utilized for creation of similar such models for different topographic conditions.

2 Study Area

Uttarkashi district in Uttaranchal has its border with China and Himachal Pradesh in India. The study area covers an area of 2054sq.km. The location map of the study area is shown in fig.1. The forest cover in the district is 38.31% of the geog. Area.(State of forest report, 2001). The common tree species in the area are either broad-lived or conifers. The broad-lived forests are chiefly composed of Oaks. The conifer trees either form pure strands or occur mixed with broad-lived species in the montane zones. The important species present are *Abies pindrow* (found above 2100m), *A. spectabilis* (above 2700m), *cedrus deodara (*above 1800m), *Cupressus torulosa* (above 2000m), *Picea smithiana* (above 2700m), *pinus roxburghii* (900-1600m), *P. wallichiana* (above 2400m) and *Taxus baccata* (above 2500m).

Pinus roxburghii covers large tracts of intermediate ranges between submontane and montane zones.

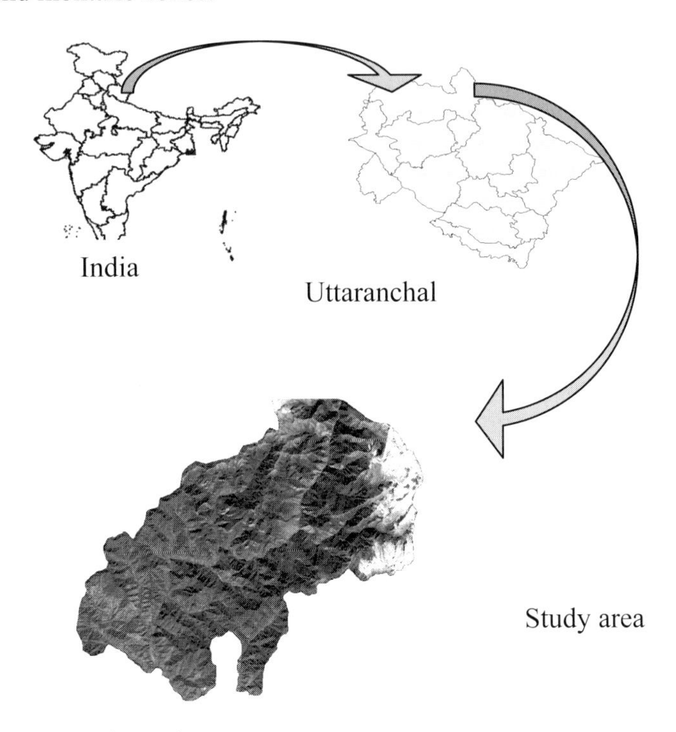

Fig. 1. Location map of the study area

3 Data Used and Methodology

Indian remote sensing satellite data(IRS- LISS-III,thematic maps(prepared from aerial photographs), forest vegetation map showing forest cover with density (FSI,2001,report) & GPS were used for analysis and data collection .The digital image analysis work was carried out using Erdas/imagine 8.6, installed on Windows-XP professional.

Indian remote sensing data of the period November 2000 was chosen for the fuel risk study. For that species type classification was done using the approach followed by champion and Seth. Contours at an interval of 40m were digitized using survey of India toposheet (scale 1:50,000) for generating the DTM (digital terrain model) (fig.2) . Based on DTM , entire study area was divided into 3 zones. Height of place between (750-1600)m, (1600-4000)m and heights above 4000m. This was done to separate one forest type from the other ,even though their tone and reflectance val ues(digital number) did not have much variance in the satellite data. Unsupervised classification technique was adopted for classifying all the three zones while taking into consideration thematic maps (based on aerial photographs), ground truth available and inventory records. Nine(9) different forest types were identified . The area covered by each forest type(table 1) and the map created is shown in (fig.3). Different forest types and the values assigned to them are shown in table 3.

The density classification was carried out using methodology developed & adopted by FSI for cover mapping of the entire country(fig.4). This was important for estimating the type and density of dry biomass which is an important factor in determining the fuel load and compactness of biomass. The area under different density classes is shown in table 2.(FSI,State of Forest report,2001). The ground truthing work was carried out on as much as 90 points. GPS was used to record the geographic location of the place affected by fire during the present and the previous year.

Fig 2: Digital elevation model **Fig 3**: Forest type map

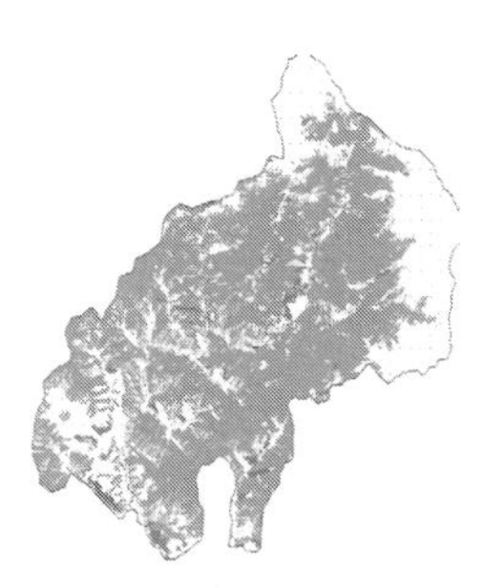

Fig. 4.: Forest Density map

Forest type	Area (in sq.km)	Color index
Pine (Dominating Chir)	380.96	
High altitude Conifers	612.43	
Broad lived	281.61	
Scrubs and Grasslands	516.37	

Table 1:Forest types and landuse distribution

Class	Area(in sq.km.)	Color index
Dense(forest cover>40%)	1050.29	
Open (forest cover between 10-40%)	224.71	
Scrubs(<10%)	38.75	
Non-Forest	477.62	

Table 2:Area under different density classes

Vegetation type	Fuel class values
Dense Pine (Chir)	4
Open Pine (Chir)	4
High altitude Conifers (Dense)	3
High altitude Conifers (Open)	2
Broad lived	2
Scrubs and grasslands	2

Table 3: Forest types for generating fuel class index map

4 Generation of DEM, Slope, Aspect & Elevation Map

Apart from the forest types & study of fuel biomass, slope, aspect & elevation of a place are quite important factors effecting the combustion of fuel and defining the rate of fire spread. Digital terrain model (DTM) has been generated using surfacing feature of the erdas/imagine s/w. The image drape function has been frequently used to visualize the surface & terrain from a point using DTM in the background.

Slope – fires usually move faster uphill than downhill , and the steeper the slope , the faster the fire will move. The degree of slope has a significant effect to the rate of spread of fire. The rate of spread is twice on a slope having slope percentage as 18 compared to a flat surface & four times on a 36 percentage slope (Anon,ITC) .Slope map of the area is given in fig.5.Based on the type of terrain, the slope was classified into four degree classes and values were assigned to each class.(table 4).

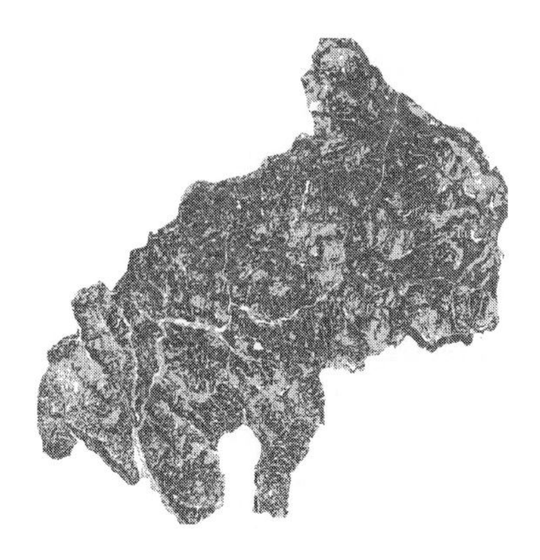

Fig. 5. Slope map

Degree of slope	Fire risk values	Color index
<10°	1	
11°-25°	2	
26°-40°	3	
>40°	4	

Table 4. Slope index

Aspect – the major effects of variation in aspect are related to the consequent variation in isolation per unit surface. Southern aspects receive more solar radiation than northern in proportion to their latitude. In higher latitude, steepness of slope cancels out the effects of lower latitude of the sun on a south aspect. Southeastern, southwestern and western slope exposures have about an equal degree of cumulative solar heating as the sun progresses to the west. Southern exposures are subjected to higher temperatures , lower humidity and rapid loss of soil moisture. The long period of solar exposure and ease of ignition causes more fires on southern slopes than northern slope. Accordingly values have been assigned & the aspect map have been reclassified (fig.6).

Fig. 6. Aspect Map

Aspect	Fire risk values	Color Index
Northern	1	
Eastern	2	
Western	3	
Southern	4	

Table 5. Aspect index

4.1 Elevation

Elevation differences have important effects on local climate, temperature , RH and vegetation. The northern Himalayas consists of varying heights ranging from 900m to more than 4000m. Based on the study area , the elevation map has been created.(Fig.7)

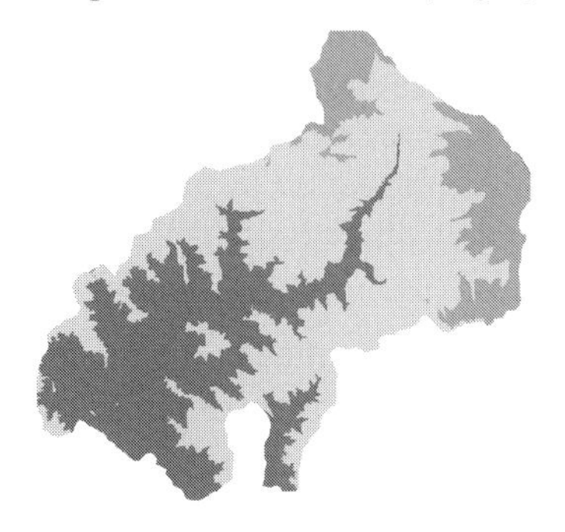

Fig 7. Elevation map

Elevation(in metres	Fire risk values	Color index
00-1600	3	
1600-4000	2	
>4000	1	

Table 6: Elevation index

4.2 Generation of Drainage Map

Rivers & Nalas not only provide a natural barrier to the spread of forest fire but inhibits or retards the risk of fire spread in an area surrounding it. Since percentage humidity around the water sources is comparatively higher than other places, incidences of fire risk are lesser than others. Though the study area has Bhagirathi (tributary of Ganga) as the main river, yet small tributaries with a span between (10m-50m) has been observed on the ground & digitized for the purpose of generating buffer of 100 meters. (fig. 9).

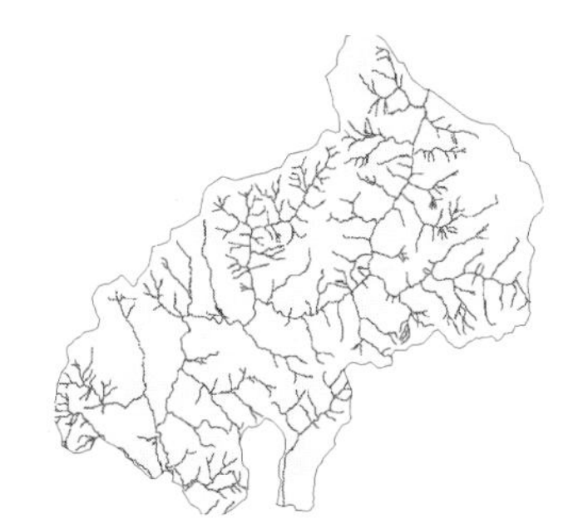

Fig. 8. Drainage map

Buffer Zone(in metres)	Fire risk values	Color index
0-100m	1	
Beyond 100m	3	

Table 7. Buffer zone along drainage

4.3 Distance to Road Map

Forest fires are ignited either by people or some natural agents such as lightning. The areas nearer to roads were considered more hazardous compared to others. Complete road network available was digitized & buffer was created for distance of 100m and 200m from the center of the road.

Fig. 9. Buffer zonation along the road

Buffer zone(in metres)	Fire risk value	Color index
0-200	4	
200-400	3	
Beyond 800	1	

Table 8. Buffer zone along roads

4.4 Habitation map

Position of settlements were marked & buffers generated around each village .(fig.10). The buffer with an radius of around 100m around the village was considered that of least risk as people will themselves control the fire as that is going to affect their field, fodder & dwellings. Buffer area beyond that was considered of higher risk. The two maps clearly indicate the position of villages more in the southern part of district compared to north.(fig.11 and 12). Presence of more settlements have been one of the major causes of fire outbreaks in the southern part of the district.

Fig. 10 Buffer zonation around habitation

Bufferzone(in metres)	Fire risk values	Color index
0-100	1	
100-400	4	
400-800	3	
Beyond 800	2	

Table 9: Buffer zones around habitation

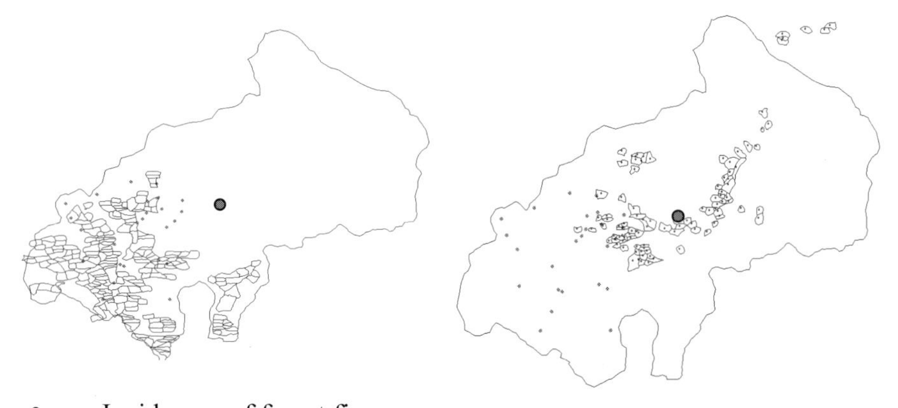

Incidences of forest fire

Uttarkashi City

Fig. 11: Villages and fire points to the south of district

Fig. 12: Villages and fire points to the north of district

Since there are usually many people in and around the forest and the probability that any of them will start a fire is very small, it is reasonable to use the Poisson distribution to model daily people caused forest fire occurrence. (Johnson and Miyanishi).

Fire risk zonation Index – Slope, aspect, drainage, accessibility, habitation, elevation and settlements were assigned different weight factors in order of their influence on occurrence of fire. Five such zones were created for the purpose of analysis. Maps of zone 2 and zone 5 were in close resemblance and fulfilled the criteria on ground. The criteria adopted was-higher the value, higher the risk. Intensive fieldwork justified the zonation on the ground. Road in some of the cases are closely following along the rivers or their tributaries, hence some areas have been misinterpreted as those of high risk even though they are close to rivers. This is because rivers have been assigned a weight factor of 2 whereas roads considered to be of higher risk to initiate fire , have been given a weight age of 8 . Post field knowledge has been applied to such areas and weigh factor assigned accordingly. Different parameters have been assigned weight ages in the order of 1-10 . This has been achieved by several such relations developed based on knowledge and experience about the study area and their verification on ground. The final fire risk zonation map(fig.13) generated has the following class and weights assigned to it (table 9).

The relation developed for Fire risk(FRZI) zonation index is:

$$FRZI= (5As+4El+5Sl+9Ft+Dr+7Hb+5Rd)/10,$$

Where abbreviations used are:
As-aspect,El-altitude,Sl-slope,Ft-foresttypes,Dr-drainage,Hb-habitation,Rd-road.

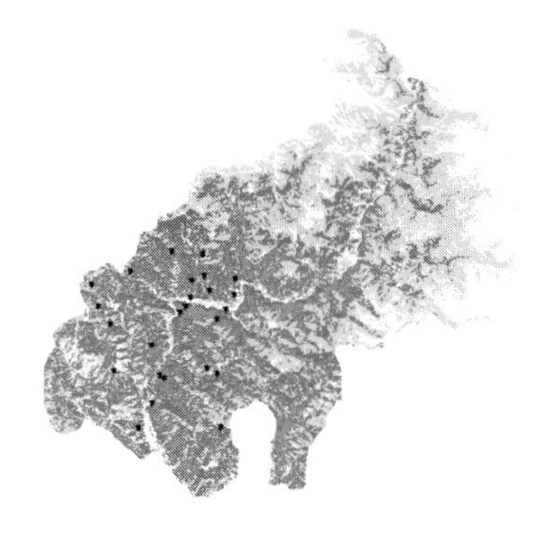

Fig. 13. Fire risk zonation map

Fire risk zonation index

No risk

Low risk

Moderate risk

High risk

★Fire points(incidences of fire reported in the previous season)

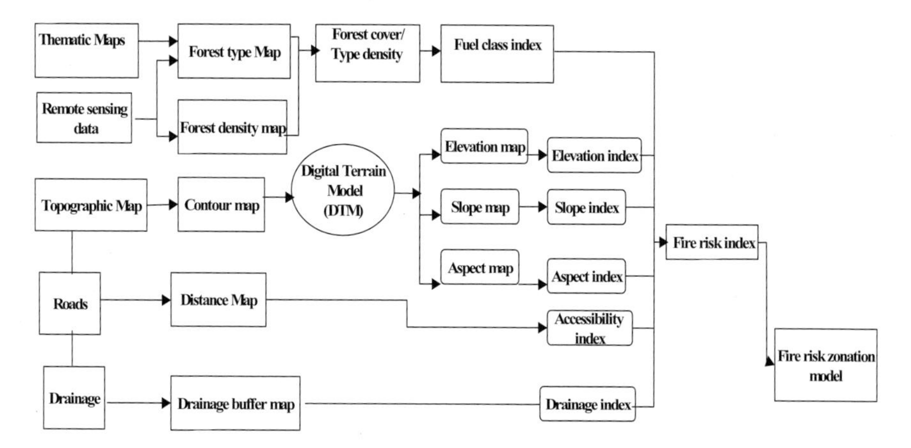

Class Name	Weight factor	Classes	Ratings	Fire sensitivity
Aspect	5	Northern	4	Low
		Eastern	5	moderate
		Western	6	high
		Southern	7	Very high
Forest type	9	Dense Pine(Chir)	10	Very high
		Open Pine(Chir)	8	Very high
		High altitude Conifers(Dense)	6	high
		High altitude Conifers(Open)	5	high
		Broad lived	4	moderate
		Scrubs and grasslands	3	moderate
Slope	5	<10°	6	Low
		11°-25°	7	Moderate
		26°-40°	8	High
		>40°	9	Very high
Altitude	4	<=2000metres	4	high
		2000-4000metres	3	moderate
		<4000metres	2	low
Road	5	within 200m	4	Very high
		200-400m	3	High
		>400m	2	moderate
Habitation	7	Within 400m	1	Low
		400-800m	6	Very high
		800-1200m	4	High
		>1200m	2	moderate
Drainage	1	Within 50m	1	Low
		>50m	2	high

Area under different Fire risk zones

Class	Area (in Sq.km)
Low	574.62
Moderate	742.88
High	490.23
Very high	246.27

Percentage area under difficult fire risk zones

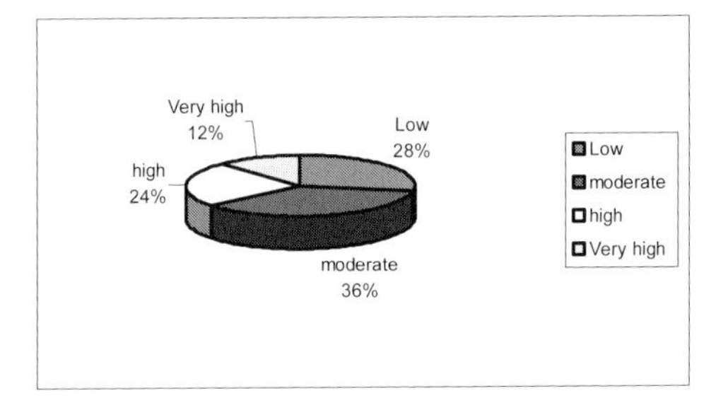

4.5 Results

The study shows that around 36% of the area is prone to forest fire. Forest types, aspect, slope and habitation have played an important role in zonation of different fire risk areas. Incidences of forest fires reported during the previous season are mostly coming under high and very high risk areas in the present study.

4.6 Implementation Mechanism

The study area is frequently under the threat of forest fire for several years. The problem is more severe during fire season. The district forest department is the main agency at the district level to take measures for prevention of fire. As a part of technical collaboration, forest survey of India(FSI) has provided the risk zonation maps to the foresters. Steps have been taken to create more watchtowers in the high-risk areas. Because of closed canopy, some of the forest fires are crown fires. Measures have been taken for generating fire lines to prevent the spread of fires to larger areas. Awareness program have been initiated at the district level so that the villagers take utmost care while visiting forest areas for fodder & firewood.

5 Conclusion

- The criteria developed helped in identifying the areas of fire risk in different categories.
- The model could be used to develop similar models for different topographic conditions, after assigning adequate weight factor.
- The integration of remote sensing data with GIS was successful in identifying zones of high fuel risk.
- iv) The incidents of fire have been observed more in the Pine(Chir) zone falling in the southern and the southwestern aspects. This is because of greater duration of insolation, more settlements and high resin value present in conifers.

References

Bahuguna VK Fire situation in India, MOEF, Govt. of India

Champion, Seth (1968) Forest types in India, Govt. of India Press, 1968, New Delhi

Forest Survey of India (2001) State of forest report,2001, MOEF, Govt. of India.

Hussin YA, Sharma N (2000) Forest fire hazard modelling using RS & GIS-a case study in Corbett national Park , ITC, Netherlands

HussinYA, Sharma N, Boon D (2000), Modelling Forest Fire Hazard using RS and GIS-a case study in Kali Konto, Indonesia. Nov., ITC, the Netherlands

Anon,Principles of Forest fire behaviour, International Institute for Geo-Sciences and Earth Observation(ITC), the Netherlands

Johnson, Miyanishi (2001) Forest fires, Academic press, USA

Lal JB (1989) India's forests, Myth & reality, Natraj Publishing House, Dehradun, pp 553-557

Wooster MJ Remote sensing of forest fires, Dept. of Geography, King's College, London

A Web GIS for Managing Post-Earthquake Emergencies

Matteo Crozi, Riccardo Galetto and Anna Spalla

DIET, University of Pavia, Via Ferrata, 1, 27100 Pavia, Italy.
Email: galetto@unipv.it

Abstract

This paper describes the results achieved within the framework of the national research project Reduction of the Seismic Vulnerability of Infrastructural Systems and Physical Environment, sponsored by the GNDT (Italian National Group for Defense Against Earthquakes), which is a body of the INGV (Italian Geophysic and Volcanology National Institute).

The project, which involved several Universities around Italy under the coordination of Prof. Michele Calvi of the University of Pavia, was carried out over three years and has been just completed in July 2004.

The aim of the project was to study the problems related to the seismic vulnerability assessment both of the physical environment and the infrastructural systems, to develop new techniques and tools, or improve the existing ones, to provide an efficient reduction of the risk.

One of the products of the project is the Web GIS, described in this paper, which has been applied to a test area in the South of Italy which in the past has suffered high seismic level events, to prove its efficiency and to examine the possibility of its implementation at national level.

1 The GNDT Project

The GNDT (National Group for Earthquake Defense) is a national body, which has the mission to coordinate the scientific research in the field of the reduction of seismic risk.

The realization of the Web GIS described in this paper has been carried out within the frame of the interdisciplinary research project *Reduction of the Seismic Vulnerability of Infrastructural Systems and Physical Envi-*

ronment, which faces the whole content of Topic 3 of the GNDT Framework Program in relation to the vulnerability of physical environment and infra-structural systems, like main road (transportation) networks, high voltage electrical lines, large industrial facilities and earth dams.

The general objectives of the project, which started in 2002 and has been completed in 2004, concern several aspects:

- the preparation of a territorial computerized data base, integrating all inventories of infra-structural environmental systems, linked with vulnerability and risk assessment computer programs.
- the development of integrated procedures for risk assessment for complex infra-structural systems, including all of the physical components, their interaction with the physical environment, the relevance of possible alternatives that can assure the system functionality, the significance of the induced damage;
- the development of relatively simple models for vulnerability assessment of system components, including an estimation of the variability of the considered parameters and of the consequent reliability of the results.
- the development and validation of vulnerability reduction methods based on innovative techniques, such as ground treatment, base isolation and advanced composite materials strengthening.
- an inventory of structures and systems examined, both in terms of networks of national relevance for which some sort of data base is available, and of case studies for which complete damage scenarios based on the outcomes of the projects will be produced.

To pursue these objectives researchers from many Universities around Italy were involved in the project; seven groups were established with the following tasks:

- *Task 1 – Seismic input and site effects*: refined definition of seismic input both in term of accelerograms and displacement and acceleration spectra, emphasis has been given to some fundamental aspects such as the structural effects of the non synchronous motion and surface effects; a final definition of the seismic input to be used in the test area has been provided;
- *Task 2 – vulnerability and physical environment*: analytical and experimental methods have been proposed to study the landslide and the retaining walls; catalogue software has been developed to manage related data;
- *Task 3 – GIS development:* realization of a Web GIS for managing all the data needed for the research activities and having specific functions

giving information to the Governmental Bodies in order to manage a post-earthquake emergencies;
- *Task 4 – Vulnerability of road networks*: deterministic and probabilistic studies to evaluate the bridges vulnerability and the impact of the bridges collapse on the road network; analytical tools have been implemented;
- *Task 5 – Vulnerability of electric networks*: analytical and experimental studies to determine the seismic vulnerability of the electrical network and develop isolation system to mitigate the risk
- *Task 6 – Vulnerability of industrial plants*: definition of the seismic vulnerability of the buildings and environmental impact due to partial or complete collapse.
- *Task 6 – Vulnerability of industrial plants:* definition of the seismic vulnerability of the buildings due to partial or complete collapse;
- *Task 7 – Vulnerability of earth dams*: assessment of the earth dams vulnerability and evaluation of the environmental impact.

In order to test the results achieved by the research group, a test area was chosen for practical application of new envisaged methodologies in the several field of study.

The chosen test area is localized at the northern margins of the Campano-Lucano Appenines in the south part of Italy. It is located between the towns of Benevento and Avellino. The centre of the zone is Pietrafusa and the radius of the zone of interest is roughly extended to all the Campania Region.

The Campano-Lucano Appenines is one of the sector of the appeninic chain with the most seismogenetic potential. The analysis of the macroseismic data associated to the main historical earthquakes and of the instrumental seismicity indicates that in the Appenines, the seismicity is concentrated in a band of nearly 30 km width, centered on the axis of the appeninic chain with a NW-SE direction. The directions of the actual stress field, determined by the analysis of the focal mechanism of the main earthquakes in the instrumental period ($M>4$, M meaning Magnitude) and of the deformation of the sections of the holes realized for oil exploration, indicate that the chain is subject to an regional and extensional stress regime of NE direction. The seismotectonic frame seems thus characterized by: a) main seismogenetic structures of NW-SE direction, with focal mechanisms of normal type, creating large earthquakes ($6<M<7$); b) secondary seismogenetic structures, localized at the margins of the main structures with an orthogonal direction with respect to the chain, with focal mechanisms of mainly strike slip type, generating moderated earthquakes ($M<6$). To the first category is associated as the source of the 1980 Irpinia earthquake (Ms=6.9), meanwhile the source responsible of the 1990 Potenza earthquake (Ms=5.4) may be associated to the second category. Moreover, as indicated by the hypocentral depths of either the main moderated/large earthquake, either the background seismicity, the appeninic seismic belt reaches 15 km depth.

In order to define the seismic source of interest for the area test, available macro seismic catalogues realized by previous GNDT projects have been consulted.

One of the main items of the project has been the determination of the seismic hazard assessment of the test area.

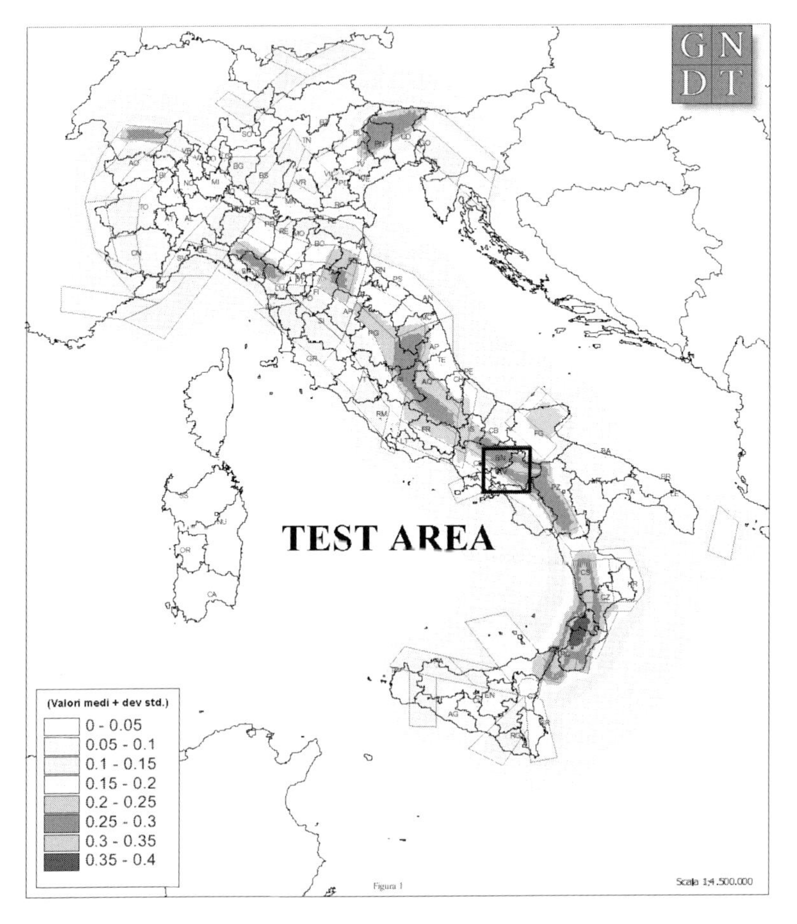

Fig. 1. Italy hazard map

The seismic hazard assessment generally is divided in two categories. The first one, known as probabilistic approach, is based on the definition of seismogenetic zones characterized by an homogeneous seismicity respect to the Gutenberg-Richter law. If an attenuation law is introduced, which gives for a magnitude/distance couple the probability to obtain some level of PGA (peak ground acceleration), the acceleration threshold, corresponding to a given probability of exceeding (generally 10%) for a given time period (generally 50 years), can be computed. A hazard map is available for the whole Italian territory.

Thus this type of approach needs limited information, but is often associated to a large standard deviation. To calibrate the method, the results of the simulation of a known earthquake (1980 Irpinia earthquake) close to the test area were compared with the observed data (fig. 1).

The second approach is deterministic. It is based on the synthetic computation of accelerograms from the definition of a kinematic source model. A large number of simulations are realized for a given site, choosing randomly the values of some source parameters, a priori unknown, as the nucleation location on the fault plane or the distribution phases of the final dislocation.

2 The Web GIS

The Web GIS, which has been realized within the frame of task 3 by the research group of Diet (Department of Building and Territory Engineering of the University of Pavia) represents the final product of the project.

Specifically, other than performing the typical functions of a Web GIS, which are to supply representations and information about a territory potentially involved in seismic events, the Web GIS has specific functions giving information to the Governmental Bodies in order to manage post-earthquake emergencies.

The Web GIS has been realized in ESRI environment based on the following components:

- Web server: IIS
- Servlet: Servletxect
- Map Server: ESRI ArcIms

The functionalities of the Web GIS are described in the following paragraphs.

2.1 Interface

The web page generated by map server is composed of 8 frames:
Top frame (1) , Tool frame (2), Map frame (3) , Toc frame (4), Text frame (5), Mode frame (6), Bottom frame (7), Post frame (hidden for the user); number enclosed in brackets refers to fig. 2.

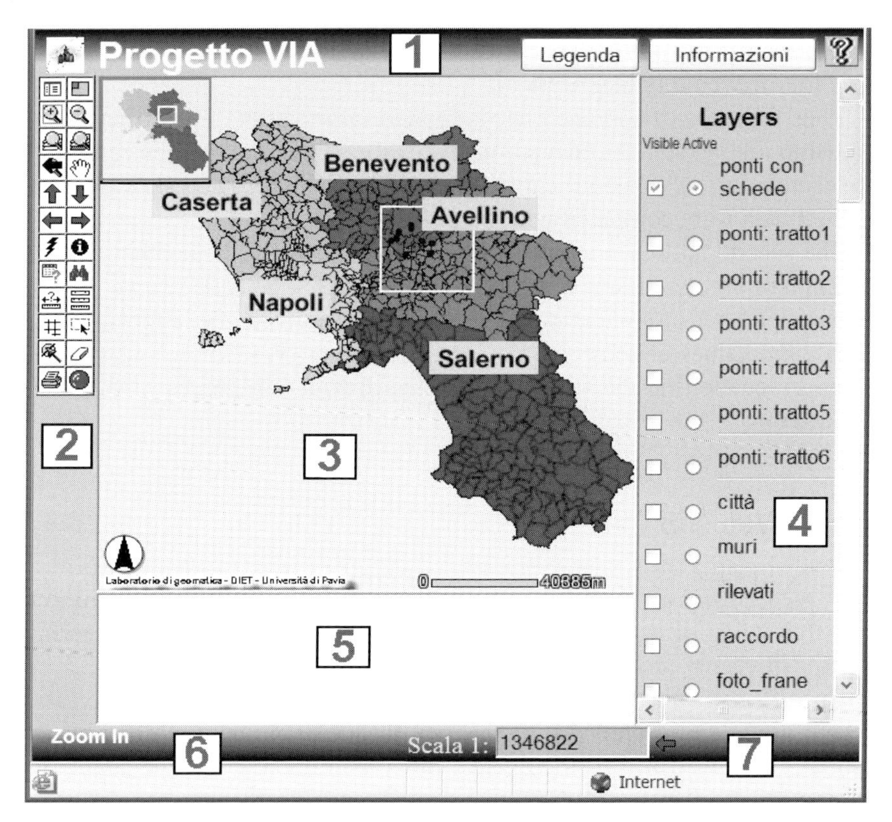

Fig. 2. Web GIS interface

Top Frame

The server for the Map Title is found in this area and next to it there is a link for its realization. Three buttons have been inserted on the right:

- *Legend*: it opens a window in which the complete legend in graphic form is shown
- *Information*: it opens, if available, an information window on the active layer

- *Help*: it opens a new window with indications regarding the basic functions of the site.

Tool Frame

It contains the instruments that characterize the functioning of Map Server; at present there are 2 columns with 24 buttons. These concern the use of the GIS functions, which are listed as follows, from top to bottom:

- *Toggle between Legend and Layer list*: alternates the visualization of the layer list (in the *Toc Frame*) between two different modes: graphic view, in which an image that shows the layers and their symbols can be inserted into the frame; module view, in which the html page inserted in the frame can be dynamically generated, permitting a choice of layers to be visualized or made active.
- *Toggle Overview Map*: alternates the visualization or not of the map overview inside the Map Frame.
- *Zoom In, Zoom Out*
- *Zoom to Full Extent*: extends the visualization of the map to comprehend all the layers present.
- *Zoom to Active Layer*: extends the visualization of the map to the extension of the layer that is active at present
- *Back to Last Extent*
- *Pan*
- *Pan to North*: predefined movement of the map in the direction indicated.
- *Pan to South*: predefined movement of the map in the direction indicated.
- *Pan to West*: predefined movement of the map in the direction indicated.
- *Pan to East*: predefined movement of the map in the direction indicated.
- *Hyperlink*: instrument concerning the visualization of information of various nature (images, text, multimedia) correlated to objects present on the map. It is a function available for the active layer and only if it has been associated to the desired material.
- *Identify*: It is a function available for the active layer and allows the visualization of those properties, in table form, that are associated to each object and are shown in the *Text Frame*.
- *Query*: this instrument is utilized to perform complex research on table data associated to the layers; once selected, a request form will appear in the *Text Frame* that must be compiled by the user: this request can be written directly in SQL format or using the wizard inserted in the form. The answer will be inserted in table form in the *Text Frame* and by se-

lecting the returned record it is possible to interact with the map (ex. zooming in on data).

- *Find*: similar to the above, but much simpler, with which it is possible to search for a word or a phrase inside the data.
- *Measure*: with this instrument it is possible to make direct measurements on the map, visualizing both simple distances and multiples of the defined unit of measure.
- *Set units*: it is possible to use this instrument to modify the predefined unit of measure of the map; and then use the measure or buffer.
- *Buffer*: function dedicated to the interaction between layers; after having selected one or more objects of the active layer (see the following two instruments) it is possible to verify the distance between these objects and the other vectorial layers. The map will be refreshed showing an area of the desired dimensions around the selected objects and in the *Text Frame* the objects of the desired layers which fall in the area of the buffer will be shown in table form.
- *Select by Rectangle*: an instrument to select the objects of the active layer, which fall in a defined rectangle defined by the user on the map. The map will be refreshed showing the selected objects with an appropriate symbol and the respective properties in table form in the *Text Frame.*
- *Select by Line/Polygon*: like the preceding instrument, but the selection of the objects can be made by their intersection with a line drawn by the user, or by their belonging to an irregular polygon defined by the user.
- *Clear Selection*: cancels the selections made using the preceding instruments and return the symbols to their original state by refreshing the map.
- *Print*: allows the creation of a print layout of the desired portion of the map, including the optional elements
- *Earthquake Simulation*: this instrument was entirely developed for this project and allows the user to simulate an earthquake and visualize the effects on the infrastructures situated in the test area. The first action visible upon selecting this instrument is a form for defining the principle characteristics of the earthquake to be simulated, in the *Text Frame*. In this form it is possible to manually insert the coordinates east, north, or as an alternative, to click on the desired point of the map with the cursor to automatically fill the 2 fields (a symbol will appear at the epicentre); the user must then insert the depth values (in meters) and the magnitude; and finally push the simulation button. At this point the server will send the answer in one of two ways: graphic or table. The first is constituted by a new map in which the constructions present in the visualization are

overlaid by an appropriate symbol that can be of three different types: one for the infrastructures that remain undamaged; a second one for the infrastructures which have sustained damage (e.g. for a bridge this means that it is passable only by vehicles under a certain tonnage); a third one for collapsed infrastructures, impeding their use (e.g. A bridge no longer passable). The answer table is constituted by a web page, in which a table gives a summary of the earthquake data and a list of the constructions present, their characteristics and the damage done. By selecting a construction on the table, the map will be refreshed, making an enlargement of the zone.

Map Frame

It contains the map image. On loading the web page, a map with an extension defined by the GIS manager is visualized. At each request of the user that involves the visualization of the map, the server refreshes this by sending a new image to the client with the desired characteristics (e.g. zoom).

Inside the frame it is possible to visualize the Overview in superimposition to the main map, the overview map may be created with the map server independently from the main one.

Toc Frame

It contains all the layers available within the selected map server. The vectorial type layers are identifiable by the 2 option buttons alongside them: the first one puts the layer into view and the second one makes it "active". Unlike the standard options, the map refresh is automatically performed when the right button of each layer is clicked. The definition of active layer is important, because most of the options offered by the map server are available only on one layer at a time, i.e. on the "active" one.

Raster layers are recognizable because they do not have a button to activate them. Some of the raster layers are not visible in all the scales, but only in specific ranges; this is to shorten the waiting time.

Text Frame

This frame is dynamically refreshed according to the tool used; for example, when using the "Identify" tool, this frame will show the information on a table that is relative to the section of the active layer selected by "a click" of the user, or it could show a form usable through the "Query" tool, etc.

Mode Frame

It shows the tool selected or the mode being selected by the pointer.

Bottom Frame

Usually used to fill and complete the web page from a graphic point of view, in this project it has been usefully modified by inserting a numeric scale control of the map into it; the value is dynamically updated and the chosen value can be inserted to show the map at the desired scale.

Post frame

This is a hidden frame, i.e. it is present with a dimension value of zero. The user cannot interact with it; its use is only for communication between the server and the client. At each request of the client, the html file contained in the frame is updated, to include every request in a special form, and sent automatically to the server in a "ArcXML" format. This frame is also used to receive the response from the server and forward the results to the right frame; for example on a Zoom command, the server will be asked for a new size for the map; it will answer by sending the new map in a graphic format which will be shown in the *Map Frame* instead of the previous one.

2.2 Layers

2.2.1 Raster Layers

All of the raster layers are linked to the related vectorial coordinate system.
1. Technical and thematic cartography
   - IGM sheets scale 1:100.000
   - IGM sheets scale 1:25.000
   - SGN sheets 1:100.000
   - 8 types of thematic maps
   - A map of accidents regarding data collected from the "Drainage Basin Authority", and/or provided by territorial organizations. This type of map offers a survey of the slide phenomena from a movement-type viewpoint.
   - Map of slide phenomena. In this map the slides are classified according to their movement type (rock fall, flow slide, etc.)

- Geo-lithological map. This map shows the geo-lithological features of the area through a colored representation
- Map of the slide intensity in function to the expected maximum velocity. This map shows a representation of the slide phenomena in the sample area.
- Map of the urban areas and infrastructures. This map shows a full view of the infrastructures and the urban areas that could be vulnerable to slide phenomena.
- Map of restricted areas, national and regional parks; this provides a representation of the restrictions that are present in an area regarding its archeological, architectural, artistic and environmental assets.
- Map of vulnerability shown by urban areas and infrastructures. This map shows the damage after a seismic event to: people, buildings, roads, railways, cultural and environmental assets, and industrial areas.
- Map of high-risk slide areas: this map indicates only the areas at high-risk of slide phenomena.

2. DTM and its elaboration
   - The DTM has been provided by the CGR enterprise of Parma; its extension is the same as that of the test area, and it has a resolution of 40m at ground level. Other maps were derived through the GIS software using this DTM data such as: aspect maps, hill shade maps, and slope maps which have also been used by other research groups.
3. Ortophoto
   - The complete set of ortophoto regarding the test area was provided by the CGR enterprise of Parma; they have a resolution of 1 m at ground level.

2.2.2 Vectorial Layers

- Borderlines of the communities of the Campania region: this layer shows the region Campania divided into its communities. This is a vectorial layer, extracted from the geographical papers provided by the ESRI enterprise using the GIS Arc Info 8 software
- Test Area: perimeter of the test area
- Road system: vectorial graph provided by Tele Atlas regarding the main roads of the Campania region
- Geological/hydro-geological layers: geological map, lithotechnical map, faults, forms of denudation by washout, forms of denudation by gravity, structural forms, slides, hydrology, geological sections

- Types of industrial plants with its layers: reservoirs, subsystems, pipelines.

2.2.3 Active Layers

Active layers are those that participate in an active way to the aim of the project, or in other words, those which allow the user to interact with the earthquake simulation tool to create various risk scenarios and to build a case study of the possible consequences of seismic events in the area.

The most significant layers in this category are contained in the ***Bridges*** layer (first layer in the *Toc Frame*), because, during this project, detailed studies were made on the vulnerability of bridges to verify the effect of an earthquake on them using strict criteria.

For bridges, specifically, due to their structure, two parameters were identified: the PGA of yielding and the PGA of collapse; these two values stand for the resistance of constructions to the stress induced by an earthquake. The earthquake discharges a force that is dependent on various factors (distance from the epicenter, geological characteristics, etc.); this force is then compared to the resistance of each construction. For every PGA value (lower, equal or higher than the reference value) a different reaction of the construction can be expected; it is represented on the map with different symbols, as previously stated under the dedicated tool.

3 The Earthquake Simulation Functionality

Two solutions of the problem have been envisaged: the first, which has been more deeply developed, refers to *deterministic* approach; the second one refers to a *probabilistic* approach.

3.1 The Deterministic Approach

The problem has been approached in the following way: given a specific area, at the moment in which the characteristics of a seismic event, that is magnitude and coordinates of the epicenter, are entered in the GIS, a devoted application allows the generation, in real time, of a series of maps and tabulations relating to the foreseen damage to the existing infrastructure in the area of interest. The effects of the simulated seismic event on the infrastructures are determined in relation to their vulnerability.

Special emphasis has been given to problem of evaluating the effects of a seismic event on the bridges and on the potential landslides in order to foreseen the situation of the road network.

The function that allows the simulation of the effects of an earthquake in a selected area works in the following way:

1. the function is activated using an specific button
2. the infrastructure layer is selected on which one wants to determine the effect of an earthquake
3. the parameters that characterize the seismic activity are then given: magnitude, depth of the epicenter, local situation of the terrain
4. choice of the planimetric position of the epicenter either by inserting its coordinates or by indicating its position with the cursor on the map displayed on the screen.

After having entered these data the software calculates the value of the PGA (peak ground acceleration) as it corresponds to each of the structures of the selected layer; for each structure this value is compared with the main damage and collapse PGA value stored in the database; the output consists in a map showing with different symbols the damaged structures and the collapsed ones (fig.3). Besides the effect of the simulated earthquake is also displayed in table format.

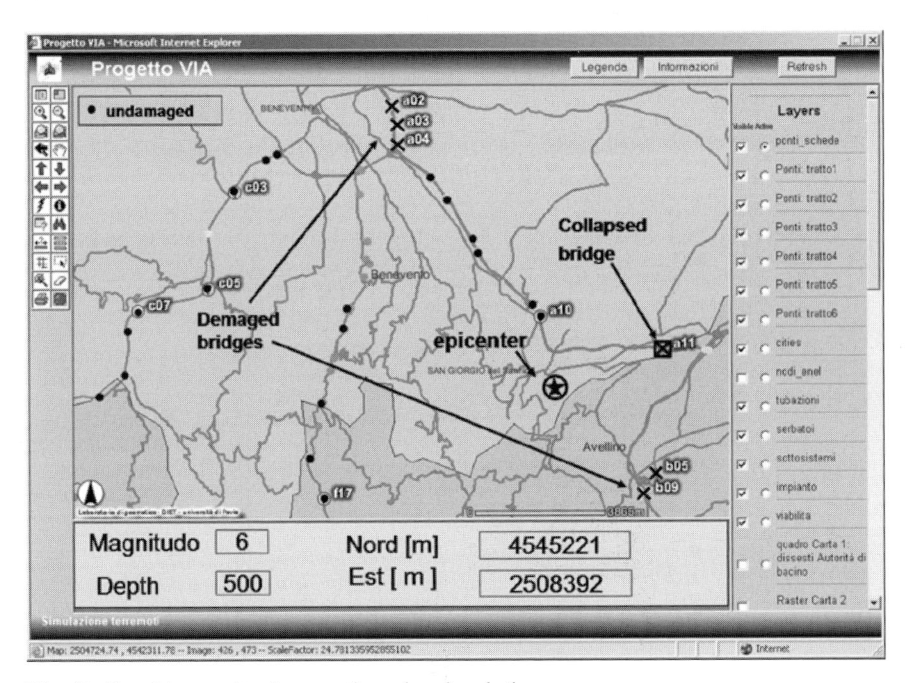

Fig. 3. Graphic result of an earthquake simulation

3.2 The Probabilistic Approach

The goal of the probabilistic approach was the following: given a road network involved in a seismic event, to compute the minimum and maximum value of the probability that two points of the road network are connected.

In order to allow the wide interactivity requested by this problem a second Web GIS has been realized in an environment different from the previous one; its characteristics are the following: Web server: Apache; Servlet: Tomcat; Map Server: Geovisio (ASI Mantova).

A functionality of the Web GIS allow the user to perform this actions:

- to select a certain number of road sections which constitute a network
- to select two points A and B in this network
- to select all the sets of minimum number of road sections which prevent the connection from A to B (cut sets)
- to compute the probability that a given seismic event, which characteristics can be entered by the user, can cause the obstruction of each cut sets
- to compute the minimum and maximum probability that A and B are connectable after the seismic event.

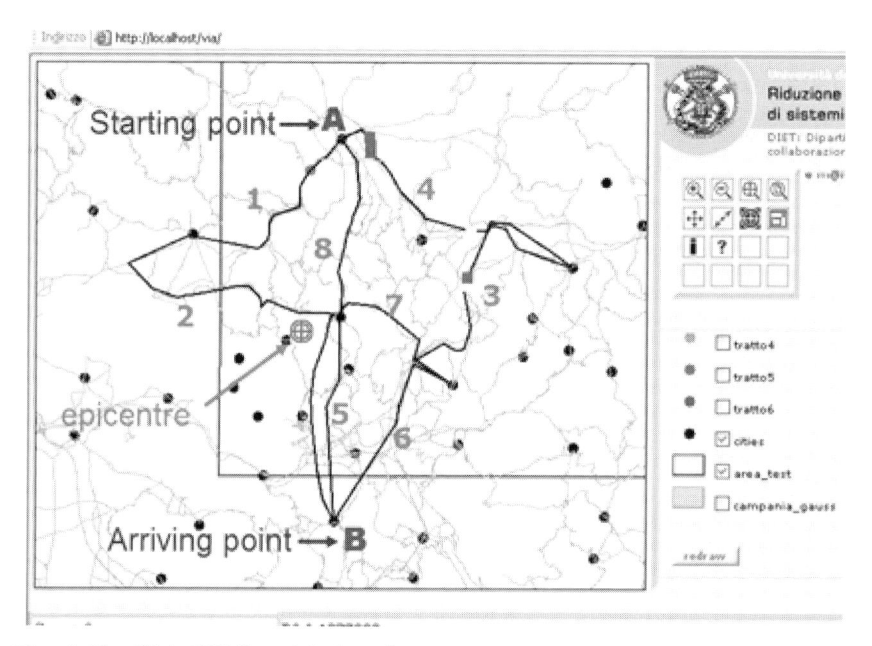

Fig. 4. The Web GIS Geovisio interface

This second Web GIS needs some implementation in order to achieve a better interactivity.

Terminology[1]

Attenuation Refers to the variation in ground motion severity with distance from fault rupture to site. Given everything else is held constant, the further the distance from fault rupture, the less ground motion intensity/severity experienced at site.

Deterministic model Refer to hazard models based on specific scenarios without explicit consideration of the probability of their occurrences.

Epicenter Refers to the earthquake's "point" source location on the earth's *surface* given in planimetric coordinates. It is vertically above the hypocenter. Earthquakes do not occur at a single point, they occur along fault ruptures, which are three-dimensional planes.

Hazard Refers to the frequency and severity of a threat inflicting losses on people, property, systems or functions. *Natural* hazards are those beyond the control of human beings, e.g., earthquakes, hurricanes, tornadoes, flood, brush fires, etc.

Hypocenter Refers to the focal "point" source location of an earthquake given in planimetric coordinates and *focal depth*. Earthquakes do not occur at a single point, they occur along fault ruptures, which are three-dimensional planes. Therefore, there are several definitions for hypocenters, the most common of which is the point of initiation of fault rupture. The hypocenter is located below the epicenter by the focal depth.

Magnitude Refers to the size of the earthquake and is a function of its energy release. Magnitude is an attribute of the earthquake itself. Earthquake magnitude is measured on many different scales; the Richter scale is most commonly used. Magnitude scales are logarithmic. A 1 increment on the magnitude scale is approximately 30 times more seismic energy release. In other words, it takes almost a *thousand* magnitude 6.0 earthquake to release the same energy of a single magnitude 8 earthquake.

PGA A measure of the ground motion severity experienced at site due to an earthquake. PGA attenuates with distance from fault rupture. PGA usually refers to the maximum *horizontal* acceleration measured at site.
A PGA of 0.2g means that the maximum *horizontal* acceleration is 20% of the earth's gravity. Since force is proportional to acceleration, which would mean that the earthquake generated horizontal forces equivalent to 20% of the structures weight *at its base*. PGA is used as a measure of input ground motion at the base of the structure.

---

[1] Courtesy of GEORISK®, Insurance, Risk Management & GIS Consulting - Santa Clara, California, USA).

Probabilistic model Refer to **hazard** models *taking into account the probabilities associated with many events affecting the region of interest.* Probabilistic models are based on simulations of hundreds or thousands of scenarios.

Vulnerability The susceptibility to losses due to exposure to hazard. Vulnerability reflects the extend of losses for any given hazard.

Risk It is a misnomer to use the phrase "Earthquake Hazard Reduction". For natural hazards, we can only attempt to reduce the risk not the hazard, either by controlling exposure to hazards or their vulnerability.

References

Bonazountas M (2001) Geographical city information via internet, GIM International, 58-61

Colombo L, Manara B (2001) WWW standardisation issues, GIM International, pp 45-47

Harder C (1998) Serving Maps on the Internet, edited by ESRI press. I e,m vol. 1 Redlands

Korte GB (1997) The GIS book, edited by Onword press. IV ed,. vol. 1 Santa fé.

Longley P, Goodchild M, Maguire D, Rhind D (1999) Geographical Information Systems. Vol. 1,2 -2 edition, John Wiley & Sons

Mooney P, Winstanley A (2001) Internet based transport maps, GIM International, pp 13-15

Pfund M (2002) 3D GIS architecture. GIM International, pp 35-37

Quintela J, Seco R, Salinas E (2001) GIS for natural environment mapping, GIM International, pp 44-47

Raper J (2000) Multidimensional geographic information science. Taylor & Francis

Van Elzakker C (2002) Cartographic visualization with ESRI's ArcGIS 8.1, GIM International, pp 40-43

Web Based Information System for Natural Hazard Analysis in an Alpine Valley

Constantin R. Gogu, Helen Freimark, Boris Stern and Lorenz Hurni

Institute of Cartography, Swiss Federal Institute of Technology ETH,
Hoenggerberg CH-8093 Zurich, Switzerland.
Email: {gogu, freimark, stern, hurni}@karto.baug.ethz.ch

Abstract

A platform for geospatial hazard and risk information system, comprising graphical and numerical geospatial data, aerial and satellite images, georeferenced thematic data, and real time monitoring feeds is being developed by the Research Network on Natural Hazards at ETH Zurich (HazNETH). It will allow researchers to build efficient systems for handling, pre-processing, and analyzing the existing large and variable datasets from different natural hazard phenomena as well as different natural environments in the Swiss region. The final product will be a geospatial hazard information system.

Three main steps will be followed in order to create this information system: the spatial database development, an integrated hazard procedure design, and a web enabled data query and visualization tool set. The geospatial database including the entire set of natural hazards phenomena occurring in an alpine valley will offer a platform to study existing hazard assessment methods and will allow the analysis and combination of various hazard parameters in relationship to phenomena. The final application based on the concept of Atlas Information Systems (AIS), will be used as an additional tool for risk and emergency assessment as well as for planning and decision making purposes.

1 Introduction

Switzerland has always been exposed to a wide variety of natural hazards happening most frequently in its alpine valleys (Fig.1). Recent events, such as those which occurred in Canton Wallis in October 2000 and included floods, debris flows and slope instabilities, or avalanches in February 1999, led to substantial loss of life and damage to property, infrastructure, cultural heritage and environment. Thus the need for an integrated natural hazard management and sustainable hazard prevention culture became obvious. Methods were developed and improved to identify areas affected by natural hazards, and parameters allowing quantifying static and dynamic impacts on structures in these areas were defined.

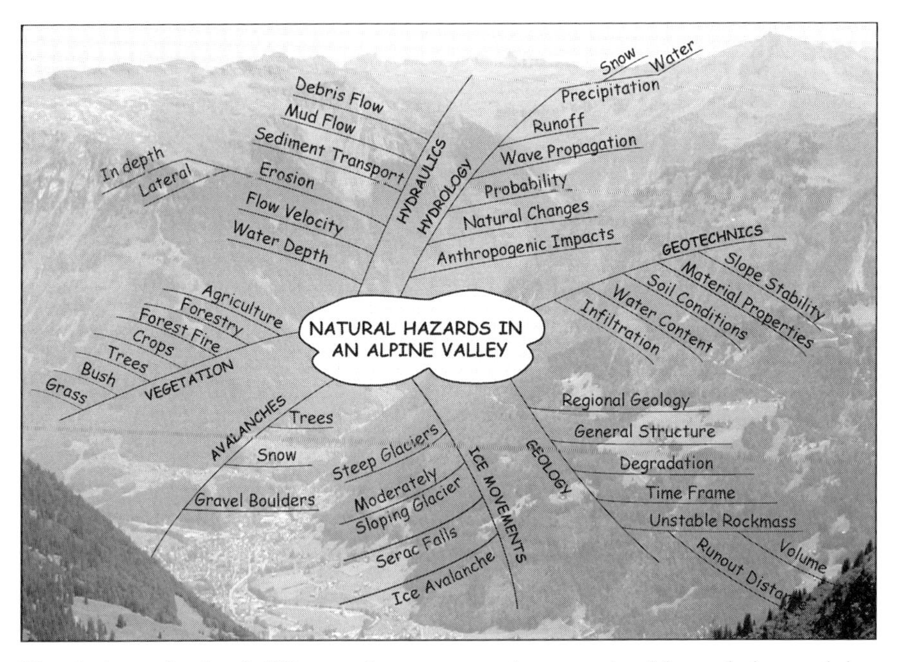

Fig. 1. A synthesis of different phenomena acting as natural hazards in an alpine valley

HazNETH is the Research Network on Natural Hazards at ETH Zurich and combines the expertise of several partner institutes: Atmospheric Physics, Climatology, Hydrology, Hydraulic Engineering, Water Management, Risk Engineering, Construction Engineering, Forest Engineering, Engineering Geology, Geotechnics, Seismology, Geodynamics, Geodesy, Cartography, Environmental Social Sciences, and Economics. HazNETH provides a platform for trans-disciplinary projects focusing on natural hazard research. It contributes to the improvement of methods and tools for integral risk management as a base for sustainable development.

The HazNETH project intends to develop a web browser enabled geospatial hazard information system (Fig. 2), which allows the project partners to share and analyze their datasets (HazTool). Apart from standard GIS functionalities, the system will provide expert analysis tools which are tailored specifically to the needs of interdisciplinary hazard research. Technically the system will be implemented as a multi-tier architecture: Internet client, application server with web server and spatial engine, and spatially enabled database.

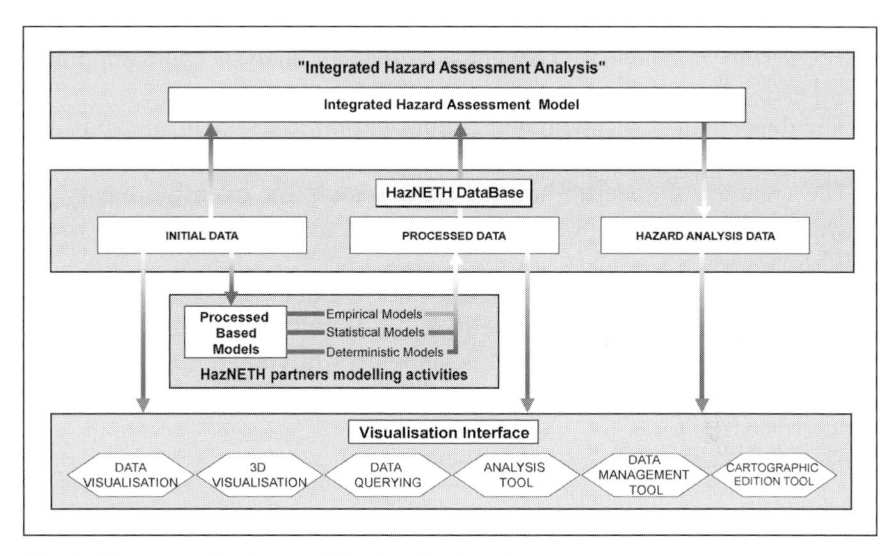

Fig. 2. The general scheme of the HazTool system

Apart from scientists, who are experts in their field, HazTool should also be used by decision makers such as emergency organizations, public authorities, and politicians. This requires a very flexible graphical user interface which lives up to the needs of the various user groups and allows the administrator to set different levels of access to the information. Experience with information systems or GIS expertise should not be a prerequisite for using HazTool as long the user is adept in the field of hazard research or management. It is the concept of the Atlas Information System (AIS), explained in Sect. 2.3, which will guide the design of such an "intuitive" user interface. Analysis results will be depicted in interactive two and three dimensional presentations and in time series, which allow the user to visually grasp correlations between different types of information. Building on the comprehensive information in the HazTool database, the project partners will be able to evaluate existing hazard assessment methods, to improve them, or to even discover new methods.

2 Geospatial System for Data Management, Modeling, Visualization, and Analysis (HazTool)

The conception and development of a web based information system for natural hazard analysis is carried out at the Institute of Cartography of the Swiss Federal Institute of Technology in Zurich (ETHZ). It is based on research in the following three areas:

1. The design of a database concept as a basis for analysis and natural hazard assessment method deduction (Sect. 2.1).
2. The development of integrated natural hazard assessment methods and data modeling tools (Sect. 2.2).
3. The creation of specific hazard analysis tools for interdisciplinary research as well as for specific research needs of particular partners (Sect. 2.3).

The following paragraphs describe the research goals and tasks that have to be completed in the above named research areas. The significance of the performed research for the development of HazTool is highlighted and the resulting technical implementation concept of HazTool (Sect. 2.4) is described.

2.1 Database Concept Design

Intensive cooperation between the project partners is a prerequisite for designing a common research platform. This implies that each partner contributes his datasets and knowledge to identify research foci. For this purpose each HazNETH partner delegated one or more responsible persons who discussed the natural hazards to be considered as relevant for Switzerland, as well as each partner's research interests and suggestions on how to map hazard related phenomena in a database. The parameters relevant to monitor each natural hazard phenomenon were identified and grouped thematically and known relationships between phenomena were documented.

Information gathered for natural hazards research is complex. Apart from managing very large interrelated datasets of different scale and spatial extent, important issues of combining hazards and their effects have to be solved. A well thought-out database concept and implementation will provide the basis for deriving new hazard assessment methods from intrinsic interactions between the studied hazards. Methods of quantifying hazard parameters as well as the uncertainty of such methods and of the datasets themselves have to be taken into account when designing the database concept.

2.1.1 Data Collection on Three Spatial Scales

Depending on the spatial extent and local distribution of natural hazard phenomena three spatial scales of data collection and analysis were identified (Fig. 3):

- General Level: Switzerland (country)
- Regional Level: Wallis (river basin)
- Local Level: (alpine valley)

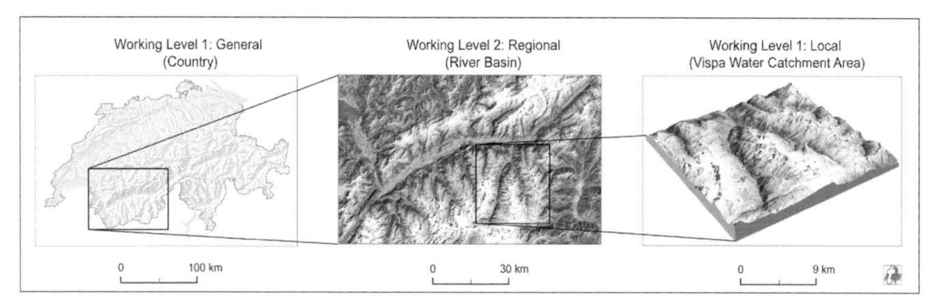

Fig. 3. Different spatial scales of data collection and analysis

The Local level (alpine valley) is targeting natural hazard phenomena occurring at a local level including landslides, torrent streams, debris flows, glaciers hazard events, etc. The datasets for this level were generally sampled at a higher resolution than for the other two levels.

The Regional Level datasets were collected for an entire hydrological system, a river basin, which roughly corresponds to the administrative boundaries of Swiss Canton Wallis plus the areas that are not part of Canton Wallis but belong to the river basin. The research focus on the regional level is directed towards natural hazards that concern the whole river basin (e.g. floods). The resolution of the collected datasets is generally lower than at the local level.

The General (country) level covers the administrative boundaries of Switzerland and to some extent the neighboring countries taking into account that natural phenomena do not respect manmade administrative boundaries. Phenomena like earthquakes or other various tectonic phenomena are observed and analyszd on such a scale.

2.1.2 Dataset Description

The information collected in the HazNETH project consists in geological, hydrological, geomorphologic, soil, climate, land use, and anthropogenic parameters and their geometry. Topological, photogrammetric, and geological information acquired from other organizations (e.g. SwissTopo, Natural Hazard Office of Canton Wallis) complement the datasets used with HazTool. The database concept requires these datasets to be organized thematically.

The datasets will be available to the project partners at different stages of processing: raw data, processed data, and project generated hazard data. Raw data is obtained through scientific measurements; processed data is received when applying different calibration and modeling procedures (process based, statistical, or empirical) of phenomena analysis to raw data. This step will be chiefly performed by each project partner with their specific modeling software. The treated datasets will then be made available to the other HazNETH partners in the database. The project generated hazard data regroups the results (maps or other output types) obtained through applying various hazard quantification methods. This group of datasets may be persistently stored and not generated by HazTool on-the-fly for reasons of system performance or lack of functionality integration into the system. At the current stage the raw data was structured and has been processed to some extent.

2.1.3 Data Models

In order to set up the database several data models were analyzed. The task was to design a data model that takes into account the special features of natural hazards of Switzerland, while permitting to direct the research focus at the beginning of the project on alpine valleys. This was a logical consequence of the completeness and detail of available datasets of the local research area (Vispertal including Saastal und Mattertal). It was also decided to direct research attention during the first two years of project duration primarily on the phenomena of torrent streams and debris flow.

Experience gathered during a similar project carried out at the Institute of Cartography of ETH Zurich (the GEOWARN project, Hurni et al. 2004) influenced the design of the database concept. Two other data models constituted sources of inspiration for the design process: ArcHydro data model developed by ESRI (Maidment 2002) for managing surface water resources and HYGES (Gogu et al. 2001) developed by University of Liege for managing groundwater resources.

The ESRI ArcHydro Data Model is used to restructure the base surface hydrology data to study torrent streams and debris flow phenomena. Using the ArcHydro data model with ESRI ArcGIS, it is possible to extract different themes from hydrological data (Maidment 2002): *Network* (showing pathways and water flow), *Drainage* (drainage areas and stream lines), *Channel* (three dimensional representations of river and channel shapes) and *Hydrography* (hydrographic features as found on topographic maps).

The ESRI ArcHydro Data Model has to be customized to the needs of the HazNETH project. Significant modifications are the adoption of a river cross section database scheme of the Swiss Federal Office for Water and Geology, the addition of a *Sedimentology* layer representing sediment volumes showing a potential to be eroded, relevant characteristics of *soil composition* (describing the soil layers, the ground covering, and the grain size distribution), and the addition of the *Ground covering layer* storing parameters like land roughness and erosion threshold.

2.2 Development of Integrated Natural Hazard Assessment Methods

Another research area deals with studying qualitative and quantitative hazard assessment for interrelated phenomena. The focus resides also with torrent streams and debris flows including related hazards such as soil and rock mass movements, and flood hazards. The main objective is to develop the scientific framework to derive new methods for integrated assessment of natural hazards in alpine valleys. Furthermore, a method of analyzing the input data uncertainty as well as the sensibility of spatial analysis reflected in hazard assessment procedures will be developed. Further details concerning this research area will be described in a different paper.

2.3 Web Based Information System for Natural Hazard Analysis

In order to create appropriate hazard analysis tools, an initial step was to organize several bilateral interviews and surveys with the various HazNETH partners. The following list describes the characteristics of the user groups identified during the interviews:

- The HazNETH partners: mostly experienced scientists with good GIS knowledge; their main interest is to combine their own information with the partners' information; they need access to all datasets with unlimited download/upload possibilities.

- The "Section dangers naturels" (Natural Hazard Office of Canton Wallis) and "Section des routes et des cours d'eau" (Division of Roads and Rivers of Canton Wallis): field specialists who work with the local council and natural hazards engineering companies for risk assessment; one of their main interest is to consult the HazTool system in order to manage their building planning permissions; they need access to all datasets with visualization/download permissions.
- Guest users: interested persons who will only be granted limited visualization permissions.

Based on the aforementioned discussions with the project partners, general objectives and priorities for the system were defined. The needs of the different user groups with none to expert experience in using computer systems need to be satisfied. Therefore, an easy to manipulate user interface should be designed to facilitate access to information, visualisation and to the use of analysis tools. The concept of the Atlas Information Systems (AIS) is suitable to guide the design of such a user-friendly system since AIS are state of the art for visualising and analysing predefined thematic collections of spatial data. They can be defined as “computerized geographic information systems related to a certain area or theme in conjunction with a given purpose – with an additional narrative faculty, in which maps play a dominant role” (Ormeling 1995, Van Elzakker 1993). The major difference to established Geographical Information Systems (GIS) is their ease of use and their cartographic quality.

Two other system priorities are on-the-fly creation of high quality web maps and integration of real-time data. Additional optional objectives are three dimensional modeling and cross-section/volume visualization of phenomena, as well as the ability to support modeling of standard decision chains of users (agencies, offices, administrations, etc). This will help to strengthen decision support systems in the field of natural hazards. The interoperability of the data, metadata, and web-services should be guaranteed through the implementation of standards (OGC and ISO).

Based on theses objectives, a set of specific natural hazard analysis tools for the HazTool system was identified, supplemented by standard GIS tools. A typical tool set could be composed by:

- panning and adaptive zooming tools on the map,
- map scaling tool,
- caption legend display,
- information queries on attributes and metadata,
- distance measurement,
- buffer zone tool,
- specific hazard queries combining interdisciplinary datasets,

- drill-tool that displays all natural hazards information available for a user specified point or surface,
- user account for saving temporarily the user defined map and settings,
- map layout and printing,
- a contextual help enabling the non-expert user to quickly understand the hang the different functionalities.

Nowadays, more and more unpleasant web based information systems are found on the Internet. Many do not respect basic cartographic rules and are therefore unattractive, unfriendly and sometimes incorrect. Visualization with the HazTool system will therefore be guided strongly by aspects of cartographic quality and graphical semiology.

2.4 Technical Implementation Concept of HazTool

With regard to the project objectives and expectations, the following decisions (Fig. 4) were taken when considering different technical implementation alternatives:

Dynamic map content, geodata, and hazard analysis services will be delivered via the web using ESRI ArcIMS. This software provides a highly scalable framework for geodata visualisation, analysis, and publishing which meets HazNETH's needs. Furthermore it supports layers, database management, client-side information entry, and geospatial information sharing. The choice of this product over an open source product or another commercial system was made along the following criteria: most of the HazNETH partners currently work with ESRI products, are therefore familiar with the software, and store their geodata in the corresponding formats. Through ETH Zurich the HazNETH partners have access to a campus licence for ESRI ArcIMS, ESRI ArcSDE, and ESRI ArcGIS products, as well as to customer support. No additional financial investment is necessary to use these products. The possibility of developing natural hazard custom applications, as well as the good reputation of the product, and its ease of use and administration made the ESRI software more attractive.

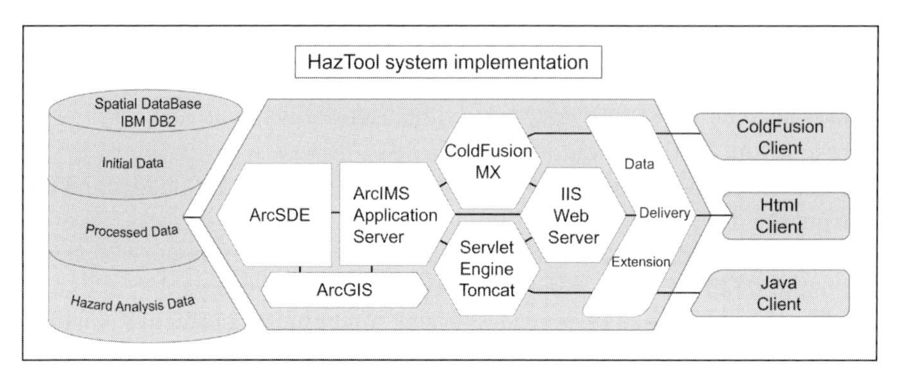

Fig. 4. HazTool system implementation

Complementary to ArcIMS, a special engine is needed to access and index geodata: ArcSDE is an ESRI server software product used to access massively large, multi-user geographic databases. Its primary roles are to provide a suite of services that enhance data management performance, but also to offer configuration flexibility, and to provide a spatial extension to the geodata.

As database management software IBM DB2® Universal Database™ with Spatial Extender was selected. This powerful multi-user geographic database was chosen according to its high performance, its capacity to store extremely large data volumes, and its compatibility with ArcSDE. Under the IBM Scholars license agreement, the database can be used free of charge for non-profit research.

In order to enable users to easily select, export, and deliver data in multiple formats and projections HazTool uses the ArcIMS Data Delivery extension. This tool allows users to upload/download geodata in 20 different spatial formats corresponding to industry standards and to project features to a variety of projections.

Each client technology proposed for ArcIMS has its respecting advantages and problems. To achieve maximum user-friendliness a benchmark will be realized in order to chose the best solution considering flexibility, performance, cartographic quality, and plug-in download necessity. At the moment, the planned client solution will probably be based on the ColdFusionServer© and SVG technologies.

3 Conclusion

Potential damage in Switzerland constantly increases because man moves into areas that are more exposed to natural hazards. The described research will provide a knowledge base and network with an easy-to-use web based client software for scientists, civil protection, and politicians when considering natural hazard prevention and protection.

The final software product will be easily adaptable to the purposes of federal and cantonal government agencies concerned with Swiss geotechnical and risk inventories, hazard assessment and disaster relief, or cantonal environmental offices. The software may eventually be made available to international and foreign national agencies, such as UNEP (United Nations Environment Program), or through Swiss foreign aid and disaster relief agencies.

This project is not only relevant for Switzerland and other countries of the European alpine region. Its results can also be transferred mountainous areas in general where similar problems with natural hazards are experienced (e.g. the Himalayas and the Andes).

References

Gogu RC, Carabin G, Hallet V, Peters V, Dassargues A (2001) GIS based hydrogeological databases and groundwater modelling. Hydrogeology Journal 9 (6), pp 555-569

Hurni L, Jenny B, Terribilini A, Freimark H, Schwandner FM, Gogu CR, Dietrich VJ (2004) GEOWARN: Ein Internetbasiertes Multimedia-Atlas-Informationssystem für vulkanologische Anwendungen, Kartographische Nachrichten, vol 54 (2), pp 67-72

Maidment DR (2002) Arc Hydro: GIS for Water Resources. ESRI Press, Redlands, California

Ormeling, F (1995) Atlas information Systems. Proceedings of the International Cartographic Conference ICA Barcelona vol 2, pp 2127-2133

Van Elzakker, CPJM (1993) The use of electronic atlases. In: Klinghammer I et al. (eds) Proceedings of the Seminar on Electronic Atlases

Using Explorative Spatial Analysis to Improve Fire and Rescue Services

Jukka M. Krisp, Kirsi Virrantaus and Ari Jolma

Helsinki University of Technology, Department of Surveying, Cartography and Geoinformatics, PO Box 1200, FIN-02015 HUT, Finland.
Email: jukka.krisp@hut.fi; kirsi.virrantaus@hut.fi; ari.jolma@hut.fi

Abstract

Within this paper we examine the use of explorative spatial analysis methods to enhance the calculation and representation of an emergency risk assessment for the Finnish fire and rescue services in the metropolitan area of the Finnish capital Helsinki (including Espoo, Vantaa and Kauniainen). Some of these methods are available in existing GIS (Geographic Information System) software. Within the strategic levels of planning and management, improvements in the identification of high or low risk areas can assist the emergency preparedness planning and resource evaluation. To enhance the determination of a risk area we visualize the spatial distribution of phenomena like population distribution, building types, workspace distribution etc. and their relevance to the emergency services. In this paper we examine the relation between population density distribution and incident density. To visualize potential variables indicating risk areas we use the third dimension. The results of the paper show that by developing explorative visualizations one can enhance the process of finding and integrating these variables into risk analysis. Within this paper we suggest innovative types of maps as tools for decision makers, which can be used to attract the public to participate in planning procedures.

1 Introduction

This paper is part of ongoing research at the Helsinki University of Technology (HUT), Institute of Cartography and Geoinformatics (ICG). The research focuses on the human vulnerability in built environments and considers issues like differences between common and rare accidents, commuting of people, and relations between accidents and networks. Natural hazard assessment and mapping is not a part of this work at this point. One case study in this research project investigates and enhances risk modeling in the case of fire and rescue services of the Helsinki Metropolitan Area. Risk model development should be an interactive and explorative process and include methods like visualization, spatial data analysis, and even more advanced spatial data mining methods.

Visualization and interaction help the user to understand the dependencies and the clustering within and between data sets. Visualization supports formulating hypotheses and answering questions about correlations between certain variables and accidents. Explorative visualization may reveal that some new variables are relevant to the model and it may also tell us more about the relevance of already used variables. If we want to analyze the correlations more deeply we should combine spatial data analysis methods with visualization.

To visually indicate clusters, if there are any, one can produce density maps. To quantify the correlations methods are needed, which estimate how strong the correlation is and how it is spatially dependent. Also the methods of geostatistics can be used as they reveal the strength of spatial autocorrelation between data sets. For example K-function has been used in spatial data analysis for analyzing the dependency of clustering between two data sets (Cressie, 1993). These methods can give answers to the questions: what kind of spatial correlation exists between given variables? We research these relationships to assist the development of a better model by finding and weighting the proper variables. In this paper we investigate correlations between population and incidents using explorative visualization of available data.

1.1 Risk Assessment in the Finnish Municipalities

In 1992 the Finnish Ministry of the Interior gave out guidelines on preparedness for municipal fire brigades stating the need to develop a systematic risk assessment practice. In these guidelines it stated that the preparedness in the fire brigades must be based on municipal risk analysis. Such a risk assessment should, according to the Guidelines, create a basis for setting the target level of the preparedness for emergencies in a municipality (Lonka, 1999). To assist the municipal fire brigades in making the risk assessments a handbook was published in 1994 in co-operation with the Ministry of the Interior and the Federation for Fire Brigade Chiefs in

Finland (Alliniemi, 1994). According to Lonka 1999 the risk is calculated for planning purposes as follows:

$$R = (L + F + P + E) * Pb$$

where

R is risk, L are the consequences for life and health, F is the rapidity of the development of accident, P are the consequences for property, E are the consequences for the environment, Pb is probability

The consequences can be deaths, injuries, property losses, interruptions and environmental damages. They are influenced by the rapidity at which the accident develops and progresses. These calculations give a very rough estimate of the risk and in practice only relative orders of magnitudes are evaluated. Application of this model is not straightforward but the distinct characteristics of each municipality make it very subjective.

In the Helsinki Metropolitan Area (which comprises the cities of Helsinki, Espoo, Vantaa and Kauniainen), the fire and rescue department focuses on protection plans for "normal" situations as well as plans for "extreme" situations (e.g. wartime). Their general interest lays in the enhancement of risk analysis for the "normal", i.e., everyday situations.

1.2 Integration of a Geographic Information System (GIS) in the Risk Analysis in the City of Espoo

To fulfill the risk assessment required, the Rescue Office of Espoo developed a model, which is now used in calculation the risk zones and associated service levels mentioned in the law (Ministry of Interior 2000). The existing model has been developed on the basis of expertise of people working in rescue services and on national statistics. No comprehensive spatio-statistical analyses were carried out about the variables of the model.

A GIS tool, developed and based on this model, can be used to determine the risk zones in any municipality. The analysis is similar and thus the results are comparable. These results can be visualized on a map and they can be used for, e.g., planning of a new fire station. The information is based on a grid map with a 250 x 250 m grid size. Each grid cell is classified into a risk level from 1 to 4. The levels define the time in which this area has to be reached by the rescue service unit (for example level 1 indicates that the location has to be reached in maximum of 6 minutes). The response time is based roughly on the behavior of a fire that develops in a building and on the maximum time for the start of resuscitation.

The GIS application in place is intended as a practical tool. The model identifies three variables that are considered in the risk analysis. Those are the population distribution data, the floor area (in square meters), and the amount of traffic. These variables are used in an automatic classification routine, which assigns a risk level to each grid cell. The result is displayed on a raster background map (Figure 1).

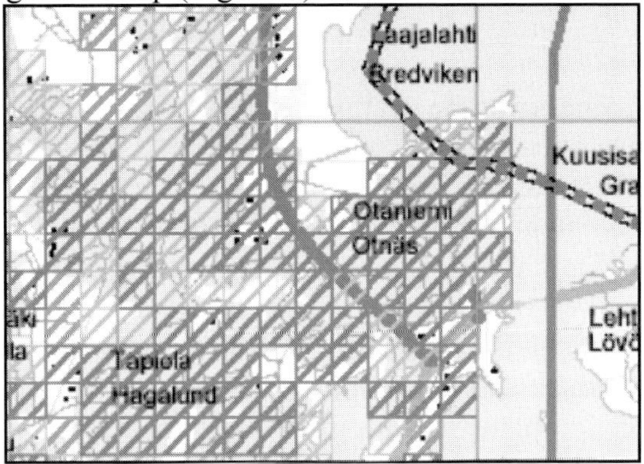

Fig. 1. Risk analysis in the city of Espoo (Ihamäki 2003)

The risk level is used in practical work so that the higher the risk level of an area the more rescue resources has to be allocated in or close to that area. In this work we try to improve the model and develop explorative visualizations to assist this process.

1.3 Explorative Visualization in Risk Model Development

As Peuquet and Kraak (Peuquet and Kraak, 2002) state, cartographic representation needs to change, as mapping has already changed. Research how maps can be best used as tools for exploring digital geographic databases and as interactive aids in experiencing the world, deriving decisions and solving spatial problems is needed. There is a need for development of tools for visually exploring spatial functional relationships.

In this paper, we concentrate on maps as visual components and separate them from the exploratory process and from the interaction between the user and the visualization. We study visualization of the relation between population density and incident density and apply this as a case where spatial datasets are combined and visualized in a three-dimensional environment. An overriding opinion amongst general users of modern visualization techniques is that much can be gained from interacting with innovative and dynamic graphical representations such as animation, three-dimensional maps or virtual worlds (Ogao, 2002). The utility and usability of visualization depend heavily on the user and the purpose of the visualization (Nielsen, 1993). Exploratory geovisualization tools have been investigated, e.g., by MacEachren, Kraak, Ogao and Miller (Kraak, 1999, Ogao, 2002, MacEachren and Kraak, 1997, Miller, 2000). Interaction and expertise are key issues for planners. An interpretation of abstract phenomena requires expertise that cannot be expected from either the general public or decision makers. Decision makers typically do not have the time for exploring situations but prefer hearing reasons for the proposed actions.

Visualizations may be animations or static images. The choice of the visualization, the orientation of the geographic space in relation to the viewer, as well as the selection of the spatial area portrayed, all similarly affect the overall message in these cases (Cosgrove, 1999).

2 Scope of this Paper

With this paper we aim to assist the emergency preparedness planning for the fire and rescue services by using a data driven approach. The first goal is to enhance the risk analysis carried out by fire and rescue services. Our methods are based on explorative visual analysis of existing data. By applying this method we try to answer questions, which arise when one is building a vulnerability model. For example, what variables are important and how important they are when? In this paper we illustrate this by using explorative visualizations of the population density and its relation to the incident data in the Helsinki region. The second goal is to provide more informative and appealing maps as tools for decision makers. Better maps can also attract the public to participate in planning procedures.

2.1 Explorative methods

Explorative methods, explorative spatial analysis methods and explorative visualization rely on interaction between a user and a computer and on the user's capability to find correlations between variables using visualizations. Explorative spatial analysis methods are needed for finding and evaluating variables for the risk model. Explorative visualizations are needed for determining the spatial relationships between these variables.

Variables can be based on a statistical analysis that considers the spatial distribution of the phenomena and its relevance to the emergency services (e.g. age distribution, building types, workspace distribution). By combining a hazard model that considers technological hazards in the build environment and vulnerability model that evaluates the build up environment (dependent on the hazard type) we aim to analyze the emergency risk. Finding and evaluating the proper variables is a difficult task and explorative spatial analysis or explorative visualization methods can be utilized to assist this. By modifying the input variables and data of the model we aim to investigate changes in the past or develop scenarios for the future.

To enhance the determination of a risk area we can visualize the spatial distribution of the phenomena and its relevance to the emergency services. By developing explorative visualizations we intend to enhance the process finding and integrating these variables into the risk analysis model. In the following we use the population distributions in relation to the incidents as an example case.

An innovative approach to be used in the explorative visualization is the use of the third dimension. Much of cartography's familiarity with three-dimensional landscapes originates from digital elevation models (DEMs), panoramas, sculpted physical models and orthographic globes. The fundamentals to design three-dimensional visualizations, using grayscale DEMs, vertical exaggeration and illumination are explained by Raper (Raper, 1989) and Petterson (Patterson, 1999) among others.

The possibilities to use the third dimension to visualize thematic data have been explored by Krisp & Fronzek (Krisp and Fronzek, 2003) among others. The use of the third dimension can aid the visualization of spatial datasets consisting of two thematic variables and allows comparing them more directly. It can also help to stress certain components of the information. Selecting and intensifying of specific parts of the information is especially important in the creation of thematic maps (Krisp and Fronzek, 2003).

2.2 Population and Incident Datasets

Generally the unequal quality of spatial data requires individual models and scales for a risk analysis in each community. Municipalities in Finland are obliged to gather register data on their population, buildings and land use plans. Due to the uniform character of the region it is important for the planners and decision makers to have reliable register data on the whole area, irrespective of these municipal boundaries. For this reason that Helsinki Metropolitan Area Council (YTV) has been working since 1997 on the production of a data package, SeutuCD, the data sets of which cover the whole metropolitan area. It is a data package gathered from the municipalities' registers. It is maintained once a year and includes register data maps for different scales and metadata software (YTV, 1999). In addition to the map of the region, the SeutuCD includes register data on buildings and land use plans as well as enterprises and agencies located within the metropolitan area. The population information is gathered by extracting the number of people registered in the individual buildings.

The Helsinki Fire & Rescue department has provided sample datasets, which contains all the fire alarm, rescue missions and also automated fire alarm systems missions within Helsinki city area for the years 2000 - 2003. The material also includes selected attribute information. That makes it possible to make different selections e.g. building block apartment fire missions, in certain time periods, in certain areas.

2.3 Determination of the Population and Incident Densities

Density analysis take known quantities of some phenomena and spreads it across the landscape based on the quantity that is measured at each location and the spatial relationship of the locations of the measured quantities (ESRI, 2004). The density of the population and accident incidents can be modeled using kernel density estimation. Kernel estimation was originally developed to obtain a smooth estimate of a univariate or multivariate probability density from a sample of observations (Bailey and Gatrell, 1995).

Given a data sample consisting of numbers of humans living in buildings and of numbers of incident reported at defined locations, the probability distribution of these variables at every location needs to be estimated. Replacing each point observation with a kernel, giving a "spatial meaning" in some sense to it, we obtain, as a sum, a continuous, smooth surface for the variables. The kernel is defined as a two-dimensional function with two parameters: the width of the kernel and the bandwidth. In practice only the bandwidth, which represents the search radius, is adjusted. The density λ at each observation point s is estimated by

$$\lambda(s) = \left\{ \sum_{i=1}^{n} K_h(s - s_i)x_i \right\}, s \in U$$

where K is the kernel and h the bandwidth (Silverman, 1986). The bandwidth (h) for the population density, as well as for the incident data was set to 100 m.

Selecting an appropriate bandwidth is a critical step in kernel estimation. The bandwidth determines the amount of smoothing and defines the radius of the circle centered on each location, which contains the observation points that contribute to the density calculation. In general, a large bandwidth will result in a large amount of smoothing and in low-density values, producing a map that is generalized in appearance. In contrast, a small bandwidth will result in less smoothing, producing a map that depicts local variations in point densities (Bailey and Gatrell, 1995). For a sample area with a scale of 1:20.000 a 100 m search radius was found to be suitable for the purpose of this study.

2.4 Results - Visualizing the Significance of Population Density in relation to the Incidents

The population and incident densities are calculated for the Helsinki city centre area. The map has a scale of 1:20.000. Figure 2 shows the population density for Helsinki in a color-coding from green, indicating a low density, to yellow, orange and red, representing a high density. Color figures can be viewed on the website:
http://www.hut.fi/Units/Cartography/research/rem/index.html. The incidents are visualized in the same way for the same sample area.

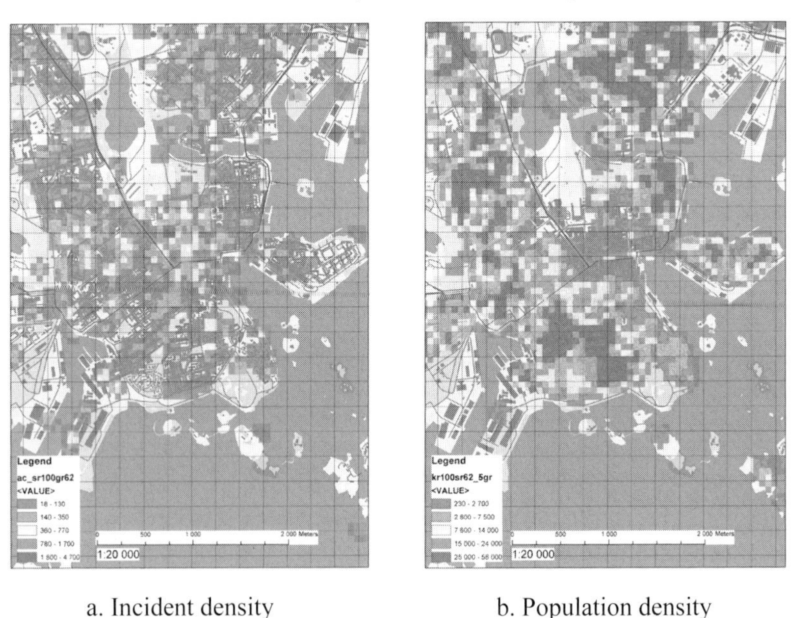

a. Incident density b. Population density

Fig. 2. Incident density (a) and Population density in Helsinki center area (b)

By comparing the to representations visually we can identify that the areas with a high population density and hot spots for incidents. The visual comparison of the two datasets indicates that areas with a high population density do in fact not correspond with areas with a high density of reported incidents.

2.5 Visualizing the Incidents Density by Using the Third Dimension

For the three-dimensional model we use the same input data as for the two dimensional maps (in Figure 2). The incidents are visualized in a similar way for the same sample area. Figure 3 illustrates the population and incident density in the third dimension.

a. Incident density

b. Population density

Fig. 3. Incident (a) and Population (b) density visualization using the third dimension

In the three dimensional maps, hills indicate a high while valleys a low density. Additionally a color coding from yellow, indicating a low density, to orange and red, representing a high density is applied. In both maps the thematic layer (population and incident density) apply a thirty percent transparency to show the underlying background map.

2.6 Explorative Four-Dimensional Visualizations

A four-dimensional visualization integrates the density information at each location using a z variable over time. We generate three-dimensional maps for all available years (2000-2003) and combine them into an interactive system illustrated in Figure 4.

Fig. 4. Animation and interactive three-dimensional map showing the incident density

The interface is easy to use, allowing the user to navigate between the different years. By clicking through the different maps for each year at various speeds the user can visually explore the changes in the incident density over time.

3 Conclusions

The case study has shown that via explorative visual analysis it is possible to review the significance of a variable in is context for the risk analysis. Population density is still considered a key variable for the vulnerability model. We don't want to give up on the hypothesis that people cause the main risk of an incident and that population is the most valuable variable in a risk analysis. The visual comparison between the population density data and the incident data shows that the connection between the population density and the incidents reported by fire and rescue services are not as strong as assumed.

Visualizing the results is a curtail point in getting the message across. A four dimensional animation can help to visually analyze the density patterns over time. Integrating the maps into an interactive system makes it easy to navigate between the different maps created for each point in time. In our example the time span from 2000 to 2003 appears to be too short to identify significant trends in the development of the incidents. Further research should aim to integrate data from a longer time period. The Fire & rescue services will continue to record incidents, so future research will be able to consider longer time spans.

By using explorative visualizations we enhance the process of transforming data into information and subsequently into knowledge. The exploration and interpretation of densities requires informative and appealing types of visualizations. Using three-dimensional maps can result in an appealing, maybe flashy visualization. However, the usefulness and usability of these combined three-dimensional maps needs to be proven.

To achieve the support of a proposed planning action it is necessary to get the attention of the public. In spatial planning very often maps are used to demonstrate the intentions of a plan. We suggest, that informative and appealing types of maps are essential tools for decision maker, to be used additionally to attract the public to participate in planning procedures (e.g. to justify the location of a new fire station).

3.1 Further Research

Until this all these methods are based on the user defined variables, but how sure we can be that our assumptions about the variables are corrects? If we want to go further in our analysis of finding the correct variables the methods of data mining and spatial data mining can be used. Data mining is the process of identifying or discovering useful and as yet undiscovered structure in the data (Fayyad and Grinstein, 2003). Spatial data mining, researched by Koperski, Han, Josslin and Ester among others(Koperski and Han, 1999, Josselin, 2003, Ester et al., 1999), is still in development stage, so no commercial tools are available. Based on the research made and the first implementations of software prototypes we can see that for example the methods like association rules can be useful. Using spatial data mining means that no hypothesis is available about the important variables that might relate to the amount of incidents. Data mining is providing knowledge about the potential variables. This knowledge can then be used for definition of the model.

The variables can be enhanced (e.g. by the integration of housing types and / or work places). Based on the risk grid it might be possible to integrate more detailed, specific target areas (e.g. stadium, airport) or events (concerts, ice hockey game times). These areas could be based on the fire inspection database of each building (which area available also nationwide).

Further research has to consider individual incidents recorded in the data and the time of each incident in relation to the time of the population density calculations. The integration of a time variable (daytimes / nighttimes) seems to be essential when relating the vulnerability model to the hazard model. More on the determination of the population density for the fire and rescue service is documented and research is still ongoing (Krisp, 2004, Krisp et al., 2005). Special events (like a hockey game) may result a temporary high risk in a certain area. Subsequently the amount of fireman on one shift can be based on a risk analysis. Up to now the shifts are based on the practical knowledge of the fire officers (e.g. more staff on new years eve). The time planning for the fireman shifts could be based on the risk analysis changing over time and simulations can integrate the spatial development over time and develop scenarios for the future (e.g. changing population density).

To acquire a better risk area model it is possible to add the individual risk evaluation for every building into the model. Subsequently the risk analysis and plans for extreme situations (e.g. war, national crises) can be linked to the "normal" risk analysis.

References

(Ministry of Interior) (2000) Kunnan Pelastustoimen Palvelutasoa Koskevat Päätökset, in Dnro SM-1999-000939/Tu-31
Alliniemi J (1994) Uhat ja mahdollisuudet – tapa tutkia onnettomuuksia ja niiden vaikutuksia
Bailey TC, Gatrell AC (1995) Interactive Spatial Data Analysis, Addison-Wesley, Reading
Cosgrove D (1999) Introduction: Mapping Meaning, in Mappings, (Ed. Cosgrove, D.) Reaction Books, London
Cressie N (1993) Statistics for Spatial Data, Wiley, NewYork
ESRI (2004) ArcGIS 9.0 Manual
Ester M, Kriegel H-P, Sander J (1999) Knowledge Discovery in Spatial Database, in Proceedings Invited Paper at 23rd German Conference on Artifical Intelligence(Ed. Institute for Computer Science, U. o. M.) Bonn, pp 13
Fayyad U, Grinstein GG (2003) Information Visualization in Data Mining and Knowledge Discovery, in San Fransisco, pp 1-17
Josselin D (2003) Spatial Data Exploratory Analysis and Usability, Data Science Journal, 2, pp 100-117
Koperski K, Han J (1999) Data Mining Methods for the Analysis of Large Geographic Databases, Simon Fraser University, Burnaby, pp 4
Kraak MJ (1999) Visualization for exploration of Spatial Data, International Journal of Geographical Information Science, 13
Krisp JM (2004) Kernel density maps for the analysis of spatial features, Helsinki University of Technology, Helsinki, pp 15
Krisp JM, Fronzek S (2003) Visualising thematical spatial data by using the third dimension, in Proceedings ScanGIS(Eds, Virrantaus, K. and Tveite, H.) Espoo, pp 157-166
Krisp JM, Henriksson R, Hilbig A (2005) Modeling and Visualizing population density for the fire and rescue servides in Helsinki, Finland, in Proceedings International Cartographic Conference A Coruna
Lonka H (1999) Risk Assessment Procedures Used in the Field of Civil Protection and Rescue Services in Different European Union Countries and Norway, SYKE, Helsinki
MacEachren AM, Kraak M-J (1997) Exploratory Cartographic Visualisation: Advancing the Agenda, Computer & Geosciences, 23, pp 335-343
Miller HJ (2000) Geographic representation in spatial analysis, Journal of Geographical Systems, pp 55-60
Nielsen J (1993) Usability Engineering, Academic Press, New York
Ogao PJ (2002) Exploratory visualization of temporal geospatial data using animation, ITC, Utrecht
Patterson T (1999) Designing 3D Landscapes, in Multimedia Cartography, (ed Gartner, G) Berlin

Peuquet DJ, Kraak, MJ (2002) Geobrowsing: creative thinking and knowledge discovery using geographic visualization, Information Visualization, 1, pp 80-91

Raper JF (1989) Key 3D modelling concepts for geoscientific analysis, in Three dimensional applications in GIS, (ed Raper) London, pp 215-232

Silverman BW (1986) Density estimation for statistics and data analysis, Chapman and Hall, London

YTV (1999) Establishments in the Metropolitan Area 1999, Helsinki Metropolitan Area Council (YTV), Helsinki

The Global Terrestrial Network for River Discharge (GTN-R): Near Real-Time Data Acquisition and Dissemination Tool for Online River Discharge and Water Level Information

Thomas Maurer

Global Runoff Data Centre (GRDC) in the Federal Institute of Hydrology (BfG), Am Mainzer Tor 1, 56068 Koblenz, Germany.
Email: thomas.maurer@bafg.de

Abstract

Today, many countries operate national near-real-time water level or river discharge transmission schemes. Increasingly, countries also publish this data online, typically by way of web sites of their national hydrological services (NHS). Though this is a major step forward, from a global perspective the diversity of data sources is still quite heterogeneous. Consequently, it still remains a tedious task to draw together all information needed for global assessments and models. Here GRDC aims at providing an additional service, called Global Terrestrial Network for River Discharge (GTN-R). The basic idea of the GTN-R project is to draw together the already available heterogeneous information on near real time river discharge data provided by individual National Hydrological Services and redistribute it in a harmonized way. GRDC has identified a priority network of 377 river discharge reference stations that constitute the first application network for GTN-R. The core of GTN-R is a software that collects near-real-time (NRT)-discharge data from distributed servers in the internet, harmonizes and summarizes it, and makes it available again in one standard format via a FTP-server. On the mid to long term, it will be necessary to overcome such tedious approaches and to arrive at internationally agreed standards for the exchange of metadata and data on measurements of geophysical and biogeochemical processes in general.

1 Introduction

An inevitable prerequisite for the sustainable management of the complex earth system respectively parts or sub-systems of it (including disasters) is unrestricted access to sound and comprehensive data and information on the state variables and fluxes of the governing processes which we try to mimic in computer models of ever growing complexity and refinement.

Besides the extension of operational monitoring and observation networks itself, there is the urgent need for the development of a more general, globally standardized data infrastructure ensuring time saving, highly automated access to the huge variety of observational data. However, authority over data and information, especially in the terrestrial domain is often scattered regionally and sectorally, resulting in highly fragmented approaches to their management and involving problems of scientific, technological, political, organizational and financial nature. Consequently, researchers and managers striving for integrated approaches including the development of indicators are on the horns of the dilemma of either spending too much of their valuable time on searching, retrieving and organizing fundamental data (which, at a large scale, is a non-trivial task for which they typically are not optimally trained) or alternatively omitting relevant information, both being unprofessional approaches that ultimately lead to stagnation in the development of suitable solutions. Though the question of how to cope with the challenges of the earth system's future stands high on the agenda of international organizations and consequently related meetings are mushrooming all around the world, yet an overarching rigorous approach aimed at tackling the fundamental data organization issue is pending, though it might evolve from recent activities such as Global Earth Observation System of Systems (GEOSS), Implementation Plan for the Global Observing System for Climate in Support of the UNFCCC (GCOS-IP, GCOS 2004a) and Infrastructure for SPatial InfoRmation in Europe (INSPIRE, EC 2004).

This is especially true for the field of hydrology and more specific for measurements of river discharge. Today, many countries operate national near-real-time water level or river discharge transmission schemes. Increasingly, countries also publish this data online, typically by way of web sites of their national hydrological services (NHS). Though this is a major step forward, from a global perspective the diversity of data sources and their management is still quite heterogeneous. Consequently, it still remains a tedious task to draw together all information needed for global assessments and models. Here, the Global Runoff Data Centre (GRDC) aims at providing an additional service, termed Global Terrestrial Network for River Discharge (GTN-R).

2 Framework

The GRDC is the digital world-wide repository of river discharge data and associated metadata, mandated by the World Meteorological Organization (WMO, cf. WMO 2001a, 2001b, GCOS 2004b). GTN-R is a GRDC contribution to the Global Terrestrial Network for Hydrology (GTN-H), which in turn is a joint effort of WMO's Hydrology and Water Resources Programme (HWRP), the Global Climate Observing System (GCOS) and the Global Terrestrial Observing System (GTOS). The GTN-H is a global hydrological "network of networks" for climate that is building on existing networks and data centers and producing value-added products through enhanced communications and shared development (cf. Fig. 1).

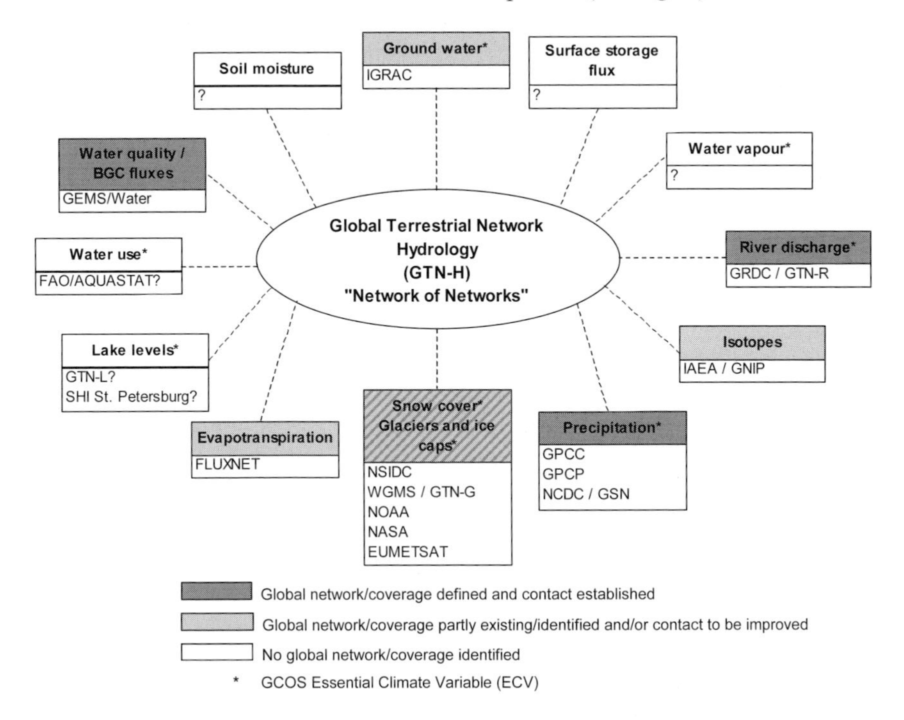

Fig. 1. The structure and status of the Global Terrestrial Network for Hydrology (GTN-H). Acronyms denote organizations that already deal with the respective hydrological variables. Question marks denote absence of a definite commitment.

The goal of the GTN-H is to meet the needs of the international science community for up-to-date hydrological data and information to address global and regional climate, water resources and environmental issues (GCOS 2000, 2002, 2003a, 2003b).

3 Problem

In general, data are used by other groups and organizations then those who collected and stored it and who are providing it. Depending on the variable the quality of the communication between these at least two groups are more or less developed. Moreover, the quality of this communication in general highly depends on the spatial scale, i.e. while there might be a well developed infrastructure in place already on a regional, national or even continental scale, this might not be the case on a larger, especially on the global scale. As illustrated in the introduction, this is especially true for the variable river discharge. Fig. 2 schematically illustrates various modes of communication between data providers and clients, i.e. data users.

In absence of any organization the situation resembles the anarchical structure as depicted in Fig. 2a. Each client has to individually contact every single data provider and, after establishing consent with regard to data exchange, has to deal with each data provider's individual interface (If) and to implement an access routine for each of them. The ineffectiveness of this approach is evident and illustrated by the $m \times n$ connecting lines in Fig. 2a. Plenty of work and communication is spend redundantly, which quickly yields frustration and stagnation as the number of involved data providers and clients increases. The problem is even worsening by the typical feature of such a setting, that data providers continue to change their interfaces without necessarily informing their clients.

A good step forward is achieved with the introduction of a network service as illustrated in Fig. 2b. While the network service still has to deal with the peculiarities of all the data providers, the clients are already relieved from the burden of communicating with every service and the total communication load is reduced by almost a factor of m (the number of clients involved). For each client, the communication load is already reduced by a factor of n. A further improvement is the introduction of a smaller number of standardized interfaces at the data provider end, as depicted in Fig. 2c. The work load of the network service reduces, which is indicated by the reduced size of the corresponding box.

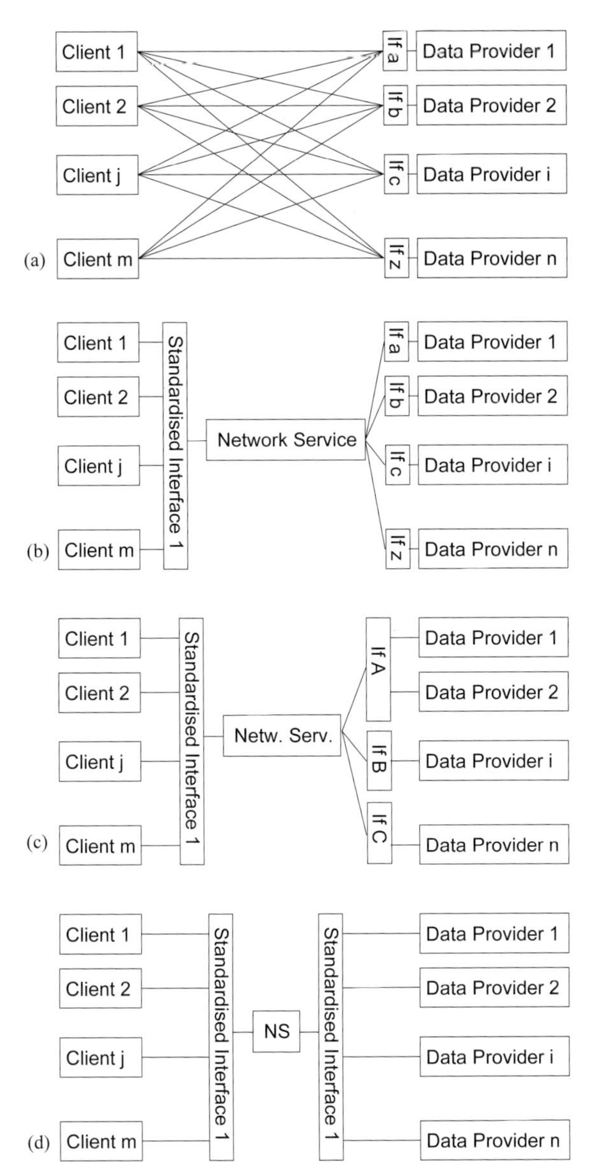

Fig. 2. Schematic diagram illustrating different types of communication network structures, (a) ad hoc communication, (b) client-side standardized intensive network service supported communication, (d) fully standardized network service communication, and (c) hybrid of (b) and (d). Further explanation in the text.

Finally, the introduction of a general international standard as shown in Fig. 2d further reduces the work load of the network service, which in this case will easily be completely automated and become an almost invisible part of the communication infrastructure, as indicated by a further size reduction of the corresponding box.

With regard to *near real-time* river discharge data on a global scale the current situation is much like depicted by the ad hoc communication model in Fig. 2a, where the *n* data providers are basically the National Hydrological Services (NHS) of the world or the large River Basin Authorities (RBA). Looking at the exchange of *historical* river discharge data, the GRDC has been acting as a Fig. 2b-type client-side standardized intensive network service since its establishment in 1988 (see also WMO 2001a, 2001b, GCOS 2004b). The long term ultimate goal for both, historical and near real-time discharge data, is certainly arriving at a fully standardized network service communication as illustrated in Fig. 2d (see also: Maurer 2004), however this will be depending on the development and universal implementation of international standards, which will take its time, though encouraging efforts are currently going on (with regard to activities as e.g. GEOSS, GCOS-IP and INSPIRE). For the time being, smaller steps have to be taken to improve the situation in the short to mid term. GTN-R is an initiative to disburden communications related to the access to near real-time river discharge data in a follow-up step by establishing a simple, robust, flexible, modular and extendible Fig. 2b-type client-side standardized intensive network service requiring minimum technical prerequisites.

4 Beneficiaries

GTN-R is an infrastructure, in general suitable for managing arbitrary networks of near real-time gauging stations. Currently, GRDC has identified a priority network of around 377 river discharge reference stations as a contribution to the Implementation Plan for the Global Observing System for Climate in Support of the UNFCCC, Action T4 (GCOS 2004a). This Network, depicted in Fig. 3, is called the "GCOS Baseline River Discharge Network" and constitutes a first application of GTN-R. It will serve an increasing number of purposes and projects in the field of climate research but also as basis for future versions of the GRDC product "Long Term Mean Annual Freshwater Surface Water Fluxes into the World Oceans" and for the estimation of biogeochemical fluxes in cooperation with the UN GEMS/Water Programme Office of UNEP/DEWA.

5 Basic Concept

The basic idea of the GTN-R project is simple: by application of an automated procedure (software), regularly draw together heterogeneously available information on near real-time river discharge data provided by the world's National Hydrological Services, harmonize and store the information in a database, and regularly redistribute the harmonized data in a standard format,. In essence this task is nothing more than a copying and reformatting routine. However, the devil is in the details.

Depending on the level of development of the national networks and data exchange policies in individual countries it is more or less demanding to upgrade the national infrastructure and to create a climate of trust and consequently cooperation. The provision of financial assistance may be required occasionally. The following levels of network development may exist at individual National Hydrological Services (from higher developed to less developed networks):

1. NHS that already provide ready to read data in near real-time via internet (though possibly in their proprietary format).
2. NHS that already publish some kind of near real-time river information via internet (but not ready to read).
3. NHS that have to create a new interface to their digital and automated national networks.
4. NHS that have to automate their inland data transfer schemes of already digitally recording gauging stations.
5. NHS that have to upgrade their non-digital gauging stations.

Within the current first phase of the GTN-R project, concentration is on NHS of level 1-3. As an organizational prerequisite in many cases individual cooperation agreements between GRDC and the individual NHS involved are necessary, clarifying issues of data policy and possibly appropriate technical capacity and infrastructure, i.e. gauging stations need to automatically transmit their data digitally to NHS in near real-time and an automated interface into the internet is required at the national level. Furthermore, GRDC has to be flexible with regard to interchange formats. Though ideally a standardized format of GRDC will be applied, for pragmatic reasons provider defined formats have to be accepted as well.

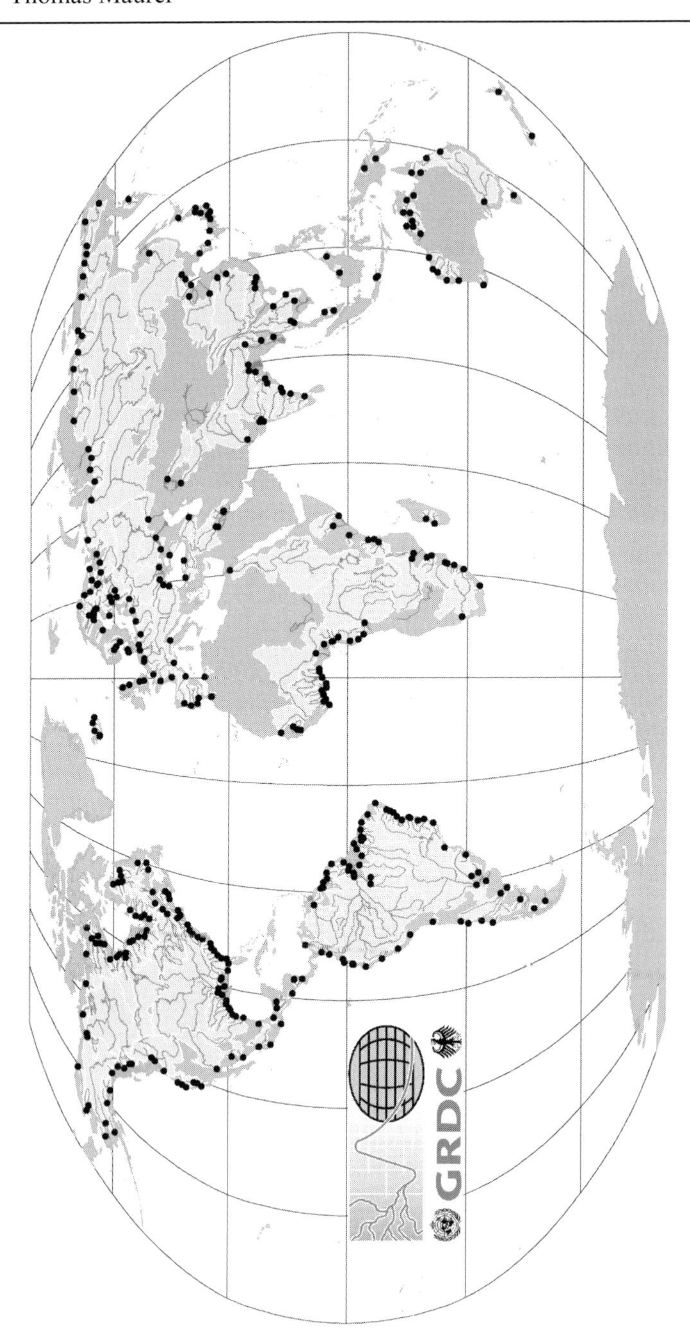

Fig. 3. GCOS Baseline River Discharge Network. The 377 stations along the continental coastlines correspond to 352 of the 358 basins depicted here (for 6 basins no GRDC station is currently available). See http://gtn-r.bafg.de for further details.

6 Software

The underlying core of the GTN-R project is a software system running at a GRDC server that allows:

- to draw together/ download near real-time (NRT) river water level or discharge data (with time steps of at least 1 hour) provided by individual NHS via Internet protocols (HTTP or FTP) in various formats,
- to transform water level data into discharge data where required
- to process and store the data in a database
- to classify the data on the background of historical data
- to check the plausibility of the data
- to redistribute/ upload all required discharge data in a harmonized way via the Internet, i.e. provision of the data in GRDC standard format at a FTP-site.

The software can be operated in two basic modes, the configuration mode and the operational mode.

The configuration environment provides the following features:

- Definition of the objects Provider, Data Source and Station, where a Provider may have multiple Data Sources and a Data Source may contain discharge data or water level data of multiple Stations.
- Each Data Source is described by its name, Unified Resource Identifier (URI), a time and rhythm of download and a code for its format. A mechanism is implemented allowing the definition of a subset of Stations potentially provided by a Data Source for regular reading.
- All information defined using the objects Provider, Data Source and Station is stored in the GRDC database, along with the discharge data read regularly from the Data Sources.
- Definition of the object Project to define subsets of all stations for regular upload for redistribution of harmonized data.
- Each Project is described by its name, URI, time and rhythm of upload.

In operational mode the software is running in the background, downloading and uploading discharge data according to the settings of Data Sources and Projects in the configuration mode. Fig. 4 illustrates the navigation through the input forms of the software while Fig. 5 shows a screenshot of the appearance of the Data Source Add/Edit Form.

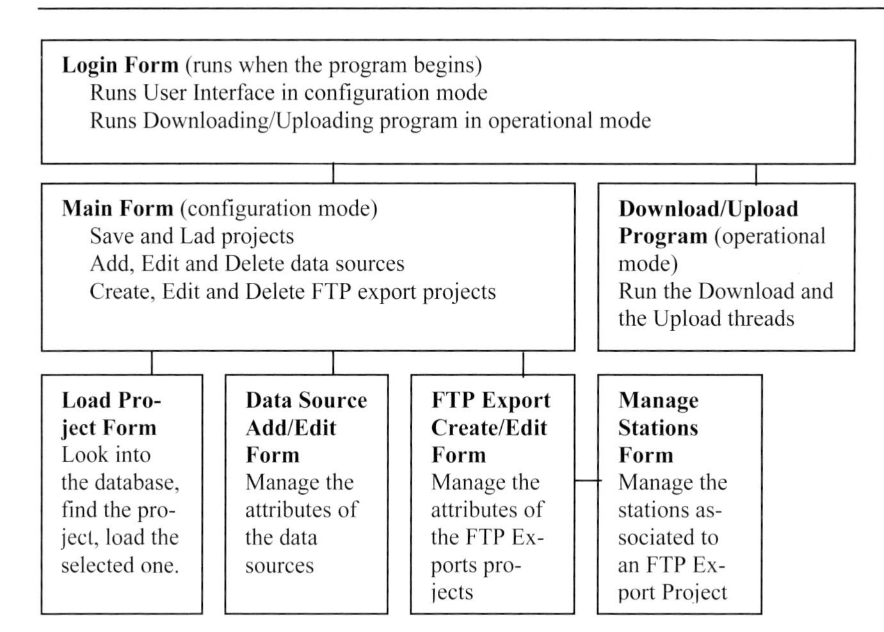
Login Form (runs when the program begins)
Runs User Interface in configuration mode
Runs Downloading/Uploading program in operational mode
Main Form (configuration mode)
Save and Lad projects
Add, Edit and Delete data sources
Create, Edit and Delete FTP export projects
Download/Upload Program (operational mode)
Run the Download and the Upload threads
Load Project Form
Look into the database, find the project, load the selected one.
Data Source Add/Edit Form
Manage the attributes of the data sources
FTP Export Create/Edit Form
Manage the attributes of the FTP Exports projects
Manage Stations Form
Manage the stations associated to an FTP Export Project

Fig. 4. Flow chart of forms in the GTN-R software

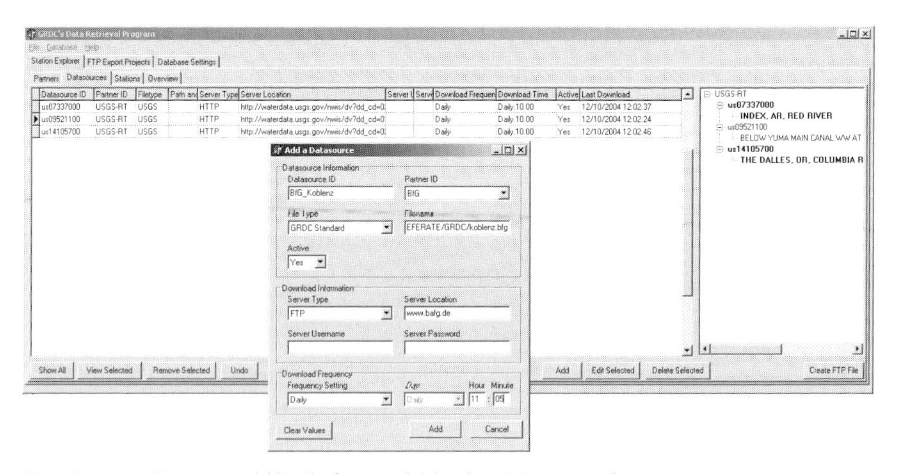
Add a Datasource
Datasource Information
Datasource ID
Partner ID
File type
Filename
Active
Download Information
Server Type
Server Location
Server Username
Server Password
Download Frequency
Frequency Setting
Hour
Minute
Clear Values
Add
Cancel
Show All
View Selected
Remove Selected
Undo
Edit Selected
Delete Selected
Create FTP File

Fig. 5. Data Source Add/Edit form within the GTN-R software

7 Add-On Modules

The regular uploaded harmonized data set is the core product of the operating GTN-R environment. It will serve as the basis for add-on products and services, build by data users as well as GRDC.

One such add-on service will be the development of a mapping application for the harmonized data product, i.e. a graphically display of the most recent collected and harmonized data of NRT-discharge stations in an interactively scaleable world map at a web page by means of an internet map server (IMS). Absolute discharge values as well as classified percentiles relative to the long term characteristics of the stations will be displayed as attributes, similar to e.g. the USGS WaterWatch.

8 Data Acquisition Issues

The success of the project is critically dependent from the willingness of NHS to cooperate. Due to the mere communication load it is not possible to advance data acquisition for all river basins at the same time. The following activities have to or may have to be carried out:

- Establishing contact to the NHS responsible for the operation of national gauging networks, convincing them to collaborate in the project.
- Receiving metadata on gauging stations available from the NHS for river basins of interest.
- Selection of stations of interest and determination of their status of operation (manual or digital, delayed or real-time, offline or online distribution).
- Negotiation of capacity building activities, their schedule and sources of funding and signing of a cooperation agreement where necessary.

For the selected and agreed stations it is necessary for GRDC to receive offline the following information for each station, once initially and thereafter every time it changes:

- Manual provision of all relevant metadata by the NHS and storage in the GRDC database, i.e.
  - National Station ID
  - Station Name
  - River Name
  - Country Name
  - Latitude in degree
  - Longitude in degree
  - Altitude in m above sea level

  - Catchment area in km^2
  - Full postal address of the organization delivering the data
  - Name, telephone, fax and email address of a focal point
- Metadata of the agreed online provision (called a Data Source in the GTN-R software), describing the download parameters, i.e.
  - Server location and filename
  - User-ID, if required
  - Password, if required
  - Frequency, time and time zone of data update at your side (e.g. daily, 09:30, UTC+8h)
  - Full postal address of the organization handling and delivering the online provision
  - Name, telephone, fax and email address of a focal point for the online provision
- Complementary water level-discharge relationship as a function or look-up table in cases where only water level data can be provided.
- Complementary longest possible record of historical daily discharge values for all selected and agreed stations in order to determine long-term statistical characteristics used to put the instantaneous/ daily values into perspective (classification and plausibility-check, add-on product development).

9 Conclusion and Outlook

The GTN-H project described here can be looked at from two completely different points of view. Having in mind *short-term* perspectives it might look complex and demanding with many tedious steps involved. However, looking at it with a more *long-term* perspective it might appear as very specialized or little generalized project, as it only cares for one variable and uses very conventional technologies. This apparent contradiction reflects the fact that the approach chosen tries to provide a solution with a *mid-term* perspective in mind: being better and more integrated than the current system but at the same time being more pragmatic than waiting for an ultimate solution, requiring to overcome current tedious approaches and to arrive at a ultimate, globally integrated data infrastructure by means of internationally agreed standards for the exchange of metadata and data on measurements of geophysical and biogeochemical processes in general (cf. Maurer 2004a, 2004b, 2003b, 2003c).

Wherever one looks, all over the world and disciplines people and organizations can be observed struggling hard to improve access to their data or the data they need, integrating some of the resources they are aware of and which they need from their current perspective for the domain(s) they currently consider. This certainly helps local/ regional respectively sectoral communities to some extend for some time. And it is the only way to proceed instantaneously and it thus will not stop.

The GRDC is one of such mechanisms currently in place, solely helping within it's limited capacity to integrate one of the many variables of interest in a bottom up fashion step by step. As long as there is no development of an overarching approach driven by something that is much more influential than GRDC, this is the only option.

However, the progress with this bottom-up approach is unsatisfying, as usually a terrible amount of manual/ personal communication is required per data acquisition. Besides contacting National Hydrological Services (NHS) directly, GRDC also scans their Web Sites and finds more and more online presentations of country's individual data holdings, but the problem is, that they all follow quite different philosophies and provide different level of detail and change from time to time without prior notice. It is thus still very tedious to retrieve data, as it involves individual treatment of each source. This paper describes the near real-time monitoring scheme for river discharge, called GTN-R, however it will provide only the basic infrastructure. The bulk of work will follow, i.e. writing interface routines for each online resource GRDC will trace down or alternatively convince data providers to deliver their data in the GRDC proposed format and with an information depth that GRDC defines. And all this has not to be done only once, but GRDC will have to follow up all changes that will be made remotely without notifying GRDC.

There are two principle ways to improve the situation:

- raising capacity by increasing manpower (the brute force approach) thus multiplying the manual/ personal acquisition activity or
- introducing automation, which inevitably requires thorough standardization to ensure machine-readability (the smart approach).

The first option is not really a sustainable approach. Though it certainly can be called integration, it will be only until a group's perspective widens that they need to revise their scheme and harmonize it again with those developments that took place elsewhere in the area/ by the communities that they now want to include. Probably this kind of comparatively uncoordinated bottom-up approach will eventually also arrive at a universal standard, but certainly at higher costs and with longer development time. However, looking at Fig. 2a, it can be easily imagined that alternatively all limited energy available is lost in friction[1]. In any case, progress will require at some point somebody who will substantiate, streamline and synchronize the process by developing and setting very general standards.

So the question is, why not spending a larger portion of the available resources for starting/ initiating rigorous standardization now, taking the rare chance of the current awareness reflected in activities as GEOSS and GCOS-IP? We urgently need to carve out over-arching standards for data representation. Only such a top-down approach will save in the mid to long term the global community a number of (costly and time consuming) subsequent iterations, that will be the consequence if domains continue to develop their individual standards in a bottom-up fashion.

It seems to be advisable to advocate for an international technical commission or coordinating body on standards for this purpose that is not specialized to any specific domain (as e.g. WMO to weather, climate and water) but rather cares for the geosphere as a whole and is associated to a more neutral organization such as a Technical Committee of ISO (recruiting of course specialists from all domains and specialized agencies), which as a side effect would also help the acceptance of the result by a wider community. Moreover, ISO has already published standards for geographic information in its ISO 19100 series of standards (Kresse and Fadaie 2004, ISO 2003). This is the ideal starting point for the search for the common denominator, as all data on any geophysical variable share geographic information associated with it. Everything defined there (in a years-long process!) could be used right away as a start for the yet to find common denominator for the representation of geophysical variables, but it needs to be extended.

In fact, encouragingly, that is what is currently happening at many places, especially in the geographic information community. But also activities such as the Framework WMO Information System (FWIS 2003), the Draft WMO Core Metadata Standard (ET-IDM 2004, WMO 2004) as well as JCOMM (2002) with its Oceans Information Technology Project (OIT) will build on it, however, care has to be taken that no unnecessary divergence takes place at a too early stage.

An impressive and encouraging example for the increasing awareness of the importance of state-of-the-art data infrastructures by governments is the afore mentioned very recent proposal for a directive of the European Parliament and of the Council establishing an Infrastructure for SPatial InfoRmation in Europe (INSPIRE).

---

[1] E.g., GRDC has encountered the situation that NHS reacted annoyed when GRDC contacted them, undertone: "Every other day someone contacts us to provide data in the format he requires, sorry we do not have the capacity".

The rational behind this proposal is given in the introduction of the Explanatory Memorandum of EC (2004). Essentially for pretty much the same reasons that were stressed in this section, EC is starting by this directive (in the field of Environmental Spatial Information) exactly such a top-down approach which needs to be initiated for all types of Earth System Science variables. This EC proposal could in fact become a template/ starting point for discussing and developing a constituting document of a yet to establish international coordinating body on standards, as it has been recommended already by the Second Report on the Adequacy of the Global Climate Observing Systems (GCOS 2003c) as well as by the GCOS-IP (GCOS 2004a).

So, let's not be too pusillanimous! This is a plea for the initiation of an international coordinating mechanism or body that guides in consultation with all major players involved (international organizations, governmental authorities, leading companies, etc.) a process of defining the fundaments for an global geosphere information system by developing standards and a data infrastructure for geophysical, biogeochemical and socioeconomic variables observed and predicted in the geosphere.

All this will certainly not happen in one single step but rather in an evolutionary process. However, it has to be the goal to minimize the number of iterations involved in this process, especially as the iterations become more and more complex as the system evolves and as resources are spent by an increasing number of organizations on harmonizing their data holdings according to the latest developments. Thus, great care has to be taken to detect "dead ends" in the evolutionary process as early as possible and prevent dissipation of energy and subsequent frustration of participants. It will require a smart strategy to cope with the task of finding an efficient way ahead. The key to success will be in the interplay of a strongly focused supervising structure (top-down) and a number of organizations developing prototypes and putting them in test beds (bottom-up), however, without falling in love with their approaches too much, as to remain able to abolish their development as soon as something else proves to better serve the purpose.

And finally it is important to recall, that striving for the ambitious goal of establishing a very general internationally standardized data infrastructure is only one task among many, in a sense of being a necessary prerequisite but not a solution in itself for the pressing question of how to manage the Earth System in order to increase societal benefits as e.g. defined by GEOSS. Without doubt, other activities, especially in the areas of monitoring and modeling must not be neglected. But they need to be better supported by information science and technology and at the same time, reciprocally, they need to better support the data infrastructure by rigorously documenting themselves. This needs to become an inescapable obligation for granting funds of both, research and operational networks.

References

EC (2004) Proposal for a directive of the European Parliament and of the Council establishing an infrastructure for spatial information in the Community (INSPIRE), Brussels, 23 July 2004, COM(2004) 516 final, 2004/0175 (COD) (Online available at http://inspire.jrc.it/proposal/COM_2004_0516_F_EN_ACTE.pdf)

ET-IDM (2004) Fourth Meeting of the Expert Team on Integrated Data Management, Geneva, 1-3 September 2004, Commission for Basic Systems, World Meteorological Organization, Geneva, Switzerland. (Online available at http://www.wmo.int/web/www/WDM/ET-IDM-4/documents.html)

FWIS (2003) Fifth Meeting of the Inter-Programme Task Team on the Future WMO Information System, Kuala Lumpur, 20-24 October 2003, Final Report. Commission For Basic Systems, World Weather Watch, World Meteorological Organization, Geneva, Switzerland. (Online available at http://www.wmo.int/web/www/FWIS/FWIS-2003-final.pdf; see also http://www.wmo.int/web/www/FWIS/documents.htm)

GCOS (2000) Establishment of a Global Hydrological Observation Network for Climate. Report of the GCOS/GTOS/HWRP Expert Meeting, Geisenheim, Germany, June 26-30, 2000. Cihlar J, Grabs W, and Landwehr J (eds), (GCOS-63; GTOS-26) (WMO/TD-No. 1047). (Online available at http://www.fao.org/gtos/doc/pub26.pdf)

GCOS (2002) Report of the GCOS/GTOS/HWRP Expert Meeting on the Implementation of a Global Terrestrial Network - Hydrology (GTN-H), Koblenz, Germany, June 21-22, 2001. Grabs W, Thomas AR (eds), (GCOS-71; GTOS-29) (WMO/TD-No. 1099).
(Online available at http://www. wmo.int/web/gcos/Publications/gcos-71.pdf)

GCOS (2003a) Global Terrestrial Network-Hydrology (GTN-H). Report of the GTN-H Coordination Panel Meeting, Toronto, Canada, 21-22 November 2002. Harvey KD, Grabs W, Thomas AR (eds), (GCOS-83; GTOS-33) (WMO/TD-No. 1155). (Online available at http://www.wmo.int/web/gcos/Publications/gcos-83.pdf)

GCOS (2003b) Report of the GCOS/GTOS/HWRP Expert Meeting on Hydrological Data for Global Studies, Toronto, Canada, 18-20 November 2002. Harvey KD, Grabs W (eds), (GCOS-84; GTOS-32) (WMO/TD-No. 1156). (Online available at http://www.wmo.int/web/gcos/Publications/gcos-84.pdf)

GCOS (2003c) Second Report on the Adequacy of the Global Climate Observing Systems. Developed on request of the United Nations Framework Convention for Climatic Change/ Subsidiary Body for Scientific and Technological Advice (UNFCCC/SBSTA). (Online available from the GCOS-homepage at http://www.wmo.int/web/gcos/gcoshome.html)

GCOS (2004a) Implementation Plan for the Global Observing System for Climate in Support of the UNFCCC. Developed on request of the United Nations Framework Convention for Climatic Change/ Subsidiary Body for Scientific and Technological Advice (UNFCCC/SBSTA). (Online available from the GCOS-homepage at http://www.wmo.int/web/gcos/gcoshome.html)

GCOS (2004b) Analysis of Data Exchange Problems in Global Atmospheric and Hydrological Networks. Initial Summary Report (for SBSTA-20). (Online available from the GCOS-homepage at http://www.wmo.int/web/gcos/gcoshome.html)

ISO (2003) ISO 19115: Geographic information – Metadata

JCOMM (2002) An Ocean Information Technology Project. Document DMCG-I/28 (20 May 2002) to the First Session of the Data Management Coordination Group (DMCG) of the Joint WMO-IOC Technical Commission for Oceanography and Marine Meteorology (JCOMM) of the Intergovernmental Oceanographic Commission (of UNESCO) (IOC) and the World Meteorological Organization (WMO), Paris, 22-25 May 2002 (Online available at http://ioc.unesco.org/oit/files/DMCG1_doc28_oit.pdf)

Kresse, W, Fadaie, K (2004) ISO Standards for Geographic Information, Springer, ISBN 3-540-20130-0

Maurer, T (2004a) Globally agreed standards for metadata and data on variables describing geophysical processes. A fundamental prerequisite to improve the management of the Earth System for our all future, GRDC Report No. 31, Global Runoff Data Centre, Koblenz, Germany. (Online available at http://grdc.bafg.de/?911)

Maurer, T (2004b) Transboundary and transdisciplinary environmental data and information integration - an essential prerequisite to sustainably manage the Earth System. INDUSTRY IDS-Water Europe 2004 Online Conference (http://www.idswater.com), 10 May-28 May 2004, 14 pp (Online available at http://grdc.bafg.de/?6413)

Maurer, T (2003a) Development of an operational internet-based near real time monitoring tool for global river discharge data. A contribution to the Global Terrestrial Network for Hydrology (GTN-H). GRDC Report No. 30, Global Runoff Data Centre, Koblenz, Germany.
(Online available at http://grdc.bafg.de/?911)

Maurer, T (2003b) Intergovernmental arrangements and problems of data sharing, In: Timmerman JG, Behrens HWA, Bernardini F, Daler D, Poss P, Van Ruiten KJM, Ward RC (eds): Information to support sustainable water management: From local to global levels, Monitoring Tailor-Made IV Conference, St. Michielsgestel, the Netherlands 15-18 September 2003, RIZA, Lelystad, The Netherlands (Online available at http://grdc.bafg.de/?3997)

Maurer, T (2003c) Challenges in transboundary and transdisciplinary environmental data integration in a highly heterogeneous and rapidly changing world - A view from the perspective of the Global Runoff Data Centre. In: Harmancioglu NB, Ozkul SD, Fistikoglu O, Geerders P (eds): Integrated Technologies for Environmental Monitoring and Information Production , Proc. NATO Advanced Research Workshop, 10 - 14 September 2001, Marmaris, Turkey.

Nato Science Series IV Volume 23, Kluwer Academic Publishers (Online available at http://grdc.bafg.de/?2535)
WMO (2001a) Exchanging hydrological data and information, WMO policy and practice, WMO brochure No. 925, World Meteorological Organization, Geneva, Switzerland.
WMO (2001b) Exchange of Hydrological Data and Products, WMO Technical Reports in Hydrology and Water Resources No.74, P. Mosley, World Meteorological Organization, Geneva, Switzerland. (WMO/TD-No. 1097). (Online available at http://www.wmo.int/web/homs/documents/TD74.pdf)
WMO (2004) WMO Core Metadata Standard (v0-2) (Online available at http://www.wmo.int/web/www/WDM/Metadata/documents.html)

Contribution of Earth Observation Data Supplied by the New Satellite Sensors to Flood Disaster Assessment and Hazard Reduction

Gheorghe Stancalie and Vasile Craciunescu

National Meteorological Administration 97, Soseaua Bucuresti-Ploiesti, Sector 1, 013686 Bucharest, Romania.
Email: gheorghe.stancalie@meteo.inmh.ro; vasile.craciunescu@meteo.inmh.ro

Abstract

The risk of flooding due to runoff is a major concern in many areas around the globe and especially in Romania. In the latest years river flooding and accompanying landslides, occurred quit frequently in Romania, some of which isolated, others-affecting wide areas of the country's territory.

The main objective of the NATO SfP project "Monitoring of extreme flood events in Romania and Hungary using EO data" is to improve the existing local operational flood hazard assessment and monitoring using the functional facilities supplied by the GIS info-layers, combined with Earth Observation (EO) data-derived information, Digital Elevation Models (DEM) and hydrological modeling. The study area is situated in the Crisul Alb - Crisul Negru - Kőrős transboundary basin, crossing the Romanian – Hungarian border.

The orbital remote sensing can provide necessary information for flood hazard and vulnerability assessment and mapping, which are directly used in the decision - making process. The EO data-derived information of the land cover/land use is important because it makes possible periodical updating and comparisons, and thus contribute to characterize the human presence and to provide elements on the vulnerability aspects, as well as the evaluation of the impact of the flooding. In order to obtain high-level thematic products the data extracted from the satellite images must be integrated with other geo-information data (topographical, pedological, meteorological data) and hydrologic/hydraulic models outputs.

The paper presents the specific methods developed for deriving satellite-based applications and products useful for flood disaster assessment and hazard reduction. An important contribution of EO derived information in the topic of managing flooding connected phenomena could be envisaged at the level of mapping aspects. Using the optical and microwave data supplied by the new satellite sensors (U.S. DMSP/Quikscat, RADARSAT, LANDSAT–7/TM, EOS-AM "TERRA"/MODIS and ASTER) different products like accurate updated digital maps of the hydrographical network and land cover/land use, mask of flooded areas, multi-temporal maps of the flood dynamics, hazard maps with the extent of the flooded areas and the affected zones, etc. have been obtained. These results, at different spatial scales, include synthesis maps easy to access and interpret, adequate to be combined with other information layouts resulted from the GIS database and to ingest rainfall-runoff models outputs.

The presented applications will contribute to preventive consideration of the extreme flood events by planning more judiciously land-use development, by elaborating plans for food mitigation, including infrastructure construction in the flood-prone areas and by optimization of the flood - related spatial information distribution facilities to end – users.

1 Introduction

Flooding remains the most widely distributed natural hazard in Europe leading to significant economic and social impacts.

In the assessment and analysis of flood risk it is important to remember that risk is entirely a human issue. Floods are part of the natural hydrological cycle and are random. The risk arises because the human use and value of the river flood plains conflicts with their natural function of conveyance of water and sediments.

Orbital remote sensing of the Earth is presently capable of making fundamental contributions towards reducing the detrimental effects of extreme floods (Brakenridge et al. 1998). Effective flood warning requires frequent radar observations of the Earth's surface through cloud cover. In contrast, both optical and radar wavelengths will increasingly be used for disaster assessment and hazard reduction. These latter tasks are accomplished, in part, by accurate mapping of flooded lands, which is commonly done over periods of several or more days. The detection of new flood events and public warnings thereof is still experimental, but making rapid progress; radar sensors are preferred due to their cloud penetrating capability as reported by Brakenridge et al. 2003. Relatively low spatial resolution, but wide-area and frequent coverage, are appropriate; the objective is to locate where within a region or watershed the flooding occurs, rather than to map the actual inundated areas. The rapid-response flood mapping and measurement provide information useful for disaster assessment, and has become a relatively common activity in many developed countries.

In the latest years river flooding and accompanying landslides, occurred quit frequently in Romania, some of which isolated, others-affecting wide areas of the country's territory. One region, which suffers from flood damages on a regular basis, is the transboundary area of the Crisul Alb and Crisul Negro rivers flowing from Romania into Hungary, where they are known as Kőrős rivers. Floods in this area typically start in the mountainous terrain of the upper parts of the basin in Romania and propagate to the plains in Hungary.

Historically, there has been a close co-operation between both countries in flood management in this area. The issues connected with the transboundary rivers crossing the Romanian – Hungarian border are covered by the bilateral Agreement for the settlement of the hydrotechnical problems, which was issued on Nov. 20, 1986. To facilitate the implementation of this agreement, working groups from the Crisuri Water Authorities in Oradea, Romania and Körös Valley District Water Authority (KOVIZIG) in Gyula, Hungary meet regularly to address the issues of mutual interest (Brakenridge et al. 2001).

The flood forecast and defense related information provided by Romania to Hungary is presently based entirely upon the ground-observed data, which are mostly collected by non-automatic hydrometeorological stations. Such data are somewhat limited in terms of spatial distribution, temporal detail, and speed of collection and transmission, and these limitations should be remedied.

Recognizing the threat of floods and the need for further improvement of flood management in this area, at the initiative of the Romanian Meteorological Administration, an international team was formed, with representatives of Hungary, Romania and USA, and proposed a project on "Monitoring of Extreme Flood Events in Romania and Hungary Using EO (Earth Observation) Data" to the NATO Science for Peace (SfP) Program. The project aims to provide to the local and river authorities as well as to other key organizations an efficient and powerful flood-monitoring tool, which is expected to significantly contribute to the improvement of the efficiency and effectiveness of the action plans for flood defense.

Some applications, for flood hazard and vulnerability assessment and mapping, developed in the framework of this NATO SfP project is presented.

2 Study Area

The study area represents the Crisul Alb/Negru/Kőrős transboundary basin spanning across the Romanian–Hungarian border, with a total area of 26,600 km2 (14,900 km2 on the Romanian territory). In Romania, the catchment (basin) comprises mountainous areas (38%), hilly areas (20%) and plains (42%). About 30% of the catchment is forested. On the Hungarian side, the catchment relief represents plains. Annual precipitation ranges from 600-800 mm/year in the plain and plateau areas to over 1200 mm/year in the mountainous areas of Romania. This precipitation distribution can be explained by the fact that humid air masses brought by fronts from the Icelandic Low frequently enter this area. The orography of the area (Apuseni Mountains) amplifies the precipitation on the western side of the mountain range. Thus, the Crisuri Rivers Basin frequently experiences large precipitation amounts in short time intervals and the frequency of such events seem to be increasing in recent years.

In terms of hydrography, there is a marked difference between high rates of mountain runoff and low rates of runoff in plains. Thus, runoff flood waves formed quickly in the Romanian part of the basin move rapidly to the plains in the Hungarian part of the basin, which is characterized by relatively slow flows and a potential for inundation. In terms of flood forecasting, the Romanian part of the basin is of greater interest with respect to flood formation, which is also reflected in this paper. The hydrography of the study area is well established. There are 62 hydrometric stations in Romania, on the Crisul Alb and Negru (and their tributaries); 7 of these stations have flow records longer than 80 years. On the Hungarian territory, the hydrometric stations at Gyula and Sarkad are particularly of interest. In Gyula, the flow was decreasing in time, but the stage was rising, in part due to the hydrotechnical structures; at Sarkad, both discharge and stage were increasing (reflects more natural conditions, without much change in the river channel geometry).

The list of significant floods includes the events of June 1974, July-August 1980, March 1981, December 1995-January 1996, March 2000, April 2000 and April 2001. The spring 2000 flood caused on the Romanian territory damages of more than $US 20 million (Munich Re, 2003) included damages to houses, roads and railways, bridges, hydraulic structures, loss of domestic animals, and business losses. On the Hungarian territory, a particularly notable was the flood of summer 1980, with total losses of $US 15 million, including destruction of farmhouses and large losses in agriculture (Munich Re 2003).

The frequency and importance of floods in the study region require further work to reduce flood damages and improve flood monitoring by the agencies in charge of flood protection, such as government agencies, civil protection authorities or municipalities. To mitigate flood impacts in the study area, structural and non-structural measures have been undertaken in the past. The Romanian area is defended by dikes along the Crisul Alb River and Crisul Negru River. These dikes were built in the 19th century for a 20-year design return period and further improved in later years. Currently, the dikes on the right bank of the Crisul Negru River and the Teuz River (43 km) are designed for a 50-year return period, and on the Crisul Alb, 67 km of dikes on the right bank and 59 km on the left bank are designed for a 100-year return period. In spite of these improvements, in April 2000, the right bank dike of the Crisul Negru broke near the village Tipari (a 130 m breach) and caused significant flooding of , and damages in, the adjacent territory. Other structural flood protection measures include permanent retention storage facilities (total volume of 34 x 106 m3) and temporary storage facilities (a total storage volume of almost 80 x 106 m3).

On the Hungarian side, in the Kőrős valley, high flood potential is recognized and exacerbated by low flood plains. Much of the area is, therefore, protected by flood dikes, of which construction started in the 18th century. More than 440 km of dikes are maintained by the Kőrős River Authority (KOVIZIG). Following the 1979 flood, construction of detention reservoirs started. Altogether, these reservoirs provide storage capacity of 188 million m3 and serve to reduce critical flood levels. The reservoirs are activated during floods, by a controlled explosion opening a protected spillway (a side weir) in flood dikes. Detained water inundates areas with lower intensity of agricultural activities and causes limited damages. Nevertheless, the reservoirs are activated only when necessary to avoid higher losses caused otherwise.

The analysis and management of floods constitute the first indispensable step towards, and a rational basis for, the development of flood protection. Where certain flood risk levels are inevitable, the affected parties must know it and be appropriately warned. To reduce the frequency and magnitude of the damages due to flooding, comprehensive, realistic and integrated strategies must be developed and implemented.

The flood forecasting and monitoring systems existing in the study area do not reflect well the spatial distribution of floods and the related phenomena (pertaining to geographic distances or patterns) in both pre- and post crises phases. To mitigate these limitations, the SfP project was initiated with emphasis on a satellite-based surveillance system connected to a dedicated GIS database that will offer a much more comprehensive evaluation of the extreme flood effects. Also, so far, the flood potential, including the risk and the vulnerability of flood-prone areas, have not been yet quantitatively assessed. An inventory of the past floods observed by the EO facilities would allow a more cost-effective design of structural and non-structural measures for flood protection and disaster relief. Finally, such data also provide important validation of the hydrological modelling-based flood risk assessment, because they show the actual extent of past flooding.

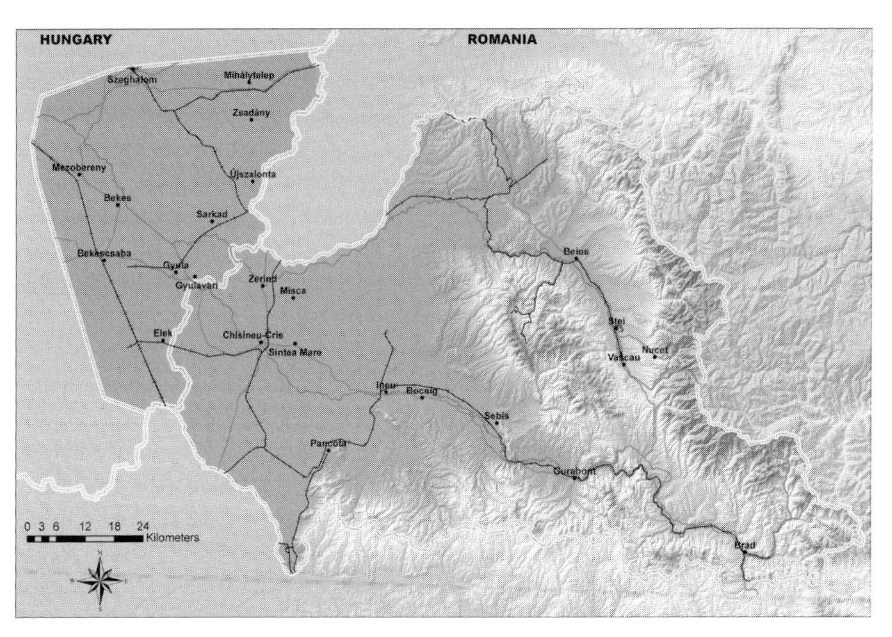

Fig. 1. Study area: the Crisul Alb - Crisul Negru - Kőrős transboundary basin, crossing the Romanian – Hungarian border

3 EO Data Used

Apart from ground information on the occurrence and evolution of the flood, NOAA/AVHRR satellite data (locally received), microwave data from U.S. DMSP and Quikscat and follow-on satellites, and the high resolution images supplied by the orbital platforms (SPOT, IRS, LANDSAT–7, RADARSAT, QUIKSCAT, EOS-AM "TERRA" and EOS–PM "AQUA"), substantially contribute to determining the flood-prone areas.

RADARSAT-1and 2, based on Synthetic Aperture Radar microwaves penetrates clouds and rain being very efficient for the flood monitoring. RADARSAT operates in different modes with selective polarization with a nominal swath width varying from 50 Km (for the fine beam mode) to 500 Km (for the ScanSAR wide beam mode). The approximate resolution varies between 10 x 9 m (for the fine beam mode) and 100 x 100m (for the ScanSAR wide beam mode). RADARSAT data with finer resolution allow the structural flood damage assessment.

SeaWinds on QuikSCAT is a new space borne Ku-band scatterometer with the resolution of about 25 km x 25 km, the swath of 1800 km (for a vertical polarization (VV) at a constant incidence angle of 54o) and of 1400 km (for a horizontal polarization (HH) at 46o). SeaWinds can provide near daily global coverage with the capability to see through clouds and darkness. It can, in principle, detect where flooding is occurring without necessarily imaging.

Especially the new American TERRA and AQUA platforms, equipped with different sensors such as MODIS and ASTER, can provide comprehensive series of flood event observations with much higher spatial resolution where available.

The Moderate Resolution Imaging Spectroradiometer (MODIS) continues the lineage of the Coastal Zone Color Scanner (CZCS), the Advanced Very High Resolution Radiometer (AVHRR), the High Resolution Infrared Spectrometer (HIRS), and the Thematic Mapper (TM). MODIS has 36 spectral bands with center wavelengths ranging from 0.412 mm to 14.235 mm. Two of the bands are imaged at a nominal resolution of 250m at nadir, five bands are imaged at 500m, and the remaining bands at 1000m.

The Advanced Spaceborne Thermal Emission and Reflection Radiometer (ASTER), an imaging instrument, acquires 14 spectral bands and can be used to obtain detailed maps of surface. Each scene covers 60 x 60 km. The ASTER spectral bands are organized into three groups: 3 bands with 15 m resolution in VNIR, 6 bands with 30m resolution in SWIR and 5 bands with 90 m resolution in TIR.

The information provided by these new sensors is of a higher quality than previously possible, and especially given the need for frequent repeat coverage while floods are underway. Information obtained from these optical and radar images have been used also for the determination of certain parameters necessary to monitor flooding: hydrographic network, water accumulation, size of flood-prone area, land cover/land use features.

A Satellite Image Database (SID) provided by different platforms and sensors has been set up. The purpose of the SID is to gather information about the raw satellite scenes available as well as of the derived products and make it available in a simple format. This information is useful to test the processing and analysis algorithms in order to establish an operational methodology for the detection, mapping and analysis of flooding. The SID was build in Microsoft Works and will be available on-line on the file server, being updated as new satellite images are acquired. Each record of the database describe the characteristics of each satellite image: platform, sensor, date and time of data acquisition, duration of pass, spectral band, coordinates of the area covered, projection, calibration, size, bits/pixel, image file format, physical location (machine, directory), origin of data, type (raw/processed), type of processing applied, algorithm used, quick-look available, cloudiness.

4 Methods for Obtaining Useful Products for Flood Risk Assessment

It is generally recognized that the management and mitigation of flood risk require a holistic, structural set of activities, approached in practice on several fronts with appropriate institutional arrangements made to deliver the agreed standard services of the community at risk (P.G. Samuels, 2004). The flood management includes pre-flood, flood emergency and flood recovery activities. Pre-flood activities are:

- Flood risk management for all causes of floods;
- Disaster contingency planning to establish evacuation routes, critical decision thresholds, public service and infrastructure requirements for emergency operations, etc;
- Construction of flood defense infrastructure, both physical defenses and implementation of forecasting and warning systems;
- Maintenance of flood defense infrastructure;
- Land-use planning and management within the whole catchment;
- Discouragement of inappropriate development within flood plains;
- Public communication and education of flood risk and actions to take in a flood emergency;
- Operational flood management includes:
- Detection of the likelihood of a flood forming;
- Forecasting of future river flow conditions from the hydrological and meteorological observations;
- Warning issued to the appropriate authorities and the public on the extent, severity and timing of the flood;
- Response to the emergency by the public and authorities.

Post-flood activities are:

- Relief for the immediate needs of those affected by the disaster;
- Reconstruction of damaged buildings, infrastructure and flood defense;
- Recovery and generation of the environment and economic activities in the flooded areas;
- Review of the flood management activities to improve the process and planning for the future events in the area affected.

Within the framework of flood surveying, optical and radar satellite images can provide up-to-date geographical information. Integrated within the GIS, flood derived and landscape descriptive information is helpful during their characteristic phases of the flood (Tholey et al., 1997):

before flooding, the image enables the description of the land cover of the studied area under normal hydrological conditions;

during flooding the image data set provide information on the inundated zones, flood map extent, flood's evolution;

after flooding, the satellite image point out the flood's effects, showing the affected areas, flood deposits and debris, with no information about the initial land cover description unless a comparison is performed with a normal land cover description map or with pre-flood data.

In order to obtain high-level thematic products the data extracted from the EO images must be integrated with other non-space ancillary data (topographical, pedological, meteorological data) and hydrologic/hydraulic models outputs. This approach may be used in different phases of establishing the sensitive areas such as: the management of the database - built up from the ensemble of the spatially geo-referenced information; the elaboration of the risk indices from morpho-hydrographical, meteorological and hydrological data; the interfacing with the models in order to improve their compatibility with input data; recovery of results and the possibility to work out scenarios; presentation of results as synthesis maps easy to access and interpret, additionally adequate to be combined with other information layouts resulted from the GIS database.

The products useful for flooding risk analysis, referred to: accurate updated maps of land cover/land use, comprehensive thematic maps at various spatial scales with the extent of the flooded areas and the affected zones, maps of the hazard prone areas.

Optical and radar satellite data have been used to perform the analysis for inventory purposes and different kind of flood related thematic information.

A series of specific processing operations for the images were performed, using the ERDAS Imagine software; geometric correction and geo-referencing in the UTM or STEREO 70 map projection system, image improvement (contrast enhancing, slicking, selective contrast, combinations between spectral bands, re-sampling operation), statistic analyses (for the characterization of classes, the selection of the instructing samples, conceiving classifications).

Optical high-resolution data have been used to perform the analysis for the inventory purposes under normal hydrological conditions as well as for determining the hydrographic network. The radiometric information contained in these images allows the derivation of both biophysical criteria and those from human activity, through supervised standard classification methods or advanced segmentation of specific thematic indices. Once extracted, these geographical information coverages were integrated within the GIS for further water crisis analysis and management.

The interpretation and analysis of remotely sensed data in order to identify, delineate and characterize flooded areas was based on relationships between physical parameters such as reflectance and emittance from feature located on the surface: reflectance and/emittance decreases when a water layer covers the ground or when the soil is humid; also reflectance and/emittance increases in the red band because of the vegetation stress cause by moisture; reflectance and/emittance changes noticeably when different temperatures, due to thick water layer are recorded.

In the microwave region the water presence could be appreciated by estimating the surface roughness, where water layers smooth surfaces dielectric constant is then heavily correlated to soil water content. In case of radar images the multi-temporal techniques was considered to identify and highlight the flooded areas. This technique uses black and white radar images of the same area taken on different dates and assigns them to the red, green and blue color channels in a false color image. The resulting multi-temporal image is able to reveals change in the ground surface by the presence of color in the image; the hue of a color indicating the date of change and the intensity of the color the degree of change. The proposed technique requires the use of a reference image from the archive, showing the « normal » situation.

4.1 The Land Cover/Land Use Mapping

The methodology for the achievement of the land cover/land use from medium and high-resolution images developed within the Remote Sensing &

GIS Laboratory of NIMH (Stancalie et al. 2000), is based on the observation of the following requirements:

- The structure of this type of information must be at the same time cartographic and statistic;
- It must be suited to be produced at various scales, so as to supply answers adapted to the different decision making levels;
- Up-dating of this piece of information must be performed fast and easily.
- The used methodology implies following the main stages below:
- Preliminary activities for data organizing and selection;
- Computer-assisted photo-interpretation and quality control of the obtained results;
- Digitization of the obtained maps (optional);
- Database validation at the level of the studied geographic area;
- Obtaining the final documents, in cartographic, statistic and tabular form.

Preliminary activities comprise collection and inventorying of the available cartographic documents and statistic data connected to the land cover: topographic, land survey, forestry, and other thematic maps at various scales.

To obtain the land cover map/land use map, satellite images with a fine geometrical resolution and rich multispectral information have to be used. In case of IRS and SPOT data the preparation stage consisted in merging data obtained from the panchromatic channel, which supply the geometric fineness (spatial resolution of 5 m for the IRS, 10 m for the SPOT), with the multispectral data (LISS for IRS, XS for SPOT), which contain the multispectral richness. In the figure 2 a flowchart for the generation of the land cover/land use maps using high resolution satellite data is presented.

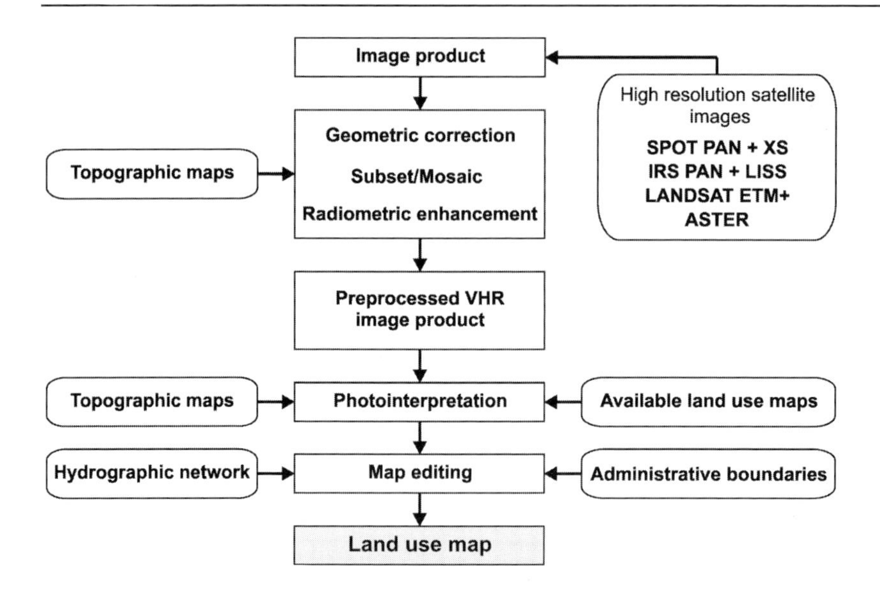

Fig. 2. Flowchart for the generation of the land cover/land use maps

For this application TERRA/ASTER data have also been used. These data proved to be suitable for detailed maps of land cover/land use, especially the visible and near infrared bands (1, 2, 3B) with 15 m resolution. The ASTER data were obtained from the Earth Observing System Data Gateway, by courtesy of Prof. Brakenridge from DFO, USA.

A series of specific image processing operations were performed with the ERDAS Imagine and ENVI software. Those operations included: geo-referencing of the data, detection of cloud and water, image improvement (through using the histogram, contrast enhancing, slicking, selective contrast, combinations between spectral bands, re-sampling operation), statistic analyses (for the characterization of classes, the selection of the instructing samples, conceiving classifications). The computer-assisted photo-interpretation finalizes in the delimitation of homogeneous areas from images, in their identification and framing within a class of interest. Discriminating and identifying the different land occupation classes rely on the classical procedures of image processing and leads to a detailed management of the land cover/land use, followed by a generalizing process, which includes:

- identification of each type of land occupation, function of the exogenous data, of the "true-land" data establishing a catalogue;
- delimitation of areas suspected to represent a certain unity of the land;

- expanding this delimitation over the ensemble of the image areas, which display resembling features.

Validation of results from photo-interpretation, mapping (by checking through on land sampling at local and regional level) and building up the database aims at knowing the reliability level and the precision obtained for the delimitation of the units and their association to the classes in the catalogue.

The satellite based cartography of the land cover/land use is important because it makes possible periodical updating and comparisons, and thus contribute to characterize the human presence and to provide elements on the vulnerability aspects, as well as the evaluation of the impact of the flooding.

The land cover maps are useful to classify the terrain function of the main types of land cover, thus allowing their characterization function of the land impermeability degree, of their absorption capacity or resilience to in-soil water infiltration.

4.2 Method for the Identification and Mapping of the Flooded Areas

The methodology for the identification, determination and mapping of the areas affected by floods is based on the different classification procedure of the optical and radar satellite images (Brakenridge et al. 1998). The advantage of the use of high resolution optical satellite images consists in the possibility to select precise spatial information upon the respective area (through merging images) and to localize and define the flooded or flooding risk areas (through classifications). The radar images can bring even during the periods with abundant rainfalls useful information regarding the flooded areas. The multi-temporal image analysis, combined with the land cover/land use information allow the identification of the area covered by water (included the permanent water bodies) and then of the flooded areas.

The figure 3 presents flowchart for the generation of the flood extent maps using satellite radar (SAR) images.

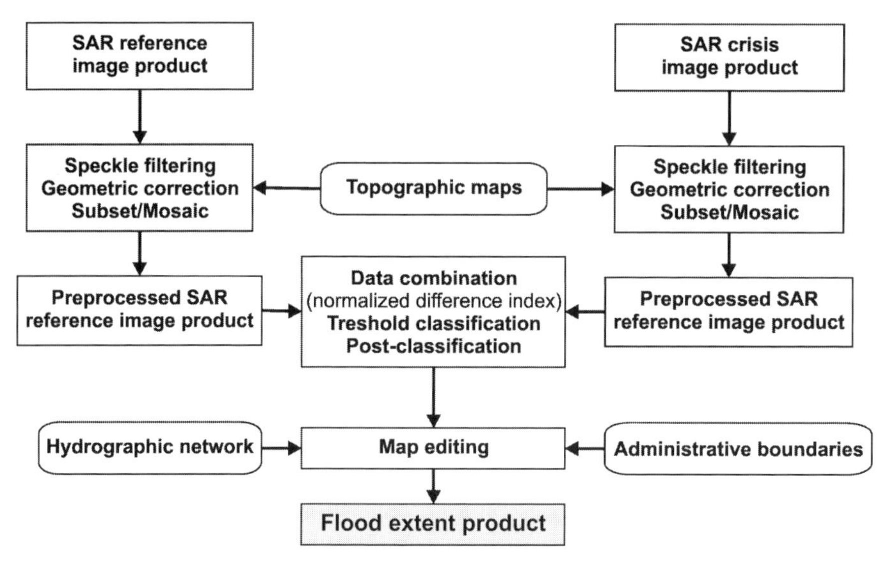

Fig. 3. Flowchart for the generation of the flood extend maps using satellite radar (SAR) images

Using the methodology for the identification and mapping of the flooded areas is possible to monitor and investigate the flood evolution during different phases. The figure 4 shows an example of the utilization of optical and radar data for the flood evolution monitoring, in the Crisul Alb basin using the RADARSAT image of 7.04.2000, during the flood event, comparing with the reference IRS image of 4.08.2000. This approach is very useful especially after the crisis for damages inventory and for recovery actions, taken to re-building destroyed or damaged facilities and adjustments of the existing infrastructure.

4.3 Method for the Flooding Risk Maps Preparation

The assessment of the flooding risk hazard requires a multidisciplinary approach; coupled with the hydrological / hydraulic modeling, the contribution of geomorphology can play an exhaustive and determining role using the GIS tools (Townsend & Walsh, 1998).

The structure of the GIS was planned in order to be used for the study, evaluation and management of information that contribute to flooding occurrence and development, as well as for the assessment of damages inflicted by flooding effects. In this regard the database represented by the spatial geo-referential information ensemble (satellite images, thematic maps and series of the meteorological and hydrological parameters, other exogenous data) is structured as a set of file-distributed quantitative and qualitative data focused on the relational structure between the info-layers. The GIS database will be connected with the hydrological database, which will allow synthetic representations of the hydrological risk, using separately or combined parameters. The figure 5 resumes the integration procedure of hydrologic/hydraulic model outputs and GIS info-layers for flooding risk maps preparation.

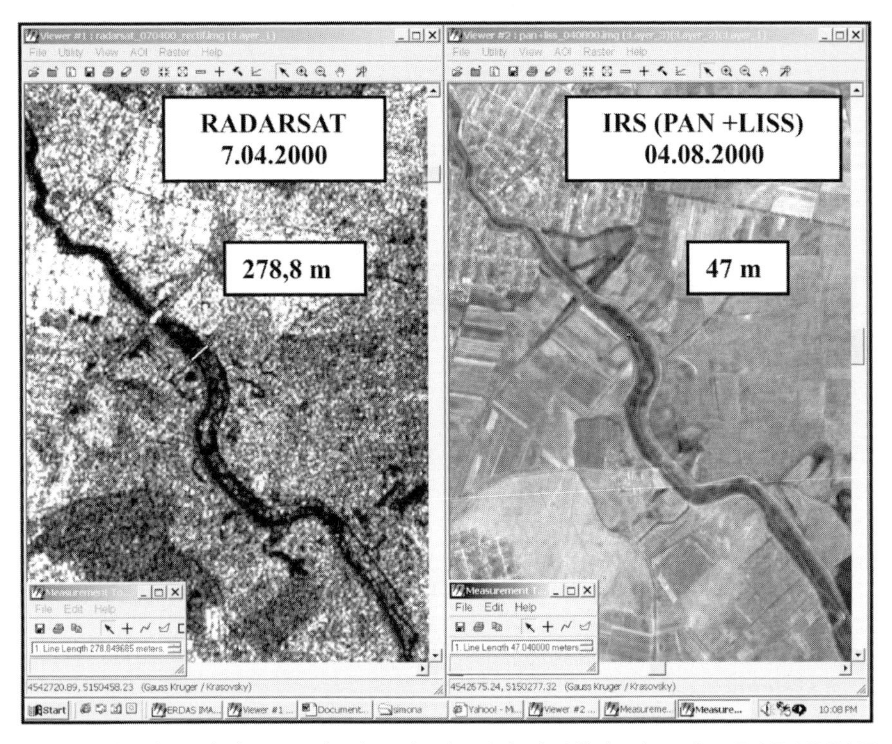

Fig. 4. Flood evolution monitoring, in the Crisul Alb basin using RADARSAT image of 7.04.2000, during the flood event, comparing with the reference IRS image of 4.08.2000

The construction of this GIS for the study area of the Crisul Alb and Crisul Negru basins was based, mainly on classical mapping documents, particularly represented by maps and topographic plans. Most of the thematic plans will be extracted from this classical mapping support. Due to the fact that, in most of the cases, the information on the maps is old-fashioned, it is imposed to update it on the basis of the recent satellite images (e.g. the hydrographic network, land cover/land use) or by field measurements (e.g. dikes and canals network). The GIS database contains the following info-layers: sub-basin and basin limits; land topography (organized in DEM); hydrographic network, dikes and canals network; communication ways network (roads, railways), localities, meteorological stations network, rain-gauging network, hydrometric stations network; land cover/land use, updated from satellite.

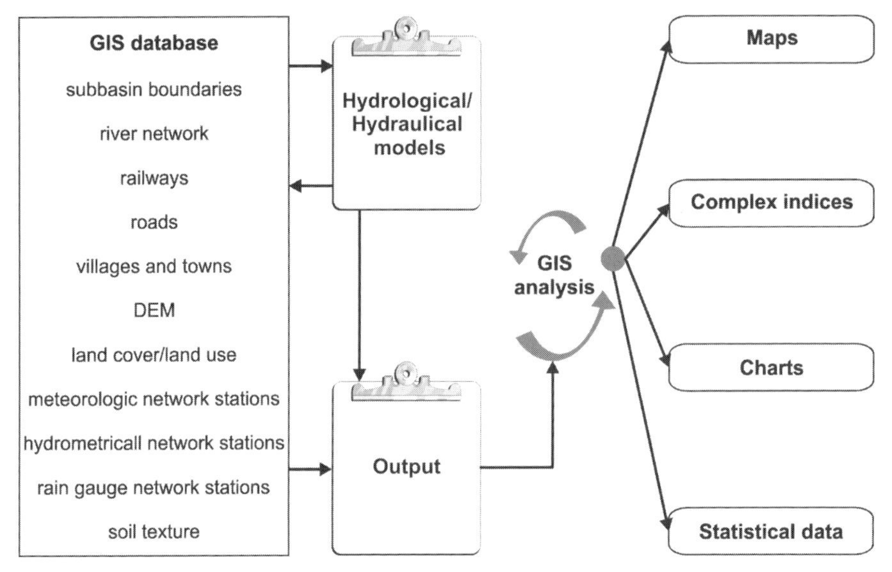

Fig. 5. Integration of hydrologic/hydraulic model outputs and GIS info-layers for flooding risk maps preparation

In order to obtain the flood risk map for the test-area, a important step refers to construction of the DTM and its integration, together with the land cover/land use maps in the GIS using a common cartographic reference system.

The DEM was realized after the following steps:

- scanning the topographic maps at 1:5,000 and 1:10,000 at 300 dpi resolution;
- geo-referencing the maps in the UTM projection;
- color separation and raster information layer extraction as a linear image (black & white without gray tones);
- vectorization of raster images.
- merging the maps;
- generating a triangulated irregular network (TIN) model.

In a TIN the point density on any part of the surface is proportional to the variation in terrain. A surface is a continuous distribution of an attribute over a two-dimensional region. TIN represents the surface as contiguous non-overlapping triangular faces. The surface value is estimated in this way for any location by simple (or polynomial) interpolation of elevation in a triangle. As elevations are irregularly sampled in a TIN, it was possible to apply a variable point density to areas where the terrain changes sharply, yielding an efficient and accurate surface model.

Various morphological criteria may be extracted from a DEM, designed using topographical maps and geodetic measurement criteria, such as altitudes, slopes, exposures, transversal profiles and thalweg locations. All these parameters are useful to evaluate the local or cumulated potential flowing into a zone of the basin. This approach is also useful to get a realistic simulation of floods, taking into account the terrain topography, the hydrological network and the water levels in different transversal profiles on the river, obtained from the hydraulical modeling.

Using the GIS database for the study area, that include the DTM, the land cover / land use maps and the vector info-layers (hydrographical, dams and canals networks, the communication and localities network etc) several simulation outputs of hydrological or hydraulic models could be superimposed in order to elaborate the risk maps to flooding.

5 Conclusions

Flood risk analysis needs to make use of and integrate many sources of information. This approach is more demanded in case of a transboundary river. The integrated flood management approach is in harmony with the recommendations of the International Strategy for Disaster Reduction and that of the EU Best Practices on Flood Prevention, Protection and Mitigation (Z. Balint, 2004).

Although satellite sensors cannot measure the hydrological parameters directly, optical and microwave satellite data supplied by the new European and American orbital platforms like the EOS-AM "Terra" and EOS–PM "Aqua", DMSP, Quikscat, SPOT, ERS, RADARSAT, Landsat7 can supply information and adequate parameters to contribute to the improvements of hydrological modeling and warning.

Considering the necessity to improve the means and methods to flood hazard and vulnerability assessment and mapping, the paper presents the capabilities offered by remotely sensed data and GIS techniques to manage flooding and the related risk. The study area is situated in the Crisul Alb - Crisul Negru - Kőrős transboundary basin, crossing the Romanian – Hungarian border.

The specific methods, developed in the framework of the NATO SfP "TIGRU" project "Monitoring of extreme flood events in Romania and

Hungary using EO data" for deriving satellite-based applications and products for flood risk mapping (maps of land cover/land use, thematic maps of the flooded areas and the affected zones, flooding risk maps) are also presented.

The satellite-based applications will contribute to preventive consideration of the extreme flood events by planning more judiciously land-use development, by elaborating plans for food mitigation, including infrastructure construction in the flood-prone areas and by optimization of the flood-related spatial information distribution facilities to end-users. In the same time the project will provide the decision-makers with updated maps of land cover/land use, hydrological network and with more accurate/comprehensive thematic maps at various spatial scales with the extent of the flooded areas and the affected zones.

References

Balint Z (2004) Flood protection in the Tisza basin. Proc. of the NATO Advanced Research Workshop – Flood Risk Management Hazards, Vulnerability, Mitigation Measures, Ostrov u Tise, Czech Republik, pp 171-183

Brakenridge GR, Tracy BT, Knox JC (1998) Orbital SAR remote sensing of a river flood wave. Int. J. Remote Sensing, vol 19 (7), pp 1439-1445

Brakenridge GR, Stancalie G, Ungureanu V, Diamandi A, Streng O, Barbos A, Lucaciu M, Kerenyi J, Szekeres J (2001) Monitoring of extreme flood events in Romania and Hungary using EO data. NATO SfP project plan, Bucharest, Romania

Brakenridge GR, Anderson E, Nghiem SV, Caquard S, Shabanch TB (2003) Flood Warnings, Flood Disaster Assessments, and Flood Hazard Reduction: The Roles of Orbital Remotite Sensing. Proc of the 30th International Symposium on Remote Sensing Environ, Honolulu, Hawaii

Munich Re Annual review of natural catastrophes (2003) Topics geo, available at http://www.munichre.com, Munich, Germany

Samuels PG (2004) An European perspective on current challenges in the analysis of inland flood risks. Proc. of the NATO Advanced Research Workshop – Flood Risk Management Hazards, Vulnerability, Mitigation Measures, Ostrov u Tise, Czech Republik, pp 3-12

Stancalie G, Alecu C, Catana S, Simota M (2000) Estimation of flooding risk indices using the Geographic Information System and remotely sensed data. Proc. of the XXth Conference of the Danubian Countries on hydrological forecasting and hydrological bases of water management, Bratislava, Slovakia

Townsend P, Walsh SJ (1998) Modeling of floodplain inundation using an integrated GIS with radar and optical remote sensing. Geomorphology, vol 21, pp 295-312

The Relationship between Settlement Density and Informal Settlement Fires: Case Study of Imizamo Yethu, Hout Bay and Joe Slovo, Cape Town Metropolis

Helen M. Smith

Department of Environmental & Geographical Science, University of Cape Town in collaboration with DiMP, Rondebosch 7701, Cape Town, South Africa.
Email: hucks@absamail.co.za

Abstract

Informal settlements have been part of the South African landscape as far back as the late 1800s. However, in the post-apartheid era, they have increasingly come under the spot light as a media concern. Housing shortages, service delivery, political tension, violence and crime, poverty and the high environmental and health risks are just some of what these settlements have to face. The problem is growing in many urban centres as in-migration continues and exacerbates the problems. In the case of Cape Town (main study context), an estimated 48 000 people are migrating into the City annually (CCC, 2004). Last year alone (2003) the City of Cape Town had a backlog of 240 000 houses (DiMP, 2004) and this year it has increased to 302 000 (CCC, 2004). Besides the huge backlog which the city faces, these areas have become the environments for accumulating disaster risk which result in repeat events that destroy the existing infrastructure and leave many people destitute and homeless. The causes for these disaster occurrences are very under researched and understanding of the complexity is just starting. The data collected shows a 120% increase in fire incidents in the last four years, as well as the fact that the frequencies of small events and the severity of disaster events are increasing disproportionately to the intensity of the hazard (MANDISA, 2004). This fact alone forms the basis of this study which investigation the relationship between settlement density and densification and the increasing patterns of fire severity over time. For this purpose two settlements, namely Imizamo Yethu (Hout Bay) and Joe Slovo (Cape Town Metropol) have been chosen.

Upgrading the infrastructure has, up until recently, been how many local authorities have dealt with informal settlements. They have started realising that the structural upgrading method used to counter the vulnerability of the community only partly solves the problem of fires. Weak community unity, limited awareness of risk, inappropriate behaviour, and irresponsible tenancy by landlords need to be addressed in parallel with structural upgrades (DiMP, 2004). GIS is the main tool used in this study to examine the relationship between dwelling density and informal settlement fires. Its main functionality revolves around analysing certain environments spatially and representing structural factors, however does not currently include non-structural mapping tools. This study will show that the socio-political and economic dynamics in informal settlements, which play an important role in the fire severity and community risk, also needs to be included when undertaking any comprehensive disaster risk assessment. The challenge posed by the complex nature of risk in informal settlements in the South African urban landscape, underlines the need for a holistic and integrated approach to GIS and Disaster Management.

1 Research Context

1.1 Introduction

In the beginning of the 1990s, an extraordinary process of political and social change started to take root in South Africa and it took only a few short years for the political party that had ruled the country for 48 years to turn its power over to a Government of National Unity. The world was in awe as from 1994, South Africa started its era of democracy and the society began to remodel itself as people became part of the post-apartheid society. Unfortunately, with the practices of the pasts, scars of epic proportions were left on the landscape and people; the new leadership was faced with the severe problems where many people, who had never been treated equally before, had to be provided for (Gawith, 1996:1).

One manifestation of the scars left on the South African landscape is informal settlements. "Informal settlements[1] are dense settlements comprising communities housed in self-constructed shelters under conditions of informal or traditional land tenure"[2] (http://www.sli.unimelb.edu.au/informal accessed 01/10/04). The landscape of developing countries illustrates these settlements among their defining features and the development and consolidation of these make-shift dwellings are often in response to an urgent need for shelter by the urban poor which the government of these countries cannot provide. A report on human settlements in general, known as the UNCHS[3] global report of 1986, pointed out that between 30% and 60% of inhabitants of most large cities in developing countries live in informal settlements. As such, the settlements are characterized by an impenetrable abundance of undersized, crude shelters built from various materials, degradation and erosion of the local ecosystem, relentless social problems, rapid, un-structured and un-planned development. Informal settlements arise when the existing land administration and planning fails to address the needs of the whole community and no low cost, basic formal housing is available (http://www.sli.unimelb.edu.au/informal accessed 01/10/04).

Internationally, settlements of this nature have become a major problem in countries of the Third World where the majority of the populations are the worlds' disadvantaged and the situation never seems to improve (Cross, et al 1996:1). This following sections will investigate the international context and then more specifically the South African context that applies to informal settlements.

1.2 The International Context

Urban migration[4] is a phenomenon found in First World and Third World countries and much research has been done in attempting to explain the processes and motivations behind their manifestation. Previously, researchers believed that the manifestation was that of total urbanization in

[1] Similar names for these types of settlements are squatter settlements/camps or shanty towns.

[2] For the purpose of this study this definition, while one of the many interpretations given to informal settlements, will be used as the primary definition.

[3] UNCHS - United Nations Centre for Human Settlements (Habitat).

[4] One of the definitions given by the Oxford Concise English Dictionary (10th Ed) is "migrate/migration (of a person) move to settle in a new area in order to find work" is the definition used in this review, however it must be noted that reasons other than finding work will be sited and used to explain the urban migration. This definition still panders to the old idea of migration.

which rural occupants left their origins in increasingly larger groups to permanently relocate to the urban centers. This would get to a point where the whole country would reach full urbanization. As explained by Cross, et al (1993), this process was often assumed to "behave according to gravity flow models", by which the movement of population was attributed to economic reasons where people from rural beginnings moved into urban areas in search of a better economic stature. This movement was predicted to remain reasonably constant. But, as further cited by Cross's article and argued by Mabin (1989 and 1990), it is questionable whether this model of a distinct and stable urban evolution actually applies in the Third World. Recently, a trend was identified in which the population flows were found to be heading towards the urban areas as well as moving back to the rural origin. This suggests that this back-and-forth movement or circulatory migration might continue for years to come rather than settling and resulting in complete urbanization. Circulatory migrations are a constant feature of migration patterns within South Africa as well as other Third World countries (Cross, et al, 1993). The complex nature of why people migrate into and out of urban centers is still very misunderstood and needs to be grasped in order for any government to handle this movement and provide for it.

1.3 The South African Context

1.3.1 Data Availability and Accuracy

There are very limited data sources profiling the South African population especially the population residing in informal settlements. Many of these areas were not even officially counted as settlements up until very recently.[5] As Hindson and McCathy (1994) point out that regardless of the national population National Census that was held in 1991, correct, featured and adequately categorized data on the composition of population within the informal settlements does not exist, as well as data on the migration patterns for the entire period prior to 1996. The character of migration and its innate intricacies and convolutions, such as the phenomenon of circular migration, exacerbate the difficulties in analyzing migration patterns in South Africa especially when the essential data records are missing (Hart 1992).

[5] One example of this is Joe Slovo in the Cape Town Metropolitan area where up until 2001, no official zoning of National Census was preformed independently for the area thus no statistics exist prior to 2001.

1.3.2 Future Population Growth

Presently, South Africa has a high rate of population growth which is believed by some to increase exponentially in the near future. However, there is reason to believe that this rate might drastically decrease due to the increase in mortality after the year 2020 as the Aids epidemic plateaus, thus causing a mass death rate to occur (MacInnis, 1997). This population boom within the urban centers as well as in-migration of the ever increasing rural communities is placing immense pressure on the cities in the form of mushrooming informal settlements (Saff 1993:235). South Africa's estimated annual population growth for urban Africans is ranging from 2.4% to 3.5% (Republic of South Africa 1994:9, South African Institute of Race Relations 1994:328,367, Barry and Mason 1997). Given the historical inequities of the past, the majority of the poor are African, as are the majority of informal settlement residents.

1.4 Disaster Science and Hazard Context

Originally, approaches to disaster events were reactive and focused on providing relief to the affected areas and households. Recent developments have seen a shift of focus to a proactive approach of decreasing the risk rather than managing the event. Risk reduction reflects a more comprehensive approach than disaster management and requires different, more inclusive development strategies than in the past.

1.4.1 Fire Focus and Occurrence

Fire occurrence is widespread across most urban South African informal settlements. The direct and indirect triggers vary from province to province along with the dynamics of the settlement. Fires have become very topical, especially as local government is directly responsible for infrastructural provision. This poses an enormous challenge with the dramatic increases in fire incidence, as not only is local government expected to provide adequate infrastructure, it is also required by law to address the level of risk. The DiMP[6] MANDISA database contains 12 500 disaster incidents recorded from 1990 to 1999 for the Cape Town Metropol. From 2000 to 2003, 11 000 fire incidents were estimated – a staggering 120% increase in four years. Most of these incidents were considered small to medium events as they did not affect more than 60 dwellings. The statistics indicate that the frequencies of small events and the severity of disaster events are

[6] DiMP - Disaster mitigation for Sustainable Livelihoods Program.

increasing disproportionately to the intensity of the hazard (DiMP; MANDISA, 2004).

1.4.2 The Historical Background to Fire Risk in Cape Town Informal Settlements

In the last decade, rapid growth in the Cape Town informal settlements has been accompanied by a high fire risk. "Between 1995 and 1999 a survey[7] of fire incidence across ten settlements[8] in the Cape Town area revealed a total of 1 612 reported informal dwelling fires which destroyed a total of **10 206** informally constructed homes. Of all dwellings destroyed, 3 227 (32%) were recorded in Langa/Joe Slovo." (DiMP, 2004) The massive increase in fire risk in the Cape Town Metropol has said to have the following contributing factors:

1. The fast rate of in-migration, policy ambiguity, with regards to the management of informal settlements, that has directed the national, provincial and local government approaches.
2. The lack of and sluggish delivery of low income housing.
3. Low cost housing decline over the past five years as a result of poor building standards, resulting in a need for necessary upgrades.
4. The provision of affordable services.
5. Utilizing all the housing allocations for the Western Cape of which 60% currently remains unspent[9].
6. Inadequate, integrated development planning in informal settlements.
7. Increases in the levels of poverty, crime and social instability within certain informal settlements.

   (DiMP, 2004)

---

[7] DiMP, 2004. Report on Informal Settlement Fire Occurrence and Loss, Cape Town Metropolitan Area. 1995-1999. Disaster Mitigation for Sustainable Livelihoods Programme. University of Cape Town.

[8] Bonteheuwel, Brown's Farm, Elsies River, Gugulethu, Imizamo Yethu, Langa, Manenberg, Nyanga, Redhill and Wallacedene.

[9] Rhoda Kadalie in *Business Day*, April 15th, 2004 as cited by DiMP, 2004.

2 Study Site Descriptions

2.1 Study Site Context

Cape Town, a major urban centre, and the location of two settlements, Imizamo Yethu and Joe Slovo, is experiencing an estimated growth of 48 000 migrants annually. This has resulted in incredible pressure on a number of informal settlements of Cape Town (DiMP, 2004). The Provincial Government and City of Cape Town have not been able to handle this influx at all with the latest figure on housing backlog as of this year (2004) is reported to be 302 000 houses. It is estimated that this figure is increasing by 27 000 annually with the local authority only able to build 23 000 homes a year (CCC[10], 2004).

The following map illustrates the widespread location of informal settlements across Cape Town, with particular focus on the two chosen study sites:

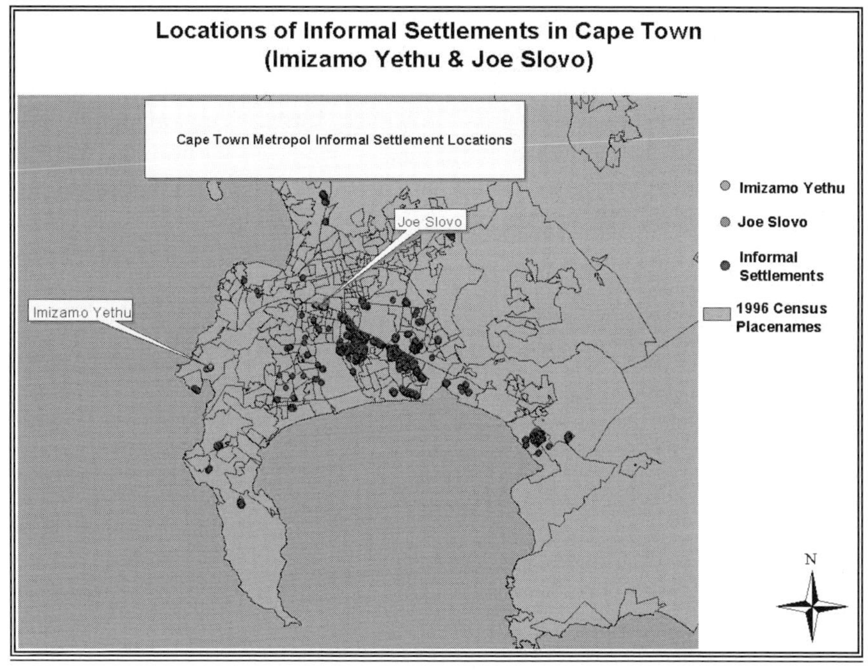

Fig. 1. Locations of Informal Settlement in the Cape Town Metropol

---

[10] CCC – Cape Town City Council

The following two settlements are found in this very context, where their history has molded their rise, formation, consolidation and composition. They were chosen on the basis of adequate data and established research prior to this study, so to have a foundation on which to work.

2.2 Study Site 1: Imizamo Yethu, Hout Bay

The following map is of Imizamo Yethu, Hout Bay and its surrounding areas.

From the 1960s Hout Bay started becoming popular as a residential suburb and its original purpose in agriculture started to phase itself out as the predominate land use form. Growth occurred in the fishing industry which provided more of a labor demand in the area. The white elite increased in the area as a result of the available real estate. This development prompted a greater need for African labor both as temporary construction workers and also on a more permanent basis in terms of domestic work and labor. The conflict between apartheid's segregationist policy and the need for African labor in the area of Hout Bay meant that evidence of early migrants can be traced to the area as far back as ±1940. To begin with, the white rate-payers and the apartheid regime managed to limit the availability of African housing in the area. However, forced removal of the coloured community to a township established near the harbour as well as strict controls and lack of provision of formal housing resulted in squatting[11] (Greene 1991).

[11] The term 'squatting' or 'squatter' refers to dwellings constructed on land which is not zoned as a residential area and thus deems any type of settlement illegal. The land chosen is usually preferable due to its location to places of work or opportunities of work. The land is also not serviced and thus slums have been known to result after a while.

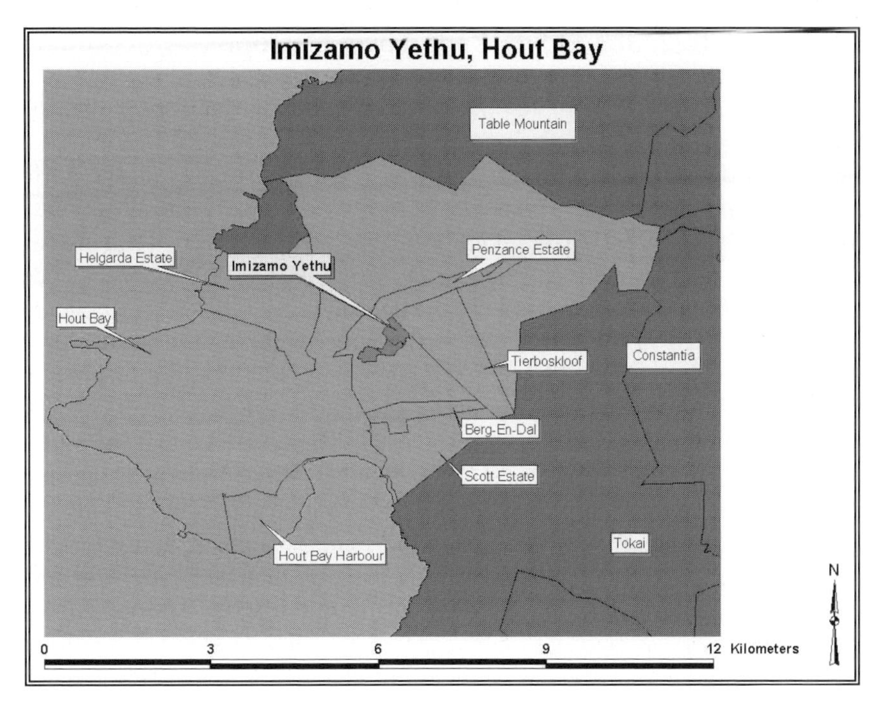

Fig. 2. Imizamo Yethu, Hout Bay and Surrounds (Data Source: 1996 National Census GIS Data)

The population in the area had grown substantially by the mid-1970s and was largely seasonal with the fishing population growing from 130 in 1944 to over 2000 and growing to as large as 5000 during seasonal Snoek catches. The fishing industry growth in the area increased the number of factories, which again, in turn resulted in a boost in the demand for African labor. Princess Bush was the name given to the area used by the Africans for settlement. After a severe fire in Princess Bush in 1990 and a long history of other fires and flood events prior to this, the local authorities realized that the close proximity of the settlement to the coast and the lack of infrastructure placed the community residing in the settlement at great risk. Thus Imizamo Yethu was founded in 1990. The area was originally forestry land which was converted into an 18 hectare, 429 plot, site and service scheme. (www.sli.unimelb.edu.au/informal/h_bay_history.html).

According to the last available figures, Imizamo Yethu comprised 3067 households (2003) with a total population of over 8062 people (2001 National Census). According to the 2001 National census the main infrastructure found in Imizamo Yethu is reasonable when compared to other informal settlements in the greater Cape Town Area, as a basic level of formal housing exists.

2.3 Study Site 2: Joe Slovo, Cape Town Metropol

Figure 3 is a macro view of Joe Slovo and it's location in Cape Town. Joe Slovo began in early 1994 as a natural and steady spill-over of people from the already dense Langa Township. It was originally situated on the buffer strip that separates Langa Township from the N2 and N7 off Washington Avenue and Vanguard Drive. This informal settlement lay on a narrow strip of land which shared its boundaries with hostels, formal houses and a 'coloured' settlement, Bonteheuwel (DiMP, 2004:20).

Langa[12] started in the 1970s and is thus one of the oldest townships in the Western Cape. It faced many challenges prior to 1994 like Influx control[13] and Pass Laws[14] but the community's fears were put to rest after the country became a democratic nation in 1994. Many believe that the consequences of democracy and easing of influx controls resulted in the consolidation and formation of Joe Slovo. The residents now occupying this tract of land are believed to have been old Langa residents whose families moved to Cape Town to join them thus causing them to search for land that would accommodate the increase in household size. The location of Joe Slovo is very attractive to new comers as it lies in close proximity to the Epping Industrial area as well as all the major transport networks which are linked to the Cape Town CBD (DiMP, 2004:21).

[12] Langa is named after Chief Langalibalele.

[13] Limited influx into urban areas.

[14] Divided residential location of people according to race.

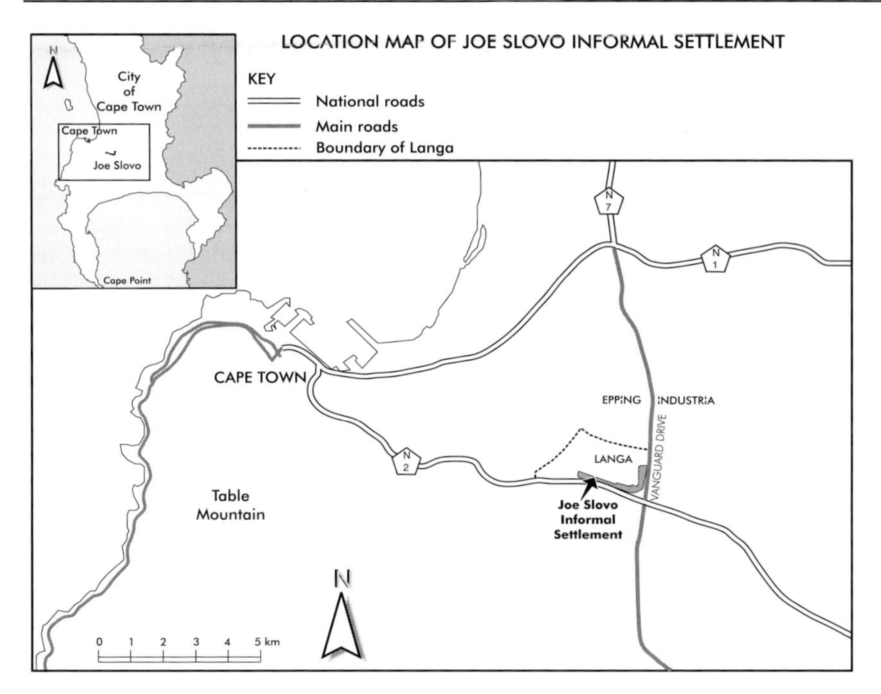

Fig. 3. Location of Joe Slovo in Cape Town (DiMP, 2004)

This area, was however, only recently officially zoned after a massive fire pressured the local authorities to dedensify the land and place fire tracks between newly established residential blocks. Presently the area is very densely populated and is in a constant growth phase (DiMP, 2004:21)

3 Methodology

3.1 Introduction

The methodology used sought to achieve the following:

1. A full analysis into the trends and patterns of dwelling densification, settlement density and fire severity in Joe Slovo and Imizamo Yethu
2. An examination and map product representing the densification in Joe Slovo and Imizamo Yethu in a time series manner.
3. A map product which reflected the density of Joe Slovo and Imizamo Yethu at a point in time in the selected years.

4. The close correlation and data overlay of the MANDISA fire severity recorded for Joe Slovo and Imizamo Yethu in a time series manner.
5. A map product reflecting these patterns and trends of fire severity in a time series manner.
6. A graphical and statistical analysis of the findings and map products as well as possible explanations for the identified trends and patterns.

This was achieved through the collection of the data from various sources, preparation of these different data sources, correcting overlay errors in spatial data, representing the different chosen study site's trends and patterns, analyzing these graphically and statistically. The full method employed by this study is available in the main thesis which is available at the University of Cape Town or on request to the author. The following is merely a skeleton outline of what steps were taken to achieve the results.

3.2 Collection of Data

A number of sources of data were collected.

3.3 EAs and Density Counts

The aerial photographs were in MrSID[15] format and thus the MrSID image extension was loaded to enable ArcView 3.3 to display the images.

3.3.1 Joe Slovo

The aerial photographs available for this settlement were: 1997, 1998, 2000, 2002 and 2003. The CCC GIS department provided 2002 and 2003 dwelling counts but in polygon format. These polygons were converted to a dot/point feature to simplify the density. This allowed for a count and total of each of the dwellings contained in each aerial photograph for each year. Density was determined using Feature Density Script.

3.3.2 EAs Use

The enumerator areas (EAs) from the 1996 Statistics SA national Census were brought in as a polygon theme layer. The EAs were incorrectly positioned and digitized on the aerial photographs and thus when transposed on top of the aerial photograph, didn't fit. The first process that took place

[15] MrSID is one of the many image formats that are used to store aerial photography

was using a script, Shape Warp[16], which enabled the warping and re-positioning of the EAs on top of the aerial photograph.

3.3.3 Imizamo Yethu

The aerial photographs available for this settlement were: 1996, 1997, 1998, 2000, 2002 and 2003. The same aerial photographs were supplied by the CCC GIS Department and they underwent the same conversion as per the settlement mentioned above.

3.4 Fire Data Preparation Method

The MANDISA database aims to record all small, medium and large scale disaster incidents in the Cape Town Metropol. It has been based on the 1996 EAs as the data sources location information consists mainly of street address. These addresses are either manually entered or automatically assigned by the database. One of the primary reasons for choosing Imizamo Yethu and Joe Slovo was that both settlements' fire data had been completed and captured.

3.5 Preparation and Production of the Density and Fire Severity Maps for Both Settlements

Once all the layers were edited, each was loaded into a view. Maps of dwelling density and fire severity were constructed and once the placing of all the map elements was completed, the maps were exported as a JPEG. This allowed for easy insertion into the report and presentations.

3.6 Data Manipulations and Analysis

The aerial analysis produced figures which gave mainly the density for each year as well as the associated fire severity. This study was done using aerial photographs where dwellings under trees were excluded and the National Census was done going from door to door throughout the different areas. Therefore discrepancies occurred throughout the overlaying of dif-

[16] Shape Warp is a script freely available on the web which can be loaded in ArcView to allow the user to warp feature themes to fit an image if required. The script has other available features which were not however made use of in this study.

ferent data sources. In 1999 and 2001 no aerials existed for either settlement and thus a linear interpolation was used to complete the time series. This was done in the following way:
E.g. in the case of 1999: 2000 dwelling density count – 1998 dwelling density count)/2 + 1998 dwelling density count.

4 Results

4.1 The Main Findings

The main aim behind this study was to enhance the understanding of the relationship between density and densification patterns of selected informal settlements and the associated fire severity that is directly related to these patterns. However, for the purpose of this symposium, the main finding involving density are only one of the many issues involved and can be further explored using the original base document found at the University of Cape Town. Using GIS allowed for an insight into the short comings of the programs in the ArcGIS series and illuminated the different areas where development needs to take place. The following diagrams presented in a time series manner illustrate some of the outputs produced in the study.

4.1.1 Imizamo Yethu & Joe Slovo

The following diagram of the Imizamo Yethu Settlement illustrates how the aerial photographs were used to as foundation to the visual analysis and the resultant data.

Figure 4 and 5 shows how two different layers, namely enumerator areas (EA) 1996 and actual area, to combine the area of the settlement to the fire and census data. Due to the change in EA, densification and subsequent increase in area of the settlement the layers do not overlay exactly. After the exclusion of the aerial photograph, one is able to see this problem more clearly.

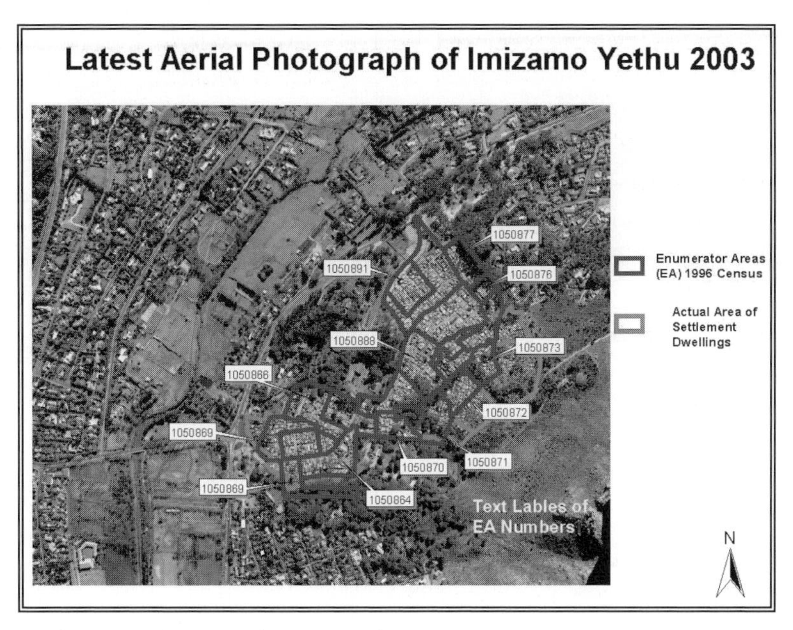

Fig. 4. Imizamo Yethu, Hout Bay (ArcView)

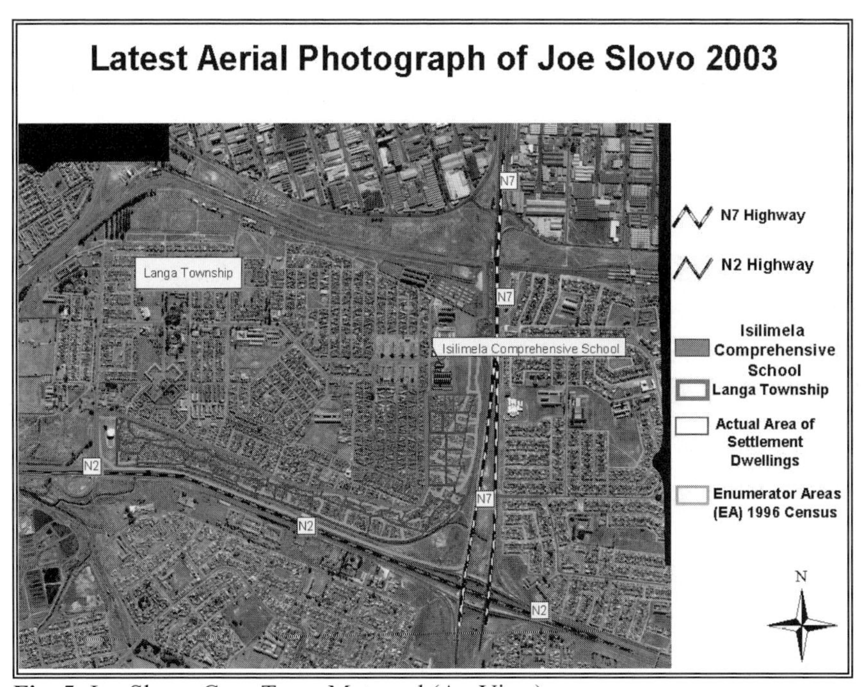

Fig. 5. Joe Slovo, Cape Town Metropol (ArcView)

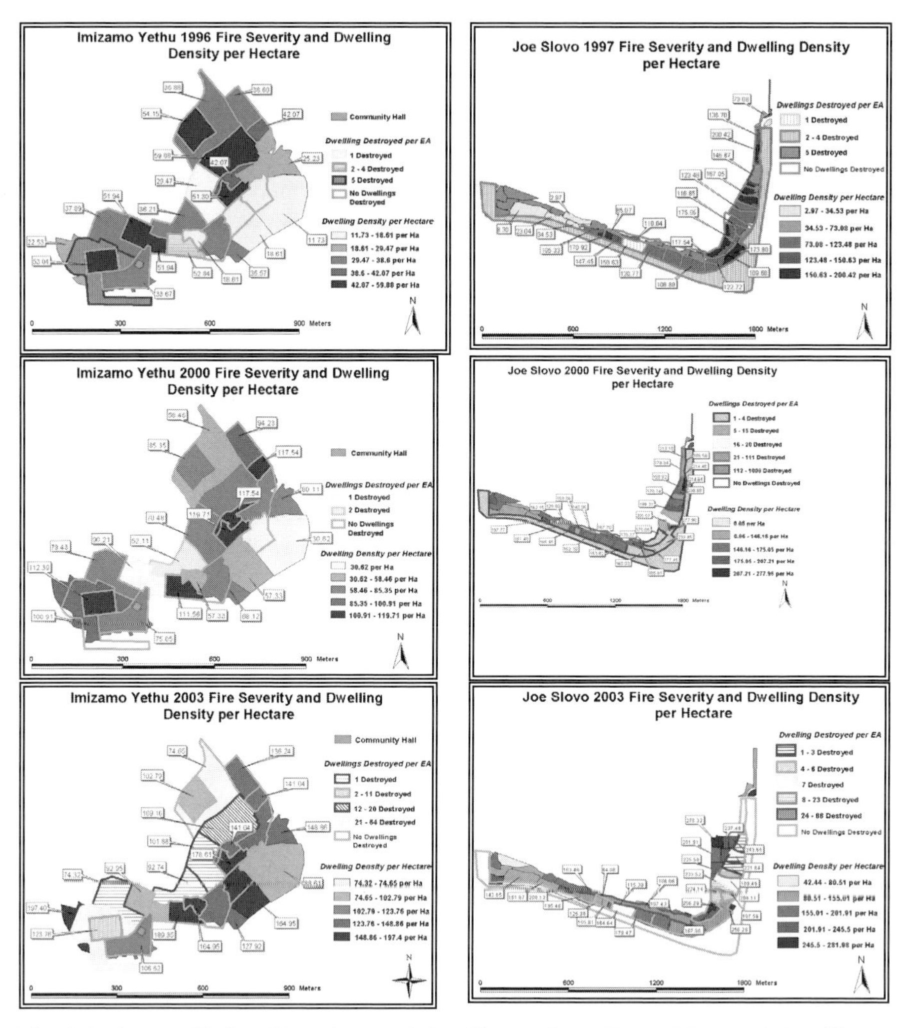

Fig 6. Imizamo Yethu, Hout Bay and Joe Slovo, Cape Town Metropol Dwelling Density and Fire Severity

In the time series above, one can see that in the case of both settlements, EAs, on which fire data was based and the actual area of the settlement did not match up and impose over one another exactly. In the case of Joe Slovo (the settlement contained by the right column in figure 6), the area changed drastically during the official zoning and recognition of the settlement in 2000. This has caused havoc with the fire incident records as they were based on the 1996 EAs and now did not correlate with the defined area, thus making the analysis difficult and increasing the error.

The findings showed that settlement density did play a part in driving the severity of the fires but not that start of the fires. The following graphically illustrate the relationship between these two variables.

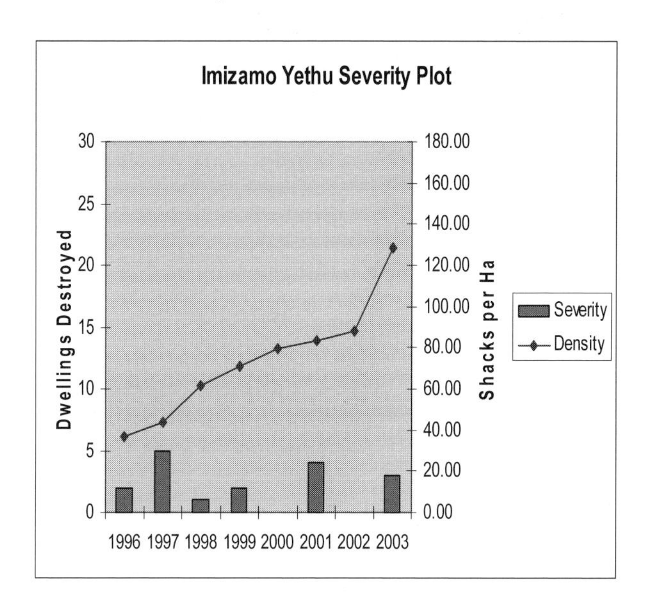

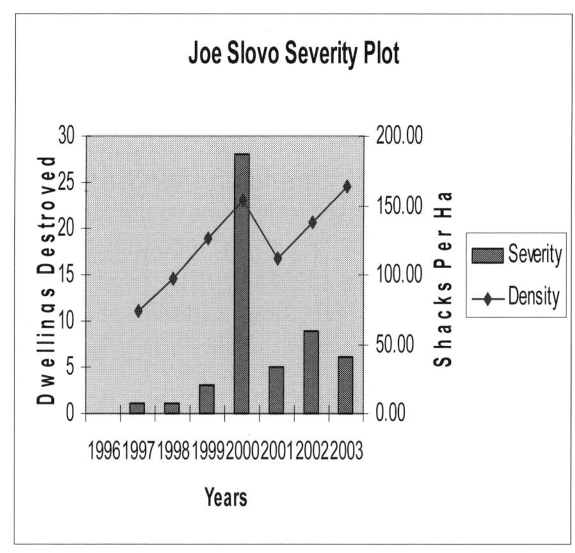

Fig. 7. Imizamo Yethu, Hout Bay and Joe Slovo, Cape Town Metropolis Dwelling Density and Fire Severity Plots

The plots show that the relationship between fire severity and dwellings density in the case of Imizamo Yethu is extremely poor as the dwelling density continues to rise; the fire severity seems haphazard in nature and shows no distinct relationship. However, in the case of Joe Slovo a slight relationship does seem to exist. This thus prompted further exploration which produced the following plot.

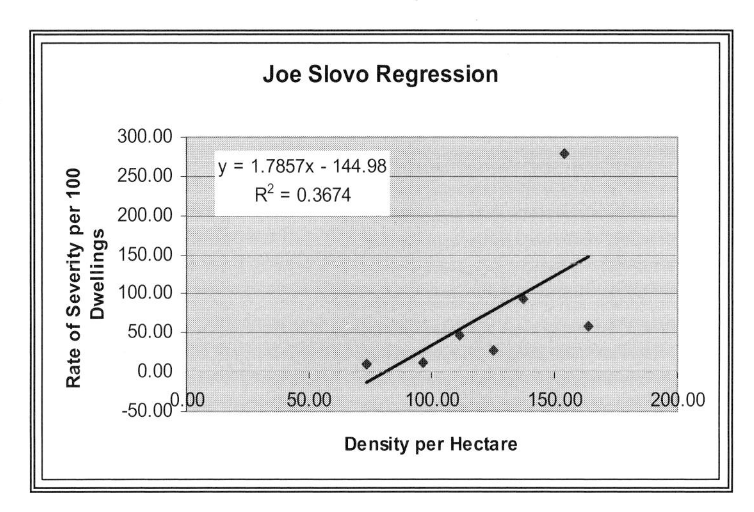

Fig. 8. Joe Slovo Regression of Rate of Severity and Dwelling Density per hectare

The graph shows a weak relationship level still as the R-squared value is only 0.3674. However, density definitely plays a role where in 2000 the zoning and de-densification of the settlement did cause a marked decrease in the fire incident and severity. This unfortunately doesn't last for long as in 2003 the fire incidents increases to just above the 2000 figure of 23. but, even in this settlement, more factors seem to be playing roles greater than that of settlement density, thus illustrating the need for further research into the environment that surrounds these settlements.

4.2 Discussion and Conclusion

When dealing with informal settlements, one must realize that the environmental context is extremely complicated with different factors playing roles which influence the risk of the settlement differently. Not enough evidence is available to imply that density primarily drives the fire severity within a settlement as too many areas in these two settlements are experiencing high fire severities yet average to below average densities. The severity is not near as bad but this is probably attributed to other factors like the climate at the time not being as hot as the climate experienced in 2000 or other socio-political factors and the density continues to climb back to its original levels experienced in 2000.

4.2.1 The Downfall of Disaster Management

When dealing with the complex environment of informal settlement fires, the complexities tend to show themselves after the technical, simplistic approach is chosen by the authorities and disaster managers. This tends to be in the form of infrastructural upgrades involving the provision of housing and basic services. Once this is done the authorities seem to wash their hands of the settlement's risk levels as they believe it has been addressed. This is fundamentally flawed as many of these settlements risks lie in their social, economic and political make-ups. When dealing with these settlements a holistic, individualistic and integrated approach needs to be taken in order to truly address the needs and risk levels of the unique communities resident in these settlements.

4.2.2 The GIS Context

GIS is problematic as it handles factors which can be easily defined and are structurally seen. If further research were to uncover that the level of community cohesion were a major factor influencing fire severity, GIS would have limited capability to plot this, as no real social mapping tools are encapsulated in the program. This is an area which needs to be explored and developed not only for the GIS applications but also for disaster risk science. The resources required for this are immense as on the ground studies would have to be preformed in all settlements across a province to get a general index that could be applied generally. What complicates it further is past factors involved in South Africa's history which clearly divides the provinces as pointed out in the literature review. A model mapping social, political and economic factors would probably have to be developed for each province individually as informal settlements in KwaZulu Natal, for example are completely different to the informal settlements found in the Western Cape. The evolution of GIS into starting to grapple with social mapping factors has to take place for the technology to stay abreast as well as act as a tool for Provincial or City disaster managers.

Databases like MANDISA are useful in their conceptual design as they could supply the right people with the right information and give a general indication of the level of fires in an area as well as the times or years where these fires are most likely to start. However, being able to use this as an accurate source of fire data as input into a GIS software program is far from occurring. Any database is only as accurate as its source data. By relying on EAs, no solid foundation exists as these EAs constantly change. It should be noted though that this is no faults of the database and its design as the fire incident records are not locating the fire according to a universal reference frame like x:y co-ordinates and thus forces the database to attempt a location based on a street address. GIS cannot function without an EA, x;y co-ordinate or some unchanging, universal reference systems and so for the purposes of this study EA were necessary. But EAs can change shape as well as location, as seen in the case of the Joe Slovo 2001 rezoning, and thus are not at all solid, unchanging reference frame. When looking at First World countries, who are streets ahead of us in terms of record and data capture, co-ordinates were their fundamental reference frame and until our fire services and other related fire response bodies start using co-ordinate readings off a GPS, a database like MANDISA is, in the long term, redundant especially in the case of informal settlements where the environment is constantly in flux.

4.2.3 Recommendations for Further Research

Informal Settlements present an environment filled with research opportunities which span across diverse fields of expertise. The disaster events that are experienced by these settlements not only incorporate fire hazards but also hazards like flooding. Researching these different hazard characteristics and their influence on the settlements risk profile could have interesting study implications. Within hazards lies another study area whereby the development of the index which represents the level of severity and occurrence could be developed. Defining whether, for example, a fire is a small, medium or large event is still very undefined and could be further refined.

In settlement density, another research direction which involves the study of the actual materials used to construct the dwelling is revealed. This study was limited in that the time limited caused the study to rely on aerial photographic information as its main base of data. However, there is evidence to suggest that fire severity is often influenced by the types of materials used in the settlement as certain materials are more flammable than others. A door to door study could be done to establish the level of each material and its influence on the fire severity of that settlement.

Another direction could be to research the other factors influencing the risk level of the settlement and see whether they are quantifiable. This type of study could mainly involve the non-structural dynamics which include social, political and economic. Although, totally dependant on the specific settlement they are found in, one could find a common pattern allowing certain assumption to be made that could be applied universally. If they are, multiple GIS technical evolutions could be developed in the form of for an example, an informal settlement fire index or early warning system.

One of the fundamental problems facing researchers wishing to choose this direction is the lack of strong, accurate, complete datasets. One study could start in the data collection arena where explorations into the costing and human resource requirement of installing systems which record the events using a universal reference frame. There are also many other data issues to be explored, like its use in the GIS software. GIS's need to develop to keep abreast of the Disaster Science field is highly necessary and could provide more available research opportunities.

The informal settlement environment has only recently been accepted as officially part of the urban landscape and thus many avenues of further research besides the ones mentioned above, in diverse fields are yet to be uncovered.

Internet Sources

http://www.pmg.org.za/docs/2004/viewminute.php?id=4312 accessed 6/10/04 Housing Portfolio Committee 18 August 2004 Statistics South Africa on Informal Settlements: briefing

http://www.durban.gov.za/eThekwini/Municipality accessed on 06/10/04

http://www.findarticles.com/p/articles/mi_m0856/is_4_30/ai_102980261/pg_2 accessed on 06/10/04 - Population reports

http://www.findarticles.com/p/articles/mi_m0856/is_2001_Summer/ai_84894273 accessed on 06/10/04 - Cities at the forefront

http://www.sacc-ct.org.za/ppu_1998.html - Accessed on 19/10/04

http://www.sli.unimelb.edu.au/informal assessed on 01/10/04

References

Cape Town City Council (CCC) and Disaster Management Departments Statistical estimates for 2004 for the City of Cape Town

Cole J (1982) Crossroads, Ravan Press, Johannesburg

Crankshaw O (1993) Squatting, Apartheid and Urbanisation on the Southern Witwatersrand. The Journal of African Affairs (1993), 92, pp 31-51

Cross C, Bekker S, Clark C (1993) Fresh Starts: Migration streams in the Southern Informal Settlements of the DRF. Natal Town and Regional Planning supplementary Report, Vol. 40, The Natal Town and Regional Planning Commission, Pietermaritzburg

DiMP (2004) or MANDISA (2004) refers to any data, findings and interpretations attributed to the Disaster Mitigation for Sustainable Livelihoods Programme, University of Cape Town and MANDISA disaster risk analysis applications as well as the following

Disaster mitigation for Sustainable Livelihoods Programme (DiMP). (2004) Evaluation of the fire prevention awareness campaign. Joe Slovo

DiMP (2004) Report on Informal Settlement Fire Occurrence and Loss, Cape Town Metropolitan Area. 1995-1999. Disaster Mitigation for Sustainable Livelihoods Programme. University of Cape Town

Du Toit B (1975) A decision-making model for the study of migration, In Migration and urbanization: models and adaptive strategies, Du Toit B, Safa H (eds.) 1975, Mouton Publishers, the Hague, Paris

Gawith M (1996) Towards a Framework for integrating environmental and community concerns into the planning and development of informal settlements: A case study of Hout Bay, Western Cape. Submitted in partial requirement for the degree of Masters of Science in the Department of Environmental and Geographical Science, University of Cape Town.

Greene C (1991) The origins and development of informal settlement in the Hout Bay area 1940 – 1986, submitted in partial requirement for the degree for Bachelor of Arts (honours) in Economic History, University of Cape Town.

Hart T (1992) Informal Housing in South Africa: an Overview, In South Africa in transition: urban and rural perspectives on squatting and informal settlement in environmental context, The University of South Africa

Hindson D, McCarthy D (1994) Here to Stay: Informal Settlements in KwaZulu Natal, Indicator Press and University of Natal

Kok P, O'Donovan M, Bouare O, Van Zyl, J (2003) Post-apartheid patterns of internal migration in South Africa, Human sciences research council, HSRC Cape Town

Li J (2000) Informal Settlement Modelling using Digital Small-format Aerial Imagery, submitted in fulfilment of the requirements for the degree of Doctor of Philosophy in Engineering, Department of Geomatics, University of Cape Town

MacInnis R (1997) What's care got to do with it? Responding to the AIDS pandemic, AIDSLINK Mar-Apr 1997 44, pp 1-3

Marshall S, Stevens L, Kola S, Kimmie, Z (2001) Upgrading Gauteng's Informal Settlements; Vol. 6: Summative Evalution of time series study at Albertina, Eatonside, Johandeo and Soshanguve South Extension 4. CASE (Community Agency for Social Enquiry) Researched for the Gauteng Housing Department

Mazur RE, Qangule VN (1995) African Migration and Appropriate Housing Responses in Metropolitan Cape Town. Western Cape Community-based Housing Trust, Draft Report

Metropolitan Spatial Development Framework (1995) A Guide for Spatial Development in the Cape Metropolitan Region. Draft for Discussion. Western Cape Economic Development Forum, Urban Development Commission

Republic of South Africa (1994), South African Institute of Race Relations (9), p. 328, p. 367, Barry and Mason 1997

Saff G (1996) Claiming a Space in a Changing South Africa: The "Squatters" of Marconi Beam, Cape Town. Annals of the Association of American Geographers, 86(2), 235-255. Blackwell, Cambridge, Massachusetts

Tomlinson R (1990) Urbanization in post-Apartheid South Africa, Unwyn Hyman Ltd. London

Yirenkyi SY (2000) Conceptual Design of a GIS-based land inventory model for urban informal settlement land management, submitted in partial fulfilment of the requirements for the degree of Masters of Science in Engineering, Department of Geomatics, University of Cape Town

AVHRR Data for Real-Time Operational Flood Forecasting in Malaysia

Lawal Billa[1], Shattri Mansor[1], Ahmad Rodzi Mahmud[2] and Abdul Halim Ghazali[2]

[1] Spatial & Numerical Modeling Laboratory, Institute of Advanced Technology &
[2] Department of Civil Engineering, Faculty of Engineering, University Putra Malaysia, 43400 Serdang, Selangor, Malaysia.
Email: biwal2000@yahoo.com

Abstract

Flash floods strike quickly and in most cases without warning. They are usually observed before any warning can be issued and usually persons and property have been affected before the warning reaches them. Such are the conditions prevalent in Malaysia's extreme monsoon weather that occasionally causes floods and results in the extensive damage to property and sometimes loss of lives. Over the years variously hydrological and structural engineering measures have been implemented for flood monitoring and forecasting. These measures have only yielded limited success as may be seen in the recurring flood situation. Yearly financial and property loss estimates have increased and an estimated cost of over 2.5 billion RM is projected for the year 2004 according to sources from the drainage and irrigation department of Malaysia. It has thus become apparent that Malaysia institutes an effective operational flood forecasting to arrest the persisting flood problem.

In this paper we will expound on current flood management and forecasting system being implemented in the country, particularly the Klang Valley that includes Kuala Lumpur where there has been tremendous urban growth and development in the last one and half decades. The paper further discusses where current flood management systems have been lacking in the absence of real-time hydro-meteorological forecasts. Where as hydrodynamic simulations and structural control measures have been emphasized in many flood management systems in Malaysia, the integration of real-time hydro-meteorological forecasts have been conspicuously absent, rendering most in-situ flood forecasts and early warnings ineffective in address the flood problem in the country.

Malaysia is a tropical country that lies along the path of the northeast and southwest monsoon. Although satellite image based NWP have proved useful for the tropical and equatorial regions of the world in flood forecasting, they have yet to be applied in Malaysia. Observations have generally shown heavy cumulonimbus clouds formation and thunderstorms precede the usual heavy monsoon rains that cause floods in the region. This makes quantitative precipitation forecast a must be input to any flood early warning design. Numerous empirical studies have determined that cloud top temperatures less that 235k in the tropics are generally expected to produce convective rainfall at the rate of 3mm/hr. In this study we thus investigate monsoon cloud formation that has the propensity to precipitate using NOAA-AVHRR data for real-time operational flood early warning in Malaysia. The AVHRR data has been preferred for its relatively high temporal resolution of at most 6/hours, its easy acquisition and cost effectiveness and its ability for automated geometric rectification when compared to GEOS and GMS data.

Cloud cover and types are processed using cloud indexing and pattern recognition techniques on the AVHRR data. The cloud indexing technique was initially developed for NOAA but was later also adapted for Geostationary satellite images. The technique assigns rainfall levels to each cloud type identifies in an image based on the relationship between cold and bright clouds top temperature and the high probability of precipitation. We discuss how visible (VIS) and infrared (IR) techniques are applied to bi-spectral cloud classification and rain areas are determined by classifying pixel clusters in the VIS/IR histogram. Precipitation probability is evaluated based on the relationship between cold and brightness temperature of clouds. The near infrared (NIR) and infrared (IR) channels 3, 4, and 5 of the data are processed for temperature and brightness. Cold clouds with temperature below 235k threshold value are taken as indication of rain. Rainfall is estimated based on the assumption that every cloud pixel has a constant unit rain-rate of $3mm^{h-1}$, which is appropriate for tropical precipitation over 2.5^{o} x 2.5^{o} areas around the equator. The paper finally discusses current developments in "nowcasting" that utilizes latest satellite observations together with numerical weather prediction models and how this system can be adapted to the needs of very short term forecast for flood early warnings in Malaysia.

1 Flood Forecasting and the Impact of Tropical Monsoon Storms in Malaysia

Every year tropical monsoon storms result in severe flooding and causes enormous economic damage, social disruption, and sometimes loss of lives. Extreme monsoon storms weather phenomena are the most destructive natural disaster afflicting Malaysia in respect of the cost, damage to property and the area extent (Keizrul and Chong, 2002). Accurate forecasting of floods induced by tropical storm is thus instrumental to the reduction of flood impacts. The damages caused are generally associated with wind damage, storm surge, and flooding. Accurate forecasting of these impending floods requires adequate meteorological inputs such as real-time rainfall, quantitative precipitation forecasts (QPF), and the cyclone landfall location. Hence, close interaction, technological development and cooperation between forecasting and meteorological technologies are of significant importance to improving flood forecasting and early warning in Malaysia.

The monsoon storm threat in Malaysia is such that flash floods strike quickly and in most cases without warning. Flooding is usually observed before any warning can be issued and usually persons and property have been affected before the warning reaches them. Variously hydrological models and structural engineering measures have been implemented over the years for flood monitoring and forecasting but have yielded only limited success. Flood forecasting model being implemented have been effective at simulating runoff and basin responses to flood based on synoptic data, however for a flood forecast to be really effective it should provide information about expected rainfall intensity well before it actually occurs. This should increase the emergency response time and provide enough time for contingencies. In the absence of an effective hydro-meteorological forecasting system, there has been sturdy and yearly increase in cost of damage due to flood. Sources from the drainage and irrigation department of Malaysia, estimate the costs of damage due flood for the year 2004 to be over 2.5 billion Ringgits. It has thus become apparent that Malaysia institutes an effective operational flood forecasting to arrest the recurring flood problem.

According to Chong (2001) the World Meteorological Organization (WMO) has helped many countries in improving their national capabilities in flood forecasting and warning under the Tropical Cyclone Program (TCP) in the last decades. The Typhoon Operation Experiment (TOPEX) from 1982 to 1983 was a classical example in which six countries in the Typhoon Committee area had improved their flood forecasting systems. It was thought necessary at the time to address the limitations and problems affecting the effectiveness of the flood forecasting and early warning systems in its model selection, model calibration and real-time operation of flood forecasting models. Experts have prioritized research areas of operational flood forecasting to be in the development of improved meteorological inputs and hydrological considerations.

2 AVHRR Data for Meteorological and Precipitation Monitoring

The availability of data from meteorological satellites such as polar orbiting NOAA series and Geostationary satellites (GMS, METEOSAT, GOES) for more than 30 years have made them the most suitable and currently the most commonly used data set in the monitoring and collection of clouds information. Meteorological satellites provide excellent aerial coverage with regard to the observation of many cloud variables and other useful information on indirect and vertically integrated quantities, like radiance at the top of the atmosphere, rather than direct measurements of model variables. The information gathered contributes to the basic research and better understanding of the earth's radiation budget, hydrological cycle and of the role of clouds in both systems.
A survey conducted by CLOUDMAP2 (Accessed 15/8/04) show the use of geostationary METEOSAT (79%) dominates the use of polar orbiting NOAA satellite (21%) worldwide. The situation is however opposite in high latitude countries such as the Scandinavian countries, where the oblique view angle of METEOSAT makes it on suitable. NOAA is most used due to its temporal coverage of these areas. The NOAA-AVHRR images come from a series of polar orbiting environmental satellite operated by the United States. They have on board the Advance Very High Resolution Radiometer (AVHRR) and are the most wildly use meteorological and regional vegetation monitoring around the world.

3 AVHRR Data for Real-Time Flood Forecasting in Malaysia

NOAA AVHRR satellite data in local area coverage (LAC) has been investigated for its meteorological application in monsoon area of the tropical regions of the world such as Malaysia and also for its appropriateness in providing better understanding into rain bearing tropical clouds.

AVHRR data is preferred because of its repeated coverage and relatively high temporal resolution of at most 6/hr daily, easy acquisition, cost effectiveness and it has the ability for automated geometric rectification when compared to GEOS and GMS data. AVHRR data has relatively high resolution data with average IFOV of 1.3 miliradians and GFOV of about 1.1km at nadir, 6+ km at the edge of the scan (Levizzani et al.2002). Another motivation for using AVHRR in Malaysia is the use of its five-channel multi-spectral observations for the purpose of discriminating tropical storm cloud properties not derivable from the primary three-channel in geostationary satellite data. The challenge is to utilize the latest observations of NOAA satellite images of monsoon cloud formation together with suitable hydro-meteorological models to improve very short forecasts (Nowcasting). Karlsson et al (1999), defined nowcasting as flood forecasting in the approximate range 0-9 hours from observation time.

Satellite-based information provided by the polar-orbital meteorological satellite series (NOAA), in the field of meteorological and climatological application are carried out to improve real-time operational monitoring of the atmosphere and hydrological environment, selection of meteorological and hydrological forecasting over short, medium and long range, as well as warning of dangerous hydro-meteorological phenomena (Karlsson,1999). This understanding should lead to improved parameterization of clouds in operation weather forecasting models based on research focuses in:

- Cloud analysis and classification
- Cloud top temperature (CTT) and cloud top height (CTH)
- Cloud mask
- Surface albedo maps
- Cloud motion derived winds
- Integration of remotely sensed information in meteorological numerical forecasting models

4 Processing of AVHRR data for Rainfall Intensity

The need for satellite- estimated precipitation arises from non dependable, poorly maintained and spatially distributed synoptic rainfall data, some as a result of inaccessible dense tropical rainforest and most importantly the need for rain fall estimates before the rain had actually fallen. Empirical methods have determined that cloud top temperatures less than 235K in the tropics are generally expected to produce stratiform rainfall at the rate of 1,5mm/half hour (Arkin and Meisner, 1987). AVHRR data provides the means of investigating and measuring cloud parameter from monsoon cloud formation that are most prevalent in Malaysia.

Multi-spectral techniques based on the relationship between cold and brightness temperature of clouds are used to evaluate precipitation probability in AVHRR data. The MIR and IR channels 3, 4, and 5 of the data are processed for temperature and brightness (T_B). A processed T_B data, typically highlights the top of the atmosphere brightness temperature for different observed features, cold clouds are high clouds and where cloud T_B falls below 235k they are identified as cumulonimbus cloud with a high probability to precipitate (Levizzani et al.2002) Lower probabilities are associated to warm but bright stratus cloud and thin cirrus cloud that are cold but dull.

According the Arkin (1979) cold clouds with temperature below 235k threshold value are indicative of rain. Rainfall is estimated based on the assumption that every cloud pixel has a constant unit rain-rate of $3mm^{h-1}$, which is appropriate for tropical precipitation over 2.5^o x 2.5^o areas around the equator. Based on storm tracing assumptions coupled with estimation techniques cloud clusters below some temperature thresholds have high probability of rainfall. Information on the radiance temperature on top of clouds in thermal infrared band can be empirically calculated into average rainfall intensity by mean of various techniques. These techniques have been detailed in (Griffith et al., 1978; Adler and Negri, 1988). Total cold cloud cover and the portion of the catchment covered by cloud determines rainfall intensity.

5 Algorithm for Processing of AVHRR Data for Rainfall Intensity

An algorithm has been developed to enhance and streamline the processing of AVHRR data for real-time tropical monsoon weather monitoring to improve quantitative precipitation forecast as input to hydrological model for flood early warning. The algorithm combines processing techniques such as the cloud indexing that assigns rain-rates to individual cloud type identified in the satellite imagery as described by Arkin (1979) and the cloud model method developed by Gruber (1973). The algorithm is best illustrated in the flow chart (Fig 1).

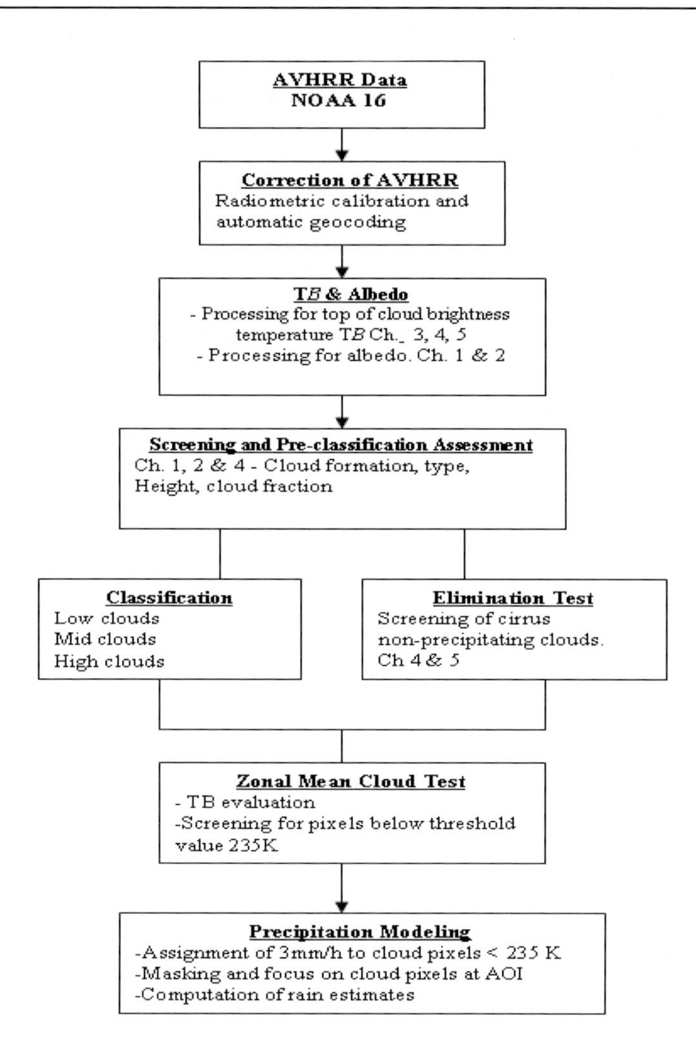

Fig. 1. Flow Chart of the Rainfall Estimation Algorithm

NOAA AVHRR data of the monsoon season covering Malaysia (Fig 1) and the South East Asian region received at the local ground receiving station of the Institute of Advanced Technology (ITMA) University Putra Malaysia is prepared for processing. The processing of the entire scene of the image involves radiometric calibration, where the visible and near infrared channels are calibrated in radiance or reflectance and the thermal IR channels in brightness temperatures. Other processing includes automatic geometric correction and rectification of the image, atmospheric correction and cloud detection.

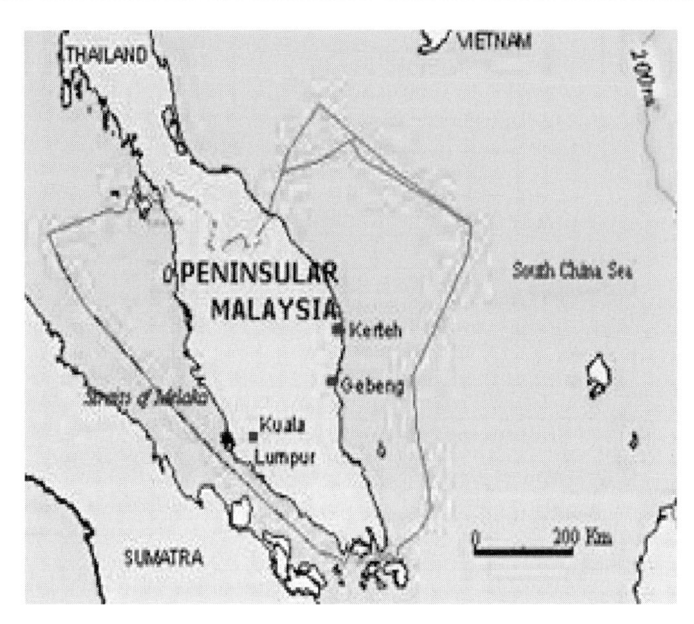

Fig.2. Peninsular Malaysia

The multispectral images provided by the NOAA satellites are essential for the top cloud temperature determination. Clouds in the image are investigated to establish cloud base height using AVHRR channel 1, 2 and 4 (Fig. 3).

Fig. 3. AVHRR data (Monsoon Period)

The first test is the brightness temperature and albedo processing. It uses the mid infrared channel 3 and the thermal channel 4 and 5 to process the entire scene for brightness temperature (Fig. 4) and channels 1 and 2 for albedo. The visible channels measure the cloud thickness and offer information about albedo, while the infrared channels measure the temperature of the cloud and offer information about thermal radiation emitted by the land and cloud top.

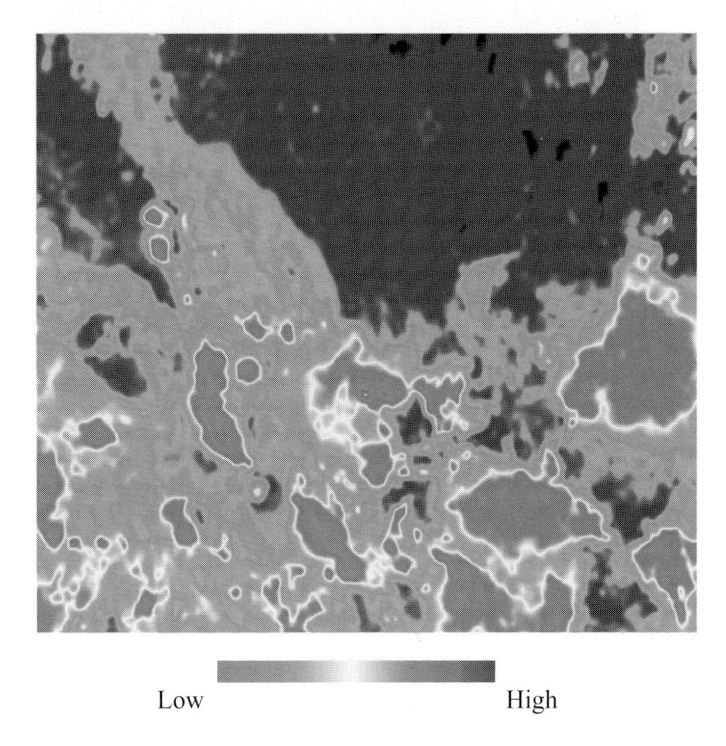

Fig. 4. Brightness Temperature (T_B in K)

Image is further processed by performing a classification to determine cloud type, fraction and height. Through this parameterization of clouds clusters, it will be possible to classify them into different precipitation level (Karlsson et al, 1999). Figure 5 shows high cloud cluster classification from which precipitation bearing clouds have been delineated (Fig.6). High precipitation cloud clusters may also be detected by means of an iterative numerical process that tests the mass, dimension and the compactness of each identified structure against the established threshold parameter (O'Sullivan et al, 1990). The screening for cirrus non-precipitation cloud is based on an established empirical discrimination of thin cirrus temperature and slope plane (Adler and Negri, 1988). Where the local minima in the IR T_B are sought and screened to eliminate thin non-precipitating cirrus.

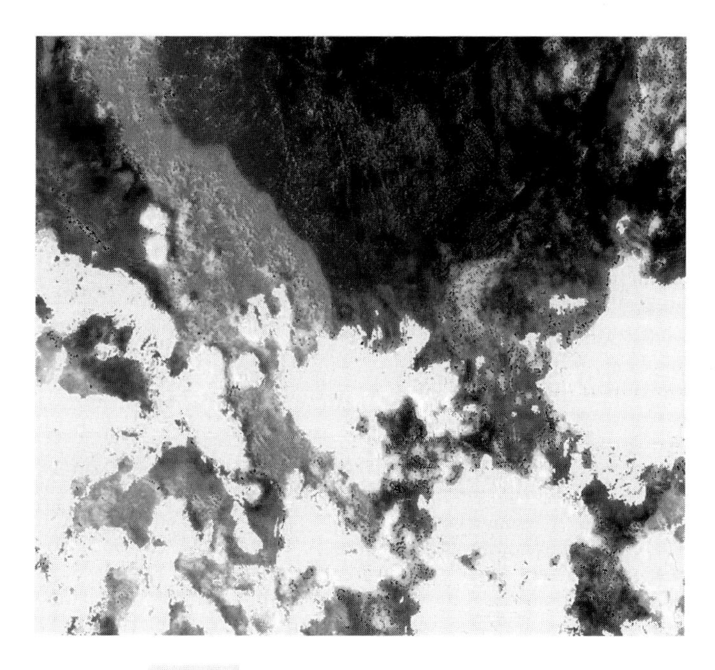

Fig.5. Classification and Delineation of High Cloud

Operational cloud top temperature (CTT) and maps are produced using AVHRR data calibrated in temperature. The CTT is created from channel 4 and 5 using a threshold method where threshold values are chosen interactively by examining the histogram of the value from the look up table (LUT) of the image processing software. The multispectral colored imagery is enhanced to visualized cloud top temperature (Fig.7). The value of temperature less than 235K is displayed on a color from red to blue as shown in the temperature map presented in figure 4. The rain parameters of 3-12 mm^{h-1} depending on monsoon rainfall intensity are finally assigned based on the 1-D cloud model that calculates maximum rain-rate as a function of maximum cloud height and minimum cloud model temperature at a threshold level of 235k.

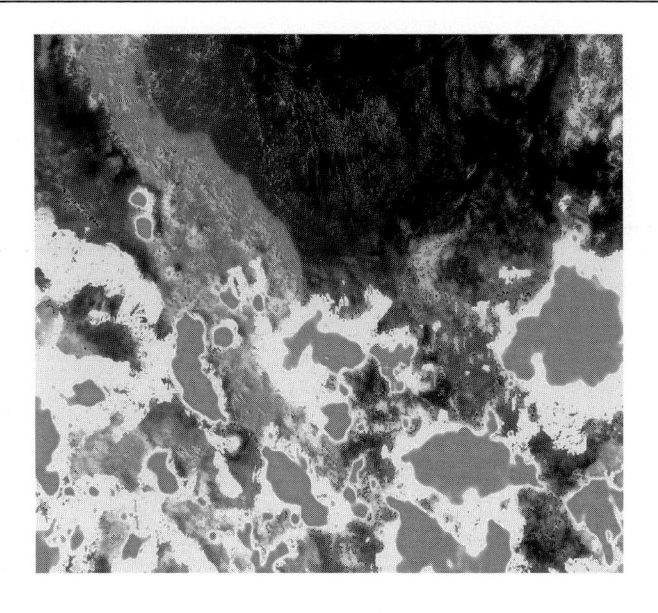

Fig. 6. Screening of Cirrus Non-precipitating Clouds

Since AVHRR data is received well in advance of rainfall and has a high temporal resolution of at least five scene coverages a day, cloud information is thus processed pre real-time and can subsequently be monitored regularly. Processed rainfall intensity is projected over catchment area or river basin in Malaysia depending on the cloud fraction and together with an integrated GIS for hydrological data processing and suitable and well calibrated hydrodynamic model for rainfall runoff and flood simulation to determine areas likely to be inundated.

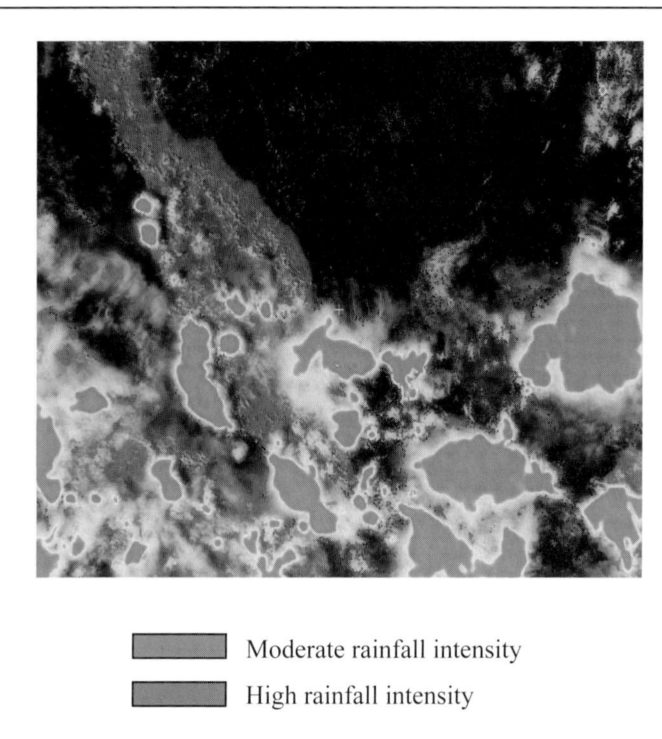

Moderate rainfall intensity

High rainfall intensity

Fig. 7. High Rainfall Intensity with T_B below 235k

6 Conclusion

Monsoon flood by all accounts is the most significant disaster affecting Malaysia in terms of costs and extent of damage. The need for improved hydro-meteorological models for real – time forecasting and early warning has been discussed in this paper in the light of an algorithm developed to improve the modeling of rainfall intensity and short term forecast using NOAA- AVHRR Data. Multi-spectral AVHRR data is excellent for the observation and modeling of many cloud variables. Cloud top brightness temperature is related to precipitation and cloud clusters with T_B below the threshold of 235k are indicative of impending rainfall.

The intensity of rainfall is shown to be depending on how low the T_B is for the cloud clusters. Through a supervised classification and T_B processing, high clouds have been identified and thin cirrus non-precipitating cloud eliminated. Rain intensity of 3-12 mm^{h-1} are assigned to the 1-D cloud model to calculates maximum rain-rate as a function of maximum cloud height and minimum cloud model temperature at a threshold level of

235k. Good and encouraging results have been observed as they provide a good aerial coverage that delineates areas likely to receive intense rainfall on a regional scale. The computed rainfall may however in most cases not have the same accuracy of the conventional observations.

AVHRR data is received in Real-time and has a high temporal resolution thus the derive cloud information reveals many application in the monitoring of top of the cloud temperature, cloud fraction and clouds type that are important for hydro-meteorological studies with consequent flood information as the determination of potential flood vulnerable river catchment area. In Malaysia where the continuous devastation of monsoon floods has become a great cause for concern, the question is how current flood forecasting system may be improved to provide information of impending flood and the likely areas to be affected well before it actually rains. The concern of using satellite-based information provided by the polar–orbiting meteorological satellite (NOAA) is based on its suitable characteristics as has been discussed and most importantly its operational applicability in hydro-meteorological and climatologically short term forecast.

References

Adler RF, Negri AJ (1988) A satellite infrared techniques to estimate tropical convective and stratiform rainfall. J. Appl. Meterology .27, pp 30-51

Arkin PA (1979) The relationship between fractional coverage of high cloud and rainfall accumulations during GATE over the B-scale array. Mon. Wea. Rev, 106, pp 1153-1171

Arkin PA and Meisner BA(1987) The relationship between large scale convective rainfall and cold cloud over the western hemisphere during 1982-84. Mon.Wea. Rev 115, pp 51-74

Chong S F (Accessed 11/29/04) Hydrological models of Precipitation. Fifth InternationalWorkshop on Tropical Cyclones, WMO/CAS/WWW http://www.aoml.noaa.gov/hrd/iwtc/Chong2-3.html. Accessed 11/29/04

CLOUDMAP2 (15/8/04) Current use of cloud data in operational meteorology and climate research. http://www-research.ge.ucl.ac.uk/cloudmap2/products/cur_use_cd.html#nwp. Accessed 15/8/04

Griffith CG, Woodley WL, Grube PG, Martin DW, Stout, Sikdar, DN (1978) Rain estimation from geosyncronous satellite imagery - visible and infrared studies Mon. Wea. Rev. 106, pp 1153 -1171

Gruber A (1973) Estimating rainfall in regions of active convection. J. Appl. Meteorology, 12, pp 110- 118.

Karlsson K, Thoss A, Dybbroe A (1999) High resolution cloud products from NOAA AVHRR and AMSU. Swedish Meteorological and Hydrological Institute (SMHI) S-601 76 Norrköping, Sweden

Keizrul BA, Chong S. F (2002) Flood forecasting and warning systems in Malaysia. DID, Hydrology and Water Resources Division, Department of Irrigation and Drainage, Malaysia.
http://agrolink.moa.my/did/papers/floods1.PDF

Levizzani V, Amorati R, Meneguzzo F (2002) A review of satellite-based rainfall estimation methods
http://www.isao.bo.cnr.it/~meteosat/papers/MUSIC-Rep-Sat-Precip-6.1.pdf. Accessed on 03/04/03

O'Sullivan F, Wash CH, Stewart M, Motell CE (1990) Rain estimation from infrared and visible GOES satellite data. J. Appl. Meteorol., 29, pp 209-223.

Supporting Flood Disaster Response by Freeware Spatial Data in Hungary

Zsofia Kugler and Arpad Barsi

Department of Photogrammetry and Geoinformatics, Budapest University of Technology and Economics, H-1111 Budapest, Muegyetem rkp. 3, Hungary.
Email: zsofia.kugler@mail.bme.hu, barsi@eik.bme.hu

Abstract

In Hungary flood disasters are the major natural hazard as a consequence of lowland topographic conditions. Due to both structural and non-structural mitigation measures from the mid 19th century, like building dikes and developing adequate warning system, no loss of life was reported since the 1950's. Nevertheless the value of damage to human property in flood-prone areas may reach millions of euros each year throughout the country. Flood forecast is performed by the national hydrological institute, and warning is transmitted to the local hydrological authorities about the forthcoming disaster. Disaster response measures are based on conventional hydrological information and action plans are developed with traditional methods. The latest technology of GIS and remote sensing is not yet tested in the country to support the spatial information need of disaster response.

The aim of our investigation was to test impact modeling using mostly freely available spatial data. This may ensure feasibility in a case of a future flood catastrophe in various geographic location. The decision of using non-commercial data was further based on the experience that data acquisition from dissimilar sources like local authorizes and the military mapping agency may last weeks or months and was found not suitable for rapid results.

Investigation was based on the recent flood disaster event in the Northern part of the country in the summer of 2004.

To analyze the elements possibly being damaged by the flood, the extent of the inundation was calculated. Flood simulation was executed using digital elevation model of the remotely sensed SRTM radar data and hydrological data from the Hungarian Water Resources Research Centre (VITUKI). Modeled flood extent was serving the basis of spatial analysis of the disaster impact. Settlements hit by the flood disaster were queried. Moreover the condition of flooded road and railway network was analyzed to support route planning of mitigation efforts.

Results of this investigation should help to speed up decision making processes during future disaster events. Moreover should help to reduce property damages and suffering of flood victims.

1 Introduction

The northern rural part of Hungary was suffering sever flooding in the summer of 2004. A cyclone was causing heavy rainfall between 26th and 30th July in the northern Carpathian mountains contributing to the occurrence of extreme runoff in the upstream drainage basin of the River Hernad. The amount of precipitation was reaching in 5 days the 150-250% of its monthly average. The river system rising in the neighboring country of Slovakia, entering Hungary in the Northeast was facing a 100-year flood disaster. Not only its downstream area in Hungary was inundated but also its upstream catchment area in Slovakia. Peak flows were reaching flow records thus the highest alert level was issued on the 26th July. Release came on the 4th August.

Flooding damaged several households; around 100 people were evacuated due to the risk of injuries and collapsing houses. Several roads were closed in the period from the 26th to the 4th August. Helicopters of the Hungarian military forces were helping the efforts of the local authorities in disaster response (VÍZÜGY 2004).

The severity of the flood disaster was not unique in the region. Hungary was more often hit by inundation in the last decade (Danube, 2002 summer, Tisza, 2001 and 2000, Tarna 2003). The frequency of 50 and 100-years probability floods are likely to increase in the region. For this reason studying the phenomenon of flood disaster is crucial to establish sustainable response strategy for the future. The aim of this investigation was to study the use of remote sensing techniques and Geographic Information Systems in supporting the relief efforts by proper spatial information. Spatial data acquisition was restricted to free satellite images and geographic data to assure feasibility disregarding spatial location. The next chapter will highlight data sources and data gathering for our study aims.

2 Spatial Data Acquisition

The last chapter gave an introduction into the investigation and a general overview of our aims. This chapter will discuss the acquisition and preprocessing of selected spatial data of dissimilar origin. As mentioned in the last chapter data was gathered from freely available sources to assure feasibility for different regions in the future.

- Remote sensing satellite imagery was planned to form a basis of flood extent derivation. The freely available highest resolution imagery of the American satellite system MODIS Aqua and Terra was obtained for flood mapping. The system has the great advantage of viewing the Earth's surface with its sweeping 2230 km wide swath every day. Thus the change in flood extent can be monitored in time sequences. For this reason MODIS images are often applied for large-scale flood mapping like at the well-know Dartmorth Floodwatch Institute too. Images were obtained from the 4th August. The remaining period of the flood event cloud cover was restricting the use of the data due to the optical characteristic of the satellite system. Obtained images were imported using ENVI/IDL routine. Lat/long coordinates were assigned to each 4th pixel in the image from the dataset (MODIS Web 2004). Thus preprocessing resulted in georeferenced spatial data for the flooded region. At this step we have to note, that preprocessing algorithms still have difficulties caused by images errors of double projecting the same area. After georeferencing these appear in form of stripes in the image mostly at image edges. Visual interpretation revealed that the spatial resolution of the data was insufficient for study aims. However a not flooded cross section of the River Hernad is around 100 m moreover the flood inundated almost 500-1000 m broad floodplains still the 250 m spatial resolution of MODIS images were not sufficient to perform flood mapping. Figure 1. illustrates a MODIS level 1B satellite image acquired on the 4th August.

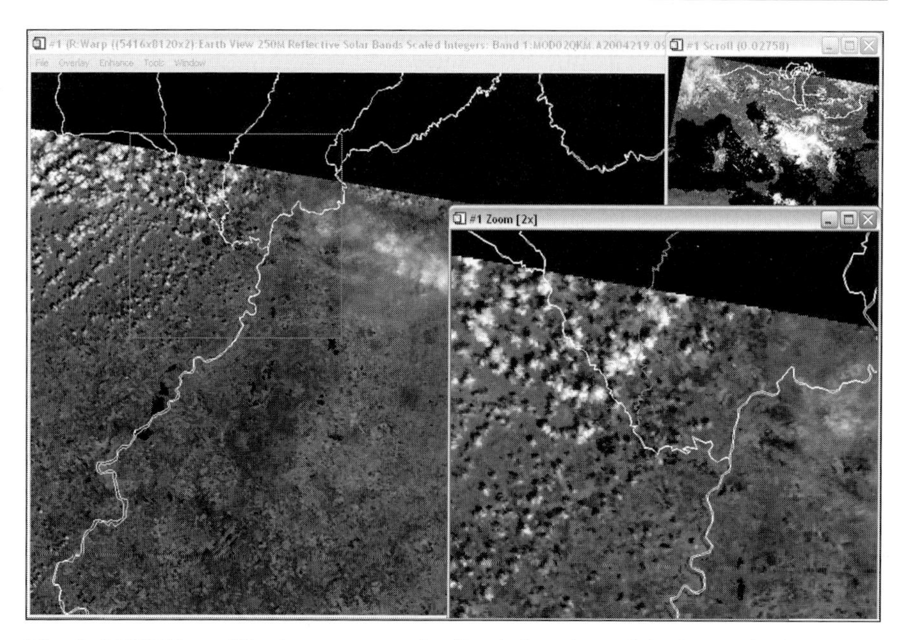

Fig. 1. MODIS satellite images over the flooded region of the Hernad River. Spatial resolution was insufficient for flood mapping aims

Since no freely available data was present for flood extent mapping we had to decide for modeling of the inundation, based on digital terrain model and hydrological parameters of the flood event.

- For this new aim we obtained spatial data of the terrain elevation. C-band radar elevation data of the SRTM mission was obtained in the resolution of 90 meters (NASA 2004). Some preprocessing was also necessary this time as well. After importing data into grid type using an ArcInfo AML script, mosaicing of grid subsets had to be performed before analyzing terrain conditions.
- For hydrological modeling channel profile and cross section elevation data had to be gathered too. The Hungarian Water Resources Research Centre (VITUKI) is responsible to acquire and store data. Their public library enables access to these data sources for research aims on a non-commercial basis. River channel geometry data was collected for the River Hernad including profile and cross sections along the river channel.
- Flood level data is published each day on several different medium like the Internet or several radio stations in the morning. Historical data is stored in two different databases accessible free of charge thought the Internet. Discharge data of the flood event on the River Hernad was collected for two hydrological stations along the river from the database

operated by National Directorate for Environment, Nature and Water in Hungary (VÍZÜGY 2004).

- Last but not least to analyze flood disaster situation basic spatial data of the area has to be collected. These are unfortunately mostly commercial datasets. Datasets were however present at out department over the region. Vector data of settlements, population, road and railway network were used to study flood situation.

After gathering above described spatial and hydrological data, hydrological modeling was executed for the River Hernad to derive flood extent and analyze flood situation. The next chapter will describe analysis steps.

3 Spatial Extent of the Flooded Areas

To improve the effectiveness of disaster management enabling rapid response, better information on the flood situation has to be provided. Reducing uncertainties in decision making processes of relief operations may help to improve public safety and lessen the cost of damages caused by a disaster events. As described above the spatial extent of the inundation had to be modeled due to the lack of non-commercial satellite imagery with sufficient spatial resolution.

Hydrological simulation was executed using the freeware modeling software HEC-RAS developed by the American Army Corps of Engineers. It performs 3D modeling of water dynamics based on geometrical data of the river channel with its proximity and hydrological data of the flood level. The software includes GIS interface enabling to import channel geometry from geographical data of many GIS software. Thus necessary cross section data of the River Hernad and profile was imported from ArcView using the data preprocessing tool of HEC-GeoRAS.

After importing channel geometry in HEC-RAS and editing hydrological parameters including channel discharge hydrological modeling can be executed. Unfortunately modeling with this sophisticated tool was too much time consuming consequently, we had to restrict modeling to river geometry to enable rapid result.

SRTM data of terrain elevation and channel geometry data of river profile was used to derive a first approach of inundated areas. Flood peak levels of the two hydrological stations of Gesztely and Hidasnémeti were interpolated according to the channel slope properties to obtain approximate flood levels every 10 km segment along the River Hernad. As a result the elevation of the different flood level was assigned to each 10 km segment. Inundation was performed in each segment with the assigned flood level using the digital terrain data of SRTM. In a last step flood mask segments were unified into one great flood map. Resulting flood mask was no output of a sophisticated hydrological model, but a good approximation of the assumed flood extent. At the time of the flow peak (31st July) around 150 km^2 was found to be inundated along the River Hernad in Hungary.

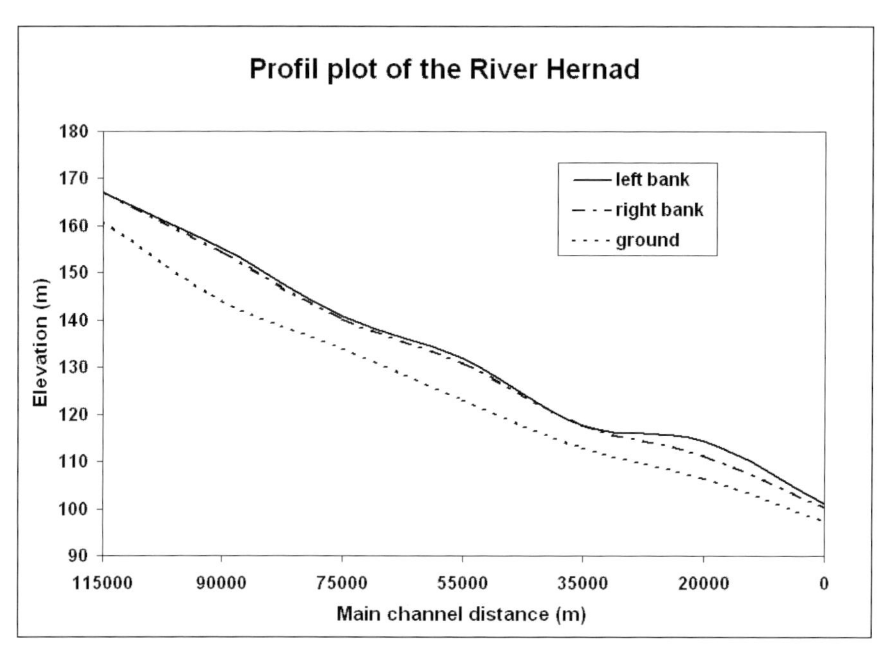

Fig. 2. Profil of the River Hernad including right and left bank dike elevation. Slope of the river channel was used to perform flood extent modeling

This chapter discussed hydrological modeling of the inundation to deliver information on the latest flood extent. Next chapter aims to describe the integration of this spatial information to perform analysis of the flood situation in order to assist geographical information need of disaster response.

4 Spatial Analysis of the Crisis Situation

Mitigation of natural disasters can be successful only when detailed knowledge is obtained on the crisis situation (Skidmore 2002). Relief planning after a disaster has occurred, needs numerous information with spatial reference like where did the catastrophe strike, how many people are affected, where are possibly people in danger and how could they be rescued? This chapter will highlight the results of the spatial analysis about the crisis situation in the Hernad River valley during the flood of 2004. Geographic data introduced in chapter 3 was used to derive spatial information on flood conditions. (see figure 3).

Natural phenomena becomes a catastrophe when it affects human activity, thus an inundation occurring in low populated regions causes less damage than in densely inhabited areas (Smith and Ward 1998). Thus spatial distribution of human population has to be analyzed in a first step impact assessment. Assumed flood extent described in the last chapter was overlaid with the geographic data on settlements including number of inhabitants and additional attribute information. Spatial query revealed that around 10 villages were hit by the flood disaster. Furthermore around 12000 inhabitants are settled in these affected settlements. As a consequence around 100 people were evacuated during the crisis event as mentioned in the introduction. The number of damaged households is also a crucial information to support mitigation efforts. Unfortunately, due to the lack of this attribute no answer could be generated from our existing data to this question.

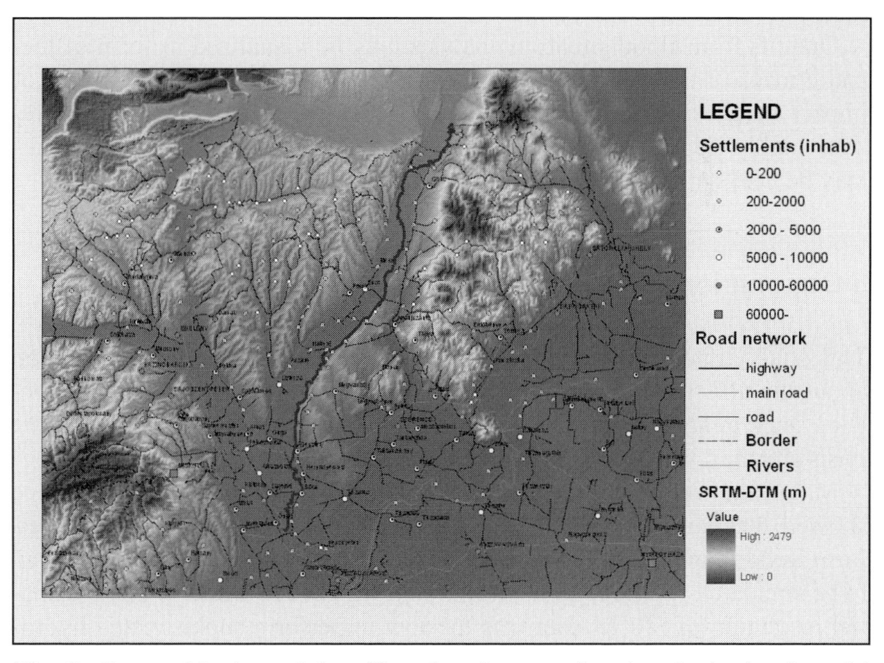

Fig. 3. Geographic data of the affected region was forming the basis of spatial analysis of the flood situation

Planning evacuation or transportation of primary gods for flood victims requires up-to-date information on road network conditions. Several roads were closed during the flood crisis furthermore bridges crossing the River got impassable. Thus in a next step the condition of road and railway network was analyzed.

Overlaying the flood mask with the road network flooded road sections were derived. The length and number of flooded road sections are listed in table 1.

	Main road	Secondary road	Total
Length (km)	5,9	38,8	44,6
Nr. of sections	10	80	90

Table 1. Condition of inundated road network. Road section are divided in two categories: main and secondary roads

Result reveals that mostly secondary roads run parallel to the river channel were inundated during the catastrophe. Furthermore 4 bridges were flooded according to our analysis. No railway or highway was affected in the flood.

Outputs from flood situation analysis may be visualized in form of thematic maps or tables of statistics. This chapter highlighted some means of impact analysis. After the flood mask is generated situation assessment results highly depend on the available spatial data inputs. Further queries may be performed if more spatial information is available over the area.

5 Conclusions

This study aimed support flood disaster response with spatial information about the affected area on the River Hernad in the summer 2004. To reduce uncertainties in problem solving and improve decision making the crisis situation was analyzed during the catastrophe.

Main aim was to test feasibility on freeware spatial data. Experience showed that non-commercial satellite images were not sufficient to perform flood mapping on the Hernad River. For this reason flood extent had to be simulated for the region. At this step we have to mention that the spatial resolution of SRTM data was too low to perform sophisticated hydrological modeling. It gave a good approximation of the possible flood extent, but higher resolution could enhance results. After simulating the spatial extent of the inundation the impact of the disaster was investigated.

Thus spatial analysis was executed on the flood situation. In this step results highly depended on the availability of input geographic information.

Our future aims are to perform hydrological modeling with more sophisticated simulation tool of HEC-RAS. Furthermore to test feasibility on commercial datasets in order to draw a comparison to free spatial data.

References

Dave T (2003) Fundamentals of Hydrology. Taylor and Francis, London and New York

MODIS Web (2004) [online] http://www.ovf.hu

NASA (2004) [online] Shuttle Radar Topography Mission (SRTM) http://www2.jpl.nasa.gov/srtm/

Skidmore A (2002) Environmental Modelling with GIS and Remote Sensing. Taylor and Francis, London and New York

Smith K, Ward R (1998) Floods, Physical Processes and Human Impacts. John Wiley&Sons Ltd.

VÍZÜGY (2004) Vízügyi adatbank. [online] http://www.vizadat.hu/

VÍZÜGY (2004) Árvíz a Hernádon 2004. nyarán - a védekezés kronológiája. [online] http://www.ovf.hu

The Use of GIS as Tool to Support Risk Assessment

Giuseppe Orlando, Francesco Selicato and Carmelo M. Torre

Department of Architecture and Planning, Bari Polytechnic,
Via Orabona 4, 7125 Bari, Italy.
Email: beppe.orlando@aliceposta.it

Abstract

Risk assessment is a complex issue aiming at evaluation of different aspects of disaster damages. Traditionally, risk analysis relies on mathematical models to establish the likelihood of a given event occurring with a given degree of intensity in a given site. The major limitation of this type of approach is the fact that the field of risk necessarily entails uncertainty and it is not therefore always possible to make realistic hypotheses about possible future scenarios.

Therefore, a new approach is required, that can take into account social, economic, cultural, and political aspects that are not generally considered in traditional assessment methods, but that serve to define the capacity of response of a territorial system to a disaster.

In this paper we present a new approach, according to which the assessment process breaks down the principal goals into a hierarchy of lower level sub-goals. Each element of the hierarchy is assigned a local weight that evaluates the importance of that element not in overall terms (i.e. referred to the principal goal), but only in relation to the supra-ordinate element with which it is compared.

The approach is tested in a case study estimating the risk of flooding in the Municipality of
Monopoli, Italy.

1 The Complex Issues Related to Risks

Although the risk of particular calamities appears in a certain sense unavoidable, the same cannot be said of the entity of the damage they may cause. Damage due to disastrous events can be classified according to the time phase when they occur:

- material damage
  This occurs at the moment of impact and involves damage to people and property. It is present principally in the zones most affected by the disaster.
- systems-organizational damage
  This occurs during the emergency assistance phase and can be caused by inefficient emergency services, lack of specialist staff and resources, inadequate aid.
- process damage
  This occurs during the phases of rebuilding and "return to normal". Such damage is closely linked to the complexity of the urban system, and has repercussions on the economic, social, political-administrative conditions. This kind of damage is more spread out and less homogeneous over time and space.

A good knowledge of the structural complexity of an urban system can serve to define its *vulnerability* (liability to suffer devastation) to a calamity and hence to limit the real damage.

2 The Need for a Different Approach

Traditionally, ever since a definition of *risk* was proposed by the UNDRO in 1979, according to which risk is the result of the product of the *hazard, vulnerability and exposure scores*, risk analysis relies on mathematical models to establish, starting from a series of experimental data, the likelihood of a given event occurring with a given degree of intensity in a given site (hazard score). By subjecting a model of the host system (vulnerable, exposed sites) to the calculated stress, the possible scenarios can be enacted and the ensuing material damage "ascertained".

One of the limitations of this type of approach, that can be described as ***quantitative***, is the fact that the field of risk necessarily entails uncertainty and it is not therefore always possible to make realistic hypotheses about the possible future scenarios. Still more importantly, such a method of analysis does not take into account all the aspects related to the *social, economic and political spheres*, that are strongly linked to the problem because they serve to describe reality in the sense of a complex system of relations.

It is due to this very complexity that any territorial system presents an intrinsic *systemic vulnerability* whereby the effects of any disaster will have repercussions, in terms of both space and time, going well beyond those observable in the zone actually physically struck by the disaster.

For this reason, a ***qualitative*** approach is required, that can take into account all those aspects (social, economic, cultural, political) that are not generally considered in traditional assessment methods, but that serve to define the capacity of response of a territorial system to a disaster.

It is necessary to define how and to what extent these aspects affect, and are affected by, a hypothetical catastrophe in order to be able to reduce systemic vulnerability and hence the damage ensuing after such an event.

Far from being merely a collection of data or an aseptic, quantitative technical evaluation, therefore, risk analysis becomes a procedure aiming to investigate how spatial, territorial, social and economic factors interact in the face of a particular hazard, or a combination of several hazards" (Menoni1997).

3 The Assessment Method

"The emergence of a systemic notion of reality, and the discovery of the inter-relations that link human beings with each other and the habitat, have caused the notion of the possibility of isolating single study elements to vacillate" (Menoni 1997).

The difficulties inherent to the search for relationships based on the links among the various aspects of the problem make it preferable to consider them separately, albeit within a logical, hierarchically structured framework.

For this reason, first of all the assessment process involves breaking down the principal goals (risk analysis) into a hierarchy of lower level sub-goals (analysis of vulnerability and hazard scores).

N.B. The concept of exposure which is traditionally referred to, is "included" in that of vulnerability, that therefore effectively represents the *vulnerable, exposed sites.* This procedure relies on a multicriteria method of support for decision-making developed by T. L. Saaty towards the end of the '70s: the AHP (Analytic Hierarchy Process) method, that "breaks down a macro-problem into more easily solved micro-problems and can manage a certain degree of incoherency in the manager within limits judged to be acceptable, so that it does not require perfect rationality" (Giangrande 2001).

Each element of the hierarchy is assigned a *weight* (local weight) that evaluates the importance of that element not in overall terms (i.e. referred to the principal goal), but only in relation to the supra-ordinate element with which it is compared. This weight is the result of the score attributed by experts on the basis of a series of paired comparisons among the various elements subordinate to the same higher level element. The global weight (referred to the goal) of each element is equal to the product of its local weight multiplied by the local weights of all the elements to which it is subordinate.

Thus, the problem is structured according to the following hierarchy:

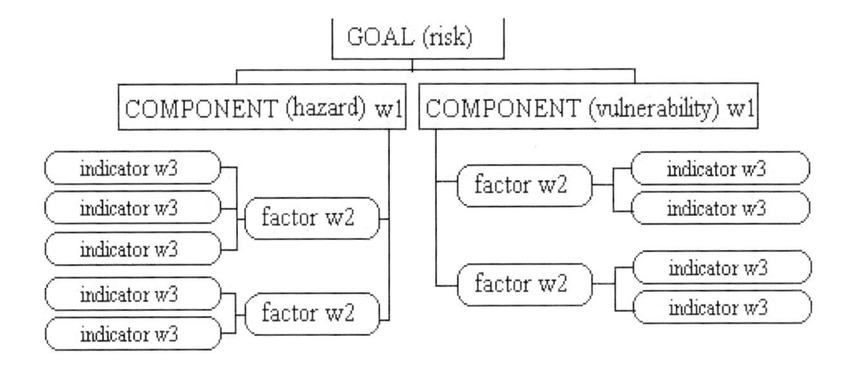

Fig. 1. Risk assessment hierarchical scheme

The vulnerability and hazard scores constitute the risk *components* and are described by *factors*, whose presence and entity on the territory are expressed by *markers*.

The global weight W of a marker is equal to

$$W = w3 * w2 * w1.$$

N.B. The sum of the weights of the elements belonging to the same level and subordinate to the same element is equal to 1.

In this way, the risk can be defined starting from an analysis of the *presence* and *entity* on the territory of those elements that constitute the risk.

4 Building the Maps

The presence and entity of the risk factors individuated is graphically illustrated by maps, and overlay of these gives rise to supra-ordinate level maps. The overlay operations are effected by GIS software, which makes it possible to associate graphic data (individuation of the presence of the factors) and numerical data (the weights associated with the single factors), in a simple, dynamic manner.

Essentially, risk assessment consists of a process of overlay mapping which yields complex information by structuring, according to causal relationships, levels of more simple data.

The vulnerability and hazard score maps are obtained by means of overlay of the maps of the respective factors going to make up each map.

Overlay of these maps yields the risk map. From the graphic overlay mapping operations, the sum of the overall weights of the single markers of hazard score and vulnerability is obtained, which expresses the overall risk index (see Eq. 4.1)

$$I_R = I_V + I_P = \sum p_{Vi}\, \delta_i + \sum p_{Pj}\, \delta_j \tag{4.1}$$

I_R : global risk index
I_V : general vulnerability index
I_P : general hazard index
p_{Vi} : local weight of vulnerability factors
p_{Pi} : local weight of hazard factors
$\delta_{i/j}$: presence index : 1= present, 0 = not present

5 Application of the Method in Field: Assessment of the Risk of Flooding in the Municipal Territory of Monopoli

A study was conducted of the risk of flooding in the Municipality of Monopoli. The territory under study is exposed to the risk of flooding above all because of the interference by the urban built-up area with the natural orographic gully system. This generates a condition of risk because it contributes to the hazard score (it can cause alterations in the natural drainage processes of the water running down from the Murgia highlands to the sea) and vulnerability score (exposing the population living near the gully system to the risk of flooding).

The hierarchical scheme adopted to assess the risk is as follows:

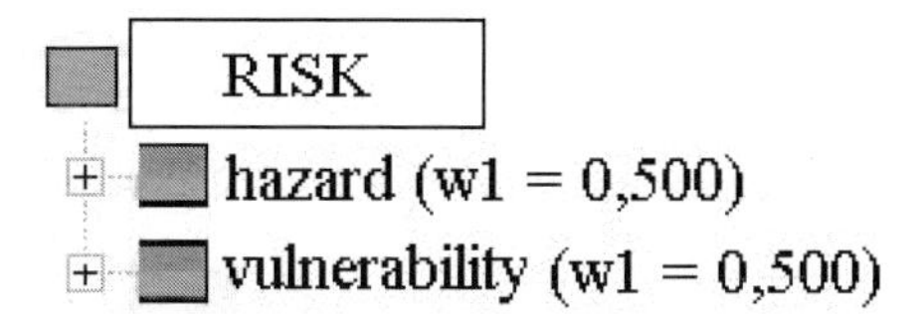

5.1 Analysis of the Hazard Score

The representative factors of the hazard score are:

- rainfall (0.278);
- slope (0.167);
- permeability (0.167);
- vegetation (0.167);
- anthropic interference with the gully system (0.222).

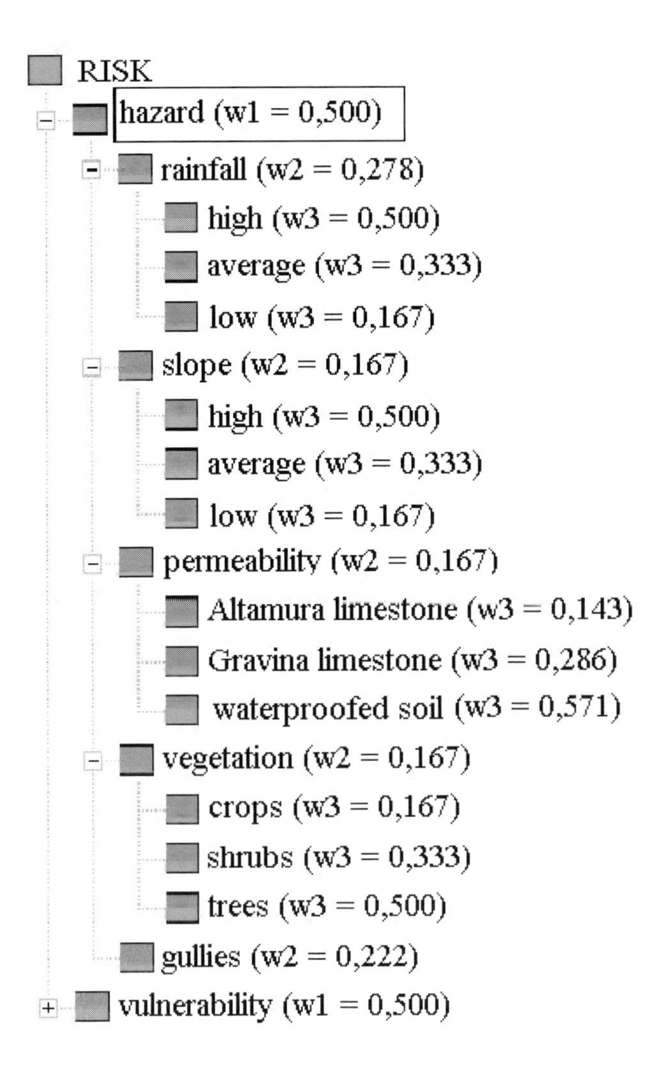

Fig. 3. Hazard assessment hierarchy

5.1.1 Rainfall

An excessive quantity of rainfall can overwhelm the natural drainage system and flood the surrounding areas. On the basis of the rainfall data recorded at the rain stations of Fasano, Polignano and Castellana, in the period 1976-1996, the table below was drawn up:

		Rain station		
		Castellana	Fasano	Polignano
1st h intensity		49	66	61
day	min	0	0	0
	max	77.4	81.0	78.4
	aver.	1.7	1.6	1.5
month	min	0	0	0
	max	274.0	231.8	225.0
	aver.	49.9	48.1	45.8
year	min	368.4	364.6	349.6
	max	849.8	914.4	808.8
	aver.	598.9	577.6	549.4

Table 1. Rainfall data recorded at the rain stations of Fasano, Polignano and Castellana

On the basis of these values, three areas with different levels of rainfall can be determined, coinciding from the geographic point of view, with the relative zones ("topoieti") (recorded by the same rainfall station) covered by the three rainfall stations, attributed three different classes of hazard score:

- high rainfall zone (Fasano): 0.500

  Hpio1= (0.500/0.500)*0.278*0.500 = 0.139
- average rainfall zone (Polignano): 0.333

  Hpio2= (0.333/0.500)*0.278*0.500 = 0.093

- low rainfall zone (Castellana): 0.167

$$Hpio3= (0.167/0.500)*0.278*0.500 = 0.046.$$

5.1.2 Slope

Exclusively sloping terrain has the effect of increasing the rush of water and reducing the quantity absorbed by the earth: this phenomenon can result in breaking down the natural and/or artificial river banks and inundating the surrounding areas. For this reason, the presence of impluvia, dolinas, gullies and depressed terrain below sloping terrain becomes extremely important.

Three grades of slope are defined, associated with different classes of hazard:

- steep slopes: 0.500

$$Hpen1= (0.500/0.500)*0.167*0.500 = 0.084$$

- average slopes: 0.333

$$Hpen2= (0.333/0.500)*0.167*0.500 = 0.056$$

- mild slopes: 0.167

$$Hpen3= (0.167/0.500)*0.167*0.500 = 0.028$$

5.1.3 Permeability

This characteristic of the terrain allows rainfall to be partly absorbed by the soil, a favorable circumstance (in this case) because it reduces the quantity of rushing water.

Basically, there are two lithotopes (Altamura limestone and Gravina limestone) in the territory of Monopoli, distributed along two strips lying parallel to the coast. These can be attributed different degrees of permeability. Another element to be considered in the context of this characteristic is that of waterproofing, that causes a reduction (partial or total) of the filter properties of the soil.

In this case, again, three classes of hazard can be defined:

- waterproofed soil: 0.571

$$Hper1= (0.571/0.571)*0.167*0.500 = 0.084$$

- Gravina limestone: 0.286

$$Hper2= (0.286/0.571)*0.167*0.500 = 0.056$$

- Altamura limestone: 0.143

$$Hper3= (0.143/0.571)*0.167*0.500 = 0.028$$

5.1.4 Vegetation

The vegetation and soil contribute to define the catchment and absorption of rainfall, and hence the quantity of rushing water. Catchment and absorption vary according to whether trees (and suchlike), shrubs (and suchlike) or crops are present on the terrain, featuring three hazard classes:

- trees: 0.500

$$Hveg1= (0.500/0.500)*0.167*0.500 = 0.084$$

- shrubs: 0.333

$$Hveg2= (0.333/0.500)*0.167*0.500 = 0.056$$

- crops: 0.167

$$Hveg3= (0.167/0.500)*0.167*0.500 = 0.028$$

5.1.5 Anthropic Interference with the Gully System

As stated above, this factor can trigger a latent hazard because it can cause overflowing of the section of natural canals (gullies) where the water runs down.

Disturbing elements can include a single raised road (that can pose a true dam), or inhabited areas built near the gully. In some cases an interference factor can actually be an advantage from the safety standpoint; this is the case of the quarry flanking the Monopoli-Castellana road: owing to interference with the Belvedere gully higher up, this deep quarry acts as a reservoir, preventing the water running down from the hills in cases of heavy rainfall from reaching the town below.

To represent the interference condition, the graphic concept of the buffer, referred to the impluvia, is used. The width of the margin of impluvia must be 50 m from the axis of the running water: the buffer zone must therefore be this wide. If there is an interference element within the buffer, a value of 1 is associated (0 if it should constitute a safety advantage); multiplied by the weight associated with this factor, the true weight is therefore:

$$Hlam= 1*0.222*0.500 = 0.111$$

5.2 Analysis of Vulnerability

The representative factors of vulnerability are:

- population: 0.600
- built-up zone: 0.400

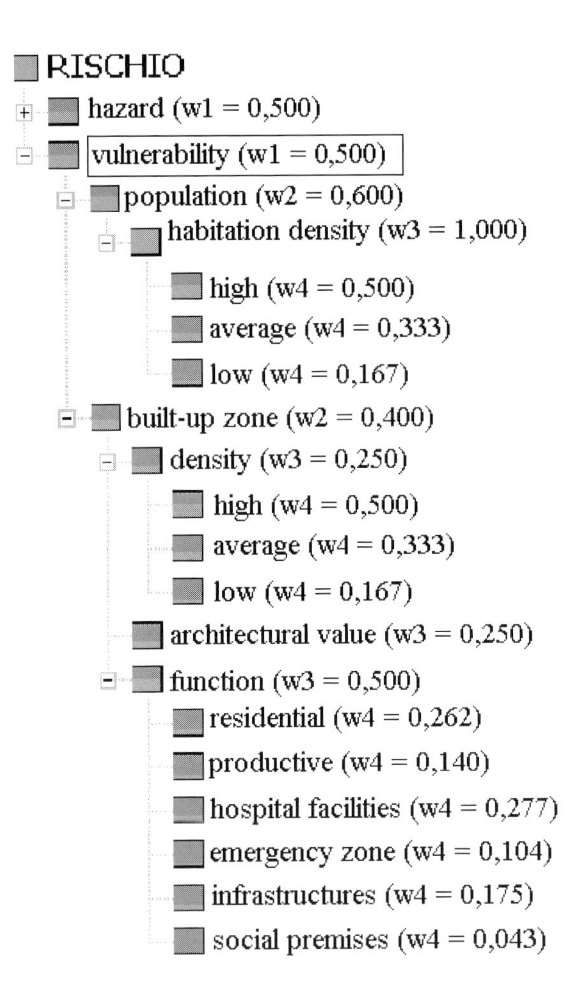

Fig. 4. Vulnerability assessment hierarchy

5.2.1 Built-Up Zone

The vulnerability of this factor depends on three markers, attributed the following weights:

- habitation density: 0.250
- architectural-historical-cultural value: 0.250

- function: 0.500

The following classes are individuated for **habitation density**:

- high density: 0.500

Hdins1= (0.500/0.500)*0.250*0.400*0.500 = 0.050

- average density: 0.333

Hdins2= (0.333/0.500)*0.250*0.400*0.500 = 0.033

- low density: 0.167

Hdins3= (0.167/0.500)*0.250*0.400*0.500 = 0.017

To indicate the **architectural-historical-cultural value** (representing a local identity value, a sense of belonging), the following overall weight is obtained:

Hintarc= 0,250*0,400*0,500 = 0,05.

For the **function** marker, the classes *residential*, *productive*, *hospital facility*, *emergency zone*, *infrastructures* and *social premises* are identified, and attributed the following classes of vulnerability:

- residential: 0.262

Hfun2= (0.262/0.277)*0.500*0.400*0.500 = 0.095

- productive: 0.140

Hfun4= (0.140/0.277)*0.500*0.400*0.500 = 0.050

- hospital facility: 0.277

Hfun1= (0.277/0.277)*0.500*0.400*0.500 = 0.1

- emergency zone: 0.104

Hfun5= (0.104/0.277)*0.500*0.400*0.500 = 0.037

- infrastructures: 0.175

Hfun3= (0.175/0.277)*0.500*0.400*0.500 = 0.063

- social premises: 0.043

Hfun6= (0.043/0.277)*0.500*0.400*0.500 = 0.015.

The hospital facilities have the highest associated value, both because of their importance in emergencies (to save human lives) and because they are highly vulnerable during the impact phase (many people with reduced motor function are present), so it is vitally important to safeguard these structures. The “emergency zones” are those areas deputed to become points of reference in the emergency phase, hosting temporary structures for aid and lodging (medium and long term) for the homeless. Such zones include schools, gymnasiums, and large sports facilities, as well as large open spaces.

Residential areas also have a high weight, because they are essential to life and loss of these areas can generate obvious practical problems (survival), and economic hardship, as well as psychological problems.

5.2.2 Population

To assess the vulnerability of this factor, the population density factor is used, individuating the three classes listed below:

- *high*: 0.500

  Hdabi1= (0.500/0.500)*1*0.600*0.500 = 0.300
- *average*: 0.333

  Hdabi1= (0.333/0.500)*1*0.600*0.500 = 0.200
- *low*: 0.167

  Hdabi1= (0.167/0.500)*1*0.600*0.500 = 0.100.

When assessing systemic vulnerability, some markers such as *mean age* of the resident population or *income* or *cultural level* may be relevant because capacity of response, in terms of action, to the calamity can be deduced from them. In fact, a population with a high level of culture and high income, and with a relatively lower mean age, clearly has a greater capacity of response (in terms of strength and economic power) than an older, less well-off population. The classification below is therefore considered.

It is evident that risk analysis cannot ignore *social* factors, because they serve to calculate the balance (long term) of the damage suffered by a community struck by a disaster.

6 The Maps

A map has been built on the basis of each of the factors highlighted, representing their presence and entity within the municipal territory of Monopoli.

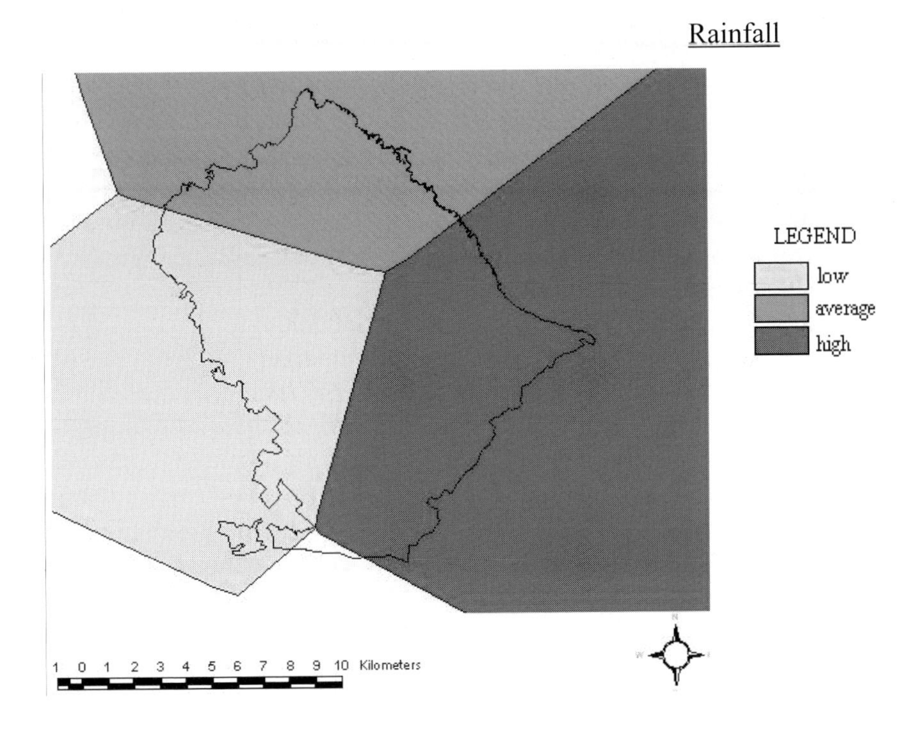

Fig. 5. Rainfall map

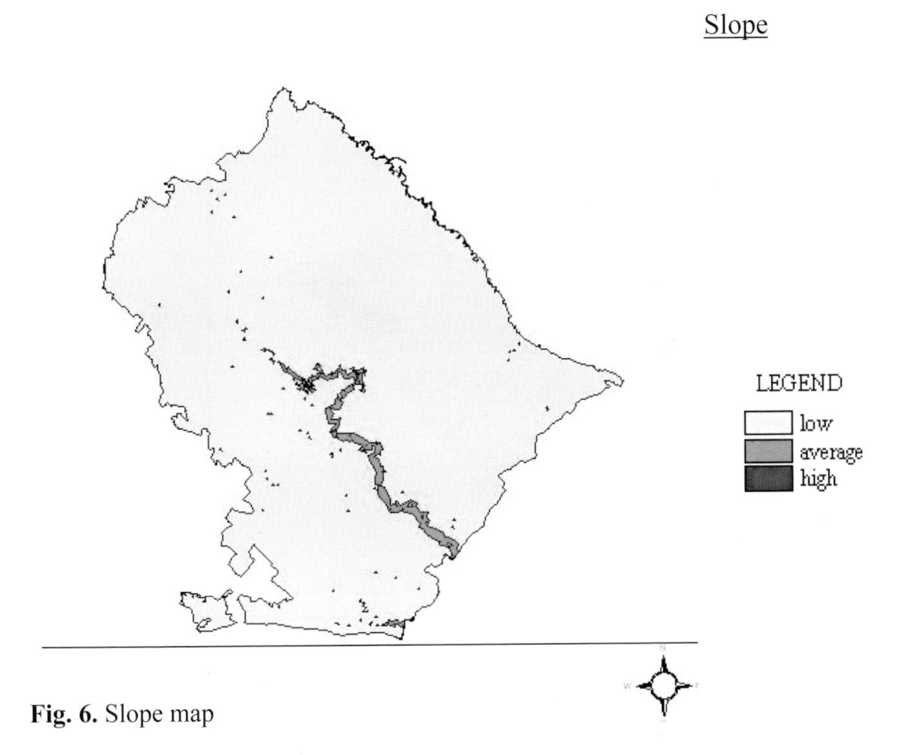

Fig. 6. Slope map

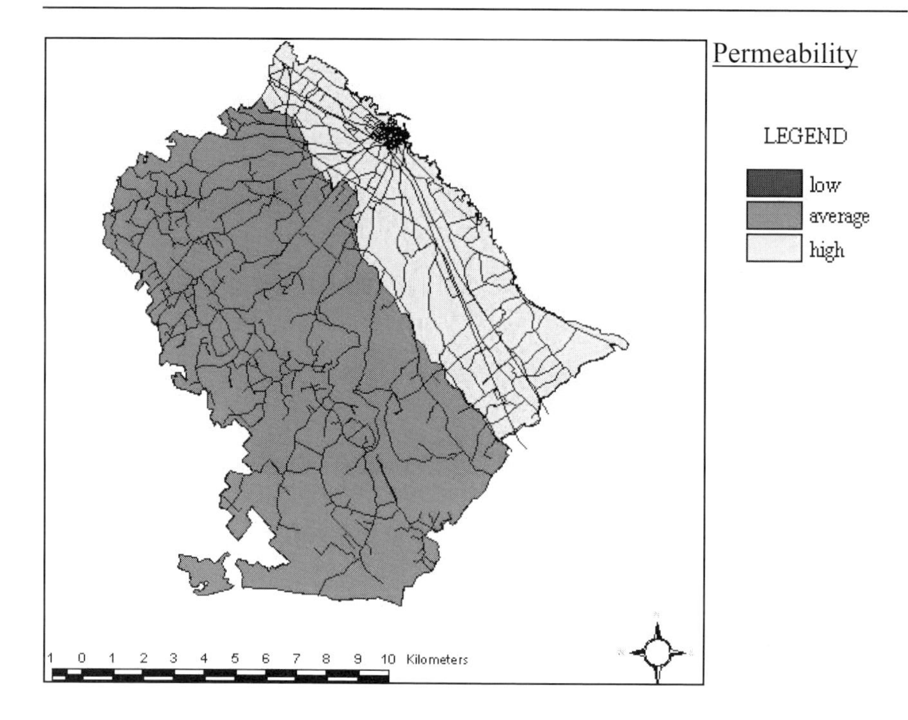

Fig. 7. Permeability map

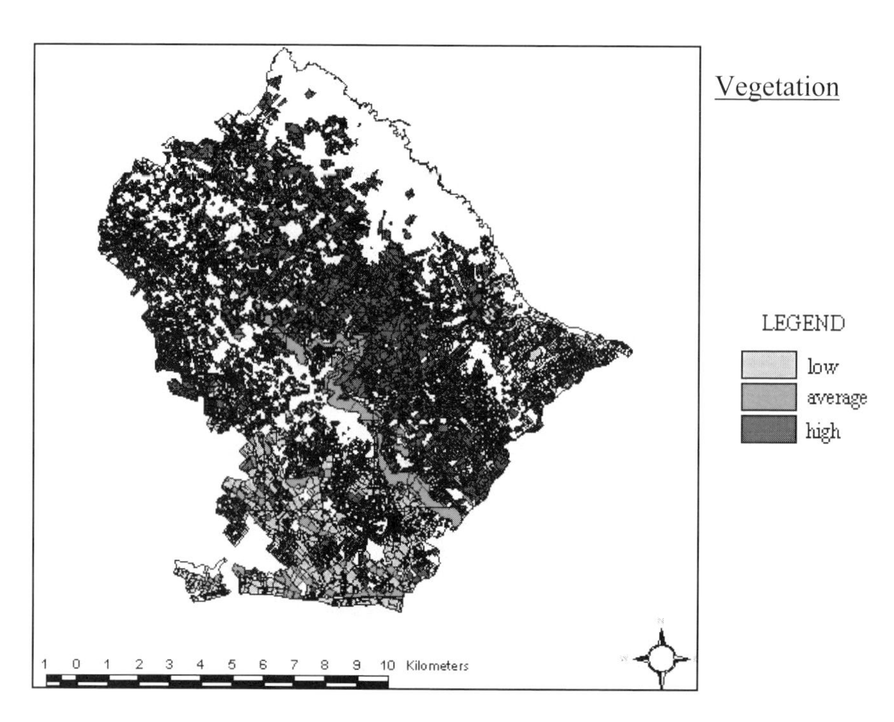

Fig. 8. Vegetation map

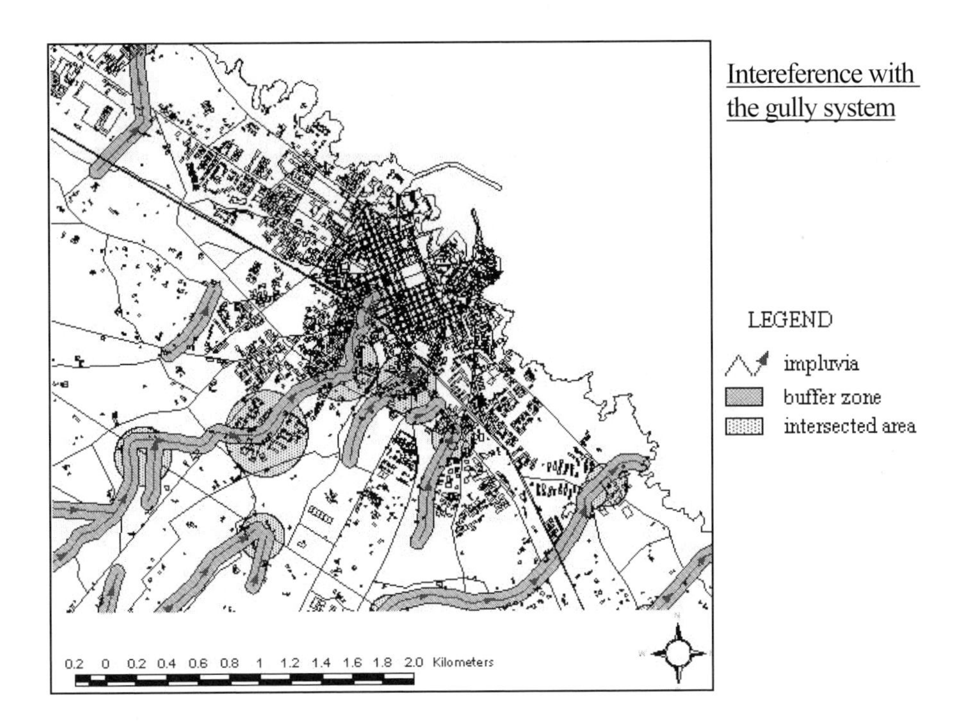

Fig. 9. Anthropic interference with the gully system map

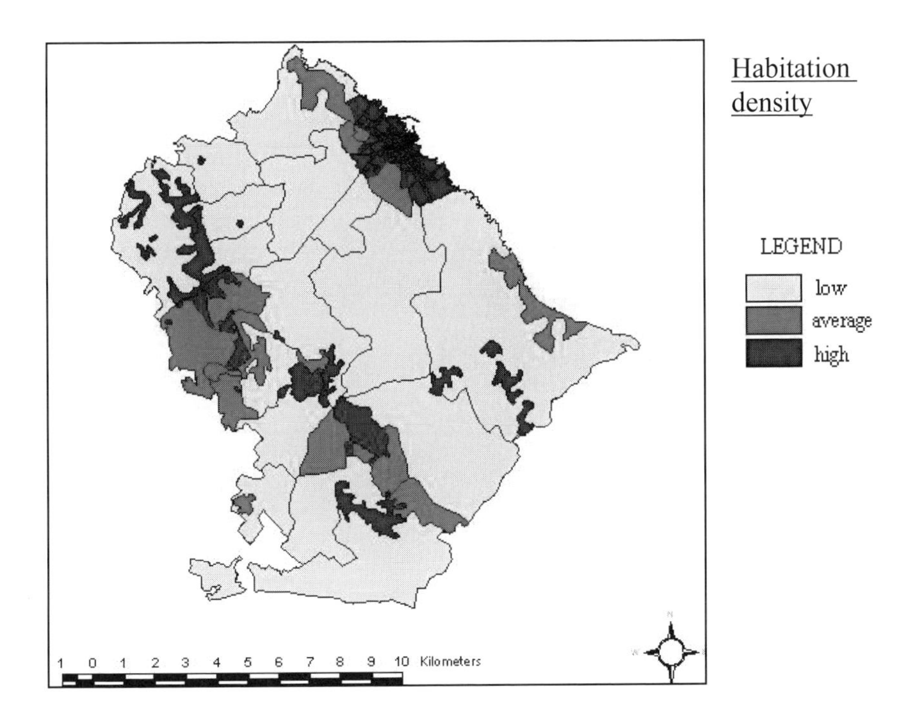

Fig. 10. Habitation density map

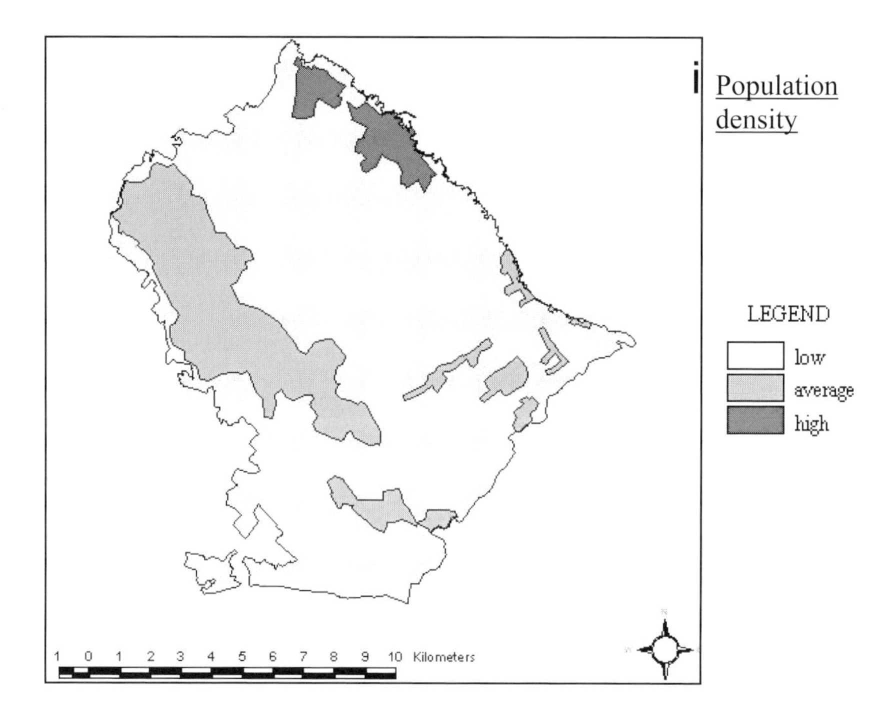

Fig. 11. Population density map

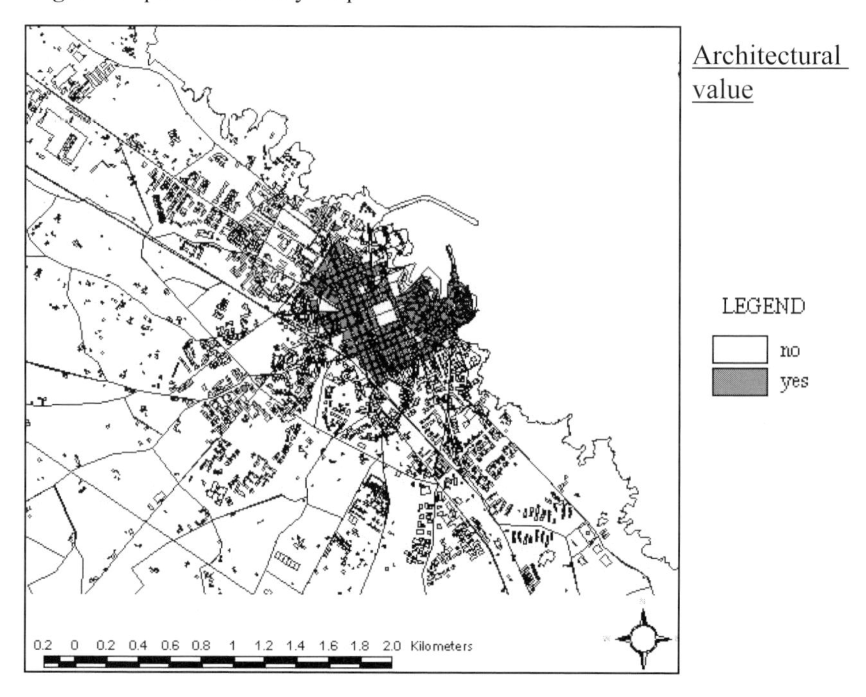

Fig. 12. Architectural value map

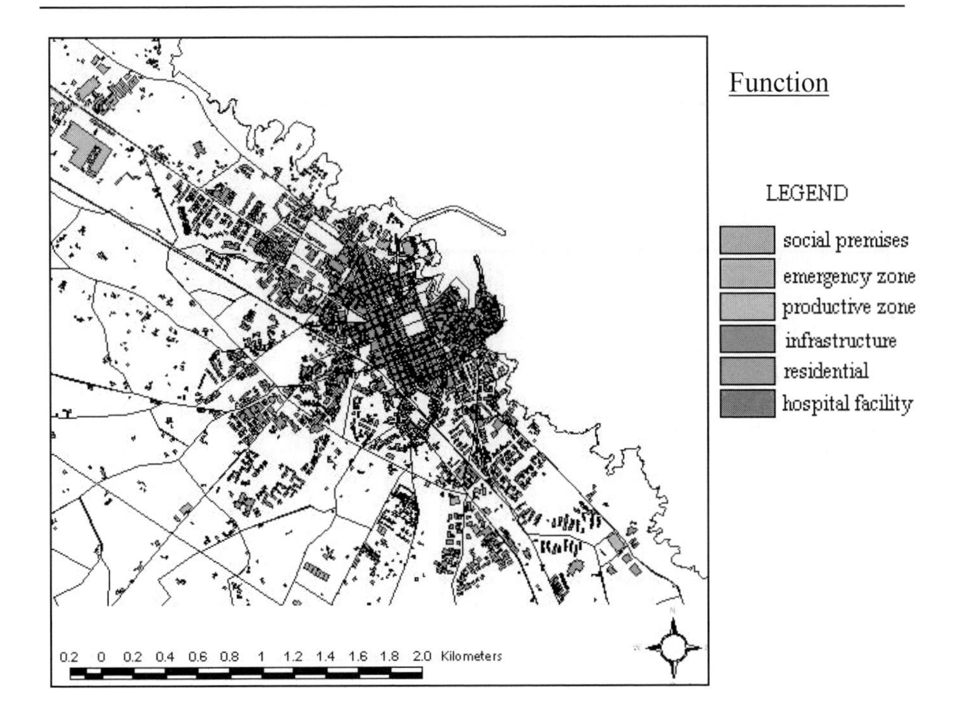

Fig. 13. Function map

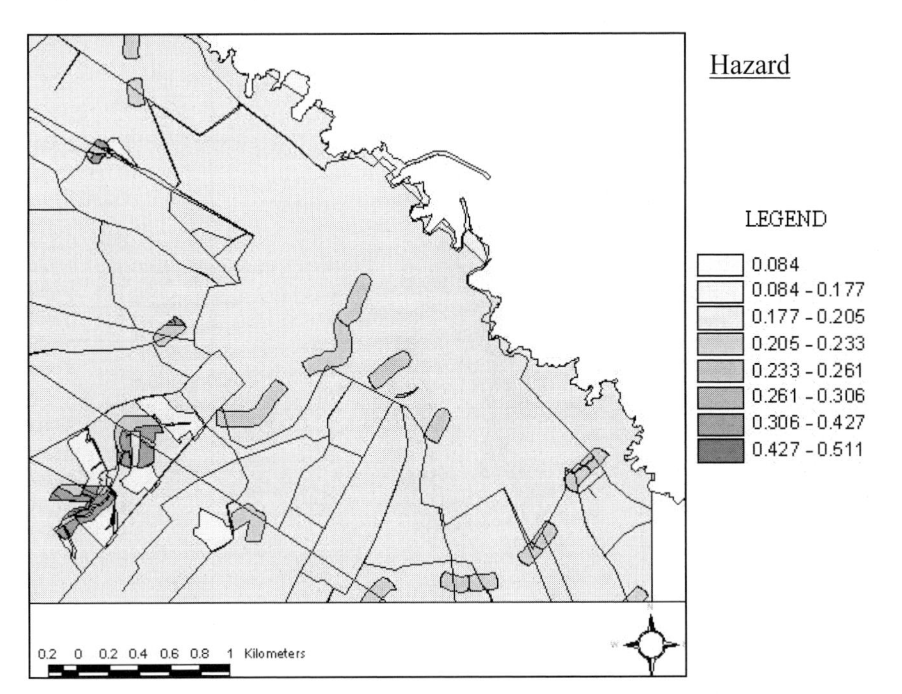

Fig. 14. Hazard map

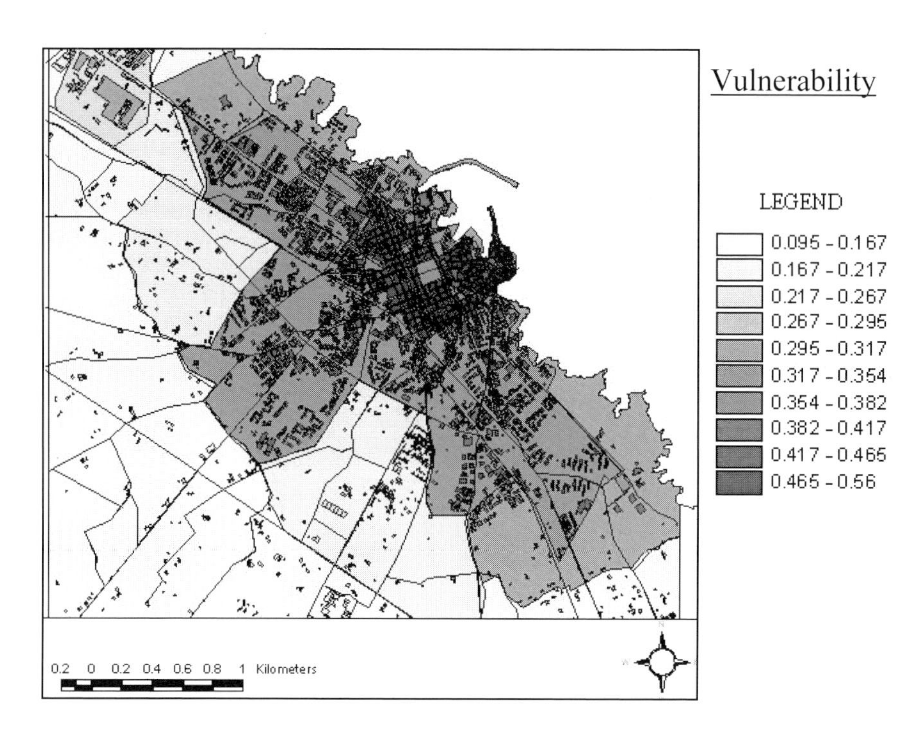

Fig. 15. Vulnerability map

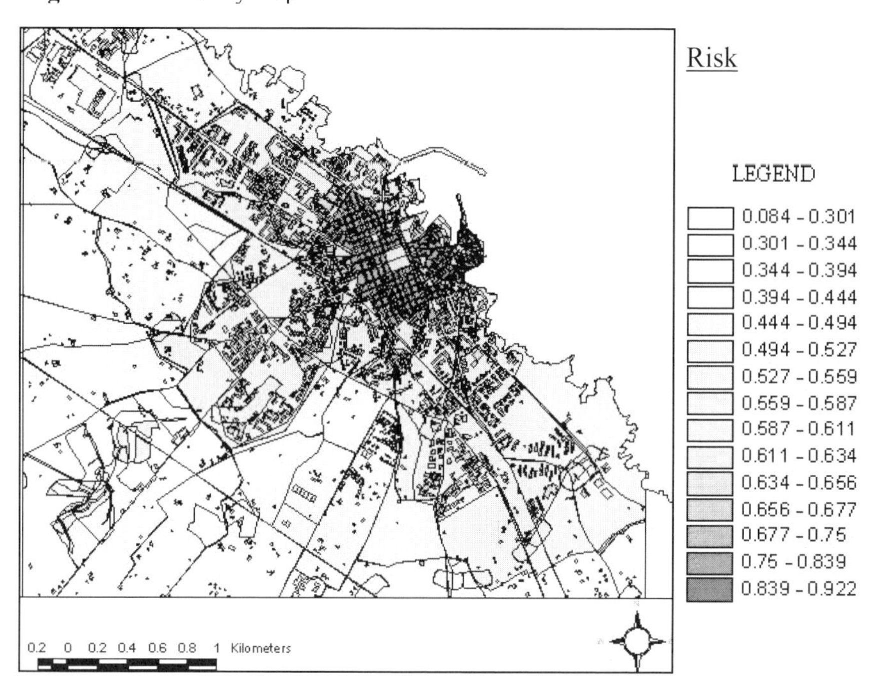

Fig. 16. Risk map

7 Conclusions

The individuation of areas at risk of flooding due to interference with the natural rainwater drainage system by the urban system has been shown to be tragically reliable. In fact, on the occasion of an intense rainfall that occurred in the zone analyzed in this study, on the 26th July 2004, some of the highlighted areas were flooded by the abundant rush of water which, as a result of the obstruction or interruption of the natural drainage canals, ran down into the inhabited centers and surrounding zones, provoking huge damage.

References

Giangrande (2001) Il metodo AHP.http://rmac.arch.uniroma3.it/Master/lezioni/giangrande/dispense/AHP.htm
Menoni S (1996) Pianificazione e incertezza. Elementi per la valutazione e la gestione dei rischi territoriali. FrancoAngeli, Milano

Ignored Devastating Disasters and Hazards: The Case of the Horn of Africa

Ambrose S. Oroda

Regional Centre for Mapping of Resources for Development (RCMRD),
PO Box 18118, 00500, Nairobi, Kenya.
Email: aoroda@rcmrd.org

Abstract

The eastern Africa sub-region is an expansive land-mass, more than 6 million km^2 comprising about 9 countries. Ecologically and environmentally the area is highly precarious, more than 65% of the sub-region being classifiable as semi-arid or arid. Some of the countries in the sub-region experience real desert like conditions with less than 250 mm annual rainfall. The sub-region experiences very frequent droughts, while rainfall distribution and intensity vary considerably, spatially and temporally. Frequent drought-related disasters include famines as a result of crop failures and lack of grazing and browse. The famines often result into human calamities such as hunger, starvation, malnutrition, mass migration of populations and in many cases death. These negative impacts of famine, resulting from environmental factors, are often compounded by several socio-economic and political factors that have over the years impacted negatively on the general production of the area. Many of these calamities due to their slow onset are ignored and often realised late, many times when their negative impacts and toll on the population have reached devastating levels. Part of the reason is lack of or inadequate Early Warning Systems (EWS) for monitoring environmental conditions as well as crop yields. This paper gives an overview of some the often ignored disasters suggesting some disaster management mechanisms that need to put into consideration in addressing and managing such calamities.

1 Introduction

1.1 Definitions

Adopted from UN, 2002 and Draft National Policy on Disaster Management in Kenya:

- Disaster refers to an overwhelming occurrence resulting into serious disruption of the functioning of a society or community. Disasters lead to widespread human, material and environmental losses exceeding the ability of the affected society or community to cope hence requiring external assistance or intervention.
- A Hazard is a dangerous situation or event that is likely to lead to or result into a disaster.
- Vulnerability is the susceptibility by an individual, community or society or property or infrastructure to hazardous conditions or to a disaster.
- Risk refers to probability or likelihood of a given population to suffer the disruption of a disaster.
- Management in this case refers to mechanisms for disaster analysis, planning, decision-making, and assignment of resources for prevention, preparation and mitigation. Management may involve recovery programmes from the effects of disasters.

1.2 Background

It is important to note that disasters know no boundaries and neither do they recognize the mightiness of nations nor their smallness (UN, 2002). Disasters vary from unexpected occurrences such as the earthquakes (that caused the December 2004 Tsunami in Asia) to more predictable seasonal floods and periodic storms. Whereas some disasters are less immediate and slow in evolving such as droughts and environmental degradation, others are spontaneous such as the Tsunamis. In the book "Living with Risks" hazards are categorised into natural, technological and environmental degradation (UN, 2002).

Natural hazards have been listed to include:

- Geological hazards such as earthquakes; tsunami; volcanic activity and emissions; and mass movements such as land slides, rock slides, rock fall, liquefaction, submarine slides. Other geological hazards include subsidence such as surface collapse and geological fault.

- Hydro-meteorological hazards which include floods, mud flows, debris; tropical cyclone, storm-surges, thunderstorms, rainstorms, windstorms; drought and desertification as well as heat waves, dust-storms and wild fires.
- Biological hazards – include outbreaks of epidemics, epizootic conditions and extensive infestation.

Technological disasters or hazards refer to danger originating from technological or industrial accidents (such as the Chernobyl atomic accident of 1986), dangerous procedures, infrastructure failure or certain human activities which may cause loss of life or injury, property damage, social or economic disruption or environmental degradation. They are sometimes referred to as anthropogenic hazards. Examples include industrial pollution, nuclear activities and radioactivity, toxic wastes, dam failures, transport, industrial or technological accidents.

Environmental degradation as a hazard refers to processes induced by man's behavior and activities (many of the time in combination with natural processes) that damage the natural resource base or destroy ecosystems.

There are, however, a number of disasters that do not fit in the above descriptions or categorizations. I am tempted to categorize such disasters as "*human disasters*" or "*human induced or human inflicted disasters*". I have in mind disasters such as the Twin-Tower disaster that was deliberately caused. Similarly are the Rwanda and Darfur genocides. These are disasters that were deliberate or intentional. Their management are equally very difficult.

2 The Horn of Africa and the Commonly Occurring disasters

2.1 The Horn of Africa

The eastern Africa sub-region also referred to as the Horn of Africa comprises Burundi, Djibouti, Eritrea, Ethiopia, Kenya, Rwanda, Somalia, Sudan, Tanzania and Uganda. Due to several other factors the Democratic Republic of Congo (DRC) tends to be included in this sub-region to some extent. Of these, seven countries (Djibouti, Eritrea, Ethiopia, Kenya, Somalia, Sudan and Uganda) form the IGAD sub-region. Human population of this sub-region is estimated at more than 200 million people (WRI, 1996), with an estimated land surface area of more than 6 million km^2. Ecologically and environmentally the area is highly precarious, more than 65% of the sub-region being classifiable as semi-arid or arid. Other countries such as Sudan experience real desert like conditions with less than 250 mm annual rainfall.

2.2 Common disasters

2.2.1 Droughts

Among the most natural disasters in the Horn of Africa are the less immediate and slowly evolving droughts and environmental degradation – desertification. Because of their less immediate and slow onset nature, they are often ignored and yet they have greater impacts – affecting far more people and have potentially greater costs for the people's future (UN, 2002). Exemplified by the records by the Kenya Government, droughts usually accompanied by land degradation (desertification) have, between 1977 and 2000 affected more than 7.7 million people compared with about 1.5 million people affected by floods and less than 20,000 people affected by fires (table 1). Edmund Barrow (Barrow, E. G. C., 1996) purports that drought, famine and relief aid now appear to be commonplace in the arid and semi-arid lands of many parts of the world, but more so in Africa. Systematic ecological and environmental degradation is becoming highly influential and this is lowering the natural resilience to disaster impact, delaying recovery. The Kenya Government in its Draft National Policy on Disaster Management recognized drought as a recurrent phenomenon that affects large areas and numbers of people in the country. According to the Centre for Research in the Epidemiology of Disaster (Blaikie et al, 1994), the frequency of occurrence of droughts has been increasing and whereas only 62 droughts were reported in the 1960s, the number rose to nearly 240 in the 1990s. However, according to Donald Wilhite (Wilhite, D. A., 2000), these figures are misleading because drought is one of the most under-reported natural disasters because the sources of most of these statistics are international aid or donor organizations. If countries afflicted by drought do not request assistance from the international community or donor governments, these episodes are not reported.

Some of the most affected regions of the world are the Sahelian and Sub-Saharan countries in which fall the eastern Africa countries. The droughts of the mid 1980s directly adversely affected more than 40 million people in Sub-Saharan Africa (Wilhite, D. A., 2000). The eastern Africa sub-region has been significantly affected as evidenced by the frequent famines and starvation that have attracted world attention in the recent past. One of the main resulting consequences of droughts has been the escalating poverty due to diminished means of economic production such as the cases where pastoralists' livestock have been wiped out during a drought (Oroda, A., 2001). Like most of the regions of Africa, the economies of all the countries of eastern Africa are highly dependent on rain-fed agriculture – both crop and livestock production, mainly for subsistence. The cumulative effects of these droughts include the erosion of assets, decreasing ability to cope with future droughts, impoverishment of rural communities and depletion of the government coffers. The effects of drought have become more pronounced in recent decades. The Government of Kenya reports that the effect of the 1991/92 drought in its arid districts led to livestock losses of up to 70% of herds and unprecedented high rates of child malnutrition of up to 50%. During this drought 1.5 million people in seventeen arid and semi-arid districts of four provinces, received relief food assistance. Rains failed again at the end of 1995 and 1996, leading to another drought situation, which affected an estimated 1.41 million people.

It is important to appreciate that besides determining the quality and quantity of agricultural production, rainfall is also a major factor determining water availability for the various socio-economic uses. The consequences of drought are, thus, manifold.

The demographic pressure, like everywhere else in the world, has meant more forest loss and more land degradation. This has meant more droughts as well as more flooding due to diminished water holding capacities. The sub-region, thus, experiences very frequent droughts as exemplified by the Kenyan data above, while at the same time rainfall distribution and amounts varying considerably, both spatially and temporally. The frequent droughts often lead to crop failures and lack of grazing and browse with frequent occurrence of famines in the sub-region (see figures 1 and 2 below). The famines often result into hunger, starvation, malnutrition, mass-migration of populations and in many cases death.

Year	Type of Disaster	Area of Coverage	Number of People Affected
1999/2000	Drought	Widespread	4.4 million
August 1998	Terrorism	Nairobi	214 killed and 5,600 injured
1997/1998	Elnino Flood	Widespread	1.5 million
1995/1996	Drought	Widespread	1.41 million
1994	Ferry accident	Mtongwe channel	270 died
1992	Train accident	Mtito Andei	31 died, 207 were injured
1991/1992	Drought	Arid and semi-arid Districts of NE, Rift, Eastern and Coast	1.5 million
1990	Fire	Lamu	20 died
1985	Floods	Nyanza/Western	10,000
1984 to 2001	HIV/AIDS	Widespread and continuing	2.2 million by 2001 700 die daily as of 2001
1983/1984	Drought	Widespread	200,000
1982	Fire	Nairobi	10,000
1982	Flood	Nyanza	4,000
1982	Fire	Lamu	4,000
1980	Drought	Widespread	40,000
1977	Drought	Widespread	20,000
1975	Drought	Widespread	16,000
1971	Drought	Widespread	150,000

Table 1. Recent History of Disasters in Kenya

These negative impacts of famine, resulting from environmental factors, are often compounded by several other social, economic and political factors that have over the years impacted negatively on the general production of the area, principally poor economic management and political instability.

Although the economic, social and environmental costs and losses associated with droughts and desertification have been increasing dramatically in the Horn of Africa, it is difficult to quantify their trends precisely because of lack of reliable historical estimates as a result of lack of systematic assessments, monitoring and recording and neither has there been systems that have specifically mapped the extent and trend of drought and desertification. This lack of reliable and operational warning and study systems has all along caused inadequacy in food supply information, a situation significantly exploited by unscrupulous and speculative traders thereby denying a reasonable population size access to food and hence compounding food insecurity situations in the subregion.

Fig. 1. Failed drought afflicted maize crop **Fig. 2.** Devastating impacts of droughts

The above scenario calls for setting up systems, Early Warning Systems (EWS), to continually provide reliable data and information on drought, desertification and the resulting consequences useful for decision-making.

The unfortunate situation is that drought has not acquired precise and universal definition. As such it is often forgotten once it ends, and everyone seems to be caught unawares again by the next drought. Some of the definitions adopted have been impact or application specific, and some of them are regional specific. The focus is, thus, often on three types of droughts viz.:- meteorological drought, agricultural drought and hydrological drought. Meteorological drought refers to deficiency of precipitation from expected or normal levels over an extended period of time while hydrological drought is defined by deficiencies in surface or sub-surface water supplies. Hydrological droughts result in lack of water for meeting normal and specific water demands. Agricultural drought is characterized by deficiency in the water availability for specific agricultural operations such as crop production.

It is predicted (UN, 2002) that drought vulnerability will increase with time due to development pressures, population increases and environmental degradation that leads to climate change.

Drought, besides its slow onset, has some other unique characteristics hence requiring different approaches in its address. Some of these characteristics are:

- Drought does not directly destroy food in storage, shelter or infrastructure,
- Its effects are cumulative,
- It is difficult to detect its onset, and,
- The impacts of drought are often widespread geographically hence difficult to quantify.

2.2.2 Land and Environmental Degradation: Desertification

Land and environmental degradation is a major hazard in the Horn of Africa where about 65% of land is arid or semi-arid, of which over 70% is classified as degraded (UN, 2002). Evidence of desert-like conditions has been on the increase and in the past the situation was described as "Sahara Desert marching southwards". The environmental degradation has been and continues to be caused by deforestation and increased agricultural activities. Other causes include inappropriate agricultural practices and increased settlement and infrastructure development.

Consequences of environmental degradation include decreased vegetation cover, decrease in soil and land stability which has resulted in land slides, and increase in soil erosion. Other consequences are decreased soil productivity resulting in poor agricultural yields and increased runoff and soil salinization whose consequences are flooding, flash floods and mud and land slides. Decrease in vegetation cover has been due to logging (for timber and other wood products), increase in demand for wood fuel and increase in human population putting more pressure on land for more food production.

Like drought, environmental degradation requires thorough studies and monitoring systems. No proper systems have, however, been established in the sub-region.

2.2.3 Human Immune Virus and Acquired Immune Deficiency Syndrome: HIV/AIDS

Using Kenya as an example, HIV/AIDS stands out as the second most severe disaster in the Horn of Africa sub-region. About 10% of the population has been affected by HIV/AIDS in Kenya. The situation was worse in Uganda in the late 1980s and early 1990s. The HIV/AIDS must have been exacerbated by the war situations in Rwanda, Burundi and the DRC.

Considered a biological hazard, HIV/AIDS has impacts on other natural disasters. For example, it reduces the labor force usually critical during the droughts (more so because it affects the most energetic members of the community). It also diminishes resources required for inputs in the management of other disasters. Like drought, HIV/AIDS has slow onset.

2.2.4 Industrial Pollution

The Kenya Government objective, according to its blue print, is to industrialize by the year 2020. The National Environment Management Authority, based on its unpublished reports, this industrialization spells a lot of doom. Lake Victoria alone receives more than 15,000 tons untreated wastes from the industries located near or within its catchment basin annually.

Again, due to its slow onset, the impacts of industrial pollution are going unnoticed hence not being realized immediately.

2.2.5 Other Hazards and Disasters

Other hazards and disasters commonly occurring but with less magnitude include floods, earthquakes, wildfires, land slides and motor accidents among others.

3 Disaster Management

Disaster Management includes a development-based set of activities aimed at reducing vulnerability within populations that are at risk to particular hazards. Disaster management, therefore, requires that adequate measures are in place to prevent the onset of a disaster and mitigate disaster effects and threats to development. Disaster mitigation activities need to overlap with development activities, as they are long-term in focus and continuous in nature.

Disaster management needs to include preparedness as precautionary measures taken in advance of an imminent threat to help people and institutions respond to and cope with the effects of a disaster. Effective disaster preparedness is supposed to be based upon a comprehensive and continuous assessment of vulnerabilities and risks. The assessment of risks will create awareness of the most likely hazards, their geographical spread, their magnitude and the elements at risk. The following are, thus, critical:

- Effectively review and updating of early warning systems with a view of improving efficiency and state preparedness. The early warning systems must be closely linked to the decision-making arms of the government as well as the international disaster response organs. These warning systems must also be capable of delivering critical information to emergency management organs as efficiently as possible and in real-time.
- Regular and independent assessment strategies and intervention policies;
- Determination of adequacy and efficient utilization of disaster resources;
- Assessment of impacts of disaster management programmes on the population, economy and the environment;
- Establishing an institutional framework that will manage disasters;
- Ensuring that disaster management and institutions involved are well coordinated and focused on both risk and vulnerability reduction. This should take cognizance of the fact that disaster management is a multi-sectoral and multi-disciplinary issue, involving governmental, non-governmental and international players;
- Promoting the linkages between disaster management and development planning;
- Fostering partnerships between the government and stakeholders at all levels, including regional and international bodies;
- Promoting disaster management culture, training, research and information dissemination, community awareness and preparedness.

4 The Role of Geo-Spatial Technologies: The Use of Space Technology and GIS in Disaster Management

Remote Sensing and GIS are increasingly being used in disaster management due to various reasons. Among the reasons are timeliness and repetitiveness of information, the synoptic coverage of the disasters and more details of remote sensing data due to their multi-spectral nature. Other benefits include free availability of certain satellite datasets and improvements in methods for application of space technology, whereas GIS has become an important tool for data integration, analysis, modeling and management.

References

Barrow EGC (1996) The Drylands of Africa, pp 1-25, pp 99-117

Blaikie P, Cannon T, Davis I, Wisner B (1994)At Risk: Natural Hazards, People's Vulnerability and Disasters, London: Routledge

Oroda A (2001) Towards Establishing an Operational Early Warning System for Food Security in the horn of Africa: Remote Sensing and Geo-Information Journal of The Netherlands Remote Sensing Board (BCRS), Nr 2, August 2001, pp 26-30

Republic of Kenya (2001) Draft National Policy on Disaster Management in Kenya, Office of President, Nairobi

UNEP (1996) Gudelines for Integrated Planning and Management of Coastal and Marine Areas in the Wider Caribbean Region

United Nations (2002) Living with Risks: A global review of disaster reduction initiatives, pp 11-78

Wilhite DA (2000) Drought: A Global Assessment Volume I, London, Routledge, pp 3-17

World Resources (1996) World Resources Information Table 8.1, World Resources Institute, Oxford University Press, New York, pp 190-191

Zschau J, Kuppers AN (2003) Early Warning Systems for Natural Disaster Reduction, pp 13-14, pp 67-69

Tight Coupling of SFlood and ArcView GIS 3.2 for Flood Risk Analysis

Shanker Kumar Sinnakaudan and Sahol Hamid Abu Bakar

Water Resources Engineering and Management Research Centre (WAREM), Faculty of Civil Engineering, MARA University of Technology, Pulau Pinang, 13500 Permatang Pauh, Penang, Malaysia.
Email: drsshan@yahoo.com

Abstract

This paper describes the development of an Arcview GIS extension namely SFlood.avx to integrate the SFlood hydraulic and sediment transport model (modified version of HEC-6 model) within ArcView GIS environment. The extension was written in an Avenue script language and Dialog Designer with a series of '*point and click*' options. The flood risk model was tested using the hydraulic and hydrological data from Pari River catchment area in Malaysia. The required sediment input parameters were obtained from field sampling. Sflood.avx has the capability of analyzing the computed water surface profiles generated from SFlood hydraulic model and producing a related flood risk map for Pari River in ArcView GIS. The user-friendly menu Graphic User Interface guides the user to understand, visualize, build query, conduct repetitious and multiple analytical tasks with SFlood model outputs. The results of this study clearly show that GIS provides an effective environment for flood risk analysis and mapping.

1 Introduction

Most computer models used in the flood risk analysis of rivers have inadequate functions in its spatial analytical capabilities and without sediment transport simulation capacity or suitable equations to represents correctly in-situ hydraulic processes (Sinnakaudan et al., 2001, Sinnakaudan, 2003). These models give detailed results on water surface elevation, erosion and sedimentation, riverbed changes and other hydraulic characteristics in huge text files. However, a great amount of time, expertise and cost are needed for visualizing the model results in presentable formats so that can be easily used by engineers, planners and decision makers. Further more, the consistent deficiencies of these models are their inability to connect the information describing the water profiles with their physical locations on the land surface. This is where a Geographic Information System (GIS) becomes a valuable tool in spatial modeling for engineers, planners and geoscientist (Burrough, 1986; Sinnakaudan et.al, 2003).

In recent years, efforts have been made to integrate hydraulic models and GIS to facilitate the manipulation of the model output for flood risk analysis. Tate (1999) introduces some of the flood risk analyzing methods by integrating HEC-RAS model with ArcView GIS. Similar attempts were also made by Jones *et al.* (1998) and Anrysiak (2000). Unfortunately, these attempts miss the important element in river modeling that is the sediment transport processes. Due to this, Sinnakaudan et al (2003) had loosely coupled ArcView GIS and HEC-6 sediment transport model by writing an integrator tool namely AVHEC-6.avx (Sinnakaudan et al., 2002[b]).

However, Sinnakaudan (2003) has confirmed that the sediment transport equation embedded in the HEC-6 model such as Ackers-White (1973) and Yang (1973) (USACE, 1991) gives less reliable water/flood level prediction incorporating sediment transport for Malaysian Rivers. As a result, the current research presents the development of a new total bed material load equation using multiple linear regression analyses that is applicable for flood risk analysis in Malaysian rivers. It was developed and embedded as a modified version of HEC-6 model and named as SFlood model which later used together with ArcView GIS for flood risk analysis (Sinnakaudan, 2003; Gee, 2003).

2 Study Area

Pari River, which is one of the main tributary of Kinta River located in Ipoh, Perak, Malaysia (Figure 1) has been chosen to quantify the flooding scenarios to meet the tasks specified in this study. A 3.0 km stretch between the gauging stations at Silibin Bridge (upstream) and Kinta River confluence (downstream) is chosen. The design main channels are rectangular in shape with an average width of 18 meter at the downstream of Tapah River and 16 meters for the rest.

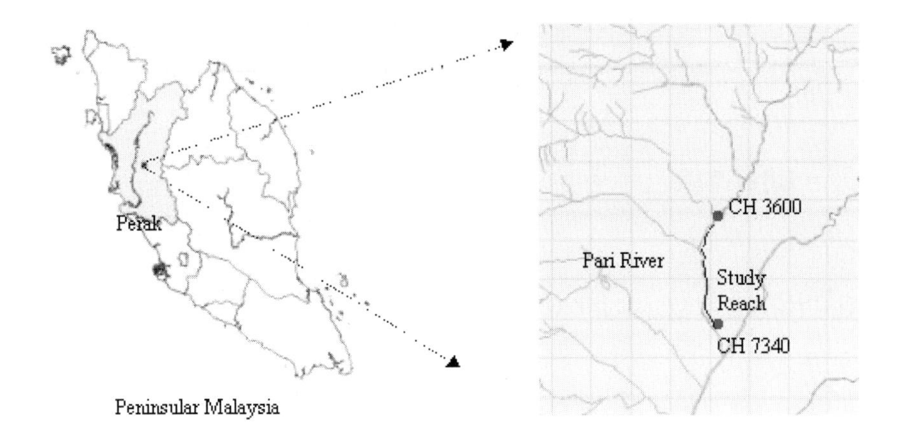

Fig. 1. Study area

3 Methodology

This study was carried out in four stages. This includes field sampling, spatial and non-spatial data collection and processing, total bed material load equation development, customization and modification of the HEC-6 model source codes using Compaq Visual FORTRAN to create SFlood sediment transport model. The modeling tool was compiled with an ArcView GIS extension and is named as *Sflood.avx.* The procedure is comprised of three elements, which are (1) a set of equations compiled in the form of SFlood geospatial model governing the hydraulic processes, (2) maps that define the study area and (3) database tables that numerically describe the study area and the model parameter. The GUI for the modeling system has been designed so that it perfectly integrates the three components as stated above.

A total of 346 reliable sediment and hydraulic database was established from recent studies (Ariffin *et al.*, 2001; Sinnakaudan *et al.*, 2003; Sinnakaudan, 2003; DID, 2003). The data sets were then divided, in which 181 data were used for analyses process (equation development), and the balance of 165 data were utilized for model validation. The validation process was further extended using a total of 987 available sediment samples and hydraulic data from rivers in the United States and Pakistan (Brownlie, 1981).

The regression technique namely multiple linear regression was used to predict sediment discharge using selected flow and sediment discharge parameters (Hair *et al,* 1995). The final regression equation derived is as follows:

$$C_v = 1.811 * 10^{-4} \left(\frac{VS_0}{\omega_s} \right)^{0.293} \left(\frac{R}{d_{50}} \right)^{1.390} \left(\frac{\sqrt{g(S_s - 1) d_{50}^{\;3}}}{VR} \right) \tag{1.0}$$

The Total Bed Material Load, T_j is derived using:

$$T_j = C_v * Q * \rho \tag{1.1}$$

Where,

C_v	=	volumetric concentration of Sediment
V	=	average velocity
ω_s	=	sediment fall velocity
S_0	=	Energy slope
g	=	acceleration due to gravity
S_s	=	specific gravity of sediment
d_{50}	=	sediment diameter where 50% of bed material are finer
R	=	hydraulic radius
T_j	=	Total bed material load
Q	=	Discharge
ρ_s	=	Specific weight of sediment

Equation 1.0 accounts for 71.51 % of the variability in analyses data and 63.63 % in the validation data. However, Yang's (1973) equation only caters for 35.91 % and 38.78 % variability in the analyses and validation data. Ackers & White (1973) predicts 12.71% for analyses data and nil for variability data. Figure 2 show the comparison between measured and estimated total bed material load and validation of Equation 1.0. The model is best suited for perennial rivers having uniform sediment size distribution with a d_{50} value within the range of 0.37 mm to 4.0 mm and performs better than the commonly used Yang, Graf and Ackers-White total bed material load equations.

Equation 1.0 was coded in FORTRAN 90 and embedded into the existing HEC-6 source codes. The modified codes of HEC-6 were compiled and named as SFlood.exe with the permission from Hydraulic Engineering Center (HEC) of United States Army Corps of Engineers (USACE) (Gee, 2003). The variable names in the new coding remained the same so that they will tally with the various existing subroutines hydraulic calculations in SFlood hydraulic model. The sample output file/header information of the SFlood is shown in Figure 3. This model later used to simulate the effect of sediment transport mechanism on flood risk.

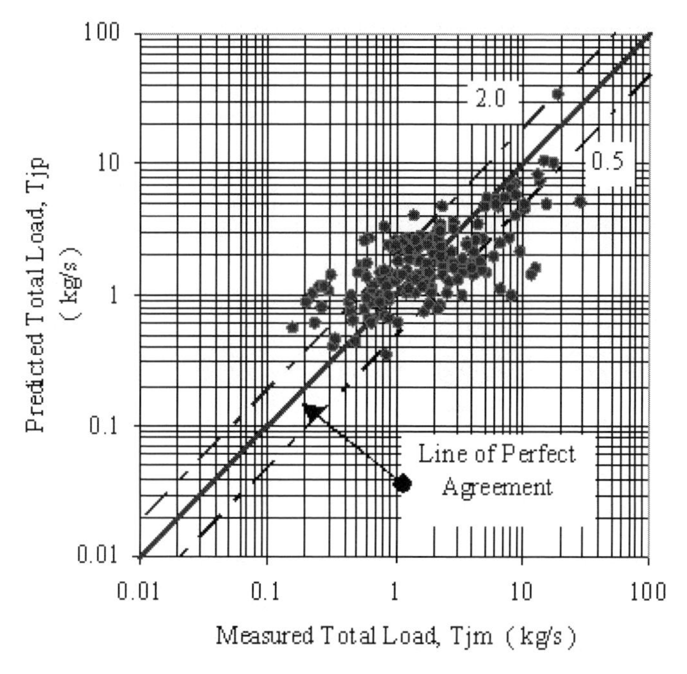

Fig. 2. Validation of predicted total bed material load using Equation 1.0

```
* SCOUR AND DEPOSITION IN RIVERS AND RESERVOIRS *
*      Version: 4.2.00 -  JUNE 2003             *
* INPUT FILE:  d30ari10_10.dat                  *
* OUTPUT FILE: d30ari10_10.out                  *
* RUN DATE: 30 SEP 04    RUN TIME: 15:11:04     *

* MODIFIED BY:                                  *
* DR. SHANKER KUMAR SINNAKAUDAN                 *
* WATER RESOURCES ENGINEERNG AND MANAGEMENT     *
* RESEARCH CENTRE (WAREM), MARA UNIVERSITI OF   *
* TECHNOLOGY, 14300 PERMATANG PAUH, PENANG      *
* MALAYSIA. TEL: +604-3822714, FAX: +604-3822812 *

* ORIGINAL SOURCE CODE: FROM HEC-6 MODEL                *
* U.S. ARMY CORPS OF ENGINEERS                          *
* HYDROLOGIC ENGINEERING CENTER, 609 SECOND STREET      *
* DAVIS, CALIFORNIA 95616-4687                          *
* (916) 756-1104                                        *

* DR.CHOONG KOK KEONG                                   *
* SCHOOL OF CIVIL ENGINEERING                           *
* UNIVERSITI SAINS MALAYSIA, SERI AMPANGAN              *
* MALAYSIA. TEL: 04-5937788 Ext.6225                    *

* MAXIMUM LIMITS FOR THIS VERSION ARE:                          *
*       10  Stream Segments (Main Stem + Tributaries)           *
*      500  Cross Sections                                      *
*      200  Elevation/Station Points per Cross Section          *
*       20  Grain Sizes                                         *
*       10  Control Points                                      *
* #### 1 NEW SED.TRANS EQS.FOR MALAYSIAN RIVER CONDITIONS       *
* #### RECIEVE SFLOOD GIS GENERATED INPUT FILE                  *
```

Fig. 3. Header information of the SFlood output File

The water surface profile through the reach was computed for ARI 10, 50 and 100 years with 30, 60 and 120 minutes design storm conditions. 4th November 1997 flood was used to calibrate and validate the simulation results. The rainfall and discharge pattern at Silibin Bridge which were used in the calibration and validation process is shown (Figure 4). The calibration and validation results are shown in Figure 5.

A user-friendly, menu-driven GUI for two and three-dimensional (2D & 3D) digital floodplain delineation was developed through ArcView GIS and SFlood tight coupling procedure by utilizing Avenue Scripting Language and Dialog Designer. This version of the model comprises user-friendly interfaces for Pre-Processor, Post Processor, SFlood Tools and SFlood buttons (Figure 6 and Figure 7). It is capable to produce quick analysis (snapshots) at any desired discharge time steps in flood risk mapping procedure. Field measurements were carried out to validate the hydraulic setting and the accuracy of model outputs.

The feasibility of simulating a flood event along a river channel and floodplain by using SFlood model was tested for Pari River catchment's area. The model calibration is focused mainly to high flows which cause floods and is related to the water level (WL) data obtained for 4th November 1997 flood (Figure 4). Flood risk analysis were conducted for the design flood events for 10, 50, 100-year Average Recurrence Interval (ARI). The design rainfall duration (D) of 30, 60 and 120 minutes for the present and future land use conditions (year 2020) were considered in the simulation scenarios.

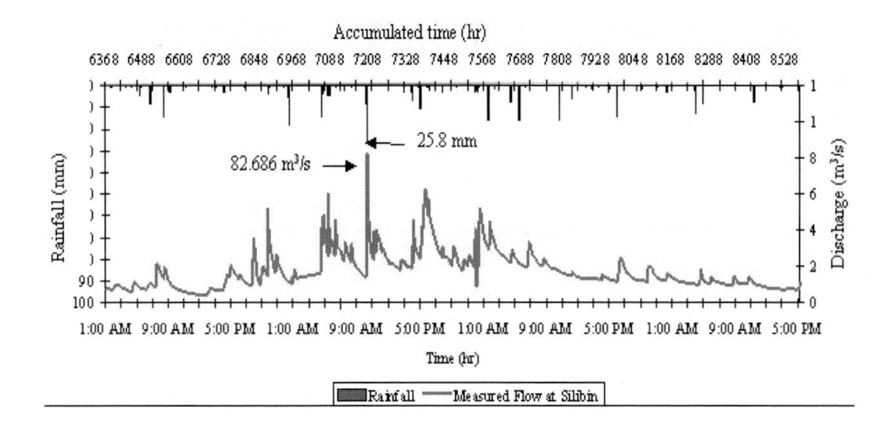

Fig. 4. 29th October 1997 – 31st December 1997 Rainfall (Station 4511111) and Discharge Hydrograph (Station 4610466) Records

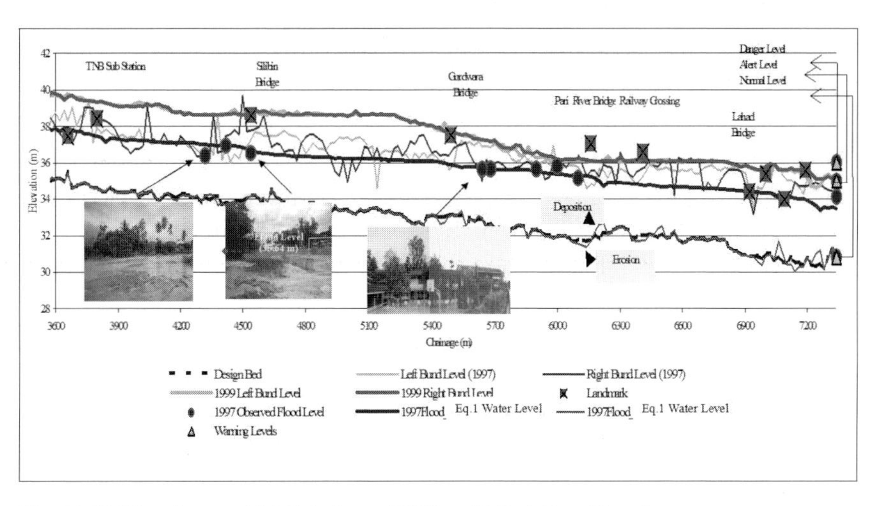

Fig. 5. Calibration and Validation of SFlood Modeling Results

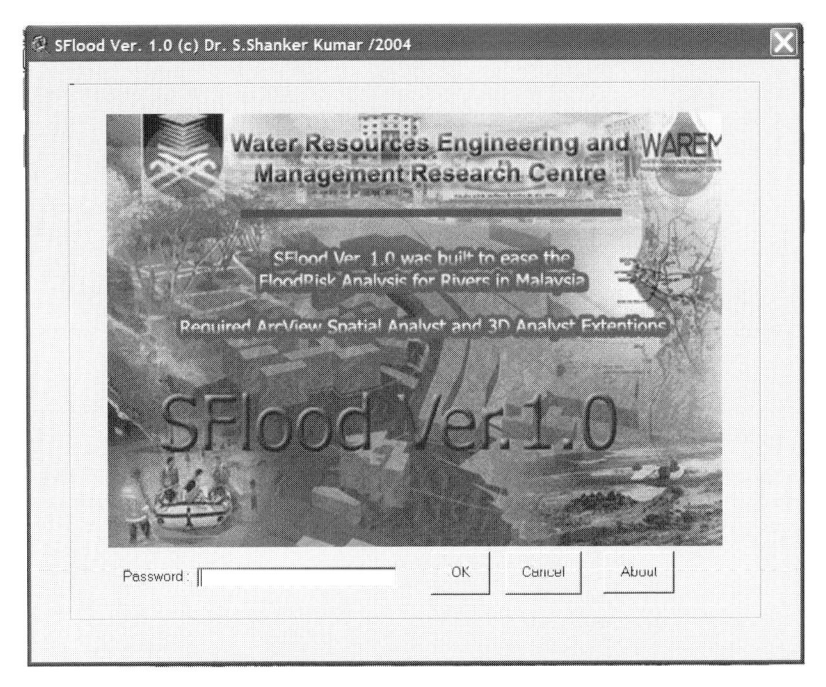

Fig. 6. Graphic User Interface (GUI) of SFlood model

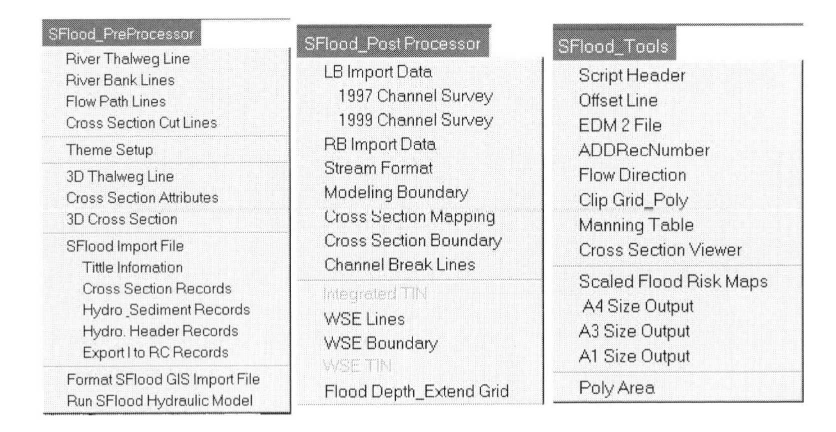

Fig. 7. SFlood Pre-Processor, Post Processor and Tools Menus

The model simulation results between Silibin Bridge (upstream) and Lahat Bridge (downstream) were analyzed. The simulation results between Chainage 3600 - 4500 and 7020 – 7300 (downstream) were treated as model stabilization sections and not used for flood risk analysis and mapping.

Wherever the flood level is greater than the bund surveyed in the year 1999, the areas are considered as flooded and validated with field observation, flood photographs and water level records as provided by Department of Irrigation and Drainage, Perak, Malaysia. The predicted flood level for ARI 10, 50 and 100 years for present and future land use conditions were draped over an Integrated TIN (Sinnakaudan et al., 2002[a]) (Figure 8) to derive the flood inundation and flood risk zone map.

Figure 9 and Figure 10 shows the flood inundation map for present and future land use conditions respectively. Figure 11 and Figure 12 shows the delineated flood risk zones based on the probability of flood event for present and future land use (year 2020) conditions.

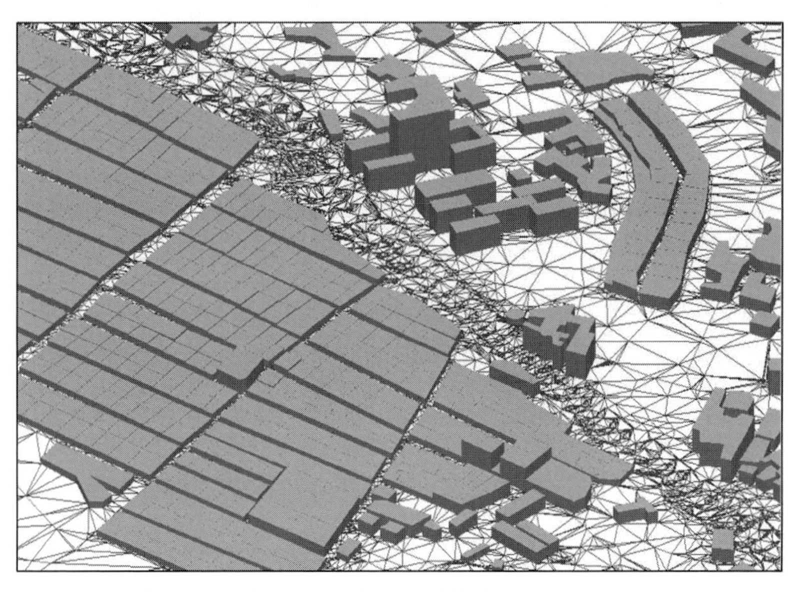

Fig. 8. Sample 3D Mesh of Integrated TIN (ITIN) for study area

The existing bund level is adequate to cater for 10 to 50 year flood for present land use conditions. However, for future land use conditions (Year 2020), the Pari River is unable to convey the excess water at the Chainage 4120 to 4520 at the upstream of Silibin Bridge and Chainage 5240 to 7340 (Figure 13).

Thus, flood-proofing measures may be considered such as raise up the existing bunding crest level above the predicted flood level (Figure 14) with an appropriate freeboard. The existing channel also may be widening up by removing the compound channels. This alternative will provide more flood conveyance for Pari River at the affected chainages. The second alternative may be implemented with the relocation of the flood plain dwellers who are most vulnerable to flood according to the flood risk zones as shown in Figure 11, 12 and 13. The third alternative is to implement “source control” oriented designs based on the Urban Stormwater Management Manual for Malaysia (2000) for new development in the Pari River catchment area. The main aim is to delay the time of the excess runoff to reach the main conveyance channel.

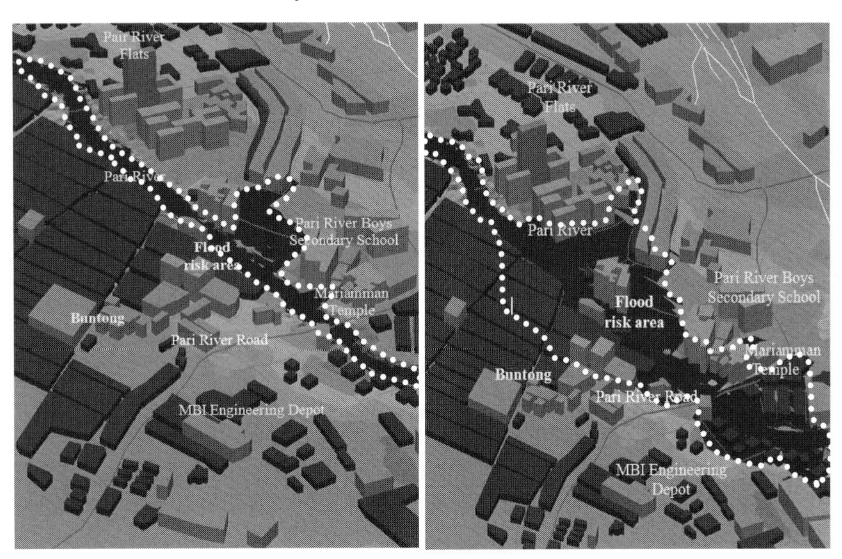

Fig. 9. Flood risk map for D120, ARI 100 years (Q = 220 m^3/s) - Present land use conditions

Fig. 10. Flood risk map for D120, ARI 100 years (Q = 343.0 m^3/s) - Year 2020 land use conditions

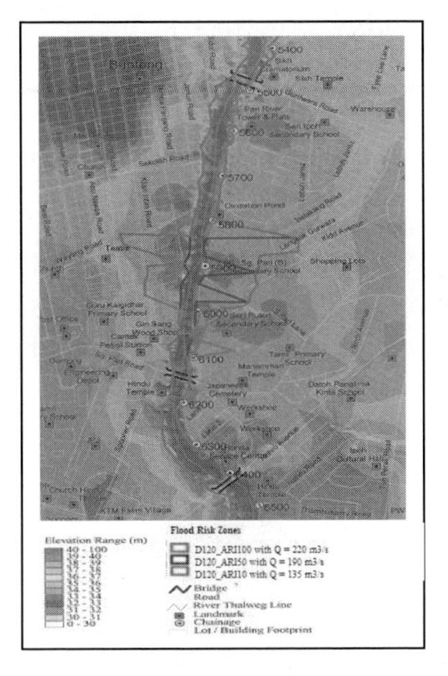

Fig. 11. Delineated flood risk zones based on the probability of flood event for present land use conditions

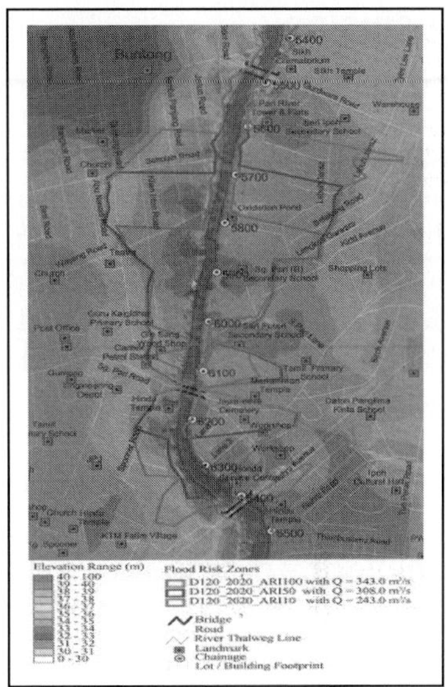

Fig. 12. Delineated flood risk zones based on the probability of flood event for future land use (year 2020) conditions

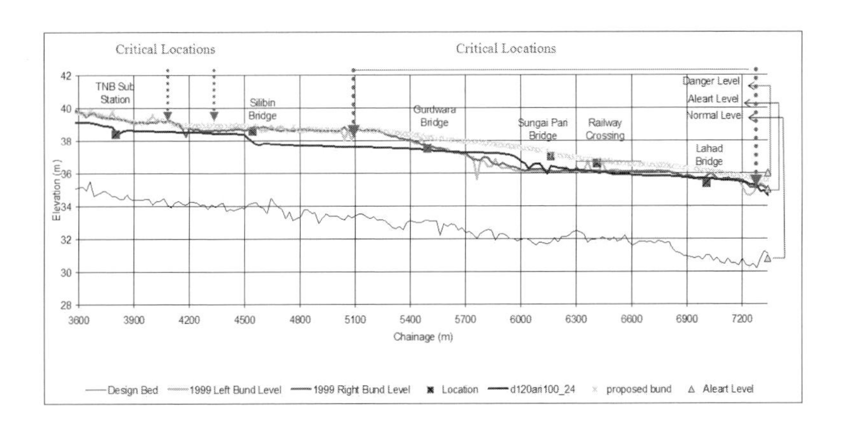

Fig. 13. Proposed bund elevations and channel improvement locations (Sinnakaudan, 2003)

4 Conclusion

The principal purpose of this research is to develop a new sediment transport equation applicable to rivers in Malaysia is achieved by the derivation of Equation 1.0. This equation successfully applied to simulate the effect of sediment transport mechanism on flood level and extend by developing SFlood modeling interface. The result of this research indicates that GIS is an effective environment for flood risk analysis and its integration with hydraulic model is not only feasible but also mutually beneficial for both GIS users and hydraulic modelers.

Acknowledgements

This paper depicts the current improvement to the PhD research that was carried out by the first author at the Universiti Sains Malaysia. The author would like to thank the Ministry of Science, Technology and Innovation (MOSTI), Malaysia for the National Science Fellowship that has resulted in this paper. An acknowledgement due to the anonymous reviewers of the published papers related to this research in Journal and conferences for their useful comments that served to strengthen this research. Many thanks are also to Department of Irrigation and Drainage, Malaysia for providing the required data on Pari River. An appreciation goes to the Faculty of Civil Engineering, Universiti Sains Malaysia on their support, advice and encouragement.

References

Anrysiak PB (2000) Visual Floodplain Modeling with Geographic Information Systems (GIS). Austin, The University Of Texas at Austin

Ariffin J, Ab Ghani A, Zakaria NA, Yahya AS, Abdul Talib S (2001) Sediment discharge on Sungai Langat and its tributaries. In Proceedings, R & D Colloquium on River Engineering and Urban Drainage (Ab Ghani A, Zakaria NA, Abdullah R, Mohd. Sidek, L (eds.), 14 –15 August, Penang, Malaysia, pp 72-78

Brownlie WR (1981) Compilation of alluvial channel data: laboratory and field. Report No. KH-R-43B. Pasadena: California Institute of Technology

Burrough PA (1986) Principles of Geographic Information Systems for Land Resources Assessment. Oxford: Clarendon Press

DID - Department of Irrigation and Drainage Malaysia (2000) Urban Stormwater Management Manual for Malaysia. Kuala Lumpur: Department of Irrigation and Drainage Malaysia

DID - Department of Irrigation and Drainage Malaysia (2003) River Sediment Data Collection and Analysis Study (Final Report). Kuala Lumpur: Department of Irrigation and Drainage Malaysia

Hair JF, Anderson RE, Tatham RL, Black W. (1995). Multivariate Data Analysis with Readings. 4th Ed. New Jersey: Prentice Hall Inc.

Gee M (2003) Hydraulic Engineering Center (HEC). HEC-6 Model source-codes. E-mail correspondence

Jones J, Haluska TL, Williamson AK, Erwin ML (1998) Updating flood maps efficiently: building on existing hydraulic information and modern elevation data with a GIS. U.S. Geological Survey Open-File Report 98-200. Reston: USGS

Sinnakaudan S Ab. Ghani A, S. Ahmad MS, Zakaria NA (2001) Integrating GIS with hydraulic and sediment transport model for flood risk analysis. In Proceedings, R & D Colloquium on River Engineering and Urban Drainage, 14 – 15 August. Penang, Malaysia, pp 47-54

Sinnakaudan S, Ab Ghani A, Chang CK, S. Ahmad MS, Zakaria NA (2002a). Integrated Triangular Irregular Network (ITIN) model for flood risk analysis. case study: Pari River, Ipoh, Malaysia. In Proceeding, Thirteenth Asia and Pacific Division Congress of International Association of Hydraulic Engineering Research on Hydraulic and Water Resources Engineering in early 21st Century, Guo J (ed), 6-8 August, Singapore: World Scientific, pp 656-660

Sinnakaudan S, Ab Ghani A, Chang CK (2002b) Flood inundation analysis using HEC-6 and ArcView GIS 3.2a. In Proceeding, 5th International Conference on Hydro - Science & Engineering, September 18-21, Warsaw University of Technology, Poland. (available in CD-Format)

Sinnakaudan S, Ab Ghani A, S. Ahmad MS, Zakaria NA (2003) Flood Risk Mapping for Pari River Incorporating Sediment Transport. Journal of Environmental Modeling and Software. 18(2), pp 119-130

Sinnakaudan S (2003) Sediment Transport Modeling And Flood Risk Mapping In Geographic Information System, PhD Thesis, Univesiti Sains Malaysia

Tate EC, Olivera F, Maidment D (1999) Floodplain Mapping Using HEC-RAS and ArcView GIS. Austin, The University of Texas at Austin

USACE - United States Army Corps of Engineers (1993) Scour and Deposition in Rivers and Reservoirs (HEC-6), User's Manual, California: Hydrologic Engineering Center

Public Participation Geographic Information Sharing Systems for Community Based Urban Disaster Mitigation

Suha Ülgen

Boğaziçi University, Kandilli Observatory and Earthquake Research Institute, Disaster Preparedness Education Program, 34342 Bebek, Istanbul, Turkey.
Email: sulgen@imagins.com

Abstract

This paper presents the efforts by Boğaziçi University, Kandilli Observatory and Earthquake Research Institute (BU-KOERI), Istanbul Community Impact Project in Istanbul, Turkey in piloting three innovative uses of geospatial information technology for community-participation in disaster mitigation. 1) ***The Neighborhood Geographic Information Sharing System*** provided volunteers with skills and tools for identification of seismic risks and response assets in their neighborhoods. Field data collection volunteers used low-cost hand-held computers and data compiled was fed into a geospatial database accessible over the Internet. Interactive thematic maps enabled discussion of mitigation measures and action alternatives. This pilot evolved into a proposal for sustained implementation with local fire stations. 2) ***The Basic Disaster Awareness Program Monitoring Facility*** involved collaboration between the Istanbul Community Impact Project and the Istanbul Education Directorate. School Basic Disaster Awareness instructors reported on dissemination of training using a web-based data entry application. The data was used to create a series of interactive Web-based thematic maps showing the geographic reach of the training program with hot links which allow the public to drill down to see local contact information. 3) ***The School Commute Contingency Pilot*** was designed to track school-bus routes in Istanbul, in order to stimulate contingency planning for commute-time emergencies when 400,000 students travel an average of 45 minutes each way on 20,000 service buses. Global Positioning System (GPS) data loggers were used to determine service bus routes displayed on printed maps highlighting nearest schools along the route. It is proposed that bus-drivers, parents and school managers be issued route maps with nearest schools that could serve as both meeting places and shelters. The system is designed to be in synchrony with the Istanbul Earthquake Early Warning and Rapid Response System housed at BU-KOERI.

1 Introduction

Following the 1999 Kocaeli and Düzce earthquakes in Turkey where 20,000 people lost their lives and 120,00 were injured, there was a concerted effort to identify the seismic risk faced by Istanbul, a mega city of 12 million inhabitants, and to develop mitigation programs. Istanbul is some 15 km. away from the fault that generated the 7.4 magnitude earthquake that hit the Kocaeli province 60 km. to the east. There is strong historical evidence that earthquake epicenters move westward on this section of the North Anatolian Fault system referred to as the Marmara Fault that is now threatening Istanbul.

One of the disaster mitigation projects undertaken by Boğaziçi University, Kandilli Observatory and Earthquake Research Institute (BU-KOERI) was the Istanbul Community Impact Project (ICIP) funded by the United States Agency for International Development, Office of Foreign Disaster Assistance. (1) ICIP focused on the development of disaster preparedness education programs and their dissemination in and around the metropolitan city of Istanbul. Within the Istanbul Community Impact Project three innovative uses of geospatial information technologies were piloted. The motivation behind these pilots was to go beyond the customary post-disaster "war-room" Geographic Information Systems (GIS) implementations by deploying bottom-up, open access and participatory, multi-purpose/user, voluntary and low-cost public geospatial information sharing applications.

2 The Three Pilot Projects

2.1 The Neighborhood Geographic Information Sharing System

There are 32 municipalities and over 650 neighborhoods in metropolitan Istanbul representing a wide range of natural, socio-economic, physical and even cultural characteristics shaped by the long history of the city. For disaster mitigation measures to be effective they need to be developed in recognition of the local differences and adopted by the active participation of each community. To assist the identification, compilation and sharing of neighborhood-specific disaster risk information, preparedness resources and the development of mitigation alternatives, a public participation GIS, named the Neighborhood Geographic Information Sharing System, was deployed in three Istanbul districts. In the process, volunteers were provided with skills and tools for identification of seismic risks and response assets in their neighborhoods. Field data collection volunteers were trained in structural awareness for seismic safety, hand-held computer and Global Positioning System (GPS) receiver usage and fieldwork protocols. Office staff were trained in data transfer and creation of thematic maps using a desktop mapping application. Data compiled was fed into a geospatial database accessible over the Internet via interactive thematic maps (Figure 1). Seeing the results of their efforts taking shape in a relatively short time and sharing them over the Internet with their neighbors was another source of satisfaction and motivation for the volunteers. The map displays helped facilitate discussion of mitigation measures and action alternatives by local stakeholders.

The Greater Municipality of Istanbul Fire Department is in the process of establishing mini fire stations throughout the city manned by a newly trained cadre of professional and voluntary fire fighters. It is suggested that the Neighborhood Geographic Information Sharing System pilot evolve into a proposal for sustained implementation with local fire stations, with a GIS technician trained in each station, and with community organization support in the first two years to forge partnerships between fire stations and local neighborhood and other civic organizations.

The pilot project assessed the potential of neighborhood organizations and volunteers to collect and utilize disaster preparedness field data using advanced technologies. It highlighted the critical importance and feasibility of compiling and fusing local data with institutional data for the information to be current. It offered a model that can be generalized for a community-based disaster mitigation and preparedness Public Participation Geographic Information Sharing System.

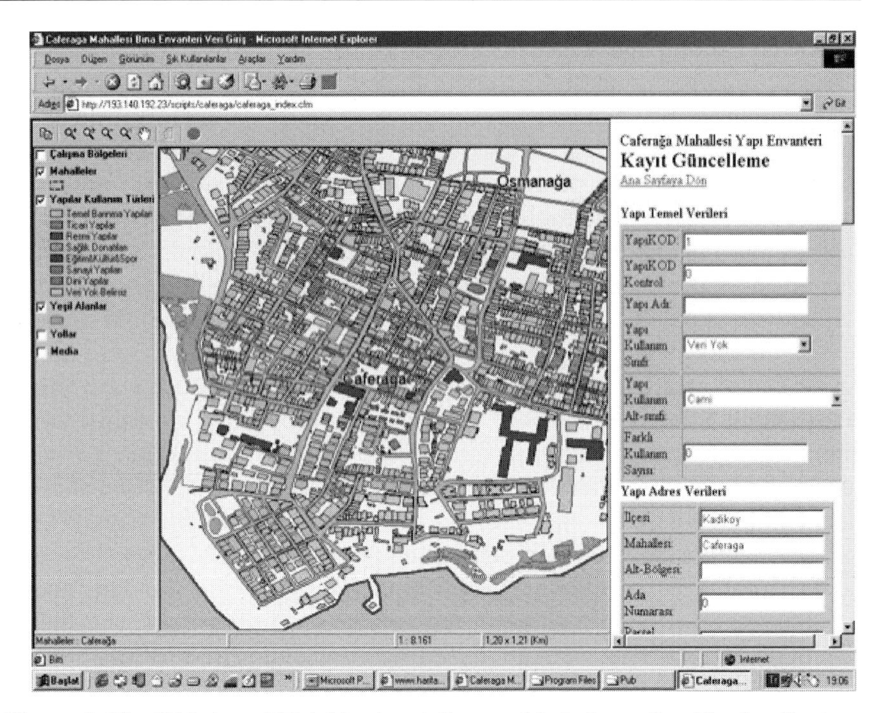

Figure 1: The Web-based Neighborhood Geographic Information Sharing System data entry and query page

2.2 The Basic Disaster Awareness Program Monitoring Facility

BU-KOERI Istanbul Community Impact Project staff collaborated with the Istanbul Education Directorate to implement a primary school-based Basic Disaster Awareness Education Program. ICIP experts trained Lead Basic Disaster Awareness trainers who, in turn, trained School Instructors. School Instructors delivered Basic Disaster Awareness Seminars in primary schools throughout the 32 districts of Istanbul reaching close to 2,000,000 students plus thousands of teachers, staff and parents.

After each school seminar, instructors reported the number of trained students, teachers, support staff and parents using a web-based data entry application. Over the course of the school year data compiled at the school, neighborhood and district levels were mapped against base population of students, teachers, support staff and estimated number of parents to display rates of population penetration and geographical diffusion of the program. An interactive map (Figure 2) of Istanbul districts was posted on the ICIP web site (www.iahep.org) where each district area constituted a hot link to another Web page which displayed program statistics and contact information for that district (Figure 3). A further click on each district page supplied the names of School Basic Disaster Awareness Instructors at the district together with their telephone numbers.

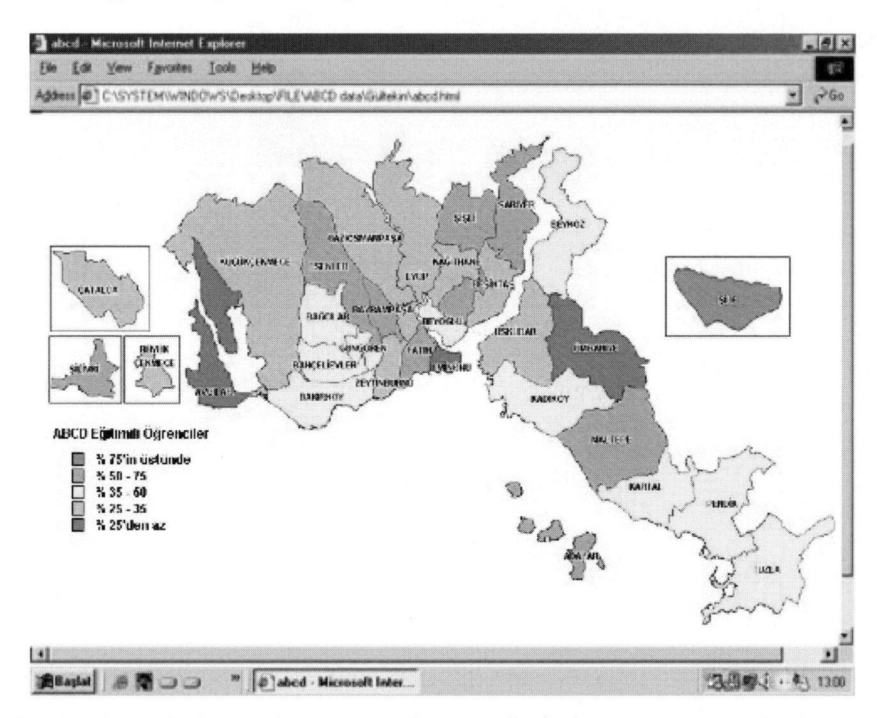

Fig. 2. Thematic interactive map Web page displaying percentage of primary school students who have received Basic Disaster Awareness Education by district. Each colored area on the map is a hot link to district-level operational statistics

The purpose of this effort is to increase accountability and civic interest and involvement in the implementation of this educational program. The Basic Disaster Awareness Program Monitoring Facility allowed the dissemination of disaster preparedness information to be tracked geographically by the program coordinator, district superintendents and school principles at the metropolitan region, district and school levels, respectively. Using the facility, parents and students were able to identify and contact School Basic Disaster Preparedness educators to ask for training or to seek answers to questions.

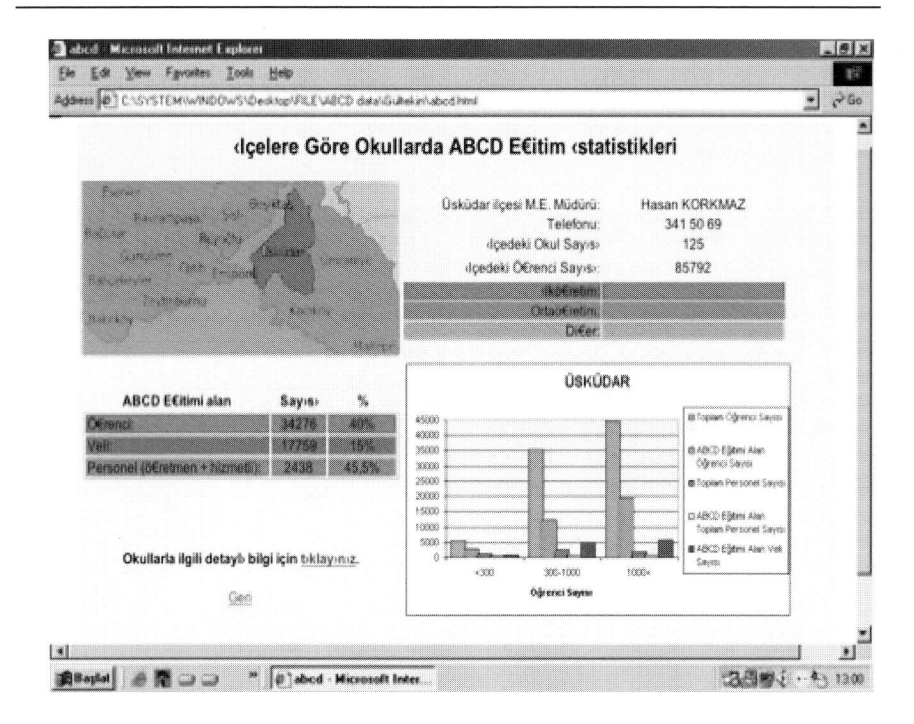

Fig. 3. Web page displaying district-level operational statistics and district superintendent contact information

2.3 The School Commute Contingency Pilot

In response to educators', administrators' and parents' concerns for student safety while in transit between home and school, the School Commute Contingency Pilot was implemented. The project involved the tracking of three school-bus routes in Istanbul, in order to stimulate contingency planning for commute-time emergencies when 400,000 students travel an average of 45 minutes each way on approximately 20,000 service buses every school day. GPS data loggers were used to record the three routes, pick-up points on each route and vehicle speed. Each route was then segmented into 5-15 minute travel segments of approximately equal length. School buildings on the route were identified as potential shelters in case of an emergency. Prototype maps highlighting the route of each service bus and the schools in proximity to the route were produced.

The idea was that should there be a major tremor or any other generalized emergency that disrupts communication for a prolonged period of time during the commute to or from the school, the driver of the service bus and/or the hostess on board would have advanced instructions to guide their passengers to the nearest school identified as a shelter. Administrators of shelter schools would refer to the table of demand estimates for shelter by school and time slots during commuting hours to have a pretty good idea about the size of the student population that would seek shelter and plan their response accordingly. Similarly, a parent or guardian who is issued the route map for the service bus his child is taking would refer to this resources to narrow the number of sites where he could meet his child instead of being completely in the dark about the child's whereabouts.

The School Commute Contingency Pilot results suggest that ICIP's School Disaster Management Planning Manual is supplemented to incorporate commute time emergency response procedures that encourage information sharing to:

1. Increase safety and security (of children) and reduce (parental) anxiety
2. Reduce transportation overload/chaos and increase access of emergency vehicles
3. Increase citizen response in the vicinity and efficient mobilization of personal resources
4. Encourage cooperation with organized responders
5. Reinforce ability of authorities to exercise legitimate control

The system is currently being redesigned to be in synchrony with the Istanbul Earthquake Rapid Response and Early Warning System housed at BU-KOERI. (2) In addition to the printed individual route maps and shelter school estimated population charts, the feasibility of near real-time service bus position reporting and shelter assignment based on load balancing among shelter schools is being explored. An emergency dispatch center software application module is being designed to manage data communication and processing, and to send automated cell phone text messages containing customized information and instructions to shelter school administrators as well as parents who subscribe to the system, Discussions are underway among a number of interested parties for the formation of a public-private partnership to make this system one of the first examples of a Disaster Response Location Based Service (LBS) in the world.